MW00446875

The Microsporidia and Microsporidiosis

The Microsporidia and Microsporidiosis

EDITOR

Murray Wittner

Division of Parasitology and Tropical Diseases
Departments of Pathology and Medicine
Albert Einstein College of Medicine
Bronx, New York

CONTRIBUTING EDITOR

Louis M. Weiss

Division of Parasitology and Tropical Diseases
Departments of Pathology and Medicine
Albert Einstein College of Medicine
Bronx, New York

ASM PRESS

WASHINGTON, D.C.

1325 Massachusetts Avenue NW
Washington, DC 20005-4171

Library of Congress Cataloging-in-Publication Data

The microsporidia and microsporidiosis / editor, Murray Wittner;
contributing editor, Louis M. Weiss.
p. cm.
Includes bibliographical references and index.
ISBN 1-55581-147-7 (hardcover)
1. Microsporidiosis. 2. Microsporidia. I. Wittner, Murray.
II. Weiss, Louis M.
[DNLM: 1. Microsporidia—physiology. 2. Microsporidia—
pathogenicity. 3. Microspora Infections—diagnosis. QX 50 M626
1999]
QR201.M73M53 1999
579.4—dc21
DNLM/DLC
for Library of Congress 98-47495
CIP

Printed in the United States of America

Cover photo: Encephalitozoon intestinalis cultured in Vero E6 green monkey cells. Reprinted from N. P. Kock, "Diagnosis of Human Pathogenic Microsporidia" (dissertation, Bernhard Nocht Institute for Tropical Medicine, Hamburg, Germany), with the permission of N. P. Kock, C. Schmertz, and J. Schottelius.

To my teacher
Richard Rokusaburo Kudo (1886–1967),
Professor of Zoology, University of Illinois,
whose unselfish devotion to the science
of protozoology and to his students
has created an indelible legacy
for all to emulate

CONTENTS

CONTRIBUTORS

Theodore G. Andreadis
Connecticut Agricultural Experiment Station, New Haven, CT 06504

James J. Becnel
USDA/ARS, Center for Medical, Agricultural, and Veterinary Entomology, Gainesville, FL 32604

G. Todd Bessinger
Department of Tropical Medicine, School of Public Health and Tropical Medicine, Tulane University Medical Center, New Orleans, LA 70112

Ralph T. Bryan
National Center for Infectious Diseases, Centers for Disease Control and Prevention, Albuquerque, NM 87110

Ann Cali
Department of Biological Sciences, Rutgers University, Smith Hall, 101 Warren Street, Newark, NJ 07102

Peter Deplazes
Institute of Parasitology, University of Zurich, Winterthurerstrasse 266A, CH-8057 Zurich, Switzerland

Elizabeth S. Didier
Department of Microbiology, Tulane Regional Primate Research Center, Tulane University Medical Center, Covington, LA 70433

V. Dolgikh
All Russian Institute for Plant Protection, Podbelskii 3, St. Petersburg-Puskin 189620, Russia

Ann M. Findley
Department of Biology, Northeast Louisiana University, Monroe, LA 71029

Dorothy Nahm Friedberg
Department of Ophthalmology, New York University School of Medicine, New York, NY 10016

Michael L. Kent
Department of Fisheries and Oceans, Pacific Biological Station, Nanaimo, British Columbia V9R 5K6, Canada

Elaine M. Keohane
Department of Clinical Laboratory Sciences, University of Medicine and Dentistry of New Jersey, Newark, NJ 07107

Donald P. Kotler
Gastrointestinal Division, Department of Medicine, St. Luke's-Roosevelt Hospital Center, and College of Physicians and Surgeons, Columbia University, New York, NY 10025

J. I. Ronny Larsson
Department of Zoology, University of Lund, Helgonavägen 3, S-223 62 Lund, Sweden

Gordon J. Leitch
Department of Physiology, Morehouse School of Medicine, Atlanta, Ga.

Hercules Moura
Division of Parasitic Diseases, MS F13, National Center for Infectious Diseases, Centers for Disease Control and Prevention, 4770 Buford Highway NE, Atlanta, GA 30341-3724, and Faculdade de Ciências Médicas, Universidade do Estado do Rio de Janeiro and Hospital Evandro Chagas, Instituto Oswaldo Cruz, FIOCRUZ, Rio de Janeiro, Brazil

Jan M. Orenstein
Department of Pathology, George Washington University School of Medicine, Washington, DC 20037

David C. Ritterband
Department of Ophthalmology, New York Eye and Ear Infirmary, New York, NY 10003

David A. Schwartz
Department of Pathology, Emory University and Grady Memorial Hospital, 80 Butler Street SE, Atlanta, GA 30335

John A. Shadduck
Heska Corporation, 1825 Sharp Point Drive, Fort Collins, CO 80525

Ross W. Shaw
Department of Zoology, University of British Columbia, Vancouver, British Columbia V6T 1Z4, Canada

Karen F. Snowden
Department of Veterinary Pathobiology, College of Veterinary Medicine, Texas A&M University, College Station, TX 77843-4467

J. Sokolova
All Russian Institute for Plant Protection, Podbelskii 3, St. Petersburg-Puskin 189620, Russia

Victor Sprague
Chesapeake Biological Laboratory, Solomons, MD 20688

Peter M. Takvorian
Department of Biological Sciences, Rutgers University, Smith Hall, 101 Warren Street, Newark, NJ 07102

Jiří Vávra
Department of Parasitology and Hydrobiology, Faculty of Science, Charles University, Prague, Czech Republic

Govinda S. Visvesvara
Division of Parasitic Diseases, MS F13, National Center for Infectious Diseases, Centers for Disease Control and Prevention, 4770 Buford Highway NE, Atlanta, GA 30341-3724

Charles R. Vossbrinck
The Connecticut Agricultural Experiment Station, 123 Huntington Street, New Haven, CT 06504, and Albert Einstein College of Medicine, 1300 Morris Park Avenue, Bronx, NY 10461

Rainer Weber
Division of Infectious Diseases and Hospital Epidemiology, Department of Internal Medicine, University Hospital, CH-8091 Zurich, Switzerland

Earl Weidner
Department of Biological Sciences, Louisiana State University, Baton Rouge, LA 70803

Louis M. Weiss
Division of Infectious Diseases and Division of Parasitology and Tropical Diseases, Albert Einstein College of Medicine, Room 504 Forchheimer, 1300 Morris Park Avenue, Bronx, NY 10461

Murray Wittner
Department of Pathology, Division of Parasitology and Tropical Diseases, Albert Einstein College of Medicine, 1300 Morris Park Avenue, Bronx, NY 10461

PREFACE

This volume on the microsporidia has been prepared in response to the need for an up-to-date and comprehensive treatment of a rapidly expanding area in parasitology and, in particular, protozoology. The flood of new information on the microsporidia that has emerged since the last compendium appeared in 1976 (J. Vávra and V. Sprague, *in* L. A. Bulla and T. C. Cheng (ed.), *Biology of the Microsporidia,* vol. 1, *Comparative Pathobiology,* Plenum Press, New York, N.Y., 1976) has been overwhelming. Two important developments since 1976 have been the emergence of this group of organisms as important human pathogens, especially in immunocompromised hosts, and the number of new genera and species that have been recognized. As pointed out by Vávra and Sprague (1976), by the end of the 19th century Labbé, in his review of the microsporidian literature, listed 3 genera, 33 named species, and 20 unnamed species. A quarter of a century later, Kudo (*Ill. Biol. Monogr.* **9:**1–268, 1924) listed 14 genera and about 170 named species. By 1976, there were 525 named species as well as 200 that had not been named. In the appendix to the current volume, Sprague and Becnel list 143 genera, and there are more than 1,200 species!

Interest in the microsporidia has extended from protozoologists and parasitologists to encompass veterinarians, entomologists, agricultural workers, and physicians specializing in such medical disciplines as infectious diseases (especially AIDS), pathology, gastroenterology, ophthalmology, and otorhinolaryngology. The microsporidia have become a favorite subject for biologists studying molecular phylogeny.

This volume is aimed as a comprehensive guide to microsporidiology. Chapter 1 attempts to trace the place of the microsporidia in biology and medicine. Chapter 2, by Vávra and Larsson, is devoted to the structure of these organisms. This chapter illustrates in detail the unique and characteristic structural features of the members of this phylum. In addition, technical notes are provided for those who wish to undertake a study of these organisms. Chapter 3, by Cali and

Takvorian, discusses microsporidian life cycles and illustrates the developmental changes of various groups of the microsporidia throughout their life stages. Chapter 4, by Weiss and Vossbrinck, provides insights into the molecular phylogeny of the microsporidia and a detailed account of their molecular biology, with special reference to the application of molecular techniques for the diagnosis of the various species infecting humans. Using the power of molecular analysis, it challenges the traditional views of the taxonomic status of the microsporidia. Chapter 5, by Weidner et al., covers the physiology and biochemistry of the microsporidia and discusses the most recent data on the subject. Chapter 6, by Keohane and Weiss, is a detailed account of that unique structure that characterizes all of the microsporidia, the polar tube, and its role in invasion is presented in detail. Studies of polar tube formation and its unique biochemical and molecular structure are described.

Chapter 7, by Didier and Bessinger, is a modern consideration of immunologic host responses to microsporidian infections and the utility of various animal models that are available for the study of microsporidiosis. In chapter 8, Kotler and Orenstein review the clinical aspects of human microsporidiosis. Particular emphasis is given to the pathogenesis of intestinal and hepatobiliary diseases. Many of their observations have emerged from their studies of clinical biopsy material of AIDS patients by both light and electron microscopy. They discuss current therapy for the various microsporidial infections. In chapter 9, Friedberg and Ritterband consider the pathogenesis, pathology, diagnosis, and treatment of ocular microsporidiosis. In chapter 10, Weber et al. describe techniques that are suitable for the diagnosis of microsporidiosis in stool, urine and other body fluids, and tissues. They also provide a detailed guide to the pathology of various microsporidial diseases. Chapter 11, by Visvesvara et al., presents methods for the culture and propagation of microsporidia. This chapter provides investigators with methods for isolation, handling, and culturing of various species of microsporidia and serves as a guide for those wishing to initiate in vitro studies.

In chapter 12, Snowden and Shadduck describe the biology of species of microsporidia of higher vertebrates other than humans. They discuss the widespread host distribution of these parasites in mammals and birds and describe in detail the life cycles, host specificity, and detection of microsporidia and the clinical presentations and pathology of microsporidian diseases of these animals. The microsporidia of fish are presented in chapter 13 by Shaw and Kent. These important infections, which are widespread among marine and freshwater fish, have a significant impact on the economics of the commercial fishing industry and may provide important clues to the epidemiology of disease transmission. In chapter 14, Becnel and Andreadis discuss the microsporidia of insects. Insects represent the type hosts in a substantial proportion of the genera of microsporidia. The chapter describes the biological and life cycle features of insect microsporidia. An appreciation of this large group of organisms may have profound implications for developing an understanding of how the microsporidia are transmitted in nature and therefore may provide the clues for the epidemiology of human microsporidial transmission. Moreover, insect microsporidia are important because it has been suggested that they be used for biological control of insect pests in agriculture.

In chapter 15, Bryan and Schwartz consider the epidemiology of microsporidian infections of humans. Although there is little firm data on this subject, some tantalizing clues to the origins of human microsporidian infections have recently come to light. The authors speculate upon these clues. We are left with the knowledge that much more information is required in order to understand the transmission and subsequent control of human microsporidian disease.

In the appendix, Sprague and Becnel provide a checklist of the known genera of the microsporidia. They have also written a glossary in which they define the specialized terms used in microsporidiology. The glossary should be particularly useful for those just entering this field.

ACKNOWLEDGMENTS

The support of the Opportunistic Infections Research Branch, Division of AIDS, National Institute of Allergy and Infectious Diseases, National Institutes of Health, is gratefully acknowledged. An unrestricted educational grant to support publication of this book was generously provided by Pfizer Inc.

HISTORIC PERSPECTIVE ON THE MICROSPORIDIA: EXPANDING HORIZONS

Murray Wittner

I

The development of protozoology as a science awaited the introduction of the microscope in the seventeenth century. Probably the first reference to protozoa in print, according to d'Archiac and Haime, is by the botanist Clusius (about 1550), who mentions that Tartars retreating from Teuton soldiers scattered coins as they fled in order to divert their pursuers. However, this tactic failed because *Nummulites,* a genus of foraminifera, caused the coins to turn to stones (Cole, 1926). Gesner's 1565 description of the foraminiferan *Vaginulina* probably represents the first description of a protozoan as an animal; although he believed this organism was a tiny mollusc. A century later, Hooke (1665) illustrated a similar organism in his *Micrographia* (Foster, 1965).

The birth of protozoology as a science can be dated to the summer of 1674 when Leeuwenhoek observed free-living ciliates in freshwater (Dobell, 1932). The science of parasitic protozoology began a few months later when, according to Dobell (1932), in the fall of 1674 Leeuwenhoek observed objects in the bile of an old rabbit which were probably oocysts of *Eimeria stediae.* In a letter to the Royal Society (London), in the same year he described them as "oval corpuscles." In 1680 Leeuwenhoek observed a motile organism in the gut of a horsefly, and in 1681, after examining his own feces, he described what was most likely *Giardia lamblia*, thereby providing the first report of a parasitic protozoan of humans. Subsequently, in 1683 he examined material from a frog's intestine and "beheld an unconceivably great number of living animalicules, and these of diverse sorts and sizes." Undoubtedly he was describing *Opalina* and *Nyctotherus* (Dobell, 1932). The word "protozoa" was first mentioned by Goldfuss in 1817 but was not formally introduced until 1820 (Cole, 1926); Goldfuss included the Cnidaria within the protozoa (Kudo, 1966), and in 1841 and 1845 Dujardin, in his important treatise *Histoire naturelle des helminthes,* described the then recognized parasitic protozoa. In 1845 Siebold defined the protozoa in essentially modern terms, "Die Thiere in welchen die verschiedenen Systeme der Organe nicht scharf ausgeschieden sind, und deren unregelmässige Forme und einfache Organization sich auf eine Zelle reduzieren lassen."

During the early part of the 19th century, an epidemic disease ravaged the silkworm industry of Europe, especially in Italy and France. Jean Louis Quatrrefages called this disease

Murray Wittner, Department of Pathology, Division of Parasitology, Albert Einstein College of Medicine, 1300 Morris Park Avenue, Bronx, NY 10461

The Microsporidia and Microsporidiosis (Murray Wittner, editor; Louis M. Weiss, contributing editor), ©1999 American Society for Microbiology, Washington, D.C.

pébrine, which in the dialect of southern France meant "pepper disease." Close examination of diseased silkworms revealed that their skin was covered with blackish blotches. During this period many Italian observers showed that the tissues of affected silkworms always contained small, shiny, oval corpuscles measuring 0.002 to 0.003 mm in diameter. A number of biologists became engaged in investigating the cause of pébrine, and in 1863 Louis Pasteur's teacher, Jean-Baptiste Dumas, a member of the Senate and a distinguished chemist, requested that Pasteur leave Paris for the south of France to study silkworm disease. By 1870 Pasteur had published his landmark studies, *Etude sur la maladie des vers à soie,* in which he described a method for controlling and preventing pébrine. This work was perhaps the first scientific study of a disease caused by a parasitic protozoan, a microsporidian parasite, that resulted in an effective practical method of control (Pasteur, 1870).

The agent causing pébrine was named *Nosema bombycis* by Nägeli (1857). However, he regarded it as a yeast and placed it in the Schizomycetes which, at the time, was a conglomeration consisting of yeast (see below) and bacteria. (A similar and important infection of honeybees that markedly diminishes the productivity of the hive is caused by *Nosema apis*.) The microsporidia have been recognized as a distinct group of unicellular organisms since 1882 when Balbiani suggested the order Microsporida to accommodate *N. bombycis,* the only microsporidian known at that time. In 1976 Sprague created the phylum Microspora, which was updated in 1992 (Sprague et al., 1992). Sprague and Becnel (1998) have proposed that Microsporidia Balbiani 1882 should be the acknowledged phylum name as well as the correct author and date.

The polar filament is the defining character of the microsporidia, and its function has attracted the attention of many investigators. Early observations to determine the stimuli for its extrusion were undertaken by Thélohan (1895), Stempell (1909), Fantham and Porter (1912), and Kudo (1913, 1916, 1918). At that time, the consensus seems to have been that changes in osmotic pressure were the cause of polar filament extrusion. While it was generally acknowledged that the polar tube functioned to conduct the sporoplasm to a host cell, it was not until the middle of the twentieth century that it was generally agreed that the sporoplasm traveled out of the spore through the tube. Oshima (1937) suggested that the sporoplasm passed from the spore through the hollow tube formed during discharge. Subsequently, many investigators confirmed these observations (Gibbs, 1953; West, 1960; Lom and Vavra, 1963; Weidner, 1976, among others).

Microsporidian species have been reported to infect nearly all of the invertebrate phyla, including such unicellular organisms as ciliates and gregarines, myxozoans, cnidarians, platyhelminths, nematodes, rotifers, annelids, molluscs, bryozoans, and arthropods, as well as all five classes of vertebrates; the greatest numbers of species infect arthropods and fish. Currently, more than 1,200 species belonging to 143 genera are recognized, and with increasing reports of new genera and species these numbers will undoubtedly increase greatly.

Microsporidia are obligate intracellular parasites of eukaryotes, the transmittable stage being a resistant spore which is usually small, possesses a thick wall, and contains a characteristic polar tube apparatus. The latter serves to transmit the genetic material to the host cell. The host range for most microsporidian species is usually restricted, although species of microsporidia with a wide host range have been described (i.e., various mammalian species). Previously regarded as pathogenic only for invertebrates, microsporidia have been reported to infect mammals as well (Trammer et al., 1997), and closely related microsporidia have been found to infect widely separated animal hosts.

Glugea anomala (Moniez, 1887), which causes subcutaneous cysts in fish (e.g., stickleback), was the first microsporidian parasite recognized to cause disease in vertebrates. Subsequently, other species such as *Pleistophora* spp and *Nucleospora (Enterocytozoon) salmonis*

Fish/Water

(Chilmonczyk et al., 1991) were implicated in sporadic but serious disease outbreaks among fish populations. These outbreaks have had an important economic impact on tropical freshwater fish, commercial fish farming, and the sport fishing industry.

Mammalian infection by microsporidia was initially reported by Wright and Craighead (1922) in the central nervous system of laboratory rabbits. This finding was also made by Levaditi et al. (1923), who believed that the observed encephalitis was caused by a microsporidian parasite which they named *Encephalitozoon cuniculi,* suggesting that it was a microsporidian. The host range of this organism is extremely broad, and it has been reported from many mammalian species including rabbits, rats, mice, various carnivores, and primates including humans (Canning and Lom, 1986).

Although the microsporidia have been recognized as pathogenic agents in a wide variety of animals, the first human case of well-documented microsporidiosis was reported by Matsubayashi et al. (1959) in a 9 year old boy suffering from recurrent fever, headache, seizures, and loss of conciousness who apparently recovered. Subsequently, disseminated nosematosis was reported by Margileth et al. (1973) in a 4-month old immunocompromised infant with thymic aplasia. Subsequent light and transmission electron microscope studies by Sprague (1974) revealed a new species of microsporidian, *Nosema connori.* Ocular microsporidiosis was first recognized by Ashton and Wirasinha (1973) in a corneal ulcer of an 11-year-old Sri Lankan boy of unknown immune status. Six years previously he had been gored by a goat, and a wound of the right upper eyelid had been sutured. In 1973 Marcus et al. reported microsporidian infection of pancreatic carcinoma cells. Subsequently, interstitial keratitis localized to the conjunctiva was described in immunologically competent patients by Shadduck et al. (1990) and Davis et al. (1990).

Aside from these occasional case reports incriminating microsporidia as a cause of human disease, they were not regarded as important or frequent agents of disease. However, during the middle and later parts of the 1970s numerous reports of severe diarrheal disease caused by several gastrointestinal bacterial and protozoan infections were reported among male homosexuals; this syndrome was called gay bowel disease. The etiology of gay bowel disease was often found to be organisms such as salmonellae, shigellae, and giardiae, and these infections were usually treated successfully. However, a subset of these patients in whom no organism could be found failed to respond to any antibiotic therapy and continued to suffer with large-volume diarrhea—so-called wasting disease—from which they usually died. In June 1981 AIDS was first recognized. It became evident that wasting disease, often chcrterized by large-volume diarrhea, was present in many AIDS patients. By 1985 several reports of AIDS patients suffering with diarrhea and wasting syndrome had been published. Desportes and colleagues (1985) first reported a new microsporidian, *Enterocytozoon bieneusi,* primarily in small intestinal enterocytes of such an AIDS patient. Interestingly, at almost the same time, Dobbins et al. (1985) and Modigliani et al. (1985) also described this organism. Subsequently, *E. bieneusi* has been found to infect biliary (Beaugerie et al., 1992) as well as tracheal, bronchial, and nasal epithelium (Eeftinck Schattenkerk et al., 1992; Hartskeerl et al., 1993), pulmonary and intestinal epithelium (Weber et al., 1992), and nonparenchymal hepatic cells (Pol et al., 1993). This organism also has been found in normal and immunocompetent individuals with self-limited diarrhea (Sandfort et al., 1994), as well as in adult pigs (Deplazes et al., 1996) and in monkeys infected with simian immunodeficiency virus (Mansfield et al., 1997).

Once it became clear that human microsporidiosis was largely a disease of immunoincompetent individuals, particularly those with AIDS, additional species were recognized that also caused disseminated infection (Canning et al., 1992). For example, several reports of disseminated microsporidiosis caused by *E. cuniculi* (Terada et al., 1987; Zender et al., 1989) were followed by numerous reports of *E. (Septata) intestinalis* (Cali et al., 1993) causing severe intractable large-volume diarrhea and was also found in urine causing interstitial nephritis as well as cholecystitis; *Encephalitozoon hellem* was recognized as the cause of keratoconjunctivitis, cystitis, ureteritis, and bronchiolitis; and *Pleistophora* sp. was found to cause myositis (Chup et al., 1993). At present, 13 species of microsporidia have been recorded that infect humans, and new microsporidian species that infect humans are being recognized with increasing frequency.

Prior to the development of molecular phylogeny, it was generally assumed that the microsporidia were primitive eukaryotes, having branched very early among the eukaryotes. Their amitochondrial state (Vávra, 1976), lack of peroxisomes, Golgi stacks, and the typical eukaryotic 5.8S ribosomal subunit fused to a large subunit RNA (Vossbrinck and Woese, 1986) further serve to confirm this notion. However, it has been shown that microsporidia possess genes for heat shock proteins of the *hsp70* class which are related to mitochondria-type *hsp70* genes (Hirt et al., 1997), suggesting that at one point in their evolution they possessed mitochondria which were lost secondarily. Interestingly, these microsporidian genes appear to be closely related to the *hsp70* genes of yeast (fungal) mitochondria. Moreover, there is accumulating evidence supporting the notion that the microsporidia have close fungal relationships. For example, thymidylate synthase and dihydrofolic acid reductase are usually found as separate epitopes on the same protein in higher plants and protists but are separate proteins in animals, fungi, and microsporidia (Vivares et al, 1996) . Furthermore, like mitochondrial *hsp70,* both the α- and β-tubulin genes of microsporidia bear similarities to genes for fungal tubulins (Edlind et al., 1996). In addition, the presence of chitin and trehalose in microsporidia lends further strength to the notion of the fungal relatedness of the microsporidia. Future studies may determine whether, like *Pneumocystis,* the microsporidia should be placed among the fungi.

It is ironic that when Nägeli first named *N. bombycis* in 1857, he placed it among the fungi (yeast). In 1998, current molecular phylogeny studies have identified several genes, including the mitochondrial *Hsp70,* α- and β-tubulin, and elongation factor genes of microsporidia that suggest their close affinity with the fungi (yeast). In addition, protozoan thymidylic synthetase and dihydrofolic acid reductase, which are present in protozoa as separate epitopes on the same protein, are found as separate proteins in the microsporidia as they are in yeast. Curiously, Nägeli's prescience that the microsporidia were fungi may soon be borne out.

ACKNOWLEDGMENTS

This work was supported by NIH grants AI RO1 31788 and AI 41398.

REFERENCES

Ashton, N., P. and A. Wirasinha. 1973. Encephalitozoonosis (nosematosis) of the cornea. *Br. J. Ophthalmol.* **60:**618–631.

Balbiani, G. 1882. Sur les microsporidies ou sporogspermies des articules. *C. R. Acad. Sci.* **95:**1168–1171.

Beaugerie, L., M. F. Teihac, A. M. Deloul, J. Fritsch, P. M. Girard, W. Rozenbaum, Y. LeQuintrec, and F. P. Chatelet. 1992. Cholangiopathy associated with microsporidia infection of the common bile duct mucosa in a patient with HIV infection. *Ann. Intern. Med.* **117:**401–402.

Cali, A., D. P. Kotler, and J. M. Orenstein. 1993. *Septata intestinalis* n. g., n. sp., an intestinal microsporidian associated with chronic diarrhea and dissemination in AIDS patients. *J. Protozool.* **40:** 101–112.

Canning, E. U., A. Curray, C. J. Lacey, and D. Fenwich. 1992. Ultrastructure of *Encephalitozoon* sp. Infecting the conjunctival, corneal and nasal epithelia of a patient with AIDS. *Eur. J. Protistol.* **28:**226–227.

Canning, E. U., and J. Lom. 1986. *The Microsporidia of Vertebrates.* Academic Press, London, United Kingdom.

Chilmonczyk, S., W. T. Cox, and R. P. Hedrick. 1991. *Enterocytozoon salmonis* n.sp. an intranuclear microsporidum from salmonid fish. *J. Protozool.* **38:**264–269.

Chup, G. L., J. Alroy, L. S. Adelman, J. C. Breen, and P. R. Skolnik. 1993. Myositis due to *Pleistophora* (microsporidia) in a patient with AIDS. *Clin. Infect. Dis.* **16:**15–21.

Cole, F. J. 1926. *The History of Protozoology.* University of London Press, London, United Kingdom.

Davis, R. M., R. L. Font, M. S. Keesler, and J. A. Shadduck. 1990. Corneal microsporidiosis: a case report including ultrastructural observations. *Ophthalmolology* **97:**958–959.

Deplazes, P., A. Mathis, C. Muller, and R. Weber. 1996. Molecular epidemiology of *Encephalitozoon cuniculi* and first detection of *Enterocytozoon bieneusi* in fecal samples of pigs. *J. Eukaryot. Microbiol.* **43:** 93S.

Desportes, I., Y. LeCharpentier, A. Galian, F. Bernard, B. Cochand-Priolett, A. Laverne, P. Ravisse, and R. Modigliani. 1985. Occurrence of a new microsporidian: *Enterocytozoon bieneusi* n. g., n. sp., in the enterocytes of a human patient with AIDS. *J. Protozool.* **32:**250–254.

Dobell, C. 1932. *Antony Van Leeuwenhoek and His "Little Animals."* Russel and Russel, New York, N.Y. [Reprint of 1932 volume.]

Dobbins, W. O., III, and W. Weinstein. 1985. Electron microscopy of the intestine and rectum in acquired immunodeficiency syndrome. *Gastroenterology* **88:**738–749.

Dujardin, F. 1845. Histoire naturelle des helminthes ou vers intestinaux. Paris, France.

Edlind, T. D., J. Li, G. S. Visvesvara, M. H. Vodkin, G. L. McLaughlin, and S. K. Katiyar. 1996. Phylogenetic analysis of β-tubulin sequences from amitochondrial protozoa. *Mol. Phylogenet. Evol.* **5:**359–367.

Eeftinck Schattenkerk, J. K., T. van Gool, L. S. Schot, M. van den Bergh Weerman, and J. Dankert. 1992. Chronic rhinosinusitis, a new clinical syndrome in HIV-infected patients with microsporidiosis. *In Workshop on Intestinal Microsporidiosis in HIV Infection.* (Abstract.)

Fantham, H. B., and A. Porter. 1912. The morphology and life history of *Nosema apis* and the significance of its various stages in the so-called "Isle of Wight" disease in bees (microsporidiosis). *Ann. Trop. Med. Parasitol.* **6:**163–196.

Foster, W. D. 1965. *A History of Parasitology.* E. S. Livingstone, Edinburgh, United Kingdom.

Gesner C. 1565. *De omne rerum fossilium genere.* Tiguri.

Gibbs, A. J. 1953. *Urleya* sp. (Microsporidia) found in the gut tissue of *Trachea secalis* (Lepidoptera). *Parasitol.* **43:**143–147.

Hartskeerl, R. A., A. R. J. Schuitema, T. van Gool, and J. Terpstra. 1993. Genetic evidence for the occurrence of extra-intestinal *Enterocytozoon bieneusi* infections. *Nucleic Acids Res.* **21:**41–50.

Hirt, R. P., D. Healy, C. R. Vossbrinck, E. U. Canning, and T. M. Embley. 1997. Identification of a mitochondrial HSP70 orthologue in *Vairimorpha necatrix:* molecular evidence that microsporidia once contained mitochondria. *Curr. Biol.* **7:**995–997.

Hooke, R. 1665. *Micrographia.* London, United Kingdom.

Kudo, R. 1913. Eine neue Methode die Sporen von *Nosema bombycis* Nägeli mit ihren ausgeschnellten Polfäden dauerhaft zu präparieren und deren Länge genauer zu bestimmen. *Zool. Anz.* **41:**368–371.

Kudo, R. 1916. Contribution to the study of parasitic protozoa. II. *Bull. Seric. Exp. Stn. Jpn.* **1:**31.

Kudo, R. 1918. Experiments on the extrusion of polar filaments of cnidosporidian spores. *J. Parasitol.* **4:**141.

Kudo, R. 1924. A biologic and taxonomic study of the Microsporidia. *Ill. Biol. Monogr.* **9:**1–268.

Kudo, R. 1966 *Protozoology,* 5th ed. Charles C Thomas, Springfield, Ill.

Ledford, D. K., M. D. Overman, A. Gonzala, A. Cali, W. Mester, and R. F. Lockey. 1985. Microsporidiosis myositis in a patient with acquired immunodeficiency syndrome. *Ann. Intern. Med.* **102:**628–630.

Levaditi, C., S. Nicolau, and R. Shoen. 1923. L'étiologie de l'encéphalite. *C.R. Acad. Sci.* **177:**985–988.

Lom, J., and J. Vavra. 1963. The mode of sporoplasm extrusion in microsporidian spores. *Acta Protozool.* **1:** 81–92.

Mansfield, K. G., A. Carville, D. Schoetz, J. Mackey, S. Tzipori, and A. Lackner. 1997. Identification of *Enterocytozoon bieneusi*-like microsporidan parasite in simian immunodeficiency virus-inoculated macaques with hepatobiliary disease. *Am. J. Pathol.* **150:**1395–1405.

Marcus, P. B., J. J. van der Watt, and P. J. Burger. 1973. Human tumor microsporidiosis. *Arch. Pathol.* **95:**341–343.

Margileth, A. M., A. J. Strano, R. Chandra, R. Neafie, M. Blum, and R. M. McCully. 1973. Disseminated nosematosis in an immunogically compromised infant. *Arch. Pathol.* **95:**145–150.

Matsubayashi, H., T. Koike, I. Mikata, H. Takei, and S. Hagiwara. 1959. A case of encephalitozoon-like body in man. *Arch. Pathol.* **67:**181–187.

Modigliani, R., C. Bories, Y. Le Charpentier, M. Salmeron, B. Messing, A. Galian, J. C. Rambaud, A. Lavergne, B. Cochand-Priollet, and I. Desportes. 1985. Diarrhea and malabsorption in acquired immunodeficiency syndrome: a study of four cases with special emphasis on opportunistic protozoan infestation. *Gut* **2G:**179–181.

Moniez, R. 1887. Observations pour la revision des Microsporidies. *C.R. Acad Sci.* **104:**1312–1315.

Nägeli, K. 1857. Ueber die neue Krankheit die Seidenraupe und verwandte Organismen. *Bot. Zeitung* **15:** 760–761.

Oshima, K. 1937. On the function of the polar filament of *Nosema bombycis. Parasitology* **29:** 220–224.

Pasteur, L. 1870. *Études sur la maladie des vers à soie.* Paris, France.

Pol, S., C. A. Romania, S. R. Richard, P. Amouyal, I. Desportes-Livage, F. Carnot, J. F. Paays, and P. Berthelot. 1993. Microsporidia infection in patients with human immunodeficiency virus and unexplained cholangitis. *N. Engl. J. Med.* **328:**95–99.

Sandfort, J., A. Hannemann, D. Stark, R. L. Owen, B. Ruf. 1994. *Enterocytozoon bieneusi* infection in an immunocompetent HIV-negative patient with acute diarrhea. *Clin. Infect. Dis.* **19:**514–516.

Shadduck, J. A., R. A. Meccoli, R. Davis, R. L. Font. 1990. Isolation of a microsporidian from a human patient. *J. Infect. Dis.* **162:**773–776.

Sprague, V. 1974. *Nosema connori* n. sp.: a microsporidian parasite of man. *Trans. Am. Microsc. Soc.* **93:**400–403.

Sprague, V. 1976. Classification and phylogeny of the Microsporidia p. 1–30. *In Comparative Pathobiology,* vol. 2. L. A. Bulla and T. C. Cheng (eds.), Plenum Press, New York, NY.

Sprague, V., J. J. Becnel, and E. L. Hazard. 1992. Taxonomy of phylum Microspora. *Crit. Rev. Microbiol.* **18:**285–395.

Sprague, V., and J. J. Becnel. 1998. Note on the name-author-date combination for the taxon microsporidies Balbiani, 1882, when ranked as a phylum. *J. Invertebr. Pathol.* **71:**91–94.

Stempell, W. 1909. Uber *Nosema bombycis* Nägeli. *Arch. Protistenkd.* **16:**281–358.

Terada, S., K. R. Reddy, L. J. Jeffers, A. Cali, and E. R. Schiff. 1987. Microsporidian hepatitis in the acquired immunodeficiency syndrome. *Ann. Intern. Med.* **107:**61–62.

Thélohan, P. 1895. Recherches sur les myxosporidiens. *Bull. Sci. France Belg.* **26:**100–394.

Trammer, T., F. Dombrowskí, M. Doehring, W. A. Maier, and H. M. Seitz. 1997. Opportunistic properties of *Nosema algerae* (Microspora), a mosquito parasite, in immunocompromised mice. *J. Eukaryot. Microbiol.* **44:**258–262.

Vávra, J. 1976. Structure of Microsporidia, pp. 1–8. *In* L. A. Bulla and T. C. Cheng (ed.), *Comparative Pathobiology,* vol. I. Plenum Press, New York, N.Y.

Vivares, C., C. Biderne, F. Duffieux, E. Peyretaillade, P. Peyret, G. Metenier, and M. Pages. 1996. Chromosomal localization of five genes in *Encephalitozoon cuniculi* (Microsporidia). *J. Eukaryot. Microbiol.* **43:**975.

Vossbrinck, C. R., and C. R. Woese. 1986. Eukaryotic ribosomes that lack a 5.8S RNA. *Nature* **320:** 287–288.

Weber, R., H. Kuster, R. Keller, T. Bächi, M. A. Spycher, J. Briner, E. Russi, and R. Lüthy. 1992. Pulmonary and intestinal microsporidiosis in a patient with acquired immunodeficiency syndrome. *Am. Rev. Respir. Dis.* **146:**1603–1605.

Weidner, E. 1976. The microsporidian spore invasion tube: the ultrastructure, isolation and characterization of the protein comprising the tube. *J. Cell Biol.* **71:**23–34.

West, A. F. Jr. 1960. The biology of a species of *Nosema* (Sporozoa, Microsporidia) in the flour beetle *Tribolium confusum*. *J. Parasitol.* **46:**747–753.

Wright, J. H., and E. M. Craighead. 1922. Infectious motor paralysis in young rabbits. *J. Exp. Med.* **36:**135–149.

Zender, H. O., E. Arrigoni, J. Eckert, and Y. Kapanci. 1989. A case of *Encephalitozoon cuniculi* peritonitis in a patient with AIDS. *Am. J. Clin. Pathol.* **92:**352–356.

STRUCTURE OF THE MICROSPORIDIA

Jiří Vávra and J. I. Ronny Larsson

2

Microsporidians are intracellular parasites ranked among the protists, which means that they are eukaryotic and unicellular. They exhibit a number of important, unique features. They have no mitochondria, no typical Golgi apparatus, and no microbodylike organelles (peroxisomes or hydrogenosomes), and their ribosomes are not of the typical eukaryotic type but resemble the ribosomes of prokaryotic organisms (Ishihara and Hayashi, 1968; Curgy et al., 1980). Further, the microsporidian spore is equipped with a unique set of organelles, the extrusion or infection apparatus, which functions when the spore content is injected into a host cell (Lom and Vávra, 1963a; Vávra, 1976a; Weidner, 1976a; Larsson, 1986b).

STRUCTURAL CHARACTERS AND CLASSIFICATION OF MICROSPORIDIA

Structural characters are the basis for the systematic ranking of microsporidia, which are acknowledged to constitute a phylum of their own, Microspora Sprague, 1977, classified in the kingdom Protozoa Goldfuss, 1818. The names and type material of new species follow the rules of the International Code of Zoological Nomenclature (1985).

However, a different interpretation of microsporidian cytology has suggested a ranking of microsporidia in the kingdom Archezoa Haeckel, 1894, resurrected by Cavalier-Smith (1983). Archezoa is an obviously polyphyletic taxon for supposedly primitive unicellular eukaryotes that have failed to acquire mitochondria and other cell organelles by endosymbiosis (Cavalier-Smith, 1983; Corliss, 1994). This ranking has never been generally accepted, however, and with new molecular biology data at hand, the validity of the kingdom Archezoa has been questioned. Although microsporidians do not have morphologically identifiable mitochondria (Vávra, 1965), molecular biology indicates that archezoans had mitochondria in the past. They possess a molecular signature of mitochondria in the form of the heat shock protein HSP70 (Germot et al., 1997; Hirt et al., 1997) (Chapter 4). Microsporidians probably acquired mitochondria but lost them when they evolved into the highly specialized intracellular parasitic life.

The unique structural organization of the microsporidian cell separates the phylum Microspora Sprague, 1977, from other protists and indicates that microsporidians constitute a

Jiří Vávra, Department of Parasitology and Hydrobiology, Faculty of Science, Charles University, Prague, Czech Republic.
J. I. Ronny Larsson, Department of Zoology, University of Lund, Helgonavägen 3, S-223 62 Lund, Sweden.

The Microsporidia and Microsporidiosis (Murray Wittner, editor; Louis M. Weiss, contributing editor), ©1999 American Society for Microbiology, Washington, D.C.

monophyletic taxon with no evident relationship to other protists. However, on the basis of new information about their molecular biology, fungal affinities have been suggested (Müller, 1997). These affinities, if they truly exist, give no indication of a new systematic position. At the present time there is no fungal taxon in which microsporidians might be ranked.

Classification of microsporidia is based on life cycle and structural (mostly ultrastructural) characters. There are divergent opinions about the classification of microsporidia below the phylum level (Sprague, 1977; Weiser, 1977; Issi, 1986; Sprague et al., 1992) (Appendix). On the basis of cytology, three groups are distinguished which also differ in life cycle characters.

Typical Microsporidia

Microsporidia having the usual cytology for these protists are by far the largest group, with more than 1,000 named species. Their hosts are found in all groups of animals, from protozoa to humans. They exhibit the most complex life cycles and produce spores in a great variety of shapes, from spherical to rodlike.

Atypical Microsporidia

Life cycle and cytology distinguish typical microsporidia from the other two groups: chytridiopsids and metchnikovellideans. Our knowledge of these groups is superficial, especially in the case of the metchnikovellideans. The two groups share an extrusion apparatus of similar, although not identical, construction that diverges from the common type of extrusion apparatus. They also share a life cycle that differs from that of typical microsporidians. Whether these similarities indicate that the two groups are closely related is unknown.

The chytridiopsids, represented by about 20 species, are parasites of terrestrial or freshwater invertebrates, mostly insects. Their spores are spherical.

The metchnikovellideans (about 25 species) are hyperparasites of gregarines, with two exceptions from marine hosts. The spores are more or less subspherical or, exceptionally, rod-shaped. Spores lack a polaroplast, and the polar filament is of a special construction.

MICROSPORIDIANS UNDER THE LIGHT MICROSCOPE

It is fairly easy to recognize a microsporidian as such if the organism has completed its life cycle and produced spores. Identification of presporal stages is difficult even for a specialist. The spore is thus the main diagnostic element.

Living spores of most microsporidians are only a few micrometers long, but the size range between the smallest and the largest species is considerable. Spores of the human parasite *Enterocytozoon bieneusi* measure about 1 µm in length, while spores of *Bacillidium filiferum,* a parasite of freshwater oligochetes, approach 40 µm. A sample of microsporidian spores is fairly uniform in shape and size, in contrast to a sample of yeast cells which can erroneously be taken for microsporidians. The uniform spore size is an important indication of the microsporidian nature of an organism.

The spore shape in most species is oval or pyriform, but rodlike, spherical, and other shapes are not unusual (Fig. 1), and fixation and staining may change the shape. For example, in the predominantly insect-parasitic *Amblyospora* spp., which produce spores in groups of eight, living spores are nearly oval in shape, while they appear barrel-shaped in a stained smear (Hazard and Oldacre, 1975).

Live microsporidian spores are refractive, and usually only a voluminous vacuole at the posterior pole of the spore is visible (Fig. 1G, J, and K). Only exceptionally are traces of the coiled, threadlike infection apparatus visible in large and living immature spores. It is not unusual for spores produced in aquatic hosts to be ornamented with filamentous projections like the characteristic spores of the blackfly parasites of the genera *Caudospora* (Fig. 1H) and *Hirsutosporos* (Batson, 1983). Very fine projections not seen in fresh preparations may be revealed by negative staining. Spores of microsporidia from aquatic invertebrate hosts often exhibit a mucous coat when released into water (Fig. 1M) (Lom and Vávra, 1963b).

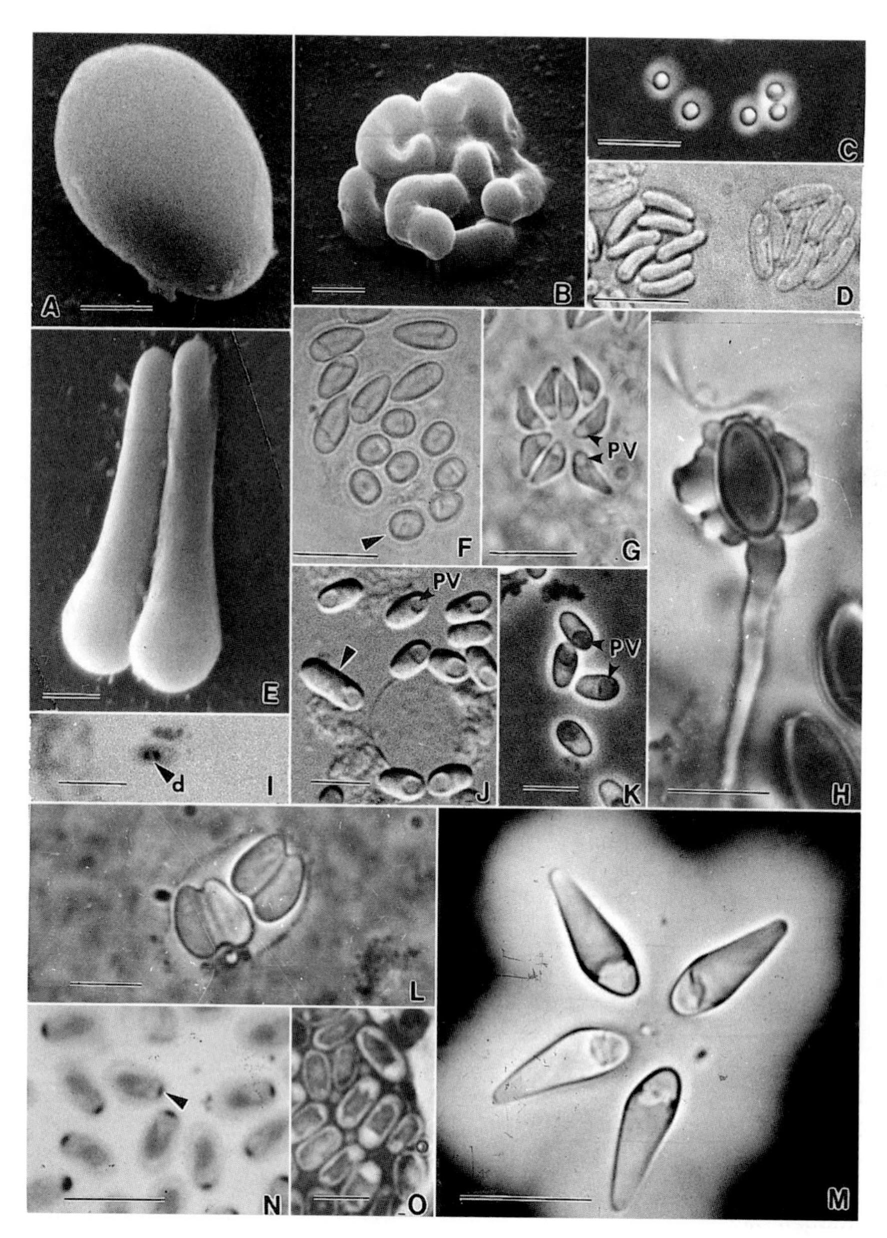

FIGURE 1 Shape and staining reactions of microsporidian spores. (A) Lightly pyriform (*Episeptum invadens*); (B) horseshoe-shaped (*Toxoglugea variabilis*); (C) spherical (*Coccospora micrococcus*); (D) rod-shaped (*Resiomeria odonatae*); (E) lageniform (*Cougourdella polycentropi*); (F) dimorphic sporogony with pyriform free spores and ovoid octospores (arrowhead) (*Amblyospora callosa*); (G) angular octospores (*Bohuslavia asterias*); (H) ovoid spore with prominent exospore projections (*Caudospora simulii*); (I) diplokaryon (*Nosema mesnili*); (J and K) live ovoid spores (*Glugea anomala*); (L) coupled spores (*Norlevinea daphniae*); (M) tetraspores with prominent mucus production (*Gurleya elegans*); (N) PAS reaction stains spores in the "polar cap," i.e., the polar sac-anchoring disk complex (arrowhead) (*Nosema lepiduri*); (O) spores in stained smear (*Nosema apis*). (A, B, and E) SEM; (C) dark field; (D and J) interference phase contrast; (F to H, and K) phase contrast; (I and O) Giemsa stain; (L and M) India ink negative staining; (N) PAS. Scale bars: A, B, and E, 1 μm; C, D, F to H, 10 μm; I to O, 5 μm. Reprinted with permission from *Arch. Protistenkd.* (A), *Protistologica* (B and D), *Eur. J. Protistol.* (E), *Zool. Anz.* (I), and *J. Protozool.* (L to N). Abbreviations: d, diplokaryon, PV, posterior vacuole.

A few more details are visible in a fixed, stained preparation. Most commonly, smears are Giemsa-stained. This is the best technique for staining nuclei of presporal stages. Mature spores exhibit a dark band across the center of the spore where the nucleus lies (Fig. 1O), a dark spot at the anterior pole (the anchor for the extrusion apparatus), and sometimes a reddish granule, the posterosome, in the posterior vacuole. In mature spores the nuclei are seldom clearly revealed after Giemsa staining (Fig. 1I) unless the staining is performed after acid hydrolysis (see "Spore Nuclei").

A number of microsporidians produce their spores in packets. The envelopes are rarely visible in light microscopic preparations, neither in fresh preparations nor in stained smears. If groups with 2, 4, 8, 16, or other regular numbers of spores are seen in carefully made smears, it might be suspected that the organism is a microsporidian (Fig. 1B, D, F, and G).

Application of specific techniques, described later in this chapter (see "Polar Filament Extrusion"), helps to determine if an organism is a microsporidian. The ultimate diagnostic test forces spores to eject the polar filament and is the classical light microscopic method for proving that an organism is a microsporidian.

REPRODUCTION AND LIFE CYCLES

Microsporidian life cycles are treated in detail in Chapter 3. However, as basic knowledge of the life cycle and of the terminology applied to the various stages is necessary for a presentation of the cytology, a brief introduction is given here. Further information on microsporidian life cycles and terminology can be found in Vávra (1976b), Vávra and Sprague (1976), Larsson (1986b, 1988c), and Sprague et al. (1992).

Basic Life Cycle

The stage in which microsporidians are normally encountered is the spore. It is the infectious stage and the only life cycle stage able to survive outside the host cell. The spore normally reaches the new host through the gut even if other routes of transmission are known to occur.

When the spore is in an appropriate host, an infectious cell, the sporoplasm, is injected from the spore into a host cell, resulting in the transmission of infection. In the new host the sporoplasm starts a more-or-less efficient reproduction, ending up with the production of spores. In most microsporidians two sequences of reproduction follow each other: an initial vegetative reproduction (merogony) yielding daughter cells (merozoites) with the potential either to repeat merogony or to enter the second reproductive phase, the production of spores (sporogony). Some microsporidians, like *Nosema* spp., reproduce weakly during sporogony, in which each mother cell yields only two spores (Brooks et al., 1985). Species with a weak sporogony reproduce more efficiently during merogony, where one or more cycles of divisions might follow each other.

In a small number of microsporidians, merogonial reproduction has never been observed and is presumed to be absent. The only reproduction known to occur is the production of spores. This situation is characteristic of the genera *Chytridiopsis* (Larsson, 1993), *Metchnikovella* (Vivier and Schrével, 1973), and a few other related genera.

Reproduction Modes

Both merogony and sporogony proceed as binary fission, plasmotomy, or schizogony depending on the species (Fig. 2). The manner in which the divisions occur is often best visible in light microscopic preparations.

When reproduction involves binary fission, two daughter cells are formed by each mother cell. Often the division products adhere to each other, forming chains of cells (Fig. 2B; see Fig. 26D). Plasmotomy and schizogony start with a series of nuclear fissions, producing a cell with numerous nuclei which is called a plasmodium. A plasmodium is usually more-or-less rounded (Fig. 2C; see Fig. 22D and 23B), but ribbonlike plasmodia with lined nuclei occur frequently (Larsson et al., 1997a). If the plasmodium splits into smaller parts by a series of successive divisions, the division process is called plasmotomy (Fig. 2C). In a small number

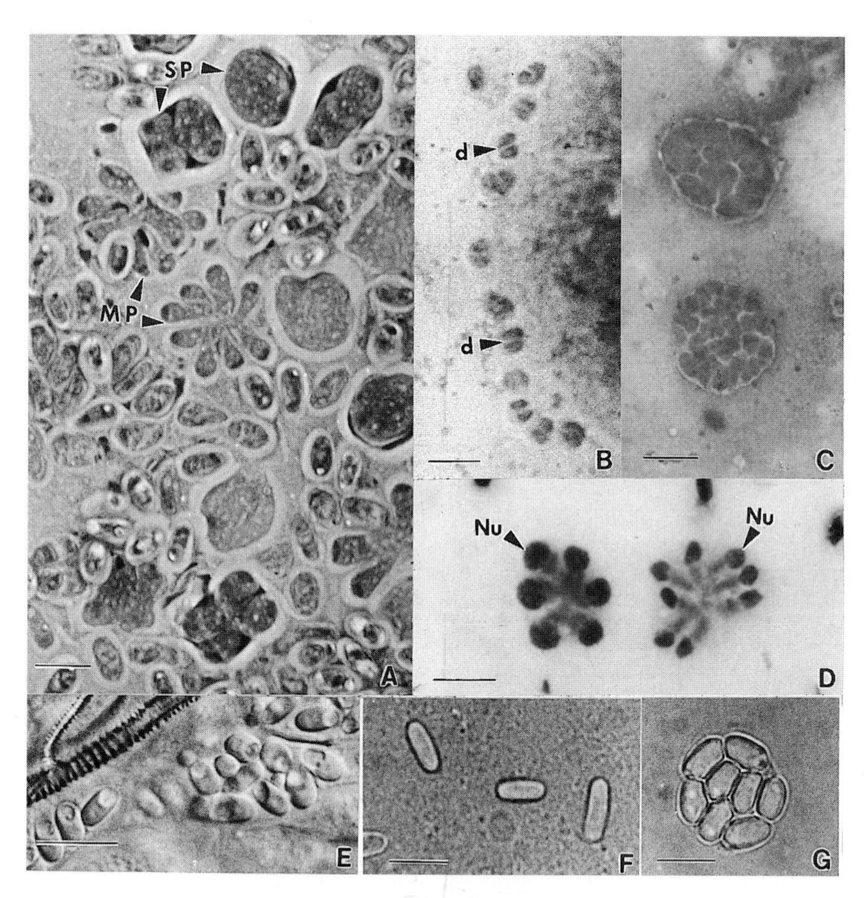

FIGURE 2 Reproduction and spore morphs. (A) Merogony, sporogony, and mature spores (*Systenostrema corethrae*); (B) chain of diplokaryotic merozoites (*Nosema* sp.); (C) two stages of plasmotomy (*Vavraia holocentropi*); (D) two stages of schizogony (*Trichotuzetia guttata*); (E to G) spore morphs of *Vairimorpha* sp. from the gypsy moth *Lymantria dispar* larvae. (E) first-generation spores for intertissue transmission; (F) *Nosema*-type binucleate spores for interhost transmission; (G) *Thelohania*-like uninucleate spores for interhost transmission. (A, B, and D) Giemsa stain; (C) hematoxylin; (E) interference phase contrast; (F and G) phase contrast. Scale bars: A, C, and D to G, 10 μm; B, 5 μm. Reprinted with permission from *Arch. Protistenkd.* (A and D). Abbreviations: d, diplokaryon; MP, merogonial plasmodium; Nu, nucleus; SP, sporogonial plasmodium.

of microsporidia the plasmodium splits endogenously, in that vacuoles are formed inside the plasmodium and the plasmodium is fragmented into cells by their successive growth. This process has been named vacuolation (Beard et al., 1990).

Schizogony proceeds in three stages: the nuclei of the multinucleate plasmodium accumulate at the periphery; lobes are formed at the periphery in the same number as there are nuclei (Fig. 2D); and finally all the lobes bud off simultaneously, each yielding one daughter cell.

Sexual Processes

Sexual reproduction has been reported to occur in a fairly small number of microsporidians. The events are not well known, as they have been described only from static observations with light and electron microscopes. The first well-documented light microscopic observation of karyogamy and meiosis at the transition from merogony to sporogony was made from a mosquito parasite of the genus *Amblyospora* (Hazard and Brookbank, 1984) (see "Microsporidian Nucleus in Mitosis and Meiosis").

This work was followed by the description of a unique type of cell with an anterior nipplelike protrusion (papilla) formed as a doubled, thickened plasma membrane occurring early in merogony (Hazard et al., 1985; Becnel, 1992, 1994; Becnel et al., 1987). The cell was interpreted to be a uninucleate gamete destined for fusion with another such cell to restore by plasmogamy (fusion of cells without nuclear fusion) the diplokaryotic state (with two apposed nuclei) of a meront. In a number of microsporidians sexuality has been deduced from more-or-less clear observations of meiotic chromosomes in ultrathin sections.

Life Cycle Types

Complete life cycles have been elucidated for only a small fraction of microsporidian species. The simplest cycles are completed in one host individual, but the most complex cycles need an alternation of hosts and involvement of more than one host generation. In such cases the respective parts of the life cycles, taking place in different tissues, hosts, or host generations, are structurally different. If two cytologically different kinds of spores (differing not only in size) are produced, the life cycles are considered dimorphic, and with three different kinds of spores, trimorphic. The sporogonial divisions and the cytology of the respective spore morphs may differ to such an extent that without knowledge of the whole cycle, some organisms in different hosts or tissues might be considered to belong to different genera (Fig. 1F and 2E to G).

ESSENTIALS OF MICROSPORIDIAN CYTOLOGY

Our presentation of the microsporidian cytology starts with a review of general cytological features. The specific structures of the merogonial and sporogonial stages and of the mature spores will be dealt with separately.

The Plasma Membrane

The plasma membrane of microsporidians is a unit membrane of classical trilaminar structure about 7 nm thick. In the merogonial stage this membrane may show evidence of active interaction with the host cell cytoplasm in the form of vesicular or tubular projections, increasing the surface area of the parasite (Vávra, 1976a). At a certain moment in the life cycle, the cell membrane becomes the site of deposition of electron-dense material which will later be incorporated into the spore wall or form various kinds of envelopes.

The Nucleus

The microsporidian nucleus is usually round or oval. The size of the nucleus varies during the life cycle. In meronts it is usually several micrometers in diameter, during sporulation its size decreases, and in spores it is usually very small (1 to 2 μm or even less).

The nucleus is of the typical eukaryotic type. It is limited by double unit membranes and separated by a perinuclear space; normally pores are visible (Fig. 3C and E; see Fig. 10F). The nucleoplasm is usually homogeneous. Condensations of chromatin, corresponding to chromosomes, have been reported occasionally. Nucleoli are seldom seen in microsporidian nuclei (Larsson, 1986b) except in stages preceding spore formation (Fig. 3C) (Vávra, 1976a).

NUCLEAR CONFIGURATION

Two types of nuclear configurations exist in microsporidians: monokaryon, an individual nucleus (Fig. 3A), and diplokaryon, two apposed nuclei (Fig. 3B; see Fig. 30B and F and 31A). Some life cycles are monokaryotic throughout, having multiple monokarya in dividing stages, while others are diplokaryotic, having multiple diplokarya in dividing stages. There are also microsporidians that shift from one nuclear type to another in different life cycle stages. In such cases diplokaryotic nuclei are typical of merogony and the transition to sporogony, while monokarya are seen in sporogony (Canning, 1988). Polymorphic species shift between the nuclear configurations when changing hosts or tissues.

Structurally the diplokaryon consists of two nuclei in a coffee bean-like association, each member of the pair having a complete nuclear membrane which, in the area where the two nuclei abut, is flat and devoid of nuclear pores (Vávra, 1976a) (Fig. 3C and D; see Fig. 10F). Both nuclei of the diplokaryon are structurally

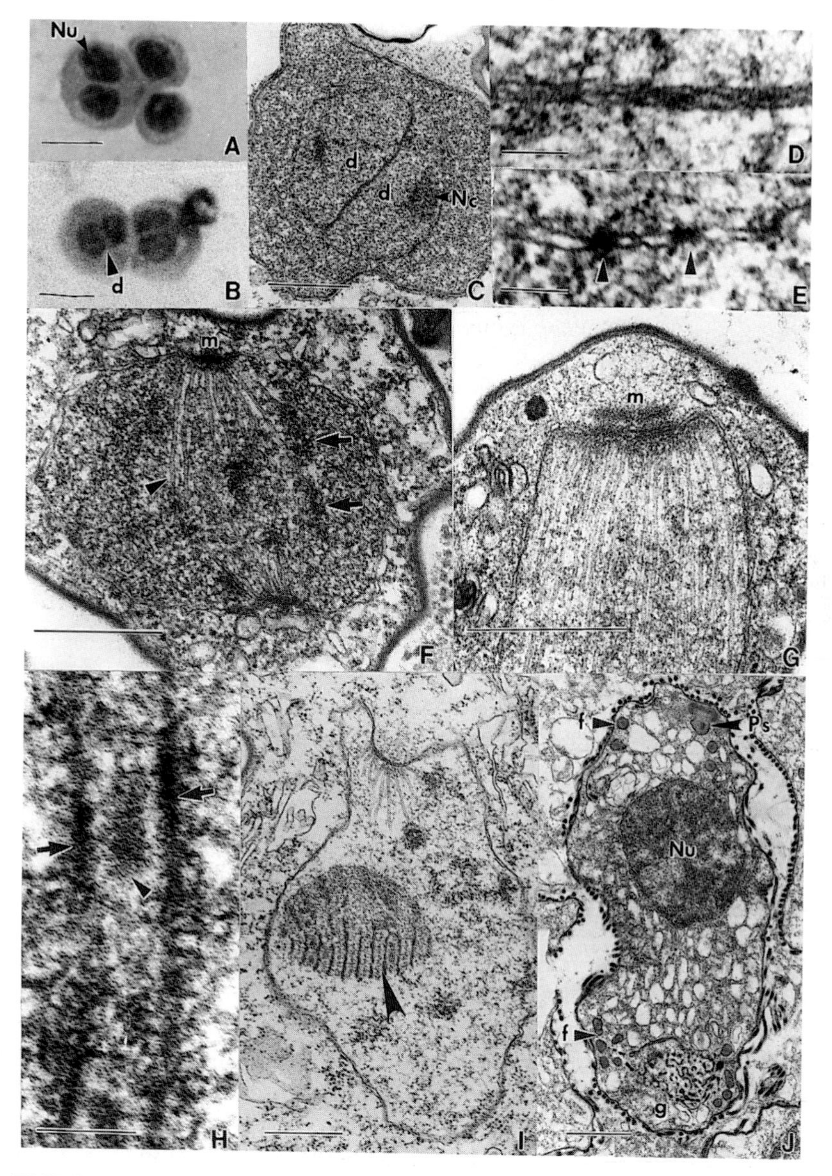

FIGURE 3 Cytology of early stages. (A) Monokaryotic meronts (*Berwaldia schaefernai*); (B) diplokaryotic sporonts (*Nosema tractabile*); (C to E) diplokaryotic sporont (*Helmichia aggregata*) with details of the zone of nuclear apposition (D) and nuclear periphery (E) (arrowheads indicate nuclear pores); (F) mitotic spindle exhibiting microtubules (arrowhead), spindle plaques, polar vesicles, and chromosomes (arrows) (*Toxoglugea variabilis*); (G) two-layered spindle plaque (*Hyalinocysta expilatoria*); (H) synaptonemal complex with lateral (arrows) and central (arrowhead) elements (*H. aggregata*); (I) synaptonemal polycomplex (arrowhead) (*Systenostrema alba*); (J) sporoblast exhibiting Golgi apparatus, primordial anchoring apparatus, and polar filament (*Janacekia debaisieuxi*). (A and B) Giemsa stain, (C to J) TEM. Scale bars: A to C and J, 1 μm; D, E, and H, 100 nm; F, G, and I, 0.5 μm. Reprinted with permission from *Eur. J. Protistol.* (A), *Protistologica* (B to F), *J. Invertebr. Pathol.*, and *Syst. Parasitol.* (I). Abbreviations: d, diplokaryon; f, polar filament; g, Golgi apparatus; m, mitotic spindle plaque; Nu, nucleus; Nc, nucleolus; Ps, polar sac; V, polar vesicle.

identical and divide synchronously (Vávra, 1976b). There is ultrastructural evidence, from some microsporidians having diplokaryotic meronts but monokaryotic sporonts and spores, that the diplokaryon originates from the association of nuclei during a sexual event (copulation of cells considered to be gametes) in the initial phase of the life cycle (Hazard et al., 1985). In some diplokaryotic microsporidians, the monokaryotic state is restored by either nuclear separation or fusion (Sprague et al., 1992). The separation is achieved by progressive dissociation of the twin nuclei (Mitchell and Cali, 1993). The fusion of the two nuclei of the diplokaryon occurs as part of meiosis.

THE MICROSPORIDIAN NUCLEUS IN MITOSIS AND MEIOSIS

Microsporidians express closed intranuclear pleuromitosis (Hollande, 1972; Raikov, 1982), which means that the nuclear membrane is preserved during mitosis and meiosis and that the mitotic spindle is intranuclear. The microtubules of the mitotic spindle attach at one end to the kinetochores of the chromosomes (Fig. 3F) and converge at the other end into a microtubule organizing center, the spindle (or centriolar) plaque (Fig. 3F and G; see Fig. 7B). Some microtubules run from one spindle plaque to the opposite one, and others radiate away from the spindle plaque into the cytoplasm (Desportes, 1976; Desportes-Livage et al., 1996). There is no centriole in the microsporidian cell, and the fine structure of the spindle plaque closely resembles that of yeasts and other ascomycetous fungi (Vávra, 1976a; Desportes and Théodoridès, 1979). Like those of yeasts (Alfa and Hyams, 1990), its components are found both internally and externally to the nuclear envelope. Externally the plaque is a lenticular electron-dense area of more-or-less stratified material situated in a depression of the nuclear envelope. Several (three to six) polar vesicles, electron-opaque vesicles with double membranes, are present close to the spindle plaque, however, their significance is not known. The internal part of the plaque, where the spindle microtubules converge, is an aggregation of electron-dense material on the internal nuclear membrane (Fig. 3G; see Fig. 7B). Some variations of the above-mentioned structure of the spindle plaque were described (e.g., in the appearance and number of strata in the lenticular area). Whether these variations are artifacts of preparation, are true structural characters of different microsporidia, or express different spindle plaque organization in different life cycle events is not known. It should be mentioned that in yeast the structure of the spindle plaque changes during mitosis and meiosis (Zickler and Olson, 1975).

Usually more than two spindle plaques are present in a nucleus during active growth and divisions of the parasite. Supernumerary plaques are ready for the next mitosis (Vávra, 1976b). The new spindle plaques arise by division of the old ones (Larsson, 1986b), and they migrate along the nuclear surface to their final site (Vávra, 1976a). It is not known if spindle plaques persist in some form in the spores, but their presence in young sporoblasts has been reported (Walker and Hinsch, 1972).

Meiosis of microsporidians has been ascertained by both light microscope cytology (Hazard and Brookbank, 1984; Chen and Barr, 1995) and electron microscopy (Loubès et al., 1976; Loubès, 1979). With the latter method, synaptonemal complexes indicating reductional division and representing coupled meiotic chromosomes were seen in a number of microsporidians representing about 20 genera (Loubès, 1979; Larsson, 1986b). Synaptonemal complexes of microsporidians are similar to those of other eukaryotes. The complex is about 100 nm wide, with three parallel dark strands (Fig. 3H): two lateral elements and one central element. Perpendicular to the longitudinal strands is a system of microfibrils.

It was originally thought that synaptonemal complexes appeared in the transition from merogony to sporogony in microsporidia shifting from diplokarya in merogony to monokaryon in the spore (Loubès et al., 1976; Loubès, 1979). Synaptonemal complexes have also been reported from species with isolated nuclei in all life cycle stages, such as *Amphiamblys bhatiellae* (Ormières et al., 1981), and from species with diplokaryotic nuclei throughout, such as *Nosema rivulogammari* (Larsson, 1983b). Of unknown significance are the synaptonemal polycomplexes (Fig. 3I) observed in three species of the genus *Systenostrema* (Larsson, 1988a, unpublished data). However, intranuclear dark material in parallel arrangement is not necessarily a synaptonemal complex, and all observations are not sufficiently documented.

Whether microsporidian meiosis is a special process, or if its course can be explained in a conventional way, has been debated (Hurst, 1993; Flegel and Pasharawipas, 1995). Apparently meiosis is not a general event in the life of microsporidians. Lack of the gene DMC1, which is a marker for meiosis, was reported in *Nosema, Encephalitozoon,* and *Spraguea* (Logsdon et al., 1997).

Endoplasmic Reticulum

Very few membranous elements are present in early merogonial cells, but their number and order of arrangement increase as the life cycle proceeds toward sporogony (Fig. 3C and G). Some of the membranous elements belong to the rough endoplasmic reticulum (ER), as evidenced by their arrangement as parallel cisternae covered by ribosomes. Numerous single-membrane-covered vesicles with lucent interiors exist in microsporidian cells other than spores. Whether they belong to the smooth type of ER or whether they are exocytic or endocytic vesicles cannot be determined with certainty.

Ribosomes

Ribosomes are a predominant component of the microsporidian cytoplasm. The large amount of these elements reveals a very high rate of protein synthesis which is also obvious from the high rate of reproduction. Within the first 48 h after the sporoplasm has reached the host cell, several rounds of division, and even spore formation, may occur.

In early developmental stages ribosomes are dispersed mostly freely in the cytoplasm, while in late merogonial and in sporogonial stages ribosomes attached to lamellae of the ER become prominent (Fig. 3C and G). In sporoblasts and young spores ribosomes arranged in a crystalline pattern, polyribosomes, are frequently observed (Fig. 13).

Ribosomes of microsporidia express prokaryotelike properties, being of the 70S sedimentation type for the monosome with subunits of 50S and 30S. They also lack 5.8S rRNA and, as in prokaryotes, have a region corresponding to part of the 5.8S rRNA in the large subunit (Ishihara and Hayashi, 1968; Curgy et al., 1980; Vossbrinck and Woese, 1986) (Chapter 4).

Golgi Apparatus

In one or several areas of the cytoplasm of presporal stages, accumulations of small, opaque vesicles, enclosed by a single membrane, are present (Fig. 3J; see Fig. 7A and 8A). The vesicles form a meshwork in which the central components are smaller than the marginal vesicles. No ribosomes are present in the area of the vesicles (Vávra, 1976a), and the vesicles are actually embedded in a matrix having greater density than the surrounding cytoplasm. The area of vesicles was called a "primitive Golgi zone" by Vávra (1965) because it does not resemble the classical Golgi apparatus of eukaryotic cells, which is composed of stacked lamellar cisternae. The idea that microsporidians might possess an atypical Golgi apparatus was at first subjected to debate (see arguments in Vávra, 1976a) but presently is universally accepted. It has been further corroborated by histochemical detection of thiamine pyrophosphatase, which is specific for the transmembranes of the Golgi apparatus (Takvorian and Cali, 1994). It is a general feature of the microsporidian Golgi apparatus that it becomes more numerous and more prominent as development progresses toward the formation of spores. In late sporonts and in sporoblasts this area occupies a progressively larger area of the cell and assumes a central role in the elaboration of various spore organelles (Vávra, 1976a) (see Fig. 7A and 14A).

STRUCTURE OF MICROSPORIDIAN DEVELOPMENTAL STAGES

Merogonial and sporogonial stages can usually be distinguished under the light microscope, but the transition from merogony to sporogony is visible only with the use of electron microscopy.

The Spore

Microsporidian development starts and ends with a spore (Fig. 4A and 5A). The spore has a thick wall, with exospore and endospore layers outside the plasma membrane and a unique infection apparatus composed of three parts: a long, threadlike polar filament, which is a multilayered structure; a polaroplast, which is a system of lamellae or sacs in the anterior half of the spore; and a posterior vac-

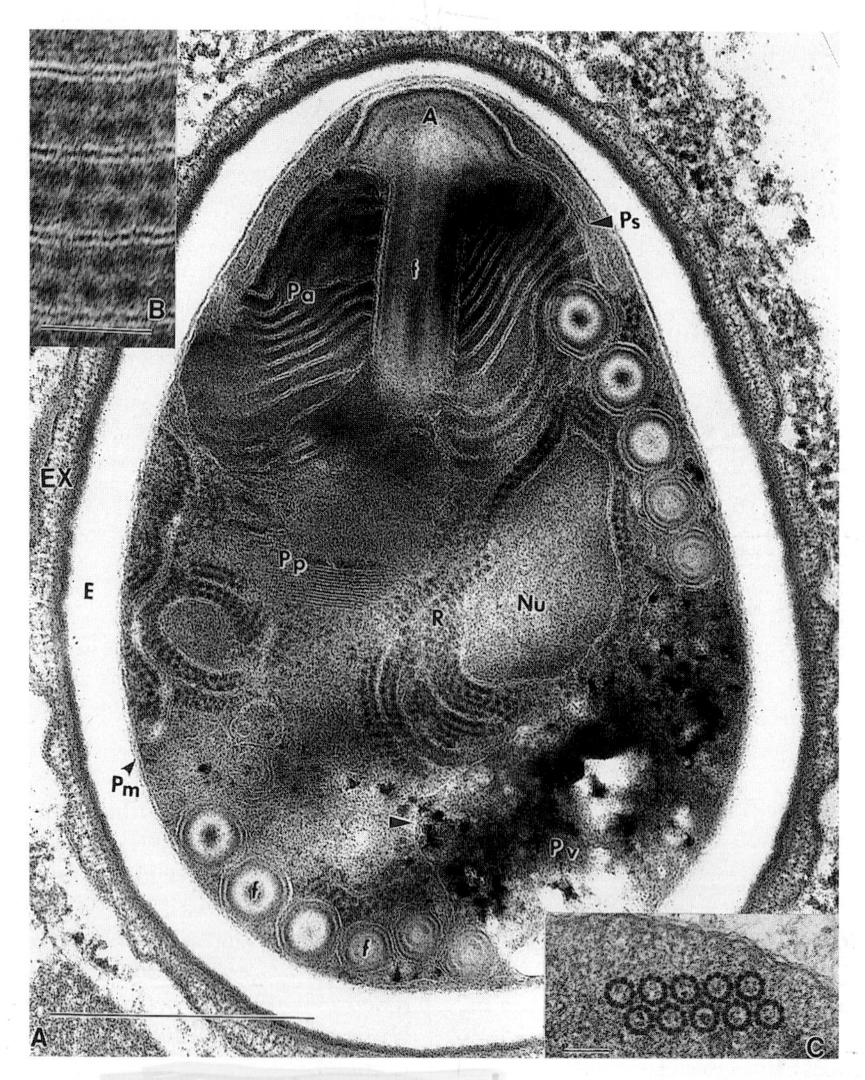

FIGURE 4 Ultrastructure of the mature spore. (A) Longitudinally sectioned spore exhibiting the characteristic organelles (arrowhead indicates vacuolar membrane) (*Episeptum inversum*); (B) polyribosomes attached to membranes (*Napamichum dispersus*); (C) circularly arranged polyribosomes (*Flabelliforma magnivora*). TEM was used for all panels. Scale bars: A, 0.5 μm; B and C, 100 nm. Reprinted with permission from *Arch. Protistenkd.* (A), *Protistologica* (B), and *Acta Protozool.* (C). Abbreviations: A, anchoring disk; E, endospore; EX, exospore; f, polar filament; Nu, nucleus; Pa, anterior polaroplast; Pm, plasma membrane; Pp, posterior polaroplast; Ps, polar sac; Pv, posterior vacuole; R, polyribosomes.

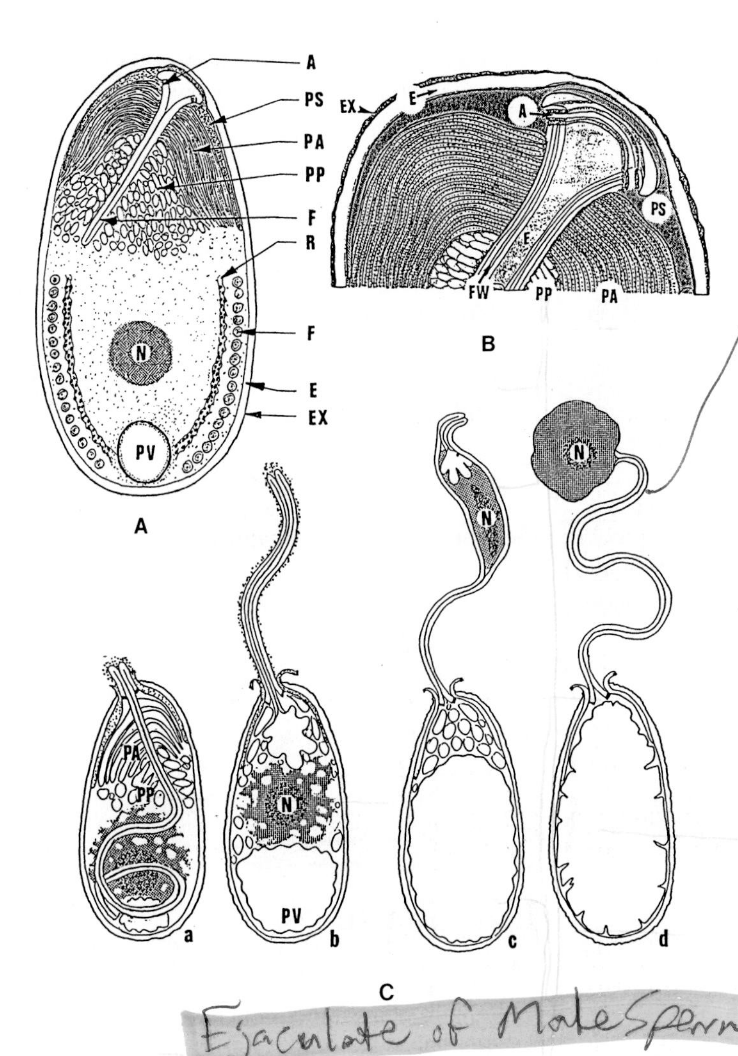

FIGURE 5 (A) Schematic representation of the construction of a microsporidian spore (*Glugea atherinae*). (B) Schematic representation of the anterior part of the spore. (C) Schematic representation of the spore germination: polar filament eversion (a and b), passage of the spore contents (sporoplasm) through the lumen of the everted polar tube (c), and exit of the sporoplasm in the form of a minute cell at the tip of the everted polar tube (d). Modified with permission from Berrebi (1978) (A to C). Abbreviations: A, anchoring disk; E, endospore; EX, exospore; F, polar filament; FW, polar filament wall; N, nucleus; PA, anterior (lamellar) polaroplast; PP, posterior (vesicular) polaroplast; PV, posterior vacuole; PS, polar sac; R, endoplasmic reticulum with (poly)ribosomes.

uole close to the posterior pole of the spore. (For details of the spore structure see "The Mature Spore.")

INFECTION BY SPORE GERMINATION

Infection starts with germination of the spore. The polar filament is triggered by an appropriate stimulus. It everts from the spore and during this process turns inside out, like the finger of a glove, and forms a tube, in an everted state called a polar tube, through which the spore contents are expelled (Fig. 5C; see Fig. 8G and 29J).

The polaroplast is active during germination. It swells, resulting in increased volume of the intralamellar matrix and disturbing the arrangement of membranous lamellae, thereby increasing the turgor within the spore. It is responsible for the initial stage of polar filament evagination (Lom and Vávra, 1963a). In the later stage of germination, when a sporoplasm is being formed, its membranes contribute the plasma membrane of the sporoplasm (Weidner et al., 1984).

The posterior vacuole expands during spore germination and pushes the spore contents into the evaginating filament (Fig. 5C). Spore germination is an osmotic event driven by the splitting of trehalose into glucose (Undeen, 1990; Undeen and Vander Meer, 1994). It seems plausible that the trehalose needed for germination is stored in the posterior vacuole, however, there is no structural proof of such a hypothesis. After the spores have emptied their contents, most of the inner space is filled by the enormously expanded vacuole, which is enclosed by a wrinkled membrane. This membrane is apposed to the plasma membrane that subtends the endospore layer of the spore wall.

The spore content is injected through the polar tube into the cytoplasm of the host cell. It appears at the tip of the tube as a minute cell, the sporoplasm, with one nucleus in some species and with a diplokaryon in other species (Fig. 5C and 6A; see Fig. 8G). Immediately after extrusion, the sporoplasm is connected to the tip of the extruded tube (Weidner, 1972), but after a few seconds it becomes detached from the tube which retracts by elasticity (Lom and Vávra, 1963a). Freshly extruded sporoplasms were reported to be surrounded by two membranes. One of them, which is continuous with the outer sheath of the extruded tube, quickly disappears (Weidner, 1972). The released sporoplasm has an external unit membrane, some vesicles, and a weakly developed ER with ribosomes and approximately equals the cytoplasmic contents of the spore. The external membrane is obtained during passage through the polar tube, evidently from the membranes of the extrusion apparatus (Weidner et al., 1984). In the light microscope the sporoplasm is a minute cell with a cytoplasm stained dark blue if Giemsa staining is used. The nucleus (or diplokaryon) is seen as one or two tiny, intensely stained, red dots.

Meronts and Merozoites

The sporoplasm injected from the spore into a suitable host cell matures into a meront which, after a period of growth and nuclear fission, divides into daughter cells, merozoites. Meronts and merozoites (Fig. 6B) are morphologically identical: rounded or slightly irregular cells covered by a unit membrane, with from one to many nuclear components which, depending on the species, are arranged as single nuclei or as diplokarya. The cytoplasm is homogeneously granular with weakly developed or even non-existent ER and numerous free ribosomes (Fig. 6D). There is one or several areas in which comparatively electron-dense Golgi vesicles are accumulated (the Golgi apparatus). In Giemsa-stained light microscopic preparations the cytoplasm of meronts and merozoites is less densely stained than the sporoplasm, and the nuclei are larger and more distinct.

If several bouts of merogony occur, the only difference in successive merogonial generations is in the increase in the lamellae of ribosome-covered ER.

Transition from Meront to Sporont

At the end of merogony patches of electron-dense material are deposited on the outer face of the plasma membrane (Fig. 6F to H). Secretion of this material is ultrastructural evidence that the parasite has entered the sporogonial phase of reproduction. When the cells are covered by the electron-dense material, they are classified as sporonts, the initial stage of sporogony. The trigger for the switch from merogony to sporogony is not known. Obviously, after a microsporidian cell has entered sporogony it cannot revert to merogony.

Sporogony

Sporogony involves division of the sporont and of its daughter cells until formation of the final division product: the sporoblast (Fig. 3J and 6C; see Fig. 24A, 25C, 26F, 27F, 30E, and 31A). This stage matures and transforms into a spore. Depending on the microsporidian species, the number of sporont divisions varies from one, bisporous sporogony, in which two sporoblasts are formed, to many, polysporous sporogony. In some species sporogony yields a regular number of spores, often eight (octosporous sporogony). Bisporous sporogony proceeds as simple binary fission, and polysporous sporogony can proceed in different ways: plasmotomy, schizogony, or vacuolation. Sporogony can take place in direct contact with the cytoplasm of the host cell, open sporogony, or inside an envelope of microsporidian origin, closed sporogony.

In a stained smear viewed under the light microscope, sporogonial stages are usually less

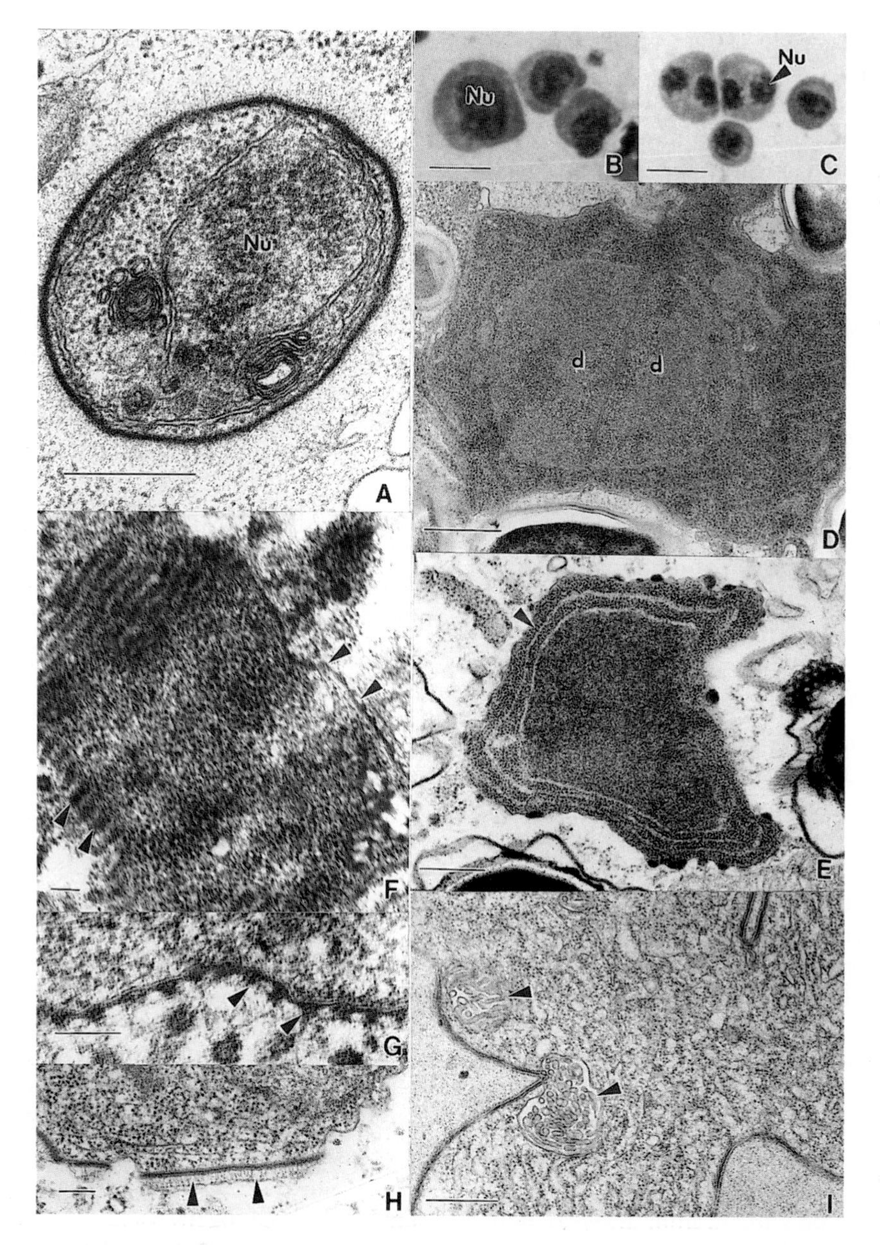

FIGURE 6 From sporoplasm to early sporogony. (A) Sporoplasm (*Ameson michaelis*); (B and C) meronts and sporonts (*Berwaldia schaefernai*); (D) diplokaryotic meront (*Helmichia aggregata*); (E) sporont with electron-dense material (arrows) accumulating spotwise (arrowhead indicates concentric layers of endoplasmic reticulum) (*B. schaefernai*); (F and G) strandlike initiation of the exospore (arrowheads) (*Nosema tractabile*); (H) wide initiation of the exospore (arrowheads) (*Episeptum inversum*); (I) paramural bodies (arrowheads) (*Microsporidium* sp. ex *Daphnia longispina*). (A and D to I) TEM; (B and C) Giemsa stain. Scale bars: B, C, and I, 1 μm; A, D, and E, 0.5 μm; F to H, 100 nm. Reprinted with permission from E. Weidner (A), *Eur. J. Protistol.* (B, C, and E), *Protistologica* (F and G), and *Arch. Protistenkd.* (H). Abbreviations: d, diplokaryon; Nu, nucleus.

intensely stained than merogonial stages, and their nuclei are smaller (Fig. 6B and C). In ultrathin sections the cytoplasm is less electron-dense and less packed with free ribosomes but contains more membranous elements in the form of vesicles and cisternae of rough ER. The cisternae are characteristically arranged in concentric layers around the nucleus (Fig. 6E). The cluster of small vesicles representing the Golgi apparatus becomes more prominent in the sporont.

THE ELECTRON-DENSE MATERIAL ON THE SPORONT SURFACE

The appearance and progressive accumulation of electron-dense material on the outer face of the plasma membrane of the sporont (Fig. 6E to H) indicates that the microsporidian cell is entering sporogony. Depending on the species, this material has the potential to develop in one of three directions.

1. It may immediately become the primordium of the exospore layer of the spore wall, which occurs frequently. The dense material on the sporont plasma membrane gradually covers the cell as a uniform layer which divides with the parasite and becomes the spore wall. The wall material is in direct contact with the cytoplasm matrix of the host cell in open sporogony. The first signs of a future exospore layer look different depending on species. In *Encephalitozoon cuniculi* (Barker, 1975), *Nosema tractabile* (Larsson, 1981c), and *Unikaryon slaptonleyi* (Canning et al., 1983), the primordial exospore material is laid down as uniform electron-dense strands in a parallel arrangement (Fig. 6F and G). More commonly the material is spread from a few foci over wide surface areas (Fig. 6H). Either the exospore primordium grows into an exospore of uniform electron-dense material (see Fig. 9A) (as, for example, in *Loma fontinalis* [Morrison and Sprague, 1981]) or it develops into layers of different electron density and thickness (see Fig. 9B and C). In some microsporidia the electron-dense material jointly envelops two spores and ensures their coupling (Fig. 1L) (Codreanu and Vávra, 1970; Larsson, 1981a; Vávra, 1984) (see "Spore Shape").

2. It may form a thin, membranous sac, a sporophorous vesicle, enclosing the sporont and its division products, including spores (see Fig. 28C and 29B, E, G, and H). In this case the exospore is formed secondarily from a second production of electron-dense material. This mode also occurs frequently (see "Sporophorous Vesicles").

3. It may form a cystlike envelope together with the plasma membrane, a sporont-derived sac (see Fig. 20). The electron-dense material of the spore wall is formed secondarily. This is the most unusual fate of the dense material (see "Sporont-Derived Sacs").

ABERRANT SPORONTS

If at the point at which the appearance and amount of electron-dense material on the sporont surface are considered, there are in a small number of microsporidia two situations in which sporonts cannot be recognized as a stage with electron-dense material on its surface:

1. In microsporidians of the family Pleistophoridae (*Pleistophora, Trachipleistophora,* and *Vavraia*), the dense material is already present on the meronts (see Fig. 18G and 28A).

2. There are also a few microsporidians, e.g., *E. bieneusi,* with a small amount and a delayed production of electron-dense material, which means that their sporogonial plasmodia have little or no reinforcement of the plasma membrane until the moment when the sporoblasts are formed (Cali and Owen, 1990) (see Fig. 22 to 24).

PARAMURAL BODIES

When a multinucleate sporogonial stage divides into unicellular daughter cells, cell separation is achieved by formation of a groove in the plasmodium which deepens progressively until complete separation of a daughter cell is achieved. The electron-dense material on the surface of the sporogonial plasmodium does not cover the bottom of the groove. At this site a whorl of plasma membrane is sometimes observed (Fig. 6I). It probably represents the site where a new plasma membrane, required to cover the increasing cell surface, is being formed. The membranous whorl was originally called a scindosome ("a body occurring at a cleaving site") (Vávra, 1975, 1976a), but this name was later replaced by paramural body ("a body occurring close to the wall") because of its structural and probably also functional resemblance to similar structures in fungi (Vávra, 1976c).

The Sporoblast

The transition of the sporont to sporoblasts is a continuous process during which the nuclei divide, electron-dense material accumulates on the plasma membrane to form a thick, continuous surface cover, and the extrusion apparatus starts to develop. The sporoblast is the cell, which gradually assumes the shape of the spore and continues to build the wall of the future spore. The specific organelles of the spore start to assume their final form in the sporoblast (Fig. 3J; see Fig. 25). In ultrastructural studies young sporoblasts appear as wrinkled cells of considerable electron density in which it is difficult to observe cell organelles. The wrinkling at a certain stage of sporoblast maturation is probably due to increased osmotic sensitivity of the cell. The sporoblast is enclosed by a plasma membrane covered by an almost continuous layer of electron-dense material which will develop into the exospore layer of the future spore. In microsporidia with a closed type of sporogony, this layer frequently forms tubular or fibrillar expansions protruding from the surface of the sporoblast.

Even if the transition from sporoblast to spore is a continuous process, for terminological reasons it is useful to agree on a point at which a cell is no longer called a sporoblast but a spore. Larsson (1986b) considers polarization of a cell the turning point: after the cell initiates polarization (cell organelles are no longer randomly distributed), it is designated a spore.

The Origin of the Extrusion Apparatus

It is more or less universally agreed that the Golgi apparatus is the organelle forming the individual parts of the extrusion apparatus (Fig. 3J and 7A). However, as the Golgi apparatus and the ER are parts of the same membrane system, some authors claim that the ER generates the extrusion apparatus (Vávra, 1976a; Desportes-Livage et al., 1996). The location of nucleoside diphosphatase, a histochemical marker for ER and, in some cells, for the outermost cis-Golgi membrane, suggests that the rough ER and the Golgi apparatus may in fact have overlapping functions in microsporidia (Takvorian and Cali, 1996).

The first traces of the extrusion apparatus may appear in the form of dilated distal ends of ER cisternae containing electron-dense material. This material is probably exported to the Golgi apparatus, and fusion of its vesicles results in formation of the extrusion apparatus (Fig. 7A and 8A).

THE POLAR FILAMENT

The first traces of the future polar filament appear in some cases in life cycle stages preceding the sporoblasts (Fig. 7A). The first sign is an accumulation of electron-dense Golgi vesicles in an area immediately adjacent to the nucleus (Fig. 7A, part a). Next, a small, round vacuole containing some dense granular material appears at the same site. The vacuole is surrounded at most of its circumference by a unit membrane, but the remaining part not covered by the membrane is in direct contact with a large reticulum of anastomosing Golgi vesicles. The vacuole represents the future polar sac (Vinckier, 1973, 1975, 1990; Vávra, 1976a) (Fig. 7A, part b). The polar filament appears initially as rings with thick, dense borders and a dark center situated at the edge of the Golgi reticulum. These rings are in fact cross sections of a coiled tube (Fig. 7A, part c). The outer limiting border (a membrane?) of the tube is continuous with the vacuole borders, and the central dense material of the tube penetrates some distance into the polar sac. This penetrating material is the primordium of the anchoring disk (Fig. 7A, part d). The reticulum of the Golgi vesicles is progressively displaced away from the nucleus toward the future posterior end of the spore, and during this process more filament coils are formed at its edges (Fig. 7A, parts d and e). When the Golgi vesicle reaches the posterior end of the future

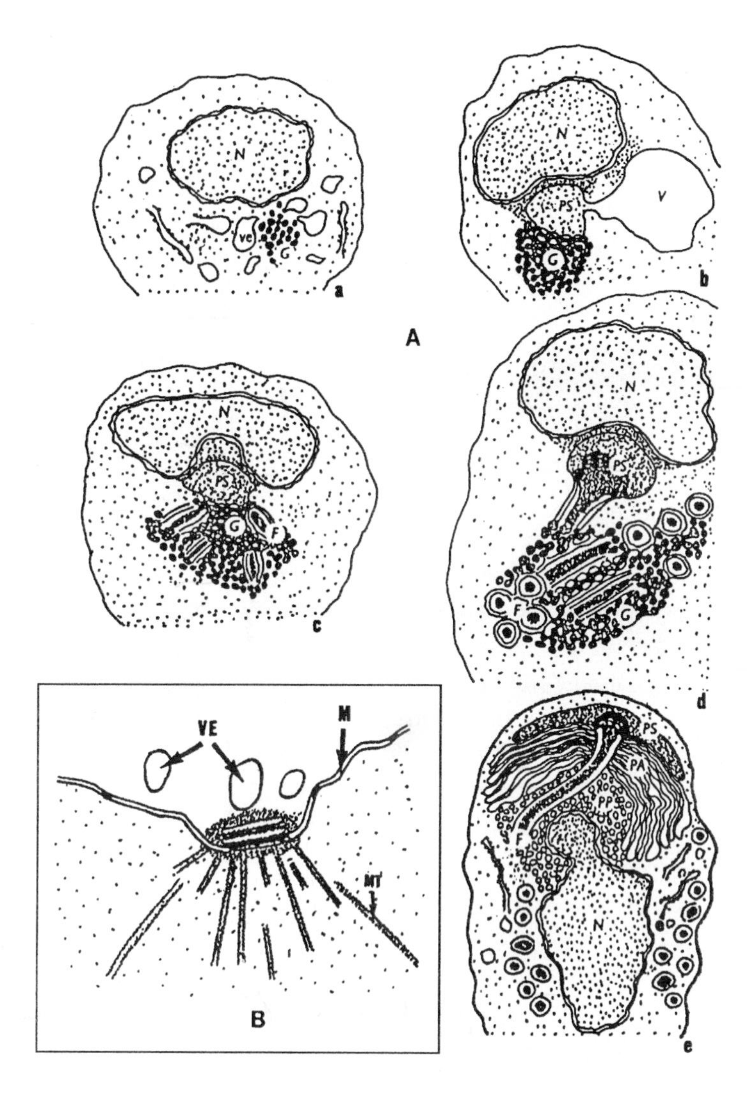

FIGURE 7 (A) Schematic representation of the formation of the extrusion apparatus during sporont-to-spore transition (*Nosemoides vivieri*). (a) A small vacuole (ve) and Golgi vesicles (G) assemble close to the nucleus; (b) the polar sac primordium (PS) and a large vacuole appear close to the nucleus, and the Golgi (G) abuts the PS; (c) coils of the polar filament (F) start to form by coalescence of Golgi vesicles; (d) the polar sac and polar filament material start to be organized into layers; (e) the polar sac migrates toward the tip of the forming spore, and the polaroplast with both parts (PA, PP) appears. (B) Schematic representation of the spindle plaque of microsporidia (*Glugea atherinae*). The plaque (an MTOC) is situated in a depression of nuclear membrane (M) and consists of a stack of flattened electron-lucent vesicles surrounded by dense granular material from which microtubules (MT) radiate. Several vesicles (VE) are situated close to the plaque on its cytoplasmic side. Modified with permission from Vinckier (1973) (A) and Berrebi (1978) (B).

spore, the polar filament formation ceases and a vacuolar space forms with the remnants of the reticulum inside. This mode of formation was confirmed by the freeze-fracture technique (Vinckier et al., 1993; Vinckier, 1990). The Golgi apparatus at this time is a large, spongiform structure occupying a considerable part of the cell volume (Fig. 14A).

In microsporidians of the genera *Enterocytozoon* (Cali and Owen, 1990; Desportes-Livage et al., 1996) and *Nucleospora* (Desportes-Livage et al., 1996), the electron-dense material of the future polar filament appears first in the lumen of cisternae of the ER and even in the perinuclear cisternae of *Enterocytozoon* from which it is probably exported to a small Golgi apparatus situated close to the vesicular terminals of the ER. This is why the ER of *Enterocytozoon* is claimed to be the origin of the polar filament (Hilmarsdottir et al., 1993). The filament material then appears as round or flattened stacks of cylinders with electron-dense walls and less dense centers. These cylinders later join to form coils (see Fig. 22D and 23B to D).

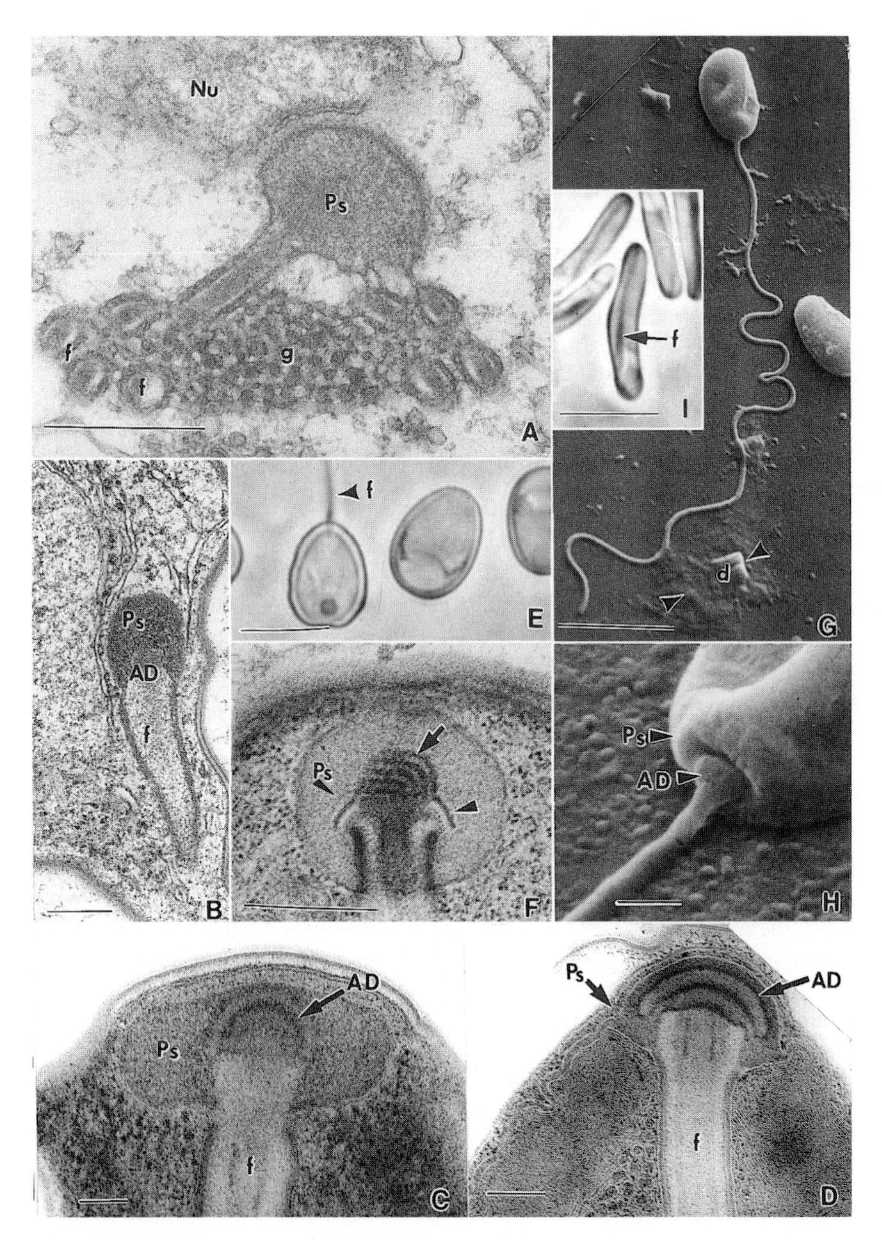

FIGURE 8 Initiation of the extrusion apparatus and ejection. (A) Initiation of the polar filament and polar sac (*Nosemoides vivieri*). Note the close association of the sac with the nucleus. (B to D) polar sac and anchoring disk in sporoblast (B), immature spore (C), and mature spore (D). (B and D) *Systenostrema alba;* (C) *S. candida.* (E) Polar filament seen in situ in one spore and ejected in the other (*Amblyospora* sp. ex *Aedes* sp.). (F) Polar sac and anchoring disk during formation (*Amblyospora culicis*). Note the layering of the nascent anchoring disk (arrow) and the presence of layered material to which the walls of the polar filament (arrowhead) abut. This structure serves as a "hinge" during filament and turns inside up. (G) Extruded polar tube and ejected sporoplasm (arrowhead) (*Nosema tractabile*). (H) Details of the anterior pole of the spore, the polar sac, and the penetrated anchoring disk are visible (*N. tractabile*). (I) polar filament of the manubrium type (*Mrazekia cyclopis*). (A to D and F) TEM; (E) phase contrast; (G and H) SEM; (I) fresh, unstained. Scale bars: A and F, 0.5 μm; B, 25 nm; C and D, 100 nm; E, G, and I, 5 μm; H, 0.5 μm. Reprinted with permission from D. Vinckier and *J. Protozool.* (A), *Syst. Parasitol.* (B to D), *Can. J. Zool.* (F), and B. S. Toguebaye and *Arch. Protistenkd.* (G and H). Abbreviations: AD, anchoring disk; d, diplokaryon; f, polar filament; g, Golgi apparatus; Nu, nucleus; Pa, anterior polaroplast; Ps, polar sac.

THE POLAROPLAST

Because the polaroplast is a delicate membranous structure, the earliest signs of its appearance are easily overlooked. There are divergent opinions about its origin, and it is possible that the time and mode of initiation vary among different microsporidia. In any case there is a close association between the polaroplast and the polar filament, which has led to speculation that both these organelles are parts of the same polar filament-polaroplast membrane complex (Lom and Corliss, 1967).

One mode of formation presumes that the polaroplast lamellae originate from progressive infolding of the membrane surrounding the vacuole, which occurs in some microsporidia (e.g., *Nosemoides vivieri*) (Vinckier, 1973, 1975) close to the nucleus during polar sac and polar filament formation (Vávra, 1976a) (Fig. 11A, parts b and c).

Other observations, e.g., from *Amblyospora culicis* (Toguebaye and Marchand, 1986) and *Bacillidium criodrili* (Larsson, 1994a), suggest that the polaroplast is initiated at a later stage in the maturing sporoblast (Fig. 8B to D). When the growing polar filament reaches the anterior pole of the spore, the first signs of the polaroplast begin to appear. The primordia are small, membrane-lined protuberances from the unit membrane cover of the filament and are first seen in the proximity of the polar sac. They widen to become membrane-lined sacs, new compartments are added in the posterior direction, and finally the compartments become packed in various arrangements.

THE POSTERIOR VACUOLE

The posterior vacuole is fully formed only in the mature spore (Fig. 4 and 5A). Its primordium appears in sporoblasts in the form of a meshwork of vesicles and tubules of the Golgi apparatus located in the area where the vacuole later appears (Vinckier, 1975). Remnants of the meshwork appear as an inclusion body, the posterosome, in the posterior vacuole of young spores of some microsporidia (Weiser and Žižka, 1974, 1975) (see "The Posterior Vacuole").

THE MATURE SPORE

Shape and size are important structural characters, as they tend to be rather constant and characteristic of respective genera and species. Other situations, however, have to be considered when the spore shape and size of an individual microsporidian are evaluated. Most microsporidians produce spores of a single type having a similar size and shape. Some microsporidians, however, produce spores of several types (Fig. 1G and 2E to G). These spores may simply differ in size (microspores, macrospores) but in other cases may differ in size, shape, and cytology. Spores are produced either within the same infected host and tissue (Pilley, 1976) or within different tissues of the same host or even in tissues of a different host (Becnel, 1994). So far, three is the maximum number of cytologically different spore types observed to be formed by a single species. Shape and size are best evaluated using a light microscope, but to be able to correctly evaluate the construction of the spore, it is necessary to make ultrathin sections and examine them by transmission electron microscopy (TEM). A longitudinal section taken through the center of the spore reveals a unique, complex organization, the principal parts of which are treated in greater detail under separate headings.

Spore Shape

The spore shape (Fig. 1) is more or less constant for all species of a genus. Spores are most commonly oval, as in the genera *Enterocytozoon* (see Fig. 22D and 24C) and *Nosema* (Fig. 1O), or pyriform, as in *Gurleya* (Fig. 1M) and *Agglomerata.* Spherical spores are, for example, found in the genera Pilosporella and *Chytridiopsis,* and rodlike spores are characteristic of several genera. Spores of *Bacillidium, Helmichia,* and *Mrazekia* are stout rods (Fig. 1D and 8I; see Fig. 11C to E), *Cylindrospora* spores are slender rods (Larsson, 1986c), and the spores of *Nadelspora* are almost filiform (Olson et al., 1994). Some genera produce spores of unique or nearly unique shape for the genus. The spores of *Toxoglugea* (Fig. 1B) and *Toxospora* are bent like horseshoes, *Campanulospora* spores are bell-shaped, and *Cougourdella* spores are lageniform (Fig. 1E).

In some microsporidian genera spores are joined in pairs, completely embedded in an amorphous substance as in *Telomyxa glugeiformis* (see Fig. 17G to J), or cemented together by patches of electron-dense material as in *Berwaldia singularis* (Larsson, 1981b), *Norlevinea daphniae* (Fig. 1L) (Vávra, 1984), and *Microsporidium fluviatilis* (Voronin, 1994). In *Telomyxa* the coupling is so intimate that the couple was for a long time thought to be a single spore (Codreanu and Vávra, 1970).

Spore Size

Even if all the spores produced belong to the same size class, a number of genera produce a small number of macrospores together with the more common microspores (see Fig. 17F). This sporogony is characteristic of the polysporoblastic genera *Pleistophora* (Canning and Nicholas, 1980) and *Vavraia* (Larsson, 1986d) and is also frequently observed in *Amblyospora* (Hazard and Oldacre, 1975). *Hyalinocysta expilatoria* produces spores of three size classes: octosporous microspores, tetraspores of intermediate size, and diplospores of the largest size class (Larsson, 1983c).

The Spore Wall

The spore wall is built in such a way that for some time it withstands osmotic pressure during spore germination but ruptures at a certain moment. The rupture occurs at the spore apex where the spore wall is thin, creating an exit for the polar filament. The wall consists of two layers external to the plasma membrane.

THE EXOSPORE

The exospore is the surface layer (Fig. 4 and 9). It is electron-dense and forms a coat of uniform thickness all around the spore. Its thickness and construction vary in different microsporidia from a thin, dense, unstratified layer about 10 nm to a complex, multilayered structure about 200 nm thick (Larsson, 1986b). High resolution by TEM must be used in order to appreciate exospore structure. For example, in the genus *Encephalitozoon,* the exospore has been described as a homogeneous layer (Barker, 1975), but high resolution shows that it actually consists of three layers of different electron density and structure (Bigliardi et al., 1996). Freeze-fracture of the exospore of *Amblyospora* spp. shows it as a granulofibrillar stratum in which short, coarse fibers are scattered among abundant granules (Vávra et al., 1986) (Fig. 10A). Bigliardi et al. (1996) recognized in freeze-fractured and freeze-etched exospores of *Encephalitozoon hellem* two layers in which 4-nm fibrils were arranged in different spatial orientations. The chemical composition of the exospore is proteinaceous, and 13 amino acids and sulfhydryl proteins were identified in this layer (Vávra, 1976a; Maurand and Loubès, 1973; Vivarès et al., 1976).

The exospore of some microsporidia differentiates rapidly, while in others it becomes more complex through a series of stages and attains its final structure only in the mature spore. The sequence of layers is a useful tool in the identification of microsporidia (Larsson, 1986b). A number of genera have exospores of a unique type, such as *Napamichum* (Fig. 9C) and *Episeptum* (Larsson, 1986e, 1996). The ultrastructure of the exospore is probably an indicator of relationships. For example, in the genera of Thelohaniidae and Amblyosporidae, the exospore of one of the spores types (the octospore) typically has a basic sequence of layers, one of the intermediate layers resembling a double membrane. In addition each genus might have more-or-less complex layering (Fig. 9B and C). The genera of the family Mrazekiidae share an exospore having a wide, uniform basal layer and an external double-membrane-like stratum (see Fig. 11 and 20A to C). In microsporidia from aquatic hosts the exospore frequently forms ornamentations in the form of appendages, fibers, spore tails, etc. (Vávra, 1963), resulting in exospores of a unique type for the genus. The exospore of *Caudospora* spp. is enormously hypertrophied

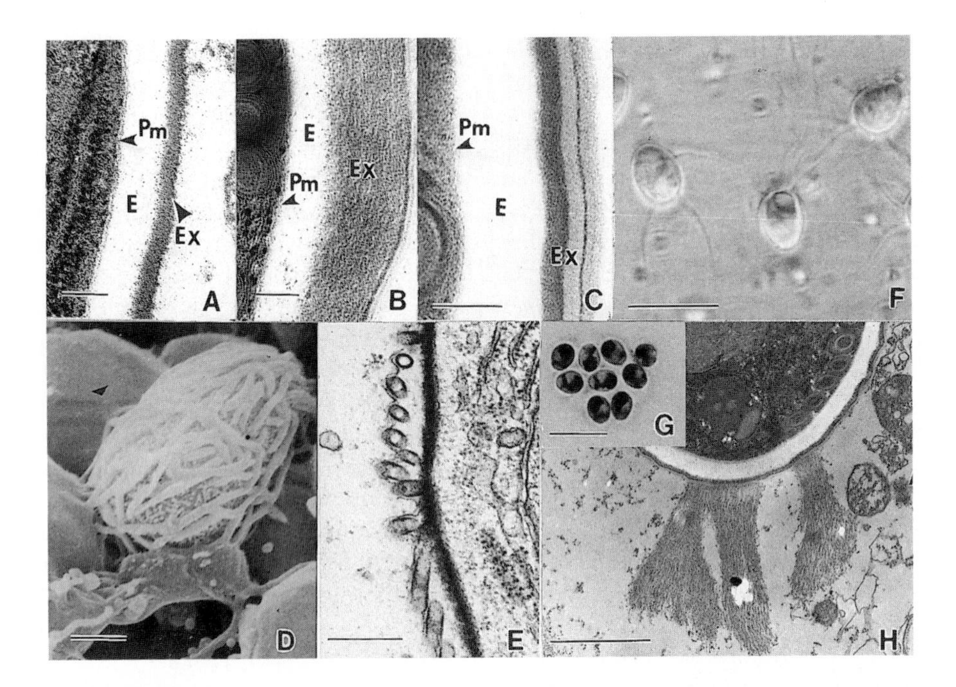

FIGURE 9 The spore wall. (A) Uniform exospore (free spore) (*Amblyospora callosa*); (B) spore wall of *Amblyospora* type (octospore) (*A. callosa*); (C) layered exospore (*Napamichum dispersus*); (D and E) exospore with tubular projections (*Janacekia debaisieuxi*); (F to H) exospore with filamentous projections, invisible in stained smears (G) (*Trichoctosporea pygopellita*). (A to C, E, and H) TEM; (D) SEM; (F) interference phase contrast; (G) Giemsa stain. Scale bars: A to C, 100 nm; D, 1 μm; E, 0.5 μm; F to H, 10 μm. Reprinted with permission from M. Rausch (D) and *Arch. Protistenkd.* (F to H). Abbreviations: E, endospore; EX, exospore; Pm, plasma membrane.

and forms a winglike and taillike expansion, giving the spore the appearance of a human sperm which can be discerned in light microscopic preparations (Fig. 1H). The often long tails of *Jirovecia* spp. are also distinct with light microscopy (Lom, 1958; Larsson, 1989b, 1990b). Tubular exospore projections make the spores of the genera *Hirsutosporos* and *Inodosporus* easy to identify. In *Hirsutosporos* the tubular projections form a girdle and a posterior tuft (Batson, 1983), whereas in *Inodosporus* they are arranged as a small anterior fork and three or four long posterior appendages (Overstreet and Weidner, 1974). The filamentous spore projections of *Trichoctosporea pygopellita* are easily seen with interference phase-contrast optics but are invisible with other light microscopic techniques (Fig. 9F to H). They are best observed by TEM, as are the tubular exospore projections of *Janacekia* spores (Vávra, 1965; Loubès and Maurand, 1976; Rausch and Grunewald, 1980) (Fig. 9D and E). In this category are mucous envelopes which are present on the spores of some aquatic microsporidia. When the spore comes in contact with water, the mucus swells and provides it with buoyancy (Lom and Vávra, 1963b). The presence of a mucous layer is revealed in the light microscope by mixing spores in suspension with India ink (see "Demonstration of Exospore Ornamentations and Mucocalyx") (Fig. 1M). In negative-stained TEM preparations the mucus has a fine, fibrous structure (Vávra, 1976a).

THE ENDOSPORE

The endospore is electron-transparent and hence appears as a structureless layer below the exospore (Fig. 4 and 9). This layer is 100 nm or

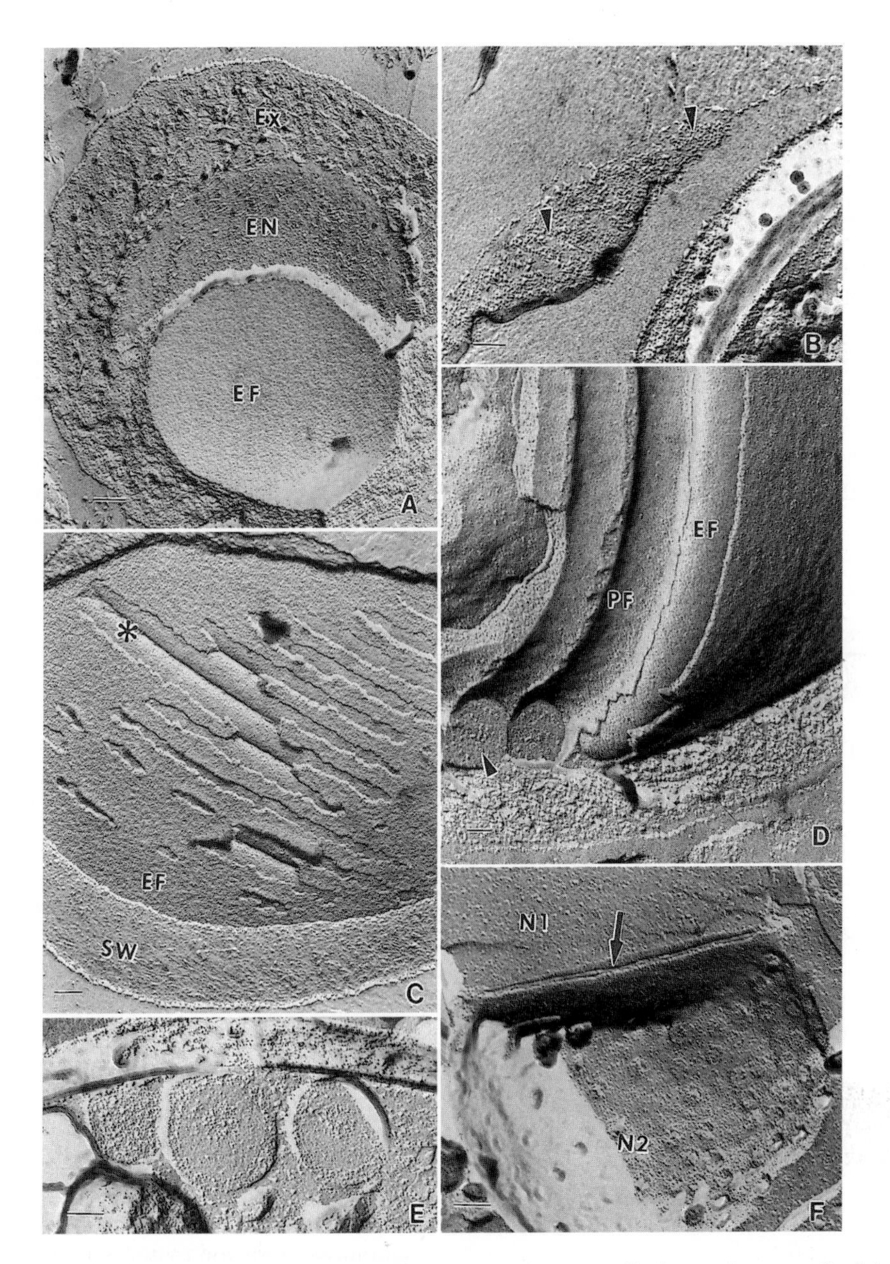

FIGURE 10 Freeze-fracture of microsporidian spores. (A) Tangentially fractured spore wall of *Amblyospora opacita,* showing the ectoplasmic face of the plasma membrane, the endospore with fibers, and the exospore with coarse fibers and granules. (B) Fracture through the sporophorous vesicle membrane of *A. opacita* (arrowheads), showing irregular granular structure demonstrating that it is not a cytomembrane. (C) Tangential fracture through immature spore of a *Tuzetia* sp., showing the ectoplasmic face of the plasma membrane covered with intramembranous particles (IP) and intramembranous particles on the ectoplasmic face of the unit membrane enveloping the polar filament (asterisk). (D) Fractured spore of *Amblyospora varians,* showing cross-fractured polar filament coils (arrowhead) and the protoplasmic and ectoplasmic faces of the unit membrane ensheathing the polar filament. In mature spores the hemimembranes of the filament are nearly devoid of intramembranous particles. (E) Cross fractured polar filament. Note its solidlike appearance with an outer ring and a central spot of fine particles. (F) Fractured diplokaryon of *Ambylospora bracteata.* One nucleus (N1) is cross-fractured. The second nucleus (N2) shows pores in the nuclear membrane. Arrow indicates the apposed nuclear membranes. Scale bars: A to F, 100 nm. Abbreviations: EN, endospore; Ex, exospore; N, nucleus, SW, spore wall; EF and PF, respective ectoplasmic and protoplasmic faces of a split unit membrane. Panels A and B are reprinted with permission from *Protistologica.*

less thick. The thickness is uniform except at the apex of the spore, where it is considerably more narrow, often half the size or less. In freeze-fracture preparations the endospore of *Amblyospora* spp. is granulofibrillar with short fibers preferentially oriented parallel to the spore surface (Vávra et al., 1986) (Fig. 8A). The material of the endospore is α-chitin, confirmed by staining reactions and physical analyses using infrared spectroscopy and X-ray diffraction (reviewed in Vávra, 1976a). The presence of chitin can be utilized for fluorescent detection of microsporidia (Vávra and Chalupsky, 1982; Van Gool et al., 1993; Vávra et al., 1993a) (for details, see "Chitin Detection").

The endospore is formed late during spore maturation, and the layer grows until the spore is completely mature. The late completion of the endospore is probably a consequence of the fact that it seals the spore, separating the microsporidian from the surrounding host cytoplasm on which it is metabolically dependent until the spore matures.

The Plasma Membrane

The internal surface of the endospore is lined with the plasma membrane (Fig. 4 and 9A to C). This membrane delimits the cytoplasmic contents of the spore against the wall and is usually considered part of it because it remains in situ in the emptied spore (Lom, 1972; Undeen and Frixione, 1991; Weidner, 1972). The sporoplasm acquires a new plasma membrane (believed to be derived from the polaroplast) when it leaves the spore.

Freeze-fracture reveals the plasma membrane to be a classical cytomembrane containing intramembranous particles (Vávra et al., 1986; Bigliardi et al., 1996). These particles are very numerous in young spores but have nearly disappeared in mature spores. The intramembranous particles are interpreted as transport proteins which can be expected to decrease during spore maturation (Vávra et al., 1986) (Fig. 10A and C; see Fig. 14D).

It must be stressed that the microsporidian spore is a single cell. The only plasma membrane present in the spore is the membrane that delimits the spore cytoplasm from the spore wall. There is no membrane-delimited "germ" inside the spore.

The Nucleus

One or two typical eukaryotic nuclei are found in the center of the spore (Fig. 4 and 5A). In spores with double nuclei they are coupled as diplokarya (Fig. 1I) with one exception. In the binucleate spores of *Binucleospora elongata,* each nucleus is enveloped in multiple membranes, disturbing the normal diplokaryotic association (Bronnvall and Larsson, 1995a). The shape of the nuclei is spherical to ovoid; exceptionally it is horseshoe-shaped like the nucleus of *Metchnikovella hovassei* (Vivier and Schrével, 1973).

Cytoplasmic Organelles

The cytoplasm of the spore contains ER often seen as strands of polyribosomes, especially around the nucleus in young spores. Either polyribosomes look like double rows of ribosomes sandwiched between membranes (Fig. 4A and B) or no membranes at all are visible. Usually there are no prominent differences between transversely and longitudinally sectioned polyribosomes. However, it has occasionally been seen, e.g., in *Heterosporis finki* (Schubert, 1969), *Trachipleistophora anthropophthera* (Vávra et al., 1998), and *Flabelliforma magnivora* (Larsson et al., 1998) (Fig. 4C), that the components of transversely sectioned polyribosomes are arranged in circular configurations with nine ribosomes in each circle. In this arrangement no membranes are visible.

The Golgi reticulum is prominent in young spores as long as the infection apparatus still has not attained its final shape.

The Polar Filament

The polar filament is the most conspicuous structure in the mature spore. Chapter 7 covers its biochemical and functional aspects, while data on its mode of formation, structure, and diversity are presented here.

POLAR FILAMENT TYPES

In most microsporidia the filament is thread-like and is usually much longer than the spore. Commonly three regions are visible: a straight anterior part (sometimes called the manubrial portion of the filament) usually about one-third of the spore length, a section where the filament bends to the side and almost touches the spore wall, and finally one or more filament coils (Fig. 4 and 5A). With few exceptions, the greatest part of the filament is coiled in the posterior half of the spore.

There are several modifications, however, in the polar filament arrangement inside the spore.

A small number of microsporidians (*Chytridiopsis* and related genera) have a short, thread-like polar filament (see Fig. 15C). There is no straight part, which means that the complete filament is arranged in a few coils at one pole of the spore.

Exceptionally, the polar filament is shorter than the spore in length, appearing as a straight rod. Such filaments are characteristic of a small number of genera such as *Baculea* (Loubès and Akbarieh, 1978), *Cylindrospora* (see Fig. 26A), *Helmichia* (Larsson, 1982) and *Scipionospora* (Bylén and Larsson, 1996). These filaments are of the usual construction, and the only difference in filaments of related genera is the shorter length.

Some microsporidia have a thickened filament. The degree of thickening is variable and can involve only the straight, apical part of the filament or also some filament coils. In some genera the thickened filaments have a very reduced coiled portion. (For details, see "Polar Filament Descriptors," item 4.)

POLAR FILAMENT DESCRIPTORS

Despite the basic similarity of its structure, there are several features of the polar filament that vary and thus are useful for characterization of individual microsporidia.

In evaluation of the polar filament it is necessary to study mature spores. The filament grows in length until the spore is completely mature, new coils are added up to that time, and immature coils are narrower than mature ones. In a nearly mature spore the last one or two coils, which are the most immature, are often narrower than the mature coils and their internal organization is less complex.

1. Polar filament types, as mentioned in "Polar filament types" above, are the basic characters used in the characterization of microsporidia at the genus or even the supra-generic level.

2. Number and arrangement of coils. Depending on the length of the filament, coils are packed in one or more layers (Fig. 4 and 5A). They are usually arranged regularly and close to each other, and it is unusual for them to form a double layer (Fig. 11A), as in *E. bieneusi* (see Fig. 24D), or an irregular stack of coils, as in *Systenostrema corethrae* (Larsson, 1986a, 1988a), or a group of coils as in *N. apis* (Scholtyseck and Daneel, 1962) and *Pleistophora hyphessobryconis* (Lom and Corliss, 1967). The number of coils in a certain species usually tends to vary in a more-or-less narrow limit in different species. The number of coils should be counted in nearly mature or mature spores because during spore maturation this number changes.

3. Tilt of polar filament coils. The coils of the polar filament are frequently tilted in relation to the longitudinal axis of the spore. The angle of this tilt has been used in discriminating species (Burges et al., 1974). The angle of tilt is difficult to measure because it depends on the plane of sectioning. The spores must be sectioned longitudinally, and the most anterior and the most posterior points of the anterior filament coil must be cut. Most species exhibit an angle of tilt between 40° and 60°, with an individual variation of 5 to 10° between different spores of the same species. In the plane perpendicular to the correct plane for measuring, the angle of tilt appears to be 90°.

4. Width of the polar filament. Two basic types of polar filaments can be identified (Weiser, 1977). One type has the same, or nearly the same, thickness along its entire length, the isofilar filament (Fig. 5A; see Fig.

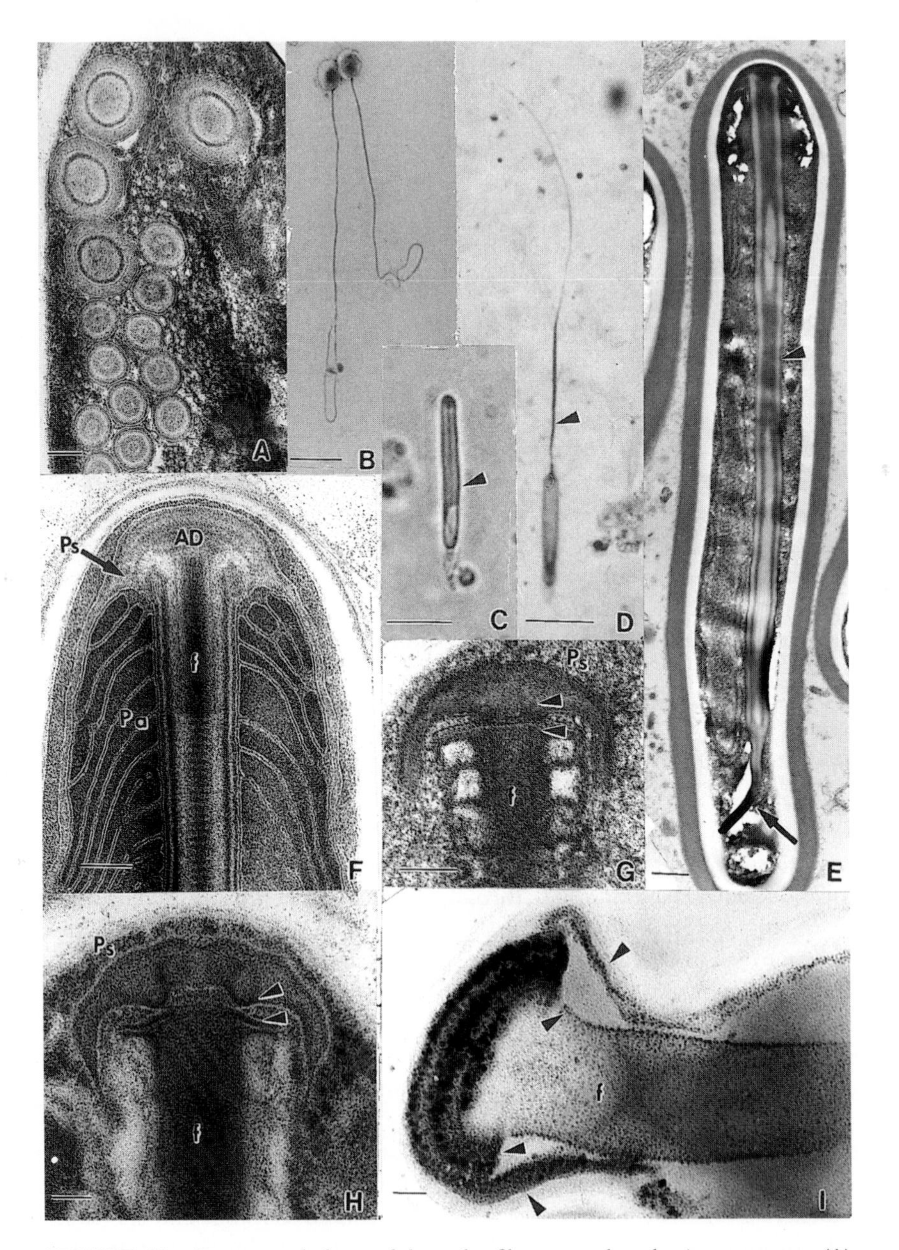

FIGURE 11 Gross morphology of the polar filament and anchoring apparatus. (A) Anisofilar filament (*Amblyospora* sp.); (B) ejected anisofilar filament (polar tube) (*Trichoctosporea pygopellita*); (C to E) spores with straight (manubroid) filament, visible in living spores (C) (*Jirovecia involuta*), ejected (D) (arrowhead indicates the straight, wide part) (*J. involuta*), and longitudinally sectioned (E) (arrowhead indicates the wide part and arrow indicates the narrow posterior part) (*Bacillidium strictum*); (F) common type of filament anchoring, with a wide polar sac, layered anchoring disk, and all layers of the filament (except for the unit membrane cover) united with the anchoring disk (*Cougourdella polycentropi*); (G and H) caplike polar sac, no visible anchoring disk, and collar-to-socket attachment of the filament (arrowheads) (G) (*Chytridiopsis trichopterae*) and (H) (*Intexta acarivora*); (I) presence of carbohydrates with α-glycol groups revealed by Thiery's reaction (arrowheads) (*Mrazekia cyclopis*). (A, and E to H) TEM; (B) Giemsa stain; (C) phase contrast; (D) hematoxylin stain; (I) PAS technique, TEM. Scale bars: A, F, and G, 100 nm; B, C, and D, 10 μm; H and I, 50 nm; E, 0.5 μm. Reprinted with permission from *Eur. J. Protistol.* (C, F, and I), *J. Eukaryot. Microbiol.* (G), and *Acta Protozool.* (H). Abbreviations: AD, anchoring disk; f, polar filament; Pa, anterior part of polaroplast; Ps, polar sac.

16C). In the other type, the anisofilar filament, the anterior part is thicker (Fig. 4A and 11A; see Fig. 13E). An abrupt constriction located in the coiled section separates the wide and the narrow parts. The anisofilar condition is visible in the ejected polar filament and can be observed with light microscopy (Fig. 11B). The most proximal zone, close to the anchoring site, is usually somewhat wider in terms of both isofilar and anisofilar filaments (Larsson, 1986c). A new term, "heterofilar," was introduced to describe a polar filament in which one or a few intermediate coils differ in size from the anterior and posterior coils (Voronin, 1989a).

A slightly different type of filament thickening is characteristic of genera of the family Mrazekiidae and apparently also of *Microfilum lutjani* (Faye et al., 1991), in which the anterior part appears as a wide, stiff, rodlike structure (Fig. 8I and 11C to E). Léger and Hesse (1916) introduced the term "manubrium" for this handlelike part of the filament. The straight section is often followed by a thin, distal part (Fig. 11E) forming one half-coil, as in *B. filiferum* (Larsson, 1989c), or more commonly a few coils. Polar filaments of microsporidia having a manubrium instead of a coiled filament are basically of the normal type and resemble anisofilar filaments. However, in contrast to the situation in a typical anisofilar filament, there is no abrupt constriction between the wide and narrow parts, but there is a zone with a successive reduction in the diameter, and the differences in width between the wide and narrow parts are greater than in normal cases of anisofilarity.

The term "manubrium" has also been used for the wide filament of the family Metchnikovellidae (see Fig. 15B and F) (Hildebrand and Vivier, 1971; Vivier, 1975; Desportes and Théodoridès, 1979; Ormières et al., 1981; Sprague et al., 1992). However, this filament is of a unique type, and therefore this term should be avoided.

The anisofilar/isofilar construction of the filament and the presence of the manubrium are characteristic features at the generic level.

5. The length of the filament is a character that must be interpreted with caution. The length calculated from electron micrographs of spores is different from the values obtained by measuring extruded filaments. Weidner (1972) calculated that the length of the polar filament in *Ameson michaelis* was 20 to 30 times the spore length when inside the spore and 60 to 100 times the spore length as an ejected tube. The diameter of 100 to 120 nm remained unchanged.

6. Internal structure of the polar filament is dealt with in detail in the section "Usual Polar Filament Structure." It is now customary to describe the polar filament structure, including the sequence of layers, when describing a new species.

7. Mode of polar filament-polar sac connection. This character divides microsporidia into two structural and probably two evolutionary groups. (See "The Polar Sac-Anchoring Disk Complex" and Fig. 12A and B.)

THE POLAR SAC-ANCHORING DISK COMPLEX

In the mature spore the polar filament terminates in a bell-shaped structure, the polar sac-anchoring disk complex, situated close to the spore apex (Fig. 4A, 8B to D, and 11F to I). At this point the spore wall is thinnest, and the rupture through which the filament emerges during spore germination is formed here. In microsporidia with oval, pyriform, or rodlike spores, the polar sac-anchoring disk complex is usually located terminally and the filament exits as a prolongation of the longitudinal axis of the spore (Fig. 4A and 11C to E). It is more unusual for the filament anchoring site to be subterminal and for the filament to emerge slightly sideways (see Fig. 29C).

The polar sac-anchoring disk complex has two components. In mature spores the polar sac surrounds the proximal part of the polaroplast in a bell-like fashion (Fig. 4A, 5A and B, and 11F). The sac is a vesicular structure enclosed by a unit membrane and filled with an electron-dense substance. The polar filament enters the polar sac and terminates in its center

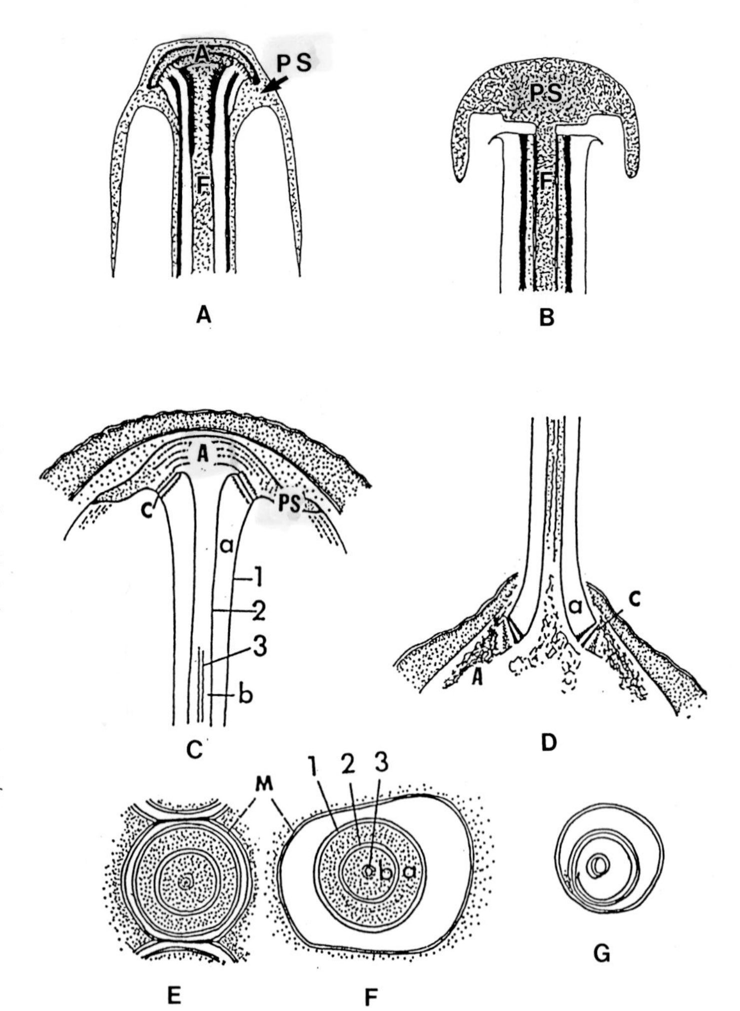

FIGURE 12 (A and B) Schematic representation of polar filament-polar sac connection in microsporidia. Typical microsporidia (A) and *Chytridiopsis*-like microsporidia (B) are shown. (C to E) Schematic representation of polar filament evagination (tip of the extrusion apparatus in a dormant spore [C]; polar sac-polar filament relationship in extruded spore [D]). Note that the structure labeled c serves as a hinge for the evaginating filament. (E) Cross section of the filament in a dormant spore. (F) Filament in an activated spore. (G) Cross section of the extruded filament. Modified with the permission of J. Lom and *Z. Parasitenkd.* Abbreviations: A, anchoring disk; a, polar filament wall; b, polar filament lumen; c, hinges; M, membrane ensheathing the polar filament; PS, polar sac; 1, 2, and 3, layers seen in the filament.

in the anchoring disk. The disk is a biconvex, often layered body which can be interpreted as a modified terminal, flattened part of the filament (Fig. 5B and 8B to D and F). In cross section the disk shows several alternating layers of more-or-less dense material. The polar sac and the anchoring disk contain carbohydrates as shown by (PAS) (periodic acid-Schiff staining) (Lillie, 1965) (Fig. 1N) and by its modification for electron microscopy (Thiéry, 1972) (Fig. 1N and 11I). Demonstration of this material by staining spores in smears or sections is a diagnostic method for microsporidians (see "Periodic Acid-Schiff Positivity").

Polar Filament-Polar Sac Relationship. Microsporidians exhibit two modes of polar filament-polar sac connections:

1. In the majority of microsporidians, the polar filament in its entire width enters the polar sac and all layers of the filament connect with the disk (Fig. 2D, 13A, and 24F). The external layers of the filament widen in a funnel-like fashion inside the sac, making the filament the same width as the anchoring disk. It has been observed in some microsporidians, e.g., *Napamichum dispersus* (Larsson, 1984) and *Amblyospora culicis* (Toguebaye and Marchand,

1985, 1986), that fibrous material connects the external layers of the filament to the anchoring disk (Fig. 8F). This fibrous material serves as "hinges" around which the polar filament turns during eversion (Lom, 1972) (see "The Polar Filament during Spore Germination").

2. The polar sac of *Chytridiopsis*-like microsporidians (Larsson, 1993) and Metchnikovellidae (Vivier and Schrével, 1973) is a rather small, cuplike body (Fig. 11G and H, 12B, and 15B). This sac is also a folded unit membrane containing electron-dense material. However, there is no visible anchoring disk. In *Chytridiopsis*-like microsporidia and obviously also in metchnikovellideans the posterior surface of the central part of the polar sac is shaped like a socket (Fig. 11G and H). The external layers of the polar filament widen anteriorly into a collarlike structure with approximately the same width as the socket (Fig. 11G and H). The unit membrane of the polar sac is continuous with the surface layer of the filament. However, the socket and collar are separated by a distinct gap. Only the center of the polar filament enters the polar sac (Fig. 12B).

FINE STRUCTURE OF THE POLAR FILAMENT

During polar filament initiation, the transversely sectioned filament usually appears as a ring limited by a thick, electron dense border and containing a fibrogranular electron-lucent matrix and a dark, circular central spot (Fig. 8A). Longitudinal sections reveal that the limiting border of the filament is continuous with the polar sac and that the dark material in the center of the filament penetrates some distance into the sac (Fig. 8C and D). The electron-dense material inside the polar sac is the first sign of the future anchoring disk. During morphogenesis of the spore the original structure of the filament is extensively reorganized, and the filament in its final form is a complex, multilayer structure. The construction of the mature polar filament is best evaluated in transverse sections, in which a number of concentric layers of different electron density and thickness are visible (Fig. 13).

Usual Polar Filament Structure. A basic filament organization is shared by the majority of microsporidian species. However, the appearance of individual layers is influenced both by the visualization techniques used and by the degree of maturity of the spore. For these reasons all layers are not always distinct, and the thickness of the respective layers varies among species. The following sequence of layers is usually discernible in the transversely sectioned filament (from the outside inward) (Fig. 13A to C): an external unit membrane (1); three layers often of approximately the same thickness, electron-dense (2), translucent (3), and dense (4); a layer resembling transversely sectioned translucent fibrils embedded in a moderately dense matrix (5); and the center (6). The center is subdivided, but the layers are often not distinct. For example, species of the genera *Episeptum* (Larsson, 1996), *Systenostrema* (Larsson, 1988a), and *Vavraia* (Larsson, 1986d) possess this kind of polar filament. Normally the wide and narrow coils of an anisofilar filament exhibit the same sequence of layers, and the differences in diameter are mainly due to a different width at the center (layer 6), as in *Vavraia holocentropi* (Larsson, 1986d) and in *Episeptum* spp. (Larsson, 1996) (Fig. 4A).

Vávra (1976a) monitored the filaments of several microsporidia from initiation to maturation, gave a more detailed account of the layering, and expressed the idea that the filament is a thick-walled tube with composite outer and inner walls consisting of three (dense, transparent, dense) layers each. Between the outer and inner walls, there is a wide electron-transparent layer. During filament formation, the thick, dark outer zone of the immature filament splits into two electron-dense concentric rings separated by a lucent layer (layers 2 to 4). These three layers represent the outer wall of the filament tube. Another three, but less thick, layers (two dense rings separated by less opaque material and probably a subdivided layer 5) represent the inner filament wall. The inner tube seems to be built from fine fibers arranged helically along the filament length. These fibers can be seen both

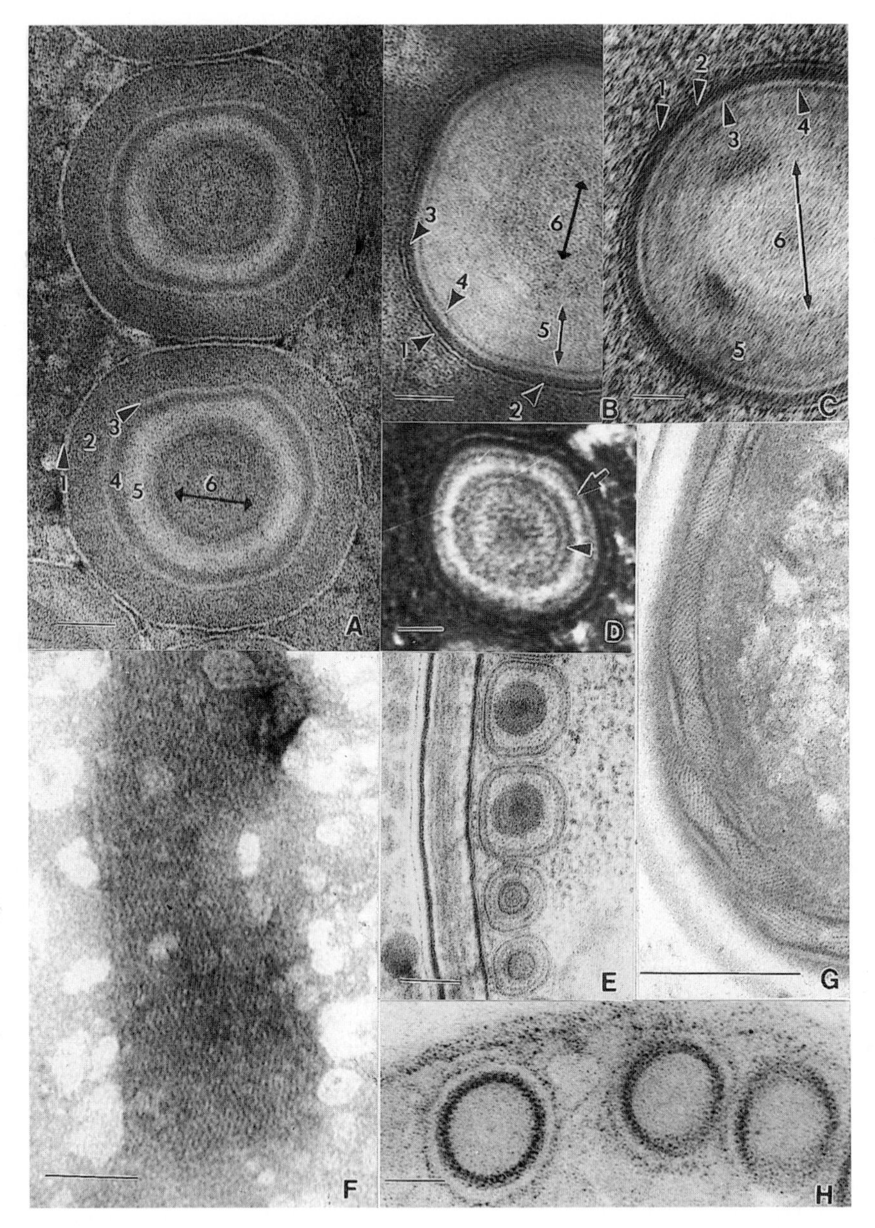

FIGURE 13 Fine structure of the polar filament. (A to C) Transversely sectioned filaments; 1 to 6 indicate the most easily observed layers in *Trichoctosporea pygopellita* (A), *Janacekia adipophila* (B), and *Jirovecia involuta* (C). (D) Layering of the filament, showing the double layers of the outer (arrow) and inner (arrowhead) filament walls (*Heterosporis finki*). (E) Layering seen in immature spores. Note the difference in polar filament layers in the anterior and posterior coils (*Amblyospora culicis*). (F and G) Fibrillar substructure of the polar filament. A *Nosema whitei* filament after tryptic digestion (F) and the fibrillar substructure of one of the filament layers revealed by Thiery's reaction for carbohydrates in *Microsporidium* sp. from *Artemia salina* (G) are shown. (H) Cross section of a filament stained by Thiery's method (*Norlevinea daphniae*). TEM was used for all panels. Scale bars: A to C, F, and H, 50 nm; D, 25 nm; E, 100nm, G, 0.5 μm. Reprinted with permission from *Arch. Protistenkd.* (A), *J. Protozool.* (B), *Eur. J. Protistol.* (C), G. Schubert (D), B. S. Toguebaye (E), Plenum Press (F and H), and E. Porchet-Henneré (E and G).

during polar filament formation and as a meshwork in negatively stained extruded filaments (Vávra, 1976a) (Fig. 13F). In between the outer and inner composite walls is an electron-transparent layer into which extend, in cartwheel fashion, the fibers from the wall of the inner tube. These fibers react positively when stained by Thiéry's (1972) reaction for carbohydrates (Fig. 13G). The lumen of the inner tube (the center layer 6) contains granular material of two electron densities: a dense central spot surrounded by a less dense matrix. In freeze-fracture preparations the filament looks like a solid rod. When a specimen is cross-fractured, an inconspicuous ring of granules close to its outer rim and another accumulation of granules in its center are revealed (Fig. 10D to E and 14A to C). The filament is enveloped by a unit type membrane as shown by the freeze-fracture technique (Vinckier et al., 1993) (Fig. 10C and D and 14C). It has been argued that the filament-ensheathing membrane is not part of the filament proper, but belongs to the polaroplast membrane system (Vávra, 1976a). The carbohydrate (glycoprotein?) material is located mainly in the inner filament wall and is much less dense than in the outer filament wall (Fig. 13H).

Polar Filaments of Unusual Structure. A minority of microsporidians have unusual types of polar filaments.

1. Microsporidians of the family Mrazekiidae produce rod-shaped spores possessing a thick polar filament of the manubrium type (Fig. 8I and 11E). The internal organization differs from the common type in that layer 5, in the straight portion is exceptionally wide and has a mottled structure (Fig. 13C). In the narrow part of the filament it is reduced to a narrow, dense stratum (Larsson, 1994a).

2. The *Chytridiopsis*-like microsporidia (species of the genera *Chytridiopsis* and *Burkea* and *Steinhausia brachynema, Nolleria pulicis, Intexta acarivora,* and *Buxtehudea scaniae*) differ from other microsporidians in several ways. The spores are small and spherical, a normal polaroplast is absent, and polar filament coils are mostly untilted. A smaller number of filament layers can be distinguished. The external dense layer of the filament is transformed into an alveolate layer in which expansions of the dense layer enclose small, round alveoli with electron-lucent content. In cross section such a filament has a stellate form (Fig. 15E and H). In tangential section a honeycomblike alveolar layer on the surface of the filament is seen in *S. brachynema* (Richards and Sheffield, 1970) (Fig. 26G and H), *Chytridiopsis* (Larsson, 1993), and *Nolleria* (Beard et al., 1990) (Fig. 26C). It is believed that the honeycomblike layer on the surface of the filament is the product of combined layers 1 to 4 of a typical filament (representing in reality the outer wall of the unextruded polar filament covered by a unit membrane in a mature spore). In the central part of the filament of *Chytridiopsis*-like microsporidia another darker ring can be distinguished, probably representing the inner wall of the polar filament tube.

As none of the *Chytridiopsis*-like microsporidia has a polaroplast, it is believed that the lucent alveoli at the filament surface represent a special type of polaroplast vesicle and that the honeycomblike layer is a primitive type of polaroplast (Vávra, 1976a).

In *I. acarivora* (Larsson et al., 1997b) the unit membrane at the polar filament surface is a smooth layer below which is a zone with winding tubules of electron-lucent material (Fig. 15D). *B. scaniae* (Larsson, 1980) and *B. gatesi* (Puytorac and Tourret, 1963) differ from the other microsporidia of this group in having a long polar filament with a distinct tilt, which gives the spores the same polarity as the normal microsporidian spore. The polar sac at one pole of the spore crowns the straight anterior part of the filament. The posterior section of the filament is arranged in numerous coils at the opposite pole. The unit membrane covering the filament of *B. scaniae* is folded, and the folds wind around the filament. The transversely sectioned filament resembles an eight-point star (Fig. 15E). The interior of the filament is similar to the interior of filaments of normal cytology, including the presence of a fibrous layer (layer 4 in the normal polar filament layering).

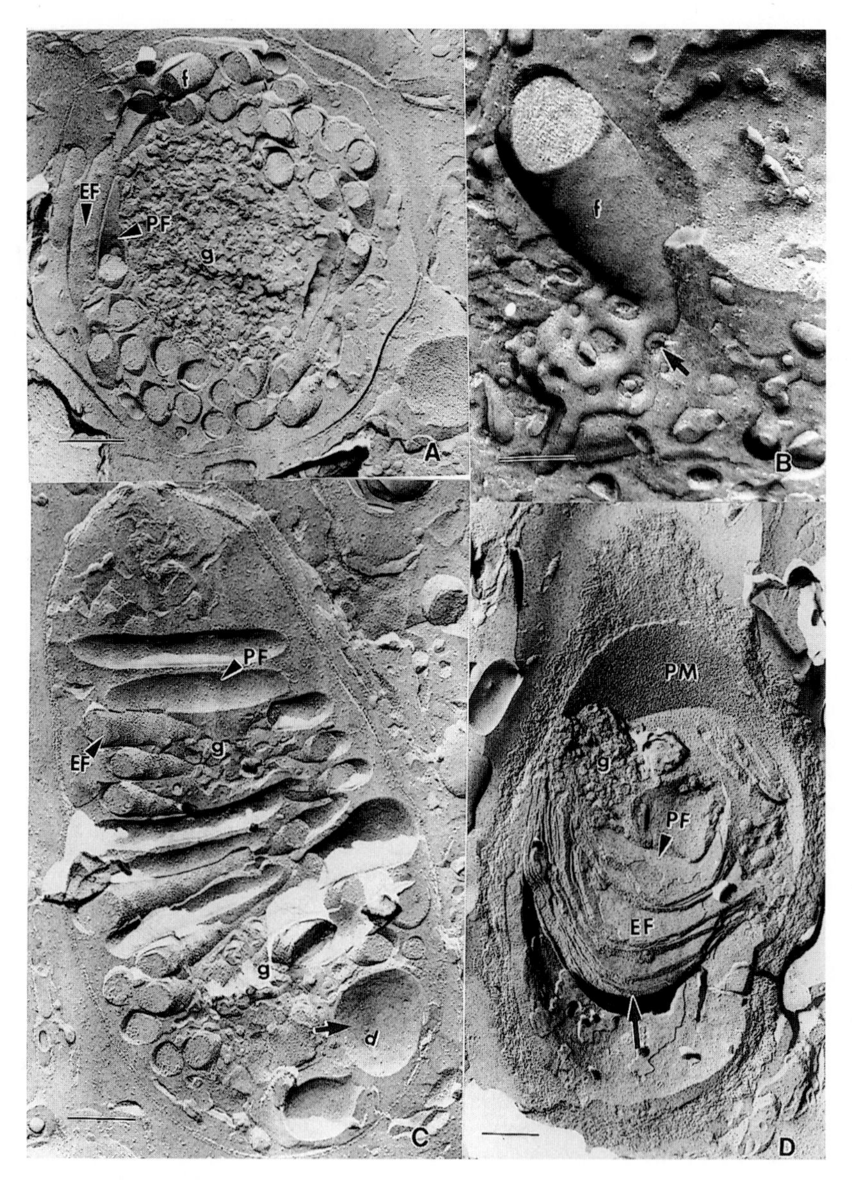

FIGURE 14 Freeze-fracture of microsporidian spores. (A) Tangential fracture of young spore of *Nosema apis,* showing the hypertrophied reticulum of the Golgi zone and cross and tangential fractures of several coils of the polar filament. Both faces (EF and PF) of the cytoplasmic membrane around the filament are shown. (B) Sporoblast of *Amblyospora varians,* showing the origin of the polar filament by coalescence of Golgi reticulum vesicles (arrow). (C) Longitudinal fracture through *N. apis* young spore, demonstrating that the polar filament is a solid cylinder enveloped by a cytoplasmic membrane with EF and PF faces; the EF face bears more numerous intramembranous particles; posterior vacuole membrane PF face is at arrow. (D) The polaroplast (arrow) in *N. apis* is revealed as a stack of membranes bearing intramembranous particles on both its hemimembrane faces. The ectoplasmic face (EF) of the spore plasma membrane (Pm) bears numerous intramembranous particles. Scale bars: A, C, and D, 0.5 μm; B, 250 nm. Reprinted with permission from *Eur. J. Protistol.* (A to C) and D. Vinckier (D). Abbreviations: g, Golgi zone; F, polar filament; EF and PF, respective ectoplasmic and protoplasmic faces of a split unit membrane; Pm, plasma membrane (ectoplasmic face).

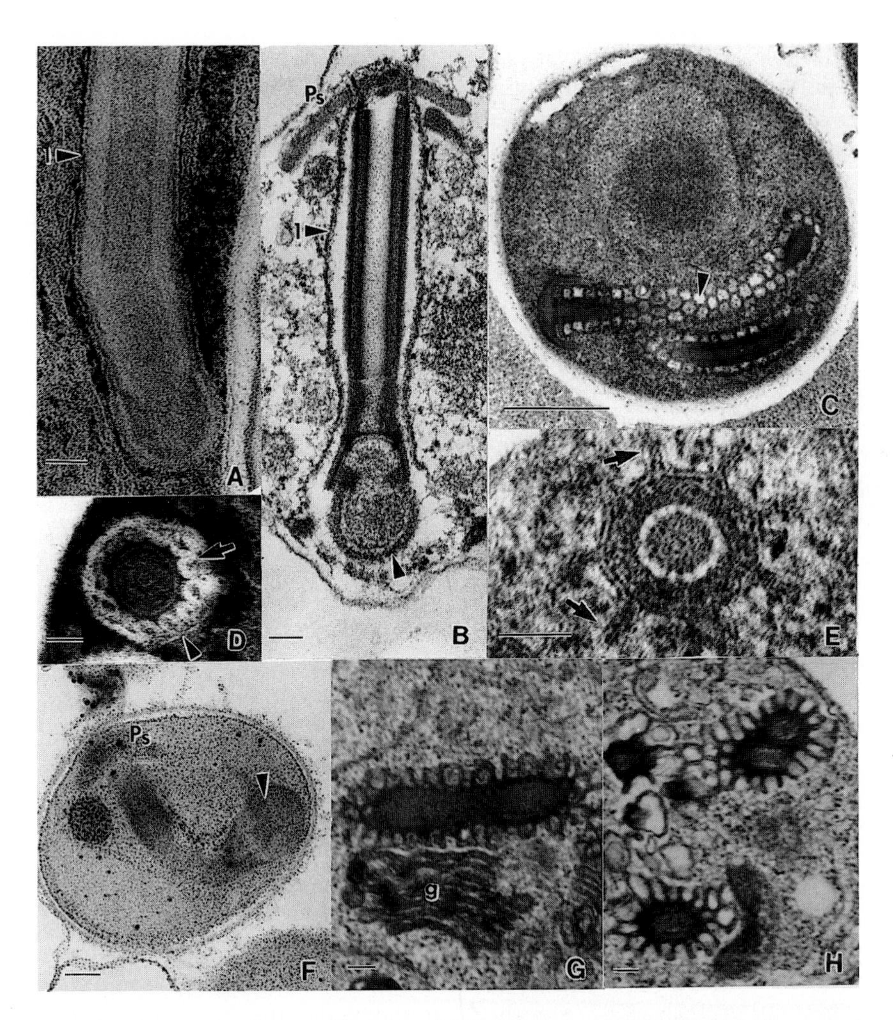

FIGURE 15 Unusual types of filaments. (A) Posterior part of a short, straight filament of normal cytology (*Cylindrospora fasciculata*); (B) straight filament of Metchnikovellidae, ending with a "gland" (arrowhead) (*Amphiacantha* sp.); (C) short and completely coiled filament of *C. trichopterae* covered by a honeycomblike surface layer (arrowhead); (D) transversely sectioned filament of *I. acarivora* (arrow indicates tubular material and arrowhead indicates the covering unit membrane); (E) transversely sectioned filament of *Buxtehudea scaniae* (arrows indicate unit membrane folds); (F) filament of *Metchnikovella hovassei* with the bulbous gland (arrowhead); (G and H) longitudinal (G) and transversal (H) sections of the filament covered with honeycomb layer (*Steinhausia brachynema*). TEM was used for all panels. Scale bars: A, D, and E, 50 nm; C and F, 0.5 μm; B, G, and H, 100 nm. Reprinted with permission from *Protistologica* (A), *J. Eukaryot. Microbiol.* (C), *Acta Protozool.* (D), J. Schrével (F), and H. G. Sheffield and C. S. Richards (G and H). Abbreviations: Nu, nucleus; Ps, polar sac, g, Golgi.

3. The polar filament of metchnikovellideans is a wide, stout structure ending with a posterior bulbous swelling (Fig. 15B and F). The filament is uncoiled in *Amphiamblys bhatiellae* (Ormières et al., 1981) and *Desportesia laubieri* (Desportes and Théodoridès, 1979) and is bent in a half-coil posteriorly in *Metchnikovella hovassei* (Vivier and Schrével, 1973). The straight filament has been called a manubrium (see "Polar Filament Descriptors"), and posterior swollen part, a gland (Vivier and Schrével, 1973). The filament has a unit membrane cover, a zone of granular material, and a distinctly organized central axis with a sequence of layers similar to the normal stratification of a polar filament. A fold of the unit membrane projects, from the posterior swollen part, creating a lamellar structure filled with granular material similar to the material in the external zone of the filament.

THE RELATION BETWEEN POLAR FILAMENT COMPOSITION AND FUNCTION

Both proteins (Weidner, 1976a) and polysaccharides have been detected in the polar filament and the anchoring apparatus. In the anchoring disk up to three layers of carbohydrates with 1,2-glycol groups occur in the anchoring disk and the whole interior of the polar sac stains positive with Thiéry's periodic acid-thiosemicarbazide-silver proteinate method (Thiéry, 1972) (Fig. 11I) (Vávra, 1972, 1976a; Larsson et al., 1993). Glycoconjugates are further present in one surface layer of the polar filament, and in one of its internal layers (Fig. 13G to H). All microsporidians examined so far with the Thiéry reaction have exhibited identical layers.

THE POLAR FILAMENT DURING SPORE GERMINATION

At present, it remains unresolved whether the filament inside the spore is a final product or a pool of materials from which a tube is formed during spore germination. Biochemical analysis revealed the presence of a protein monomer with a molecular weight of 23 kD as the main component of the filament (Weidner, 1976a; Weidner et al., 1995), and it was proposed that the filament actually grows and forms a tube by polymerization at the filament tip during eversion (Weidner, 1982). There is, however, no compelling reason not to believe that the filament simply turns inside out (as a finger of a glove does) during eversion as Lom and Vávra argued (1963a) and as corroborated by light microscope and ultrastructural observations (Lom, 1972; Vávra, 1976a). According to this model the filament exit proceeds as follows. In the germinating spore, the narrow space between the unit membrane ensheathing the filament and the filament proper increases first (Fig. 12E and F) as a consequence of swelling of the polaroplast. Later, the intrasporal pressure squeezes the base of the filament (at the site of its attachment to the anchoring disk) through the thin part of the spore wall (Fig. 5C). The filament turns inside out during the exit, its basal part turning around the "hingelike" fibrous structure (Fig. 8F and 12D) fastening the wall of the filament tube to the anchoring disk (Fig. 12C and D). The extruded filament shows several concentric membranous layers (Lom, 1972) (Fig. 12G).

The "simple eversion model" is certainly more complex than presented above. In any case, the presence in the filament of fibrillar structures and structures containing carbohydrate in both its unextruded and extruded states (Vávra, 1976a) argue against simplification of filament structure and function based on biochemical characterization of one of its components.

The Polaroplast

The polaroplast is a system of membrane-limited cavities in the anterior part of the spore (Fig. 4A, 5A and B, and 16). In most microsporidians it is a voluminous structure surrounding the straight part of the polar filament and ending at the level of the first filament coils. One-third to one-half of the spore volume is occupied by the polaroplast, and only rarely is it rudimentary or nonexistent. In immature spores the developing polaroplast has an irregular texture, while the polaroplast of mature spores is a distinctly organized structure with compartments that vary in shape and are often lamellar.

CONDITIONS FOR POLAROPLAST STRUCTURE EVALUATION

In evaluating the construction of the polaroplast it is necessary to study the polaroplast of mature spores. All members of a genus share a polaroplast of nearly identical construction, and construction is thus a useful classification

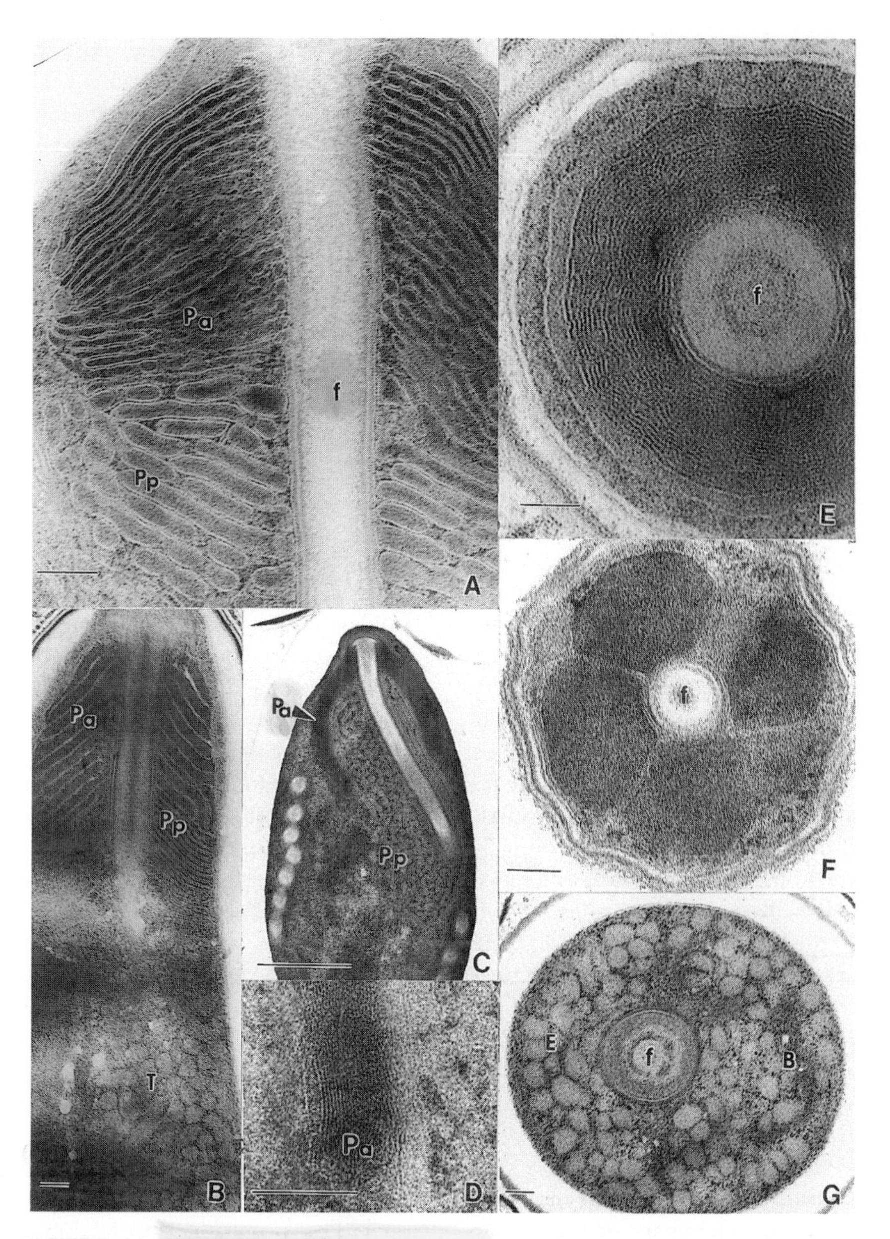

FIGURE 16 Fine structure of the polaroplast. (A) The most common type of bilaminar polaroplast, with more closely packed anterior lamellae (*Vavraia holocentropi*); (B) polaroplast with three regions (wide lamellae, narrow lamellae, and tubules) (*Agglomerata sidae*); (C and D) bilamellar polaroplast with exceptionally closely packed anterior lamellae ("cavum") (*Nosema artemiae*); (E to G) three aspects of the transversly sectioned polaroplast, concentric lamellae (E) (*Cylindrospora fasciculata*), petallike arrangement (F) (*Cougourdella polycentropi*), and tubules (G) (*Jirovecia involuta*). TEM was used for all panels. Scale bars: A, B, D, F, and G, 100 nm; E, 50 nm; C, 0.5 μm. Reprinted with permission from *Protistologica* (A to E), *Arch. Protistenkd.* (B), and *Eur. J. Protistol.* (F and G). Abbreviations: Pa, anterior polaroplast; Pp, posterior polaroplast; T, tubules, f, polar filament.

character. As an organelle built of membranous compartments, the polaroplast is sensitive to fixation and embedding procedures. Because of this sensitivity, its structure is easily influenced by the handling of the spores during preparation. In well-preserved specimens and in spores fixed in hypertonic fixatives, the polaroplast lamellae are tightly packed and regularly spaced. Hypotonic fixatives cause the polaroplast to swell (Vávra, 1976a; Vinckier et al., 1993). Good preservation of unit membranes, both of the polaroplast and the polar filament, shows their trilaminar nature, which is a prerequisite for correct interpretation of polaroplast structure. Further, it is important to note how the polaroplast has been sectioned. A perfect longitudinal section gives a different picture of the polaroplast than an oblique section. The images also differ when a longitudinal section taken close to the midline of the spore is compared to a section taken at the periphery. Some of the aberrant polaroplast types described are illustrated only by oblique sections, and there are reasons to doubt some of the interpretations. Even if two lamellar polaroplasts look identical when sectioned longitudinally, transverse sections might reveal a difference, because the lamellae might be arranged in two ways. Either the lamellae are radially complete, which means that each one surrounds the polar filament completely (Fig. 16E), or they are more narrow, arranged like the petals of a flower (Fig. 16F).

FINE STRUCTURE OF THE POLAROPLAST

The compartments of the polaroplast are shaped like lamellae, chambers, or tubules delimited by approximately 5-nm thick unit membranes. The unit membrane character of the polaroplast was confirmed by the freeze-fracture technique which revealed the presence of intramembranous particles on the faces of split polaroplast membranes (Vinckier et al., 1993) (Fig. 14D). The polar sac, the compartments of the polaroplast, and the unit membrane covering the polar filament belong to the same membrane system, and the membranes of the polaroplast form anastomoses to the unit membrane of the polar filament. The polaroplast is most commonly divided into a shorter anterior part and a more voluminous posterior section, but there are several variations on the theme, including subdivision into up to four different regions.

1. The most commonly observed type of polaroplast has two lamellar parts. Anterior lamellae are closely packed and regularly arranged, while posterior lamellae are wider and less regularly organized (Fig. 5A and B and 16A; see Fig. 21A). This polaroplast is typical of *Encephalitozoon* (Barker, 1975), *Pleistophora* (Canning and Nicholas, 1980), and numerous other genera. The two regions are referred to as the lamellar polaroplast and the vesicular polaroplast, respectively. A polaroplast with inverted construction, with two regions of regularly arranged lamellae, where the anterior lamellae are wide and the posterior ones narrow, is characteristic of the genera *Episeptum* (Fig. 4A) and *Trichotuzetia* (Vávra et al., 1997).

2. Certain microsporidia, e.g., *Ameson michaelis* (Sprague et al., 1968), *Spraguea lophii* (Loubès et al., 1979), and *Helmichia aggregata* (Larsson, 1982), have an electron-dense zone anterior to the uniform lamellar polaroplast. Also, the dense zone is lamellar, and the lamellae are unusually electron-dense and compressed (Fig. 16C and D). Originally this zone was believed to be a cavity, not a lamellar organelle, and the term "cavum" was used to denote such a region (Sprague et al., 1968).

3. Variations in the posterior part of the two-region lamellar polaroplast are not unusual. The posterior lamellae of some *Amblyospora* spp. (Andreadis, 1983; Dickson and Barr, 1990) are so wide and irregularly shaped that they are best described as chambers. In *Napamichum cellatum*, the posterior compartments are almost tubular (Bylén and Larsson, 1994).

4. Two variations of a polaroplast with three regions have been described. One type has narrow lamellae anteriorly, a median section of wide lamellae, and a posterior region with tubules. This is the polaroplast of *Binucleospora* *elongata* (Bronnvall and Larsson, 1995a). In the second type of three-region polaroplast, a section with wide lamellae is followed by one with narrow, closely packed lamellae and a final portion with tubules. This type of polaroplast was originally observed in the genus *Agglomerata* (Fig. 16B), but is also known to occur in *Lanatospora tubulifera* (Bronnvall and Larsson, 1995b).

5. A polaroplast with four regions was observed in *Microsporidium fluviatilis* (Voronin, 1994). In the anteroposterior direction the following regions are visible: loosely arranged chambers, closely packed chambers, large globular compartments, and posterior lamellae.

6. A few genera have uniform lamellar polaroplasts. The polaroplast of *Cystosporogenes deliaradicae* is lamellar in the normal way, protruding sideward from the polar filament (Larsson et al., 1995). The lamellae in the polaroplasts of *Cylindrospora fasciculata* (Larsson, 1986c) and *Ordospora colligata* (Larsson et al., 1997a) are folded, one outside the other.

7. In the trimorphic *Amblyospora* spp., which alternates between mosquitoes and copepods, the elongate type of spore produced in the copepod exhibits a different type of uniform polaroplast composed entirely of globular sacs (Sweeney et al., 1985; Becnel and Sweeney, 1990). It was originally considered unique to *Amblyospora,* but new studies have revealed that a similar spore with a polaroplast of the same type is also produced by the dimorphic or trimorphic mosquito parasites of the genera *Culicospora* (Becnel et al., 1987), *Culicosporella* (Becnel and Fukuda, 1991), and *Edhazardia* (Becnel et al., 1989).

8. *Bohuslavia asterias* has a polaroplast lacking distinct regions (Larsson, 1985). The shape of the compartments changes successively from wide, irregular chambers anteriorly to closely packed lamellae posteriorly.

9. A minority of microsporidians have in their spores membranous or tubular elements with no resemblance to the polaroplasts of other microsporidia. These structures are sometimes interpreted as being primitive or rudimentary polaroplasts. *Baculea daphniae* has such a rudimentary polaroplast in the form of a vacuole with membranous septa (Loubès and Akbarieh, 1978), however, *Chytridiopsis-* and *Metchnikovella*-like microsporidia lack a distinct polaroplast (Fig. 15B, C, and F). The surface structures of the uniquely constructed polar filaments have been interpreted as being homologous to a polaroplast (Vávra, 1976a).

The Posterior Vacuole

The posterior vacuole is the third component of the extrusion apparatus of the spore (Fig. 4A and 5A). It is a membrane-lined area with a clear or spongy content located in the posterior part of the spore. It looks like an empty vacuole when observed in the light microscope, and the size and shape are quite variable. At one extreme some microsporidians apparently do not have a posterior vacuole, and in other cases it looks like an inconspicuous slit. Typically, however, the vacuole is readily visible, and in some species it is large and occupies more than half of the spore volume. Fish microsporidia frequently have such large vacuoles.

The vacuole is enclosed by a unit type membrane as shown by the freeze-fracture technique (Vinckier et al., 1993) (Fig. 14C). Like other parts of the extrusion apparatus, the vacuole is evidently a product of the Golgi vesicles. This is corroborated by the presence in the posterior vacuole of immature spores of a spongy reticulum of electron-dense vesicles, similar to and probably identical to the Golgi reticulum forming the polar filament. This structure was called a posterosome by Weiser and Žižka (1974, 1975), a term replacing earlier names such as "posterior body" and "inclusion body".

The mutual spatial relationship between the polar filament and the posterior vacuole is not well known. There are divergent opinions as to whether the filament enters the vacuole and terminates there. Weidner (1972) postulated that the polar filament enters the vacuole but is separated from it by a membrane. However, freeze-fracture technique have failed to prove that the filament enters the vacuole (Vinckier et al., 1993). On the other hand, Lom and Vávra (1963a, unpublished data) have noticed that during spore germination and polar filament unwinding, the posterosome vigorously spins inside the vacuole. This observation is interpreted as proof that the filament ends inside the vacuole and that in immature spores the posterosome is identical to the Golgi material at the distal end of the filament.

ENVELOPES OF MICROSPORIDIAN ORIGIN

The whole development of the microsporidian takes place in intimate relationship with the host cell cytoplasm. This contact is maintained until the moment the spores are nearly mature and reflects the fact that the microsporidian is fully dependent on the metabolism of the host cell.

The sporoplasm is inoculated into the host cell, which ensures that the microsporidian enters the host cell unrecognized and protected from host defense reactions. The parasite plasma membrane is in direct contact with the host cell cytoplasmic matrix during merogony, and in many microsporidia also during the sporogony. The host cell envelopes the microsporidian with a cisterna of ER with ribosomes. Less frequent is the case in which the parasite induces the formation of a membranous boundary, delimiting a vacuole of presumed host cell origin within the parasite develops (see "The Parasitophorous Vacuole"). Many microsporidia, however, have a closed type of sporogony (see "Sporogony") and produce various vesiclelike structures inside which the spores are formed: sporophorous vesicles of two kinds, exospore-derived envelopes, and cystlike structures. The various types of envelopes provide a barrier between the host fluids and the spores. If envelopes are used as taxonomic criteria, it is crucial to discriminate between those produced by the host cell in reaction to a microsporidian infection and those derived from microsporidia, as well to determine the manner in which the microsporidia-derived envelopes are produced.

Sporophorous Vesicles

The sporophorous vesicle (Fig. 17) is known in older literature as a pansporoblast, an obsolete and not recommended term for microsporidia (Vávra and Sprague, 1976; Canning and Nicholas, 1980).

The sporophorous vesicle originates as a layer of secretions generated by the organism. It loses contact with the plasma membrane and forms an envelope. The volume of the sporophorous vesicle between the envelope and the microsporidian cell is called the episporontal space (Vávra, 1984). A vesicle produced by the sporont is called a sporontogenetic sporophorous vesicle, and when produced by the meront, a merontogenetic sporophorous vesicle (Canning and Hazard, 1982). In some microsporidians the vesicle is a transient structure: it remains until the spores mature, and then rapid degeneration ensues. Other microsporidians sporulate in more or less persistent vesicles. As most envelopes are delicate structures, light microscopy is rarely sufficient to confirm the presence of sporophorous vesicles.

Sporophorous vesicles are normally spherical or ovoid, as in *Amblyospora, Parathelohania, Vavraia,* and *Trachipleistophora* (Fig. 17A; see Fig. 28C) (Hazard and Oldacre, 1975; Vávra et al., 1981). Microsporidians belonging to the genera *Chapmanium* (Hazard and Oldacre, 1975), *Napamichum* (Larsson, 1984), and *Ormieresia* (Vivarès et al., 1977) sporulate in fusiform sporophorous vesicles.

The spherical sporophorous vesicles of *Trichoduboscqia epeori,* a parasite of dayflies, are ornamented with usually four, more than 20 μm long, needlelike appendages (Batson, 1982). They have an internal core of material similar to collagen. The number of projections varies from two to four, correlated with the number of spores enclosed (Léger, 1926; Weiser, 1961; Batson, 1982). Sporophorous vesicles are released into water from dead hosts, and the projections are likely to be floating devices.

The sporophorous vesicle of *Telomyxa glugeiformis* is an oval body (Fig. 17G to J) in

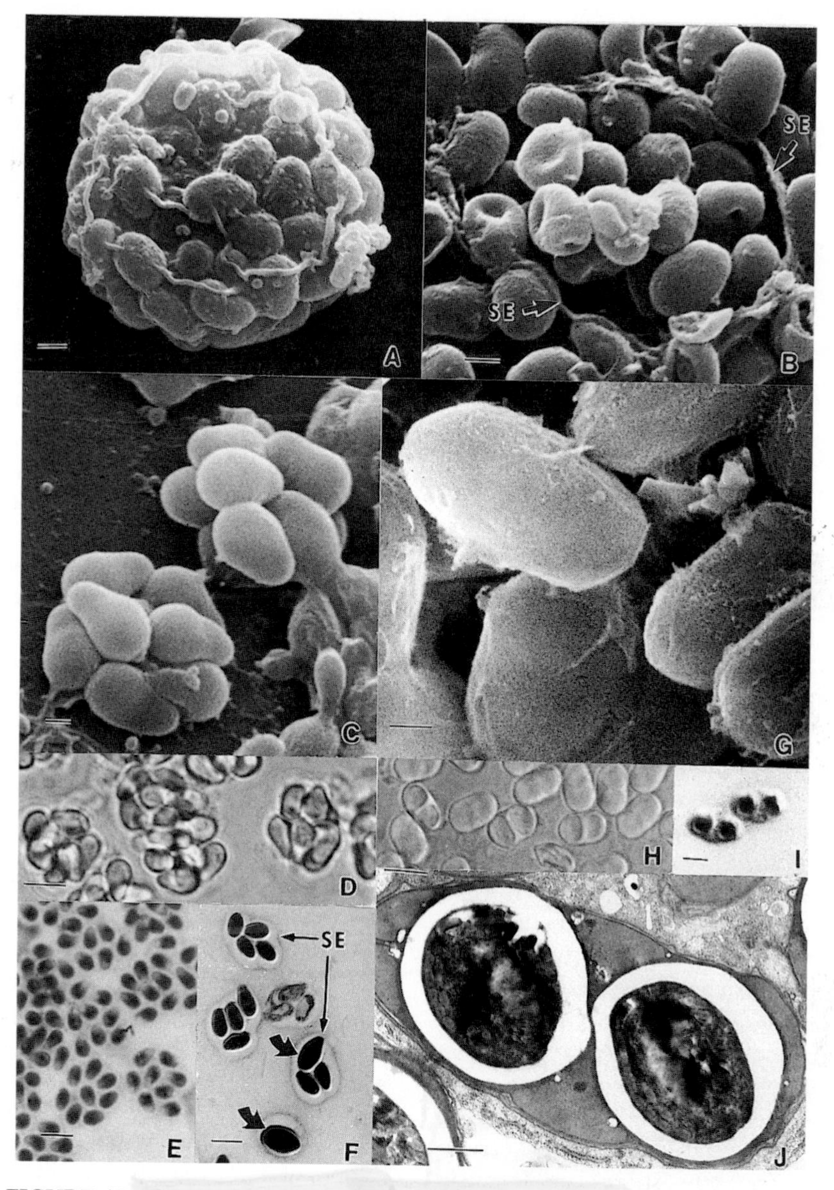

FIGURE 17 Gross morphology of the sporophorous vesicle. (A and B) The merontogenetic sporophorous vesicle of *Vavraia holocentropi,* complete and sectioned; (C to E) three aspects of the octosporous sporophorous vesicle of *Systenostrema candida;* (F) tetrasporous sporophorous vesicles of *Gurleya dorisae* with two vesicles exhibiting macrospores (large arrows) and a reduced number of spores; (G to J) four aspects of the disporous vesicle of *Telomyxa glugeiformis.* (A to C and G) SEM; (D and H) phase contrast; (E and F) hematoxylin stain; (I) Giemsa stain; (J) TEM. Scale bars: A to C, 1 μm; D, E, F, and H, 5 μm; J, 0.5 μm; G and I, 1 μm. Reprinted with permission from *Syst. Parasitol.* (C and D), *Acta Protozool.* (F), and *Zool. Anz.* (J). Abbreviation: SE, sporophorous vesicle envelope.

which two spores are embedded in an electron-dense material (Léger and Hesse, 1910; Codreanu and Vávra, 1970; Larsson, 1981a). Its shape resembles that of a microsporidian spore, and it was originally thought to be a single spore (Léger and Hesse, 1910).

THE SPORONTOGENETIC SPOROPHOROUS VESICLE

The first deposited layer of electron-dense secretions on the surface of the sporont builds the vesicle, usually as a thin, electron-dense structure which progressively detaches from the plasma membrane, forming an envelope (Fig. 18A to C). Two kinds of sporontogenetic sporophorous vesicles exist:

1. A multicell sporophorous vesicle, the most frequently occurring type, collects all sporoblasts inside a common envelope. Multicell vesicles are found in many microsporidia, such as the genera *Amblyospora* (Hazard and Oldacre, 1975) and *Episeptum* (Larsson, 1996). Multicell vesicles are usually prominent under the light microscope, appearing as more or less well-defined packets of spores (Fig. 1B, D, and F and 17C to F; see Fig. 28C and 29E). The number of spores in such vesicles can be between two and many. In some microsporidian genera the number is constant, and in others it is variable. The number of spores per vesicle is a useful character in identifying microsporidians at levels above the species. Caution must be exercised when sporophorous vesicles with many spores are seen in the intestinal tissues of a host. Spore-loaded intestinal epithelium cells shed into the gut lumen can easily be mistaken for multisporous sporophorous vesicles (see Fig. 21A and B).

2. If the sporophorous vesicle envelope divides simultaneously with the plasmodium inside, finally enclosing each spore in a vesicle of its own, single cell sporophorous vesicles are formed (Fig. 18D and E). Single-cell vesicles are characteristic of a small number of genera, *Tuzetia, Janacekia, Alfvenia, Nelliemelba* (Maurand et al., 1971; Larsson, 1983d), and *Lanatospora* (Bronnvall and Larsson, 1995b), and can be seen only with TEM.

Sporophorous Vesicle Structure. The primordium of a sporophorous vesicle either develops simultaneously all over the sporont and a complete envelope is released, as in the genus *Cystosporogenes* (Fig. 18A and B), or vesicle primordia appear spotwise over the surface of the sporont (Fig. 6E). The latter appears to be the more common occurrence. The envelope of the future sporophorous vesicle is then released as blisters (Fig. 18C) which later join together to form a complete vesicle. This initiation is typical of *Tuzetia*-like microsporidia. These blisters are filled with electron-dense material (Fig. 6E) easily stained with hematoxylin stains, which makes the sporogonial stages appear spotted in stained smears (Fig. 18E).

The envelope of a sporophorous vesicle is normally a thin layer, measuring only a few nanometers, as in *Cystosporogenes deliaradicae* (Fig. 18A). The layer sometimes has the appearance of a two-layered membrane, but in other species it is clearly a uniform layer. Freeze-fracture studies on vesicles of *Tuzetia* and *Amblyospora* spp. have revealed that it is not a unit type membrane (Vávra et al., 1986) (Fig. 10B).

Aberrant Types of Sporophorous Vesicles. Only a few genera produce aberrant types of sporontogenetic sporophorous vesicles. Microsporidia of the genus *Berwaldia* have individual vesicle envelopes composed of a membranelike coat supported by tubules. The sporophorous vesicle appears folded and is attached at intervals to the exospore (Larsson, 1981b; Vávra and Larsson, 1994). The envelopes of *T. glugeiformis* (Fig. 18H), *Cryptosporina brachyfila* (Fig. 18I), and *Pegmatheca lamellata* (Larsson, 1987) are thicker structures measuring up to about 30 nm. The envelope of *T. glugeiformis* has a layered texture different from that of the electron-dense exospore of the mature spore (Codreanu and Vávra, 1970; Larsson, 1981a), while the layered envelope of *C. brachyfila* is constructed like the exospore (Hazard and Oldacre, 1975). In this species stratified exospore material is produced twice, the first layer of secretions yielding the

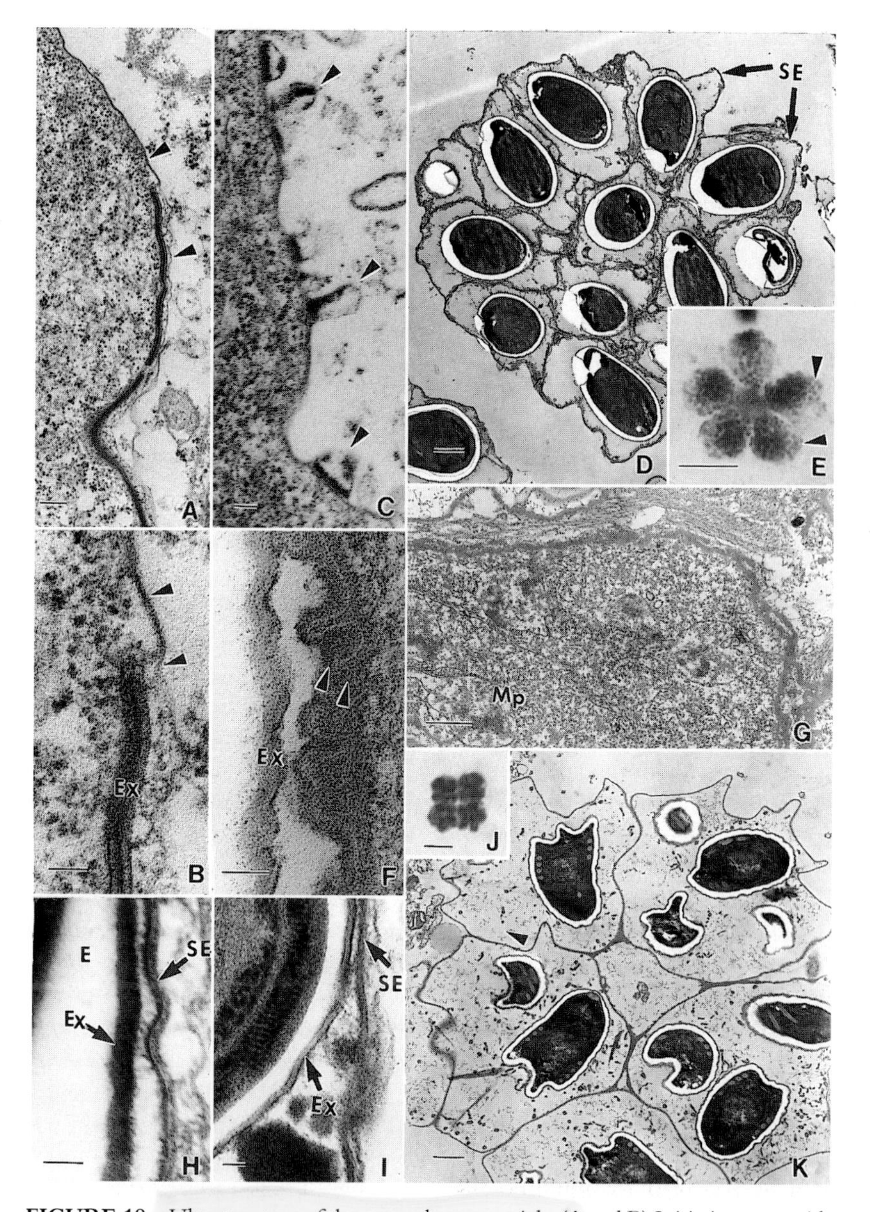

FIGURE 18 Ultrastructure of the sporophorous vesicle. (A and B) Initiation over wide areas of the sporont surface (arrowheads indicate envelope) (*Cystosporogenes deliaradicae*); (C) blisterlike initiation (arrowheads) (*Tardivesicula duplicata*); (D) mature spores enclosed in individual sporophorous vesicles (*Janacekia undinarum*); (E) blisterlike protuberances (arrowheads) of the individual vesicles stain darkly on the rosettelike dividing sporont (*Janacekia adipophila*); (F and G) layered (arrowheads) merontogenetic sporophorous vesicle of *Vavraia holocentropi* showing surrounding mature spores (F) and initiation (G); (H and I) unusual types of thick sporophorous vesicles from *Telomyxa glugeiformis* (H) and *Cryptosporina brachyfila* (I) in which the envelope is identical to the exospore; (J and K) connected sporophorous vesicles (*Pegmatheca lamellata*). (A to D, F to I, and K) TEM; (E and J) hematoxylin stain. Scale bars: A, 100 nm; B, C, F, H, and I, 50 nm; D, 1 μm; E, 10 μm; G and K, 0.5 μm; J, 5 μm. Reprinted with permission from *Eur. J. Protistol.* (A to C), *Protistologica* (D and G), *J. Protozool.* (E), and *Zool. Anz.* (J to K). Abbreviations: E, endospore; Ex, exospore; Mp, merogonial plasmodium; SE, sporophorous vesicle envelope.

sporophorous vesicle. After this layer has been released, a second production of exospore material creates the exospore (Fig. 18I). The vesicles of *Pegmatheca* spp. are even more unusual in that all sporonts generated by the last merogonic division remain together and the sporophorous vesicles adhere, even when containing mature spores (Hazard and Oldacre, 1975; Larsson, 1987) (Fig. 18J and K). Another unusual condition is seen in *Polydispyrenia* (Canning and Hazard, 1982). The multinucleate meront does not release merozoites but remains undivided when it enters sporogony. A sporophorous vesicle is formed at the onset of sporogony originating as a typical sporontogenetic sporophorous vesicle but functioning as a merontogenetic vesicle in collecting all daughter cells produced by the meront.

THE MERONTOGENETIC SPOROPHOROUS VESICLE

In a small number of microsporidia, the dense material building the sporophorous vesicle appears during merogony (Canning and Hazard, 1982). Great quantities of electron-dense material are deposited on the external surface of the plasma membrane of the meront (Fig. 18G; see Fig 28A). The result is an up to 165-nm-thick (*Vavraia culicis*), often amorphous cell wall (Fig. 21F) permeated by membranous channels and forming a surface labyrinth intermeshed with the host cell cytoplasm. At the time of sporogony, the sporogonial plasmodium retracts and membranous channels and labyrinthlike protrusions disappear, leaving the amorphous layer as a persistent envelope collecting all daughter cells originating from the meront. Finally, it is filled with mature spores (Fig. 17A and B). Characteristically the envelope of a merontogenetic sporophorous vesicle remains thick. This kind of envelope is produced by microsporidians of the genera *Pleistophora, Vavraia,* and *Trachipleistophora* (Fig. 18F and G; see Fig 31B and C). Each genus builds envelopes of characteristic structure: two or three layers of different electron density in *Pleistophora* (Canning and Nicholas, 1980) and two thick, dense strata separated by a thin layer of lucent material in *Vavraia* (Canning and Hazard, 1982) and *Trachipleistophora anthropophthera* (Vávra et al., 1998) (Fig. 18F). There is an uniformly dense layer in *Trachipleistophora hominis* (Hollister et al., 1996). In the genus *Trachipleistophora* the envelope forms a tuft of 25- to 40-nm fibers or tubules extending into the host cell cytoplasm (Weidner et al., 1997; Vávra et al., 1998) (see also "*Trachipleistophora* spp.").

THE EPISPORONTAL SPACE AND INCLUSIONS

The episporontal space is defined as the cavity between the envelope of the sporophorous vesicle and the sporont wall (Vávra, 1984). At the onset of sporogony the episporontal space is small and may not even be continuous over the whole surface of the parasite. The episporontal space, however, becomes important as the primordial envelope of the sporophorous vesicle detaches more and more from the sporont, or its daughter cells, and the volume of the vesicle increases.

The episporontal space is not an empty space because it contains a variety of electron dense inclusions. These structures are more numerous in the early phase of sporogony, with a peak when sporoblasts are released. Simultaneously with maturation of the spores, the quantity of inclusions decreases, and sporophorous vesicles with ripe spores usually have only remnants of this material. Exceptionally, this material disappears almost completely, as in *Hyalinocysta* spp. (Hazard and Oldacre, 1975; Larsson, 1989a).

The inclusions are of three kinds—fibrous, tubular, or crystalline (granular)—but in many cases these individual forms are combined. Inclusions are also dynamic structures changing in quantity and appearance during sporophorous vesicle maturation.

Thin, fibrous material is a regular component of sporophorous vesicles, as in *Amblyospora capillata* (see Larsson, 1983a) and *Thelohania maenadis* (Vivarès, 1980). The fibrils often traverse the episporontal space from the surface of the sporont to the envelope of the sporophorous vesicle. The fibrils of *Trichotuzetia guttata* have a distinct substructure and are arranged in a regular manner (Fig. 19A and B).

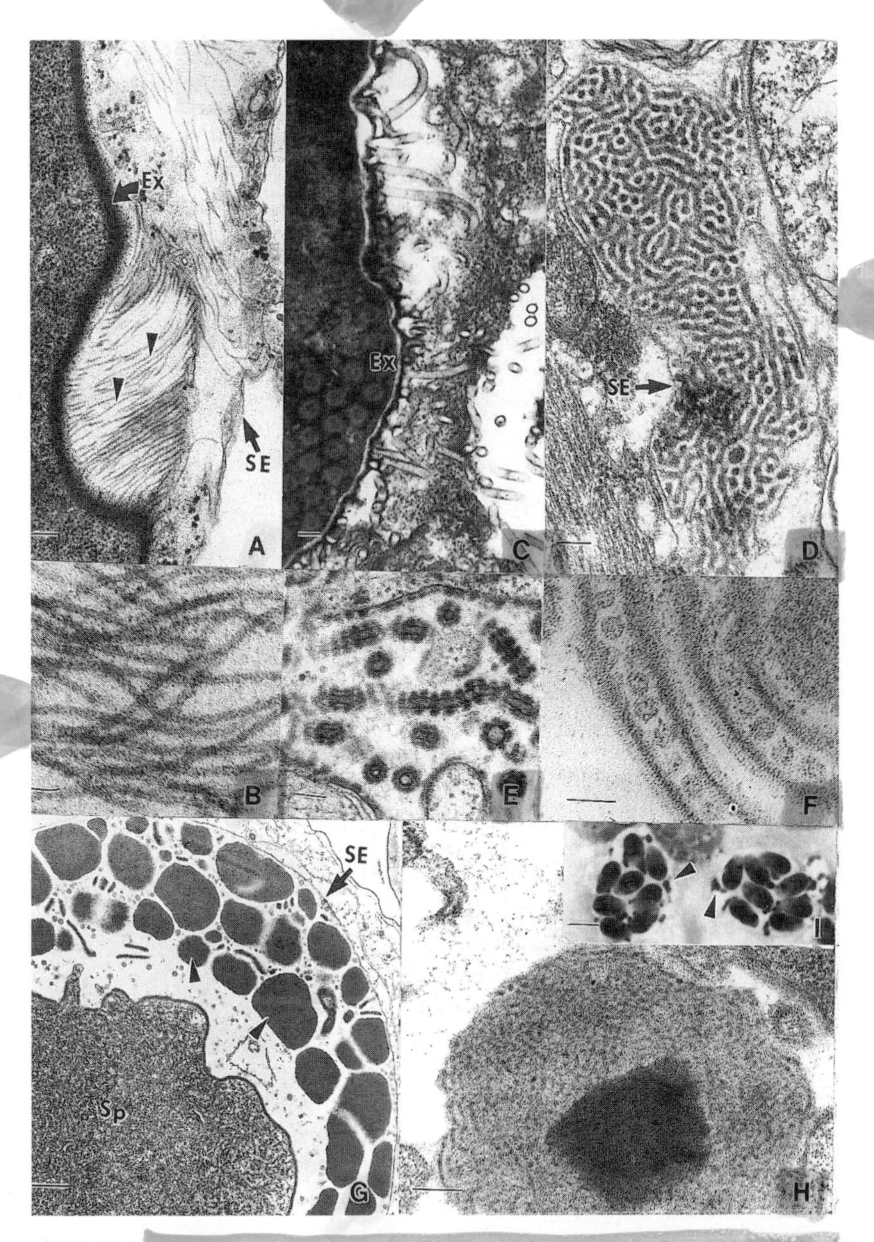

FIGURE 19 Exospore projections and inclusions of the episporontal space. (A and B) Fibrous material (arrowheads) connecting the exospore with the envelope (*Trichotuzetia guttata*); (C) wide tubules from exospore to envelope (*Janacekia debaisieuxi*); (D) labyrinthlike tubular material (arrowheads) (*Toxoglugea variabilis*); (E) tubular exospore-derived material (*Agmasoma penaei*); (F) septate, exospore-derived tubules (*Systenostrema candida*); (G) uniform amorphous crystallike material (arrowheads) (*Amblyospora* sp.); (H and I) two-layered crystals (arrowheads) (*Napamichum dispersus*). (A to H) TEM; (I) hematoxylin stain. Scale bars: A, C to E, and H, 100 nm; B and F, 50 nm; G, 0.5 µm; I, 5 µm. Reprinted with permission from *Arch. Protistenkd.* (A and B), *Eur. J. Protistol.* (E), *Syst. Parasitol.* (F), and *Protistologica* (H). Abbreviations: Ex, exospore; SE, sporophorous vesicle envelope; Sp, sporont.

In *Trichoctosporea pygopellita,* fibrous material, persisting as exospore projections from the mature spore (Fig. 9F to H) occurs together with a unique type of inclusion appearing as arrays of globular chambers (Larsson, 1994b).

The episporontal space of *Tuzetia*-like microsporidia is traversed by nonseptate tubules, which are narrow in *Tuzetia* (Fig. 9D and E and 19C) and *Lanatospora* (Voronin, 1989) and wide in *Janacekia* (Loubès and Maurand, 1976; Weiser and Žižka, 1975). In *Berwaldia* spp., tubular material is so closely associated with the envelope of the sporophorous vesicle that it appears two-layered (Larsson, 1981b). *Thelohania*-like microsporidia have tubular inclusions of two different kinds. The lumen of the narrow tubules is often filled with electron-dense material, and the tubular nature is not always obvious. The wide tubules, which are projections from the sporoblast wall, are septate and have a wall constructed like the exospore layer (Fig. 19E and F). *Cystosporogenes deliaradicae* (Larsson et al., 1995) shares characteristic tubules with *Agmasoma penaei* (Fig. 19E) (Hazard and Oldacre, 1975). In *Glugea stephani* three different kinds of tubular structures have been defined and a numbered system was proposed (Takvorian and Cali, 1983).

Inclusions in the form of parallel thick-walled tubules forming a kind of labyrinth (Fig. 19D) within the episporontal space are characteristic of sporophorous vesicles of the genus *Vairimorpha* (Weiser and Purrini, 1985; Moore and Brooks, 1992, 1994).

Crystalline or granular material is a prominent feature of newly formed sporophorous vesicles of *Amblyospora* spp., also visible in light microscopic preparations (Fig. 19G). The crystals are successively reduced during maturation of the spores, but a small number of crystals are normally also present in old vesicles. Still greater production of crystalline material is seen in *Cryptosporina brachyfila,* in which spores are more or less obscured by the crystals (Hazard and Oldacre, 1975). At the ultrastructural level crystals are composed either of a uniform electron-dense material, as in *Amblyospora* (Fig. 19G) (Hazard and Oldacre, 1975) and *C. brachyfila* (Fig. 21I), or of material of variable electron density and texture, as in *Napamichum* spp. (Fig. 19H and I).

Little is known about the chemical nature of the secretory material of the episporontal space, but the close connection, and sometimes structural similarity, to the exospore suggest that the exospore and the inclusions have a the same or similar chemical composition. Weidner and collaborators identified protein fibers in the episporontal space of an unidentified *Thelohania* sp. as cytokeratin intermediates and desmosomal analogs (Weidner et al., 1990).

Various functions have been attributed to episporontal secretions. An older term for these structures, "metabolic granules," vaguely suggested some role in the metabolism of the organism (Hazard and Oldacre, 1975). More specifically, it has been proposed that the secretions are surplus material from the production of sporoblasts which is more or less used up during the formation of spores (Larsson, 1986b). The observation that some types of secretory material are structurally identical to exospore material (Fig. 19E) supports this idea. The other interpretation is based on the tendency of this material to form tubules, channels, or strands sometimes connecting the outer layer of sporoblasts or nascent spores with the wall of the sporophorous vesicle, e.g., as in *Chapmanium cirritus* (Hazard and Oldacre, 1975) (Fig. 19A to C). This finding suggests a conductive function, which was experimentally demonstrated by Overstreet and Weidner (1974) in *Inodosporus spraguei*. For more information on different kinds of sporophorous vesicles and secretory inclusions, see Larsson (1986b).

Exospore-Derived Envelopes

Some members of the families Mrazekiidae and Tuzetiidae enclose their spores in exospore-derived envelopes which can be mistaken for individual sporophorous vesicles of the *Tuzetia* type.

1. Mrazekiidae. The sporoblasts of Mrazekiidae generate a thick, stratified exospore in which the basal layer is wide and uniform and the surface layer resembles a double membrane. The two-layered exospore remains unchanged in mature spores of *Bacillidium strictum* (Larsson, 1992), *Jirovecia caudata* (Larsson, 1990b), and probably also *Hrabyeia xerkophora* (Lom and Dyková, 1990). In *Bacillidium filiferum* the surface layer is released in the shape of filiform projections (Larsson, 1989c), while in *Jirovecia brevicauda* the surface layer is transformed into tubular projections (Larsson and Götz, 1996). In both species only the wide basal layer remains as the exospore of the mature spore. In *Bacillidium criodrili* (Larsson, 1994a), *Jirovecia involuta* (Larsson, 1989b), and *Rectispora reticulata* (Larsson, 1990d), the surface component of the exospore of the sporoblast is completely released from the basal layer, which remains as the exospore of the mature spore. The surface layer forms a complete sac around the mature spore (Fig. 20A to C), making it indistinguishable from an individual sporophorous vesicle.

2. Tuzetiidae. Spores of microsporidians of the family Tuzetiidae are enclosed in individual sporophorous vesicles generated by the sporont (Fig. 6E and 18D). In two of the species, *Alfvenia nuda* (Larsson, 1983d, 1986b) and *Nelliemelba boeckella* (Milner and Mayer, 1982), a second envelope is formed internally to the sporophorous vesicle when surface material from the layered exospore of the sporoblast is released. At least in *A. nuda,* part of the sporoblast retains the complete exospore, with the result that spores with double and single envelopes occur together.

Sporont-Derived Sacs

The dense material remains as a cover on the plasma membrane of the sporont, and together with it forms a persistent cystlike sac which collects the spores (Fig. 20D to G). This is typical of *Chytridiopsis-* and *Metchnikovella*-like microsporidians, the spores of which are produced endogenously (Desportes and Théodoridès, 1979; Purrini and Weiser, 1984; Beard et al., 1990; Larsson et al., 1997b). In this process vacuoles appear within the sporogonial plasmodium, separating each nucleus and a surrounding zone of cytoplasm from each other. Each of these units matures into a sporoblast, and the vacuole membranes are incorporated as the plasma membrane of the sporoblast. Thus the plasma membrane of the sporogonial plasmodium is not incorporated into the spore wall but remains as the internal layer of a complex envelope (Fig. 20E). In both *Metchnikovella-* and *Chytridiopsis*-like microsporidia, enveloped sporogony in sporont-derived sacs occurs together with sporogony in which spores may be collected in a parasitophorous vacuole of host cell origin (Fig. 20F and H) (Vivier and Schrével, 1973; Larsson, 1993).

The sporont-derived sacs of *Metchnikovella*-like microsporidians are long, fusiform, or nearly cylindrical structures with blunt or threadlike poles (Fig. 20H) (Vivier and Schrével, 1973; Ormières et al., 1981; Desportes and Théodoridès, 1979). *Chytridiopsis*-like microsporidia sporulate in spherical sacs (Fig. 20F and G).

CYTOLOGICAL ANOMALIES IN MICROSPORIDIA

Cytological anomalies, reported or visible in micrographs, are not unusual in microsporidians. Many of them are obviously artifacts caused by the techniques used or can be explained by external influences on the host. Others stem from internal conditions within the microsporidian itself. A third kind of anomaly involves the presence of foreign inclusions in the microsporidian cell.

1. External influences. Teratological development of microsporidians is, for example, known to be caused by drugs administered to the host (Liu and Myrick, 1989; Ditrich et al., 1994). Contact with another microorganism is the probable explanation for a number of anomalies seen in *Cystosporogenes deliaradicae* (Larsson et al., 1995). The anomalies were observed in host specimens simultaneously infected by the fungus *Strongwellsea castrans,* and teratological spores were especially frequent in the proximity of hyphae.

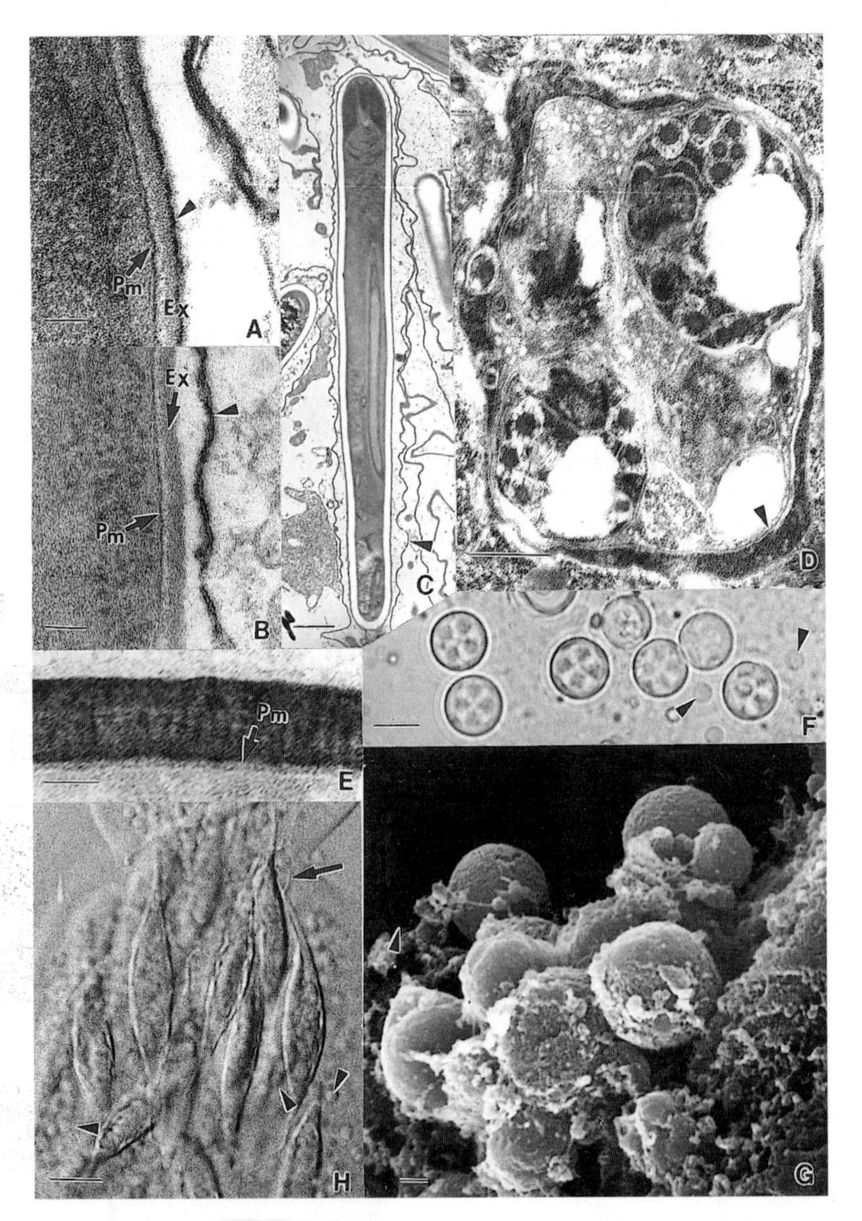

FIGURE 20 Exospore-derived envelopes and sporont-derived sacs. (A to C) Two steps in the initiation of an exospore-derived envelope. Arrowheads indicate the surface layer that forms the envelope and an envelope surrounding a mature spore (*Jirovecia involuta*). (D) External part of a sporont which is transferred into an envelope (arrowheads) (*Intexta acarivora*). (E to G) Thick-walled cystlike sacs of *Chytridiopsis trichopterae,* showing mature envelope (E), live cysts enclosing spores and free spores as indicated by arrowheads (F), and mature cysts (G). (H) Live fusiform sacs of an *Amphiacantha* sp. (arrows indicate threadlike projections, and arrowheads indicate spores in the sacs and free in the cytoplasm. (A to E) TEM; (F and H) phase contrast; (G) SEM. Scale bars: A, B, and E, 50 nm; C and G, 1 µm; D, 0.5 µm; F and H, 5 µm. Reprinted with permission from *Eur. J. Protistol.* (C) and *J. Eukaryot. Microbiol.* (E to G). Abbreviations: Ex, exospore; Pm, plasma membrane.

2. Internal conditions. Frequent anomalies include incompletely separated sporoblasts, which are especially common in microsporidians with rod-shaped spores, and disorganization of the polar filament of mature spores (Becnel et al., 1989; Vávra, 1962; Larsson, 1989c, 1990d, 1995). Anomalies in the polar filament are visible as disturbed coiling, supernumerary and undifferentiated polar filament coils, or coils with an increased number of centers. There are also frequent reports of abortive sporogony, e. g. in *Edhazardia aedis* (Becnel et al., 1989) and *Hyalinocysta expilatoria* (Larsson, 1983c), in which a number of sporoblasts fail to mature to spores.

Some observations of aberrant cytology can probably be explained as reductions. At least two species of microsporidians, *Amblyospora capillata* (Larsson, 1983a) and *Napamichum aequifilum* (Larsson, 1990a), have polar filaments that appear to be isofilar. However, it is more likely that the filaments are anisofilar and have had the posterior narrow part reduced. Both species belong to genera with distinct cytological characteristics and anisofilar polar filaments. The absence of sporophorous vesicles in the sporogony of *Nudispora biformis* is probably another example of reduction (Larsson, 1990c). This microsporidian is cytologically similar to *Thelohania*-like microsporidia, which sporulate in sporophorous vesicles, and it behaves identically in both merogony and sporogony, producing episporontal space inclusions of the *Thelohania* type.

3. A different kind of anomaly is the presence of cytological inclusions of foreign origin. Three reports classified the foreign bodies as virus particles. The earliest observation was made by Liu (1984) who described particles resembling bee viruses in the cytoplasm of lysed spores of the honeybee parasite *N. apis*. Others have described intranuclear particles, which were spherical and aggregated, having a lucent periphery and an electron-dense center and measuring 20 to 25 nm in diameter. The first observation was in two unidentified species of the family Thelohaniidae (Larsson, 1988b), and the second in *Trichotuzetia guttata* (Vávra et al., 1997).

MICROSPORIDIAN-INDUCED EFFECTS ON HOST CYTOLOGY

The host cell is more or less visibly affected by the development of a parasite, but a clear pathogenic effect caused by a microsporidian is apparent only during parasite sporogony (Fig. 21). In earlier stages the microsporidian behaves more like a cellular symbiont with whom the host cell is cooperating.

Association with Host Cell Organelles

On rare occasions microsporidians establish close associations with specific organelles of the host cell. Species of the genera *Chytridiopsis* (Sprague et al., 1972) and *Nolleria pulicis* (Beard et al., 1990) develop in such a close association with the host nucleus that the sporulating microsporidian rests in an invagination of the nuclear envelope (Fig. 21E). A location close to the host nucleus is also frequent in *E. hellem* (see Fig. 27F), as is the association of microsporidia with host cell mitochondria (see Fig. 22D and 23E). For example, *Buxtehudea scaniae* provokes the mitochondria of the host cell to aggregate closely around sporulating microsporidians, creating a framelike structure (Fig. 21D). *Nucleospora salmonis* is found within the host cell nucleus.

The Parasitophorous Vacuole

Some microsporidia, such as *Encephalitozoon cuniculi* (Petri, 1969) and *Glugoides intestinalis* (Larsson et al., 1996), induce the production of a boundary in the host cell, separating the microsporidian from the cytoplasm of the host cell. The membrane-lined cavity thus formed around the microsporidian is called a parasitophorous vacuole (Fig. 21C; see Fig 25C, 26E, and 27F). The mechanism by which this vacuole is formed is not known. Ultrastructural observations show that the membrane limiting the vacuole appears in the early developmental stages of the microsporidian in the form of a membrane tightly applied to the cell plasma membrane (see Fig. 25A and B). In the case of *E. intestinalis,* it has been suggested that the parasitophorous vacuole surrounding the internalized sporoplasm originates from the invaginated cell membrane. Invagination occurs

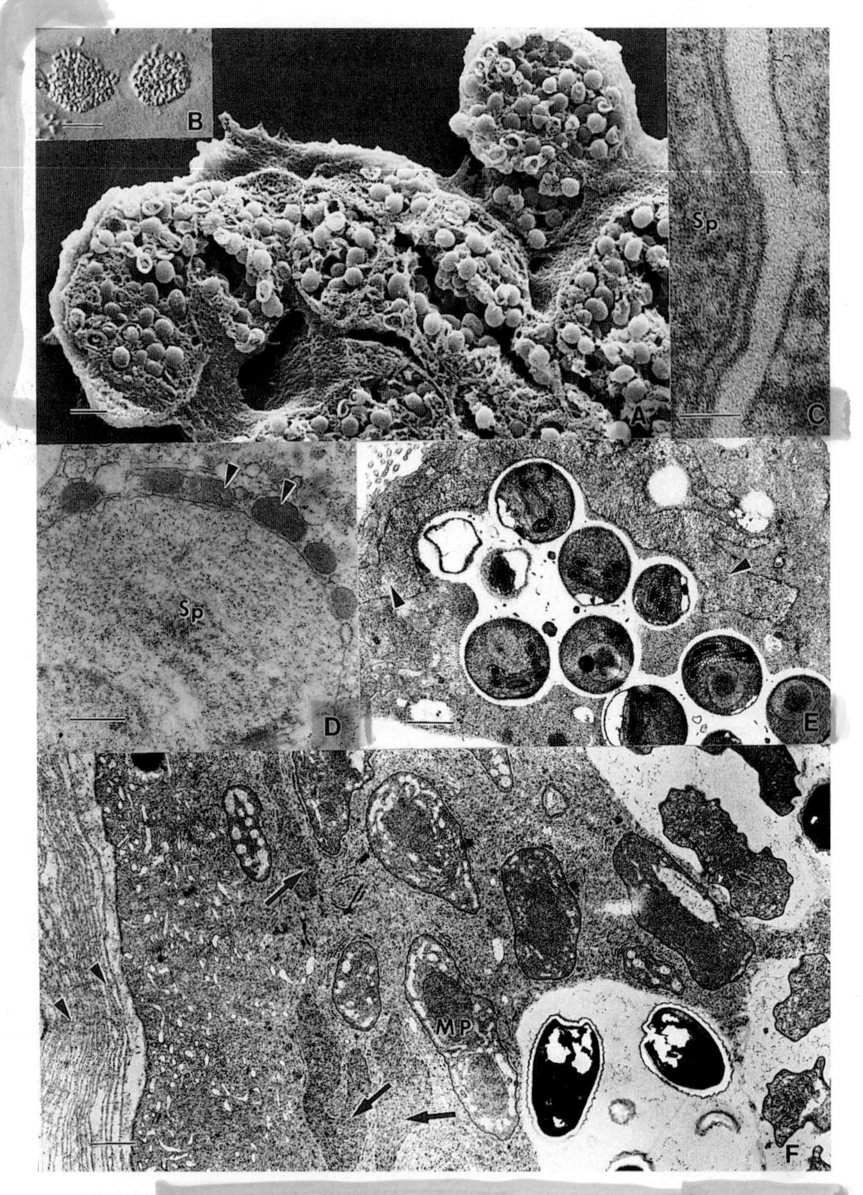

FIGURE 21 Microsporidian-induced effects on the host cytology. (A) Gut epithelium cells filled with microsporidian spores (*N. apis*); (B) epithelium cells released into the gut lumen which can be mistaken for multisporous sporophorous vesicles (*N. apis*); (C) parasitophorous vacuole (*Glugoides intestinalis*); (D) host cell mitochondria (arrowheads) accumulated at the surface of a sporont (*Buxtehudea scaniae*); (E) host nucleus (arrowheads) enveloping spores of *Chytridiopsis trichopterae;* (F) the *Glugea* xenoma with multilayered wall (arrowheads) and with the most immature stages of the microsporidian at the periphery (arrows indicate host nuclei) (*Glugea anomala*). (A) SEM; (B) interference phase contrast; (C to F) TEM. Scale bars: A, 5 μm; B, 25 μm; C, 50 nm; D, 0.5 μm; E and F, 1 μm. Panel C is reprinted with permission from *Eur. J. Protistol.* Abbreviations: MP, merogonial plasmodium; PM, plasma membrane; S, spore; SN, sporont; Sp, sporogonial plasmodium.

at the site where the apical part of the spore contacts the host cell, and the spore then discharges its filament into this invagination (Magaud et al., 1997). This observation requires verification because it is difficult to distinguish a spore extruding into a phagocytic cell invagination from an extruded spore that has been secondarily phagocytized.

The parasitophorous vacuole border appears as a unit membrane closely applied to the plasma membrane of merogonial stages (Barker, 1975) (see Fig. 25A and B). Presporal stages are seen adhering to the vacuolar membrane in a manner similar to a tight junction attachment (Weidner, 1976b). During sporogony the membrane becomes detached from the parasite surface, and sporoblasts and spores lie freely in the center of the vacuole (see Fig. 26C to F). The free membrane bordering the vacuole forms a rich canopy of blebs protruding into the vacuolar lumen (Weidner, 1976b) (see Fig. 27A). Other microsporidia, such as *Janacekia debaisieuxi*, induce lysis (Maurand, 1973; Weiser, 1976a; Larsson, 1983d), whereby the walls of infected cells are destroyed and the infected tissue is transformed into a syncytium, a cell soup in which thousands of microsporidian cells are dispersed among nuclei and organelles of the host cells.

Host Cell Enlargement

A common effect of microsporidiosis is hypertrophic enlargement of the host cell, including the nucleus (Weissenberg, 1976). In insects infected by microsporidians the effects are especially visible on the nuclei of the fat cells (Martins and Perondini, 1974). Not only the number of nuclei but also the volume of chromosomes is increased (Pavan et al., 1969). Comparing freeze-etched uninfected cells with microsporidia-infected cells, Liu (1972) observed larger particles on the nuclear envelopes of infected cells and found the nuclear pores of infected cells were larger in diameter. It is unusual for the nuclei of host cells to be invaded, as in infections caused by *Nosema bombycis* (Takizawa et al., 1973), *Nosemoides simocephali* (Loubès and Akbarieh, 1977), and *Nucleospora salmonis* (Hedrick et al., 1991).

The most striking example of hypertrophic growth is the formation of giant cells filled with microsporidian spores (a xenoma) as is frequently encountered in fish but also in other hosts. The classical example is the infection by *Glugea anomala* (Canning et al., 1982), a common parasite of epidermal cells of sticklebacks (*Gasterosteus* and *Pungitius* spp.), provoking a complex host-parasite association. The invaded cell is ruled by the parasite, and the host nucleus is forced to divide. The result is a xenoma, a giant epithelial cell containing thousands of host nuclei and millions of microsporidia (Fig. 21F). It is stratified, with mature spores in the center and immature stages at the periphery. The xenoma is externally covered by a multilayered coat into which fibroblasts and epithelial fragments are incorporated.

Infections in the Digestive Tract Epithelium

In the epithelium of the digestive tract the high regenerative capacity compensates to some extent for destruction, and microsporidia-filled epithelial cells are shed into the gut lumen (Fig. 21B). Such spore-filled cells can easily be mistaken for multisporous sporophorous vesicles.

COMMENTS ON THE STRUCTURE OF SOME HUMAN OPPORTUNISTIC MICROSPORIDIA

Only human microsporidia about which enough structural data are available to allow comparison with other microsporidia are considered here.

Enterocytozoon bieneusi

E. bieneusi Desportes, Le Charpentier, Galian, Bernard, Cochand-Priollet, Laverne, Ravisse and Modigliani, 1985, occurs relatively frequently in enterocytes of the small intestine of severely immunodeficient humans, and rarely in immunocompetent individuals. The ultrastructure was described by Desportes et al., (1985), Cali and Owen (1990), Desportes-Livage et al.,

(1991, 1996), Hilmarsdottir et al., (1993), and Orenstein (1991).

All data on the fine structure of the organism have been based on biopsy materials. The organism is apparently sensitive to fixation procedures, and some of its recognized ultrastructural characters (the rare presence of membranes, the paleness of the cytoplasm in early stages, and the extremely dilated perinuclear cisternae) are evidently due to handling of the material. When ferriosmium fixation is used, better preservation of cytoplasmic elements is achieved (Hilmarsdottir et al., 1993; Desportes-Livage et al., 1996).

E. bieneusi is monomorphic and has single nuclei throughout its life cycle (Cali and Owen, 1990). The nuclei in diplokaryotic arrangement reported in the original description were probably closely adjacent nuclei following karyokinesis. The sausagelike nuclei encountered in early plasmodia are evidently nuclei ready for rapid sequential mitoses (Fig. 22D), a prerequisite for an organism living in short-lived enterocytes. Spindle plaques were seen positioned across the short axis of the nucleus, suggesting that splitting of chromosome chromatids occurs first without physical separation into individual nuclei (Cali and Owen, 1990). Nuclear divisions evidently take place in rapid sequence, as the transition between long, sausagelike nuclei and round nuclei is rarely seen and shows deeply lobed nuclei (Orenstein, unpublished data). Actual karyokinesis, however, was observed only once, with the spindle plaques positioned across the short axis of the sausagelike nucleus in an invagination of the nuclear membrane, giving the elongate nucleus a dumbbell form (Spycher, unpublished data) (Fig.22E). Meiosis probably does not occur in *E. bieneusi,* as synaptonemal complexes were not seen in its nuclei. All stages occur in direct contact with the cytoplasm of the enterocytes in the zone between the host cell nucleus and the brush border. The parasites are often wedged into a shallow depression of the enterocyte nucleus (Fig. 22F) and surrounded by host cell mitochondria adhering to the parasite plasma membrane (Fig. 22D and 23B). Merogony in its classical sense (as a multiplicative stage) probably does not occur in *Enterocytozoon* or is very limited. Two adjacent parasite cells abutted without intervening host cell cytoplasm were seen (Cali and Owen, 1990; Orenstein, unpublished data), but the actual division was not observed. The isolated locations of individual *Enterocytozoon* cells in enterocytes suggest that all developmental stages are represented by a single cell, first uninucleate (Fig. 22C) and later multinucleate (plasmodium), in which organelles of future spores differentiate (Fig. 23B). The only cell division observed so far was that taking place during sporoblast formation (see Fig. 24A).

Conventional splitting of the microsporidian life cycle into merogony and sporogony, on the basis of secretion of electron-dense material on the cell plasma membrane, cannot be applied to *Enterocytozoon.* The parasite plasma membrane remains as a naked unit membrane until the sporoblast stage, a long time after the extrusion apparatus elements have formed in the plasmodium. Because this formation starts very early, it is impossible to define what stage is a meront or a sporont. Cali and Owen (1990) circumvented this problem by using the term "early proliferative stages," but it is doubtful that these stages really proliferate (divide). Terms such as "early uninucleate stages," "early plasmodia," "advanced plasmodia," "sporoblasts," and "spores" seem to describe the *E. bieneusi* developmental stages sufficiently.

Early uninucleate stages are small cells (about 1 μm) enveloped by a unit membrane, and having a single nucleus and cytoplasm with ribosomes and traces of ER (Fig. 22C). Early plasmodia are round or oval cells with several elongate, sausagelike nuclei and a few lamellae of ER. Expanded regions of the nuclear envelope (expanded perinuclear cisterna) or of rough ER form electron-lucent clefts with a dense border consisting of a multilamellar material of phospholipid composition (Desportes-Livage et al., 1993) (Fig. 22D). Small aggregates of Golgi vesicles are located at the distal ends of some ER lamellae and are identical with the "vesicular precursors of the

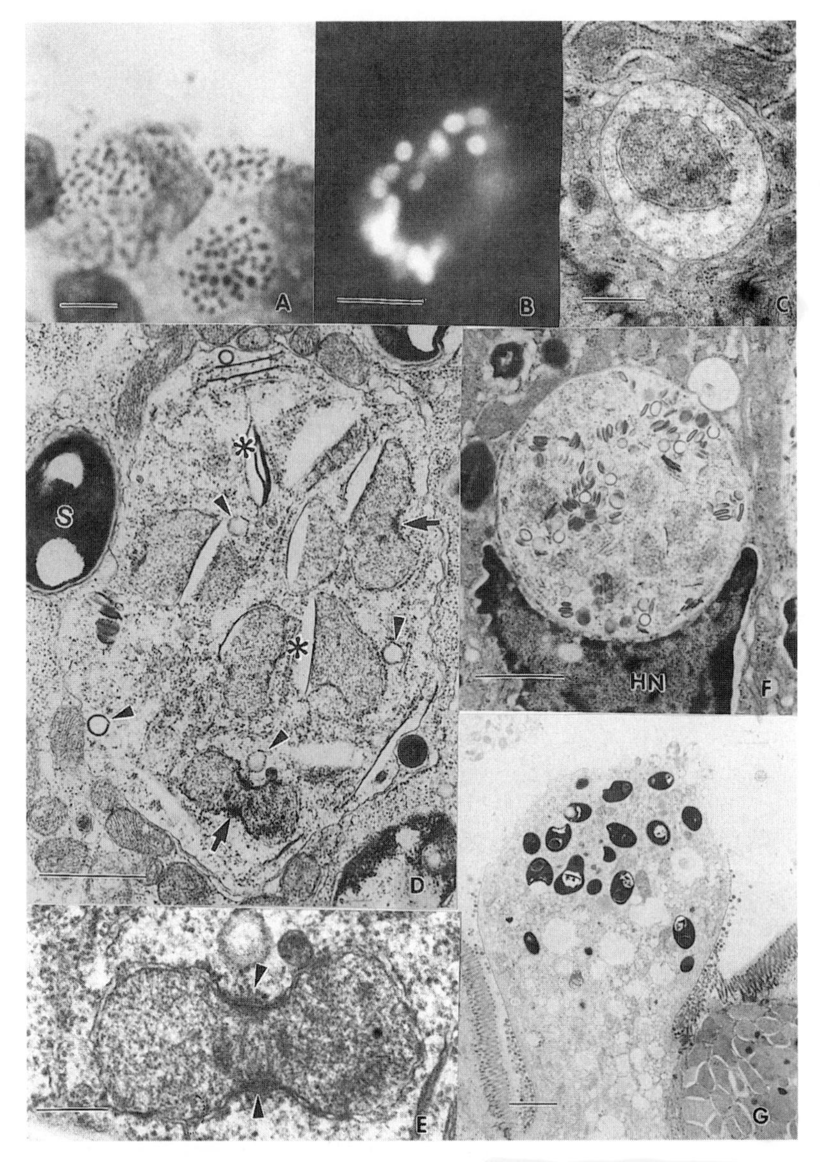

FIGURE 22 *Enterocytozoon bieneusi*. (A) Groups of spores originating from an individual plasmodium (Giemsa stain). (B) Spores of a sporogonial plasmodium in intestinal biopsy (fluorescence, Calcofluor staining). (C) Uninucleate, initial stage. (D) Multinucleate plasmodium with elongate nuclei, some of which are dividing (arrows). Expanded perinuclear cisternae contain dense material (asterisks). First primordia of the extrusion apparatus (possible polar sac) appear as dense rings (arrowheads). (E) Dividing nucleus as in (D). Spindle plaques (arrowheads) and the spindle are situated across the shorter axis of the nucleus. (F) Association of a multinucleate plasmodium with the nucleus of the host enterocyte. (G) Infected enterocyte with spores is released into the intestine lumen. Scale bars: A and B, 5 μm; C, 0.5 μm; D, 1 μm; E, 250 nm; F and G, 2 μm. (C to G) TEM. Reprinted with permission from M. A. Spycher (D and E) and J. M. Orenstein (F and G). Abbreviations: HN, host cell nucleus; S, spore.

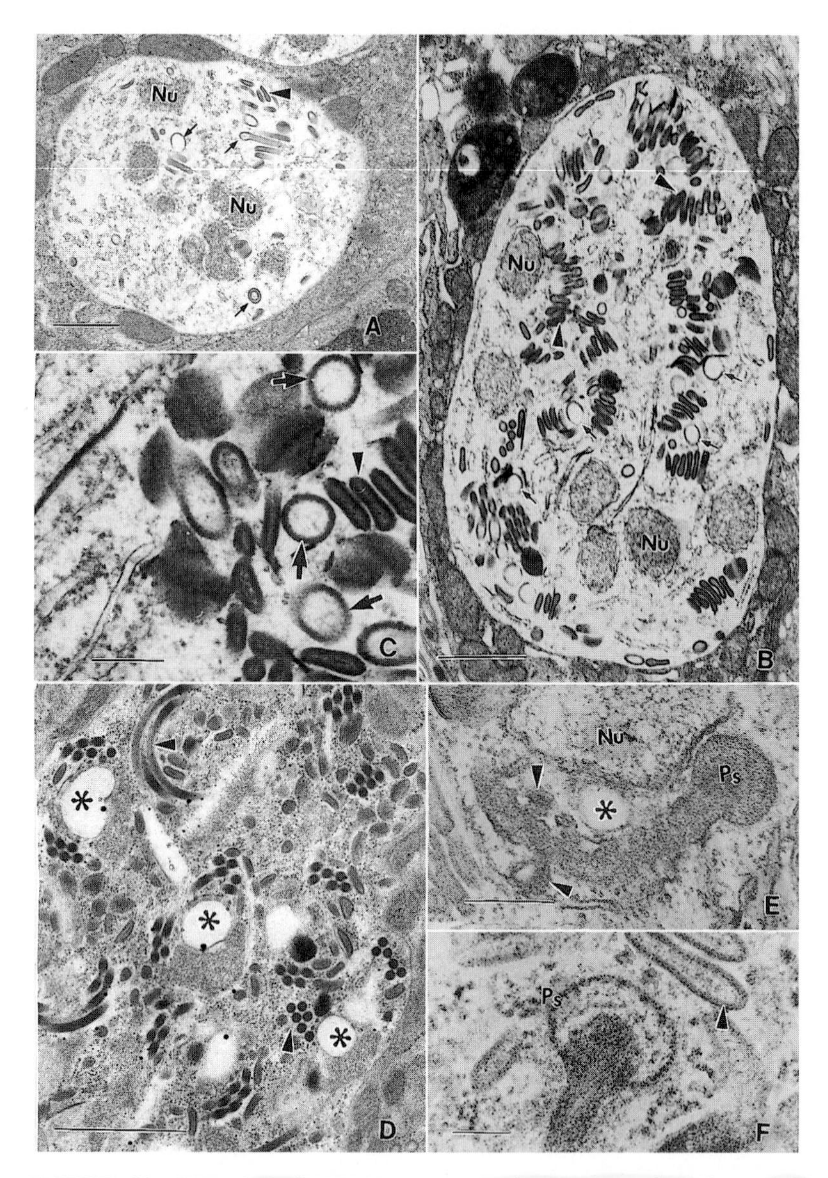

FIGURE 23 Formation of extrusion apparatus elements in multinucleate plasmodia of *E. bieneusi*. Polar sac (?) primordia are shown by arrows, and polar filament primordia are shown by arrowheads. (A) Initial stage. (B) Advanced sporogonial plasmodium. (C) Detail at high magnification. (D) More advanced plasmodium. A vacuole representing the primordium of the polaroplast and of the future posterior vacuole is associated with each nucleus (asterisk). (E) Both the polar sac and the vacuole (asterisk) are associated with the nucleus. Vesicular outgrowth of the polar filament probably represents formation of the polaroplast (arrowheads). (F) Detail of the polar sac. TEM was used for all panels. Scale bars: A, B, and D, 1 μm; C and E, 250 nm; F, 100 nm. Reprinted with permission from J. M. Orenstein (A and D), M. A. Spycher (B and C), and *J. Eukaryot. Microbiol.* and I. Desportes-Livage (E and F). Abbreviations: Nu, nucleus; Ps, polar sac.

polaroplast" described by Desportes-Livage et al., (1991) and Hilmarsdottir et al., (1993).

Advanced plasmodia are identified by the presence of many round nuclei and of several structures indicating the entry of the microsporidian into sporogony. Although these structures are described separately, they are probably parts of the same vesiculomembranous system forming the extrusion apparatus.

1. First to appear in the cytoplasm are several ovoid or spherical "empty-looking vesicles" (ELV) about 200 to 250 nm in diameter and limited by a thick, dense, but sometimes incomplete border (Fig. 23A and C). These structures are identical with the "polar tube precursors" of Hilmarsdottir et al. (1993) and of Desportes-Livage et al. (1996). These vesicles originate in association with the ER and are often seen lining the edges of ER cisternae and in contact with the electron dense lamellar inclusions in the ER (Cali and Owen, 1990) (Fig. 22D). The destiny of these vesicles is not clear. They remain identifiable until the time the electron dense disks (see item 2 below) begin to organize into stacks representing future polar filament coils. Each stack of disks, which will form a polar filament of one future spore, assembles around an ELV (Spycher, unpublished data) (Fig. 23B). It is hypothesized here that the ELV are primordia of polar sacs.

2. The second structure is represented by "electron-dense disk-like (EDD) structures" (Cali and Owen, 1990) appearing in sections as sausagelike formations with very dense borders and less dense centers (Fig. 23B and C). Ferriosmium fixation and higher resolution show that the borders of EDD structures are limited by a membrane and are multilayered (Hilmarsdottir et al., 1993; Desportes-Livage et al., 1996). Individual EDD structures seem to be bound together by membrane bridges (Ditrich et al., 1994). In addition to the ELVs, EED structures appear in the cytoplasm at the borders of the electron-lucent clefts (expanded ER lamella) and in the vicinity of Golgi vesicles. The disks accumulate first in stacks and are reported to fuse later to form arcs and complete coils of the polar filament (Fig. 23B and D).

3. Each nucleus becomes associated with a membrane-bound lucent vacuole (perinuclear vacuole, PNV) (150 to 200 nm) situated in a shallow depression of the nucleus (Fig. 23D and E). The PNV is probably identical with the "electron-lucent inclusions" in Fig. 3 of Cali and Owen (1990). The PNV increases in size during development and is often seen either incompletely or completely divided into two compartments (Fig. 23D). During this period the vacuole is about 400 to 500 nm in size. It is hypothesized here that one of the compartments gives rise to the polaroplast and the other to the posterior vacuole.

4. The polar sac is formed as a dense vesicle with a membranous border and is closely associated with the nucleus and partly invaginated in its membrane. The sac is located close to the PNV, and at this site the perinuclear cisterna contains electron-dense material (Fig. 23E). The polar filament appears as a filamentous extension of the polar sac (Fig. 23E). It has a dark core surrounded by a less dense layer bounded externally by a double membrane, the same membrane that encloses the polar sac. The dense core material penetrates a certain distance into the sac and inside the sac forms a cupola-like heap of dense material (Fig. 23F). During the advanced stage of sporogenesis, the polar sac flattens, the coils of the polar filament merge into a continuous thread, and small vesicles of the vesicular polaroplast appear as outgrowths of the membrane enclosing the straight part of the filament (Fig. 23E).

Finally, the complete set of extrusion apparatus organelles belonging to each nucleus is formed in the plasmodium (Fig. 24B). The individual polar sacs, each with its extrusion apparatus complement, assemble close to the plasma membrane of the plasmodium (Orenstein, 1991) (Fig. 24C). Invagination of the plasma membrane, concurrently thickened by addition of the surface coat, separates the uninucleate sporoblasts, each with a complete set of spore organelles (Fig. 24A). Sporogony is

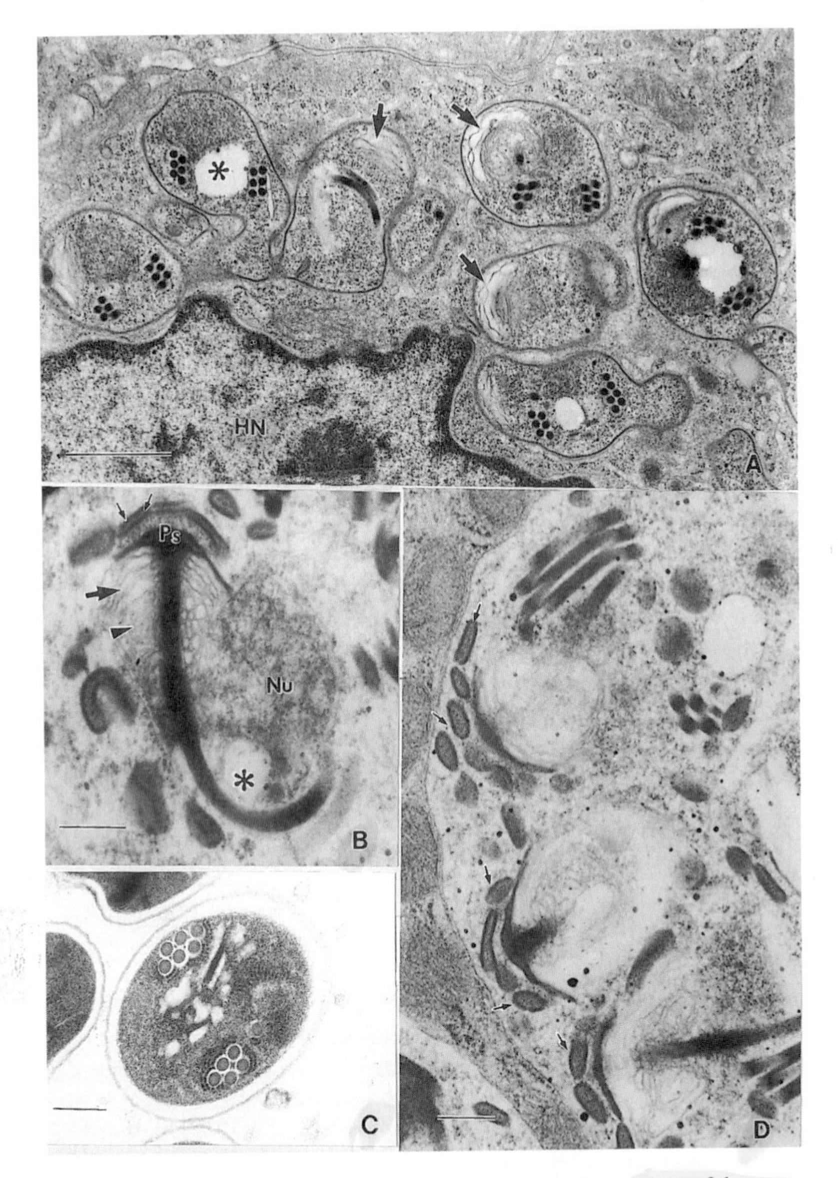

FIGURE 24 Spore formation in *Enterocytozoon bieneusi*. (A) Origin of the sporoblasts. In the sporoblast the former vacuole is split into an anterior vacuolar space, in which a lamellar polaroplast is formed (arrows) and a posterior part, the future posterior vacuole (asterisk). Polar filament coils start to be organized in the final form. (B) Detailed view of the polar sac, polar filament, lamellar polaroplast (large arrow), vesicular polaroplast (arrowhead), nucleus, and future posterior vacuole (asterisk). Note several dense vesicular structures above the polar sac and the opaque material between these vesicles and the polar sac (small arrows). (C) Nearly mature spore with a very thin exospore, relatively thick endospore, and characteristically coiled polar filament forming two rows. (D) Beginning of sporoblast formation. The anterior parts of the extrusion apparatus assemble near the plasmodium surface. Note that each polar sac is exteriorly lined with elongated vesicles (small arrows). TEM was used for all panels. Scale bars: A, 1 μm; B to D, 250 nm. Reprinted with permission from J. M. Orenstein and *J. Parasitol.* (C and D). Abbreviations: Nu, nucleus; HN, host cell nucleus; Ps, polar sac.

thus polysporoblastic (up to about 60 spores arising from one sporogonial plasmodium) (Vávra, unpublished data) (Fig. 22A and B) with precocious spore organelle formation. Spore maturation involves formation of the posterior vacuole and of the endospore layer beneath the surface coat (exospore).

Spores, 1.5 by 0.9 μm, are ellipsoid and have a single nucleus. Despite many observations, their fine structure is poorly known, as most observed spores were not completely mature. Fully mature spores have a relatively thick (40-nm) endospore and a thin (13-nm) single layer exospore (Orenstein, 1991). The polaroplast has a thin anterior lamellar part and a posterior vesicular part. Five to six isofilar coils of the polar filament are arranged in two rows (Fig. 24D). Teratoid sporogenesis probably caused by the AIDS drug azidothymidine was described by Ditrich et al. (1994). It included the formation of large, deeply lobed sporogonial plasmodia, incompletely separated sporoblasts, and unusually large spores (2.5 μm) with 10 coils of the polar filament.

The unique nature of this microsporidian resides primarily in the precocious development of spore organelles within the plasmodium and late initiation of formation of the electron-dense coat on the sporogonial plasmodium outer face. This order of events is probably due to the necessity for the organism to complete its life cycle within a short lived enterocyte (4 to 5 days) (Desportes-Livage et al., 1991) with the spores being shed into the intestinal lumen (Orenstein, 1991) (Fig. 22G). The dense, multilamellar phospholipid material occurring early in the ER lamellae evidently serves as a pool of material for the polar filament and membranous structures of the future spore. The outer coat of dense material is formed late, probably because the relatively large plasmodium, with its many nuclei, synthesizing extrusion apparatus organelles must have a very high metabolic rate and transmembrane transport must not be hindered.

If the development and structure of *E. bieneusi* are compared with that of other microsporidia, its closest counterpart is *Nucleospora salmonis* (previously classified in the genus *Enterocytozoon*), a parasite of nuclei of immature blood cells of salmonid fish. In both species the extrusion apparatus develops early, before fission of the sporogonial plasmodium into sporoblasts (Desportes-Livage et al., 1996). In principle, however, both microsporidia share characteristics common to the "typical microsporidia" (see "Typical Microsporidia").

Encephalitozoon spp.

Three species of the genus *Encephalitozoon, E. cuniculi* Levaditi, Nicolau, and Schoen, 1923, *E. hellem* Didier, Didier, Friedberg, Stenson, Orenstein, Yee, Tio, Davis, Vossbrinck, Millichamp, and Shadduck, 1991, and *E.* (formerly *Septata*) *intestinalis* Cali, Kotler and Orenstein, 1993, are known to infect humans. These species are morphologically similar, and only *E. intestinalis* can be distinguished from the other two species by structural characters. Whether *E. intestinalis* rightfully belongs to the genus *Encephalitozoon* or should be retained in its original genus, *Septata,* is discussed by Cali et al. (1996b). The genus *Encephalitozoon* is monomorphic, and single nuclei occur in all life cycle stages. Development typically takes place in a parasitophorous vacuole presumably of host origin. Reports of direct development in host cell cytoplasm are discussed below. The limiting membrane of the parasitophorous vacuole appears first as a unit membrane so closely applied to the plasma membrane that the meront seems to have two unit membranes at its surface (Barker, 1975) (Fig. 25A and B). Later, the outer membrane surrounding the parasite starts to detach, and small electron-lucent areas appear between the parasite plasma membrane and the vacuolar membrane (Fig. 25C and 26E). As the parasite grows and divides, a single vacuolar space develops in which parasite cells are situated (Fig. 25D, 26B and C, and 27E and F). The vacuolar space contains meronts adhering to the vacuolar membrane (Fig. 26C and F) by an attachment resembling a tight junction (Weidner, 1975, 1976b). Meronts divide by repeated binary fission, and

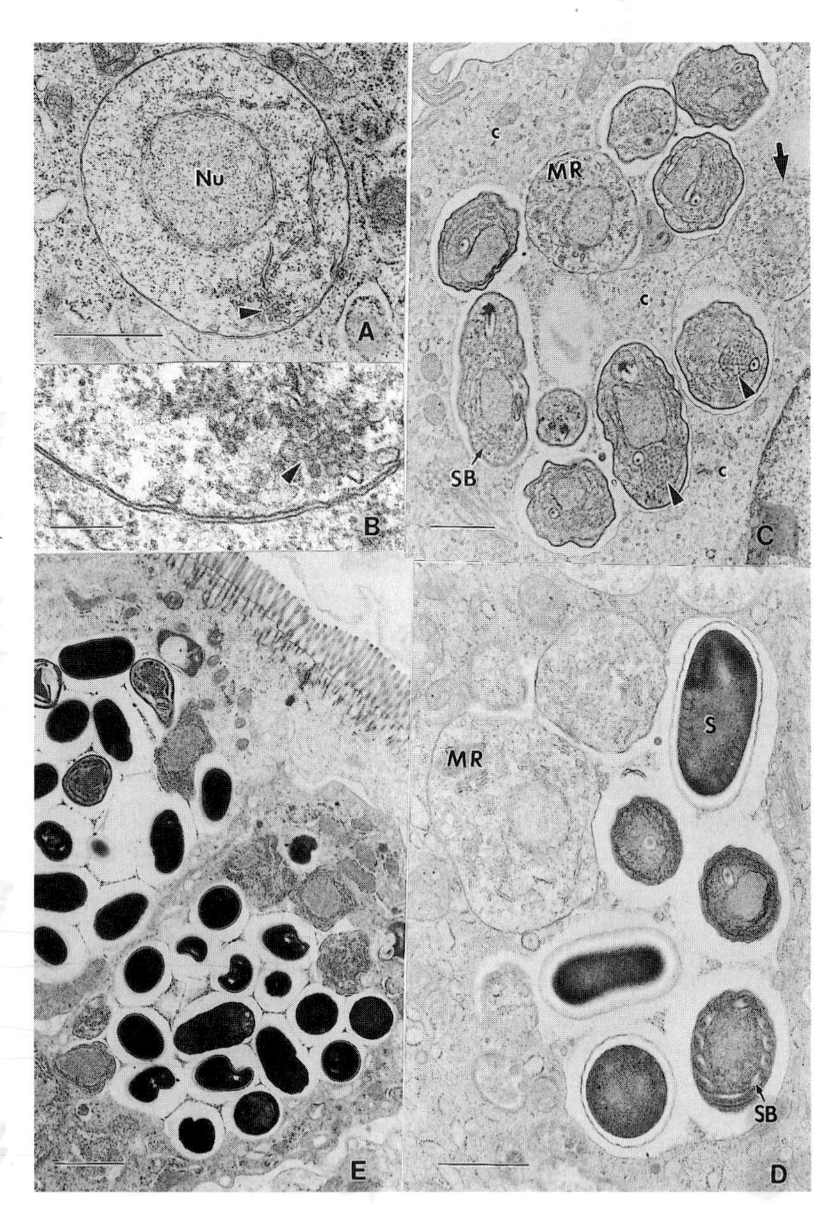

FIGURE 25 *Encephalitozoon* (syn. *Septata*) *intestinalis* (Golgi is shown at arrowhead). (A) Initial meront surrounded by snugly adhering membrane of the future parasitophorous vacuole. (B) Detailed view of the membranous complex of the meront in panel A. (C and D) A large parasitophorous vacuole originates by progressive confluence of individual vacuoles around dividing parasites. A meront (arrow) is fully embedded in the host cell cytoplasm. (E) Fully formed vacuole with septa forming incomplete chambers around individual spores. TEM was used for all panels. Scale bars: A, C, and D, 1.0 μm; B, 250 nm, E, 2 μm. Reprinted with permission from M. A. Spycher (A to D) and J. M. Orenstein (E). Abbreviations: c, host cell cytoplasm; MR, meronts; Nu, nucleus; SB, sporoblasts; S, spore.

because of delayed cytokinesis, small chains of meronts are formed (Fig. 26C). Sporonts are characterized by a layer of dense material on their plasma membrane, first deposited in strands and later as a homogeneous layer (Barker, 1975). Sporonts detach from the vacuole membrane (Fig. 26F). Sporogony takes place in the lumen of the vacuole (Fig. 26C and F) and is thought to be disporoblastic and occasionally tetrasporoblastic. The border of the parasitophorous vacuole that is not in direct contact with parasite cells forms a canopy of blebs extending into the vacuolar space (Weidner, 1975, 1976b) (Fig. 27A).

The parasitophorous vacuole of *E. intestinalis* has a special character, being prominently lobed and containing septa consisting of a granular material (Fig. 25C). The septa form incomplete chambers enclosing individual parasites within the vacuole and the material

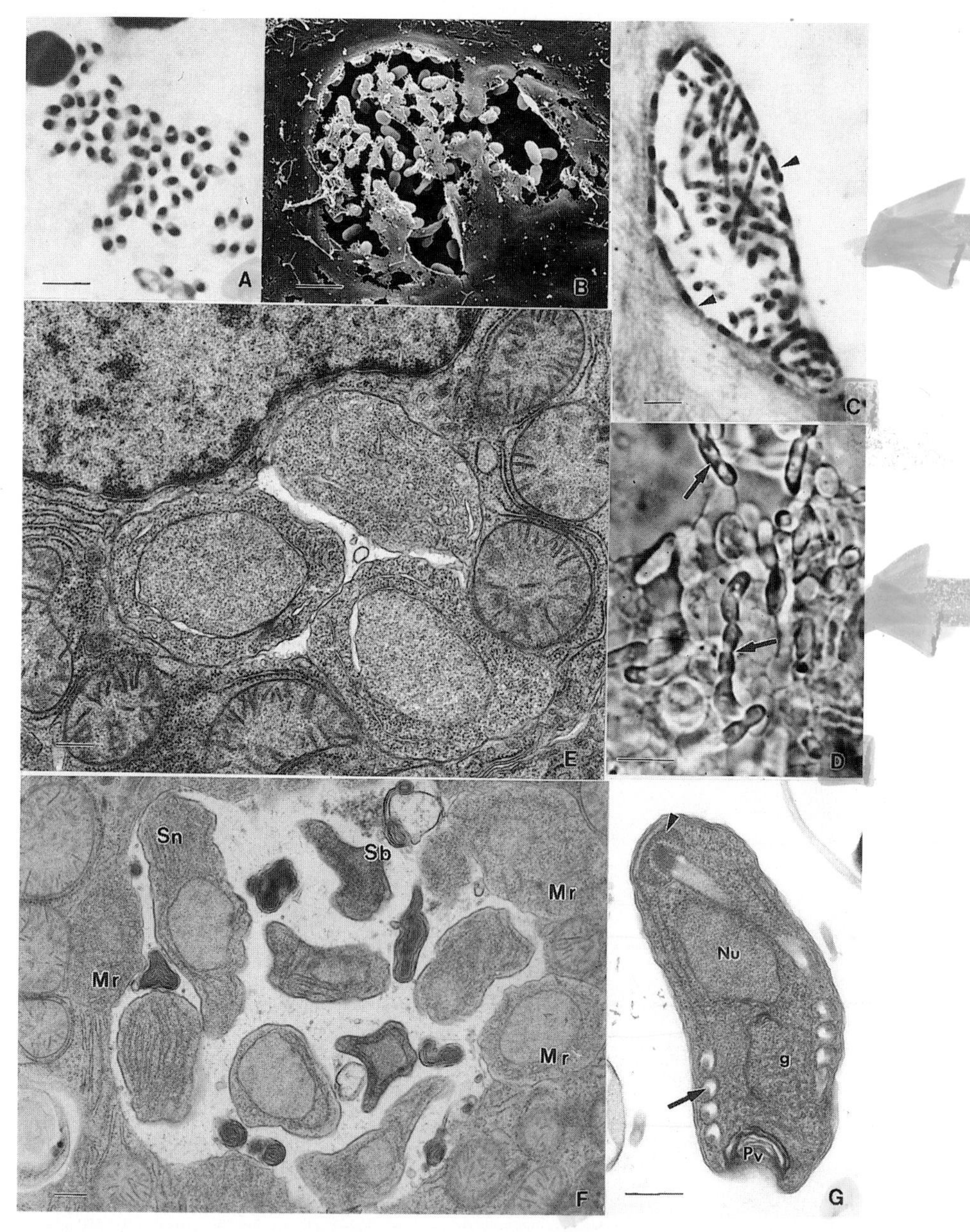

FIGURE 26 *Encephalitozoon cuniculi.* (A) Spores on a smear (Giemsa stain); (B) SEM view of a broken parasitophorous vacuole with spores; (C) parasitophorous vacuole with meronts lining the vacuole borders (arrowheads) and sporogonial stages and spores in center of vacuole; (D) chains of sporoblasts inside the parasitophorous vacuole (arrows) (Giemsa stain); (E) early parasitophorous vacuole originating around three meronts (phase contrast); (F) fully formed parasitophorous vacuole with meronts adhering to vacuole border, sporonts, and sporoblasts detached to the vacuole interior; (G) young spore with organelles; arrow indicates polar filament early form; arrowhead indicates polar sac. (A) Mouse peritoneal exudate; (B and E to G) liver of SCID mouse; (C and D) rabbit chorioid plexus cell tissue culture. (E to G) TEM. Scale bars: A to D, 5 μm; E to G, 0.5 μm. Reprinted with permission from B. Koudela (B and E) and *Folia Parasitol.* (C and D). Abbreviations: g, Golgi reticulum; Mr, meront; Nu, nucleus; Sb, sporoblast; Sn, sporont; Pv, posterior vacuole (collapsed).

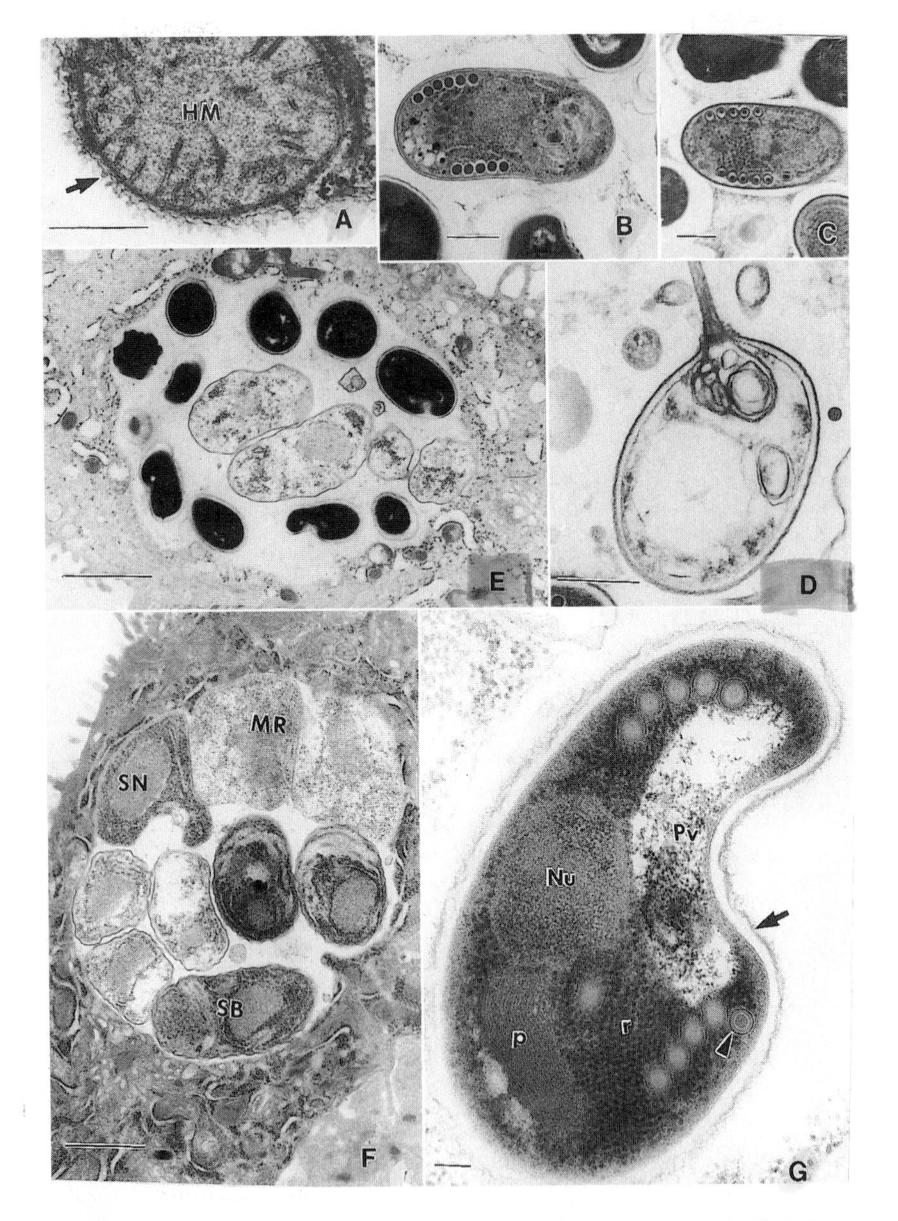

FIGURE 27 *Encephalitozoon* spp. (A) Border of a parasitophorous vacuole. The canopy of blebs protruding into the vacuole is shown at arrow. (B to D) *E. intestinalis*. (B and C) Young spores with five filament coils; (D) extruded spore. (E to G) *E. hellem*. (E) Fully formed parasitophorous vacuole; (F) earlier vacuole with meronts, sporont, and sporoblasts transforming into spores; (G) spore with organelles. A polar filament with a substructure of concentric rings is shown at the arrowhead. The three layers of the spore wall, the plasma membrane, and the endo- and exospore are indicated by the arrow. TEM was used for all panels. Scale bars: A to D, 0.5 µm; E, 2 µm; F, 1 µm; G, 100 nm. Reprinted with permission from J. M. Orenstein (B to D) and A. Curry (E to G). Abbreviations: HM, host cell mitochondrion; MR, meront; Nu, nucleus; P, polaroplast lamellae; r, area with polyribosomes; SB, sporoblast; SN, sporont; Pv, posterior vacuole (distorted).

forming the septa is believed to be secreted by the parasites (Cali et al., 1993). Canning et al. (1994) believe that the septa originate from compression of slightly electron opaque, loosely dispersed reticulate material present in *Encephalitozoon* spp. parasitophorous vacuoles that disappears during sporogony in *E. cuniculi* and *E. hellem*. Evidently, the interior of the parasitophorous vacuoles in *Encephalitozoon* is more structured than the loose contents seem to indicate. The fact that the sporoblasts and spores of *E. intestinalis* have a seemingly loose halo of a rather constant width (about 0.3 μm), separating them from the septa (Fig. 25E), favors this idea. The lobed appearance of the parasitophorous vacuole of *E. intestinalis* is a result of the fusion of vacuoles surrounding individual parasitic cells accompanied by host cell cytoplasmic debris caught as vacuoles coalesce (Fig. 25D). The fusion of several larger vacuoles was also described (Canning et al., 1994).

Spores, 2.5 by 1.5 μm, are ellipsoid and uninucleate (Fig. 26A and B) with a rugose exospore and a moderately thick endospore (Fig. 25D and 27G). At low magnifications the exospore looks like a single layer about 30 nm thick, but high-resolution TEM shows that it has three layers: an outer electron-dense "spiny layer" (12 nm), an intermediate electron-lucent lamina (3 nm), and an inner dense fibrous layer (15 nm) (Bigliardi et al., 1996). The polaroplast is lamellar, and the polar filament is isofilar with three to eight coils in a single row (Fig. 26G and 27B, C, and G). A posterior vacuole is sometimes present (Fig. 26G and 27G). When released from the parasitophorous vacuole, the spores frequently extrude their polar filaments (Fig. 27D), a prerequisite for efficient intertissue transmission of the parasite.

Despite a number of reports on the subject, certain points of *Encephalitozoon* spp. structure are still not clear. Whether the parasite can grow in direct contact with host cell cytoplasm as claimed by several authors (Cali et al., 1996b; Desser et al., 1992) should be reinvestigated. The origin of the parasitophorous vacuole is intriguing. It has a similar configuration in many types of infected cells, appearing as an additional unit membrane surrounding the parasite. It has been proposed that this membrane might be the host cell plasma membrane originally invaginated by the extruding polar tube (Magaud et al., 1997). This proposal implies that the manner of entry of *Encephalitozoon* into host cells is different from that of other microsporidia which pierce the host cell with the evaginating polar tube. The published micrograph of an *E. intestinalis* polar tube invaginating the plasma membrane of a macrophage (Magaud et al., 1997) is not persuasive in this respect, as it can also be interpreted as a macrophage pseudopodium engulfing a spore with a previously evaginated polar tube.

Whether *Encephalitozoon* sporogony is disporoblastic or tetrasporoblastic has been argued (Cali et al., 1993; Canning et al., 1994; Cali et al., 1996b). This feature may depend on the relationship of the speed of growth to the cytokinesis of individual cells and on conditions under which the parasite has been grown. In tissue culture rather long chains of sporonts (Vávra et al., 1972; Desser et al., 1992) are formed (Fig. 26D).

Trachipleistophora spp.

Two known species of the genus *Trachipleistophora, T. hominis* Hollister, Canning, Weidner, Field, Kench and Marriott, 1996, and *T. anthropophthera* Vávra, Yachnis, Shadduck, and Orenstein, 1998, were found in corneal scrapings, in muscle biopsy specimens, and in systemic infections of AIDS patients, respectively.

T. hominis, the type species of the genus, was described as monomorphic and having a single nucleus. The meronts have a dense coat on their plasma membrane, forming branched processes into the host cell cytoplasm (Fig. 28A). This coat is the precursor of the sporophorous vesicle (of the merontogenetic type) and divides with the body of the meront. Merogony is probably limited to binary fission, as no chains or multinucleate merogonic plasmodia were observed. Sporogony begins with retraction of the sporont plasma membrane within the surface coat, which becomes a rather thick-walled envelope of the

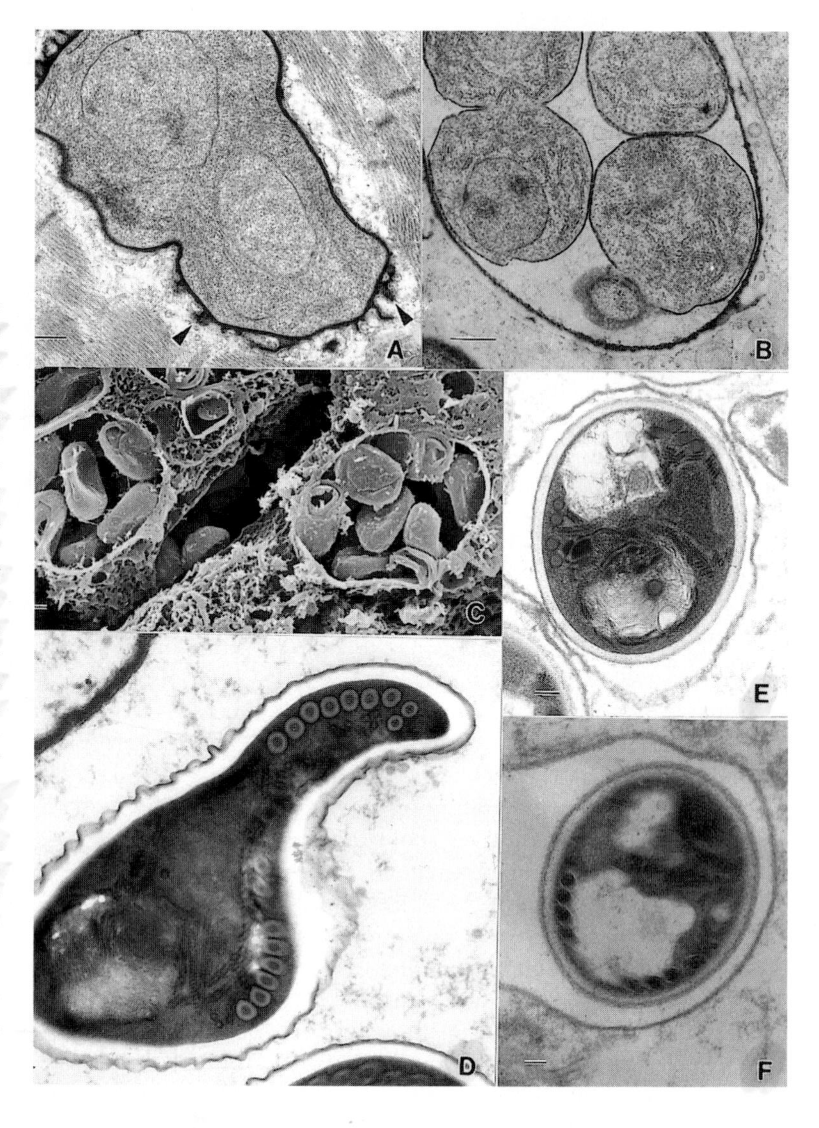

FIGURE 28 *Trachipleistophora* spp. (A) Binucleate meront in skeletal muscle cell, showing the plasmalemma covered by dense material forming connections between meronts (arrow) and protuberances into the host cell cytoplasm (arrowheads); (B) sporophorous vesicle with four presporoblast cells; (C) sporophorous vesicles with type I spores; (D) thick-walled type I spore with seven thick and two thin coils; (E and F) thin-walled type II spores with four isofilar coils. (A and B) TEM of *T. hominis* in SCID mouse skeletal muscle (A) and RK-13 cell tissue culture (B); (C to F) *T. anthropophthera* in human brain (SEM [C] and TEM [D to F]). Scale bars: A, 0.5 μm; B and C, 1 μm; D to F, 100 nm. Panel B is reprinted with permission from E. U. Canning.

sporophorous vesicle (Fig. 28B). The sporont divides by a series of binary fissions into uninucleate sporoblasts and spores, 2 to 32 per vesicle (Fig. 29B). Spores are pyriform, 4.0 by 2.4 μm (fresh) (Fig. 29A to C), and have an anterolateral anchoring disk, a lamellar polaroplast with two different types of lamellar spacing, and about 11 coils of polar filament in a single row. Observations demonstrate that the polar filament is anisofilar, usually with three thinner posterior coils (Canning, unpublished data; Vávra, unpublished data).

T. anthropophthera is similar to the type species of the genus but differs by being dimorphic; it forms two types of sporophorous vesicles, each enclosing spores of a different type. Sporophorous vesicles of type I are similar to those of the type species of the genus and usually contain eight or more thick-walled spores (3.7 by 2.0 μm, fixed) (Fig. 28C and 29D, E, and G). Type I spores are similar in organization to those of the type species of the genus but have an anisofilar polar filament usually with seven thicker and two thinner coils (Fig. 28D).

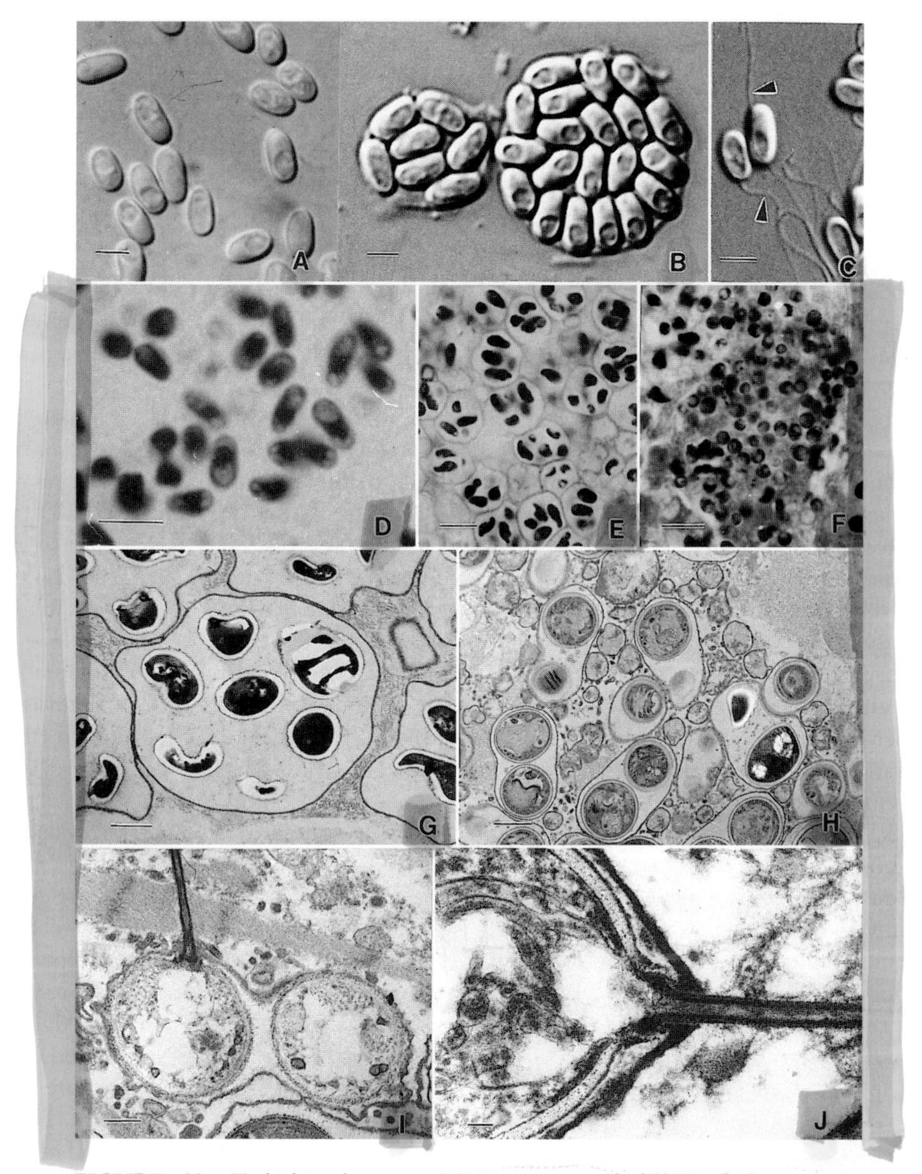

FIGURE 29 *Trachipleistophora* spp. (A) Free spores (fresh); (B) fresh spores in sporophorous vesicles; (C) extruded spore; the polar tube extrudes subapically (arrowheads); (D) stained spores (Goodpasture's carbolfuchsin stain); (E) type I spores in sporophorous vesicles (semithin plastic section, methylene blue-azure II-basic fuchsin staining); (F) aggregate of type II spores in histology section (hematoxylin and eosin stain); (G) type I spores in a multisporous sporophorous vesicle; (H) type II spores in bisporous sporophorous vesicles; (I) extruded type II spore; (J) detail of tip of extruded type II spore, showing the tubular aspect of the polar tube. (A to C) *T. hominis,* skeletal muscle homogenate of SCID mice. (D to J) *T. anthropophthera,* human brain (D to H) and human heart muscle (I and J). (A to C) Interference phase contrast; (G to J) TEM. Scale bars: A to F, 5 μm; G and H, 1 μm; I, 0.5 μm; J, 100 nm. Reprinted with permission from B. Koudela (B and C) and J. M. Orenstein (J).

Sporophorous vesicles of type II are formed in the same tissues and sometimes even in the same cell as type I vesicles (Fig. 29F). Type II sporophorous vesicles are thin-walled and contain only two type II spores (Fig. 29H), which are small, nearly round (2.2 to 2.5 by 1.8 to 2.0 µm, fixed), and thin-walled and have four or five isofilar polar filament coils (Fig. 28E and F). It is believed that type II spores serve in the dissemination of infection within the host tissues as they are frequently found with their polar tubes extruded (Orenstein, unpublished data) (Fig. 29I and J).

Vittaforma corneae

V. corneae (Shadduck, Meccoli, Davis, and Font, 1990) Silveira and Canning, 1994, was isolated from human corneal stroma. The organism was also found in the urinary tract of an AIDS patient (Deplazes et al., 1997). This parasite is a monomorphic species having diplokarya throughout its life cycle. Synaptonemal complexes were not observed in nuclei. The nuclear configuration was the reason why the organism was originally described as *Nosema corneum* Shadduck, Meccoli, Davis, and Font, 1990.

The host-parasite interface is peculiar and appears to involve three parallel, closely apposed unit membranes. This arrangement can be explained by the parasite being surrounded by the host ER cisterna: the innermost membrane is the parasite plasma membrane (PP), the next outer membrane (so closely applied that it appears to be a second parasite membrane) is presumed to be the ribosome-free component of the host cell ER membrane (HERM1), and the outermost membrane is studded with ribosomes and represents the other part of the host ER cisterna (HERM2). This peculiar membranous arrangement occurs in all stages including spores and represents the main distinguishing character of the genus *Vittaforma* (Fig. 30B to G). Merogony proceeds as binary fission of diplokaryotic stages, and both the HERM1 and the HERM2 divide with the meront. "Clefts" in the form of an enlarged part of the perinuclear cisterna in meronts have been described (Silveira and Canning, 1995). Dark material, similar in appearance to that present in perinuclear clefts of *E. bieneusi*, was seen in clefts of *Vittaforma* meronts, but it is probable that these clefts were fixation artifacts.

During sporogony, elongate, ribbon-shaped sporonts (plasmodia) are formed as a result of delayed cytokinesis (Fig. 30F). The sporonts are marked by deposition of electron-dense material on the PP, first in small patches appearing as curved, lenticular structures which progressively form a uniform dense coat on the sporont. The division is effected by means of deep clefts perpendicular to the long axis of the ribbonlike sporogonial plasmodium. The PP, together with the electron-dense coat and the inner face of the host ER (HERM1), invaginate and separate the plasmodial ribbon into up to eight linear arrays of sporoblasts. Membranous whorls resembling paramural bodies are formed at the bottom of the invaginating furrows. During sporoblast individualization, the HERM2 looses its close contact with the surface of the parasite but, as a loose sheath still completely envelops, each sporoblast and spore (Silveira and Canning, 1995).

The spores (in tissue culture) are polymorphic, typically appearing as elongate cylinders (Fig. 30A). Small, ovoid spores or spores resembling long rods are also present. Spore size is reported to be 3.7 by 1.0 µm (Shadduck et al., 1990), or 3.8 by 1.02 µm (Silveira and Canning, 1995). However, when fresh, the spores measure only 2.8 by 1.1 µm (typical form). The ovoid forms are half the indicated size, and long spores are twice as large as the typical form (image-splitting eyepiece measurements) (Vávra, unpublished data). The spores are diplokaryotic. The polar filament is isofilar with about six coils (Fig. 30H), and the exospore is about half as thick as the endospore and is surrounded by loose membranes (Fig. 30G). The polaroplast is composed of tightly packed lamellae situated around the straight portion of the polar filament and a few posterior lamellae in groups of four (Silveira and Canning, 1995).

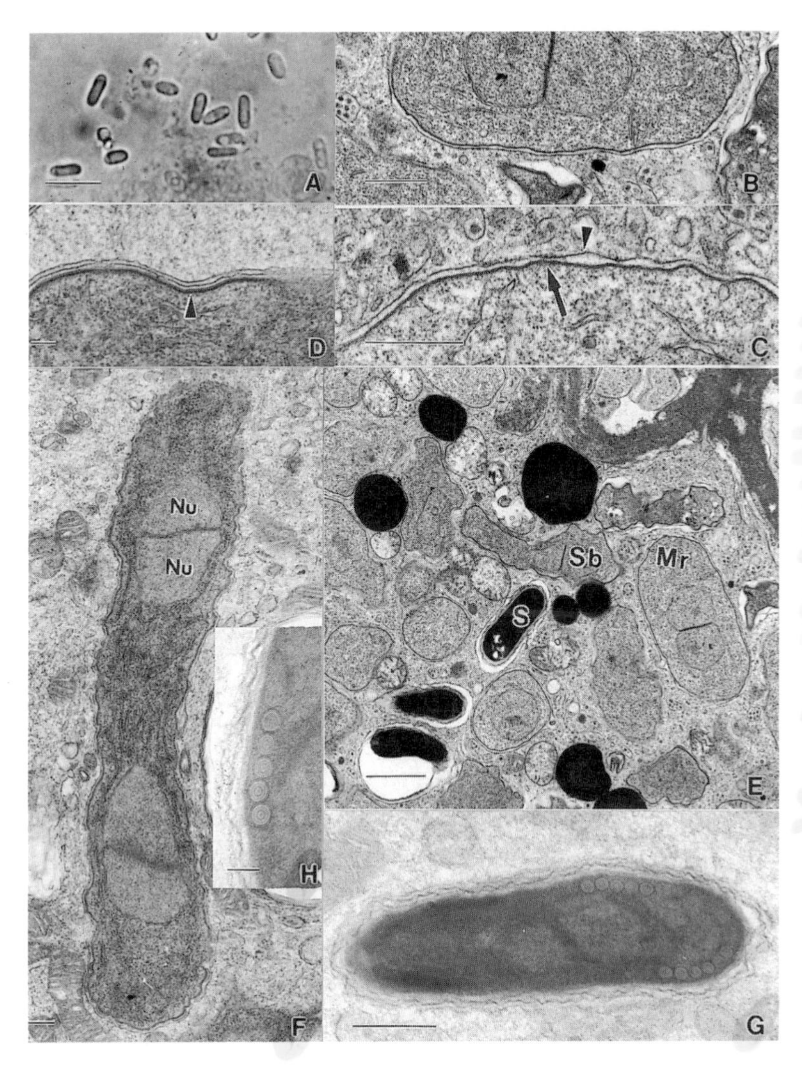

FIGURE 30 *Vittaforma corneae.* (A) Fresh spores (phase contrast); (B) meront; (C) meront enclosed by an inner membranous complex (arrow) of two adhering membranes (the inner one supposedly belonging to the parasite and the outer one belonging to the host) and one outer host membrane (arrowhead); (D) cell boundary of the sporont, showing electron-dense material deposited on the innermost (parasite) membrane (arrowhead); (E) low-magnification view of infected mouse liver; (F) dividing diplokaryotic sporont; (G) mature spore enclosed in several membranous layers; (H) polar filament coils with substructure. (B to H) TEM. Scale bars: A, 5 μm; B and C, F, and G, 0.5 μm; D and H, 100 nm; E, 1 μm. Reprinted with permission from J. A. Shadduck (D and F to H). Abbreviations: Mr, meront; Nu, nucleus; Sb, sporoblast; S, spore.

This microsporidian is unique in having a complex of three unit membranes covering the organism, the external two of which are presumed to be of host origin. The persistence of this membranous complex during the entire life cycle, including the spore stage, suggests that these membranes are transformed in some way by the microsporidian and cannot simply be considered host cell membranes. This situation is in contrast to that in a number of microsporidia that are encased in host cell ER cisterna, but only during merogony (for more details see "Techniques in Microsporidial Research" below and Silveira and Canning, 1995). The situation in *Vittaforma* resembles that in *Endoreticulatus* spp. in which meronts are tightly enveloped by the host ER. However, in *Endoreticulatus,* sporogonic divisions occur in a vacuolar space bordered by the host ER, and in contrast to the arrangement in *Vittaforma* the spores are not enclosed in individual sacs of host cell origin (Brooks et al., 1988; Cali and El Garhy, 1991).

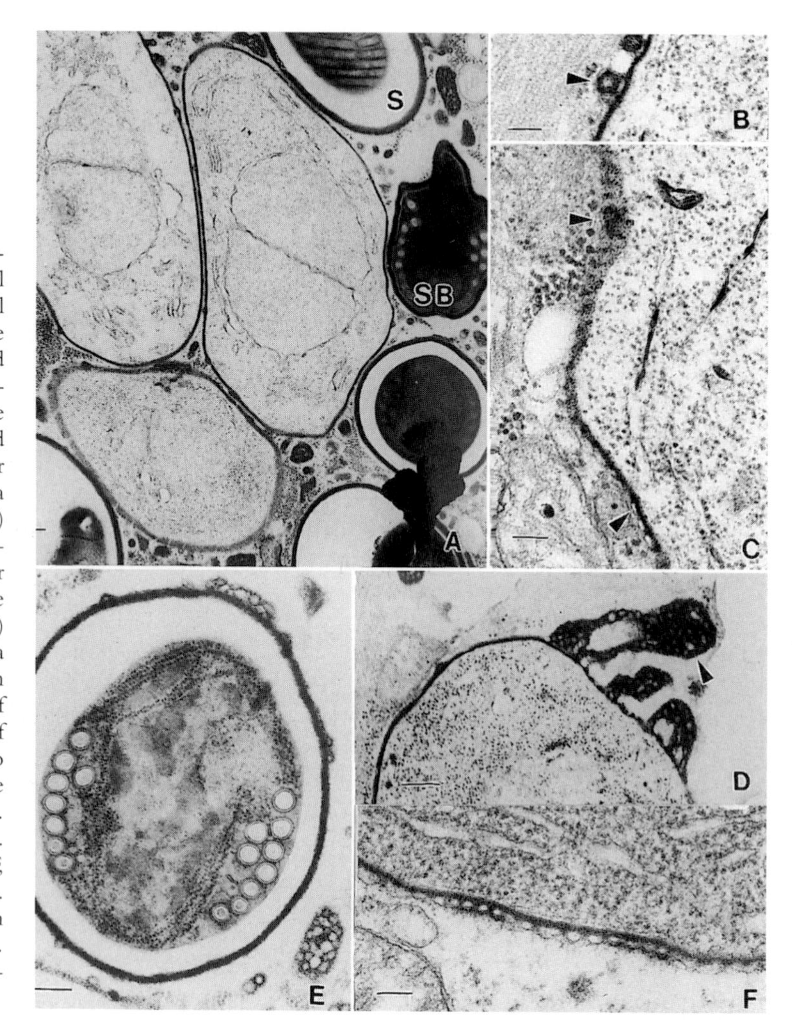

FIGURE 31 *Brachiola vesicularum* from human skeletal muscle. (A) Developmental stages (meronts?), showing the diplokaryotic nuclei; (B and C) details of the electron-dense material deposits in the form of membrane-bound tubules (B) or dense tubular strands (C) at the plasma membrane (arrowhead); (D) parasite cell showing large expansion of the vesiculotubular electron-dense material at the cell surface (arrowhead); (E) spore with PF arranged in a double coil; (F) construction of the meront surface of *Nosema algerae* parasite of *Anopheles* mosquito larvae is to a certain degree similar to the meront surface of *Brachiola*. TEM was used for all panels. Scale bars: A, 1 μm; B, C, and F, 100 nm; D, 250 nm; E, 0.5 μm. Reprinted with permission from A. Cali and *J. Eukaryot. Microbiol.* (A to E). Abbreviations: SB, sporoblast; S, spore.

Brachiola vesicularum

B. vesicularum Cali, Takvorian, Lewin, Rendel, Sian, Wittner, Tanowitz, Keohane and Weiss, 1998, was described from a skeletal muscle biopsy specimen from an AIDS patient (Cali et al., 1996a, 1998). It is a monomorphic, diplokaryotic species developing in direct contact with the host cell cytoplasm. Cells with a single diplokaryon or dividing cells with two diplokarya were seen, but it cannot be established whether they were meronts or sporonts, as in all stages the plasma membrane was covered by a thick coat of electron-dense material deposited in tubulovesicular strands (Fig. 31A to C) and sometimes extending a certain distance from the parasite surface (Fig. 31D). Sporogony is disporoblastic, producing spores 2.9 by 2.0 μm in size with 8 to 10 coils of the polar filament arranged in one to three, usually two, rows (Fig. 10E). The polar filament is anisofilar, 2 or 3 of the last coils being thinner.

In its life cycle and structure, this microsporidian is similar to the genus *Nosema*. However, it differs from typical *Nosema* spp. by having on its plasma membrane a thick layer of electron-dense material forming extensive vesiculotubular appendages. In this respect it is similar to *Nosema algerae* Vávra and Undeen, 1970, a mosquito parasite, which, as molecular biology indicates, is phylogenetically distant from typical representatives of the genus *Nosema* (Baker, Vossbrinck, Maddox, and Undeen, 1994) (Fig. 13F). In contrast to *N. algerae,* however, *B. vesicularum* grows at mammalian body temperature and forms more extensive vesiculotubular appendages.

TECHNIQUES IN MICROSPORIDIAL RESEARCH

Techniques useful in studying microsporidian structure have been summarized by Vávra and Maddox (1976), Hazard et al. (1981), and Undeen and Vávra (1997). The most important methods are reported here.

Diagnosis

In most cases the spores are the stage leading to diagnosis of a microsporidial infection. Developmental stages of microsporidia other than spores are inconspicuous and are not resistant to environmental conditions after the death of the host. The spores are resistant and persist for more than a year in dead hosts and tissues (Weiser, 1961), making them excellent for diagnostic purposes. The diagnosis of human microsporidian infections is treated in detail in Chapter 10.

The Appearance of Infected Hosts and Tissues and the Making of Smears

Spore-filled tissues of a more-or-less transparent host may exhibit a milky appearance caused by light being scattered off the spores. The milky appearance together with the background coloring of the host can cause a milky-yellow, green, or rosy appearance in the infected host or organs.

The parasites are normally studied in smears from infected tissues. Thus, in preparing a smear, it is important to use a technique that preserves the dividing stages in a way that allows modes of division to remain visible and avoids destruction of sporophorous vesicles. If the smear is prepared like a bacteriological smear, the spore groups, which are the visible sign of sporogony in sporophorous vesicles, are lost unless the vesicle envelope is unusually resistant. A squash preparation preserves these features better. Thus, a small piece of microsporidia-filled tissue is squashed in a small drop of water or saline between two microscope slides. When the slides are separated, it is important to avoid gliding one slide over the other (Larsson, 1988c).

RECOGNIZING MICROSPORIDIAN SPORES

Microsporidian spores are in most cases similar in size and shape to spores of other organisms, such as other protists, bacteria, and fungi. They may even resemble metazoan sperm, such as those of mites. There are several techniques that allow the investigator to determine that spores are from microsporidia.

Gram Positivity. Microsporidian spores are thick-walled structures and are able to retain stain complexes during destaining. They are positive by Gram staining and also by many other methods based on this technique. The Goodpasture-Perrin method for staining encephalitozoa gives the same result as classical Gram staining and is recommended for both smears and paraffin sections of embedded material (Vávra and Maddox, 1976).

Goodpasture-Perrin Method for Staining Encephalitozoa (Lillie, 1965) (Modified). (1) Fix a dry smear from infected tissue or a piece of an organ in any type of fixative. In the latter case proceed by standard techniques (paraffin embedding, sectioning) in order to obtain deparaffinized sections on a slide. (2) Place the smear or tissue section in distilled water by routine techniques. (3) Stain with Goodpasture's aniline-carbolfuchsin (0.39 g of basic fuchsin, 1 ml of aniline, 1 g of crystalline phenol, 100 ml of 30 % ethanol) for 5 min at 60° to 70° C. (4) Rinse in distilled water. (5) Destain in 40% formaldehyde for 10 to 20 min. (6) Rinse in distilled water. (7) Counterstain in a saturated solution of picric acid in water. (8) Wash in 95% ethanol. (9) Dehydrate in 100% ethanol followed by xylene. (10) Mount in a neutral mounting medium (e.g., Permount, Histoclad, DPX).

Result: Spores are deep red on a slightly yellow background.

Periodic Acid-Schiff Positivity. When spores are subjected to oxidation by periodic acid and then stained by Schiff's reagent, a minute red granule (a polar cap or a polar sac-anchoring disk complex) stains at one pole of the spore. This method is specific for distinguishing microsporidian spores from similar-appearing spores of other organisms (Vávra, 1959). The staining reveals polysaccharides with 1, 2-glycol groups located in the anchoring apparatus of the filament. The oxidation of glycol groups provides an aldehyde that is demonstrated by Schiff's leukofuchsin reagent (Hotchkiss, 1948).

Staining of the Polar Cap by the PAS Reaction (Lillie, 1965). This method can be applied both to smears and to deparaffinized tissue sections. Any fixative can be used, but aldehyde fixatives may give nonspecific staining. (1) Place the material in distilled water. (2) Allow oxidation in a 0.5% to 1.0% solution of periodic acid in distilled water for 8 to 10 min. (3) Allow reduction for 20 s in a mixture of 100 ml of distilled water, 2.0 g of potassium iodide, 2.0 g of crystalline sodium thiosulfate, and 20 ml of 1 N HCl. (4) Wash in distilled water for 2 min. (5) Stain with Schiff's reagent (leukofuchsin) for 15 min. (For preparation, see any histology manual.) (6) Wash twice in water containing SO_2 (5 ml of 1 N HCl, 6 ml of 10% sodium bisulfite, and 100 ml of H_2O distilled). The first quick rinse serves to wash out most of the Schiff's reagent, and the second rinse, for 4 min, washes out the remaining traces of the reagent. (7) Wash well in tap water. (8) Stain in Harris's hematoxylin (or another hematoxylin stain) and counterstain with 1% light green SF in distilled water. (9) Dehydrate quickly in ascending ethanol and mount.

Result: Spores show a minute red granulum at one spore pole (use oil immersion for observation). A short staining time should be used in the hematoxylin step in order not to mask the PAS positive structure.

Chitin Detection. Microsporidian spores have chitin in the endospore layer of the spore wall (see The Endospore). Although this substance is not limited to microsporidian spores, detection of its presence by optical brighteners (e.g., Calcofluor or Uvitex) offers a useful diagnostic technique. Optical brighteners are organic compounds with aromatic rings as part of their structure which emit light in the visible spectrum upon ultraviolet or short-wave light excitation. The theoretical basis of chitin labeling with brighteners is presented in Vávra et al. (1993a).

Labeling of Microsporidian Spores with Calcofluor White M2R (Vávra et al., 1993). 1. Fresh material. Mix the spore-containing material with a 0.01% solution of Calcofluor White M2R (American Cyanamid Co.) or Fluorescence Brightener 28 (Sigma) in 0.1 M phosphate buffer, pH 8.0. Alternatively, flood a dry smear with the same solution. Apply a coverslip and observe by fluorescence microscopy with a violet or blue excitation filter. The spore wall fluoresces strongly. Fixed (but not picric acid containing fixatives) or nonfixed material can be equally effective. The same result will be obtained if the Calcofluor solution is washed off after a brief application and replaced by buffered glycerol or another mounting medium suitable for fluorescence. Buffered glycerol is obtained by mixing 0.25 M sodium carbonate buffer, pH 9.5, with glycerol in a proportion of 1:9. 2. If fresh material (e.g., spores) has been stored for a long period or the material has been fixed, it is suggested that a 0.001% Calcofluor White M2R solution in 1 N NaOH be used. This solution is also recommended for rapid examination of tissue biopsy specimens in which the dye achieves fast penetration of tissues crushed between the slide and the coverslip. Fluorescence detection of microsporidia with this method can be completed in minutes (Vávra et al., 1993b).

Remark: A number of optical brighteners that give results similar to those obtained with Calcofluor are the active substances of several commercial chitin detection kits used by mycologists. Van Gool et al. (1993) use Uvitex B (Ciba-Geigy, Basel, Switzerland) with success in the clinical diagnosis of microsporidiosis (Chapter 10).

Fluorescence Screening of Microsporidia in Histological Sections (Weir and Sullivan, 1989) (Modified). (1) Fix material in Carnoy's fixative. Do not use Bouin's fixative or any other fixative containing picric acid. (2) Process the material routinely into paraffin. (3) Stain deparaffinized sections with a hematoxylin stain and do not counterstain with eosin. (4) Wash off the stain and flood the slides with 1% Uvitex 2B (Ciba-Geigy) or a 0.01% solution of Calcofluor White M2R in 0.1 M phosphate-buffered saline, pH 7.0 to 8.0. The staining time is not critical since binding of the fluorescent agent is nearly instantaneous. Wash well in water and either observe wet under a coverslip or mount in buffered glycerol. Observe under fluorescent microscopy with a violet or blue excitation filter.

Polar Filament Extrusion. Definite proof that an organism is a microsporidian is obtained when the spore extrudes its filament. Polar filament extrusion is, however, a capricious event. Some microsporidia extrude filaments easily without any special stimulation, whereas others are quite unresponsive to mechanical and chemical stimuli. No single reliable method can be recommended for filament extrusion, and experiments may be necessary for each species. Useful hints for initiating spore germination can be found in Undeen and Vávra (1997). The following methods may be successful, but in any case not all spores are extruded. Many spores may also be incompletely extruded.

1. Pressure. Apply thumb pressure on the coverslip of a wet preparation with spore material (for safety reasons cover the slide with a piece of filter paper).

2. Hydrogen peroxide. Mix a suspension of material containing spores in a 1:1 proportion with 3 to 6% H_2O_2 or with the same hydrogen peroxide concentration in 0.1 M KCl.

3. Alkaline priming. Mix the material to be examined (spore suspension) in a 1:1 proportion with 0.2 M KOH and allow to stand for about 30 min. Centrifuge and discard the supernatant. On a slide, mix a very small portion of the sediment with a large drop of Dulbecco's phosphate-buffered saline, pH 7.3 (without Mg/Ca), containing 0.93 g of KCl in 100 ml, apply a coverslip, and observe. (Modified method of Yasunaga et al., 1995).

4. Rewetting of spores. Try to flood with tap water a freshly dried smear of material with spores.

5. To a drop of material with spores in water on a slide add a few small crystals of KCl. Cover and observe. Convection currents during salt dissolution will move the spores through different concentrations of KCl; some spores will be primed and will be extruded.

Transmission Electron Microscopy. Another unambiguous method for proving that an organism is a microsporidian is TEM. The recommended handling of microsporidia for TEM has been summarized by Becnel (1997). Routine techniques are used for fixation (two fixation times are recommended: normal for presporal stages, and extended periods up to 24 h or more for spores; spores can be stored in glutaraldehyde for long periods of time) and embedding, but spores require longer infiltration times in the resin. Useful information on the structure of spores can be obtained by reprocessing histologic materials embedded in paraffin. Even materials on stained slides provide information on the polar filament structure and the spore wall when processed for electron microscopy. For such purposes, after the mounting medium is removed by soaking in xylene, sections are rehydrated to 50% ethanol and transferred via 0.1 M cacodylate buffer into 1% OsO_4 in a buffer. After they are washed and dehydrated in ethanol, they are transferred to propylene oxide, to mixtures of 70:30 and 30:70 propylene oxide:epoxy embedding resin, and finally to resin alone. A prepolymerized block of resin is placed on top of each section, and the slide is incubated overnight at 65°C to allow polymerization of the new resin. The slide is then placed on a hot plate, and the section, now embedded in resin, is snapped off. Sections are cut from the exposed face of the block (Curry, personal communication).

OBSERVATION AND MEASURING OF MICROSPORIDIAN SPORES

Although microsporidian spores seem to be quite uniform in appearance (spores of most species being oval or pyriform), they are valuable for diagnosis provided that their shape and size are observed in vivo and properly described. Especially in the case of microsporidia from water-dwelling invertebrate hosts, the spores bear ornamentations or mucous layers on their surfaces that can be used as taxonomic characters (see "Microsporidians under the Light Microscope" and "The Exospore").

Staining of Microsporidia. Giemsa staining of methanol-fixed smears is the standard technique for microsporidia. When preparing the smear, it is important to apply the technique described in "The Appearance of Infected Hosts and Tissues and the Making of Smears."

Sections of paraffin-embedded material can be stained by a variety of methods, although the Giemsa staining (Short and Cooper, 1948) and Gomori's trichrome staining methods are recommended (Vávra, 1976c). Excellent results are obtained with hematoxylin stains. Modified Heidenhain's iron hematoxylin stain is relatively fast and also gives excellent results:

Iron Hematoxylin Staining (Sprague, 1981). (1) Place sections in distilled water. (2) Mordant in Lang's solution (300 ml of 4% ferric ammonium sulfate, 5.0 ml of acetic acid, and 0.6 ml of sulfuric acid) for 10 to 30 min or longer (time is not critical). (3) Wash in running tap water for 5 min. (4) Rinse in distilled water. (5) Stain in 0.5% aqueous hematoxylin (for preparation, see any histology manual) about one or two times as long as the time in mordant (time is not critical). (6) Destain in a saturated picric acid solution in distilled water for 10 to 60 min (control the amount of destaining under the microscope). (7) Wash for 1 h in running tap water. (8) Counterstain with eosin, light green, etc. (9) Dehydrate, clear, and mount.

Immobilization of Spores. The microscopist is confronted with the fact that in fresh mounts the spores are moving as a result of Brownian movement, making it difficult to determine shape and size. Mounting the spores on an agar monolayer prevents Brownian movement and allows an exact measurement of size and a recording of shape.

Spore Monolayers for Size and Shape Recording (Vávra, 1964). (1) Prepare a 1.5% solution of agar melted in hot, distilled water. (2) Pour a thin layer of agar on a slide and let it cool and gel (the surface of the agar layer should be a little convex and smooth). (3) Put a *small* drop of a *thick* spore suspension on a coverslip. (4) Invert the coverslip and place it on the agar layer in a single motion. Do not move it any further.

Result: There will be areas where the spores are immobilized in a single layer between the agar and the coverslip. Although the preparation can be observed immediately, sometimes it improves after several hours after the agar layer starts to dry down and the spores are pressed more firmly into the agar layer. The preparation can be stored (in a wet petri dish or sealed with paraffin) for no more than several days. An alternative to this method is the technique described by Hostounský and Žižka (1979) in which agar-coated slides are prepared in advance and are dried and stored in the cold. A coverslip with spores is then put on the dried agar, which absorbs the water, swells to some degree, and immobilizes the spores.

Recording of Shape. A photomicrograph of fresh spores provides more information than the best verbal description. It is surprising that very few descriptions of new microsporidian taxa include good photomicrographs showing the important features of the spores in question. Good photographs not only show the shape and size of the spores but also provide information on the size and position of the posterior vacuole.

For photographic recording use high-contrast negative material (e.g., material for reproduction of line drawings). Do not close the condenser aperture of the microscope excessively, and do not use phase contrast because the outline of spore contours is distorted by diffraction lines. The entire photographic process should be aimed at obtaining maximum print contrast.

Measuring of Spore Size. The size of spores is an important structural character, but its exact recording is difficult, especially in microsporidia with very small spores. Current ocular micrometers do not provide enough accuracy since it is difficult to superimpose the scale exactly with the refractive spore edge. The use of a special measuring eyepiece (Image Splitting Eyepiece; Vickers Instruments, Ltd., York, United Kingdom) is strongly recommended (Kramer, 1964). A more modern method is the use of a computerized image analysis program. Measuring spores on photographs, made with a carefully calibrated microscope and photographic enlarger, is a suitable alternative if a special eyepiece or image analyzer is not available. Fixation causes spores to shrink up to 25%, which means that measurements of spore samples can be compared only if the samples have been treated in an identical way.

Demonstration of Exospore Ornamentations and the Mucocalyx. Spores of microsporidia infecting aquatic organisms often bear appendages in the form of tails, fibers, mucous layers, etc. (see "The Exospore"). Relatively thick elements of this kind can be shown by negative staining methods, i.e., techniques in which the stain does not penetrate the spore structures and they therefore appear unstained on a stained background. Fine ornamentations, however, can be seen only with electron microscopy.

Mucocalyx. Mix a large drop of a fresh spore suspension in water with a small drop of India ink. Apply a coverslip and observe. If the spores have a mucocalyx, it will appear as a transparent area, i.e., a halo, around the spore (Lom and Vávra, 1963b) (Fig. 1M).

Spore appendages. Mix a spore suspension in water with a small drop of bacteriological ink (either original Burri ink for bacteriology or a 5 to 10% solution of water-soluble nigrosine in water employing the modification of Deflandre (1923). Make a smear and let the preparation dry. Spore appendages are revealed as white fibers on a dark background. The thickness of the smear and the amount of dye are critical. This method is one of the simplest ways to demonstrate spores in a smear (Vávra, 1963).

Spore Nuclei. Nuclear configuration is an important structural marker. Current staining methods (e.g., Giemsa and hematoxylin), fail to reveal nuclei in spores properly since the spore contents stain too deeply. Also, the Feulgen reaction, which is DNA-specific, stains spore nuclei indistinctly (Jírovec, 1932). The best method is Giemsa staining following hydrolysis, which decreases the stainability of the cytoplasm surrounding the nucleus.

Giemsa Staining after Hydrolysis for Microsporidian Nuclei (Piekarski, 1937) (Modified). (1) Fix the smear of spores in any conventional fixative. (2) Wash in distilled water. (3) Hydrolyze in 1 N

HCl at 60°C for 2 to 10 min. Time is critical and should be adjusted for each species. (4) Wash well in several changes of tap water followed by distilled water to remove all traces of the acid. (5) Stain in a conventional way with Giemsa stain.

Result: The nuclei appear bright red in an almost colorless spore.

Remark: This technique is practically the only staining method that shows microsporidian nuclei satisfactorily. Do not mistake nuclei for posterosomes (see "The Posterior Vacuole"), which also stain in the posterior portion of young spores.

A simple variation of this technique, which provides less control of the process, involves heating a dried spore smear covered with a large drop of acid over a small flame of a Bunsen burner until the first fumes appear. Further washing and staining should then proceed as stated above (Weiser, 1976b).

ACKNOWLEDGMENTS

We express our sincere thanks to the following colleagues who kindly contributed to the chapter by providing photomicrographs, material, information, or criticism: J. J. Becnel (USDA, ARS, Gainesville, Fla.), A. Cali (Rutgers University, Newark, N.J.), E. U. Canning (Imperial College, London, United Kingdom), A. Curry (Whittington Hospital, Manchester, United Kingdom), I. Desportes-Livage (INSERM, Paris, France), I. Fries (Swedish University of Agricultural Sciences, Uppsala, Sweden), D. Koudela (Institute of Parasitology, České Budějovice, Czech Republic), J. M. Orenstein (George Washington University, Washington, D.C.), E. Porchet-Henneré (Université de Lille, Lille, France), J. Schrével (Musée d' Histoire Naturelle, Paris, France), J. A. Shadduck (Heska Corp., Fort Collins, Colo.), K. Snowden (Texas A & M University, College Station, Tex.), M. A. Spycher (University Hospital, Zurich, Switzerland), B. S. Toguebaye (Université Cheik Antal Diop, Dakar, Senegal), D. Vinckier (Université de Lille, Lille, France), R. Weber (University Hospital, Zurich, Switzerland), E. Weidner (Louisiana State University, Baton Rouge, La.), J. Weiser (Institute of Entomology,, České Budějovice, Czech Republic). L. Gefors and I. Norling (Lund, Sweden), and H. Kulikova (Prague, Czech Republic) are thanked for excellent technical assistance. We express our sincere thanks to P. M. Takvorian (Rutgers University, Newark, N.J.) and L. M. Weiss (Albert Einstein College of Medicine, Bronx, N.Y.) for their assistance in preparation of the figures for this chapter.

The Larsen Foundation (Lund, Sweden), the Swedish Natural Science Research Council, Charles University (Prague, Czech Republic; grant 280/1996/B BIO), and the National Institutes of Health (grant AI31788) provided economic support.

REFERENCES

Alfa, C. E., and J. S. Hyams. 1990. Many routes lead to the pole. *Nature* **348:**484.

Andreadis, T. G. 1983. Life cycle and epizotiology of *Amblyospora* sp. (Microspora: Amblyosporidae) in the mosquito, *Aedes cantator. J. Protozool.* **30:** 509–518.

Baker, M. D., C. R. Vossbrinck, J. V. Maddox, and A. H. Undeen. 1994. Phylogenetic relationships among *Vairimorpha* and *Nosema* species (Microspora) based on ribosomal RNA sequence data. *J. Invertebr. Pathol.* **64:**100–106.

Barker, R. J. 1975. Ultrastructural observation on *Encephalitozoon cuniculi* Levaditi, Nicolau et Schoen, 1922, from mouse peritoneal macrophages. *Folia Parasitol.* **22:**1–9.

Batson, B. S. 1982. A light and electron microscopical study of *Trichoduboscqia epeori* Léger, 1926 (Microspora, Duboscqiidae). *J. Protozool.* **29:**202–212.

Batson, B. S. 1983. A light and electron microscope study of *Hirsutosporos austrosimulii* gen. n., sp. n. (Microspora: Nosematidae), a parasite of *Austrosimulium* sp. (Diptera: Simuliidae) in New Zealand. *Protistologica* **19:** 263–280.

Beard, C. B., J. F. Butler, and J. J. Becnel. 1990. *Nolleria pulicis* n. gen., n. sp. (Microsporida: Chytridiopsidae), a microsporidian parasite of the cat flea, *Ctenocephalides felis* (Siphonaptera: Pulicidae). *J. Protozool.* **37:**90–99.

Becnel, J. J. 1992. Horizontal transmission and subsequent development of *Amblyospora californica* (Microsporida, Amblyosporidae) in the intermediate and definitive hosts. *Dis. Aquat. Org.* **13:**17–28.

Becnel, J. 1994. Life cycles and host-parasite relationships of Microsporidia in culicine mosquitoes. *Folia Parasitol.* **41:**91–96.

Becnel, J. J. 1997. Complementary techniques: preparations of entomopathogens and diseased specimens for more detailed study using microscopy, p. 335–365. *In* L. Lacey (ed.), *Manual of Techniques in Insect Pathology.* Academic Press, San Diego, Calif.

Becnel, J. J., and T. Fukuda. 1991. Ultrastructure of *Culicosporella lunata* (Microsporida: Culicosporellidae fam. n.) in the mosquito *Culex pilosus* (Diptera: Culicidae) with new information on the developmental cycle. *Eur. J. Protistol.* **26:**319–329.

Becnel, J. J., E. I. Hazard, T. Fukuda, and V. Sprague. 1987. Life cycle of *Culicospora magna* (Kudo, 1920) (Microsporida: Culicosporidae) in *Culex restuans* Theobald with special reference to sexuality. *J. Protozool.* **34:**313–322.

Becnel, J., V. Sprague, T. Fukuda, and E. I. Hazard. 1989. Development of *Edhazardia aedis* (Kudo, 1930) n. g., n. comb. (Microsporida: Amblyosporidae) in the mosquito *Aedes aegypti* (L.) (Diptera: Culicidae). *J. Protozool.* **36:**119–130.

Becnel, J. J., and A. W. Sweeney. 1990. *Amblyospora trinus* n. sp. (Microsporida: Amblyosporidae) in the Australian mosquito *Culex halifaxi* (Diptera: Culicidae). *J. Protozool.* **37:**584–592.

Berrebi, P. 1978. Contribution a l'étude biologique des zones saumâtres du littoral mediterranéen français. Biologie d'une microsporidie: *Glugea atherini* n. sp. parasite de l'atherine: *Atherina boyeri* Risso, 1810 (Poisson-Teleostéen) des étangs côtiers. Ph.D. Thesis, Académie de Montpellier, Université des Sciences et Techniques du Languedoc, Montpellier, France.

Bigliardi, E., M. G. Selmi, P. Lupetti, S. Corona, S. Gatti, M. Scaglia, and L. Sacchi. 1996. Microsporidian spore wall: ultrastructural findings on *Encephalitozoon hellem* exospore. *J. Eukaryot. Microbiol.* **43:**181–186.

Bronnvall, A., and J. I. R. Larsson. 1995a. Description of *Binucleospora elongata* gen. et sp. nov. (Microspora, Caudosporidae), a microsporidian parasite of ostracods of the genus *Candona* (Crustacea, Cyprididae) in Sweden. *Eur. J. Protistol.* **31:** 63–72.

Bronnvall, A., and J. I. R. Larsson. 1995b. Description of *Lanatospora tubulifera* sp. n. (Microspora, Tuzetiidae) with emended diagnosis and new systematic position for the genus *Lanatospora*. *Arch. Protistenkd.* **146:**69–78.

Brooks, W. M., J. J. Becnel, and G. G. Kennedy. 1988. Establishment of *Endoreticulatus* n. g. for *Pleistophora fidelis* (Hostounský and Weiser 1985) (Microsporida: Pleistophoridae) based on the ultrastructure of a microsporidium in the Colorado potato beetle *Leptinotarsa decemlineata* (Say) (Coleoptera: Chrysomelidae). *J. Protozool.* **35:** 481–488.

Brooks, W. M., E. I. Hazard, and J. Becnel. 1985. Two new species of *Nosema* (Microsporida: Nosematidae) from the Mexican bean beetle *Epilachna varivestis* (Coleoptera: Coccinellidae). *J. Protozool.* **32:**525–535.

Burges, H. D., E. U. Canning, and I. K. Hulls. 1974. Ultrastructure of *Nosema oryzaephili* and the taxonomic value of the polar filament. *J. Invertebr. Pathol.* **23:**135–139.

Bylén, E. K. C., and J. I. R. Larsson. 1994. *Napamichium cellatum* n. sp. (Microspora, Thelohaniidae), a new parasite of midge larvae of the genus *Endochironomus* (Diptera, Chironomidae) in Sweden. *J. Eukaryot. Microbiol.* **41:**450–457.

Bylén, E. K. C. and J. I. R. Larsson. 1996. Ultrastructural study and description of *Mrazekia tetraspora* Léger & Hesse, 1922 and transfer to a new genus *Scipionospora* n. g. (Microspora, Caudosporidae). *Eur. J. Protistol.* **32:**104–115.

Cali, A., and M. El Garhy. 1991. Ultrastructural study of the development of *Pleistophora schubergi* Zwoelfer, 1927 (Protozoa: Microsporida) in larvae of the spruce budworm, *Choristoneura fumiferana* and its subsequent taxonomic change to the genus *Endoreticulatus*. *J. Protozool.* **38:**271–278.

Cali, A., D. P. Kotler, and J. M. Orenstein. 1993. *Septata intestinalis* n. g., n. sp., an intestinal microsporidian associated with chronic diarrhea and dissemination in AIDS patients. *J. Eukaryot. Microbiol.* **40:**101–112.

Cali, A., and R. L. Owen. 1990. Intracellular development of *Enterocytozoon,* a unique microsporidian found in the intestine of AIDS patients. *J. Protozool.* **37:**145–155.

Cali, A., P. M. Takvorian, S. Lewin, M. Rendel, C. Sian, M. Wittner, H. B. Tanowitz, E. Keohane, and L. M. Weiss. 1998. *Brachiola vesicularum,* n. g., n. sp., a new microsporidium associated with AIDS and myositis. *J. Eukaryot. Microbiol.* **45:**240–251.

Cali, A., P. M. Takvorian, S. Lewin, M. Rendel, C. Sian, M. Wittner, and L. M. Weiss. 1996a. Identification of a new *Nosema*-like microsporidian associated with myositis in an AIDS patient. *J. Eukaryot. Microbiol.* **43:**108S.

Cali, A., L. M. Weiss, and P. M. Takvorian. 1996b. Microsporidian taxonomy and the status of *Septata intestinalis*. *J. Eukaryot. Microbiol.* **43:**106S-107S.

Canning, E. U. Unpublished data.

Canning, E. U. 1988. Nuclear division and chromosome cycle in microsporidia. *BioSystems* **21:**333–340.

Canning, E. U., R. J. Barker, J. C. Hammond, and J. P. Nicholas. 1983. *Unikaryon slaptonleyi* sp. nov. (Microspora: Unikaryonidae) isolated from echinostome and strigeid larvae from *Lymnaea peregra:* observations on its morphology, transmission and pathogenicity. *Parasitology* **87:**175–184.

Canning, E. U., A. S. Field, M. C. Hing, and D. J. Mariott. 1994. Further observations on the ultrastructure of *Septata intestinalis* Cali, Kotler et Orenstein, 1993. *Eur. J. Protistol.* **30:**414–422.

Canning, E. U., and E. I. Hazard. 1982. Genus *Pleistophora* Gurley, 1893: an assemblage of at least three genera. *J. Protozool.* **29:**38–49.

Canning, E. U., J. Lom, and J. P. Nicholas. 1982. Genus *Glugea* Thelohan, 1891 (Phylum Microspora): redescription of the type species *Glugea anomala* (Moniez, 1887) and recognition of its sporogonic development within sporophorous vesicles (pansporoblastic membranes). *Protistologica* **18:**193–210.

Canning, E.U., and P. Nicholas. 1980. Genus *Pleistophora* (Phylum Microspora): redescription of the type species, *Pleistophora typicalis* Gurley, 1893 and ultrastructural characterization of the genus. *J. Fish Dis.* **3:**317–338.

Cavalier-Smith, T. 1983. A 6-kingdom classification and a unified phylogeny, p. 265–279. *In* W. Schwemmler and H. E. A. Schenn (ed.), *Endocytobiology II.* de Gruyter, Berlin, Germany.

Chen, W. J., and A. R. Barr. 1995. Chromosomal evidence on the sporogony of *Amblyospora californica* (Microspora: Amblyosporidae) in *Culex tarsalis* (Diptera: Culicidae). *J. Eukaryot. Microbiol.* **42:**103–108.

Codreanu, R., and J. Vávra. 1970. The structure and ultrastructure of the microsporidian *Telomyxa glugeiformis* Léger and Hesse, 1910, parasite of *Ephemera danica* (Mull.) nymphs. *J. Protozool.* **17:** 374–384.

Corliss, J. O. 1994. An interim utilitarian ("user-friendly") hierarchical classification and characterization of the protists. *Acta Protozool.* **33:**1–51.

Curgy, J. J., J. Vávra and C. Vivarès. 1980. Presence of ribosomal RNAs with prokaryotic properties in Microsporidia, eukaryotic organisms. *Biol. Cell.* **38:**49–52.

Curry, A. Personal communication.

Deflandre, G. 1923. Emploi de la nigrosine dans l'étude des algues inférieures. *Bull. Soc. Bot. Fr.* **70:** 738–741.

Deplazes, P., M. Van Saanen, A. Iten, A. Mathis, R. Keller, I. Tanner, R. Weber, and E. U. Canning. 1997. Double infection with *Vittaforma corneae* and *Encephalitozoon hellem* in an AIDS patient, p. 75. Program and Abstr. 10th Int. Congr. of Protozool. 1997. Business Meetings and Incentives, Sydney, Australia.

de Puytorac, P., and M. Tourret. 1963. Études de kystes d'origine parasitaire (Microsporidie ou Grégarines) sur la paroi interne du corps des vers Megascolecidae. *Ann. Parasitol. Hum. Comp.* **38:** 861–874.

Desportes, I. 1976. Ultrastructure de *Stempellia mutabilis* Léger & Hesse, microsporidie parasite de l'ephemère *Ephemera vulgata* L. *Protistologica* **12:** 121–150.

Desportes, I., Y. Le Charpentier, A. Galian, F. Bernard, B. Cochan-Priollet, A. Lavergne, P. Ravisse, and R. Modigliani. 1985. Occurrence of a new microsporidian: *Enterocytozoon bieneusi* n. g., n. sp., in the enterocytes of a patient with AIDS. *J. Protozool.* **32:**250–254.

Desportes, I., and J. Théodoridès. 1979. Étude ultrastructurale d'*Amphiamblys laubieri* n. sp. (Microsporidie, Metchnikovellidae) parasite d'une Grégarine (*Lecudina* sp.) d'un Echiurien abyssal. *Protistologica* **15:** 435–457.

Desportes-Livage, I., S. Chilmonczyk, R. Hedrick, C. Ombrouck, D. Monge, I. Maiga, and M. Gentilini. 1996. Comparative development of two microsporidian species: *Enterocytozoon bieneusi* and *Enterocytozoon salmonis,* reported in AIDS patients and salmonid fish respectively. *J. Eukaryot. Microbiol.* **43:**49–60.

Desportes-Livage, I., I. Hilmarsdottir, C. Romana, S. Tanguy, A. Datry, and M. Gentilini. 1991. Characteristics of the microsporidian *Enterocytozoon bieneusi:* a consequence of its development within short-living enterocytes. *J. Protozool.* **38:**111S-113S.

Desportes-Livage, I., F. Harper, I. Hilmarsdottir, Y. Benhamou, C. Ombrouck and M. Gentilini. 1993. The phospholipids in *Enterocytozoon bieneusi:* an electron spectroscopic imaging study. *Folia Parasitol.* **40:**275–278.

Desser, S. S., H. Hong and Y. J. Yang. 1992. Ultrastructure of the development of a species of *Encephalitozoon* cultured from the eye of an AIDS patient. *Parasitol. Res.* **78:**677–683.

Dickson, D. L., and A. R. Barr. 1990. Development of *Amblyospora campbelli* (Microsporida: Amblyosporidae) in the mosquito *Culiseta incidens* (Thomson). *J. Protozool.* **37:**71–78.

Didier, E. S., P. J. Didier, D. N. Friedberg, S. M. Stenson, J. M. Orenstein, R. W. Yee, F. O. Tio, R. M. Davis, C. Vossbrinck, N. Millichamp, and J. A. Shadduck. 1991. Isolation and characterization of a new human microsporidian *Encephalitozoon hellem* (n. sp.), from three AIDS patients with keratoconjunctivitis. *J. Infect. Dis.* **163:**617–621.

Ditrich, O, J. Lom, I. Dyková, and J. Vávra. 1994 First case of *Enterocytozoon bieneusi* infection in the Czech Republic: comments on the ultrastructure and teratoid sporogenesis of the parasites. *J. Eukaryot. Microbiol.* **41:**35S-36S.

Faye, N., B. S. Toguebaye, and G. Bouix. 1991. *Microfilum lutjani* n. g., sp. n. (Protozoa, Microsporidia), a gill parasite of the golden African snapper *Lutjanus fulgens* (Valenciennes, 1830) (Teleost Lutjanidae): developmental cycle and ultrastructure. *J. Protozool.* **38:**30–40.

Flegel, T. W., and T. Pasharawipas. 1995. A proposal for typical eukaryotic meiosis in microsporidians. *Can. J. Microbiol.* **41:**1–11.

Germot, A., H. Philippe, and H. Le Guyader. 1997. Evidence for loss of mitochondria in microsporidia from a mitochondrial-type HSP70 in *Nosema locustae. Mol. Biochem. Parasitol.* **87:** 159–168.

Hazard, E. I., and J. W. Brookbank. 1984. Karyogamy and meiosis in an *Amblyospora* sp. (Microspora) in the mosquito *Culex salinarius. J. Invertebr. Pathol.* **44:**3–11.

Hazard, E. I., E. A. Ellis, and D. J. Joslyn. 1981. Identification of microsporidia, p.163–182. *In* H.D. Burges (ed.), *Microbial Control of Plant Diseases 1970–1980.* Academic Press, London, United Kingdom.

Hazard, E. I., T. Fukuda, and J. J. Becnel. 1985. Gametogenesis and plasmogamy in certain species of Microspora. *J. Invertebr. Pathol.* **46:**63–69.

Hazard, E. I., and S. W. Oldacre. 1975. Revision of Microsporida (Protozoa) close to *Thelohania,* with descriptions of one new family, eight new genera, and thirteen new species. *U.S. Dept. Agric. Tech. Bull.* 1530. U.S. Department of Agriculture, Washington, D.C.

Hedrick, R. P., J. M. Groff, and D. V. Baxa. 1991. Experimental infections with *Nucleospora salmonis* n. g., n. sp.: an intranuclear microsporidium from chinook salmon (*Oncorhynchus tshawitscha*). *Fish Health Sect./Am. Fish. Soc. Newsl.* **19:**5.

Hildebrand, H., and E. Vivier. 1971. Observations ultrastructurales sur le sporoblaste de *Metchnikovella wohlfarthi* n. sp. (Microsporidies), parasite de la grégarine *Lecudina tuzetae. Protistologica* **7:** 131–139.

Hilmarsdottir, I., I. Desportes-Livage, A. Datry, and M. Gentilini. 1993. Morphogenesis of the polaroplast in *Enterocytozoon bieneusi* Desportes et al., 1985, a microsporidian parasite of HIV infected patients. *Eur. J. Protistol.* **29:**88–97.

Hirt, R. P., B. Healy, C. R. Vossbrinck, E. U. Canning, and T. M. Embley. 1979. A mitochondrial *Hsp70* orthologue in *Vairimorpha necatrix:* molecular evidence that microsporidia once contained mitochondria. *Curr. Biol.* **7:**995–998.

Hollande, A. 1972. Le déroulement de la cryptomitose et les modalités de la ségrégation des chromatides dans quelques groupes de Protozoaires. *Ann. Biol.* **11:**427–466.

Hollister, W. S., E. U. Canning, E. Weidner, A. S. Field, J. Kench, and D. J. Marriott. 1996. Development and ultrastructure of *Trachipleistophora hominis* n. g., n. sp., after in vitro isolation from an AIDS patient and inoculation into athymic mice. *Parasitology* **112:**143–154.

Hostounský, Z., and Z. Žižka. 1979. A modification of the "agar cushion method" for observation and photographic recording microsporidian spores. *J. Protozool.* **26:**41A–42A.

Hotchkiss, R. D. 1948. A microchemical reaction resulting in the staining of polysaccharide structures in fixed tissue preparation. *Arch. Biochem.* **16:** 131–141.

Hurst, L. D. 1993. Drunken walk of the diploid. *Nature* **365:**206–207.

International Trust for Zoological Nomenclature. 1985. *International Code of Zoological Nomenclature.* International Trust for Zoological Nomenclature, London, United Kingdom.

Ishihara, R., and Y. Hayashi. 1968. Some properties of ribosomes from the sporoplasm of *Nosema bombycis. J. Invertebr. Pathol.* **11:**377–385.

Issi, I. V. 1986. Microsporidia as a phylum of parasitic protozoa, p. 6–136 *In* T.V. Beyer, and I.V. Issi (eds.), *Protozoology,* vol. 10. Nauka, Leningrad, USSR. (In Russian with English summary.)

Jírovec, O. 1932. Ergebnisse der Nuclealfärbung an den Sporen den Microsporidien nebst einigen Bemerkungen über *Lymphocystis. Arch. Protistenk.* **77:**379–390.

Kramer, J. P. 1964. *Nosema kingi* sp. n., a microsporidian from *Drosophila willistoni* Sturtevant, and its infectivity for other muscoids. *J. Insect Pathol.* **6:**491–499.

Larsson, R. Unpublished data.

Larsson, R. 1980. Insect pathological investigations on Swedish Thysanura. II. A new microsporidian parasite of *Petrobius brevistylis* (Microcoryphia, Machilidae); description of the species and creation of two new genera and a new family. *Protistologica* **16:**85–101.

Larsson, R. 1981a. The ultrastructure of the spore and sporogonic stages of *Telomyxa glugeiformis* Léger and Hesse 1910 (Microsporida: Telomyxidae). *Zool. Anz.* 20**6:**137–153.

Larsson, R. 1981b. A new microsporidium *Berwaldia singularis* gen. et sp. nov. from *Daphnia pulex* and a survey of microsporidia described from Cladocera. *Parasitology* **83:**325–342.

Larsson, R. 1981c. Description of *Nosema tractabile* n. sp. (Microspora, Nosematidae), a parasite of the leech *Helobdella stagnalis* (L.) (Hirudinea, Glossiphoniidae). *Protistologica* **17:** 407–422.

Larsson, R. 1982. Cytology and taxonomy of *Helmichia aggregata* gen. et sp. nov. (Microspora,

Thelohaniidae), a parasite of *Endochironomus* larvae (Diptera, Chironomidae). *Protistologica* **18:**355–370.

Larsson, R. 1983a. *Thelohania capillata* n. sp. (Microspora, Thelohaniidae)—an ultrastructural study with remarks on the taxonomy of the genus *Thelohania* Henneguy, 1892. *Arch. Protistenkd.* **127:** 21–46.

Larsson, R. 1983b. On two microsporidia of the amphipod *Rivulogammarus pulex:* light microscopical and ultrastructural observations on *Thelohania muelleri* (Pfeiffer, 1895) and *Nosema rivulogammari* n. sp. (Microspora, Thelohaniidae and Nosematidae). *Zool. Anz.* **211:**299–323.

Larsson, R. 1983c. Description of *Hyalinocysta expilatoria* n. sp., a microsporidian parasite of the blackfly *Odagmia ornata. J. Invertebr. Pathol.* **42:**348–356.

Larsson, R. 1983d. A revisionary study of the taxon *Tuzetia* Maurand, Fize, Fenwick and Michel, 1971, and related forms (Microspora, Tuzetiidae). *Protistologica* **19:**323–355.

Larsson, R. 1984. Ultrastructural study and description of *Chapmanium dispersus* sp. n. (Microspora, Thelohaniidae), a microsporidian parasite of *Endochironomus* larvae (Diptera, Chironomidae). *Protistologica* **20:**547–563.

Larsson, R. 1985. On the cytology, development and systematic position of *Thelohania asterias* Weiser, 1963, with creation of the new genus *Bohuslavia* (Microspora, Thelohaniidae). *Protistologica* **21:** 235–248.

Larsson, R. 1986a. Development, ultracytology and taxonomical position of *Thelohania corethrae* Schuberg and Rodriguez, 1915 (Microspora, Thelohaniidae). *Arch. Protistenkd.* **132:**245–264.

Larsson, R. 1986b. Ultrastructure, function and classification of microsporidia, p. 325–390. *In* J. O. Corliss and D. J. Patterson (ed.), *Progress in Protistology,* vol. 1. Biopress, Ltd., Bristol, United Kingdom.

Larsson, J. I. R. 1986c. Ultrastructural investigation of two microsporidia with rod-shaped spores, with descriptions of *Cylindrospora fasciculata* sp. nov. and *Resiomeria odonatae* gen. et sp. nov. (Microspora, Thelohaniidae). *Protistologica* **22:**379–398.

Larsson, R. 1986d. Ultrastructural study and description of *Vavraia holocentropi* n. sp. (Microspora, Pleistophoridae), a parasite of larvae of the caddisfly *Holocentropus dubius* (Trichoptera, Polycentropodidae). *Protistologica* **22:** 441–452.

Larsson, J. I. R. 1986e. Ultracytology of a tetrasporoblastic microsporidium of the caddisfly *Holocentropus picicornis* (Trichoptera, Polycentropodidae), with description of *Episeptum inversum* gen. et sp. nov. (Microspora, Gurleyidae). *Arch. Protistenkd.* **131:**257–279.

Larsson, J. I. R. 1987. Ultrastructure and development of a microsporidium of the genus *Pegmatheca* with description of *P. lamellata* sp. nov. (Microspora, Thelohaniidae). *Zool. Anz.* **218:**304–320.

Larsson, R. 1988a. On the taxonomy of the genus *Systenostrema* Hazard & Oldacre, 1975 (Microspora, Thelohaniidae), with description of two new species. *Syst. Parasitol.* **11:**3–17.

Larsson, R. 1988b. Isometric viruslike particles in spores of two microsporidia belonging to the Thelohaniidae. *J. Invertebr. Pathol.* **51:**163–165.

Larsson, J. I. R. 1988c. Identification of microsporidian genera (Protozoa, Microspora)—a guide with comments on the taxonomy. *Arch. Protistenkd.* **136:**1–37.

Larsson, J. I. R. 1989a. On the ultrastructure of *Hyalinocysta expilatoria* (Microspora, Thelohaniidae) and the taxonomy of the species. *J. Invertebr. Pathol.* **54:**213–223.

Larsson, J. I. R. 1989b. Light and electron microscope studies on *Jírovecia involuta* sp. nov. (Microspora, Bacillidiidae), a new microsporidian parasite of oligochaetes in Sweden. *Eur. J. Protistol.* **25:**172–181.

Larsson, J. I. R. 1989c. The light and electron microscopic cytology of *Bacillidium filiferum* sp. nov. (Microspora, Bacillidiidae). *Arch. Protistenkd.* **137:**345–355.

Larsson, J. I. R. 1990a. Description of a new microsporidium of the water mite *Limnochares aquatica* and establishment of the new genus *Napamichum* (Microspora, Thelohaniidae). *J. Invertebr. Pathol.* **55:**152–161.

Larsson, J. I. R. 1990b. On the cytology of *Jirovecia caudata* (Léger and Hesse, 1916) (Microspora, Bacillidiidae). *Eur. J. Protistol.* **25:**321–330.

Larsson, J. I. R. 1990c. On the cytology and taxonomic position of *Nudispora biformis* n. g., n. sp. (Microspora, Thelohaniidae), a microsporidian parasite of the dragon fly *Coenagrion hastulatum* in Sweden. *J. Protozool.* **37:** 310–318.

Larsson, R. 1990d. *Rectispora reticulata* gen. et sp. nov. (Microspora, Bacillidiidae), a new microsporidian parasite of *Pomatothrix hammoniensis* (Michaelsen, 1901) (Oligochaeta, Tubificidae) *Eur. J. Protistol.* **26:**55–64.

Larsson, R. 1992. The ultrastructural cytology of *Bacillidium strictum* (Léger and Hesse, 1916)

Jírovec, 1936 (Microspora, Bacillidiidae). *Eur. J. Protistol.* **28:**175–183.

Larsson, J. I. R. 1993. Description of *Chytridiopsis trichopterae*, n. sp. (Microspora, Chytridiopsidae), a microsporidian parasite of the caddis fly *Polycentropus flavomaculatus* (Trichoptera, Polycentropodidae), with comments on relationships between the families Chytridiopsidae and Metchnikovellidae. *J. Eukaryot. Microbiol.* **40:**37–48.

Larsson, J. I. R. 1994a. Characteristics of the genus *Bacillidium* Janda, 1928 (Microspora, Mrazekiidae): reinvestigation of the type species *B. criodrili* and improved diagnosis of the genus. *Eur. J. Protistol.* **30:**85–96.

Larsson, J. I. R. 1994b. *Trichoctosporea pygopellita* gen. et sp. nov. (Microspora: Thelohaniidae), a microsporidian parasite of the mosquito *Aedes vexans* (Diptera, Culicidae). *Arch. Protistenk.* **144:**147–161.

Larsson, J. I. R. 1995. A light microscopic and ultrastructural study of *Gurleya legeri* sensu Mackinnon (1911)—with establishment of the new species *Gurleya dorisae* sp. n. (Microspora, Gurleyidae). *Acta Protozool.* **34:** 45–56.

Larsson, J. I. R. 1996. Two tetrasporoblastic microsporidian parasites of caddis flies (Trichoptera) with description of the new species *Episeptum invadens* sp. n. and *Episeptum circumscriptum* sp. n. (Microspora, Gurleyidae). *Arch. Protistenk.* **146:** 349–362.

Larsson, J. I. R., D. Ebert, K. L. Mangin, and J. Vávra. 1998. Ultrastructural study and description of *Flabelliforma magnivora* sp. n. (Microspora, Duboscqiidae), a microsporidian parasite of *Daphnia magna* (Crustacea: Cladocera: Daphniidae). *Acta Protozool.* **37:**41–52.

Larsson, J. I. R., D. Ebert, and J. Vávra. 1997a. Ultrastructural study and description of *Ordospora colligata* gen. et sp. nov. (Microspora, Ordosporidae fam. nov.), a new microsporidian parasite of *Daphnia magna* (Crustacea, Cladocera). *Eur. J. Protistol.* **33:**432–443.

Larsson, J. I. R., D. Ebert, J. Vávra, and V. N. Voronin. 1996. Redescription of *Pleistophora intestinalis* Chatton, 1907, a microsporidian parasite of *Daphnia magna* and *Daphnia pulex*, with establishment of the new genus *Glugoides* (Microspora, Glugeidae). *Eur. J. Protistol.* **32:**251–261.

Larsson, J. I. R., J. Eilenberg, and J. Bresciani. 1995. Ultrastructural study and description of *Cystosporogenes deliaradicae* n. sp. (Microspora, Glugeidae), a microsporidian parasite of the cabbage root fly *Delia radicum* (Linnaeus, 1758) (Diptera, Anthomyidae). *Eur. J. Protistol.* **31:**275–285.

Larsson, J. I. R., and P. Götz. 1996. Ultrastructural evidence for the systematic position of *Jirovecia brevicauda* (Léger and Hesse, 1916) Weiser, 1985 (Microspora, Mrazekiidae). *Eur. J. Protistol.* **32:** 366–371.

Larsson, J. I. R., M. Steiner, and S. Bjornson. 1997b. *Intexta acarivora* gen. et sp. n. (Microspora: Chytridiopsidae)—ultrastructural study and description of a new microsporidian parasite of the forage mite *Tyrophagus putrescentiae* (Acari: Acaridae). *Acta Protozool.* **36:**295–304.

Larsson, J. I. R., J. Vávra, and J. Schrével. 1993. *Bacillidium cyclopis* Vávra, 1962. Description of the ultrastructural cytology and transfer to the genus *Mrazekia* Léger and Hesse, 1916 (Microspora, Mrazekiidae). *Eur. J. Protistol.* **29:**49–60.

Léger, L. 1926. Une microsporidie nouvelle à spores épineux. *C.R. Acad. Sci.* **182:**727–729.

Léger, L., and E. Hesse. 1910. Cnidosporidies des larves d'Éphémères. *C. R. Acad. Sci.* **150:**411–414.

Léger, L., and E. Hesse. 1916. *Mrazekia,* genre nouveau de microsporidies à spores tubuleuses. *C.R. Soc. Biol.* **79:**345–348.

Levaditi, C., S. Nicolau, and R. Schoen. 1923. L'agent étiologique de l'encéphalite épizootique du lapin (*Encephalitozoon cuniculi*). *C. R. Acad. Sci. Biol.* **89:**984–986.

Lillie, R. D. 1965. *Histopathological Techniques and Practical Histochemistry.* McGraw-Hill Book Co., New York.

Liu, T. P. 1972. Ultrastructural changes in the nuclear envelope of larval fat body cells of *Simulium vittatum* (Diptera) induced by microsporidian infection of *Thelohania bracteata. Tissue Cell* **4:**493–502.

Liu, T. P. 1984. Virus-like cytoplasmic particles associated with lysed spores of *Nosema apis. J. Invertebr. Pathol.* **44:**103–105.

Liu, T. P., and G. R. Myrick. 1989. Deformities in the spore of *Nosema apis* as induced by intraconazole. *Parasitol. Res.* **75:**498–502.

Logsdon, J. M., M. Dorey, and W. Ford Doolittle. 1997. Molecular evolution of recA-homologous genes in protists, p. 135. Program Abstr. 10th Int. Congr. Protozool. 1997. Business Meetings and Incentives, Sydney, Australia.

Lom, J. 1958. Contribution to the development of *Mrazekia caudata* Léger et Hesse 1916. *Cesk. Parasitol.* **5:**147–151. (In Czech with English summary.)

Lom, J. 1972. On the structure of the extruded microsporidian polar filament. *Z. Parasitenkd.* **38:** 200–213.

Lom, J., and J. O. Corliss. 1967. Ultrastructure observations on the development of the microsporidian protozoon *Plistophora hyphessobryconis* Schaeperclaus. *J. Protozool.* **14:**141–152.

Lom, J., and I. Dyková. 1990. *Hrabyeia xerkophora* n. gen. n. sp., a new microsporidian with tailed spores from the oligochaete *Nais christinae* Kasprzak, 1973. *Eur. J. Protistol.* **25:**243–248.

Lom, J., and J. Vávra. 1963a. The mode of sporoplasm extrusion in microsporidian spores. *Acta Protozool.* **1:**81–89.

Lom, J., and J. Vávra. 1963b. Mucous envelopes of spores of the subphylum Cnidospora (Doflein 1901). *Vest. Cesk. Spol. Zool.* **27:**4–6.

Loubès, C. 1979. Recherches sur la meiose chez les Microsporidies: Consequences sur les cycles biologiques. *J. Protozool.* **26:**200–208.

Loubès, C., and M. Akbarieh. 1977. Étude ultrastructurale de *Nosemoides simocephali* n. sp. (Microsporidie), parasite intestinal de la Daphnie *Simocephalus vetulus* (Muller, 1776). *Z. Parasitenk.* **54:**125–137.

Loubès, C., and M. Akbarieh. 1978. Étude ultrastructurale de la Microsporidie *Baculea daphniae* n. g, n. sp., parasite de l'épithélium intestinal de *Daphnia pulex* Leydig, 1860 (Crustacé, Cladocère). *Protistologica* **14:**23–38.

Loubès, C., and J. Maurand. 1976. Étude ultrastructurale de *Pleistophora debaisieuxi* Jírovec, 1943 (Microsporida): son transfert dans le genre *Tuzetia* Maurand, Fize, Michel et Fenwick, 1971 et rémarques sur la structure et la genèse du filament polaire. *Protistologica* **12:** 577–591.

Loubès, C., J. Maurand, and R. Ormières. 1979. Étude ultrastructurale de *Spraguea lophii* (Doflein, 1898), Microsporidie parasite de la Baudroi: éssai d'interpretation du dimorphisme sporal. *Protistologica* **15:**43–54.

Loubès, C., J. Maurand, and V. Rousset-Galangau. 1976. Présence de complexes synaptonématiques dans le cycle biologique de *Gurleya chironomi* Loubès et Maurand, 1975: un argument en faveur d'une sexualité chez les Microsporidies? *C. R. Acad. Sci.* **282:**1025–1027.

Magaud, A., A. Achbarou, and I. Desportes-Livage. 1997. Cell invasion by the microsporidium *Encephalitozoon intestinalis. J. Eukaryot. Microbiol.* **44:**81S.

Martins, R. R., and A. L. P. Perondini. 1974. Hipertrofia nuclear e sintese de RNA em celulas de *Rhynchosciara* infectadas por miccrosporideos. *Cien. e Cult.* **27:**961–968.

Maurand, J. 1973. Recherches biologiques sur les Microsporidies des larves des Simulies. Ph.D. thesis. Université des Sciences et Techniques du Languedoc, Académie de Montpellier, Montpellier, France.

Maurand, J., A. Fize, B. Fenwick, and R. Michel. 1971. Étude au microscope électronique de *Nosema infirmum* Kudo 1921, microsporidie parasite d'un copépode cyclopoïde; création du genre nouveau *Tuzetia* à propos de cette éspêce. *Protistologica* **7:**221–225.

Maurand, J., and C. Loubès. 1973. Recherches cytochemiques sur quelques microsporidies. *Bull. Soc. Zool. Fr.* **98:**373–381.

Milner, R. J., and J. A. Mayer. 1982. *Tuzetia boeckella* sp. nov. (Protozoa: Microsporida), a parasite of *Boeckella triarticulata* (Copepoda: Calanoidea) in Australia. *J. Invertebr. Pathol.* **39:**174–184.

Mitchell, M. J., and A. Cali. 1993. Ultrastructural study of the development of *Vairimorpha necatrix* (Kramer, 1965) (Protozoa, Microsporida) in larvae of the corn earworm, *Heliothis zea* (Boddie) (Lepidoptera, Noctuidae) with emphasis on sporogony. *J. Eukaryot. Microbiol.* **40:** 701–710.

Moore, C. B., and W. M. Brooks. 1992. An ultrastructural study of *Vairimorpha necatrix* (Microspora, Microsporida) with particular reference to episporontal inclusions during octosporogony. *J. Protozool.* **39:**392–398.

Moore, C. B., and W. M. Brooks. 1994. An ultrastructural study of the episporontal inclusions produced during octosporogony by five species/isolates of *Vairimorpha* (Microspora: Microsporida). *J. Invertebr. Pathol.* **63:**197–206.

Morrison, C. M., and V. Sprague. 1981. Electron microscopical study of a new genus and new species of microsporida in the gills of Atlantic cod *Gadus morhua. J. Fish Dis.* **4:**15–32.

Müller, M. 1997. What are microsporidia? *Parasitol. Today* **13:**455–456.

Olson, R. E., K. L. Tiekotter, and P. W. Reno. 1994. *Nadelspora canceri* n. g., n. sp., an unusual microsporidian parasite of the Dungeness crab, *Cancer magister. J. Eukaryot. Microbiol.* **41:**349–359.

Orenstein, J. M. 1991. Microsporidiosis in the acquired immunodeficiency syndrome. *J. Parasitol.* **77:**843–864.

Orenstein, J. M. Personal communication.

Ormières, R., C. Loubès, and J. Maurand. 1981. *Amphiamblys bhatiellae* n. sp., microsporidie parasite de *Bhatiella marphysae* Setna, 1931, eugrégarine d'annélide polychète. *Protistologica* **17:**273–280.

Overstreet, R. M., and E. Weidner. 1974. Differentiation of microsporidian spore-tails in *Inodosporus spraguei* gen. et sp. n. *Z. Parasitenkd.* **44:**169–186.

Pavan, C., A. L. P. Perondini, and T. Picard. 1969. Changes in chromosomes and in development of cells of *Sciara ocellaris* induced by microsporidian infections. *Chromosoma* **28:**328–345.

Petri, M. 1969. Studies on *Nosema cuniculi* found in transplantable ascites tumours with a survey of microsporidiosis in mammals. *Acta Pathol. Microbiol. Scand. Suppl.* **204:**1–92.

Piekarski, G. 1937. Cytologische Untersuchungen an Paratyphus und Colibakterien. *Arch. Mikrobiol.* **8:**428–438.

Pilley, B. M. 1976. A new genus *Vairimorpha* (Protozoa: Microsporida) for *Nosema necatrix* Kramer 1965: pathogenicity and life cycle in *Spodoptera exempta* (Lepidoptera: Noctuidae). *J. Invertebr. Pathol.* **28:**177–183.

Purrini, K., and. J. Weiser. 1984. Light and electron microscopic studies of *Chytridiopsis typographi* (Weiser, 1954) Weiser, 1970 (Microspora) parasitizing the barkbeetle *Hylastes cunicularius* Er. *Zool. Anz.* **212:**369–376.

Raikov, I. 1982. *Cell Biology Monographs,* vol. 9. *The Protozoan Nucleus, Morphology and Evolution.* Springer-Verlag, Vienna.

Rausch, M., and J. Grunewald. 1980. Light and stereoscan electron microscopic observations on some microsporidian parasites (Cnidosporidia, Microsporidia) of blackfly larvae (Diptera, Simuliidae). *Z. Parasitenkd.* **63:**1–11.

Richards, C. S., and H. G. Sheffield. 1970. Unique host relations and ultrastructure of a new microsporidian of the genus *Coccospora* infecting *Biomphalaria glabrata,* p. 439–452. *In Proceedings of the IVth International Colloquium on Insect Pathology.* Society for Invertebrate Pathology, Gainesville, Fla.

Scholtyseck, E., and R. Daneel. 1962. Ueber die Feinstruktur der Spore von *Nosema apis. Dtsch. Entomol. Z.* **9:**471–476.

Schubert, G. 1969. Ultracytologische Untersuchungen an der Spore der Mikrosporidienart, *Heterosporis finki,* gen. et sp. n. *Z. Parasitenkd.* **32:**59–79.

Shadduck, J. A., R. A. Meccoli, R. Davis, and R. L. Font. 1990. Isolation of a microsporidian from a human patient. *J. Infect. Dis.* **162:**773–776.

Short, H. E., and W. Cooper. 1948. Staining of microscopical sections containing protozoal parasites by modification of McNamara's method. *Trans. R. Soc. Trop. Med. Hyg.* **41:**427–428.

Silveira, H., and E. U. Canning. 1995. *Vittaforma corneae* n. comb. for the human microsporidium *Nosema corneum* Shadduck, Meccoli, Davis and Font, 1990, based on its ultrastructure in the liver of experimentally infected athymic mice. *J. Eukaryot. Microbiol.* **42:**158–165.

Sprague, V. 1977. Systematics of the Microsporidia, p. 1–510. *In* L. A. Bulla, Jr., and T. C. Cheng (ed.), *Comparative Pathobiology,* vol. 2. Plenum Press, New York.

Sprague, V. 1981. Iron haematoxylin staining, p. 271–272. *In* C. Clark (ed.), *Staining Procedures.* William and Wilkins, Baltimore, Md.

Sprague, V., J. J. Becnel, and E. I. Hazard. 1992. Taxonomy of Phylum Microspora. *Crit. Rev. Microbiol.* **18:**285–395.

Sprague, V., R. Ormières, and J.-F. Manier. 1972. Creation of a new genus and a new family in the Microsporida. *J. Invertebr. Pathol.* **20:**228–231.

Sprague, V., S. H. Vernick, and B. J. Lloyd, Jr. 1968. The fine structure of *Nosema* sp. Sprague, 1965 (Microsporida, Nosematidae) with particular reference to stages in sporogony. *J. Invertebr. Pathol.* **12:**105–117.

Spycher, M. A. Unpublished data.

Sweeney, A. W., E. I. Hazard, and M. F. Graham. 1985. Intermediate host for an *Amblyospora* sp. (Microspora) infecting the mosquito, *Culex annulirostris. J. Invertebr. Pathol.* **46:**98–102.

Takizawa, H., E. Vivier, and A. Petitprez. 1973. Développement intranucléaire de la microsporidie *Nosema bombycis* dans les cellules de vers à soie après infestation expérimentale. *C. R. Acad. Sci.* **277:**1769–1772.

Takvorian, P. M., and A. Cali. 1983. Appendages associated with *Glugea stephani,* a microsporidian found in flounder. *J. Protozool.* **30:**251–256.

Takvorian, P. M., and A. Cali. 1994. Enzyme histochemical identification of the Golgi apparatus in the microsporidian, *Glugea stephani. J. Eukaryot. Microbiol.* **41:**63S-64S.

Takvorian, P. M., and A. Cali. 1996. Polar tube formation and nucleoside diphosphatase activity in the microsporidian, *Glugea stephani. J. Eukaryot. Microbiol.* **43:**102S-103S.

Thiéry, J. P. 1972. Mise en évidence des polysaccharides sur coupes fines en microscopie électronique. *J. Microsc.* **6:**987–1018.

Toguebaye, B. S., and B. Marchand. 1985. Pathogénie, cycle de développement et ultrastructure d'*Amblyospora culicis* n. sp. (Protozoa: Microspora), parasite du moustique *Culex quinquefasciatus* Say, 1823 (Diptera, Culicidae). *Can. J. Zool.* **63:**1197–1809.

Toguebaye, B. S., and B. Marchand. 1986. Genèse des différents organites de la spore uninucléée d'*Amblyospora culicis* (Protozoa, Microspora). *Arch. Protistenkd.* **132:**231–244.

Undeen, A. H. 1990. A proposed mechanism for the germination of microsporidian (Protozoa: Microspora) spores. *J. Theor. Biol.* **142:**223–235.

Undeen, A. H., and E. Frixione. 1991. Structural alteration of plasma membrane in spores of the microsporidium *Nosema algerae* on germination. *J. Protozool.* **38:**511–518.

Undeen, A. H., and R. K. Vander Meer. 1994. Conversion of intrasporal trehalose into reducing sugars during germination of *Nosema algerae* (Protista: Microspora) spores: a quantitative study. *J. Eukaryot. Microbiol.* **41:**129–132.

Undeen, A. H., and J. Vávra. 1997. Research methods for entomopathogenic protozoa, p. 117–151. *In* L. Lacey (ed.), *Manual of Techniques in Insect Pathology.* Academic Press, San Diego, Calif.

Van Gool, T., F. Snijders, P. Reiss, J. K. M. Eeftinck Schattenkerk, M. A. van den Bergh Weerman, J. F. W. M. Bartelsman, J. J. M. Bruins, E. U. Canning, and J. Dankert. 1993. Diagnosis of intestinal and disseminated microsporidial infections in patients with HIV by a new rapid fluorescence technique. *J. Clin. Pathol.* **46:**694–699.

Vávra, J. Unpublished data.

Vávra, J. 1959. Beitrag zur Cytologie einiger Mikrosporidien. *Vest. Cesk. Spol. Zool.* **23:**347–350.

Vávra, J. 1962. *Bacillidium cyclopis* n. sp. (Cnidospora, Microsporidia), a new parasite of copepods. *Vest. Cesk. Spol. Zool.* **26:**295–299.

Vávra, J. 1963. Spore projection in microsporidia. *Acta Protozool.* **1:**153–155.

Vávra, J. 1964. Recording microsporidian spores. *J. Insect Pathol.* **6:**258–260.

Vávra, J. 1965. Étude au microscope électronique de la morphologie et du développement de quelques microsporidies. *C. R. Acad. Sci.* **261:**3467–3470.

Vávra, J. 1972. Detection of polysaccharides in microsporidian spores by means of periodic acid-thiosemicarbazide-silver proteinate test. *J. Microsc.* **14:** 357–360.

Vávra, J. 1975. Scindosome—a new microsporidian organelle. *J. Protozool.* **22**(Suppl.)**:**69A.

Vávra, J. 1976a. Structure of the microsporidia, p. 1–85. *In* L. A. Bulla, Jr., and T. C. Cheng (ed.), *Comparative Pathobiology,* vol. 1. Plenum Press, New York, N.Y.

Vávra, J. 1976b. The development of microsporidia, p. 87–110. *In* L. A. Bulla, Jr., and T. C. Cheng (ed.), *Comparative Pathobiology,* vol. 1. Plenum Press, New York, N.Y.

Vávra, J. 1976c. The occurrence of paramural bodies in microsporidia. *J. Protozool.* **23:**21A.

Vávra, J. 1984. *Norlevinea* n. g., a new genus for *Glugea daphniae* (Protozoa: Microspora), a parasite of *Daphnia longispina* (Crustacea: Phyllopoda). *J. Protozool.* **31:**508–513.

Vávra, J., R. J. Barker, and C. P. Vivarès. 1981. A scanning electron microscope study of a microsporidian with pansporoblast: *Thelohania maenadis. J. Invertebr. Pathol.* **37:**47–53.

Vávra, J., P. Bedrník, and J. Cinátl. 1972. Isolation and in vitro cultivation of the mammalian microsporidian *Encephalitozoon cuniculi. Folia Parasitol.* **19:**349–354.

Vávra, J., and J. Chalupský. 1982. Fluorescence staining of microsporidian spores with the brightener "Calcofluor White M2R." *J. Protozool.* **29:**503.

Vávra, J., R. Dahbiová, W. S. Hollister, and E. U. Canning. 1993a. Staining of microsporidian spores by optical brighteners with remarks on the use of brighteners for the diagnosis of AIDS associated human microsporidioses. *Folia Parasitol.* **40:**267–272.

Vávra, J., and J. I. R. Larsson. 1994. *Berwaldia schaefernai* (Jírovec, 1937) comb. n. (Protozoa, Microsporida), fine structure, life cycle and relationship to *Berwaldia singularis* Larsson, 1981. *Eur. J. Protistol.* **30:**45–54.

Vávra, J., J. I. R. Larsson, and M. D. Baker. 1997. Light and electron microscopic study of *Trichotuzetia guttata* gen. et sp. n. (Microspora, Tuzetiidae), a microsporidian parasite of *Cyclops vicinus* Uljanin, 1875 (Crustacea, Copepoda). *Arch. Protistenkd.* **147:**293–306.

Vávra, J., and J. V. Maddox. 1976. Methods in microsporidiology, p. 281–319. *In* L. A. Bulla, Jr., and T. C. Cheng (ed.), *Comparative Pathobiology,* vol. 1. Plenum Press, New York, N.Y.

Vávra, J., E. Nohynková, L. Machala, and J. Spála. 1993b. An extremely rapid method for detection of microsporidia in biopsy materials from AIDS patients. *Folia Parasitol.* **40:**273–274.

Vávra, J., and V. Sprague. 1976. Glossary for the microsporidia, p. 341–363. *In* L. A. Bulla, Jr., and T. C. Cheng (ed.), *Comparative Pathobiology,* vol. 1. Plenum Press, New York, N.Y.

Vávra, J., and A. H. Undeen. 1970. *Nosema algerae* n. sp. (Cnidospora: Microsporida) a pathogen in a laboratory colony of *Anopheles stephensi* Liston (Diptera, Culicidae). *J. Protozool.* **17:**240–249.

Vávra, J., D. Vinckier, G. Torpier, E. Porchet, and E. Vivier. 1986. A freeze-fracture study of microsporidia (Protozoa: Microspora). I. The sporophorous vesicle, the spore wall, the spore plasma membrane. *Protistologica* **22:**143–154.

Vávra, J., A. T. Yachnis, J. A. Shadduck, and J. M. Orenstein. 1998. Microsporidia of the genus *Trachipleistophora*—causative agents of human microsporidiosis: description of *Trachipleistophora anthropophthera* n. sp. (Protozoa: Microsporidia). *J. Eukaryot. Microbiol.* **45:**273–283.

Vinckier, D. 1973. Étude des cycles et de l'ultrastructure de *Lecudina linei* n. sp. (Grégarine parasite de

Nemerte) et de son parasite, la microsporidie *Nosema vivieri* (V. D et P.). Thesis. Université des Sciences et Techniques de Lille, Lille, France.

Vinckier, D. 1975. *Nosemoides* gen. n., *N. viveri* (Vinckier, Devauchelle & Prensier, 1970) comb. nov. (Microsporidie): étude de la différenciacion sporoblastique et genèse des differentes structures de la spore. *J. Protozool.* **22:**170–184.

Vinckier, D. 1990. Différenciation cellulaire chez les microsporidies: étude ultrastructurale du développment de *Nosemoides vivieri* V.D. et P. et *Nosema apis* Zander et mise en évidence de l'évolution des systemes membranaires par la technique de cryofracture. Thesis. Université des Sciences et Techniques de Lille Flandres Artois, Lille, France.

Vinckier, D., E. Porchet, E. Vivier, J. Vávra, and G. A. Torpier. 1993. A freeze-fracture study of microsporidia (Protozoa: Microspora). II. The extrusion apparatus: polar filament, polaroplast, posterior vacuole. *Eur. J. Protistol.* **29:**370–380.

Vivarès, C. P. 1980. Étude ultrastructurale de *Thelohania maenadis* Pérez (Microspora: Microsporida) et donnés nouvelles sur le genre *Thelohania* Henneguy. *Arch. Protistenkd.* **123:**44–60.

Vivarès, C. P., G. Bouix, and J. F. Manier. 1977. *Ormieresia carcini,* gen. n., sp.n., microsporidie du crabe mediterranéen, *Carcinus mediterraneus* Czerniavsky, 1884: cycle évolutif et étude ultrastructurale. *J. Protozool.* **24:** 83–94.

Vivarès, C. P., C. Loubès, and G. Bouix. 1976. Recherches cytochimiques approfondies sur les Microsporidies parasites du crabe vert de la Méditerranée, *Carcinus mediterraneus* Czerniavsky, 1884. *Ann. Parasitol. Hum. Comp.* **51:**1–14.

Vivier, E. 1975. The microsporidia of the Protozoa. *Protistologica* **11:**345–361.

Vivier, E., and J. Schrével. 1973. Étude en microscopie photonique et électronique de differents stades du cycle de *Metchnikovella hovassei* et observations sur la position systématique des Metchnikovellidae. *Protistologica* **9:**95–118.

Voronin, V. N. 1989a. Classification of the polar filaments of microsporidia on the base of their ultrastructure. *Citologia* **31:**1010–1015. (In Russian with English summary.)

Voronin, V. N. 1989b. The ultrastructure of *Lanatospora macrocyclopis* (Protozoa: Microsporida) from the cyclope *Macrocyclops albidus* (Jur.) (Crustacea, Copepoda). *Arch. Protistenkd.* **137:**357–366.

Voronin, V. N. 1994. *Microsporidium fluviatilis* sp. n. (Protozoa: Microsporidia) from the cyclope *Eucyclops serratulus* (Fisch.) with description of a new type of polaroplast. *Parazitologiya* **28:**48–51. (In Russian with English summary.)

Vossbrinck, C. R., and C. R. Woese. 1986. Eukaryotic ribosomes that lack a 5.8 S RNA. *Nature* **320:**287–288.

Walker, M. H., and G. W. Hinsch. 1972. Ultrastructural observations of a microsporidian protozoan parasite in *Libnia dubia* (Decapoda). I. Early spore development. *Z. Parasitenkd.* **39:**17–26.

Weidner, E. 1972. Ultrastructural study of microsporidian invasion into cells. *Z. Parasitenkd.* **40:**227–242.

Weidner, E. 1975. Interactions between *Encephalitozoon cuniculi* and macrophages. *Z. Parasitenkd.* **47:**1–9.

Weidner, E. 1976a. The microsporidian spore invasion tube. The ultrastructure, isolation, and characterization of the protein comprising the tube. *J. Cell Biol.* **71:**23–34.

Weidner, E. 1976b. Some aspects of microsporidian physiology, p. 111–126. *In* L. A. Bulla, Jr., and T. C. Cheng (ed.), *Comparative Pathobiology,* vol. 1. Plenum Press, New York.

Weidner, E. 1982. The microsporidian spore invasion tube. III. Tube extrusion and assembly. *J. Cell Biol.* **93:**976–979.

Weidner, E., W. Byrd, A. Scarborough, J. Pleshinger, and D. Sibley. 1984. Microsporidian spore discharge and the transfer of polaroplast organelle into plasma membrane. *J. Protozool.* **31:**195–198.

Weidner, E., E. U. Canning., and W. S. Hollister. 1997. The plaque matrix (PQM) and tubules at the surface of intramuscular parasite *Trachipleistophora hominis. J. Eukaryot. Microbiol.* **44:**359–365.

Weidner, E., S. B. Manale, S. K. Halonen, and J. W. Lynn. 1995. Protein-membrane interaction is essential to normal assembly of the microsporidian spore invasion tube. *Biol. Bull.* **188:** 128–135.

Weidner, E., R. M. Overstreet, B. Tedeschi, and J. Fuseler. 1990. Cytokeratin and desmoplakin analogues within an intracellular parasite. *Biol. Bull.* **179:**237–242.

Weir, G. O., and J. T. Sullivan. 1989. A fluorescence screening technique for microsporidia in histological sections. *Trans. Am. Microsc. Soc.* **108:**208–210.

Weiser, J. 1961. *Die Mikrosporidien als Parasiten der Insekten.* Monographien zur Angewandte Entomologie, vol. 17. Paul Parey Verlag, Hamburg, Germany.

Weiser, J. 1976a. The *Pleistophora debaisieuxi* xenoma. *Z. Parasitenk.* **48:**263–270.

Weiser, J. 1976b. Staining of the nuclei of microsporidian spores. *J. Invertebr. Pathol.* **28:**147–149.

Weiser, J. 1977. Contribution to the classification of microsporidia. *Vest. Cesk. Spol. Zool.* **41:**308–320.

Weiser, J., and K. Purrini. 1985. Light- and electron-microscopic studies on the microsporidian *Vairimorpha ephestiae* (Mattes) (Protozoa, Microsporidia) in the meal moth *Ephestia kuehniella. Arch. Protistenkd.* **130:**179–189.

Weiser, J., and Z. Žižka. 1974. Stages in sporogony of *Pleistophora debaisieuxi* (Microsporidia). *J. Protozool.* **21:**477, (abstr.).

Weiser, J., and Z. Žižka. 1975. Stages in sporogony of *Plistophora debaisieuxi* Jírovec (Microsporidia). *Acta Protozool.* **14:**185–194.

Weissenberg, R. 1976. Microsporidian interactions with host cells, p. 203–237. *In* L. A. Bulla, Jr., and T. C. Cheng (eds.), *Comparative Pathobiology.* Vol. 1. Plenum Press, New York.

Yasunaga, C., I. Shino, M. Funakoshi, T. Kawarabata, and S. Hayasaka. 1995. A new method for inoculation of poor germinator *Nosema* sp. NIS M11 (Microsporida: Nosematidae) into an insect cell culture. *J. Eukaryot. Microbiol.* **42:**191–195.

Zickler, D., and L. W. Olson. 1975. The synaptonemal complex and the spindle plaque during meiosis in yeast. *Chromosoma* (Berlin) **50:** 1–23.

DEVELOPMENTAL MORPHOLOGY AND LIFE CYCLES OF THE MICROSPORIDIA

Ann Cali and Peter M. Takvorian

3

This chapter reviews the morphological features and development of the microsporidia. Since there are over a hundred genera and hundreds of species of microsporidia, the life cycle and the nature of the diseases they produce are quite variable. The host range is extremely wide, extending from other protists to invertebrates (especially insects) to vertebrates (particularly fish and mammals), including humans. Consequently, some features are general while others are specific, related to only particular genera or species. We will discuss and demonstrate the more typical microsporidial features in detail, as well as some of the morphological variations, and present examples of these features.

While the phylum Microsporidia Balbiani, 1882 (Issi, 1986), or Microspora Sprague, 1977, represents a diverse group of obligate intracellular parasitic protists, they share a specialized means of initiating infection (Fig. 1). This mechanism is the one feature they all have in common: the production of a unique spore containing a polar filament with an anterior attachment complex (Fig. 2 and 3).

Ann Cali and Peter Takvorian, Department of Biological Sciences, Rutgers University, 101 Warren Street, Smith Hall Newark, NJ 07102.

LIGHT MICROSCOPIC IDENTIFICATION

The spores of microsporidia are generally small, oval or pyriform, resistant structures that vary in length from approximately 1 to about 10 to 12 μm, with a few needlelike spores as long as 20 μm (Cepede, 1924; Sprague, 1977; Canning and Lom, 1986; Olson et al., 1994). In mammals spores are generally 1 to 4 μm in length (Bryan et al., 1991; Weber et al., 1994). The presence of a microsporidial infection is most often diagnosed by the detection of spores. Fresh spores are extremely refractile when viewed by phase contrast microscopy (Fig. 4). When fixed, wax embedded, and stained for light microscopy, they do not stain well with routine hematoxylin and eosin but can be visualized if one looks for birefringence or uses any one of a number of special stains, including Grocott methenamine-silver, acid-fast, Giemsa, or Heidenhain iron hematoxylin stain (Fig. 5) (Gray et al., 1969; Strano et al., 1976). The periodic acid-Schiff (PAS) stain technique produces a small, red-stained granule located anteriorly (Fig. 6) and is a light microscopic diagnostic technique for identification of microsporidian spores at least 4 μm long (Strano et al., 1976). Unfortunately, many human-infecting spores are too small (1 to 2 μm) for detection of this granule. Stains for fixed

The Microsporidia and Microsporidiosis (Murray Wittner, editor; Louis M. Weiss, contributing editor), ©1999 American Society for Microbiology, Washington, D.C.

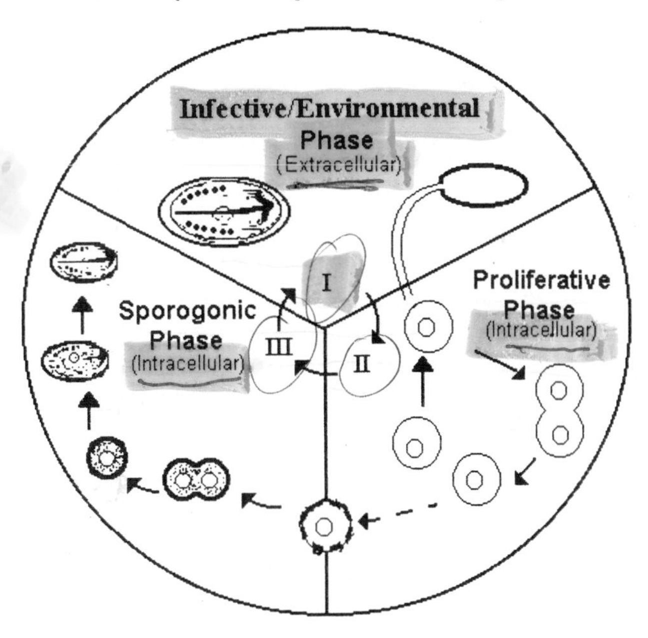

FIGURE 1 Diagram of a typical developmental cycle of the microsporidia. The three regions represent the three phases of the microsporidian life cycle. Phase I is the infective/environmental phase, the extracellular phase of the cycle. It contains the mature spores in the environment. Under appropriate conditions, the spore is activated (e.g., if the spore is ingested by an appropriate host, it is activated by the gut environment) and triggered to evert its polar filament (which becomes a hollow tubule). If the polar tubule pierces a susceptible host cell and injects the sporoplasm into it, phase II begins. Phase II is the proliferative phase, the first phase of intracellular development. During the proliferative part of the microsporidian life cycle, organisms are usually in direct contact with the host cell cytoplasm and increase in number. The transition to phase III, the sporogonic phase, represents the organism's commitment to spore formation. In many life cycles this stage is indicated morphologically by parasite secretions through the plasmalemma producing the thickened membrane. The number of cell divisions that follow varies depending on the genus in question, and the result is spore production.

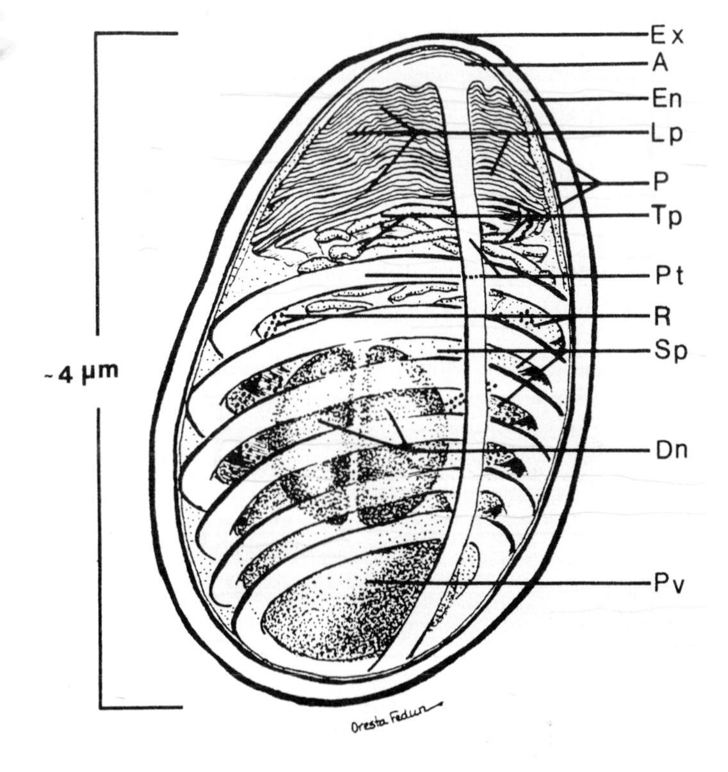

FIGURE 2 Diagram of the internal structure of a microsporidian spore. The spore coat has an outer electron-dense region called the exospore (Ex) and an inner thicker electron-lucent region, the endospore (En). A unit membrane (P) separates the spore coat from the spore contents. The extrusion apparatus, anchoring disk (A), polar tubule (Pt), lamellar polaroplast (Lp), and tubular polaroplast (Tp), dominates the spore contents and is diagnostic for microsporidian identification. The posterior vacuole (Pv) is a membrane-bound vesicle which sometimes contains a "membrane whirl," a "glomerularlike" structure, flocculent material, or some combination of these structures. The spore cytoplasm is dense and contains ribosomes (R) in a tightly coiled helical array. The nucleation may consist of a single nucleus or a pair of abutted nuclei, a diplokaryon (D). The size of the spore depends on the particular species and can vary from less than 1 to more than 10 µm. The number of polar tubule coils also varies from a few to 30 or more, again depending on the species observed. (Reprinted with permission from Cali and Owen, 1988.)

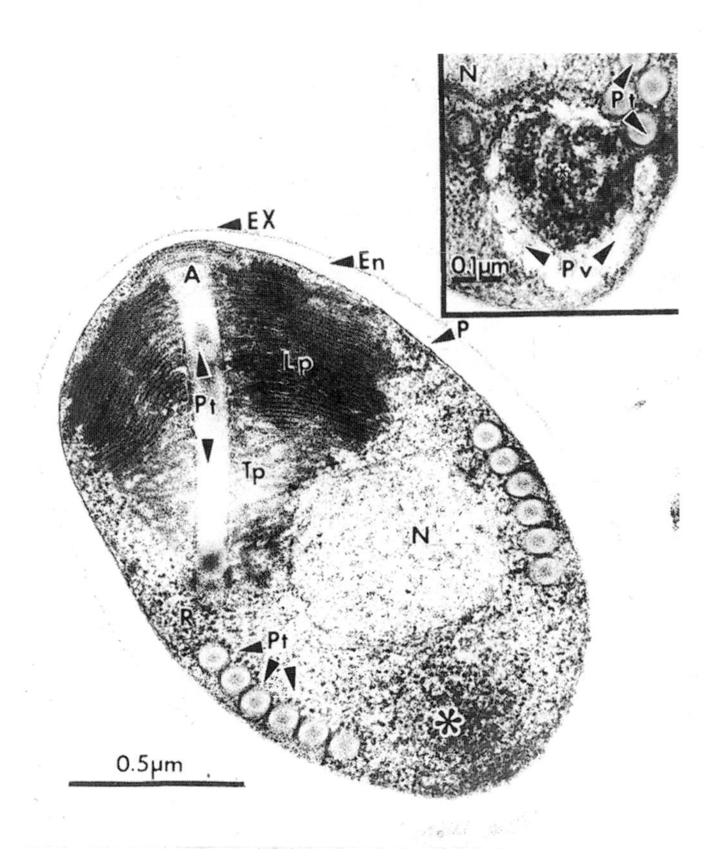

FIGURE 3 Electron micrograph of *Glugea americanus* spore from the angler fish, *Lophius americanus,* depicting the structure of a typical uninucleate spore. The anterior end is filled with the anchoring disk (A), polar tube (Pt), and the various lamellar (Lp) and tubular (Tp) polaroplast membranes of the extrusion apparatus. Ribosomes (R), polar tube cross sections (Pt), and a single nucleus (N) occupy the mid-portion of the spore. The posterior end (asterisk) contains a vacuole that is not evident in this oblique section. (Inset) The posterior vacuole (Pv), often contains a dense body composed of tubular material which may appear "glomerular" (asterisk). The posterior vacuole is often abutted to the coils of the polar tube (Pt). (Reprinted with permission from Takvorian and Cali, 1986.)

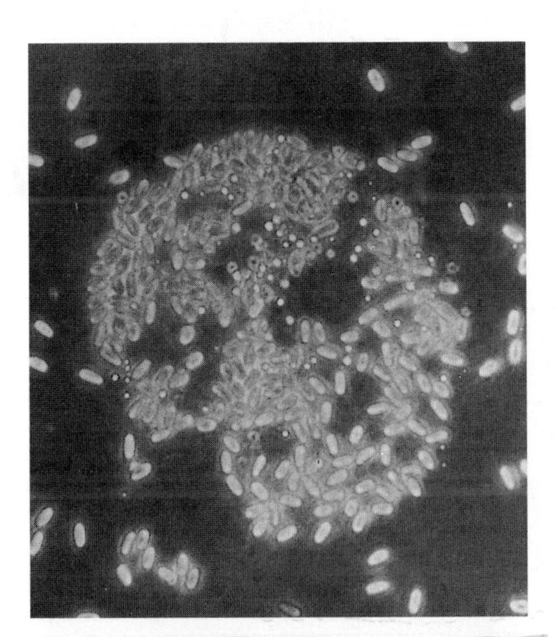

FIGURE 4 *Nosema apis* spores in an intestinal epithelial cell from the honeybee. The spores appear highly refractile when observed in a fresh squash preparation by phase contrast microscopy. Spores are 4 by 2 μm. (Reprinted with permission from Cali and Owen, 1988.)

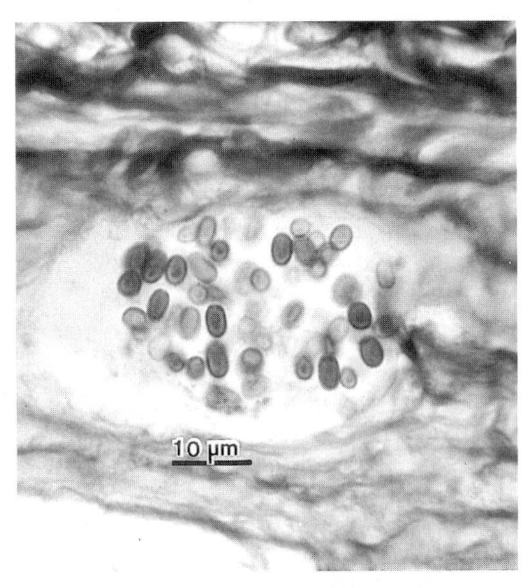

FIGURE 5 *Brachiola connori* (Sprague, 1974) Cali et al. 1998, formerly *Nosema connori*. Spores in the muscularis of the jejunum are shown. The spore coats are well illustrated by Grocott-methenamine-silver stain AFIP# 71-5887. (Reprinted with permission from Strano et al., 1976.)

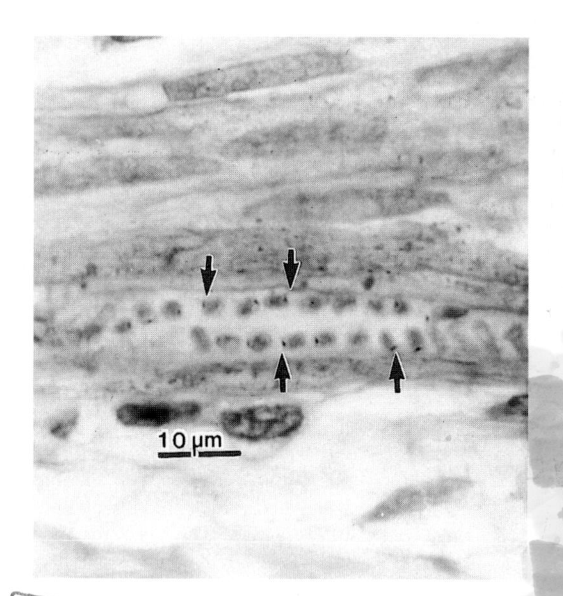

FIGURE 6 *B. connori* spores in the muscularis. The anterior end of the spores contains a PAS-positive granule (arrows). AFIP # 71-5883. (Reprinted with permission from Strano et al., 1976.)

smear preparations include Giemsa, Gram, Gram-chromotrope, and fluorescent (Strano et al., 1976; Moura et al., 1997; Schwartz et al., 1992, van Gool et al., 1993). Additionally, "thick" sections from plastic tissue blocks embedded for electron microscopy can be mounted on slides and stained with toluidine blue for light microscopic observation (Fig. 7) (Orenstein et al., 1992).

LIFE CYCLE PATTERN OF THE MICROSPORIDIA

The general life cycle pattern of the microsporidia can be divided into three phases: the infective or environmental phase, the proliferative phase, and the sporogony or spore forming phase (Fig. 1). The infective phase includes the liberation of spores, their travels in the environment, and the environmental factors necessary for "germination." The parasite cells of the proliferative and sporogonic phases develop intracellularly within infected host cells. (Most microsporidia develop in the host cell cytoplasm; however, the genus *Nucleospora* develops in the nucleoplasm of its host cells [Docker et al., 1997] [see Chapter 13 for details].) Although parasite development is variable depending on the genus of the microsporidian, all intracellular cycles produce sporonts and sporoblasts and terminate in the production of more spores (Fig. 8). (See Issi [1986], Larsson [1986], or Sprague et al. [1992] for descriptions of each genus and species and the higher taxonomy.) In order for successful transmission of infection to occur, spores must be liberated from the infected host, maintain

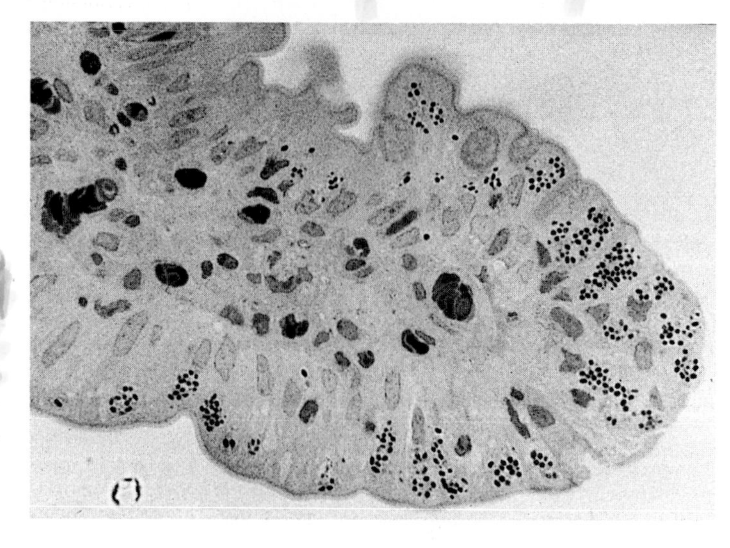

FIGURE 7 *Encephalitozoon* (*Septata*) *intestinalis* in biopsy tissue from human intestine. A light micrograph of densely staining spores concentrated in the cytoplasm of enterocytes at the tip of a distorted villus is shown (semithin plastic section, toluidine blue stain). (Reprinted with permission from Orenstein et al., 1992.)

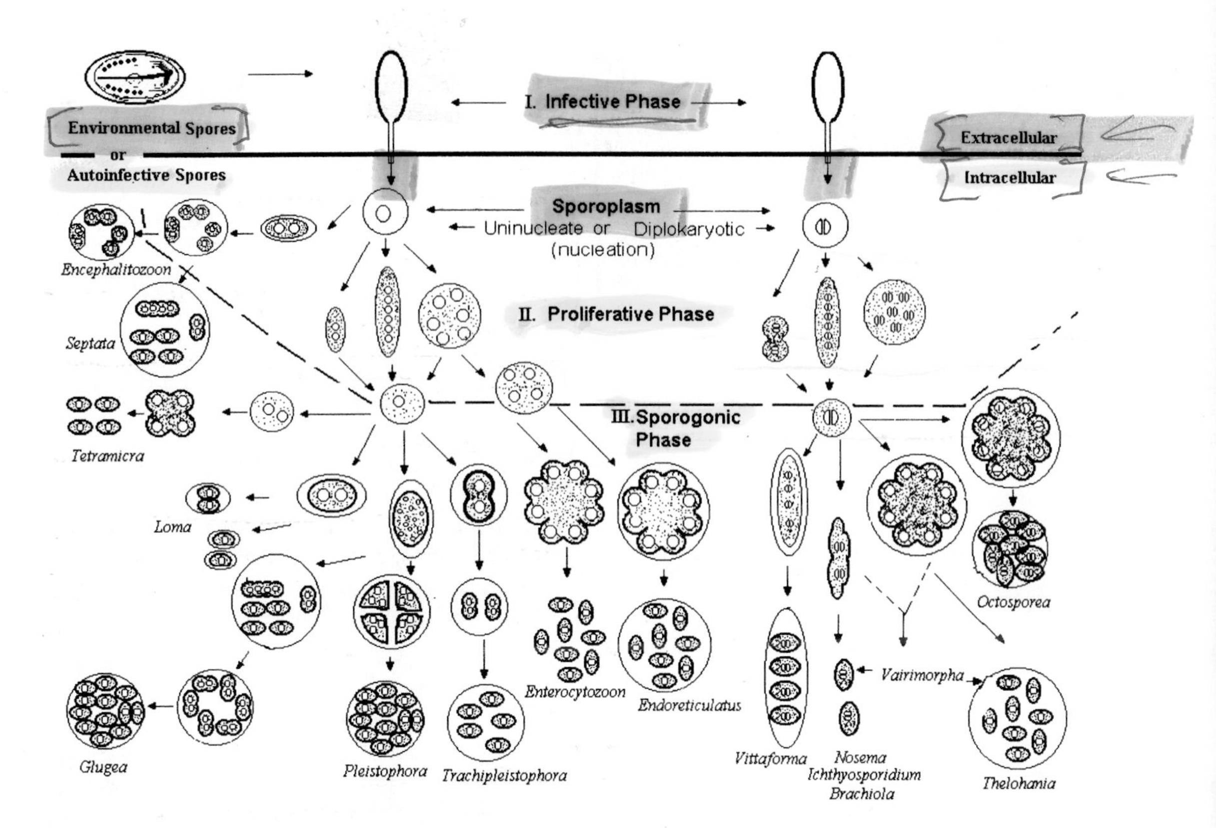

FIGURE 8 Diagram of the life cycles of several horizontally transmitted genera of microsporidia, illustrating developmental diversity. In the infective phase, the proper environmental conditions are required to activate mature spores, resulting in polar tubule extrusion. The polar tubule of each spore is shown piercing the host cell plasmalemma, represented by the solid black line. Below it is the intracellular cytoplasmic area. The sporoplasms travel through the everted polar tubules and are deposited inside the host cells, initiating the proliferative phase of development. The sporoplasm on the left is uninucleate, and the cells produced from it represent the developmental patterns of several microsporidia with isolated nuclei. The sporoplasm on the right is diplokaryotic, and it similarly produces the various diplokaryotic developmental patterns. Cells containing either type of nucleation produce one of three basic developmental forms. Some cycles have cells that divide by binary fission immediately after karyokinesis (e.g., *Brachiola*). A second type forms elongated moniliform multinucleate cells that divide by multiple fission (e.g., some *Nosema* spp.). The third type forms rounded plasmodial multinucleate cells that divide by plasmotomy (e.g., *Endoreticulatus*). Cells may repeat their division cycles one to several times in the proliferative phase. The intracellular stages in this phase are in direct contact with the host cell cytoplasm or closely abutted to the host ER. There are two types of exceptions: (1) the proliferative cells of *Encephalitozoon* and *Septata* are surrounded by a host formed parasitophorous vacuole throughout their development (possibly Tetramicra); and (2) the proliferative plasmodium of the genus *Pleistophora* is surrounded by a thick layer of parasite secretions in the proliferative phase that separates and becomes the sporophorous vesicle in the sporogonic phase. Below the dashed line are the stages of the sporogonic phase. A few cycles maintain direct contact with the host cell cytoplasm in the sporogonic phase: *Nosema, Ichthyosporidium, Brachiola, Tetramicra,* and *Enterocytozoon*. The remaining genera form a sporophorous vesicle as illustrated by the circles around developing sporogonial stages. Note that in the *Thelohania* cycle and the *Thelohania*-like part of the *Vairimorpha* cycle, the diplokarya separate and continue their development as cells with isolated nuclei.

their viability in the environment, encounter a new susceptible host, and gain entry to host cells that can support growth (multiplication) of the parasite. All these factors are hurdles that must be overcome for the parasite to succeed. A means of bypassing many of these problems is vertical (transovarial) transmission, a process that has developed in some insect- and fish-infecting microsporidia. (Details are presented in other chapters.) Additionally, transplacental transmission has been demonstrated in the mammalian microsporidium *Encephalitozoon cuniculi* (Hunt et al., 1972). This chapter will deal with typical horizontally transmitted infections.

PHASE I OF THE MICROSPORIDIAN LIFE CYCLE

The Infective or Environmental Phase

Many environmental factors related to phase I have been identified; some are detrimental to the spores while others are necessary to trigger or activate the spores of a particular microsporidian species. It is the infective phase of the life cycle that covers these factors, which are often the most elusive. Usually, when life cycles are described, only the intracellular developmental phases of the cycle are included, leaving the environmental phase to the epidemiologist.

Spore Survival in the Environment

With the exception of those undergoing transovarial or transovum transmission (Kellen et al., 1965), most microsporidian spores spend at least some part of the infective phase in the environment outside the host. Kramer described this portion of the spores' existence as "extracorporeal" (Kramer, 1976).

Regardless of the method by which spores enter the extracorporeal environment, they must survive for some period of time under harsh conditions until they gain access to a new susceptible host. Environmental factors such as temperature, moisture, ionizing radiation, and the material(s) in which spores are deposited appear to have an impact on their ability to remain viable outside their host. Reports indicate that the material in which spores exit the host body also influences spore survival.

Spores become extracorporeal by several routes, depending on the host they are infecting and the site of infection. Infected intestinal epithelial cells generally slough off and degrade within the digestive system, allowing spores to be released with the feces (Orenstein et al., 1992). Renal infections produce spores that are excreted in the urine (Orenstein et al., 1992), while spores produced from sinus and respiratory infections can exit the body via sputum (Schwartz et al., 1992). In some lepidopterous insects, regurgitation of spores is a means of expulsion (Thomson, 1958). Infected animals that die provide spores that enter the environment following decomposition of the body, cannibalism, scavenging, or physical abrasion or tearing due to environmental action (wind and/or wave action).

Kramer (1970) demonstrated variations in spore longevity to be dependent on environmental factors. When placed on dry surfaces at room temperature, spores from insects remained infective for up to 1 year. He also reported that spores in fecal pellets or dried cadavers lasted less than 6 months under the same conditions but retained infectivity for over a year in cold aqueous media. *Octosporea muscaedomesticae* spores survived 16 months in dried feces when maintained at 5°C with 50% humidity, but only 8 months at room temperature (Kramer, 1970). *Nosema destrutor* retained viability after a 20-min. exposure at 55°C but was inactivated at 70°C (Maddox, 1973). Similar effects were observed in *N. necatrix*, which lost viability after 30 min. at 60°C, 90 minutes at 55°C, and 5 h at 50°C but remained infective for 3 weeks at 40°C (Maddox, 1973). These reports indicate that elevated temperatures severely reduce infectivity, while cold seems to lengthen spore survival (Undeen et al., 1993). In addition to temperature, exposure to water also has variable effects. According to Kramer (1970), spores that retained infectivity for only 6 months when dry were still infective for as long as 10 years when stored in cold distilled water. He also reported that *Octosporea* spores, which retain infectivity for 2 years when stored in water, lose their infectivity in 2 months when rewet after drying (Kramer, 1970).

The effects of sunlight on spores appear to be species dependent. Maddox (1973) reported that *Nosema apis* spores remained viable after 24 h of exposure to sunlight, while *N. necatrix* spores lost viability after only 5 h of exposure. *Nosema algerae* spores showed no change in infectivity after being exposed to bright sunlight for up to 4 h, but 8 min of exposure to an artificial ultraviolet (UV) light source decreased infectivity and infection intensity (number of spores produced in the insects infected) by 99.9% (Kelly and Anthony, 1979). Additional experiments showed that infectivity was reduced by 48% after 1 min of UV exposure and 76% after 2 min, while intensity of infection decreased by 91% after only 2 min of exposure (Kelly and Anthony, 1979). Undeen and Vander Meer (1990) utilized UV light to stimulate germination in *N. algerae*, reporting that short exposures to low doses of UV (254 nm) produced some germination but that 20 min of exposure destroyed the ability to germinate in all the spores. They related the inability to germinate to a decrease in the carbohydrate trehalose, which appears to be necessary for establishing the osmotic potential of the spore during activation (Undeen and Vander Meer, 1990). Olsen et al. (1986) reported similar effects on *N. apis* when they exposed spores to artificial UV light.

Laboratory Conditions for Spore Storage

Methods of spore storage in the laboratory vary with the type of organism, length of storage necessary, and type of equipment available. Spores have remained viable when stored in refrigerated distilled water (Fuxa and Brooks, 1979), stored in refrigerated seawater (Cali et al., 1986), kept as frozen water suspensions (Henry and Oma, 1974), and lyophilized (Bailey, 1972). Microsporidian spores that had been stored for up to 25 years in liquid nitrogen were thawed and tested for their ability to infect various hosts (Maddox and Solter, 1996); spores from terrestrial hosts retained their ability to germinate and produce infections. *N. algerae* spores, which are from the aquatic stage of mosquitoes, did not survive this method of storage, however, they can maintain viability for years when stored in distilled water and refrigerated at a few degrees above freezing (Undeen and Vander Meer, 1994).

Spore Germination (Triggering)

Microsporidian spores transfer their infective sporoplasm to a susceptible host cell following a series of complex events which include environmental changes necessary to activate spore-triggering mechanisms (Undeen, 1990). These changes may be chemical and/or physical, affect spore permeability, and result in eversion of the polar filament, the process that results in the filament becoming tubular (literally turning inside out) as it comes out of the spore. It is emitted with sufficient force to pierce the plasmalemma of a host cell if one is encountered. The sporoplasm is then transmitted from the spore to the host cell cytoplasm through the everted polar tubule, providing a unique means of inoculation (Fig. 9). As a result, the parasite sporoplasm enters the host cell without interacting with the host cell plasmalemma, thus bypassing some of the host's signaling mechanisms. It is at this point that the host-parasite interface becomes important and involves several variable factors that will be discussed later in this chapter.

The requirements necessary for initial activation of the extrusion apparatus have been studied in several genera. Most studies indicate that the initial activation requirements are variable and depend on the organism being studied (Undeen and Avery, 1984; Undeen, 1990). Experimentation with several physical and chemical factors in the spores' environment has led to some interesting hypotheses explaining the mechanisms involved in polar filament extrusion. Some studies on physical change involve subjecting spores to high pressure (20,000 lb in a French press) (Weidner, 1982), air drying followed by rewetting (Olsen et al., 1986), and

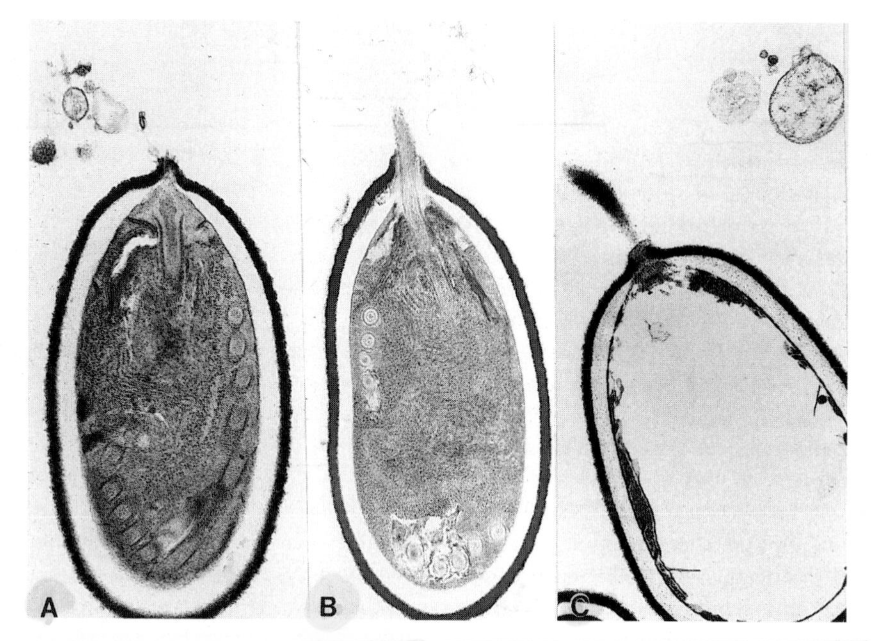

FIGURE 9 *Nosema algerae.* Mature spores were incubated in a germination medium and fixed for EM. (A) Activated spore just about to evert its polar filament; (B) activated spore with polar tube in the process of everting as it passes through the apical portion of the spore coat; (C) empty spore shell with the polar tube still attached.

mild sonication (Frixione et al., 1994). Success with these factors has been moderate.

The majority of microsporidian infections are initiated in the susceptible host gut; consequently, its chemical environment seems a logical place to look for specific stimuli. A comprehensive study of the chemical environment in which *N. algerae* spores germinated in various susceptible mosquito hosts was conducted. It was concluded that there was a correlation between variations in the chemical environment throughout the alimentary canal and the stimulation, activation, and extrusion of spores in the mosquito gut (Jaronski, 1979). "The germination pattern was consistent for each host species but varied among the different species studied." It was concluded that "the spores responded to one or more stimuli: pH, ion concentration, osmolarity, digestive enzymes, redox potential, and digestive products. Several lines of evidence suggest that sodium and potassium ions, within a limited pH range, act as primary germination stimuli for *N. algerae*" (Jaronski, 1979).

Spore germination has also been reported to involve a change in the concentration of calcium ions accompanied by a pH shift. During studies on germination in *Encephalitozoon hellem*, removal of calcium ions from the germination solution resulted in a decrease in polar filament extrusion (Leitch et al., 1995). Additionally, Leitch et al. (1995) suggest that the parasitophorous vacuole in which *E. hellem* develops maintains a lower calcium concentration, thus preventing germination. They hypothesize that during cell death and the accompanying rupture of the host cell, the levels of calcium rise dramatically and stimulate germination (Leitch et al., 1995). The role of calcium changes inside the spore during germination has also been studied. In *Spraguea lophii*, a calcium ion influx into the extrusion apparatus is reported to come from the spore wall or from "plasma membrane pools" (Pleshinger and Weidner, 1985), while in *Glugea hertwigi* the influx is reported in association with the membranous polaroplast (Weidner and Byrd, 1982).

The effect of pH on spore extrusion has been related to "priming" the spores. For *Ameson (Nosema) michaelis* pH 10 has been reported (Weidner, 1972). While *G. hertwigi* (Scarborough-Bull and Weidner, 1985) and *E. hellem* respond to pH 9.0 (Leitch et al., 1995), some organisms (e.g., *Vavraia culicis*), require a neutral or acidic pH to activate germination (Undeen, 1983).

The role of monovalent ions (Na^+, K^+) during spore germination has been the subject of numerous studies on *N. algerae*. In this organism, the ions appear to initiate a series of events that result in increased intrasporal hydrostatic pressure. The activity of these ions is enhanced by the presence of Cl^- and an alkaline pH (8 to 10). The influx of these monovalent cations into the spore is believed to redistribute Ca^{2+} within the spore and somehow trigger activation (Frixione et al., 1994). Undeen and Vander Meer (1994), investigating the same organism under similar germination conditions (NaCl at pH 9.5), also reported high rates of spore activation. In addition, they were able to link the role of the ions to activation or release of the enzyme trehalase, which cleaves the disaccharide trehalose into smaller molecules. The rapid increase in solute concentration resulting from this enzymatic action is thought to increase the intrasporal hydrostatic pressure (believed to be as high as 79 atm), providing the force for germination (Undeen and Vander Meer, 1994).

Undeen and Solter (1997) observed two different buoyant densities (1.198 and 1.150) for mature spores of *Vairimorpha necatrix*. The higher density spores germinated at a greater rate than the lower density spores. These authors attribute the difference in density to the carbohydrate concentration in the spores. Chemical analysis of the denser spores indicated that 88% of the increase in spore weight was due to an increase in carbohydrate level. They reported similar results with *N. algerae* and suggested that "sugar acquisition appears to occur during the attainment of the final spore density and perhaps signals spore maturation" (Undeen and Solter, 1997).

In 1997 Frixione et al. examined several of the above features and provided a plausible hypothesis combining the interaction of ions (Na^+, K^+, Cl^-), alkaline pH, large quantities of sugars in the spores, enzymatic activity causing the breakdown of these disaccharides, and the role of water in germination. They suggest that spore germination is dependent on the hydration state of the plasma membrane-wall complex, stimulant ions, and the flow of water into spores through transmembrane pathways analogous to water channels in highly permeable cells (Frixione et al., 1997).

In summary, the interaction of several factors seems to be responsible for triggering a series of sequential events that can produce sufficient hydrostatic pressure inside the spore to initiate polar filament ejection, and this pressure varies among different microsporidian genera and species. Ions such as sodium and/or potassium, in the presence of chlorine ions in an alkaline environment, can enter the spore. These ions in turn trigger two events within the spore: a calcium ion flux and activation of the enzyme trehalase. As a result, the disaccharide trehalose is cleaved into small molecules, increasing the osmotic pressure potential. Water then enters the spore through transmembrane pathways and interacts with these molecules. This interaction results in a dramatic increase in intrasporal pressure, which provides the force for the rupture of the spore coat and polar filament eversion into a polar tubule. (Details of the polar tubule are dealt with in Chapter 6.) The polar tubule acts as the transfer vehicle through which the sporoplasm travels. When the polar tubule pierces a compatible host cell, the proliferative phase begins.

INTRACELLULAR PHASES OF DEVELOPMENT

General Features of Intracellular Developing Stages of the Microsporidia

Before describing the phases of intracellular development, it is important to review some basic microsporidial cellular features. As protists, microsporidia are eukaryotic, but they lack mitochondria and centrioles, features typically ascribed to such a group. Their ribosomes resemble those of prokaryotes in size (70S) (Ishihara and Hayashi, 1968; Curgy et al., 1980), but the 16S rRNAs of microsporidia that have been sequenced are significantly shorter than both eukaryotic and prokaryotic small subunit rRNAs (Vossbrinck et al., 1987; Weiss et al., 1992). Phylogenetic analysis of the rRNA of *V. necatrix* has suggested that the microsporidia are ancient organisms that branched off very early in the eukaryotic line (Vossbrinck et al., 1987), and a subsequent reclassification groups them with the Archezoa (Cavalier-Smith, 1987). However, research on microsporidia demonstrating the presence of Golgi and endoplasmic reticulum (ER) enzymes (Takvorian and Cali, 1994, 1996) and the presence of heat shock proteins, normally associated with mitochondria (Germot et al., 1997), suggest that they may be very "evolved" as parasites and thus degenerate. Additionally, tubulin gene analysis suggests a taxonomic relationship with the fungi (Edlind et al., 1996). This is quite interesting when one looks at the fact that the nuclear envelope remains intact during karyokinesis in both the microsporidia and the fungi, which also lack centrioles.

Nucleation and Nuclear Activity

Nucleation and nuclear activity vary considerably among the microsporidia. Isolated nuclei and a nuclear arrangement called diplokarya (paired abutted nuclei) are found in the microsporidia (Fig. 10 and 11). Diplokarya can occur throughout some microsporidian developmental cycles and in parts of other cycles and can be completely absent in still others (Fig. 8). Diplokarya have been reported to multiply, to fuse, or to separate from each other (Fig. 12 to 14). Synaptonemal complexes, indicating meiotic activity, have been demonstrated in some insect microsporidia (see Chapter 14 for details) and are absent in others, including all those studied in mammals. In *Enterocytozoon*, elongated nuclei have been demonstrated followed by the production of many nuclei with no transitional stages observed (Fig. 15). In this microsporidian, it is possible that chromatin duplication occurs several times before karyokinesis (Cali and Owen, 1990). Additionally, greatly elongated nuclei have been reported in the spore stages of *N. canceri* (Olson et al., 1994). The nature of their nuclear activity is not known.

Nuclear Division

In general, nuclear division is indicated by the presence of dense-staining material (centriolar plaques) on the nuclear envelope, which has microtubules attached to it intranuclearly. Dense bodies of condensed chromatin associated with the microtubules and invagination of the nuclear envelope during anaphase are typically seen in association with karyogamy in microsporidia (Fig. 12). In some microsporidia, karyokinesis and cytokinesis are linked, and in others they are not (Fig. 8).

Cytokinesis

When cytokinesis immediately follows nuclear division, the cell divides by binary fission (Fig. 13). In many genera, nuclear division occurs several times without cytokinesis, resulting in the production of multinucleate cells. They may form large, rounded multinucleate cells, plasmodia, which eventually divide by plasmotomy (Fig. 16). Alternatively, they form elongated ribbonlike (moniliform) multinucleate stages (Fig. 17) which divide by progressive fragmentation into smaller multinucleate cells which ultimately divide (by multiple fission) into cells each containing one nucleus or diplokaryon.

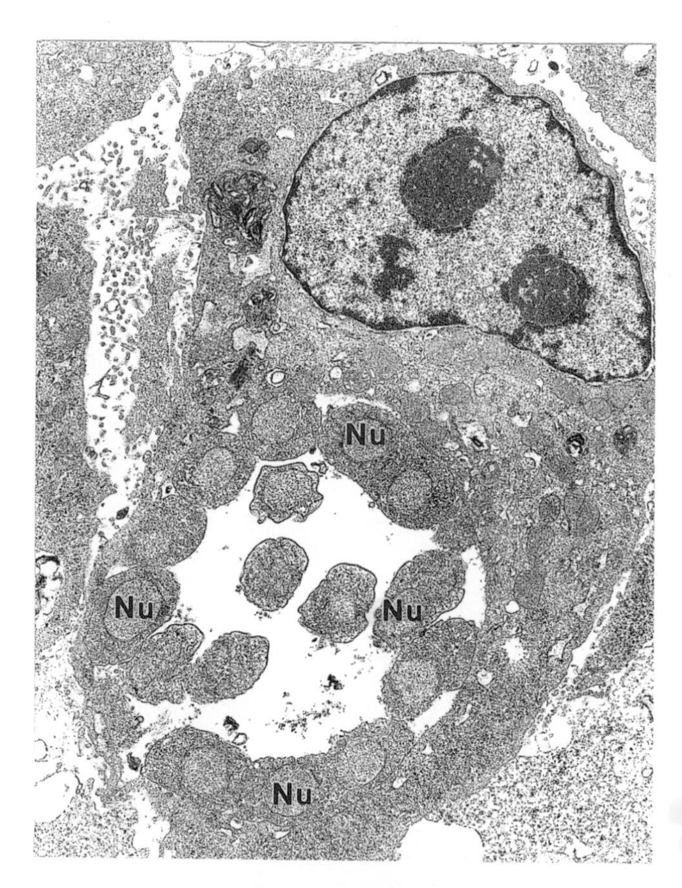

FIGURE 10 *Encephalitozoon cuniculi.* A parasitophorous vacuole containing proliferative cells with large, round isolated nuclei (Nu) is shown.

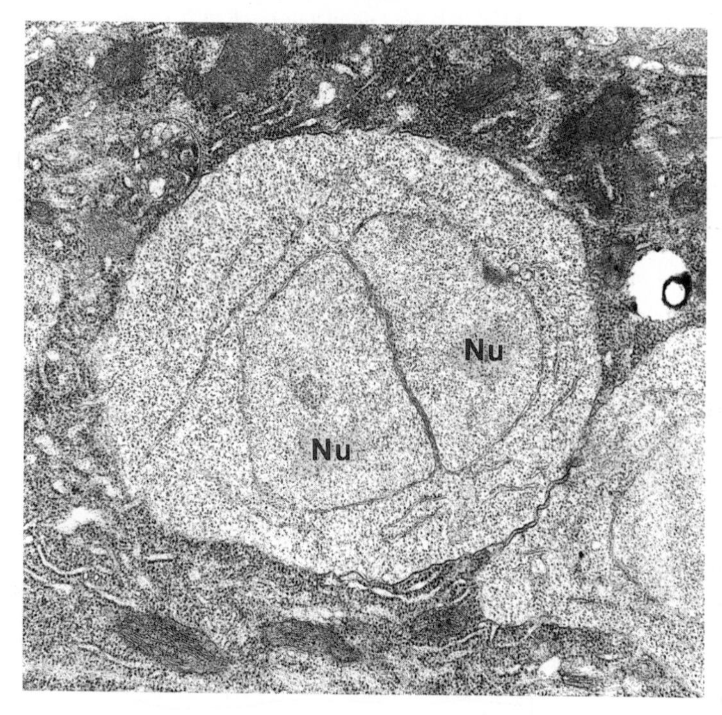

FIGURE 11 *Nosema bombycis.* A proliferative cell containing nuclei (Nu) in diplokaryotic arrangement is shown. The parasite cells are in direct contact with the host cell cytoplasm.

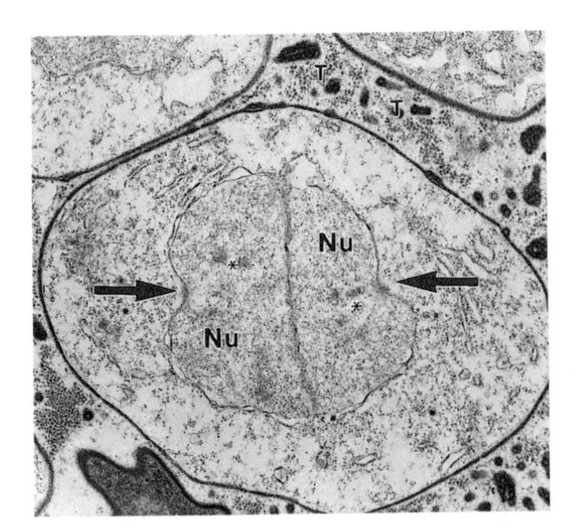

FIGURE 12 Early proliferative cell of *Brachiola vesicularum* with diplokaryotic nuclear pair (Nu) undergoing karyokinesis. Nuclear membrane invaginations containing spindle plaques on each nuclear envelope (arrows) of the diplokaryotic pair and chromosomes (asterisk) within each respective nucleoplasm are present. Note the presence of vesiculotubular material (T) in the host cytoplasm and thickened plasmalemma. (Reprinted with permission from Cali et al.,1998.)

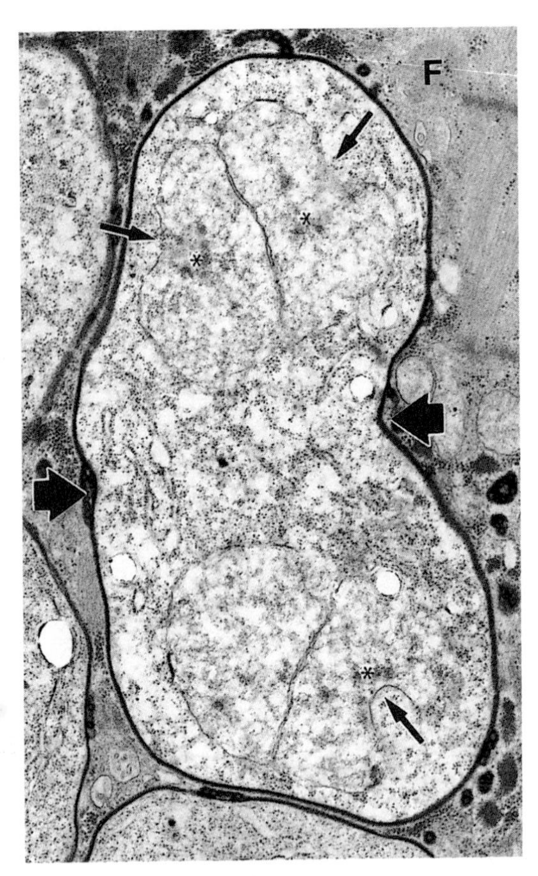

FIGURE 13 Proliferative cell of *B. vesicularum* containing two diplokaryotic nuclear pairs, both with nuclei in late anaphase. Note chromosomes (asterisk) in close association with spindle plaque (arrows). Plasmalemmal invagination (broad arrowheads) indicates that cytokinesis has begun before the nuclei enter interphase, thus linking the two processes. The parasite cells are in direct contact with the muscle cell cytoplasm with no intervening parasitophorous vacuole as evidenced by the proximity of the myofilaments (F). (Reprinted with permission from Cali et al.,1998.)

Host-Parasite Interface during Parasite Development

Since microsporidia are obligate intracellular parasites, no description of development and life cycles would be complete without a discussion of the host-parasite interface. Like most features of the microsporidia, this aspect is quite variable depending on the genus or family observed.

Before we provide a detailed description of the host-parasite interfaces, a few terms need to be defined. A parasite is either in direct contact with the host cell cytoplasm or isolated from it. According to Sprague et al. (1992), all types of isolation involve an interfacial envelope defined as "a structure of any kind or origin that encloses the parasite and thereby prevents direct contact between the parasite plasmalemma and the host cell hyaloplasm." This definition is quite useful because sometimes the origin or nature of the envelope is not obvious.

Several other terms have been used with varying definitions as knowledge of parasite development has evolved (Sprague et al., 1992), but we will limit the terminology to the two terms most commonly used. The term for a host-derived interfacial envelope is "parasitophorous vacuole," and the term for a parasite-derived interfacial envelope is "sporo-

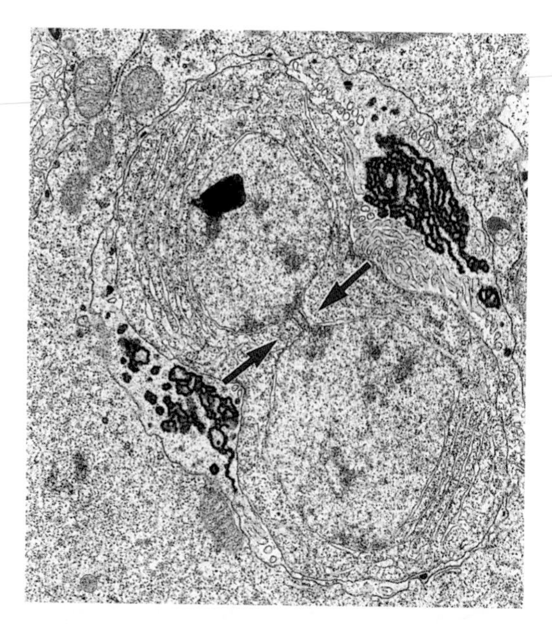

FIGURE 14 *Vairimorpha necatrix*. Nuclear "isthmus" (arrow) between separating diplokaryotic nuclei of a sporont and invagination of cytoplasm. (Reprinted with permission from Mitchell and Cali, 1993.)

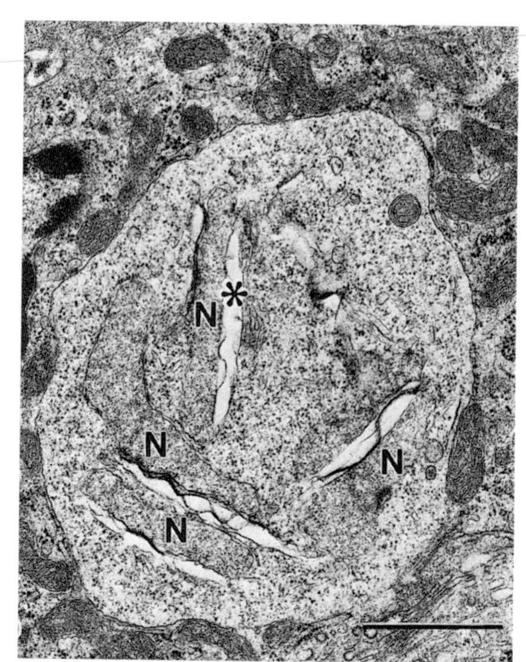

FIGURE 15 Proliferative plasmodial stage of *Enterocytozoon bieneusi*. The cytoplasm is relatively simple, containing only ribosomes and small amounts of membrane. Adjacent to electron-lucent inclusions (asterisk) are multiple elongated nuclei (N), none of which are abutting in diplokaryon form. Note the close approximation of the host nucleus and mitochondria to the parasite plasmalemma. (Reprinted with permission from Cali and Owen, 1990.)

phorous vesicle" (SPOV of Canning and Lom, 1986), regardless of the phase in the life cycle in which it was produced or the nature of the envelope (secretions or membranes).

Our knowledge of host-parasite interfacial relationships continues to develop as we observe new organisms or additional features of previously described organisms. At present, organisms can be divided into four categories (Cali, 1986; Cali and Owen, 1988): those in which there is direct contact of the parasite plasmalemma with the host cell cytoplasm (hyaloplasm), parasite-produced indirect contact, host-produced indirect contact, and host-parasite produced indirect contact (Table 1). With the application of electron microscopy to the study of microsporidial development, these relationships have been given more significance (Cali, 1971; Tuzet et al., 1971) and have been found to reflect different taxonomic groupings (Larsson, 1986; Issi, 1986; Sprague et al., 1992). In general, interfacial relationships are family characteristics, however, we have refrained from using family taxons because they vary with the classification scheme used. Instead, we will cite representative genera to illustrate relationships.

Most microsporidia spend all or part of the proliferative phase in direct contact with the host cell cytoplasm before interfacial envelopes form. Thus, the developmental stages of a given parasite may occur in one or more of these relationships with the host cell cytoplasm during their intracellular development. It is not surprising that parasites can form one relationship with the host cell cytoplasm during the proliferative phase and a different relationship during the sporogonic phase since the needs of the parasite probably change dramatically from one phase to the other. In an effort to systematically describe distinct types of

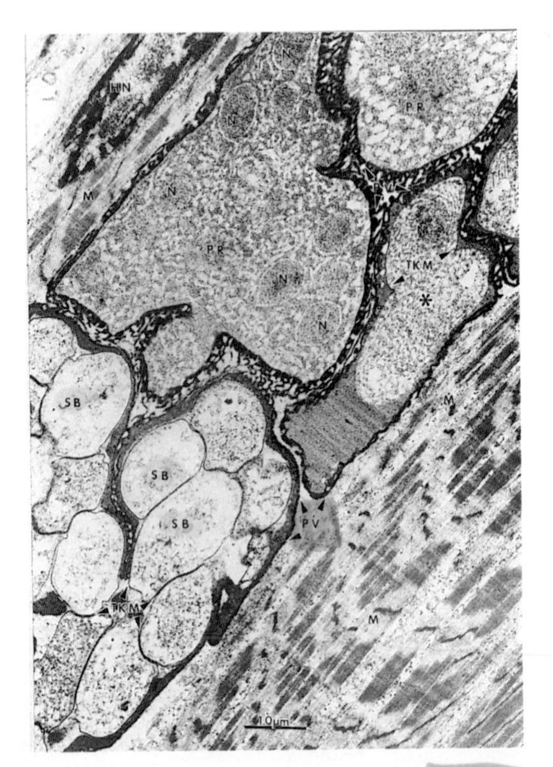

FIGURE 16 Electron micrograph of a *Pleistophora* sp. developing in the skeletal muscle (M) of a patient with AIDS. The proliferative stages (PR) and sporoblasts (SB) are enclosed in thick walled sporophorous vacuoles (PV). The proliferative forms are multinucleated, with only isolated nuclei (N). An early sporont has a plasmalemma pulling away from the sporophorous vacuole (PV), and the plasmalemma has started to thicken (TKM). (Reprinted with permission from Cali and Owen, 1988.)

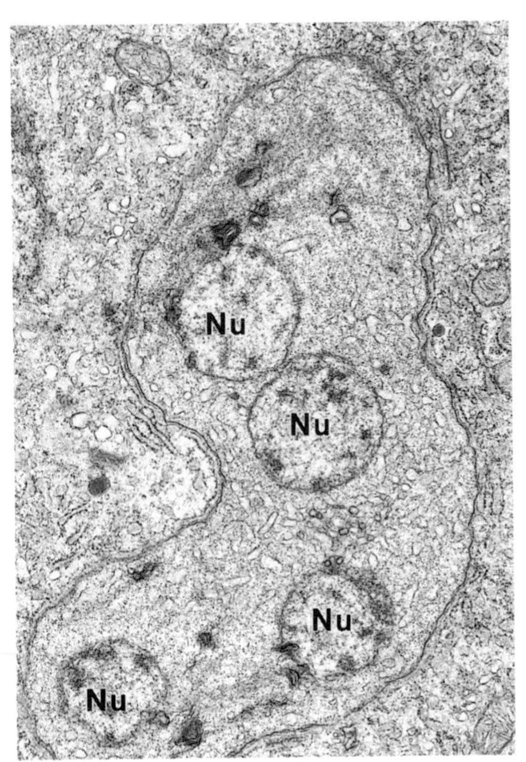

FIGURE 17 Elongated tetranucleate cell of *Glugea stephani* tightly abutted by host ER in the xenoma periphery. Nu, nucleus.

SPOVs, Larsson (1986) distinguished five morphological arrangements. While interfacial relationships may not be well understood, it is obvious that they have important physiological and nutritional significance (Overstreet and Weidner, 1974; Weidner, 1975; Becnel et al., 1986; Cali and Owen, 1988, 1990). Furthermore, while the host-parasite interface varies from the parasite plasmalemma being in direct contact with the host cell cytoplasm to isolation, a variety of appendages form and may allow communication between the two.

DIRECT CONTACT BETWEEN HOST AND PARASITE

Microsporidia having direct contact are those that maintain direct contact between their plasmalemma and the host cell cytoplasm (hyaloplasm) throughout the developmental cycle (lacking any type of interfacial envelope). Genera as widely divergent in their development as *Nosema* and *Enterocytozoon* (Fig. 18 and 19) share the same type of interface (Cali, 1971; Desportes et al., 1985; Cali and Owen, 1990). Additional examples include *Hirsutusporos, Issia, Unikaryon,* and *Ameson*.

In these genera, the parasite's plasmalemma thickens as a result of deposition of material on its surface at the commencement of sporogony. Examples are *Nosema* (Cali, 1971), *Unikaryon* (Canning and Nicholas, 1974), *Ameson* (Weid-

TABLE 1 Interfacial relationships of the microsporidia

Type I: direct contact

The parasite plasmalemma is in direct contact with the host-cell cytoplasm (hyaloplasm). Examples: *Nosema* and *Enterocytozoon*.

Type II: indirect contact by host produced isolation

A parasitophorous vacuole is a host-formed single membrane surrounding the developing parasite cell cluster. This vacuole is present during both the proliferative phase and the sporogonic phase, however, the relationship between it and the parasite changes. Developing parasite cells maintain a very close relationship with this envelope until they develop the thickened sporont plasmalemma; then they appear loose within the vacuole. Example: *E. cuniculi*.

A host ER double membrane surrounds the parasite cells throughout development. In the proliferative phase the host ER double membranes follow the plasmalemma of the dividing cells so that no obvious vacuole is formed. In sporogony, the host ER does not divide with the sporonts but instead forms a double membraned parasitophorous vacuole surrounding the cluster of organisms formed in sporogony. Example: *Endoreticulatus*.

Type III: indirect contact by parasite-produced isolation

A parasite-secreted envelope surrounds the parasite cells throughout development. It becomes a sporophorous vesicle (SPOV) in sporogony when the parasite plasmalemma pulls away from the secreted envelope and then the plasmalemma thickens following additional secretions. Example: *Pleistophora*.

The parasite develops in direct contact with the host cell cytoplasm during early development, but a parasite-formed membrane then isolates the sporonts, sporoblasts, and spores from host cytoplasmic contact. Examples: *Thelohania*, *Octosporea*, and *Vairimorpha*.

Type IV: indirect contact by host- and parasite-produced isolation

Host ER closely abuts the parasite plasmalemma in the proliferative phase. Then the parasite produces blisters arising from the plasmalemma to form the interfacial envelope. Thus, a SPOV is formed in sporogony. It may also contain tubules. Examples: *Loma* and *Glugea*.

The host and parasite contribute to the formation of a thick interfacial envelope that surrounds all stages of parasite cells. Example: *Trachipleistophora*.

The host forms a parasitophorous vacuole that surrounds the parasite cluster, and parasite-secreted material surrounds each parasite cell inside the vacuole. Example: *Septata*.

---

ner, 1972) and *Enterocytozoon* (Desportes et al., 1985; Cali and Owen, 1990). While a few of these microsporidia exhibit precocious formation of the thickened membrane, in *Nosema algerae, Anncaliia varivestis,* and *Brachiola vesicularum* direct contact is still maintained (Avery and Anthony, 1983; Brooks et al., 1985; Issi et al., 1993; Cali et al., 1998). All these microsporidia avoid the host mechanism by which phagosomes are formed around them and instead are found among the cytoplasmic organelles as though they were a natural part of this environment. Some of these microsporidia have been reported as possibly interacting with host organelles such as the mitochondria, nuclei, or ribosomes. Host distribution varies depending on the genus. For example, *Nosema* is a genus that seems to be ubiquitous but is most abundant in insects, *Unikaryon* is found in trematodes (Canning and Nicholas, 1974) and insects (Toguebaye and Marchand, 1984), *Ameson* is found in crabs (Weidner, 1970), and *Enterocytozoon* (Desportes et al., 1985) and *Brachiola* (Cali et al., 1997; Cali et al., 1998) are found in humans.

HOST-PRODUCED INTERFACIAL ENVELOPE

Microsporidia with host-produced interfacial envelopes are those that are separated from the hyaloplasm by host-produced membranes during proliferation through spore formation. This host-formed envelope may be a single or a double membrane.

The developing stages are enclosed in a host-produced, phagosomelike limiting vacuole (parasitophorous vacuole). Although it is typical for cells to surround any foreign material coming into their cytoplasm with a phagosome membrane, microsporidia of most genera seem capable of eluding this mechanism. The

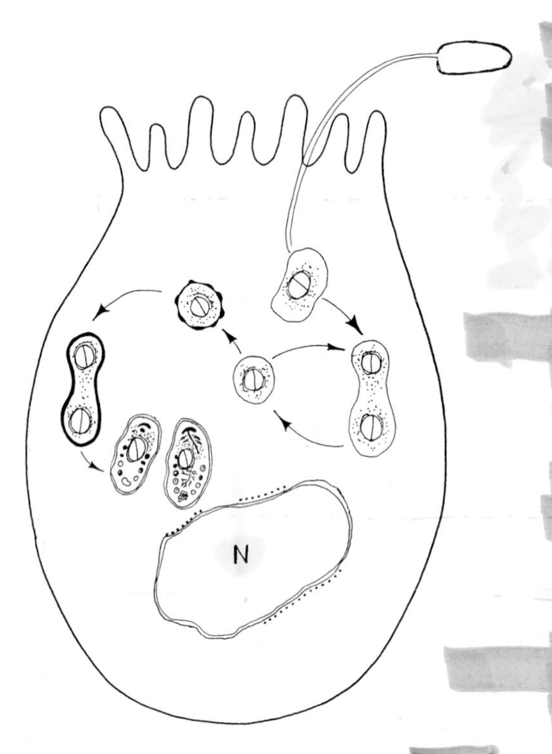

FIGURE 18 Diagram of the life cycle of *Nosema*. A triggered spore injects its diplokaryotic sporoplasm into the host cell cytoplasm beginning the proliferative phase. The organisms multiply by binary fission or multiple fission of ribbonlike moniliform multinucleate cells. The transition to the sporogonic phase is morphologically indicated by parasite secretions through the plasmalemma, producing the thickened membrane. In the genus *Nosema*, each sporont produces two sporoblasts, which results in the production of two spores. This entire cycle takes place in direct contact with the host cell cytoplasm (type I interfacial relationship). N, nucleus.

genus *Encephalitozoon* is an exception (Cali, 1971; Pakes et al., 1975; Sprague and Vernick, 1971). The proliferative stages are tightly abutted to the host-formed interfacial envelope (parasitophorous vacuole) until they develop the thickened plasmalemmal membrane of the sporogonic stages. Then they detach and are observed in the lumen of the vacuole (Fig. 10 and 20).

The interfacial envelope is formed by double-membraned host ER. In the genus *Endoreticulatus,* host ER surrounds the proliferative cells in very close proximity, forming an individual vacuole around each organism (Fig. 21). When sporogony begins and the parasite plasmalemma thickens, the surrounding ER no longer adheres to the organism. Instead of invaginating between the dividing cells, a vacuolar space forms, limited by a double membrane. As this vacuole becomes filled with sporoblasts and then spores, it begins to break down (subpersistent parasitophorous vacuole) (Fig. 21, inset) (Brooks et al., 1988; Cali and El Garhy, 1991).

PARASITE-PRODUCED INTERFACIAL ENVELOPE

Microsporidia with parasite-produced interfacial envelopes are those that are separated from the hyaloplasm by parasite-produced secretions or membranes. These types of organisms maintain interfacial envelopes throughout development, while membranous interfacial envelopes do not form until the beginning of sporogony (Sprague et al., 1992).

Some microsporidia begin secreting material during early proliferative development. It becomes a thick, amorphous coat that persists throughout development. In the genus *Pleistophora,* proliferative multinucleate plasmodia are formed. They are surrounded by amorphous material (Fig. 22) which becomes modified by further secretions. At the onset of sporogony this material separates from the plasmalemma and forms the SPOV (Canning and Hazard, 1982). Subsequently, the plasmodial plasmalemma secretes additional material, resulting in a thickened membrane. This plasmodium divides, forming sporoblasts and eventually spores all inside the SPOV (Fig. 16 and 23).

The genus *Pleistophora* has been reported from many hosts, both vertebrate and invertebrate. Since 1980, however, when Canning and Nicholas redescribed the type species, none of the invertebrate *Pleistophora* spp. have survived ultrastructural scrutiny. For example, *Endoreticulatus fidelis* and *E. schubergi* were *Pleistophora fidelis* (Brooks et al., 1988) and *P. schubergi* (Cali

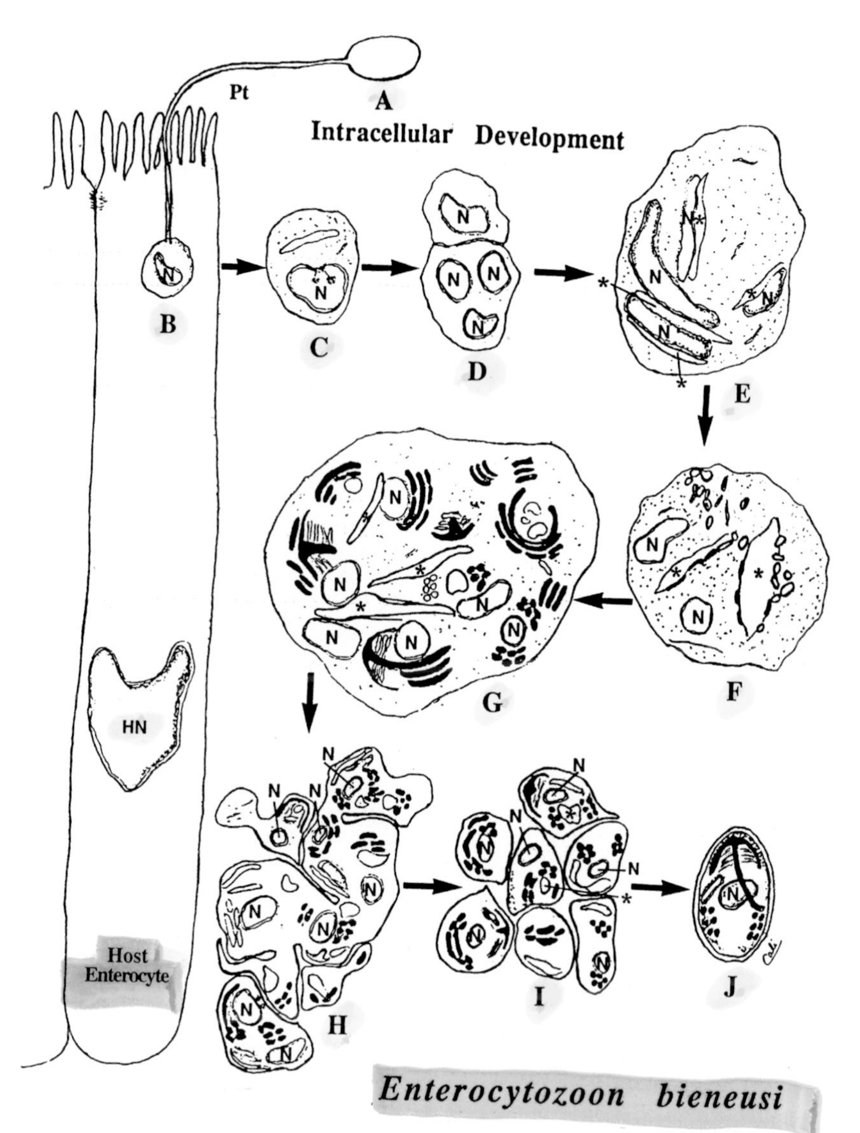

FIGURE 19 Diagram of the life cycle of *Enterocytozoon bieneusi*. (A) An empty spore with its polar tubule extruded is injected into the cytoplasm of a host intestinal enterocyte cell and has a sporoplasm (B) at the end. (B to J) Intracellular developmental stages. (C) An early proliferative cell containing a single dividing nucleus. (D) Abutted proliferative plasmodial cells with no intervening host cytoplasmic organelles suggest recent cytokinetic division. (E) A proliferative plasmodial cell containing multiple elongated nuclei (N) and electron-lucent inclusions (asterisk). (F) Early sporogonial plasmodium showing electron-dense disks forming at the surface of the electron-lucent inclusions. (G) A late sporogonial plasmodium filled with electron-dense disks, some in stacks and some fused in arcs in advanced stages of polar tubule formation. Nuclei are now round and dense and usually associated with electron-dense disk complexes and electron-lucent inclusions. Polar tubule attachment complexes (umbrella-shaped dense structures) appear at this stage. (H) During sporogonial division the plasmalemma thickens and invaginates, segregating individual nuclei with polar tubule complexes. (I) Several newly formed sporoblast cells which are irregularly shaped and possess a thickened plasmalemma and five to six coils of polar tube. (J) Mature spores characterized by the presence of a small electron-lucent inclusion in the sporoplasm, a single nucleus, a polar tubule forming about six coils in a double row, an anterior attachment complex extending down around the polaroplast, and a thick electron-lucent spore coat. (Reprinted with permission from Cali, 1993.)

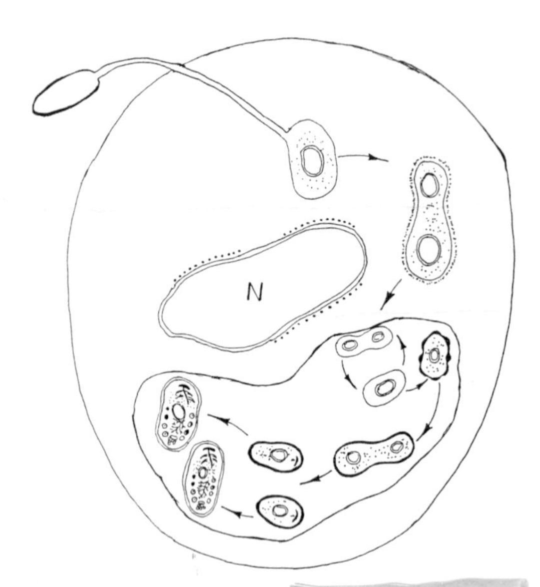

FIGURE 20 Diagram of the life cycle of *Encephalitozoon*. A triggered spore injects its unikaryotic sporoplasm into the host cell cytoplasm, beginning the proliferative phase. This genus develops in a host-formed parasitophorous vacuole surrounding all the developing stages. The organisms multiply by binary fission or multiple fission of ribbonlike moniliform multinucleate cells. The transition to the sporogonic phase is morphologically indicated by parasite secretions through the plasmalemma, producing the thickened membrane of the sporont. In this genus the number of sporoblasts produced depends on the species in question. *E. cuniculi* produces two sporoblasts, while *E. hellem* may produce more. All sporoblasts undergo a metamorphosis into spores. This entire cycle takes place in a parasitophorous vacuole in the host cell cytoplasm (type II interfacial relationship). N, nucleus.

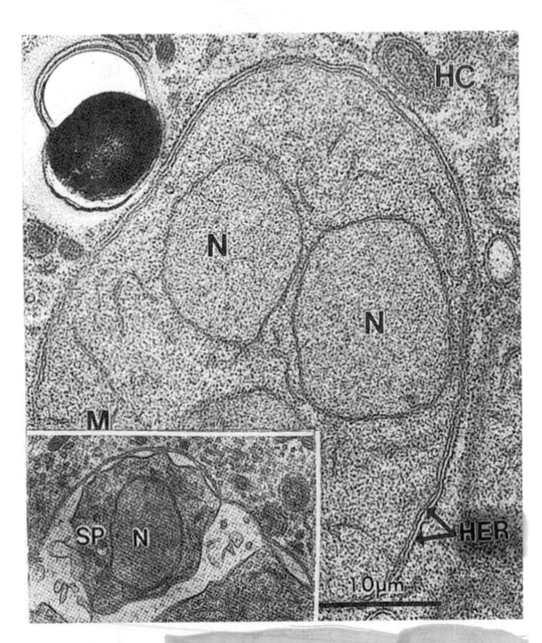

FIGURE 21 *Endoreticulatus schubergi*. The double membrane of the host endoplasmic reticulum (HER) completely surrounds the multinucleated plasmodium (M) in the host cytoplasm (HC) with at least three nuclei (N). (Inset) The sporogonial plasmodium divides by plasmotomy, producing several uninucleate sporonts. A portion of the double-membraned parasitophorous vacuole contains single nucleated sporonts (SP). (Reprinted with permission from Cali and El Garhy, 1993.)

and El Garhy, 1991). *Pleistophora* is found in the muscle tissue of fish, amphibians, and humans (Canning and Lom, 1986; Ledford et al., 1985).

The proliferative cells are in the hyaloplasm and at the end of the proliferative phase; they then develop "blisters" on the plasmalemma which enlarge and pull away, forming a second membrane which separates from the plasmalemma. These organisms produce an interfacial envelope that surrounds them as they commence sporogony, a membranous (SPOV), a typical thin unit membrane that may or may not persist. The plasmalemma of the sporogo-

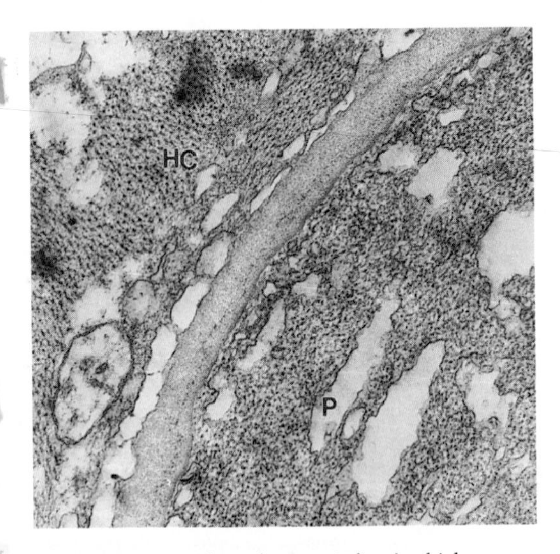

FIGURE 22 *Pleistophora typicalis*. A thick amorphous coat surrounding a proliferative plasmodium is shown. (Reprinted with permission from Canning and Hazard, 1982.)

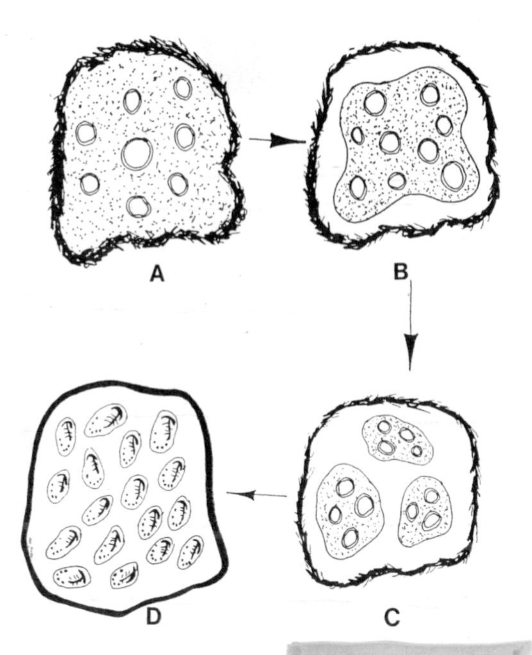

FIGURE 23 Diagram of the *Pleistophora*-secreted SPOV. The proliferative phase includes formation of a plasmodium by multiple nuclear divisions of isolated nuclei (A). The proliferative plasmodium secretes a thick, electron-dense amorphous coat which is retained and becomes the SPOV when the parasite plasmalemma retracts away from it (B). The plasmalemma thickens as it retreats from the SPOV. The sporogonial plasmodium divides by plasmotomy, producing smaller multinucleate cells (C). This division process continues until all the cells are uninucleate. These cells are sporoblasts and all undergo a metamorphosis into spores. The entire cycle takes place in the SPOV in the host cell cytoplasm (type III interfacial relationship).

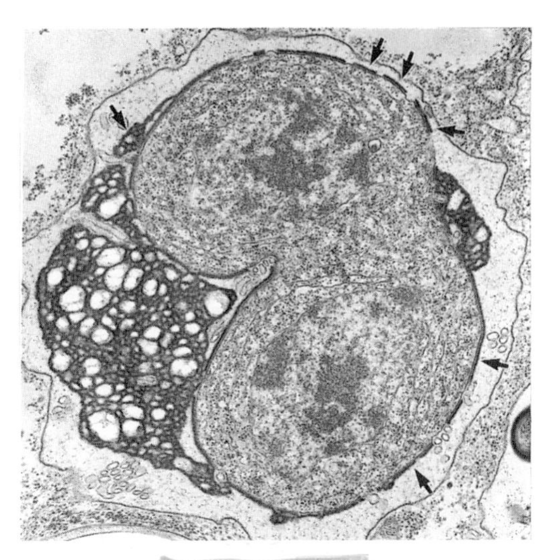

FIGURE 24 *Vairimorpha necatrix.* Plasmodia forming electron-dense coats from the material in the episporontal space, which appears to be connected to the parasite (arrows). (Reprinted with permission from Mitchell and Cali, 1993.)

nial plasmodium, now inside the SPOV, thickens as a result of progressive deposition of secretions until the entire membrane becomes uniformly thick (Fig. 24). These genera include *Thelohania, Octosporea,* and *Vairimorpha.*

The genus *Vairimorpha* is unique in that it has a *Nosema*-like proliferative phase followed by two sporogonic cycles (Fig. 25). One cycle remains *Nosema*-like throughout sporogony, while the other cycle is *Thelohania*-like and forms an interfacial envelope isolating the remaining developmental stages from the host cytoplasm (Pilley, 1976; Mitchell and Cali, 1993). These microsporidia are primarily parasites of insects, and the only member of this group found in mammals is *Thelohania apodemi,* reported from the brain of a field mouse (Doby et al., 1963).

INDIRECT CONTACT BETWEEN HOST AND PARASITE

Microsporidia having indirect contact between host- and parasite-produced interfacial envelopes are separated from the hyaloplasm by membranes and/or secretions produced by both the host and the parasite.

The host ER closely abuts the parasite plasmalemma during proliferative development. The parasite then produces blisters arising from the plasmalemma to form the interfacial envelope. Thus, a SPOV is formed in sporogony.

This type of development is seen in *Glugea.* (Canning et al., 1982). During proliferative development, the parasite plasmalemma is tightly abutted by host ER, forming an interfacial envelope (Fig. 26). Additionally, at the onset of sporogony, the parasite plasmalemma develops externally produced blisters. As they enlarge and separate from the plasmalemma, the SPOV is formed (Fig. 27) (Canning and Lom, 1986). The

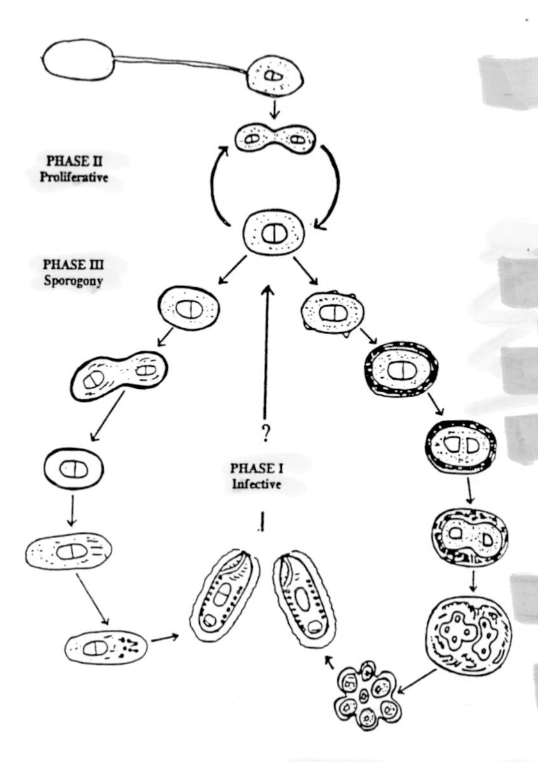

FIGURE 25 Diagram of the proposed life cycle of *V. necatrix,* showing the two patterns of development, disporoblastic (*Nosema*-like) and octosporoblastic (*Thelohania*-like). During the sporogonic phase of octosporoblastic development, the parasite appears to develop as two plasmodia in a single sporophorous vesicle. (Reprinted with permission from Mitchell, 1993.)

sporogonial plasmodium membrane thickens following the deposition of secretions, forming a scalloped surface that develops into a uniform coat (thick membrane). Similar development has been observed in the genus *Loma* (Canning and Lom, 1986).

The host and the parasite contribute to the formation of a thick interfacial envelope that surrounds all stages of parasite cells. *Trachipleistophora,* a disseminating parasite found only in humans, is an example of this type of development (Weidner et al., 1997). They describe a surface plaque matrix (PQM) associated with tubular elements, both of parasite origin. "The PQM surfaces are characteristically overlain by host membrane which appears to be endoplasmic reticulum since ribosomes are often attached. This host cell endoplasmic reticulum surrounding the PQM and tubules is thought to be fixed in position . . . at the host parasite interface" (Fig. 28) (Weidner et al., 1997).

A host-formed parasitophorous vacuole surrounds each cluster of developing parasites. Additionally, parasite-secreted material surrounds each organism inside the vacuole, partitioning them from each other and resulting in a septate (honeycomb) appearance. This type of interface is characteristic of the genus *Septata* (Cali et al., 1993). Both proliferative and sporogonic stages of this organism continuously secrete an electron-dense fibrillar material that forms a network around the individual parasite cells developing within a parasitophorous vacuole (Fig. 29 and 30).

Appendages

An additional feature relating to the host-parasite interface and present in many of the above-mentioned interfacial relationships, is appendages. In 1983 Takvorian and Cali attempted to classify the tubules, and subsequently other researchers have described various appendages (Table 2). There are several distinct types (Fig. 31 to 34) designated types I to IV (Takvorian and Cali, 1983; Moore and Brooks, 1992). Tubules have been reported in a diversity of organisms (Larsson, 1986; Sprague et al., 1992) and are usually associated with sporogony, often appearing in conjunction with interfacial envelopes. They have been observed in both parasitophorous vacuoles, e.g., in *Septata* (Cali et al., 1993), and SPOVs, e.g., in *Glugea* (Takvorian and Cali, 1983) and *Vairimorpha* (Moore and Brooks, 1992). However, they have also been identified in the proliferative phase and in the hyaloplasm of *B. vesicularum* (Cali et al., 1998) and *N. algerae* (Avery and Anthony, 1983), two organisms exhibiting a precocious thickening of the plasmalemma.

Another group of appendages includes filaments, fibers, bristles, mucous coverings, and ridges, which have been described as various types of surface structures (Vavra, 1963; Takvo-

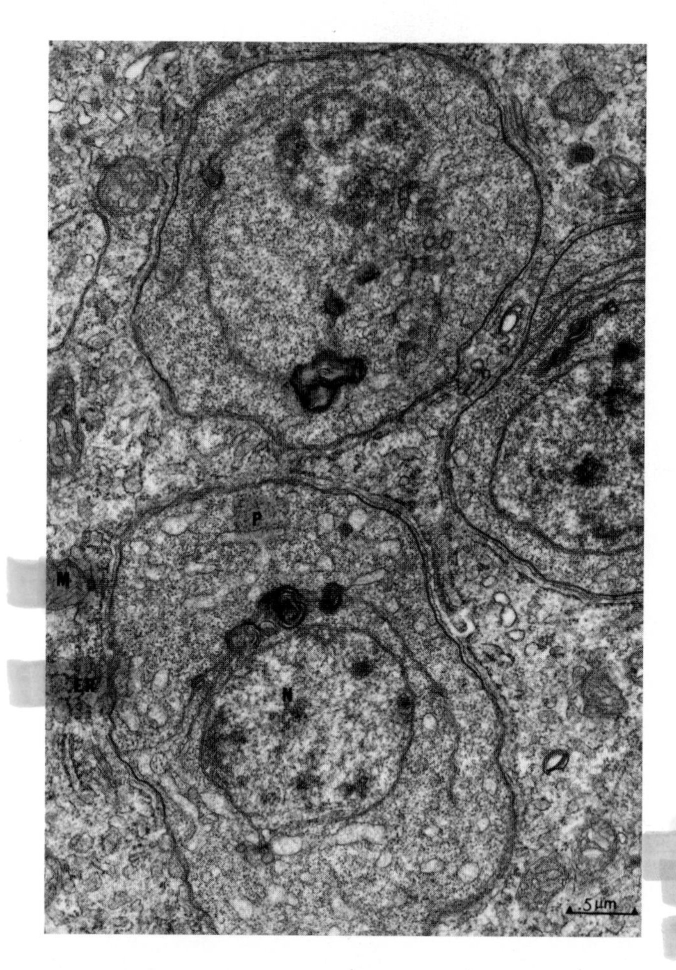

FIGURE 26 *Glugea stephani*. Early proliferative parasites in the periphery of a xenoma surrounded by host endoplasmic reticulum (ER) and host mitochondria (M) are shown. Note the diffuse parasite cytoplasm devoid of any mitochondria.

rian and Cali, 1983; Morrison and Marryaatt, 1986; Weidner et al., 1990). Weidner illustrated an elaborate array of filaments on the spores of *Ameson* (Fig. 35) (Weidner, 1970). The functions of all these appendages are not known, but they probably serve to facilitate host-parasite interchange of materials.

PHASE II OF THE MICROSPORIDIAN LIFE CYCLE

The proliferative phase includes all cell growth and division from the sporoplasm through the parasite's commitment to spore formation. This phase has been referred to as schizogonic and merogonic by some authors, however, different authors have assigned different types of nuclear activity to the terms merogony and schizogony with respect to the microsporidia. Schizogony and merogony have been considered synonymous by some (Vavra and Sprague, 1976); in the 1986 book by Canning and Lom, the term meronts was used. They defined meronts as having isolated nuclei or diplokarya that may divide repeatedly by binary fission or multiple fission or by plasmotomy (Canning and Lom, 1986). Other authors use both terms and define them

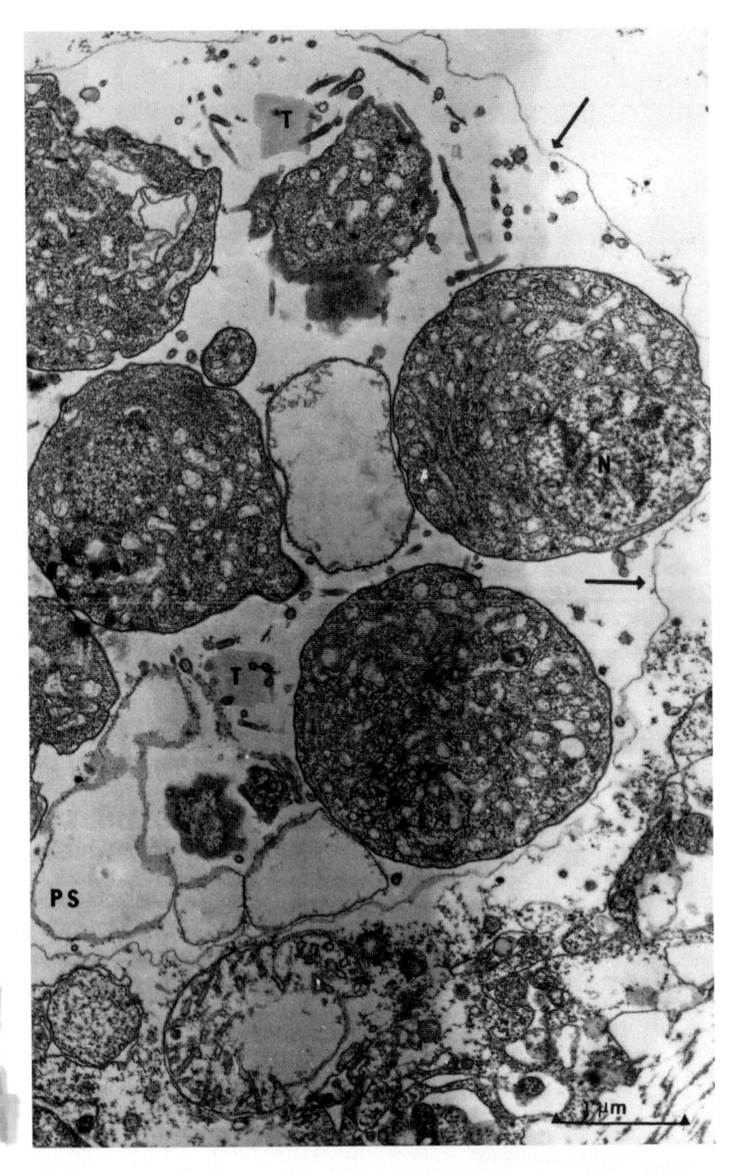

FIGURE 27 *G. stephani* sporoblasts in a SPOV. Note the thickened plasmalemma, the increased cytoplasmic density of sporoblasts, and the presence of tubules (T) in the SPOV (arrows).

as representing differing types of nucleation, merogony applying only to diplokaryotic cells, and schizogony applying to division of cells with unpaired nuclei (Sprague et al., 1992). Larsson (1986) uses the apicocomplex approach and defines schizogony as a type of plasmodial cell division that may occur in the proliferative or the sporogonic phase of development. Additionally, the microsporidia are both diverse and enigmatic in regard to ploidy numbers, the presence or absence of meiotic divisions, and the interaction of the nuclei of diplokaryotic pairs. Because of the diversity of these organisms and the interpretation of terms, it seems better to refer to this phase of development as the proliferative phase and describe what is known of the nuclear activity of the particular microsporidian.

Proliferative cells in general have a large nuclear region and a simple cytoplasm. Nucleation is represented by isolated rounded (Fig. 10) or double D-shaped diplokarya (Fig. 11). An atypical or rarely occurring nuclear form is the long,

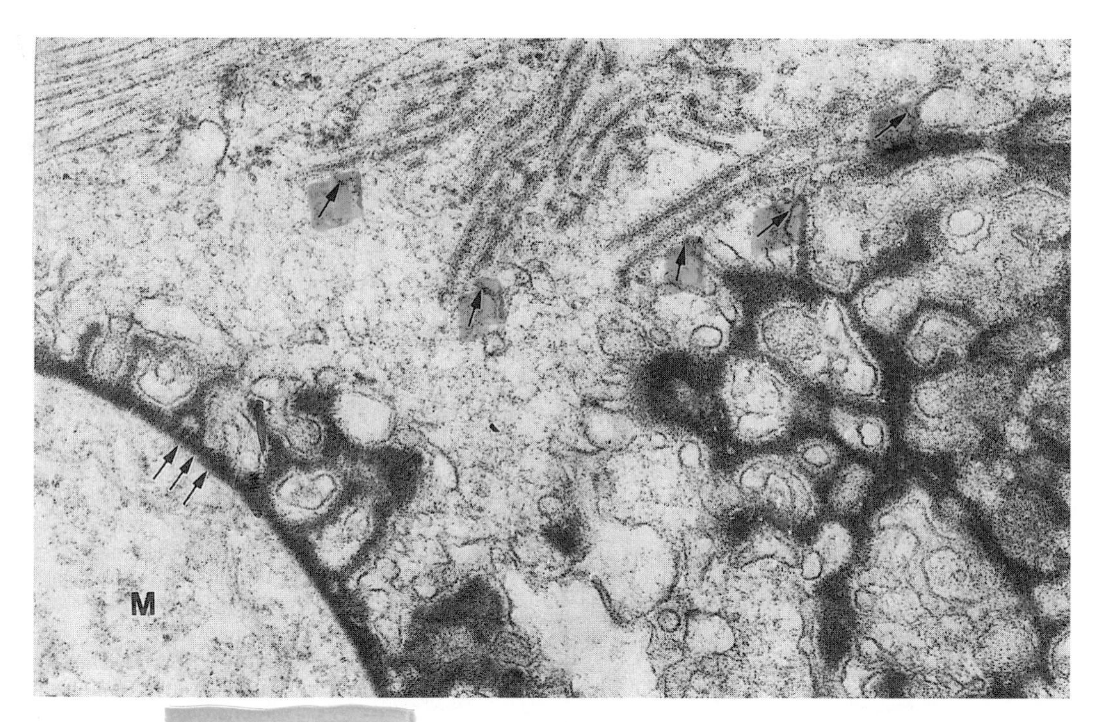

FIGURE 28 *Trachipleistophora hominis.* An electron micrograph of an anastomosing complex of PQM with tubules adjoining the meront (M) is shown. The position of the meront plasma membrane is indicated by three arrows. Abundant ER vesicles are aligned with the PQM surface. Single arrows indicate the membrane-surrounded tubules and the points at which the tubule-enveloped membrane is confluent with ER membrane around PQM. (Reprinted with permission from Weidner et al., 1997.)

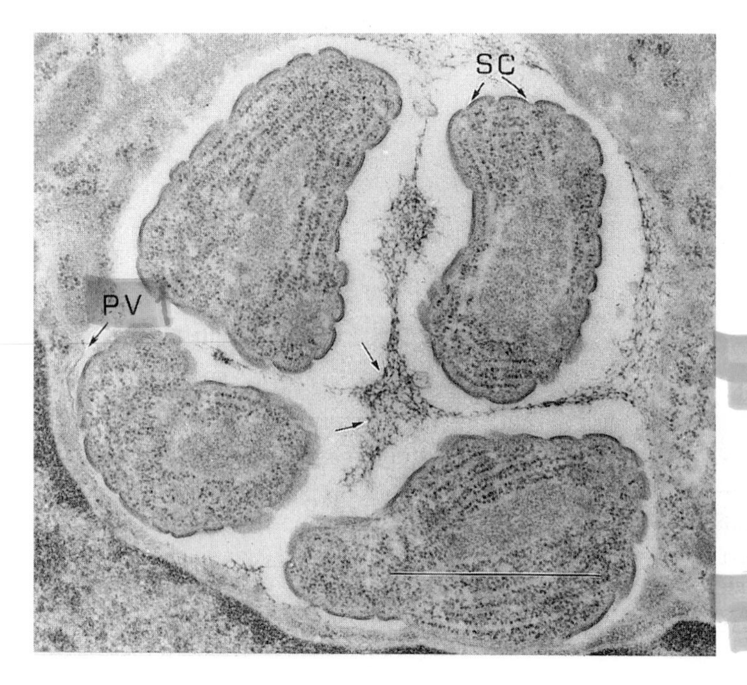

FIGURE 29 Commencement of sporogony and formation of sporonts in *E. intestinalis.* The first sign of sporogony is the deposition of secretory material on the plasmalemmal surface (SC) of single nucleated cells now called sporonts. As deposition of this material continues (SC), the sporont cell surface takes on a scalloped appearance and pulls away from the fibrillar lamina (arrows). It is at this stage that the presence of a parasitophorous vacuole membrane becomes evident (PV). (Reprinted with permission from Cali et al., 1993.)

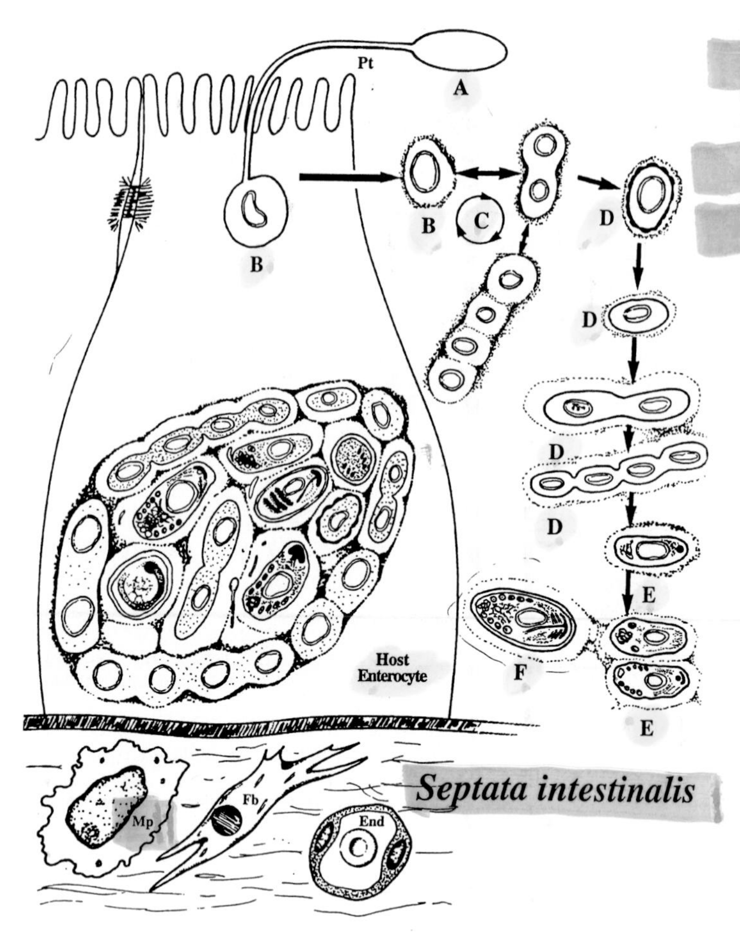

FIGURE 30 Diagram of the life cycle of *E.* (*S.*) *intestinalis.* (A) An empty spore with its polar tubule extruded is injected into the cytoplasm of a host intestinal enterocyte cell. It has a sporoplasm (B) at the end. (B to G) Intracelluar developmental stages. (C) Proliferation cells may be uninucleated, binucleated, or tetranuleated. Multinucleate cells are elongated and divide by simple fission. The plasmalemma of these cells is a unit membrane which secretes droplets that become a fibrillar matrix in which the parasites are embedded. (D) Sporont cells with thick plasmalemmal membranes may be uninucleated, binucleated, or tetranucleated and continue to secrete the fibrillar matrix. The sporont cell surface appears to pull away from the fibrillar matrix, making the cells appear to be in individual chambers. A parasitophorous vacuole membrane becomes evident, and long tubular appendages form at the surface of the sporonts and are scattered between the parasite cells. (E) After the last cell division, sporoblast metamorphosis begins. These single nucleated, oval cells contain a vesicular Golgi-like mass associated with polar tubule formation. (F) Spores measure 2 by 1 μm and are characterized by a thick electron-lucent spore coat, a single nucleus, a polar tubule that forms about five coils in a single row and its anterior attachment complex, and a posterior vacuole. (G) A parasite cluster, illustrating the asynchronous nature of development, the separation of individual cells by the fibrillar matrix, and the presence of type I tubular appendages. Other cells in the lamina propria that may become infected are fibroblastic (Fb), endothelial (End), and macrophage (Mp) cells. (Reprinted with permission from Cali, 1993.)

TABLE 2 Classification of tubular appendages[a]

Type I: a single tubule that emanates from the sporont surface and usually terminates in a bulblike structure (Fig. 31).
Type II: a cluster of uniformly thin tubules arranged in cablelike appearance (Fig. 32).
Type III: a single uniform-diameter tubule that may be branched and contains electron-dense, regularly spaced particles (Fig. 33).
Type IV: an elaborate branching network of tubules surrounding sporogonial plasmodia (Fig. 34). The tubules appear either electron-dense or electron-lucent depending on their contents.

[a]Types I to III have been described by Takvorian and Cali (1986); type IV has been described by Moore and Brooks (1992).

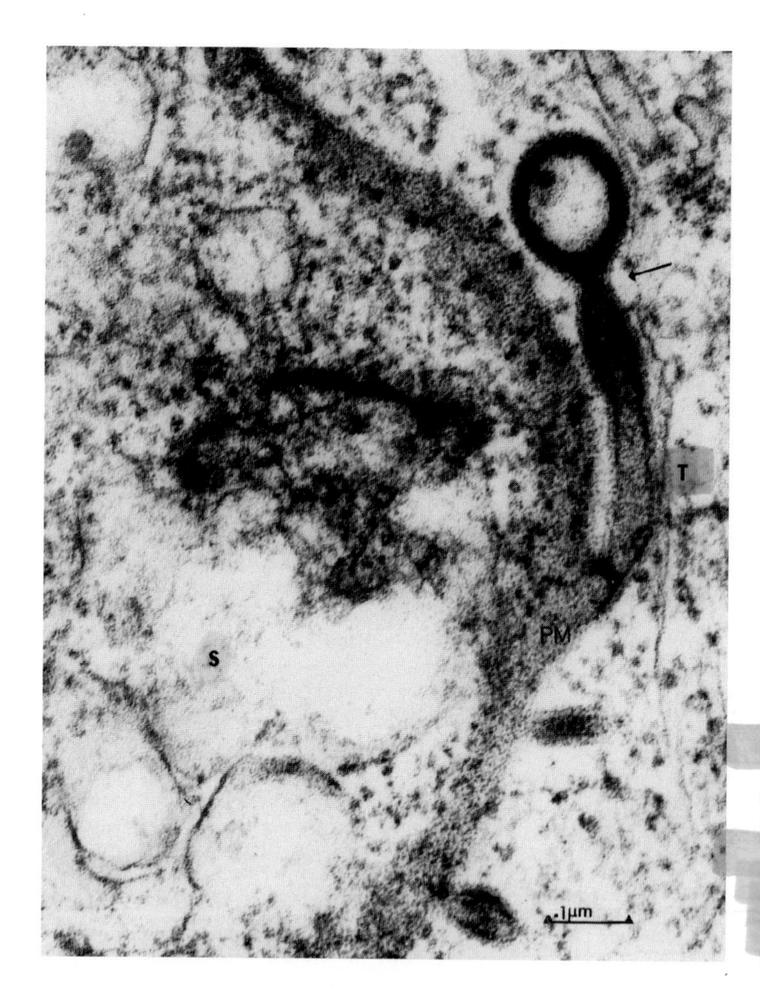

FIGURE 31 Example of a type I tubule (T) in *G. stephani*. The tubule is continuous with the sporoblast (S) plasmalemma (PM), and its distal end constricts (arrow) before terminating in a bulbous structure. Note the granular and filamentous material covering the bulb. (Reprinted with permission from Takvorian and Cali, 1983.)

sausage-shaped nuclei seen in some early stages of *Enterocytozoon bieneusi* (Fig. 15) (Cali and Owen, 1990). The ribosomes and ER are relatively sparse, and the Golgi complex generally is not apparent until the beginning of sporogony. The plasmalemma of most genera appears as a typical unit membrane and may be in direct contact with the host cell cytoplasm or have a variety of interfacial separations, depending on the parasite genus (Table 1).

Division of the Proliferative Cells

Proliferative cells may contain one or more nuclei. When more than one nucleus is present, they are isolated or are in a diplokaryon arrangement. Diplokaryotic nuclei divide as a pair, producing new pairs. As nuclear division occurs, the cytokinetic process may be linked so that the cells divide by binary fission (Fig. 36), or the cells may continue to grow without cytokinesis, producing larger multinucleate cells. If these cells are rounded, plasmodia are produced and the cells divide by plasmotomy (Fig. 21). Alternatively, the cytoplasm surrounding the nuclei may elongate into a ribbonlike form which eventually divides by forming indentations of the plasmalemma between the nuclei, resulting in multiple fission (Fig. 17). Different genera follow different patterns (Fig. 8).

In some genera, the proliferative phase includes cells that divide continuously, producing hundreds of organisms (*Glugea*) (Fig. 37), while in other genera relatively few cells are produced in this phase, most of their parasite replication occurring later in sporogony (*Pleistophora*) (Fig. 16).

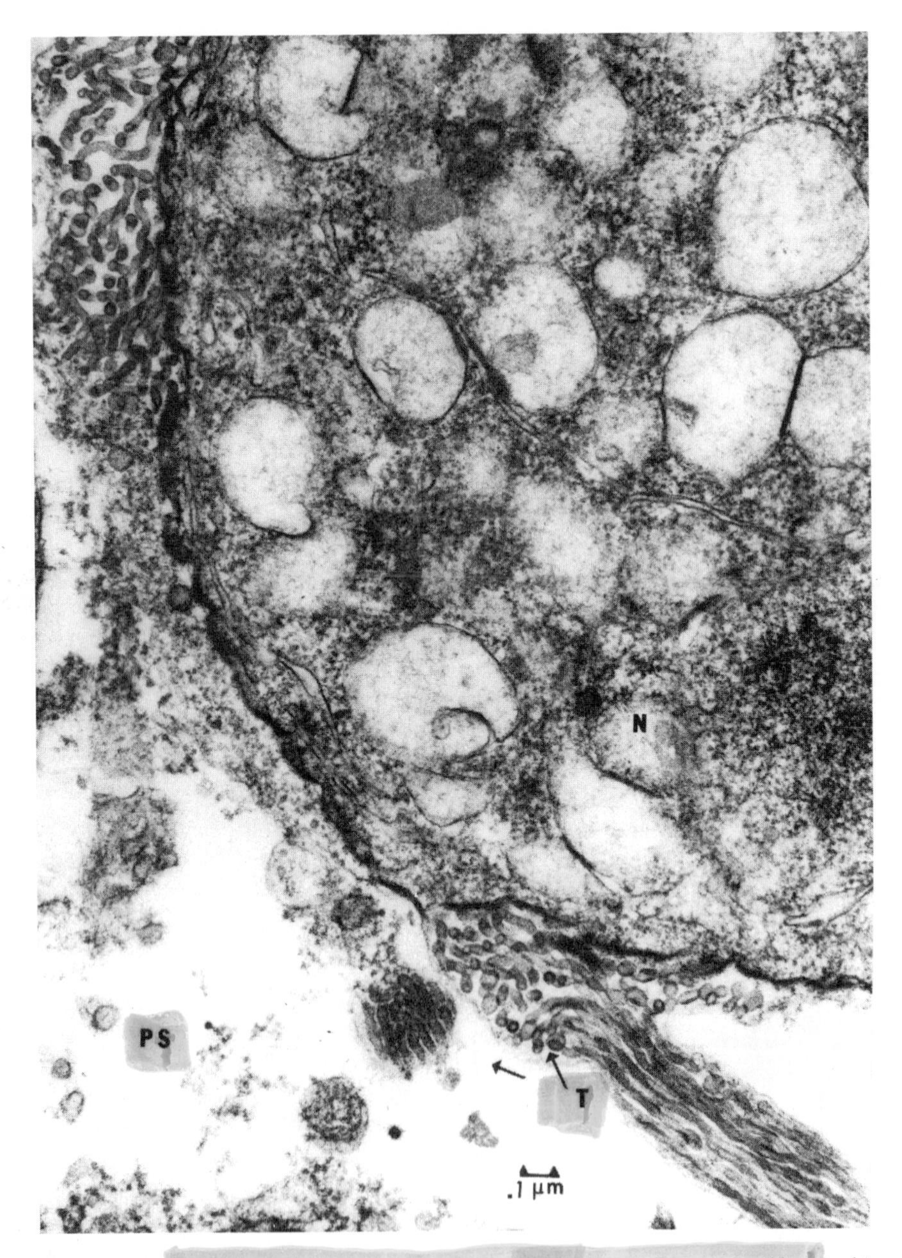

FIGURE 32 *G. stephani.* A cluster of type II tubules (T) projects from a sporoblast. At this magnification, the constrictions along the length of the tubules are apparent. PS indicates a SPOV. (Reprinted with permission from Takvorian and Cali, 1983.)

PHASE III OF THE MICROSPORIDIAN LIFE CYCLE

The sporogony phase involves sporonts (cells that produce two to several sporoblasts), sporoblasts (cells that undergo a metamorphosis into spores), and spores.

Sporonts

As the initial cells in this phase of development, sporonts are also the cells that make the transition into the sporogonic phase of development. In microsporidia that have meiosis, its occurrence is the point of transition to the sporogonic

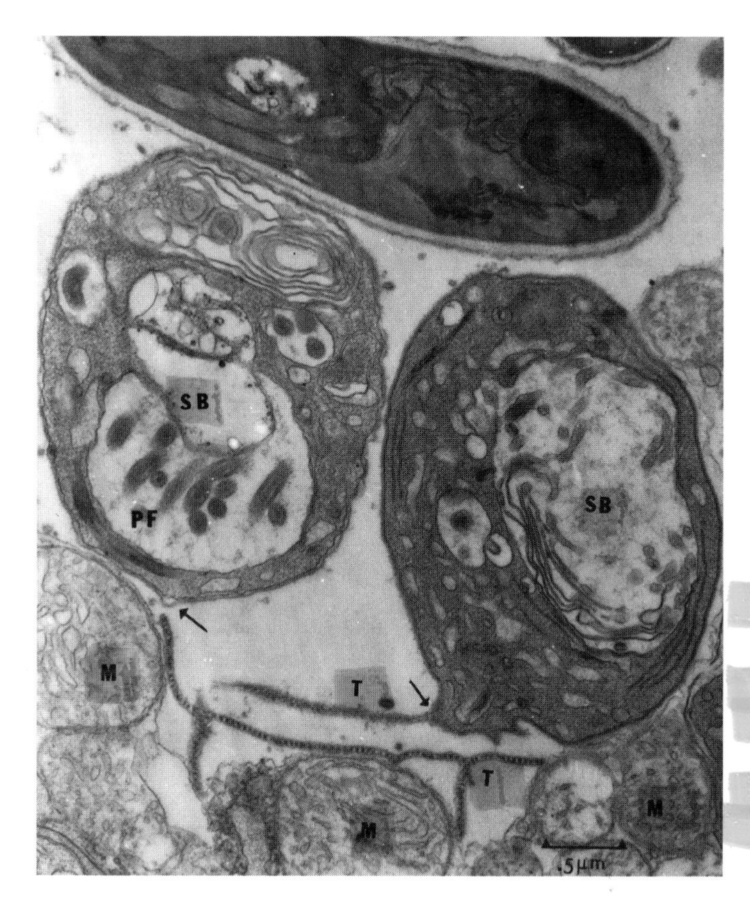

FIGURE 33 *G. stephani* sporoblasts (SB), identified by the presence of a developing polar filament (PF), have type III tubules (T) projecting from their plasmalemma (arrows). The tubules contain regularly spaced electron-dense particles inside the tubule lumen. Note the tubule abutting the host mitochondrion (M). (Reprinted with permission from Takvorian and Cali, 1983.)

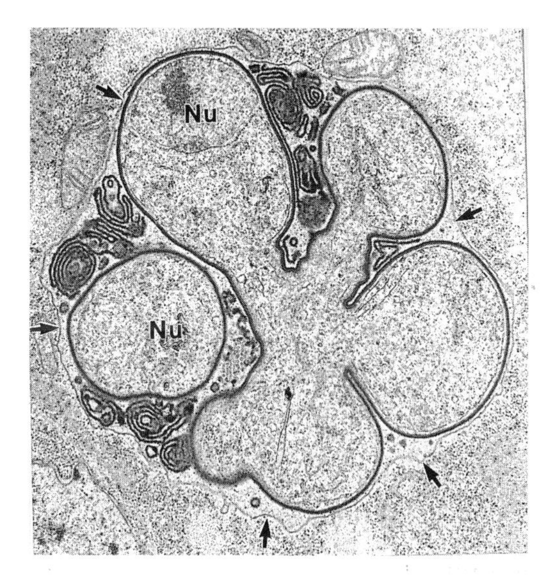

FIGURE 34 *V. necatrix*. Shown is an electron micrograph of the *Thelohania*-like pattern of development during sporogony in an SPOV which contains multinucleate plasmodium surrounded by type IV tubules in the episporontal space. Nu, nucleus. (Reprinted with permission from Mitchell and Cali, 1993.)

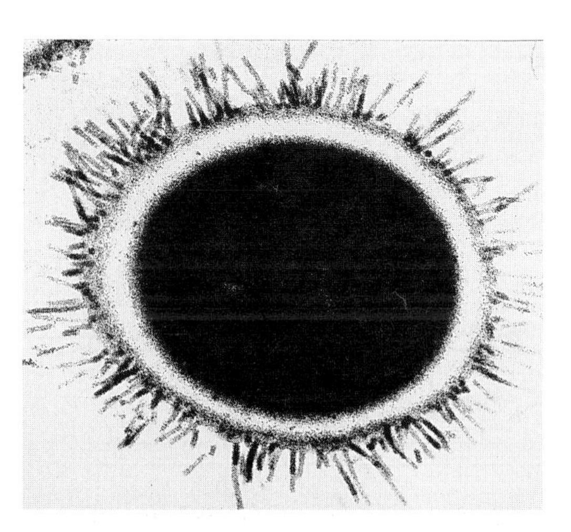

FIGURE 35 *Ameson michaelis.* Electron micrograph of a spore surface with projections. (Reprinted with permission of E. Weidner.)

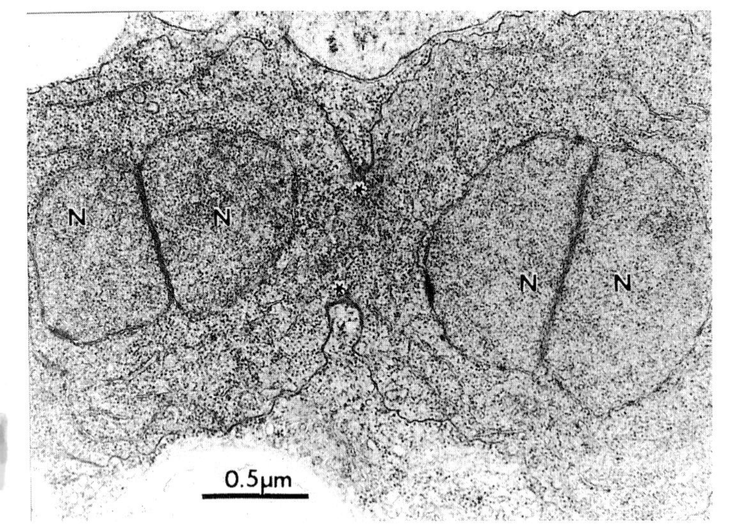

FIGURE 36 Electron micrograph of *N. bombycis* proliferative stage undergoing division (asterisk). The nuclei (N) are in the diplokaryon arrangement (paired and abutted) typical of this genus. Note the direct contact between the parasite and the host cytoplasm. (Reprinted with permission from Cali, 1971.)

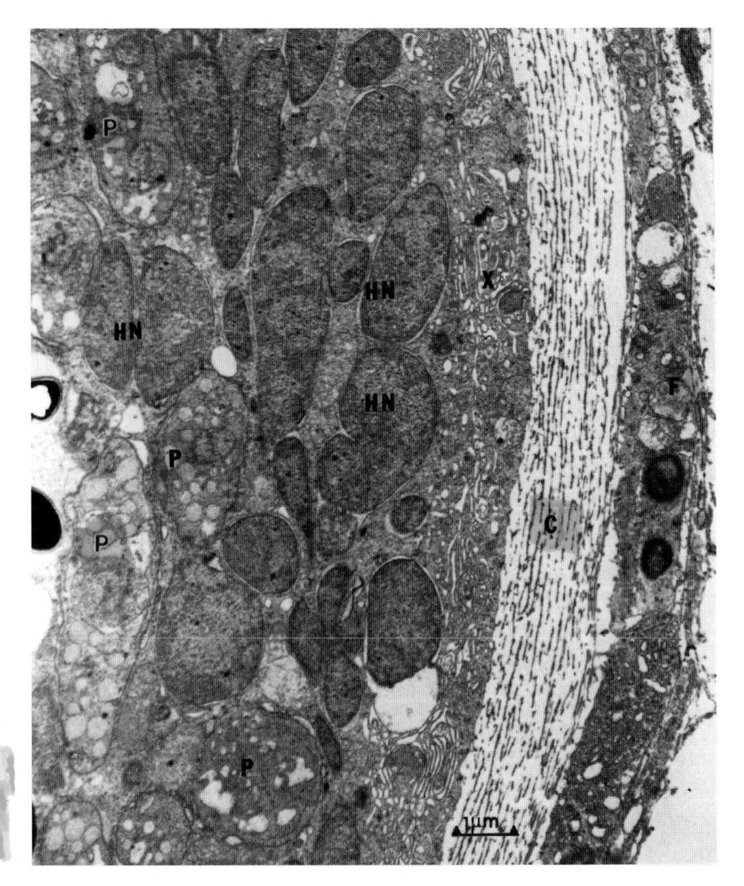

FIGURE 37 *G. stephani*. The periphery of the xenoma (X) containing proliferating parasites (P) distributed among host nuclei (HN) is shown. Collagen (C) and fibroblasts (F) encapsulate the xenoma.

phase. However, in most microsporidia meiosis is not an essential feature of sporogony. In genera in which diplokaryotic nuclei separate, this event is considered the beginning of sporogony. Morphological markers indicating the transition to sporogony include a general increase in cytoplasmic density due to increased ER and ribosomes and a change in the appearance of the parasite plasmalemma (cell-limiting membrane). In the development of many microsporidia, a cell becomes a sporont with the onset of plasmalemmal thickening. Material secreted by the parasite forms an electron-dense addition to the plasmalemma. Being part of a progressive process, the deposited material usually forms patches, producing a scalloped appearance (Fig. 29). But it may be the result of irregular additions of secretions (Fig. 24), creating large and small patches of material. This thickening progresses to a uniformly dense covering (Fig. 38) which may be smooth, ridged, or ornamented and is often referred to as the "thickened membrane" of the sporont. This electron-dense layer becomes the exospore coat of the spore stage (Fig. 2 and 3).

Membrane thickening was initially demonstrated in the genus *Nosema* in 1971 and has been reported in many other genera since then (Cali, 1971). A few exceptions are known, the most notable being *N. algerae,* in which a precocious thickened membrane is present from the sporoplasm stage on, being present in all the proliferative developmental stages (Avery and Anthony, 1983). Additionally, this situation is a characteristic of two species in two other genera, *A. varivestes* (Brooks et al., 1985; Issi et al., 1993) and *B. vesicularum* (Cali et al., 1998). The reverse has also been observed, and in *E. bieneusi,* plasmalemmal thickening does not occur until just prior to spore formation (Fig. 39).

Once the sporont's thickened membrane is formed, the number of nuclear and cell divisions that follow vary depending on the genus. Three types of division are observed, similar to those seen in the proliferative phase. Karyokinesis linked to cytokinesis, also referred to as binary fission of binucleates, results in the production of two sporoblasts. In genera such as *Nosema* (diplokaryotic) and *Encephalitozoon* (isolated nuclei), each sporont cell undergoes one nuclear division process, then the cell divides by binary fission, producing two sporoblast cells (Fig. 40 and 41). Each sporoblast cell undergoes metamorphosis to produce a spore; hence these genera are referred to as disporous genera.

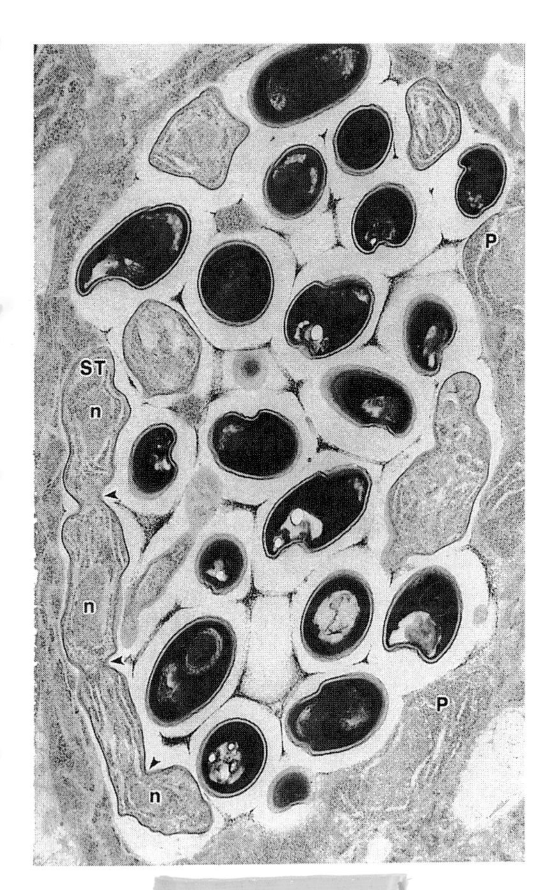

FIGURE 38 *E. intestinalis* sporogony. Deposition of material results in a uniformly thick plasmalemma surrounding the sporont cells. These sporonts continue to secrete the fibrillar lamina. The sporont (ST) is a tetranucleate (n) elongated cell in the process of cytokinesis (arrowhead). This cluster of parasite cells also contains many mature electron-dense spores, proliferative cells (P), and a dense fibrillar lamina separating the individual parasite cells. (Reprinted with permission from Cali et al., 1993.)

If cytokinesis is not linked to karyokinesis, nuclear division is repeated within the cell and four nuclei are produced. These cells may elongate, forming varying lengths of ribbon-like, moniliform sporonts or a sporogonial plasmodium containing four nuclei. In genera

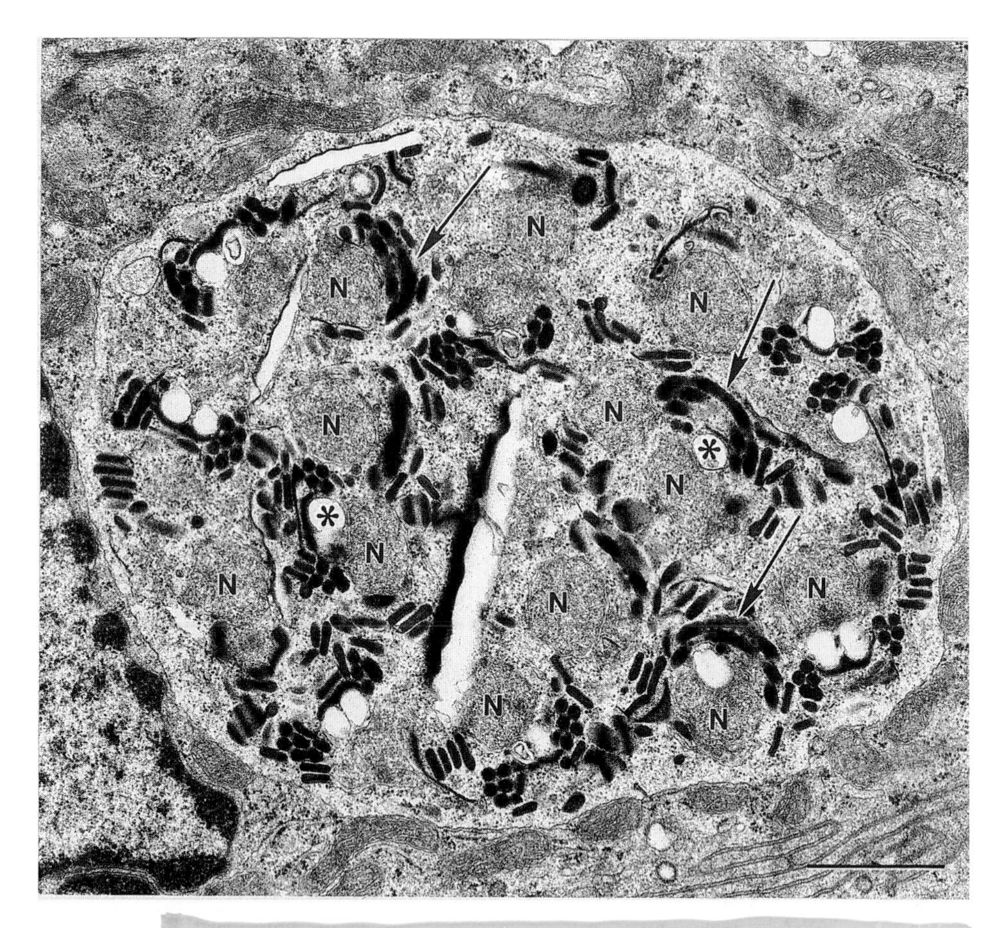

FIGURE 39 *E. bieneusi* plasmodium with multiple developing polar filaments and many nuclei. A sporogonial plasmodium containing at least 12 nuclei (N) in a single plane of the section is shown. The round, dense nuclei are each associated with electron-dense disk complexes and electron-lucent inclusions (asterisk). Electron-dense disks fuse into arcs, forming polar filament coils. Despite the advanced maturation and organelle separation associated with each nucleus, there is not yet any evidence of cytokinesis or plasmalemmal thickening. (Reprinted with permission from Cali and Owen, 1990.)

such as *Tetramicra* and *Septata,* four nuclei are produced and a postkaryokinetic cytokinesis results in four sporoblasts (Fig. 29 and 30). When nuclear division is repeated many times within the cell, the cell enlarges by becoming a rounded multinucleate cell, called a sporogonial plasmodium, and divides by plasmotomy. In the genera *Thelohania* (isolated nuclei) and *Octosporea* (diplokaryotic nuclei) and in the plasmodial form of *Vairimorpha,* the sporont develops into a sporogonial plasmodium containing eight nuclei which divides by plasmotomy and produces eight spores (Fig. 34). In other polysporous genera, such as *Pleistophora,* the majority of cell divisions may occur during sporogony, sometimes producing as many as 100 or more spores from a sporogonial plasmodium.

Sporoblasts

Sporoblasts are cells that are derived from the final division of the sporont. They are the products of the last cell division of the entire intracellular development and are the cells

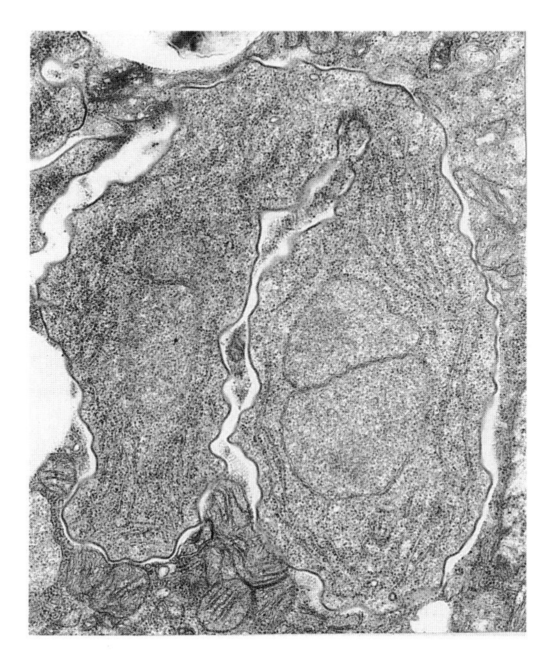

FIGURE 40 *N. bombycis* cell after the diplokaryon has completed its division in a cell possessing a thickened membrane. Cytokinesis has commenced, but a connection between the two diplokaryotic parts of the cell is still present. (Reprinted with permission from Cali, 1971.)

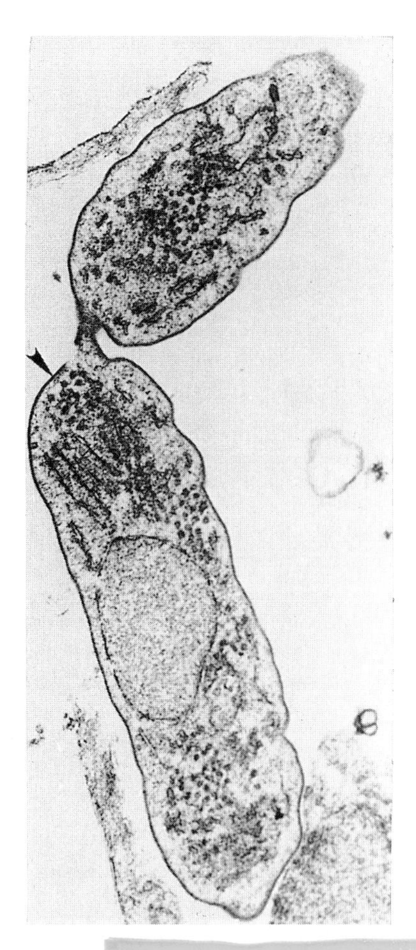

FIGURE 41 Electron micrograph of *E. cuniculi* sporont undergoing cytokinesis. Note the thickened membrane (arrowhead) and single nucleus. It is the occurrence of cell division at this stage of development that makes this parasite disporous. (Reprinted with permission from Pakes et al., 1975.)

that undergo morphogenesis into spores. They form the extrusion apparatus which consists of the polar filament, its anchoring disk complex, the polaroplast membranes and/or tubules, and the posterior vacuole. In addition, large quantities of ribosomes are produced. As sporoblasts continue their maturation, they crenate and decrease in size and their cytoplasm becomes more electron-dense (Fig. 42). This increase in cytoplasmic density and complexity is a result of the rapid production of ER and ribosomes, enlargement of the Golgi complex, and the presence of both numerous electron-dense bodies and portions of the forming polar filament complex.

Early in sporoblast morphogenesis, a cluster of vesicles, generally referred to as a "primitive Golgi" or a Golgi-like complex, appears. As early as 1965, Vavra described this "area of vesicles" as a "primitive Golgi zone" (Vavra, 1965). Subsequently, the term "Golgi" has been used to describe a variety of structures in the sporoblast. Jensen and Wellings (1972) describe saccules in *Glugea stephani* as Golgi, calling the area of vesicles particulate material. Sprague and Vernick (1969) describe the PAS-positive region in the anterior polaroplast area as Golgi. However, it is the area of vesicles associated with the forming polar filament that has been cytochemically demonstrated to have thiamine pyrophosphatase (TPPase) activity (Takvorian

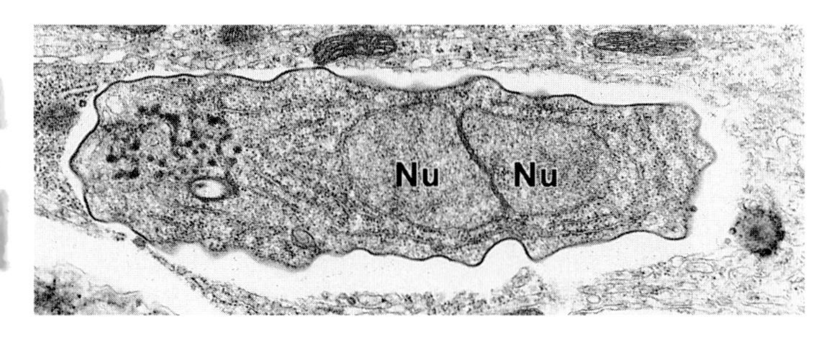

FIGURE 42 *N. bombycis* sporoblast. The cell-limiting membrane is thick, a diplokaryon is present, and polar filament formation has commenced. The morphogenic process of this cell into a spore has begun. Nu, nucleus. (Reprinted with permission from Cali, 1971.)

and Cali, 1994), a known marker for distinguishing the Golgi apparatus (Fujita and Okamoto, 1979). The enzyme activity of TPPase was used in the study of sporoblast development in *G. stephani* (Takvorian and Cali, 1994). Membrane-bound vesicular and lamellar structures associated with the forming polar filament contained increasing amounts of reaction product (Fig. 43), indicating that true Golgi activity is present and associated with polar filament formation in the microsporidia and that the vesicular structure is a Golgi complex (Takvorian and Cali, 1994). From the early 1970s on, a number of authors have illustrated and/or described the area of vesicles in many genera of the microsporidia (Cali, 1971; Walker and Hinsch, 1972; Youssef and Hammond, 1971).

Additionally, nucleoside diphosphatase (NDPase), a marker of ER and, in some cells, of the outermost *cis*-Golgi membrane (Goldfischer, 1982; Goldfischer et al., 1971; Novikoff, 1976), was used in a histochemical study on *G. stephani* (Takvorian and Cali, 1996). The NDPase reaction product was present in the posterior end of early sporoblasts, in tubules, and on the forming polar filament. The reaction product appeared to "bleb" off multilobed bodies adjacent to the forming polar filament. At the end of polar filament formation, reaction product was visible on lamellar bodies appearing in the area of the posterior vacuole, in polaroplast membranes, and on the polar filament sheath. These enzyme histochemical studies suggest that both the ER and the Golgi apparatus are responsible for producing the polar filament complex.

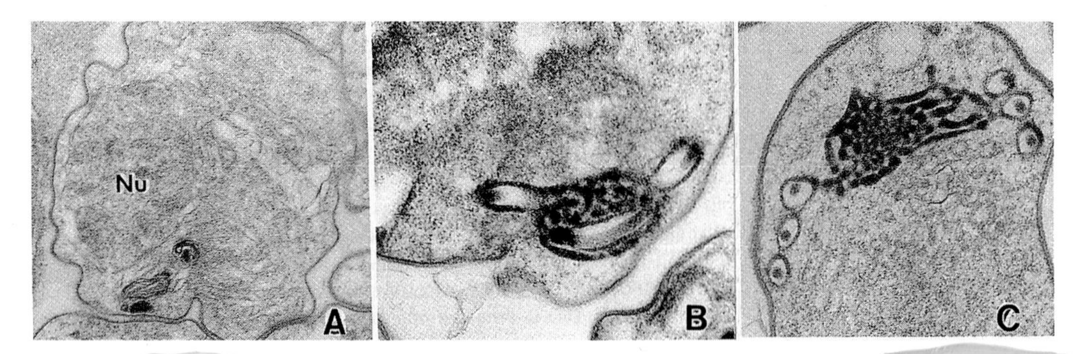

FIGURE 43 *G. stephani* sporoblasts histochemically treated to demonstrate the presence of the Golgi-associated enzyme TPPase, which produces an electron-dense reaction product (RP). (A) The TPPase activity in membranes and a dense body (forming polar filament). (B) As maturation of the sporoblast continues, vesicular and lamellar structures associated with the forming polar filament contain to increase the amount of RP. (C) RP is visible in the forming polar filament coil cross sections and the associated vesicular and lamellar structures. No other sporoblast structures contained TPPase activity. (Reprinted with permission from Takvorian and Cali, 1994.)

The Polar Filament Complex

The anchoring disk, a mushroom- or umbrella-shaped structure, forms in the anterior most portion of these cells. Within the anchoring disk center is an electron-dense area which appears to be the point at which the straight or manubroid portion of the polar filament emerges. In many microsporidia, a membrane, the polar sac, covers the anchoring disk and is the limiting membrane of this portion of the polar filament (Fig. 2 and 3) (Vavra, 1976; Larsson, 1986).

The polaroplast, a series of multilayered lamellae and/or tubular structures, develops around the manubroid portion of the polar filament shaft. This structure ranges from being absent (*Hessea squamosa*) or poorly defined (*Buxtehudea scaniae*) in a few microsporidian genera to being a "voluminous" component of the spore in most genera (Larsson, 1986). It is adjacent to the underside of the anchoring disk cap and emanates from the polar sac perpendicular to the manubroid portion of the polar filament (Fig. 47). The polaroplast is formed from vesicular material that widens to become sacs. These sacs elongate and become closely packed together, producing a series of electron-dense lamellae with spaces between them (Vavra, 1976; Larsson, 1986; Takvorian and Cali, 1986). Although the polaroplast morphology varies depending on the species, it generally develops into two different types of sacs present in different regions along the polar filament.

The coils of the polar filament are generally present from below the forming polaroplast to the posterior of the cell. The coiling follows the periphery of the sporoblast, encircling what will become the sporoplasm. These cells are significantly more electron-dense than the early sporoblasts, are usually smaller, and appear crenated as they shrink away from the surrounding tissue.

Finally, as the sporoblast nears maturation into a spore, after the spore shape and the polar filament have become well established (Fig. 44), the posterior vacuole and the endospore coat structures are produced. The posterior

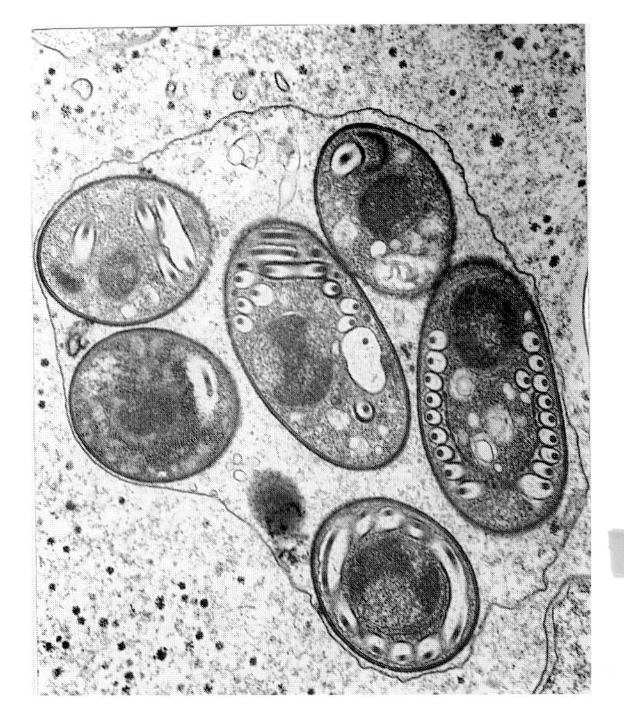

FIGURE 44 *V. necatrix* maturing sporoblasts in a sporophorous vesicle. Note the advanced stage of the developing polar filaments, the absence of the electron-lucent endospore, and the decreased amount of electron-dense material in the episporontal space. (Reprinted with permission from Mitchell and Cali, 1993.)

vacuole is a membrane-bound space believed to be a remnant of the Golgi apparatus (Sprague and Vernick, 1969). The endospore coat is an electron-lucent layer of secretions which forms and becomes thicker as the sporoblast becomes a mature spore. It is located between the plasmalemma and the electron-dense, previously secreted coat.

Spores

The spores are diagnostic for the microsporidia (Fig. 2 and 3). The spore possesses a thick, resistant coat divided into three regions. The outermost is the electron-dense exospore that originated from parasite secretions at the beginning of sporogony. This layer is from two to several times thicker than a membrane and may be smooth or ornamented with ridges, filaments (including intermediate filaments) (Weidner et al., 1990; Weidner, 1992), tubular structures (Batson, 1983), or a mucous coating containing filaments (Vavra and Sprague, 1976) (Fig. 35). The next region is the electron-lucent endospore. It is a protein-chitin complex with variable thickness but is usually many times thicker than the exospore. This electron-lucent layer is thinnest in the region of the anterior anchoring disk complex (Fig. 45 and 46). The innermost layer of the spore coat is the membrane surrounding the entire spore contents, the extrusion apparatus, a sporoplasm (the nucleus and the cytoplasm of the spore), and the posterior vacuole. The combined spore coats provide a great deal of protection for the spore's survival in the environment.

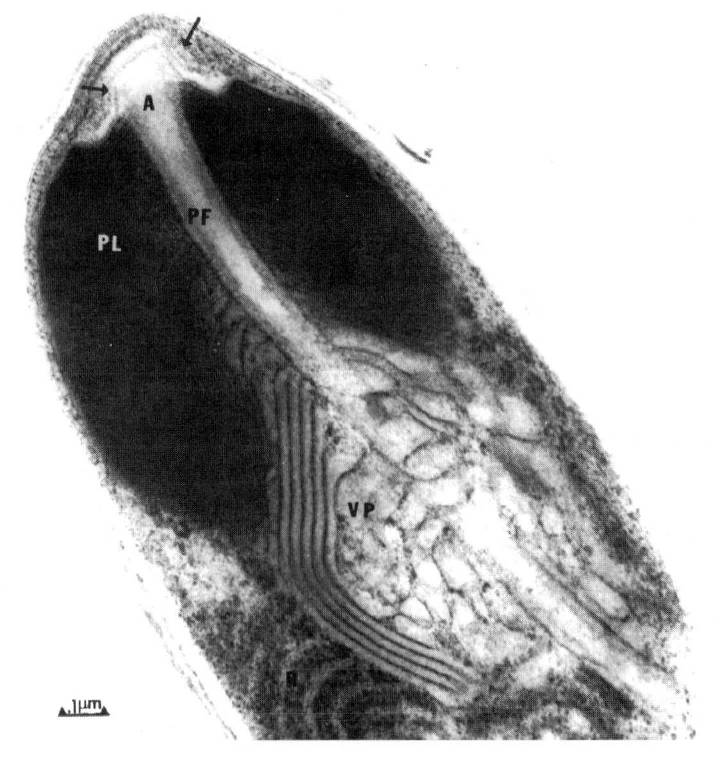

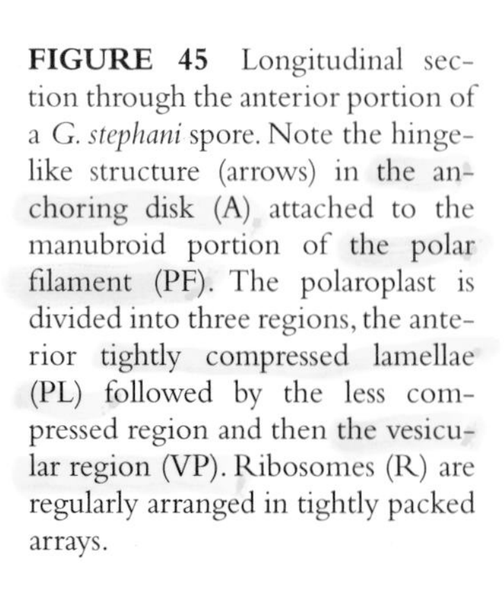
FIGURE 45 Longitudinal section through the anterior portion of a *G. stephani* spore. Note the hinge-like structure (arrows) in the anchoring disk (A) attached to the manubroid portion of the polar filament (PF). The polaroplast is divided into three regions, the anterior tightly compressed lamellae (PL) followed by the less compressed region and then the vesicular region (VP). Ribosomes (R) are regularly arranged in tightly packed arrays.

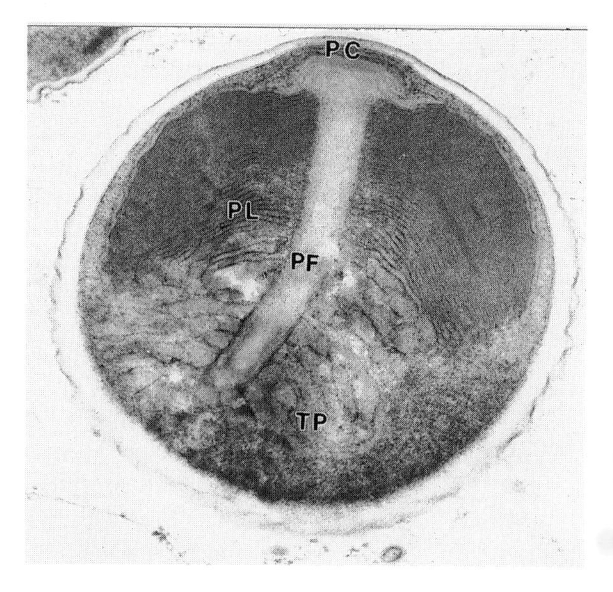

FIGURE 46 *G. stephani.* The polar cap (PC) is well illustrated in this oblique section through the anterior portion of a spore. The tubular nature of the third region of the polaroplast (TP) is well illustrated in this section of the spore. Note the thinning of the spore wall in the region of the polar cap. PL, pleated "compressed" lamellae; PF, polar filament. (Reprinted with permission from Takvorian and Cali, 1981.)

The extrusion apparatus consists of several parts that formed in the sporoblast. The most obvious is the long, coiled polar filament also known as the polar tubule. Both terms are used for this structure because when it is in the spore it appears solid and when it is extruded it everts and becomes tubular. It is attached inside the anchoring disk complex, and the polaroplast surrounds it, filling the anterior third to half of the spore. In a mature spore, the polar filament appears multilayered, composed of a number of concentric rings of varying thickness and electron density, and the entire structure is surrounded by a membrane sheath (Vavra, 1976; Takvorian and Cali, 1981, 1986; Larsson, 1986). Because of its unique structure and function in initiating infection, the development of the polar filament, the extrusion apparatus, and the organelle(s) that produces them have always been of great interest to microsporidiologists.

The anteriormost region of the polaroplast abuts the anchoring disk with membrane sacs that generally appear tightly compressed into regular arrays of lamellae, while the more posterior sacs are less compressed and more irregular, sometimes having a tubular or vesicular appearance (Fig. 45 and 46) (Vavra, 1976; Takvorian and Cali, 1981, 1986). These regions have been called the lamellar polaroplast, the tubular polaroplast, and the vesicular polaroplast. Recognizing the diversity of polaroplast morphology and organization in various species of microsporidia, Larsson (1986) described five types of polaroplast structural arrangements. While the polarblast is variable in volume and morphology, it seems to consistently terminate before the coiled region of the polar filament. The polar filament coils around the contents of the remainder of the spore, the sporoplasm and the posterior vacuole (Fig. 2).

The polar filament is variable in length, arrangement, and diameter. The number of coil cross sections varies from one organism to another but is usually fairly consistent in any given species. As few as a half-dozen (*E. bieneusi* and *Encephalitozoon cuniculi*) (Fig. 47) to as many as three to four dozen (*Pleistophora macrozoarcidis*) coils may be formed (Fig. 48) (El

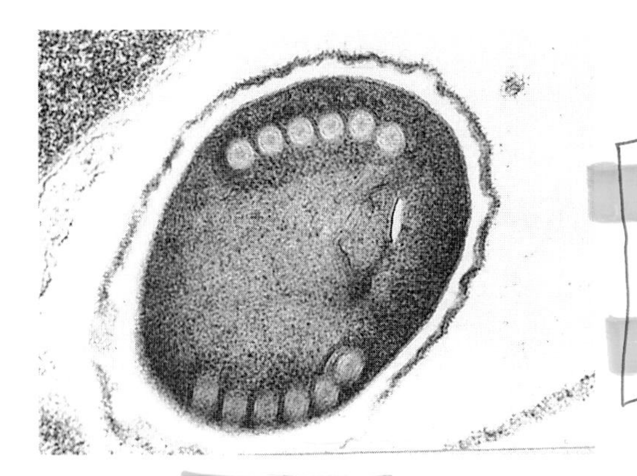

FIGURE 47 *E. cuniculi* spore containing six cross sections of the isofilar polar filament.

Garhy, 1993). The arrangement can be variable or consistent in a given species. In *E. bieneusi* the polar filament always forms six coils in a double row (Cali and Owen, 1990), and in *E. cuniculi* it forms six coils of the polar filament in a single row (Pakes et al., 1975). In *B. vesicularum,* the polar filament typically forms two rows, but single and triple rows of coils also occur (Cali et al., 1998). The polar filament diameter may be isofilar, uniform throughout its length (Fig. 3) (most microsporidia), or anisofilar, a portion of the filament being smaller in diameter than the rest (Fig. 49).

The sporoplasm consists of nuclear and cytoplasmic material. The nuclear material may be in the form of a single nucleus or a diplokaryon, depending on the genus of the microsporidian (Fig. 8). The surrounding cytoplasm is extremely dense and rich in ribosomes, which are usually arranged in a regular helical array or in tightly packed sheets (Fig. 49).

In the most posterior region of the spore is the posterior vacuole, which often appears irregular in shape. This membrane-limited structure has been described as an empty vacuole or one containing flocculent, globular, vesicular, glomerular, or lamellar material in concentric whorls or fine particulate material (Vavra, 1976; Takvorian and Cali, 1986).

The spores are produced intracellularly within the infected host cells. When the infected host cell dies, the spores are released and may pass out of the body in waste products, or when the host organism dies, the spores may gain access to the external environment. The resistant spore coat enables them to survive until they are consumed or find the proper envi-

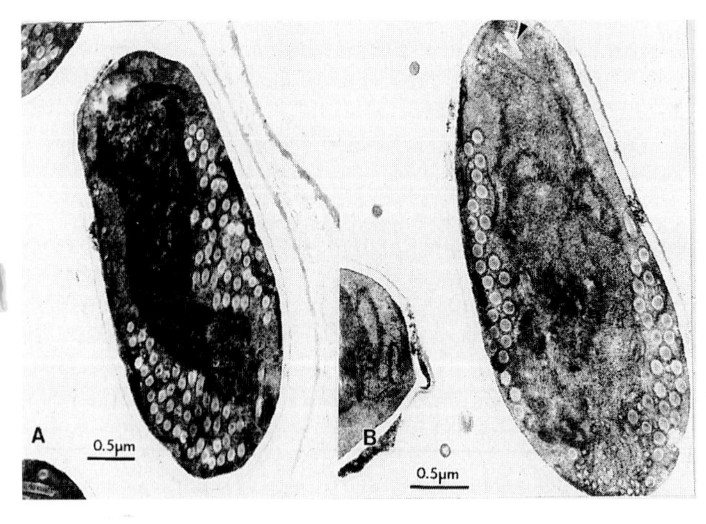

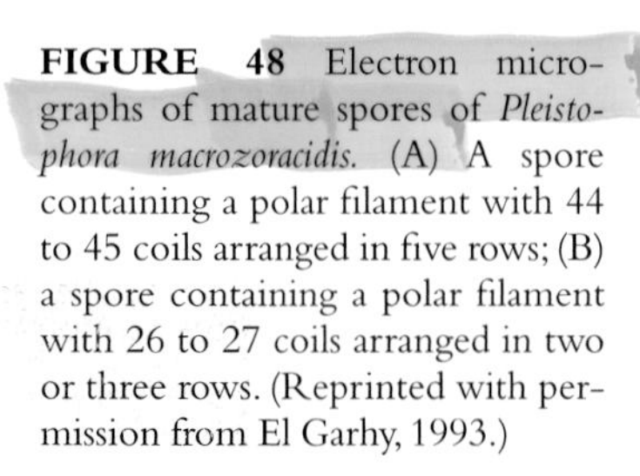

FIGURE 48 Electron micrographs of mature spores of *Pleistophora macrozoracidis.* (A) A spore containing a polar filament with 44 to 45 coils arranged in five rows; (B) a spore containing a polar filament with 26 to 27 coils arranged in two or three rows. (Reprinted with permission from El Garhy, 1993.)

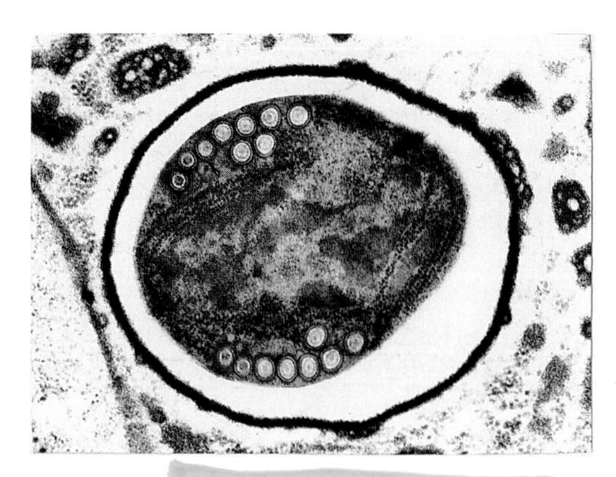

FIGURE 49 *Brachiola vesicularum* mature spore containing a fully developed electron-lucent endospore coat. The exospore surface has several vesiculotubular structures on it. Note the presence of nine polar filament cross sections arranged in two rows. This polar filament is anisofilar, and the last two or three cross sections are smaller in diameter than the others. (Reprinted with permission from Cali et al., 1998.)

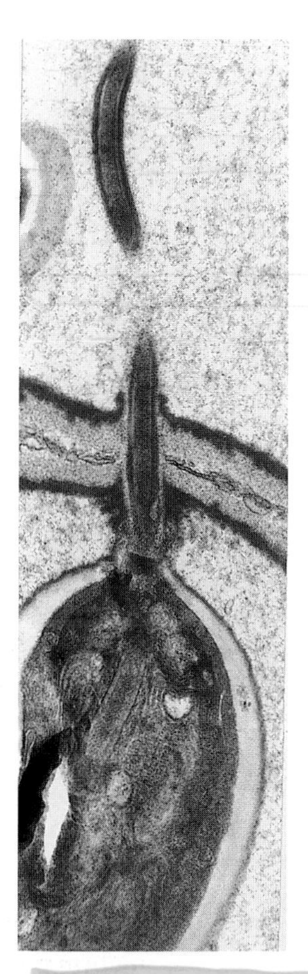

FIGURE 50 *Pleistophora macrozoracidis* autoinfective spore with extruded polar tubule. Note its ability to pierce through the thick SPOV wall surrounding it. A portion of the tubule is visible in the adjacent cytoplasm where the sporoplasm has been injected. (Reprinted with permission from El Garhy, 1993.)

ronment for their polar filament eversion system to be stimulated. These spores can be referred to as environmental spores.

In addition to environmental spores, some microsporidia produce autoinfective spores which become activated and germinate within the same host in which they were just produced. They evert their polar filament, piercing through whatever cells are in their path. The sporoplasm is deposited within the cell pierced by the end of the polar tubule, thus other host cells are inoculated and the intracellular developmental cycle is repeated in newly infected cells of the same host (Fig. 50). These spores are called autoinfective spores because they promote the spread of infection within the host in which they were produced. Microsporidia that produce both often have two different numbers of polar filament coils within their spores (Iwano and Ishihara, 1989, 1991), the autoinfective ones having shorter filaments (Iwano and Kurtti, 1995), probably because an autoinfective spore does not have to deposit its contents very far from its location in order to reach an adjacent uninfected cell. In our experience, an organism with a short polar filament, such as *E. cuniculi* (about six coils), has only one polar filament size (Fig. 47) and produces both types of spores (Cali and Takvorian, unpublished observations).

Environmental spores, spores that transmit infection from host to host, require the proper stimuli to be triggered. The majority of microsporidial infections can be initiated by oral inoculation, the digestive tract providing the proper environmental stimuli that result in polar tubule eversion and injection of the sporoplasm into the surrounding host tissues. This event takes place in the intestine of the host after the spores have been consumed, and gut conditions initiate extrusion. The polar tubule everts with a great deal of force, enabling it to pierce adjacent host cells and deposit the sporoplasm. If the parasite is to be successful, the infective sporoplasm must be deposited within host cells that can support growth of the parasite and continue the life cycle. Infection of mucosal surfaces such as the intestinal wall of the host are easy to explain as being caused by direct inoculation from the polar tubule into the host cell. However, tissues far removed from the intestine become infected by a less direct and more elusive process that is often not well understood (Cali et al., 1991a; Cali et al., 1991b). (Some microsporidia are transmitted transovarily [Andreadis and Hall, 1979].) In some organisms, such as *Septata intestinalis* and *Nosema bombycis,* there are several cells in addition to those forming the epithelial lining that support the growth of the parasite, including such wandering cells as macrophages and endothelial and fibroblastic cells (Cali, 1993; Cali et al., 1993). These cells allow the infection to be disseminated to other parts of the host body. With other microsporidia such as *E. bieneusi* and *N. apis,* the parasite seems limited to the epithelial surface cells associated with the digestive tract (Pol et. al., 1993; Gray et al., 1969). The parasites may spread from cell to cell via these surface cells, but the connective tissue below does not seem to support the infection. The reverse may also be true. The fish microsporidian, *G. stephani,* can infect by oral inoculation; however, the gut epithelial cells are not infected; instead the infection is located in connective tissue below the epithelial lining (Cali et al., 1986).

VARIATIONS WITHIN A SPECIES

As previously stated, some organisms form two populations of spores: those that are autoinfective and those that are environmental. In addition, a third type of spore, a macrospore, occurs rarely in some species such as *Pleistophora* (Canning and Hazard, 1982). It is called a macrospore because it is about two times the size of the spores typical of the species. A few microsporidia of the genus *Amblyospora* have two host cycles with morphologically different spores in each and different development in the males and females of the same host species (Andreadis, 1985, 1988; Sweeney et al., 1985, 1988). The genus *Vairimorpha* has two cycles and forms two different types of spores in the same host and tissues at the same time (Fig. 25) (Moore and Brooks, 1992; Mitchell and Cali, 1993). These life cycles are covered in detail in Chapter 13.

Developmental Variations due to Temperature

In homothermic hosts, temperature is a factor in the success or failure of parasite development. Most poikilotherm-infecting microsporidia develop at temperatures typical of their hosts' temperature range. For example, *G. stephani* infects a variety of flatfish that develop in relatively cold water. When the fish are exposed to water at or above 16°C, parasite development occurs. Parasites continue to multiply as long as the fish is in this water temperature range, however, when the temperature goes below 15°C, parasite development is arrested. Reexposure to the higher temperature triggers continued parasite development (Olsen, 1981). Most insect microsporidia develop at ambient temperatures for their environmental range. A notable exception is *N. algerae,* which has been experimentally inoculated into rodents in which successful infection was accomplished in the footpads and tail only, the temperature presumably being the limiting factor (Trammer et al., 1997). Since the mammalian temperature is 37°C, this seems to be a rather fortunate barrier and probably represents physiological differences between mammalian and poikilothermic parasites.

Another effect of temperature is morphological. The genus *Vairimorpha* was created for a microsporidian that develops two different morphological patterns (Fig. 25). Depending on the temperature of its poikilothermic host, it can develop a *Nosema*-like morphology or a *Thelohania*-like morphology, usually a mixture of both since temperature varies and morphologies seem to overlap (Pilley, 1976, Mitchell and Cali, 1993).

SUMMARY

While there seems to be a diversity of microsporidia already identified, new ones are reported every day, giving new meaning to the word diversity. It makes one think that we have seen only the tip of the iceberg. With the new interest in microsporidia generated from AIDS infections and the new technologies being applied to them, we have had an explosion of new information in the past 10 years. It will be interesting to see how our knowledge of this group grows in the next 10 years.

ACKNOWLEDGMENTS

We thank Robert Shimony and Charlene Gallo for their suggestions and for proofreading the manuscript. We also thank Paul Lowman for assistance with the organization of the references.

REFERENCES

Andreadis, T. G. 1985. Life cycle, epizootiology, and horizontal transmission of *Amblyospora* (Microspora: Amblyosporidae) in a univoltine mosquito, *Aedes stimulans*. *J. Invertebr. Pathol.* **46:**31–46.

Andreadis, T. G. 1988. Comparative susceptibility of the copepod *Acanthocyclops vernalis* to a microsporidian parasite, *Amblyospora connecticus*, from the mosquito *Aedes cantator*. *J. Invertebr. Pathol.* **52:**73–77.

Andreadis, T. G., and D. W. Hall. 1979. Significance of transovarial infections of *Amblyospora* sp. (Microspora: Thelohaniidae) in relation to parasite maintenance in the mosquito *Culex salinarius*. *J. Invertebr. Pathol.* **34:**152–157.

Avery, S. W., and D. W. Anthony. 1983. Ultrastructural study of early development of *Nosema algerae* in *Anopheles albimanus*. *J. Invertebr. Pathol.* **42:**87–95.

Bailey, L. 1972. The preservation of infective microsporidan spores. *J. Invertebr. Pathol.* **20:**252–254.

Batson, B. S. 1983. A light and electron microscopic study of *Hirusutusporos austrosimulii* gen. n., sp. n. (Microspora: Nosematidae), a parasite of *Austrosimulium* sp. (Diptera: Simuliidae) in New Zealand. *Protistologica* **19:**263–280.

Becnel, J. J., E. I. Hazard, and T. Fukuda. 1986. Fine structure and development of *Pilosporella chapmani* (Microspora: Thelohaniidae) in the mosquito, *Aedes triseriatus* (Say). *J. Protozool.* **33:**60–66.

Brooks, W. M., J. J. Becnel, and G. G. Kennedy. 1988. Establishment of *Endoreticulatus* n. g. for *Pleistophora fidelis* (Hostounsky & Weiser, 1975) (Microsporida: Pleistophoridae) based on the ultrastructure of a microsporidium in the Colorado potato beetle, *Leptinotarsa decemlineata* (Say) (Coleoptera: Chrysomelidae). *J. Protozool.* **35:** 481–488.

Brooks, W. M., E. I. Hazard, and J. Becnel. 1985. Two new species of *Nosema* (Microsporida, Nosematidae) from the Mexican bean beetle *Epilachna varivestis* (Coleoptera, Coccinellidae). *J. Protozool.* **32:**525–535.

Bryan, R. T., A. Cali, R. L. Owen, and H. C. Spencer. 1991. Microsporidia: opportunistic pathogens in patients with AIDS, p. 1–26. *In* T. Sun (ed.), *Progress in Clinical Parasitology,* vol. 2. Field and Wood Medical Publishers, New York, N.Y.

Cali, A. 1971. Morphogenesis in the genus *Nosema*. Proceedings of the IVth International Colloquium on Insect Patholology, College Park, Md.

Cali, A. 1986. Comparison of the biology and pathology of Microsporida from different host groups, p. 356–359. *In* R. A. Samson, J. M. Black, and D. Peters (ed.), *Fundamental and Applied Aspects of Invertebrate Pathology.* Foundation for the 4th International Colloquium on Invertebrate Pathology, Wageningen, The Netherlands.

Cali, A. 1993. Cytological and taxonomical comparison of two intestinal disseminating microsporidioses. *AIDS* **7:**S12–S16.

Cali, A., R. T. Bryan, R. L. Owen, and H. C. Spencer. 1991a. Microsporidia: opportunistic pathogens in patients with AIDS. *Prog. Clin. Parasitol.* **2:**1–26.

Cali, A., and M. El Garhy. 1991. Ultrastructural study of the development of *Pleistophora schubergi* Zwolfer, 1927 (Protozoa, Microsporida) in larvae of the spruce budworm, *Choristoneura fumiferana,* and its subsequent taxonomic change to the genus *Endoreticulatus*. *J. Protozool.* **38:**271–278.

Cali, A., D. P. Kotler, and J. M. Orenstein. 1993. *Septata intestinalis* n. g., n. sp., an intestinal microsporidian associated with chronic diarrhea and dissemination in AIDS patients. *J. Eukaryot. Microbiol.* **40:**101–112.

Cali, A., D. M. Meisler, C. Y. Lowder, R. Lembach, L. Ayers, P. M. Takvorian, I. Rutherford, D. L. Longworth, J. McMahon, and R. T. Bryan. 1991b. Corneal microsporidioses: characterization and identification. *J. Protozool.* **38:**215–217S.

Cali, A., and R. L. Owen. 1988. Microsporidiosis, p. 929–950. *In* A. Balows, W. J. Hausler, M. Ohashi, and A. Turano (eds.), *Laboratory Diagnosis of Infectious Diseases: Principles and Practice* vol. 1. Springer-Verlag, New York, N.Y.

Cali, A., and R. L. Owen. 1990. Intracellular development of *Enterocytozoon,* a unique microsporidian found in the intestine of AIDS patients. *J. Protozool.* **37:**145–155.

Cali, A., and P. M. Takvorian. Unpublished observations.

Cali, A., P. M. Takvorian, E. Keohane, and L. M. Weiss. 1997. Opportunistic microsporidian infections associated with myositis. *J. Eukaryot. Microbiol.* **44:**86S.

Cali, A., P. M. Takvorian, S. Lewin, M. Rendel, C. S. Sian, M. Wittner, H. B. Tanowitz, E. Keohane, and L. M. Weiss. 1998. *Brachiola vesicularum,* n. g., n. sp., a new microsporidium associated with AIDS and myositis. *J. Eukaryot. Microbiol.* **45:**240–251.

Cali, A., P. M. Takvorian, J. J. Ziskowski, and T. K. Sawyer. 1986. Experimental infection of American winter flounder (*Pseudopleuronectes americanus*) with *Glugea stephani* (Microsporida). *J. Fish Biol.* **28:**199–206.

Canning, E. U., and E. I. Hazard. 1982. Genus *Pleistophora* Gurley, 1893: an assemblage of at least three genera. *J. Protozool.* **29:**39–49.

Canning, E. U., and J. Lom. 1986. *The Microsporidia of Vertebrates.* Academic Press, London, United Kingdom.

Canning, E. U., J. Lom, and J. P. Nicholas. 1982. Genus *Glugea* Thelohan 1891 (Phylum Microspora): Redescription of the type species *Glugea anomala* (Moniez, 1887) and recognition of its sporogonic development within sporophorous vesicles (pansporoblastic membranes). *Protistologica* **18:**193–210.

Canning, E. U., and J. P. Nicholas. 1974. Light and electron microscope observation on *Unikaryon legeri* (Microsporida, Nosematidae), a parasite of the metacercaria of *Meigymnophallus minutus* in *Cardium edule. J. Invertebr. Pathol.* **23:** 92–100.

Canning, E. U., and J. P. Nicholas. 1980. Genus Pleistophora (Phylum Microspora): redescription of the type species, *Pleistophora typicalis* Gurley, 1893 and ultrastructural characterization of the genus. *J. Fish Dis.* **3:**317–338.

Canning, E. U., and R. E. Sinden. 1973. Ultrastructural observations on the development of *Nosema algerae* Vavra and Undeen (Microporida, Nosematidae) in the mosquito *Anopheles stephensi* Liston. *Protistologica* **9:**405–415.

Cavalier-Smith, T. 1987. Eukaryotes with no mitochondria. *Nature* **326:**352–353.

Cepede, C. 1924. *Mrazekia piscicola* n.sp. microsporidie parasite du Merlan (*Gadus merlangus*). *Bull. Soc. Zool. Fr.* **49:**109–113.

Curgy, J., J. Vavara, and C. Vivares. 1980. Presence of ribosomal RNAs with prokaryotic properties in microsporidia: eukaryotic organisms. *Biol. Cell.* **38:**49–52.

Desportes, I., Y. Le Charpentier, A. Galian, F. Bernard, B. Cochand-Priollet, A. Lavergne, F. Ravisse, and R. Modigliani. 1985. Occurrence of a new microsporidian: *Enterocytozoon bieneusi* n. g., n. sp., in the enterocytes of a human patient with AIDS. *J. Protozool.* **32:**250–254.

Doby, J. M., A. Jeannes, and B. Rault. 1963. *Thelohania apodemi* n. sp., première microsporidie du genre *Thelohania* observée chez un mammifère. *C. R. Acad. Sci.* **257:**248–251.

Docker, M. F., M. L. Kent, D. M. Hervio, J. S. Khattra, L. M. Weiss, A. Cali, and R. H. Devlin. 1997. Ribosomal DNA sequence of *Nucleospora salmonis* Hedrick, Groff and Baxa, 1991 (Microsporea: *Enterocytozoonidae*): implications for phylogeny and nomenclature. *J. Eukaryot. Microbiol.* **44:**55–60.

Edlind, T. D., J. Li, G. S. Visvesvara, M. H. Vodkin, G. L. McLaughlin, and S. K. Katiyar. 1996. Phylogenetic analysis of beta-tubulin sequences from amitochondrial protozoa. *Mol. Phylogenet. Evol.* **5:**359–367.

El Garhy, M. F. 1993. *Comparison of the Development of Two Microsporidian Species* Pleistophora schubergi *Zwolfer, 1927 in Spruce Budworm larvae* (Choristoneura fumiferana) *and* Pleistophora macrozoarcidis *Nigrelli, 1946 in Ocean Pout* (Macrozoarces americanus) *and the Evaluation of Their Taxonomic Status.* Ph.D. thesis. Rutgers, The State University of New Jersey, Newark.

Frixione, E., L. Ruiz, J. Cerbon, and A. H. Undeen. 1997. Germination of *Nosema algerae* (Microspora) spores: - conditional inhibition by D_2O, ethanol and Hg^{2+} suggests dependence of water influx upon membrane hydration and specific transmembrane pathways. *J. Eukaryot. Microbiol.* **44:**109–116.

Frixione, E., L. Ruiz, and A. H. Undeen. 1994. Monovalent cations induce microsporidian spore germination in vitro. *J. Eukaryot. Microbiol.* **41:** 464–468.

Fujita, H., and H. Okamoto. 1979. Fine structural localization of thiamine pyrophosphatase and acid phosphatase activities in the mouse pancreatic acinar cell. *Histochemistry* **64:**287–295.

Fuxa, J. R., and W. M. Brooks. 1979. Mass production and storage of *Vairimorpha necatrix* (Protozoa: Microsporida). *J. Invertebr. Pathol.* **33:**86–94.

Germot, A., H. Philippe, and H. Leguyader. 1997. Evidence for loss of mitochondria in microsporidia from a mitochondrial-type HSP70 in *Nosema locustae. Mol. Biochem. Parasitol.* **87:**159–168.

Goldfischer, S. 1982. The internal reticular apparatus of Camillo Golgi: a complex, heterogeneous organelle, enriched in acid, neutral, and alkaline phosphatases, and involved in glycosylation, secretion, membrane flow, lysosyme formation, and intracellular digestion. *J. Histochem. Cytochem.* **30:** 717–733.

Goldfischer, S., E. Essner, and B. Schiller. 1971. Nucleoside diphosphatase and thiamine pyrophosphatase activities in the endoplasmic reticulum and Golgi apparatus. *J. Histochem. Cytochem.* **19:**349.

Gray, F. H., A. Cali, and J. D. Briggs. 1969. Intracellular stages in the life cycle of the microsporidan *Nosema apis. J. Invertebr. Pathol.* **14:** 391–394.

Henry, J. E., and E. A. Oma. 1974. Effect of prolonged storage of spores on field applications of *Nosema locustae* (Microsporida: Nosematidae) against grasshoppers. *J. Invertebr. Pathol.* **23:**371–377.

Hunt, R. D., N. W. King, and H. L. Foster. 1972. Encephalitozoonosis: evidence for vertical transmission. *J. Infect. Dis.* **126:**212–214.

Ishihara, R., and Y. Hayashi. 1968. Some properties of ribosomes from the sporoplasm of *Nosema bombycis. J. Invertebr. Pathol.* **11:**377–385.

Issi, I. V. 1986. Microsporidia as a phylum of parasitic protozoa. *Protozoology* **10:**1–136.

Issi, I. V., S. V. Krylova, and V. M. Nicolaeva. 1993. The ultrastructure of the Microsporidium *Nosema meligethi* and establishment of the new genus *Anncaliia. Parazitologiya* **27:**127–133.

Iwano, H., and R. Ishihara. 1989. Intracellular germination of spores of a *Nosema* sp. immediately after their formation in cultured cell. *J. Invertebr. Pathol.* **54:**125–127.

Iwano, H., and R. Ishihara. 1991. Dimorphism of spores of *Nosema* spp. in cultured cell. *J. Invertebr. Pathol.* **57:**211–219.

Iwano, H., and T. J. Kurtti. 1995. Identification and isolation of dimorphic spores from *Nosema furnacalis* (Microspora: Nosematidae). *J. Invertebr. Pathol.* **65:**230–236.

Jaronski, S. T. 1979. *Role of the Larval Mosquito Midgut in Determining Host Susceptibility to* Nosema algerae *(Microsporida).* Ph.D. thesis. Cornell University, Ithaca, N.Y.

Jensen, H. M., and S. R. Wellings. 1972. Development of the polar filament-polaroplast complex in a microsporidian parasite. *J. Protozool.* **19:**297–305.

Kellen, W. R., H. C. Chapman, T. B. Clark, and J. E. Lindegren. 1965. Host-parasite relationships of some *Thelohania* from mosquitoes (Nosematidae: Microsporidia). *J. Invertebr. Pathol.* **7:**161–166.

Kelly, J. F., and D. W. Anthony. 1979. Susceptibility of spores of the microsporidian *Nosema algerae* to sunlight and germicidal ultraviolet radiation. *J. Invertebr. Pathol.* **34:**164–169.

Kramer, J. P. 1970. Longevity of microsporidian spores with special reference to *Octosporea muscaedomesticae* Flu. *Acta Protozool.* **8:**217–224.

Kramer, J. P. 1976. The extra-corporeal ecology of microsporidia, p. 127–135. *In* L. A. Bulla, Jr., and T. C. Cheng (ed.), *Comparative Pathobiology of the Microsporidia,* vol. 1. Plenum Press, New York, N.Y.

Larsson, R. 1986. Ultrastructure, function, and classification of microsporidia. *Prog. Protistol.* **1:**325–390.

Ledford, D. K., M. D. Overman, A. Gonzalvo, A. Cali, S. W. Mester, and R. F. Lockey. 1985. Microsporidiosis myositis in a patient with the acquired immunodeficiency syndrome. *Ann. Intern. Med.* **102:**628–630.

Leitch, G. J., M. Scanlon, G. S. Visvesvara, and S. Wallace. 1995. Calcium and hydrogen ion concentrations in the parasitophorous vacuoles of epithelial cells infected with the microsporidian *Encephalitozoon hellem. J. Eukaryot. Microbiol.* **42:** 445–451.

Maddox, J. V. 1973. The persistence of the microsporida in the environment. *Misc. Publ. Entomol. Soc. Am.* **9:**99–106.

Maddox, J. V., and L. F. Solter. 1996. Long-term storage of infective microsporidian spores in liquid nitrogen. *J. Eukaryot. Microbiol.* **43:**221–225.

Mitchell, M. 1993. *Morphology and Morphometry of* Vairimorpha necatrix *in* Heliothis zea. Ph.D. thesis. Rutgers, The State University of New Jersey, Newark.

Mitchell, M. J., and A. Cali. 1993. Ultrastructural study of the development of *Vairimorpha necatrix* (Kramer, 1965) (Protozoa, Microsporida) in larvae of the corn earworm, *Heliothis zea* (Boddie) (Lepidoptera, Noctuidae) with emphasis on sporogony. *J. Eukaryot. Microbiol.* **40:**701–710.

Moore, C. B., and W. M. Brooks. 1992. An ultrastructural study of *Vairimorpha necatrix* (Microspora, Microsporida) with particular reference to episporontal inclusions during octosporogony. *J. Protozool.* **39:**392–398.

Morrison, C. M., and V. Marryaatt. 1986. Further observations on *Loma morhua* Morrison & Sprague, 1981. *J. Fish. Dis.* **9:**63–67.

Morrison, C. M., and V. Sprague. 1981. Electron microscopical study of a new genus and new species of Microsporida in the gills of Atlantic cod *Gadus morhua* L. *J. Fish Dis.* **4:**15–32.

Moura, H., D. A. Schwartz, F. Bornay-Llinares, F. C. Sodre, S. Wallace, and G. S. Visvesvara. 1997. A new and improved quick-hot Gram-chromotrope technique that differentially stains microsporidian spores in clinical samples, including paraffin-embedded tissue sections. *Arch. Pathol. Lab. Med.* **121:**888–893.

Novikoff, A. B. 1976. The endoplasmic reticulum: a cytochemist's view (a review). *Proc. Natl. Acad. Sci. USA* **73:**2781–2787.

Olsen, P. E., W. A. Rice, and T. P. Liu. 1986. In vitro germination of *Nosema apis* spores under conditions favorable for the generation and maintenance of sporoplasms. *J. Invertebr. Pathol.* **47:**65–73.

Olson, R. E. 1981. Effects of low temperature on the development of the microsporidan *Glugea stephani* in English sole (*Parophrys vetulus*). *J. Wildl. Dis.* **17:**559–563.

Olson, R. E., K. L. Tiekotter, and P. W. Reno. 1994. *Nadelspora canceri* n. g., n. sp., an unusual microsporidian parasite of the Dungeness crab, *Cancer magister. J. Protozool.* **41:**349–359.

Orenstein, J. M., D. T. Dieterich, and D. P. Kotler. 1992. Systemic dissemination by a newly recognized intestinal microsporidia species in AIDS. *AIDS* **6:**1143–1150.

Orenstein, J. M., M. Tenner, A. Cali, and D. P. Kotler. 1992. A microsporidian previously undescribed in humans, infecting enterocytes and macrophages, and associated with diarrhea in an acquired immunodeficiency syndrome patient. *Hum. Pathol.* **23:**722–728.

Overstreet, R. M., and E. Weidner. 1974. Differentiation of microsporidian spore tails in *Inodosporus spraguei* gen. et sp. n. *Z. Parasitenkd.* **44:**169–186.

Pakes, S. P., J. A. Shadduck, and A. Cali. 1975. Fine structure of *Encephalitozoon cuniculi* from rabbits, mice and hamsters. *J. Protozool.* **22:**481–488.

Pilley, B. M. 1976. A new genus, *Vairimorpha* (Protozoa: Microsporida), for *Nosema necatrix* Kramer 1965: pathogenicity and life cycle in *Spodoptera exempta* (Lepidoptera: Noctuidae). *J. Invertebr. Pathol.* **28:**177–183.

Pleshinger, J., and E. Weidner. 1985. The microsporidian spore invasion tube. IV. Discharge activation begins with pH-triggered calcium influx. *J. Cell Biol.* **100:**1834–1838.

Pol, S., C. A. Romana, S. Richard, P. Amouyal, I. Desportes-Livage, F. Carnot, J. F. Pays, and P. Berthelot. 1993. Microsporidia infection in patients with the human immunodeficiency virus and unexplained cholangitis. *N. Engl. J. Med.* **328:**95–99.

Scarborough-Bull, A., and E. Weidner. 1985. Some properties of discharged *Glugea hertwigi* (Microsporida) sporoplasms. *J. Protozool.* **32:**284–289.

Schwartz, D. A., R. T. Bryan, K. O. Hewan-Lowe, G. S. Visvesvara, R. Weber, A. Cali, and P. Angritt. 1992. Disseminated microsporidiosis (*Encephalitozoon hellem*) and acquired immunodeficiency syndrome: autopsy evidence for respiratory acquisition. *Arch. Pathol. Lab. Med.* **116:**660–668.

Sprague, V., J. J. Becnel, and E. I. Hazard. 1992. Taxonomy of phylum Microspora. *Crit. Rev. Microbiol.* **18:**285–395.

Sprague, V. 1977. Systematics of the microsporidia, p. 1–510. *In* L. A. Bulla, Jr., and T. C. Cheng (ed.), *Comparative Pathobiology,* vol. 2. Plenum Press, New York, N.Y.

Sprague, V., and S. Vernick. 1969. Light and electron microscope observations on *Nosema nelsoni* Sprague, 1950 (Microsporida, Nosematidae) with particular reference to its Golgi complex. *J. Protozool.* **16:**264–271.

Sprague, V., and S. H. Vernick. 1971. The ultrastructure of *Encephalitozoon cuniculi* (Microsporida, Nosematidae) and its taxonomic significance. *J. Protozool.* **18:**560–569.

Strano, A. J., A. Cali, and R. C. Neafie. 1976. Microsporidiosis, p. 336–339. *In* C. H. Binford and D. H. Connor (ed.), *Pathology of Tropical and Extraordinary Diseases*, vol. 1. Armed Forces Institute of Pathology, Washington D.C.

Sweeney, A. W., E. I. Hazard, and M. F. Graham. 1985. Intermediate host for an *Amblyospora* sp. (Microspora) infecting the mosquito *Culex annulirostris. J. Invertebr. Pathol.* **46:**98–102.

Sweeney, A. W., E. I. Hazard, and M. F. Graham. 1988. Life cycle of *Amblyospora dyxenoides* sp. nov. in the mosquito *Culex annulirostris* and the copepod *Mesocyclops albicans. J. Invertebr. Pathol.* **51:** 46–57.

Takvorian, P. M., and A. Cali. 1981. The occurrence of *Glugea stephani* (Hagenmuller, 1899) in American winter flounder, *Pseudopleuronectes americanus* (Walbaum) from the New York-New Jersey lower bay complex. *J. Fish Biol.* **18:**491–501.

Takvorian, P. M., and A. Cali. 1983. Appendages associated with *Glugea stephani,* a microsporidian found in flounder. *J. Protozool.* **30:**251–256.

Takvorian, P. M., and A. Cali. 1986. The ultrastructure of spores (Protozoa: Microsporida) from *Lophius americanus,* the angler fish. *J. Protozool.* **33:** 570–575.

Takvorian, P. M., and A. Cali. 1994. Enzyme histochemical identification of the Golgi apparatus in the microsporidian, *Glugea stephani. J. Eukaryot. Microbiol.* **41:**63S-64S.

Takvorian, P. M., and A. Cali. 1996. Polar tube formation and nucleoside diphosphatase activity in the microsporidian, *Glugea stephani. J. Eukaryot. Microbiol.* **43:**102S-103S.

Thomson, H. M. 1958. Some aspects of the epidemiology of a microsporidian parasite of the spruce budworm, *Choristoneura fumiferana* (Clem.). *Can. J. Zool.* **36:**309–316.

Toguebaye, B. S., and B. Marchand. 1984. Ultrastructural study of developmental stages and sporogonic mitosis of *Nosema henosepilachnae* new species (Microsporida: Nosematidae): parasitic in *Henosepilachna elaterii* (Coleoptera: Coccinellidae). *Protistologica* **20:**165–180.

Trammer, T., F. Dombrowski, M. Doehring, W. A. Maier, and H. M. Seitz. 1997. Opportunistic properties of *Nosema algerae* (Microspora), a mosquito parasite, in immunocompromised mice. *J. Eukaryot. Microbiol.* **44:** 258–262.

Tuzet, O., J. Maurand, A. Fize, R. Michel, and B. Fenwick. 1971. Proposition d'un nouveau cadre systematique' pourles generes de microsporidies. *C. R. Acad. Sci.* **272:**1268–1271.

Undeen, A. H. 1983. The germination of *Vavraia culicis* spores. *J. of Protozool.* **30:**274–277.

Undeen, A. H. 1990. A proposed mechanism for the germination of microsporidian (Protozoa, Microspora) spores. *J. Theor. Biol.* **142:**223–235.

Undeen, A. H., and S. W. Avery. 1984. Germination of experimentally nontransmissible microsporidia. *J. Invertebr. Pathol.* **43:**299–301.

Undeen, A. H., M. A. Johnson, and J. J. Becnel. 1993. The effects of temperature on the survival of *Edhazardia aedis* (Microspora, Amblyosporidae), a pathogen of *Aedes aegypti. J. Invertebr. Pathol.* **61:** 303–307.

Undeen, A. H., and L. F. Solter. 1997. Sugar acquisition during the development of microsporidian (Microspora, Nosematidae) spores. *J. Invertebr. Pathol.* **70:**106–112.

Undeen, A. H., and R. K. Vander Meer. 1990. The effect of ultraviolet radiation on the germination of *Nosema algerae* Vavra and Undeen (Microsporida: Nosematidae) spores. *J. Protozool.* **37:**194–199.

Undeen, A. H., and R. K. Vander Meer. 1994. Conversion of intrasporal trehalose into reducing sugars during germination of *Nosema algerae* (Protista: Microspora) spores: a quantitative study. *J. Eukaryot. Microbiol.* **41:**129–132.

van Gool, T., F. Snijders, P. Reiss, J. K. Eeftinck Schattenkerk, M. A. van den Bergh Weerman, J. F. Bartelsman, J. J. Bruins, E. U. Canning, and J. Dankert. 1993. Diagnosis of intestinal and disseminated microsporidial infections in patients with HIV by a new rapid fluorescence technique. *J. Clin. Pathol.* **46:**694–699.

Vavra, J. 1963. Spore projections in microsporidia. *Acta Protozool.* **1:**153–156.

Vavra, J. 1965. Etude au microscope electronique de la morphologie et du developpement de quelques microsporidies. *C. R. Acad. Soc.* **261:** 3467–3470.

Vavra, J. 1976. Biology of the microsporidia, p. 371. *In* L. A. Bulla and T. C. Cheng (ed.), *Comparative Pathobiology,* vol. 1. Plenum Press, New York.

Vossbrinck, C. R., J. V. Maddox, S. Friedman, B. A. Debrunner-Vossbrinck, and C. R. Woese. 1987. Ribosomal RNA sequence suggests microsporidia are extremely ancient eukaryotes. *Nature* **326:**411–414.

Walker, M. H., and G. W. Hinsch. 1972. Ultrastructural observations of a microsporidian protozoan parasite in *Libinia dubia* (Decapoda). I. Early spore development. *Z. Parasitenkd.* **39:**17–26.

Weber, R., R. T. Bryan, D. A. Schwartz, and R. L. Owen. 1994. Human microsporidial infections. *Clin. Microbiol. Rev.* **7:**426–461.

Weidner, E. 1970. Ultrastructural study of microsporidian development. *Z. Zellforsch.* **105:**33–54.

Weidner, E. 1972. Ultrastructural study of microsporidian invasion into cells. *Z. Parasitenkd.* **40:** 227–242.

Weidner, E. 1975. Interactions between *Encephalitozoon cuniculi* and macrophages. *Z. Parasitenkd.* **47:** 1–9.

Weidner, E. 1982. The microsporidian spore invasion tube. III. Tube extrusion and assembly. *J. Cell Biol.* **93:**976–979.

Weidner, E. 1992. Cytoskeletal proteins expressed by microsporidian parasites. *Subcell. Biochem.* **18:**385–399.

Weidner, E., and W. Byrd. 1982. The microsporidian spore invasion tube. II. Role of calcium in the activation of invasion tube discharge. *J. Cell Biol.* **93:**970–975.

Weidner, E., E. U. Canning, and W. S. Hollister. 1997. The plaque matrix (PQM) and tubules at the surface of an intramuscular parasite, *Trachipleistophora hominis. J. Eukaryot. Microbiol.* **44:** 359–365.

Weidner, E., R. M. Overstreet, B. Tedeschi, and J. Fuseler. 1990. Cytokeratin and desmoplakin analogues within an intracellular parasite. *Biol. Bull.* **179:**237–242.

Weiss, L. M., E. Keohane, X. Zhu, H. Tanowitz, M. Wittner, and A. Cali. 1992b. PCR identification of microsporidia by rRNA. *In* Workshop on Intestinal Microsporidia in HIV Infection. (Abstract.)

Youssef, N., and D. M. Hammond. 1971. The fine structure of the developmental stages of the microsporidian *Nosema apis* (Zander). *Tissue Cell* **3:** 283–294.

MOLECULAR BIOLOGY, MOLECULAR PHYLOGENY, AND MOLECULAR DIAGNOSTIC APPROACHES TO THE MICROSPORIDIA

Louis M. Weiss and Charles R. Vossbrinck

4

The term "microsporidia" refers to a ubiquitous group of eukaryotic obligate intracellular protozoan parasites which were recognized over 100 years ago with the description of *Nosema bombycis*, a parasite of silkworms. Microsporidia infect all major animal groups most often as gastrointestinal pathogens, however, infections of the reproductive, respiratory, muscle, excretory, and nervous systems are well documented (Canning and Lom, 1986; Sprague and Vávra, 1977; Weber et al., 1994a; Wittner et al., 1993). In fact, microsporidia have been reported from every tissue and organ, and their spores are common in environmental sources such as ditch water. Of the more than 100 genera in the phylum Microspora, several have been demonstrated in human disease (Weber et al., 1994a, Wittner et al., 1993; Canning and Hollister, 1992): *Nosema, Vittaforma* (Silveira and Canning, 1995), *Brachiola* (Cali et al., 1998), *Pleistophora, Encephalitozoon, Enterocytozoon* (Desportes et al., 1985), *Septata* (Cali et al., 1993), reclassified as *Encephalitozoon* (Hartskeerl et al., 1995; Baker et al., 1995), and *Trachipleistophora* reported from AIDS patients (Field et al., 1996). The genus *Microsporidium* has been used to designate microsporidia of uncertain taxonomic status. *Pleistophora, Nosema, Brachiola*, and *Trachipleistophora* have been associated with myositis (Weber et al., 1994a; Wittner et al., 1993). The Encephalitozoonidae (*Encephalitozoon hellem, E. cuniculi*, and *E.* [*Septata*] *intestinalis*) have been linked with disseminated disease as well as with keratoconjunctivitis (Rastrelli et al., 1994), sinusitis, respiratory disease, prostatic abscesses, and intestinal infection (Weber et al., 1994a; Wittner et al., 1993). *Nosema, Vittaforma*, and *Microsporidium* have been associated with stromal keratitis following trauma in immunocompetent hosts (Rastrelli et al., 1994). *Enterocytozoon bieneusi*, first reported in 1985 (Desportes et al., 1985), is linked with malabsorption and diarrhea and has been described only in humans.

The recognition of microsporidia as opportunistic pathogens in humans has led to increased interest in their molecular biology. Much of the recent work has focused on determination of the nucleotide sequences for rRNA genes. These sequences have been used in the development of diagnostic tools for species identification. In addition, such rRNA gene sequences have facilitated the development of a molecular phylogeny of these organisms. These microsporidian rRNA sequences,

Louis M. Weiss, Albert Einstein College of Medicine, 1300 Morris Park Avenue, Room 504 Forchheimer, Bronx, NY 10461. *Charles R. Vossbrinck*, The Connecticut Agricultural Experiment Station, 123 Huntington Street, New Haven, CT 06504, and Albert Einstein College of Medicine, 1300 Morris Park Avenue, Bronx, NY 10461.

The Microsporidia and Microsporidiosis (Murray Wittner, editor; Louis M. Weiss, contributing editor), ©1999 American Society for Microbiology, Washington, D.C.

as well as recent sequence data on other conserved structural genes, have also proved invaluable in the investigation of theories of early eukaryotic evolution. Given the increasing evidence of the importance of the microsporidia as both human and agricultural pathogens, recent work has focused on the identification of microsporidial genes that could serve as potential therapeutic targets.

MOLECULAR BIOLOGY

General Organization and Characteristics

Microsporidia are obligate intracellular pathogens whose biological nature remains enigmatic and which display a number of characteristics that are unusual for eukaryotic organisms. They lack mitochondria and centrioles and possess prokaryote-size ribosomes (70S, consisting of a large [23S] subunit and a small [16S] subunit) (Curgy et al., 1990; Ishihara and Hayashi, 1968). The microsporidian *Vairimorpha necatrix* was found to lack a 5.8S ribosome subunit but had sequences homologous to the 5.8S region at the beginning of the 23S subunit (Vossbrinck and Woese, 1986). The absence of an internal transcribed spacer 2 (ITS2) region resulting in fusion of the 5.8S region with the large-subunit rRNA gene is a feature usually found in bacteria and has not been reported in eukaryotes other than the microsporidia. Polyadenylation occurs on mRNA in microsporidia, as in all other eukaryotes (Vossbrinck and Weiss, unpublished observations). In a microsporidian polar tube gene, polar tube protein (PTP) Eh_{55}, a putative polyadenylation signal sequence (AATAAA) is present within 5 bp of the stop codon (Keohane et al., 1998). U2 and U6 snRNA genes have been identified in *Nosema locustae,* suggesting the presence of a functional spliceosome (Fast et al., 1998). *V. necatrix* has also been found to possess a unique U2 RNA homolog (DiMaria et al., 1996). The cap structures of the U2 small nuclear RNAs and the mRNAs are neither 2,2,7- nor 7-methylguanosine and can be added to the list of unusual characters that set microsporidia apart from other eukaryotes. Identification of this unusual cap structure may well prove to be of importance from both a basic molecular biological and an antimicrosporidial drug development viewpoint. To date, the only intron found in a microsporidian gene is a 28-bp intron at the beginning of an *E. cuniculi* large-subunit ribosomal protein (Biderre et al., 1998). The intron frequency in microsporidia is probably below 0.2 per kb.

The genome size of several different microsporidia has been determined and varies from 2.9 to 19.5 Mb (Munderloh et al., 1990; Malone and McIvor, 1993; Biderre et al., 1994, 1995; Viveres et al., 1995). The genomic size of *E. cuniculi* has been demonstrated to be 2.9 Mb (Biderre et al., 1995), and that of *E. intestinalis* is 2.3 Mb (Peyretaillade et al., 1998a), which suggests that they are the smallest eukaryotic nuclear genomes so far identified. This value is also about half the size of the smallest previously known microsporidian genome of 5.3 Mb in *N. locustae* (Streett, 1994; Biderre et al., 1995). By using a lacto-aceto-orcein stain, an *Amblyospora* sp. was estimated to have seven chromosomes (Hazard et al., 1979). The karyotype of a few members of the phylum Microspora has also been determined by pulsed field electrophoresis (Munderloh et al., 1990; Vivares et al., 1995). *Nosema pyrausta* and *N. furnacalis* have 13 chromosome bands ranging in size from 130 to 440 kb (Munderloh et al., 1990), *N. costelytrae* has 8 bands of 290 to 1,810 kb (Malone and McIvor, 1993), *N. locustae* has 18 bands of 139 to 651 kb, *Glugea atherinae* has 16 bands of 420 to 2,700 kb, *Spragea lophii* has 12 bands of 230 to 980 kb (Biderre et al., 1994), *Vairimorpha oncoperae* (isolated from grass grubs [*Costelytra zealandica*]) has 14 bands of 130 to 1,930 kb (Malone and McIvor, 1993), *V. oncoperae* (isolated from porcina caterpillars [*Wiseana* sp.]) has 16 bands of 140 to 1,830 kb (Malone and McIvor, 1993), a *Vairimorpha* sp. (isolated from *Pieris rapae*) has 8 bands of 720 to 1,790 kb (Malone and McIvor, 1993), *E. cuniculi* has 11 bands of 217 to 315 kb, *E. hellem* has 12 bands of 175 to 315 kb, and *E. intestinalis* has 11 bands of 190 to 280 kb (Peyretaillade et al.,

1998a). Thus, for currently studied microsporidia the chromosome number has varied from 7 to 16.

The small microsporidian genomic size indicates that these organisms may have developed strategies for packing genetic information tightly into the genome (e.g., viruslike genomic organization) or that they may have lost genetic information for metabolic pathways and depend on host cell sources for these compounds. A study on chromosome 1 of *E. cuniculi* has demonstrated that in a 6-kb region four protein-coding genes only 195 bp apart were identified with open reading frames (ORFs) in the same orientation (Vivaras et al., 1996). Partial gene clones for cdc2 and aminopeptidase have been localized to chromosome 8 (272 kb), and aminopeptidase, thymidylate synthase, serine hydroxymethyl transferase, and dihydrofolate reductase to chromosome 1 (217 kb) of *E. cuniculi* (Vivares et al., 1996). These genes had a strong bias (75%) for a C or G residue in the third codon position (Vivares et al., 1996). Examination of the β-tubulin genes of *E. hellem* and *E. intestinalis* demonstrates an identical ORF, encoding an unknown protein, 162 bp upstream of the start codon for the β-tubulin genes on the strand opposite the β-tubulin-coding strand (Li et al., 1996). In a polar tube gene from *E. hellem* (PTP Eh_{55}) an ORF encoding an unknown protein is present 54 bp upstream of the putative PTP start codon on the strand opposite the PTP coding strand (Keohane et al., 1998). A series of anonymous single probes have been developed and mapped to the different chromosomes of *E. cuniculi* (Biderre et al., 1997). These probes should provide useful markers for the localization of genes on these chromosomes and facilitate the development of a chromosomal map for *E. cuniculi*. Consistent with its small genome size, *E. cuniculi* has a low abundance of repetitive DNA elements such as microsatellites (about 9 to 10%) (Biderre et al., 1997). Expression screening of genomic libraries has been feasible in these organisms (Keohane et al., 1998; Weiss, unpublished observations). Currently, no techniques exist for the transfection and manipulation of genes in the microsporidia. Such methods are needed to facilitate studies on gene function and structure in these organisms.

rRNA Genes

Microsporidian 16S rRNA diverges greatly from the small subunit rRNA (16S) sequences of other eukaryotes. In *V. necatrix*, for example, the small-subunit rRNA sequence is far shorter than a typical eukaryotic sequence (1,244 bp) and shares little or no homology with other eukaryotes (Vossbrinck et al., 1987). The rRNA genes of other microsporidia have also been found to be significantly shorter than both eukaryotic and prokaryotic small-subunit rRNA (Vossbrinck et al., 1987; Weiss et al., 1994). Moreover, microsporidian rRNA genes lack regions of the small-subunit sequence considered "eukaryotic." *V. necatrix* lacks regions present in both eukaryote and prokaryotic sequences, such as positions 180 to 225 and 590 to 650 (numbered according to the standard *Escherichia coli* rRNA sequence), where little homology exists between eukaryotes and prokaryotes. This divergence has been used as evidence that microsporidia are an early branch in eukaryotic evolution. This divergence, however, could also be caused by specialization (i.e., evolution) of the rRNA genes of these obligate intracellular pathogens due to unknown selective pressures. Recent data suggest that microsporidian rRNA genes may be reduced to a universal core (Peyretaillade et al., 1998a).

The small-subunit rRNA sequence as well as the ITS and adjoining large-subunit rRNA genes of many microsporidia has been determined (Table 1). With the use of conserved (i.e., universal) rRNA sequences and sequences specific to the microsporidia, a set of primers which are useful in cloning microsporidian rRNA genes from spores or clinical material has been designed (Table 2) (Fig. 1). These data have proven useful in the construction of a molecular phylogeny for this phylum (see "Phylogeny"). Riboprinting (restriction mapping of amplified rRNA genes)

TABLE 1 Microsporidian genes in GenBank[a]

Type and accession no.	Organism	Reference
Small-subunit (16S rRNA) rRNA genes		
U68474	*Amblyospora* sp. (*Amblyospora salinaria*)	Baker et al., 1997
U68473	*Amblyospora californica*	Baker et al.,1998
AF027685	*Amblyospora stimuli*	Baker et al., 1998
L15741	*Ameson michaelis*	Zhu et al., 1994
AF027683	*Culicosporella lunata*	Baker et al., 1998
AF027684	*Edhazardia aedis*	Baker et al., 1998
X98469	*Encephalitozoon cuniculi* (Stewart strain)	Hollister et al., 1996
X98470	*Encephalitozoon cuniculi* (Donovan strain)	Hollister et al., 1996
X98467	*Encephalitozoon cuniculi* (D. Owen strain)	Hollister et al., 1996
L39107	*Encephalitozoon cuniculi*	Baker et al., 1995
L17072	*Encephalitozoon cuniculi*	Visvesvara et al., 1994b
L13295	*Encephalitozoon cuniculi* partial *cds*	Vossbrinck et al., 1993
Z19563	*Encephalitozoon cuniculi*	Zhu et al., 1993b
L07255	*Encephalitozoon cuniculi*	Hartskeerl et al., 1993
L13393	*Encephalitozoon hellem*	Vossbrinck et al., 1993
L39108	*Encephalitozoon hellem*	Baker et al., 1995
L19070	*Encephalitozoon hellem*	Visvesvara et al., 1994b
U39297	*Encephalitozoon intestinalis* partial *cds*	Franzen et al., 1996
U09929	*Encephalitozoon intestinalis*	Visvesvara et al., 1995
L39113	*Encephalitozoon intestinalis*	Baker et al., 1995
L19567	*Encephalitozoon intestinalis*	Zhu et al., 1993a
L16867	*Encephalitozoon* sp.	Hartskeerl, unpublished data
L16866	*Encephalitozoon* sp.	Hartskeerl et al., 1993
L39109	*Endoreticulatus shubergi*	Baker et al., 1995
L16868	*Enterocytozoon bieneusi*	Hartskeerl et al., 1993
L07123	*Enterocytozoon bieneusi*	Zhu et al., 1993c
U15987	*Glugea atherinae*	Pieniazek, unpublished data
L13293	*Icthyosporidium giganteum* partial *cds*	Vossbrinck et al., 1993
L39110	*Icthyosporidium* sp.	Baker et al., 1995
U26534	*Nosema apis*	Pieniazek et al., 1996
X73894	*Nosema apis*	Malone et al., 1995
U26158	*Nosema bombi* partial *cds*	Malone and McIvor, 1996
D85504	*Nosema bombycis* spore	Inoue, unpublished data
D14632	*Nosema bombycis* partial *cds* pseudoribosomal gene	Kawakami et al., 1994
D85503	*Nosema bombycis* spore	Inoue, 1997
U26157	*Nosema bombycis* partial *cds*	Malone and McIvor, 1996
L39111	*Nosema bombycis*	Baker et al., 1995
U26533	*Nosema ceranae*	Fries et al., 1996
U26532	*Nosema furnacalis*	Pieniazek et al., 1996
U11051	*Nosema necatrix* (ATCC 30460)	Fries et al., 1996
U27359	*Nosema oulemate*	Pieniazek et al., 1996
U09282	*Nosema tricholusiae*	Pieniazek et al., 1996
U09283	*Nosema trichoplusiae*	Pieniazek et al., 1996
U11047	*Nosema vespula*	Pieniazek et al., 1996
L31842	*Nosema vespula*	Ninham, unpublished data
D85501	*Nosema* sp. spore	Inoue, unpublished data
U10883	*Nucleospora* (*Enterocytozoon*) *salmonis*	Barlough et al., 1995
U78176	*Nucleospora* (*Enterocytozoon*) *salmonis*	Kent et al., 1996
AF027682	*Parathelohania anophelis*	Baker et al., 1997
D85500	*Pleistophora* sp. spore	Inoue, 1997
U47052	*Pleistophora anguillarum*	Huang et al., 1998
U10342	*Pleistophora* sp. (ATCC 500400)	Pieniazek, unpublished data

(Continued)

TABLE 1 *(Continued)*

Type and accession no.	Organism	Reference
AF033197	*Spraguea lophii*	Hinkle et al., 1997
AF031538	*Thelohania solenopsis*	Moser et al., 1998
AF031537	*Thelohania* sp.	Moser et al., 1998
AJ002605	*Trachipleistophora hominis*	Cheney, 1998
L13294	*Vairimorpha lymantriae* partial *cds*	
M24612	*Vairimorpha necatrix*	Vossbrinck et al., 1987
U26159	*Vairimorpha* sp.	Malone and McIvor, 1996
Y00266	*Vairimorpha necatrix*	Vossbrinck et al., 1987
AF031539	*Vairimorpha* sp. (from *S. richteri*)	Moser et al., 1998
L39114	*Vairimorpha* sp	Baker et al., 1995
L28977	*Vairimorpha* sp.	Baker et al., 1994
L28976	*Vairimorpha* sp.	Baker et al., 1994
D85502	*Varimorpha* sp. spore	Inoue, 1997
X74112	*Vavraia oncopertae*	Malone et al., 1995
L39112	*Vittaforma corneae*	Baker et al., 1995
U11046	*Vittaforma corneae*	Da Silva et al., 1997
3'-end small-subunit ITS and 5'-end large-subunit rRNA genes		
L28960	*Amblyospora* sp.	Baker et al., 1994
L20293	*Ameson michaelis*	Zhu et al., 1994
X98466	*Encephalitozoon cuniculi* (Donovan strain)	Hollister and Canning, 1996
X98468	*Encephalitozoon cuniculi* (D. Owen strain)	Hollister and Canning, 1996
L29560	*Encephalitozoon cuniculi*	Katiyar et al., 1995
L13332	*Encephalitozoon cuniculi*	Vossbrinck et al., 1993
AJ00581	*Encephalitozoon cuniculi* rDNA unit	Peyretaillade et al., 1998a
L13331	*Encephalitozoon hellem*	Vossbrinck et al., 1993
L29557	*Encephalitozoon hellem*	Katiyar et al., 1995
L20292	*Encephalitozoon intestinalis*	Zhu et al., 1994
Y11611	*Encephalitozoon intestinalis*	Schnittger et al., 1997
L20290	*Enterocytozoon bieneusi*	Zhu et al., 1994
U61180	*Enterocytozoon bieneusi* ITS region	Deplazes et al., 1996
AF023245	*Enterocytozoon bieneusi*	Mansfield et al., 1998
AF044391	*Glugea anomala*	Nilsen et al., 1998
L13430	*Icthyosporidium giganteum*	Vossbrinck et al., 1993
U78815	*Loma embiotocia*	Shaw et al., 1997
U78736	*Loma salmonae*	Docker et al., 1997
L28961	*Nosema algerae*	Baker et al., 1994
U76706	*Nosema apis* including complete 23S sequence	Rice, 1998
U97150	*Nosema apis* complete RNA gene sequence	Gatehouse and Malone, 1998
L28962	*Nosema bombycis*	Baker et al., 1994
D14631	*Nosema bombycis* partial *cds*	Kawakami et al., 1992
L28963	*Nosema distriae*	Baker et al., 1994
L28964	*Nosema epilachnae*	Baker et al., 1994
L28965	*Nosema heliothidis*	Baker et al., 1994
L28967	*Nosema locustae*	Baker et al., 1994
L28966	*Nosema kingi*	Baker et al., 1994
L28968	*Nosema pyrausta*	Baker et al., 1994
U78176	*Nucleospora (Enterocytozoon) salmonis*	Kent et al., 1996; Docker et al., 1997
AF044394	*Pleistophora* sp. 1	Nilsen et al., 1998
AF044389	*Pleistophora* sp. 2	Nilsen et al., 1998
AF044390	*Pleistophora* sp. 3	Nilsen et al., 1998
AF044392	*Pleistophora ehrenbaumi*	Nilsen et al., 1998

(Continued)

TABLE 1 *(Continued)*

Type and accession no.	Organism	Reference
AF044393	*Pleistophora finisterrensis*	Nilsen et al., 1998
AF044388	*Pleistophora hippoglossoideos*	Nilsen et al., 1998
AF044387	*Pleistophora typicalis*	Nilsen et al., 1998
L28969	*Parathelohania anophelis*	Baker et al., 1994
L28972	*Vairimorpha ephistiae*	Baker et al., 1994
L28973	*Vairimorpha heterosporum*	Baker et al., 1994
L13330	*Vairimorpha lymantriae*	Vossbrinck et al., 1993
L28974	*Vairimorpha lymantriae*	Baker et al., 1994
L28975	*Vairimorpha necatrix*	Baker et al., 1994
L28971	*Vairimorpha* sp.	Baker et al., 1994
L28970	*Vairimorpha* sp.	Baker et al., 1994
Tubulin genes		
L31807	*Encephalitozoon cuniculi* β-tubulin partial *cds*	Edlind et al., 1994
L47271	*Encephalitozoon hellem* β-tubulin full *cds*	Li et al., 1996
L31808	*Encephalitozoon hellem* β-tubulin partial *cds*	Edlind et al., 1994
L47274	*Encephalitozoon intestinalis* β-tubulin partial *cds*	Li et al., 1996
U66908	*Encephalitozoon hellem* α-tubulin partial *cds*	Keeling and Doolittle, 1996
L47272	*Nosema locustae* (ATCC 30860) α-tubulin partial *cds*	Li et al., 1996
U66907	*Nosema locustae* α-tubulin partial *cds*	Keeling and Doolittle 1996
L47273	*Nosema locustae* β-tubulin partial *cds*	Li et al., 1996
U66906	*Spraguea lophii* α-tubulin partial *cds*	Keeling and Doolittle, 1996
Other gene sequences		
AJ005666	*Encephalitozoon cuniculi* PTP gene	Delbac et al., 1998
AF044915	*Encephalitozoon hellem* PTP gene PTP Eh_{55}	Keohane et al., 1998
AF031701	*Encephalitozoon hellem actin* (ACT1) partial *cds*	Edlind, 1997
AF054829	*Encephalitozoon cuniculi* L27a homolog	Biderre et al., 1998
AJ005582	*Encephalitozoon cuniculi* 5S rRNA	Peyretaillade et al., 1998a
AJ012470	*Encephalitozoon cuniculi* hsp70 homolog	Peyretaillade et al., 1998b
AJ005644	*Encephalitozoon cuniculi* dihydrofolate reductase locus	Duffieux et al., 1998
AJ006824	*Encephalitozoon cuniculi* thymidine kinase	Duffieux, unpublished
AJ006823	*Encephalitozoon cuniculi* elongation factor 2	Duffieux, unpublished
AJ006825	*Encephalitozoon cuniculi* trehalose 60-phosphate phosphatase	Duffieux, unpublished
D32139	*Glugea plecoglossi* EF-1α	Kamaishi et al., 1996a
D84253	*Glugea plecoglossi* peptide EF-1α	Kamaishi et al., 1996b
D79220	*Glugea plecoglossi* peptide EF-2α	Kamaishi et al., 1996b
U28045	*Nosema bombycis* DNA fragment (unknown gene)	Malone et al., 1995
U28046	*Nosema costelyatae* DNA fragment (unknown gene)	Malone et al., 1995
L37097	*Nosema locustae* isoleucyl-tRNA synthetase partial *cds*	Brown and Doolittle, 1995
AF005490	*Nosema locustae* glutamyl-tRNA synthetase gene	Brown and Doolittle, 1997
AF005489	*Nosema locustae* glutaminyl-tRNA synthetase gene	Brown and Doolittle, 1997
AF031702	*Nosema locustae* actin (ACT1) partial *cds*	Edlind, 1997
AF053589	*Nosema locustae* U2 snRNA	Fast et al., 1998
AF053588	*Nosema locustae* U6 snRNA	Fast et al., 1998
U97520	*Nosema locustae* mitochondrial-type *hsp70* gene	Germot et al., 1997
AF019227	*Spraguea lophii* RNA polymerase	Hinkle et al., 1997
AF019228	*Spraguea lophii* chitin synthase	Hinkle et al., 1997
AF019229	*Spraguea lophii* reverse transcriptase	Hinkle et al., 1997
AF008215	*Vairimorpha necatrix* mitocondrial *hsp70* homolog	Hirt et al., 1997
Z50072	*Vairimorpha necatrix* DNA for U2 snRNA gene	DiMaria et al., 1996

[a]Adapted from Weiss and Vossbrinck (1998) with permission.

TABLE 2 Primers for the identification and sequencing of microsporidian rDNA1[a]

Primer[b]	Sequence
ss18f[c]	CACCAGGTTGATTCTGCC
ss18sf	GTTGATTCTGCCTGACGT
ss350f	CCAAGGA(T/C)GGCAGCAGGCGCGAAA
ss350r	TTTCGCGCCTGCTGCC(G/A)TCCTTG
ss530f	GTGCCAGC(C/A)GCCGCGG
ss530r	CCGCGG(T/G)GCTGGCAC
ss1047r	AACGGCCATGCACCAC
ss1061f	GGTGGTGCATGGCCG
ss1492r	GGTTACCTTGTTACGACTT (universal primer)
ss1537	TTATGATCCTGCTAATGGTTC
ls212r1	GTT(G/A)GTTTCTTTTCCTC
ls212r2	AATCC(G/A/T/C)(G/A)GTT(G/A)GTTTCTTTTCCTC
ls580r	GGTCCGTGTTTCAAGACGG

[a]Primers 18f and 1492r amplify most of the small-subunit rRNA of the microsporidia. Primers 530f and 212r1 or 212r2 are used to amplify the small-subunit rRNA and the ITS region. The remaining primers are used to sequence, with overlap, the forward and reverse strands of the entire small-subunit rRNA and ITS region. ls580r amplifies a variable region of the 5′ end of the large subunit rRNA gene of many microsporidia (e.g., *Nosema* and *Vairimorpha*), but it does not work on all microsporidia. ss1537 allows sequencing closer to the 3′ end of the small-subunit rRNA of many but not all microsporidia. ss350f and ss350r may not be needed for sequencing reactions if 18f and 530r provide sufficient overlap to obtain clear sequence data. The table is adapted from Weiss and Vossbrinck (1998) with permission.

[b]ss, primers in the small-subunit rRNA gene; ls, primers in the large subunit rRNA gene; f, forward primer (positive strand); r, reverse primer (negative strand).

[c]Similar to V1 primer (Zhu et al., 1993a, 1993b; Weiss et al., 1994).

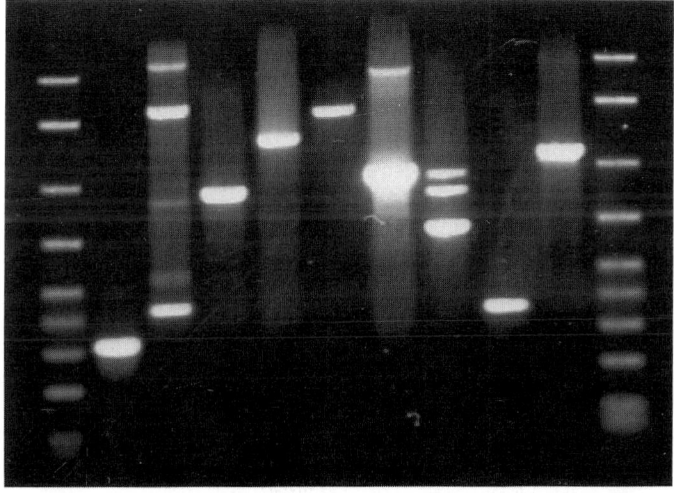

FIGURE 1 PCR of *E. cuniculi* with the microsporidian rRNA primers described in Table 2. Lane S, standard (50, 100, 200, 300, 400, 500, 700, 1,000, 1,500, and 2,000 bp); lane 1, 18f::350r; lane 2, 18f-530r; lane 3, 18f::1047r; lane 4, 18f::1492r; lane 5, 18f-212r1; lane 6, 350f-1492r; lane 7, 530f::1492r; lane 8, 1061f-1492r; lane 9, 530f-212r1. Products of the predicted sizes are present with all the primer sets. In cases in which more than one band is present, cloning of the amplicon of the correct size usually yields the rRNA fragment of interest. Additional bands are due to hybridization to unrelated genes in the samples.

has also been utilized to investigate the evolutionary relationships among the microsporidia (Pomport-Castillin et al., 1997). Amitochondrial protozoa, such as the Microsporida, display relatively short ITS regions as well as 47-47′ regions (Katiyar et al., 1995). In addition, all microsporidian rRNA gene sequences lack ITS2 regions and display a bacterium-like fusion of the homologous 5.8S RNA sequence with the beginning of the large-subunit rRNA gene sequence. In *Vairimorpha lymantriae*, *V. necatrix*, *E. cuniculi*, and *E. hellem* the rRNA sequence demonstrates a small hairpin loop analogous to the 9-9′ structure that defines the 5.8S large-subunit junction (Katiyar et al., 1995). This hairpin loop may have functional significance, as it is also conserved in bacterial and plastid large subunit rRNA sequences (Gutell et al., 1993). In many protozoa the aminoglycoside paromomycin inhibits protein synthesis, presumably by binding to ribosomes at the base of the 47-47′ hairpin loop (De Stasio and Dahlberg, 1990). Because of the absence of this binding site for paromomycin in the microsporidian rRNA sequence, it was suggested that microsporidia are not sensitive to this drug (Katiyar et al., 1995), and in vitro testing has confirmed that *E. cuniculi* is resistant to paromomycin (Beauvais et al., 1994). One copy of the rDNA gene has been found on each chromosome of *E. cuniculi* and *E. intestinalis*, but in *N. bombycis*, an insect microsporidium with a genome of 15.3 Mb, the rRNA gene(s) is found only on one chromosome (a 760-kb chromosome) (Kawakami et al., 1994; Vivares et al., 1996; Peyretaillade et al., 1998a).

Sequence data for rRNA from several microsporidia infecting humans, *E. cuniculi*, *E. hellem*, *E. bieneusi*, and *E. intestinalis* (Baker et al., 1995; Didier et al., 1995, 1996a, 1996b; Hartskeerl et al., 1993, 1995; Katiyar, 1995; Weiss et al., 1994; Visvesvara et al., 1994a; Vossbrinck et al., 1993; Zhu et al., 1993a, 1993b, 1993c, 1994), have been useful in the evaluation of phylogenetic relationships among these microsporidia as well as in the development of PCR primers (see "Molecular Diagnostic Techniques") for diagnosis and epidemiologic studies (see "Phylogeny"). In *E. cuniculi* a set of tetranucleotide repeats (5′GTTT3′) in the ITS has been found to vary among isolates from different hosts, resulting in the definition of three isotypes (types I, II, and III) for this parasite (Didier et al., 1995, 1996c). This ITS region heterogeneity may be useful in examining the epidemiology of human infection with *E. cuniculi*. In a similar fashion heterogeneity in the ITS region has been described in *E. bieneusi* isolates (Rinder et al., 1997). In the case of *E. bieneusi*, however, the heterogeneity is due to nucleotide substitutions at nine independent sites in the ITS region, generating three distinct rRNA ITS groups. At this point it is unknown if these groups will correlate with as yet unidentified animal reservoirs of *E. bieneusi* or if this heterogeneity merely represents variation in a multicopy gene within the same organism.

Tubulin Genes

The antimicrosporidial activity of many benzimidazoles (particularly albendazole), which act by binding to β-tubulin, has lead to an interest in tubulin genes in the microsporidia. Subsequently, several microsporidian β-tubulin genes have been cloned and sequenced (Edlind et al., 1994, 1996; Li et al., 1996; Weiss, unpublished observations). The tubulin genes identified and sequenced for several microsporidia, all of which have been shown to be sensitive to benzimidazoles, display the predicted amino acids (Cys_{165}, Phe_{167}, Glu_{198}, Phe_{200}, Arg_{241}, $His_{6)}$) associated with this phenotype in other eukaryotes (Edlind et al., 1994, 1996; Katiyar et al., 1994; Li et al., 1996). Interestingly, phylogenetic analysis of the β-tubulin genes of the microsporidia (Edlind et al., 1996) suggests that these organisms are related to the fungi (see "Phylogeny"). The β-tubulin gene of *E. bieneusi* has not been cloned but would be of particular interest given the poor efficacy of albendazole in the treatment of this microsporidian infection in humans. In addition to microsporidian β-tubulin genes, several α-tubulin genes (Li et al., 1996; Keeling and

Doolittle, 1996), as well as actin genes (Edlind, unpublished observations) have been cloned from various microsporidia and are in the GenBank database (Fig. 1).

The β-tubulin gene of *E. cuniculi* appears to be present in two copies localized to chromosomes 2 (235 kb) and 3 (241 kb) as determined by hybridization to pulsed field electrophoresis preparations (Vivares et al., 1996). However, Southern hybridization to restriction endonuclease digestions of genomic DNA of both *E. hellem* and *E. intestinalis* suggests that the β-tubulin gene is a single-copy gene (Li et al., 1996). The *E. hellem* and *E. intestinalis* 5′ flanking regions do not have recognizable canonical promoter elements. There is a 90% identity in the 50-bp stretch preceding the start codon of these two genes (compared with only a 50% identity in the −50 to −110-bp region) (Li et al., 1996). Additionally, two 7 to 8-bp AT-rich regions are present in this 50-bp segment, and palindromic elements 6 to 10 bp long are centered 80 bp upstream of the start codon (Li et al., 1996). This area may represent a promoter region, but analysis of additional protein-coding genes of microsporidia will be necessary for a consensus to emerge.

Other Genes

In addition to those for rRNA and β-tubulin, the following genes have been cloned from microsporidia for use in phylogenetic anaysis: α-tubulin (Li et al., 1996; Keeling and Doolittle, 1996), isoleucyl-tRNA synthetase (Brown and Doolittle, 1996), *hsp70* (Hirt et al., 1997; Germot et al., 1997; Peyretaillade et al., 1998b), actin (Edlin, unpublished results), elongation factor 1α (Kamiashi et al., 1996a, 1996b; Hashimoto et al., 1996), and elongation factor 2 (Kamiashi et al., 1996b). Because of the small genomic size of microsporidia, as well as interest in the phylogenetic placement of these organisms, several groups have started genome sequencing projects on microsporidia. Currently, research is under way involving a *Nosema* sp., *E. cuniculi*, and *S. lophii*. From *E. cuniculi* partial gene clones for cdc2 and aminopeptidase have been localized to chromosome 8 (272 kbp), and aminopeptidase, thymidylate synthase, serine hydroxymethyl transferase, and dihydrofolate reductase localized to chromosome 1 (217 kbp) of *E. cuniculi* (Vivares et al., 1996; Biderre et al., 1997; Duffieux et al., 1998). Dihydrofolate reductase and thymidylate synthase appear to occur as distinct transcription units on chromosome 1. The dual chromosome localization of aminopeptidase is probably related to the existence of two enzymatic types, and partial sequence data provide evidence of a Zn^{2+} binding region in this microsporidian enzyme (Biderre et al., 1997). In *S. lophii* genes have been identified for reverse transcriptase, DNA-directed DNA polymerase, and chitin synthase in preliminary studies involving a "shotgun" (i.e., genomic dBEST) genomic analysis (Hinkle et al., 1997). While the microsporidian chitin synthase was homologous with that of the ascomycete *Emericella nidulans*, it was divergent from fungal chitin synthase sequences (Hinkle et al., 1997). The identification of a reverse transcriptase most commonly associated with retrotransposons suggests that mobile genetic elements may be present in this organism (Hinkle et al., 1997).

Microsporidia are defined by the production of a small spore containing a single polar filament that coils around the interior of the spore. When appropriately stimulated, the polar filament rapidly discharges from the spore, forming a hollow polar tube which remains attached to the spore. Sporoplasm passes through this tube, facilitating transmission of the parasite into a new host cell. PTP genes have been identified from *E. cuniculi* (PTP-Ec) (Delbac et al., 1998) and *E. hellem* ($PTP\text{-}Eh_{55}$) (Keohane et al., 1998, submitted). Both these sequences have an N-terminal signal sequence that is cleaved during processing of the protein and is predicted to be a signal sequence with standard computer algorithms. The 5′ untranslated regions share similarities in sequence suggestive of a similar promoter region in these two PTP genes. Interestingly, as in tubulin genes, there are AT-rich regions immediately upstream of the start codon of these genes. In the *S. lophii* genomic project it has also been observed that microsporidian genes are relatively GC-rich compared with intergene regions (Hinkle et al., 1997). This skew may prove useful in both shotgun cloning of protein-coding regions (through the use of restriction enzymes that recognize GC-rich regions), as well as in the identification of putative start codons.

PHYLOGENY

Currently, taxonomy and species classification is based on ultrastructural and ecological features, including the size and morphology of the spores, the number of coils in the polar tube, the developmental life cycle, the host-parasite interface, and the developmental cycle in the host. A brief review of modern classification schemes based on these characters is presented below (see also Sprague and Vávra, 1977; Levine et al., 1980; Sprague et al., 1992). Molecular analysis of rDNA is changing our view of the taxonomic significance of these structural and ecological characters. As more molecular characters for more microsporidia are collected and analyzed, it is likely that the classification scheme for the microsporidia will be altered (Baker et al., 1995).

Placement of Microsporidia among the Eukaryotes

Prior to the advent of comparative analysis of molecular data, the classification of eukaryotes was based on morphological, ecological, and physiological characteristics. Speculation based on this information is problematic because the homology among such characters is often not discernible and characters that could be considered were thought to be either shared or derived depending on the underlying hypothesis used in construction of the phylogeny. Still, eukaryotic phylogeny as an intellectual pursuit is very important because this information has important ramifications in understanding evolution as well as in choosing organisms for a comparative approach in order to broaden our understanding of molecular and biochemical processes.

Collection of ultrastructural data began with the first electron micrograph of a biological specimen published by Marton in 1934. The first reported electron microscopic studies on a microsporidia were those of Krieg (1955), who reported on a *Pleistophora* sp. from white grubs, and Weiser (1959), who presented electron microscopic images of *Nosema laphygmae* from *Laphygma (Spodoptera) frugiperda*. Unfortunately, because of the early fixation techniques used with these spores, these images were not clear and add little to our knowledge of spore structure. The first study that demonstrated ultrastructural spore structure was that of Huger (1960). Images of the laminar structure of the "polaroplast" (Huger proposed the term in his paper), the inner and outer spore coat, the nuclei, polar granules, and clear cross sections of the polar filament were seen for the first time. Huger also presented an accurate schematic diagram of the polar filament coiled around the inside of the periphery of the spore. Thus, ultrastructural characters were available for comparative phylogenetic purposes both to study the relationship of the microsporidia among eukaryotes and to generate phylogenies within the microsporidia. Studies by Lom and Vavra (1961), Kudo and Daniels (1963), and others soon followed.

The presence of nematocystlike structures in the Myxosporidia and Actinomyxida resulted in the placement of these two groups with the Microsporidia early in the systematics of the Protozoa. Doflein (1901) groups the suborder Microsporidia with the suborders Myxosporidia and Actinomyxida in the order Cnidospora based on the presence of a spore-containing polar filament. The order Cnidosporidia is placed in the subclass Neospora, in the class Sporozoa. Lom and Vavra (1962) concurred with Dolflein's classification but noted that the only character common to the subphylum Cnidospora was the polar filament. They further indicated that the Myxosporidia and Actinomyxida are much more closely related to each other than to the microsporidia because in both the Myxosporidia and the Actinomyxida (class Heteronucleida), "during development the nuclei are twice differentiated: 1) into vegetative nuclei of the 'plasmodium' and into 'sporogenous' nuclei, which 2) give rise to nuclei of shell valves, polar capsules, and sporoplasm. The spores thus originating are polycellular ones." Consistent with this observation, it is now thought that the Myxosporidia are diminutive coelenterates (Siddall et al., 1995). The microsporidia, on the other hand, were placed in the class Isonucleida and, as indicated by Lom and Vavra, "During the development no differentiation of the nuclei into vegetative and sporogenous ones occurs. The spores are monocellular ones. The germ is always liberated from the spore on passing through extruding filament. There is always developed a McManus positive structure closing the pore of the spore." Kudo (1966) adopted the same scheme in his book and placed the Microsporida as a separate order in the class Cnidospora along with the orders Myxosporidia, Actinomyxida, and Helicosporida. Vavra (1966) indicated that the microsporidia are in no way related to the other protozoan groups and stated, "Phylogenetic note: if we compare what we know about the spores of Microsporidia, Myxosporidia and Actinomyxida we see that (a) Microsporidia are a very homogenous group, (b) there is no relation between Microsporidia and Myxo- and Actinomyxida, and (c) the microsporidia are a very isolated group—we do not (know) anything about their phylogenetic relationships."

The collection of molecular sequence data began in 1951 with Sanger's and Tuppy's report entitled "The Amino-acid Sequence in the Phenylalanyl Chain of Insulin." Soon many proteins were being sequenced, and this work was followed by the development of molecular phylogenies for a variety of organisms (Dayhoff et al., 1972) with several proteins and nucleic acids, including the 5S rRNA. The beginning of DNA sequencing dates to 1977, when papers were published by Sanger and Coulson (1977) describing chain-terminating (dideoxy) inhibitors and by Maxam and Gilbert (1977) reporting differential cleavage. Phylogenies were soon developed based on the small-subunit rRNA sequences of various organisms. The first molecular analysis including a microsporidian species was performed by Vossbrinck et al. (1987) who demonstrated, based on comparative rRNA data, a high degree of divergence between *V. necatrix* and other eukaryotes. These authors proposed that since microsporidia lack mitochondria, they may have diverged from other eukaryotic life forms before the production of oxygen and its introduction into the earth's atmosphere by blue-green algae and before the symbiotic relationship developed between mitochondria (α-purple bacteria according to small-subunit rRNA analysis) (Yang et al., 1985) and eukaryotes. The production of oxygen in the earth's atmosphere is based on the geological evidence of a layer of iron oxide estimated to have formed 2.5 billion to 2.8 billion years ago (Walker, 1983). In addition to the lack of mitochondria (Vavra, 1976), other unique characteristics of the microsporidia are the lack of a 5.8S rRNA (Vossbrinck and Woese, 1986) and what has been described as a "primitive " form of meiosis (Raikov, 1982).

On the basis of *V. necatrix* analysis, Cavalier-Smith (1987) proposed that other amitochondriate protozoa also may have diverged early and also may be primitively lacking mitochondria. Consistent with Cavalier-Smith's (1987) proposal, based on rDNA analysis, *Giardia lamblia*, an amitochondriate parasitic diplomonad, is also a highly divergent protistan (Sogin et al., 1989). On the basis of distance methods, *G. lamblia* is seen as the first organism to diverge from the eukaryotic tree of life, with *V. necatrix* second. According to maximum parsimony analysis of the rRNA sequences, *V. necatrix* branched first from the eukaryotic lineage, however, because of its long branch length, it is believed that it is changing at an evolutionary rate faster than that of the other organisms in the analysis and therefore that maximum parsimony analysis is not as reliable as distance methods in these circumstances (Sogin et al., 1989). Sogin et al. (1989) addressed the issues of the high G + C content of *G. lamblia* (75%) and the low G + C content of *V. necatrix* (35%) by performing additional phylogenetic analyses excluding *Sulfolobus solfataricus* (67% G + C content) as the outgroup. In doing this work they obtained similar results with *G. lamblia* and *V. necatrix* as early diverging organisms from the eukaryotic tree of life. It could not be determined from this analysis if eukaryotes are as ancient as prokaryotes or if eukaryotes are evolving faster, yet from their analysis these authors stated conclusively that *V. necatrix* has an "accelerated rate of evolutionary change" (Sogin et al., 1989).

A more recent analysis of small-subunit rRNA (Fig. 2) by Leipe et al. (1993), which included additional protozoan groups, also supports early divergence of amitochondriate protozoa. These authors, however, speculate that various factors may have an effect on altering the true tree topology. These factors include the parasitic nature of the amitochondriate protozoa, the high G + C content of *G. lamblia* (75% G + C), the low G + C content of *V. necatrix* (35% G + C), and the relative shortness of some of the sequences examined. Leipe et al. (1993) included the small-subunit rDNA sequence for *Hexamita inflata*, a free-living diplomonad with a G + C content of 51%, to address bias due to the G + C content and to changes in evolutionary rates due to parasitism. Their results indicated that when prokaryotes with a high G + C content were eliminated, *V. necatrix* was the most divergent organism, and that *Giardia* and *Hexamita* branched together, implying that the diplomonads were monophyletic. When pro-

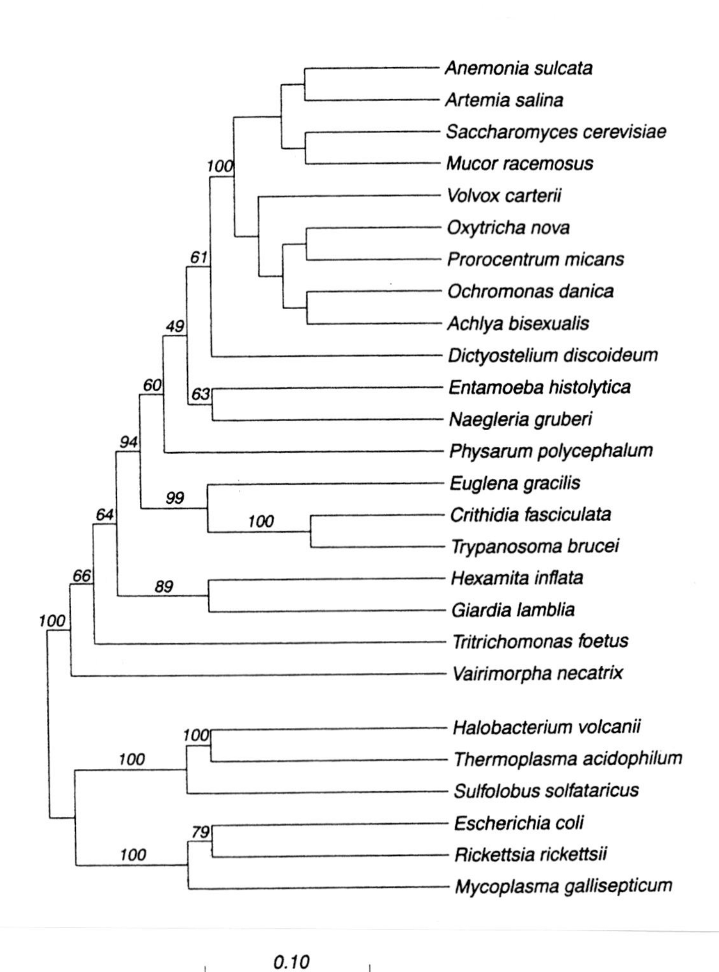

FIGURE 2 Multikingdom phylogeny inferred from 16S rRNA sequences. The phylogenetic tree was inferred by using unambiguously aligned sequences with the neighbor joining method in the PHYLIP program. Bootstrap support for topological elements in the tree is based on 200 resamplings. Horizontal distances between nodes of the tree represent relative evolutionary distances. The bar scale corresponds to 10 changes per 100 positions. (Reprinted with permission from Leipe et al., 1993.)

karyotes with a high G + C content were used in the analysis, *G. lamblia* was positioned at the base of the tree, but the diplomonads were split and were considered a paraphyletic group. While this is strong evidence that using outgroups with similarly skewed nucleotide compositions biases the outcome of the analysis (separating organisms within the same taxonomic group because of low and high G + C contents), the low G + C content could also bias the outcome for *V. necatrix* to branch first, leaving the lowest branching eukaryote in question. They therefore speculated that microsporidia may have diverged first from the eukaryotic line but concluded that because of its low G + C content the position of *V. necatrix* was still ambiguous.

Studies by Hashimoto and Hasegawa (1996) and by Kamaishi et al. (1996a, 1996b) (Fig. 3) focus on the maximum likelihood analysis of elongation factor 1α (EF-1α) and its eubacterial homolog EF-Tu and elongation factor 2 (EF-2) and its eubacterial homolog EF-G to understand the relationships of eukaryotes to each other and to the eubacteria and archaea. EF-1α/Tu is a GTP-dependent enzyme responsible for binding aminoacyl-tRNA to the A site of ribosomes in a codon-dependent manner, and EF-2/G is responsible for the GTP-dependent translocation of peptidyl-tRNA from the A site to the P site on the ribosome. These authors provided a rigorous analysis of the amino acid sequence data and tested various models of amino acid substitution. They addressed the concern that the analysis of microsporidia could be biased by the low G + C content of *V. necatrix* and instead used the microsporidian parasite of the ayu fish, *Glugea plecoglossi*, which has a G + C content of about 50%. Noting previous concerns about the G + C bias, these authors proposed that phylogenies based on the analysis of amino acid sequences of proteins are not biased by genomic G + C content, providing a "more

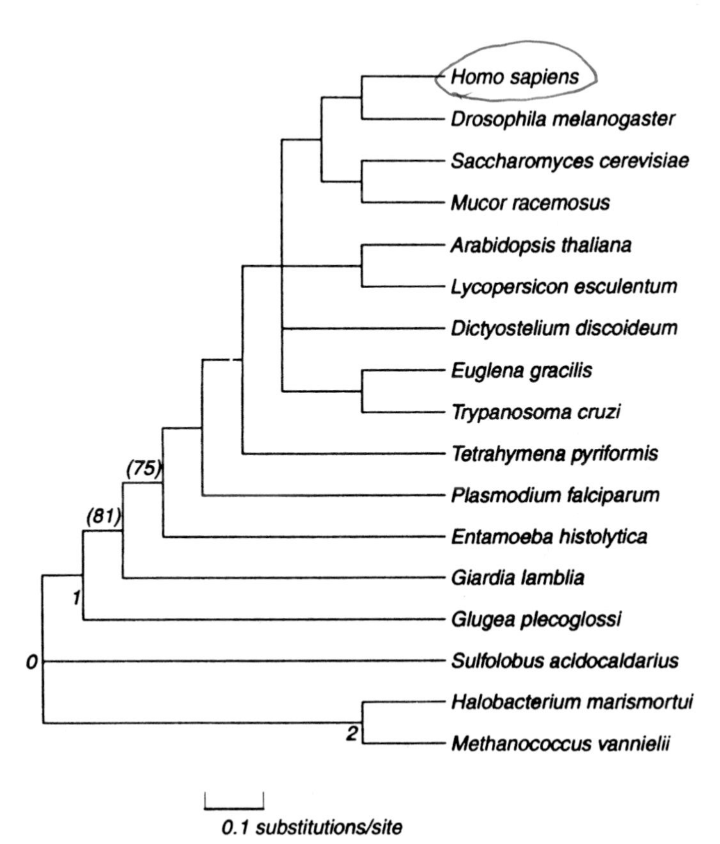

FIGURE 3 EF-1α tree of eukaryotes with archaebacteria as an outgroup. The phylogenetic tree was constructed by the maximun likelihood method based on the JTT model. The horizontal length of each branch is proportional to the estimated number of substitutions. Bootstrap probabilities are shown in parentheses. (Reprinted with permission from Kamaishi et al., 1996b.)

robust estimation of the early divergence of eukaryotes" (Hashimoto and Hasegawa, 1996; Kamaishi et al., 1996a, 1996b).

Analogous to the small-subunit rRNA analysis of *V. necatrix*, the EF-1α analysis of *G. plecoglossi* demonstrated a long branch length. Kamaishi et al. (1996a) proposed that maximum likelihood analysis is a good method when such a long branch length exists. On the basis of the JTT model of the EF-1α data, Kamaishi et al. (1996a) concluded that the microsporidia are the outgroup of the eukaryotic tree followed by *Giardia* (the diplomonads) and then by *Entamoeba histolytica* (Fig. 3). They reiterate the hypotheses of Vossbrinck et al. (1987) and Cavalier-Smith (1989) that these mitochondria-lacking protozoa might have diverged before occurrence of the symbiotic event. Further investigation with maximum likelihood analysis and the JTT-F model (Kamaishi et al., 1996b) including the EF-2 gene demonstrated that *G. plecoglossi* branched first from the eukaryotic tree followed by *G. lamblia* and *E. histolytica*. Further evaluation of the EF-1α and EF-2 analyses was accomplished by adding the estimated log likelihood for each tree topology. This analysis favored the relationship of the amitochondriate protozoa seen in the EF-1α analysis. The major difference between the EF-1α/EF-2 phylogeny and results from previous studies with rRNA (Leipe et al., 1993; Vossbrinck et al., 1987) is in the position of the flagellates *Euglena* and *Trypanosoma* as the first mitochondria-containing organisms to diverge from the eukaryotic line (rRNA analysis) or as a branch after the ciliate genus *Tetrahymena* (EF-1α analysis). Additionally, in the rRNA phylogenies *Entamoeba* (which has no mitochondria) and *Naegleria* (which has mitochondria) are sister taxa, implying that these two sarcodines are related and that the mitochondria were secondarily lost from *Entamoeba*. In contrast, the EF-1α tree places *Entamoeba* much lower phylogenetically.

Microsporidia as Derived Fungi

A much different view of the phylogenetic placement of the Microspora has been suggested based on analysis of β-tubulin genes (Katiyar et al., 1994; Edlind et al., 1994, 1996; Li et al., 1996). Some perceived advantages of using β-tubulin genes for phylogenetic analysis are that (1) they are unique to eukaryotes and therefore may more accurately parallel those of other eukaryotic-specific structures such as the nucleus; (2) they are major components of the cytoskeleton, the 9 + 2 axonemes, and most importantly the mitotic spindle and therefore may have paralleled the nucleus; and (3) they are easier to align unambiguously. Both α- and γ-tubulins are used as outgroups in these studies. Maximum parsimony analysis and distance analysis of the β-tubulin amino acid sequence data (Edlind et al., 1994, 1996; Li et al., 1996; Katiyar et al., 1994) place *E. histolytica* as the most divergent of the eukaryotes and relegates the microsporidia (in this case *E. hellem*) to the fungi (Fig. 4). Both parsimony and distance analyses (PRODIST and NEIGHBOR [Felsenstein, 1978, 1988]) gave the same results. The unusual rRNA and the absence of typical fungal structures (cell walls, mitochondria, ergosterol-containing membranes) are postulated to reflect secondary loss during evolution rather than primitive characteristics (Edlind et al., 1996). It should be noted, however, that when the yeast *Candida albicans* was added to the β-tubulin distance analysis, it changed the position of the yeasts in relation to the fungi, and the yeasts were shown to branch before *Trichomonas vaginalis*.

Edlind et al. (1996) explain differences between the rRNA and β-tubulin phylogenies with respect to the microsporidia by postulating that deletions in microsporidial rRNA cause changes in adjacent and possibly in distant rRNA sequences, making it less reliable for phylogenetic analysis. Such changes may also have led to alterations in other genes involved in protein translation such as EF-1α and EF-2, isoleucyl-tRNA (Brown and Doolittle, 1998, in press) as well as the lack of 5.8S RNA. With regard to differences between β-tubulin and rRNA phylogenies for *E. histolytica*, Edlind et al. (1996) state, "It is more difficult to rationalize the dramatic difference in the placement of *Entamoeba histolytica* within the rRNA and tubulin (both α and β) phylogenies." A number of factors also point to the possibility that simi-

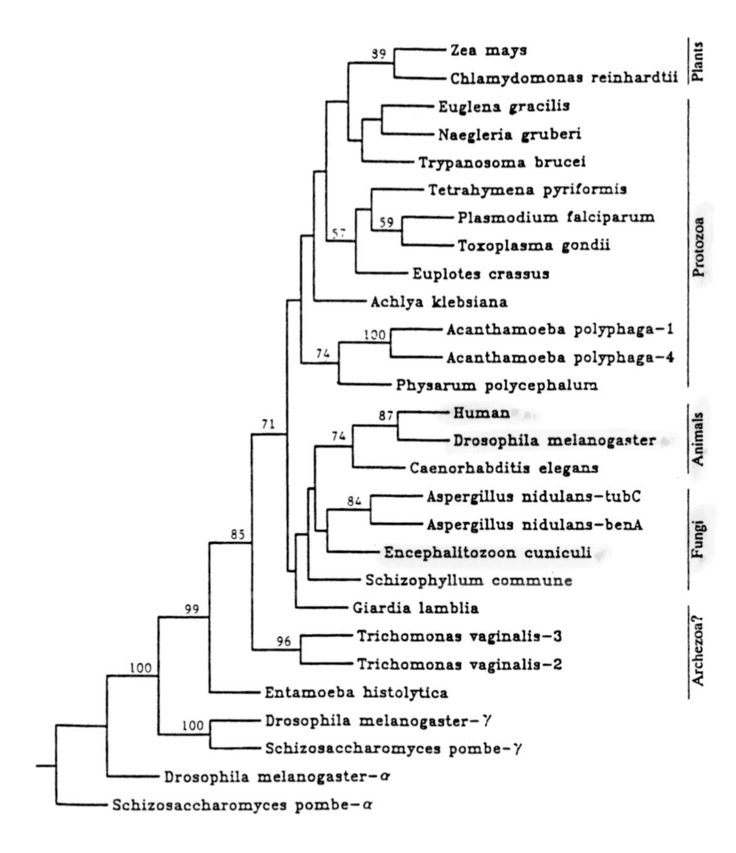

FIGURE 4 Phylogenetic tree of representative partial β-tubulin sequences (residues 108 to 259). Selected α-tubulin and γ-tubulin sequences are included, with *S. pombe* α-tubulin as the designated outgroup. Parsimony analysis (100 bootstrap resamplings) was used for the consensus tree. Numbers are percentages of trees in which the group of species to the right of that branch was found. A nearly identical tree, except that *E. cuniculi* and *S. commune* branches were reversed, was obtained with a distance matrix method (500 resamplings). (Reprinted with permission from Edlind et al., 1996.)

larities among taxa in the β-tubulin phylogeny also represent a case of convergent evolution due to selection pressure from naturally occurring antimicrobial compounds. The facts that some microsporidia and helminths are sensitive to albendazole while other microsporidia are not, and that some fungi are sensitive to the related compound nocodazole while others are not, imply that benzimidazole sensitivity is either plesiomorphic or polyphyletic in origin. Edlind et al. (1996) report the identification of a nocodazole-resistant *Saccharomyces cerevisiae* isolate. By comparative analysis with *E. hellem* and *Aspergillus nidulans*, the resistance correlates with a mutation from Val to Gly at amino acid residue 268, confirming that resistance to benzimidazole can be achieved by a single mutation. These observations lend credence to the idea that β-tubulin may be a target for antimicrobial activity and that over the evolutionary time scale organisms exposed to similar naturally occurring antimicrobial compounds may have developed similar mutations.

There is additional evidence for the view that microsporidia may be related to the fungi (Muller, 1997; Keeling and McFadden, 1998). The *E. cuniculi* genes for thymidylate synthase and dihydrofolate reductase are separate genes although they are on the same chromosome (Vivares et al., 1996). In the EF-1α sequence of *G. plecoglossi* there is an insertion found only in fungi (Kamaishi et al., 1996a). Microsporidia also display similarities to the fungi in mitosis (closed mitosis and spindle pole bodies) and meiosis (Desportes, 1976; Canning, 1988; Haig, 1993; Flegel and Pasharawipas, 1995). Mitochondrial heat shock protein 70 (*hsp70*) genes have been identified in *V. necatrix, E. cuniculi, E. hellem,* and *N. locustae*, and on phylogenetic analysis these genes cluster with the mitochondrial *hsp70* genes of fungi (Hirt et al., 1997; Germot et al., 1997; Peyretaillade et al., 1998b;

Hashimoto, unpublished). Phylogenetic analysis of both α-tubulin (Keeling and Doolittle, 1996; Li et al., 1996) and RNA polymerase (Embley, personal communication) genes also supports placement of the microsporidia with the fungi. In addition, a repeat analysis of the EF-1α and EF-2 data with removal of the outgroups also places the microsporidia with the fungi (Embley, personal communication).

The identification of prokaryotic homologs of *cpn60* and *hsp70* in all three groups lacking mitochondria—Diplomonada as detemined by *G. lamblia cpn 60* (Soltys and Gupta, 1994; Roger et al., 1998); Trichomonada as determined by *T. vaginalis cpn60* (Bui et al., 1996; Horner et al., 1996; Roger et al., 1996) and *hsp70* (Bui et al., 1996; Germot et al., 1996); and Microspora as determined by *N. locustae hsp70* (Germot et al., 1997), *V. necatrix hsp70* (Hirt et al., 1997), *E. cuniculi* (Peyretaillade et al., 1998b), and *E. hellem hsp70* (Hashimoto, unpublished data) (Fig. 5)—suggests at one point the presence of mitochondria in the evolution of these organisms. In addition, mitochondria-type *cpn60* has also been identified in *E. histolytica* (Clark and Roger, 1995). The lack of mitochondria is thus probably an apomorphic rather than a pleisomorphic character. This finding is not in itself proof that these organisms once had mitochondria, because all life forms appear to have this gene. However, the fact that phylogenetic analysis places the eukaryotic *hsp70* gene with the α-Proteobacteria (Germot et al., 1997; Hirt et al., 1997), as does mitochondrial rRNA analysis (Yang et al., 1985), provides convincing evidence for the idea that these organisms once had mitochondria. Further support for this belief is the recognition that trichomonad hydrogenosomes were derived from mitochondria (Embley et al., 1997).

Maximum likelihood analysis provides relatively strong evidence that the prokaryotic *hsp70* identified in the microsporidia (Germot et al., 1997; Hirt et al., 1997) belongs to the mitochondrial group and demonstrated weak support for the fungal clade (Fig. 5). This analysis provides further evidence for the hypothesis that the microsporidia are related to the fungi. In addition to the mitochondrial *hsp70* and *cpn60*, valyl-tRNA synthetase genes consistent with the secondary absence of mitochondria have been found in *T. vaginalis* (Brown and Dolittle, 1995; Hashimoto et al., 1998), in *G. lamblia* (Hashimoto et al., 1998), and in the microsporidia (Hashimoto, unpublished data). These data also imply that the microsporidia did not diverge from other eukaryotes before the advent of mitochondrial symbiosis.

Prior to molecular analysis of the eukaryotes, very little to nothing was stated with certainty about the relationships among the most distantly related taxa based on morphological and ecological characters. Molecular phylogenies do not by any means resolve all eukaryotic relationships but do make an important addition to our knowledge. As more and more molecules are analyzed, a consensus will develop concerning the best guess relationship. The discrepancies among the analyses of various molecules may be explained by the hypothesis that it is more likely that a group of unrelated molecules diverged rapidly to show a similar divergent "nonrelationship" and less likely that unrelated genes repeatably converged making unrelated organisms appear related. Therefore, analysis of separate molecules leading to the same divergence or unrelatedness may be less significant than molecular analysis of separate molecules leading to the same relationship. Placement of the microsporidia as the most divergent organism may simply mean that they have changed very rapidly and being considered an outgroup may reflect traits that have changed too rapidly to reflect evolutionary relationships. There are now a number of lines of evidence supporting the view that the microsporidia are divergent fungi. The rather unique evolution of the polar filament for the infection of hosts may explain the great success (rapid divergence) microsporidia have attained. Currently, it is probably in poor judgment to use "traditional" morphological and ecological characters to substantiate or deny molecular phylogenies. It is more productive (and it allows independent evaluation) to document on the molecular phylogeny the traditional characters (Baker et al., 1995).

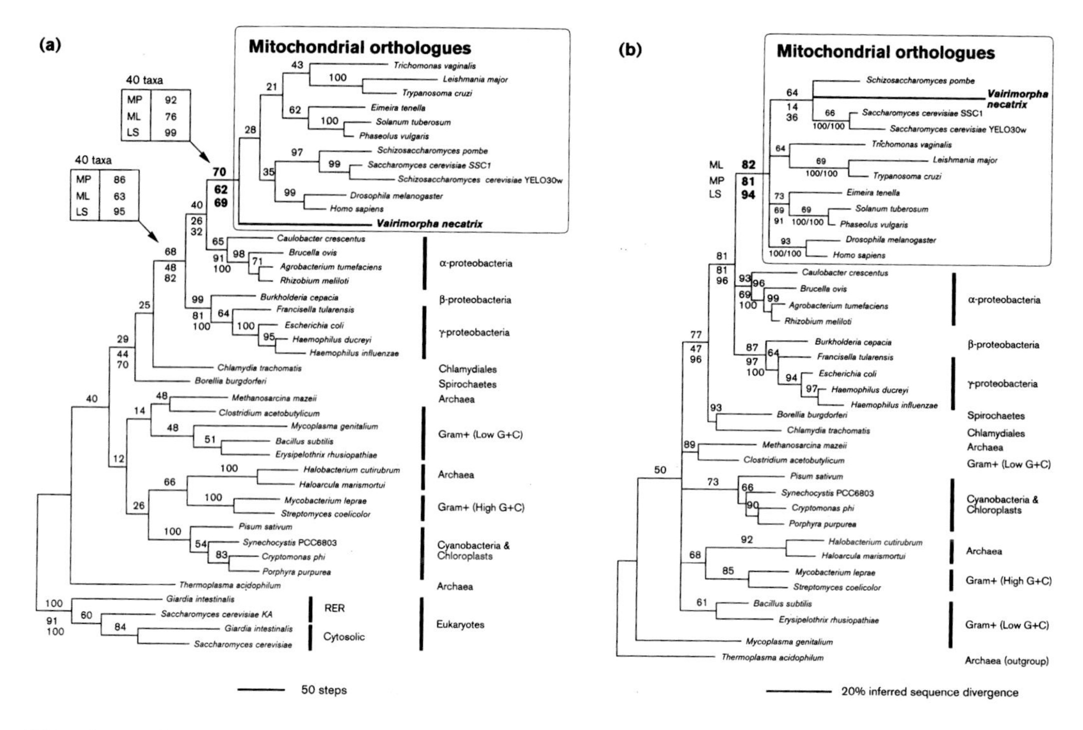

FIGURE 5 *hsp70* phylogenetic analysis. (a) Maximum parsimony (MP) bootstrap (500 replicates) consensus tree for 41 *hsp70* sequences; (b) maximum likelihood (ML) consensus tree (using the PUZZLE 3.0 program with 1,000 puzzling steps) for 37 *hsp70* sequences. Both trees demonstrate that the *V. necatrix* sequence is within the mitochondrial clade. Support values above branches correspond to the analysis shown. Support values below relevant branches are from analyses of the same data set by different methods: (a) an ML and a least squares (LS) distance (100 bootstrap replicates); (b), an MP and an LS distance analyses. Boxed values in panel b correspond to support values with the *V. necatrix* sequence removed from the analysis. (Reprinted with permission from Hirt et al., 1997.)

There is little mention in the protozoan literature about the possibility that discerning the correct phylogeny may not be possible. Figure 6 shows a divergence of taxon B from taxon AC. From a cladistic view the only characters linking taxa A and C and separating them from B are those that changed after the divergence of B from AC and before A and C split (period D). If period D is short, then statistically only the most variable sites will change. There is a good chance that these characters will change again during the period from the split of A and C to the present. Phylogenetic analysis, therefore, has a limited resolution even if all nucleotides (the entire genome) are sequenced for all organisms being analyzed. The fact that a phylogeny can be obtained for a group of organisms does not guarantee that it reflects evolutionary history. It has been shown for parsimony analysis and compatibility methods that when an incorrect tree is obtained, including additional data sometimes verifies the incorrect phylogeny (Felsenstein, 1978). The phylogeny of the insect orders, the animal phyla, and the protozoa all include rapid radiation a relatively long time ago. Much effort has been made by systematists to obtain solutions to these highly prized

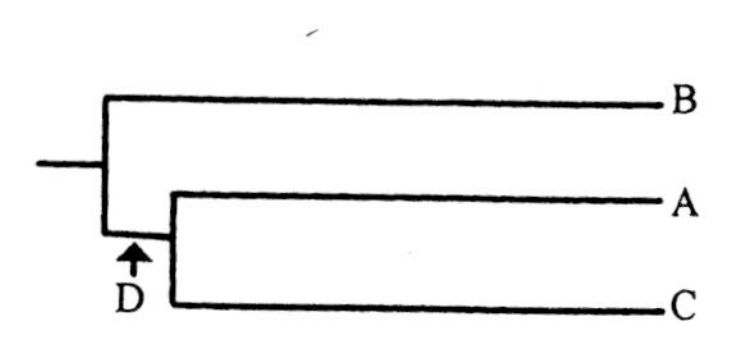

FIGURE 6 Hypothetical phylogenetic tree.

phylogenetic questions, and they may be impossible to answer. Yet scores of research papers have been written on protozoan phylogeny based on morphological and biochemical characters that at the outset had no chance of resolving these problems. Several review articles address these ideas (Maley and Marshall, 1998; Philippe and Adoutte, 1996). The small-subunit ribosomal data were for a while thought to hold the answer to the placement of the microsporidia (Vossbrinck et al., 1987). Now more genes are being added, and we may finally have the illusion that we "know" the answers to unanswerable questions. Analysis of two different molecules suggests that microsporidia have a long evolutionary branch length; however, this could be due to rapid evolution based in part on assumption of the correct outgroup. One must remember that there is no outgroup to the tree of life and that the three primary kingdoms (archaebacteria, bacteria, and eukaryotes) are all equally likely to contain characters present in their common ancestor. For example, without fossil data it is not clear whether the common ancestor of all entant life had a nucleus and that for purposes of rapid reproduction the nucleus as well as introns were lost in bacterial forms.

Phylogeny of the Phylum Microspora: Morphological and Life Cycle Characteristics

Taxonomic studies on the microsporidia depend on separating taxa based on available ultrastructural and ecological characters (Chapter 3). Traditional descriptions, obtained using the light microscope, are based primarily on spore characteristics such as size, shape, and number of nuclei per spore, and on the host. Weiser (1982) lists 70 characters for the identification of microsporidia that can be distinguished by light microscopy. The use of electron microscopy for taxonomic purposes provides an additional, more discriminating set of characters. These features include the structure of the polaroplast and the pansporoblastic membrane, the shape and number of coils in the polar filament, and details of the exospore structure (for review, see Larsson, 1986). In addition to morphological studies on spores from what Larsson (1986) refers to as "spontaneously infected hosts," life cycle studies yield important information about vegetative growth and development (merogony and sporogeny) and development in alternate hosts (Andreadis, 1983).

For discussion purposes the microsporidia can be divided into three basic groups: (1) "primitive" microsporidia (Metchnikovellidae), hyperparasites of gregarines in annelids, which are separated from the other microsporidia by the presence of a rudimentary polar filament (a short, thick, manubrium-shaped tube) and a spore lacking a polaroplast; (2) the Chytridiopsidae, Hesseidae, and Burkeidae, which can be seen as "intermediates" described as having a short polar filament and minimal development of the polaroplast and endospore; and (3) "higher" microsporidia, which have a well-developed polar filament, polaroplast, and posterior vacuole. Differences among modern classifications of microsporidia center on the arrangement of these three groups and, most significantly, the characters used to divide the third group, the higher microsporidia, into subgroups.

Tuzet et al. (1971) separated suborders into Apansporoblastina and Pansporoblastina based on the presence of a membrane surrounding the sporoblast. The next divisions were based on whether a sporogonial plasmodium is present in the Apansporoblastina as well as on the number of spores produced in the pansporoblast in the Pansporoblastina.

Classification of the Microsporidia according to Tuzet et al. (1971)

- Class Microsporidea Corliss and Levine 1963
  - Order Microsporida Balbiani 1882
    - Suborder Apansporoblastina
      - Family Caudosporidae Weiser 1958
        - Genus *Caudospora*
      - Family Nosematidae Labbe 1899
    - Suborder Pansporoblastina
      - Family Monosporidae
        - Genus *Tuzetia*

Family Telomyxidae Leger and Hess 1922
  Genus *Telomyxa*
Family Polysporidae
  Genera *Glugea, Gurleya, Thelohania, Heterosporis, Duboscqia, Trichoduboscqia, Plistophora, Weiseria, Pyrotheca.*

Sprague and Vavra (1977), like Tuzet, separated the higher microsporidia based on the presence of a pansporoblastic membrane. The suborders were further broken down into families according to the details of sporogony and the nuclear condition. In this monumental work, the taxa, including the genera, are defined, and information, including host and site, vegetative stages, sporulation stages, spores, and locality, are given for each species. Sprague discusses these characters in relation to phylogeny and qualifies his comments at the outset as "sheer speculation."

Classification of the Microsporidia according to Sprague and Vavra (1977)
- Phylum Microspora
  - Class Rudimicrosporea
    - Order Metchnikovellidae
      - Family Metchnikovellidae
        - Genera *Metchnikovella, Amphiacantha, Ambliamblys*
  - Class Microsporea
    - Order Chytridopsida
      - Family Chytridiopsidae
        - Genera *Chytridiopsis, Steinhausia*
      - Family Hesseidae
        - Genus *Hessea*
      - Family Burkeidae
        - Genus *Burkea*
    - Order Microsporida
      - Suborder Pansporoblastina
        - Family Pleistophoridae
        - Family Pseudopleistophoridae
        - Family Duboscquiidae
        - Family Thelohaniidae
        - Family Gurleyidae
        - Family Telomyxidae
        - Family Tuzetiidae
      - Suborder Apansporoblastina
        - Family Glugeidae
        - Family Unikaryonidae
        - Family Caudosporidae
        - Family Nosematidae
        - Family Mrazekiidae

Weiser (1977) placed *Chytridiopsis* and *Hessea* with the metchnikovellids based on the presence of a rudimentary polar filament with spherical spores enclosed in persistent thick-walled pansporoblasts. He then divided the rest of the class Microsporididea into the order Pleistophoridida (sporogony and spores uninuclear) and the order Nosematidida (sporogony and spores diplokaryotic). Levine et al. (1980) used Sprague's 1977 divisions based on Tuzet's Apansporoblastina and Pansporoblastina designations as the primary divisions of the higher microsporidia. Issi (1986) listed 11 characters visible at the light and electron microscopic levels and included a chart listing the states of these characters for 68 genera of microsporida. She separated the microsporidia into four subclasses: Metchinikovellidea, Chytridiopsidea, Cylindrosporidea, and Nosematidea. The phylogenetic tree Issi proposes places *Vairimorpha* close to *Nosema*, which is in agreement with molecular phylogenetic analysis. Larsson (1986) presented a similar chart with 11 characters describing the character states for 66 species from 51 genera. The important morphological characters used for taxonomic purposes include spore shape; number of sporoblasts per sporont; shape, number, and location of the nuclei; structure of the polaroplast; structure of the parasitophorous vacuole (host origin); structure of the sporophorous vesicle (parasite origin); shape and number of coils in the polar filament; and details of the exospore. Larsson (1988) presented another important work on the identification of microsporidian genera. Microsporidial characters are detailed at both the light and electron microscopic levels. A key to the genera is given, and a character matrix is shown listing the character states for the available genera.

Sprague et al. (1992) present a review of the taxonomy of the microsporidia in which they separate out the metchnikovellids (including *Hessea*) as *incertae sedis* taxa. The higher-level divisions of the microsporidia are based on what is referred to as the "chromosome cycle." The class Dihaplophasea has diplokaryotic stages and undergo a pairing of gametes which then proliferate and end by undergoing haplosis to produce gametes again. Haplosis can occur either by meiosis (order Meiodihaplophasida) or by nuclear dissociation (order Dissociodihaplophasida). The other class of microsporidia, the Haplophasea, are entirely haplophasic. Further taxonomic divisions and definitions are based on the presence of an interfacial envelope (pansporoblastic membrane), number of nuclei, number of spores in a sporophorous vesicle, and other details of the life cycle. This work also includes a glossary defining microsporidial terms and their classification with and without definitions of taxa. An abbreviated form (without family and generic levels) of their classification is as follows:

Classification of the Microsporidia according to Sprague et al. (1992)

Phylum Microspora
- Class 1. Diphaplophasea
  - Order 1. Meiodihaplophasida
  - Order 2. Dissociodihaplophasida
- Class 2. Haplophasea

The above-mentioned references (Tuzet et al., 1971; Sprague and Vavra, 1977; Issi, 1986; Larsson, 1986, 1988; Sprague et al., 1992) provide overviews of the history, ultrastructural and structural characteristics, and life cycle differences among taxa of microsporidia. These works are key references for many microsporidiologists and are important for anyone interested in learning about the biology and ultrastructure of these organisms.

Present methods for phylogenetic analysis generally involve a listing of characters (i.e., spore coat and polar filament) that have two or more character states (i.e., present or absent and isofilar or anisofilar). It is important to know the character states for the same characters for all organisms. Missing characters are acceptable for various taxa as long as there are enough characters common to all taxa examined. As stated previously, the only characters that help yield the correct phylogeny are those that changed when these organisms diverged and have not changed subsequently. All other characters will either result in no information or give misleading results. The inclusion of fossil data can be key in determining the minimum age of a particular characteristic and in concluding whether character states are primitive (pleisomorphic) or derived (apomorphic). Unfortunately there are no fossil data for the microsporidia, and the morphological, life cycle, and ecological characters are too few to allow determination of phylogenetic relationships at higher taxonomic levels (above the family level) and thus to establish the significance of each character and which characters are derived.

Phylogeny of the Microsporidia: Molecular Methods

Cloning of many microsporidian rRNA genes has been accomplished (see Table 1 for a list of genes and Table 2 for cloning and sequencing primers). The first use of comparative analysis of rRNA sequence data was presented as an unrooted tree of five microsporidial species (Vossbrinck et al., 1993) based on partial sequence data from the small and large subunits. *E. hellem* (Didier et al., 1991) had previously been characterized by Western blot and sodium dodecyl sulfate-polyacrylamide gel electrophoresis (SDS-PAGE) analysis from three AIDS patients and demonstrated differences from *E. cuniculi* with these techniques. Ultrastructurally *E. hellem* and *E. cuniculi*, however, were almost identical (Didier et al., 1991). At that time it was not clear if *E. hellem* was simply an isolate of *E. cuniculi*. Analysis of *E. hellem*, *E. cuniculi*, *V. necatrix*, *V. lymantriae*, and *Ichthyosporidium giganteum* revealed that *E. hellem* and *E. cuniculi* were most likely distinct yet closely related species. In this analysis *V. necatrix* and *V. lymantriae* were chosen because it was known that they are distinct yet closely related parasites of *Lepidoptera*, and *I. giganteum* was chosen as an outgroup species (Vossbrinck et al., 1993). A similar analysis demonstrated that *E. intestinalis* was also a distinct organism (Zhu et al., 1993a).

Analysis of a variety of microsporidia (Fig. 7) (Baker et al., 1995), including five species isolated from AIDS patients, highlights the polyphyletic nature of AIDS-related microsporidia and brings into further doubt the use of any single character for developing higher taxonomic groupings. Figures 7 and 8 demonstrate that *E. intestinalis*, *E. cuniculi*, and *E. hellem* are three distinct yet closely related species. *E. hellem* and *E. cuniculi* are almost indistinguishable at the ultrastructural level, while *E. intestinalis* can be distinguished by the presence of an extracellular matrix surrounding the sporoblasts and spores (Cali et al., 1993). On the basis of a comparative rDNA analysis, *E. (Septata) intestinalis* and *E. cuniculi* are more similar to each other than to *E. hellem*. A more recent phylogenetic tree that includes most of the rRNA sequences available is presented in Fig. 8. This phylogenetic tree is similar to one reported by Keeling and McFadden (1998). This analysis of the microsporidia confirms the polyphyletic nature of the human microsporidia and the close relationship among the three *Encephalitozoon* spp. as outlined above.

As indicated in Fig. 9, by rRNA phylogeny *Nosema bombycis*, the type genus for *Nosema* and a parasite of the silk moth (*Bombyx mori*), is

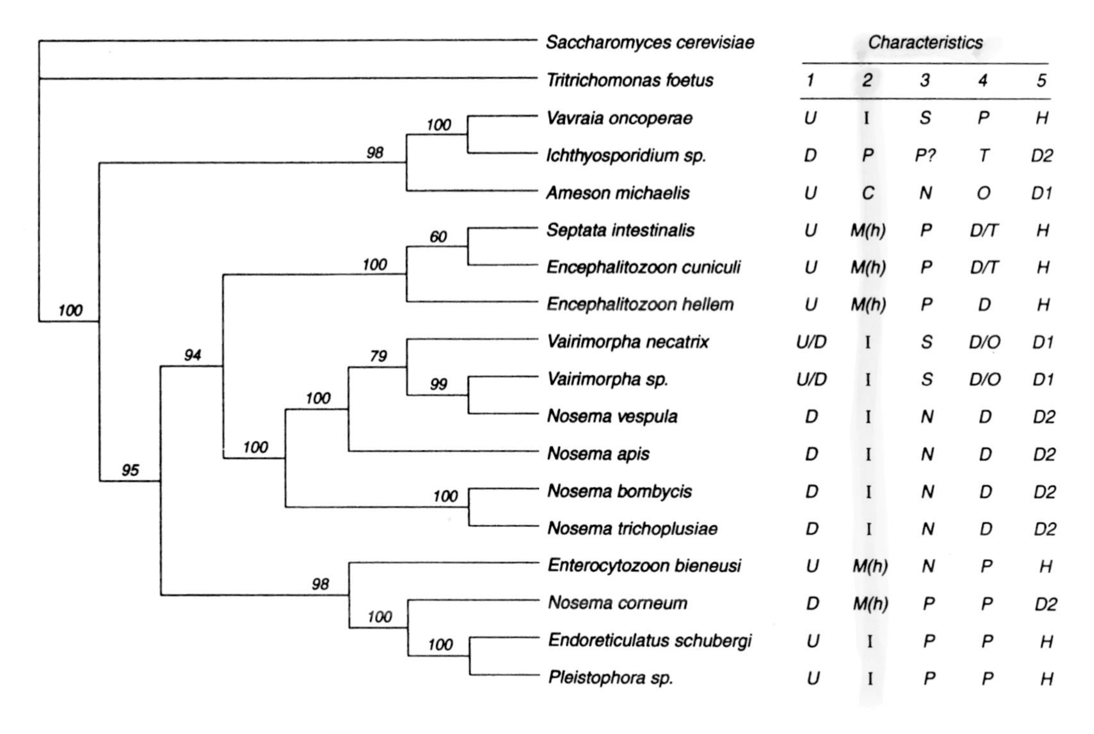

FIGURE 7 Microsporidian phylogeny: molecular and morphologic characteristics. Bootstrap analysis (400 replicates) of the most parsimonious tree. Numbers represent percentages of bootstrap replicates. Characters represented are as follows. (1) Nuclear condition of the spore: D, diplokaryotic; U, uninucleate. (2) Host: C, Crustacea; I, Insecta; M, Mammalia (h, human); P, Pisces. (3) Membrane surrounding spores: N, none; P, parasitophorous vacuole; S, sporophorous vesicle. (4) Sporogony: D, disporous; O, octosporous; P, polysporous; T, tetrasporous. (5) Chromosome cycle: D1, dihaplophasic/haplophasic with meiosis; D2, dihaplophasic/haplophasic with nuclear dissociation; H, haplophasic only. *Nosema* sp. characters are based on genomic placement. (Reprinted with permission from Baker et al., 1995.)

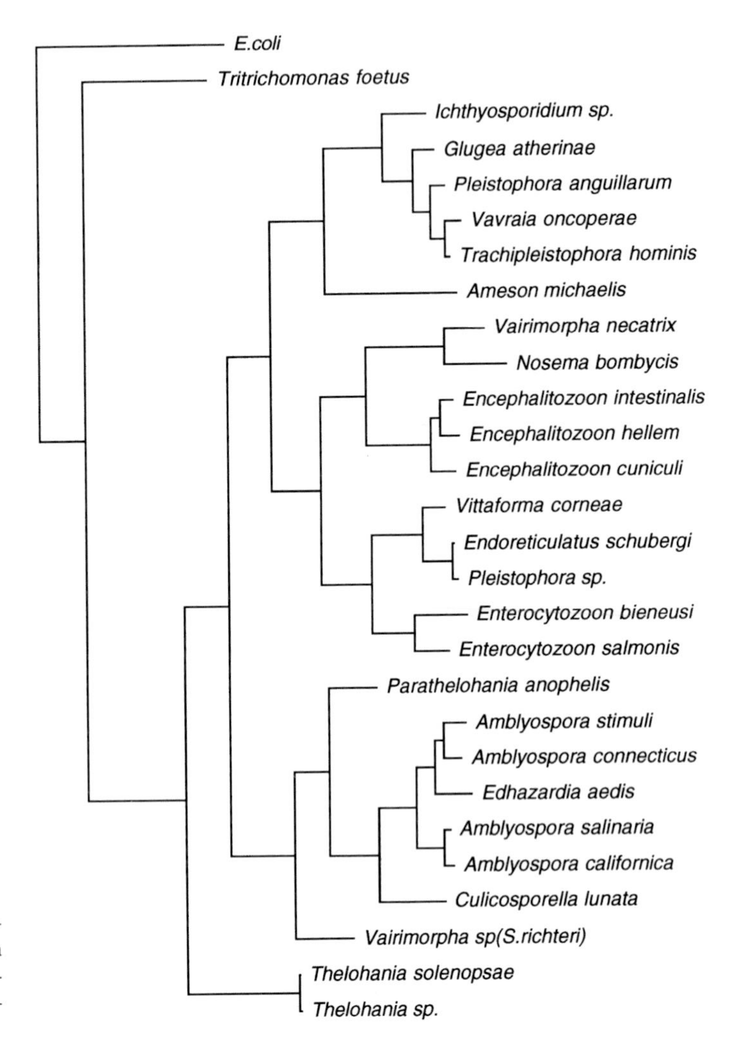

FIGURE 8 Microsporidian small-subunit rRNA phylogeny. This is a bootstrap analysis of the most parsimonious tree containing rRNA sequences from GenBank (Table 1).

closely related to *Vairimorpha* spp. which are also parasites of *Lepidoptera*. There are two distinct groups in Fig. 9, one with *N. bombycis*, the type species for *Nosema*, and one with *V. necatrix*, the type species for *Vairimorpha*, and there are distinct sequence characteristics separating the two groups. Similar results were obtained by Baker et al. (1994) who, by direct reverse transcriptase dideoxy sequencing of the 580 reverse region of the rRNA, demonstrated the close relationship between several *Vairimorpha* spp. and *N. bombycis*. In addition to highlighting the importance of the host as an important taxonomic character, this study showed that an organism believed to be diplokaryotic throughout its life cycle was closely related to an organism that produces both isolated diplokaryotic spores and uninucleate octospores in packets surrounded by a pansporoblastic membrane. Several of the classification schemes mentioned above based on traditional characters have divided *Vairimorpha* and *Nosema* into distantly related taxa because of these differences. Molecular analysis demonstrates that *N. bombycis* is much more distantly related to other "*Nosema*" spp., *N. kingi*, *N. locustae*, and *N. algerae*. Thus neither the distinction of being diplokaryotic throughout the life cycle nor the presence of a pansporoblast can be

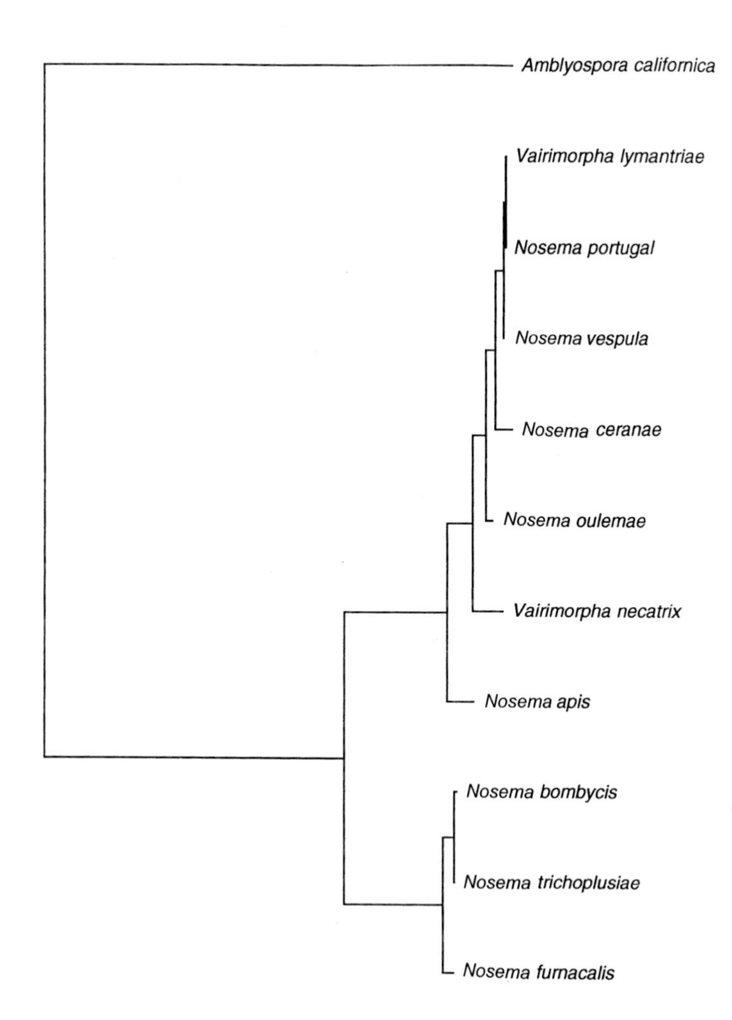

FIGURE 9 *Nosema/Vairimorpha* phylogeny. This is an analysis of *Nosema* and *Vairimorpha* rRNA sequences in GenBank (Table 1). As can be seen, despite the difference in nuclear number, there is overlap in these two genera at the molecular level of analysis.

used alone to group microsporidia at higher levels. These characteristics may be changing too rapidly to be of use in classification at higher taxonomic levels. The genus *Nosema*, defined as being diplokaryotic throughout its life cycle, is in all likelihood a polyphyletic group of unrelated taxa (Sprague and Vavra, 1992). Evidence of this is that *N. corneum* (syn. *Vittaforma corneae*) is more closely related to *Endoreticulatus schubergi* than to the true *Nosema* spp. Since *N. corneum* shares only the diplokaryotic condition with other *Nosema* spp., it was renamed *V. corneae* based on the ribbon-shaped (i.e., *Vittaforma*) sporonts (Silveira and Canning, 1995). Further evidence of this is found in a report that a newly described microsporidian, *Nosema portugal* from the gypsy moth (*Lymantria dispar*), has only a single nucleotide difference from *V. lymantriae* in its rRNA coding region (by rDNA sequencing) and only a two nucleotide difference in its rRNA spacer (ITS) region (Maddox et al., 1998). However, the authors feel compelled to classify these two taxa in separate genera because *Nosema* is defined taxonomically as being diplokaryotic throughout its life cycle, whereas *Vairimorpha* is defined as having a uninucleate octosporous stage during its development.

In a comparative rDNA analysis of three microsporidia from fire ants (Moser et al., 1998) two microsporidia classified as *Thelohania* and one classified as a *Vairimorpha* sp. demonstrated divergence from the others. Our analysis (Fig. 8) demonstrated a slightly different relationship among these microsporidia than that reported by Moser et al. (1998) which could be due to alignment differences or to the use of a different outgroup. Interestingly, Moser et al. (1998) found that the two *Thelohania* spp. were most closely related to a microsporidian classified as a *Vairimorpha* sp. that was unrelated to the lepidopteran *Vairimorpha* and to other fire ant microsporidia. Thus, neither host type nor production of octospore packets linked these two groups.

A report (Baker et al., 1995) on the taxonomic position of *Amblyospora* demonstrated that *Amblyospora californica* and an *Amblyospora* sp. (now *A. salinaria* [Becnel and Andreadis, 1998]) were outgroups to the remainder of the microsporidia analyzed in that study. *Amblyospora* is polymorphic (heterosporous), having three distinct spore types with both uninucleate and diplokaryotic stages in both a primary and an intermediate host (Chapter 14). This observation makes a case that the ability to develop various morphologies and karyomorphs is ancestral (pleisomorphic) in the microsporidia and is consistent with the idea that various taxa may have lost or may simply be repressing characters present in *Amblyospora*. More recently Baker et al. (1995) have given an indication of host-parasite cospeciation among the *Amblyospora*, demonstrating that *A. californica* and *Amblyospora* sp., both from *Culex* mosquitoes, are related and that *Edhazardia aedis* and *A. stimuli*, both from *Aedes* mosquitoes, are also related. Figure 8 shows the same relationship with the addition of a sequence from *A. connecticus* (Vossbrinck et al., 1998) from *Aedes cantator*, in agreement with the cospeciation hypothesis. The genus *Edhazardia* has been defined as lacking an intermediate host but has many morphological similarities to *Amblyospora* and may need redefining or renaming in the future. Vossbrinck et al. (1998) demonstrated that by sequencing the microsporidia from a single *Acanthocyclops vernalis* copepod, detection of the intermediate host for a microsporidian species is possible. This technique eliminates the difficult process of laboratory life cycle studies and is more conclusive.

The most prevalent human microsporidial parasite is *E. bieneusi*. Comparative rDNA analysis (Baker et al., 1995) shows a close relationship between *E. bieneusi* and *Nucleospora salmonis* (syn. *Enterocytozoon salmonis*), a parasite of salmonid fish (Chilmonczyk et al., 1991) (Fig. 8). Ultrastructural similarities include precocious development of the polar tube before division of the sporogonial plasmodium into sporoblasts and the lack of a pansporoblastic membrane (growth of all stages of the parasite in direct contact with the host). The primary distinguishing feature is growth of *N. salmonis* in the nucleus of the host cell rather than in the cytoplasm as seen in *E. bieneusi*. *N. salmonis* has proven useful as an animal model in screening drugs for activity against the Enterocytozoonidae (Coyle et al., 1998; Weiss and Kent, unpublished data). Molecular data (Fig. 8) also indicate that *Vittaforma corneae* is related to the family Enterocytozoonidae. As *E. bieneusi* cannot be cultivated continuously in vitro, *V. corneae* has been used for in vitro screening of drugs for activity against *E. bieneusi* (Didier, 1997).

Only one case of infection with *V. corneae* has been reported from the corneal stroma of a nonimmunocompromised individual (Shadduck et al., 1990). rDNA analysis reveals a relatively close relationship to *E. schubergi* and a "*Pleistophora*" sp. (ATCC 50040), both parasites of insects. It should be pointed out that the *Pleistophora* sp. was placed in the American Type Culture Collection (ATCC) before it was realized that *Pleistophora* is a composite genus based simply on the formation of spore packets (by a pansporoblastic membrane) containing more than 16 spores per packet. The genus was subdivided into three genera: *Pleistophora typicalis* from the fish *Myoxocephalus scorpius*, *Vavraia culicis* from the mosquito *Anopheles quadrimaculatus*, and *Polydispyrenia simulii* from the blackfly *Simulium tuberosum* (Canning and Hazzard, 1982). Brooks et al. (1988) established the genus *Endoreticulatus* for *Pleistophora fidelis* from the Colorado potato beetle, which is the genus to which the *Pleistophora* sp. in the ATCC should be assigned. Thus, this is a case in which the molecular data resolved a systematic polyphyletic grouping that had been previously resolved at the ultrastructural level. *Trachipleistophora hominis*, a human pathogen that disseminates, is related to *V. oncoperae* and *Pleistophora anguillarum*.

Molecular relationships may be useful in suggesting the environmental reservoir for the microsporidia found in humans. Encephalitozoonidae appear to be found in many animals, and infections may stem from contact with other humans, pets, or food sources. *E. bieneusi* may be a water- or food-borne pathogen related to fish. *V. corneae* infection may represent a random opportunistic event initiated by direct contact with spores in the environment (which normally infect insects). The microsporidia most similar to *V. corneae* according to rRNA phylogeny (Fig. 8) have been isolated from lepidoptera. *T. hominis*, the most recently reported human microsporidian infection to be sequenced (Hollister et al., 1996), was isolated from the skeletal muscle of an AIDS patient. The sequence is very similar to that of *V. oncoperae*, which is in the same group as *P. anguillarum* (Fig. 8). This finding suggests that the source of this infection may be through contact with freshwater or contact with insects. Many of the human-pathogenic microsporidia have been identified in water sources by PCR (Dowd et al., 1998).

E. cuniculi, first isolated from rabbits, has since been isolated from a number of mammals (Canning and Lom, 1986; Didier et al., 1995; Deplazes et al., 1996). Examination of the ITS region of several *E. cuniculi* isolates (Didier et al., 1995) reveals differences in the number of GTTT repeats in mouse (two repeats), rabbit (three repeats), and dog (four repeats) isolates (Fig. 10). ITS region sequences of additional *E. cuniculi* isolates (two from mouse, two from rabbit, and one from dog) show a correspondence between host and number of GTTT repeats. Sequencing of the ITS region of *E. cuniculi* from AIDS patients in two independent studies (Didier et al., 1996c; Hollister et al., 1996) shows the four GTTT repeats as seen in the dog strains. Eleven isolates of *E. cuniculi* from rabbits in Switzerland (Deplazes et al., 1996) all had three GTTT repeats, adding more evidence of a correlation between the number of repeats and the host. The same study reported six isolates from humans having three GTTT repeats as found in rabbits, rather than four GTTT repeats as found previously in dogs. This observation may imply different sources of *E. cuniculi* in AIDS patients. It has also been appreciated that differences exist in the ITS region of different *E. bieneusi* isolates (Rinder et al., 1997). In the case of *E. bieneusi*, however, the heterogeneity is due to nucleotide substitutions at nine independent sites in the ITS region, generating three distinct rRNA ITS groups. At this point it is unknown if these groups will correlate with as yet unidentified animal reservoirs of *E. bieneusi* or if this heterogeneity merely represents variation in a multicopy gene within this organism.

```
Mouse     5'TGTTGTTGTGTTTTGATGGATGTTTGTTT--------GTGG 3'
Rabbit    5'TGTTGTTGTGTTTTGATGGATGTTTGTTTGTTT----GTGG 3'
Dog       5'TGTTGTTGTGTTTTGATGGATGTTTGTTTGTTTGTTTGTGG 3'
```

FIGURE 10 ITS of *E. cuniculi* isolates. This analysis demonstrates the variation in the number of GTTT repeats in the ITS region of the various *E. cuniculi* isolates and the relationship to host species.

The Usefulness of Molecular Data in Relation to Morphological and Ecological Characters in the Study of Microsporidia

With the advent of molecular sequencing techniques the relationships among these different sets of data and what should comprise a species description may be called into question. The answer lies in understanding what information each data set provides. The molecular data present an excellent means of identifying a species and provide an excellent data set for proposing evolutionary relatedness through phylogenetic analysis. The rDNA sequence data provide a set of characters that are usually unique to an organism. Therefore, the species description should, if possible, contain a complete rRNA sequence including the ITS region to provide a means of unequivocally confirming whether the same species is isolated subsequently. Comparison of this sequence data should have greater reliability than comparison of electron micrographs. This molecular information is particularly useful in studying aquatic microsporidia and in identifying intermediate hosts of the same species and has obvious importance in the diagnosis of microsporidiosis at the species level. Ultrastructural analysis (electron microscopy) also provides information that is often unique to a species, although some closely related species may be indistinguishable by electron microscopy. Ultrastructural characters, however, do not provide information for producing accurate phylogenies, particularly at the higher taxonomic levels. It appears that characters such as the number of nuclei, the number of spores or sporonts, the length and structure of the polar filament, the presence of a pansporoblast, and the details of the life cycle are attributes that microsporidia change relatively rapidly in adapting to different hosts and tissues types during the speciation process. Detailed life cycle and ecological studies as well as ultrastructural changes observed in different hosts, provide additional characters for the comparison of microsporidia. Such detailed information is sometimes supplied with the ultrastructural images in species descriptions (Maddox et al., 1998) but for practical purposes these descriptions are often based on ultrastructural observations of a single infected host. This situation may be the reason for the large number of monotypic genera described in the microsporidia (Larsson, 1986, 1988). Future considerations of species descriptions should be based on the information that is most important in furthering our understanding of these organisms. In the meantime the comparative phylogenetic analysis of rDNA sequence data is facilitating determination of the environmental source of the microsporidia pathogenic to humans by demonstrating which of the sequenced microsporidia are most closely related to these human pathogens.

MOLECULAR DIAGNOSTIC TECHNIQUES

The diagnosis of microsporidial infections has been steadily improving over the past few years with the development of noninvasive diagnostic techniques. While serology has not proven useful for the diagnosis of microsporidiosis, these organisms can be identified in stained specimens by light or electron microscopy (Weber et al., 1994a; Weber et al., 1994b; Wittner et al., 1993). The specific staining methods and procedures are reviewed in Chapter 10. Several PCR methods based on the amplification of rRNA gene fragments have been published for use in the diagnosis and species identification of microsporidia infecting humans. Molecular techniques for the diagnosis of microsporidiosis appear to have high sensitivity and specificity. These methods have proven extremely useful both in the identification of animal models as well as in investigations of the epidemiology of microsporidia pathogenic to humans. A large international multicenter study comparing these PCR tests to those involving conventional microscopy was recently completed (Rinder et al., 1998). It found that interlaboratory variability accounted for most of the differences between the reported sensitivity and specificity of microscopy and PCR methods of diagnosis. Overall, the sensitivity of the different PCR methods was 89% for clinical specimens with a specificity of 98%, compared with a sensitivity of 80% and a specificity of 95% for microscopic diagnosis (Rinder et al., 1998). In spiked stool specimens, it was also found that PCR methods were more sensitive for low concentrations of spores ($\leq 10^4$ per g of stool) (Fig. 11), whereas light microscopy was more robust for high concentrations of spores ($\geq 10^6$ per g of stool) (Rinder et al., 1998). The limit of detection by light microscopy appears to be between 10^3 and 10^4 spores per g of stool (Rinder et al., 1998), which suggests that many cases of microsporidiosis are being overlooked when light microscopy is used as the only means of diagnosis.

The small-subunit rRNA sequences of many microsporidia, both those that are pathogenic in humans and environmental organisms, have been determined (Table 1) and have been useful in the study of phylogenetic relationships (see "Phylogeny"). Homology PCR cloning of the rRNA genes of microsporidia

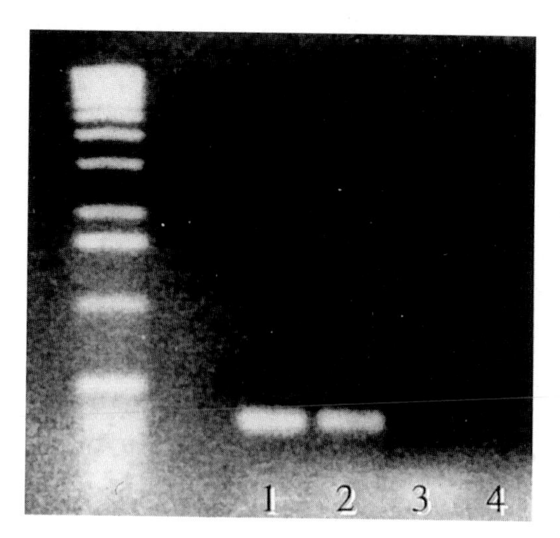

FIGURE 11 PCR for *E. intestinalis* in stool specimens. Amplification of a band of the correct size (375 bp) is demonstrated by PCR in stool specimens from two patients with primer set V1::SI500 to the small-subunit rRNA gene. Lanes 1 and 2, stool specimens from two patients with *E. intestinalis* infection; lane 3, stool from a patient with *E. bieneusi* infection; lane 4, negative control (no DNA).

has been accomplished with primers complementary to conserved sequences in *V. necatrix* (Vossbrinck et al., 1987), as well as with other phylogenetically conserved primers (Table 2). These primers are the basis of a molecular toolbox that allows the cloning of rRNA genes from uncharacterized microsporidia. This cloning can be done on archived or fresh material. Placement of these cloned rRNA genes into the microsporidian molecular phylogeny provides additional evidence for the assignment of a microsporidian to a specific genus as well as an ability to distinguish morphologically similar microsporidia at the species level. It has been suggested that such molecular characterization be included, whenever possible, in the description of new microsporidia. Such analysis has revealed that some microsporidia thought to be separate species, i.e., *N. bombycis* and *N. tichoplusiae* (Pieniazek et al., 1996), are the same and has confirmed that other organisms, i.e., *E. hellem* and *E. intestinalis*, are indeed separate species (Zhu et al., 1994). These cloned rRNA genes can then be used to define PCR primers for studies on the diagnosis and epidemiology of microsporidia. A recent use of this technology has confirmed that different morphological forms of a microsporidian seen in different host species are the same organism (Vossbrinck et al., 1998).

The small-subunit rRNA gene sequences for the human pathogens *E. cuniculi*, *E. hellem*, *E. intestinalis*, *E. bieneusi*, and *V. corneae* (Baker et al., 1995; Didier et al., 1995a, 1996a, 1996b; Hartskeerl et al., 1993a, 1993b, 1995; Katiyar et al., 1995; Weiss et al., 1994; Visvesvara et al., 1994a; Vossbrinck et al., 1993; Zhu et al., 1993a, 1993b, 1993c, 1994) have been determined and are available in the GenBank data base (Table 1). It has been possible to design PCR primers for these small-subunit rRNA genes to identify microsporidia in clinical samples (Table 3). It is also possible by this method to identify microsporidia at the species level without ultrastructural examination. PCR primer sets based on rRNA genes have also been described for the diagnosis of microsporidia pathogenic to vertebrates other than humans, but these primers will not be reviewed in this chapter.

Sample Preparation

Microsporidian nucleic acids can be readily extracted from clinical samples such as tissue biopsies, corneal scrapings, duodenal aspirations, and urine specimens, as well as from in vitro cultures, with routine procedures such as proteinase K digestion followed by phenol chloroform extraction and ethanol precipitation (Ausubel et al., 1997) or with commercial kits for the extraction of DNA such as QIAmp (Qiagen, Santa Clarita, Calif.) or Wizard Preps (Promega Corp., Madison, Wis.) (Hartskeerl et al., 1993a, 1993b; Zhu et al., 1993a, 1993b, 1993c; Coyle et al., 1996; Franzen et al., 1995, 1996a, 1996c; Schuitema et al., 1993). Microsporidian nucleic acid can also be isolated from paraffin-embedded material with standard methods (Innis et al., 1990) or with commercial kits such as DexPAT (Takera Biochemical, Inc., Berkeley, Calif.) (Cali et al., 1998).

TABLE 3 Diagnostic PCR primers for microsporidia pathogenic in humans[a]

Microsporidia amplified	Primer	Name[b]	ΔT^{c}	Amplicon[d]	Reference(s)
Encephalitozoonidae and *Enterocytozoon bieneusi*	5′CACCAGGTTGATTCTGCCTGAC3′ 5′CCTCTCCGGAACCAAACCTG3′	PMP1 (V1) PMP2	60	Eb 250 Ec 268 Ei 270 Eh 279[e]	Fedorko et al., 1995
	5′TGAATG(G/T)GTCCCTGT3′ 5′TCACTCGCCGCTACT3′ 5′GTTCATCGCACTACT3′ 5′GGAATTCACACCGCCCGTC(A/G)(C/T) TAT3′ 5′CCAAGCTTATGCTTAAGT(C/T)(A/C) AA(A/G)GGGT3′ 5′CCAAGCTTATGCTTAAGTCCAGGGAG3′	MSP1 MSP2A MSP2B MSP3 MSP4A MSP4B	58	Eb 508 Ec 289 Ei 305	Katzwinkel-Wladarsch et al., 1996
	5′CCAGGUTGATUCTGCCUGACG3′ 5′TUACCGCGGCUGCUGGCAC3′ 5′AAGGAGCCTGAGAGATGGCT3′ 5′CAATTGCTTCACCCTAAGGTC3′ 5′GACCCCTTTGCACTCGCACAC3′ 5′TGCCCTCCAGTAAATCACAAC3′ 5′CCTCCAATCAATCTCGACTC3′	Mic3U Mic421U Mic266 Eb379 Ec378 Eh410 Ei395	65/62	Eb 132 Ec 113 Eh 134 Ei 128	Kock et al., 1997
	5′CACCAGGTTGATTCTGCC3′ 5′GTGACGGGCGGTGTCTAC3′	C1 (V1) C2	56	Eb 1170 Ec 1190 Eh 1205 Ei 1186[f]	Raynaud et al., 1998
Encephalitozoonidae	5′TGCAGTTAAAATGTCCGTAGT3′ 5′TTTCACTCGCCGCTACTCAG3′	int530f int580r	40	1000	Didier et al., 1996a
Encephalitozoon intestinalis	5′CACCAGGTTGATTCTGCCTGAC3′ 5′CTCGCTCCTTTACACTCGAA3′	V1 Si500	58	375	Weiss et al., 1994

	5′GGGGGTAGGAGTGTTTTTG3′ 5′CAGCAGGCTCCCTCGCCATC3′	3 3	65	930	Schuitema et al., 1993; David et al., 1996
	5′TTTCGAGTGTAAGGAGTCGA3′ 5′CCGTCCTCGTTCTCCTGCCCG3′	SINTF1 SINTR	55	520	DaSilva et al., 1997; Visvesvara et al., 1995
Encephalitozoon cuniculi	5′ATGAGAAGTGATGTGTGTGCG3′ 5′TGCCATGCACTCACAGGCATC3′	ECUNF ECUNR	55	549	Visvesvara et al., 1994; DeGroote et al., 1995
Encephalitozoon hellem	5′TGAGAAGTAAGATGTTTAGCA3′ 5′GTAAAAAGACTCTCACACTCA3′	EHELF EHELR	55	547	Visvesvara et al., 1994
Enterocytozoon bieneusi	5′GAAACTTGTCCACTCCTTACG3′ 5′CCATGCACCACTCCTGCCATT3′	EBIEF1 EBIER1	55	607	DaSilva et al., 1996
	5′CACCAGGTTGATTCTGCCTGAC3′ 5′ACTCAGGTGTTATACTCACGTC3′	V1 EB450	48	353	Zhu et al., 1993; Coyle et al., 1996
	5′CACCAGGTTGATTCTGCCTGAC3′ 5′CAGCATCCACCATAGACAC3′	V1 Mic3	54	446	Mansfield et al., 1997; Carville et al., 1997
	5′TCAGTTTTGGGTGTGGTATCGG3′ 5′GCTACCCATACACACATCATTC3′	Eb.gc Eb.gt	49	210	Velasquez et al., 1996
	5′GCCTGACGTAGATGCTAGTC3′ 5′ATGGTTCTCCAACTGAAACC3′	2 2	55	1265	David et al., 1996
Vittaforma corneae	5′TGAGACGTGAAGATGAGTATC3′ 5′TCCCTGCCCACTGTCTCCAAT3′	NCORF1 NCORR1	55	375	Pieniazek et al., 1998

[a]Adapted from Weiss and Vossbrinck (1998) with permission.
[b]Designation of primers in references.
[c]Annealing temperature in PCR.
[d]Size of amplified fragment in base pairs. Eb, *E. bieneusi*; Ec, *E. cuniculi*; Ei, *E. intestinalis*; Eh, *E. hellem*.
[e]*PstI* and *Hae*III restriction analysis differentiates these amplicons.
[f]*Hind*III and *Hin*fI restriction analysis differentiates these amplicons.

The isolation of microsporidian nucleic acids from spores or stool specimens is more difficult. Mechanical disruption of spores with 500-μm glass beads (Mini-Bead Beater, Biospec Products, Inc., Bartlesville, Okla.) in combination with proteinase K is often effective in the isolation of DNA (Zhu et al., 1993b; Weiss et al., 1994; Fedorko et al., 1995; Didier et al., 1996a; Raynaud et al., 1998). Alternatively, many types of microsporidian spores can be germinated by incubation in H_2O_2 (0.5 to 3%), and nucleic acids purified by routine extraction methods (Keohane et al., 1998). For stool specimens mechanical disruption and/or harsh extraction conditions have generally been required to obtain reproducible amplification. Successful methods for obtaining amplification of different pathogens from stool have included the use of 0.5% sodium hypochlorite (Fedorko et al., 1995), chitinase (Fedorko et al., 1995), lyticase (Raynaud et al., 1998), guanidine thiocyanate (Boom et al., 1990; Kock et al., 1997; Talal et al., 1998), 10% formalin or 1 M KOH (Fedorko et al., 1995, Katzwinkel-Wladarsch et al., 1996; Liguory et al., 1997), dithiothreitol (DTT) (Katzwinkel-Wladarsch et al., 1996) or hexadecyltrimethylammonium bromide (Carville et al., 1997) or boiling the samples (Ombrouck et al., 1996a, 1996b, 1997). Inhibitors of polymerase enzymes are often a problem in stool specimens and reportedly be removed by many of the methods described above. Furthermore, it has been shown that inhibition can be overcome by simple dilution of the samples (Ombrouck et al., 1997). Guanidine thiocyanate (GITT) (Boom et al., 1990) extraction, however, may be the most useful method in that it can be carried out in less than 1 h and removes most polymerase inhibitors present in stool specimens.

PCR Primers That Identify Several Microsporidian Species

Two main approaches have been employed in the construction of PCR primers for microsporidia: the design of universal panmicrosporidian primers and the design of species-specific primer pairs. The primer set V1(18f)::1492r amplifies the small-subunit rRNA gene sequence of many microsporidia, producing an amplicon of 1,200 to 1,500 bp (Weiss et al., 1994). A second primer set, 530f::580r, has been used to amplify part of the small-subunit rRNA gene, the ITS, and part of the large-subunit rRNA gene of many microsporidia (Baker et al., 1994, 1995; Vossbrinck et al., 1993; Weiss et al., 1994; Zhu et al., 1993c). These "universal" primer pairs are extremely useful in obtaining rRNA sequence data for recently identified microsporidia, thereby facilitating their classification by molecular means (Table 2). Unfortunately these primers are not useful for diagnosis, as the large size of the product (a 1.3-kb amplicon) limits their sensitivity. This problem is particularly evident in formalin-fixed tissue where reliable amplification is best obtained when primers are separated by 100 to 400 bp. In fresh tissue, however, primers separated by 700 to 1,000 bp can still provide adequate sensitivity.

A panmicrosporidian primer set, PMP1::PMP2 (Fedorko et al., 1995), has been designed that can amplify the Encephalitozoonidae (*E. cuniculi*, *E. hellem*, and *E. intestinalis*), *E. bieneusi* and *V. corneae* (Fedorko and Hijazi, 1996). PMP1 is similar to the universal primer V1(18f). This primer set yields a 250- to 279-bp product depending on the species of microsporidia involved. Unique restriction products of the amplicons are produced when they are digested by *Pst*I or *Hae*III, allowing species identification; however, these restriction fragments differ by 2 to 11 bp. The *E. bieneusi* amplicon does not have a *Pst*I site, but the amplicon from any of the Encephalitozoonidae is cut into two fragments. Nevertheless, because of the small difference in restriction fragment size, known microsporidian controls should be employed and acrylamide electrophoresis is useful. This technique is especially important in distinguishing between *E. intestinalis* and *E. hellem*. This primer set has been utilized successfully with formalin-fixed stool specimens treated with sodium hypochlorite, as well as with in vitro cultures. A second panmicrosporidian primer set, C1::C2 (Raynaud et al., 1998), which can also amplify the Encephalitozoonidae (*E. cuniculi*, *E. hellem*, and *E. intestinalis*), as well as *E. bieneusi*, has been identified. C1 is similar to the universal primer V1(18f). This primer set yields an approximately 1,200-bp amplicon from these microsporidia pathogenic to humans. Unique restriction products of the amplicons are produced by digestion with *Hin*dIII or *Hin*fI, facilitating species identification. The *E. bieneusi* amplicon is cut into two fragments (784 and 386 bp) by *Hin*dIII, whereas all the Encephalitozoonidae amplicons are not cut. The Encephalitozoonidae amplicons are distinguished by *Hin*fI digestion: *E. cuniculi* yields two fragments (830 and 350 bp), *E. hellem* yields three fragments (580, 350, and 260 bp), and *E. intestinalis* yields four fragments (460, 350, 250, and 120 bp). This method has been reported to work with stool specimens and has identified *E. intestinalis* in travelers and *E. bieneusi* in AIDS patients with diarrhea (Raynaud et al., 1998). The detection limit of *E. cuniculi* spiked stool specimens varied from 20 to 100 spores per 0.1 g sample with this method (Raynaud et al., 1998).

A nested PCR has been described for the detection of microsporidia pathogenic for humans from stool specimens (Katzwinkel-Wladarsch et al., 1996). This nested PCR produces an amplicon across the small-subunit rRNA gene, ITS region, and large-subunit rRNA gene region. This PCR can detect and discriminate *E. bieneusi*, *E. cuniculi*, and *E. intestinalis* with primers MSP1::MSP2A and MSP2B for the first PCR and MSP3::MSP4A and MSP4B for the second (i.e., nested). With these primers *E. bieneusi* yields a 508-bp amplicon, and the Encephalitozoonidae a 300-bp amplicon. *Mnl*I restriction digestion of the amplicon can be used to identify the microsporidian species present (*E. cuniculi*, 289 bp; *E. intestinalis*, 305 bp; no data provided for *E. hellem*). This nested PCR was able to detect 3 to 100 *E. cuniculi* spores in 0.1 g of stool (Katzwinkel-Wladarsch et al., 1996). Comparison of this PCR assay with light microscopy involving 34 specimens from 31 HIV-infected patients produced identical results in 82% (28 of 34) of specimens (Katzwinkel-Wladarsch et al., 1997). Four samples were microscopy negative and PCR positive, and two samples microscopy positive and PCR negative. Species identification had a 100% concordance in all cases. This PCR also confirmed the identification of *E. bieneusi* in the feces of pigs (Deplazes et al., 1996; Katzwinkel-Wladarsch et al., 1996). The initial primer pair, MSP1::MSP2A/MSP2B, also amplifies DNA from *V. necatrix*, *V. lymantriae*, *Ameson michaelis*, and *I. giganteum*.

Another nested PCR has also been described that identifies the same pathogenic microsporidia (Kock et al., 1997). In this procedure an initial PCR (annealing temperature 65°C) with primers Mic3U::Mic421U yields amplicons of 410 to 433 bp depending on the species of microsporidia. This step is followed by four species-specific PCRs (annealing temperature 62°C) with Mic266::Eb379 (132-bp amplicon), Mic266::Ec378 (113-bp amplicon), Mic266::Eh410 (134-bp amplicon), or Mic266::Ei395 (128-bp amplicon), which allows identification of the species of microsporidia in the sample. This PCR was tested with culture-derived spores of *E. hellem* and *E. cuniculi*, as well as with stools and tissue biopsies from patients with *E. bieneusi* and *E. intestinalis*. The nested PCR was able to detect all these samples and was reported to detect as few as 200 spores per g in spiked fecal samples (20 *E. cuniculi* spores per 0.1 g).

A pan-Encephalitozoonidae primer set, int530f::int580r, that amplifies a 1,000-bp amplicon only from microsporidia belonging to the family Encephalitozoonidae has been described (Schuitema et al., 1993; Didier et al., 1996a, 1996b). No amplification is seen with DNA from *E. bieneusi* or *V. corneae*. Species-specific diagnosis of the amplicon can be obtained by using either a set of species-specific oligonucleotides for Southern blotting (*E. cuniculi*, 5′TAGCGGCTGACGAAGCTGC3′; *E. hellem*, 5′TGAGTGTGAGAGTGTTTT-TACAT3′; *E. intestinalis*, 5′CGGGCAAGGA-GAACGAGGACGG3′), *Fok*I digestion to create species-specific restriction patterns, or heteroduplex mobility shift analysis of the amplicon (Didier et al., 1996a). This primer set has been used to successfully amplify microsporidia from urine, conjunctiva, and tissue culture. With this primer set and a second set of specfic primers developed for the diagnosis of *E. bieneusi* and *E. intestinalis* (David et al., 1996; Liguory et al., 1997), about 10 copies of recombinant plasmids containing rRNA genes from *E. bieneusi* and *E. intestinalis* were detected. These primers allowed 26 out of 28 intestinal biopsy specimens with transmission electron microscopy (TEM)-confirmed infection with either *E. bieneusi* or *E. intestinalis* to be identified with no false positives in 84 TEM-negative biopsies (David et al., 1996).

Species-Specific PCR Primers

PRIMERS FOR THE ENCEPHALITOZOONIDAE

Primer sets specific to *E. hellem* (Visvesvara, 1994), *E. cuniculi* (De Groote et al., 1995), and *E. intestinalis* (Weiss et al., 1994; Schuitema et al., 1993; David et al., 1996) have been described. The *E. hellem* primer pair ECUNF::ECUNR has been used to amplify a 547-bp product from *E. hellem* (CDC:0291:V213) spores obtained from tissue culture. The *E. cuniculi* primer pair EHELF::EHELR was used to amplify a 549-bp product from *E. cuniculi* spores obtained from tissue culture (Visvesvara et al., 1994, 1995; De Groote et al., 1995). Both these primer pairs are species-specific. A set of primers originally designed for the identification of *Echinococcus multilocularis* small-subunit rRNA genes were also found to amplify *E. cuniculi* rRNA, and it has been demonstrated that the small-subunit rRNA sequence of *E. cuniculi* contains regions regarded as sequence-specific for the genus *Echinococcus* and the species *Echinococcus multilocularis* (Furuya et al., 1995; Nagano et al., 1996).

The *E. intestinalis* primer set V1::SI500 has been used to amplify a 375-bp amplicon from intestinal biopsies, duodenal fluid, and formalin-fixed stool (Fig. 11) from patients with *E. intestinalis* (Weiss et al., 1994; Franzen et al., 1995, 1996a; Coyle et al., 1996; Ombrouck et al., 1996a, 1996b; and Weiss, unpublished data). In a case of disseminated *E. intestinalis* infection this PCR also detected *E. intestinalis* in a blood specimen (Franzen et al., 1996b). This primer set did not amplify DNA from human duodenal tissue, *E. cuniculi*, *E. hellem*, *Glugea stephani*, *N. locustae*, *N. bombycis*, *Pleistophora*, *Escherichia coli*, and *Saccharomyces cerevisiae*. An internal 30-bp oligomer (Ombrouck et al., 1997) and a 18-bp oligomer SI60 (5′TGTTGATGAACCTTGTGG3′) have been used to confirm the amplicon by Southern hybridization (Weiss et al., 1994; Franzen et al., 1995, 1996a). In a study of 46 human immunodeficiency virus (HIV)-infected patients with diarrhea, PCR with V1::SI500 identified 10 patients infected with *E. intestinalis* (Franzen et al., 1996a). Five of these 10 patients also were infected by *E. bieneusi* as demonstrated by PCR (Franzen et al., 1996a). Very few data are available on the sensitivity and specificity of these primers. In a study with the *E. intestinalis* primer set V1::SI500, amplification occurred for all five TEM-confirmed cases (Coyle et al., 1996), and it has been reported that this primer set can detect as few as 100 spores per g (10 spores per 0.1 g) of stool (Ombrouck et al., 1996b, 1997). The *E. intestinalis* primer set 3, which amplifies a 930-bp region, has also been used in tissue biopsies and with duodenal fluid and stool specimens (Schuitema et al., 1993; David et al., 1996; Liguory et al., 1997). This primer pair detected *E. intestinalis* in three biopsy specimens from TEM-confirmed cases (David et al., 1996), and three stool specimens with *E. intestinalis* (Liguory et al., 1997) and was negative in 25 cases of TEM-proven *E. bieneusi* biopsies (David et al., 1996). Primer set SINTF1::SINTFR also detected *E. intestinalis* from tissue culture samples and duodenal fluid from one patient (DaSilva et al., 1997). Another *E. intestinalis* primer pair, V1::Sep1 (5′CCTGCCGCTTCAGAACC3′), was used to exclude *E. intestinalis* infection in simian immunodeficiency virus (SIV)-infected rhesus monkeys with experimental *E. bieneusi* infection but has not been reported for clinical specimens (Tzipori et al., 1997).

ENTEROCYTOZOON BIENEUSI PRIMERS

Several specific primer sets have been developed for the diagnosis of *E. bieneusi* infections. Primer set V1::EB450 has been used to amplify a 353-bp amplicon from *E. bieneusi*-infected tissue, duodenal fluid, and stool (Zhu et al., 1993c; Franzen et al., 1995, 1996a, 1996b; Ombrouck et al., 1996a, 1997; Coyle et al., 1996). This primer is specific for *E. bieneusi* and does not amplify *E. hellem*, *E. cuniculi*, *V. necatrix*, *N. locustae*, *N. bombycis*, *Pleistophora*, *Toxoplasma gondii*, *E. coli*, *S. cerevisiae*, or *Trypanosoma cruzi* (Coyle et al., 1996). The original report demonstrating a weak reaction with *E. hellem* (Zhu et al., 1993c) appeared to be due to contamination of *E. hellem* cultures with a plasmid containing *E. bieneusi* rRNA (Coyle et al., 1996). The small-subunit rRNA sequence of *E. hellem* does not suggest that V1::EB450 would result in an amplicon. An internal 30-bp oligomer, EB150 (5′TGTTGCGGTAATTTGGTCTCTGTGTGTAAA3′), complementary to a region of the V1::EB450 amplicon, has been used for Southern blot analysis (Zhu et al., 1993; Weiss et al., 1994, Franzen et al., 1995, 1996c; Coyle et al., 1996; Ombrouck et al., 1997). This primer set reliably amplifies *E. bieneusi* DNA from gastrointestinal biopsy specimens with infections confirmed by light and electron microscopy (TEM) (Franzen et al., 1995, 1996; Coyle et al., 1996). In one study involving TEM-confirmed cases of *E. bieneusi*, all 25 cases were identified by this primer pair (Coyle et al., 1996). The sensitivity of PCR is probably higher than that of TEM, as patients initially identified only by PCR (i.e., negative by TEM) were subsequently demonstrated on follow-up pathology to have this parasite (thus, the positive PCR was not a false positive reaction) (Coyle et al., 1996). Primer set V1::EB450 has also been used to identify *E. bieneusi* in stool specimens (Ombrouck et al., 1997; Talal et al., 1998). In addition, this primer set amplified *E. bieneusi* from stool specimens in experimental *E. bieneusi* infection in SIV-infected rhesus monkeys (Tzipori et al., 1997). There is about a 2% difference between the sequence for *E. bieneusi* on which these primers were based and other *E. bieneusi* rRNA sequences in GenBank. This difference may represent true variation or a sequencing error. On the basis of the sequence difference a modified EB450 primer with a C instead of a G in position 9 has been suggested (DaSilva et al., 1996). This explanation could account for the failure of primer set V1::EB450 in one study to amplify *E. bieneusi* from short-term in vitro cultures and one bile sample (DaSilva et al., 1996). However, one study reported that the suggested modified EB450 primer (with C at position 9) had decreased sensitivity compaired to the original primer set V1::EB450 in the examination of stool specimens (Talal et al., 1998).

The primer set EB1EF1::EB1ER1 amplified a 607-bp fragment from duodenal fluid and bile fluid from a patient with *E. bieneusi* and from culture supernatant from a short-term *E. bieneusi* in vitro culture (DaSilva et al., 1996). Specificity was confirmed by testing for amplification with 13 species of microsporidia as well as human DNA, and amplification occurred only with *E. bieneusi*. Primer set EB1EF1::EB1ER1 may have a higher sensitivity than V1::EB450 (DaSilva, 1996).

An *E. bieneusi*-like rDNA sequence was identified in the stool of SIV-infected macaques by EB1EF1::EB1ER1. The primer set V1::Mic3 amplifies a 446-bp product from *E. bieneusi* (Mansfield et al., 1997), and this amplicon has been used to demonstrate *E. bieneusi* by in situ hybridization. Primer set V1::Mic3 amplified *E. bieneusi* from the stool of SIV-infected macaques with an experimental infection (Tzipori et al., 1997) and in formalin-fixed stool of SIV-infected macaques with a spontaneous infection (Mansfield et al., 1997). In a study of formalin-fixed stool specimens from 74 HIV-positive Zimbabwean patients, 45 of whom had diarrhea, and 6 biopsy specimens from biopsy TEM-confirmed *E. bieneusi*-infected patients, primer set V1::Mic3 was positive in all 6 TEM-positive patients and in 34 of the 74 other patients (Carville et al., 1997). This PCR method detected twice as many cases (40 out of 80) as the modified trichrome stain (Weber et al., 1992) for microsporidia (19 of the 80 patients tested positive) (Carville et al., 1997).

Primer set 2 amplifies a 1,265-bp DNA from *E. bieneusi* and has been successful in the identification of *E. bieneusi* in clinical specimens (tissue and stool) (Liguory et al., 1997; David et al., 1996). Examination of 31 stool specimens from 26 patients with intestinal *E. bieneusi* infection, 6 stool specimens from 3 patients with *E. intestinalis*, and 61 stool specimens from 45 patients without intestinal microsporidiosis with primer set 2 demonstrated a sensitivity and specificity of 100% (Liguory et al., 1997). For gastrointestinal biopsy specimens this primer set confirmed that 23 out of 23 samples were positive for *E. bieneusi*, and no amplification occurred in two biopsy specimens from *E. intestinalis* (David et al., 1996).

The ITS was utilized in the design of primer set Eb.gc::Eb.gt, which amplifies a 210-bp fragment from *E. bieneusi* (Velasquez et al., 1996). This primer set has identified *E. bieneusi* in two stool specimens and did not amplify four *E. intestinalis* stool specimens or DNA from *Isosopora belli*, *Giardia lamblia*, or *Cryptosporidia*. Primer set Eb.gc::Eb.gt was also successful in the identification of *E. bieneusi* in gastrointestinal biopsy specimens and duodenal aspirates.

The use of PCR in the diagnosis of microsporidiosis shows great promise. It is likely that it will be adapted for large-scale epidemiological studies on these pathogens once better techniques are developed to eliminate PCR inhibitors and decrease the amount of manipulation currently required to process stool specimens. These techniques have already been applied in investigations on animal reservoirs of *E. bieneusi*, and both SIV-infected macaques (Mansfield et al., 1997) and pigs (Deplazes et al., 1996) have been identified as being infected with *E. bieneusi*. Primer sets have also been described for the other Enterocytozoonidae, *N. (Enterocytozoon) salmonis*, which is an important pathogen in fish aquaculture (Barlough et al., 1995; Docker et al., 1997). Sequencing of the rRNA genes of the *E. bieneusi*-like organisms identified in pig and macaque studies will confirm if these are identical to the human pathogen or a closely related organism. The identification of strain differences (Rinder et al., 1997) in the ITS region of *E. bieneusi*, as demonstrated with *E. cuniculi* (Didier et al., 1996c), may allow further epidemiological characterization of such isolates.

VITTAFORMA CORNEAE (SYN. *NOSEMA CORNEUM*) PRIMERS

Primers based on the rDNA sequence of *V. corneae* have also been described (Pieniazek et al., submitted). This primer set, NCORF1::NCORR1, amplifies a 375-bp product from *V. corneae* and does not amplify any of the other microsporidia seen in humans. In addition, no amplification was seen with DNA from 20 other unrelated microsporidia.

BRACHIOLA VESICULARUM PRIMERS

B. vesicularum was identified as etiological in a case of myositis in an HIV-1-infected individual (Cali et al., 1998). Because of the similarities between this organism and *N. algerae*, a set of primers, NAGf (5′GCCGTTTGGCAAGTTGG3′) and NAG178r (5′ATATCGACGGGACTCTCACC3′), were designed that amplifed a 192-bp fragment from DNA prepared from formalin-fixed *N. algerae* spores. These primers, however, did not amplify *B. vesicularum* DNA prepared from a formalin-fixed muscle biopsy. Currently, no molecular diagnostic techniques exist for this organism.

In Situ Hybridization

In situ hybridization has been used only occasionally for the diagnosis of microsporidian infections. A 443-bp amplicon withV1 and Mic3 from *E. bieneusi* was used to identify microsporidia in the tissues of SIV-infected rhesus monkeys after experimental infection with human *E. bieneusi* (Tzipori et al., 1997) and in spontaneously occurring infection with an *E. bieneusi*-like organism (Mansfield et al., 1997). Examination of histologic sections by in situ hybridization revealed characteristic supranuclear staining in villus tip epithelium (Fig. 12) which was confirmed by electron microscopy (Tzipori et al., 1997; Mansfield et al., 1997). Cleftlike spaces and intracytoplasmic spores could be visualized by this technique, as well as organisms in the billiary tree (Mansfield et al., 1997). The V1::Mic3 probe was also able to detect infection by in situ hybridization in duodenal biopsies from six TEM-proven cases of *E. bieneusi* infection (Carville et al., 1997).

SUMMARY

Knowledge of the phylum Microspora and its relationship to humans is in a period of rapid growth. In the last few years noninvasive diagnostic techniques for investigating microsporidia have been developed and are being applied to both clinical and epidemiological studies. As techniques have improved, new microsporidian pathogens and reservoir hosts of the microsporidia infecting humans are being identified. Studies on the molecular phylogeny of the microsporidia have demonstrated that they display a large evolutionary distance from most eukaryotes, and as more molecular phylogenies are developed, their place in the "tree of life" may be clarified. Work has recently begun on the molecular biology of these organisms and is beginning to suggest that their genomic organization may be different from that of other eukaryotes.

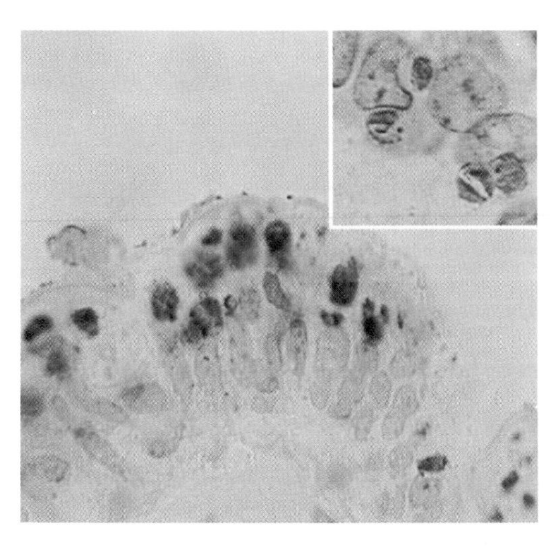

FIGURE 12 In situ hybridization for *E. bieneusi*, demonstrating the detection of *E. bieneusi* in an intestinal biopsy specimen using an rRNA hybridization probe. (Reprinted with permission from Carville et al., 1997.)

ACKNOWLEDGMENTS

This work was supported by NIH grant AI31788.

We thank Saul Tzipori for pictures and T. Hashimoto and T. Edlind for figures and conversations on phylogeny.

REFERENCES

Andreadis, T. G., 1983. Life cycle and epizootiology of *Amblyospora* sp. (Microspora, Amblyosporidae) in the mosquito *Aedes cantator*. *J. Protozool.* **30:** 509–518.

Ausubel, F. M., R. Brent, R. E. Kingston, D. D. Moore, J. G. Seidman, J. A. Smith, and K. Struhl (ed.) 1997. *Current Protocols in Molecular Biology.* John Wiley and Sons, New York, N.Y.

Baker, M. D., et al. 1997. Unpublished data.

Baker, M. D., C. R. Vossbrinck, J. J. Becnel, and T. G. Andreadis. 1998. Phylogeny of *Amplyospora* (Microsporidia: Amblyosporidae) and related genera based on small subunit ribosomal DNA data: a possible example of host parasite co-speciation. *J. Eukaryot. Microbiol.* **71:**199–206.

Baker, M. D., C. R. Vossbrinck, E. S. Didier, J. V. Maddox, and J. A. Shadduck. 1995. Small subunit ribosomal DNA phylogeny of various microsporidia with emphasis on AIDS related forms. *J. Eukaryot. Microbiol.* **42:**564–570.

Baker, M. D., C. R. Vossbrinck, J. V. Maddox, and A. H. Undeen. 1994. Phylogenetic relationships among *Vairimorpha* and *Nosema* species (Microspora) based on ribosomal RNA sequence data. *J. Invertebr. Pathol.* **64:**100–106.

Barlough, J. E., T. S. McDowell, A. Milani, L. Bigornia, S. B. Slemenda, N. J. Pieniazek, and R. P. Hedrick. 1995. Nested polymerase chain reaction for detection of *Enterocytozoon salmonis* genomic DNA in chinook salmon *Oncorynchus tshawytscha*. *Dis. Aquat. Org.* **23:**17–23.

Beauvais, B., C. Sarfati, S. Challier, and F. Derouin. 1994. In vitro model to assess the effect of antimicrobial agents on *Encephalitozoon cuniculi*. *Antimicrob. Agents Chemother.* **38:**2440–2448.

Becnel, J. J., and T. G. Andreadis. 1998. *Amblyospora salinaria* n. sp. (Microsporidia: Amblyosporidae): parasite of *Culex salinarius* (Diptera: Culicidae), its life cycle stages and an intermediate host and establishment as a new species. *J. Inverteb. Pathol.* **71:**258–262.

Biderre, C., F. Duffieux, E. Peyretaillade, P. Glaser, P. Peyret, A. Danchin, M. Pages, G. Metenier, and C. P. Vivares. 1997. Mapping of repetitive and non-repetitive DNA probes to chromosomes of the microsporidian *Encephalitozoon cuniculi*. *Gene* **191**:39–45.

Biderre, C., G. Metenier, and C. P. Vivares. 1998. A small spliceosome-type intron occurs in a ribosomal protein gene of the microsporidian *Encephalitozoon cuniculi*. *Mol. Biochem. Parasitol.* **94:** 283–286.

Biderre, C., M. Pages, G. Metenier, E. U. Canning, and C. P. Vivares. 1995. Evidence for the smallest nuclear genome (2.9 Mb) in the microsporidium *Encephalitozoon cuniculi*. *Mol. Biochem. Parasitol.* **74**:229–231.

Biderre, C., M. Pages, G. Metenier, D. David, J. Bata, G. Prensier, and C. P. Vivares. 1994. On small genomes in eukaryotic organisms: molecular karyotypes of two microsporidian species (Protozoa) parasites of vertebrates. *C. R. Acad. Sci. III* **317**:399–404.

Boom, R., C. J. A. Sol, M. M. M. Salimans, C. L. Jansen , P. M. E. Wertheim-vanDillen, and J. Van Der Noordaa. 1990. Rapid and simple method for purification of nucleic acids. *J. Clin. Microbiol.* **28**:495–503.

Brooks, W. M., J. J. Becnel, and G. G. Kennedy. 1988. Establishment of *Endoreticulatus* n.g. for *Pleistophora fidelis* (Hostounsky & Weiser, 1975) (Microsporidia: Pleistophoridae) based on the ultrastructure of a microsporidium in the Colorado potato beetle, *Leptinotarsa decemlineata* (Say) (Coleopera: Chrysomelidae). *J. Protozool.* **35**:481–488.

Brown, J. R., and W. F. Doolittle. 1996. Root of the universal tree of life based on ancient aminoacyl-tRNA synthetase gene duplications. *Proc. Natl. Acad. Sci. USA* **92**:2441–2445.

Brown, J. R., and W. F. Doolittle. Gene descent, duplication, and horizontal transfer in the evolution of glutamyl-tRNA and glutamyl-tRNA synthetases. *J. Mol. Evol.*, in press.

Bui, E. T., P. J. Bradley, and P. J. Johnson. 1996. A common evolutionary origin for mitochondria and hydrogenosomes. *Proc. Natl. Acad. Sci. USA* **93**:9651–9656.

Cali, A., D. P. Kotler, and J. M. Orenstein. 1993. *Septata intestinalis* n. g., n. sp., an intestinal microsporidian associated with chronic diarrhea and dissemination in AIDS patients. *J. Eukaryot. Microbiol.* **40**:101–112.

Cali, A., and R. L. Owen. 1988. Microsporidiosis, p. 929–947. *In* A. Ballows, W. J. Hausler, Jr., M. Ohashi, and H. Turano (ed.), *Laboratory Diagnosis of Infectious Diseases: Principles and Practice*, vol 1. Springer-Verlag, New York.

Cali, A., P. M. Takvorian, S. Lewin, M. Rendel, C. Sian, M. Wittner, H. B. Tanowitz, E. Keohane, and L. M. Weiss. 1998. *Brachiola vesicularum*, a new AIDS microsporidian. *J. Eukaryot. Microbiol.* **45**:240–251.

Canning, E. U. 1988. Nuclear division and chromosome cycle in microsporidia. *BioSystems* **21**:333–340.

Canning, E. U., and E. I. Hazard. 1982. Genus *Pleistophora* Gurley, 1893: an assemblage of at least three genera. *J. Protozool.* **29**:39–49.

Canning, E. U., and W. S. Hollister. 1992. Human infections with microsporidia. *Rev. Med. Microbiol.* **3**:35–42.

Canning, E. U., and J. Lom. 1986. *The Microsporidia of Vertebrates.* Academic Press, New York.

Carville, A., K. Mansfield, G. Widner, A. Lackner, D. Kotler, P. Wiest, T. Gumbo, S. Sarbah, and S. Tzipori. 1997. Development and application of genetic probes for the detection of *Enterocytozoon bieneusi* in formalin-fixed stools and in intestinal biopsies of infected patients. *Clin. Diag. Lab. Immunol.* **4**:405–408.

Cavalier-Smith, T. 1987. Eukaryotes with no mitochondria. *Nature* **326**:332–333.

Cheney, S. A. 1998. Unpublished data.

Chilmonczyk, S., W. T. Cox, and R. P. Hedrick. 1991. *Enterocytozoon salmonis* n.sp.: an intranuclear microsporidium from salmonid fish. *J. Protozool.* **38**:264–269.

Clark, C. G., and A. J. Roger. 1995. Direct evidence for secondary loss of mitochondria in *Entamoeba histolytica*. *Proc. Natl. Acad. Sci. USA* **92**:6518–6521.

Coyle, C. M., M. Kent, H. B. Tanowitz, M. Wittner, and L. M. Weiss. 1998. TNP-470 is an effective antimicrosporidial agent. *J. Infect. Dis.* **177**:515–518.

Coyle, C. M., M. Wittner, D. Kotler, C. Noyer, J. M. Orenstein, H. B. Tanowitz, and L. M. Weiss. 1996. Prevalence of microsporidiosis due to *Enterocytozoon bieneusi* and *Encephalitozoon (Septata) intestinalis* among patients with AIDS-related diarrhea: determinatin by polymerase chain reaction to the microsporidian small subunit-rRNA gene. *Clin. Infect. Dis.* **23**:1002–1006.

Curgy, J. J., J. Vavra, and C. Vivares. 1990. Presence of ribosomal RNAs with prokaryotic properties in microsporidia, eukaryotic organisms. *Biol. Cell.* **38**:49–52.

Da Silva, A. J. et al. 1997. Unpublished data.

Da Silva, A. J., D. A. Schwartz, G. S. Visvesvara, H. deMoura, S. B. Slemenda, and N. J. Pieniazek. 1996. Sensitive PCR diagnosis of infections by *Enterocytozoon bieneusi* (microsporidia) using primers based on the region coding for small-subunit rRNA. *J. Clin. Microbiol.* **34**:986–987.

Da Silva, A. J., S. B. Slemenda, G. S. Visvesvara, D. A. Schwartz, C. Mel-Wilcox, S. Wallace, and N. J. Pieniazek. 1997. Detection of *Septata intestinalis* (Microsporidia) Cali et al. 1993 using polymerase chain reaction primers targeting the small subunit ribosomal RNA coding region. *Mol. Diagn.* **2**:47–52.

David, F., A. R. G. Schuitema, C. Sarfati, O. Liguory, R. A. Hartskeerl, F. Derouin, and J. Molina. 1996. Detection and species identification of intestinal microsporidia by polymerase chain reaction in duodenal biopsies from human immunodeficiency virus-infected patients. *J. Infect. Dis.* **174:**874–877.

Dayhoff, M. O., R.V. Eck, and C. M. Park. 1972. A model of evolutionary change in proteins, p. 89–99. *In* M. O. Dayhoff (ed.), *Atlas of Protein Sequence and Structure*, vol. 5. National Biomedical Research Foundation, Washington D.C.

De Groote, M. A., G. Visvesvara, M. L. Wilson, N. J. Pieniazek, S. B. Slemenda, A. J. daSilva, G. L. Leitch, R. T. Bryan, and R. Reves. 1995. Polymerase chain reaction and culture confirmation of disseminated *Encephalitozoon cuniculi* in a patient with AIDS: successful therapy with albendazole. *J. Infect. Dis.* **171:**1375–1378.

Delbac, F., P. Peyret, G. Metenier, D. Danielle, A. Danchin, and C. P. Vivares. 1998. On protein of the microsporidian invasion apparatus: complete sequence of a polar tube protein in *Encephalitozoon cuniculi*. *Mol. Microbiol.* **29:**825–834.

Deplazes, P., A. Mathis, C. Muller, and R. Weber. 1996. Molecular epidemiology of *Encephalitozoon cuniculi* and first detection of *Enterocytozoon bieneusi* in fecal samples of pigs. *J. Eukaryot. Microbiol.* **43:**93S.

Desportes, I. 1976. Ultrastructure de *Stempellia mutabilis* Leger et Hess, microsporidie parasite de l'ephemere *Ephemera vulgata* L. *Protistologica* **12:** 121–150.

Desportes, I., Y. Le Charpentier, A. Calian, F. Bernard, B. Cochand-Priollet, A. Lavergne, P. Ravisse, and R. Modigliani. 1985. Occurrence of a new microsporidian: *Enterocytozoon bieneusi* n.g., n. sp., in the enterocytes of a human patient with AIDS. *J. Protozool.* **26:**179–187.

De Stasio, E. A., and A. E. Dahlberg. 1990. Effects of mutagenesis of a conserved base pair site near the decoding region of *Escherichia coli* 16S ribosomal RNA. *J. Mol. Biol.* **212:**127–133.

Didier, E. S. 1997. Effects of albendazole, fumagillin and TNP-470 on microsporidial replication in vitro. *Antimicrob. Agents Chemother.* **41:**1541–1546.

Didier, E. S., P. J. Didier, D. N. Friedberg, S. M. Stenson, J. M. Orenstein, R. W. Yee, F. W. Tio, R. M. Davis, C. Vossbrinck, N. Millichamp, and J. A. Shadduck. 1991. Isolation and characterization of a new human microsporidian, *Encephalitozoon hellem* (n.sp.) from three AIDS patients with keratoconjuctivitis. *J. Infect. Dis.* **163:** 617–621.

Didier, E. S., L. B. Rogers, A. D. Brush, S. Wong, V. Traina-Dorge, and D. Bertucci. 1996a. Diagnosis of disseminated microsporidian *Encephaltizoon hellem* infection by PCR-Southern analysis and successful treatment with albendazole and fumagillin. *J. Clin. Microbiol.* **34:**947–952.

Didier, E. S., L. B. Rogers, J. M. Orenstein, M. D. Baker, C. R. Vossbrinck, T. Van Gool, R. Hartskeerl, R. Soave, and L. M. Beaudet. 1996b. Characterization of *Encephalitozoon (Septata) intestinalis* isolates cultured from nasal mucosa and bronchoalveolar lavage fluids of two AIDS patients. *J. Eukaryot. Microbiol.* **43:**34–43.

Didier, E. S., G. S. Visvesvara, M. D. Baker, L. B. Rogers, D. C. Bertucci, M. A. DeGroote, and C. R. Vossbrinck. 1996c. A microsporidian isolated from an AIDS patient corresponds to *Encephalitozoon cuniculi* III, originally isolated from domestic dogs. *J. Clin. Microbiol.* **34:**2835–2837.

Didier, E. S., C. R. Vossbrinck, M. D. Baker, L. B. Rogers, D. C. Bertucci, and J. A. Shadduck. 1995. Identification and characterization of three *Encephalitozoon cuniculi* strains. *Parasitology* **111:** 411–421.

DiMaria, P., B. Palic, B. A. Debrunner-Vossbrinck, J. Lapp , and C. R. Vossbrinck. 1996. Characterization of the highly divergent U2 RNA homolog in the microsporidian *Vairimorpha necatrix*. *Nucleic Acids Res.* **24:**515–522

Docker, M. F., R. H. Devlin, J. Richard, and M. L. Kent. 1997. Sensitive and specific polymerase chain reaction assay for detection of *Loma salmonae* (Microsporea). *Dis. Aquat. Org.* **29:**41–48.

Docker, M. F., M. L. Kent, D. L. Hervio, J. S. Khattra, L. M. Weiss, A. Cali, and R. H. Devlin. 1997. Ribosomal DNA sequence of *Nucleospora salmonis* Hedrick, Groff and Baxa, 1991 (Microsporea: Enterocytozoonidae): implications for phylogeny and nomenclature. *J. Eukaryot. Microbiol.* **44:**55–60.

Doflein, F. 1901. Die Protozoen als Parasiten und Krankheitserreger nach biologischen Gesichtspunkten dargestellt. Verlag von Gustav Fischer.

Dowd, S. E., C. P. Gerba, and I. L. Pepper. 1998. Confirmation of the human-pathogenic microsporidia *Enterocytozoon bieneusi, Encephalitozoon intestinalis,* and *Vittaforma corneae* in water. *Appl. Environ. Microbiol.* **64:**3332–3335.

Duffieux, F. 1998. Unpublished data.

Duffieux, F., P. Peyret, B. A. Roe, and C. P. Vivares. 1998. First report of the systematic sequencing of the small genome of *Encephalitozoon cuniculi* (Microspora, Protozoa): gene organization of a 4.3 kbp region on chromosome I. *Microb. Comp. Genomics* **3:**1–11.

Edlind, T. 1997. Unpublished data.

Edlind, T., G. Visvesvara, J. Li, and S. Katiyar. 1994. Cryptosporidium and microsporidial betatubulin sequences: predictions of benzimidazole sensitivity and phylogeny. *J. Eukaryot. Microbiol.* **41:**38S.

Edlind, T. D., J. Li, G. S. Visvesvara, M. H. Vodkin, G. L. McLaughlin, and S. K. Katiyar. 1996. Phylogenetic analysis of beta-tubulin sequences from amitochondrial protozoa. *Mol. Phylogenet. Evol.* **5:**359–367.

Embley, M. Personal communication.

Embley, T. M., L. J. Horner, and R. P. Hirt. 1997. Anaerobic eukaryote evolution: hydrogenosomes as biochemically modified mitochondria? TREE **12:**437–441.

Fast, N. M., A. J. Roger, C. A. Richardson, and W. F. Doolittle. 1998. U2 and U6 snRNA genes in the microsporidian *Nosema locustae*: evidence for a functional spliceosome. *Nucleic Acids Res.* **26:** 3202–3207.

Fedorko, D. P., and Y. M. Hijazi. 1996. Application of molecular techniques to the diagnosis of microsporidial infection. *Emerg. Infect. Dis.* **2:**183–191.

Fedorko, D. P., N. A. Nelson, and C. P. Cartwright. 1995. Identification of microsporidia in stool specimens by using PCR and restriction endonucleases. *J. Clin. Microbiol.* **33:** 1739–1741.

Felsenstein, J. 1978. Cases in which parsimony or compatibility methods will be positively misleading. *Syst. Zool* **27:**401–410.

Felsenstein, J. 1988. Phylogenies from molecular sequences: inference and reliability. *Annu. Rev. Genet.* **22:**521–65.

Field, A. D., D. J. Marriott, S. T. Milliken, B. J. Brew, E. U. Canning, J. G. Kench, P. Darveniza, and J. L. Harkness. 1996. Myositis associated with a newly described microsporidian, *Trachipleistophora hominis*, in a patient with AIDS. *J. Clin. Microbiol.* **34:**2803–2811.

Flegel, T. W., and T. Pasharawipas. 1995. A proposal for typical eukaryotic meiosis in microsporidians. *Can. J. Microbiol.* **41:**1–11.

Franzen, C., A. Muller, P. Hegener, B. Salzberger, P. Hartmann, G. Fatkenheuer, V. Diehl, and M. Schrappe. 1995. Detection of microsporidia (*Enterocytozoon bieneusi*) in intestinal biopsy specimens from human immunodeficiency virus-infected patients by PCR. *J. Clin. Microbiol.* **33:** 2294–2296.

Franzen, C., R. Kuppers, A. Muller, B. Salzberger, G. Fatkenheuer, B. Vettern, V. Diehl, and M. Schrappe. 1996a. Genetic evidence for latent *Septata intestinalis* infection in human immunodeficiency virus-infected patients with intestinal microsporidiosis. *J. Infect. Dis.* **173:** 1038–1040.

Franzen, C., A. Muller, M. Hartmann, V. Kochanek, V. Diehl, and G. Fatkenheuer. 1996c. Disseminated *Encephalitozoon (Septata) intestinalis* infection in a patient with AIDS. *N. Engl. J. Med.* **335:**110–1611.

Franzen, C., A. Muller, P. Hegener, P. Hartmann, B. Salzberger, B. Franzen, V. Diehl, and G. Fatkenheuer. 1996b. Polymerase chain reaction for microsporidian DNA in gastrointestinal biopsy specimens of HIV-infected patients. *AIDS* **10:** F23–F27.

Fries, I. M., F. Feng, A. J. daSilva, S. B. Slemenda, and N. J. Pieniazek. 1996. *Nosema ceranae* n. sp. (Microsporidia, Nosematidae), morphological and molecular characterization of a microsporidian parasite of the Asian honey bee *Apis cerana* (Hymenoptera, Apidae) *Eur. J. Protistol* **32:**356–365.

Furuya, K., H. Nagano, and C. Satoh. 1995. Primers designed for the amplification of *Echinococcus multilocularis* DNA amplify the DNA of *Encephalitozoon*-like spores in the polymerase chain reaction. *J. Eukaryot. Microbiol.* **42:**526–528.

Gatehouse, H. S., and L. A. Malone. 1998. The ribosomal RNA gene of *Nosema apis* (Microspora): DNA sequence for small and large subunit rRNA genes and evidence of a large tandom repeat size. *J. Invertebr. Pathol.* **71:**97–105.

Germot, A., H. Philippe, and H. LeGuyader. 1996. Presence of a mitochondrial-type 70-kDa heat shock protein in *Trichomonas vaginalis* suggests a very early mitochondrial endosymbiosis in eukaryotes. *Proc. Natl. Acad. Sci. USA* **93:**14614–14617.

Germot, A., H. Philippe, and H. LeGuyader. 1997. Evidence for loss of mitochondria from a mitochondrial-type HSP 70 in *Nosema locustae. Mol. Biochem. Parasitol.* **87:**159–168.

Gupta, R. S., and B. Singh. 1994. Phylogenetic analysis of 70 kD heat shock protein sequences suggests a chimeric origin for the eukaryotic cell nucleus. *Curr. Biol.* **4:**1104–1114.

Gutell, R. R., M. W. Gray, and M. N. Schnare. 1993. A compilation of large subunit (23 S and 23 S-like) ribosomal RNA structures. *Nucleic Acids Res.* **21:**3055–3074.

Haig, D. 1993. Alternatives to meiosis: the unusual genetics of red algae, microsporidia, and others. *J. Theor. Biol.* **163:**15–31.

Hartskeerl, R. A. Unpublished data.

Hartskeerl, R. A., A. R. Schuitema, and R. deWachter. 1993a. Secondary structure of the small subunit ribosomal RNA sequence of the microsporidium *Encephalitozoon cuniculi. Nucleic Acids Res.* **21:**1489.

Hartskeerl, R. A., A. R. Schuitema, T. van Gool, and W. J. Terpstra. 1993b. Genetic evidence for the occurrence of extra-intestinal *Enterocytozoon bieneusi* infections. *Nucleic Acids Res.* **21:**4150.

Hartskeerl, R. A., T. Van Gool, A. R. Schuitema, E. S. Didier, and W. J. Terpstra. 1995. Genetic and immunological characterization of the microsporidian *Septata intestinalis* Cali, Kotler and Orenstein, 1993: reclassification to *Encephalitozoon intestinalis*. *Parasitology* **110:**277–285.

Hashimoto, T. Unpublished data.

Hashimoto, T., and M. Hasegawa. 1996. Origin and early evolution of eukaryotes inferred from the amino acid sequences of translation elongation factors 1α/Tu and 2/G. *Adv. Biophys.* **32:** 73–120.

Hashimoto, T., L. B. Sanches, T. Shirakura, M. Muller, and M. Hasegawa. 1998. Secondary absence of mitochondria in *Giardia lamblia* and *Trichomonas vaginalis* revealed by valyl-tRNA synthetase phylogeny. *Proc. Natl. Acad. Sci. USA*, in press.

Hazzard, E. I., T. G. Andreadis, D. J. Joslyn, and E. A. Ellis. 1979. Meiosis and its implications in the life cycles of *Ambylospora* and *Paratheohania* (Microspora). *J. Parasitol.* **65:** 117–122.

Hinkle, G., et al. 1997. Unpublished data.

Hinkle, G., H. G. Morrison, and M. J. Sogin. 1997. Genes coding for reverse transcriptase, DNA-directed RNA polymerase and chitin synthase form the microsporidian *Spraguea lophii*. *Biol. Bull.* **193:**250–251.

Hirt, R. P., D. Healy, C. R. Vossbrinck, E. U. Canning, and T. M. Embley. 1997. Identification of a mitochondrial Hsp70 orthologue in *Vairimorpha necatrix:* molecular evidence that microsporidia once contained mitochondria. *Curr. Biol.* **7:**995–998.

Hollister, W. S., E. U. Canning, and C. L. Anderson. 1996. Identification of microsporidia causing human disease. *J. Eukaryot. Microbiol.* **43:**104S–105S.

Horner, D. S., R. P. Hirt, S. Kilvington, D. Lloyd, and T. M. Embley. 1996. Molecular data suggest an early acquisition of the mitochondrion endosymbiont. *Proc. R. Soc. Lond. B Biol. Sci.* **263:** 1053–1059.

Huang, H. W., C. F. Lo, C. C. Tseng, S. E. Peng, C. M. Chou, and G. H. Kou. 1998. The small subunit ribosomal RNA gene sequence of *Pleistophora anguillarum* and the use of PCR primers for diagnostic detection of the parasite. *J. Eukaryot. Microbiol.* **45:**556–560.

Huger, A. 1960. Electron microscope study on the cytology of a microsporidian spore by means of ultrathin sectioning. *J. Insect Pathol.* **2:**84–105.

Innis, M. A., D. H. Gelfand, J. J. Sninsky, and T. J. White. 1990. *PCR Protocols: A Guide to Methods and Applications.* Academic Press, Inc., San Diego, CA.

Inoue, T. 1997. Unpublished data.

Ishihara, R., and Y. Hayashi. 1968. Some properties of ribosomes from the sporoplasm of *Nosema bombycis*. *J. Invertebr. Pathol.* **11:**377–385.

Issi, I. V. 1986. Microsporidia as a phylum of parasitic protozoa. *Acad. Sci. USSR* (*Leningrad*) **10:** 6–136.

Kamaishi, T., T. Hashimoto, Y. Nakamura, Y. Masuda, F. Nakamura, K. Okamoto, M. Shimizu, and M. Hasegawa. 1996b. Complete nucleotide sequence of the genes encoding translation elongation factors 1α and 2 from a microsporidian parasite, *Glugea plecoglossi:* implications for the deepest branching of eukaryotes. *J. Biochem.* **120:**1095–1103.

Kamaishi, T., T. Hashimoto, Y. Nakamura, F. Nakamura, S. Murata, N. Okada, D. Okamoto, M. Shimizu, and M. Hasegawa. 1996a. Protein phylogeny of translation elongation factor EF-1 alpha suggests microsporidians are extremely ancient eukaryotes. *J. Mol. Evol.* **42:**257–263.

Katiyar, S. K., V. R. Gordon, G. L. McLaughlin, and T. D. Edlind. 1994. Antiprotozoal activities of benzimidazoles and correlations with β-tubulin sequence. *Antimicrob. Agents Chemother.* **38:**2086–2090.

Katiyar, S. K., G. S. Visvesvara, and T. D. Edlind. 1995. Comparisons of ribosomal RNA sequences from amitochondrial protozoa: implications for processing, mRNA binding and paromomycin susceptibility. *Gene* **152:**27–33.

Katzwinkel-Wladarsch, S., M. Lieb, W. Helse, T. Loscher, and H. Rinder. 1996. Direct amplification and species determination of microsporidian DNA from stool specimens. *Trop. Med. Int. Health* **1:**373–378.

Kawakami, Y., T. Inoue, K. Ito, K. Kitamizu, C. Hanawa, T. Ando, H. Iwano, and R. Ishihara. 1994. Identification of a chromosome harboring the small subunit ribosomal RNA gene of *Nosema bombycis*. *J. Invertebr. Pathol.* **64:**147–148.

Kawakami, Y., T. Inoue, M. Kikuchi, M. Takayanagi, M. Sunairi, T. Ando, and R. Ishihara. 1992. Primary and secondary structures of 5S ribosomal RNA of *Nosema bombycis* (Nosematidae, Microsporidia). *J. Seric. Sci. Jpn.* **61:**321–327.

Keeling, P. J., and W. F. Doolittle. 1996. Alpha-tubulin from early diverging eukaryotic lineages and the evolution of the tubulin family. *Mol. Biol. Evol.* **13:**1297–1305.

Keeling, P. J., and G. I. McFadden. 1998. Origins of microsporidia. *Trends Microbiol.* **6:**19–23.

Kent, M. L., D. M. Hervio, M. F. Docker, and R. H. Devlin. 1996. Taxonomy studies and diagnostic tests for myxosporean and microsporidian pathogens of salmonid fishes utilizing ribosomal DNA sequence. *J. Eukaryot. Microbiol.* **43:**98S–99S.

Keohane, E. M., G. A. Orr, H. S. Zhang, P. M. Takvorian, A. Cali, H. B. Tanowitz, M. Wittner, and L. M. Weiss. 1998. The molecular characterization of the major polar tube protein gene from *Encephalitozoon hellem*, a microsporidian parasite of humans. *Mol. Biochem. Parasitol.* **94:** 227–236.

Kock, N. P., H. Petersen, T. Fenner, I. Sobottka, C. Schmetz, P. Deplazes, N. J. Pieniazek, H. Albrecht, and J. Schottelius. 1997. Species-specific identification of microsporidia in stool and intestinal biopsy specimens by the polymerase chain reaction. *Eur. J. Clin. Microbiol. Infect. Dis.* 16, 369–376.

Krieg, A. 1955. Ueber Infektionskrankheiten bei Engerlingen von Melolontha sp. unter besonderer Berucksichtigung einer Mikrosporidien-Erkrankung. *Zentbl. Bakteriol. Parasitenkd. Abt. 2* **108:**533–538.

Kudo, R. R. 1966. *Protozoology,* 5th ed. Thomas, Springfield, Ill.

Kudo, R. R., and E. W. Daniels. 1963. An electron microscope study of the spore of a microsporidian, *Thelohania californica. J. Protozool.* **10:**112–120.

Larsson, J. I. R. 1986. Ultrastructure, function, and classification of microsporidia. *Prog. Protistol.* **1:** 325–390.

Larsson, J. I. R. 1988. Identification of microsporidian genera (Protozoa, Microspora): a guide with comments on taxonomy. *Arch. Protistenkd.* **136:**1–37.

Leipe, D. D., J. H. Gunderson, T. A. Nerad, and M. L. Sogin. 1993. Small subunit ribosomal RNA of *Hexamita inflata* and the quest for the first branch in the eukaryotic tree. *Mol. Biochem. Parasitol.* **59:**41–48.

Levine, N. D., J. O. Corliss, F. E. Cox, G. Deroux, J. Grain, B. M. Honigberg, G. F. Leedale, A. R. Loeblich, III, J. Lom, D. Lynn, E. G. Merinfeld, F. C. Page, G. Poljansky, V. Sprague, J. Vavra, and F. G. Wallace. 1980. A newly revised classification of the protozoa. *J. Protozool.* **27:**37–58.

Li, J., S. K. Katiyar, A. Hamelin, G. S. Visvesvara, and T. D. Edlind. 1996. Tubulin genes from AIDS-associated microsporidia and implications for phylogeny and benzimidazole sensitivity. *Mol. Biochem. Parasitol.* **78:**289–295.

Liguory, O., F. David, C. Sarfati, A. R. Schuitema, R. A. Hartskeerl, F. Derouin, J. Modai, and J. M. Molina. 1997. Diagnosis of infections caused by *Enterocytozoon bieneusi* and *Encephalitozoon intestinalis* using polymerase chain reaction in stool specimens. *AIDS* **11:**723–726.

Lom, J., and J. Vavra. 1961a. Contribution to the Knowledge of microsporidian spore. I. Electron Microscopy. II. The Sporoplasm Extrusion. Abstracts of Papers Presented at the International Conference of Protozoologists, Prague, 1961, p. 259–260.

Lom, J., and J. Vavra. 1961b. Niektore wyniki baden nad ultrastruktura spor posozyta ryb Pleistophora hyphessobrycornis (Microsporidia). *Waidomosci Parazytol.* **7:**828–832.

Lom, J., and J. Vavra. 1962. A proposal to the classification within the subphylum cnidospora. *Syst. Zool.* **11:**172–175.

Maddox, J. V., M. D. Baker, M. R. Jeffords, M. Kuras, A. Linde, L. F. Solter, M. L. McManus, J. Vavra, and C. R. Vossbrinck. *Nosema portugal,* n. sp., isolated from gypsy moths (*Lymantria dispar* L.) collected in Portugal. *J. Invertebr. Pathol.*, in press.

Malone, L. A., A. H. Broadwell, E. T. Lindridge, C. A. McIvor, and J. A. Ninham. 1995. DNA probes for two Microsporidia, *Nosema bombycis* and *Nosema costelytrae. J. Invertebr. Pathol.* **65:**269–273.

Malone, L. A., and C. A. McIvor. 1993. Pulsed-field gel electrophoresis of DNA from four microsporidian isolates. *J. Invertebr. Pathol.* **61:**203–205.

Malone, L. A., and C. A. McIvor. 1994. Ribosomal RNA genes of two microsporidia, *Nosema apis* and *Vavraia oncoperae*, are very variable. *J. Invertebr. Pathol.* **64:**151–152.

Malone, L. A., and C. A. McIvor. 1996. Use of nucleotide sequence data to identify a microsporidian pathogen of *Pieris rapae* (Lepidoptera, Pieridae) *J. Invertebr. Pathol.* **68:**231–238.

Maley, L. E., and C. R. Marshall. 1998. The coming of age of molecular systematics. *Science* **279:**505–506.

Mansfield, K. G., et al. 1998. Unpublished data.

Mansfield, K. G., A. Carville, D. Shvetz, J. MacKey, S. Tzipori, and A. A. Lackner. 1997. Identification of *Enterocytozoon bieneusi*-like microsporidian parasite in simian immunodeficiency virus-inoculated macaques with hepatobillary disease. *Am. J. Pathol.* **150:**1395–1405.

Marton, L. 1934. Electron microscopy of biological objects. *Nature* **133:**911.

Maxam, A. M., and W. Gilbert. 1977. A new method for sequencing DNA. *Proc. Natl. Acad. Sci. USA* **74(2):**560–564.

Moser, B., J. J. Becnel, J. Maruniak, and R. Patterson. Analysis of the ribosomal DNA sequences of the microsporidia *Thelohania* and *Vairimorpha* of fire ants. *J. Invertebr. Pathol.*, in press.

Muller, M. 1997. What are the microsporidia? *Parasitol. Today* **13:**455–456.

Munderloh, U. G., T. J. Kurtti, and S. Ross. 1990. Electrophoretic characterization of chromosomal DNA from two microsporidia. *J. Invertebr. Pathol.* **56:**243–248.

Nagano, H., C. Satoh, and K. Furuya. 1996. Nucleotide sequences of DNA fragments of *Encephalitozoon cuniculi* amplified by polymerase chain reaction with primers considered specific for *Echinococcus. J. Eukaryot. Microbiol.* **43:**218–221.

Nilsen, F., C. Endresen, and I. Hordvik. 1998. Molecular phylogeny of microsporidians with particular reference to species that infect the muscles of fish. *J. Eukaryot. Microbiol.* **45:**535–543.

Ninham, J. A. Unpublished data.

Ombrouck, C., L. Ciceron, S. Biligui, S. Brown, P. Marechal, T. van Gool, A. Datry, M. Danis, and I. Desportes-Livage. 1997. Specific PCR assay for direct detection of intestinal microsporidia *Enterocytozoon bieneusi* and *Encephalitozoon intestinalis* in fecal specimens from human immunodeficiency virus-infected patients. *J. Clin. Microbiol.* **35:**653–655.

Ombrouck, C., L. Ciceron, and I. Desportes-Livage. 1996a. Specific and rapid detection of Microsporidia in stool specimens from AIDS patients by PCR. *Parasite* **3:**85–86.

Ombrouck, C., I. Desportes-Livage, A. Achbarou, and M. Gentilini. 1996b. Specific detection of microsporidia *Encephalitozoon intestinalis* in AIDS patients. *C.R. Acad. Sci.* **319:**39–43.

Peyretaillade, E., C. Biderre, P. Peyret, F. Duffieux, G. Metenier, M. Gouy, B. Michot, and C. P. Vivares. 1998a. Microsporidian *Encephalitozoon cuniculi*, a unicellular eukaryote with an unusual chromosomal dispersion of ribosomal genes and a LSU rRNA reduced to the universal core. *Nucleic Acids Res.* **26:**3513–3520.

Peyretaillade, E., V. Broussolle, P. Peyret, G. Metenier, M. Gouy, and C. P. Vivares. 1998b. Microsporidia, amitochondrial protists, possess a 70-kDa heat shock protein gene of mitochondrial evolutionary origin. *Mol. Biol. Evol.* **15:**683–689.

Philippe, H. and A. Adoutte. 1996. How far can we trust the molecular phylogeny of protists? *Verh. Dtsch. Zool. Ges.* **89.2:**49–62.

Pieniazek, N. J. Unpublished data.

Pieniazek, N. J., et al. Unpublished data.

Pieniazek, N. J., A. J. da Silva, S. B. Slemenda, and G. S. Visvesvara. 1998. Molecular and morphological characterization of a human microsporidian pathogen *Vittaforma corneae* (syn. *Nosema corneum*). Submitted for publication.

Pieniazek, N. J., A. J. da Silva, S. B. Slemenda, G. S. Visvesvara, T. J. Kurtti, and C. Yasunaga. 1996. *Nosema trichoplusiae* is a synonym of *Nosema bombycis* based on the sequence of the small subunit ribosomal RNA coding region. *J Invertebr. Pathol.* **67:**316–317.

Pomport-Castillon, C., B. Romestand, and J. F. De Jonckheere. 1997. Identification and phylogenetic relationships of microsporidia by riboprinting. *J. Eukaryot. Microbiol.* **44:**540–547.

Raikov, I. B. 1982. *The Protozoan Nucleus: Morphology and Evolution*, p. 124–129. Springer-Verlag, Berlin.

Rastrelli, P. D., E. Didier, and R. W. Yee. 1994. Microsporidial keratitis. *Opthalmol. Clin. N. Am.* **7:**617–633.

Raynaud, L., F. Delbac, V. Broussolle, M. Rabodonirina, V. Girault, M. Wallon, G. Cozon, C. P. Vivares, and F. Peyron. 1998. Identification of *Encephalitozoon intestinalis* in travelers with chronic diarrhea by specific PCR amplification. *J. Clin. Microbiol.* **36:**37–40.

Rice, R. N. 1998. Unpublished data.

Rinder, H., K. Janitschke, H. Aspöck, A. J. Da Silva, P. Deplazes, D. P. Fedorko, C. Franzen, U. Futh, F. Hünger, A. Lehmacher, C. G. Meyer, J.-M. Molina, J. Sandfort, R. Weber, T. Löscher, and the Diagnostic Multicenter Study Group on Microsporidia. 1998. Blinded, externally controlled multicenter evaluation of light microscopy and PCR for detection of microsporidia in stool specimens. *J. Clin. Microbiol.* **36:**1814–1818.

Rinder, H., S. Katzwinkel-Wladarsch, and T. Loscher. 1997. Evidence for the existence of genetically distinct strains of *Enterocytozoon bieneusi. Parasitol. Res.* **83:**670–672.

Roger, A. J., C. G. Clark, and W. F. Doolittle. 1996. A possible mitochondrial gene in the early-branching amitochondriate protist *Trichomonas vaginalis. Proc. Natl. Acad. Sci. USA* **93:**14618–14622.

Roger, A. J., S. G. Svard, J. Tovar, C. G. Clark, M. W. Smith, F. D. Gillin, and M. L. Sogin. 1998. A mitochondrial-like chaperonin 60 gene in *Giardia lamblia:* evidence that diplomonads once harbored an endosymbiont related to the progenitor of mitochondria. *Proc. Natl. Acad. Sci. USA* **95:** 229–234.

Sanger, F., and A. R. Coulson. 1977. DNA sequencing with chain-terminating inhibitors. *Proc. Natl. Acad. Sci. USA* **74:**5463–5467.

Sanger, F., and H. Tuppy. 1951. The amino-acid sequence in the phenylalanyl chain of insulin. 1. The identification of lower peptides from partial hydrolysates. *Biochemistry* **49:**463–481.

Schnittger, L., et al. 1997. Unpublished data.

Schuitema, A. R. J., R. A. Hartskeerl, T. van Gool, R. Laxminarayan, and W. J. Terpstra. 1993. Application of the polymerase chain reaction for the diagnosis of microsporidiosis. *AIDS* **7**(Suppl. 3)**:**S57–S61.

Shadduck, H. A., R. A. Meccoli, R. Davis, and R. L. Font. 1990. Isolation of a microsporidian from a human patient. *J. Infect. Dis.* **162:**773–776.

Shaw, R. W., M. L. Kent, M. F. Docker, A. M. V. Brown, R. H. Devlin, and M. L. Adamson. 1997. A new *Loma* species (Microsporidia) in shiner perch (*Cymatogaster aggregata*). *J. Parasitol.* **83:**296–302.

Siddall, M. E., D. S. Martin, D. Bridge, S. S. Desser, and D. K. Cone. 1995. The demise of a phylum of protists: phylogeny of myxozoa and other parasitic cnidaria. *J. Parasitol.* **81:**961–967.

Silveira, H., and E. U. Canning. 1995. *Vittaforma corneae* n. comb. of the human microsporidium *Nosema corneum* Shadduck, Meccoli, Davis and Font, 1990, based on its ultrastructure in the liver of experimentally infected athymic mice. *J. Eukaryot. Microbiol.* **42:**158–165.

Sogin, M. L., J. H. Cunderson, H. J. Elwood, R. A. Alonso, and D. A. Peattie. 1989. Phylogenetic meaning of the kingdom concept: an unusual RNA from *Giardia lamblia*. *Science* **243:**75–77.

Soltys, B. J., and R. S. Gupta. 1994. Presence and cellular distribution of a 60-kDa protein related to mitochondrial HSP 60 in *Giardia lamblia*. *J. Parasitol.* **80:**580–590.

Sprague, V., and J. Vavra. 1977. Systematics of the microsporidia, p. 31–335. *In* L. A. Bulla and T. C. Cheng (ed.), *Comparative Pathobiology*, vol. 2. Plenum Press, New York, N.Y.

Sprague, V., J. J. Becnel, and E. I. Hazard. 1992. Taxonomy of the phylum Microspora. *Crit. Rev. Microbiol.* **18:**285–395.

Streett, D. A. 1994. Analysis of *Nosema locustae* (Microsporidia: Nosematidae) chromosomal DNA with pulsed-field gel electrophoresis. *J. Invertebr. Pathol.* **63:**301–303.

Talal, A., D. P. Kotler, J. Orenstein, and L. M. Weiss. 1998. Detection of *Enterocytozoon bieneusi* in stool specimens by the polymerase chain reaction. *Clin. Infect. Dis.* **26:**673–675.

Tuzet, O., J. Maurand, A. Fize, R. Michel, and B. Fenwich. 1971. Proposition d'un nouveau cadre systematique por les genres de Microsporidies. *C.R. Acad. Sci.* **272:**1268–1271.

Tzipori, S., A. Carville, G. Widner, D. Kotler, D. Mansfield, and A. Lackner. 1997. Transmission and establishment of a persistent infection of *Enterocytozoon bieneusi*, derived from a human with AIDS, in simian immunodeficiency virus-infected rhesus monkeys. *J. Infect. Dis.* **175:**1016–1020.

Vavra, J. 1966. Some recent advances in the study of microsporidian spores, p. 443–444. *In Proceedings of the First International Congress of Parasitology.* Pergamon Press, New York.

Vavra, J. 1976. Structure of microsporidia, p. 1–86. *In* L. A. Bulla and T. C. Cheng (ed.), *Comparative Pathobiology*, vol. 1. Plenum Press, New York.

Velasquez, J. N., S. Carnevale, E. A. Guarnera, J. H. Labbe, A. Chertcoff, M. G. Cabrera, and M. I. Rodriguez. 1996. Detection of the microsporidian parasite *Enterocytozoon bieneusi* in specimens from patients with AIDS by PCR. *J. Clin. Microbiol.* **34:**3230–3232.

Visvesvara, G. S., A. J. da Silva, G. P. Croppo, N. J. Pieniazek, G. J. Leitch, D. Ferguson, H. de Moura, S. Wallace, S. B. Slemenda, I. Tyrrell, and J. Meador. 1995. In vitro culture and serologic and molecular identification of *Septata intestinalis* isolated from urine of a patient with AIDS. *J. Clin. Microbiol.* **33:**930–936.

Visvesvara, G. S., G. J. Leitch, A. J. Da Silva, G. P. Croppo, H. Moura, S. Wallace, S. B. Slemenda, D. A. Schwartz, D. Moss, R. T. Bryan, and N. J. Pieniazek. 1994. Polyclonal and monoclonal antibody and PCR-amplified small-subunit rRNA identification of a microsporidian, *Encephalitozoon hellem*, isolated from an AIDS patient with disseminated infection. *J. Clin. Microbiol.* **32:**2760–2768.

Vivares, C., C. Biderre, F. Duffieux, E. Peyretaillade, P. Peyret, G. Metenier, and M. Pages. 1996. Chromosomal localization of five genes in *Encephalitozoon cuniculi* (microsporidia). *J. Eukaryot. Microbiol.* **43:**97S.

Vossbrinck, C. R., and L. M. Weiss. Unpublished data.

Vossbrinck, C. R., and C. R. Woese. 1986. Eukaryotic ribosomes that lack a 5.8S RNA. *Nature* **320:**287–288.

Vossbrinck, C. R., T. G. Andreadis, and B. A. Debrunner-Vossbrinck. 1998. Verification of intermediate hosts in the life cycles of microsporidia by small subunit rDNA sequencing. *J. Eukaryot. Microbiol.* **45:**290–292.

Vossbrinck, C. R., M. D. Baker, E. S. Didier, B. A. Debrunner-Vossbrinck, and J. A. Shadduck. 1993. Ribosomal DNA sequences of *Encephalitozoon hellem* and *Encephalitozoon cuniculi*: species identification and phyogenetic construction. *J. Eukaryot. Microbiol.* **40:**354–362.

Vossbrinck, C. R., J. V. Maddox, S. Friedman, B. A. Debrunner-Vossbrinck, and C. R. Woese. 1987. Ribosomal RNA sequence suggests microsporidia are extremely ancient eukaryotes. *Nature* **326:**411–414.

Walker, C. G. 1983. p. 280–289. *In* J. W. Schopf (ed.), *Earth's Earliest Biosphere.* Princeton University Press, Princeton, N.J.

Weber, R., R. T. Bryan, R. L. Owen, C. M. Wilcox, L. Gorelkin, and G. S. Visvesvara. 1992. Improved light-microscopical detection of microsporidial spores in stool and duodenal aspiriates: the Enteric Opportunistic Infections Working Group. *N. Engl. J. Med.* **326:**161–166.

Weber, R., R. T. Bryan, D. A. Schwartz, and R. L. Owen. 1994a. Human microsporidial infections. *Clin. Microbiol. Rev.* **7:**426–461.

Weber, R., B. Sauer, M. A. Spycher, P. Deplazes, R. Keller, R. Ammann, J. Briner, and R. Luthy. 1994b. Detection of *Septata intestinalis* in stool specimens and coprodiagnostic monitoring of successful treatment with albendazole. *Clin. Infect. Dis.* **19:**342–345.

Weiser, J. 1959. *Nosema laphygmae* n. sp., and the internal structure of the microsporidian spore (not). *J. Insect Pathol.* **1:**52–59.

Weiser, J. 1977. Contribution to the classification of microsporidia. *Vestn. Cesk. Spol. Zool.* **41:**308–320.

Weiser, J. 1982. Methods of identification of microsporidia with vector control potential, p. 393–399. *In Proceedings of the IIIrd International Colloquium on Invertebrate Pathology.*

Weiss, L. M. Unpublished data.

Weiss, L. M., and M. Kent. Unpublished data.

Weiss, L. M., and C. R. Vossbrinck. 1998. Microsporidiosis: molecular and diagnostic aspects. *Adv. Parasitol.* **40:** 351–395.

Weiss, L. M., X. Zhu, A. Cali, H. B. Tanowitz, and M. Wittner. 1994. Utility of microsporidian rRNA in diagnosis and phylogeny: a review. *Folia Parasitol.* (Prague) **41:**81–90.

Wittner, M., H. B. Tanowitz, and L. M. Weiss. 1993. Parasitic infection in AIDS patients: cryptosporidiosis, isosporiasis, microsporidiosis, cyclosporiasis. *Infect. Dis. Clin. N. Am.* **7:**569–586.

Yang, D., Y. Oyaizu, H. Oyaizu, G. J. Olsen, and C. R. Woese. 1985. Mitochondrial origins. *Proc. Natl. Acad. Sci. USA* **82:**4443–4447.

Zhu, X., M. Wittner, H. B. Tanowitz, A. Cali, and L. M. Weiss. 1993a. Small subunit rRNA sequence of *Septata intestinalis. Nucleic Acids Res.* **21:** 48–46.

Zhu, X., M. Wittner, H. B. Tanowitz, A. Cali, and L. M. Weiss. 1993b. Nucleotide sequence of the small ribosomal RNA of *Encephalitozoon cuniculi. Nucleic Acids Res.* **21:**1315.

Zhu, X., M. Wittner, H. B. Tanowitz, A. Cali, and L. M. Weiss. 1994. Ribosomal RNA sequences of *Enterocytozoon bieneusi, Septata intestinalis* and *Ameson michaelis:* phylogenetic construction and structural correspondence. *J. Eukaryot. Microbiol.* **41:**204–209.

Zhu, X., M. Wittner, H. B. Tanowitz, D. Kotler, A. Cali, and L. M. Weiss. 1993c. Small subunit rRNA sequence of *Enterocytozoon bieneusi* and its potential diagnostic role with use of the polymerase chain reaction. *J. Infect. Dis.* **168:**1570–1575.

MICROSPORIDIAN BIOCHEMISTRY AND PHYSIOLOGY

Earl Weidner, Ann M. Findley, V. Dolgikh, and J. Sokolova

5

The microsporidia are a large group of highly specialized obligate intracellular protozoan parasites. Although they are best known as parasites of arthropods and fish, microsporidians are the etiological agents of various pathologies evidenced in amphibians, reptiles, birds, and a variety of mammals, including humans (Cali et al., 1993; Canning, 1977; Canning and Hollister, 1991; Orenstein et al., 1990; Sinderman, 1970; Weiser, 1976). Despite the wide distribution of these parasites in nature and the devastating effects they inflict on host animals, relatively little is known about the physiological characteristics of the microsporidia or the infections they produce. Indeed, previous studies concerning these parasites have dealt almost exclusively with ultrastructural descriptions of their functional morphology and life cycle details.

Microsporidians are known to possess a number of unusual cytological and molecular peculiarities that point to their probable position as a very early branch of eukaryotic evolution (Baker et al., 1995; Olsen and Woese, 1993; Philippe and Adoutte, 1995; Vossbrinck et al., 1987, 1993; Vossbrinck and Woese, 1986; Zhu et al., 1994). The prokaryotic structure of the microsporidian ribosome as well as sequence data for 16S-like rRNA reveal greater similarity to the archaebacteria than to other eukaryotes. The finding of a mitochondrial-type HSP70 in the cytoplasm of *Nosema locustae* (Germot et al., 1997) adds to the evidence in favor of an ancient origin for this group. It is further assumed that an extremely long period of coevolution for microsporidian-host systems has lead to the establishment of excellent and (most likely) unique mechanisms of metabolic interaction between these parasites and their host cells. In this context it is particularly remarkable that practically no information exists on the metabolism of microsporidia.

ENERGY METABOLISM OF MICROSPORIDIA

Microsporidia-Host Cell Interactions: *Ameson michaelis* in the Blue Crab (*Callinectes sapidus*)

Disturbances in the biochemical composition of tissues infected by intracellular parasites are of interest because such infections often significantly alter the electrolyte, carbohydrate,

Earl Weidner, Department of Biological Sciences, Louisiana State University, Baton Rouge, LA 70803. *Ann M. Findley,* Department of Biology, Northeast Louisiana University, Monroe, LA 71209. *V. Dolgikh and J. Sokolova,* All Russian Institute for Plant Protection, Podbelskii 3, St. Petersburg-Puskin 189620, Russia.

The Microsporidia and Microsporidiosis (Murray Wittner, editor; Louis M. Weiss, contributing editor), ©1999 American Society for Microbiology, Washington, D.C.

protein, and free amino acid pools of host cells (von Brand, 1973). Of equal importance is the fact that the resultant host cell environment alterations almost certainly influence the metabolic activities of an intracellular parasite. In a preliminary study, we investigated the effects of microsporidian infection on the biochemical composition of host tissues (Findley et al., 1981). *A. michaelis* invades the blue crab (*C. sapidus*), causing widespread muscle damage which is ultimately reflected in biochemical changes in the host blood. Sporogenesis proceeds rapidly within the blue crab sarcoplasm. After entering myofibrils, schizonts differentiate into sporonts and then multiply into sporoblasts (Weidner, 1970). The massive numbers of sporoblast colonies in the sarcoplasm result in disorientation and eventual loss of the highly organized myofibrillar structure of the host muscle (Weidner, 1970). As the infection proceeds to its terminal stages, parasite spores largely replace the host skeletal musculature (up to 10^9 parasites/g of infected muscle).

This study (Findley et al., 1981) clearly illustrated that the interaction of *A. michaelis* interaction with its blue crab host significantly altered the biochemical constituents of host tissues. Modifications in skeletal muscle protein and carbohydrate metabolism were reflected in substantial variations in hemolymph composition. Blood osmolality and Cl^- and Na^+ levels decreased with heavy parasitic invasion, while levels of K^+ and ninhydrin-positive substances (NPS) increased in both light and heavy infections. Microsporidiosis resulted in a general increase in the amount of all hemolymph free amino acids detected except glutamic acid. These data suggest dilution of the extracellular fluid compartment with intracellular ions and amino acids and are consistent with the widespread host cell damage observed.

The effects of *A. michaelis* sporogenesis were observed by comparing the biochemical composition of thoracic and cheliped skeletal musculature. (Thoracic muscle is invaded first as parasites leave the gut wall via the open vascular system of the crab. Cheliped musculature, on the other hand, becomes infected later, and significant damage is seen here only in the heavy, terminal stages of the disease.) Protein and carbohydrate levels were lower in infected thoracic muscle (Fig. 1), but the opposite trend was observed for tissue-free amino acids (NPS). The concentration of 9 of the 16 amino acids detected remained unchanged with infection. Skeletal muscle glutamate, proline, glycine,

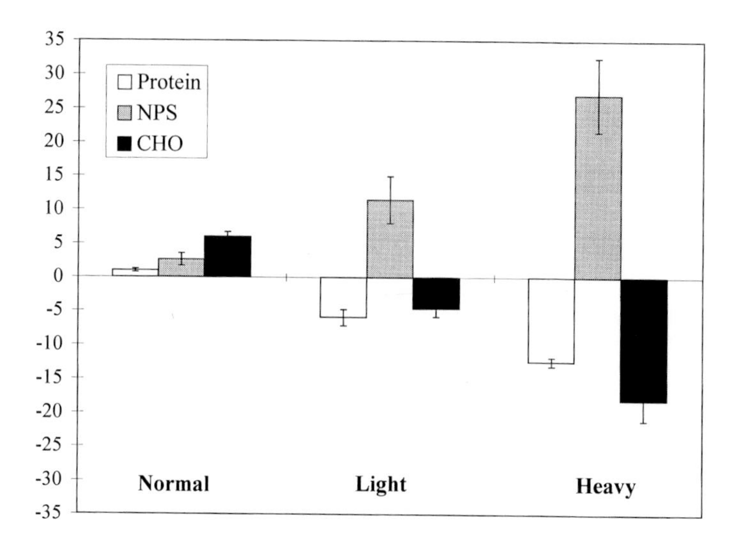

FIGURE 1 Effect of *Ameson michaelis* infection on the skeletal muscle composition of the blue crab (*Callinectes sapidus*). Data are presented as the percentage difference in protein, free amino acid (NPS), and carbohydrate (CHO) levels of thoracic versus cheliped skeletal muscle. Bar direction indicates whether values for thoracic muscle were higher (+) or lower (-) than those for cheliped muscle (mean ± 95% confidence interval).

alanine, and arginine levels declined, while taurine and tyrosine levels increased (Findley et al., 1981).

Parasitic infection may also result in host tissues being deprived of oxygen (von Brand, 1973). Oxygen insufficiency can result from the increased metabolic demand of host tissues and/or developing parasites or from an impaired oxygen delivery system. Under these conditions, anaerobic metabolism becomes increasingly important as a means of energy production (Burke, 1979). In crustaceans, glycolysis is a major functional anaerobic pathway. During hypoxia, significant quantities of pyruvate are converted to lactic acid in crustacean skeletal muscle (Burke, 1979; Dedinger and Schatzlein, 1973; Schatzlein et al., 1973). Lactic acid accumulated in the hemolymph, thoracic muscle, and hepatopancreas of parasitized blue crabs (Fig. 2). Lactate concentrations reached six to seven times their normal levels in hemolymph and skeletal muscle and four times the control value in the hepatopancreas. Since the amount of lactic acid is always higher in the thoracic muscle than in the hemolymph, it is reasonable to assume that this muscle is the ultimate source of lactate production (Phillips et al., 1977). Blood glucose levels declined during the terminal stages of microsporidian infection. Reduced hemolymph glucose may result from increased metabolic demands of host cells and/or developing parasites.

The detailed characterization of the blue crab host cell environment during *A. michaelis* infection provides critical background information for metabolic studies on isolated extracellularly maintained microsporidian parasites. Attempts to characterize the metabolic and biochemical properties of obligate intracellular protozoan parasites are subject to several formidable problems. (1) As unicellular organisms, parasitic protozoans necessarily possess biochemical capabilities that are more complex than the specialized activities of metazoan cells. (2) By virtue of the obligate intracellular existence of these parasites, their metabolic activities are intimately associated with those of their host cell environment. (3) The intricate life cycles of these parasites often involve multiple hosts and/or various specialized tissues within a single host. (4) Metabolic studies conducted on different developmental stages of the same parasite may produce dissimilar and often conflicting data concerning biochemical characteristics. (For an early review of the problems associated with metabolic studies on intracellular parasites, see Trager [1974].) In any case, caution must be exercised in the interpretation of data derived from in vitro manipulation of intracellular parasites. Since culture conditions cannot faithfully approximate the

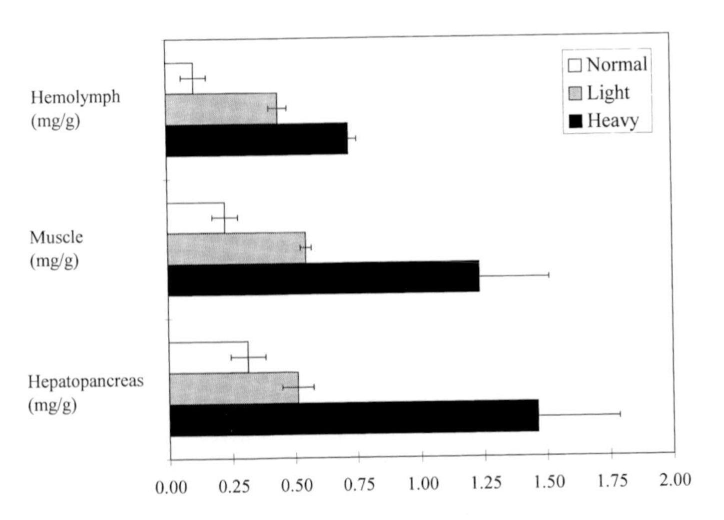

FIGURE 2 Lactate concentrations in the hemolymph, thoracic skeletal muscle, and hepatopancreas of normal, lightly infected, and heavily infected blue crabs (mean ± 95% confidence interval).

host cell environment, it is quite possible that biochemical potentialities are being studied rather than metabolic realities as they occur in vivo (von Brand, 1973).

Until recently, biochemical investigations of the metabolic processes of the microsporidia have suffered because of insufficient numbers of the different parasite stages and inadequate methods for the in vitro cultivation and maintenance of these organisms. However, microsporidian infection in the blue crab, for example, results in the production of large numbers of parasites. Since *A. michaelis* spores can be easily recovered and induced to hatch by certain osmotic and ionic shifts (Weidner, 1972, 1976), the opportunity exists to investigate pure populations of these intracellular parasites during brief periods of extracellular maintenance. Viable sporoplasms subjected to tissue culture media supplemented with ATP, coenzyme A, and pyruvic acid retain their ultrastructural integrity for up to 4 h (Weidner and Trager, 1973). The addition of ATP to the incubation medium was prompted by the apparent lack of mitochondria in all microsporidian growth stages (Weidner, 1970) and the natural conclusion that this necessary energy currency may be supplied to developing parasites by the host cell environment (approximated here by the culture media). It is important to note that there is no indication of sporoplasm growth or differentiation during the incubation period. This situation appears to be a natural one, however, since sporoplasms of several microsporidian species injected into host cells in vitro remain quiescent for some time (up to 24 h) before displaying obvious evidence of development (Shadduck, 1969; Undeen, 1975).

This initial "static" period may actually be a rather dynamic time when during which several essential transport and metabolic processes become functionally competent prior to the growth and multiplicative phases of parasite development. As a first step in characterizing the metabolic potential of the early sporoplasm stage, we examined the carbohydrate activity of sporoplasm populations during brief periods of extracellular maintenance (Findley and Weidner, unpublished observations). *A. michaelis* sporoplasms were found to readily utilize glucose when incubated in medium 199 supplemented with 3 mM ATP. A decline in exogenous glucose was followed by the appearance of substantial quantities of both lactate and, to a lesser extent, pyruvic acid (Fig. 3).

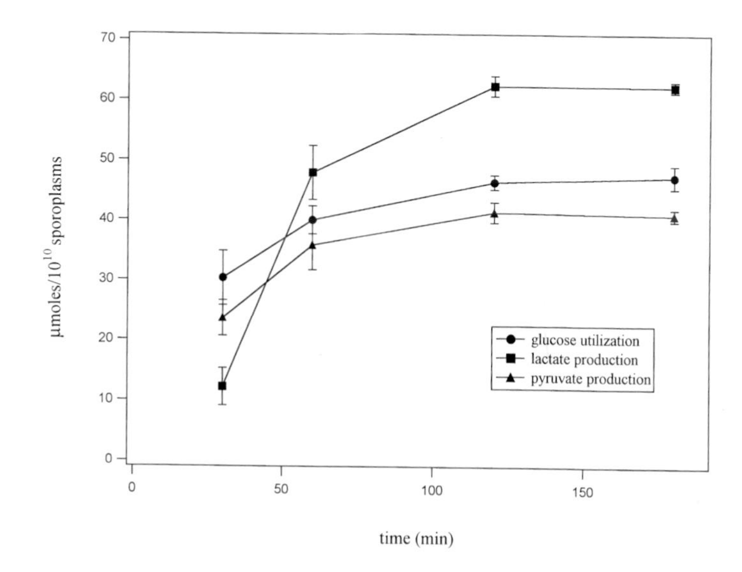

FIGURE 3 Glucose utilization, lactate production, and pyruvate production by isolated *A. michaelis* sporoplasms incubated in medium 199 containing 5.5 mM glucose. Data are presented as (cumulative) micromoles of solute per 10^{10} sporoplasms (mean ± 95% confidence interval).

Therefore, the production of lactic acid by isolated parasites may represent a significant contribution to the accumulation of this metabolite in the thoracic skeletal muscle of infected animals.

Cells deprived of ATP readily lost their ultrastructural integrity, did not consume exogenous glucose, and did not produce significant quantities of lactate or pyruvate. In the presence of 10 mM NaF, a potent glycolytic inhibitor, sporoplasms did not utilize the glucose supplied in the medium nor did they evolve substantial quantities of glycolytic end products.

The rate of glucose transport into microsporidian cells proceeded more rapidly at low (0.5 mM) rather than high (5.5 mM) substrate concentrations (Fig. 4). Glucose uptake by *A. michaelis* sporoplasms displayed sensitivity toward known inhibitors (i.e., 1 mM ouabain and 0.5 mM amiloride) of sodium transport. The relative importance of Na^+-dependent glucose transport appeared to vary with substrate concentration.

Energy Metabolism: Possible Ways of Coupling with Host Cell Energy Systems

SPORES

No storage nutrients have been detected in spores, but it is well known that spore viability can be preserved for years. Hence the internal source of energy sustaining this viability must exist within spores. High concentrations of trehalose and trehalase activity have been demonstrated in the spores of *Nosema apis* (Vander Meer and Gochnauer, 1969, 1971) and *N. algerae* (Undeen et al., 1987; Undeen and Vander Meer, 1994). Trehalase catalyzes the conversion of trehalose to glucose. It is likely that: (1) this disaccharide serves as the main fuel source in spores, and (2) sufficient amounts of ATP can be obtained from glucose catabolism via glycolysis. Starch gel electrophoresis analysis with subsequent histochemical staining revealed the presence of the glycolytic enzyme phosphoglucose isomerase (PGI) in *N. heterosporum* spores (Hazard et al., 1981). This result agrees with analo-

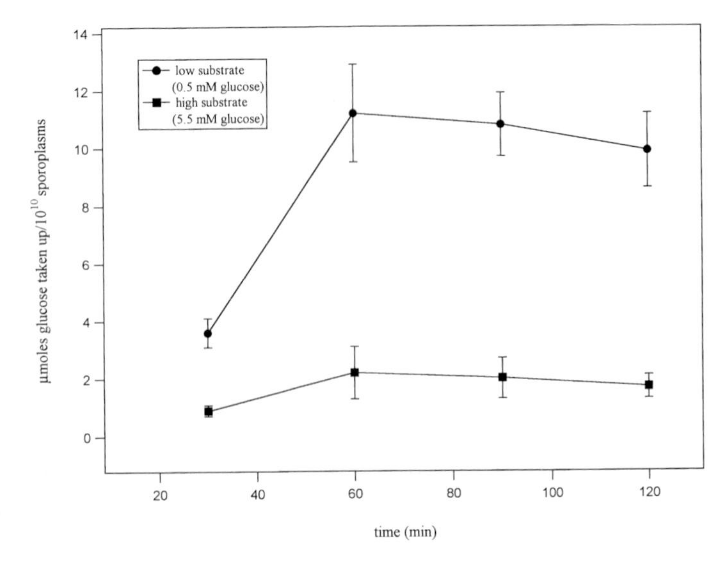

FIGURE 4 Uptake of U-^{14}C-D-glucose by isolated *A. michaelis* sporoplasms. Data are presented as micromoles of glucose taken up per 10^{10} sporoplasms. Incubations in low-substrate (minimum essential medium, 0.5 mM glucose) and high-substrate (medium 199, 5.5 mM glucose) media are compared (mean ± 95% confidence interval).

gous data obtained from *N. grylli*. Five enzymes of glycolysis, PGI, fructose 6-phosphate kinase, aldolase, 3-phosphoglycerate kinase, and pyruvate kinase (PK), were detected in spores of *N. grylli* (Table 1) (Fig. 5). Despite extremely high PGI activities, we were unable to measure any hexokinase (HK)activity (Dolgikh et al., 1997). However, minute concentrations of HK may be present in spores (indeed, all other parasites appear to possess this enzyme) sufficient to support the low rate of metabolism necessary to sustain long-term spore survival. In this situation, high PGI activity provides fast entrance of small concentrations of glucose 6-phosphate (from the HK reaction) into glycolysis and furnishes successful concurrent competition with other glucose-6P pathways. The activities of malate dehydrogenase, malic enzyme, lactate dehydrogenase, alcohol dehydrogenase, and succinate dehydrogenase in *N. grylli* spores were also below the sensitivity of our methods (Dolgikh et al., 1997). Hence it appears likely that glycerol 3-phosphate dehydrogenase (G3PDH) is utilized in microsporidians to reoxidize the NADH generated during glycolysis. Such a mechanism has been described in trypanosomes (Opperdoes, 1987) and trichomonads (Steinbuchel and Muller, 1986a; 1986b).

Quantitative analysis of the trehalose-glucose balance in dormant and germinated spores of *N. algerae* indicates that glucose catabolism occurs during normal spore germination (Undeen et al., 1987). Although fluoride, a glycolytic inhibitor, failed to prevent germination, the inability to do so may be attributable to the impermeability of the spores to this inhibitor (Undeen, 1990).

The fact that phosphoglucomutase (PGM) and glucose 6-phosphate dehydrogenase (G6PDH) were detected in *N. grylli* spores suggests that microsporidia may use glucose 6-phosphate in trehalose synthesis (PGM) and/or in the pentose phosphate pathway (G6PDH). However, the rates of these processes (at least that of the pentose phosphate sequence) are not thought to be significant in spores. Our inability to measure the very next enzyme of the pentose phosphate pathway (6-phosphate-gluconate dehydrogenase) in *N. grylli* spores appears to confirm this. PGM has also been detected in *N. heterosporum* and *Vairimorpha necatrix* spores (Hazard et al., 1981).

It is very probable that oxidative processes are not involved in the energy metabolism of spores. This assumption is supported by the following considerations: (1) the lack of mitochondria, (2) spore germination in the absence of oxygen, and (3) the failure of inhibitors of oxidative metabolism to prevent spore germination (Undeen, 1990).

TABLE 1 Activities of enzymes in microsporidian *N. grylli* spores

Enzyme	Activity (nmol/min × mg of protein, mean ± SE)[a]
Glucose-6-phosphate dehydrogenase	15 ± 1 (7)
6-Phosphogluconate dehydrogenase	<1 (4)
Phosphoglucomutase	7 ± 1 (7)
Hexokinase	<0.2 (8)
Phosphoglucoisomerase	1,549 ± 255 (5)
Fructose 6-phosphate kinase	10 ± 1 (4)
Aldolase	5 ± 1 (4)
3-Phosphoglycerate kinase	16 ± 4 (4)
Pyruvate kinase	6 ± 1 (6)
Glycerol-3-phosphate dehydrogenase	16 ± 2 (7)
Malate dehydrogenase	<0.2 (9)
Malic enzyme	<0.2 (4)
Lactate dehydrogenase	<0.2 (6)
Alcohol dehydrogenase	<0.2 (4)
Succinate dehydrogenase	<0.2 (4)

[a]Figure in parentheses indicates the number of separate spore purification procedures.

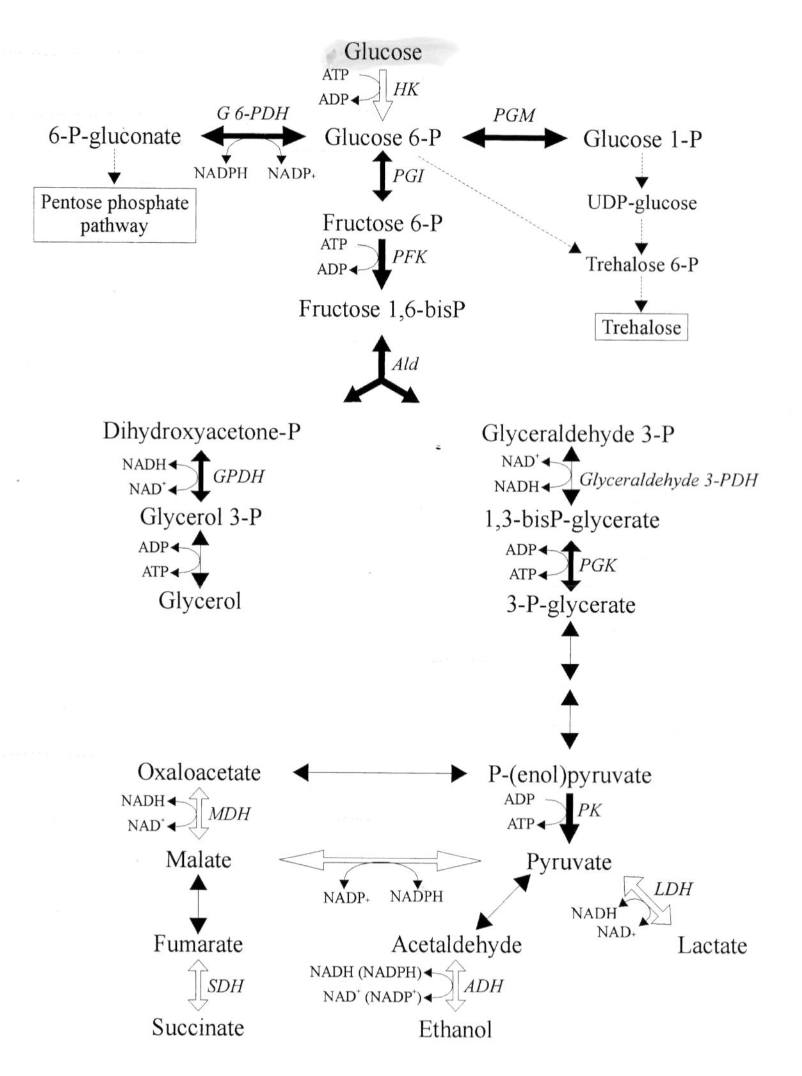

FIGURE 5 A proposed scheme of energy metabolism in microsporidian spores.

INTRACELLULAR STAGES

Like the spore, the intracellular stages of the microsporidian life cycle (meronts and sporonts) lack mitochondria, plastids, hydrogenosomes, and glycosomes and apparently do not possess any nutrient storage within granules. Inside host cells the growth of microsporidians causes the essential rearrangement of cytoskeletal elements and results in the replacement of host cell organelles, especially the mitochondria and endoplasmic reticulum usually concentrated around developing parasites, with spore material (Cali and Owen, 1990; de Graaf et al., 1994; Sokolova et al., 1988, 1994). Indeed, close association of the parasite membrane system with the outer membrane of host cell mitochondria is often observed in thin sections of infected cells (Dufort et al., 1987; Sokolova et al., 1988). Most microsporidian species (including all *Nosema* spp.) develop in direct contact with the host cell cytoplasm and without formation of the complex wall structure of a parasitophorous vacuole. This intimate association indicates that the parasites may require some external energy supply.

Microsporidian intracellular stages are thought to utilize host-derived ATP for their energy needs. Other intracellular parasites, e.g., *Rickettsia prowazekii* (Winkler, 1976), *Plasmodium falciparum* (Choi and Mikkelsen, 1990), and probably *Toxoplasma gondii* (Sorensen et al., 1987), have been shown to possess ATP/ADP translocases within their plasma membrane similar to those found in the mitochondrial membrane system. Microsporidia may have also acquired such a mechanism for harvesting host cell energy. In support of this hypothesis are the following observations: (1) Addition of ATP to the extracellular culture medium helped maintain the structural integrity of *Nosema michaelis* sporoplasms (Weidner and Trager, 1973). (2) In isolated intracellular stages of *N. grylli* (Selesnjov et al., 1996) the activities of PK and G3PDH were no higher than in spores, hence the rate of glycolysis in metabolically active intracellular stages may be comparable to that in dormant spores. (3) Fat reserves decreased in insect fat bodies with microsporidian infection (Canning, 1962; Darwish et al., 1989; Dolgikh et al., 1996; Selesnjov et al., 1996), and this observation cannot be explained by parasite lipid utilization since microsporidia lack mitochondria. (4) An increase in the activities of enzymes participating in the interconversion of C_3 substrates (G3PDH, PK, LDH) was detected in infected host cells (Dolgikh et al., 1995; Kucera and Weiser, 1975) (Table 2). These enzymes may take part in the utilization of glycerol released from fat body triglycerides. (5) Elevated oxygen consumption by infected insects (Lewis et al., 1971), as well as by mitochondria isolated from host tissues, was observed during the course of microsporidian development (Baburina et al., 1989). A significant increase in free oxygen uptake by mitochondria isolated from heavily infected larvae of *Agrotis segetum* (Lepidoptera) was evidenced during parasite merogony (1 to 6 days after invasion). This initial increase was followed by an abrupt drop in O_2 consumption during sporogony and spore formation (Fig. 6). The increase in O_2 is likely a result of the intensive oxidation of fats by host cells during early stages of parasite development. Clearly, the last three points indicate a strong enhancement of oxidative metabolism by infected cells. The resultant energy can be harnessed by the parasite in the form of ATP in addition to being used by the host cell for synthetic and repair processes.

Microsporidians probably utilize host cell carbohydrates for energy production, and ATP can be produced via glycolysis. Various glycolytic enzymes have been detected in spores (and sporoplasms) and consequently should also be present in other intracellular stages. We have observed the disappearance of glycogen granules in the host cell cytoplasm during the early stages of parasite development without any concomitant change in the size or quantity of lipid droplets. For many intracellular parasites, glucose has been shown to be a major source of energy. However, failure to detect the

TABLE 2 Activities of enzymes in the fat body of control and infected crickets (*Gryllus bimaculatus*)

Enzyme	Activity (nmol/min × mg of protein, mean ± SE)[a]	
	Control	Infected
Lactate dehydrogenase (lactate → pyruvate)	2 ± 0.4 (4)[a]	10 ± 2 (7)
Lactate dehydrogenase (pyruvate → lactate)	2 ± 0.4 (14)	14 ± 4 (10)
Glycerol-3-phosphate dehydrogenase	4 ± 1 (19)	24 ± 3 (19)
Pyruvate kinase	14 ± 3 (18)	162 ± 13 (23)

[a]Figure in parentheses indicates the number of independently examined crickets.

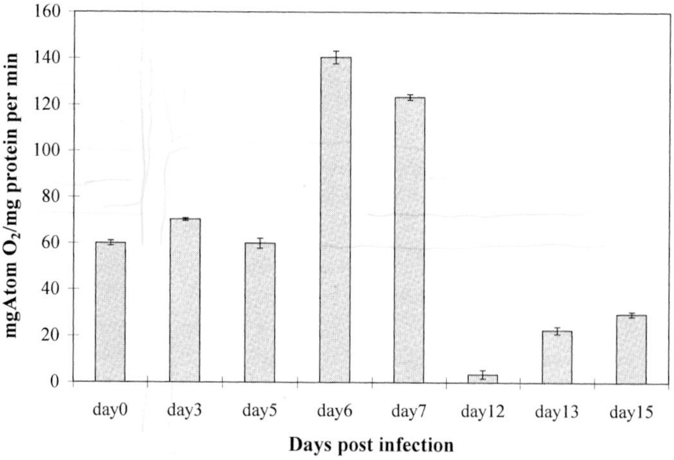

FIGURE 6 Polarographic measurement of oxygen consumption by mitochondria isolated from a *Agrotis segetum* (Lepidoptera: Noctuidae) fat body infected with *Vairimorpha antheraeae.*

presence of HK together with the high PGI activity observed in spores (Dolgikh et al., 1997; Hazard et al., 1981) leads us to speculate that without a significant HK reaction, microsporidians may be able to directly utilize phosphohexoses from the host cell environment. High PGI activity may be necessary in obtaining an advantage over competing mechanisms that may be operating within the host cell for phosphohexose utilization. In addition, significant conversion of glycogen or free glucose into trehalose (a transport disaccharide) is an important function of the insect fat body. Trehalose is formed from glucose 6-phosphate and UDP-glucose, which is in turn produced from UTP and glucose 1-phosphate. Glycogen can be formed from UDP-glucose, and glucose 6-phosphate can be generated during gluconeogenesis as well. In summary, then, phosphohexoses seem to be actively produced in the cells of the insect fat body, and microsporidia may capitalize on their presence as a source of energy production.

It is also of interest to speculate on the ultimate outcome of the end products of anaerobic carbohydrate catabolism. Parasites may release substantial quantities of pyruvate, lactate, glycerol, glycerol 3-phosphate, etc., into the host cytoplasm. Their further utilization (e.g., in the citric acid cycle) may be the cause of observed increases in the activities of host enzymes participating in the interconversion of C_3 substrates (Table 2).

SIGNAL TRANSDUCTION PATHWAYS

An External Signal Pathway for Spore Activation

The microsporidian species *Spraguea lophii* is a model for investigating externally mediated signal transduction and subsequent activation of the internal signal pathway for triggering of a missile cell, the microsporidian spore. The general order for the *S. lophii* signal pathway, as proposed over 10 years ago, begins with a pH shift followed by a Ca^{2+} influx into the spore. Once the signal reaches the spore interior, it is mediated in part by calmodulin (CAM), and the activation continues (Pleshinger and Weidner, 1985). Studies have identified specific components in the external signal pathway partly because this molecular assemblage is segregated from the internal spore wall and therefore can be easily recovered for analysis (Manale, 1996). A number of proteins from this signal pathway are bound in the outer membrane envelope (MEV) bordering the tough outer spore wall (Manale, 1996; Weidner and Halonen, 1993) (Fig. 7). These proteins reside in invaginated fold domains in this membrane.

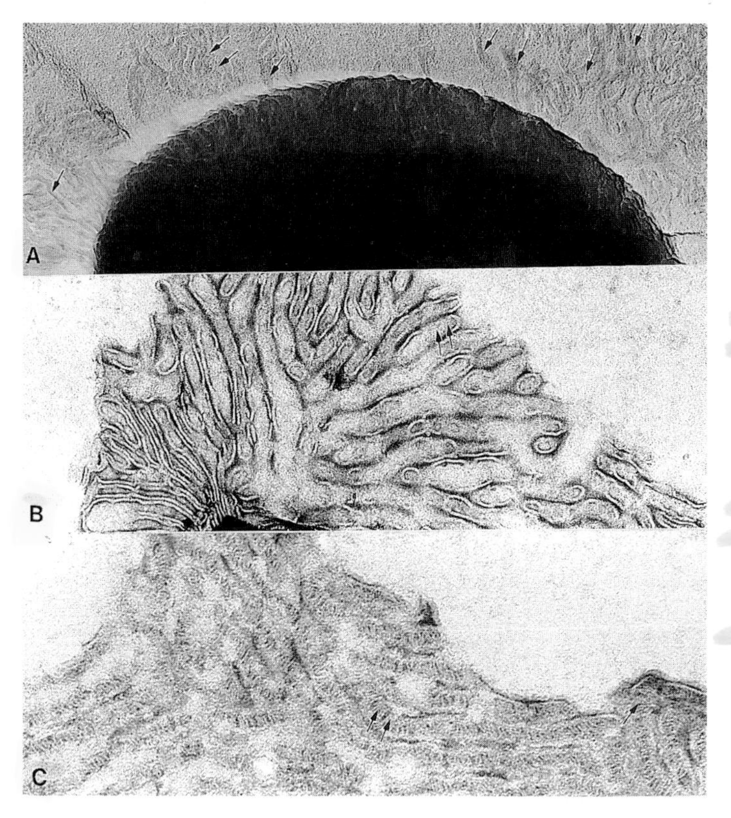

FIGURE 7 Surface topography of the outer envelope surrounding *S. lophii* spores. (A) Platinum-shadowed spore with the outer envelope partially dispersed after preincubation in urea and a rinse in distilled water. Arrows indicate envelope is compartmentalized. (B) Isolated *S. lophii* spore envelope stained with phosphotungstic acid. Arrows indicate compartments with internal matrix which probably buffer proteins within these domains. (C) Isolated *S. lophii* spore envelope stained with uranyl acetate. Visualized proteins (arrows) are weakly stained within the compartments so clearly delineated in panel B. The compartmental channels are 75 to 100 nm in width.

The fold order of the MEV onto the spore wall appears to be maintained in part by intermediate filaments (Weidner and Halonen, 1993). The laterally aligned 50 to 100-nm-wide folds contain compartments permeable to larger molecular probes although the proteins within these domains appear to be protected from some hydrolytic enzymes (Weidner, unpublished observations). The MEV compartments on *S. lophii* spores bear clathrin, heterotrimer G proteins, CAM, nucleotidase(s) and a CAM kinase (Manale, 1996). A study by Alaqui et al. (1997) indicated that *S. lophii* spores also contained sphingomyelin and cholesterol, although it was not determined whether these components were from the MEV domain. These molecules are important elements in caveolae, essential signal transduction centers characteristic of many eukaryotic cells (Anderson, 1993; Koleske et al., 1995). The standard view of membrane signaling is of a ligand-driven process with random interactions featuring G proteins operating near a cell surface. However, work on caveolae complexes indicates that they are centers for signal transduction in plasma membranes, and observation of their presence is beginning to change the traditional view of this process. An emerging model indicates that signal cascade pathways are compartmentalized and ordered, particularly in highly responsive cells (Anderson, 1993; Koleske et al., 1995; Liu et al., 1996). Since microsporidian spores may be thought of as highly responsive cells, the external signal system should be particularly well developed in this group. The following proteins have been recovered from the

MEV fold domains of *S. lophii* spores, and these identifications have been made.

G PROTEINS

G proteins are central to transducing external signals. Since they are highly conserved housekeeping proteins (Yokoyama and Starmer, 1992), we tested rat antipeptide antisera directed to consensus sequences from G protein α-subunits in order to identify the involvement of heterotrimer G proteins. The antisera tested react with G proteins across the different phyletic categories and react with *Toxoplasma* G proteins (Halonen et al., 1996). The MEV fold proteins were tested against rat brain membrane proteins as an internal standard. Immunoreactivity was detected in the MEV lysates with antibodies GC/2, GO/1, and RM/1 (Halonen et al., 1996), which recognize Go- and Gs-type α-subunits (Fig. 8). Within the MEV fold, G proteins are believed to bind to ion channels or to an enzyme. In the present case, there appears to be a Ca^{2+} pathway and a mediator molecule that releases the calcium, resulting in an ion influx through the spore polar aperture which appears to be the site on which the MEV has some orientation.

CAM AND CAM KINASE

Immunolabeling and Western blot analyses indicate that the MEV folds on *S. lophii* spores have several peptides reactive to CAM and to a CAM kinase (Manale, 1996; Weidner and Halonen, 1993). This evidence supports earlier work reporting that CAM inhibitors chlorpromazine and trifluoroperazine block *S. lophii* spore activation (Pleshinger and Weidner, 1985). It is unlikely that these drugs block an internal signal pathway since their molecular weight is nearly 500 and they therefore should be too big for diffusion passage. However, CAM inhibitors can enter the surface compartments of the MEV; indeed, evidence indicates that after the drug is removed by washing, the spores regain full capacity for spore activation.

Early evidence for a surface kinase on *S. lophii* was indicated when outer spore envelope proteins became phosphorylated in the presence of gamma-labeled ATP at the time of spore activation (Manale, 1996; Weidner and Halonen, 1993). Manale (1996) subsequently isolated the *S. lophii* MEV proteins and characterized a CAM kinase by Western blot analyses. This CAM kinase shows some resemblance to mammalian CAM kinase II since it can be resolved into three peptide bands. This similarity is expected since CAM kinase II is a multifunctional enzyme characteristically associated with calcium cascade signal pathways (Braun and Schulman, 1995; Chapman, 1995).

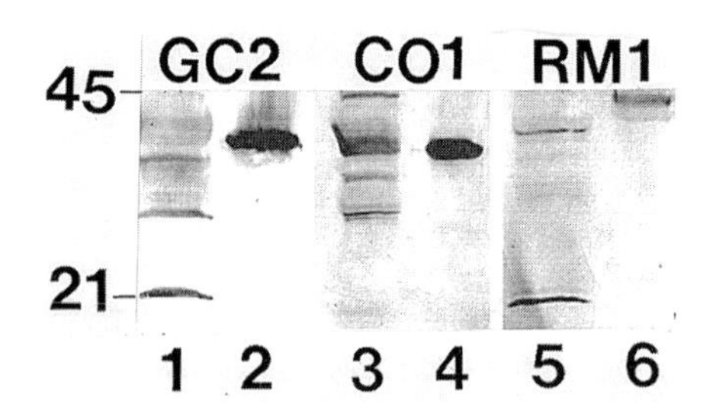

FIGURE 8 Immunoblot analyses of *S. lophii* spore envelope proteins specific for G protein (α-subunits). Antisera GC/2 recognizes Goα (lanes 1 and 2); CO/1 recognizes Goα and Giα 3 (lanes 3 and 4) and RM/1 is specific for Gsα (lanes 5 and 6). For controls, rat brain membrane proteins are in lanes 2, 4 and 6, and *S. lophii* envelope peptides are in lanes 1, 3 and 5.

CLATHRIN IN MEVS

Western blot and immunofluorescence data indicate that clathrin is present in the MEV folds of *S. lophii* spores. Manale (1996) has provided some evidence for a 180-kDa heavy chain and a single 50-kDa light chain. Immunolabeling with a peroxidase probe indicates that clathrin is localized to the folds in the MEV (Fig. 9). Clathrin may be more of a multifunctional protein than originally believed since the light chains have binding sites or domains on intermediate filaments, CAM, and HSP70 (Georgatos et al., 1989; Jackson et al., 1987).

THE EXTERNAL SIGNAL

Induced activation of the signal cascade in *S. lophii* spores requires Ca^{2+}, a pH shift to the alkaline, and the addition of a polyanionic mole-

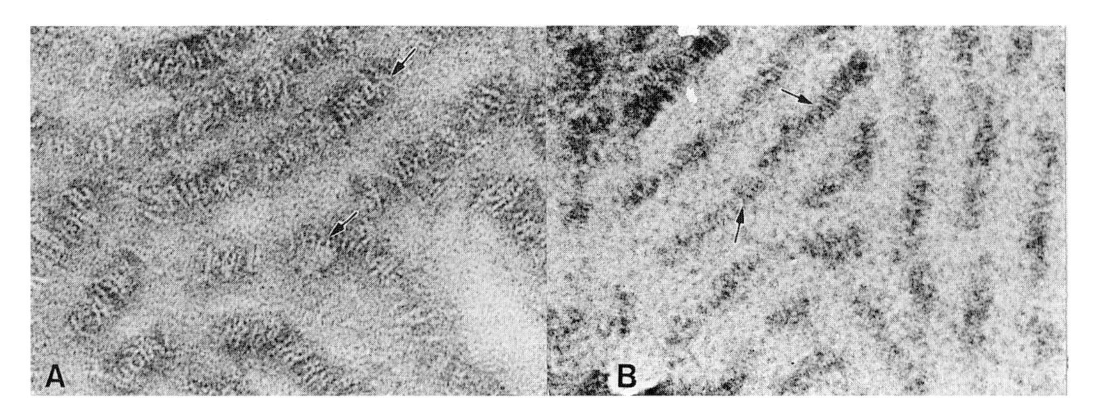

FIGURE 9 *S. lophii* spore envelope isolate. (A) Uranyl acetate-stained proteins partially isolated from spore envelope but still in the native linear arrangement (arrows). (B) Isolate of *S. lophii* envelope with proteins tested against anticlathrin and reacted with a second antibody coupled to peroxidase. Arrows indicate peroxidase reaction is limited to threadlike elements. Magnification, ×200,000.

cule such as mucin (Pleshinger and Weidner, 1985). It is thought that at the onset of the cascade CAM is the mobile calcium machine or ratchet wheel for effecting a Ca^{2+} wave through the MEV folds to the polar aperture rather than directed into the thick, relatively impermeable spore wall. Early observations of a Ca^{2+} influx through channels was suggested since verapamil effectively blocked *S. lophii* activation (Pleshinger and Weidner, 1985). Since verapamil has a molecular weight of 500, any channel to be blocked would probably have to be close to the surface. However, the polar aperture appears to be composed of plates near the surface, and this assemblage may be the target of verapamil (Fig. 10). Once the calcium ions have passed internally, the internal signal pathway begins.

The Internal Signal Cascade

The signal pathway associated with spore discharge may be similar to those found in exocytic events: an influx of external Ca^{2+}, G protein regulation or control, kinase events, and an effector molecule(s). Calcium is known to have profound effects on major families of biologically important molecules (da Silva and Reinach, 1991; McPhalen et al., 1991; Strynadka and Jans, 1991), and this is most apparent in trichocyst protein expression during discharge (Satir, 1989). In microsporidians, Ca^{2+} influx appears to affect polaroplast membranes and the matrix inside (Weidner and Byrd, 1982). However, Ca^{2+} also affects the polar tube and seems to change its viscosity and conformation (Weidner et al., 1995). Since this protein comprises a significant percentage of the total spore volume, its calcium-induced

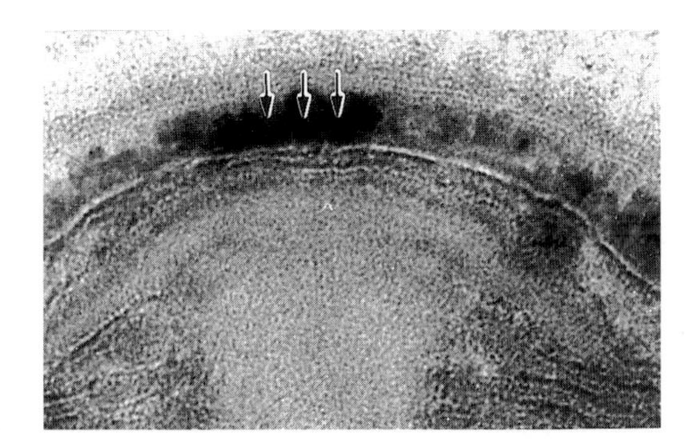

FIGURE 10 *S. lophii* spore polar aperture area. Arrows indicated precipitate that frequently accumulated near spore aperture after $CaCl_2$ incubation. Since precipitation was confined to this site, one assumption is that the calcium salt accumulates here because of the channel proteins at this aperture (where the Ca^{2+} influx is believed to occur). Magnification, ×200,000.

conformational change may account in part for the initial thrust during spore discharge. But it is also conceivable that other ion influxes may set in motion the trigger cascade for spore activation. Undeen and his associates have provided convincing evidence that sodium, chloride, or even fluoride ions can affect the discharge event in some microsporidian species (Frixione et al., 1997; Undeen and Avery, 1988a, 1988b).

Microsporidian Spore Extrusion

The extrusion of sporoplasm from the spore takes place through an everting discharge tube in a manner similar to that described a few decades ago by Lom and Vávra (1963). The extrusion apparatus in unfired spores has cytoplasm and a nucleus to the exterior. Video imaging, with appropriate probes, indicates that the polaroplast membrane moves out into the tube early during its formation (Weidner et al., 1995). Since the intrasporular sporoplasm is not surrounded by a second membrane and the original plasma membrane of the spore is discarded during discharge, the polaroplast is thought to provide part or all of the new plasma membrane of the discharged sporoplasm. Rhodamine 123 binds to and labels the polaroplast membrane in unfired spores; during discharge the fluorescence is transferred to the envelope of the extruding sporoplasm. This result is particularly compelling since rhodamine 123 is selective only for certain charged states on membranes such as those of mitochondria, presynaptic nerve terminals, and sperm (Haugland, 1996). That the sporoplasm is free within the spore and during the extrusion event is indicated by video imaging which occasionally shows sporoplasms emerging and sliding off the tube prematurely while the 4', 6' diamidino-2-phenylindole (DAPI)-stained nucleus is still traversing down the tube (Weidner et al., 1995). This situation would occur only rarely if the sporoplasm were compartmentalized within a membrane during the extrusion. Whether the polar tube protein changes its fluidity during extrusion is unresolved; however, it is apparent that the newly assembled tube is much more stable than the polar tube matrix moving down the tube during its formation (Weidner, 1982).

ORIGIN OF THE FORCE FOR SPOROPLASM EXPULSION

There are three possible sources of force generation: (1) First, a conformational shift in the polar filament protein significantly increases internal pressure by the swelling action of the protein matrix within the spore space, and it fires out the aperture. Since the movement of the tube outward involves a membrane sliding on a membrane, there is physically little resistance to flow. By capillarity, the sporoplasm moves down the tube, and the momentum of this motion allows it to traverse the entire distance of the tube. (2) When activated, the posterior vacuole generates a swelling action. Unfortunately, ths vacuole is essentially uncharacterized; however, Lom and Vávra (1963) have indicated that this compartment appears to swell noticeably at the time of spore activation. (3) Finally, Undeen and associates have clearly shown that there is a rapid shift in trehalose-to-glucose conversion at the time of spore activation. The assumption is that the osmotic increase effected by this conversion can establish sufficient intrasporular pressure to initiate discharge and extrude the sporoplasm (Undeen et al., 1987; Undeen and Vander Meer, 1994).

MOLECULAR MOTORS AND THE MERONT STAGES

Microsporidian meront stages appear to have actin-myosin and kinesin-associated molecular motors. The number of molecular species of kinesins present in cells dictates the number of possible functions carried out since specific functions each appear to be associated with a particular light chain (Muresan et al., 1997). For example, the vesicle delivery system has a distinct kinesin, as does the apparatus associated with chromosome movement during mitosis (Wordeman and Mitchison, 1995). There is much interest in the kinesins of microsporidians since kinesin motor elements have specific binding sites for intermediate filaments (Liao and Gunderson, 1997); and a number of species of microsporidians have distinct associations with intermediate filaments, particularly at the host-parasite interface. In addition to the kinesin-associated motor, there are myosin-actin motors. Endocytosis, exocytosis, and cytokinesis are under actin and myosin control. The cytokinesis ring is particularly apparent in the division of a sporont into individual sporoblasts (Canning and Lom, 1986; Xie and Canning, 1986).

BIOCHEMICALS OF PARTICULAR INTEREST IN MICROSPORIDIANS

Nucleotides and Nucleotidases

Preliminary observations indicate that *S. lophii* have surface nucleotidase(s) for dephosphorylating nucleotides to nucleosides or to purines and ribose for subsequent entry. All parasitic protozoans investigated so far are purine-dependent on the host, and microsporidians appear to have a similar requirement (Aronow et al., 1987; Sibley et al., 1994; Trager, 1986). Several methods have been used in preliminary investigations on nucleotidase activity in microsporidians. We tested a method developed by Robinson and Karnovsky (1983) in which cerium is a capture agent for liberated phosphate at sites where the enzyme was working. Our results indicated that nucleotidase was present in the MEV fold compartments at the surface of *S. lophii* spores (Fig. 11). As an alternative method, substrate *p*-nitrophenyl phosphate was used. When the terminal phosphate is cleaved from this substrate, the nitrophenyl provides a measurable color shift (Weidner, unpublished observations). Levamisole (2.0 mM), a blocking agent for ATPase, effectively inhibited the reduction of substrate nitrophenyl phosphate (Robinson and Karnovsky, 1983).

In vitro studies on isolated *S. lophii* sporoplasms maintained extracellularly show that cells required an exogenous source of nucleotides. In the absence of ATP and GTP, extracellular sporoplasms disintegrated rapidly (Weidner and Trager, 1973).

Polyamines

Attention has also been given to polyamine levels in microsporidians (Coyle, 1996). Since polyamines can be secreted by cells (McCormack, personal communication), there is interest in whether host cells can be affected by polyamines secreted by meront colonies located

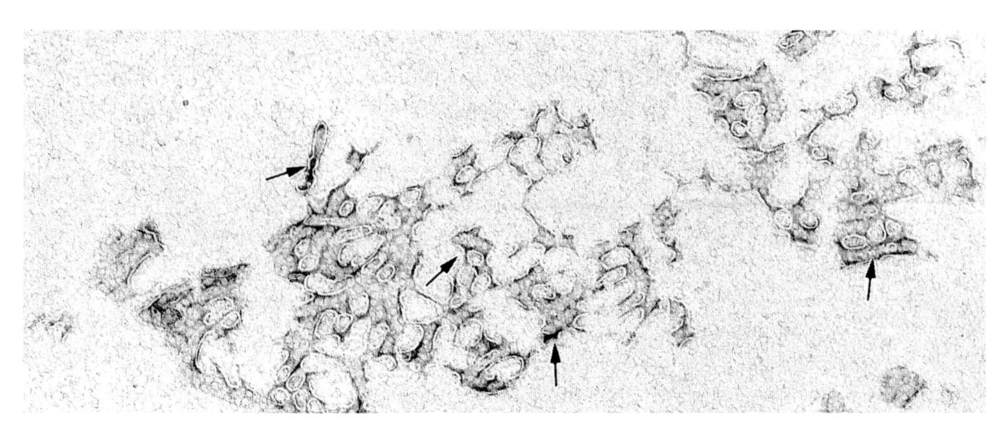

FIGURE 11 Nucleotidase activity in the isolated spore envelope of *S. lophii*. The envelopes were incubated in nitrophenyl phosphate or ATP (1 mM), and the substrate was incubated in medium with the capture agent cerium. The cytochemical medium contained 0.1 M Tris-maleate (pH 7.4), $CeCl_3$, ATP, or nitrophenyl phosphate (Robinson and Karnovsky, 1983). Reaction was confined to the envelope channels. Magnification, ×35,000.

in the host cytoplasm. The evidence indicates that elevated polyamine levels significantly affect growth and differentiation in cells (Casero, 1995). Polyamines are by nature involved in a diverse set of actions, including gene expression, ion regulation, and biosynthetic control (Pegg, 1986). Polyamine metabolism is often regulated by specific enzymes such as ornithine decarboxylase (ODC). Thus, an overexpression of ODC can affect cell growth (Shawtz and Pegg, 1994). Microsporidians do express ODC, as pointed out by Coyle et al. (1996) for *Encephalitozoon cuniculi*. There is presently no evidence that host cell perturbations are the result of (1) the secretion of polyamines into the cell by microsporidians or (2) the effect of microsporidians on host regulation of polyamine synthesis. Microsporidians do synthesize polyamines. For example, *E. cuniculi* contains spermidine, spermine, and putrescine; furthermore, *E. cuniculi* has a high ratio of spermidine and spermine to putrescine. This ratio indicates that polyamine interconversion likely occurs. Since polyamines can transfer out of cells by secretion, it is possible that secreting microsporidian meronts might have a profound effect on the activities of the host cell.

Mitochondrial (mt) HSP70 Chaperonin in Microsporidians

The HSP70 group of chaperonins, like actin, have a structural motif that consists of two lobes for binding nucleotide in the middle. mtHSP70, which is encoded in the nucleus, resides in mitochondria and can function as part of an import machine (Matouschek et al., 1997). mtHSP 70 is thought to be an ATP-driven motor associated with protein import. Horst et al. (1997) believe mtHSP70 requires several associate proteins in order to function in the import process. But the evidence indicates that mtHSP70 can serve as a mechanochemical enzyme that literally pulls proteins across membranes. However, although microsporidians lack mitochondria, they possess mtHSP70 (Germot et al., 1997). There are several likely sites where mtHSP70 can locate: First, the rhodamine 123-sensitive polaroplast domain is a likely position since it is believed to be transferred in part to the outer membrane domain of discharged sporoplasms. Conceivably, mtHSP70 might function in association with this site. However, if mtHSP70 is not located within the extrusion apparatus of the spore, it can be located only within the posterior vacuole or in the nucleus itself. The HSP70 are a family of molecular species each with a designated functional adaptation. Yeast cells have 10 types of HSP70. *Drosophila* and mammals have 8, and archaebacteria and yellow sugarbeet virus have at least one HSP70 analog. Functions include protein transport, folding, assembly or disassembly, and Ca^{2+} regulation or activation (Matouschek et al., 1997). While there are no data suggesting that mtHSP70 is part of a transport apparatus for moving some parasite proteins into the host cell domain, there is evidence that keratin intermediate filament proteins are in the host cell cytoplasm and have been transported there from the parasite without exocytosis.

KERATIN INTERMEDIATE FILAMENT CYTOSKELETONS

During the past decade it has become apparent that while intermediate filaments appear to be primarily structural proteins, they may also function as dynamic polymers. For example, their assembly and disassembly occurs within a small time frame in mitosis (Vikstrom et al., 1989) or during lipid pool growth during lipogenesis in 3T3 cells (Franke et al., 1987). Keratin intermediate filaments, although generally believed to be proteins associated with higher eukaryotes (Albers and Fuchs, 1992), appear to be part of the makeup of some microsporidian species, including *Thelohania butleri*, a *Thelohania* sp. in *Pandalus jordani*, a *Thelohania* sp. in *C. sapidus*, an *Amblyospora* sp., *Trachipleistophora*, and *A. michaelis* (Weidner, 1992). Here we will focus on keratin filaments found in two domains: (1) within the spore stage, the microsporidian sporophorous vesicle, and (2) keratin in the host cell cytoplasm domain but situated at its interface with the parasite.

Keratin with Sporophorous Vesicles

Many microsporidian species have a sporophorous vesicle (SPV) stage in which the developing spores are confined within a compartment (Canning and Lom, 1986). In a *Thelohania* sp. of the blue crab (*C. sapidus*), the keratin and desmosomal analogs are associated with envelopes surrounding the spores in the SPV (Fig. 12) (Weidner et al., 1990). Generally, vertebrate keratins have diversified significantly, although these proteins have a number of conserved domains. And yet, antibodies to vertebrate keratin exhibit some cross-reactivity with *Thelohania*-associated keratin. It is probable that these proteins are from *Thelohania* since the arthropod host species does not express keratin (Bartnik and Weber, 1989). The desmosome-keratin analogs and the SPV increase the survival of the spores inside. Preliminary data indicate that spores remain viable longer when stored within SPVs than when they are removed from SPVs (Weidner, unpublished observations). However, the intermediate filaments within the SPV domain may serve as a biochemical scaffolding since they have receptor sites for functional elements such as ATPases (Jackson et al., 1987), kinases (Jackson et al., 1987), plaque proteins (Vikstrom et al., 1989), and ankyrin (Georgatos and Blobel, 1987).

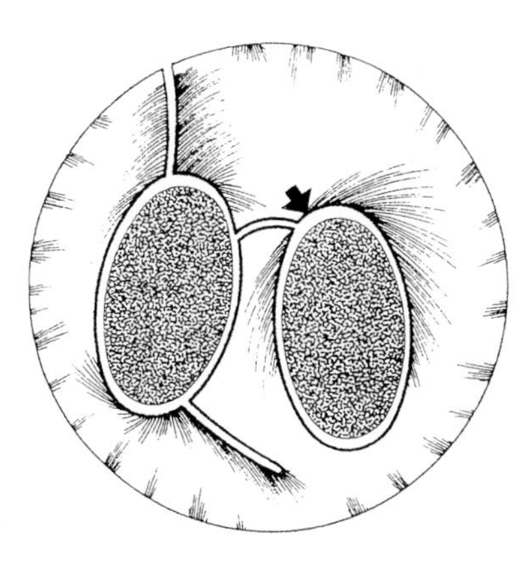

FIGURE 12 The presumed anatomy of a *Thelohania* sporophorous vesicle with keratin and plaque proteins. The arrow indicates the position of the jacket assemblage (bearing the keratin) that envelops the spore. The evidence for channel continuity with the vesicle exterior is indicated by the permeation of lanthanum and dyes into the area between the jacket and the spores; no probe material penetrated the primary space between the jacket and the vesicle envelope.

Keratin Filaments at the Host-Parasite Interface

Keratin intermediate filament networks form immediately exterior to meronts and spore-forming elements of *S. lophii*, *Trachipleistophora hominis*, and *A. michaelis*. The filament networks are so extensive around *S. lophii* that entire colonies can be compartmentalized in a mat which partially segregates from the host cell domain (Fig. 13). While it has not been proven that the intermediate filaments within the host cell cytoplasm are of parasite origin, this is a possibility in *A. michaelis* since it resides in an arthropod host that does not appear to

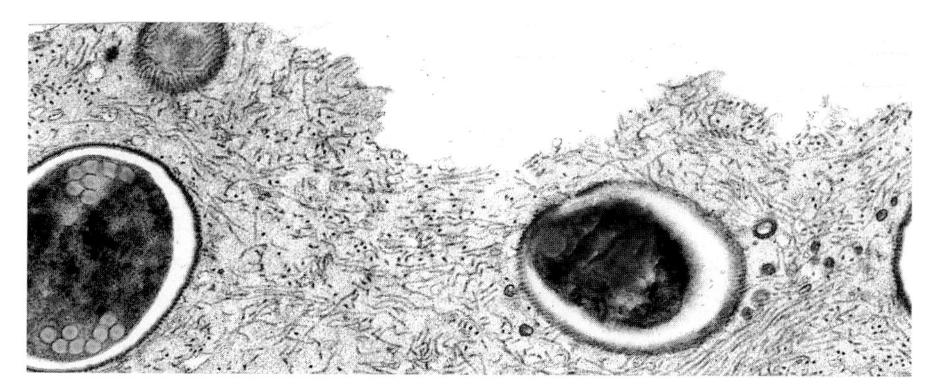

FIGURE 13 Electron micrograph showing spore colony removed from host neuron after 6 h in culture medium. Filament bundles still persist and comprise the matrix between the spores or meronts and hold these cells together. Magnification, ×30,000.

have the genes for cytoplasmic intermediate filament expression (Bartnik and Weber, 1989). Similarly, in *S. lophii*-infected neurons, the keratin filaments may be of parasite origin since neurons do not normally express keratin. Whatever the origin, the interface cytoskeleton is likely an important adaptation by the parasite. In *S. lophii,* the parasites are confined in these mat domains and therefore are restricted. This confinement appears to have some survival value for the host cell since infected neurons show some action potential and ultrastructural integrity years after the initial infection (Weidner, unpublished observations).

NUCLEIC ACID RESEARCH

Microsporidians have been reported to have the smallest genome size of any eukaryote (6 to 6.2 Mbp), and the Comparative Molecular Biology Center at the MBL is currently sequencing a partial genome of one species (Hinkle et al., 1997). Preliminary investigations indicate that microsporidians (1) lack introns, (2) have some gene relatedness to certain fungi, and (3) have an interesting reverse transcriptase that is normally associated with retrotransposons. The last observation is most intriguing because it indicates either that some species have a retrovirus or that they may utilize the reverse transcriptase gene to assimilate into the host's genome (or vice versa). The data on a reverse transcriptase were reported for *S. lophii,* a microsporidian that may have the capacity to express some vertebratelike keratins. If these data are confirmed, it will indicate that host genome assimilation may be a real possibility.

The organization of microsporidian rRNAs indicates that there has been a fusion of the 5.8S and 23S rRNAs (Biderre et al., 1997). Generally, eukaryotic rRNA genes are arranged head to tail in tandem repeats with precursor regions separated by spacers. But in *E. cuniculi* there is but one rRNA gene per chromosome and no tandem arrangements as in other amitochondriates such as *Trichomonas.* This small set of rRNA genes contrasts with those in organisms such as *Giardia* which have approximately 300 copies of rRNA in tandem (Biderre et al., 1997). Microsporidians also appear to have small nuclear RNAs associated with ribonucleoprotein complexes referred to as spliceosomes (Dimaria et al., 1996). This indicates that the pre-mRNA splicing gene apparatus evolved at the outset of eukaryotic evolution. While these amitochondriate protists appear to have spliceosomes, the amitochondriate *Giardia* do not, suggesting that spliceosome gene complexes developed before the acquisition of mitochondria, or perhaps that microsporidians may have lost their mitochondria. The presence of a mtHSP70 would indicate that they may have had mitochondria at one time.

With reference to the overall microsporidian nuclear apparatus, it appears to be that of a standard protistan eukaryote: a conventional mitotic apparatus, chromosomes, kinetochores, centriolar plaques, and a nuclear envelope with nuclear pores. However, many species have the persistent stable arrangements of paired haploid nuclei. Dikaryons that retain two separate genomes may be genetically equivalent to diploid cells; however, this condition is found elsewhere among the eukaryotes only in some higher fungi (Raper and Flexer, 1970).

IN VITRO INVESTIGATIONS ON ISOLATED EXTRACELLULAR SPOROPLASMS AND MERONTS

Since microsporidians are intracellular parasites for most of their life cycle, they have not been directly amenable to biochemical analysis. Nearly all studies have been confined to the spore stage. Recently, however, some success has been achieved in isolating and maintaining meronts and discharged sporoplasms in extracellular support medium. Developing a simple in vitro model will be useful for many subsequent biochemical analyses.

Isolation of Discharged Microsporidian Sporoplasms with Concanavalin A

The isolation of discharged sporoplasms has been achieved for *S. lophii.* Spore activation was initiated with preincubation in 100 mM

N-2-hydroxyethylpiperazine-*N*'-2-ethanesulfonic acid (HEPES) at pH 7 with 50 mM Ca^{2+} for 1 to 6 h. Subsequently, 10^4 spores in 40 μl of buffer were spread onto each 0-1 18 × 18 mm coverslip. After 30 s, 50 μl of filtered mucus (fresh human, bovine, or pig saliva prepassaged through a 0.45-μm-pore-size filter) was added over the spore film followed within seconds by the addition of 10 to 20 μl of HEPES buffered at pH 10 (without Ca^{2+}). Spores hatched within 2 min, and discharged sporoplasms attached to the coverslip. To remove the spores, concanavalin A (ConA) (0.1%) in HEPES at pH 7 was applied to each coverslip, and with some flow action with a pipet, the spores were lifted off the surface, clumped, and washed free of the attached cover, leaving the glass-attached sporoplasms. After several rinses in axenic media, the coverslips were transferred to cultivation chambers.

Maintenance of Extracellular Sporoplasms

Extracellular microsporidians were first observed to be significantly stabilized with the addition of nucleotides to culture medium (Weidner and Trager, 1973). *A. michaelis* yielded a high percentage of sporoplasms however, little success was achieved in separating the sporoplasms from the spores. Scarborough found that *Glugea hertwigi* sporoplasms showed improved structural integrity when placed in culture medium containing ATP and 3% gelatin. Although the sporoplasms were difficult to work with while in the gelatin matrix, the protein may have provided the cells with some stability (Scarborough-Bull and Weidner, 1985). However, *G. hertwigi* is a good candidate for subsequent studies since the spores are easily induced to germinate with a high yield of sporoplasms. But the recovery of isolated populations of attached *S. lophii* sporoplasms with the use of ConA may be a step forward since these cells can be transferred quickly from one medium to another and there appears to be a potential for quantitative and qualitative biochemical analyses (Fig. 14). One of the useful features of having attached sporoplasms is that molecular trafficking, cell replication, or autoradiographic analysis can take place. Cultures of *S. lophii* sporoplasms have shown some stability and indications of initiating cell division when placed in RMPI 1640 supplemented with KCl, HEPES buffer, bovine serum albumin (BSA), NAD, ATP, GTP, glucose, ribose, and a thin overlay of Matrigel (Weidner, unpublished observations). The cells of *S. lophii* appear to be more stable in the presence of some oxygen rather than no oxygen, which indicates that microsporidians may be somewhat like *Plasmodium*. This is not a surprising finding since a number of microsporidians take up residence in tissues, such as fish gills, where higher levels of oxygen can be present (Canning and Lom, 1986).

Indications of an Endocytotic Mechanism in Newly Discharged Sporoplasms

Recent evidence indicates that the polaroplast may provide part or all of the outer envelope of discharged sporoplasms since Nile red or rhodamine 123-labeled polaroplasts in the spores were reestablished as fluorescence on the outer envelope of discharged sporoplasms (Weidner, unpublished observations). To determine whether the discharged sporoplasms have an endocytotic capability, Scarborough-Bull and Weidner (1985) exposed sporoplasms to ferritin particles, and the results indicated some uptake. *S. lophii* sporoplasms were also incubated in rhodamine conjugated to albumin or dextran and the results showed some binding to the sporoplasm surface and limited uptake into vesicles (Fig. 14).

In Vitro Studies on Extracellular Meronts

S. lophii meronts grow in tightly grouped colonies within ganglionic neurons of *Lophius americanus* (Weissenberg, 1968, 1976). Although they are free within the host cytoplasm, they are confined to a specific domain by a network of intermediate filaments (Fig. 9). The parasite colonies are so bound within the intermediate filament mat that they can be

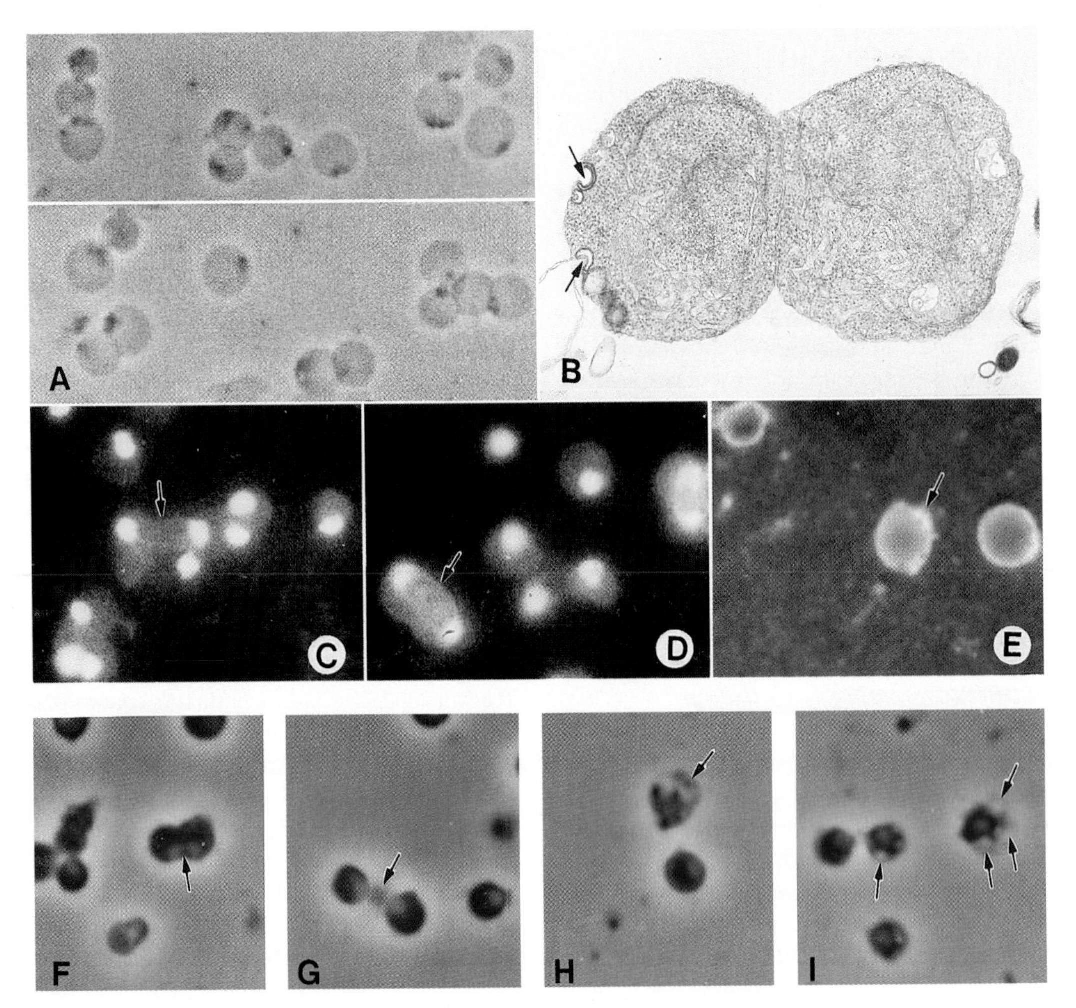

FIGURE 14 Images of *S. lophii* sporoplasms in extracellular culture media. (A) Phase light micrograph of sporoplasms after 12 h in medium. (B) Electron micrograph of sporoplasms. Arrows indicate apparent interiorizing on surface of cell. (C and D) DAPI-stained sporoplasms after 24 h in culture. Arrows indicate presumptive divider cells. (E) Sporoplasms after 10-min incubation in rhodamine-albumin with some accumulation of label in small fluorescent domains at the surface (arrows). (F and G) DAPI-stained sporoplasms after 24 h in culture. Arrows show a cell in apparent division. (H and I) Sporoplasms incubated for 20 min in dextran. Arrows point to vacuoles as possible sites of dextran endocytosis. Sporoplasms are 2 μm in diameter.

readily liberated from the host neuron and transferred directly to culture medium. Colonies of meronts were stabilized in RPMI 1640 supplemented with KCl, HEPES buffer, BSA, NAD, and 0.1 mg/ml of ATP and GTP. In addition, the culture medium was supplemented with Matrigel (Becton Dickinson) (50μl/ml of medium). Preliminary results show that the intermediate filaments remained stable in this medium, and the meronts appeared to be in better condition ultrastructurally at the interface of the colony and the medium (Fig. 15). This particular species may be of some use in biochemical studies since much of the host cytoplasm appears to be excluded from the isolated parasite colonies.

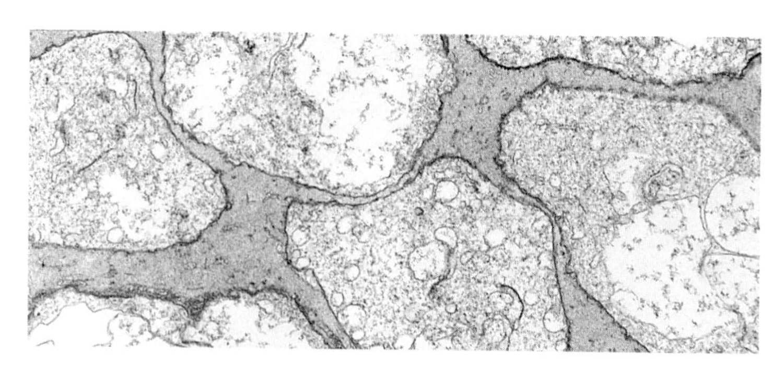

FIGURE 15 The periphery of an isolated extracellular *S. lophii* meront colony after 6 h in culture medium. Cells at the border of the colony (seen here) were apparently in better physical condition than interior cells of the colony, indicating that the dense matrix between meronts may affect nutrient infiltration. Magnification, ×25,000.

ACKNOWLEDGMENT

We thank C. M. Evans for her assistance in the preparation of several of the figures.

REFERENCES

Alaqui, H. E., J. Bata, D. Bauchart, and C. P. Vivares. 1997. Sporal lipid analysis of two microsporidian species parasites of fishes. *J. Eukaryot. Microbiol.* **44:**74S.

Albers, K., and E. Fuchs. 1992. The molecular biology of intermediate filament proteins. *Int. Rev. Cytol.* **134:**243–279.

Anderson, R. G. W. 1993. Caveolae: where incoming and outgoing messengers meet. *Proc. Natl. Acad. Sci. USA* **90:**10909–10913.

Aronow, B., K. Kaur, K. McCartan, and B. Ullman. 1987. Two high affinity nucleoside transporters in *Leishmania donovani*. *Mol. Biochem. Parasitol.* **22:**29–37.

Baburina, G. N., T. M. Ephimenko, I. V. Issi, M. C. Klyanvinsh, E. G. Rappoport, and Y. Y. Sokolova. 1989. Respiration activity of mitochondria isolated from winter moth larvae infected with microsporidia. *Bull. All-Union Inst. Plant Protection.* **73:**7–10.

Baker, M. D., C. R. Vossbrinck, E. S. Didier, J. V. Maddox, and J. A. Shadduck. 1995. Small subunit ribosomal DNA phylogeny of various Microsporidia with emphasis on AIDS related forms. *J. Eukaryot. Microbiol.* **42:**564–570.

Bartnik, E., and K. Weber. 1989. Widespread occurrence of intermediate filaments in invertebrates: common principles and aspects of diversion. *Eur. J. Cell Biol.* **50:**17–33.

Biderre, C., E. Peyretaillade, F. Duffleux, P. Peyret, G. Metenier, and C. Vivares. 1997. The rDNA unit of *Encephalitozoon*: complete 23S sequence and copy number. *J. Eukaryot. Microbiol.* **44:**76S.

Braun, A. P., and H. Schulman. 1995. The multifunctional Ca^{2+}/calmodulin-dependent protein kinase: from form to function. *Annu. Rev. Physiol.* **57:**417–445.

Burke, E. M. 1979. Aerobic and anaerobic metabolism during activity and hypoxia in two species of intertidal crabs. *Biol. Bull.* **156:**157–168.

Cali, A., D. P. Kotler, and J. M. Orenstein. 1993. *Septata intestinalis* n. g., n. sp., intestinal microsporidian associated with chronic diarrhea and dissemination in AIDS patients. *J. Eukaryot. Microbiol.* **40:**101–112.

Cali, A., and R. L. Owen. 1990. Intracellular development of *Enterocytozoon,* a unique microsporidian found in the intestine of AIDS patients. *J. Protozool.* **37:**145–155.

Canning, E. U. 1962. The pathogenicity of *Nosema locustae* Canning. *J. Insect Pathol.* **4:**248–256.

Canning, E. U. 1977. Microsporidia, p. 155–196. *In* J. P. Kreier (ed.), *Parasitic Protozoa,* vol. IV. Academic Press, New York.

Canning, E. U., and W. S. Hollister. 1991. In vitro and in vivo investigations of human Microsporidia. *J. Protozool.* **38:**631–635.

Canning, E. U., and J. Lom. 1986. *Microsporidia of Vertebrates.* Academic Press, New York.

Casero, R. A. 1995. Pages 1–3. *In* R. A. Casero (ed.), *Polyamines: Regulation and Molecular Interaction.* Springer-Verlag, New York.

Chapman, P. F. 1995. The alpha-Ca^{2+}/calmodulin kinase. II. A bidirectional modulator of presynaptic plasticity. *Neuron* **14:**491–497.

Choi, I., and R. Mikkelsen. 1990. *Plasmodium falciparum:* ATP/ADP transport across the parasitophorous vacuolar and plasma membranes. *Exp. Parasitol.* **71:**452–462.

Coyle, C., C. Bacchi, N. Yarlett, H. B. Tanowitz, M. Wittner, and L. M. Weiss. 1996. Polyamine metabolism as a therapeutic target for microsporidia. *J. Eukaryot. Microbiol.* **43:**96A.

Darwish, A., E. Weidner, and J. Fuxa. 1989. *Vairimorpha necatrix* in adipose cells of *Trichoplusia ni*. *J. Protozool.* **36:**308–311.

da Silva, A. C. R., and F. C. Reinach. 1991. Calcium binding induces conformational changes in muscle regulation proteins. *Trends Biochem. Sci.* **16:**53–57.

Dedinger, J. E., and F. C. Schatzlein. 1973. Carbohydrate metabolism in the shore crab, *Pachygrapsus crassipes.* II. Glycolytic rates of muscle, gill and hepatopancreas. *Comp. Biochem. Physiol.* **45B:**699–708.

de Graaf, D. C., H. Raes, G. Sabbe, P. H. de Rycke, and F. J. Jacobs. 1994. Early development of *Nosema apis* (Microspora: Nosematidae) in the midgut epithelium of the honey bee (*Apis mellifera*). *J. Invertebr. Pathol.* **63:**74–81.

Dimaria, P., B. Palic, B. A. Debrunner-Vossbrinck, J. Lapp, and C. R. Vossbrinck. 1996. Characterization of the highly divergent 42 RNA homolog in the microsporidian *Vairimorpha necatrix*. *Nucleic Acid Res.* **24:**515–522.

Dolgikh, V. V., M. V. Grigoryev, Y. Y. Sokolova, and I. V. Issi. 1995. An influence of infection with microsporidian *Nosema grylli* and coccidia *Adelina* sp. on an activity and isozyme pattern of lactate dehydrogenase in a fat body of the cricket *Gryllus bimaculatus. Parasitologiya* **26:**520–524.

Dolgikh, V. V., M. V. Grigoryev, Y. Y. Sokolova, and I. V. Issi. 1996. Influence of the microsporidian *Nosema grylli* and the coccidia *Adelina* sp. on the ovary development and on the activities of three dehydrogenases in fat body of female crickets *Gryllus bimaculatus. Parasitologiya* **30:**70–75.

Dolgikh, V., J. Sokolova, and I. Issi. 1997. Activities of enzymes of carbohydrate and energy metabolism of the spores of the microsporidian *Nosema grylii. J. Eukaryot. Microbiol.* **44:**246–249.

Dufort, M., Y. Valero, and M. Poguet. 1987. Particular distribution de la mitochondrias de *Mytilicola intestinalis* (Crustacea, Copepoda) en celulas parasitadas por *Unikaryon mytilicolae* (Microspora, Unikaryonidae). *Rev. Iber. Parasitol.* **Vol. Extraord:** 1–11.

Findley, A. M., E. W. Blakeney, Jr., and E. H. Weidner. 1981. *Ameson michaelis* (Microspora) in the blue crab, *Callinectes sapidus*: parasite-induced alterations in the biochemical composition of host tissues. *Biol. Bull.* **161:**115–125.

Findley, A. M., and E. H. Weidner. Unpublished observations.

Franke, W. W., M. Hergt, and C. Grund. 1987. Rearrangement of vimentin cytoskeleton during adipose conversion: formation of an intermediate filament cage around lipid globules. *Cell* **49:**131–141.

Frixione, E., L. Ruiz, J. Certon, and A. H. Undeen. 1997. Germination of *Nosema algerae* (Microspora) spores: conditional inhibition by D_2O, ethanol and Hg^+ suggests dependence of water influx upon membrane hydration and specific transmembrane pathways. *J. Eukaryot. Microbiol.* **44:** 109–116.

Georgatos, S. D., and G. Blobel. 1987. Two distinct attachment sites for vimentin along the plasma membrane and nuclear envelope in avian erythrocytes: a basis for vectorial assembly of intermediate filaments. *J. Cell. Biol.* **105:**105–115.

Georgatos, S. D., G. Blobel, and M. Chirico. 1989. Molecular interactions between intermediate filament proteins in heat shock proteins. *J. Cell Biol.* **109:**1417A.

Germot, A., H. Philipp, and H. LeGuyader. 1997. Evidence for loss of mitochondria in microsporidia from a mitochondria type HSP70 in *Nosema locustae. Mol. Biochem. Parasitol.* **87:**159–168.

Halonen, S., E. Weidner, and J. F. Siebenaller. 1996. Evidence of heterotrimer GTP-binding protein in *Toxoplasma gondii. J. Eukaryot. Microbiol.* **43:**187–193.

Haugland, R. P. 1996. *In* M. Spence (ed.), *Handbook of Fluorescent Probes and Research Chemicals.* Molecular Probes, Eugene, Oreg.

Hazard, E. I., E. A. Ellis, and D. J. Joslyn. 1981. Identification of Microsporidia, p. 163–182. *In*: H. D. Burgess (ed.), *Microbial Control of Pests and Plant Diseases 1970–1980.* Academic Press, New York.

Hinkle, G., H. L. Morrison, and M. L. Sogin. 1997. Genes coding for reverse transcriptase, DNA-directed RNA polymerase and chitin synthase for the microsporidian *Spraguea lophii. Biol. Bull.* **193:**250–251.

Horst, M., A. Azem, G. Schatz, and T. B. Glick. 1997. What's the driving force for import into mitochondria? *Biochim. Biophys. Acta* **1318:**71–78.

Jackson, A. P., H. Seow, N. J. Holmes, K. Drickamar, and P. Parham. 1987. Clathrin light chains contain brain-specific insertion sequences and a region of homology with intermediate filaments. *Nature* **336:**154–159.

Koleske, A., J. D. Baltimore, and M. P. Lisanti. 1995. Reduction of caveolin and caveolae in oncogenically transformed cells. *Proc. Natl. Acad. Sci. USA* **92:**1381–1385.

Kucera, M., and J. Weiser. 1975. The different course of lactate and glutamate dehydrogenases activity in the larvae of *Barathra brassicae* (Lepidoptera) during microsporidian infection. *Acta ntomol. Behemoslov.* **72:**370–373.

Lewis, L. C., J. A. Mutchmor, and R. E. Lynch. 1971. Effect of *Perezia pyraustae* on oxygen consumption by the European corn borer, *Ostrinia nubilalis. J. Insect Physiol.* **17:**2457–2468.

Liao, G., and G. G. Gunderson. 1997. Kinesin is a candidate for crossbridging microtubules and intermediate filaments: selectively binding of kinesin to detyrosinated tubulin and vimentin. *Mol. Biol. Cell* **8:**379A.

Liu, P., Y. Ying, Y. G. Ko, and R. G. W. Anderson. 1996. Localization of platelet-derived growth factor-stimulated phosphorylation cascade to caveolae. *J. Biol. Chem.* **271:**10299–10303.

Lom, J., and J. Vávra. 1963. The mode of sporoplasm extrusion in microsporidian spores. *Acta Protozool.* **1:**81–92.

Manale, S. B. 1996. *Spore envelope proteins of* Spraguea lophii. Ph.D. thesis. Louisiana State University, Baton Rouge.

Matouschek, A., A. Azem., K. Ratliff, B. S. Glick, K. Schmid, and G. Schatz. 1997. mtHSP70 can act as an ATP-driven force generating motor during protein import. *EMBO J.* **16:**6727–6736.

McCormack, S. Personal communication.

McPhalen, C. A., N. C. J. Strynadka, and M. N. G. James. 1991. Ca^{2+}-binding sites in proteins: a structural perspective. *Adv. Protein Chem.* **42:**77–144.

Muresan, V., P. Zerfas, A. Lyass, T. Reese, and B. J. Schnapp. 1997. Different vesicle populations in axoplasm contain different kinesin motors. *Mol. Biol. Cell* **8:**379A.

Olsen, G. J., and C. R. Woese. 1993. Ribosomal RNA: a key to phylogeny. *FASEB J.* **7:**113–123.

Opperdoes, F. R. 1987. Compartmentation of carbohydrate metabolism in trypanosomes. *Ann. Rev. Microbiol.* **41:**127–151.

Orenstein, J. M., J. Chiang, W. Steinberg, P. D. Smith, H. Rotterdam, and D. P. Kotler. 1990. Intestinal microsporidiosis as a cause of diarrhea in human immunodeficiency virus-infected patients: a report of 20 cases. *Hum. Pathol.* **21:**475–481.

Pegg, A. G. 1986. Recent advances in the biochemistry of polyamines in eukaryotes. *Biochem. J.* **234:**249–262.

Philippe, H., and A. Adoutte. 1995. How reliable is our current view of eukaryotic phylogeny? p. 17–33. *In* G. Brugerolle and J. P. Mignot (ed.), *Protistological Actualities. Proceedings of the Second European Congress of Protistology and Eighth European Conference on Ciliate Biology.* Clermont-Ferrand, France.

Phillips, J. W., J. W. McKinney, F. J. R. Hird, and D. L. MacMillan. 1977. Lactic acid formation in crustaceans and the liver function of the midgut gland questioned. *Comp. Biochem. Physiol.* **56B:** 427–433.

Pleshinger, J., and E. Weidner. 1985. The microsporidian spore invasion tube. IV. Discharge activation begins with pH-triggered Ca^{2+} influx. *J. Cell Biol.* **100:**1834–1838.

Raper, J. R., and H. S. Flexer. 1970. The road to diploidy with emphasis on a detour, p. 401–432. *In Symposium of the Society for General Microbiology.* Cambridge University Press, Cambridge.

Robinson, J. M., and M. J. Karnovsky. 1983. Ultrastructural localization of nucleotidase in guinea pig neutrophils based on the use of cerium as a capture agent. *J. Histochem. Cytochem.* **31:**1190–1196.

Satir, B. H. 1989. Signal transduction events associated with exocytosis in ciliates. *J. Protozool.* **36:** 382–389.

Scarborough-Bull, A., and E. Weidner. 1985. Some properties of discharged *Glugea hertwigi* (Microspora) sporoplasms. *J. Protozool.* **32:**284–289.

Schatzlein, F. C., H. M. Carpenter, M. R. Rodgers, and J. L. Sutko. 1973. Carbohydrate metabolism in the striped shore crab, *Pachygrapsus crassipes.* I. The glycolytic enzymes of gill, hepatopancreas, heart and leg muscle. *Comp. Biochem. Physiol.* **45B:**393–405.

Selesnjov, K., I. Issi, V. Dolgikh, G. Belostotskaya, O. Antonova, and J. Sokolova. 1995. Fractionation of different life cycle stages of Microsporidia *Nosema grylli* from crickets *Gryllus bimaculatus* by centrifugation in Percoll density gradient for biochemical research. *J. Eukaryot. Microbiol.* **42:**288–292.

Selesnjov, K. V., A. O. Antonova, and I. V. Issi. 1996. The microsporidiosis of the crickets *Gryllus bimaculatus* (Gryllidae) caused by the Microsporidia *Nosema grylli* (Nosematidae). *Parasitologiya* **30:**250–262.

Shadduck, J. A. 1969. *Nosema cuniculi*: in vitro isolation. *Science* **166:**516–517.

Shawtz, L. M., and A. E. Pegg. 1994. Overproduction of ornithine decarboxylase caused by relief of translational repression is associated with neoplastic transformation. *Cancer Res.* **54:**2313–2316.

Sibley, L. D., T. R. Niesman, T. Asai, and T. Takeuchi. 1994. *Toxoplasma gondii*: secretion of a potent nucleoside triphosphate hydrolase at the parasitophorous vacuole. *Exp. Parasitol.* **79:**301–311.

Sinderman, C. J. 1970. *Principal Diseases of Marine Fish and Shellfish.* Academic Press, New York.

Sokolova, Y. Y., K. V. Selesnjov, V. V. Dolgikh, and I. V. Issi. 1994. Microsporidia *Nosema grylli* n.sp. from the cricket *Gryllus bimaculatus. Parasitologiya* **28:**488–493.

Sokolova, Y. Y., S. A. Timoshenko, and I. V. Issi. 1988. Morphogenesis and ultrastructure of life cycle stages of *Nosema mesnili* Paillot Microsporidia, Nosematidae). *Citologiya* **30:**26–33.

Sorensen, W., A. J. Billington, S. A. Norris, J. E. Briggs, M. T. Reding, and G. A. Filice. 1987. *Toxoplasma gondii*: metabolism of intracellular tachyzoites is affected by host cell ATP production. *Exp. Parasitol.* **85:**101–104.

Steinbuchel, A., and M. Muller. 1986a. Glycerol, a metabolic end product of *Trichomonas vaginalis* and *Tritrichomonas foetus*. *Mol. Biochem. Parasitol.* **20:** 45–55.

Steinbuchel, A., and M. Muller. 1986b. Anaerobic pyruvate metabolism of *Tritrichomonas foetus* and *Trichomonas vaginalis* hydrogenosomes. *Mol. Biochem. Parasitol.* **20:**57–65.

Strynadka, N. C., and M. N. G. Jans. 1991. Towards an understanding of the effects of calcium on protein structure and function. *Curr. Opin. Struct. Biol.* **1:**905–914.

Trager, W. 1974. Some aspects of intracellular parasitism. *Science* **183:**269–273.

Trager, W. 1986. *Living Together: The Biology of Animal Parasites.* Plenum Press, New York.

Undeen, A. H. 1975. Growth of *Nosema algerae* in pig kidney cell cultures. *J. Protozool.* **22:**107–110.

Undeen, A. H. 1990. A proposed mechanism for the germination of microsporidian (Protozoa: Microspora) spores. *J. Theor. Biol.* **142:**223–235.

Undeen, A. H., and S. W. Avery. 1988a. Ammonium chloride inhibition of germination of spores of *Nosema algerae* (Microspora: Nosematidae). *J. Invertebr. Pathol.* **52:**326–334.

Undeen, A. H., and S. W. Avery. 1988b. Effects of anions on the germination of *Nosema algerae* (Microspora: Nosematidae). *J. Invertebr. Pathol.* **52:**84–89.

Undeen, A. H., L. M. Elgazzar, R. K. Vander Meer, and S. Narang. 1987. Trehalose levels and trehalase activity in germinated and ungerminated spores of *Nosema algerae* (Microspora: Nosematidae). *J. Invertebr. Pathol.* **50:**230–237.

Undeen, A. H., and R. K. Vander Meer. 1994. Conversion of intrasporal trehalose into reducing sugars during germination of *Nosema algerae* (Protista: Microspora) spores: a quantitative study. *J. Eukaryot. Microbiol.* **41:** 129–132.

Vander Meer, J. W., and T. A. Gochnauer. 1969. Some effects of sublethal heat on spores of *Nosema apis*. *J. Invert. Pathol.* **13**:442–446.

Vander Meer, J. W., and T. A. Gochnauer. 1971. Trehalase activity associated with spores of *Nosema apis*. *J. Invertebr. Pathol.* **17:**38–41.

Vikstrom, K. L., G. G. Borisy, and R. D. Goldman. 1989. Dynamic aspects of intermediate filaments networks in BHK-21 cells. *Proc. Natl. Acad. Sci. USA* **86:**549–553.

von Brand, T. 1973. *Biochemistry of Parasites.* Academic Press, New York.

Vossbrinck, C. R., M. D. Baker, E. S. Didier, B. A. Debrunner-Vossbrinck, and J. A. Shadduck. 1993. Ribosomal DNA sequences of *Encephalitozoon hellem* and *Encephalitozoon cuniculi*: species identification and phylogetic construction. *J. Eukaryot. Microbiol.* **40:**354–362.

Vossbrinck, C. R., J. V. Maddox, S. Friedman, B. A. Debrunner-Vossbrinck, and C. R. Woese. 1987. Ribosomal RNA sequence suggests microsporidia are extremely ancient eukaryotes. *Nature* **326:**411–414.

Vossbrinck, C. R., and C. R. Woese. 1986. Eukaryotic ribosomes that lack a 5.8S RNA. *Nature* **320:**287–288.

Weidner, E. 1970. Ultrastructural study of microsporidian development. I. Nosema sp. Sprague, 1965 in *Callinectes sapidus* Rathbun. *Z. Zellforsch.* **105:**33–54.

Weidner, E. 1972. Ultrastructural study of microsporidian invasion into cells. *Z. Parasitenkd.* **40:**227–242.

Weidner, E. 1976. The microsporidian spore invasion tube. *J. Cell Biol.* **71:**23–34.

Weidner, E. 1982. The microsporidian spore invasion tube. III. Tube extrusion and assembly. *J. Cell Biol.* **93:**976–979.

Weidner, E. 1992. Cytoskeletal proteins expressed by microsporidian cells. *Subcell. Biochem.* **18:**385–399.

Weidner, E. Unpublished observations.

Weidner, E., and W. Byrd. 1982. Microsporidian spore invasion tube. II. Role of Ca^{2+} in the activation of the invasion tube discharge. *J. Cell Biol.* **93:**970–975.

Weidner, E., and S. K. Halonen. 1993. Microsporidian spore envelope keratins phosphorylate and disassemble during spore activation. *J. Eukaryot. Microbiol.* **40:**783–788.

Weidner, E., S. B. Manale, S. K. Halonen, and J. W. Lynn. 1995. Protein-membrane interactions essential to normal assembly of microsporidian spore invasion tube. *Biol. Bull.* **188:** 128–135.

Weidner, E., R. M. Overstreet, B. Tedeschi, and J. Fuseler. 1990. Cytokeratin and desmoplakin analogues within an intracellular parasite. *Biol. Bull.* **179:**237–243.

Weidner, E., and W. Trager. 1973. Adenosine triphosphate in the extracellular survival of an intracellular parasite (*Nosema michaelis,* Microsporidia). *J. Cell Biol.* **57:**586–591.

Weiser, J. 1976. Microsporidia in invertebrates: Host-parasite relations at the organismal level, p. 163–201. *In* L. A. Bulla, Jr., and T. C. Cheng (ed.), *Comparative Pathobiology,* vol. 1. Plenum Press, New York.

Weissenberg, R. 1968. Intracellular development of the microsporidian *Glugea anomala* Moniez in hypertrophied migrating cells of the fish *Gasterosteus aculeatus* L., an example of formation of xenoma tumors. *J. Protozool.* **15:**44–57.

Weissenberg, R. 1976. Microsporidian interaction with host cells, p. 203–237. *In* L. A. Bulla, Jr., and T. C. Cheng (ed.), *Comparative Pathobiology,* vol. 1. Plenum Press, New York.

Winkler, H. H. 1976. Rickettsial permeability: an ADP-ATP transport system. *J. Biol. Chem.* **251:**389–396.

Wordeman, L., and J. Mitchison. 1995. Identification and partial characterization of mitotic centromere-associated kinesin, a kinesin-related protein that associates with centromeres during mitosis. *J. Cell Biol.* **128:**95–104.

Xie, W. D., and E. U. Canning. 1986. *Pyrothecas hydropsycheae* n.sp. a microsporidian parasite of caddis fly larvae. *J. Protozool.* **33:**462–467.

Yokoyama, S., and W. Starmer. 1992. Phylogeny and evolutionary rates of G protein alpha subunit genes. *J. Mol. Evol.* **35:**230–238.

Zhu, X., M. Wittner, H. B. Tanowitz, A. Cali, and L. M. Weiss. 1994. Ribosomal RNA sequences of *Enterocytozoon bieneusi, Septata intestinalis* and *Ameson michaelis*: phylogenetic construction and structural correspondence. *J. Eukaryot. Microbiol.* **41:**204–209.

THE STRUCTURE, FUNCTION, AND COMPOSITION OF THE MICROSPORIDIAN POLAR TUBE

Elaine M. Keohane and Louis M. Weiss

6

The spores of microsporidia possess a unique, highly specialized structure, the polar tube, which is used to inject the parasite from the spore into a new host cell. In the spore, the polar tube is connected at the anterior end and then coils around the sporoplasm. On appropriate environmental stimulation, the polar tube is rapidly discharged out of the spore with sufficient force to pierce a cell membrane, forming a 50- to 500-μm hollow tube that serves as a conduit for sporoplasm passage into the new host cell. The polar tube is often referred to as the polar filament while inside the spore since it contains fine, particulate, electron-dense material prior to discharge (Lom and Vávra, 1963, Lom, 1972; Weidner, 1972, 1976). Because of the small size of the spores (approximately 1 to 10 μm), the small diameter of the polar tube (0.1 to 0.2 μm), and the rapidity of the polar tube discharge and sporoplasm passage (<2 s), study of the polar filament extrusion process has been difficult.

Elaine M. Keohane, Department of Clinical Laboratory Sciences, University of Medicine and Dentistry of New Jersey, Newark, NJ 07107. *Louis M. Weiss,* Division of Infectious Diseases and Division of Parasitology, Albert Einstein College of Medicine, Bronx, NY 10461.

It has been over 100 years since Thelohan accurately described the microsporidian polar filament and the triggering of its discharge by nitric acid (Thelohan, 1892, 1894). In 1894 he wrote, "Enfin, dans un assez grand nombre de spores, on constate la sortie d'un filament qui attient trois ou quatre fois la longueur primitive de la spore, soit 12 à 14 μ. En rapport avec cette sortie du filament on trouve la capsule diminuée de volume et, surtout, beaucoup moins réfringente" (Thelohan, 1894). This observation on the polar tube extrusion process was later confirmed by Stempell (1909), Korke (1916), and Kudo (1918). Schuberg (1910), Leger and Hesse (1916), and Kudo (1916, 1920) noted that the polar filament was coiled in the spore, and drawings rendered by Kudo in 1920 depicted a spore with a coiled polar tube partially extruded to a length of 230 μm (Kudo, 1920) (Fig. 1).

Korke (1916) described the opening of the polar capsule at the anterior end of the spore, noted the extrusion of the polar filament, and identified a droplet or "amoebula-like" body at the end of the filament as the parasite. He also proposed that the function of the filament may be to "conduct the sporozoite (amoebula) to a distant part of the tissue and thus ensure an advance of the parasite into fresh areas" (Korke, 1916). Although many investigators

The Microsporidia and Microsporidiosis (Murray Wittner, editor; Louis M. Weiss, contributing editor), ©1999 American Society for Microbiology, Washington, D.C.

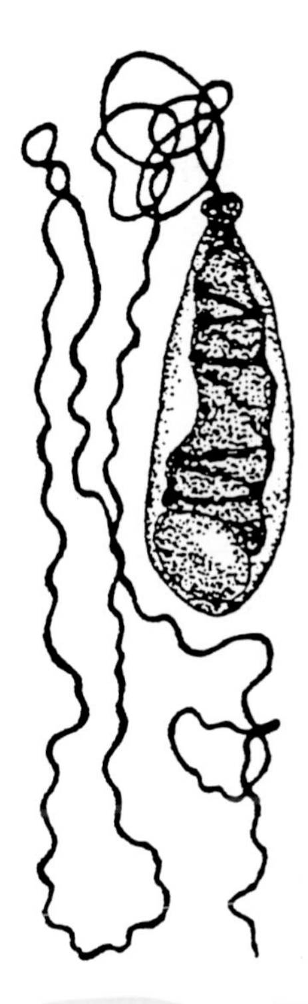

FIGURE 1 Polar tube extrusion. A spore of the microsporidium *Thelohania magna* after polar tube extrusion by mechanical pressure is shown (Kudo, 1920).

subsequently observed the drop of fluid at the end of the extruded polar filament, it was not initially clear if the drop was sporoplasm and how it arrived at the tip of the filament. Many investigators thought it highly unlikely that an entire sporoplasm could pass through such a long, narrow polar tube.

Several theories have been proposed regarding the method by which the sporoplasm exits the spore and on the function of the polar filament or tube in this process. One theory proposed that the drop of viscous fluid at the end of the polar filament was used for attachment and that the function of the filament was to fix the spore to the target tissues (Oshima, 1927). It was believed that after extrusion the sporoplasm either flowed out of the spore at the base of the filament (Fantham and Porter, 1914; Kudo, 1916; Zwolfer, 1926) or flowed out a hole in the spore that formed after the filament disconnected (Connet, 1932; Hall, 1952; Weiser, 1958). Others proposed that a solid filament was discharged from the spore in a "jack-in-the-box fashion" (Oshima, 1927), dragging the sporoplasm out of the spore (Dissanaike, 1955, Dissanaike and Canning, 1957). Stempell (1909) and Strickland (1913) thought that the sporoplasm poured out an evaginated polar filament formed from the projection of a chitinous spore covering. In 1937 Oshima, and later others, proposed evagination of the filament during discharge to form a hollow tube that remained attached to the spore, as well as passage of the sporoplasm from the spore through this tube (Oshima, 1937, 1966; Gibbs, 1953; Bailey, 1955; Walters, 1958; West, 1960; Kramer, 1960; Lom and Vávra, 1963). Evagination of the polar filament has been likened to "reversing a finger of a glove" (Lom and Vávra, 1963; Lom, 1972).

Oshima (1937) and Trager (1937) observed nuclei in the droplet at the end of the filament, and Oshima correctly interpreted this to be the organism emerging from the tube (Oshima, 1937). He also theorized that by using the tube to inject the sporoplasm into a host cell, the parasite had a mechanism that protected the sporoplasm from the digestive fluid in the gut of insects (Oshima, 1937). Further evidence of sporoplasm passage through the tube came from Gibbs (1953) who described sporoplasm flowing out of discharged polar tubes, as well as from West (1960) and Kramer (1960) who demonstrated the sporoplasm nucleus in partially extruded stained polar tubes. Lom and Vávra (1963) also detected the sporoplasm nucleus in partially extruded stained polar tubes and further reported a widening of the polar tube as the sporoplasm passed. With electron microscopy, Lom (1972) and Weidner (1976) observed elongated sporoplasm in sections of extruded polar tube, as well as the polar tube piercing membranes (Weidner, 1972) (Fig. 2 and 3). Although it is now accepted that the sporoplasm flows through the discharged polar tube and into the host cell, the mechanism of activation and tube formation during discharge still remains unclear.

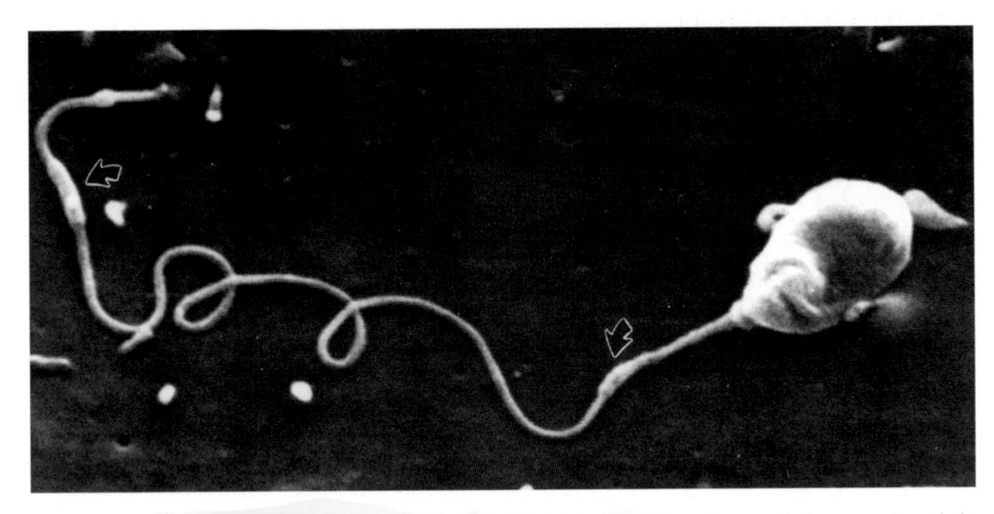

FIGURE 2 Scanning electron micrograph of sporoplasm passage. The micrograph shows an *Encephalitozoon intestinalis* spore with an extruded polar tube. Arrows indicate sporoplasm passage through the polar tube. (Reprinted with permission from Kock, 1998.)

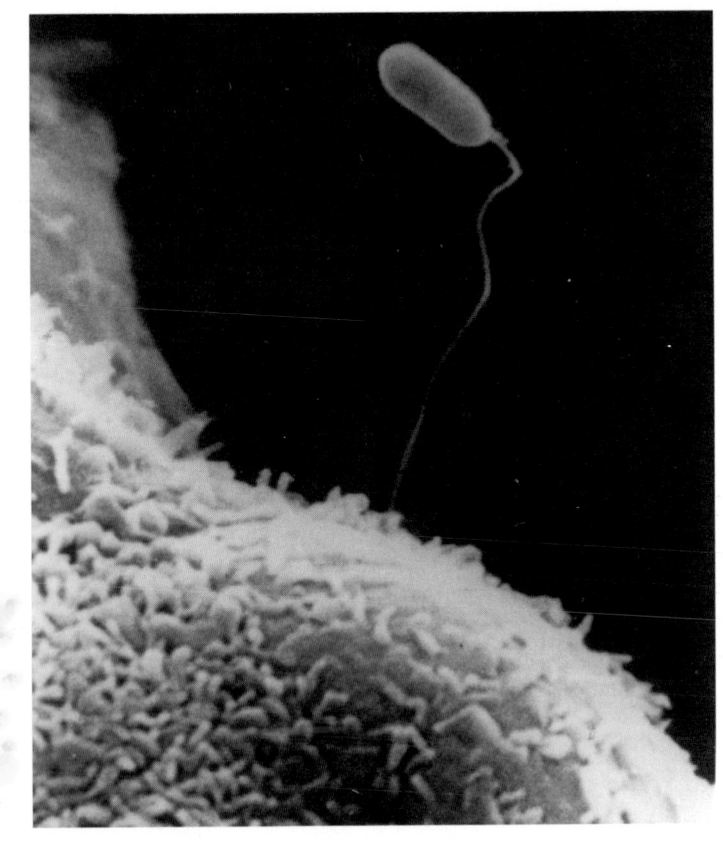

FIGURE 3 Scanning electron micrograph of a microsporidian infection of a host cell. The micrograph shows an extruded polar tube of a spore of *E. intestinalis* piercing and infecting Vero E6 green monkey kidney cells in tissue culture. (Reprinted with permission from Kock, 1998.)

MORPHOLOGY OF THE POLAR FILAMENT OR POLAR TUBE

The spores of the microsporidia vary in size from approximately 1 to 10 μm (Cali and Owen, 1988). Spores have three general features: the spore coat, the sporoplasm, and the extrusion apparatus (Vávra, 1976) (Fig. 4 and 5). The environmentally resistant spore coat consists of three layers observed by transmission electron microscopy (TEM): an electron-dense, proteinaceous exospore, an electron-lucent endospore, and a plasma membrane or plasmalemma (Vávra, 1976; Canning and Lom, 1986; Cali and Owen, 1988). The exospore may be corrugated, with ridges, filaments, or tubular structures (Vávra, 1976; Cali and Owen, 1988), but appendages on the outer surface have been found only in microsporidia of aquatic hosts (Vávra, 1976). The endospore layer is composed of chitin and protein (Kudo, 1921; Dissanaike and Canning, 1957; Vávra, 1976) and is thinnest at the anterior end of the spore where rupture occurs during discharge of the polar tube. The sporoplasm contains a nucleus which may be single or a diplokaryon (two abutted nuclei) depending on the genus (Vávra, 1976; Cali and Owen, 1988). Numerous ribosomes have also been observed in helical coils or in sheets (Vávra, 1976; Cali and Owen, 1988; Chioralia et al., 1998). The sporoplasm in the dormant spore is electron-dense, and during activation it loses this density (Lom, 1972; Chioralia et al., 1998). This can also be observed with phase-contrast microscopy in which the spore loses refractility after extrusion (Vávra, 1976). When stained with 0.2% trypan blue, extruded spores are blue and intact spores are colorless (de Graaf et al., 1993).

The spore has an extrusion apparatus consisting of a long polar filament connected to a mushroom-shaped anchoring disk at the anterior end of the spore, the polaroplast, and the posterior vacuole (Vávra, 1976) (Fig. 4 and 5). The filament is divided into two regions: the manubroid or straight portion connected to the anchoring disk, and the posterior region that forms from 4 to more than 30 coils around the sporoplasm depending on the species (Huger,

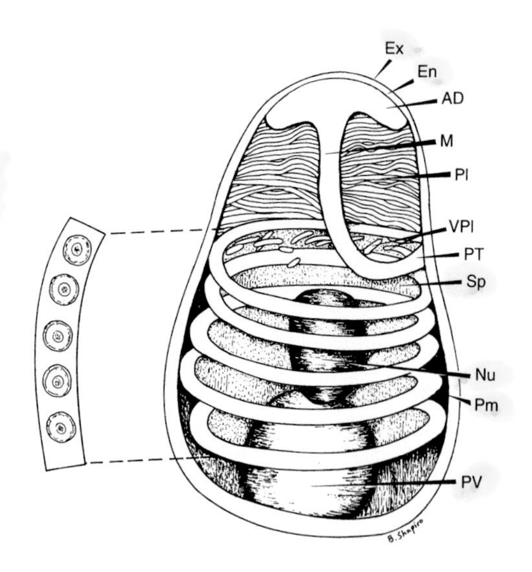

FIGURE 4 Diagram of a microsporidian spore. Spores range in size from 1 to 10 μm. The spore coat consists of an electron-dense exospore (Ex), electron-lucent endospore (En), and plasma membrane (Pm). It is thinner at the anterior end of the spore. The sporoplasm (Sp) contains a single nucleus (Nu), the posterior vacuole (PV), and ribosomes. The polar filament is attached to the anterior end of the spore by an anchoring disk (AD) and is divided into two regions: the manubroid or straight portion (M) and the posterior region forming five coils (PT) around the sporoplasm. The manubroid polar filament is surrounded by the lamellar polaroplast (Pl) and vesicular polaroplast (VPl). The inset depicts a cross section of the polar tube coils (five coils in this spore), demonstating the various concentric layers of different electron density and electron-dense core present in such cross sections.

1960; Vávra, 1976; Cali and Owen, 1988). The coils can occur in one to several rows (Vávra, 1976) (Fig. 4 and 5). The manubroid polar filament and the coils closest to the anterior end have a larger diameter (0.16 to 0.18 μm) than the distal coils (0.11 to 0.12 μm) (Kudo and Daniels, 1963; Sinden and Canning, 1974; Takvorian and Cali, 1986; Chioralia et al., 1998). Although it has been generally observed that the manubroid portion of the polar tube extends to the medial portion of the spore and then coils (Vávra, 1976; Chioralia et al., 1998), Takvorian and Cali (1986) showed that the

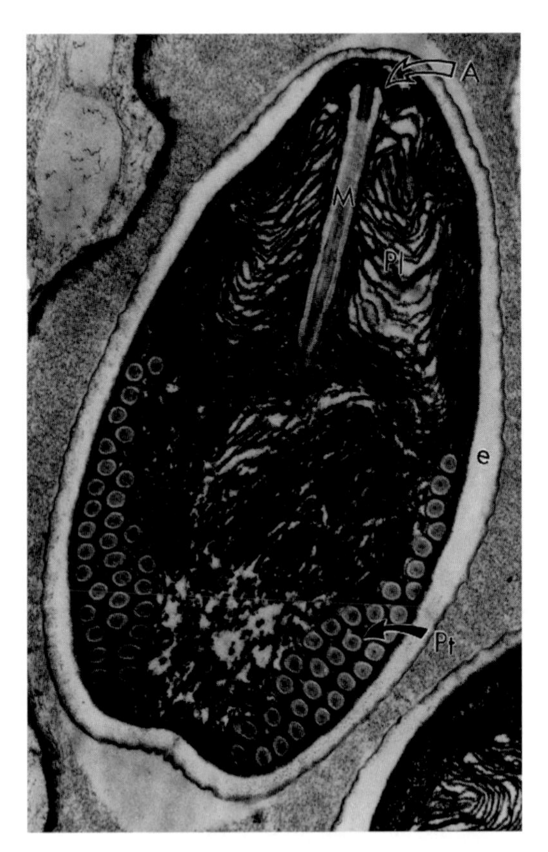

FIGURE 5 Transmission electron micrograph of a longitudinal section of a spore of *Plistophora hyphessobryconis.* The manubroid polar filament (M) is attached to the anterior end of the spore by an anchoring disk (open arrow) and then forms 33 coils (Pt and solid arrow) around the sporoplasm in one to three rows. Pl, lamellar polaroplast; e, electron-lucent endospore. Note that the endopsore layer is thinner at the apical end and that the polar tube contains concentric layers of varying electron density with a dense core. The spore measures 2.8 by 4.6 μm. (Reprinted with permission from Lom and Corliss, 1967.)

manubroid polar filament of *Glugea americanus* extends the entire length of the spore, turns anteriorly, and then begins to coil. The distal end of the polar filament has not been demonstrated, and it remains unclear if it is blunt (i.e., closed) or open-ended inside the spore (Erickson et al., 1968; Lom, 1972; Vávra, 1976; Chioralia et al., 1998). On longitudinal section a limiting membrane surrounding the polar filament can be seen (Vávra, 1976; Weidner, 1972, 1976), and it is composed of three layers of different electron density (Lom, 1972; Chioralia et al., 1998). Sinden and Canning (1974) suggested that it was a double membrane, the outer membrane being contiguous with the polar sac or membrane surrounding the anchoring disk.

The anterior portion of the manubroid polar filament is surrounded by membranous structures called the lamellar polaroplast and the tubular or vesicular polaroplast (Huger, 1960; Vávra, 1976; Takvorian and Cali, 1986; Chioralia et al., 1988). The lamellar polaroplast membranes are more anteriorly located and consist of tightly packed, flattened membranes, stacked one upon another, which abut the manubroid filament and anchoring disk and are perpendicular to the filament (Takvorian and Cali, 1986; Chioralia et al., 1988). The vesicular polaroplast is more tubular in shape and located toward the center of the spore. It has been suggested that the polaroplast membrane is contiguous with the limiting membrane around the polar filament (Weidner, 1972; Weidner et al., 1995). Also present in the sporoplasm is a membrane-bound posterior vacuole that can contain flocculent, granular material, membrane whirls, or glomerulus-like structures and often abuts the coils of the polar tube (Weidner, 1972; Lom and Corliss, 1967; Cali and Owen, 1988).

In cross section, the polar filament inside the spore is composed of electron-dense and electron-lucent concentric layers (Fig. 4 and 5) (Huger, 1960; Vávra, 1976). The number of layers observed has been variable, ranging from as few as 3 to 6 layers (Vávra et al., 1966; Lom, 1972; Chioralia et al., 1998) to as many as 11 to 20 different layers in a cross section of polar tube (Sinden and Canning, 1974; Vávra, 1976). In mature spores Chioralia et al. (1998) noted six layers in the coiled part of the polar tube and only three layers in the straight part (manubroid). It appears that the thickness of the layers varies along the polar filament (Vávra et al., 1966), while number of layers varies with spore maturity (Vávra, 1976; Chioralia et al.,

1998). Moreover, a different pattern of layers has been observed before, during, and after extrusion (Lom, 1972; Weidner, 1972; Chioralia et al., 1998). Electron-dense, particulate material fills the center of the filament (Kudo and Daniels, 1963; Lom and Vávra, 1963; Vávra, 1976) and undergoes changes during the eversion process (Lom and Corliss, 1967). Weidner proposed that this material was unpolymerized polar tube protein (PTP) (Weidner, 1972, 1976).

Spores with completely discharged tubes appear empty except for an increased posterior vacuole and folded plasma membrane (plasmalemma) (Lom, 1972). During spore discharge, the anchoring disk ruptures, the polar tube everts, and a collarlike structure in the area of the anchoring disk holds the filament in place during extrusion (Lom, 1972) (Fig. 6A and B). On the basis of ultrastructural observations, the eversion of the polar tube has been likened to a tube sliding within a tube (Weidner, 1982; Weidner et al., 1995; Chioralia et al., 1998). Further details on spore activation and discharge are presented later in this chapter.

Extruded polar tubes appear to be hollow and made of membranes (Lom, 1972). They range from 0.1 to 0.2 μm in diameter and 50 to 150 μm in length (Kudo and Daniels, 1963; Weidner, 1976; Frixione et al., 1992), although tubes as long as 300 to 500 μm have been reported in some microsporidia (Lom and Corliss, 1967; Hashimoto et al., 1976; Olsen et al., 1986). The polar tube has considerable elasticity and flexibility in that it shows variation in diameter from 0.1 to 0.25 μm during discharge (Scarborough-Bull and Weidner, 1985), its diameter can increase to 0.6 μm during sporoplasm passage (Lom and Vávra, 1963; Ishihara, 1968; Weidner, 1972, 1976; Olsen et al., 1986), and its length shortens by 5 to 10% after sporoplasm passage (Frixione et al., 1992).

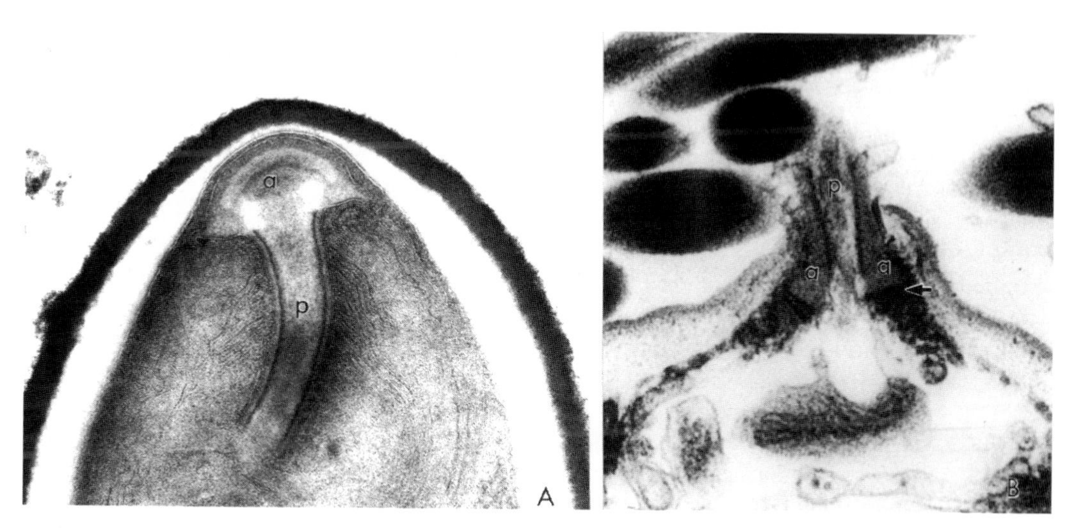

FIGURE 6 Morphology of anchoring disk. (A) Electron micrograph of the anterior portion of a longitudinal section of a spore of *Nosema algerae* showing the mushroom-shaped anchoring disk (a) connected to the manubroid polar filament (p). The polar filament and anchoring disk are surrounded by a limiting membrane. Note the thin endospore area above the anchoring disk. (Reprinted with permission from Peter M. Takvorian.) (B) Electron micrograph of the anterior portion of a longitudinal section of a spore of a *Nosema* sp. showing polar filament extrusion. Collarlike structures (a) formed from the anchoring disk hold the polar tube (p) in place during extrusion. This figure also demonstrates eversion of the polar tube. Note that the hinge region (dark lines in the anchoring disk) and the anchoring disk have rotated 90° during extrusion of the polar tube. (Reprinted with permission from Lom, 1972.)

POLAR TUBE MORPHOGENESIS

In most microsporidia, synthesis of the polar filament begins in the early sporoblast as an oval body of membranes and dense material (Takvorian and Cali, 1996). Subsequently a Golgi-like structure appears with cisternae, small vesicles, and sacs (Sprague and Vernick, 1969). The anchoring disk originates from a membrane-limited vesicle or polar sac (Vávra, 1976). Initially it begins near the nucleus and Golgi-like structure and then moves to the anterior portion of the spore to form the anchoring disk (Takvorian and Cali, 1986). Then vacuolelike structures transform into tubules forming hollow rings which pinch off, eventually forming the polar filament. As maturation proceeds in mature sporoblasts, vesicles fuse to form large, multilobed bodies closely associated with the developing polar filament (Takvorian and Cali, 1996). The core and envelope of the polar filament appear first, followed by additional layers inside the filament (Vávra, 1976). The number of layers in a polar filament coil depends on its maturity in that fewer layers are found in the coils of immature spores than in mature spores (Vávra, 1976; Chioralia et al., 1998).

Several observers have suggested that the polar filament originates from the coalescence of vacuoles of the Golgi complex, forming a tube with a limiting membrane (Vávra, 1976; Vávra and Undeen, 1970). Weidner (1970) proposed that the central core of the filament arises from the Golgi-like saccules, and the outer envelope from the endoplasmic reticulum. Sprague and Vernick (1969) concluded that the outer membrane originated from Golgi-like membranes but that the middle layer was of mitotic spindle origin. However, Jensen and Wellings (1972) believed that the base of the filament was derived from transformed nuclear material, the anterior part of the tube from cisternae and endoplasmic reticulum and the posterior part from the Golgi complex. In *Glugea stephani* thiamine pyrophosphatase was present on membranes and dense material that formed the polar filament, suggesting a trans-Golgi association (Takvorian and Cali, 1994). However, staining of the polar filament core, its outer sheath, and its originating vacuoles was also seen with nucleoside disphosphatase (NDPase), a marker for endoplasmic reticulum as well as cis-Golgi membrane (Takvorian and Cali, 1996). The polaroplast, polar sac, and posterior vacuole have also been reported to be derived from the Golgi complex (Sprague and Vernick, 1969; Jensen and Wellings, 1972). Lom and Corliss (1967) and Weidner (1970), however, provided evidence that the polaroplast was elaborated from the endoplasmic reticulum.

The morphogenesis of the extrusion apparatus is different in the Enterocytozoonidae than in other microsporidia. Polar tube formation occurs in the sporogonial plasmodium stage, before evidence of sporoblast development. In *E. bieneusi,* a large, electron-dense body associated with vesiculotubular networks forms electron-dense disks (Desportes-Livage et al., 1996) which then fuse into arcs, forming the polar filament coils (Cali and Owen, 1990). Multiple anchoring disks and polaroplasts also arise at this stage. When sporoblasts form, each one contains five to six coils of the preformed polar filament with the anchoring disk positioned at the anterior end. The extrusion apparatus then completes its development in the sporoblast. It is unclear whether the Golgi complex forms the polar filament in Enterocytozoonidae since the nucleus, mediated by the vesicular polaroplast, is associated with a forming polar tube (Desportes-Livage et al., 1996). The vesicular polaroplast appears to originate from vesicles enclosed within rough endoplasmic reticulum cisternae, while the nuclear envelope and endoplasmic reticulum cisternae form the lamellar polaroplast. In *E. salmonis* (*Nucleospora salmonis*), polar tubes do not form from electron-dense disks but from cylinders or tubular structures associated with ribosomes that line up along the endoplasmic reticulum cisternae and fuse end to end, forming the outer layer of the polar filament (Chilmonczyk et al., 1991; Desportes-Livage et al., 1996). Polaroplast precursors are small aggregates of vesicles (Desportes-Livage et al., 1996).

POLAR TUBE PROTEINS

Because of the unique function of the polar tube, elucidation of its chemical nature has been of interest to researchers. Table 1 summarizes the PTPs identified to date. Kudo (1921) observed that polar tubes were insoluble in water and saliva. However, they were found to be completely digested by trypsin in 24 h (Zwolfer, 1926; Weidner, 1972) and rapidly digested after extrusion in digestive fluid or in the midgut of insects (Oshima, 1927, 1937; Undeen, 1976; Undeen and Epsky, 1990). Polar tubes are visible by light microscopy when stained with Thiery's periodic acid thiosemicarbazide-silver proteinate method (Vávra, 1976) and the periodic acid-Schiff reaction (Vávra, 1976), and on electron microscopy they bind ferritin-conjugated concanavalin A (Con A) (Weidner, 1972). It is likely, therefore, that glycoproteins are present in the polar tube.

Polar tubes resist dissociation in 1 to 3% sodium dodecyl sulfate (SDS), 1% Triton X-100, 1 to 10% H_2O_2, 5 to 8 N H_2SO_4, 1 to 2 N HCl, chloroform, 1% guanidine HCl, 0.1 M proteinase K, 8 to 10 M urea, 50 mM $NaCO_3$, and 50 mM $MgCl_2$ (Weidner, 1972, 1976, 1982). They dissociate, however, in various concentrations of 2-mercaptoethanol (2-ME), dithiothreitol (DTT), 6% urea with 0.1 M proteinase K and 1% guanidine HCl in 0.1 M proteinase K (Weidner, 1976, 1982; Keohane et al., 1996c). The polar filament inside the spore has been found to react to reducing agents and detergents in the same manner as everted polar tubes (Weidner, 1976, 1982). It is the solubility properties of PTPs that allow their separation from other proteins in the spore. Since the spore coat and polar tube are resistant to concentrations of detergents and acids that solubilize most other proteins, treatment of spores with these agents can be used to isolate PTPs.

Weidner (1976) purified a putative PTP from *Ameson michaelis* by utilizing the unusual solubility properties of polar tubes. After spore discharge was triggered or the spore wall was disrupted with 8 N sulfuric acid, spores were washed with 3% SDS to remove sporoplasm and soluble spore coat proteins. Subsequent treatment of the spores with 1% DTT or 50% 2-ME selectively solubilized the polar tubes, leaving the spore coats intact. Since both DTT and 2-ME readily dissociate the polar tube, it is probable that disulfide bridging is important in stabilizing polar tube structure. SDS-polyacrylamide gel electrophoresis (PAGE) of the DTT or 2-ME supernatants containing the solubilized polar tubes revealed a single 23-kDa protein band, presumably representing a PTP (Weidner, 1976). Amino acid analysis of this protein demonstrated the presence of multiple cysteine residues, consistent with the hypothesis that disulfide bridging is important in PTPs (Weidner, 1976).

With a similar extraction procedure, a 43-kDa PTP was isolated from the mechanically disrupted spores of the fish microsporidium *G. americanus* (Keohane et al., 1994). After sequential extraction of glass bead-disrupted spores with 1% SDS and 9 M urea, followed by solubilization of the residual polar tubes in 2% DTT (Keohane et al., 1996c) (Fig. 7), the DTT-solubilized material contained four protein bands of 23, 27,

TABLE 1 Polar tube proteins

Organism	Molecular Mass (kDa)	Reference(s)
Ameson michaelis	23	Weidner, 1976
Encephalitozoon cuniculi	45	Keohane et al., 1996b
	35, 52 or 55[a], 150	Delbac et al., 1996, 1998a, 1998b
Encephalitozoon hellem	55[b]	Keohane et al., 1996b, 1998
Encephalitozoon intestinalis	45	Keohane et al., 1996b
	60, 120	Beckers et al., 1996
Glugea americanus	43	Keohane et al., 1994, 1996a, 1996c
Glugea atherinae	34, 75, 170	Delbac et al., 1996, 1998a

[a]Cloned gene (Delbac et al., 1998b).
[b]Cloned gene (Keohane et al., 1998a).

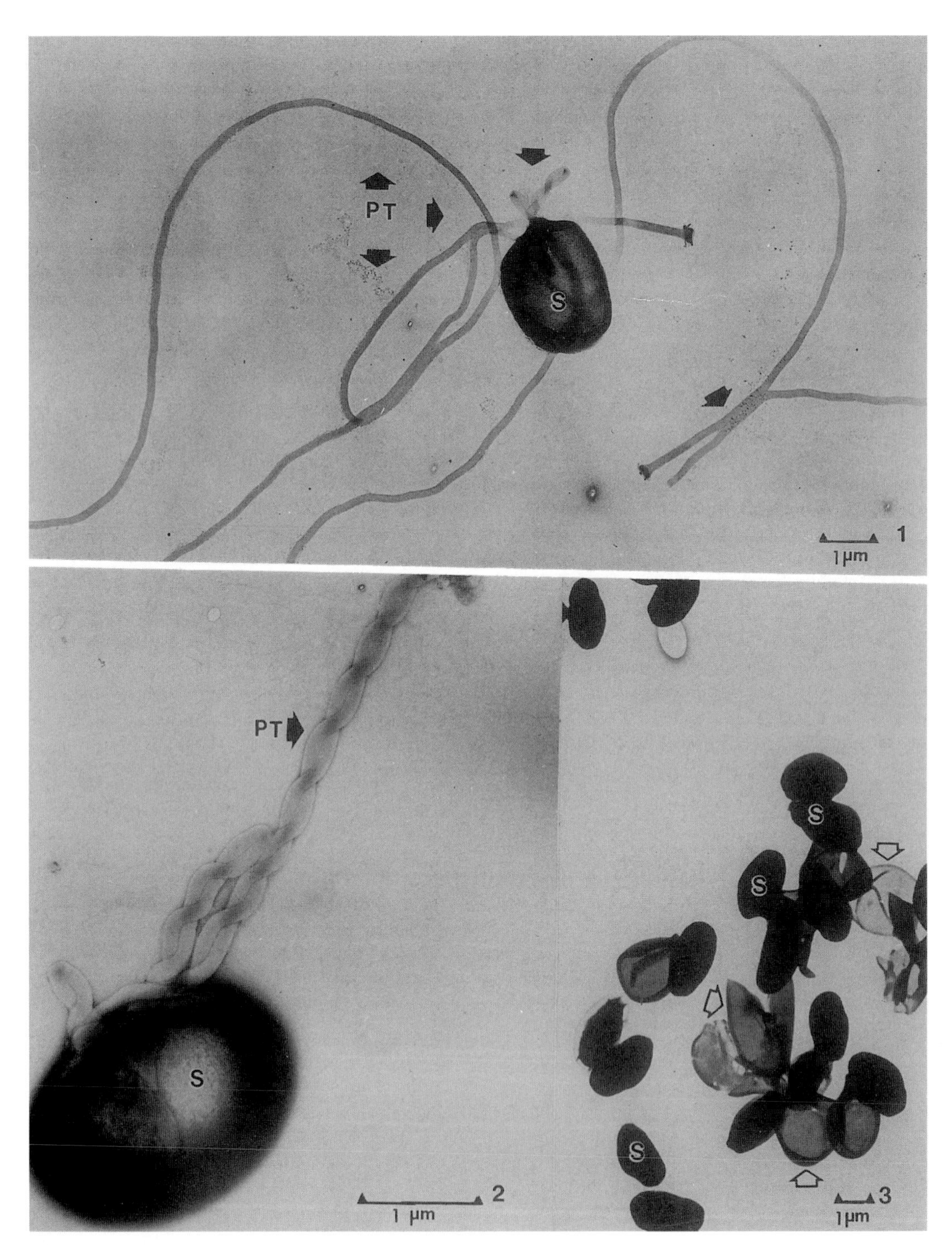

FIGURE 7 Polar tube solubility. Shown are negative stain transmission electron micrographs of spores of *Glugea americanus* disrupted with 0.5 µm acid washed glass beads in a Mini Beadbeater (Biospec Products, Bartlesville, Okla.). (Panels 1 and 2) Disrupted spores after being extracted five times with 1% SDS and once with 9 M urea. Note broken spores (S) and straight and twisted polar tubes (PT and closed arrows). (Panel 3) Disrupted spores after being washed five times with 1% SDS and once with 9 M urea and incubated 2 h with 2% DTT. Note broken spores (S), lack of spore contents (open arrows), and absence of polar tubes. (Reprinted with permission from Keohane et al., 1996c.)

34, and 43 kDa. These putative PTPs displayed no immunoblot reactivity with commercial anti-actin, anti-α-tubulin, or anti-β-tubulin selected for its reactivity across phyla. Monoclonal antibodies (mAbs) to all these proteins were produced. The mAb that was reactive to the 43-kDa protein by immunoblotting (mAb 3C8.23.1) localized to the polar tubes of *G. americanus* spores by immunogold electron microscopy (Fig. 8). This observation indicated that the 43-kDa protein was clearly of polar tube origin. Subsequently, the 43-kDa PTP in the DTT-solubilized material was purified to homogeneity by reverse phase high-performance liquid chromatography (HPLC) (Keohane et al., 1996a) (Fig. 9). This HPLC-purified protein migrated at 43 kDa by SDS-PAGE and reacted with polar tube mAb 3C8.23.1 by immunoblotting (Keohane et al., 1996a). Polyclonal sera from mice immunized with the purified 43-kDa PTP reacted to a 43-kDa protein in *G. americanus* spore lysate by immunoblotting (see Fig. 9 inset, lane C), and intrasporal and extruded polar tubes of *G. americanus* by immunogold electron microscopy. Amino acid analysis revealed this PTP to be proline-rich.

With the same purification protocol, PTPs have been purified from several microsporidia of the genus *Encephalitozoon* associated with human infections (Keohane et al., 1996b). The polar tubes from the Encephalitozoonidae (*Encephalitozoon intestinalis, E. hellem,* and *E. cuniculi*) were found to have the same solubility properties as those of *G. americanus,* that is, they were insoluble in 1% SDS and 9 M urea and soluble in 2% DTT. The PTP from all three members of the Encephalitozoonidae demonstrated an ultraviolet-absorbing peak at a retention time similar to that of *G. americanus* PTP, indicating similar hydrophobicity (Fig. 10). On SDS-PAGE and silver staining, the HPLC-purified PTP of *E. hellem* migrated at 55 kDa, while those of *E. cuniculi* and *E. intestinalis* migrated at 45 kDa. Polyclonal rabbit antibody raised to the HPLC-purified PTP of *E. hellem* (anti-PTP Eh_{55}) localized by immunogold electron microscopy to intrasporal and extruded polar tubes. This antiserum also reacted by immunoblotting with the purified PTPs of *E. hellem, E. cuniculi,* and *E. intestinalis,* as well as with *G. americanus.* Thus, PTPs identified by this purification method display similar charac-

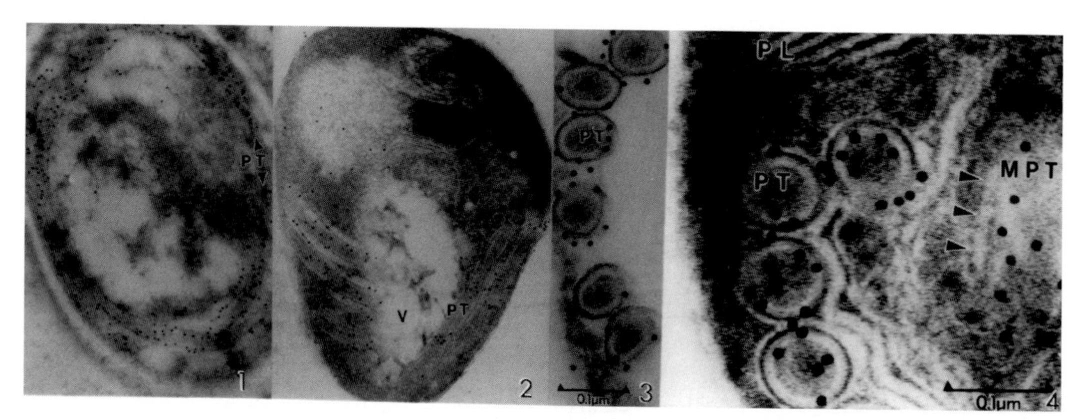

FIGURE 8 Immunogold electron microscopy of a polar tube-specific monoclonal antibody. Shown is an immunogold electron micrograph of a *G. americanus* spore with mAb 3C8.23.1 (Keohane et al., 1994), and a secondary antibody labeled with 12-nm colloidal gold, stained with 1% uranyl acetate. (Panels 1 to 3) Note gold localization on longitudinal, transverse, and cross sections of polar tubes (PT). (Panel 4) four polar tube (PT) cross sections, a portion of the lamellar polaroplast (PL), and a sagittal cut through the anterior straight portion (manubroid) of the polar tube (MPT) indicated by arrowheads. Note the localization of the gold on the "sheath" or outer portion, the "dense" core or center, and the medium-dense material of the polar tube. (Reprinted with permission from Keohane et al., 1996c).

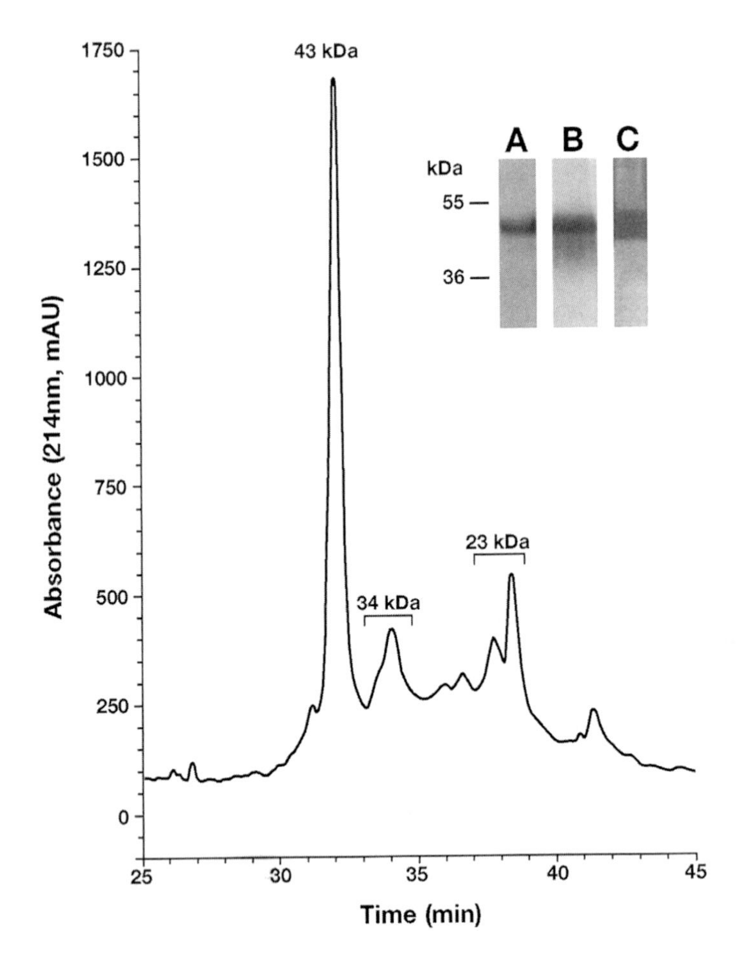

FIGURE 9 HPLC purification of *G. americanus* polar tube components. Reverse-phase HPLC of the DTT-solubilized PTPs of *G. americanus* obtained from SDS-urea extracted spores was performed. Spores of *G. americanus* were disrupted by glass beads and sequentially extracted according to a previously published protocol (Keohane et al., 1994, 1996c). The proteins were subjected to reductive alkylation by 4-vinylpyridine, followed by reverse-phase HPLC with a linear gradient of H_2O and acetonitrile containing 0.1% trifluoroacetic acid (Keohane et al., 1996a). (Inset, lane A) SDS-PAGE (10% polyacrylamide) silver stain of the major ultraviolet-absorbing peak, demonstrating a 43-kDa protein. Peaks corresponding to the previously reported 23- and 34-kDa proteins in the DTT-solubilized material were also identified. (Inset, lane B) The purified 43-kDa PTP demonstrated strong immunoblot activity with polar tube specific mAb 3C8.23.1. (Inset, lane C) A polyclonal mouse antiserum to this 43-kDa protein (anti-Ga PTP_{43}) reacted with a 43-kDa antigen in *G. americanus* spore lysate. Anti-Ga PTP_{43} also reacted with extruded and intrasporal polar tubes of *G. americanus* spores by immunogold electron microscopy. (Reprinted with permission from Keohane et al., 1996c.)

teristics of solubility, hydrophobicity, mass, proline content, and immunologic epitopes.

Amino acid analysis of PTPs from the Encephalitozoonidae and *G. americanus* demonstrated that proline was a significant component of these proteins (Keohane et al., 1996a, 1996b). The *E. hellem* 55-kDa PTP (Keohane et al., 1998) and an *E. cuniculi* PTP (Delbac et al., 1998b) have been cloned and have a predicted proline content similar to that determined by amino acid analysis of the purified native proteins. Proline is a hydrophobic amino acid, and because of its ring structure, it forms a fixed kink in a polypeptide, resulting in chain rigidity. A high proline content is a feature of several structural proteins, including collagen and elastin. Both these proteins are known for their high tensile strength, and elastin is also noted for its elasticity and ability to recoil. In a similar fashion, a microfilarial sheath protein has also been found to contain a high concentration of proline (Bardehle et al., 1992; Zahner et al., 1995), and the microfilarial sheath is also a flexible, baglike structure. Properties such as high tensile strength and elasticity appear to be important in the discharge and passage of sporo-

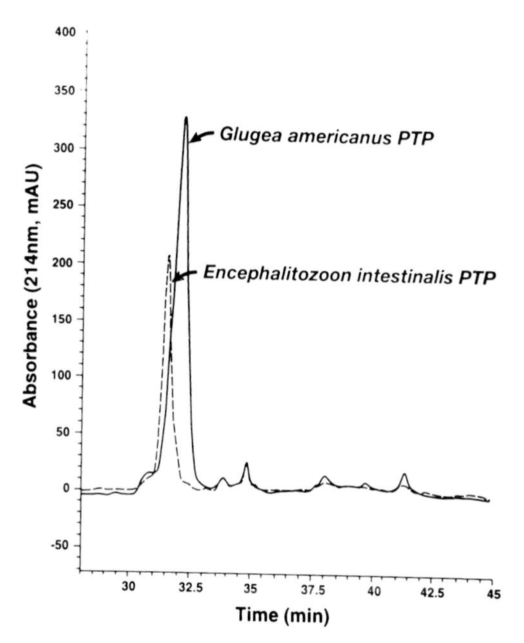

FIGURE 10 Comparison of *G. americanus* and *E. intestinalis* PTPs. Reverse-phase HPLC of the major DTT-solubilized PTP of *G. americanus* and *E. intestinalis* was performed. The method of purification is the same as that shown in Fig. 9. Note the similarity in the retention times of the PTPs of these two microsporidia.

plasm through the polar tube. The high proline content of PTPs is consistent with these functions.

PTPs appear to be highly immunogenic in both experimental and natural infections. In a large serosurvey, antibodies reacting to *E. cuniculi* were present in 5% of pregnant French women and in 8% of Dutch blood donors (van Gool et al., 1997). These positive human sera demonstrated reactivity to the polar tube by immunofluorescence techniques (van Gool et al., 1997). In the process of eversion of the polar tube, unique immunologic epitopes may be exposed. mAb Si91 is specific for extruded polar tubes of *E. intestinalis* by immunofluorescence and immunogold electron microscopy. It recognizes 60- and 120-kDa bands by immunoblotting (Beckers et al., 1996). This mAb does not react with polar tubes within the spore. Further evidence for the high specificity of some PTP epitopes is the absence of cross-reactivity of mAb Si91 in extruded polar tubes of *E. hellem* (Beckers et al., 1996).

In addition to the mAbs to PTPs described above, studies have demonstrated that polyclonal antibodies raised to whole spore lysates in experimental animals usually react with the polar tube (Schwartz et al., 1993; Zierdt et al., 1993; Weiss, unpublished data). In a study on the immunologic response to spore antigens of *Glugea atherinae* and *E. cuniculi,* several candidate PTPs were identified (Delbac et al., 1996, 1998a). Proteins of 34, 75, and 170 kDa in *G. atherinae* and 35, 52 or 55, and 150 kDa in *E. cuniculi* were localized to the polar tube by immunofluorescence and immunogold electron microscopy. It is conceivable that the 52- or 55-kDa protein identified by this technique (Delbac et al., 1996, 1998a) is the same as the 45-kDa PTP of *E. cuniculi* reported by Keohane et al., (1996b). The antibody to the 35-kDa *E. cuniculi* protein reacted with the *G. atherinae* polar tube by immunogold electron microscopy (Delbac et al., 1996, 1998a). In addition, *E. cuniculi* antibodies reacted with the polar tube of *Spraguea lophii* (*G. americanus*) by immunofluorescence (Delbac et al., 1996). These findings suggest that shared PTP epitopes must exist.

A study of the assembly properties of an isolated PTP demonstrated that SDS-washed tubes of PTP reduced by 2-ME, unalkylated, and dialyzed against an alkaline buffer remained dissociated. On acidification, however, the PTP reassembled into sheets or shells, appearing more fluid than PTP of discharged spores (Weidner, 1976). Reassembly was not observed to occur when PTP was alkylated after 2-ME treatment nor after reduction by 1% DTT and subsequent removal of the DTT (Weidner, 1976). We found that DTT-solubilized PTP aggregated when DTT was removed by dialysis, but that reductive alkylation of cysteine residues with 4-vinylpyridine prevented such aggregation (Weiss, unpublished data; Keohane et al., 1996a). Polar tubes have been reported to show branches and to coalesce into networks when suspended in 0.05 to 0.1 M $CaCl_2$ (Weidner, 1982). A putative 52-kDa PTP of *E. cuniculi* has been demonstrated to bind $^{45}Ca^{2+}$ (Delbac et al., 1996). These properties suggest that both disulfide bonds and calcium may play a role in the assembly or function of PTPs.

MOLECULAR BIOLOGY OF PTPs

Information has been obtained on the molecular biology of several PTPs. Two complementary strategies have been utilized in the cloning and characterization of PTP genes. One involves expression screening of a genomic library with a polar tube-specific antibody (Keohane et al., 1998) and the other involves the polymerase chain reaction (PCR) utilizing degenerate primers to the amino acid sequence of peptide fragments from a protein purified by two-dimensional electrophoresis (Delbac et al., 1998b).

Expression screening has proven to be a viable strategy for obtaining gene sequences due to the rarity of introns in the microsporidian genome (Weiss and Vossbrinck, 1998). In addition, the small genome size of the microsporidia (2.9 Mb for *E. cuniculi* [Biderre et al., 1995]) implies that gene density is high. A polyclonal antibody to HPLC-purified *E. hellem* 55-kDa PTP localized to intrasporal polar filaments and extrasporal polar tubes of *E. hellem* by immunogold electron microscopy. This antibody (anti-PTP Eh_{55}) was subsequently utilized by Keohane et al. (1998) to screen a partial *Sau*3A digest expression genomic library of *E. hellem* resulting in identification of the corresponding PTP gene. A clone containing a 2.3-kb insert contained the entire gene, including the sequences coding for the N-terminal sequence of the purified Eh_{55} PTP (GenBank accession number AF044915). The gene product was expressed in *E.coli,* and it reacted by immunoblotting with the polar tube specific anti-PTP Eh_{55}.

Another successful strategy for cloning and sequencing a PTP gene involved microsequencing of digested PTP fragments purified by two-dimensional electrophoresis (Delbac et al., 1998b). This procedure was followed by the use of degenerate primers for a polymerase chain reaction (PCR) to clone the corresponding gene. Proteins were extracted from the spores of *E. cuniculi* and separated by two-dimensional electrophoresis (Delbac et al., 1998b). A 55-kDa protein that reacted with a polar tube-specific antibody by immunoblotting was isolated and digested with endoprotease-Lys C. Peptide fragments were then separated by HPLC, and two of these fragments were sequenced. Degenerate primers were constructed with each peptide sequence and were used in a PCR that amplified a 1-kb fragment which was cloned into a plasmid vector and sequenced. The 5′ and 3′ ends of the gene were then obtained by rapid amplification of cDNA ends (RACE) PCR techniques (GenBank accession AJ005666). Serum from mice, immunized with the gene product expressed in *Escherichia coli,* localized to the polar tube by immunofluorescence and immunogold electron microscopy.

Although the PTP gene and its encoded protein are different for *E. hellem* and *E. cuniculi,* they have striking similarities (Table 2). The PTP gene from *E. hellem* is 1,362 bp long and encodes a protein of 453 amino acids (Keohane et al., 1998), while that of *E. cuniculi* is 1,188 bp long and encodes a protein of 395 amino acids (Delbac et al., 1998b). Both genes contain no introns and reveal no homology with other known genes. Restriction enzyme digestion of genomic DNA is consistent with these PTP genes being single-copy genes. *E. cuniculi* PTP gene localizes to chromosome 6 (Delbac et al., 1998b). A putative signal sequence of 22 amino acids predicted to target these proteins to the endoplasmic reticulum is present in both *E. hellem* and *E. cuniculi.* In the *E. hellem* PTP, after cleavage of the putative signal, the remaining coding sequence translates into a mature protein of 431 amino acids with a predicted molecular mass of 43 kDa and an N terminus corresponding to the N-terminal sequence of HPLC-purified native PTP Eh_{55} (Keohane et al., 1998). A polyadenylation sequence (AATAAA) was observed at the end of the coding region. The predicted *E. cuniculi* PTP molecular mass is 37 kDa (373 amino

TABLE 2 Comparison of PTP genes and predicted proteins

	Encephalitozoon hellem[a,b]	*Encephalitozoon cuniculi*[a,c]
PTP gene length (bp)	1,362	1,188
Introns	None	None
Copy no. per haploid genome	1	1
No. of amino acids in translated protein	453	395
22 amino acid signal peptide	Present	Present
No. of amino acids in mature protein after cleavage of signal peptide	431	373
Predicted molecular mass (kDa)	43	37
SDS-PAGE molecular mass (kDa)	55	55
Amino acid composition (%)		
Proline	14.6	13.4
Glycine	13	11.8
Arginine	0	0
Tryptophan	0	0
Phenylalanine	0	0
Central core repeats	6 (20 amino acids)	3 (26 amino acids) 1 partial repeat
Sequence of repeat region	GSNQTIPGIVYPCQPGQGGS	PGQQQILSGTLPPGATLCQGQAMPST
	GSNQTIPGVISPCQPGQGGS	PGQQQILSGTLPPGATLCQGQAMPST
	GSNQTIPGIVYPCQPGQGGS	PGQQQILSGTLPPGATLCQGQAMPST
	GSNQTIPGVISPCQPGQGGS	PGQQQ**V**LSGTLLPGATLCQ**D**Q**C**MP**G**T
	GSNQTIPGIVYPCQPGQ**NGD**	
	GSNQTIPGVISPCQPGQGG**N**	
GenBank no.	AF044915	AJ005666

[a]Underlined amino acids in *E. hellem* PTP are an alternating motif. Amino acids in a boldface type are substitutions in the repeat sequences.
[b]Keohane et al., 1998.
[c]Delbac et al., 1998b.

acids) after cleavage of the signal sequence. The major amino acids coded by the *E. hellem* and *E. cuniculi* PTP genes were proline (14.6 and 13.4%, respectively) and glycine (13.0 and 11.8%, respectively). Both encoded proteins lacked arginine, tryptophan, and phenylalanine, which is consistent with amino acid analysis of Encephalitozoonidae PTPs (Keohane et al., 1996b; Weiss, unpublished data) and *G. americanus* PTP (Keohane et al., 1996a, 1996c) previously purified by HPLC. This amino acid analysis differs from a previously reported analysis of a 23-kDa PTP of *A. michaelis* in which the proline content was 3.5%, arginine 3.5%, and phenylalanine 3% (Weidner, 1976). The difference between the molecular mass determined from the PTP gene sequences (*E. cuniculi* 37 kDa and *E. hellem* 43 kDa) and that from SDS-PAGE (55 kDa for each protein) may be the result of an aberrant migration in SDS-PAGE due to the high proline content of the protein or to posttranslation modification of the proteins as a result of N-glycosylation or O-glycosylation.

E. hellem PTP contained a central core of six tandem repeats of 20 amino acids, with the motif isoleucine-valine-tyrosine alternating with valine-isoleucine-serine in the repeats (Table 2). The repeat in the core was predominantly hydrophilic and contained 14 polar and 6 hydrophobic amino acids. The protein contained nine putative N-glycosylation sites (NXT/S), six of which were in the central domain. Also, numerous putative O-glycosylation sites were predicted for the N-terminus of the protein, with virtually none occurring in the tandem repeat area. This arrangement is consistent with the reported presence by staining of glycoproteins in the polar tube (Weidner, 1972; Vávra, 1976). *E. cuniculi* PTP also contained a predominately hydrophilic central core with three tandem repeats of 26 amino acids followed by one partial repeat, also with predicted O- and N-glycosylation sites. In addition, the central region was bounded by two repeats of nine residues, which was not observed in *E. hellem* PTP.

E. hellem PTP had a high antigenicity index in the core and in the N-terminus and a predicted β-pleated sheet structure with only two small α-helix regions outside the central core and one small α-helix at the N-terminal signal sequence. The core also lacked lysine/arginine, methionine, aspartic acid-proline bonds, and glutamic acid and consequently was resistant to digestion by trypsin, cyanogen bromide, formic acid, and endoproteinase Glu-C, respectively (Keohane et al., 1998). Cysteine comprised 4.9% of the *E. hellem* and 4.6% of *E. cuniculi* PTP amino acids and was present in the core repeat sequences of both genes. It is likely that disulfide linkages between cysteine residues are important in the structure of the polar tube in that they solubilize in the presence of thiol reducing agents such as DTT and 2-ME.

SPORE GERMINATION: ACTIVATION

Thelohan (1894) was the first to note swelling of the spore prior to extrusion. Although Stempell (1909) and Kudo (1918) proposed that extrusion of the polar filament was due to osmotic pressure, Oshima (1927, 1937) was the first to experimentally demonstrate the relationship of osmotic pressure and extrusion. By using H_2O_2 and various concentrations of NaCl, he decreased the rate of spore extrusion with a high concentration of saline. Lom and Vávra (1963) found polaroplast and posterior vacuole swelling before extrusion, and West (1960) observed contraction of the filament and indicated its importance in sporoplasm ejection. The polar tube discharges from the anterior pole of the spore in an explosive reaction occurring in less than 2 s (Oshima, 1937; Lom and Vávra, 1963; Vávra et al., 1966; Weidner, 1972; Frixione et al., 1992) and is thought to form a hollow tube by a process of eversion, similar to everting the finger of a glove (Oshima, 1937, 1966; Gibbs, 1953; Lom and Vávra, 1963; Ishihara, 1968; Lom, 1972; Weidner, 1982).

Spore discharge is generally believed to occur in several phases: (1) activation, (2) breakdown of sporoplasm compartments, (3) increase in intrasporal osmotic pressure, (4) discharge of the polar tube, and (5) passage of sporoplasm through the polar tube. The exact mechanism of spore discharge is not understood.

There is no universal stimulant for spore activation. Conditions that activate spores vary widely among species, presumably reflecting an organism's adaptation to its host and external environment (Undeen and Epsky, 1990). Since microsporidia are found in a wide range of terrestrial and aquatic hosts, different species may require unique activation conditions for spore discharge. These specific conditions are probably important in preventing accidental discharge into the environment (Undeen and Avery, 1988a; Undeen and Epsky, 1990) and may contribute to host specificity. Table 3 is a summary of various methods utilized to induce spore discharge experimentally.

Various pH conditions have been shown to promote spore discharge, including incubation at an alkaline pH (Ishihara, 1967; Undeen, 1978; Undeen and Avery, 1984), or an acidic pH (Korke, 1916; Hashimoto et al., 1976; Undeen, 1978, 1983; de Graaf et al., 1993), or a pH shift from acid or neutral to alkaline (Pleshinger and Weidner, 1985; Weidner et al., 1995) or from alkaline to less alkaline or neutral (Oshima, 1937, 1964a; Weidner, 1972; Undeen, 1978; Malone, 1984, 1990). Several microsporidia are less dependent on the pH of the environment and have demonstrated spore discharge under both acidic and alkaline conditions (Hashimoto et al., 1976; Undeen, 1983; Undeen and Avery, 1988a). In some

TABLE 3 Reported conditions for activation and discharge of polar tubes

Organism	In vitro method of polar tube discharge	Reference(s)
Amblyospora sp.	1.6 M sucrose plus 0.2 M KCl, pH 9	Undeen and Avery, 1984
Edhazardia aedis	0.1 M KCl, pH 10.5	Undeen et al., 1993
Encephalitozoon hellem	140 mM NaCl, 5 mM KCl, 1 mM $CaCl_2$, 1 mM $MgCl_2$, pH 9.5 or 7.5, with and without 5% H_2O_2	Leitch et al., 1993; He et al., 1996
Encephalitozoon intestinalis	140 mM NaCl, 5 mM KCl, 1 mM $CaCl_2$, 1 mM $MgCl_2$, pH 9.5 or 7.5, with and without 5% H_2O_2	He et al., 1996
	Spores from urine resuspended in 0.025 N NaOH in phosphate-buffered saline	Beckers et al., 1996
Encephalitozoonidae *Encephalitozoon cuniculi* *Encephalitozoon hellem* *Encephalitozoon intestinalis*	0.3% H_2O_2 at 37°C for 12 h	Weiss, unpublished data
Glugea fumiferanae	Chlorides of alkali metal ions at pH 10.8: CsCl, RbCl, KCl, NaCl, or LiCl	Ishihara, 1967
Glugea hertwigi	Calcium ionophore A23187	Weidner and Byrd, 1982
	pH shift from neutral (7.0) to alkaline (9.5) in 150 mM phosphate buffer	
	50 mM sodium citrate in 100 mM glycylglycine buffer, pH 9.5	
	150 mM phosphate buffer in 100 mM glycylglycine buffer, pH 9.5	
Gurleya sp.	Desiccation followed by rehydration with normal saline	Gibbs, 1953
Nosema sp.	3% 40-volume H_2O_2	Walters, 1958
Nosema algerae	$KHCO_3$-K_2CO_3 buffer, pH 8.8	Vávra and Undeen, 1970
	KCl, NaCl, RbCl, CsCl, or NaF, pH 9.5; $KHCO_3$, pH 9.0 (0.1- to 0.3-M solutions), requires pretreatment in distilled H_2O	Undeen, 1978
	0.05 M halogen anion Br^-, Cl^-, or I^- in combination with Na^+ or K^+, pH 9.5; or 0.05 M F^- in combination with Na^+ or K^+, pH 5.5	Undeen and Avery, 1988a
	0.1 M NaCl buffered at pH 9.5 with 20 mM glycine-NaOH or borate-NaOH	Undeen and Avery, 1988b; Undeen and Frixione, 1991
	0.1 M NaCl buffered at pH 9.5 with 20 mM Tris-borate	Frixione et al., 1992
	Alkali metal cations in 0.1 M NaCl or KCl, pH 9.5, or 0.1 M $NaNO_2$, pH 9.5, or Na^+ ionophore monesin in 0.04 M NaCl, pH 9.5	Frixione et al., 1994
Nosema apis	Dehydration in air followed by rehydration with neutral distilled H_2O	Kramer, 1960
	Dehydration in air, followed by rehydration in phosphate-buffered saline, pH 7.1	Olsen et al., 1986
	0.5 M NaCl with 0.5 M $NaHCO_3$, pH 6	De Graaf et al., 1993
Nosema bombycis	30% H_2O_2 or 30% H_2O_2 with 1% $NaHCO_3$	Kudo, 1918
	Boiled digestive fluid of silkworm or 3% H_2O_2	Oshima, 1927
	Digestive fluid of silkworm or liver extract medium, pH > 8.0	Trager, 1937
	NaOH (N/10 to N/160), pH 11 to 13 neutralized with HCl to pH 6.0 to 9.0	Oshima, 1937

(Continued)

TABLE 3 *(Continued)*

Organism	In vitro method of polar tube discharge	Reference(s)
Nosema bombycis *(Continued)*	KOH (N/7 to N/640) neutralized with HCl to pH 6.5–8.0	Oshima, 1964a
	0.375 M KCl, 0.05 M glycine, 0.05 M KOH, pH 9.4–10.0	Oshima, 1964b
	1.5 to 3% H_2O_2	Oshima, 1966
	0.1 N KOH followed by preheated silkworm hemolymph	Ishihara, 1968
Nosema costelytrae	Pretreatment with 0.2 M KCl, pH 12, followed by 0.2 M KCl, pH 7	Malone, 1990
Nosema heliothidis	Pretreatment with 0.15 M cation (K, Na, Li, Rb, or Cs), pH 11, followed by 0.15 M cation (K), pH 7	Undeen, 1978
Nosema helminthorum	Mechanical pressure	Dissanaike, 1955
Nosema locustae	Dehydration with 2.5 M sucrose or 5% polyethylene glycol followed by 0.1 M Tris-HCl, 0.1 M NaCl, or 0.1 M glycine-NaOH, 0.1 M NaCl, pH 9–10	Undeen and Epsky, 1990
	Dehydration in air followed by rehydration in 0.1 M Tris-HCl, pH 9.2, 37°C	Whitlock and Johnson, 1990
Nosema michaelis	Pretreatment in veronal acetate buffer, pH 10, followed by tissue culture medium 199	Weidner, 1972
Nosema pulicis	Weak acetic acid-iodine water	Korke, 1916
Nosema whitei	Dehydration in air followed by rehydration with neutral distilled H_2O	Kramer, 1960
Perezia pyraustae	Dehydration in air followed by rehydration with neutral distilled H_2O	Kramer, 1960
Plistophora anguillarum	0.1 M potassium citrate-HCl, pH 3 to 4, or 0.01 M $KHCO_3$-K_2CO_3, pH 10, or 0.5 to 50% H_2O_2	Hashimoto et al., 1976
Plistophora hyphessobryconis	5% H_2O_2	Lom and Vávra, 1963
Spraguea lophii[a]	pH shift from acid/neutral to alkaline, (pH 9.0), in 0.5 M glycylglycine or 0.5 M carbonate buffer containing 2% mucin or 0.5 M poly-D-glutamic acid	Pleshinger and Weidner, 1985
	Calcium ionophore A23187	Pleshinger and Weidner, 1985
	Phosphate buffered saline, pH 8.5–9.0, containing 0.1-0.5% porcine mucin	Weidner et al., 1984
	Storage in 0.05 M HEPES, pretreatment in 10^{-5} M Ca^{2+} pH 7, followed by HEPES, pH 9.5, containing 2% mucin	Weidner et al., 1995
Thelohania californica	Mechanical pressure	Kudo and Daniels, 1963
Thelohania magna	Mechanical pressure	Kudo, 1916, 1920
Vairimorpha necatrix	Pretreatment with 0.15 M cation (K, Li, Rb or Cs), pH 10.5, followed by 0.15 M cation (Na or K), pH 9.4	Undeen, 1978
Vairimorpha plodiae	Pretreatment with 0.1 or 1 M KCl, pH 11, followed by 0.1 or 1 M KCl, pH 8.0	Malone, 1984
Vavraia culicis	0.2 M KCl, pH 6.5 (one isolate), pH 7.0-9.0 (another isolate)	Undeen, 1983
Vavraia oncoperae	Pretreatment with 3 mM EDTA followed by 0.2 M KCl pH 11	Malone, 1990

[a]Also known as *Glugea americanus*.

species dehydration by drying or hyperosmotic solutions followed by rehydration has been effective in promoting spore discharge (Gibbs, 1953; Kramer, 1960; Olsen et al., 1986). In other microsporidia, dehydration followed by rehydration at an alkaline pH was effective (Undeen, 1978; Undeen and Avery, 1984; Undeen and Epsky, 1990; Whitlock and Johnson, 1990).

Various cations, including potassium, lithium, sodium, cesium, and rubidium (Oshima, 1964b; Ishihara, 1967; Undeen, 1978, 1983; Malone, 1984, 1990; Undeen and Epsky, 1990; Whitlock and Johnson, 1990; de Graaf et al., 1993; Frixione et al., 1994) and anions such as bromide, chloride, iodide, and fluoride (Undeen and Avery, 1988a) have been used to promote discharge. Sodium or potassium cations at an alkaline pH have also been reported to stimulate germination (Oshima, 1964b; Ishihara, 1967; Undeen and Avery, 1988b; Undeen and Frixione, 1991; Frixione et al., 1994). It is apparent from these studies that both cations and anions enter the spore passively. While the spore wall forms a barrier to larger molecules (i.e., functions as a molecular sieve), it appears that alkali metal cations pass freely through the spore wall and plasma membrane (plasmalemma). These cations are required for the germination process, and to some extent smaller cations are more effective (Frixione et al., 1994). Mucin or polyanions (Pleshinger and Weidner, 1985; Weidner et al., 1995), hydrogen peroxide (Kudo, 1918; Lom and Vávra, 1963; Hashimoto et al., 1976; Leitch et al., 1993; He et al., 1996), low-dose ultraviolet radiation (Undeen and Vander Meer, 1990), and a Na^+ ionophore (monensin) (Frixione et al., 1994) have also been used to trigger discharge. Inhibitors of spore discharge include concentrated alcohols (Kudo, 1918), 0.01 to 0.1 M magnesium chloride (Malone, 1984), ammonium chloride (Undeen, 1978; Undeen and Avery, 1988c; Undeen and Epsky, 1990), low salt concentrations (10 to 50 mM) (Undeen, 1978), sodium fluoride (Undeen and Avery, 1988a), 1.8 M sucrose (Undeen, 1978), silver ions (Ishihara, 1967), gamma radiation (Undeen et al., 1984), ultraviolet light (Whitlock and Johnson, 1990), temperatures greater than 40°C (Whitlock and Johnson, 1990), a microfilament disrupter (cytochalasin D), a microtubule disrupter (demecolcine), itraconazole (Leitch et al., 1993), metronidazole (He et al., 1996) and the nitric oxide donors *S*-nitroso-*N*-acetylpenicillamine and sodium nitroprusside (He et al., 1996).

Calcium chloride (0.001 to 0.1 M) has been found to inhibit spore germination in some studies (Oshima, 1964a, 1964b; Ishihara, 1967; Undeen, 1978, 1983; Weidner and Byrd, 1982; Malone, 1984), while 0.2 M $CaCl_2$ at pH 9.0 (Pleshinger and Weidner, 1985) and 1 mM $CaCl_2$ (Leitch et al., 1993; He et al., 1996), as well as the calcium ionophore A23187 (Weidner and Byrd, 1982; Pleshinger and Weidner, 1985), promoted discharge in other studies. Ethylene glycol-bis(β-aminoethyl ether)-*N,N,N′,N′*-tetraacetic acid (EGTA) in the presence of calcium also promoted spore discharge in one study (Malone, 1984) and inhibited discharge in another (Pleshinger and Weidner, 1985). The calcium channel antagonists lanthanum, verapamil, and nifedipine, as well as the calmodulin inhibitors chlorpromazine and trifluroperazine, have also been reported to inhibit spore discharge (Pleshinger and Weidner, 1985; Leitch et al., 1993; He et al., 1996). Removal of clathrin and calmodulin from the intermediate filament cage assembly that envelopes the spores of *S. lophii* (*G. americanus*) resulted in irreversible inactivation of spore discharge (Weidner, 1992). Calcium has been been proposed to be involved directly in spore discharge, in which its displacement from the polaroplast membrane either activates a contractile mechanism or combines with the polaroplast matrix and causes polaroplast swelling (Weidner and Byrd, 1982). In support of this concept the calcium ionophore A23187 was found to trigger polaroplast swelling and polar tube discharge, while calcium chloride inhibited the reaction (Weidner, 1982; Weidner and Byrd, 1982). These studies suggest that calcium may have a pivotal role in the germination process.

It has been theorized that microsporidia exhibit the same response to germination stimuli regardless of the mode of activation, which is to increase the intrasporal osmotic pressure (Kudo, 1918; Oshima, 1937; Lom and Vávra, 1963; Undeen, 1990; Undeen and Frixione, 1990; Frixione et al., 1992). This increase in osmotic pressure results in an influx of water into the spore accompanied by swelling of the polaroplasts and posterior vacuole prior to spore discharge (Huger, 1960; Lom and Vávra, 1963; Weidner and Byrd, 1982; Frixione et al., 1992). Indirect evidence for the osmotic pressure theory is provided by the observation that in hyperosmolar solutions polar tube discharge is inhibited or slowed down (Oshima, 1937; Lom and Vávra, 1963; Undeen and Frixione, 1990; Frixione et al., 1992) and sporoplasm passage does not occur (Weidner, 1976; Frixione et al., 1992). It is postulated that the osmotic pressure forces eversion of the polar tube and subsequent expulsion of the sporoplasm (Undeen, 1990). The spore wall provides structural resistance, as well as elasticity and flexibility for this process (Undeen and Frixione, 1990, 1991).

Water flow across the spore wall and plasma membrane is a clear requirement in osmotic theories of spore discharge. A study involving D_2O has demonstrated that water influx into spores occurs through a specific transmembrane pathway (i.e., an aquaporin) sensitive to $HgCl_2$ and the hydration state of the membrane (Frixione et al., 1997). D_2O has been demonstrated to inhibit spore discharge in a concentration dependent manner (Frixione et al., 1997). This inhibition of germination by D_2O could be overcome by manipulations associated with a decrease in water structure related to an increase in temperature or to the ionic strength of the solution (Frixione et al., 1997). These observations support the presence and participation of aquaporins in spore germination (Frixione et al., 1997). Further supporting evidence for the presence of an aquaporin in spore plasma membranes is the observation of 7- to 10-nm intramembrane particles of uniform appearance similar to monomeric aquaporins in freeze-fracture electron micrographic studies on the spore plasma membrane (Liu and Davis, 1973; Vávra et al., 1986; Undeen and Frixione, 1991). The presence of such aquaporins is consistent with reported observations of the effects of cations and anions on spore germination (i.e., the larger the hydration sphere of the ion, the less effective the ion is in activation via this pathway).

Several theories have been proposed regarding the mechanism for increasing osmotic pressure in the spore. One of the earliest explanations was that activation simply increases the permeability of the spore coat to water (Lom and Vávra, 1963), however, data suggest that while the spore coat functions as a molecular sieve, it is freely permeable to water. In addition, aquaporins in other systems are also freely permeable to water. Dall (1983) proposed that spore germination was due to the creation of a proton gradient by the alkaline environment surrounding the spore. He hypothesized that during germination a proton gradient drives a proton-cation exchange mechanism consisting of a carboxylic acid ionophore. As protons are depleted in the sporoplasm, the increase in alkalinity triggers the same mechanism in the membranes of organelles, particularly the polaroplast and posterior vacuole. Water then flows into the spore because of the generalized osmotic imbalance, resulting in an increase in intrasporal pressure (Dall, 1983). It should be noted, however, that not all microsporidia require an alkaline pH for spore discharge and that the cationic flux described does not appear sufficient for the osmotic pressure required for germination.

Microsporidian spores have been reported to contain trehalose and trehalase (Wood et al., 1970; Vandermeer and Gochnauer, 1971). On the basis of a finding of decreased trehalose levels in discharged spores as compared to those in undischarged spores of *Nosema algerae,* an alternative theory for spore discharge has been proposed (Undeen et al., 1987; Undeen and Vander Meer, 1994). In this mechanism, activation causes changes in the spore, such as a breakdown in compartmentalizaton, bringing trehalose in contact with the enzyme trehalase (Undeen, 1990). Trehalose is degraded into a larger number of small molecules with a resultant increase in osmotic pressure. The subsequent flow of water into the spore causes an increase in intrasporal pressure and spore discharge (Undeen, 1990). Ultrastructural evidence exists for the breakdown of cellular compartments prior to polar filament discharge in that there is a decrease in electron density beginning at the posterior end of the spore and disappearance of polaroplast membranes and polyribosomes (Lom, 1972; Chioralia et al., 1998). However, in *Nosema apis,* de Graaf et al. (1993) found little difference in the trehalose/glucose ratio before and after germination, which suggests that this mechanism may not apply to all microsporidia.

Spore discharge is a complex process requiring several steps before the polar tube begins to evert and transfer of sporoplasm to its host cell can occur. One possible synthesis of the current data (Keohane and Weiss, 1998) is that ion flux into the spore occurs in response to external stimuli (such as a pH change) and that this ionic flux results in the displacement of calcium from intramembranous compartments. This calcium flux may be tied to a subsequent loss of compartmentalization of the spore (perhaps as a result of effects on microfilaments). Displacement of calcium would also cause activation of an enzyme (with a specific pH requirement), such as trehalase, that amplifies the hydrostatic force driving water into the spore by breaking down a complex molecule (trehalose) into its components (glucose and its subsequent metabolites). This would result in an increase in the number of osmotically active molecules in the spore. The subsequent influx of water through aquaporins causes swelling of the posterior vacuole and polaroplast, a decrease in sporoplasm density, and eventually rupture of the spore through the anterior attachment complex with eversion of the polar tube. Interfering with this process at any one of these steps may result in inhibition of spore germination. For example, an alteration in spore pH caused by ammonium chloride could inhibit ion flux and/or enzyme activity. Stimulation by low-dose ultraviolet radiation or hydrogen peroxide may result from damage to intracellular compartments allowing enzyme and substrate to interact and bypassing the calcium activation mechanism.

SPORE GERMINATION: POLAR TUBE DISCHARGE

The polar filament coils around the internal contents of the spore and is surrounded by a series of polaroplast membranes at its anterior end. On spore activation, an increase in intrasporal osmotic pressure and a swelling of these membranes occur (Lom and Vávra, 1963). The first sign of spore discharge is a visible protrusion at the anterior end of the spore at the polar cap (Lom and Vávra, 1963; Kudo and Daniels, 1963; Frixione et al., 1992). The polar cap ruptures, and the polar filament rapidly emerges from the spore in a helicoidal fashion along nearly a straight line, forming the hollow polar tube (Frixione et al., 1992) (Fig. 11). Full discharge of the polar tube requires less than 2 s with a maximum velocity of 105 μm/s (Frixione et al., 1992). The velocity of this extrusion increases as the polar tube reaches its maximum length, supporting the concept of the "everting finger of a glove" model (Frixione et al., 1992). Polar tubes range from 50 to 500 μm in length and 0.1 to 0.2 μm in diameter (Kudo and Daniels, 1963; Lom and Corliss, 1967; Weidner, 1976; Frixione et al., 1992). The same polar tube thickness has been observed before and after discharge, including discharge into media of various viscosities (Weidner, 1976). Inside the spore the polar tube is filled with electron-dense material, while the discharged tubes appear as hollow cylinders (Weidner, 1976, 1982). An incompletely discharged tube resembles a cylinder within a cylinder (i.e., a double-membrane cylinder) at its distal end (Weidner, 1982; Weidner et al., 1994, 1995). Spores can be broken by mechanical methods such as pressure (Kudo, 1918; Kudo and Daniels, 1963; Weidner, 1982; Dall, 1983; Undeen, 1990) or by glass bead disruption (Connor, 1970; Langley et al., 1987; Keohane et al., 1996c), releasing polar tubes from the sides of the spores.

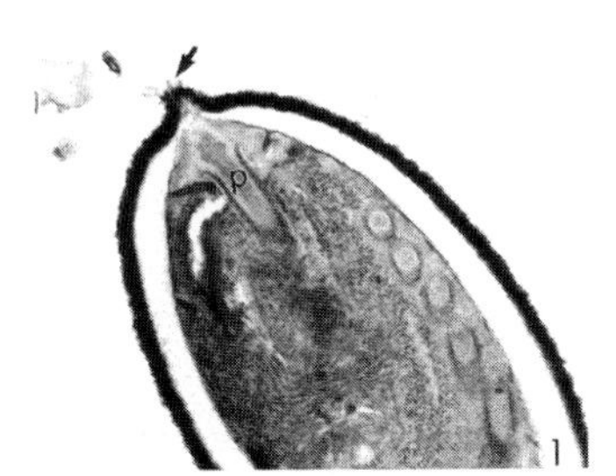

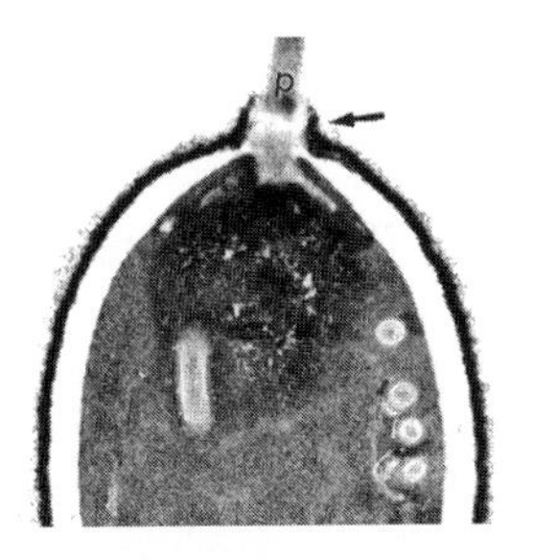

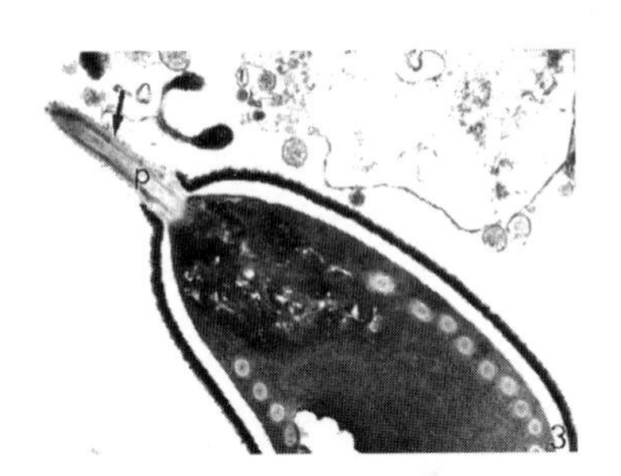

FIGURE 11 Transmission electron micrographs of polar tube discharge. Electron micrographs of anterior portion of longitudinal sections of the spores of *N. algerae* are shown. (Panel 1) Swelling of the anterior end of the spore and disruption of the anchoring disk. Note protein filaments already present above the anterior end of the spore. (Panel 2) Eversion of polar tube membrane (arrow) and release of material (presumably PTP) from the polar tube (p). (Panel 3) Further formation of the polar tube (p). Note polar tube membrane and deposition of PTP coat on the outside of the membrane (arrow). (Reprinted with permission from Peter M. Takvorian.)

It has long been observed that during discharge the portion of the polar tube already everted remains unchanged while the tube elongates and even changes direction at the growing tip (Kramer, 1960; West, 1960; Weidner, 1982; Frixione et al., 1992). It is also known that the length of the discharged polar tube appears to be about two to three times that of the coiled tube inside the spore (Lom and Corliss, 1967; Weidner, 1972). Some investigators believe that during extrusion the tube simply stretches to this increased length because of its elasticity (Lom and Corliss, 1967; Vávra, 1976). Another explanation that has been proposed is that the polaroplast membranes contribute to lengthening of the tube during eversion (Weidner, 1972; Weidner et al., 1995), however, this theory has not been accepted by all investigators (Sinden and Canning, 1974; Vávra, 1976; Toguebaye and Marchand, 1987). Furthermore, immunogold electron microscopy suggests that the polar tube and polaroplast are antigenically distinct (Keohane et al., 1994, 1996a, 1996b, 1996c; Delbac et al., 1996; Beckers et al., 1996).

Chioralia et al. (1998) proposed a theory involving the unfolding of a fully formed polar filament. In the polar filament, they described six concentric layers in the coiled part and only three layers in the straight part (manubroid) and in the extruded tube. They interpret this finding as demonstration of a formed tube within a tube in the coiled section of the polar filament (thus this section has twice the number of membranes as the manubroid), partially everted into itself and with the distal end of the coil open to the sporoplasm. In this model, the tubular polaroplast expands during activation and pushes the lamellar polaroplast into the open distal end of the polar filament. This event triggers the manubroid portion of the filament to evert. The sporoplasm flows into the filament, and the filled filament uncoils and slides in a rotating or screwlike motion through the everted part (Chioralia et al., 1998). Thus, the polar filament is fully formed prior to eversion, and the portion of the tube within a tube accounts for the apparent increase in length of the extruded polar tube.

Other evidence suggests the tube is lengthened by the deposition of PTPs at the growing tip. In this model, unpolymerized PTPs in the dense core of the polar filament are deposited on the outside of the polar tube during eversion and are incorporated at the growing tip (Weidner, 1982; Toguebaye and Marchand, 1987; Frixione et al., 1992; Weidner et al., 1984, 1995). This theory is supported by observations involving pulse labeling with latex particles (Weidner, 1982) and by video interference-contrast microscopy (Frixione et al., 1992; Weidner et al., 1994, 1995).

We believe the evidence supports the view that after rupture of the anchoring disk, the polar tube forms from an eversion of the polar tube membrane beginning at its anterior end (Fig. 12 and 13). Swelling of the polaroplast exerts pressure at this anterior point and initiates this eversion, while the collarlike structure in the anchoring disk turns and holds the tube in place. Ultrastructural evidence for eversion is seen in Fig. 6B and 11 in which the polar tube membranes at the anchoring disk are folding out (i.e., the tube is everting). Simultaneously, unpolymerized PTPs are released from the polar filament core and polymerize on the outside of the membrane scaffolding as it is being everted, creating a polar tube that increases in length at the tip (Fig. 12 and 13). Figures 6, 11, and 14 demonstrate release of material from the polar filament and deposition of this material (Fig. 14) on the outside of the polar tube membrane. In this model, the lipophilic regions of the PTPs adhere to the membrane and the hydrophilic regions are oriented into the aqueous extrasporal environment. The two Encephalitozoonidae PTPs that have been cloned display hydrophilic and hydrophobic regions consistent with this model. It is also probable that a change in redox potential between the inside of the polar filament and the outside of the spore could trigger the formation of disulfide bonds in PTPs favoring the polymerization process along these membranes. The tube would then continue to form until the supply of unpolymerized PTP and membrane is exhausted.

After eversion, the polar tube remains attached to the anterior end of the spore by a collarlike structure formed from the anchoring disk (Lom, 1972) (Fig. 6B). After complete discharge of the tube, the sporoplasm flows

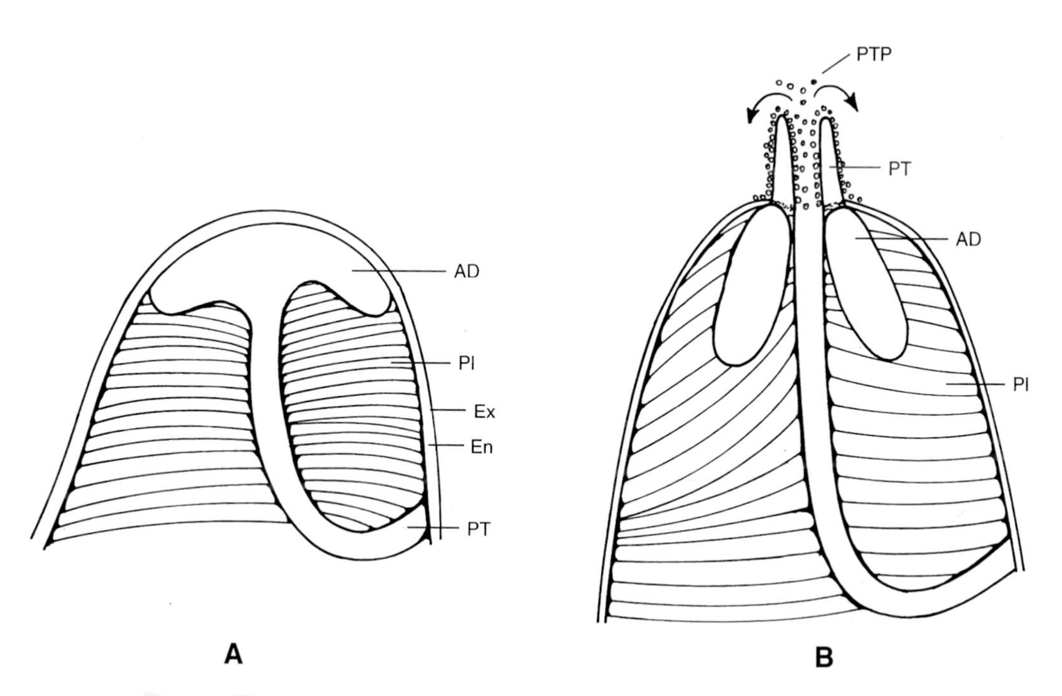

FIGURE 12 Model of polar tube discharge. (A) A resting spore. (B) Initial eversion of the polar filament. Note that the anchoring disk has everted or rotated to form a collar and that the polaroplast membranes have swollen. Unpolymerized PTP is released from the polar tube core and polymerizes on the outside of the membrane scaffolding as it is being everted. AD, anchoring disk, Pl, polaroplast membranes: Ex, exospore; En, endospore; PT, polar tube; PTP, polar tube protein(s).

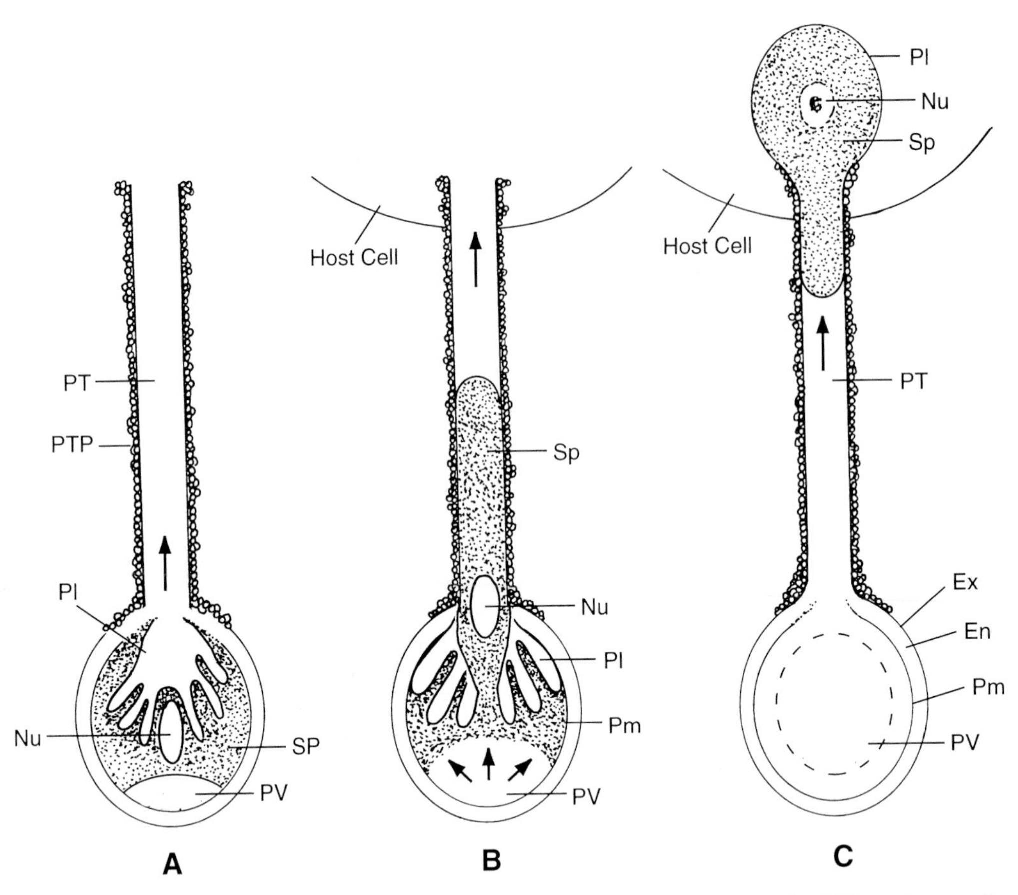

FIGURE 13 Model of sporoplasm exit from microsporidian spores. (A) As the polar tube (PT) continues to form with polymerization of the polar tube protein(s) (PTP), the polaroplast membrane begins to enter the hollow portion of the tube. The posterior vacuole (PV) starts to enlarge. (B) Sporoplasm (SP) and nucleus (Nu) flow into the tube surrounded by the polaroplast membrane (Pl), leaving behind the plasma membrane (Pm) (still attached to the spore coat [Ex and En]). The polar tube has penetrated a host cell, and the posterior vacuole (PV) has swollen and fills the space vacated by the sporoplasm. The swelling of the posterior vacuole generates the osmotic pressure driving the extrusion of the contents of the spore. (C) After the polar tube has pierced the host cell membrane, the sporoplasm and nucleus now surrounded by the polaroplast membrane emerge from the tip of the hollow polar tube inside the host cell. The spore contains the plasma membrane (Pm), posterior vacuole (PV), and spore coat (exospore [Ex] and endospore [En]) which are left behind. The limiting membrane of the sporoplasm in the host cell is provided by the polaroplast membranes (Pl).

through the polar tube and appears as a droplet at its distal end (Fig. 13 and 15) (Oshima, 1937; Gibbs, 1953; Lom, 1972; Weidner, 1972; Frixione et al., 1992). It appears that the polar tube must be completely discharged for sporoplasm passage, as it has not been observed in partially discharged tubes (Gibbs, 1953; Weidner, 1972; Frixione et al., 1992). It has been reported that the polar tube diameter can increase from 0.1 μm to 6 μm during sporoplasm passage (Lom and Vávra, 1963; Oshima, 1966; Ishihara, 1968; Weidner, 1972, 1976; Olsen et al., 1986). When video-enhanced contrast microscopy is used, there is a time delay of about 15 to 500 ms between completion of discharge and appearance of the droplet (Frixione et al., 1992). It has

FIGURE 14 Extruded polar tube, demonstrating membranes and polar tube protein coat. A transmission electron micrograph of an extruded polar tube of *N. algerae* demonstrating a polar tube membrane (curved arrow) and the coating of PTP over the membrane (straight arrow) is shown. (Reprinted with permission from Peter M. Takvorian.)

been suggested that the delay might be due to eversion of a blind-ended tube which needs to be opened by some mechanism prior to sporoplasm release (West, 1960; Erickson et al., 1968; Frixione et al., 1992). While in contact with the tip of the polar tube, the sporoplasm droplet enlarges to a volume in excess of what might be expected from the size of the spore (Frixione et al., 1992). This enlargement is most likely due to the movement of water into the sporoplasm secondary to an osmotic gradient between the sporoplasm and the environment (Frixione et al., 1992).

The spore contents emerge from the tip of the tube as a membrane-bound structure (Weidner, 1972; Frixione et al., 1992). Since sporoplasm membrane remains in the empty spore coat after germination, it is believed that the polaroplast membranes form the new limiting membrane of the emerging nucleoplasm (Weidner, 1976; Weidner et al., 1984, 1994, 1995; Undeen and Frixione, 1991). If the polar tube is discharged next to a cell, it can pierce the cell and transfer (i.e., inject) the sporoplasm into that cell (Fig. 3) (Oshima, 1937; Gibbs, 1953; Bailey, 1955; Walters, 1958; Lom and Vávra, 1963; Oshima, 1966; Ishihara, 1968; Weidner, 1972; Iwano and Ishihara, 1989). If there are no adjacent cells, the droplet of sporoplasm remains

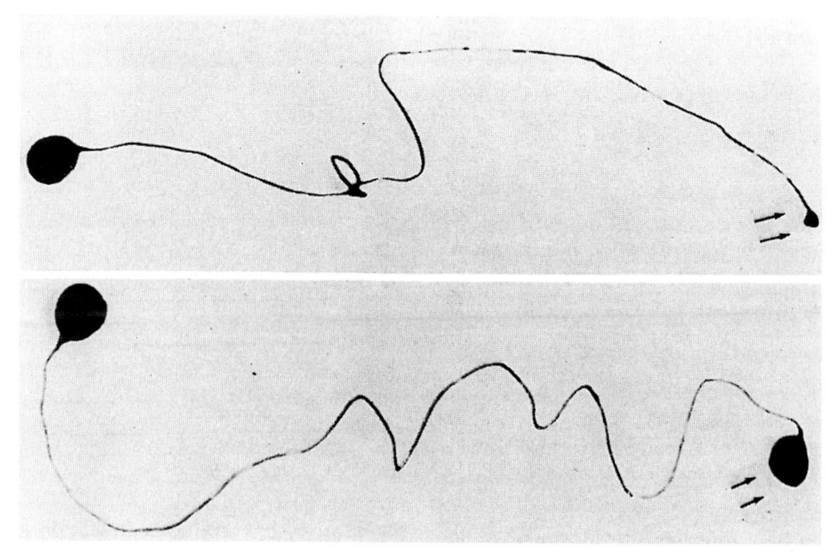

FIGURE 15 Micrograph of sporoplasm passage through the polar tube. A negative stain of transmission electron microscopy of discharged spores of *Nosema michaelis* (*A. michaelis*) shows extruded polar tubes. Arrows indicate appearance of sporoplasm at the end of the discharged polar tube. (Reprinted with permission from Weidner, 1972.)

attached to the polar tube for a period of time. The mechanism by which the polar tube penetrates host cells is unknown and it is also not known if the polar tube binds to a specific receptor on the host cell.

SUMMARY

While the polar tube was first described as a unique microsporidian structure over 100 years ago (Thelohan, 1892, 1894), its biochemical components and the mechanism of germination of spores remain to be definitively determined. Application of the techniques of modern biology (both immunologic and molecular) has resulted in the identification of several PTPs although the interactions and functional significance of these proteins remains to be determined.

Experimental evidence suggests that osmotic pressure generated in the spore prior to polar tube eversion is a key event in germination. The mechanisms by which this osmotic gradient is generated and the significance of both alkali cations and anions, as well as of calcium, in the initiation of this process remain to be resolved. Work on *N. algerae* has established that trehalose may be a key compound in germination (Undeen et al., 1990). Studies of this nature need to be extended to microsporidia found in noninsect hosts to determine whether these findings can be generalized to other microsporidia.

The polar tube serves as an unique vehicle for transmission of infection by piercing an adjacent host cell, thereby inoculating the sporoplasm directly into the cell and functioning essentially as a hypodermic needle. Additional characterization of the early events in the rupture of the anterior attachment complex and in eversion of the polar tube, as well as of the mechanism of host cell attachment and penetration is needed. The invasion organelle of the microsporidia (polar tube, polaroplast, and posterior vacuole) has successfully served this diverse phylum, resulting in a group of obligate intracellular organisms capable of infecting almost any type of cell. Further study may lead to novel strategies for control of these important parasitic protozoa.

ACKNOWLEDGMENTS

This work was supported by NIH grant AI31788.

We thank George A. Orr for advice and assistance on protein purification methods, Ann Cali and Peter M. Takvorian for advice and assistance with immunogold electron microscopy, Jiří Lom, Peter M. Takvorian, Earl Weidner, and Nico Paul Kock for photographs, and Jiří Vávra for discussions on the mechanics of polar tube extrusion.

REFERENCES

Bailey, L. 1955. The infection of the ventriculus of the adult honeybee by *Nosema apis* Zander. *Parasitology* **45:**86–94.

Bardehle, G., A. Jepp-Libutzki, D. Linder, K. Moehnle, H. H. Schott, H. Zahner, U. Zahringer, and S. Stirm. 1992. Chemical composition of *Litomosoides carinii* microfilarial sheaths. *Acta Trop.* **50:**237–247.

Beckers, P. J. A., G. J. M. M. Derks, T. van Gool, F. J. R. Rietveld, and R. W. Sauerwien. 1996. *Encephalitozoon intestinalis*-specific monoclonal antibodies for laboratory diagnosis of microsporidiosis. *J. Clin. Microbiol.* **34:**282–285.

Biderre, C., M. Pages, G. Metenier, E. U. Canning, and C. P. Vivares. 1995. Evidence for the smallest nuclear genome (2.9 Mb) in the microsporidium *Encephalitozoon cuniculi. Mol. Biochem. Parasitol.* **74:**229–231.

Cali, A., and R. L. Owen. 1988. Microsporidiosis, p. 929–950. *In* A. Ballows, W. J. Hausler Jr., M. Ohashi, and H. Turano (eds.), *Laboratory Diagnosis of Infectious Diseases: Principles and Practice,* vol. 1. Springer-Verlag, New York.

Cali, A., and R. L. Owen. 1990. Intracellular development of *Enterocytozoon,* a unique microsporidian found in the intestine of AIDS patients. *J. Protozool.,* **37:**145–155.

Canning, E. U., and J. Lom. 1986. *The Microsporidia of Vertebrates,* p. 1–16. Academic Press, New York.

Chilmonczyk, S., W. T. Cox, and R. P. Hedrick. 1991. *Enterocytozoon salmonis* n. sp.: an intranuclear microsporidium from salmonid fish. *J. Protozool.* **38:**264–269.

Chioralia, G., T. Trammer, W. A. Maier, and H. M. Seitz. 1998. Morphologic changes in *Nosema algerae* (Microspora) during extrusion. *Parasitol. Res.* **84:**123–131.

Connet, A. 1932. Le cycle évolutif du *Plistophora chironomi. Cellule* **41:**181–202.

Connor, R. M. 1970. Disruption of microsporidian spores for serologic studies. *J. Invertebr. Pathol.* **15:**138.

Dall, D. J. 1983. A theory for the mechanism of polar filament extrusion in the Microspora. *J. Theor. Biol.* **105:**647–659.

de Graaf, D. C., G. Masschelein, F. Vandergeynst, H. F. De Brabander, and F. J. Jacobs. 1993. *In vitro* germination of *Nosema apis* (Microspora: Nosematidae) spores and its effect on their α-trehalose/d-glucose ratio. *J. Invertebr. Pathol.* **62:**220–225.

Delbac, F., F. Duffieux, D. David, G. Metenier, and C. P. Vivares. 1998a. Immunocytochemical identification of spore proteins in two microsporidia, with emphasis on extrusion apparatus. *J. Eukaryot. Microbiol.* **45:**224–231.

Delbac, F., F. Duffieux, P. Peyret, D. David, G. Metenier, and C. Vivares. 1996. Identification of sporal proteins in two microsporidian species: an immunoblotting and immunocytochemical study. *J. Eukaryot. Microbiol.* **43:**101S.

Delbac, F., P. Peyret, G. Metenier, D. David, A. Danchin, and C. P. Vivares. 1998b. On proteins of the microsporidian invasion apparatus: complete sequence of a polar tube protein in *Encephalitozoon cuniculi*. *Mol. Microbiol.* **29:**825–834.

Desportes-Livage, I., S. Chilmonczyk, R. Hedrick, C. Ombrouck, D. Monge, I. Maiga, and M. Gentilini. 1996. Comparative development of two microsporidian species: *Enterocytozoon bieneusi* and *Enterocytozoon salmonis,* reported in AIDS patients and salmonid fish, respectively. *J. Eukaryot. Microbiol.* **43:**49–60.

Dissanaike, A. S. 1955. Emergence of the sporoplasm in *Nosema helminthorum*. *Nature* **174:**1002–1003.

Dissanaike, A. S., and E. U. Canning. 1957. The mode of emergence of the sporoplasm in Microsporidia and its relation to the structure of the spore. *Parasitology* **47:**92–99.

Erickson, B. W., Jr., S. H. Vernick, and V. Sprague. 1968. Electron microscope study of the everted polar filament of *Glugea weissenbergi* (Microsporida, Nosematidae). *J. Protozool.* **15:**758–761.

Fantham, H. B., and A. Porter. 1914. The morphology, biology, and economic importance of *Nosema bombi* n. sp. parasitic in various bumblebees (*Bombus* spp.) *Ann. Trop. Med. Parasitol.* **8:** 623–638.

Frixione, E., L. Ruiz, J. Cerbon, and A. H. Undeen. 1997. Germination of *Nosema algerae* (Microspora) spores: conditional inhibition by D_2O, ethanol and Hg^{2+} suggests dependence of water influx upon membrane hydration and specific transmembrane pathways. *J. Eukaryot. Microbiol.* **44:**109–116.

Frixione, E., L. Ruiz, M. Santillan, L. V. de Vargas, J. M. Tejero, and A. H. Undeen. 1992. Dynamics of polar filament discharge and sporoplasm expulsion by microsporidian spores. *Cell Motil. Cytoskeleton* **22:**38–50.

Frixione, E., L. Ruiz, and A. H. Undeen. 1994. Monovalent cations induce microsporidian spore germination in vitro. *J. Eukaryot. Microbiol.* **41:** 464–468.

Gibbs, A. J. 1953. *Gurleya* sp. (Microsporidia) found in the gut tissue of *Trachea secalis* (Lepidoptera). *Parasitology* **43:**143–147.

Hall, I. M. 1952. A new species of Microsporidia from the fawn-colored lawn moth, *Crambus bonifatellus* (Hulst) Lepidoptera, Crambidae. *J. Parasitol.* **38:** 487–491.

Hashimoto, K., Y. Sasaki, and K. Takinami. 1976. Conditions for extrusion of the polar filament of the spore of *Plistophora anguillarum,* a microsporidian parasite in *Anguilla japonica*. *Bull. Jpn. Soc. Sci. Fish.* **42:** 837–845.

He, Q., G. J. Leitch, G. S. Visvesvara, and S. Wallace. 1996. Effects of nifedipine, metronidazole and nitric oxide donors on spore germination and cell culture infection of the microsporidia *Encephalitozoon hellem* and *Encephalitozoon intestinalis.* *Antimicrob. Agents Chemother.* **40:**179–185.

Huger, A. 1960. Electron microscope study on the cytology of a microsporidian spore by means of ultrathin sectioning. *J. Insect Pathol.* **2:**84–105.

Ishihara, R. 1967. Stimuli causing extrusion of polar filaments of *Glugea fumiferanae* spores. *Can. J. Microbiol.* **13:**1321–1332.

Ishihara, R. 1968. Some observations on the fine structure of sporoplasm discharged from spores of a microsporidian, *Nosema bombycis.* *J. Invertebr. Pathol.* **12:**245–258.

Iwano, H., and R. Ishihara. 1989. Intracellular germination of spores of a *Nosema* sp. immediately after their formation in cultured cell. *J. Invertebr. Pathol.* **54:**125–127.

Jensen, H. M., and S. R. Wellings. 1972. Development of the polar filament-polaroplast complex in a microsporidian parasite. *J. Protozool.* **19:**297–305.

Keohane, E. M., G. A. Orr, P. M. Takvorian, A. Cali, H. B. Tanowitz, M. Wittner, and L. M. Weiss. 1996a. Purification and characterization of a microsporidian polar tube protein. *Mol. Biochem. Parasitol.* **79:**255–259.

Keohane, E. M., G. A. Orr, P. M. Takvorian, A. Cali, H. B. Tanowitz, M. Wittner, and L. M. Weiss. 1996b. Purification and characterization of human microsporidian polar tube proteins. *J. Eukaryot. Microbiol.* **43:**100S.

Keohane, E. M., G. A. Orr, H. S. Zhang, P. M. Takvorian, A. Cali, H. B. Tanowitz, M. Wittner, and L. M. Weiss. 1998. The molecular characterization of the major polar tube protein gene from *Encephalitozoon hellem,* a microsporidian parasite of humans. *Mol. Biochem. Parasitol.* **94:**227–236.

Keohane, E., P. M. Takvorian, A. Cali, H. B. Tanowitz, M. Wittner, and L. M. Weiss. 1994.

The identification and characterization of a polar tube reactive monoclonal antibody. *J. Eukaryot. Microbiol.* **41:**48S.

Keohane, E. M., P. M. Takvorian, A. Cali, H. B. Tanowitz, M. Wittner, and L. M. Weiss. 1996c. Identification of a microsporidian polar tube protein reactive monoclonal antibody. *J. Eukaryot. Microbiol.* **43:**26–31.

Keohane, E. M., and L. M. Weiss. 1998. Characterization and function of the microsporidian polar tube: a review. *Folia Parasitol.* **45:**117–127.

Kock, N. P. 1998. *Diagnosis of human pathogen microsporidia.* Ph.D. thesis. Bernard Nocht Institute for Tropical Medicine, Hamburg, Germany.

Korke, V. T. 1916. On a *Nosema* (*Nosema pulicis* n.s.) parasitic in the dog flea (*Ctenocephalus felis*). *Ind. J. Med. Res.* **3:**725–730.

Kramer, J. P. 1960. Observations on the emergence of the microsporidian sporoplasm. *J. Insect Pathol.* **2:**433–439.

Kudo, R. 1916. Contributions to the study of parasitic protozoa. I. On the structure and life history of *Nosema bombycis* Naegeli. *Bull. Imp. Seric. Exp. Stn. Jpn.* **1:**31–51.

Kudo, R. 1918. Experiments on the extrusion of polar filaments of cnidosporidian spores. *J. Parasitol.* **4:**141–147.

Kudo, R. 1920. On the structure of some microsporidian spores. *J. Parasitol.* **6:**178–182.

Kudo, R. 1921. On the nature of structures characteristic of cnidosporidian spores. *Trans. Am. Microsc. Soc.* **40:**59–74.

Kudo, R. R., and E.W. Daniels. 1963. An electron microscope study of the spore of a microsporidian, *Thelohania californica*. *J. Protozool.* **10:**112–120.

Langley, R. C., Jr., A. Cali, and E. W. Somberg. 1987. Two-dimensional electrophoretic analysis of spore proteins of microsporidia. *J. Parasitol.* **73:** 910–918.

Leger, L., and E. Hesse. 1916. Sur la structure de la spore des Microsporidies. *C.R. Soc. Biol.* **79:**1049–1053.

Leitch, G. J., Q. He, S. Wallace, and G. S. Visvesvara. 1993. Inhibition of spore polar filament extrusion of the microsporidium, *Encephalitozoon hellem,* isolated from an AIDS patient. *J. Eukaryot. Microbiol.* **40:**711–717.

Liu, T. P., and D. M. Davies. 1973. Ultrastructural architecture and organization of the spore envelope during development in *Thelohania bracteata* (Strickland, 1913) after freeze-etching. *J. Protozool.* **20:**622–630.

Lom, J. 1972. On the structure of the extruded microsporidian polar filament. *Z. Parasitenk.* **38:** 200–213.

Lom, J., and J. O. Corliss. 1967. Ultrastructural observations on the development of the microsporidian protozoon, *Plistophora hyphessobryconis* Schaperclaus. *J. Protozool.* **14:**141–152.

Lom, J., and J. Vávra. 1963. The mode of sporoplasm extrusion in microsporidian spores. *Acta Protozool.* **1:**81–92.

Malone, L. A. 1984. Factors controlling *in vitro* hatching of *Vairimorpha plodiae* (Microspora) spores and their infectivity to *Plodia interpunctella, Heliothis virescens* and *Pieris brassicae*. *J. Invertebr. Pathol.* **44:**192–197.

Malone, L. A. 1990. *In vitro* spore hatching of two microsporidia, *Nosema costelytrae* and *Vávraia oncoperae* from New Zealand pasture insects. *J. Invertebr. Pathol.* **55:**441–443.

Olsen, P. E., W. A. Rice, and T. P Liu. 1986. *In vitro* germination of *Nosema apis* spores under conditions favorable for the generation and maintenance of sporoplasms. *J. Invertebr. Pathol.* **47:**65–73.

Oshima, K. 1927. A preliminary note on the structure of the polar filament of *Nosema bombycis* and its functional significance. *Annot. Zool. Jpn.* **11:** 235–243.

Oshima, K. 1937. On the function of the polar filament of *Nosema bombycis*. *Parasitology* **29:**220–224.

Oshima, K. 1964a. Effect of potassium ion on filament evagination of spores of *Nosema bombycis* as studied by neutralization method. *Annot. Zool. Jpn.* **37:**102–103.

Oshima, K. 1964b. Stimulative or inhibitive substance to evaginate the filament *Nosema bombycis* Nageli: I. The case of artificial buffer solution. *Jpn. J. Zool.* **14:**209–229.

Oshima, K. 1966. Emergence mechanism of sporoplasm from the spore of *Nosema bombycis* and the action of filament during evagination. *Jpn. J. Zool.* **15:**203–220.

Pleshinger, J., and E. Weidner. 1985. The microsporidian spore invasion tube. IV. Discharge activation begins with pH-triggered Ca^{+2} influx. *J. Cell Biol.* **100:**1834–1838.

Scarborough-Bull, A., and E. Weidner. 1985. Some properties of discharged *Glugea hertwigi* (Microsporida) sporoplasms. *J. Protozool.* **32:**284–289.

Schuberg, A. 1910. Ueber Mikrosporidien aus dem Hoden der barbe und durch sie verursachte Hypertrophie der Kerne. *Arb. Kais. Gesundh.* **33:**401–434.

Schwartz, D. A., G. S. Visvesvara, G. J. Leitch, L. Tashjian, M. Pollack, J. Holden, and R. T. Bryan. 1993. Pathology of symptomatic microsporal (*Encephalitozoon hellem*) bronchiolitis in the acquired immunodeficiency syndrome: a new respiratory pathogen diagnosed from lung biopsy, brochoalveolar lavage, sputum and tissue culture. *Hum. Pathol.* **24:**937–943.

Sinden, R. E., and E. U. Canning. 1974. The ultrastructure of the spore of *Nosema algerae* (Protozoa, Microsporida) in relation to the hatching mechanism of microsporidian spores. *J. Gen. Microbiol.* **85:**350–357.

Sprague, V., and S. H. Vernick. 1969. Light and electron microscope observations on *Nosema nelsoni* Sprague, 1950 (Microsporida, Nosematidae) with particular reference to its Golgi complex. *J. Protozool.* **16:**264–271.

Stempell, W. 1909. Über *Nosema bombycis* Naegeli nebst Bemerkungen über Mikrophotographie mit gewõhnlichem und ultraviolettem Licht. *Arch. Protistenk.* **16:**281–358.

Strickland, E. H. 1913. Further observations on the parasites of *Simulium* larvae. *J. Morph.* **24:**43–102.

Takvorian, P. M., and A. Cali. 1986. The ultrastructure of spores (Protozoa: Microsporida) from *Lophius americanus,* the angler fish. *J. Protozool.* **33:**570–575.

Takvorian, P. M., and A. Cali. 1994. Enzyme histochemical identification of the golgi apparatus in the microsporidian, *Glugea stephani. J. Eukaryot. Microbiol.* **41:**63S–64S.

Takvorian, P. M., and A. Cali. 1996. Polar tube formation and nucleoside diphosphatase activitiy in the microsporidian, *Glugea stephani. J. Eukaryot. Microbiol.* **43:**102S–103S.

Thelohan, P. 1892. Observations sur les Myxosporidies et essai de classification de ces organismes. *Bull. Soc. Philom.* **4:**165–172.

Thelohan, P. 1894. Sur la présence d'une capsule à filament dans les spores des microsporidies. *C. R. Acad. Sci.* **118:**1425–1427.

Toguebaye, B. S., and B. Marchand. 1987. Intracellular emergence of the microsporidian sporoplasm as revealed by electron microscopy in *Nosema couilloudi* (Microspora, Nosematidae) *Arch. Protistenk.* **134:**397–407.

Trager, W. 1937. The hatching of spores of *Nosema bombycis* Nageli and the partial development of the organism in tissue cultures. *J. Parasitol.* **23:**226–227.

Undeen, A. H. 1976. *In vivo* germination and host specificity of *Nosema algerae* in mosquitos. *J. Invertebr. Pathol.* **27:**343–347.

Undeen, A. H. 1978. Spore hatching processes in some *Nosema* species with particular reference to *N. algerae* (Vávra and Undeen), p. 29–49. *In* W. M. Brooks (ed.), *Selected Topics on the Genus* Nosema (*Microsporida*). Entomological Society of America, College Park, Md.

Undeen, A. H. 1983. The germination of *Vávraia culicis. J. Protozool.* **30:**274–277.

Undeen, A. H. 1990. A proposed mechanism for the germination of microsporidian (Protozoa: Microspora) spores. *J. Theor. Biol.* **142:**223–235.

Undeen, A. H., and S. W. Avery. 1984. Germination of experimentally nontransmissible microsporidia. *J. Invertebr. Pathol.* **43:**299–301.

Undeen, A. H., and S. W. Avery. 1988a. Effect of anions on the germination of *Nosema algerae* (Microspora: Nosematidae) spores. *J. Invertebr. Pathol.* **52:**84–89.

Undeen, A. H., and S. W. Avery. 1988b. Spectrophotometric measurement of *Nosema algerae* (Microspora: Nosematidae) spore germination rate. *J. Invertebr. Pathol.* **52:**253–258.

Undeen, A. H., and S. W. Avery. 1988c. Ammonium chloride inhibition of the germination of spores of *Nosema algerae* (Microspora: Nosematidae). *J. Invertebr. Pathol.* **52:**326–334.

Undeen, A. H., L. M. El Gazzar, R. K. Vander Meer, and S. Narang. 1987. Trehalose levels and trehalase activity in germinated and ungerminated spores of *Nosema algerae* (Microspora: Nosematidae). *J. Invertebr. Pathol.* **50:**230–237.

Undeen, A. H., and N. D. Epsky. 1990. *In vitro* and *in vivo* germination of *Nosema locustae* (Microspora: Nosematidae) spores. *J. Invertebr. Pathol.* **56:**371–379.

Undeen, A. H., and E. Frixione. 1990. The role of osmotic pressure in the germination of *Nosema algerae* spores. *J. Protozool.* **37:**561–567.

Undeen, A. H., and E. Frixione. 1991. Structural alteration of the plasma membrane in spores of the microsporidium *Nosema algerae* on germination. *J. Protozool.* **38:**511–518.

Undeen, A. H., and R. K. Vander Meer. 1990. The effect of ultraviolet radiation on the germination of *Nosema algerae* Vávra and Undeen (Microsporida: Nosematidae) spores. *J. Protozool.* **37:**194–199.

Undeen, A. H., and R. K. Vander Meer. 1994. Conversion of intrasporal trehalose into reducing sugars during germination of *Nosema algerae* (Protista: Microspora) spores: a quantitative study. *J. Eukaryot. Microbiol.* **41:**129–132.

Undeen, A. H., R. K. Vander Meer, B. J. Smittle, and S. W. Avery. 1984. The effect of gamma radiation on *Nosema algerae* (Microspora: Nosematidae) spore viability, germination and carbohydrates. *J. Protozool.* **31:**479–482.

Vandermeer, J. W., and T. A. Gochnauer. 1971. Trehalase activity associated with spores of *Nosema apis. J. Invertebr. Pathol.* **17:**38–41.

Van Gool, T., J. C. M. Vetter, B. Weinmayr, A. Van Dam, F. Derouin, and J. Dankert. 1997. High seroprevalence of *Encephalitozoon* species in immunocompetent subjects. *J. Infect. Dis.* **175:** 1020–1024.

Vávra, J. 1976. Structure of microsporidia, p. 1–86. *In* L. A. Bulla and T. C. Cheng (ed.), *Comparative Pathobiology,* vol. 1. Plenum Press, New York.

Vávra, J., L. Joyon, and P. de Puytorac. 1966. Observation sur l'ultrastructure du filament polaire des microsporidies. *Protistologica* **2:**109–112.

Vávra, J., and A. H. Undeen. 1970. *Nosema algerae* n.sp. (Cnidospora, Microsporida), a pathogen in a laboratory colony of *Anopheles stephensi* Liston (Diptera, Culicidae). *J. Protozool.* **17:**240–249.

Vávra, J., D. Vinckier, G. Torpier, E. Porchet, and E. Vivier. 1986. A freeze-fracture study of microsporidia I (Protozoa: Microspora). The sporophorous vesicle, the spore wall and the spore plasma membrane. *Protistologica* **22:**143–154.

Walters, V. A. 1958. Structure, hatching and size variation of the spores in a species of *Nosema* (Microsporidia) found in *Hyalophora cecropsia* (Lepidoptera). *Parasitology* **48:**113–120.

Weidner, E. 1970. Ultrastructural study of microsporidian development. *Z. Zellforsch.* **105:**33–54.

Weidner, E. 1972. Ultrastructural study of microsporidian invasion into cells. *Z. Parasitenkd.* **40:**227–242.

Weidner, E. 1976. The microsporidian spore invasion tube: the ultrastructure, isolation, and characterization of the protein comprising the tube. *J. Cell Biol.* **71:**23–34.

Weidner, E. 1982. The microsporidian spore invasion tube. III. Tube extrusion and assembly. *J. Cell Biol.* **93:**976–979.

Weidner, E. 1992. Cytoskeletal proteins expressed by microsporidian parasites. *Subcell. Biochem.* **18:**385–399.

Weidner, E., and W. Byrd. 1982. The microsporidian spore invasion tube. II. Role of calcium in the activation of invasion tube discharge. *J. Cell Biol.* **93:**970–975.

Weidner, E., W. Byrd, A. Scarborough, J. Pleshinger, and D. Sibley. 1984. Microsporidian spore discharge and the transfer of polaroplast organelle membrane into plasma membrane. *J. Protozool.* **31:**195–198.

Weidner, E., S. B. Manale, S. K. Halonen, and J. W. Lynn. 1994. Microsporidian spore invasion tubes as revealed by fluorescent probes. *Biol. Bull.* **187:**255–256.

Weidner, E., S. B. Manale, S. K. Halonen, and J. W. Lynn. 1995. Protein-membrane interaction is essential to normal assembly of the microsporidian spore invasion tube. *Biol. Bull.* **188:** 128–135.

Weiser, J. 1958. *Nosema laphygmae* n.sp. and the internal structure of microsporidian spore. *J. Insect Pathol.* **1:**52–59.

Weiss, L. M. Unpublished data.

Weiss, L. M., and C. R. Vossbrinck. 1998. Microsporidiosis: molecular and diagnostic aspects. *Adv. Parasitol.* **40:**351–395.

West, A. F., Jr. 1960. The biology of a species of *Nosema* (Sporozoa, Microsporidia) parasitic in the flour beetle *Tribolium confusum*. *J. Parasitol.* **46:**747–753.

Whitlock, V. H., and S. Johnson. 1990. Stimuli for the *in vitro* germination and inhibition of *Nosema locusta* (Microspora: Nosematidae) spores. *J. Invertebr. Pathol.* **56:**57–62.

Wood, P. J., I. R. Siddiqui, J. W. Vandermeer, and T. A. Gochnauer. 1970. Carbohydrates of *Nosema apis* spores. *Carbohydr. Res.* **15:**154–158.

Zahner, H., G. Hobom, and S. Stirm. 1995. The microfilarial sheath and its proteins. *Parasitol. Today* **11:**116–120.

Zierdt, C. H., V. J. Gill, and W.S. Zierdt. 1993. Detection of microsporidian spores in clinical samples by indirect fluorescent-antibody assay using whole-cell antisera to *Encephalitozoon cuniculi* and *Encephalitozon hellem*. *J. Clin. Microbiol.* **31:**3071–3074.

Zwolfer, W. 1926. *Pleistophora blochmanni,* eine neue Microsporidie aus *Gammarus pulex L*. *Arch. Protistenkd.* **54:**261–340.

HOST-PARASITE RELATIONSHIPS IN MICROSPORIDIOSIS: ANIMAL MODELS AND IMMUNOLOGY

Elizabeth S. Didier and G. Todd Bessinger

7

Microsporidia are single-celled, spore-forming, obligately intracellular parasites that belong to the phylum Microspora. The microsporidia are considered emerging pathogens because new species have been identified as causes of disease in humans during the last 20 years. In addition, the incidence of microsporidiosis in humans has increased during the last 20 years and is expected to continue to rise. Because species of microsporidia that were recognized causes of disease in animals are now causing infections in humans, microsporidia also are considered reemerging pathogens. Generally, microsporidia are recognized as causing opportunistic infections in immunologically compromised individuals such as AIDS patients and organ transplant recipients undergoing immunosuppressive therapy, but these organisms are being recognized in immunologically competent individuals as well (Bryan, 1995; Bryan et al., 1991; Hautvast et al., 1997; Orenstein, 1991; Sandfort et al., 1994; Shadduck, 1989; Shadduck and Greeley, 1989; Shadduck et al., 1996; Sobottka et al., 1995; Wanke et al., 1996; Weber and Bryan, 1994; Weber et al., 1994). Studies on natural and experimental infections in laboratory animals have provided insights about innate barriers of resistance, early induced responses, and adaptive immune-mediated responses expressed by the host to control microsporidia. In immunologically compromised hosts, microsporidia replicate with little restraint to cause potentially lethal infections. In immunologically competent hosts, microsporidia may cause asymptomatic yet persistent infections. The purpose of this chapter is to describe the host-parasite relationships in animals with microsporidiosis and how they relate to microsporidian infections in humans.

Elizabeth S. Didier, Department of Microbiology, Tulane Regional Primate Research Center, Tulane University Medical Center, Covington, LA 70433. *G. Todd Bessinger,* Department of Tropical Medicine, School of Public Health and Tropical Medicine, Tulane University Medical Center, New Orleans, LA 70112.

ANIMAL MODELS OF MICROSPORIDIA THAT INFECT HUMANS

Microsporidia have been recognized as etiologic agents of disease in animals since the early 1920s but were not routinely recognized as pathogens in humans until the AIDS pandemic began in the 1980s. There are more than 1,000 species of microsporidia, of which 13 are presently known to infect humans (Canning and Hollister, 1992; Canning et al., 1986; Didier et al., 1998; Hollister et al., 1996a; Schwartz and Bryan, 1997; Weber et al., 1994).

The Microsporidia and Microsporidiosis (Murray Wittner, editor; Louis M. Weiss, contributing editor), ©1999 American Society for Microbiology, Washington, D.C.

Encephalitozoon cuniculi

E. cuniculi Levaditi, Nicolau, and Schoen, 1923, is the best-studied microsporidian species of mammals for several reasons. As the first microsporidian identified in a mammal, *E. cuniculi* was detected by Wright and Craighead (1922) in the brain, spinal cord, and kidneys of a rabbit with motor paralysis and was named in 1923 by Levaditi (Levaditi et al., 1923). *E. cuniculi* also was the first microsporidian successfully isolated from a mammalian host (i.e., a rabbit) for long-term culture (Shadduck, 1969), thereby providing a source of organisms for developing diagnostic methods and for studies on experimental infections in laboratory animals. Of epidemiological interest is the fact that *E. cuniculi* has an extraordinarily broad host range among mammals, and spontaneous (or natural) infections have been reported in hosts that include (but are not limited to) rodents, lagomorphs, carnivores, and ruminants, as well as human and nonhuman primates (Canning et al., 1986; Didier et al., 1998; Weber et al., 1994).

Experimental infections with *E. cuniculi* have been transmitted to many mammalian hosts, although murine models with euthymic, athymic, and severe combined immunodeficient (SCID) mice have been most extensively used in studying the host-parasite relationships of microsporidiosis (Didier et al., 1994; Gannon, 1980b; Hermanek et al., 1993; Koudela et al., 1993; Schmidt and Shadduck, 1983). In addition, a xenograft model utilizing fetal rabbit intestinal tissue grafted subcutaneously onto athymic or SCID mice has been developed for establishing extraintestinal infections with *E. cuniculi* (as well as the other *Encephalitozoon spp.*) (Wasson et al., 1998). The murine models for *E. cuniculi* have been particularly useful because mice are natural hosts of *E. cuniculi*, dissemination of the organism in mice is similar to that in humans, and immune responses can be easily studied in mice through the availability of inbred strains of mice and immunological reagents. In addition, the clinical manifestations of *E. cuniculi* infections in immunologically competent and immunodeficient mice parallel those seen in humans infected with *E. cuniculi*.

Three strains of *E. cuniculi* have been identified on the basis of biochemical, immunological, and molecular variations (Table 1), and these differences are of epidemiological interest in defining the potential zoonotic sources of *E. cuniculi* infections in humans (Deplazes et al., 1996a, 1996b; Didier et al., 1995; Katiyar et al., 1995; Mathis et al., 1997). *E. cuniculi* strain I, which was originally isolated and cultured from rabbits, has been identified in human immunodeficiency virus (HIV)-infected individuals in Switzerland (Deplazes et al., 1996a, 1996b), and *E. cuniculi* strain III, originally isolated and cultured from domestic dogs, has been identified in AIDS patients in the United States (DeGroote et al., 1995; Didier et al., 1996b). *E. cuniculi* strain II was originally isolated for culture from a mouse, and has been found to naturally infect blue foxes but has not been reported to infect humans (Didier et al., 1995; Mathis et al., 1996). Experimentally, these *E. cuniculi* strains were not host-specific (Deplazes et al., 1996a, 1996b; Didier et al., 1995; Pang and Shadduck, 1985; Shadduck and Geroulo, 1979; Van Dellen et al., 1989), and the biological and epidemiological significance of strain variations among *E. cuniculi* isolates awaits further study.

Encephalitozoon hellem

E. hellem Didier et al., 1991, was first identified in three AIDS patients with keratoconjunctivitis (Friedberg et al., 1990; Yee et al., 1991). Morphologically, *E. hellem* is nearly identical to *E. cuniculi* (Didier et al., 1991a, 1991b) and it is likely that some of the earlier human cases of *E. cuniculi* infection may in fact have been due to *E. hellem*. *E. hellem* infections have not been identified in any other mammals but have been reported in psittacine birds (Black et al., 1997; Pulparampil et al., 1998). Experimental *E. hellem* infections have been established in euthymic and athymic mice and in rhesus macaques (*Macaca mulatta*) (Didier et al., 1994; Hollister et al., 1993). In addition, *E. hellem* has

TABLE 1 Microsporidia of humans

Species	Examples of nonhuman hosts	In vitro culture	Sites of infection	Comment(s)
Commonly detected				
Encephalitozoon cuniculi	Wide range of mammals		Disseminated	Wide mammalian host range
Strain I	Rabbits	Yes		Identified in humans in Switzerland
Strain II	Mice, blue foxes	Yes		Not identified in humans
Strain III	Domestic dogs	Yes		Identified in humans in the United States
Encephalitozoon hellem	Psittacine birds	Yes	Disseminated	
Encephalitozoon intestinalis	Dogs, donkeys, pigs, cows, goats	Yes	Small intestine, disseminated	Originally named *Septata intestinalis*
Enterocytozoon bieneusi	Pigs, nonhuman primates	No	Small intestine, gallbladder, liver	Short-term culture established Three strains identified but not named
Less commonly detected				
Brachiola vesicularum	None known	Not attempted	Corneal stroma, skeletal muscle	Originally called a *Nosema*-like species
Microsporidium africanum	None known	Not attempted	Corneum	Catch-all genus for microsporidia that could not be classified
Microsporidium ceylonensis	None known	Not attempted	Corneum	The infected individual was gored by a goat 6 years earlier.
Nosema connori	None known	Not attempted	Disseminated	Nosema spp. often infect insects. Proposed new classification is *Brachiola connori*.
Nosema ocularum	None known	Not attempted	Corneal stroma	
Pleistophora species	None known	Not attempted	Skeletal muscle	*Pleistophora* spp. often infect fish.
Trachipleistophora antropophtera	None known	Not attempted	Brain, heart, kidney	
Trachipleistophora hominis	None known	Yes	Skeletal muscle, nasal sinuses	
Vittaforma corneae	None known	Yes	Corneal stroma	Originally named *Nosema corneum*

been isolated from AIDS patients for long-term culture (Didier et al., 1991a; Hollister et al., 1993; Scaglia et al., 1994; Visvesvara et al., 1991, 1994).

Encephalitozoon intestinalis

First reported in AIDS patients with chronic diarrhea (Blanshard et al., 1992; Orenstein et al., 1992a, 1992b), *E. intestinalis* was originally named *Septata intestinalis* Cali, Orenstein, and Kotler, 1993, because of the presence of extracellular matrix material that "septated" the parasitophorous vacuole into compartments. As additional biochemical, immunological, and molecular information became available, this organism was reclassified as *E. intestinalis* (Baker et al., 1995; Hartskeerl et al., 1995). Recently, *E. intestinalis* spores were identified in the feces of dogs, donkeys, pigs, cows, and goats by PCR (Bornay-Llinares et al., 1998). Experimentally, mice (Achbarou et al., 1996; Bessinger et al., 1997; Enriquez, 1997) and rhesus macaques (Didier, 1998, unpublished observations) have been infected with *E. intestinalis.* This organism disseminated in experimentally infected animals, as in humans, but variably caused diarrhea in infected mice. *E. intestinalis* has been placed in long-term culture (Didier et al., 1996a; Doultree et al., 1995; Van Dellen et al., 1989; VanGool et al., 1994; Visvesvara et al., 1995a).

Enterocytozoon bieneusi

E. bieneusi Desportes et al., 1985, is the most frequently reported microsporidian that infects and causes chronic diarrhea in AIDS patients (Orenstein, 1991; Orenstein et al., 1990, 1997; Weber et al., 1994). In addition to humans, natural hosts of *E. bieneusi* include pigs (Deplazes et al., 1996b) and several species of nonhuman primates (Mansfield et al., 1997). Natural infections in pigs and nonhuman primates appeared to be similar to those in humans in that they were associated with diarrhea and were primarily localized in the small intestine, often with biliary tract involvement. As also reported for *E. cuniculi,* there appear to be at least three strains of *E. bieneusi* (Rinder et al., 1997). Experimentally, *E. bieneusi* has been passaged to rhesus macaques (Tzipori et al., 1997). No successful infections have been reported in mice, but hydrocortisone acetate-suppressed rats inoculated orally with *E. bieneusi* developed asymptomatic infections and shed low but consistent numbers of spores in their feces (Accoceberry et al., 1997). Although *E. bieneusi* has been grown in short-term culture (Visvesvara et al., 1995b), long-term culture has not been possible, hampering studies on this important microsporidian.

Trachipleistophora hominis

T. hominis Field et al., 1996, was first described to occur in an AIDS patient and caused severe myositis and sinusitis. This organism has been described only recently and has no other known natural hosts. When inoculated intraperitoneally into athymic mice, *T. hominis* was observed to infect the skeletal muscle as in spontaneous human infections (Hollister et al., 1996b). In addition, *T. hominis* can be grown in long-term culture, thus generating organisms for further study (Hollister et al., 1996b).

Vittaforma corneae

V. corneae was first named *Nosema corneum* Shadduck et al., 1990, after being identified in the corneal stroma of an otherwise healthy individual who was not infected with HIV (Davis et al., 1990). Observations from subsequent ultrastructural studies warranted the change to *V. corneae* Silveira and Canning, 1995, and small-subunit rDNA sequence data further corroborated this change to a new genus (Baker et al., 1995). Spontaneous (natural) infections with *V. corneae* have not been reported in nonhuman hosts. Experimental *V. corneae* infections have been transmitted to athymic mice, and after intraperitoneal inoculation, *V. corneae* organisms disseminated but infected the retina rather than the corneal stroma of the eye (Silveira et al., 1993). Since *V. corneae* infections are so rare in humans, it is difficult to determine if murine infections simulate human infections. *V. corneae* can be cultured to generate organisms for experimental investigations (Shadduck et al., 1990).

Other Species

Several additional microsporidia have been reported to infect humans. *Nosema connori* Sprague, 1974 (*Brachiola connori* [Cali et al., 1998]), caused a disseminated infection in an infant with SCID (Margileth et al., 1973). *Nosema ocularum* Cali et al., 1991, and a *Nosema*-like microsporidian (later named *Brachiola vesicularum*) (Cali et al., 1996, 1998) caused infections of the corneal stroma of HIV-seronegative individuals, and the latter was found in skeletal muscle as well. *Pleistophora* spp. have been identified as the cause of myositis in HIV-seronegative and HIV-infected individuals (Chupp et al., 1993; Ledford et al., 1985). *Trachipleistophora antropophtera* is the most recently described microsporidian in humans and was found in the brain tissues of two AIDS patients at autopsy (Vávra et al., 1997; Yachnis et al., 1996). Two microsporidia placed in a catch-all genus, *Microsporidium ceylonensis* and *M. africanum,* were reported to cause corneal ulcers in HIV-seronegative individuals (Ashton and Wirasinha, 1973; Pinnolis et al., 1981). Most of these microsporidia were identified at autopsy, and so live organisms were not available for passage into new hosts for experimental transmission of the infection or for possible long-term culture. Natural infections with these species have not been reported in nonhuman hosts.

HOST-PARASITE RELATIONSHIPS

Chronic Infections in Immunologically Competent Hosts

The majority of immunologically competent hosts who became infected with microsporidia (e.g., rabbits and mice) developed chronic and persistent infections with few clinical signs of disease. This condition has been described as a balanced host-parasite relationship because the host survives and the parasite persists. *E. cuniculi*-infected euthymic mice often developed ascites which resolved within a few weeks (Canning et al., 1986; Nelson, 1967; Perrin, 1943), and in rare cases, infected rabbits developed signs of cerebral dysfunction (Shadduck and Pakes, 1971). Typically, shedding of *E. cuniculi* spores in the urine of infected hosts was sporadic, and parasite replication appeared to be under control of the host's immune system. Conversely, mice with chronic microsporidiosis developed ascites with parasites after treatment with corticosteroids, which caused generalized immunosuppression (Bismanis, 1970; Innes, 1970).

On the basis of serological studies, it is likely that immunologically competent humans have been chronically (or persistently) infected with microsporidia (see section Acquired Antibody Responses). In addition, a few cases of microsporidiosis in HIV-seronegative individuals that resolved clinically after a few weeks have been reported (Sandfort et al., 1994; Silverstein et al. 1997; Sobottka et al., 1995; Weber and Bryan, 1994). As improved sensitive diagnostic methods become more widely used (e.g., polymerase chain reaction), it should be possible to verify if humans commonly carry subclinical microsporidial infections (Fedorko and Hijazi, 1996).

Disease in Immunologically Compromised Hosts

In the absence of a competent immune system, microsporidia may replicate without control, resulting in disease and often the death of the host. Immunodeficient athymic and SCID mice experimentally infected with *E. cuniculi* or *V. corneae* developed lethal disease, usually associated with severe ascites containing microsporidia (Didier et al., 1994; Gannon, 1980b; Hermanek et al., 1993; Koudela et al., 1993; Schmidt and Shadduck, 1983, 1984; Silveira et al., 1993), and athymic mice inoculated with *T. hominis* developed severe skeletal muscle disease (Hollister et al., 1996b). Carnivores, squirrel monkeys, and horses infected transplacentally with *E. cuniculi* often died, presumably because of the immature immune system in these hosts (Anver et al., 1972; Botha et al., 1979, 1986a, 1986b; McCully et al., 1978; Mohn and Nordstoga, 1982; Mohn et al., 1982a; Mohn et al., 1982b; Nordstoga, 1972; Nordstoga and Mohn, 1978; Shadduck and Orenstein, 1993; Shadduck et al., 1978; Van Dellen et al., 1978; van Rensburg et al., 1991; Zeman and Baskin, 1985).

Microsporidiosis in humans was first seen in children with impaired immune systems (Margileth et al., 1973; Matsubayashi et al., 1959) and has been increasingly recognized during the AIDS pandemic, particularly in individuals with fewer than 100 CD4$^+$ T lymphocytes per μl of blood (Asmuth et al., 1994; Orenstein, 1991; Weber et al., 1994). In addition, microsporidiosis has been recognized in organ transplant recipients undergoing immunosuppression treatments (Bryan, 1995; Bryan et al., 1996; Kelkar et al., 1997; Rabodonirina et al., 1996; Sax et al., 1995).

Immune-Mediated Disease in Immunologically Competent Hosts (Carnivores)

Carnivores such as domestic dogs, foxes, and mink become infected with *E. cuniculi* more frequently by transplacental transmission than do other animals. These hosts often died, but carnivores that survived the acute infection generally developed chronic infections associated with hypergammaglobulinemia that progressed to renal disease via type III hypersensitivity immune responses (Arneson and Nordstoga, 1977; Botha et al., 1979, 1986a, 1986b; McCully et al., 1978; McInnes and Stewart, 1991; Mohn, 1982a, 1982b; Mohn and Nordstoga, 1975, 1982; Mohn and Odegaard, 1977; Mohn et al., 1974, 1982a, 1982b; Nordstoga, 1972, 1976; Nordstoga and Westbye, 1976; Shadduck and Orenstein, 1993; Stewart et al., 1988; Zhou and Nordstoga, 1993). Perivascular granulomatous lesions were typically observed in infected carnivores, and granular deposits containing IgM and IgG were detected in the glomerular basement membranes in *E. cuniculi*-infected mink (Zhou and Nordstoga, 1993). It is unknown, however, if the immune complexes were generated between *E. cuniculi* antigens and antibody specific to *E. cuniculi* or resulted from autoimmune responses against host antigens.

HOST RESPONSES TO MICROSPORIDIA

Microsporidia are ubiquitous organisms and therefore are encountered frequently by mammals, yet infections and clinical manifestations associated with microsporidiosis are relatively rare. At least three levels or stages of defense mechanisms are expressed to prevent or control infection with most microorganisms. These are innate resistance, early induced responses, and adapted immune responses (Janeway and Travers, 1996).

Innate Resistance

Mechanisms of innate resistance involve host specificity, mechanical and chemical barriers, and defense responses expressed by the host within a few hours of exposure to a microorganism. Innate resistance is probably important in preventing infection when one considers that of the more than 1,000 species of microsporidia found in nature only 13 are presently known to infect humans. In one example, McDougall and colleagues reported the presence of a *Nosema* sp. in the feces of an AIDS patient but considered it an incidental finding since the individual was not otherwise infected with this microsporidian (McDougall et al., 1993). This observation suggested that innate and early induced responses may be important in preventing many microsporidian species from establishing infections in humans.

HOST SPECIFICITY

The microsporidia species that infect humans demonstrate a great deal of variability in host range. The most frequently reported microsporidian that infects humans, *E. bieneusi,* also infects pigs and nonhuman primates but has not been successfully transmitted to mice, including immunodeficient strains of mice. *E. intestinalis,* the next most common microsporidian to infect humans, was successfully transmitted to mice and monkeys (Achbarou et al., 1996; Bessinger et al., 1997; Didier, 1998, unpublished observations; Enriquez, 1997) and appears to infect dogs, donkeys, pigs, cows, and goats (Bornay-Llinares et al., 1998). *E. cuniculi* also can infect a wide range of mammals (Canning et al., 1986; Didier et al., 1998; Shadduck et al., 1979b; Weber et al., 1994), whereas *E. hellem,* which was also transmitted to mice and monkeys (Didier et al., 1994; Hollister et al., 1993), has been found to infect only birds and AIDS patients under natural conditions (Black et al., 1997; Friedberg et al., 1990; Pulparampil et al., 1998; Rastrelli et al., 1994; Yee et al., 1991). Most of the microsporidian species reported in humans, including *V. corneae, Pleistophora, Trachipleistophora,* and *Nosema* spp., have been described in only a handful of case reports, suggesting that some of these species are found only incidentally in humans. These examples of natural and experimental infections, however, suggested that host specificity played some role in limiting the susceptibility and resistance to some species of microsporidia. Two factors believed to affect host specificity are the genetic makeup of the host and the host's body temperature.

Genetic Background. It is not known if specific receptors or adhesion factors mediate infection by microsporidia. However, studies utilizing inbred strains of mice indicated that the genetic background of a host could influence progression of an established microsporidian infection. Niederkorn et al. (1981) infected several strains of mice with *E. cuniculi* intraperitoneally and compared their level of resistance or susceptibility on the basis of percentage of infected peritoneal macrophages (PMPs) 2 weeks later. Relatively resistant strains of mice with fewer than 1% infected PMPs included BALB/c, J/J, C57BL/6 × BALB/c F_1, and SJL mice. Relatively susceptible mice included C57BL/6, DBA/1, and 129/J mice with ≥ 15% infected PMPs. Strains with intermediate levels of infected PMPs included C57BL/10, DBA/2, and AKR mice. These variations were not found to be major histocompatibility complex (MHC)-restricted since BALB/c mice (H-2^d) were relatively resistant while DBA/2 mice (also H-2^d) displayed intermediate levels of susceptibility, and C57BL/6 mice (H-2^b) were found to be highly susceptible to *E. cuniculi* infection, while intermediate levels of infected PMPs were observed in C57BL/10 mice (also H-2^b). Furthermore, C57BL/6 × BALB/c F_1 mice were as resistant to *E. cuniculi* infections as the parent BALB/c mice, suggesting that resistance was controlled by one or more dominant genes outside the MHC.

Resistance and susceptibility also were not affected by specific MHC-restricted receptors since fibroblasts from all strains of mice were equally susceptible to infection with *E. cuniculi* in vitro (Niederkorn et al., 1981). To date, however, no receptors have been described that might be utilized by microsporidia to infect a host cell, although some tissue tropism exists among species of microsporidia (see "Tissue Specificity").

Body Temperature of the Host. Temperature is an important limiting factor that restricts the host range and host site(s) of infection in two species of microsporidia. *N. algerae* Vávra and Undeen, 1970, a microsporidian of mosquitoes, infected porcine kidney cell cultures but did not survive beyond 3 days at 37°C. *N. algerae,* however, continued to grow indefinitely in porcine kidney cells and MDCK cells maintained at room temperatures of 25 to 27°C (Didier et al., 1991a; Undeen, 1975). Intravenous, peroral, or intranasal inoculations of *N. algerae* into immunologically competent euthymic mice failed to produce persistent infections (Undeen and Maddox, 1973), whereas subcutaneous inoculations into the ears, tail, or footpads of such mice, when temperatures were slightly lower than 37°C, resulted in localized infections and replicating stages could be observed through day 10 (Undeen and Alger, 1976). In athymic mice inoculated with *N. algerae* subcutaneously in the tail and footpads, organisms were seen to disseminate to the connective tissues and muscle fibers (Trammer et al., 1997). Inoculation of *N. algerae* into athymic mice by an intravenous, peroral, or intranasal route however, failed to establish infection. The authors suggested that immunodeficient hosts may be susceptible to insect microsporidia, such as *N. algerae,* if the initial infections are established prior to dissemination in an extremity in which temperatures are slightly lower than core body temperature (Trammer et al., 1997).

Among the microsporidia first identified in humans, temperature has also been demonstrated to affect *E. hellem* infectivity. Experimentally, *E. hellem* infected athymic mice but was slower to cause death when compared with *E. cuniculi* and *V. corneae* (previously named *N. corneum*) infections in athymic mice (Didier et al., 1994). At the time, this finding was thought to be related to the observation that at 37°C, *E. hellem* replicated more slowly than *E. cuniculi* and *V. corneae* in vitro. *E. hellem* was recognized as a cause of disease in psittacine birds which have a core body temperature of 41°C (Black et al., 1997; Pulparampil et al., 1998), and *E. hellem* was subsequently observed to replicate more rapidly in vitro at 41°C than at 37°C (Snowden, personal communication).

These observations suggested that the host body temperature was a predisposing factor for the infectivity of at least two microsporidians, *N. algerae* and *E. hellem,* and may limit some microsporidia from infecting humans, as well. Such findings are also of epidemiological interest since several AIDS patients who became infected with *E. hellem* also reported having pet parakeets (Friedberg et al., 1990), adding support to the hypothesis that *E. hellem* infections may be zoonotic. The risk of exposure to insects by immunodeficient hosts, however, is still debatable since no insect microsporidia have been reported to cause infections in AIDS patients.

PHYSICAL AND CHEMICAL BARRIERS

Microsporidia bypass the skin and primarily infect mammalian hosts via ingestion or inhalation (Canning et al., 1986; Orenstein, 1991; Schwartz and Bryan, 1997; Schwartz et al., 1993b; Weber et al., 1994). Although the infectious dose of microsporidia administered orally has not been defined in mammals, the acidic pH of the stomach is likely to reduce the infectivity of some microsporidia. For example, Shadduck and Polley (1978) reported that the viability of *E. cuniculi* decreased significantly after incubation in a pH 4 buffer, although a few viable organisms remained. The effects of lysozyme or other mucosal (intestinal) enzymes on microsporidian viability or infectivity are unknown. It is also unknown whether the normal gut flora contributes to the inhibition of microsporidial infections.

SERUM FACTORS

If microsporidia infect epithelial cells of the intestine or respiratory tract, the early defense mechanisms of the host may come into play via activation of serum factors such as opsonins or components of the alternate pathway of complement. It is unknown if microsporidia induce the alternate pathway of complement or can be opsonized by nonspecific serum factors in vivo, but in vitro data imply that these factors contribute to innate resistance. Incubation of unstimulated or thioglycollate-induced peritoneal lagomorph macrophages with heat-inactivated normal rabbit serum or normal IgG did not lead to a significant reduction in the number of *E. cuniculi* parasites compared with the number of organisms recovered from cultures of macrophages incubated with complete medium. Addition of guinea pig complement to macrophage cultures containing heat-inactivated normal IgG or normal serum, however, resulted in some reduction in the number of parasites (Niederkorn and Shadduck, 1980). Further evidence of inhibition of microsporidia by complement or other serum factors was described in a murine macrophage system (Schmidt and Shadduck, 1984). *E. cuniculi* spores were incubated with heat-inactivated normal mouse serum, hyperimmune serum, baby rabbit complement, or combinations of serum and complement. After incubation for 45 min at 37°C, parasites were counted and added to fibroblast cultures to compare the infectivity of treated versus nontreated microsporidia. The infectivity of microsporidia was significantly reduced when *E. cuniculi* were treated with immune serum alone, complement alone, normal mouse serum alone, or immune serum plus complement compared with the infectivity of nontreated microsporidia. Serum factors therefore appeared to inhibit the infectivity of microsporidia, but it is unknown which factors, in addition to complement contributed to the reduction in parasite infectivity.

PHAGOCYTES

Macrophages pose the first line of cellular defense when microorganisms break or circumvent epithelial barriers. These cells phagocytize microorganisms, release cytokines to initiate early induced responses of defense, and process antigens for presentation on their surfaces to initiate specific immune responses (Janeway and Travers, 1996). *E. bieneusi* organisms primarily replicate within epithelial cells lining the small intestine, do not replicate in macrophages, and appear to be degraded by macrophages released into intestinal tissues (Kotler et al., 1993; Orenstein, 1991; Orenstein et al., 1990; Tzipori et al., 1997). Several other species of microsporidia that infect mammals, such as the *Encephalitozoon* spp., typically disseminate and encounter macrophages (Canning and Hollister, 1991, 1992). Although many *Encephalitozoon* spores probably are destroyed by macrophages, some survive and replicate in macrophages because the lysosomes fail to fuse with the phagosomes (Weidner, 1975). Weidner and Sibley (1985) observed that the phagosomes with one microsporidian failed to acidify and that these organisms were more likely to survive and replicate than phagosomes containing more than one organism (see "Host Cell-Parasite Interface" below). The route of entry into the macrophage also appeared to determine the fate of microsporidia in macrophages. In in vitro studies, Couzinet et al. (1997) observed that of 85% of *E. cuniculi* spores found within human monocyte-derived macrophages, only 0.03% had discharged their polar tubules. The majority of the microsporidian spores appeared to be destroyed after phagocytosis, although some microsporidia taken up by phagocytosis had extruded their polar filaments into the cytoplasm of the macrophage (Couzinet et al., 1997; Ditrich et al., 1997; Markova and Koudela, 1997). The authors thus speculated that only those few spores that discharged their polar filaments and escaped the lysosomes were able to replicate.

TISSUE SPECIFICITY

In experimental and naturally occurring infections, *E. bieneusi* organisms remained primarily localized to the small intestine, *Nosema* spp. mainly infected the corneal stroma, and *Pleistophora* spp. and *T. hominis* infected the skeletal muscle of their hosts. Conversely, *Encephalitozoon* spp. typically disseminated and thereby expressed significantly less tissue specificity (Canning and Hollister, 1992; Didier et al., 1994; Schwartz and Bryan, 1997; Weber et al., 1994). It is unknown, however, if the tissue specificity observed for some species of microsporidia was due to parasite selection (or tropism), parasite evasion of immune responses at immunologically privileged sites, or pressure exerted by the host that determined where a microsporidian can persist.

Early Induced Responses

When the innate defense barriers are evaded, overwhelmed, or circumvented, the host employs a series of early induced responses in an attempt to destroy or at least control microorganisms until immune responses are generated (Janeway and Travers, 1996). Early induced responses are triggered by receptor-signal transduction pathways leading to a limited array of responses. These responses (1) are identical or invariant for each exposure to a given microorganism (i.e., no secondary response occurs), (2) are maintained for the same length of time after each antigen (immunogen) exposure, and (3) are not antigen (immunogen)-specific. Inflammation is the principal early induced response the host employs and involves the recruitment of phagocytic cells to the site of infection and the secretion of mediators that influence the type of immune response that subsequently develops.

EARLY CELLULAR RESPONSES

During early induced responses, macrophages that first encounter a microorganism typically release cytokines such as tumor necrosis factor alpha (TNF-α), interleukin-1 (IL-1), IL-6, IL-8, and IL-12 which cause fever and recruit additional macrophages and other effector cells such as neutrophils, natural killer (NK) cells, and lymphocytes to express the acute phase response. One clinical manifestation of an early response to microsporidiosis was transient ascites which developed in mice after intraperitoneal inoculation with *E. cuniculi* (Arison et al., 1966; Morris et al., 1958; Petri 1965, 1969). The ascites consisted primarily of macrophages and, in euthymic mice, resolved after approximately 2 to 3 weeks, presumably when the specific immune responses came into play. Splenomegaly also was observed in mice infected with *E. cuniculi* (Niederkorn et al., 1983), as the result of an increase in cellular proliferation or recruitment of cells to this lymphoid organ. The spleen weight of BALB/c mice given *E. cuniculi* intraperitoneally peaked at 1 week with a spleen weight index of 2.87 and began to subside by 2 weeks with a spleen weight index of 1.97 (Didier and Shadduck, 1988). Indirect evidence from in vitro studies also supported the likelihood that early induced responses could be generated against microsporidia whereby addition of live, but not dead, *E. cuniculi* generated an oxidative burst as determined by increased superoxide generation in murine peritoneal macrophages (Ditrich et al., 1997).

The strongest evidence that early induced responses occurred during microsporidiosis was from reports by Niederkorn and colleagues (Niederkorn 1985; Niederkorn et al., 1983) who demonstrated increased splenic and pulmonary NK activity in *E. cuniculi*-infected euthymic and athymic mice. The enhanced NK activity in euthymic BALB/c and C57BL/6 mice peaked approximately 3 to 4 days after inoculation with viable *E. cuniculi* and returned to baseline levels by day 21. No enhancement was seen when the mice were inoculated with dead microsporidia. Interestingly, NK-defective (beige) mice did not die of infection with *E. cuniculi,* whereas athymic BALB/c mice, which expressed the highest NK activity after inoculation with viable organisms, failed to survive the infection (Niederkorn et al., 1983). These results indicated that enhanced NK activity probably helped control microsporidiosis during early induced responses and that additional responses contributed to protection against lethal disease.

In humans, early induced responses appeared to be generated because inflammatory responses developed in the absence of functional T cell-mediated responses in AIDS patients. Lesions consisting of poorly formed granulomas with macrophages and plasmacytoid cells were observed in AIDS patients with negligible numbers of T cells who were ill-equipped to generate specific immune responses. Fever, an expression of an early induced response, was often a clinical symptom associated with microsporidiosis in AIDS patients as well (Kotler et al., 1990, 1993; Kotler and Orenstein, 1994; Modigliani et al., 1985; Orenstein, 1991; Schwartz and Bryan, 1997; Schwartz et al., 1996).

EARLY CYTOKINE RESPONSES

Indirect evidence exists that microsporidia induced early cytokine responses that were expressed prior to specific adapted immune responses. Armstrong et al. (1973) reported that an interferonlike substance was generated in rabbit kidney cells infected with *E. cuniculi* in vitro which became resistant to infection with vesicular stomatitis virus or eastern equine encephalitis virus in vitro. In addition, *Nucleospora salmonis,* a cousin of *E. bieneusi* that infects chinook salmon, induced a secretion of factors that stimulated lymphocyte proliferation in vitro (Wongtavatchai et al., 1995a, 1995b). Activated NK cells are known to release IFN-γ, and it is likely that the NK cells activated during the early responses to microsporidia in *E. cuniculi*-infected mice generated IFN-γ as well (Niederkorn, 1985; Niederkorn et al., 1983).

The cytokine TNF also appeared to be generated during microsporidiosis. Elevated fecal TNF-α levels were detected in AIDS patients with microsporidiosis and a barely detectable level of $CD4^+$ T cells (a range of 0 to 80 cells per µl of blood and a median of 27 $\pm$ 25 cells per µl of blood), suggesting that the elevated TNF-α level was not generated via antigen-specific immune cell activation of macrophages (Sharpstone et al., 1997). Indirect evidence also suggested that TNF-α may contribute to resistance during the early stages of infection since incubation of murine peritoneal macrophages with TNF-α (1,000 u/ml) inhibited the replication of *E. cuniculi* for at least 48 h (i.e., was parasitistatic) even though no reduction in organisms (i.e., parasiticidal activity) was observed (Didier, 1995; Didier and Shadduck, 1994). The TNF-α expressed during the later stages of microsporidiosis in AIDS patients is believed to contribute to the pathogenesis of these infections (Sharpstone et al., 1997), but its role during the early stages is not well understood.

Adapted Immune Responses

Whereas innate and early induced mechanisms of resistance are immediate or rapidly expressed, respectively, the development of adapted immune responses requires several days to allow for clonal expansion of the number of B and T lymphocytes. Characteristics that distinguish immune responses from early induced and innate mechanisms include (1) adaptation (i.e., development of immune responsiveness via clonal expansion of the number of lymphocytes), (2) antigen specificity, and, (3) development of anamnestic responses or immunological memory (Janeway and Travers, 1996). Classically, immune responses are separated into humoral immunity and cell-mediated immunity. Both types of responses generally develop after exposure to a microorganism, although one type may predominate or is more important than the other in clearing or controlling the pathogen. If immune responses fail to develop because of immune suppression or deficiency, microsporidia may replicate without restraint and cause disease. Conversely, unregulated or dysfunctional immune responses appear to contribute to disease, as expressed by immune complex-mediated type III hypersensitivity responses or other autoimmune pathways.

HUMORAL IMMUNE RESPONSES

Humoral immunity is classically defined as the production of antibodies by clonally expanding the number of B lymphocytes and plasma cells in response to immunogens. Immunologically competent hosts experimentally infected with microsporidia generally expressed an early serum IgM response followed by expression of serum IgG and IgA. In most immunologically competent hosts infected with microsporidia, the IgG response persisted indefinitely, usually for the life of the host, because the inducing immunogens (i.e., microsporidia) persisted. Little is known about the IgA responses in hosts infected with microsporidia, however.

Humoral Antibody Functions. Antibodies appear to contribute to resistance but are insufficient to prevent lethal microsporidiosis. Passive transfer of hyperimmune serum from BALB/c mice failed to protect BALB/c (*nu/nu*) athymic mice inoculated with *E. cuniculi* (Schmidt and Shadduck, 1983). However, maternal antibodies passively transferred from *E. cuniculi*-infected dams were observed in newborn rabbits, and rabbits necropsied at less than 1 day of age or at 11 days of age displayed no *E. cuniculi*-associated lesions (Bywater and Kellett, 1978a, 1978b, 1979; Bywater et al., 1980). Although it was possible that 11 days was insufficient for the development of *E. cuniculi*-associated lesions, the presence of the maternally transferred antibodies may have reduced the infectivity of *E. cuniculi* in the young rabbits. Most of the data supporting a role for antibody in protection against microsporidiosis, as well as information about the functions of these antibodies, are based on in vitro studies.

Opsonizing Effects of Antibody. Studies examining immune responses to microsporidia have primarily utilized *E. cuniculi*. In addition to replicating within epithelial and endothelial cells, *E. cuniculi* can survive and replicate within parasitophorous vacuoles in macrophages because of an absence of phagosome-lysosome fusion (Weidner, 1975). However, when *E. cuniculi* spores were treated with rabbit antiserum and added to rabbit peritoneal macrophages, lysosomes were observed to fuse with the parasitophorous vacuoles (Niederkorn and Shadduck, 1980). In addition, a reduction in the number of *E. cuniculi* organisms was seen in cultures of murine peritoneal macrophages incubated with *E. cuniculi* that had been opsonized (i.e., pretreated with heat-inactivated hyperimmune serum) compared with the number of spores recovered from macrophages incubated with normal mouse serum-treated organisms (Schmidt and Shadduck, 1984). Conversely, no opsonization effect was observed when spores of *Tetramicra brevilfilum*, a microsporidian of turbot (*Scophthalamus maximus*), were incubated with immune serum and turbot splenic macrophages (Leiro et al., 1996), indicating that there are probably differences between mammals and fish regarding opsonization functions of antibodies.

Complement Fixation. One of the earliest serological diagnostic tests for microsporidiosis in rabbits was the complement fixation test, which indicated that rabbits with *E. cuniculi* infection expressed complement-fixing antibodies (Wosu et al., 1977a, 1977b). The function of these complement fixing antibodies in resistance in vivo is unknown, and it is unlikely that the spore coat of a mature microsporidian can be penetrated by complement fixation although less mature stages may be susceptible to complement-mediated lysis. In vitro studies, however, suggested that antibody-mediated complement fixation reduced the infectivity of microsporidia. Niederkorn and Shadduck (1980) demonstrated that macrophages killed a significant number of *E. cuniculi* spores that had been pretreated with heat-inactivated immune serum. Addition of guinea pig complement to these cultures further reduced the number of recovered *E. cuniculi,* suggesting that complement fixation further assisted in the opsonization of microsporidia. Similar results were observed in a murine system. Fewer infectious *E. cuniculi* were recovered from BALB/c murine peritoneal macrophage cultures when the organisms had been preincubated with hyperimmune serum plus complement than when the *E. cuniculi* had been preincubated with only immune serum (Schmidt and Shadduck, 1984).

Neutralizing Antibodies. In one study, antibodies were found to inhibit microsporidial infection of nonphagocytic cells. Enriquez (1997) reported that incubation of *E. intestinalis, E. hellem,* or *E. cuniculi* spores with a monoclonal antibody (MAb 3B6) that reacts against all three *Encephalitozoon* spp. inhibited in vitro infection of Vero cells in the absence of complement. No information has been published about receptor-mediated internalization of microsporidia into host cells, but neutralization studies suggested that parasite or host cell receptors may contribute to parasite infectivity.

Acquired Antibody Responses. The time until specific antibodies were detected after exposure to microsporidia under experimental situations generally varied with the route of inoculation. In studies comparing routes of *E. cuniculi* inoculation in rabbits, serum IgG was first detected by the India ink reaction or an indirect immunofluorescent antibody test (IFAT) by 7 days after intravenous inoculation, 8 days after intracerebral inoculation, 10 days after subcutaneous inoculation, and approximately 21 days after oral inoculation (Cox and Gallichio, 1978; Cox et al., 1979; Kunstry et al., 1986; Waller et al., 1978; Waller and Bergquist, 1982; Wosu et al., 1977a, 1977b). *E. cuniculi*-specific IgM responses preceded IgG responses, and IgG antibodies persisted for at least 400 days. In addition, serum IgA was detected in rabbits inoculated intrarectally (Wicher et al., 1991), but IgA responses were not examined in the other studies.

A similar time course of antibody expression was observed in euthymic mice. As detected by IFAT, 7 days after intraperitoneal inoculation of *E. cuniculi,* BALB/c mice generated specific IgG responses which remained high for the life of the mice (Schmidt and Shadduck, 1983). C57BL mice given *E. cuniculi* by oral inoculation, however, generated specific IgM and IgG, as detected by IFAT, 22 days after inoculation (Gannon, 1980b). IgM levels fell to baseline 2 months after inoculation, while IgG levels remained high 17 months later. A similar time course of antibody expression, as detected by enzyme-linked immunosorbent assay (ELISA) and Western blot immunodetection, was observed in C57BL mice inoculated orally with *E. intestinalis* (Bessinger et al., 1997).

Rhesus macaques (*M. mulatta*) inoculated intravenously with 1×10^9 *E. cuniculi, E. hellem,* or *E. intestinalis* organisms first expressed serum IgM, as detected by ELISA and Western blot immunodetection, approximately 14 days later (Didier, 1998, unpublished observations; Didier et al., 1992, 1994). These IgM responses peaked between days 28 and 42 after inoculation and fell to baseline levels by 70 to 98 days after inoculation. The simian serum IgG responses were first detected between days 14 and 28 by ELISA and Western blot, peaked at approximately 42 days, and remained high for at least 3 years after inoculation.

These results indicated that serum antibodies developed more rapidly after intravenous, intraperitoneal, or subcutaneous inoculation than after oral or intratracheal inoculations and that IgM responses preceded IgG responses. Observations that IgG antibodies persisted for the life of these experimentally inoculated animals support the hypothesis that immunologically competent hosts that generated persistent antibody responses were probably persistently infected with microsporidia.

Serological survey data further support the hypothesis that the persistence of microsporidia-specific antibodies is due to the persistence of a chronic microsporidial infection. From 0 to 100% of guinea pigs, hamsters, rabbits, mice, and squirrel monkeys in laboratory colonies have been found to be seropositive for *E. cuniculi* and at necropsy a correlation was found between the presence of antibody and histologic evidence of infection (Bywater and Kellett, 1978a, 1978b, 1979; Chalupsky et al., 1973, 1979; Gannon, 1980a; Lyngset, 1980; Shadduck and Baskin, 1989; Zeman and Baskin, 1985). Culling of seropositive rabbits from laboratory animal colonies resulted in an absence of infected seropositive animals from these colonies (Bywater and Kellett, 1978b; Cox et al., 1977). Furthermore, inoculation of mice with killed *E. cuniculi* resulted in transient antibody responses and ELISA titers that soon fell to baseline levels (Liu and Shadduck, 1988), suggesting that antibody levels remained positive (i.e., ELISA titer ≥1:800) only if the parasites persisted. An exception to this hypothesis may exist in regard to nonhuman primates. A small number of squirrel monkeys were observed to seroconvert from positive to negative to positive, or conversely, to seroconvert from negative to positive to negative serum IgG titers as measured by IFAT against *E. cuniculi* (Shadduck and Baskin, 1989). It is possible that these monkeys were able to clear their *E. cuniculi* infection, thereby decreasing their production of specific IgG, and that reexposure to microsporidia may have led to the regeneration of an antibody response. Whether such a scenario occurs in humans is under investigation, however.

Data concerning the development or acquired nature of antibody responsiveness in humans after exposure to microsporidia are less definitive. Early serological surveys utilized *E. cuniculi* as the antigen because it was the only mammalian microsporidian that could be cultured. Relatively high numbers of individuals in various human populations expressed positive levels of specific IgG. Singh et al. (1982) found that serum antibodies positive to *E. cuniculi* were detected by IFAT in 36% (33 of 92) of patients in Ghana with malaria, 43% (38 of 89) of tuberculosis patients in Nigeria, 19% (13 of 70) of filaria-infected individuals in Malaysia, and 4.6% (11 of 263) of individuals with typhoid fever, leprosy, toxoplasmosis, or toxocariasis. Bergquist and colleagues detected anti-*E. cuniculi* IgG in sera of 33% (10 of 30) of healthy Swedish homosexual men and 12% (14 of 105) of Swedes returning from the tropics (Bergquist et al., 1984a, 1984b). Using ELISA and Western blot immunodetection methods, Canning and colleagues found that 6.8% (36 of 545) of individuals in Jali, The Gambia, and 8.6% (23 of 286) of individuals in psychiatric wards and 6.1% (10 of 163) of children with neurological disorders in the United Kingdom expressed positive serum IgG titers for *E. cuniculi* (Hollister and Canning, 1987; Hollister et al., 1991). In addition, they found that 5.5% (33 of 598) of malaria patients and 10.3% (18 of 175) of schistosomiasis patients also expressed *E. cuniculi*-specific IgG. Given that many parasitic infections cause aberrant antibody responses and that new species of microsporidia have been discovered in humans, these results are difficult to interpret regarding the relationship between developing antibody responses and true microsporidial infections. Even when the newly cultured human microsporidian species *E. hellem* and *V. corneae*, were used along with *E. cuniculi* as antigens, variable results were obtained. Eight of 8 sera from HIV seronegative homosexual individuals with no history of clinical microsporidiosis expressed positive ELISA IgG titers (≥1:800) against all three microsporidial antigens, while sera from 7 of 12 HIV-seronegative heterosexuals were positive for IgG against *E. cuniculi* and *E. hellem* and 5 of these 12 sera also contained IgG for *V. corneae* (Didier et al., 1993). The most recent and most stringent serological study involving ELISA, IFAT, and counterimmunoelectrophoresis was reported by Van Gool and colleagues who found that 8% (24 of 300) of Dutch blood donors and 5% (13 of 276) of pregnant French women expressed *E. intestinalis*-specific serum IgG (Van Gool et al., 1997).

Further confounding the interpretation of serological studies in humans is the cross-reactivity between sera and species of microsporidia. IFAT detected weak cross-reactivity between microsporidia originally from fish, insects, and mammals (Niederkorn et al., 1980), as well as between *E. bieneusi* and *Encephalitozoon* spp. assayed with experimentally induced immune sera or patient sera by IFAT, Western blot, and ELISA (Aldras et al., 1994; Didier et al., 1991a, 1991b, 1993; Hartskeerl et al., 1995; Weiss et al., 1992). Variable cross reactions also were expressed by *E. bieneusi*-infected AIDS patients against *Glugea* spp. on Western blots (Ombrouck et al., 1995). Little is known about the natural course of antibody production in humans with naturally acquired microsporidiosis. The presence of microsporidia-specific antibodies in many immunologically competent human populations suggests that people are at least exposed to, and possibly infected with, microsporidia. More sensitive methods such as polymerose chain reaction, however, will be required to assess the putative correlation between seropositivity and persistent microsporidiosis (Fedorko and Hijazi, 1996).

Antibody responses in immunologically naive (i.e., immature) and immunodeficient hosts also were variable depending on the state of immune competence at the time of exposure to microsporidia. Rabbits born to seropositive dams expressed maternal antibodies during the first 2 to 4 weeks, after which the pups became seronegative between 6 and 8 weeks of age (Bywater et al., 1978a, 1978b; 1979; Bywater et al., 1980; Lyngset, 1980). Many of the young rabbits began to seroconvert between 8 and 14 weeks of age as detected by the India ink (carbon) reaction for *E. cuniculi*-specific serum IgG. In another experiment in which pups were separated from seropositive dams, rabbits that failed to become seropositive by 16 weeks of age were unlikely to seroconvert thereafter and presumably were not infected with *E. cuniculi* (Cox and Gallichio, 1977). Neonatal mice inoculated intraperitoneally with *E. cuniculi* expressed lower *E. cuniculi*-specific serum IgG responses than adult-inoculated mice during the initial 20 weeks but subsequently expressed serum IgG at levels similar to those of mice infected as adults, as detected by ELISA (Liu et al., 1988). In addition, a slight delay in antibody expression was observed in kittens inoculated with *E. cuniculi* (Pang and Shadduck, 1985). Specific serum IgG was not detected until approximately 3 to 4 weeks after intraperitoneal inoculation of 3- and 4-day-old kittens, whereas cats who were 2 weeks of age or 2 1/2 months of age at the time of intraperitoneal inoculation expressed specific serum IgG 2 weeks later. All cats, however, expressed similarly high levels of IgG as determined by IFA on termination of the experiment. These results suggested that hosts who were infected at a young, immunologically immature age were slower to produce antibodies but eventually expressed them at the same levels as those expressed by hosts who became infected as adults.

Antibody responses in carnivores such as blue foxes and domestic dogs are of particular interest because two extremes of immune responses have been observed. Transplacentally infected carnivores usually developed renal disease and often died (Mohn, 1982a, 1982b; Mohn and Nordstoga, 1975, 1982; Mohn and Odegaard, 1977; Mohn et al., 1982a, 1982b; Nordstoga and Mohn, 1978; Shadduck and Orenstein, 1993). Blue foxes infected with *E. cuniculi* transplacentally were observed to express specific IgG at 3 weeks of age, which probably represented passively transferred maternal antibodies. The animals that survived transplacental infection with *E. cuniculi,* or became infected after birth, subsequently developed renal disease which appeared to be associated with polyarteritis nodosa via immune complex formation and deposition. By 3 to 5 months of age, blue foxes with clinical signs of microsporidiosis expressed specific serum IgG to *E. cuniculi,* along with significantly higher than normal gamma globulin protein levels (Mohn and Nordstoga, 1975). Some blue foxes that resolved their clinical symptoms expressed a corresponding reduction in serum gamma globulin protein levels. Blue foxes inoculated with *E. cuniculi* orally at 2 to 14 days of age developed mild clinical disease, and specific serum IgG was detected at 30 days and remained at a high level, while blue foxes infected at an older age with *E. cuniculi* expressed few if any clinical signs of disease (Mohn, 1982a, 1982b).

Similar results have been described for domestic dogs infected with *E. cuniculi*. Again, transplacentally infected puppies often died, but those that survived expressed specific serum antibodies at 7 days and thereafter (Shadduck et al., 1978; Stewart et al., 1986). It was likely that the antibodies detected in puppies less than 1 month old were passively transferred maternal antibodies. Intraperitoneally inoculated 2-day old puppies expressed a level of IgM 10 days later which plateaued at 45 days, and IgG was expressed by 10 days (Szabo and Shadduck, 1987). *E. cuniculi*-specific antibodies continued to increase, and the total IgG levels in these dogs were approximately threefold higher than in seronegative control dogs (Szabo and Shadduck, 1987). Adult dogs inoculated orally, intravenously, or intraperitoneally with *E. cuniculi* expressed transient IgM responses from 6 to 12 weeks after inoculation, and IgG responses were detected by IFAT approximately 6 weeks after inoculation and continued to increase during the following several months (Stewart et al., 1979; Szabo and Shadduck, 1988). Hypergammaglobulinemia was a consistent finding in *E. cuniculi* seropositive dogs (Botha et al., 1979, 1986a, 1986b; Hollister et al., 1989; Stewart et al., 1986, 1997). Thus, in carnivores, the age of the host at the time of infection with *E. cuniculi* affected the development of antibodies and the progression of disease that appeared to be mediated by the antibodies.

Immunodeficient hosts, on the other hand, failed to express specific *E. cuniculi* IgG responses. For example, athymic and SCID mice generated no detectable antibody responses to *E. cuniculi* or *E. intestinalis* (Enriquez, 1997; Gannon, 1980b; Hermanek et al., 1993). SCID mice reconstituted with immune lymphocytes from immunocompetent syngeneic donors, however, expressed serum IgG 12 to 28 days after infection with *E. cuniculi* or *E. intestinalis,* and expressed serum IgA 17 days after infection. Simian immunodeficiency virus (SIV)-infected rhesus monkeys with $CD4^{+}CD29^{+}$ T cell levels below 10% at the time of *E. cuniculi, E. hellem,* or *E. intestinalis* inoculation mounted no detectable specific antibody (Didier, 1998, unpublished observations; Didier et al., 1992, 1994). However, SIV-infected monkeys whose peripheral blood $CD4^{+}CD29^{+}$ T-cell levels were still above 10% at the time of intravenous microsporidian inoculation were able to mount antibody responses. Lower ELISA IgM and IgG titers were detected in these monkeys, and a delay in peak IgG expression occurred approximately 98 to 126 days after inoculation of

SIV-infected monkeys, compared with the peak IgG response observed approximately 42 days after microsporidian inoculation in healthy (non-SIV-infected) monkeys. Interestingly, the IgM responses, which subsided by 98 days after inoculation of healthy monkeys, remained above baseline levels for 154 to 248 days after inoculation in the SIV-infected monkeys. In most cases, in the SIV-infected monkeys microsporidia-specific antibody levels fell dramatically about 1 month prior to death.

As expected, HIV-infected individuals expressed variable antibody responses to microsporidia. Two of four HIV-infected individuals with diagnosed *E. hellem* infections failed to express ≥1:800 (positive) IgG titers to *E. hellem* (as measured by ELISA), and one of two individuals with *E. intestinalis* failed to express detectable antibodies (Didier et al., 1991c, 1993). No serological studies have been presented that followed the progession of antibody responses over time or in concert with declining $CD4^+$ T cell levels in HIV-infected persons suspected of having microsporidiosis. On the basis of the serology of SIV-infected monkeys and immunodeficient mice experimentally infected with microsporidia, it is likely that the variability in antimicrosporidia Ig immunoglobulin responses in humans with HIV and microsporidia were probably related to the immune status of the individuals at the time of microsporidial infection.

Anamnestic Antibody Responses. Anamnestic antibody responses were found to occur in euthymic mice inoculated with *E. cuniculi* intraperitoneally followed by a secondary challenge 14 to 16 weeks later, as observed by IgG responses that increased dramatically with little lag time (Liu et al., 1988; Schmidt and Shadduck, 1983). Many mammalian hosts inoculated with *E. cuniculi* continued to express antibody responses, probably because the parasite persisted and continually induced immune responsiveness. In addition, inoculation of mice with dead organisms resulted in transient antibody responses that fell to baseline levels after inoculation with *E. cuniculi* (Liu and Shadduck, 1988). This finding suggested that after infection with *E. cuniculi,* "secondary" challenges continued as the microsporidia replicated during the course of infection.

The level of microsporidian-specific serum IgG responses expressed by mammals other than carnivores against microsporidia remained consistently high but did not continue to increase indefinitely. In other words IgG responses in these hosts seemed to plateau and therefore must have been under some regulatory control. Carnivores, however, expressed *E. cuniculi*-specific IgG and gamma globulin serum protein responses that continued to increase and did not appear to be as well regulated, as evidenced by the hypergammaglobulinemia and immune complex disease that developed. These host differences in the regulation of secondary or anamnestic antibody responses are not well characterized, however.

Specificity of Humoral Antibody Responses. Specific antibody responses typically were generated in spontaneously and experimentally infected or immunized hosts. Cross-reactivity was not detected between microsporidia and other protozoans such as *Toxoplasma gondii, Trypanosoma cruzi, Trypanosoma congolese,* and *Eimeria* spp. (Wosu et al., 1977a, 1977b). Mice and rhesus macaques inoculated with a microsporidian species such as *E. cuniculi, E. hellem, E. intestinalis, V. corneae,* or *N. algerae* (a mosquito microsporidian) expressed the strongest antibody responses, as detected by Western blot immunodetection, against the homologous microsporidian species used in the inoculation (Didier et al., 1991a, 1993, 1994; Hollister et al., 1993, 1995). In humans infected with *E. hellem, E. cuniculi, E. intestinalis,* or *V. corneae,* the strongest responses by ELISA or Western blot also were strongest against the species that caused the infection (Didier et al., 1991a, 1991c, 1993; Schwartz et al. 1993a, 1993b, 1996; Visvesvara et al., 1994, 1995a).

Cross-reactivity has been observed between some species of microsporidia, indicating the existence of shared immunogens (e.g., polar filament proteins). Cross-reacting antibodies were observed between *Encephalitozoon* spp. (Croppo et al., 1994, 1997; Didier et al., 1991a, 1991c, 1993; Enriquez et al., 1997; Hartskeerl et al., 1995), and antisera generated against *Encephalitozoon* spp. weakly cross-reacted with *E. bieneusi* (Aldras et al., 1994; Beckers et al., 1996; Weiss et al., 1992; Zierdt et al., 1993). Whereas cross-reactive antibodies also were demonstrated between microsporidia that infect fish, insects, and mammals (Niederkorn et al., 1980; Ombrouck et al., 1995), species-specific antibodies could be selected out of polyclonal antisera by absorption to generate antiserum for specifically identifying *E. hellem* (Schwartz et al., 1993a, 1993b) and *E. cuniculi* (Mertens et al., 1997) infections in AIDS patients. In addition, specific antibodies for defining specific antigens among species and strains of microsporidia have been detected by Western blot immunodetection (Croppo et al., 1994, 1997; Didier et al., 1991a, 1991c, 1993, 1995; Hartskeerl et al., 1995; Visvesvara et al., 1991, 1994, 1995a).

CELL-MEDIATED IMMUNE RESPONSES

Cell-mediated immune responses are regulated and expressed by T lymphocytes. Whereas T lymphocytes help B lymphocytes expand clonally in number and differentiate for antibody production after exposure to immunogens, cell-mediated immune responses are typically characterized by T-lymphocyte functions that include the release of cytokines (e.g., IFN-γ) to activate and induce cells such as macrophages and cytotoxic T cells. Whereas antibody-mediated responses are generally required for protection against extracellular microorganisms (e.g., bacteria), T cell-mediated responses are typically required for resistance against intracellular microorganisms (Janeway and Travers, 1996).

T-Lymphocyte Functions. Resistance to lethal microsporidiosis was found to be dependent on functional T-cell responses. Experimentally, athymic mice, which lack functional mature T cells, and SCID mice, which lack functional mature T and B lymphocytes, died after inoculation with *E. cuniculi*, *E. hellem*, or *V. corneae* (Didier et al., 1994; Gannon, 1980b; Koudela et al., 1993; Schmidt and Shadduck, 1983; Silveira et al., 1993). Transfer of sensitized syngeneic T-cell enriched splenic cells protected *E. cuniculi*-infected athymic mice from lethal disease, whereas transfer of B lymphocytes or macrophages from sensitized donor mice failed to prolong survival or protect infected athymic mice (Enriquez, 1997; Hermanek et al., 1993; Schmidt and Shadduck, 1983). Furthermore, latent microsporidiosis could be seen in mice following injection with cortisone, a suppressor of T-lymphocyte function (Bismanis, 1970; Innes, 1970). Microsporidiosis in AIDS patients is clinically symptomatic in patients with $CD4^+$ T-cell levels below 100 per μl of blood (Asmuth et al., 1994) and, conversely, remission of microsporidiosis was observed in AIDS patients given combination antiretroviral therapy who showed improved $CD4^+$ T-cell levels (Goguel et al., 1997; Carr et al., 1998).

Cytokine Production for Macrophage Activation. T-lymphocytes contributed to resistance by releasing cytokines that activate macrophages to kill microsporidia. Supernatants collected from splenic or peripheral blood lymphocytes of *E. cuniculi*-infected mice (BALB/c or C57/BL) or domestic dogs that were incubated with *E. cuniculi* in vitro, activated thioglycolate-induced murine peritoneal macrophages or peripheral blood canine monocytes, respectively, to kill 60 to 90% of the *E. cuniculi* spores added to the cultures (Schmidt and Shadduck, 1984; Szabo and Shadduck, 1988). Incubation of thioglycollate-induced BALB/c murine peritoneal macrophages with recombinant murine TNF-α (1,000 U/ml) or murine recombinant IFN-γ (10 or 100 U/ml) inhibited the replication of *E. cuniculi* in vitro, whereas addition of lipopolysaccharide (LPS) (1.0 or 10.0 ng/ml) alone failed to inhibit *E. cuniculi* replication in murine macrophages (Didier and Shadduck, 1994; Didier, 1995). The addition of both LPS (1.0 ng/ml) and IFN-γ

(100 U/ml), however, activated the murine macrophages to kill *E. cuniculi* within 24 h by mechanisms that implicated nitrogen intermediates (Didier, 1995). Addition of the L-arginine inhibitor N^G-monomethyl-L-arginine to LPS- plus IFN-activated murine macrophages at a dose of 50, 100, or 250 μM significantly inhibited nitric oxide synthesis and macrophage-mediated killing of *E. cuniculi*. The addition of exogenous L-arginine to these cultures reversed the inhibition of the N^G-monomethyl-L-arginine, resulting in macrophage-mediated killing of the microsporidia. Inducible nitric oxide synthase-deficient C57BL mice infected with *E. cuniculi,* however, failed to develop lethal disease (Khan, personal communication). In addition, Achbarou et al. (1996) observed that parasite shedding in feces of IFN-γ receptor knockout mice infected with *E. intestinalis* was greater than in wild-type control mice but that *E. intestinalis* infections were not lethal to the knockout mice. This finding suggested that neither IFN-γ nor nitric oxide alone was sufficient for resistance to lethal microsporidiosis and that several immune response mechanisms contributed to resistance in these mice.

Less work has been published on the functions of cytokines in resisting microsporidia in humans. Cytokine mRNA expression in intestinal biopsies was evaluated by Snijders et al. (1995), but no significant differences were seen for TNF, IL-1β, IL-6, IL-8, or IL-10 in six HIV-infected individuals with microsporidiosis and diarrhea compared with seven HIV-infected individuals with unknown causes of diarrhea or HIV-seronegative individuals. The authors suggested that immunotherapy probably would not relieve clinical manifestations associated with microsporidiosis. However, no comparisons were made between HIV-infected individuals with microsporidiosis and HIV-seronegative individuals infected with microsporidia.

Cytotoxicity. $CD8^+$ T lymphocytes release secretory granules containing perforin, granzymes (or fragmentins), and TNF-β to kill cells expressing specific antigen and MHC class I determinants (Janeway and Travers, 1996). Khan and colleagues observed that β2 microglobulin knockout mice, which were unable to express MHC class I determinants, as well as CD8 knockout mice, died within 18 days of intraperitoneal injections with *E. cuniculi* (Khan, personal communication). Cytotoxic splenic T-cell activity was detected 17 or 24 days after euthymic mice were inoculated intraperitoneally with *E. cuniculi* using *E. cuniculi*-infected murine peritoneal macrophage target cells in vitro.

Acquired Cell-Mediated Immune Responses. In adoptive transfer experiments on BALB/c athymic mice, approximately 7 to 10 days were required for T cells to undergo clonal expansion prior to expressing effector responses that controlled microsporidia. Transfer of naive (i.e., nonsensitized) syngeneic T-cell enriched splenic lymphocytes or T-cell enriched splenic lymphocytes from donors inoculated with *E. cuniculi* 4 or 7 days earlier failed to protect *E. cuniculi*-infected athymic mice. Syngeneic T-cell enriched splenic lymphocytes from donors inoculated with *E. cuniculi* 10 or 14 days earlier, however, were able to adoptively transfer resistance to *E. cuniculi*-infected athymic mice, and these cells, when incubated with *E. cuniculi* in vitro, released cytokines that activated macrophages to kill *E. cuniculi* (Schmidt and Shadduck, 1983). Cytotoxic T-lymphocyte activity was not detected in the spleen cells recovered 14 days after the infection of donor mice with *E. cuniculi* (Khan, pesonal communication; Schmidt and Shadduck, 1984) but was observed 17 or 24 days after infection (Khan, personal communication). In addition, migration inhibition of leukocytes, a measure of cell-mediated immune responsiveness, was observed by peripheral blood leukocytes obtained from domestic dogs infected with *E. cuniculi* 13 days earlier and incubated with *E. cuniculi* antigen in vitro (Stewart et al., 1986). This expression of cell-mediated immunity peaked approximately 36 days after infection and lasted through day 97. Collectively, these results suggested that cytokine production, as determined by induction of macrophage-mediated killing of microsporidia or leukocyte migration inhibition, occurred between 10 and 13 days after infection with *E. cuniculi* while cytotoxic T-lymphocyte activities were not detected until 17 days after infection. The implications are that $CD4^+$ T cells can function during the acute phase of microsporidiosis and that $CD8^+$ cytotoxic T cells can function to control or regulate the chronic phase of microsporidiosis.

Anamnestic Cell-Mediated Immune Responses. Several measurements of anamnestic or secondary cell-mediated immune responses, including delayed-type hypersensitivity (DTH) responses, lymphocyte blastogenesis, and cytokine production, have been demonstrated in animals infected with microsporidia. The ability of rabbits to generate a positive intradermal skin test developed approximately 1 week after infection, while the DTH reaction in infected rabbits occurred between 24 and 72 h after intradermal inoculation with *E. cuniculi* antigen, indicative of a secondary response (Pakes et al., 1972, 1984). Responsiveness to intradermal skin tests correlated best with the detection of lesions associated with microsporidiosis at necropsy. Secondary DTH responses were also demonstrated after inoculation of *E. cuniculi* antigen into the footpads of infected mice (Schmidt and Shadduck, 1983). No differences in severity of footpad swelling were observed in adult compared with neonatal mice infected with *E. cuniculi* 2 weeks earlier (Liu et al., 1988).

Another indication of secondary responsiveness in hosts infected with microsporidiosis was the observation that blastogenic responses of sensitized murine lymphocytes to microsporidian antigens were higher than those of splenic lymphocytes from naive mice (Didier and Shadduck, 1988; Liu et al., 1989). Furthermore, the blasting lymphocytes from infected mice and dogs released cytokines that activated macrophages in vitro to kill microsporidia, while naive lymphocytes incubated with *E. cuniculi* failed to release macrophage-activating cytokines (Liu et al., 1989; Schmidt and Shadduck, 1984; Szabo and Shadduck, 1988).

Specificity of Cell-Mediated Immune Responses. The induction of secondary cell-mediated immune responses to microsporidia has been shown to be specific in vivo and in vitro. For example, the DTH response generated from the intradermal skin test in *E. cuniculi*-infected rabbits developed only when *E. cuniculi* was used as the antigen and did not develop against *T. gondii* (Pakes et al., 1972, 1984). In addition, footpad swelling in *E. cuniculi*-infected BALB/c mice occurred in response to *E. cuniculi* antigen but not in response to sheep erythrocytes (Schmidt and Shadduck, 1983). Specific antigens of 11, 23, 120, 133, and 165 kDa from *E. intestinalis* induced the strongest blastogenesis of sensitized murine (BALB/c) splenic lymphocytes (Didier and Bertucci, 1996). Specific antigens that induced protective versus nonprotective cytokines have not been defined, but the induction of cytokine release by blasting lymphocytes was observed to be antigen-specific. Splenic lymphocytes from *E. cuniculi*-infected mice incubated with *E. cuniculi* generated greater levels of macrophage-activating cytokines than those incubated with *Listeria monocytogenes* (Schmidt and Shadduck, 1984). This specificity, however, could be circumvented with mitogens, such as concanavalin A, which induced generalized blasting of murine and canine lymphocytes and secretion of macrophage-activating factors (Schmidt and Shadduck, 1984; Szabo and Shadduck, 1988).

MICROSPORIDIAL SURVIVAL

Microsporidia that infect mammals often persist in the face of innate resistance, early induced response, and adapted immune response mechanisms. The various intracellular environments in which microsporidia replicate appear to provide some level of protection. Furthermore, microsporidia have been observed to affect host cell physiology as well as alter immune responsiveness of the host to other immunogens.

Host Cell-Parasite Interface

Microsporidia vary with regard to the intracellular sites where the organisms replicate. One type of host cell-parasite interface is the host cell-derived membrane-bound parasitophorous vacuole in which *Encephalitozoon* spp. replicate. The parasitophorous vacuole membranes in an *E. cuniculi*-infected macrophage were described by Weidner (1975) as having blebs that appeared to be interiorizing cytoplasmic contents into the parasitophorous vacuole. The host cell cytoplasm mass was seen to decrease in proportion to the increased mass of the parasitophorous vacuole as the microsporidia continued to replicate. Although macrophages contain a well-developed lysosomal network, Weidner (1975) observed that there was an absence of fusion between the lysosomes and the parasitophorous vacuoles containing the microsporidia. In later studies with the microsporidian *Glugea hertwigi*, which also can replicate within parasitophorous vacuoles in macrophages, Weidner and Sibley (1985) reported that simultaneous infection of a macrophage with several *G. hertwigi* resulted in acidification and phagosome-lysosome fusion. Phagosomes containing a single microsporidian, though, failed to acidify, resulting in an absence of lysosomal fusion that allowed parasite replication.

The failure of phagosome acidification may have been due to the characteristics of the parasitophorous vacuole membrane. Leitch and colleagues (1995) observed that the membranes of parasitophorous vacuoles containing *E. hellem* were permeable to calcium and hydrogen ions because an equilibrium existed between these ions in the parasitophorous vacuole and the host cell cytoplasm without a concentration gradient. Induced changes in the concentrations of these ions were rapidly equilibrated. In addition, the parasitophorous vacuole membrane was found to be permeable to the relatively large anion calcein, suggesting that it was also permeable to nutrients. How microsporidia specifically affected the parasitophorous vacuole membranes, however, is unknown. No secretory factors have been described for most microsporidia that infect mammals, although *E. intestinalis* secretes an extracellular matrix substance that appeared to compartmentalize the parasitophorous vacuole (Cali et al., 1993; Orenstein, 1991; Orenstein et al., 1992a, 1992b). The purpose or function of this matrix material is unknown.

E. bieneusi replicates as multinucleated plasmodia, and each plasmodium is bound by a plasma membrane that lies in direct contact with the host cell cytoplasm. Desportes and colleagues described a close association between the parasite's plasma membrane and cisternae of the host cell endoplasmic reticulum, which may protect the parasite from host cell lysosomal recognition (Desportes et al., 1985, 1996). In addition, *E. bieneusi* infected epithelial cells but not macrophages or other phagocytic cells, suggesting that these organisms may have avoided cells rich in lysosomes, preferentially selected nonphagocytic cells, or survived only in nonphagocytic cells such as epithelial cells.

Like *E. bieneusi*, *V. corneae* replicates free in the cytoplasm but undergoes binary division with each of the organisms being surrounded by "both layers of a cisterna of endoplasmic reticulum" (Silveira and Canning, 1995). Silveira and Canning postulated that the endoplasmic reticulum may disguise the parasites to prevent lysosomal fusion with the organisms (Silveira and Canning, 1995).

T. hominis and *Pleistophora* spp. typically infect skeletal muscle cells, although the former also infects macrophages. Each organism developed an electron-dense coat outside or external to the meront's plasma membrane and generated branching projections that extended into the host cell cytoplasm (Field et al., 1996; Hollister et al., 1996b). The membranelike coats were believed to be flattened vesicles derived from host cell cytoplasm. As sporonts developed, their electron-dense coats detached and the sporonts secreted new surface coats to form sporophorous vesicles (or a pansporoblastic membrane) (Hollister et al., 1996b). The projections from *T. hominis* organisms were described by Weidner et al. (1997) as generating a surface plaque matrix with a tubular arrangement of filaments that could be stained with antibodies generated against the intermediate filament, keratin. These authors speculated that these structures could have affected molecular traffic within the infected host cell. In addition, it was possible that the electron-dense coat surrounding the meronts and the sporophorous vesicles protected the parasites from recognition by host cell lysosomes.

Effects of Microsporidia on the Host Cell

Cells infected with microsporidia such as *E. bieneusi, V. corneae,* or *Encephalitozoon* spp. morphologically seemed to display an increase in cell ribosomes and endoplasmic reticulum, suggesting that these microsporidia induced an increase in host cell metabolism (Gourley and Swedo, 1988; Orenstein, 1991; Silveira and Canning, 1995). Host cell mitochondria seemed to increase in number and were found in juxtaposition to the replicating microsporidia (Canning and Hollister, 1992; Kotler et al., 1990, 1993; Kotler and Orenstein, 1994; Shadduck and Pakes, 1971). Host cells infected with *V. corneae* became multinucleated in vitro suggesting that some microsporidia were able to arrest host cell cytokinesis (Silveira and Canning, 1995; Silveira et al., 1993). Microsporidia also affected the synthesis of host cell factors. Armstrong et al. (1973) observed that rabbit kidney cell cultures contaminated with *E. cuniculi* expressed an increased resistance to eastern equine encephalitis virus and vesicular stomatitis virus, which they suggested was due to the induction of interferon (IFN) synthesis by the microsporidia. In addition, *N. salmonis,* a microsporidian related to *E. bieneusi,* caused the secretion of factors by T lymphocytes that induced lymphocyte proliferation in vitro (Wongtavatchai et al., 1995a, 1995b).

Effects of Microsporidia on the Host

In immunologically compromised hosts, microsporidia often caused lethal disease because the parasites were able to continue replicating without restraint. However, in otherwise healthy mammalian hosts, microsporidia survived and also affected the immune system of the host in response to other immunogens. The initial observations of Armstrong and colleagues, who suggested that *E. cuniculi* induced an IFN response in vitro, also may account for the increased tumor resistance in *E. cuniculi*-infected laboratory animals (Armstrong et al., 1973). Rats inoculated with a lethal dose of Yoshida sarcoma cells that survived were later found by Petri (1965, 1966) to be concomitantly infected with *E. cuniculi.* He speculated that the *E. cuniculi* infection induced a "cancerolytic" effect because the ascites typically induced by the tumor cells resolved in rats bearing the microsporidia. In addition, the majority of tumor cells in the ascites were not infected with microsporidia, and so it was unlikely that the parasites killed the tumor cells directly. Arison et al. (1966) reported that mice inoculated with the Ehrlich tumor cell line typically died, whereas *E. cuniculi*-infected mice survived an otherwise lethal inoculation. It was possible that the *E. cuniculi* infections activated NK cells and IFN responses that prevented the tumors from killing the host. C57BL mice infected with *E. cuniculi* developed fewer pulmonary melanomas after inoculation with the lung-homing B16F10 melanoma than mice not previously infected with *E. cuniculi* (Niederkorn, 1985). These *E. cuniculi*-infected mice expressed enhanced NK cell activation. In addition, rabbits with latent microsporidiosis developed more severe inflammatory responses to neural devices coated with biomaterials than rabbits without microsporidiosis did (Ansbacher et al., 1988).

Whereas *E. cuniculi* infection improved the resistance of mice and rats to some tumors, microsporidia also transiently suppressed the immune systems of infected hosts. Rabbits naturally infected with *E. cuniculi* expressed depressed IgG responses to *Brucella abortus* 5 weeks later, while enhanced IgM responses were detected 8 weeks after inoculation with *B. abortus* (Cox, 1977). Similarly, antibody responses to sheep red blood cells (SRBCs) were significantly lower in BALB/c mice infected 1 week earlier with *E. cuniculi* compared with those of noninfected mice, but mice infected 2 weeks earlier expressed normal anti-SRBC antibody responses (Didier and Shadduck, 1988). Genetic differences among strains of mice also appeared to contribute to the modulated immune responses in *E. cuniculi*-infected mice. Anti-SRBC antibody responses were lower in C57BL mice infected with *E. cuniculi* than in infected BALB/c mice, and the suppressed antibody responsiveness to SRBCs lasted approximately 2 weeks longer in the C57BL than in the BALB/c mice infected with *E. cuniculi* (Niederkorn et al., 1981).

Lymphocyte blastogenesis responses also were affected by *E. cuniculi* infection. Splenic lymphocytes from C57BL and BALB/c mice infected 1 week earlier with *E. cuniculi* exhibited significantly lower proliferative responses to the T-cell mitogen conconavalin A or phytohemagglutinin than lymphocytes from uninfected mice did (Didier and Shadduck, 1988; Niederkorn et al., 1981). This transiently suppressed lymphoproliferative response was not observed against the B-lymphocyte mitogen pokeweed mitogen or against LPS (Didier and Shadduck, 1988). Two weeks after *E. cuniculi* infection, the splenic lymphoproliferative responses of BALB/c mice to ConA were either normal or substantially higher than those of controls, whereas responses by C57BL splenic lymphocytes obtained 6 weeks after *E. cuniculi* infection were significantly lower than in noninfected mice (Didier and Shadduck, 1988; Niederkorn et al., 1981). Since the transiently suppressed mitogenic responses were observed against T-cell mitogens and not B-cell mitogens, it was believed that suppressor T cells may have been induced during the early stages of microsporidiosis. However, mixing experiments failed to identify a suppressor T cell mechanism (Didier and Shadduck, 1988).

There appeared to be a genetic influence on the immunosupression observed in mice infected with *E. cuniculi*. Liu et al. (1989) reported that cytokine secretion occurred later in *E. cuniculi*-infected C57BL mice than in infected BALB/c mice. They further reported that partial removal of C57BL adherent splenic macrophages restored the blastogenic response to mitogens. Treatment of splenic lymphocytes from *E. cuniculi*-infected C57BL mice with indomethacin, a prostaglandin inhibitor, enhanced the blastogenic response, suggesting that *E. cuniculi* induced the transient immunosuppression in C57BL mice via prostaglandin synthesis in macrophages. It is unknown how microsporidia caused this immunosuppression, and no toxins or secretory antigens of microsporidia have been identified.

The mechanisms by which microsporidia induced hypersensitivity responses leading to disease in blue foxes and domestic dogs also are not well understood. To address this issue, Liu and Shadduck (1988) infected autoimmune-prone MRL/MPJ-LPR mice with *E. cuniculi* in an attempt to develop a hypersensitivity model for microsporidiosis. MRL/MPJ-LPR mice bear the *lpr* gene which results in an age-related development of hypergammaglobulinemia, antinuclear antibodies, and circulating immune complexes that cause autoimmune disease (Murphy and Roths, 1978). They found, however, that the *lpr* gene did not appear to play a role in progression of microsporidiosis-associated disease in these mice.

Infection of Immunologically Privileged Sites

Some microsporidia preferentially infected or survived only at immunologically privileged host sites and were thereby protected from immune responses that otherwise would have controlled replication of the microsporidia. Examples of immunologically privileged sites include the uterus, testes, brain, cornea, and fetus (Janeway and Travers, 1996). Immunogens introduced at these sites stimulated systemic immune responses, but the effector mechanisms were not able to reach the privileged sites. It is also believed that some level of tolerance may be induced against such immunogens and that autoimmune responses will develop if tolerance wanes or if damage occurs to the immunologically privileged site (Janeway and Travers, 1996).

Microsporidian infections of the lens or cornea have been described in rats (Perrin, 1943), rabbits (Ashton and Wirasinha, 1973; Ashton et al., 1976; Stiles et al., 1997; Wolfer et al., 1993), cats (Buyukmihci et al., 1977), blue foxes (Arneson and Nordstoga, 1977), and humans (Cali et al., 1991a, 1991b, 1994; Davis et al., 1990; Friedberg et al., 1990; Lowder et al., 1990, 1996; Metcalfe et al., 1992; Pinnolis et al., 1981; Rastrelli et al., 1994; Silverstein et al., 1997; Yee et al., 1991) attributable to *Encephalitozoon* spp., *V. corneae, Nosema ocularum,* and unspeciated *Nosema* organisms. Clinically, some infections progressed to perforation of corneal ulcers and spontaneous rupture of the lens (Davis et al., 1990; Pinnolis et al., 1981; Stiles et al., 1997; Wolfer et al., 1993). Interestingly, the cataract described in blue foxes infected with *E. cuniculi* was attributed by the authors to an autoimmune-type response (Arneson and Nordstoga, 1977).

Transplacental transmission of microsporidiosis has been reported in rabbits (Hunt et al., 1972), horses (Van Rensburg et al., 1991), and nonhuman primates (Anver et al., 1972; Zeman and Baskin, 1985) but has occurred most frequently in carnivores (Botha et al., 1979; Canning and Hollister, 1992; Canning et al., 1986; Mohn, 1982a, 1982b; Mohn and Nordstoga, 1975, 1982; Mohn and Odegaard, 1977; Mohn et al., 1974, 1982a, 1982b; Nordstoga, 1972, 1976; Nordstoga and Mohn, 1978; Nordstoga and Westbye, 1976; Shadduck and Orenstein, 1993; Shadduck and Pakes, 1971; Van Dellen et al., 1978; Zhou and Nordstoga, 1993). In addition, carnivores with microsporidiosis were frequently described as expressing hypergammaglobulinemia, immune complex-mediated renal disease, and polyarteritis nodosa. Since the uterus is an immunologically privileged site and the fetus is immunologically immature, it was possible that the microsporidia replicated without eliciting effector immune responses. In surviving hosts, tolerance may have waned as the host matured and the immune system developed, resulting in autoimmune disease. Transplacental transmission in rabbits is rare and difficult to induce experimentally (Owen and Gannon, 1980), but rabbits that became infected transplacentally, however, did not develop autoimmunity (Hunt et al., 1972), suggesting that differences exist between carnivores and other mammals regarding the development of autoimmunity.

CONCLUDING REMARKS

Relatively little is known about the immunology of microsporidiosis because it is a recently recognized infection in immunologically compromised people. The clinical manifestations of microsporidiosis in animals, however, have been highly predictive of the disease in humans, and a competent immune system has been shown to be important in preventing lethal disease. Athymic mice died from microsporidiosis, and it was only after the AIDS pandemic developed that microsporidiosis emerged in the human population. Adoptive transfer of T lymphocytes protected infected athymic mice, and remission of microsporidiosis has been observed in AIDS patients who exhibited improvement in their $CD4^+$ T-lymphocyte levels after combination antiretroviral therapy (Goguel et al., 1997; Carr et al., 1998). A balanced host-parasite relationship developed in many animals with microsporidiosis that displayed few clinical signs of disease yet carried persistent infections. However, there is little epidemiological information as to whether humans carry persistent subclinical microsporidial infections. A better understanding of the immune responses that establish a balanced host-parasite relationship in animals will likely prove beneficial in understanding the immunology of microsporidiosis in humans and in developing immunotherapeutic and chemotherapeutic strategies that can be applied to microsporidiosis.

ACKNOWLEDGMENTS

We acknowledge John Shadduck, Jan Orenstein, Elizabeth Canning, and Pete Didier for their support, helpful discussions, and continued encouragement, and we express our thanks to Howard Gelberg, Imtiaz Khan, Karen Snowden, and Kate Wasson for sharing their prepublished data. We also thank Donna Bertucci and Linda Rogers for their many years of excellent technical assistance, and the National Institutes of Health (grants RR00164 and AI39968) for financial support.

REFERENCES

Accoceberry, I., J. Carriere, M. Thellier, S. Biligui, M. Danis, and A. Datry. 1997. Rat model for the human intestinal microsporidian *Enterocytozoon bieneusi. J. Eukaryot. Microbiol.* **44:**83S.

Achbarou, A., C. Ombrouck, T. Gneragbe, F. Charlotte, L. Renia, I. Desportes-Livage, and D. Mazier. 1996. Experimental model for human intestinal microsporidiosis in interferon gamma receptor knockout mice infected by *Encephalitozoon intestinalis. Parasite Immunol.* **18:**387–392.

Aldras, A. M., J. M. Orenstein, D. P. Kotler. J. A. Shadduck, and E. S. Didier. 1994. Detection of microsporidia by indirect immunofluorescence antibody test using polyclonal and monoclonal antibodies. *J. Clin. Microbiol.* **32:**608–612.

Ansbacher, L., M. F. Nichols, and A. W. Hahn. 1988. The influence of *Encephalitozoon cuniculi* on neural tissue responses to implanted biomaterials in the rabbit. *Lab. Anim. Sci.* **38:**689–695.

Anver, M. R., N. W. King, and R. D. Hunt. 1972. Congenital encephalitozoonosis in a squirrel monkey (*Saimiri sciureus*). *Vet. Pathol.* **9:**475–480.

Arison, R. N., J. A. Cassaro, and M. P. Pruss. 1966. Studies on a murine ascites-producing agent and its effect on tumor development. *Cancer Res.* **26:**1915–1920.

Armstrong, J. A., Y. H. Ke, M. C. Breinig, and L. Ople. 1973. Virus resistance in rabbit kidney cell cultures contaminated by a protozoan resembling *Encephalitozoon cuniculi. Proc. Soc. Exp. Biol. Med.* **142:**1205–1208.

Arnesen, K., and K. Nordstoga. 1977. Ocular encephalitozoonosis (nosematosis) in blue foxes: polyarteritis nodosa and cataract. *Acta. Ophthalmol.* **55:**641–651.

Ashton, N., C. Cook, and F. Clegg. 1976. Encephalitozoonosis (nosematosis) causing bilateral cataract in a rabbit. *Br. J. Ophthalmol.* **60:**618–631.

Ashton, N., and P. A. Wirasinha. 1973. Encephalitozoonosis (nosematosis) of the cornea. *Br. J. Ophthalmol.* **57:**669–674.

Asmuth, D. M., P. C. DeGirolami, M. Federman, C. R. Ezratty, D. K. Pleskow, G. Desai, and C. A. Wanke. 1994. Clinical features of microsporidiosis in patients with AIDS. *Clin. Infect. Dis.* **18:**819–825.

Baker, M. D., C. R. Vossbrinck, E. S. Didier, J. V. Maddox, and J. A. Shadduck. 1995. Small subunit ribosomal DNA phylogeny of various microsporidia with emphasis on AIDS related forms. *J. Eukaryot. Microbiol.* **42:**564–570.

Beckers, P. J., G. J. Derks, T. van Gool, F. J. Rietveld, and R. W. Sauerwein. 1996. *Encephalocytozoon intestinalis:* specific monoclonal antibodies for laboratory diagnosis of microsporidiosis. *J. Clin. Microbiol.* **34:**282–285.

Bergquist, R., L. Morfeldt-Mansson, P. O. Pehrson, B. Petrini, and J. Wasserman. 1984a. Antibody against *Encephalitozoon cuniculi* in Swedish homosexual men. *Scand. J. Infect. Dis.* **16:**389–391.

Bergquist, N. R., G. Stintzing, L. Smedman, T. Waller, and T. Andersson. 1984b. Diagnosis of encephalitozoonosis in man by serological tests. *Br. Med. J.* **288:**902.

Bessinger, G. T., H. H. Stibbs, M. F. Wiser, K. Buchanan, and E. S. Didier. 1997. Murine model for *Encephalitozoon intestinalis* infection. *Amer. J. Trop. Med. Hyg.* **57:**313.

Bismanis, J.E. 1970. Detection of latent murine nosematosis and growth of *Nosema cuniculi* in cell cultures. *Can. J. Microbiol.* **16:**237–242.

Black, S. S., L. A. Steinohrt, D. C. Bertucci, L. B. Rogers, and E. S. Didier. 1997. *Encephalitozoon hellem* in budgerigars (*Melopsittacus undulatus*). *Vet. Pathol.* **34:**189–198.

Blanshard, C., W. S. Hollister, C. S. Peacock, D. G. Tovey, D. S. Ellis, E. U. Canning, and B. G. Gazzard. 1992. Simultaneous infection with two types of intestinal microsporidia in a patient with AIDS. *Gut* **33:**418–420.

Bornay-Llinares, F. J., A. J. da Silva, H. Moura, D. A. Schwartz, G. S. Visvesvara, N. J. Pieniazek, A. Cruz-Lopez, P. Hernandez-Jauregui, J. Guerrero, and F. J. Enriquez. 1998. Immunologic, microscopic, and molecular evidence of *Encephalitozoon intestinalis* (*Septata intestinalis*) infection in mammals other than humans. *J. Infect. Dis.* **178:**820–826.

Botha, W. S., I. C. Dormehl, and D. J. Goosen. 1986a. Evaluation of kidney function in dogs suffering from canine encephalitozoonosis by standard clinical pathological and radiopharmaceutical techniques. *J. S. Afr. Vet. Assoc.* **57:**79–86.

Botha, W. S., C. G. Stewart, and A. F. van Dellen. 1986b. Observations on the pathology of experimental encephalitozoonosis in dogs. *J. S. Afr. Vet. Assoc.* **57:**17–24.

Botha, W. S., A. F. van Dellen, and C. G. Stewart. 1979. Canine encephalitozoonosis in South Africa. *J. S. Afr. Vet. Assoc.* **50:**135–144.

Bryan, R. T. 1995. Microsporidiosis as an AIDS-related opportunistic infection. *Clin. Infect. Dis.* **21**(suppl. 1):S62–S65.

Bryan, R. T., A. Cali, R. L. Owen, and H. C. Spencer. 1991. Microsporidia: opportunistic pathogens in patients with AIDS. *Prog. Clin. Parasitol.* **2:**1–26.

Bryan, R. T., R. Weber, and D. A. Schwartz. 1996. Microsporidiosis in patients who are not infected with human immunodeficiency virus. *Clin. Infect. Dis.* **23:**114–117.

Buyukmihci, N., R. W. Bellhorn, J. Hunziker, and J. Clinton. 1977. *Encephalitozoon (Nosema)* infection of the cornea in a cat. *J. Amer. Vet. Med. Assoc.* **171:**355–357.

Bywater, J. E., and B. S. Kellett. 1978a. *Encephalitozoon cuniculi* antibodies in a specific-pathogen-free rabbit unit. *Infect. Immun.* **21:**360–364.

Bywater, J. E., and B. S. Kellett. 1978b. The eradication of *Encephalitozoon cuniculi* from a specific pathogen-free rabbit colony. *Lab. Anim. Sci.* **28:**402–404.

Bywater, J. E., and B. S. Kellett. 1979. Humoral immune response to natural infection with *Encephalitozoon cuniculi* in rabbits. *Lab. Anim.* **13:**293–297.

Bywater, J. E., B. S. Kellett, and T. Waller. 1980. *Encephalitozoon cuniculi* antibodies in commercially available rabbit antisera and serum reagents. *Lab. Anim.* **14:**86–89.

Cali, A., D. P. Kotler, and J. M. Orenstein. 1993. *Septata intestinalis* n. g., n. sp., an intestinal microsporidian associated with chronic diarrhea and dissemination in AIDS patients. *J. Eukaryot. Microbiol.* **40:**101–112.

Cali, A., D. M. Meisler, C. Y. Lowder, R. Lembach, L. Ayers, P. M. Takvorian, I. Rutherford, D. L. Longworth, J. McMahon, and R. T. Bryan. 1991a. Corneal microsporidioses: characterization and identification. *J. Protozool.* **38:**215S–217S.

Cali, A., D. M. Meisler, I. Rutherford, C. Y. Lowder, J .T. McMahon, D. L. Longworth, and R. T. Bryan. 1991b. Corneal microsporidiosis in a patient with AIDS. *Am. J. Trop. Med. Hyg.* **44:**463–468.

Cali, A., P. M. Takvorian, S. Lewin, M. Rendel, C. Sian, M. Wittner, and L. M. Weiss. 1996. Identification of a new *Nosema*-like microsporidian associated with myositis in an AIDS patient. *J. Eukaryot. Microbiol.* **43:**108S.

Cali, A., P. M. Takvorian, S. Lewin, M. Rendel, C. S. Sian, M. Wittner, H. B. Tanowitz, E. Keohane, and L. M. Weiss. 1998. *Brachiola vesicularum,* n. g., n. sp., a new microsporidium associated with AIDS and myositis. *J. Eukaryot. Microbiol.* **45:**240–251.

Cali, A., L. M. Weiss, P. M. Takvorian, H. Tanowitz, and M. Wittner. 1994. Ultrastructural identification of AIDS associated microsporidiosis. *J. Eukaryot. Microbiol.* **41:**24S.

Canning, E. U., and W. S. Hollister. 1991. *In vitro* and *in vivo* investigations of human microsporidia. *J. Protozool.* **38:**631–635.

Canning, E. U., and W. S. Hollister. 1992. Human infections with microsporidia. *Rev. Med. Microbiol.* **3:**35–42.

Canning, E. U., J. Lom, and I. Dykova. 1986. *The Microsporidia of Vertebrates.* Academic Press, New York.

Carr, A., D. Marriott, A. Field, E. Vasak, and D. A. Cooper. 1998. Treatment of HIV-1-associated microsporidiosis and cryptosporidiosis with combination antiretroviral therapy. *Lancet* **351:**256–261.

Chalupsky, J., J. Vavra, and P. Bedrnik. 1973. Detection of antibodies to *Encephalitozoon cuniculi* in rabbits by the indirect immunofluorescent antibody test. *Folia Parasitol.* (Prague) **20:**281–284.

Chalupsky, J., J. Vavra, and P. Bedrnik. 1979. Encephalitozoonosis in laboratory animals: a serological survey. *Folia Parasitol.* (Prague) **26:**1–8.

Chupp, G. L., J. Alroy, L. S. Adelman, J. C. Breen, P. R. Skolnik. 1993. Myositis due to *Pleistophora* (Microsporidia) in a patient with AIDS. *Clin. Infect. Dis.* **16:**15–21.

Cox, J. C. 1977. Altered immune responsiveness associated with *Encephalitozoon cuniculi* infection in rabbits. *Infect. Immun.* **15:**392–395.

Cox, J. C., and H. A. Gallichio. 1977. An evaluation of indirect immunofluorescence in the serological diagnosis of *Nosema cuniculi* infection. *Res. Vet. Sci.* **22:**50–52.

Cox, J. C., and H. A. Gallichio. 1978. Serological and histological studies on adult rabbits with recent, naturally acquired encephalitozoonosis. *Res. Vet. Sci.* **24:**260–261.

Cox, J. C., H. A. Gallichio, D. Pye, and N. B. Walden. 1977. Application of immunofluorescence to the establishment of an *Encephalitozoon cuniculi*-free rabbit colony. *Lab. Anim. Sci.* **27:**204–209.

Cox, J. C., R. C. Hamilton, and H. D. Attwood. 1979. An investigation of the route and progression of *Encephalitozoon cuniculi* infection in adult rabbits. *J. Protozool.* **26:**260–265.

Croppo, G. P., G. J. Leitch, S. Wallace, and G. S. Visvesvara. 1994. Immunofluorescence and Western blot analysis of microsporidia using anti-*Encephalitozoon hellem* immunoglobulin G monoclonal antibodies. *J. Eukaryot. Microbiol.* **41:**31S.

Croppo, G. P., G. S. Visvesvara, G. J. Leitch, S. Wallace, and M. A. DeGroote. 1997. Western blot and immunofluorescence analysis of a human isolate of *Encephalitozoon cuniculi* established in culture from the urine of a patient with AIDS. *J. Parasitol.* **83:**66–69.

Couzinet, S., P. Deplazes, R. Weber, and S. Zimmerli. 1997. Interactions between human macrophages and microsporidia, p. 18. *In*

Abstr. 2nd Workshop Microsporidiosis Cryptosporidiosis Immunodeficient Patients. České Budějovice, Czech Republic.

Davis, R. M., R. L. Font, M. S. Keisler, and J. A. Shadduck. 1990. Corneal microsporidiosis: a case report including ultrastructural observations. *Ophthalmology* **97:**953–957.

DeGroote, M. A., G. S. Visvesvara, M. L. Wilson, N. J. Pieniazek, S. B. Slemenda, A. J. DaSilva, G. J. Leitch, R. T. Bryan, and R. Reves. 1995. Polymerase chain reaction and culture confirmation of disseminated *Encephalitozoon cuniculi* in a patient with AIDS: successful therapy with albendazole. *J. Infect. Dis.* **171:**1375–1378.

Deplazes, P., A. Mathis, R. Baumgartner, I. Tanner, and R. Weber. 1996a. Immunologic and molecular characteristics of *Encephalitozoon*-like microsporidia isolated from humans and rabbits indicate that *Encephalitozoon cuniculi* is a zoonotic parasite. *Clin. Infect. Dis.* **22:**557–559.

Deplazes, P., A. Mathis, C. Muller, and R. Weber. 1996b. Molecular epidemiology of *Encephalitozoon cuniculi* and first detection of *Enterocytozoon bieneusi* in faecal samples of pigs. *J. Eukaryot. Microbiol.* **43:**93S.

Desportes-Livage, I., S. Chilmonczyk, R. Hedrick, C. Ombrouck, D. Monge, I. Maiga, and M. Gentilini. 1996. Comparative development of two microsporidian species: *Enterocytozoon bieneusi* and *Enterocytozoon salmonis,* reported in AIDS patients, and salmonid fish, respectively. *J. Eukaryot. Microbiol.* **43:**49–60.

Desportes, I., Y. Le Charpentier, A. Galian, F. Bernard, B. Cochand-Priollet, A. Lavergne, P. Ravisse, and R. Modigliani. 1985. Occurrence of a new microsporidan: *Enterocytozoon bieneusi* n. g., n. sp., in the enterocytes of a human patient with AIDS. *J. Protozool.* **32:**250–254.

Didier, E.S. 1995. Reactive nitrogen intermediates implicated in the inhibition of *Encephalitozoon cuniculi* (phylum Microspora) replication in murine peritoneal macrophages. *Parasite Immunol.* **17:**406–412.

Didier, E. S. 1998. Unpublished observations.

Didier, E. S., A. Aldras, and J. A. Shadduck. 1992. Serological studies in simian microsporidiosis. *Am. J. Trop. Med. Hyg.* **47:**168–169.

Didier, E. S., and D. C. Bertucci. 1996. Identification of *Encephalitozoon intestinalis* proteins that induce proliferation of sensitized murine spleen cells. *J. Eukaryot. Microbiol.* **43:**92S.

Didier, E. S., P .J. Didier, D. N. Friedberg, S. M. Stenson, J. M. Orenstein, R. W. Yee, F. O. Tio, R .M. Davis, C. Vossbrinck, N. Millichamp, and J. A. Shadduck. 1991a. Isolation and characterization of a new human microsporidian, *Encephalitozoon hellem* (n. sp.), from three AIDS patients with keratoconjunctivitis. *J. Infect. Dis.* **163:**617–621.

Didier, P. J., E. S. Didier, J. M. Orenstein, and J. A. Shadduck. 1991b. Fine structure of a new human microsporidian, *Encephalitozoon hellem,* in culture. *J. Protozool.* **38:**502–507.

Didier, E. S., D. P. Kotler, D. T. Dieterich, J. M. Orenstein, A. M. Aldras, R. Davis, D. N. Friedberg, W. K. Gourley, R. Lembach, C. Y. Lowder, D. M. Meisler, I. Rutherford, R. W. Yee, and J. A. Shadduck. 1993. Serological studies in human microsporidia infections. *AIDS* **7:**S8–S11.

Didier, E. S., L. B. Rogers, J. M. Orenstein, M. D. Baker, C. R. Vossbrinck, T. Van Gool, R. Hartskeerl, R. Soave, and L. M. Beaudet. 1996a. Characterization of *Encephalitozoon (Septata) intestinalis* isolates cultured from nasal mucosa and bronchoalveolar lavage fluids of two AIDS patients. *J. Euk. Microbiol.* **43:**34–43.

Didier, E. S., and J. A. Shadduck. 1988. Modulated immune responsiveness associated with experimental *Encephalitozoon cuniculi* infection in BALB/c mice. *Lab. Anim. Sci.* **38:**680–684.

Didier, E. S., and J. A. Shadduck. 1994. IFN-gamma and LPS induce murine macrophages to kill *Encephalitozoon cuniculi in vitro*. *J. Eukaryot. Microbiol.* **41:**34S.

Didier, E. S., J. A. Shadduck, P. J. Didier, N. Millichamp, and C. R. Vossbrinck. 1991c. Studies on ocular microsporidia. *J. Protozool.* **38:**635–638.

Didier, E. S., K. A. Snowden, J. A. Shadduck. 1998. The biology of microsporidian species infecting mammals. *Adv. Parasitol.* **40:**279–316.

Didier, E. S., P. W. Varner, P. J. Didier, A. M. Aldras, N. J. Millichamp, M. Murphey-Corb, R. Bohm, and J. A. Shadduck. 1994. Experimental microsporidiosis in immunocompetent and immunodeficient mice and monkeys. *Folia Parasitol.* (Prague) **41:**1–11.

Didier, E. S., G. S. Visvesvera, M. D. Baker, L. B. Rogers, D. C. Bertucci, M.A. DeGroote, and C. R. Vossbrinck. 1996b. A microsporidian isolated from an AIDS patient corresponds to *Encephalitozoon cuniculi* III, originally isolated from domestic dogs. *J. Clin. Microbiol.* **34:**2835–2837.

Didier, E. S., C. R. Vossbrinck, M. D. Baker, L. B. Roger, D. C. Bertucci, and J. A. Shadduck. 1995. Identification and characterization of three *Encephalitozoon cuniculi* strains. *Parasitology* **111:** 411–421.

Ditrich, O., M. F. Cross, J. Jones, J. Hensel, and F. J. Enriquez. 1997. Strategies of microsporidial evasion of macrophage killing, p. 19. *In* Abstr. 2nd Workshop Microsporidiosis Cryptosporidiosis

Immunodeficient Patients. České Budějovice, Czech Republic.

Doultree, J. C., A. L. Maerz, N. J. Ryan, R. W. Baird, E. Wright, S. M. Crowe, and J. A. Marshall. 1995. *In vitro* growth of the microsporidian *Septata intestinalis* from an AIDS patient with disseminated illness. *J. Clin. Microbiol.* **33:**463–470.

Enriquez, F. J. 1997. Microsporidia: immunity and immunodiagnosis, p. 16. *In* Abstr. 2nd Workshop Microsporidiosis and Cryptosporidiosis Immunodeficient Patients. Ceské Budĕjovice, Czech Republic.

Enriquez, F. J., O. Ditrich, J. D. Palting, and K. Smith. 1997. Simple diagnosis of *Encephalitozoon* sp. microsporidial infections by using a panspecific antiexospore monoclonal antibody. *J. Clin. Microbiol.* **35:**724–729.

Fedorko, D. P., and Y. M. Hijazi. 1996. Application of molecular techniques to the diagnosis of microsporidial infection. *Emerg. Infect. Dis.* **2:**183–91.

Field, A. S., D. J. Marriott, S. T. Milliken, B. J. Brew, E. U. Canning, J. G. Kench, P. Darveniza, and J. L. Harkness. 1996. Myositis associated with a newly described microsporidian, *Trachipleistophora hominis,* in a patient with AIDS. *J. Clin. Microbiol.* **34:**2803–2811.

Friedberg, D. N., S. M. Stenson, J. M. Orenstein, P. M. Tierno, and N. C. Charles. 1990. Microsporidial keratoconjunctivitis in acquired immunodeficiency syndrome. *Arch Ophthalmol.* **108:**504–508.

Gannon, J. 1980a. A survey of *Encephalitozoon cuniculi* in laboratory animal colonies in the United Kingdom. *Lab. Anim.* **14:**91–94.

Gannon, J. 1980b. The course of infection of *Encephalitozoon cuniculi* in immunodeficient and immunocompetent mice. *Lab. Anim.* **14:**189–192.

Goguel, J., C. Katlama, C. Sarfati, C. Maslo, C. Leport, and J. M. Molina. 1997. Remission of AIDS-associated intestinal microsporidiosis with highly active antiretroviral therapy. *AIDS* **11:** 1658–1659.

Gourley, W. K., and J. L. Swedo. 1988. Intestinal infection by microsporidia *Enterocytozoon bieneusi* of patients with AIDS: an ultrastructural study of the use of human mitochondria by a protozoa. *Lab. Invest.* **58:**35A.

Hartskeerl, R. A., T. Van Gool, A. R. Schuitema, E. S. Didier, and W. J. Terpstra. 1995. Genetic and immunological characterization of the microsporidian *Septata intestinalis* Cali, Kotler and Orenstein, 1993: reclassification to *Encephalitozoon intestinalis. Parasitology.* **110:**277–285.

Hautvast, J. L., J. J. Tolboom, T. J. Derks, P. Beckers, and R. W. Sauerwein. 1997. Asymptomatic intestinal microsporidiosis in a human immunodeficiency virus-seronegative, immunocompetent Zambian child. *Pediatr. Infect. Dis. J.* **16:**415–416.

Hermanek, J., B. Koudela, Z. Kucerova, O. Ditrich, and J. Travnicek. 1993. Prophylactic and therapeutic immune reconstitution of SCID mice infected with *Encephalitozoon cuniculi. Folia Parasitol* (Prague) **40:**287–291.

Hollister, W. S., and E. U. Canning. 1987. An enzyme-linked immunosorbent assay (ELISA) for detection of antibodies to *Encephalitozoon cuniculi* and its use in determination of infections in man. *Parasitology* **94:**209–219.

Hollister, W. S., E. U. Canning, and C. L. Anderson. 1996a. Identification of microsporidia causing human disease. *J Eukaryot. Microbiol.* **43:** 104S–105S.

Hollister, W. S., E. U. Canning, N. I. Colbourn, and E. J. Aarons. 1995. *Encephalitozoon cuniculi* isolated from the urine of an AIDS patient, which differs from canine and murine isolates. *J. Eukaryot. Microbiol.* **42:**367–372.

Hollister, W. S., E. U. Canning, N. I. Colbourn, A. Curry A., and C. J. Lacey. 1993. Characterization of *Encephalitozoon hellem* (Microspora) isolated from the nasal mucosa of a patient with AIDS. *Parasitology* **107:**351–358.

Hollister, W. S., E. U. Canning, and M. Viney. 1989. Prevalence of antibodies to *Encephalitozoon cuniculi* in stray dogs as determined by an ELISA. *Vet. Rec.* **124:**332–336.

Hollister, W. S., E. U. Canning, E. Weidner, A. S. Field, J. Kench, and D. J. Marriott. 1996b. Development and ultrastructure of *Trachipleistophora hominis* n. g., n. sp. after *in vitro* isolation from an AIDS patient and inoculation into athymic mice. *Parasitology* **112:**143–154.

Hollister, W. S., E. U. Canning, and A. Willcox. 1991. Evidence for widespread occurrence of antibodies to *Encephalitozoon cuniculi* (Microspora) in man provided by ELISA and other serological tests. *Parasitology* **102:**33–43.

Hunt, R. D., N. W. King, and H. L. Foster. 1972. Encephalitozoonosis: evidence for vertical transmission. *J. Infect. Dis.* **126:**212–214.

Innes, J. R. M. 1970. Parasitic infections of the nervous system of animals. *Ann. N. Y. Acad. Sci.* **174:**1042–1047.

Janeway, C. A., and P. Travers. 1996. *Immunobiology,* 2nd ed. Garland Publishing, New York, N.Y.

Katiyar, S. K., G. Visvesvara, and T. D. Edlind. 1995. Comparisons of ribosomal RNA sequences from amitochondrial protozoa: implications for processing, mRNA binding, and paromomycin susceptibilitiy. *Gene* **152:**27–33.

Kelkar, R., P. S. Sastry, S. S. Kulkarni, T. K. Saikia, P. M. Parikh, and S. H. Advani. 1997.

Pulmonary microsporidial infection in a patient with CML undergoing allogeneic marrow transplant. *Bone Marrow Transplant.* **19:**179–182.

Khan, I. 1998. Personal communication.

Kotler, D. P., A. Francisco, F. Clayton, J. V. Scholes, and J. M. Orenstein. 1990. Small intestinal injury and parasitic diseases in AIDS. *Ann. Intern. Med.* **113:**444–449.

Kotler, D. P., and J. M. Orenstein. 1994. Prevalence of intestinal microsporidiosis in HIV-infected individuals referred for gastroenterological evaluation. *Am. J. Gastroenterol.* **89:**1998–2002.

Kotler, D. P., S. Reka, K. Chow, and J. M. Orenstein. 1993. Effects of enteric parasitoses and HIV infection upon small intestinal structure and function in patients with AIDS. *J Clin Gastroenterol.* **16:**10–15.

Koudela, B., J. Vitovec, Z. Kucerova, O. Ditrich, and J. Travnicek. 1993. The severe combined immunodeficient mouse as a model for *Encephalitozoon cuniculi* microsporidiosis. *Folia Parasitol.* (Prague) **40:**279–286.

Kunstyr, I., L. Lev, and S. Naumann. 1986. Humoral antibody response of rabbits to experimental infection with *Encephalitozoon cuniculi. Vet. Parasitol.* **21:**223–232.

Ledford, D. K., M. D. Overman, A. Gonzalvo, A. Cali, S. W. Mester, and R. F. Lockey. 1985. Microsporidiosis myositis in a patient with the acquired immunodeficiency syndrome. *Ann. Intern. Med.* **102:**628–630.

Leiro, J., M. Ortega, J. Estevez, F. M. Ubeira, and M. L. Sanmartin. 1996. The role of opsonization by antibody and complement in *in vitro* phagocytosis of microsporidian parasites by turbot spleen cells. *Vet. Immunol. Immunopathol.* **51:**201–210.

Leitch, G. J., M. Scanlon, G. S. Visvesvara, and S. Wallace. 1995. Calcium and hydrogen ion concentrations in the parasitophorous vacuoles of epithelial cells infected with the microsporidian *Encephalitozoon hellem. J. Eukaryot. Microbiol.* **42:** 445–451.

Levaditi, C., S. Nicolau, and R. Schoen. 1923. L'étiogie de l'encéphalite Çpizootique du lapin, dans ses rapports avec l'étude expérimentale de l'encéphalite léthargique *Encephalitozoon cuniculi* (nov. spec.). *Ann. Inst. Pasteur* (Paris) **38:**651–711.

Liu, J. J., E. H. Greeley, and J. A. Shadduck. 1988. Murine encephalitozoonosis: the effect of age and mode of transmission on occurrence of infection. *Lab. Anim. Sci.* **38:**675–679.

Liu, J. J., E. H. Greeley, and J. A. Shadduck. 1989. Mechanisms of resistance/susceptibility to murine encephalitozoonosis. *Parasite Immunol.* **11:**241–256.

Liu, J. J., and J. A. Shadduck. 1988. *Encephalitozoon cuniculi* infection in MRL/MPJ-LPR (lymphoproliferation) mice. *Lab. Anim. Sci.* **38:**685–688.

Lowder, C. Y., J. T. McMahon, D. M. Meisler, E. M. Dodds, L. H. Calabrese, E. S. Didier, and A. Cali. 1996. Microsporidial keratoconjunctivitis caused by *Septata intestinalis* in a patient with acquired immunodeficiency syndrome. *Am. J. Ophthalmol.* **121:**715–717.

Lowder, C. Y., D. M. Meisler, J. T. McMahon, D. L. Longworth, and I. Rutherford. 1990. Microsporidia infection of the cornea in a man seropositive for human immunodeficiency virus. *Am. J. Ophthalmol.* **109:**242–244.

Lyngset, A. 1980. A survey of serum antibodies to *Encephalitozoon cuniculi* in breeding rabbits and their young. *Lab. Anim. Sci.* **30:**558–561.

Mansfield, K. G., A. Carville, D. Shvetz, J. MacKey, S. Tzipori, and A. A. Lackner. 1997. Identification of an *Enterocytozoon bieneusi*-like microsporidian parasite in simian-immunodeficiency-virus-inoculated macaques with hepatobiliary disease. *Amer. J. Pathol.* **150:**1395–1405.

Margileth, A. M., A. J. Strano, R. Chandra, R. Neafie, M. Blum, and R. M. McCully. 1973. Disseminated nosematosis in an immunologically compromised infant. *Arch. Pathol.* **95:**145–150.

Markova, P., and B. Koudela. 1997. Interaction bewteen mice peritoneal macrophages and *E. cuniculi* spores *in vivo* and *in vitro,* p. 20. *In* Abstr. 2nd Workshop Microsporidiosis Cryptosporidiosis Immunodeficient Patients. České Budějovice, Czech Republic.

Mathis, A., J. Akerstedt, J. Tharaldsen, O. Odegaard and P. Deplazes. 1996. Isolates of *Encephalitozoon cuniculi* from farmed blue foxes (*Alopex lagopus*) from Norway differ from isolates from Swiss domestic rabbits (*Oryctolagus cuniculus*). *Parasitol. Res.* **82:**727–730.

Mathis, A., M. Michel, H. Kuster, C. Muller, R. Weber, and P. Deplazes. 1997. Two *Encephalitozoon cuniculi* strains of human origin are infectious to rabbits. *Parasitology* **114:**29–35.

Matsubayashi, H., T. Koike, T. Mikata, and S. Hagiwara. 1959. A case of *Encephalitozoon*-like body infection in man. *Arch. Pathol.* **67:**181–187.

McCully, R. M., A. F. Van Dellen, P. A. Basson, and J. Lawrence. 1978. Observations on the pathology of canine microsporidiosis. *Onderstepoort. J. Vet. Res.* **45:**75–91.

McDougall, R. J., M. W. Tandy, R. E. Boreham, D. J. Stenzel, and P. J. O'Donoghue. 1993. Incidental finding of a microsporidian parasite from an AIDS patient. *J. Clin. Microbiol.* **31:**436–439.

McInnes, E. F., and C. G. Stewart. 1991. The pathology of subclinical infection of *Encephalitozoon cuniculi* in canine dams producing pups with overt encephalitozoonosis. *J. S. Afr. Vet. Assoc.* **62:**51–54.

Mertens, R. B., E. S. Didier, M. C. Fishbein, D. C. Bertucci, L. B. Rogers, and J. M. Orenstein. 1997. *Encephalitozoon cuniculi* microsporidiosis: infection of the brain, heart, kidneys, trachea, adrenal glands, and urinary bladder in a patient with AIDS. *Mod. Pathol.* **10:**68–77.

Metcalfe, T. W., R. M. Doran, P. L. Rowlands, A. Curry, and C. J. Lacey. 1992. Microsporidial keratoconjunctivitis in a patient with AIDS. *Br. J. Ophthalmol.* **76:**177–178.

Modigliani, R., C. Bories, Y. Le Charpentier, M. Salmeron, B. Messing, A. Galian, J. C. Rambaud, A. Lavergne, B. Cochand-Priollet, and I. Desportes. 1985. Diarrhoea and malabsorption in acquired immune deficiency syndrome: a study of four cases with special emphasis on opportunistic protozoan infestations. *Gut* **26:**179–187.

Mohn, S. F. 1982a. Experimental encephalitozoonosis in the blue fox: clinical and serological examinations of affected pups. *Acta Vet. Scand.* **23:**503–514.

Mohn, S. F. 1982b. Encephalitozoonosis in the blue fox: comparison between the india-ink immunoreaction and the indirect fluorescent antibody test in detecting *Encephalitozoon cuniculi* antibodies. *Acta Vet. Scand.* **23:**99–106.

Mohn, S. F., and K. Nordstoga. 1975. Electrophoretic patterns of serum proteins in blue foxes with special reference to changes associated with nosematosis. *Acta Vet. Scand.* **16:**297–306.

Mohn, S. F., and K. Nordstoga. 1982. Experimental encephalitozoonosis in the blue fox: neonatal exposure to the parasite. *Acta Vet. Scand.* **23:**344–360.

Mohn, S. F., K. Nordstoga, and I. W. Dishington. 1982a. Experimental encephalitozoonosis in the blue fox: clinical, serological and pathological examinations of vixens after oral and intrauterine inoculation. *Acta Vet. Scand.* **23:**490–502.

Mohn, S. F., K. Nordstoga, J. Krogsrud, and A. Helgebostad. 1974. Transplacental transmission of *Nosema cuniculi* in the blue fox (*Alopex lagopus*). *Acta Pathol. Microbiol. Scand. Sect. B* Microbiol Immunol. **82:**299–300.

Mohn, S. F., K. Nordstoga, and O. M. Moller. 1982b. Experimental encephalitozoonosis in the blue fox: transplacental transmission of the parasite. *Acta Vet. Scand.* **23:**211–220.

Mohn, S. F., and O. A. Odegaard. 1977. The indirect fluorescent antibody test (IFAT) for the detection of *Nosema cuniculi* antibodies in the blue fox (*Alopex lagopus*). *Acta Vet. Scand.* **18:**290–292.

Morris, J. A., J. M. McCown, and R. E. Blount. 1958. Ascites and hepatosplenomegaly in mice associated with protozoan-like cytoplasmic structures. *J. Infect. Dis.* **98:**306–311.

Murphy, E. D., and J. B. Roths. 1978. Autoimmunity and lymphoproliferation: induction by mutant gene lpr and acceleration by a male-associated factor in strain BXSB mice, p. 207–220. *In* N. R. Rose, P. E. Bigazzi, and N. L. Warner (ed.), *Genetic Control of Autoimmune Disease.* Elsevier/North-Holland, Amsterdam, The Netherlands.

Nelson, J. B. 1967. Experimental transmission of a murine microsporidian in Swiss mice. *J. Bacteriol.* **94:**1340–1345.

Niederkorn, J. Y. 1985. Enhanced pulmonary natural killer cell activity during murine encephalitozoonosis. *J. Parasitol.* **71:**70–74.

Niederkorn, J. Y., J. K. Brieland, and E. Mayhew. 1983. Enhanced natural killer cell activity in experimental murine encephalitozoonosis. *Infect. Immun.* **41:**302–307.

Niederkorn, J. Y., and J. A. Shadduck. 1980. Role of antibody and complement in the control of *Encephalitozoon cuniculi* infections by rabbit macrophages. *Infect. Immun.* **27:**995–1002.

Niederkorn, J. Y., J. A. Shadduck, and E. C. Schmidt. 1981. Susceptibility of selected inbred strains of mice to *Encephalitozoon cuniculi*. *J. Infect. Dis.* **144:**249–253.

Niederkorn, J. Y., J. A. Shadduck, and E. Weidner. 1980. Antigenic cross-reactivity among different microsporidian spores as determined by immunofluorescence. *J. Parasitol.* **66:**675–677.

Nordstoga, K. 1972. Nosematosis in blue foxes. *Nord. Vet. Med.* **24:**21–24.

Nordstoga, K. 1976. Polyarteritis nodosa: general aspects and occurrence in domestic animals, particularly in association with nosematosis in blue foxes. *Nord. Vet. Med.* **28:**51–58.

Nordstoga, K., and S. F. Mohn. 1978. Nosematosis (encephalitozoonosis) in a litter of blue foxes after intrauterine injection of *Nosema* spores. *Acta Vet. Scand.* **19:**150–152.

Nordstoga, K., and K. Westbye. 1976. Polyarteritis nodosa associated with nosematosis in blue foxes. *Acta Pathol. Microbiol. Scand. Sect. A* **84:** 291–296.

Ombrouck, C., B. Romestand, J. M. da Costa, I. Desportes-Livage, A. Datry, F. Coste, G. Bouix, and M. Gentilini. 1995. Use of cross-reactive antigens of the microsporidian *Glugea atherinae* for the possible detection of *Enterocytozoon bieneusi* by Western blot. *Am. J. Trop. Med. Hyg.* **52:**89–93.

Orenstein, J. M. 1991. Microsporidiosis in the acquired immunodeficiency syndrome. *J. Parasitol.* **77:**843–864.

Orenstein, J. M., J. Chiang, W. Steinberg, P. D. Smith, H. Rotterdam, and D. P. Kotler. 1990. Intestinal microsporidiosis as a cause of diarrhea in human immunodeficiency virus-infected patients: a report of 20 cases. *Hum. Pathol.* **21:**475–481.

Orenstein, J. M., D. T. Dieterich, and D. P. Kotler. 1992a. Systemic dissemination by a newly recognized intestinal microsporidia species in AIDS. *AIDS.* **6:**1143–1150.

Orenstein, J. M., H. P. Gaetz, A. T. Yachnis, S. S. Frankel, R. B. Mertens, and E. S. Didier. 1997. Disseminated microsporidiosis in AIDS: are any organs spared? *AIDS.* **11:**385–386.

Orenstein, J. M., M. Tenner, A. Cali, and D. P. Kotler. 1992b. A microsporidian previously undescribed in humans, infecting enterocytes and macrophages, and associated with diarrhea in an acquired immunodeficiency syndrome patient. *Hum. Pathol.* **23:**722–728.

Owen, D. G., and J. Gannon. 1980. Investigation into the transplacental transmission of *Encephalitozoon cuniculi* in rabbits. *Lab. Anim.* **14:**35–38.

Pakes, S. P., J. A. Shadduck, D. B. Feldman, and J. A. Moore. 1984. Comparison of tests for the diagnosis of spontaneous encephalitozoonosis in rabbits. *Lab. Anim. Sci.* **34:**356–359.

Pakes, S. P., J. A. Shadduck, and R. G. Olsen. 1972. A diagnostic skin test for encephalitozoonosis (nosematosis) in rabbits. *Lab. Anim. Sci.* **22:**870–877.

Pang, V. F., and J. A. Shadduck. 1985. Susceptibility of cats, sheep, and swine to a rabbit isolate of *Encephalitozoon cuniculi. Am. J. Vet. Res.* **46:**1071–1077.

Perrin, T. L. 1943. Spontaneous and experimental *Encephalitozoon* infection in laboratory animals. *Arch. Pathol.* **36:**559–563.

Petri, M. 1965. A cytolytic parasite in the cells of transplantable tumours. *Nature* **205:**302–303.

Petri, M. 1966. The occurrence of *Nosema cuniculi* (*Encephalitozoon cuniculi*) in the cells of transplantable, malignant ascites tumours and its effect upon tumour and host. *Acta Pathol. Microbiol. Scand.* **66:**13–30.

Petri, M. 1969. Studies on *Nosema cuniculi* found in transplantable ascites tumours with a survey of microsporidiosis in mammals. *Acta Pathol. Microbiol. Scand. Suppl.* **204:**1–91.

Pinnolis, M., P. R. Egbert, R. L. Font, and F. C. Winter. 1981. Nosematosis of the cornea: case report, including electron microscopic studies. *Arch. Ophthalmol.* **99:**1044–1047.

Pulparampil, N., D. Graham, D. Phalen, and K. Snowden. 1998. *Encephalitozoon hellem* in two eclectus parrots (*Eclectus roratus*): identification from archival tissues. *J. Eukaryot. Microbiol.* **45:**651–655.

Rabodonirina, M., M. Bertocchi, I. Desportes-Livage, L. Cotte, H. Levrey, M. A. Piens, G. Monneret, M. Celard, J. F. Mornex, and M. Mojon. 1996. *Enterocytozoon bieneusi* as a cause of chronic diarrhea in a heart-lung transplant recipient who was seronegative for human immunodeficiency virus. *Clin. Infect. Dis.* **23:**114–117.

Rastrelli, P. D., E. S. Didier, and R. W. Yee. 1994. Microsporidial keratitis. *Ophthalmol. Clin. N. Am.* **7:**617–633.

Rinder, H., S. Katzwinkel-Wladarsch, and T. Loescher. 1997. Evidence for the existence of genetically distinct strains of *Enterocytozoon bieneusi. Parasitol. Res.* **83:**670–672.

Sandfort, J., A. Hannemann, H. Gelderblom, K. Stark, R. L. Owen, and B. Ruf. 1994. *Enterocytozoon bieneusi* infection in an immunocompetent patient who had acute diarrhea and who was not infected with the human immunodeficiency virus. *Clin. Infect. Dis.* **19:**514–516.

Sax, P. E., J. D. Rich, W. S. Pieciak, and Y. M. Trnka. 1995. Intestinal microsporidiosis occurring in a liver transplant recipient. *Transplantation.* **60:**617–618.

Scaglia, M., L. Sacchi, S. Gatti, A. M. Bernuzzi, P. dePolver, I. Piacentini, E. Concia, G. P. Croppo, A. J. da Silva, and N. J. Pieniazek. 1994. Isolation and identification of *Encephalitozoon hellem* from an Italian AIDS patient with disseminated microsporidiosis. *APMIS* **102:**817–827.

Schmidt, E. C., and J. A. Shadduck. 1983. Murine encephalitozoonosis model for studying the host-parasite relationship of a chronic infection. *Infect. Immun.* **40:**936–942.

Schmidt, E. C., and J. A. Shadduck. 1984. Mechanisms of resistance to the intracellular protozoan *Encephalitozoon cuniculi* in mice. *J. Immunol.* **133:** 2712–2719.

Schottelius, J., Y. Lo, and C. Schmetz. 1995. *Septata intestinalis* and *Encephalitozoon cuniculi:* cross-reactivity between two microsporidian species. *Folia Parasitol.* (Prague) **42:**169–172.

Schwartz, D. A., and R. T. Bryan. 1997. Microsporidia, p. 61–94. *In* C. R. Horsburgh, Jr., and A. M. Nelson (ed.), *Pathology of Emerging Infections.* American Society for Microbiology, Washington, D.C.

Schwartz, D. A., I. Sobottka, G. J. Leitch, A. Cali, and G. S. Visvesvara. 1996. Pathology of microsporidiosis: emerging parasitic infections in patients with acquired immunodeficiency syndrome. *Arch. Pathol. Lab. Med.* **120:**173–188.

Schwartz, D. A., G. S. Visvesvara, M. C. Diesenhouse, R. Weber, R. L. Font, L. A. Wilson, G. Corrent, D. F. Serdarevic, D. F. Rosberger, P. C. Keenen, E. Grossniklaus, K. Hewan-Lowe, and R. T. Bryan. 1993a. Pathologic features and immunofluorescent antibody demonstration of ocular microsporidiosis (*Encephalitozoon hellem*) in seven patients with acquired immunodeficiency syndrome. *Am. J. Ophthalmol.* **115:**285–292.

Schwartz, D. A., G. S. Visvesvara, G. J. Leitch, L. Tashjian, M. Pollack, J. Holden, and R. T. Bryan. 1993b. Pathology of symptomatic microsporidial (*Encephalitozoon hellem*) bronchiolitis in the acquired immunodeficiency syndrome: a new respiratory pathogen diagnosed from lung biopsy, bronchoalveolar lavage, sputum, and tissue culture. *Hum. Pathol.* **24:**935–943.

Shadduck, J. A. 1969. *Nosema cuniculi: in vitro* isolation. *Science* **166:**516–517.

Shadduck, J. A. 1989. Human microsporidiosis and AIDS. *Rev. Infect. Dis.* **11:**203–207.

Shadduck, J. A., and G. B. Baskin. 1989. Serologic evidence of *Encephalitozoon cuniculi* infection in a colony of squirrel monkeys (*Saimiri sciureus*). *Lab. Anim. Sci.* **39:**328–330.

Shadduck, J. A., R. Bendele, G. T. Robinson. 1978. Isolation of the causative organism of canine encephalitozoonosis. *Vet. Pathol.* **15:**449–460.

Shadduck, J. A., and M. J. Geroulo. 1979. A simple method for the detection of antibodies to *Encephalitozoon cuniculi* in rabbits. *Lab. Anim. Sci.* **29:**330–334.

Shadduck, J. A., and E. Greeley. 1989. Microsporidia and human infections. *Clin. Microbiol. Rev.* **2:**158–165.

Shadduck, J. A., G. Kelsoe, and R. J. Helmke. 1979a. A microsporidan contaminant of a nonhuman primate cell culture: ultrastructural comparison with *Nosema connori*. *J. Parasitol.* **65:**185–188.

Shadduck, J. A., R. A. Meccoli, R. Davis, and R. L. Font. 1990. Isolation of a microsporidian from a human patient. *J. Infect. Dis.* **162:**773–776.

Shadduck, J. A., and J. M. Orenstein. 1993. Comparative pathology of microsporidiosis. *Arch. Pathol. Lab. Med.* **117:**1215–1219.

Shadduck, J. A., and S. P. Pakes. 1971. Encephalitozoonosis (nosematosis) and toxoplasmosis. *Am. J. Pathol.* **64:**657–671.

Shadduck, J. A., and M. B. Polley. 1978. Some factors influencing the *in vitro* infectivity and replication of *Encephalitozoon cuniculi*. *J. Protozool.* **25:** 491–496.

Shadduck, J. A., R. Storts, and L. G. Adams. 1996. Selected examples of emerging and reemerging infectious diseases in animals. *ASM News* **62:**586–588.

Shadduck, J. A., W. T. Watson, S. P. Pakes, and A. Cali. 1979b. Animal infectivity of *Encephalitozoon cuniculi*. *J. Parasitol.* **65:**123–129.

Sharpstone, D., A. Rowbottom, N. Francis, G. Tovey, D. Ellis, M. Barrett, and B. Gazzard. 1997. Thalidomide: a novel therapy for microsporidiosis. *Gastroenterology.* **112:**1823–1829.

Silveira, H., and E. U. Canning. 1995. *Vittaforma corneae* n. comb. for the human microsporidium *Nosema corneum* Shadduck, Meccoli, Davis & Font, 1990, based on its ultrastructure in the liver of experimentally infected athymic mice. *J. Eukaryot. Microbiol.* **42:**158–165.

Silveira, H., E. U. Canning, and J. A. Shadduck. 1993. Experimental infection of athymic mice with the human microsporidian *Nosema corneum*. *Parasitology* **107:**489–496.

Silverstein, B. E., E. T. Cunningham, T. P. Margolis, V. Cevallos, and I. G. Wong. 1997. Microsporidial keratoconjunctivitis in a patient without human immunodeficiency virus infection. *Am. J. Ophthalmol.* **124:**395–396.

Singh, M., G. J. Kane, L. Mackinlay, I. Quaki, E. H. Yap, B. C. Ho, L. C. Ho, and K. C. Lim. 1982. Detection of antibodies to *Nosema cuniculi* (Protozoa: Microscoporidia) in human and animal sera by the indirect fluorescent antibody technique. *Southeast Asian J. Trop. Med. Public Health* **13:**110–113.

Snijders, F., S. J. van Deventer, J. F. Bartelsman, P. den Otter, J. Jansen, M. L. Mevissen, T. van Gool, S. A. Danner, and P. Reiss. 1995. Diarrhoea in HIV-infected patients: no evidence of cytokine-mediated inflammation in jejunal mucosa. *AIDS* **9:**367–373.

Snowden, K. 1998. Personal communication.

Sobottka, I., H. Albrecht, J. Schottelius, C. Schmetz, M. Bentfeld, R. Laufs, and D. A. Schwartz. 1995. Self-limited traveller's diarrhea due to a dual infection with *Enterocytozoon bieneusi* and *Cryptosporidium parvum* in an immunocompetent HIV-negative child. *Eur. J. Clin. Microbiol. Infect. Dis.* **14:**919–920.

Sprague, V. 1974. *Nosema connori* n. sp., a microsporidian parasite of man. *Trans. Am. Microsc. Soc.* **93:**400–403.

Stewart, C. G., W. S. Botha, and A. F. van Dellen. 1979. The prevalence of *Encephalitozoon* antibodies in dogs and an evaluation of the indirect fluorescent antibody test. *J. S. Afr. Vet. Assoc.* **50:**169–172.

Stewart, C. G., M. G. Collett, and H. Snyman. 1986. The immune response in a dog to *Encephalitozoon cuniculi* infection. *Onderstepoort. J. Vet. Res.* **53:**35–37.

Stewart, C. G., F. Reyers, and H. Snyman. 1988. The relationship in dogs between primary renal disease and antibodies to *Encephalitozoon cuniculi*. *J. S. Afr. Vet. Assoc.* **59:**19–21.

Stewart, C. G., A. F. van Dellen, and W. S. Botha. 1979. Canine encephalitozoonosis in kennels and the isolation of *Encephalitozoon* in tissue culture. *J. S. Afr. Vet. Assoc.* **50:**165–168.

Stiles, J., E. S. Didier, B. Ritchie, C. Greenacre, M. Willis, and C. Martin. 1997. *Encephalitozoon cuniculi* in the lens of a rabbit with phacoclastic uveitis: confirmation and treatment. *Vet. Comp. Ophthalmol.* **7:**233–238.

Szabo, J. R., and J. A. Shadduck. 1987. Experimental encephalitozoonosis in neonatal dogs. *Vet. Pathol.* **24:**99–108.

Szabo, J. R., and J. A. Shadduck. 1988. Immunologic and clinicopathologic evaluation of adult dogs inoculated with *Encephalitozoon cuniculi*. *J. Clin. Microbiol.* **26:**557–563.

Trammer, T., F. Dombrowski, M. Doehring, W. A. Maier, and H. M. Seitz. 1997. Opportunistic properties of *Nosema algerae* (Microspora), a mosquito parasite, in immunocompromised mice. *J. Eukaryot. Microbiol.* **44:**258–262.

Tzipori, S., A. Carville, G. Widmer, D. Kotler, K. Mansfield, A. Lackner. 1997. Transmission and establishment of a persistent infection of *Enterocytozoon bieneusi,* derived from a human with AIDS, in simian immunodeficiency virus-infected rhesus monkeys. *J. Infect. Dis.* **175:**1016–1020.

Undeen, A. H. 1975. Growth of *Nosema algerae* in pig kidney cell cultures. *J. Protozool.* **22:**107–110.

Undeen, A. H., and N. E. Alger. 1976. *Nosema algerae:* infection of the white mouse by a mosquito parasite. *Exp. Parasitol.* **40:**86–88.

Undeen, A. H., and J. V. Maddox. 1973. The infection of nonmosquito hosts by injection with spores of the microsporidan *Nosema algerae*. *J. Invertebr. Pathol.* **22:**258–265.

van Dellen, A. F., W. S. Botha, J. Boomker, and W. E. Warnes. 1978. Light and electron microscopical studies on canine encephalitozoonosis: cerebral vasculitis. *Onderstepoort J. Vet. Res.* **45:**165–186.

Van Dellen, A. F., C. G. Stewart, and W. S. Botha. 1989. Studies of encephalitozoonosis in vervet monkeys (*Cercopithecus pygerythrus*) orally inoculated with spores of *Encephalitozoon cuniculi* isolated from dogs (*Canis familiaris*). *Onderstepoort J. Vet. Res.* **56:**1–22.

Van Gool, T., E. U. Canning, H. Gilis, M. A. Van Den Bergh Weerman, J. K. M. Eeftinck Schattenkerk, and J. D. Dankert. 1994. *Septata intestinalis* frequently isolated from stool of AIDS patients with a new cultivation method. *Parasitology* **109:**281–289.

Van Gool, T., V. C. Vetter, B. Weinmayr, A. Van Dam, F. Derouin, and J. Dankert. 1997. High seroprevalence of *Encephalitozoon* species in immunocompetent subjects. *J. Infect. Dis.* **175:**1020–1024.

van Rensburg, I. B., D. H. Volkmann, J. T. Soley, and C. G. Stewart. 1991. *Encephalitozoon* infection in a still-born foal. *J. S. Afr. Vet. Assoc.* **62:**130–132.

Vavra, J., and A. H. Undeen. 1970. *Nosema algerae* n.sp. (Cnidospora, Microsporida): a pathogen in a colony of *Anopheles stephensi* Lison (Dipetera, Culicidae). *J. Protozool.* **17:**240–249.

Vavra, J, A. T. Yachnis, J. A. Shadduck, J. M. Orenstein, E. U. Curry, and A. Curry. 1997. A *Trachipleistophora*-like microsporidium of man: its dimorphic nature and relationship to *Thelohania apodemi,* p. 11. *In* Abstr. 2nd Workshop Microsporidiosis Cryptosporidiosis Immunodeficient Patients. České Budějovice, Czech Republic.

Visvesvara, G. S., A. J. DaSilva, G. P. Croppo, N. J. Pieniazek, G. J. Leitch, D. Ferguson, H. de Moura, S. Wallace, S. B. Slemenda, I. Tyrrell, D. F. Moore, and J. Meador. 1995a. *In vitro* culture and serologic and molecular identification of *Septata intestinalis* isolated from urine of a patient with AIDS. *J. Clin. Microbiol.* **33:**930–936.

Visvesvara, G. S., G. J. Leitch, A. J. DaSilva, G. P. Croppo, H. Moura, S. Wallace, S. B. Slemenda, D. A. Schwartz, D. Moss, R. T. Bryan, and N. J. Pieniazek. 1994. Polyclonal and monoclonal antibody and PCR-amplified small-subunit rRNA identification of a microsporidian, *Encephalitozoon hellem,* isolated from an AIDS patient with disseminated infection. *J. Clin. Microbiol.* **32:**2760–2768.

Visvesvara, G. S., G. J. Leitch, H. Moura, S. Wallace, R. Weber, and R. T. Bryan. 1991. Culture, electron microscopy, and immunoblot studies on a microsporidian parasite isolated from the urine of a patient with AIDS. *J. Protozool.* **38:**105S–111S.

Visvesvara, G., G. J. Leitch, N. J. Pieniazek, A. J. Da Silva, S. Wallace, S. B. Slemenda, R. Weber, D. A. Schwartz, L. Gorelkin, C. M. Wilcox, and R. T. Bryan. 1995b. Short-term *in vitro* culture and molecular analysis of the microsporidian, *Enterocytozoon bieneusi*. *J. Eukaryot. Microbiol.* **42:**506–510.

Waller, T., and N. R. Bergquist. 1982. Rapid simultaneous diagnosis of toxoplasmosis and encephalitozoonosis in rabbits by carbon immunoassay. *Lab. Anim. Sci.* **32:**515–517.

Waller, T., B. Morein, and E. Fabiansson. 1978. Humoral immune response to infection with *Encephalitozoon cuniculi* in rabbits. *Lab. Anim.* **12:** 145–148.

Wanke, C. A., P. DeGirolami, and M. Federman. 1996. *Enterocytozoon bieneusi* infection and diarrheal disease in patients who were not infected with human immunodeficiency virus: case report and review. *Clin. Infect. Dis.* **23:**816–818.

Wanke, C. A., D. Pleskow, P. C. Degirolami, B. B. Lambl, K. Merkel, and S. Akrabawi. 1996. A medium chain triglyceride-based diet in patients with HIV and chronic diarrhea reduces diarrhea and malabsorption: a prospective, controlled trial. *Nutrition* **12:**766–771.

Wasson, K., K. Snowden, E. Didier, J. Shadduck, and H. Gelberg. Intestinal xenograft model for human microsporidiosis. Submitted for publication.

Weber, R., and R. T. Bryan. 1994. Microsporidial infections in immunodeficient and immunocompetent patients. *Clin. Infect. Dis.* **19:**517–521.

Weber, R., R. T. Bryan, D. A. Schwartz, and R. L. Owen. 1994. Human microsporidial infections. *Clin Microbiol. Rev.* **7:**426–461.

Weidner, E. 1975. Interactions between *Encephalitozoon cuniculi* and macrophages. Parasitophorous vacuole growth and the absence of lysosomal fusion. *Z. Parasitenkd.* **47:**1–9.

Weidner, E., E. U. Canning, and W. S. Hollister. 1997. The plaque matrix (PQM) and tubules at the surface of intramuscular parasite, *Trachipleistophora hominis. J. Eukaryot. Microbiol.* **44:**359–365.

Weidner, E., and L. D. Sibley. 1985. Phagocytized intracellular microsporidian blocks phagosome acidification and phagosome-lysosome fusion. *J. Protozool.* **32:**311–317.

Weiss, L. M., A. Cali, E. Levee, D. LaPlace, H. Tanowitz, D. Simon, and M. Wittner. 1992. Diagnosis of *Encephalitozoon cuniculi* infection by Western blot and the use of cross-reactive antigens for the possible detection of microsporidiosis in humans. *Am. J. Trop. Med. Hyg.* **47:**456–462.

Wicher, V., R. E., Baughn, C. Fuentealba, J. A. Shadduck, F. Abbruscato, and K. Wicher 1991. Enteric infection with an obligate intracellular parasite, *Encephalitozoon cuniculi,* in an experimental model. *Infect. Immun.* **59:**2225–2231.

Wolfer, J., B. Grahn, B. Wilcock, and D. Percy. 1993. Phacoclastic uveitis in the rabbit. *Prog. Vet. Comp. Ophthalmol.* **3:**92–96.

Wongtavatchai, J., P. A. Conrad, and R. P. Hedrick. 1995a. *In vitro* characteristics of the microsporidian: *Enterocytozoon salmonis. J. Eukaryot. Microbiol.* **42:**401–405.

Wongtavatchai, J., P. A. Conrad, and R. P. Hedrick. 1995b. Effect of the microsporidian *Enterocytozoon salmonis* on the immune response of chinook salmon. *Vet. Immunol. Immunopathol.* **48:**367–374.

Wosu, N. J., R. Olsen, J. A. Shadduck, A. Koestner, and S. P. Pakes. 1977a. Diagnosis of experimental encephalitozoonosis in rabbits by complement fixation. *J. Infect. Dis.* **135:**944–948.

Wosu, N. J., J. A. Shadduck , S. P. Pakes, J. K. Frenkel, K. S. Todd, Jr., and J. D. Conroy. 1977b. Diagnosis of encephalitozoonosis in experimentally infected rabbits by intradermal and immunofluorescence tests. *Lab. Anim. Sci.* **27:** 210–216.

Wright, J. H., and E. M. Craighead. 1922. Infectious motor paralysis in young rabbits. *J. Exp. Med.* **36:**135–140.

Yachnis, A. T., J. Berg, A. Martinez-Salazar, B. S. Bender, L. Diaz, A. M. Rojiani, T. A. Eskin, and J. M. Orenstein. 1996. Disseminated microsporidiosis especially infecting the brain, heart, and kidneys: report of a newly recognized pansporoblastic species in two symptomatic AIDS patients. *Am. J. Clin. Pathol.* **106:**535–543.

Yee, R. W., F. O. Tio, J. A. Martinez, K. S. Held, J. A. Shadduck, and E. S. Didier. 1991. Resolution of microsporidial epithelial keratopathy in a patient with AIDS. *Ophthalmology* **98:**196–201.

Zeman, D. H., and G. B. Baskin. 1985. Encephalitozoonosis in squirrel monkeys (*Saimiri sciureus*). *Vet. Pathol.* **22:**24–31.

Zhou, Z., and K. Nordstoga. 1993. Mesangioproliferative glomerulonephritis in mink with encephalitozoonosis. *Acta Vet. Scand.* **34:**69–76.

Zierdt, C. H., V. J. Gill, and W. S. Zierdt. 1993. Detection of microsporidian spores in clinical samples by indirect fluorescent-antibody assay using whole-cell antisera to *Encephalitozoon cuniculi* and *Encephalitozoon hellem. J. Clin. Microbiol.* **31:**3071–3074.

CLINICAL SYNDROMES ASSOCIATED WITH MICROSPORIDIOSIS

Donald P. Kotler and Jan M. Orenstein

8

Microsporidial parasites are ubiquitous in nature (Vossbrinck et al., 1987), and so it should come as no surprise that they can infect humans and cause clinical disease. Microsporidia were recognized by light microscopy of mammalian tissue samples more than 70 years ago (Levaditi et al., 1923) and linked to human disease almost 40 years ago (Matsubayashi et al., 1959). However, reports of clinical disease related to microsporidia in humans were rare before 1985. In mammals, clinically apparent disease due to microsporidia occurs mostly in young animals or in association with immune deficiencies. *Encephalitozoon cuniculi,* the first microsporidian species shown to cause infection in humans, has a relatively broad host range in mammals (Bergquist et al., 1983). To date, seven genera of microsporidia have been identified as potential pathogens in humans, three of which are related to species causing disease in animals (Table 1).

The emergence of AIDS has brought about an increased awareness of the broad spectrum of potential pathogens facing humans, in part because of the large number of "new" pathogens producing serious disease in AIDS patients and also because of the realization that immune suppression is a common phenomenon in modern medicine, including iatrogenic immune suppression during treatment of allergic and severe inflammatory diseases, after transplantation, and during chemotherapy. Indeed, immune suppression is an important component of protein energy malnutrition (Chandra, 1991) and has been a common problem throughout recorded history. Its adverse effects on survival in children in the developing world are well documented. The increased susceptibility of AIDS patients to develop chronic enteric infections as well as other complications has sensitized public health authorities and the general public to sources of potential disease within our environment. Especially in the case of cryptosporidiosis, the identification of large-scale as well as small outbreaks of disease has exposed limitations in the ability to protect municipal water systems. In this context, a report has documented the first apparent outbreak of microsporidiosis (Cotte et al., 1998).

Reports of diarrheal syndromes associated with microsporidiosis and human immunodeficiency virus (HIV) infection were first pub-

Donald P. Kotler, Gastrointestinal Division, Department of Medicine, St. Luke's-Roosevelt Hospital Center, and College of Physicians and Surgeons, Columbia University, New York, NY 10025. *Jan M. Orenstein,* Department of Pathology, George Washington University School of Medicine, Washington, DC 20037.

The Microsporidia and Microsporidiosis (Murray Wittner, editor; Louis M. Weiss, contributing editor), ©1999 American Society for Microbiology, Washington, D.C.

TABLE 1 Microsporidia identified as pathogenic to humans

Identified in patients with AIDS	Identified in other patients
Encephalitozoon cuniculi	*Encephalitozoon cuniculi*
Encephalitozoon hellem	
Encephalitozoon intestinalis	
Enterocytozoon bieneusi	*Enterocytozoon bieneusi*
Trachipleistophora hominis	
Trachipleistophora anthropopthera	
Pleistophora sp.	*Pleistophora* sp.
Brachiola vesiculatum	*Brachiola* (syn. *Nosema*) *connori*
	Nosema ocularum
	Vittaforma corneae
	Microsporidium africanus
	Microsporidium ceylonensis

lished in 1985 (Desportes et al., 1985; Dobbins and Weinstein, 1985), and the number of articles describing human disease increased rapidly after 1990. While most papers have dealt with gastrointestinal disease in HIV infection, other clinical syndromes associated with microsporidiosis have been reported, including keratoconjunctivitis, sinusitis, tracheobronchitis, encephalitis, tubulointerstitial nephritis, hepatitis, cholecystitis, cholangitis, osteomyelitis, and myositis. The spectrum of affected individuals also has spread beyond those infected with HIV.

The aim of this chapter is to describe the clinical and pathogenic features of microsporidiosis. Most of the discussion will be centered around intestinal disease since it is the most prevalent and most well studied.

INTESTINAL DISEASE DUE TO MICROSPORIDIOSIS

In retrospect, microsporidiosis was present in the initial AIDS patient seen by the author (D.P.K.) in the summer of 1981. *Enterocytozoon bieneusi* was first observed by transmission electron microscopy (TEM) in 1982 in villus epithelial cells in small intestinal biopsies from AIDS patients in Texas and in Washington, D.C. (Orenstein et al., 1991). The first literature reports of intestinal microsporidiosis in AIDS patients appeared almost simultaneously in 1985 from the United States and France (Desportes et al., 1985; Dobbins and Weinstein, 1985). Several case reports and small case series followed (Modigliani et al., 1985; Curry et al., 1988; Bernard et al., 1991; Michiels et al., 1991; Simon et al., 1991; Ullrich et al., 1991) as did one larger case series (Orenstein et al., 1990), the latter suggesting that intestinal microsporidiosis was not rare in AIDS patients.

These studies demonstrated characteristics different from those of all other microsporidia known at the time, allowing the identification of a new genus and species. The characteristic features of *E. bieneusi* are its development as a multinucleate plasmodium in intimate contact with the cell cytoplasm (pansporoblastic), the presence of cleftlike structures called electron-lucent inclusions, and the presence of electron-dense disks that develop into the polar tubule.

Other distinctive features include precocious development of the injection apparatus within the intact plasmodium, six turns of the polar tubule arranged in two layers, and mononuclear spores (1 by 1.5 µm) that are the smallest of all known microsporidia. Since that time, other species of *E. (Nucleospora)* have been identified (Chilmonczyk et al., 1991). A microsporidian species that differed ultrastructurally from *E. bieneusi* was identified by TEM in an intestinal biopsy from an AIDS patient with diarrhea in 1988 (Orenstein et al., 1992a). Two years later, three additional cases were discovered. Since then, many reports have described cases of *Encephalitozoon intestinalis* (*Septata intestinalis*) (Cali et al., 1993) in the United States, Europe, and Australia. In contrast to infection with *E. bieneusi, E. intestinalis* infection is not limited to small intestinal and biliary epithelia but involves macrophages and disseminates widely (Orenstein et al., 1992a). A single case report of *E. cuniculi* infection of duodenal mucosa has been reported in a patient without gastrointestinal symptoms (Franzen et al., 1995a).

Epidemiology of Intestinal Microsporidiosis

DISTRIBUTION

The initial case series of microsporidiosis involved patients from the United States and western Europe. Later case series also documented cases in Africa (Drobniewski et al., 1995) and Australia (Field et al., 1990). More recent reports have identified microsporidiosis in Spain (Subirats et al., 1996), Italy (Voglino et al., 1996; Marangi et al., 1995), Brazil (Brasil et al., 1996), Chile (Weitz et al., 1995), Mali (Maiga et al., 1997), New Zealand (Everts et al., 1997), and Thailand (Pitisuttithum et al., 1995). Thus, microsporidiosis appears to have worldwide distribution.

PREVALENCE

Prevalence rates of microsporidiosis among AIDS patients with chronic diarrhea have varied between 2 and 50% (Canning and Hollister, 1990; Greenson et al., 1991; Orenstein, 1991; Simon et al., 1991; Swenson et al., 1993; Cotte et al., 1993; Rabeneck et al., 1993; Molina et al., 1993; Kotler et al., 1994b; Drobniewski et al., 1995; Bernard et al., 1995), depending on the study population and methods of diagnosis. Prevalence rates of microsporidiosis are generally comparable to those of other AIDS-associated opportunistic enteric infections such as cryptosporidiosis. Our own prospective sequential series of gastrointestinal evaluations in AIDS patients between 1991 and 1994 revealed an incidence of 33% in AIDS patients with chronic diarrhea (Kotler et al., 1994b). The incidence of clinically significant infection with *E. bieneusi* is about 10 times greater than that of *E. intestinalis,* at least in New York City. The prevalence rate for microsporidiosis has fallen dramatically since the advent of potent combination antiretroviral agent therapy.

The mode of transmission of enteric microsporidiosis has not been studied nor have the natural reservoirs of *E. bieneusi* been defined. Most investigators suspect the parasite is acquired by ingestion of contaminated food or water. A single report identified a cluster of cases in France over a specific time interval (Cotte et al., 1998). Most but not all of the affected individuals were HIV-infected, however, the exact source of the outbreak and its mode of transmission were not identified in the study. Tzipori et al. (1997) were able to establish a persistent infection in simian immunodeficiency virus (SIV)-infected rhesus monkeys by gavage using microsporidial spores isolated from symptomatic human AIDS patients by intestinal lavage. Infection was documented by shedding of spores in stool and by in situ hybridization studies on intestinal mucosa. The infected monkeys had relatively well preserved $CD4^+$ lymphocyte counts and no clinical disease. In contrast, the investigators were unable to transmit a persistent infection to an SIV-negative immunocompetent monkey.

Most reports of intestinal microsporidiosis have involved HIV-infected individuals. Although the majority of cases have been diagnosed in homosexual males, diagnoses also have been made in heterosexual women and in children. *E. bieneusi* has been identified in patients with other immune deficiencies (Sax et al., 1995), as well as in immunocompetent individuals who are asymptomatic or have a self-limited diarrheal illness (Sandfort et al., 1994; Maiga et al., 1997; Hautvast et al., 1997; Bryan et al., 1997; Cotte, 1998).

CAUSE-EFFECT RELATIONSHIP BETWEEN MICROSPORIDIOSIS AND DIARRHEA

The pathogenicity of microsporidiosis has been assumed by most authors because of the pathologic changes seen on intestinal biopsy (Orenstein et al., 1990; Greenson et al., 1991; Eeftinck Schattenkerk et al., 1991; Kotler et al., 1993). However, simply observing a putative pathogen in an area of disease does not necessarily prove that the organism is the cause of the disease. In fact, the reverse is true of certain clinical conditions, e.g., colonization of a previously damaged tissue compartment by an organism. The problem has been compounded in AIDS since the disease initially was undefined and clinicians were faced with a large number of novel agents. Organisms such as cytomegalovirus and *Mycobacterium avium* complex were originally suspected as being commensal rather than true pathogens. For this reason, formal analysis of cause and effect is appropriate for microsporidia.

Establishing a cause-effect relationship is difficult in situations where animal models for a specific disease do not exist. Previous generations of investigators were faced with the same problem, especially near the end of the last century when histopathologic and culture techniques had identified a large number of potential pathogens. Several sets of analytical criteria were proposed, the most well recognized being those of Robert Koch (Fredericks and Relman, 1996), which were based on his clinical observations in patients with tuberculosis and anthrax. There were three criteria: (1) The microbe can be isolated in all cases of the disease under circumstances that can account for the clinical course and pathologic features; (2) the microbe does not also occur as a nonpathogenic agent; (3) introduction of the microbe to a new organism after its isolation and propagation in pure culture reproduces the disease. A corollary criterion is that effective treatment or other recovery from the disease is associated with a disappearance of the microbe. It is obvious that these criteria are appropriate for agents that can be cultured. They are less helpful for infections for which culture techniques or suitable animal models are not available and for noninfectious causes of disease.

Other criteria, such as the Bradford-Hill criteria, have been proposed that are more appropriate for analyzing the cause-effect relationships of microsporidiosis (Bradford-Hill, 1965). These are statistical criteria and include the strength, consistency, and specificity of association between the detection of microsporidia and clinical disease, the temporality of association, the biological gradient between stimulus and response, the biological plausibility of a cause-effect relationship, coherence among different studies, reversibility, and analogy with other situations. None of these epidemiological criteria is absolute.

Microsporidia were originally believed to cause intestinal disease on the basis of their identification in abnormal small intestinal mucosa of severely immunosuppressed AIDS patients with chronic diarrhea and weight loss. In order to more precisely define the effect of microsporidiosis on intestinal mucosa, we studied a series of AIDS patients with and without microsporidiosis and cryptosporidiosis, as well as control AIDS patients and healthy controls. The AIDS groups were selected on the basis of TEM of small intestinal biopsies obtained during the evaluation of diarrhea or other gastrointestinal symptoms (Kotler, 1993). Quantitative histopathology was performed, specific activities of intestinal disaccharidases were determined, and qualitative and quantitative measurements of HIV RNA and protein antigens in jejunal mucosa were made. Biopsies from patients with microsporidiosis and cryptosporidiosis showed partial villus atrophy and crypt hyperplasia, while biopsies from AIDS patients without enteric pathogens had values similar to those of normal controls. The biopsies from cryptosporidiosis and microsporidiosis showed changes similar to those in biopsies from patients with tropical sprue or partially treated celiac disease except for the presence of the enteric pathogen. The results suggested that small intestinal injury in an AIDS patient is related to the presence of an epithelial cell infection and is not a nonspecific effect of immune deficiency, though a few patients in this series had evidence of intestinal injury without an identifiable enteric pathogen. The general mucosal histologic changes were similar in patients with cryptosporidiosis and microsporidiosis. The specific activities of sucrase, lactase, and maltase were low in patients with microsporidiosis and cryptosporidiosis, while the results in AIDS patients without enteric pathogens were similar to those in normal controls. No difference in the expression of HIV RNA or p24 antigen was found in AIDS patients with or without intestinal parasites. These results showed an association of small intestinal damage and enteric infections as well as a pattern of intestinal injury consistent with the proposed mechanism of epithelial cell injury, thus satisfying the plausibility and analogy (tropical sprue) criteria of Bradford-Hill.

However, this study, which found associations between intestinal damage and the presence of microsporidia, did not prove specificity of association since patients without intestinal symptoms were not evaluated by small intestinal biopsy. Studies from another laboratory found quite different results in an evaluation of HIV-infected volunteer subjects not referred for gastrointestinal evaluation. A prospective study of HIV-infected subjects with and without chronic diarrhea, defined as two or more loose stools per day for more than 30 days, showed *E. bieneusi* prevalence rates of higher than 20% for both groups. The results were unrelated to peripheral blood CD4^{+} lymphocyte counts, coinfections, or other clinical parameters (Rabeneck et al., 1993). Small intestinal injury was not found in all patients with microsporidiosis. On follow-up studies, diarrhea or weight loss did not invariably develop in patients with microsporidiosis (Rabeneck et al., 1995). The authors concluded that *E. bieneusi* infection was not associated with clinical disease in all cases and suggested that microsporidia may not be a true enteric pathogen. However, they did not distinguish between the presence of microsporidia as a commensal organism and the possibility that microsporidia exists in a clinically latent state. Their data suggest a possible biological gradient between parasite burden and the risk of having clinical symptoms (Rabeneck et al., 1995), suggesting that other factors, e.g., immune function (CD4^{+} lymphocyte count), might influence the development of clinically overt disease.

The specificity of the association between diarrhea and intestinal microsporidiosis was analyzed by a PCR technique and TEM to detect microsporidia in intestinal biopsies. Comparisons were made in a large series of AIDS patients with and without diarrhea, the former group with and without microsporidiosis (Coyle et al., 1996). All patients with microsporidiosis studied by TEM except one complained of chronic diarrhea. The other patient with microsporidiosis had objective evidence of small intestinal injury on biopsy and by D-xylose absorption testing. Microsporidia were detected by PCR in all but one patient with microsporidia detected by TEM and in none of the controls. These results demonstrated both the strength and specificity of the association, though they do not fully address the deficiencies of the former study since the patients all had been referred for gastroenterologic evaluation and may not have been representative of all HIV-infected individuals.

Studies that have identified animal models of *E. bieneusi* also support the pathogenic potential of this microsporidian parasite. *E. bieneusi* has been reported in pigs (Deplazes et al., 1996), and *E. bieneusi*-like organisms in simian immunodeficiency virus (SIV)-infected rhesus monkeys (Mansfield et al., 1997) Experimental infection of SIV-infected rhesus monkeys with *E. bieneusi* from human tissue has been demonstrated (Tzipori et al., 1997). In SIV-infected monkeys the presence of *E. bieneusi* is associated with hepatic damage and cholangiopathy (Mansfield et al., 1997). A 30% prevalence of hepatic and small intestinal microsporidiosis was found in an autopsy series of SIV-infected monkeys, whereas no microsporidia were seen in SIV-infected monkeys without hepatic histopathology (Mansfield, 1997). Infection of immunosuppressed gnotobiotic piglets with *E. bieneusi* resulted in the shedding of spores and a self-limited diarrheal syndrome (Kondova et al., 1998).

E. bieneusi has been identified as a cause of self-limited diarrhea in immunocompetent hosts (Bryan et al., 1997; Weber et al., 1994; Hautvast et al., 1997; Sandfort et al., 1994), in an epidemiologic study on *E. bieneusi* in 1% of African children with diarrhea (Bertange et al., 1993), as well as in patients undergoing liver, heart-lung, and bone marrow transplantation (Weber et al., 1994; Kelkar et al., 1997; Bryan et al., 1997; Rabodonirina et al., 1996). The elimination of *E. bieneusi* has been associated with the resolution of symptoms, such as diarrhea, in both immunocompetent and immunocompromised hosts. For example, the elimination of *E. bieneusi* from stool was associated with the resolution of diarrhea in four patients treated with fumagillin (Molina et al., 1997). Several studies have also documented the reversal of renal failure and improvement in intestinal function during albendazole therapy for *E. intestinalis* infection (Orenstein et al., 1993; Aarons et al., 1994; Sobottka et al., 1995; Dore et al., 1995; Franzen et al., 1995b; DeGroote et al., 1995; Molina et al., 1995; Joste et al., 1996), thus fulfilling the criterion for reversibility for this microsporidian parasite.

On the basis of published studies, one can conclude that microsporidia are likely to be true enteric pathogens and may cause chronic symptomatic disease in patients with AIDS and other immune deficiencies. Symptomatic infection may occur as a self-limited disease in the absence of immune deficiencies. Since infection is not always associated with intestinal symptoms (Rabeneck et al., 1993; Hauvast et al., 1997; Bryan et al., 1997), other factors may help determine the clinical course. It is possible that chronic latent infections occur, with clinical activation related to progressive immune deficiency or other issues (Franzen et al., 1995a, 1995b, 1995c, 1995d, 1996). The probability of developing clinical disease may be modified by other factors. Most studies found that microsporidiosis typically is diagnosed in patients with a severe depletion of $CD4^+$ lymphocytes (15 to 30 cells per mm^3) (Field et al., 1990; Eeftinck Schattenkerk et al., 1991; Molina et al., 1993; Kotler et al., 1994b). A study finding similar prevalence rates in patients with and without diarrhea showed much higher $CD4^+$ lymphocyte counts in those without chronic diarrhea (113 and 192 in patients with and without diarrhea, respectively) (Rabeneck et al., 1993). Other authors also have reported finding relatively well-preserved $CD4^+$ lymphocyte counts in some patients with microsporidiosis (Sowerby et al., 1995). SIV-infected monkeys infected by the oral route had no observable diarrhea and had relatively well-preserved $CD4^+$ lymphocyte counts (Tzipori et al., 1997). These reports suggest that the level of immune function might influence the probability of developing clinically overt or chronic disease due to microsporidia.

Pathogenesis of Intestinal Injury

As discussed above, it is likely that transmission of microsporidia occurs via the oral route. The life cycle of the microsporidia outside the body has not been studied, and the factors that might influence infectivity, such as gastric acid, etc., have not been studied. The intraluminal signals that regulate excystation also are uncertain. The polar filament might be extruded in response to the higher pH in the small intestinal lumen or to other local factors (Canning and Hollister, 1987). Spores with extruded polar filaments may be seen in upper intestinal lavage specimens (Kotler, unpublished observations). The polar filament infects an epithelial cell by directly penetrating the cell and injecting its sporoplasm. The speed of microsporidial replication is unknown, but it likely is rapid. The life cycles of other microsporidia have been measured in vitro and may be complete within a few hours (Canning and Lom, 1986). It is important to remember that the villus epithelial cell is the primary target of infection for *E. bieneusi* and the initial target of *E. intestinalis.* The villus epithelial cell has a limited life span under normal circumstances, about 48 to 72 h in humans, after the cell leaves the crypt. The life span of villus epithelial cells falls markedly in the presence of crypt hyperplasia with partial villus atrophy, which is the pathologic lesion associated with microsporidiosis. For this reason, the life cycle must be rapid unless final maturation can occur after epithelial cell extrusion into the lumen. A pathologic correlate is that the intensity of infection is greatest near the villus tip.

Extrusion of spores into the lumen or sloughing of infected cells returns the spores to the intestinal lumen, presumably to reinfect other cells or to be passed in the stool. It is uncertain if spores infect contiguous cells or more distant sites. If the former occurs, the efficiency of successful reinfection is greater if the spores travel beneath mucous layer rather than in the bulk luminal contents. It is possible that signals for extrusion of the polar filament are related to the proximity to the epithelium since excystation into the bulk phase would be futile.

FEATURES OF INTESTINAL INJURY ASSOCIATED WITH MICROSPORIDIOSIS

The cardinal feature of intestinal microsporidiosis is injury to the small intestinal epithelium, leading to malabsorption. Two aspects of intestinal injury are relevant to malabsorption, decreased mucosal surface area and functional immaturity of the villus epithelial cells, since both may affect absorption. The absorption rate for any nutrient crossing a membrane is related to mucosal surface area, diffusion distance to the absorbing membrane, and concentration gradient across the membrane. Mucosal surface area is a function of the number of villus epithelial cells, which in turn depends on the rates of cell production and loss. These rates are tightly regulated under normal circumstances in order to maintain homeostasis (Johnson, 1988). Histopathologic studies suggest that the mucosal surface area is diminished as a result of microsporidial infection. The decrease in epithelial surface area in patients with microsporidiosis is reflected in subnormal serum concentrations of D-xylose following an oral test dose (Kotler et al., 1990b, 1993), implying a decrease in mucosal surface area. Studies on HIV-infected patients with small intestinal enteropathies, including microsporidiosis, demonstrated decreased absorption rates for mannitol, a passively absorbed monosaccharide (Lima et al., 1997). The excess destruction of cells associated with microsporidial infection perturbs the steady-state mechanisms that optimize nutrient absorption and leads to adaptations in epithelial proliferation. Under conditions of increased cell loss, such as those due to microsporidial infection, compensatory crypt hyperplasia occurs and returns villus architecture toward normal, resulting in partial villus atrophy and crypt hyperplasia. Intestinal dysfunction in *E. bieneusi* or *E. intestinalis* infections resembles that in patients with tropical sprue and celiac disease, two diseases also characterized by excess losses of villus enterocytes (Brunner et al., 1970). The specific mediators of crypt hyperplasia in microsporidial infection, as well as in other diseases, are not completely understood.

The average survival of an epithelial cell during its migration along the villus is 48 to 72 h, during which time functional maturation and senescence occur. Variable amounts of time are required for the expression of different enterocyte enzymes. For example, the specific activities of the brush border disaccharidases, maltase and sucrase, are similar along the length of the villus, while the specific activities of lactase and enzymes of lipid metabolism are absent in the lower villus and expressed only in the upper villus (Shiau et al., 1979; Boyle et al., 1980). If the migration rates of newly formed enterocytes are increased, insufficient time may be available for complete functional maturation. Clinically, there are greater deficits in lactose and fat absorption than in starch and sucrose absorption in diseases producing villus atrophy and crypt hyperplasia. In the study cited above (Kotler et al., 1993), the relative depletion of lactase was much greater than that of maltase, as would be expected in a state of rapid cell migration.

Intestinal Immune Response to Microsporidia

Little is known about immunity to *E. bieneusi* and *E. intestinalis.* Antibodies to microsporidia have been determined in serologic studies, but the specificity of the antibody response is uncertain (Didier et al., 1993; Aldras et al., 1994). There is limited published information about cell-mediated immunity in the control of microsporidiosis (Didier et al., 1994), though clinical specimens typically show increased numbers of intraepithelial lymphocytes in areas containing *E. bieneusi.* Symptomatic illness in AIDS patients is usually associated with severe depletion of $CD4^+$ lymphocytes, suggesting a crucial role for cell-mediated immunity. A prominent role for cell-mediated immunity is likely, given the high prevalence of clinical remission in patients treated with combinations of potent antiretroviral therapy (Goguel et al., 1997; Carr et al., 1998). Experimental studies with interferon gamma knockout mice demonstrated prolonged shedding of spores after infection with *E. intestinalis* (Achbarou et al., 1996).

Clinical Illness

The clinical illness in patients with microsporidiosis varies considerably. The clinical syndromes associated with infection by *E. intestinalis* and *E. bieneusi* are similar and are typical of malabsorption (Kotler, 1991). A small percentage of patients do not have diarrhea, though objective evidence of malabsorption may be present. Patients typically have 3 to 10 nonbloody bowel movements per day not associated with fever or night sweats. The volume of each bowel movement may vary widely, and the diarrhea may be described as being "pure water." The episodes of diarrhea occur at irregular intervals during the day, often accompanied by excessive flatus. There may be an alteration in the odor of feces and flatus, which is putrid in nature. Patients with mild disease may describe a specific intolerance for foods containing lactose and fat, whereas those with more severe symptoms are affected by almost all food intake. In severe cases, there is associated dehydration and electrolyte abnormalities, predominantly hypokalemia, hypomagnesemia, and decreased serum bicarbonate concentrations.

A single case report has documented a small intestinal perforation associated with *E. intestinalis* infection (Soule et al., 1997).

Some patients voluntarily eat less food to avoid symptoms, while others note that appetite is preserved. Objectively, calorie counts often reveal inadequate intake. The problem may be a prolonged satiety phase after a meal, similar to that seen in other clinical and experimental malabsorption syndromes (Sclafani et al., 1978). This alteration in appetite behavior has been related to the presence of unabsorbed nutrients in the lower intestine in other situations, though studies have not been performed in patients with microsporidiosis. Associated problems include dry mouth, hypochlorhydria, slowed gastric emptying, decreased pancreatic secretion, and prolonged intestinal transit time (Burn-Murdoch et al., 1978; Owyang et al., 1983; Spiller et al., 1984), which have also been associated with distal intestinal contact with unabsorbed nutrients in other conditions. Weight loss occurs slowly, and its rate may diminish over time. Rapid changes in weight usually reflect changes in hydration status associated with variations in the intensity of diarrhea. Studies on hydration status have shown that patients with microsporidiosis are chronically dehydrated compared to normal subjects and AIDS patients without malabsorption (Babameto et al., 1994).

Few quantitative studies on malabsorption in AIDS patients have been published. Malabsorption of sugars and fats can be detected in AIDS patients with microsporidiosis (Kotler et al., 1990a, 1990b, 1990c; Kotler, 1993; Asmuth et al., 1994). Quantitatively, stool losses may account for 20 to 25% of the calories from sugars and fats (Lambl et al., 1996; Kotler, unpublished observations). Energy expenditure in patients with microsporidiosis is felt to be low, as in any HIV-infected patient with a negative caloric balance. No studies on resting energy expenditure in patients with microsporidiosis have been reported, although we observed subnormal resting energy expenditure in malnourished AIDS patients with malabsorption irrespective of cause (Kotler et al., 1990a, 1990b, 1990c).

DIAGNOSIS

The ability to diagnose microsporidiosis has evolved greatly since its first identification in 1985. Initially, the diagnosis could be made only by TEM since microsporidial forms do not take up the standard hematoxylin or eosin stain. Thus, the ability to make the diagnosis was limited to the few centers employing this technique. Working backward from the ultrastructural changes, several investigators described suggestive light microscopic changes (Lucas et al., 1989; Peacock et al., 1991; Simon et al., 1991). The application of special light microscopic stains capable of detecting spores or other forms (Rijpstra et al., 1988; Giang et al., 1993; Bryan and Weber, 1993; Field et al., 1993; Kotler et al., 1994a; Franzen et al., 1995c; Weber et al., 1992a) allowed the diagnosis to be made with greater confidence by light microscopy, though with less certainty for speciation. The application of techniques for identifying microsporidial spores in stool specimens greatly expanded the number of centers capable of rendering a diagnosis of microsporidiosis and allowed the performance of epidemiologic studies (van Gool et al., 1990, 1993; Orenstein et al., 1991; Verre et al., 1992; Weber et al., 1992, 1994; DeGirolami et al., 1995; Clarridge et al., 1996). The development of molecular techniques carries the promise of greatly increased diagnostic sensitivity and specificity for *E. bieneusi* and *E. intestinalis* and could facilitate epidemiologic studies if methods of mass screening could be developed (Zhu et al., 1993a, 1993b; Weiss et al., 1994b; Franzen et al., 1995c, 1995d; Fedorko et al., 1995). These techniques also would permit the detection of subclinical infections (Franzen et al., 1996a, 1996c).

Ultrastructural Features of Intestinal Microsporidiosis

The various stages in the life cycle of *E. bieneusi* and *E. intestinalis* have been characterized by TEM in several laboratories (Orenstein et al., 1990, 1991; Cali and Owen, 1990). Since *E. bieneusi* cannot be propagated in vitro, the various developmental stages have been characterized in clinical specimens and their sequence inferred from studies on other microsporidia.

ENTEROCYTOZOON BIENEUSI

E. bieneusi infection is limited to epithelial cells in the small intestine and the hepatobiliary tree and pancreas. Spores of *E. bieneusi* have been detected in the lamina propria (Schwartz et al., 1995), probably resulting from cell lysis. A single infected enterocyte may contain microsporidial forms in different stages of development. Whether this occurs as a result of multiple infections in the same cell, binary or multiple fission of plasmodial forms, or variable rates of maturation of infectious sporoplasm is unclear.

The earliest stage identified in clinical material is the proliferating plasmodium, or meront (Fig. 1), which is a small (1 mm in diameter), oval, membrane-bound inclusion, usually seen in the apical cytoplasm and usually associated with cellular mitochondria. The meront is more electron-lucent than the surrounding cytoplasm and contains free ribosomes but no other recognizable structures. A nucleus develops next (Fig. 2) and undergoes multiple divisions. At about this time, a rough endoplasmic reticulum emerges and electron-lucent inclusions develop. The electron-lucent inclusions are lined by electron-dense material (best demonstrated by ferric osmium staining) which appears to be polar tubule precursor material (Desportes et al., 1985). Later in this stage, electron-dense disklike structures begin to develop from the electron-lucent inclusions as the parasite enters sporogony. The electron-

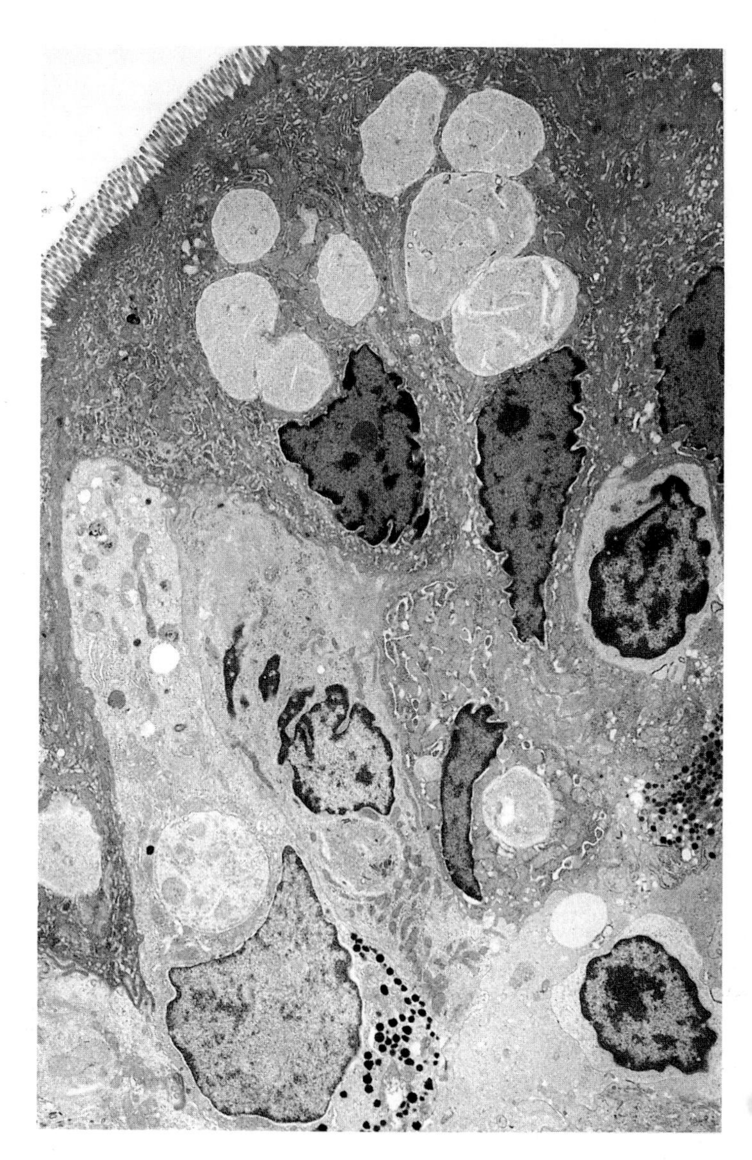

FIGURE 1 Transmision electron micrograph of a jejunal biopsy from a patient with *E. bieneusi* infection, demonstrating the plasmodial stage of development. Note their location in the supranuclear Golgi region, the nuclear cupping, and the intraepithelial lymphocytes. Magnification, ×4,525.

dense disks, whose appearance indicates the beginning of sporogony, increase in number and size. In some sections, the disks appear ringlike, while they are arranged as flat stacks in others (Fig. 3). Interconnections develop, both end to end and in a syncytiumlike pattern, leading to the coiled tubule. The coiled tube contains six turns in *E. bieneusi* and is organized into two tiers of three turns each. Furthermore, the two tiers are consistently out of register by about 45° (Fig. 4). Spores form by progressive development and association of the polar tube with its anchoring plate, polaroplast membrane, nucleus, and posterior vacuole. Through a complicated process of membrane invaginations, the plasmodium divides into multiple sporonts. With development of the endospore and ectospore, first the sporoblast and then the mature electron-dense, egg-shaped spore (1 by 1.5 μm) is formed.

Mature spores have been caught in the act of erupting through the enterocyte brush

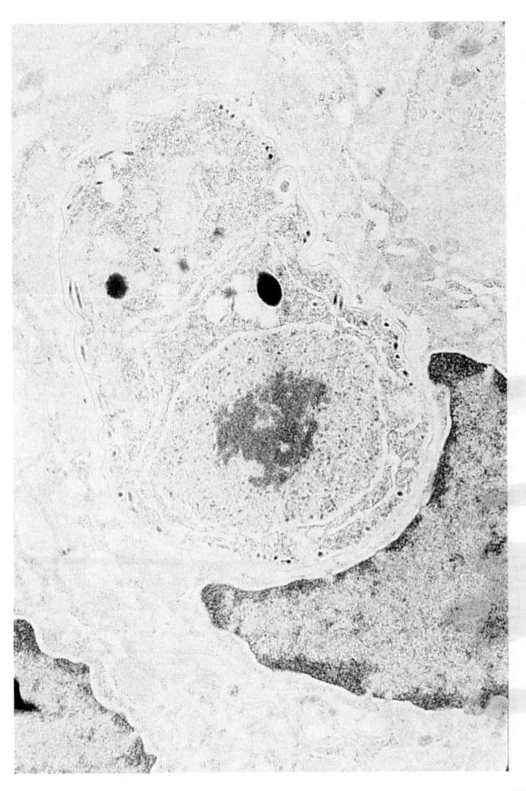

FIGURE 2 Transmission electron micrograph of a single meront with a developing nucleus. Magnification, ×10,000.

border membrane, and spores are also seen in whole cells that have sloughed from the villus (Fig. 5). Still other sections demonstrate disruption of epithelial cells containing spores and plasmodia. It is possible that cells can retain viability for a period of time after sloughing. If so, it is possible that spore maturation can proceed after the enterocyte has been extruded into the intestinal lumen.

ENCEPHALITOZOON INTESTINALIS

Distinctive morphologic features of *E. intestinalis* include a unique development of individual spores within separate chambers of a parasitophorous vacuole (Fig. 6a). The developing organisms are separated by fibrillar septa of parasite origin (Fig. 6b). Unlike those of *E. bieneusi*, *E. intestinalis* plasmodia do not contain electron-lucent inclusions or electron-dense disks. Characteristically, *E. intestinalis* completes merogony prior to development of the polar tubule, the polar tubule has a single tier of six or seven turns, and the spores are 2.2 by 1.2 μm in size (Fig. 7).

Light Microscopic Characteristics of Intestinal Microsporidiosis

ENTEROCYTOZOON BIENEUSI

Intestinal microsporidiosis affects the topography of the small intestinal surface. Close examination at endoscopy demonstrates scalloping of the valvulae conniventes. Villus fusion may be apparent (Fig. 8). There may be a mucoid haze to the mucosa or a punctate whitish color to the mucosa, the latter representing lipid accumulation in epithelial cells because of their immaturity.

Microsporidia produce a characteristic pattern of tissue injury and cytopathology visible by light microscopy. *E. bieneusi* infection results in variable degrees of villus blunting and crypt hyperplasia (Fig. 9). In some cases, villus height is nearly normal and marked crypt hyperplasia is present, but in other cases, significant villus atrophy is associated with less marked crypt hyperplasia. The variation may reflect differences in parasite burden or endogenous factors such as immune function and nutritional status. Biopsies containing long, slender villi and short crypts are unlikely to harbor *E. bieneusi*. The degree of injury is usually similar in different biopsies obtained from the same area. There is a gradient in the parasite burden along the small intestine, the distal duodenum and proximal jejunum having higher burdens than the proximal duodenum (Orenstein et al., 1992b). Diagnosis can be made by ileal biopsy (Weber, et al., 1992b, 1994), but the organism is rarely seen in the colon.

E. bieneusi is most often observed in the upper third of the villus and not in the crypt. There is no acute inflammation. In contrast, the numbers of intraepithelial lymphocytes are usually greater in areas of active infection. Affected sites show epithelial disarray and excess nuclear debris. There may be crowding of

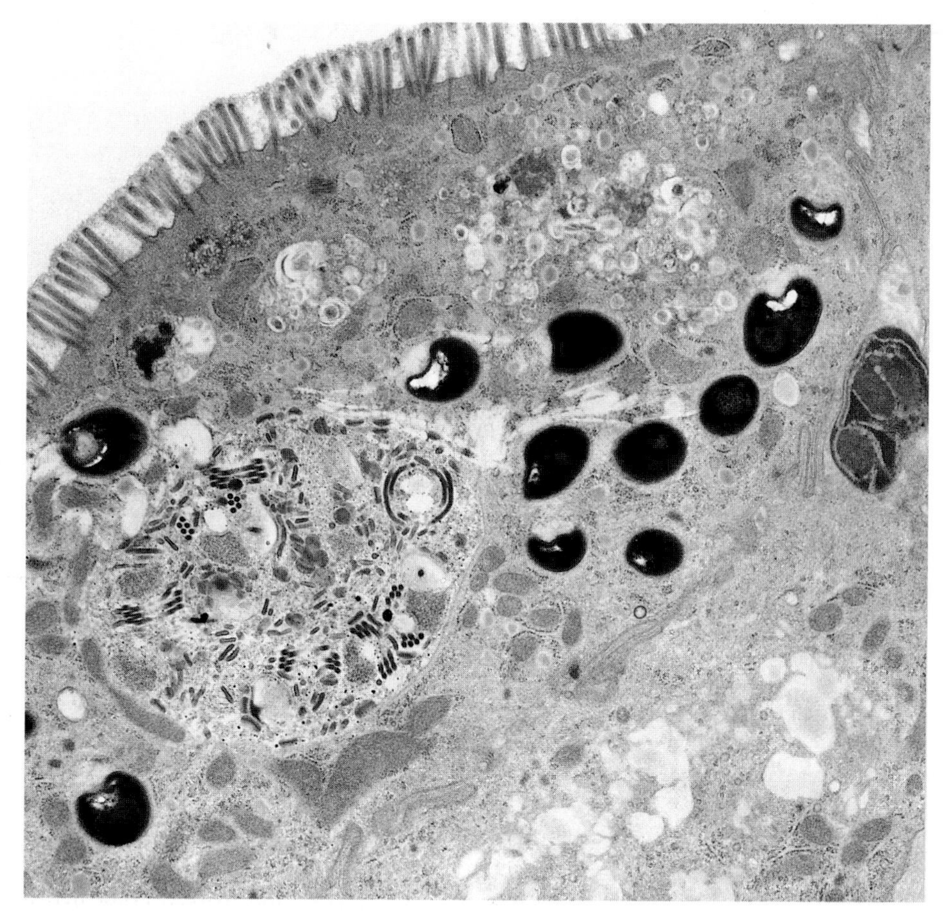

FIGURE 3 Transmision electron micrograph of a jejunal biopsy from a patient with *Enterocytozoon bieneusi* infection, demonstrating a degenerating (vesiculated) enterocyte containing many mature spores and one sporogonial plasmodium. Magnification, ×9,813.

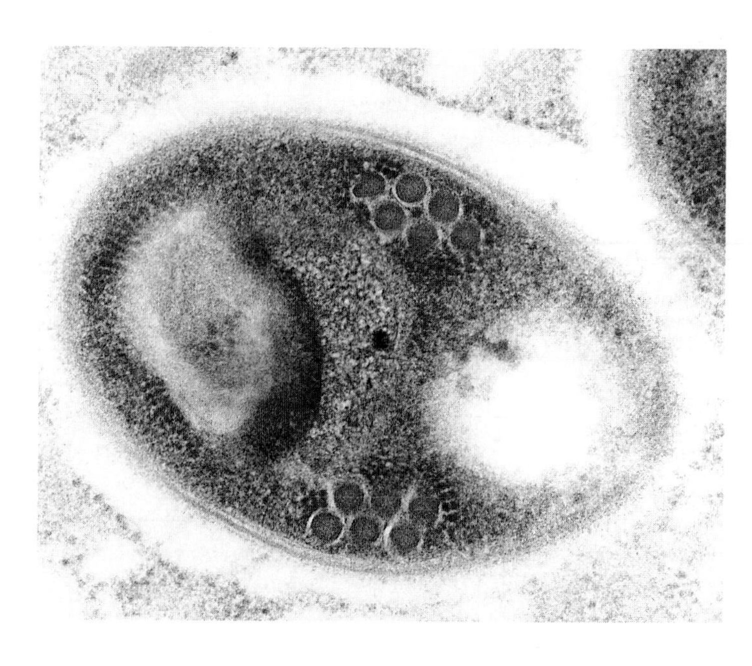

FIGURE 4 A spore of *E. bieneusi,* demonstrating the characteristic six turns of the polar tubule, which are organized into two tiers of three turns each and which are out of register by 45°. Magnification, ×83,000.

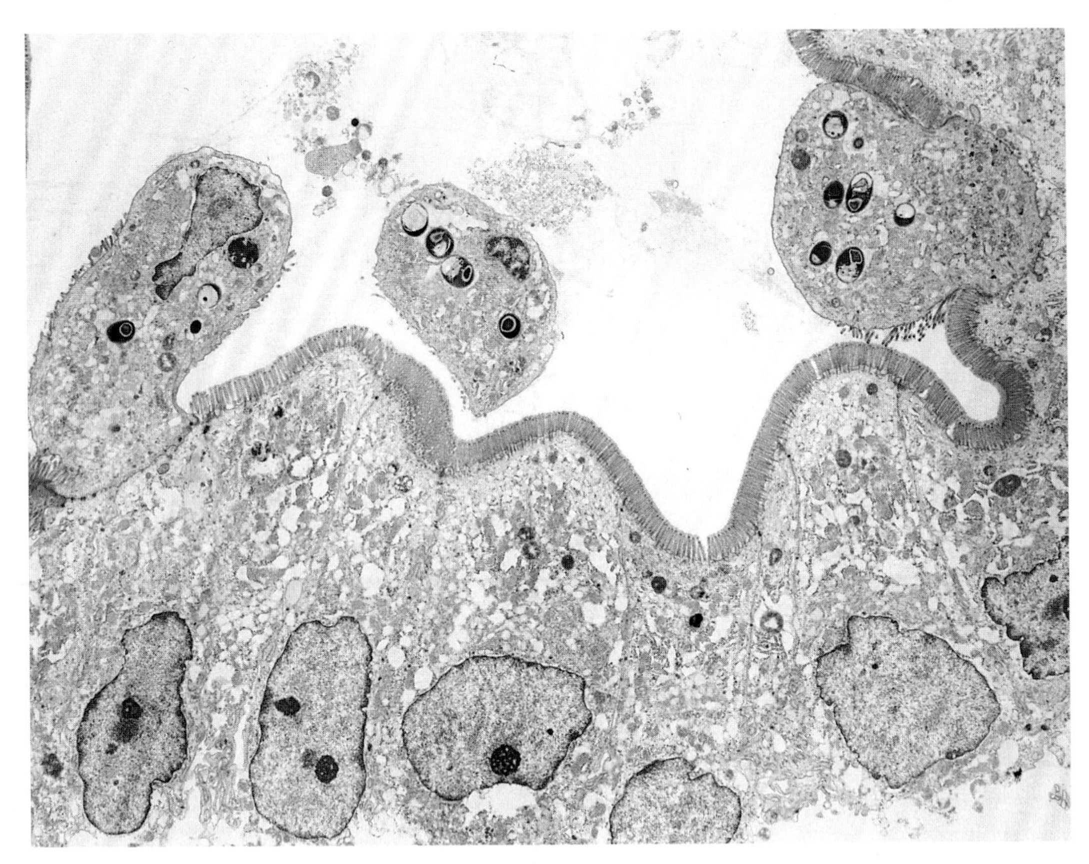

FIGURE 5 Transmission electron micrograph of a jejunal biopsy from a patient with *E. bieneusi* infection, demonstrating cells containing spores that have sloughed or are in the process of sloughing from the epithelial surface. Magnification, ×3,423.

nuclei at the villus tip, with large numbers or strips of epithelial cells in the process of sloughing. Individual cells can be seen in the process of sloughing and appear as teardrop-shaped cells, invariably containing refractile bodies, corresponding to spores (Fig. 10). This finding is quite characteristic of *E. bieneusi*. Villus enterocytes demonstrate cytopathic changes, including a cuboidal shape or pleomorphism, hyperchromatic nuclei, and loss of the basal orientation of nuclei. Cytoplasmic changes include vesiculation, vacuolization, and occasional lipid accumulation. The lamina propria may contain increased numbers of plasma cells and macrophages but no neutrophils.

Since the organisms develop in the supranuclear and apical cytoplasm, they may affect nuclear shape, producing a flattening or cupping of the apical pole (Figs. 1 and 2). The electron-lucent clefts also may be visible, especially when flanked by hematoxylin-stained material, with a cat's-eye appearance. Rarely, spores can be detected as clusters of negatively staining or refractile granular material in the cytoplasm.

ENCEPHALITOZOON INTESTINALIS

E. intestinalis is diagnosed by the same techniques as *E. bieneusi*. This organism is usually easier to detect because of its larger size, more bluish color on hematoxylin and eosin staining, greater refractivity, strong birefringence, and

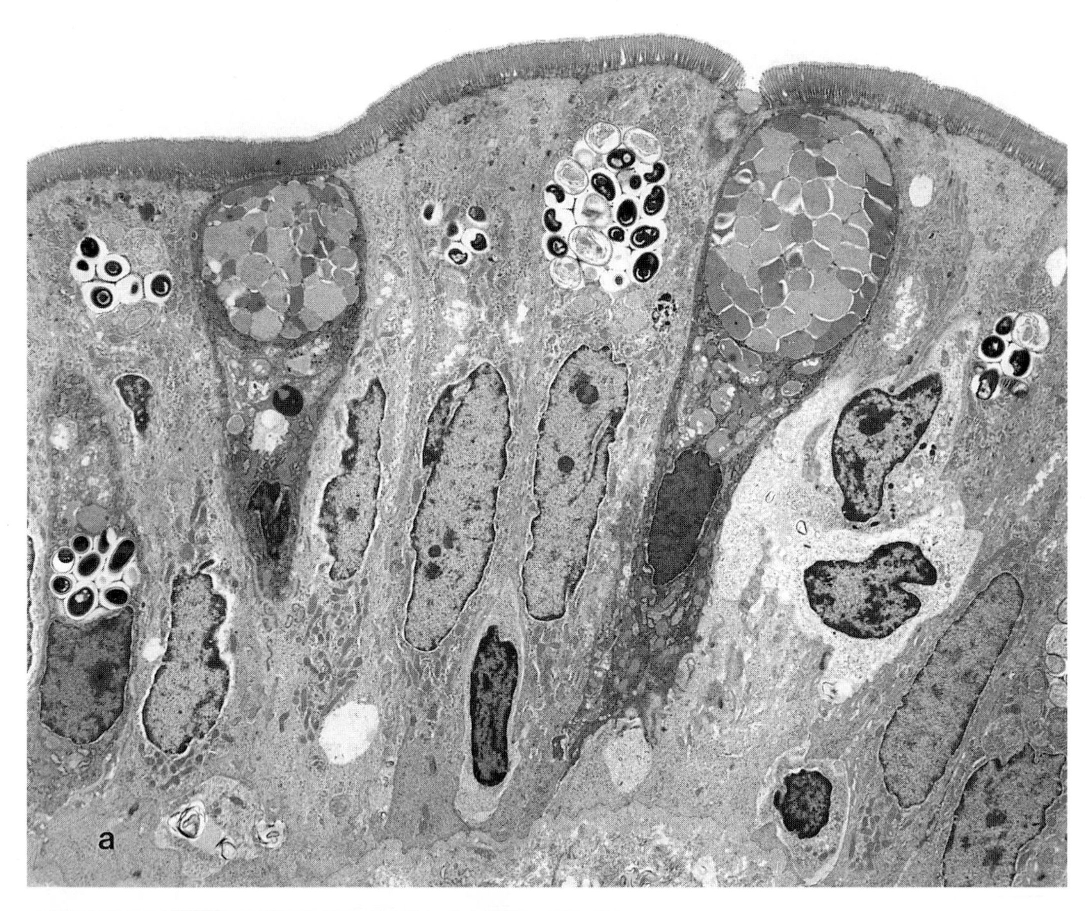

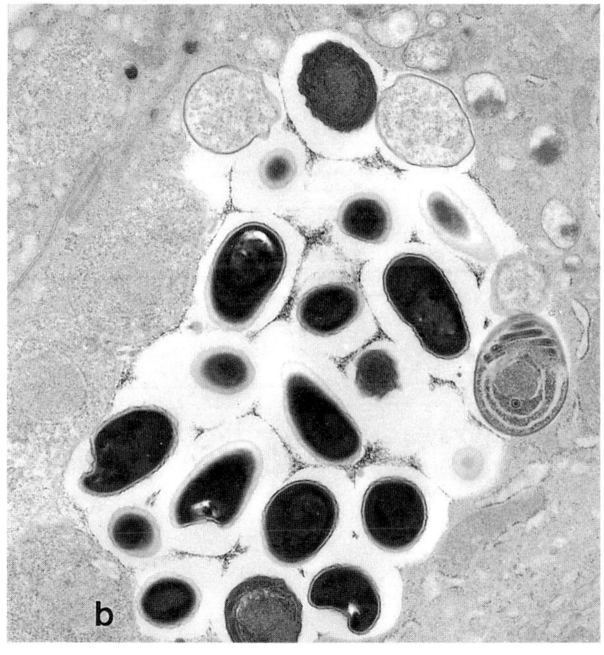

FIGURE 6 (A) Transmission electron micrograph of a jejunal biopsy, demonstrating numerous septated parasitophorous vacuoles of *Encephalitozoon intestinalis,* which are located in the Golgi-rich supranuclear cytoplasm. Note the intraepithelial lymphocytes. Magnification, ×3,270. (B) From meronts to mature spores, *E. intestinalis* development in a septated parasitophorous vacuole. Magnification, ×10,584.

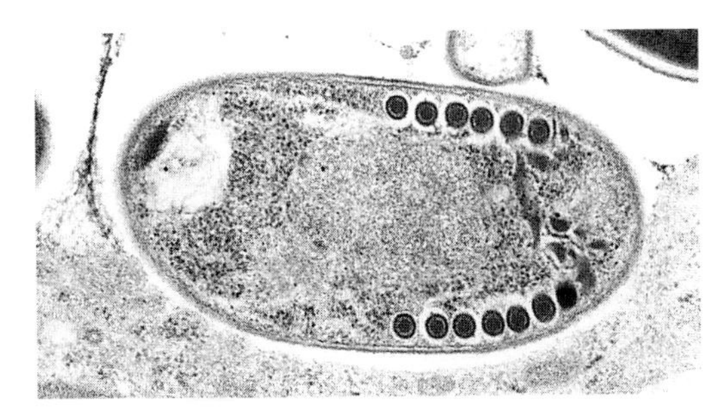

FIGURE 7 A spore of *E. intestinalis,* demonstrating the characteristic six to seven turns of the polar tubule. Magnification, ×25,000.

enhanced staining. The parasite burden is often much higher for this organism (Fig. 11). Macrophages (even endothelial cells and fibroblasts) containing spores can be seen in the lamina propria. Free spores and cells containing spores also can be found in urinary sediment (Fig. 12). In addition, there may be greater pathology, including necrosis and mucosal erosions (Fig. 13).

Special Light Microscopic Preparations and Procedures

In a suspected infection, special stains allow confirmation, especially when TEM is not readily available (Fig. 14). Useful stains include Gram, Giemsa, acid-fast, Warthin-Starry, and a modified tissue trichrome or chromotrope 2R stain as described for the examination of fecal specimens, in paraffin-embedded tissues, and Giemsa-stained touch preparations of fresh mucosal biopsies (Simon et al., 1991; Giang et al., 1993; Field et al., 1993; Kotler et al., 1994a; Franzen et al., 1995c). Fluorescence techniques have been applied to biopsy specimens, with staining for chitin, an important component of the spore (Conteas et al., 1996; Franzen et al., 1995c). Examination under polarized light is another effective means of detecting microsporidia, as the spores are birefringent, particularly in Gram- and Warthin-Starry-stained sections. The larger spores of *E. intestinalis* are brightly birefringent and stain much more intensely than *E. bieneusi* spores.

A

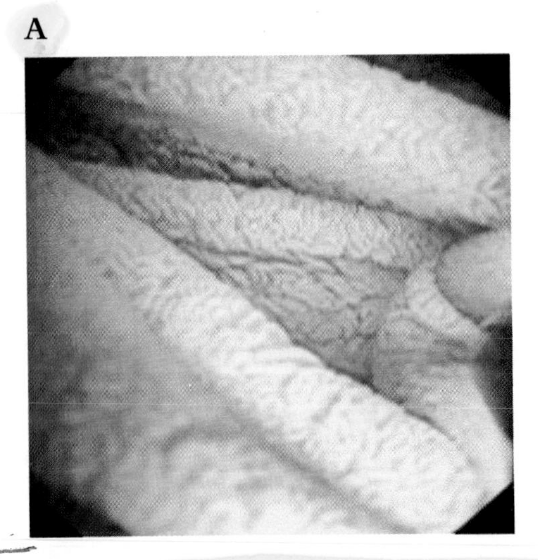

B

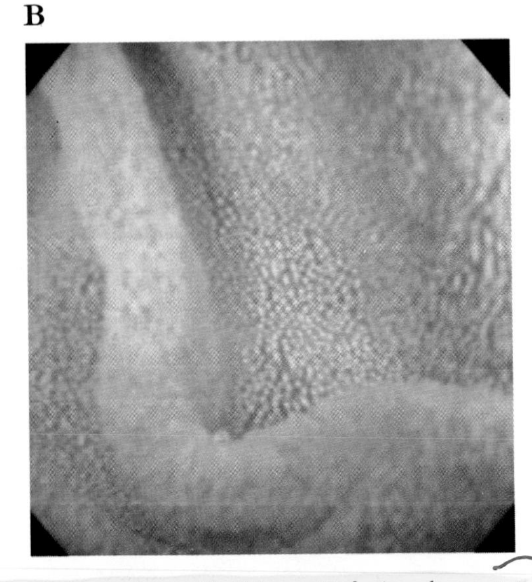

FIGURE 8 (A) Endoscopic photograph of jejunal mucosa from a patient with *E. bieneusi* infection, demonstating villus fusion. This photograph was obtained after immersing the intestinal lumen in saline. (B) Endoscopic photograph of jejunal mucosa from an uninfected subject, demonstrating normal jejunal villi.

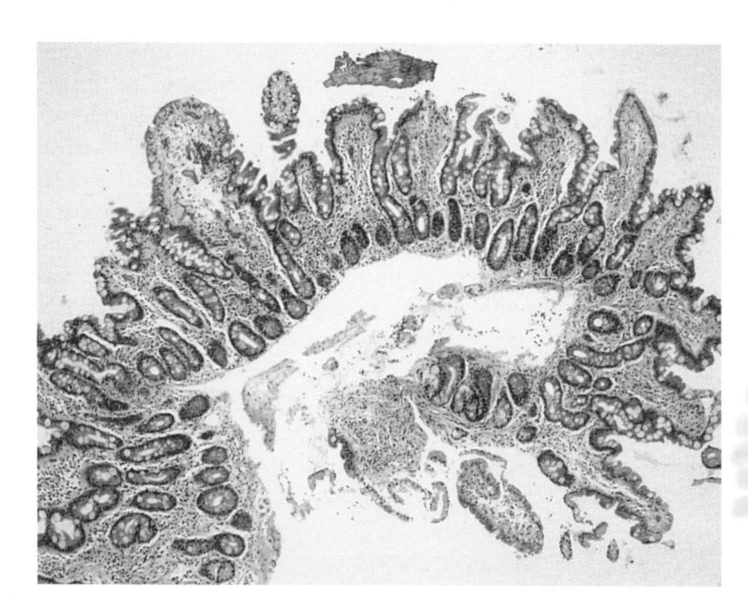

FIGURE 9 Low magnification of small bowel, showing partial villus atrophy and crypt hyperplasia. These changes are typical of microsporidial infection, although they can also be seen with cryptosporidiosis. The sample was stained with hematoxylin and eosin. Magnification, ×125.

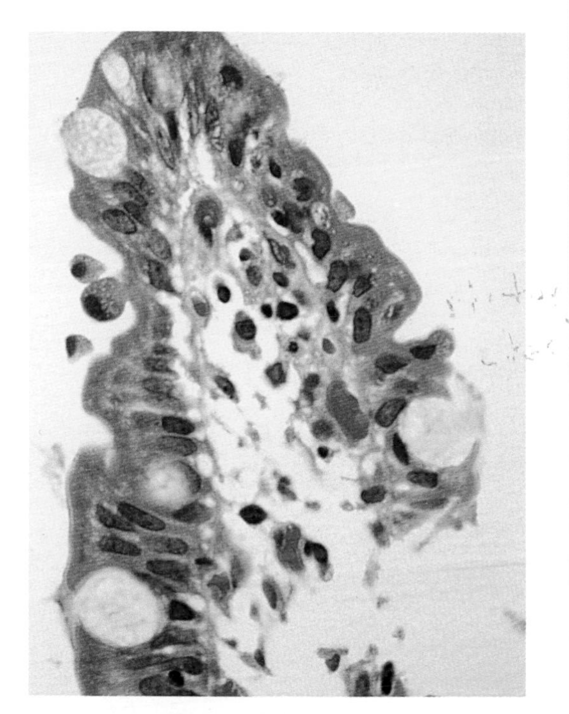

FIGURE 10 Light microscopy showing supranuclear plasmodia of *E. bieneusi* and shedding enterocytes containing mature spores. The sample was stained with hematoxylin and eosin. Magnification, ×250.

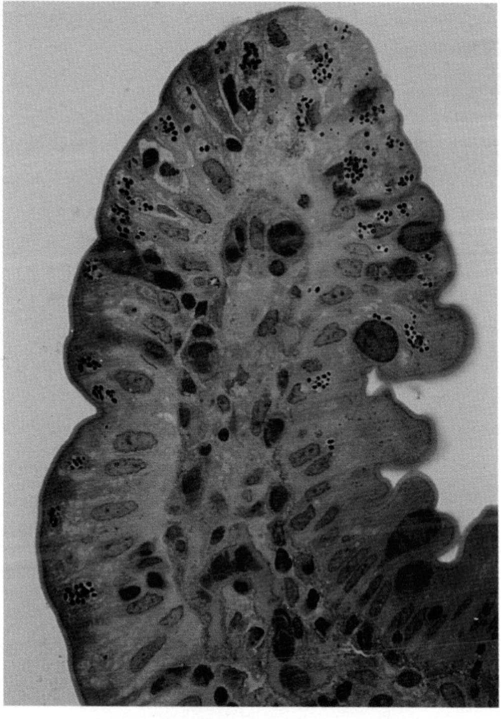

FIGURE 11 Light micrograph of villus tip in a plastic section of a patient with *E. intestinalis* infection demonstrating spores in nearly every epithelial cell near the villus tip. The sample was stained with methylene blue-azure II-fuchsin. Magnification, ×250.

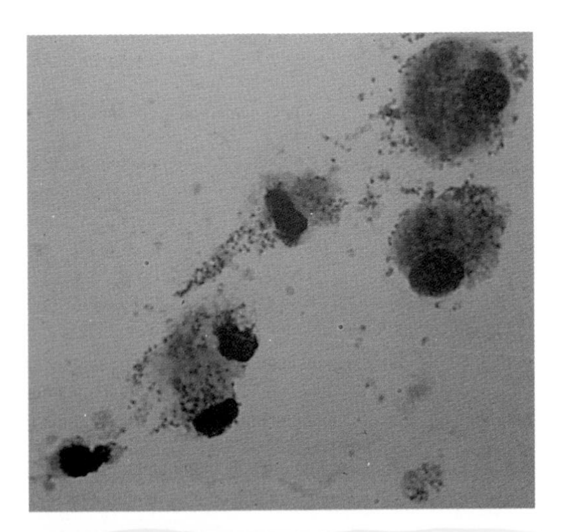

FIGURE 12 Urine sediment stained with Diff-Quik, demonstating spores of *E. intestinalis* within epithelial cells. Magnification, ×250.

We evaluated the sensitivities, specificities and predictive values for several special stains as compared to TEM methods (Kotler et al., 1994a). The sensitivities and negative predictive values of the different techniques ranged from about 60 to 90%, while the specificities and positive predictive values ranged from about 95 to 100%. A technique for quantitating the parasite burden was devised. Compared to the results with TEM, the parasite burden was lower in the false negative than in the true positive cases, suggesting that a false negative diagnosis is related to a low parasite burden. Gram and Giemsa stain spores and some other intracellular forms, whereas the chromotrope stain and acid-fast stain reveal only spores. There was a very high sensitivity for diagnosing *E. bieneusi* in hematoxylin- and eosin-stained sections in this study. The ability to correctly diagnose microsporidiosis by hematoxylin- and eosin-stained sections has been noted by several investigators (Rijpstra et al., 1988; Lucas et al., 1989; Peacock et al., 1991; Simon et al., 1991; Weber et al., 1992a, 1992b, 1994).

Samples of intraluminal fluid (DeGirolami et al., 1995) and cytobrush preparations (Orenstein et al., 1995) also are effective in the detection of microsporidia with Giemsa, Gram, Diff-Quik, and modified trichrome stains. These tests have the advantage of allowing rapid diagnosis. The chromotrope 2R stain permits differentiation between microsporidial spores and bacteria since bacteria do not stain.

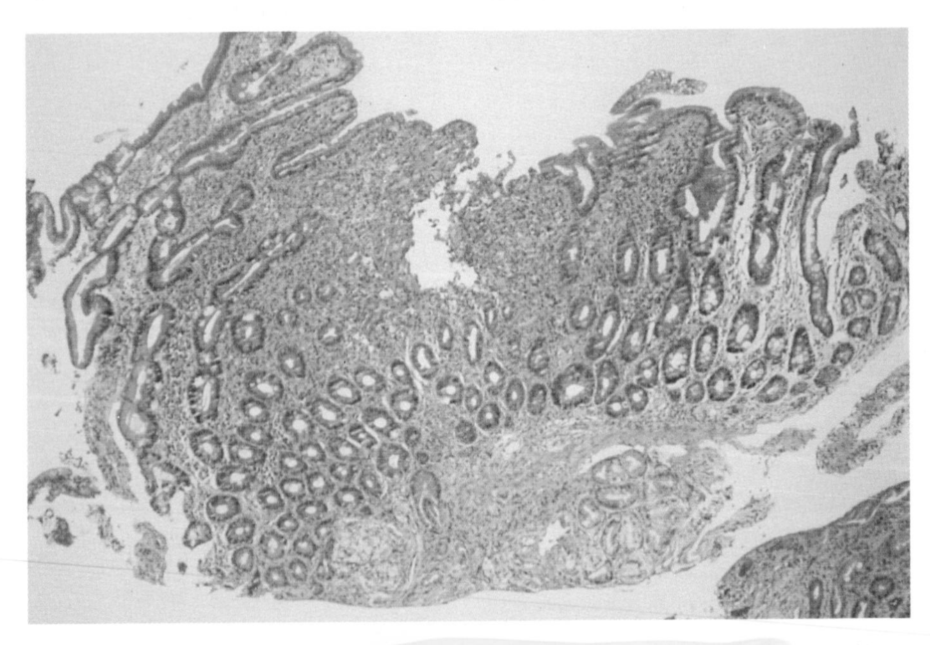

FIGURE 13 An area of ulcerated small bowel due to *E. intestinalis* infection. The sample was stained with hematoxylin and eosin. Magnification, ×125.

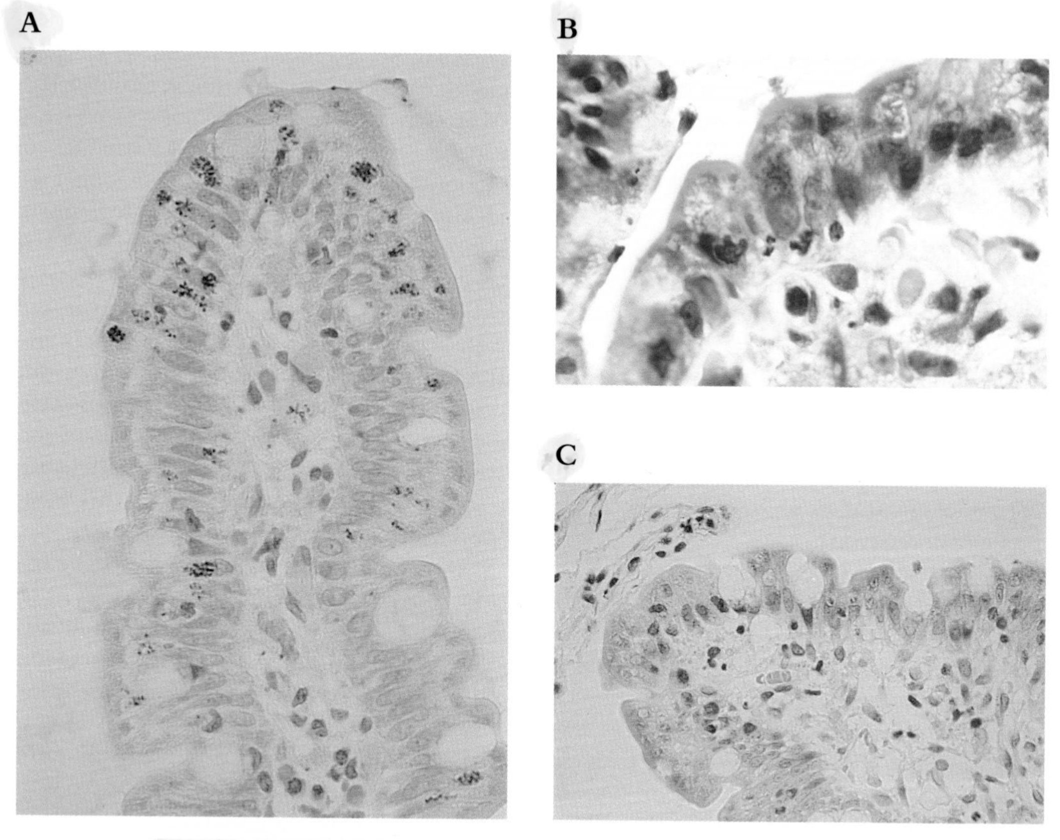

FIGURE 14 (A) Gram stain of a jejunal villus tip from a patient with *E. intestinalis* infection demonstrating gram-positive spores. Magnification, ×125. (B) Acid-fast stain of a jejunal biopsy, showing spores of *E. bieneusi*. Magnification, ×400. (C) Giemsa stain of a villus tip from a patient with *E. bieneusi* infection, showing multiple intracellular forms of *E. bieneusi*. Magnification, ×125.

Stool Examination

The need for an intestinal biopsy to diagnose microsporidiosis has limited its routine application in clinical situations and severely confounded most attempts to study the organism itself because of its inherently invasive nature. The ability to accurately diagnose microsporidiosis would greatly facilitate work in the field, and several techniques have been applied successfully. Quantitative estimation of spore excretion may be possible (Clarridge et al., 1996). Giemsa staining of a fecal preparation that was homogenized, sieved, centrifuged, and extensively washed revealed *E. bieneusi* spores (van Gool et al., 1990). TEM of stool detected spores (Orenstein et al., 1991). The chromotrope 2R modified trichrome stain also has been used to detect spores in stool (Weber et al., 1992a). Occasional yeast forms or bacteria take up the chromotrope stain but they differ from *E. bieneusi* in size and shape. The advantage of this stain is that stool specimens without special preparation or formalin fixation can be used. Staining at 56°C decreases the incubation time significantly (Bryan et al., 1991). An alternative method, using the fluorochrome Uvitex 2B or Calcifluor, is effective in staining chitin, which is a component of the wall of the microsporidial spore (van Gool et al., 1993; Franzen et al., 1995a). The modified trichrome stain and the Uvitex 2B stain were found to be approximately equivalent for the detection of microsporidia in low numbers (Ignatius et al., 1997).

Molecular Techniques for the Detection of Microsporidiosis

Several laboratories have successfully identified microsporidia in stool and other clinical specimens with a variety of techniques, such as in situ hybridization and PCR (Zhu et al., 1993a, 1993b; Weiss et al., 1994b; Franzen et al., 1995b; Coyle et al., 1996; David et al., 1996; Velasquez et al., 1996; Ombrouck et al., 1997; Liguory et al., 1997; Kock et al., 1997; Talal et al., 1998; Visvesvara et al., 1994). In addition, separate primers are available to distinguish among the different species of microsporidia. The availability of these techniques may aid in the diagnosis of microsporidiosis and facilitate epidemiologic studies on both prevalence and nonhuman reservoirs.

HEPATOBILIARY MICROSPORIDIOSIS

The epithelium of the hepatobiliary tree is vulnerable to infection with microsporidia, as well as with other protozoa and opportunistic pathogens. Autopsy studies on monkeys that died of simian AIDS have noted the frequent presence of hepatic histopathology. In one series a 30% prevalence of hepatic and small intestinal microsporidiosis was found in those animals with hepatic histopathology, however, no microsporidia were noted in animals with normal livers (Mansfield et al., 1997). The authors concluded that infection with an *E. bieneusi*-like organism occurs naturally in monkeys. The organism found very strongly resembled *E. bieneusi* by both ultrastructural and molecular analysis. It appears that hepatobiliary disease is the predominant clinical form of microsporidiosis in monkeys, while small intestinal disease is usually more significant in humans.

Bile duct injury in patients with AIDS was first described in 1983 and shows many similarities to the syndrome of sclerosing cholangitis in non-AIDS patients. Several associated infections have been identified, including cytomegalovirus, *Cryptosporidium parvum,* mycobacteria, and *Isospora belli,* as well as malignancies (Cello et al., 1989; McWhinney et al., 1991; Beaugerie et al., 1992; Orenstein et al., 1992b; Pol et al., 1993). As many as one-half of all cases do not have an associated infection. Both *E. bieneusi* and *E. intestinalis* have been reported in AIDS-related cholangitis and in acalculous cholecystitis, occurring in less than 10% of cases in most series. Pol reported a series of patients with AIDS-related cholangitis in whom no opportunistic infection could be identified but who were subsequently found to have microsporidiosis (Pol et al., 1993). Microsporidiosis is associated with AIDS-related cholangitis in about one-third of cases in our experience (Kotler, unpublished observations). Given the past difficulties in establishing the diagnosis of microsporidiosis at many centers and the finding of frequent hepatobiliary microsporidiosis in simian AIDS, it is possible that a sizable proportion of patients with AIDS-related cholangitis have microsporidiosis.

The pathogenesis of AIDS-related cholangitis is not known. Many authors believe that sclerosing cholangitis in HIV infection and other clinical situations represents a stereotypic pattern of injury in the bile duct. Suggested mechanisms include injury related to desquamation of the epithelium and exposure of the lamina propria to bile salts and other compounds in the bile, which produces an inflammatory reaction and concentric fibrosis.

The major clinical findings are abnormal liver function tests and abdominal pain suggestive of subacute cholecystitis (French et al., 1995; Ramon-Barcia and Sadowinski-Pine, 1996; Knapp et al., 1996) or biliary colic. Some patients are asymptomatic. There is no fever. Laboratory abnormalities include progressive elevation of serum alkaline phosphatase and gamma glutamyl transpeptidase levels; the bilirubin concentration is usually normal, and transaminase levels are only mildly affected. The presence of abdominal pain appears to be related to the finding of abnormalities at the ampulla of Vater, such as papillitis or papillary stenosis.

The diagnosis of AIDS-related cholangitis is based on typical findings on cholangiography (Chen and Goldberg, 1984) (Fig. 15). Cello (1989) described four distinctive patterns of AIDS-related cholangitis: papillary stenosis (15 to 20%), focal strictures and dilatation of the intra- and extrahepatic ducts (sclerosing cholangitis) (20%), combined papillary stenosis

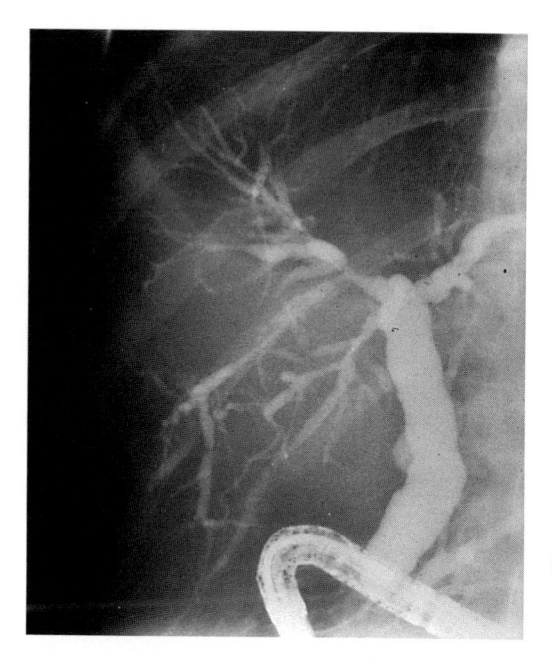

FIGURE 15 Endoscopic retrograde cholangiogram from a patient with *E. bieneusi* infection, demonstrating diffuse dilatation of the common bile duct with irregular walls, plus areas of narrowing and dilatation of the intrahepatic bile ducts.

and sclerosing cholangitis (50%), and long extrahepatic bile duct strictures (15%). Liver biopsy findings in AIDS-related cholangitis are nonspecific and may be normal in some patients. Portal fibrosis and nonspecific inflammatory cell infiltrates have been described.

There is no curative therapy for AIDS-associated cholangitis. The major therapeutic intervention is endoscopic sphincterotomy, which is associated with pain relief in a majority of patients. In other patients with findings consistent with cholecystitis, cholecystectomy by the open or laparoscopic route may bring relief of symptoms. However, the underlying process of sclerosing cholangitis is progressive. The prognosis in patients with AIDS-related cholangitis is poor, as it is a manifestation of late-stage AIDS. Short survivals have been reported and the specific causes of death have been unrelated to the cholangitis. However, prolonged survival for up to 32 months has been seen in a few patients who died either of liver failure or cachexia associated with refractory bacterial cholangitis (Kotler, unpublished observations).

Painful chronic pancreatitis was seen to develop in a pediatric patient with *E. bieneusi* infection, although histologic examinations of the pancreas or pancreatic ducts were not performed. As noted above, immunodeficient monkeys with microsporidiosis may have prominent injury of both intrahepatic and extrahepatic biliary epithelia, as well as small intestinal enterocytes, gallbladder epithelium, and pancreatic ductal epithelium.

DISSEMINATED DISEASE DUE TO *ENCEPHALITOZOON* SPECIES AND OTHER MICROSPORIDIA

Most authors agree that *E. bieneusi* infection is localized to epithelial cells in the intestine and hepatobiliary tree, though spores may occasionally be seen in the lamina propria and have been isolated from pulmonary secretions (del Aguila et al., 1997). On the other hand, patients with *E. intestinalis* infections tend to develop disseminated disease (Orenstein et al., 1992c; Cali et al., 1993; Aarons et al., 1994; Molina et al., 1995; Dore et al., 1995; Gunnarsson et al., 1995; Cowley et al., 1997). At its worst, the process produces fulminating multiorgan failure resembling the hyperinfection syndrome seen with *Strongyloides stercoralis*.

The kidney is a prominent site of injury in *E. intestinalis* infections. Although patients may be asymptomatic, some have flank pain and symptoms of urethritis (Corcoran et al., 1996; Soule et al., 1997). Renal failure has been ascribed to microsporidiosis (Aarons et al., 1994). Renal involvement has been confirmed by the identification of spores, both intracellular and free, in urinary sediment. A tubulointerstitial nephritis is present by light microscopy, but glomeruli are not involved (Fig. 16). Ultrastructural analysis shows infection of both tubule cells (containing microvilli) and transitional cells (containing cytokeratin). Spores have also been identified in most other organs, including nasal and sinus mucosae, brain (pituitary), liver, spleen, and bone (mandible), as well as in a rectal ulcer.

Several case reports and case series have described pulmonary, nasal, corneal, conjunctival, and central nervous system (CNS) infections with other microsporidia, including *E. cuniculi* and the morphologically identical *E. hellem* (Davis et al., 1990; Cali et al., 1991; Lacey et al., 1992; Schwartz et al., 1992, 1993; Weber et al., 1992a, 1993, 1997; Remadi et al., 1995; Franzen et al., 1995a, 1996b; Rossi et al., 1996; Belcher et al., 1997; Scaglia et al., 1997; Mertens et al., 1997; Orenstein et al., 1997). It is possible that *E. hellem* can be acquired by inhalation. The cerebral lesions may produce headache, vomiting, and visual and cognitive impairment. In imaging studies, multiple hypodense lesions that enhance with contrast are observed, similar to those seen in CNS toxoplasmosis. Conjunctival biopsies show cytopathic changes in infected epithelial cells plus infiltration of polymorphonuclear leukocytes. Corneal disease, which can also be caused by *Vittaforma corneae* may result in keratitis and iritis. Granulomatous hepatitis (Terada et al., 1987) and fulminant hepatic failure (Sheth et al., 1997) associated with *E. cuniculi* have been reported. Isolated cases of peritonitis and myositis also have been reported (Ledford et al., 1985; Zender et al., 1989; Hollister et al., 1996).

Disseminated infections in two patients with serious multisystem involvement including the brain, were found to be due to a *Pleistophora*-like organism in one case (Yachnis et al., 1996) which was subsequently identified as *Trachipleistophora anthropopthera* (Orenstein, unpublished observations). Both patients developed seizures and decreased mental status. Imaging studies revealed ring-enhancing lesions suggestive of CNS toxoplasmosis. The infection was located in the gray matter and was characterized by necrosis (Fig. 17A). Widespread disease involving heart, kidney, pancreas, thyroid, parathyroid, liver, bone marrow, lymph nodes, and spleen was also noted. The most heavily infected cells were epithelia, cardiac myocytes, and astrocytes (Fig. 17B). The spores were easily identified with a variety of special stains (see above). Ultrastructurally, the organism was characterized by pansporoblastic development within thick-walled sporophorous vacuoles of parasite origin. The spores were 2.0 by 2.8 μm in size and were oval to cigar-shaped. The polar tubules were arranged in a single row of six or seven thick coils and two thin coils. A similar organism, named *Trachipleistophora hominis,* has been isolated from

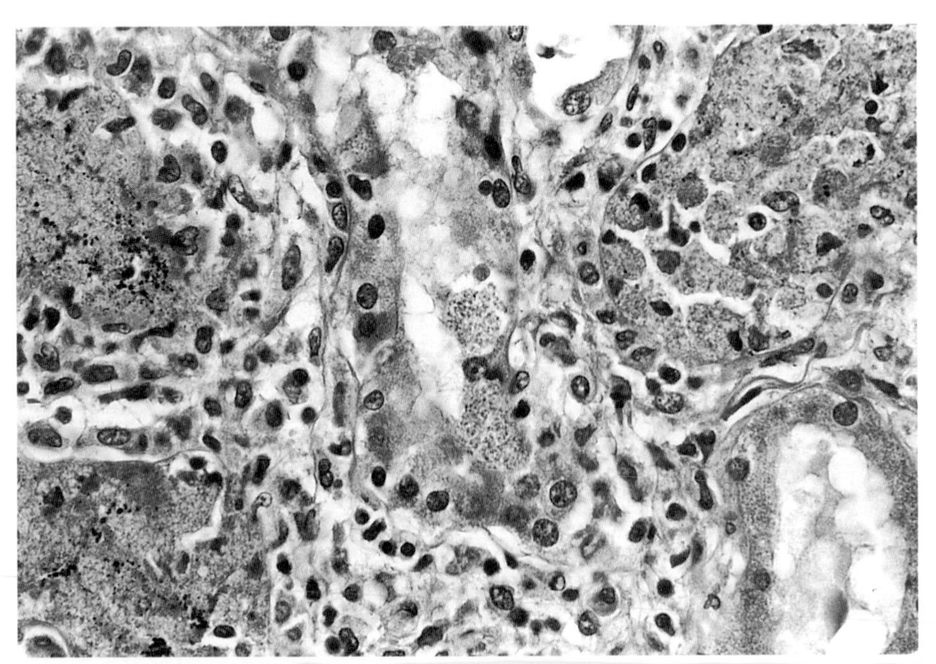

FIGURE 16 Renal tubular epithelium infected with *E. intestinalis* lyse and shed in the lumen to be subsequently found in the urinary sediment. The sample was stained with hematoxylin and eosin. Magnification, ×400.

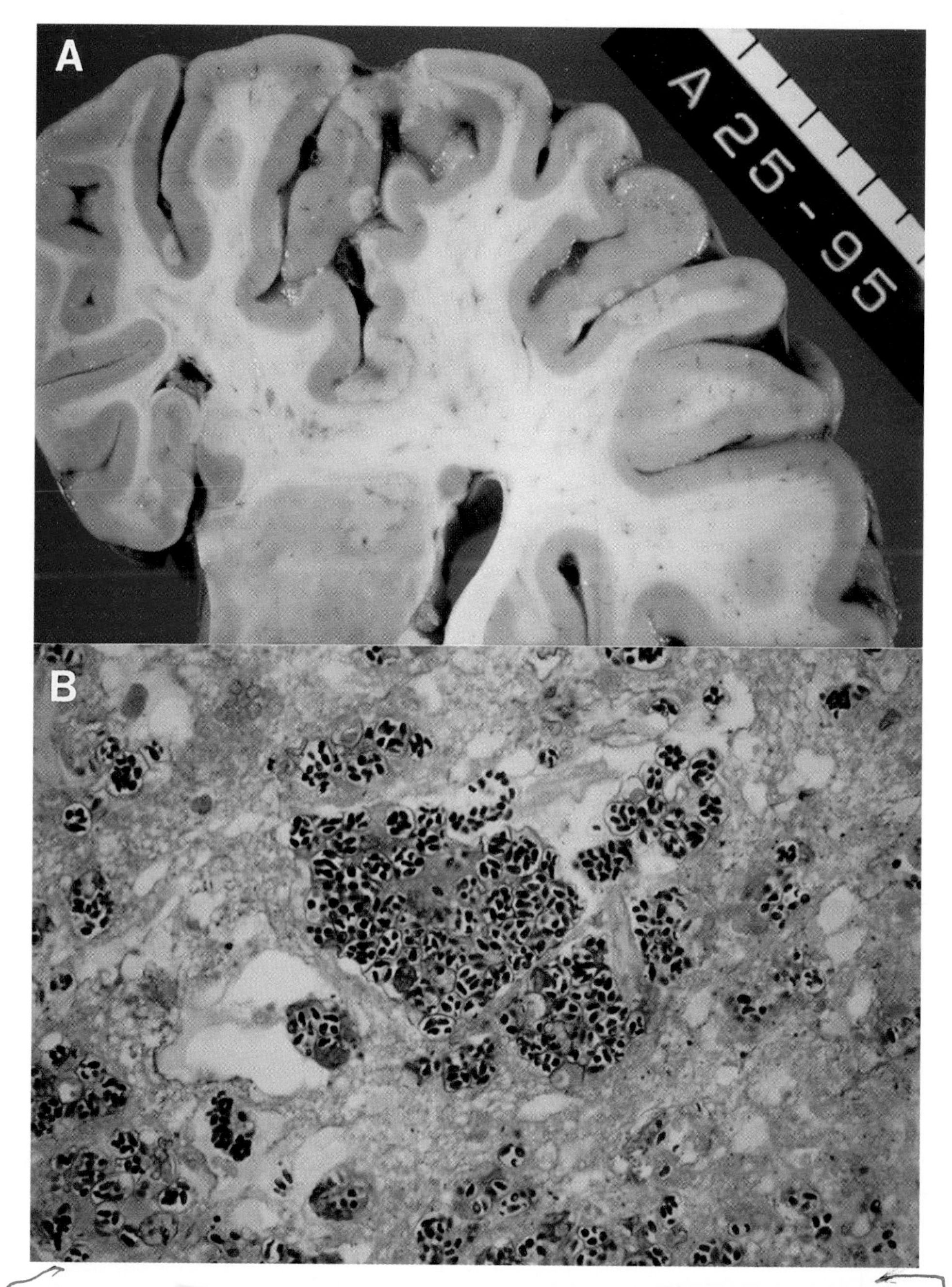

FIGURE 17 (A) Gross picture of a brain from a patient with *Trachipleistophora anthropopthera* infection demonstrating multiple necrotic lesions in the gray matter. (B) Light microscopic section from the brain shown in panel A, demonstrating spores in astrocytes and other cells. The sample was stained with Gomori methenamine-silver. Magnification, ×400.

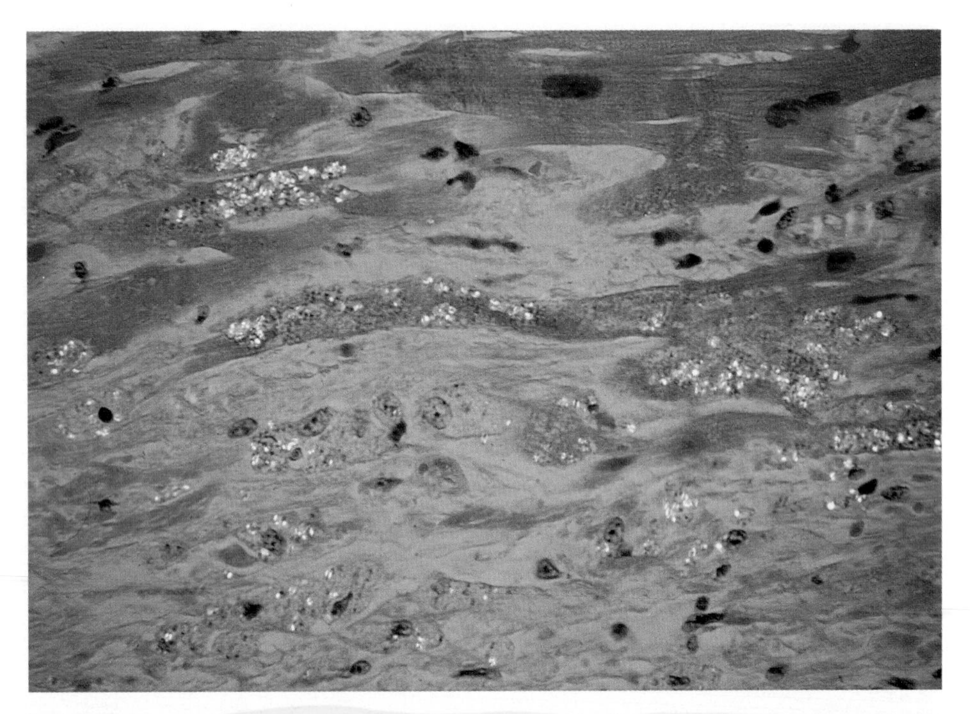

FIGURE 18 Light micrograph, under polarized light, of cardiac muscle from a patient with *Trachipleistophora* infection, demonstrating atrophy and fibrosis of the cardiac myocytes. The spores of the organism are birefringent. The sample was stained with hematoxylin and eosin. Magnification, ×250.

skeletal muscle (Fig. 18), in which infection produced degeneration, atrophy, scarring, and intense inflammation (Hollister et al., 1996).

A *Nosema*-like microsporidian organism classified as *Brachiola vesiculatum* on the basis of TEM studies was described as the etiologic agent in a case of myositis in an HIV-1 infected patient (Cali et al., 1996, 1998). Cases of myositis due to *Pleistophora* sp. have also been reported in AIDS patients (Chupp et al., 1993; Grau et al., 1996).

TREATMENT: GENERAL CONSIDERATIONS

Two agents, fumagillin and albendazole, have been shown to have activity against microsporidia both in vitro and in vivo (Table 2) (Molina et al., 1995; Dieterich et al., 1994; Gunnarsson et al., 1995; Weber et al., 1994; Dore et al., 1995; Rosenberger et al., 1993; Haque et al., 1993; Beauvais et al., 1994; Ditrich et al., 1994; Katznelson and Jamieson, 1952; Diesenhouse et al., 1993; Blanshard et al., 1993; Shadduck, 1980; Franssen and Jamieson, 1995; Lecuit et al., 1994; DiCandilo et al., 1993; Weiss et al., 1994a). Despite initial reports of favorable treatment with metronidazole for intestinal infection with *E. bieneusi,* this drug has not been effective in other studies and has no in vitro activity against *E. cuniculi* (Beauvais et al., 1994). Other agents used without success in the treatment of gastrointestinal microsporidiosis are azithromycin, paromomycin (microsporidia lack a binding site for this drug) (Katiyar et al., 1995), and quinacrine. Atovaquone has been reported to have limited efficacy in patients, however, no in vitro activity has been demonstrated (Beauvais et al., 1994; Weiss, unpublished observations). Sparfloxacin and chloroquine have shown in vitro activity against Encephalitozoonidae but have not been used clinically for these organisms (Beauvais et al., 1994).

TABLE 2 Therapy for microsporidiosis

Organism	Drug(s)	Dosage[a]
Enterocytozoon bieneusi[b]	No effective treatment; albendazole has resulted in clinical improvement in some patients; oral fumagillin is promising.	
Encephalitozoonidae (systemic infection) *Encephalitozoon cuniculi* *Encephalitozoon hellem* *Encephalitozoon intestinalis*	Albendazole	400 mg BID
Encephalitozoonidae (keratoconjunctivitis)	Fumagillin solution (Fumidil B, 3 mg/ml); patients may also need albendazole if systemic infection is present.	2 drops every 2 h for 4 days; then 2 drops 4 times a day
Trachipleistophora hominis	Albendazole	400 mg BID
Brachiola vesiculatum	Albendazole and itraconazole	400 mg BID and 400 mg QD, respectively

[a] BID, twice a day; QD, once a day.
[b] Antiretroviral treatment, which results in improvement of immune function, can result in the resolution of symptoms and spontaneous elimination of this organism by the host.

Albendazole, a benzimidazole that binds to tubulin, has activity against microsporidiosis. Albendazole is effective in inhibiting the growth of *Nosema bombycis* both in vitro in *Spodoptera frugiperda* cells and in vivo in *Heliocoverpa zea* larvae and pupae (Haque et al., 1993) In vitro, albendazole has activity against all the Encephalitozoonidae (*E. hellem, E. cuniculi* and *E. intestinalis*) at concentrations of less than 0.1 μg/ml (Haque et al., 1993; Beauvais et al., 1994; Ditrich et al., 1994; Weiss et al., 1994a; Coyle et al., 1998) and is efficacious in animal models. Other benzimidazoles such as mebendazole, fenbendazole, thiabendazole, parbendazole, and albendazole sulfoxide also are inhibitory to microsporidia at concentrations of 1 to 10 ng/ml (Katiyar et al., 1997; Weiss, unpublished observations). Other tubulin-binding drugs, such as trifluralin and oryzalin, have not shown activity in vitro (Weiss, unpublished observations). Data on the sequence of the tubulin genes of the Encephalitozoonidae demonstrate an amino acid sequence usually associated with sensitivity to benzimidazoles such as albendazole (Edlind et al., 1994, 1996).

Albendazole is poorly absorbed 2 h after an oral dose, peak levels in serum are 0.2 to 0.94 μg/ml; however, absorption is increased if the medication is taken with food containing relatively high concentrations of fat. It is protein-bound (70%); distributed into blood, bile, and cerebrospinal fluid; and eliminated by the kidneys. After oral administration albendazole is converted to albendazole sulfoxide which is detectable in the systemic circulation. While side effects are rare, the following adverse reactions have been reported: hypersensitivity (rash, pruritis, fever), neutropenia (reversible), CNS effects (dizziness, headache), gastrointestinal disturbances (abdominal pain, diarrhea [Shah et al., 1996], nausea, vomiting), hair loss (reversible), and elevated hepatic enzymes (reversible). Albendazole is not believed to be carcinogenic or mutagenic. In animals (rats and rabbits), at dosages of 30 mg/kg, it was embryotoxic and teratogenic. Thus, albendazole is not recommended for use during pregnancy. Well-controlled studies in human pregnancy have not been performed.

In vitro fumagillin inhibits the growth of *E. cuniculi, E. hellem,* and *E. intestinalis* as well as many other microsporidia (Beauvais et al., 1994; Weiss et al., 1994a; Coyle et al., 1998; Didier et al., 1997). In addition, the fumagillin analog TNP470 is active against both the Encephalitozoonidae and *V. corneae* in vitro (Didier et al., 1997; Coyle et al., 1998) and against *E. cuniculi* and *Nucleospora (Enterocytozoon) salmonis* in vivo (Coyle et al., 1998; Higgins et al., 1998). The mechanism of the action of fumagillin on microsporidia is unknown, but in endothelial cells, fumagillin binds to methionine aminopeptidase type 2 (Griffith et al., 1997). Of concern is that this agent is microsporidiostatic and not microsporidiocidal. In vitro, when fumagillin is discontinued organisms start to grow and return to pretreatment levels. In patients with *E. bieneusi* infection, oral fumagillin was reported to both decrease diarrhea and result in clearance of the organism; however, patients developed thrombocytopenia (Molina et al., 1997).

Enterocytozoon bieneusi

As noted above, the prevalence rates for microsporidiosis have fallen dramatically since the advent of combination antiretroviral agent therapy. In addition, two reports have noted the frequent development of clinical remission in patients with microsporidiosis who are started on combination antiretroviral therapy (Goguel et al., 1997; Carr et al., 1998). Thus, the current primary treatment of choice for a symptomatic patient with *E. bieneusi* infection is with combination antiretroviral agents, if available.

Relatively few drug therapy studies on *E. bieneusi* have been published. Limited symptomatic improvement during treatment with antibiotics, such as metronidazole, has been reported, although histologic evidence of infection persisted (Field et al., 1992). Other antimicrobial agents such as paromomycin and trimethoprim-sulfamethoxazole, among others, have been used (Dieterich et al., 1993; Dionisio et al., 1995). After a preliminary report on the clinical efficacy of albendazole (Blanshard et al., 1992), a prospective open-label study of 66 patients with *E. bieneusi* confirmed that albendazole therapy provided some symptomatic improvement and weight stabilization (Dieterich et al., 1994). However, follow-up studies performed in a subgroup of these patients showed that they had evidence of infection and abnormal D-xylose absorption. A double-blind, placebo-controlled trial has also been completed.

Other potentially effective therapies for microsporidiosis have been reported, including atovaquone (Anwar-Bruni et al., 1996), furazolidine (Dionisio et al., 1995), and fumagillin (Molina et al., 1997). In a study involving 10 drug regimens (albendazole plus metronidazole, sulfadiazine plus pyrimethamine, atovaquone, doxycycline plus nifuroxazide, itraconazole, flubendazole, chloroquine, paromomycin, sparfloxacin, or fumagillin) tested orally for three consecutive weeks only fumagillin was able to clear *E. bieneusi* from both stools and intestinal biopsies (Molina et al., 1997). Thalidomide (Sharpstone et al., 1995, 1997) and octreotide (Cello et al., 1991) have both been reported to decrease diarrhea in patients with microsporidiosis, probably secondary to their antisecretory effects on enterocytes.

Encephalitozoon intestinalis

In contrast to *E. bieneusi, E. intestinalis* has a uniformly excellent response to albendazole therapy (Orenstein et al., 1993; Aarons et al., 1994; Weber et al., 1994; Molina et al., 1995; Dore et al., 1995; DeGroote et al., 1995; Franzen et al., 1995b). Follow-up biopsies have shown disappearance of spores and only ghosts of spores within macrophage lysosomes. An improvement in D-xylose absorption has been noted (Orenstein et al., 1993). Autopsy studies have documented clearance of the organism after therapy (Sobottka et al., 1995; Joste et al., 1996). However, albendazole therapy has been associated with the development of pseudomembranous colitis (Shah et al., 1996).

Other Microsporidia

In cases of chronic sinusitis and disseminated infection due to *E. hellem,* treatment with 400 mg of albendazole twice daily resulted in the resolution of symptoms and clearance of the organism (Lecuit et al., 1994; Visvesvara et al., 1994). Patients with disseminated *E. cuniculi* infection of the CNS, conjunctiva, sinuses, kidney, and lungs improved clinically with albendazole therapy (Weber et al., 1997; Gordon et al., 1986). In patients with myositis caused by *T. hominis,* albendazole (400 mg

BID) resulted in clinical improvement (Field et al., 1996). In a patient with myositis due to a *Nosema*-like microsporidian (Cali et al., 1996) subsequently identified as *B. vesiculatum* (Cali et al., 1998), administration of albendazole (400 mg BID) and itraconazole (400 mg daily) resulted in clinical improvement and disappearance of the organism on subsequent muscle biopsy (Cali et al, 1998).

Supportive Therapies

The major morbidity due to intestinal microsporidiosis is chronic progressive malnutrition and chronic dehydration. Deficits in electrolytes and minerals, particularly K^+, Ca^{2+}, and Mg^{2+}, may be severe. Diet modification may be helpful in a few patients with mild to moderate disease but often does not significantly affect the clinical course. A recent study demonstrated stabilization of weight and body cell mass in patients receiving an oral semielemental diet or total parenteral nutrition (Kotler et al., 1998). Parenteral nutritional therapy resulted in nutritional repletion in some patients (Kotler et al., 1990; Melchior et al., 1996). Opiates such as diphenoxylate, paregoric, and tincture of opium may be effective, although the dose required sometimes causes excessive sedation.

CONCLUSION

Despite the growing recognition of microsporidia as a cause of disease in HIV-infected individuals and others, and the increasing number of specific microsporidial pathogens, there are major gaps in our knowledge of the disease. Information about the natural reservoirs, mode of infection, and immune response to these organisms is very limited. Except in the case of infections caused by Encephalitozoonidae, treatment efficacy is poor. At present, the brightest hope for HIV-infected patients is an improvement in immune function as a result of effective antiretroviral therapy.

ACKNOWLEDGMENTS

This work was supported by NIH grant AI21414 and by the Joyce Mertz Gilmore Foundation.

REFERENCES

Aarons, E. J., D. Woodrow, W. S. Hollister, E. U. Canning, N. Francis, and B. G. Gazzard. 1994. Reversible renal failure caused by a microsporidial infection. *AIDS* **8:**1119–1121.

Achbarou, A., D. Mazier, I. Desportes-Livage, L. Renia, F. Charlotte, T. Gneragbe, and C. Ombrouck. 1996. Experimental model for human intestinal microsporidiosis in interferon gamma receptor knockout mice infected by *Encephalitozoon intestinalis. Parasite Immunol.* **18:** 387–392.

Aldras, A. M., J. M. Orenstein, D. P. Kotler, J. A. Shadduck, and E. S. Didier. 1994. Detection of microsporidia by indirect immunofluorescence antibody test using polyclonal and monoclonal antibodies. *J. Clin. Microbiol.* **32:**608–612.

Anwar-Bruni, D. M., J. L. Lennox, R. T. Bryan, C. M. Wilcox, D. A. Schwartz, and S. E. Hogan. 1996. Atovaquone is effective treatment for the symptoms of gastrointestinal microsporidiosis in HIV-1-infected patients. *AIDS* **10:**619–623.

Asmuth, D. M., P. C. DeGirolami, M. Federman, C. R. Ezratty, D. K. Pleskow, G. Desai, and C. A. Wanke. 1994. Clinical features of microsporidiosis in patients with AIDS. *Clin. Infect. Dis.* **18:**819–825.

Babameto, G., D. P. Kotler, S. Burastero, J. Wang, and R. N. Pierson. 1994. Alterations in hydration in HIV-infected individuals. *Clin. Res.* **42:** 279A, abstr.

Beaugerie, L., M.-F. Teilhac, A.-M. Deluol, J. Fritsch, P.-M. Girard, W. Rozenbaum, Y. Le Quintrec, and F.-P. Chatelet. 1992. Cholangiopathy associated with *Microsporidia* infection of the common bile duct mucosa in a patient with HIV infection. *Ann. Intern. Med.* **117:**401–402.

Beauvais, B., C. Sarfati, S. Challier, and F. Derouin. 1994. In vitro model to assess effect of antimicrobial agents on *Encephalitozoon cuniculi. Antimicrob. Agents Chemother.* **38:**2440.

Belcher, J. W., Jr., B. M. Schmookler, and S. A. Guttenberg. 1997. Microsporidiosis of the mandible in a patient with acquired immunodeficiency syndrome. *J. Oral Maxillofac. Surg.* **55:**424–426.

Bergquist, N. R., T. Waller, S. Mravak, and U. Meyer. 1983. Report of two recent cases of human microsporidiosis. Annual Meeting, American Society of Tropical Medicine and Hygeine, San Antonio, Texas, abstr.

Bernard, E., P. Dellamonica, Y. Le Fichoux, J. F. Michiels, V. Mondain, X. Hebuterne, P. M. Roger, C. Boissy, C. Pradier, and M. Carles. 1995. Persistent diarrhea in HIV infected patients: role of *Enterocytozoon bieneusi. Presse Med.* **24:**671–674.

Bernard, E., J. F. Michiels, J. Durant, P. Hoffman, F. Desalvador, R. Loubiere, Y. Le Fichoua, and P. Dellamonica. 1991. Intestinal microsporidiosis due to *Enterocytozoon bieneusi:* a new case report in an AIDS patient. *AIDS* **5:**606–607.

Bertagne, S., F. Foulet, W. Alkassoum, J. Fleury-Feith, and M. Develoux. 1993. Prévalence des spores d'*Enterocytozoon bieneusi* dans les selles de patients sidéens et d'enfants africains non infectés par le VIH. *Bull. Soc. Pathol. Exot.* **86:**351.

Blanshard, C., D. S. Ellis, S. P. Dowell, G. Tovey, and B. G. Gazzard. 1993. Electron microscopic changes in *Enterocytozoon bieneusi* following treatment with albendazole. *J. Clin. Pathol.* **46:**898.

Blanshard, C., D. S. Ellis, D. G. Tovey, S. Dowell, and B. G. Gazzard. 1992. Treatment of intestinal microsporidiosis with albendazole in patients with AIDS. *AIDS* **6:**311–313.

Boyle, J. T., P. Celano, and O. Koldovsky. 1980. Demonstration of a difference in expression of maximal lactase and sucrase activity along the villus in the adult rat jejunum. *Gastroenterology* **79:**503–507.

Bradford-Hill, A. 1965. The environment and disease: association or causation? *Proc. R. Soc. Med.* **58:**295–300.

Brasil, P., H. Moura, H. Mattos, M.C. Gutierrez, T. Cuzzi-Maya, and F. C. Sodre. 1996. Intestinal microsporidiosis in HIV-positive patients with chronic unexplained diarrhea in Rio de Janeiro, Brazil: diagnosis, clinical presentation and follow-up. *Rev. Inst. Med. Trop. Sao Paulo* **38:**97–102.

Brunner, O., S. Edelman, and F. A. Klipstein. 1970. Intestinal morphology of rural Haitians: a comparison between overt tropical sprue and asymptomatic subjects. *Gastroenterology* **58:**655–672.

Bryan, R. T., D. A. Schwartz, and R. Weber. 1997. Microsporidiosis in patients who are not infected with human immunodeficiency virus. *Clin. Infect. Dis.* **24:**534–535. (Letter.)

Bryan, R. T., and R. Weber. 1993. Microsporidia: emerging pathogens in immunodeficient persons. *Arch. Pathol. Lab. Med.* **117:**1243–1245. (Editorial.)

Bryan, R. T., R. Weber, J. M. Stewart, P. Angritt, and G. S. Visvesvara. 1991. New manifestations and simplified diagnosis of human microsporidiosis. *Am. J. Trop. Med. Hyg.* **45:**133–134.

Burn-Murdoch, R. A., M. Fischer, and J. N. Hunt. 1978. The slowing of gastric emptying by proteins in test meals. *J. Physiol.* **274:**477–485.

Cali, A., D. P. Kotler, and J. M. Orenstein. 1993. *Septata intestinalis,* n. g., n. sp., an intestinal microsporidian associated with chronic diarrhea and dissemination in AIDS patients. *J. Eukaryot. Microbiol.* ***40:***101–112.

Cali, A., D. M. Meisler, I. Rutherford, C. Y. Lowder, J. T. McMahon, D. L. Longworth, and R. T. Bryan. 1991. Corneal microsporidiosis in a patient with AIDS. *Am. J. Trop. Med. Hyg.* **44:**463–468.

Cali, A., and R. L. Owen. 1990. Intracellular development of *Enterocytozoon,* a unique microsporidian found in the intestine of AIDS patients. *J. Protozool.* **37:**145–155.

Cali, A., P. M. Takvorian, S. Lewin, M. Rendel, C. Sian, M. Wittner, and L. M. Weiss. 1996. Identification of a new *Nosema*-like microsporidian associated with myositis in an AIDS patient. *J. Eukaryot. Microbiol.* **43:**108S.

Cali, A., P. M. Takvorian, S. Lewin, M. Tendel, C. S. Sian, M. Wittner, H. B. Tanowitz, E. Keohane, and L. M. Weiss. 1998. *Brachiola vesicularum,* n. g., n. sp., a new microsporidian associated with AIDS and myositis. *J. Eukaryot. Microbiol.* **45:**250–251.

Canning, E. U., and J. Lom. 1986. *The Microsporidia of Vertebrates.* Academic Press, Orlando, Fla.

Canning, E. U., and W. S. Hollister. 1987. Microsporidia of mammals: widespread pathogens or opportunistic curiosities? *Parasitol. Today* **3:**267–273.

Canning, E. U., and W. S. Hollister. 1990. *Enterocytozoon bieneusi* (Microspora): prevalence and pathogenicity in AIDS patients. *Trans. R. Soc. Trop. Med.* **84:**181–186.

Carr, A., D. Marriott, A. Field, E. Vasak, and D. A. Cooper. 1998. Treatment of HIV-1-associated microsporidiosis and cryptosporidiosis with combination antiretroviral therapy. *Lancet* **351:**256–261.

Cello, J. P., J. H. Grendell, P. Basuk, D. Simon, L. Weiss, M. Wittner, R. P. Rood, M. Wilcox, C. E. Forsmark, A. Read, J. A. Satow, C. S. Weikel, and R. N. Beaumont. 1991. Effect of octreotide on refractory AIDS-associated diarrhea. *Ann. Intern. Med.* **115:**705.

Cello, J. 1989. Acquired immunodeficiency syndrome cholangiopathy: spectrum of disease. *Am. J. Med.* **86:**539–546.

Chandra, R. K. 1991. 1990 McCollum Award lecture: Nutrition and immunity—lessons from the past and new insights into the future. *Am. J. Clin. Nutr.* **53:**1087–1101.

Chen, L. Y., and H. Goldberg. 1984. Sclerosing cholangitis: broad spectrum of radiographic features. *Gastrointest. Radiol.* **9:**39–46.

Chilmonczyk, S., W. T. Cox, and R. P. Hedrick. 1991. *Enterocytozoon salmonis* n. sp.: an intranuclear microsporidium from salmonid fish. *J. Protozool.* **38:**264–269.

Chupp, G. L., J. Alroy, L. S. Adelman, J. C. Breen, and P. R. Skolnik. 1993. Myositis due to

Pleistophora (Microsporidia) in a patient with AIDS. *Clin. Infect. Dis.* **16:**15.

Clarridge, J. E., III, S. Karkhanis, L. Rabeneck, B. Marino, and L. W. Foote. 1996. Quantitative light microscopic detection of *Enterocytozoon bieneusi* in stool specimens: a longitudinal study of human immunodeficiency virus-infected microsporidiosis patients. *J. Clin. Microbiol.* **34:**520–523.

Conteas, C. N., J. M. Orenstein, M. LaRiviere, J. Donovan, R. Porschen, A. Nguyen, F. Dahlan, G. W. Berlin, and T. Sowerby. 1996. Fluorescence techniques for diagnosing intestinal microsporidiosis in stool, enteric fluid, and biopsy specimens from acquired immunodeficiency syndrome patients with chronic diarrhea. *Arch. Pathol. Lab. Med.* **120:**847–853.

Corcoran, G. D., P. L. Chiodini, C. Daniels, and J. R. Isaacson. 1996. Urethritis associated with disseminated microsporidiosis: clinical response to albendazole. *Clin. Infect. Dis.* **22:**592–593.

Cotte, L., M. Rabodonirina, A. Piens, M. Perreard, M. Mofon, and C. Trepo. 1993. Prevalence of intestinal protozoons in French patients infected with HIV. *J. Acquir. Immune Defic. Syndr. Hum. Retroviral.* **6:**1024–1029.

Cotte, L., M. Rabodonirina, C. Raynal, H. Chapius, M. A. Piens, and C. Trepo. 1998. Outbreak of intestinal microsporidiosis in HIV-infected and non-infected patients, abstract 483. In *Fifth National Conference on Human Retroviruses and Opportunistic Infections.*

Cowley, G. P., S. B. Lucas, E. U. Canning, L. Papadaki, and R. F. Miller. 1997. Disseminated microsporidiosis in a patient with acquired immunodeficiency syndrome. *Histopathology* **30:**386–389.

Coyle, C. M., J. M. Orenstein, M. Wittner, D. P. Kotler, C. Noyer, H. Tanowitz, and L. M. Weiss. 1996. Prevalence of microsporidiosis in AIDS related diarrhea as determined by polymerase chain reaction to microsporidian ribosomal RNA. *Clin. Infect. Dis.* **23:**1002–1006.

Coyle, C, M. Dent, H. B. Tanowitz, M. Wittner, and L. M. Weiss. 1998. TNP-470 is an effective antimicrosporidial agent. *J. Infect. Dis.* **177:**515–518.

Curry, A., L. J. McWilliam, N. Y. Haboubi, and B. K. Mandal. 1988. Microsporidiosis in a British patient with AIDS. *Br. Med. J.* **41:**477–478.

David, F., J. M. Molina, F. Derouin, R. A. Hartskeerl, O. Liguory, C. Sarfati, and A. R. Schuitema. 1996. Detection and species identification of intestinal microsporidia by polymerase chain reaction in duodenal biopsies from human immunodeficiency virus-infected patients. *J. Infect. Dis.* **174:**874–877.

Davis, R. M., R. L. Font, M. S. Keisler, and J. A. Shadduck. 1990. Corneal microsporidiosis: a case report including ultrastructural observations. *Ophthalmology* **97:**953–957.

DeGirolami, P. C., C. R. Ezratty, G. Desal, A. McCullough, D. Asmuth, C. Wanke, and M. Federman. 1995. Diagnosis of intestinal microsporidiosis by examination of stool and duodenal aspirate with Weber's modified trichrome and Unitex 2B stains. *J. Clin. Microbiol.* **33:**805–810.

DeGroote, M. A., R. Reves, R. T. Bryan, G. J. Leitch, A. J. daSilva, S. B. Slemenda, N. J. Pieniazek, M. L. Wilson, and G. Visvesvara. 1995. Polymerase chain reaction and culture confirmation of disseminated *Encephalitozoon cuniculi* in a patient with AIDS: successful therapy with albendazole. *J. Infect. Dis.* **171:**1375–1378.

del Aguila, C., N. J. Pieniazek, A. J. Da Silva, G. P. Croppo, G. S. Visvesvara, R. Navajas, J. Cobo, C. Turrientes, S. Fenoy, and R. Lopez-Velez. 1997. Identification of *Enterocytozoon bieneusi* spores in respiratory samples from an AIDS patient with a 2-year history of intestinal microsporidiosis. *J. Clin. Microbiol.* **35:**1862–1866.

Deplazes, P., A. Mathis, C. Muller, and R. Weber. 1996. Molecular epidemiology of *Encephalitozoon cuniculi* and first detection of *Enterocytozoon bieneusi* in fecal samples of pigs. *J. Eukaryot. Microbiol.* **43:**93S.

Desportes, I., Y. Le Charpentier, A. Galian, F. Bernard, B. Cochand-Priollet, A. Lavergne, P. Ravisse, and R. Modigliani. 1985. Occurrence of a new microsporidan: *Enterocytozoon bieneusi* n. g., n. sp., in the enterocytes of a human patient with AIDS. *J. Protozool.* **32:**250–254.

DiCandilo, F., G. Bassotti, M. Marroni, F. Baldelli, and A. Morelli. 1993. *Enterocytozoon bieneusi* detection in a patient with human immunodeficiency virus infection and chronic diarrhea: response to albendazole treatment. *Ital. J. Gastroenterol.* **25:**321.

Didier, E. S., D. P. Kotler, D. T. Dieterich, J. M. Orenstein, A. M. Aldras, R. Davis, D. N. Friedberg, W. K. Gourley, R. Lembach, C. Y. Lowder, D. M. Meisler, I. Rutherford, R. W. Yee, and J. A. Shadduck. 1993. Serologic studies in human microsporidiosis. *AIDS* **7**(Suppl. 3):S8–11.

Didier, E. S., P. W. Varner, P. J. Didier, A. M. Aldras, N. J. Millichamp, M. Murphy-Korb, R. Bohm, and J. A. Shadduck. 1994. Experimental microsporidiosis in immunocompetent and immunodeficient mice and monkeys. *Folia Parasitol.* **41:**1–11.

Diesenhouse, M. C., L. A. Wilson, G. F. Corrent, G. S. Visvesvara, H. E. Grossniklaus, and R. T.

Bryan. 1993. Treatment of microsporidial keratoconjunctivitis with topical fumagillin. *Am. J. Opthalmol.* **115:**293.

Dieterich, D. T., E. Lew, D. P. Kotler, M. Poles, and J. M. Orenstein. 1993. Divergence between clinical and histologic responses during treatment of *Enterocytozoon bieneusi* infection with albendazole: prospective study and review of the literature. *AIDS* **7**(Suppl. 3)**:**S43–S44.

Dieterich, D. T., E. Lew, D. P. Kotler, M. Poles, and J. M. Orenstein. 1994. Treatment with albendazole for intestinal disease due to *Enterocytozoon bieneusi* in patients with AIDS. *J. Infect. Dis.* **169:**173–183.

Dionisio, D., G. Sterrantino, M. Meli, M. Trotta, D. Milo, and F. Leoncini. 1995. Use of furazolidone for the treatment of microsporidiosis due to *Enterocytozoon bieneusi* in patients with AIDS. *Rec. Progr. Med.* **86:**394–397.

Ditrich, O., Z. Kucerova, and B. Koudela. 1994. In vitro sensitivity of *Encephalitzoon cuniculi* and *E. hellem* to albendazole *J. Eukaryot. Microbiol.* **41:**37S.

Dobbins, W., and W. M. Weinstein. 1985. Electron microscopy of the intestine and rectum in acquired immunodeficiency syndrome. *Gastroenterology* **88:**738–749.

Dore, G. J., D. J. Marriott, M. C. Hing, J. L. Harkness, and A. S. Field. 1995. Disseminated microsporidiosis due to *Septata intestinalis* in nine patients infected with the human immunodeficiency virus: response to therapy with albendazole. *Clin. Infect. Dis.* **21:**70–76.

Drobniewski, F., P. Kelly, A. Carew, B. Ngwenya, N. Luo, C. Pankhurst, and M. Farthing. 1995. Human microsporidiosis in African AIDS patients with chronic diarrhea. *J. Infect. Dis.* **171:**515–516.

Edlind, T., G. Visvesvara, J. Li, and S. Katiyar. 1994. Cryptosporidium and microsporidial beta-tubulin sequences: predictions of benzimidazole sensitivity and phylogeny. *J. Eukaryot. Microbiol.* **41:**38S.

Edlind, T. D., J. Li, G. S. Visvesvara, M. H. Vodkin, G. L. McLaughlin, and S. K. Katiyar. 1996. Phylogenetic analysis of beta-tubulin sequences from amitochondrial protozoa. *Mol. Phylogenet. Evol.* **5:**359.

Eeftinck Schattenkerk, J. K. M., T. van Gool, R. J. van Ketel, J. F. W. M. Bartelsman, C. Kuiken, W. J. Terpstra, and P. Reiss. 1991. Clinical significance of small-intestinal microsporidiosis in HIV-1 infected individuals. *Lancet* **337:**895–898.

Everts, R., C. Newhook, G. Paltridge, and S. T. Chambers. 1997. Microsporidiosis in New Zealand. *N. Z. Med. J.* **110:**83. (Letter.)

Fedorko, D. P., N. A. Nelson, and C. P. Cartwright. 1995. Identification of microsporidia in stool specimens by using PCR and restriction endonucleases. *J. Clin. Microbiol.* **33:** 1739–1741.

Field, A. S., A. Harkness, and D. Marriott. 1992. Enteric microsporidiosis: incidence and response to albendazole or metronidazole. VII International Conference on AIDS, Amsterdam, abstr. PoB 3344.

Field, A. S., M. Hing, S. T. Milliken, and D. J. Marriott. 1990. Microsporidia in the small intestine of HIV-infected patients. *Med. J. Aust.* **158:**390–394.

Field, A. S., D. J. Marriott, and M. C. Hing. 1993. The Warthin-Starry stain in the diagnosis of small intestinal microsporidiosis in HIV-infected patients. *Folia Parasitol.* **40:**261–266.

Field, A. D., D. J. Marriott, S. T. Milliken, B. J. Brew, E. U. Canning, J. G. Kench, P. Darveniza, and J. L. Harkness. 1996. Myositis associated with a newly described microsporidian, *Trachipleistophora hominis,* in a patient with AIDS. *J. Clin. Microbid.* **34:**2803.

Franssen, F. F. J., J. T. Lumeij, and F. van Knapen. 1995. Susceptibility of *Encephalitozoon cuniculi* to several drugs in vitro. *Antimicrob. Agents. Chemother.* **39:**1265–1268.

Franzen, C., V. Diehl, G. Mahrle, P. Hartmann, G. Fatkenheuer, B. Salzberger, A. Schwenk, A. Muller, G. S. Visvesvara, and D. A. Schwartz. 1995a. Immunologically confirmed disseminated, asymptomatic *Encephalitozoon cuniculi* infection of the gastrointestinal tract in a patient with AIDS. *Clin. Infect. Dis.* **21:**1480–1484.

Franzen, C., M. Schrappe, V. Diehl, G. Mahrle, G. Fatkenheuer, B. Salzberger, A. Schwenk, and A. Muller. 1995b. Intestinal microsporidiosis with *Septata intestinalis* in a patient with AIDS: response to albendazole. *J. Infect.* **31:**237–239.

Franzen, C., A. Muller, B. Salzberger, G. Fatkenheuer, S. Eidt, G. Mahrle, V. Diehl, and M. Schrappe. 1995c. Tissue diagnosis of intestinal microsporidiosis using a fluorescent stain with Uvitex 2B. *J. Clin. Pathol.* **48:**1009–1010.

Franzen, C., A. Muller, P. Hegener, B. Salzberger, P. Hartmann, G. Fatkenheuer, V. Diehl, and M. Schrappe. 1995d. Detection of microsporidia (*Enterocytozoon bieneusi*) in intestinal biopsy specimens from human immunodeficiency virus-infected patients by PCR. *J. Clin. Microbiol.* **33:**2294–2296.

Franzen, C., M. Schrappe, V. Diehl, B. Vetten, G. Fatkenheuer, B. Salzberger, A. Muller, and R. Kuppers. 1996a. Genetic evidence for latent *Septata intestinalis* infection in human immunodeficiency virus-infected patients with intestinal microsporidiosis. *J. Infect. Dis.* **173:**1038–1040.

Franzen, C., A. Muller, B. Salzberger, G. Fatkenheuer, V. Diehl, and M. Schrappe. 1996b.

Chronic rhinosinusitis in patients with AIDS: potential role of microsporidia. *AIDS* **10:**687–688.

Franzen, C., G. Fatkenheuer, V. Diehl, B. Franzen, B. Salzberger, P. Hartmann, P. Hegener, and A. Muller. 1996c. Polymerase chain reaction for microsporidian DNA in gastrointestinal biopsy specimens of HIV-infected patients. *AIDS* **10:**323–327.

Fredericks, D. N., and D. A. Relman. 1996. Sequence-based identification of microbial pathogens: a reconsideration of Koch's postulates. *Clin. Microbiol. Rev.* **9:**18–31.

French, A. L., L. M. Beaudet, D. A. Benator, C. S. Levy, M. Kass, and J. M. Orenstein. 1995. Cholecystectomy in patients with AIDS: clinicopathologic correlations in 107 cases. *Clin. Infect. Dis.* **21:**852–858.

Giang, T., D. P. Kotler, M. L. Garro, and J. M. Orenstein. 1993. Tissue diagnosis of intestinal microsporidiosis using the chromotrope-2R trichrome stain. *J. Clin. Pathol.* **117:**1249–1253.

Goguel, J., C. Katlama, C. Sarfati, C. Maslo, C. Leport, and J. C. Molina. 1997. Remission of AIDS-associated intestinal microsporidiosis with highly active antiretroviral therapy. *AIDS* **11:** 1638–1641.

Gordon, S.C., K. R. Reddy, E. E. Gould, R. McFadden, C. O'Brien, M. DeMedina, L. J. Jeffers, and E. R. Schiff. 1986. The spectrum of liver disease in the acquired immunodeficiency syndrome. *J. Hepatol.* **2:**475.

Grau, A., M. E. Valls, J. E. Willaina, D. S. Ellis, M. J. Muntaine, and C. Nadal. 1996. Myositis caused by *Pleistophora* in a patient with AIDS. *Med. Clin.* (Barcelona) **107:**779.

Greenson, J., P. Belitsos, J. Yardley, and J. Bartlett. 1991. AIDS enteropathy: occult enteric infections and duodenal mucosal alterations in chronic diarrhea. *Ann. Intern. Med.* **114:**366–372.

Griffith, E. C., S. Su, B. E. Turk, S. Chen, Y. H. Chang, Z. Wu, K. Biemann, and J. O. Liu. 1997. Methionine aminopeptidase (type 2) is the common target for angiogenesis inhibitors AGM-1470 and ovalicin. *Chem. Biol.* **4:**461–471.

Gunnarsson, G., D. Hurlbut, P. C. DiGirolami, M. Federman, and C. Wanke. 1995. Multiorgan microsporidiosis: report of five cases and review. *Clin. Infect. Dis.* **21:**37–44.

Haque, M. A., W. S. Hollister, A. Willcox, and E. U. Canning. 1993. The antimicrosporidial activity of albendazole. *J. Invertebr. Pathol.* **62:**171.

Hautvast, J. L., R.W. Sauerwein, P. Beckers, T. J. Derks, and J. J. Tolboom. 1997. Asymptomatic intestinal microsporidiosis in a human immunodeficiency virus-seronegative, immunocompetent Zambian child. *Pediatr. Infect. Dis. J.* **16:**415–416.

Higgins, M. J., M. L. Kent, J. D. Noran, and L. M. Weiss. 1998. Efficacy of the fumagillin analog TNP-470 for *Nucleospora salmonis* and *Loma salmonae* infections in chinook salmon (*Oncorhynchus tshawytscha*). *Dis. Aquat. Org.* **34:**45–49.

Hollister, W. S., E. U. Canning, E. Wiedner, A. S. Field, J. Kench, and D. J. Marriott. 1996. Development and ultrastructure of *Trachipleistophora hominis* n. g. n. sp. after *in vitro* isolation from an AIDS patient and inoculation into athymic mice. *Parasitology* **112:**143–154.

Ignatius, R., R. Ullrich, H. Hahn, E. O. Riecken, U. Futh, W. Heise, T. Schneider, S. Koppe, W. Schmidt, U. Mansmann, O. Liesenfeld, and S. Henschel. 1997. Comparative evaluation of modified trichrome and Uvitex 2B stains for detection of low numbers of microsporidial spores in stool specimens. *J. Clin. Microbiol.* **35:**2266–2269.

Johnson, L.R. 1988. Regulation of gastrointestinal mucosal growth. *Physiol. Rev.* **68:**456–469.

Joste, N. E., D. A. Schwartz, K. J. Busam, and J. D. Rich. 1996. Autopsy verification of *Encephalitozoon intestinalis* (microsporidiosis) eradication following albendazole therapy. *Arch. Pathol. Lab. Med.* **120:**199–203.

Katiyar, S. K., and T. D. Edlind. 1997. In vitro susceptibilities of the AIDS-associated microsporidian *Encephalitozoon intestinalis* to albendazole, its sulfoxide metabolite, and 12 additional benzimidazole derivatives. *Antimicrob. Agents Chemother.* **41:**2729–2732.

Katiyar, S. K., G. S. Visvesvara, and T. D. Edlind. 1995. Comparisons of ribosomal RNA sequences from amitochondrial protozoa: implications for processing, mRNA binding, and paromomycin susceptibility. *Gene* **152:**27.

Katznelson, H., and C. A. Jamieson. 1952. Control of *Nosema* disease of honeybees with fumagillin. *Science* **115:**70.

Kelkar, R., P. S. Sastry, S. S. Kulkarni, T. K. Saikia, P. M. Parikh, and S. H. Advani. 1997. Pulmonary microsporidial infection in a patient with CML undergoing allogeneic marrow transplant. *Bone Marrow Transplant.* **19:**170.

Knapp, P. E., P. Saltzman, and P. Fairchild. 1996. Acalculous cholecystitis associated with microsporidial infection in a patient with AIDS. *Clin. Infect. Dis.* **22:**195–196.

Kock, N. P., J. Schottelius, H. Albrecht, N. J. Pieniazek, P. Deplazes, C. Schmetz, I. Sobottka, T. Fenner, and H. Petersen. 1997. Species-specific identification of microsporidia in stool and intestinal biopsy specimens by the polymerase chain reaction. *Eur. J. Clin. Microbiol. Infect. Dis.* **16:**369–376.

Kondova, I., K. Mansfield, M. A. Buckholt, B. Stein, G. Widmer, A. Carville, A. Lackner,

and S. Tzipori. 1998. Transmission and serial propagation of *Enterocytozoon bieneusi* from humans and rhesus macaques in gnotobiotic piglets. *Infect. Immun.* **66:**5515–5519.

Kotler, D. P. Unpublished observations.

Kotler, D. P., A. R. Tierney, S. K. Brenner, S. Couture, J. Wang, and R. N. Pierson, Jr. 1990a. Preservation of short-term energy balance in clinically stable patients with AIDS. *Am. J. Clin. Nutr.* **57:**7–13.

Kotler, D. P., A. Francisco, F. Clayton, J. Scholes, and J. M. Orenstein. 1990b. Small intestinal injury and parasitic disease in AIDS. *Ann. Intern. Med.* **113:**444–449.

Kotler, D. P., A. R. Tierney, J. Wang, and R. N. Pierson, Jr. 1990c. Effect of home total parenteral nutrition upon body composition in AIDS. *J. Parenteral Enteral Nutr.* **14:**454–458.

Kotler, D. P. 1991. Gastrointestinal complications of the acquired immunodeficiency syndrome, p. 86–103. *In* T. Yamada (ed.), *Textbook of Gastroenterology.* Lippincott, Philadelphia, Pa.

Kotler, D. P., S. Reka, K. Chow, and J. M. Orenstein. 1993. Effects of enteric parasitoses and HIV infection upon small intestinal structure and function in patients with AIDS. *J. Clin. Gastroenterol.* **6:**10–15.

Kotler, D. P., T. T. Giang, M. L. Garro, and J. M. Orenstein. 1994a. Light microscopic diagnosis of microsporidiosis in patients with AIDS. *Am. J. Gastroenterol.* **89:**540–544.

Kotler, D. P., and J. M. Orenstein. 1994b. Prevalence of enteric pathogens in HIV-infected individuals referred for gastrointestinal evaluation. *Am. J. Gastroenterol.* **89:**1998–2002.

Kotler, D. P., L. Fogleman, and A. R. Tierney. 1998. Comparison of total parenteral nutrition and an oral, semielemental diet on body composition, physical function, and nutrition-related costs in patients with malabsorption due to acquired immunodeficiency syndrome. *J. Parenter. Enteral Nutr.* **22:**120–126.

Lacey, C. J. N., A. M. T. Clarke, P. Fraser, T. Metcalfe, G. Bonsor, and A. Curry. 1992. Chronic microsporidian infection of the nasal mucosae, sinuses and conjunctivae in HIV disease. *Genitourin. Med.* **68:**179–181.

Lambl, B. B., M. Federman, D. Pleskow, and C. A. Wanke. 1996. Malabsorption and wasting in AIDS patients with microsporidia and pathogen-negative diarrhea. *AIDS* **10:**739–44.

Lecuit, M., E. Oksenhendler, and C. Sarfati. 1994. Use of albendazole for disseminated microsporidian infection in a patient with AIDS. *Clin. Infect. Dis.* **19:**332.

Ledford, D. M., M. D. Overman, A. Gonzalvo, A. Cali, S. W. Mester, and R. F. Lockey. 1985. Microsporidiosis myositis in a patient with the acquired immunodeficiency syndrome. *Ann. Intern. Med.* **102:**628–629.

Levaditi, C., S. Nicolau, and R. Schoen. 1923. L'étiologie de l'encéphalite. *C. R. Acad. Sci.* **177:**985–988.

Liguory, O., J. M. Molina, J. Modai, F. Derouin, R. A. Hartskeerl, A. R. Schuitema, C. Sarfati, and F. David. 1997. Diagnosis of infections caused by *Enterocytozoon bieneusi* and *Encephalitozoon intestinalis* using polymerase chain reaction in stool specimens. *AIDS* **11:**723–726.

Lima, A. A., R. L. Guerrant, D. P. Fedorko, J. W. Fox, Y. Bao, I. T. McAuliffe, L. J. Barrett, A. M. Gifoni, and T. M. Silva. 1997. Mucosal injury and disruption of intestinal barrier function in HIV-infected individuals with and without diarrhea and cryptosporidiosis in northeast Brazil. *Am. J. Gastroenterol.* **92:**1861–1866.

Lucas, S. B., L. Papadaki, C. Conlon, N. Sewankambo, R. Goodgame, and D. Serwadda. 1989. Diagnosis of intestinal microsporidiosis in patients with AIDS. *J. Clin. Pathol.* **42:**885–887.

Maiga, I., E. Pichard, M. Gentilini, A. Datry, Y. el Fakhry, L. Kassambara, L. Maiga, E. Giboyau, I. Hilmarsdottir, I. Desportes-Livage, H. Traore, M. Dembele, and O. Doumbo. 1997. Human intestinal microsporidiosis in Bamako (Mali): the presence of *Enterocytozoon bieneusi* in HIV seropositive patients. *Sante* **7:**257–262.

Mansfield, K., A. Carville, D. Shvetz, J. MacKay, S. Tzipori, and A. Lackner. 1997. Identification of an *Enterocytozoon bieneusi*-like microsporidian parasite in simian immunodeficiency virus-inoculated macaques with hepatobiliary disease. *Am. J. Pathol.* **150:**1395–1405.

Marangi, A., O. Brandonisio, C. Romanelli, S. Lisi, G. Pastore, G. Angarano, M. A. Panaro, and P. Maggi. 1995. Intestinal microsporidiosis in AIDS patients with diarrhoeal illness in Apulia (south Italy). *New Microbiol.* **18:**435–439.

Matsubayashi, H., T. Koike, I. Mikata, H. Takei, and S. Hagiwara. 1959. A case of *Encephalitozoon*-like infection in man. *Arch. Pathol.* **67:**181–187.

McWhinney, P. H. M., D. Nathwani, S. T. Green, J. F. Boyd, and J. A. H. Forrest. 1991. Microsporidiosis detected in association with AIDS-related sclerosing cholangitis. *AIDS* **5:**1394–1395.

Melchior, J. C., C. Chastang, P. Gelas, F. Carbonnel, J. F. Zazzo, A. Boulier, J. Cosnes, B. Bouletreau, and B. Messing. 1996. Efficacy of 2-month total parenteral nutrition in AIDS patients: a controlled randomized prospective trial. *AIDS* **10:**379–384.

Mertens, R. B., E. S. Didier, M. C. Fishbein, D. C. Bertucci, L. B. Rogers, and J. M. Orenstein. 1997. *Encephalitozoon cuniculi* microsporidiosis: infection of the brain, heart, kidneys, trachea, adrenal glands, and urinary bladder in a patient with AIDS. *Mod. Pathol.* **10:**68–77.

Michiels, J. F., P. Hofman, M. C. Saint Paul, V. Giorsetti, E. Bernard, H. Vinti, and R. Loubiere. 1991. Microsporidiose intestinale: 3 cas chez des sujets seropositifs pour le VIH. *Ann. Pathol.* **11:**169–175.

Modigliani, R., C. Bories, Y. Le Charpentier, M. Salmeron, B. Messing, A. Galian, J.C. Rambaud, A. Lavergne, B. Cochand-Priollet, and I. Desportes. 1985. Diarrhoea and malabsorption in acquired immune deficiency syndrome: a study of four cases with special emphasis on opportunistic protozoan infestations. *Gut* **26:**179–187.

Molina, J. M., C. Sarfati, B. Beauvais, M. Lemann, A. Lesourd, F. Ferchal, I. Casin, P. Lagrange, R. Modigliani, F. Derouin, and J. Modai. 1993. Intestinal microsporidiosis in human immunodeficiency virus-infected patients with chronic unexplained diarrhea: prevalence and clinical and biologic features. *J. Infect. Dis.* **167:**217–221.

Molina, J. M., J. Modai, F. Derouin, A. Jaccard, C. Sarfati, B. Beauvais, and E. Oksenhendler. 1995. Disseminated microsporidiosis due to *Septata intestinalis* in patients with AIDS: clinical features and response to albendazole therapy. *J. Infect. Dis.* **171:**245–249.

Molina, J. M., J. Goguel, C. Sarfati, C. Chastang, I. Desportes-Livage, J. F. Michiels, C. Maslo, C. Katlama, L. Cotte, C. Leport, F. Raffi, F. Derouin, and J. Modai. 1997. Potential efficacy of fumagillin in intestinal microsporidiosis due to *Enterocytozoon bieneusi* in patients with HIV infection: results of a drug screening study. *AIDS* **11:**1603–1610.

Ombrouck, C., I. Desportes-Livage, M. Danis, A. Datry, T. van Gool, P. Marechal, S. Brown, S. Biligui, and L. Ciceron. 1997. Specific PCR assay for direct detection of intestinal microsporidia *Enterocytozoon bieneusi* and *Encephalitozoon intestinalis* in fecal specimens from human immunodeficiency virus-infected patients. *J. Clin. Microbiol.* **35:**652–655.

Orenstein, J. M. Unpublished observations.

Orenstein, J. M. 1991. Microsporidiosis in the acquired immunodeficiency syndrome. *J. Parasitol.* **77:**843–864.

Orenstein, J. M., J. Chiang, W. Steinberg, P. D. Smith, H. Rotterdam, and D. P. Kotler. 1990. Intestinal microsporidiosis as a cause of diarrhea in human immunodeficiency virus-infected patients: a report of 20 cases. *Hum. Pathol.* **21:**475–481.

Orenstein, J. M., D. T. Dieterich, E. Lew, and D. P. Kotler. 1993. Albendazole as a treatment for disseminated microsporidosis due to *Septata intestinalis* in AIDS patients. *AIDS* **7**(Suppl. 3)**:**S40–S42.

Orenstein, J. M., D. Dieterich, M. A. Poles, and E. Lew. 1995. The endoscopic brush cytology specimen in the diagnosis of intestinal microsporidiosis. *AIDS* **9:**1199–1201. (Letter.)

Orenstein, J. M., E. S. Didier, R. B. Mertens, S. S. Frankel, A. T. Yachnis, and H. P. Gaetz. 1997. Disseminated microsporidiosis in AIDS: are any organs spared? *AIDS* **11:**385–386. (Letter.)

Orenstein, J. M., M. Tenner, A. Cali, and D. P. Kotler. 1992a. A microsporidian previously undescribed in humans, infecting enterocytes and macrophages and associated with diarrhea in an AIDS patient. *Hum. Pathol.* **23:**722–728.

Orenstein, J. M., M. Tenner, and D. P. Kotler. 1992b. Localization of infection by the microsporidian *Enterocytozoon bieneusi* in the gastrointestinal tract of AIDS patients with diarrhea. *AIDS* **6:**195–197.

Orenstein, J. M., D. T. Dieterich, and D. P. Kotler. 1992c. Systemic dissemination by a newly recognized microsporidia species in AIDS. *AIDS* **6:** 1143–1150.

Orenstein, J. M., W. Zierdt, C. Zierdt, and D. P. Kotler. 1991. Identification of spores of the Microspora, *Enterocytozoon bieneusi* in stool and duodenal fluid from AIDS patients with diarrhea. *Lancet* **336:**1127–1128.

Owyang, C., L. Green, and D. Rader. 1983. Colonic inhibition of pancreatic and biliary secretion. *Gastroenterology* **84:**470–475.

Peacock, C. S., C. Blanshard, D. G. Tovey, D. S. Ellis, and B. G. Gazzard. 1991. Histological diagnosis of intestinal microsporidiosis in patients with AIDS. *J. Clin. Pathol.* **44:**558–563.

Pitisuttithum, P., S. Vanijanond, S. Leelasuphasri, B. Punpoonwong, D. Chindanond, and B. Phiboonnakit. 1995. Intestinal microsporidiosis: first reported case in Thailand. *Southeast Asian J. Trop. Med. Public Health* **26:**378–380.

Pol, S., C. A. Romana, S. Richard, P. Amouyal, I. Desportes-Livage, F. Carnot, J. Pays, and P. Berthelot. 1993. Microsporidia infection in patients with the human immunodeficiency virus and unexplained cholangitis. *N. Engl. J. Med.* **328:** 95–99.

Rabeneck, L., R. M. Genta, F. Gyorkey, J. E. Clarridge, P. Gyorkey, and L. W. Foote. 1995. Observations on the pathological spectrum and clinical course of microsporidiosis in men infected with the human immunodeficiency virus: follow-up study. *Clin. Infect. Dis.* **20:**1229–1235.

Rabeneck, L., F. Gyorkey, R. Genta, P. Gyorkey, L. Foote, and J. M. H. Risser. 1993. The role of microsporidia in the pathogenesis of HIV-related chronic diarrhea. *Ann. Intern. Med.* **119:** 895–899.

Rabodonirina, M., M. Bertocchi, I. Desportes-Livage, L. Cotte, H. Levrey, M. A. Piens, G. Monneret, M. Celard, J. F. Mornex, and M. Mojon. 1996. *Enterocytozoon bieneusi* as a cause of chronic diarrhea in a heart-lung transplant recipient who was seronegative for human immune deficiency virus. *Clin. Infect. Dis.* **23:**114–117.

Ramon-Garcia, G., and S. Sadowinski-Pine. 1996. Acalculous cholecystitis and microsporidiosis in a patient with AIDS. *Clin. Infect. Dis.* **23:**664. (Letter.)

Remadi, S., J. Dumais, K. Wafa, and W. MacGee. 1995. Pulmonary microsporidiosis in a patient with the acquired immunodeficiency syndrome: a case report. *Acta Cytol.* **39:**1112–1116.

Rijpstra, A. C., E. U. Canning, R. J. Van Ketel, J. K. M. Eeftinck Schattenkerk, and J. J. Laarman. 1988. Use of light microscopy to diagnose small-intestinal microsporidiosis in patients with AIDS. *J. Infect. Dis.* **157:**827–831.

Rosenberger, D. F., O. N. Serdarevic, R. A. Erlandson, R. T. Bryan, D. A. Schwartz, G. S. Visvesvara, and P. C. Keenan. 1993. Successful treatment of microsporidial keratoconjunctivitis with topical fumagillin in a patient with AIDS. *Cornea* **12:**261.

Rossi, R. M., M. Federman, and C. Wanke. 1996. Microsporidian sinusitis in patients with the acquired immunodeficiency syndrome. *Laryngoscope* **106:**966–971.

Sandfort, J., A. Hannamen, H. Gelderblom, D. Stark, R. L. Owen, and B. Ruf. 1994. *Enterocytozoon bieneusi* in an immunocompetent patient who had acute diarrhea and was not infected with the human immunodeficiency virus. *Clin. Infect. Dis.* **19:**514–516.

Sax, P. E., J. D. Rich, W. S. Pieciak, and Y. M. Trnka. 1995. Intestinal microsporidiosis occuring in a liver transplant recipient. *Transplantation* **60:**617–618.

Scaglia, M., G. S. Visvesvara, S. Wallace, S. B. Slemenda, N. J. Pieniazek, A. M. Bernuzzi, A. Orani, S. Corona, S. Gatti, A. da Silva, G. P. Croppo, and L. Sacchi. 1997. Pulmonary microsporidiosis due to *Encephalitozoon hellem* in a patient with AIDS. *J. Infect.* **34:**119–126.

Schwartz, D. A., A. Abou-Elella, C. M. Wilcox, L. Gorelkin, G. S. Visvesvara, S. E. Thompson, R. Weber, and R. T. Bryan. 1995. The presence of *Enterocytozoon bieneusi* spores in the lamina propria of small bowel biopsies with no evidence of disseminated microsporidiosis. Enteric Opportunistic Infections Working Group. *Arch. Pathol. Lab. Med.* **119:**424–428.

Schwartz, D. A., R. T. Bryan, K. O. Hewan-owe, G. S. Visvesvara, R. Weber, A. Cali, and P. Angritt. 1992. Disseminated microsporidiosis (*Encephalitozoon hellem*) and acquired immunodeficiency syndrome: autopsy evidence for respiratory acquisition. *Arch. Pathol. Lab. Med.* **116:**660–668.

Schwartz, D. A., G. S. Visvesvara, M. C. Diesenhouse, R. Weber, R. L. Font, L. A. Wilson, G. Corrent, O. N. Serdarevic, D. F. Rosberger, and P. C. Keenen. 1993. Pathologic features and immunofluorescent antibody demonstration of ocular microsporidiosis (*Encephalitozoon hellem*) in seven patients with acquired immunodeficiency syndrome. *Am. J. Ophthalmol.* **115:**285–292.

Sclafani, A., H. S. Koopmans, J. Vasselli, and M. Reichman. 1978. Effects of intestinal bypass surgery on appetite, food intake, and body weight in obese and lean rats. *Am. J. Physiol.* **234:**E389–E398.

Shadduck, J. A. 1980. Effect of fumagillin on *in vitro* multiplication of *Encephalitozoon cuniculi*. *J. Protozool.* **27:**202.

Shah, V., F. L. Altice, and C. Marino. 1996. Albendazole-induced pseudomembranous colitis. *Am. J. Gastroenterol.* **91:**1453–1454.

Sharpstone, D., B. Gazzard, M. Barrett, D. Ellis, G. Tovey, N. Francis, and A. Rowbottom. 1997. Thalidomide: a novel therapy for microsporidiosis. *Gastroenterology* **112:**1823–1829.

Sharpstone, D., A. Rowbotton, M. Nelson, and B. Gazzard. 1995. The treatment of microsporidial diarrhea with thalidomide. *AIDS* **9:**658.

Sheth, S. G., C. Bates, M. Federman, and S. Chopra. 1997. Fulminant hepatic failure caused by microsporidian infection in a patient with AIDS. *AIDS* **11:**553–554.

Shiau, Y. F., D. P. Kotler, and G. M. Levine. 1979. Can normal small bowel morphology be equated with normal function? *Gastroenterology* **76:**1246, abstr.

Simon, D., L. Weiss, H. Tanowitz, A. Cali, J. Jones, and M. Wittner. 1991. Light microscope diagnosis of human microsporidiosis and variable response to octreotide. *Gastroenterology* **100:**271–273.

Simon, D., L. Weiss, M. Wittner, J. Cello, P. Basuk, R. Rood, and A. Cali. 1991. Prevalence of microsporidia in AIDS patients with refractory diarrhea. *Am. J. Gastroenterol.* **86:**1348, abstr.

Sobottka, I., D. A. Schwartz, R. Laufs, G. S. Visvesvara, J. Schottelius, H. Schafer, and H. Albrecht. 1995. Disseminated *Encephalitozoon (Septata) intestinalis* infection in a patient with AIDS: novel diagnostic approaches and autopsy-confirmed parasitological cure following treatment with albendazole. *J. Clin. Microbiol.* **33:**2948–2952.

Soule, J. B., J. M. Orenstein, M. C. Pistole, R. B. Becker, and A. L. Halverson. 1997. A patient with acquired immunodeficiency syndrome and untreated *Encephalitozoon (Septata) intestinalis* microsporidiosis leading to small bowel perforation:

response to albendazole. *Arch. Pathol. Lab. Med.* **121:**880–887.

Sowerby, T. M., C. N. Conteas, O. G. W. Berlin, and J. Donovan. 1995. Microsporidiosis in patients with relatively preserved CD4 counts. *AIDS* **9:**975.

Spiller, R. C., I. F. Trotman, B. E. Higgins, M. A. Ghatei, G. K. Rimble, Y. C. Lee, S. R. Bloom, J. J. Misiewicz, and D. B. A. Silk. 1984. The ileal brake: inhibition of jejunal motility after ileal fat perfusion in man. *Gut* **25:**365–374.

Subirats, M., C. del Aguila, M. Baquero, J. Verdejo, G. Visvesvara, A. Moody, O. Aguilera, and G. Gonzalez-Castelao. 1996. Diagnosis of 4 cases of intestinal microsporidiosis in AIDS patients. *Enferm. Infect. Microbiol. Clin.* **14:**533–537.

Swenson, J., J. D. MacLean, E. Kokoskin-Nelson, J. Szabo, J. Lough, and M. J. Gill. 1993. Microsporidiosis in AIDS patients. *Can. Commun. Dis. Rep.* **19:**13–15.

Talal, A. H., D. P. Kotler, J. M. Orenstein, and L. M. Weiss. 1998. Detection of *Enterocytozoon bieneusi* by PCR using primers to the small-subunit rRNA. *Clin. Infect. Dis.* **26:**673–675.

Terada, S., K. R. Reddy, L. J. Jeffers, A. Cali, and R. Schiffer. 1987. Microsporidian hepatitis in the acquired immunodeficiency syndrome. *Ann. Intern. Med.* **107:**61–62.

Tzipori, S. Unpublished data.

Tzipori, S., A. Carville, G. Widmer, D. Kotler, K. Mansfield, and A. Lackner. 1997. Transmission and establishment of a persistent infection of *Enterocytozoon bieneusi* derived from a human with AIDS in SIV-infected rhesus monkeys. *J. Infect. Dis.* **175:**1016–1020.

Ullrich, R., M. Zeitz, C. Bergs, K. Janitschke, and E. O. Riecken. 1991. Intestinal microsporidiosis in a German patient with AIDS. *Klin. Wochenschr.* **69:**443–445.

van Gool, T., W. S. Hollister, J. E. Schattenkerk, M. A. van den Bergh Weerman, W. J. Terpstra, R. J. van Ketel, P. Reiss, and E. U. Canning. 1990. Diagnosis of *Enterocytozoon bieneusi* microsporidiosis in AIDS patients by recovery of spores from faeces. *Lancet* **2:**697–698.

van Gool, T., F. Snijders, P. Reiss, J. K. M. Eeftinck Schattenkerk, M. A. van den Bergh Beerman, J. F. W. M. Bartlesman, J. J. M. Bruins, E. U. Canning, and J. Dankert. 1993. Diagnosis of intestinal and disseminated microsporidial infections in patients with HIV by a new rapid fluorescence technique. *J. Clin. Pathol.* **46:**694–699.

Velásquez, J. N., S. Carnevale, E. A. Guarnera, J. H. Labbé, A. Chertcoff, and M. I. Rodríguez. 1996. Detection of the microsporidian parasite *Enterocytozoon bieneusi* in specimens from patients with AIDS by PCR. *J. Clin. Microbiol.* **34:**3230–3232.

Verre, J., D. Marriott, M. Hing, A. Field, and J. Harkness. 1992. Evaluation of light microscopic detection of microsporidial spores in faeces from HIV infected patients. In: Workshop on Intestinal Microsporidia in HIV Infection, December 15–16, 1992, Paris, abstr.

Visvesvara, G. S., G. J. Leitch, A. J. Da Silva, G. P. Croppo, H. Moura, S. Wallace, S. B. Slemenda, D. A. Schwartz, D. Moss, R. T. Bryan, and N. J. Pieniazek. 1994. Polyclonal and monoclonal antibody and PCR-amplified small-subunit rRNA identification of a microsporidian, *Encephalitozoon hellem,* isolated from an AIDS patient with disseminated infection. *J. Clin. Microbiol.* **32:**2760–2768.

Voglino, M. C., E. Pozio, R. Paloscia, F. Goffredo, V. Rinaldi, A. Ludovisi, P. Rossi, and G. Donelli. 1996. Intestinal microsporidiosis in Italian individuals with AIDS. *Ital. J. Gastroenterol.* **28:**381–386.

Vossbrinck, C. R., J. V. Maddox, S. Friedman, B. A. Debrunner-Vossbrinck, and C. R. Woese. 1987. Ribosomal RNA sequence suggests microsporidia are extremely ancient eukaryotes. *Nature* **326:**411–414.

Weber, R., and R. T. Bryan. 1994. Microsporidial infections in immunodeficient and immunocompetent patients. *Clin. Infect. Dis.* **19:**517.

Weber, R., R. T. Bryan, R. L. Owen, C. M. Wilcox, L. Gorelkin, and G. S. Visvesvara. 1992a. Improved light-microscopical detection of microsporidia spores in stool and duodenal aspirates. *N. Engl. J. Med.* **326:**161–166.

Weber, R., H. Kuster, R. Keller, T. Bachi, M. A. Spycher, J. Briner, E. Russi, and R. Luthy. 1992b. Pulmonary and intestinal microsporidiosis in a patient with the acquired immunodeficiency syndrome. *Am. Rev. Respir. Dis.* **146:**1603–1605.

Weber, R., H. Kuster, G. S. Visvesvara, R. T. Bryan, D. A. Schwartz, and R. Luthy. 1993. Disseminated microsporidiosis due to *Encephalitozoon hellem:* pulmonary colonization, microhematuria, and mild conjunctivitis in a patient with AIDS. *Clin. Infect. Dis.* **17:**415–419.

Weber, R., R. Luthy, H. Kuster, B. Sauer, R. Baumann, A. Mathis, M. Flepp, and P. Deplazes. 1997. Cerebral microsporidiosis due to *Encephalitozoon cuniculi* in a patient with human immunodeficiency virus infection. *N. Engl. J. Med.* **336:** 474–478.

Weber, R., A. Muller, M. A. Spycher, M. Opravil, R. Ammann, and J. Briner. 1992c. Intestinal *Enterocytozoon bieneusi* microsporidiosis in an

HIV-infected patient: diagnosis by ileo-colonoscopic biopsies and long-term follow up. *Clin. Invest.* **70:**1019–1023.

Weber, R., B. Sauer, M. A. Spycher, P. Deplazes, R. Keller, R. Ammann, J. Briner, and R. Luthy. 1994. Detection of *Septata intestinalis* in stool specimens and coprodiagnostic monitoring of successful treatment with albendazole. *Clin. Infect. Dis.* **19:**342–345.

Weiss, L. M., E. Michalakakis, C. Coyle, H. B. Tanowitz, and M. Wittner. 1994a. The in vitro activity of albendazole against *Encephalitozoon cuniculi. J. Eukary. Microbiol.* **41:**65S.

Weiss, L. M., X. Zhu, A. Cali, H. B. Tanowitz, and M. Wittner. 1994b. Utility of microsporidian rRNA in diagnosis and phylogeny: a review. *Folia Parasitol.* **41:**81–90.

Weitz, J. C., R. Bryan, and R. Botehlo. 1995. Microsporidiosis in patients with chronic diarrhea and AIDS, in HIV asymptomatic patients and in patients with acute diarrhea. *Rev. Med. Chil.* **123:**849–856.

Yachnis, A. T., J. Berg, A. Martinez-Salazar, B. S. Bender, L. Diaz, A. M. Rojiani, T. A. Eskin, and J. M. Orenstein. 1996. Disseminated microsporidiosis especially infecting the brain, heart, and kidneys. *Am. J. Clin. Pathol.* **106:**535–543.

Zender, H. O., E. Arrigoni, J. Eckert, and Y. Kapanci. 1989. A case of *Encephalitozoon cuniculi* peritonitis in a patient with AIDS. *Am. J. Clin. Pathol.* **92:**352–356.

Zhu, X., M. Wittner, H. B. Tanowitz, A. Cali, and L. M. Weiss. 1993a. Nucleotide sequence of the small subunit rRNA of *Septata intestinalis. Nucleic Acids Res.* **20:**4846.

Zhu, X., M. Wittner, H. Tanowitz, D. P. Kotler, A. Cali, and L. M. Weiss. 1993b. Small subunit rRNA sequence of *Enterocytozoon bieneusi* and its potential diagnostic role with use of the polymerase chain reaction. *J. Infect. Dis.* **168:**1570–1575.

OCULAR MICROSPORIDIOSIS

Dorothy Nahm Friedberg and David C. Ritterband

9

INTRODUCTION AND HISTORY

Human infection with microsporidia was rare prior to the AIDS epidemic. However, some of the earliest reported cases involved the eye. In 1937 Wolf and Cowan reported a case of chorioretinitis and encephalitis which they attributed to microsporidia (*Encephalitozoon cuniculi*), although there is doubt as to whether it actually was a microsporidial infection (Wolf and Cowen, 1937). In 1973 Ashton and Wirasinha described the first well documented case of ocular involvement in humans (Ashton and Wirasinha, 1973). An 11-year-old Tamil boy had been gored by a goat 6 years prior to being examined. A corneal transplant was performed, and histopathology of the scarred and vascularized cornea revealed parasites deep in the stroma. Numerous prominent pathologists reviewed the slides, and after initially confusing the organisms with the Leishman-Donovan bodies seen in *Leishmania,* agreed that they were microsporidial spores (*E. cuniculi*). Electron microscopic examination was not performed.

Dorothy Nahm Friedberg, Department of Ophthalmology, New York University School of Medicine, New York, NY 10016. *David C. Ritterband,* Department of Ophthalmology, New York Eye and Ear Infirmary, New York, NY 10003.

In 1981 Pinnolis and colleagues reported the case of a 26-year-old African woman with a perforated corneal ulcer. There was no antecedent history of trauma. On histopathology organisms were found just anterior to Decemet's membrane, deep in the corneal stroma. Epithelioid granulomas were seen throughout the stroma, as were multinucleated giant cells. The organisms were classified as *Nosema* rather than *Encephalitozoon* because of their size (2.5 to 3 by 4.5 to 5 μm), the absence of a phagocytic vacuole, and the presence of 11 to 13 coils in the polar tubule. Two additional cases of deep stromal keratitis have been described (Bryan et al., 1990; Davis et al., 1990; Cali et al., 1991).

In 1990 there were several cases of superficial keratitis in AIDS patients, which differed in presentation and clinical course from the earlier cases seen in immunocompetent individuals (Friedberg et al., 1990; Lowder et al., 1990; Centers for Disease Control, 1990). These patients had a superficial epithelial keratopathy. Organisms were present in the conjunctiva as well as in the cornea. Vision was generally good except in the case of secondary infection. Organisms were initially identified as *E. cuniculi* because of their morphology on transmission electron microscopy (TEM) (Friedberg et al., 1990). However, further studies which included

The Microsporidia and Microsporidiosis (Murray Wittner, editor; Louis M. Weiss, contributing editor), ©1999 American Society for Microbiology, Washington, D.C.

examination of organisms grown in tissue culture and further analyzed by sodium dodecyl sulfate-polyacrylamide gel electrophoresis (SDS-PAGE) technique showed them to be immunologically different from *E. cuniculi* (Friedberg et al., 1990; Didier et al., 1991a; Yee et al., 1991). These isolates represented a new species of microsporidian, *Encephalitozoon hellem*. The name is a combination of portions of the names of two of the patients from whom the organisms were isolated. Although initially thought to represent only ocular disease, *E. hellem* has been found in the sinuses, urine, and nasal mucosa, indicating that dissemination can occur (Lacey et al., 1992; Didier et al., 1996a). Superficial epithelial keratitis has also been described in a patient with systemic *Encephalitozoon (Septata) intestinalis* (Lowder et al., 1996). A case of superficial epithelial keratitis in a non-human immunodeficiency virus (HIV)-infected patient on chronic prednisone treatment for asthma has also been reported (Silverstein et al., 1997).

The first patients were treated symptomatically with topical lubricants (Friedberg et al., 1990). More recently topical fumagillin has been successful in resolving superficial keratitis, and systemic albendazole has relieved constitutional symptoms (Diesenhouse et al., 1993; Rosberger et al., 1993; Lecuit et al., 1994; Didier et al., 1996a).

Ocular infection with microsporidia has been described in animals. Stromal keratitis in a cat thought to be due to *Encephalitozoon (Nosema)* was cured by superficial keratectomy. The organisms were present throughout the corneal stroma (Buyukmihci et al., 1977). In the blue fox, a systemic *E. cuniculi* infection can be accompanied by ocular disease, mainly cataract and vascular lesions similar to polyarteritis nodosa (Arnesen and Nordstoga, 1977). A case of bilateral cataract in a rabbit thought to be caused by *E. cuniculi* has been reported; TEM was suboptimal for speciation (Ashton et al., 1976). A small focus of microsporidial infection was described in the retina of an infected rat (Perrin, 1943).

EPIDEMIOLOGY

One of the difficulties in understanding the epidemiology of ocular microsporidiosis is the constantly changing nomenclature and the difficulty in speciation. Initially, species identification depended on identification of morphological characteristics on TEM. However, differences in immune characteristics and sequences of ribosomal DNA (rDNA) have lead to more sophisticated identifications (Davis et al., 1990; Hollister et al., 1993; Vossbrinck et al., 1993). For example, *E. cuniculi*-like organisms were identified by electron microscopy (EM) morphology in early cases of superficial keratopathy in AIDS patients (Friedberg et al., 1990). Further immunological studies differentiated these organisms as *E. hellem* (Didier et al., 1991a; Friedberg et al., 1993). *Nosema corneum (Vittaforma corneae)* (Silveira and Canning, 1995b), sometimes called *E. cuniculi* or *Nosema cuniculi* (Ashton and Wirasinha, 1973), produces a completely different infection pathologically with different morphology on TEM than the organism initially identified as *E. cuniculi,* but later proven to be *E. hellem*. The accompanying tables attempt to clarify published accounts of ocular microsporidia in both immunocompetent (Table 1) and immunosuppressed (Table 2) individuals.

Ocular microsporidial infection is rare but is more common in immunosuppressed individuals. In the case of *E. hellem* and *E. (Septata) intestinalis* there is evidence that the ocular findings are part of a systemic infection which in *E. hellem* may involve the sinuses, nasal mucosa (Lacey et al., 1992), respiratory tract, and kidneys (Schwartz et al., 1992), and in *S. intestinalis,* may affect the gastrointestinal tract (Lowder et al., 1996). There has been no large-scale seroepidemiologic screening of AIDS patients to identify the prevalence of antibody positivity to various strains of microsporidia; nor is the true number of cases, including reported and unreported cases, known. A study of antibody responses to *E. cuniculi, E. hellem, N. corneum,* and *N. algerae* evaluated six patients with known ocular microsporidiosis (four HIV positive) and eight patients who were HIV seronegative and

TABLE 1 Cases of ocular microsporidial infection in immunocompetent patients

Reference(s) for initial clinical description	Organism(s)	No. of cases	Ocular finding(s)	Systemic infection	Treatment
Wolf and Cowen (1937)	*E. cuniculi* (probably an error)	1	Chorioretinitis	+	None (patient died)
Ashton and Wirasinha (1973)	*Nosema, M. celonensis*	1	Deep stroma, history of trauma	–	Enucleation
Pinnolis et al. (1981)	*Nosema, M. africanum*	1	Deep stroma	–	Keratoplasty
Davis et al. (1990), Shadduck et al. (1990)	*N. corneum, Vittaforma corneae*	1	Deep stroma	– –	Biopsy, keratoplasty
Cali et al. (1991), Bryan et al. (1990)	*Nosema ocularum*	1	Deep stroma	–	Corneal biopsy

TABLE 2 Cases of ocular microsporidial infection in immunosuppressed patients

Reference(s) for initial clinical description	Organism(s)[a]	No. of new cases	Superficial keratitis	Systemic (documented)	Treatment
Lowder et al. (1990), Cali et al. (1991)	*E. cuniculi*-like	1	+	–	Topical antibiotics
Friedberg et al. (1990), Didier et al. (1991)	*E. hellem* (2), N.S. (1)	3	+	–	Lubrication and sulfa drops
Centers for Disease Control (1990), Lowder et al. (1990), Friedberg et al. (1990), Yee et al. (1991)	N.S.	0 (5[b])	+		
Yee et al. (1991), Didier et al. (1991)	*E. hellem*	1	+	–	Scraping/ itraconazole
Lacey et al. (1992) and Metcalfe et al. (1992), Hollister et al. (1993)	*E. hellem*	1	+	+	Brolene/ albendazole
Diesenhouse et al. (1993)	*E. hellem*	2	+	+	Fumagillin
Rosberger et al. (1993)	*E. hellem*	1	+	+	Fumagillin
Schwartz et al. (1992, 1993), Diesenhouse et al. (1993), Rosberger et al. (1993)	*E. hellem*	3 (4[b])	+ (5/7)	+ (4/7)	Fumagillin
McCluskey et al. (1993)	*Encephalitozoon*	1	+	–	Brolene
Lecuit et al. (1994)	*E. hellem*/ *E. intestinalis*	1	+	+	Albendazole
Wilkins et al. (1994)	N.S.	1	+	–	Fumagillin
Garvey et al. (1995)	N.S.	1	+	–	Fumagillin
Didier et al. (1996a)	*E. hellem*	1	+	+	Albendazole/ fumagillin
Shah et al. (1996)	N.S.	1	+	–	Fumagillin
Lowder et al. (1996)	*E. intestinalis*	1	+	+	Fumagillin
Silverstein et al. (1997)	*E. hellem*/ *Septata intestinalis*	1	+	+	Albendazole
Gritz et al. (1997)	N.S.	1	+	+	Albendazole/ fumagillin

[a]N.S., not stated.
[b]Number of cases included but previously reported.

who had no history of microsporidial infection, demonstrated variable results. The authors concluded that enzyme-linked immunosorbent assay (ELISA) titers and Western blot tests may not be powerful tools in the diagnosis of microsporidial infections, particularly in AIDS patients (Didier et al., 1991b).

PATHOGENESIS AND PATHOLOGY

The clinical and histopathologic appearance of ocular microsporidiosis can be divided into two distinct patterns of infection, one involving the deep corneal stroma and seen in immunocompetent patients, and the other a superficial epithelial infection seen in immunodeficient individuals. Four cases of deep corneal stromal disease have been reported, and in two of these cases penetrating keratoplasty was required for visual rehabilitation. In the first reported case, the responsible organism was renamed *Microsporidium celonensis* (Canning and Lom, 1986) although in the original article it was referred to both as *Nosema* and *Encephalitozoon* (Ashton and Wirasinha, 1973). The genus *Microsporidium* contains microsporidia for which the taxonomic status cannot be identified (Weiss, 1995). The penetrating keratoplasty specimen showed central corneal necrosis with acute inflammatory cells and some giant cells. Organisms were observed in the deep stroma just anterior to Decemet's membrane. Some organisms were observed in macrophages, but they were mainly between the corneal lamellae. They stained weakly with hematoxylin and eosin, and periodic acid-Schiff (PAS) but were well delineated with Giemsa stain. TEM was not performed.

The patient described by Pinnolis and colleagues underwent enucleation of a blind, painful eye (Pinnolis et al., 1981). The organism was identified as *Nosema,* and the species was not determined. However, it is now called *Microsporidium africanum* (Canning and Lom, 1986). Light microscopy (LM) of the specimen revealed a perforated corneal ulcer with infiltration of neutrophils and mononuclear cells, epithelioid granulomas, and some foreign body giant cells. Organisms were mainly deep in the corneal stroma just anterior to Decemet's membrane and measured 2.5 to 3 by 4.5 to 5 μm. They were oval and stained faintly with hematoxylin and eosin and PAS and were refractile by phase contrast microscopy. A birefringent, rodlike polar filament was seen with polarizing light. Organisms were strongly acid-fast and positive to Grocott methenamine-silver stain. They were found both free and in clusters within the cytoplasm of histiocytes. No organisms were seen elsewhere in the globe. TEM analysis revealed spores with an electron-lucent capsule and a polar filament with 11 to 13 coils. The nucleus appeared single, but fine details were not seen. No phagocytic vacuole was observed.

In the case described by Davis and colleagues, tissue from a corneal biopsy and later from a penetrating keratoplasty specimen was examined (Davis et al., 1990). The difference in staining in the two samples was thought to relate to the possible degeneration of spores in the later specimen. Organisms were 3.5 to 4 by 1.5 μm and were located mainly in keratocytes, although some were seen in extracellular collagenous lamellae. Inflammation was absent, probably because of prolonged steroid use (Figs. 1 to 6). TEM revealed spores with a thick capsule and a translucent cell wall, diplokarya, and a polar tubule with five or six coils. The organism was initially called *N. corneum* but is now known as *V. corneae* (Silveira and Canning, 1995b). The growth of these organisms in tissue culture represents the first in vitro isolation and propagation of microsporidia from a human (Shadduck et al., 1990).

Cali described a case of stromal keratitis and corneal ulceration due to *Nosema ocularum* (Bryan et al., 1990; Cali et al., 1991). The spores were 3 by 5 μm with paired abutted nuclei (diplokaryon) and 9 to 12 coils of polar tubule. The parasites were in direct contact with the host cytoplasm and were distributed throughout the cytoplasm.

A single case of superficial epithelial keratitis in a non-HIV-infected individual has been reported; however, this patient had been taking

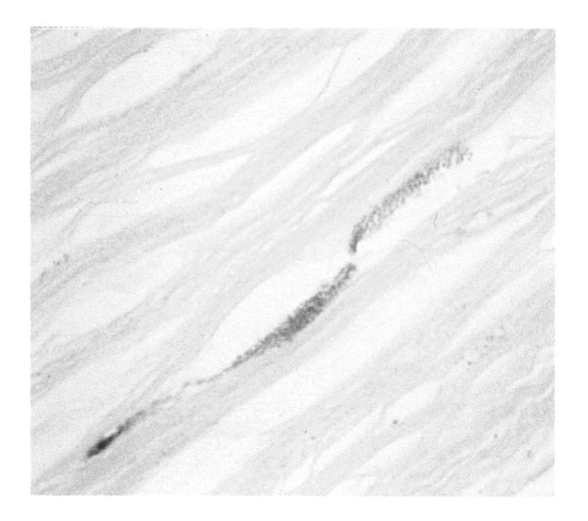

FIGURE 1 Several keratocytes with ill-defined basophilic spores of *V. corneae.* The sample was stained with hematoxylin and eosin. Original magnification, ×160.) (Reprinted with permission from *Ophthalmology* [Davis et al., 1990].)

FIGURE 2 Same field as Fig. 1 with numerous birefringent linear structures probably corresponding to portions of polar filament. The sample was viewed by polarized light. Original magnification, ×160. (Reprinted with permission from *Ophthalmology* [Davis et al., 1990].)

systemic steroids chronically for asthma. Characteristic organisms were seen on corneal scraping, and the patient responded to albendazole treatment. Further identification was not performed (Silverstein et al., 1997). There have been a number of cases of superficial epithelial keratitis involving patients with AIDS. The responsible organism in most cases was *E. hellem.* This species cannot be differentiated from *E. cuniculi* by TEM; immunofluorescent antibody staining (Schwartz et al., 1993; 1996), and SDS-PAGE analysis (Didier et al., 1991a)

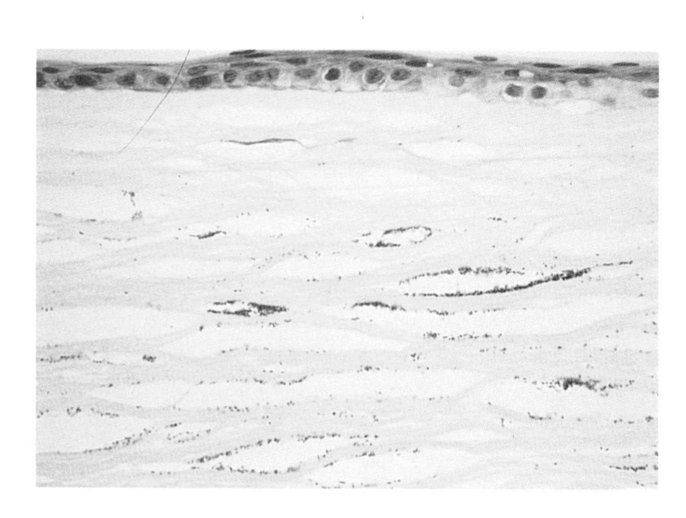

FIGURE 3 Anterior corneal stroma. Many keratocytes contain gram positive spores, and some are extracellular. Corneal epithelium appears slightly thinned, and Bowman's layer appears intact. The sample was treated with Brown and Hopps stain. Original magnification, ×100. (Reprinted with permission from *Ophthalmology* [Davis et al., 1990].)

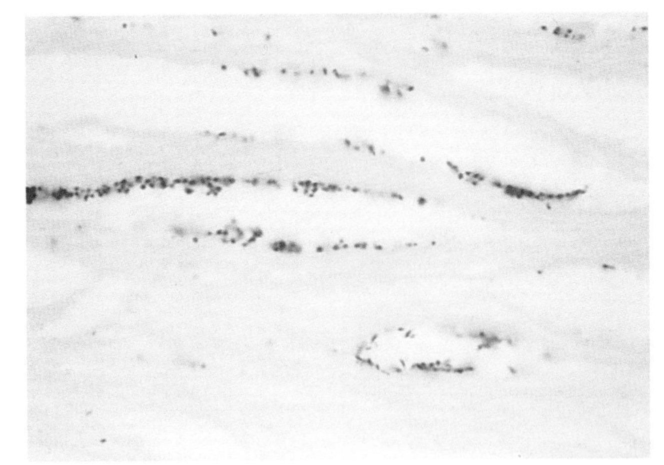

FIGURE 4 High-power view of area in Fig. 3 showing both intracellular and extracellular organisms. The sample was treated with Brown and Hopps stain. Original magnification, ×252. (Reprinted with permission from *Ophthalmology* [Davis et al., 1990].)

can differentiate the two species. At least one case of *E.* (*Septata*) *intestinalis* has been identified as causing a superficial keratitis (Lowder et al., 1996).

Histology of conjunctival scrapings from patients with microsporidial superficial epithelial keratitis shows numerous ovoid organisms mainly within epithelial cells within clear vacuoles with scant neutrophilic infiltration. The organisms are gram-positive and stain with Giemsa stain (Fig. 7 and 8). In corneal scrapings organisms are seen in the superficial epithelium generally within cells. Corneal and conjunctival biopsy specimens reveal infection confined to the superficial epithelial layers of the cornea, sparing the stroma (Fig. 9). Inflammatory cells can be seen intraepithelially and subepithelially. TEM shows uninucleate spores within parasitophorous vacuoles. They are 1 by 2.5 μm and have polar tubules with up to seven

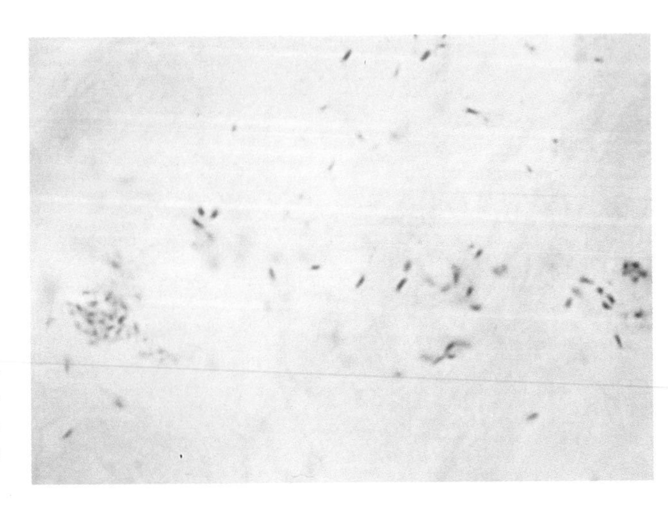

FIGURE 5 Organisms are acid fast variable. The sample was treated with Ziehl-Neelsen stain. Original magnification ×252. (Reprinted with permission from *Ophthalmology* [Davis et al., 1990].)

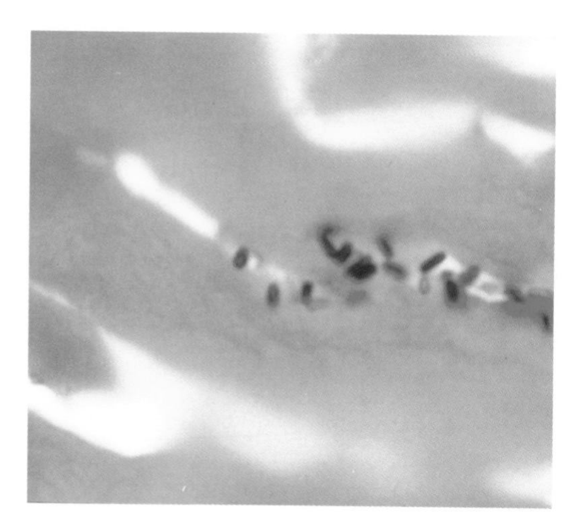

FIGURE 6 Oval to elongated extracellular spores are outlined within the corneal stroma. The sample was treated with Grocott methenamine-silver stain. Original magnification, ×403. (Reprinted with permission from *Ophthalmology* [Davis et al., 1990].)

coils (Fig. 10). The deeper epithelial layers may have fewer organisms, as well as organisms in earlier stages of development. Immature spores and meronts may also be seen (Fig. 11) (Friedberg et al., 1990; Schwartz et al., 1993).

The mechanism of ocular microsporidial infection has not been definitively determined. In the case of deep stromal disease, one of the patients recalled antecedent trauma, but others did not. Organisms were observed only in the deep corneal stroma in the enucleated globe, and no other ocular tissues were involved. In cases in which penetrating keratoplasty was performed, there was no indication that disease recurred in the grafted corneas. This implies that there was no conjunctival reservoir of infectious organisms.

Initial descriptions of the superficial keratitis recognized conjunctival involvement but did not appreciate the extent of systemic disease. An early report suggested that exposure to animals might be responsible for the infection (Friedberg et al., 1990). Subsequently it has become clear that *E. hellem* infection is systemic, involving the urinary tract, nasal mucosa, and sinuses, as well as the respiratory epithelium (Lacey et al., 1992; Metcalfe et al., 1992; Schwartz et al., 1992). Schwartz suggested that the extensive involvement of the respiratory tree in a patient with *E. hellem* infection points to this area as the initial site of infection (Schwartz et al., 1992). Infection of the urinary tract could occur hematogenously, and eye

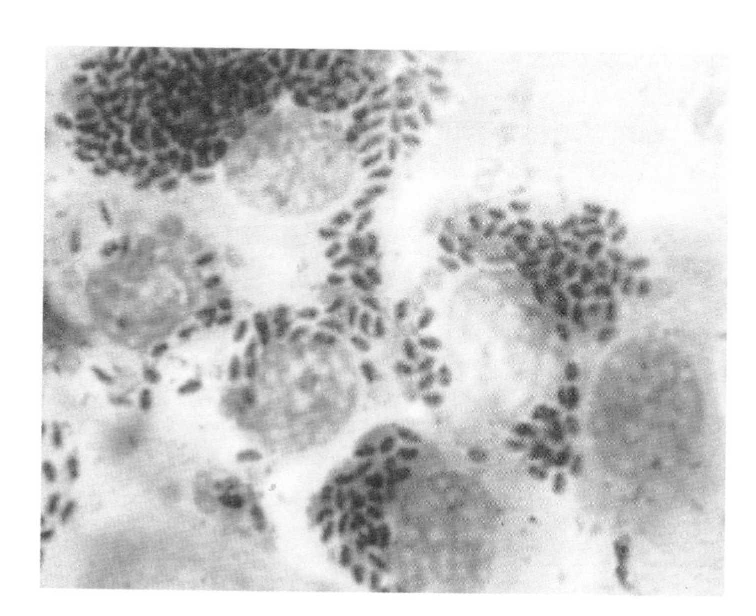

FIGURE 7 Conjunctival scraping with numerous large ovoid gram positive organisms within conjunctival epithelial cells. Original magnification, ×500. (Reprinted with permission from *Archives of Ophthalmology* [Friedberg et al., 1990]. Figure copyright 1990, American Medical Association.)

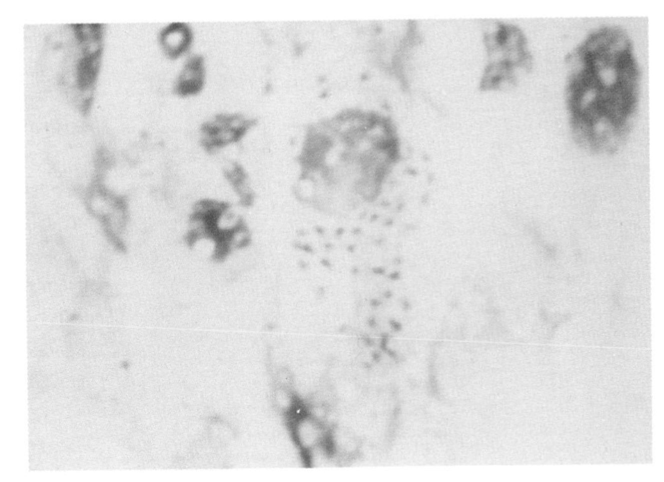

FIGURE 8 Conjunctival scraping showing organisms staining with Giemsa stain. Original magnification, ×500.

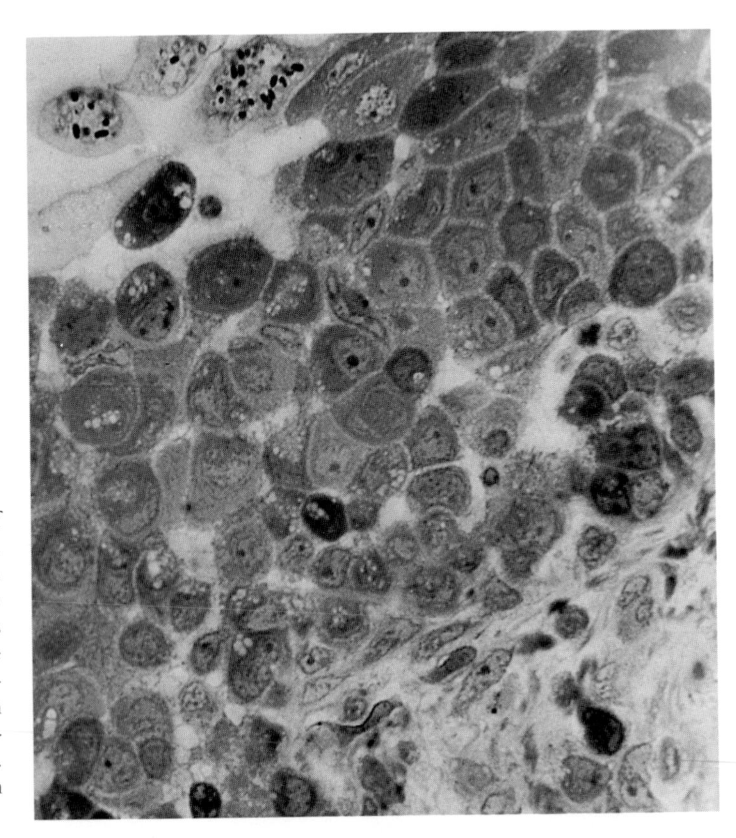

FIGURE 9 Plastic section of conjunctival biopsy specimen. Sloughing superficial cells contain parasite-laden vacuoles. Mature spores stain densely. The sample was treated with methylene blue-azure II-basic fuchsin stain. Original magnification, ×360. (Reprinted with permission from *Archives of Ophthalmology* [Friedberg et al., 1990]. Figure copyright 1990, American Medical Association.)

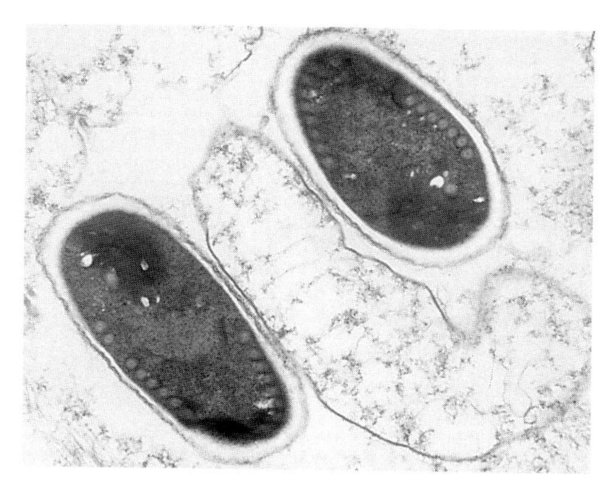

FIGURE 10 Electron micrograph from conjunctival biopsy. A mature electron-dense uninucleate spore has seven turns of the polar tubule. Original magnification, ×46,000. (Reprinted with permission from *Archives of Ophthalmology* [Friedberg et al., 1990]. Figure copyright 1990, American Medical Association.)

infection could occur via the fingers which had been in contact with urine. The case of *E. (Septata) intestinalis* had involvement of the gastrointestinal tract. Direct inoculation into the ocular surface could be the mechanism of spread (Lowder et al., 1996).

CLINICAL PRESENTATION

There are two clinical presentations of ocular microsporidial infections: corneal stromal keratitis occurring in immunocompetent patients and an epithelial keratopathy and conjunctivitis seen in immunosuppressed patients. The early cases of ocular microsporidia differed significantly in presentation and clinical course from the cases seen in immunosuppressed individuals (Centers for Disease Control, 1990). The two initial cases had deep stromal keratitis with vascularization as end stage disease, but the clinical course was not described (Ashton and Wirasinha, 1973; Pinnolis et al., 1981). The microsporidial etiology of the two cases was realized only after tissue was obtained for histopathologic studies using LM and TEM. Two additional cases were reported in 1990 and 1991, both in healthy individuals (Bryan et al., 1990; Davis et al., 1990; Cali et al., 1991). Pathologically these cases resembled the two earlier ones in that the deep corneal stroma was laden with microsporidia between the lamellae and within histiocytes. The case reported by Davis began insidiously and mimicked a progressive herpes disciform keratitis with recurrent stromal infiltration and uveitis. Diagnosis was made on corneal biopsy but penetrating keratoplasty was

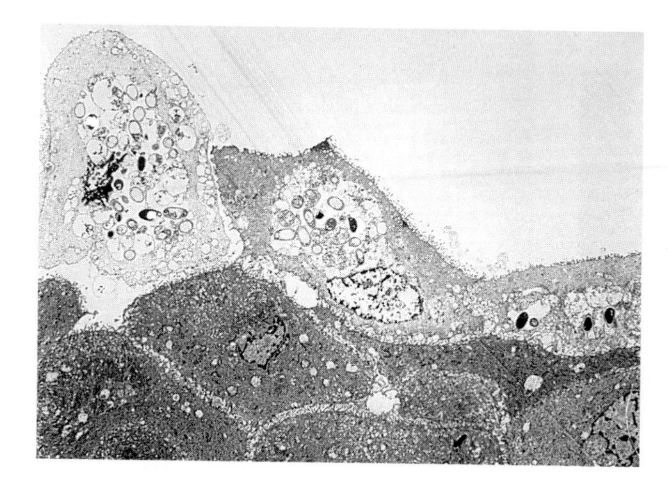

FIGURE 11 Electron micrograph from conjunctival biopsy. The superficial cell has clear parasitophorous vacuoles with parasites at various stages of development. Deeper cells show meronts in vacuole indenting the nucleus (lower right). Original magnification, ×2,700. (Reprinted with permission from *Archives of Ophthalmology* [Friedberg et al., 1990]. Figure copyright 1990, American Medical Association.)

necessary for visual rehabilitation (Figs. 12 and 13). Patients with deep stromal keratitis all suffered a marked reduction in visual acuity from the infection.

In 1990, several cases of superficial keratitis in AIDS patients were reported that differed significantly in presentation and clinical course from the earlier cases of deep stromal disease. Clinical manifestations included one or more of the following ocular signs: foreign body sensation, conjunctivitis, dry eyes, blurred vision, or photophobia. On slit lamp examination, the infection was characterized either by mild papillary conjunctivitis with or without coarse punctate epitheliopathy (Friedberg et al., 1990; Lowder et al., 1990; Centers for Disease Control, 1990; Yee et al., 1991) (Fig. 14). The eyes were often inflamed, sometimes intermittently, with periods of quiescence (Lowder et al., 1990; Schwartz et al., 1993). The keratitis was commonly bilateral and chronic. Some of the lesions stained with fluorescein, while others did not. The density of the epithelial spots varied from mild scattered lesions to a dense concentration involving the entire cornea, which looked as if salt had been shaken on it (Fig. 15 and 16). Anterior uveitis was not present unless the patient had concurrent cytomegalovirus (CMV) retinitis. None of the patients had deep stromal involvement except one in whom a secondary pseudomonas corneal ulcer with sclerokeratitis developed (Friedberg et al., 1990). Generally, vision was mildly affected but, if left untreated, in some cases acuity was compromised. Subsequent case reports have echoed these findings (Diesenhouse et al., 1993; Rosberger et al., 1993; Schwartz et al., 1993; Lowder et al., 1996; Shah et al., 1996).

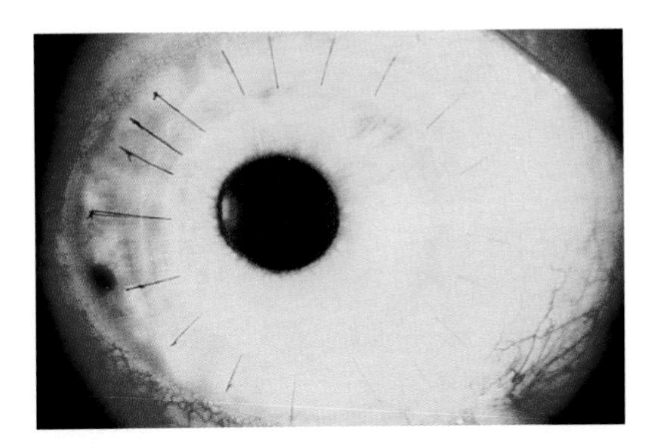

FIGURE 13 Eye in Fig. 12 six months after corneal transplantation. There is no evidence of recurrent microsporidial keratitis. (Reprinted with permission from *Ophthalmology* [Davis et al., 1990].)

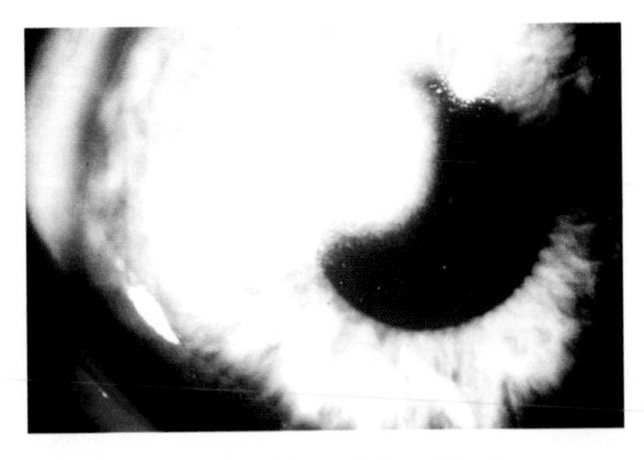

FIGURE 12 Stromal keratitis from *Vittaforma corneae*. (Reprinted with permission from *Ophthalmology* [Davis et al., 1990].)

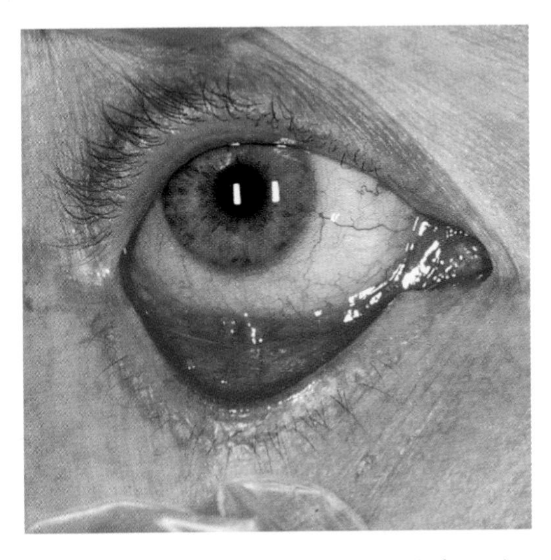

FIGURE 14 Minimal bulbar conjunctival reaction with moderate hyperemia of inferior palpebral conjunctiva with erythema and fusiform edema in the inferior fornix of an AIDS patient infected with *Encephalitozoon hellem*. (Reprinted with permission from *Archives of Ophthalmology* [Friedberg et al., 1990]. Figure copyright 1990, American Medical Association.)

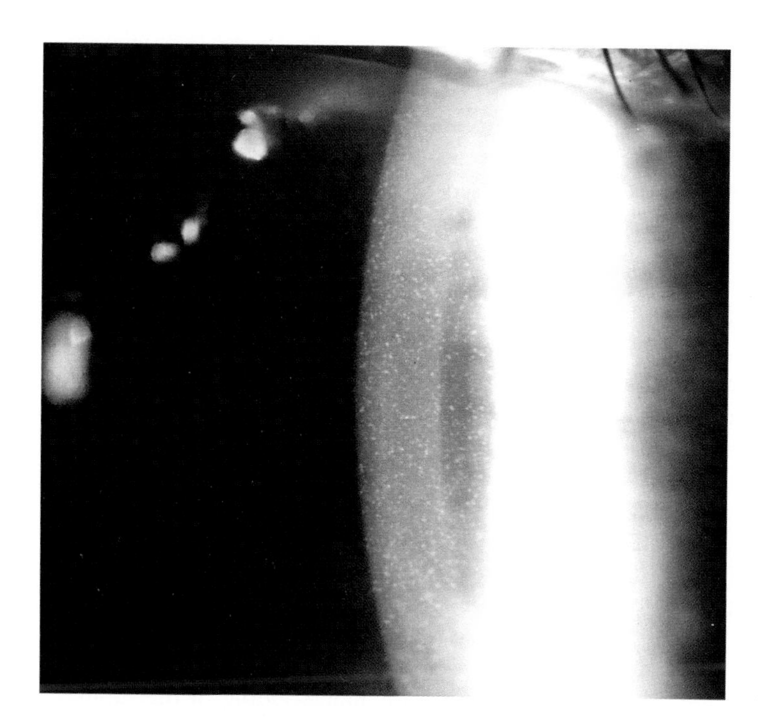

FIGURE 15 Mild superficial epithelial keratitis in AIDS patient infected with *E. hellem*. (Reprinted with permission from *Archives of Ophthalmology* [Friedberg et al., 1990]. Figure copyright 1990, American Medical Association.)

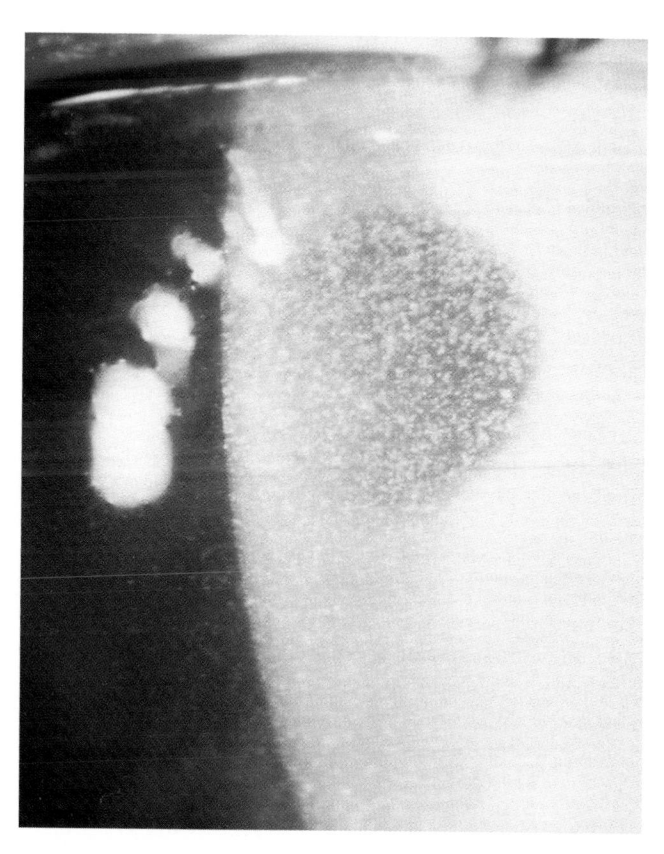

FIGURE 16 Diffuse, coarse, superficial epithelial keratitis in an AIDS patient infected with *E. hellem*. (Reprinted with permission from *Archives of Ophthalmology* [Friedberg et al., 1990]. Figure copyright 1990, American Medical Association.)

Schwartz and colleagues have suggested that ocular microsporidiosis occurs in advanced AIDS. Profoundly depressed $CD4^+$ lymphocyte levels were a common feature reported in their seven clinical cases (Schwartz et al., 1993).

DIFFERENTIAL DIAGNOSIS

Both deep stromal and superficial epithelial keratitides due to ocular microsporidiosis are rare but must be considered in situations where these entities do not respond to conventional therapy. Ocular microsporidiosis has emerged as the cause of bilateral keratoconjunctivitis in immunosuppressed individuals. The differential diagnosis includes bacterial, chlamydial, and viral keratoconjunctivitis, and dry eye syndrome. Often the diagnosis is suspected on clinical grounds, especially in an HIV-positive individual. A clinical history of chronic diarrhea or sinusitis raises the suspicion of systemic microsporidial disease. Many patients relate a history of treatment with lubricants, topical antibiotics, and/or antiviral agents without response. If initial cultures or gram stains are negative, repeat conjunctival biopsy or scrapings and clinical suspicion often lead the clinician to the diagnosis. To physicians familiar with the ocular history, signs, and symptoms, the clinical presentation may be diagnostic.

The differential diagnosis of microsporidial infection causing stromal disease is broader; diagnosis is unlikely based on clinical examination. In all reported cases of microsporidia causing stromal disease, diagnosis was made only after histopathologic tissue examination (Ashton and Wirasinha, 1973; Pinnolis et al., 1981; Davis et al., 1990; Cali et al., 1991). The differential diagnosis includes a broad list of agents causing suppurative and nonsuppurative inflammation and vascularization of the cornea. Nonsuppurative causes include congenital and acquired syphilis, tuberculosis, leprosy, onchocerciasis, herpes simplex, herpes zoster, Epstein-Barr virus, lymphogranuloma venereum, mumps, leishmaniasis, and Cogan's syndrome. Suppurative causes include infectious keratitis caused by bacteria, fungi, or acanthamoebas.

LABORATORY DIAGNOSIS

Microsporidia are sufficiently unique to be classified as a separate phylum (Canning and Lom, 1986). Within the phylum are dozens of genera and more than 1,000 species. Traditionally, the classification of microsporidia in humans depended on histochemical procedures and TEM (Sprague, 1977). The presence of a polar tubule is diagnostic and helps classify an organism as a member of the phylum Microspora (Levine et al., 1980). The classification of genera is based on spore structure, nuclear arrangement, and developmental life cycle. While morphologic studies are able to distinguish between genera of microsporidia, TEM may not always distinguish species within a genus (Didier et al., 1991a). Further advances in molecular biology and immunology have enhanced our understanding of the subtleties between species of these organisms. In the past decade numerous taxonomic changes have taken place.

Microsporidia causing ocular disease have played a significant role in our ability to isolate and speciate these organisms. Proper laboratory identification requires careful attention to the preparative technique. Difficulties ensue if the organisms are inadequately centrifuged from specimens, improperly stained, or mistaken for bacteria (Shadduck, 1989). Gram, Giemsa, acid-fast, and PAS stains have all been useful in the evaluation of body fluid sediments (Canning and Lom, 1986). Giemsa-stained organisms are easier to see in histologic specimens than those stained with hematoxylin and eosin (Shadduck and Greeley, 1989). In addition, fresh tissue squash preparations made from biopsy specimens can be viewed directly by phase-contrast microscopy in which the spores are refractile and birefringent (Rijpstra et al., 1988; Shadduck and Greeley, 1989).

Histopathologic examination of a corneal button was instrumental in documenting the first case of ocular microsporidia and invaluable in recording the LM findings (Ashton and Wirasinha, 1973). Examination of the corneal button revealed numerous refractile oval bodies measuring an average of 3.5 by 1.5 μm not only within macrophages but also between corneal lamellae. Smears from the cornea stained weakly positive with hematoxylin and eosin, PAS, and Gram stain, and intensely with Giemsa stain. Methylene blue staining showed an internal structure consisting of a prominent polar vacuole and a central banded nucleus. Review of the slides led to the diagnosis of microsporidiosis.

The early LM findings of ocular microsporidiosis were instrumental in localizing and recognizing these organisms in other tissues. Subsequently, it was recognized that the polar PAS-staining granule at the anterior end of mature spores is diagnostic for microsporidia (Cali and Owen, 1988).

Classification of microsporidia was initially based on size, nuclear arrangement (mono- or diplokaryotic), mode of division, and association of proliferative forms within the host cell (Canning and Lom, 1986). More recently the "gold standard" in the classification of microsporidia has involved spore structure and developmental life cycle as shown by TEM (Sprague, 1977). Ocular microsporidia speciation has required TEM in order to assess ultrastructural features such as the arrangement and complexity of the polar tubule, the number and size of the nuclei, the location and relationship of the organism to the cytoplasm (Bryan 1990; Didier et al., 1996a), and the number and type of divisions occurring during merogony and sporogeny (Silveira and Canning, 1995b). For diagnostic purposes, however, EM is time-consuming, requires significant expertise, and has proved relatively insensitive in identifying all microsporidia at the species level (Didier et al., 1993; Didier et al., 1996a). Newer advances in molecular biology have aided in speciation and will be discussed further below.

Patients with enteric infections have usually provided the highest diagnostic yield of microsporidia. However, ocular specimens have played a crucial role in our understanding of these organisms. Human microsporidian isolates, including those cultured from the eye, were propitious in providing a source of antigen for developing serologic tests for microsporidia. However, serum ELISA titers in AIDS patients with documented microsporidian infections have been highly variable for different reasons, including progressive immunodeficiency associated with the HIV virus, possible cross-reactivity with other microsporidian species, and the severity of the immunodeficiency at the time of infection (Didier et al., 1991c; Didier et al., 1996a). Growth of microsporidia in tissue culture has been important in the development of diagnostic tests, drug studies, and animal model studies. *V. corneae* (*N. corneum*), the first human microsporidian grown in culture, was recovered from the corneal stroma of a HIV-seronegative patient with keratitis (Shadduck et al., 1990). Shadduck et al. used pieces of corneal tissue obtained from an infected corneal button removed after penetrating keratoplasty and inoculated two cell lines with the infected tissue. Examination of the inoculated cell lines and media for organisms weeks later demonstrated microsporidial organisms in all stages of development. Ultrastructural studies indicated that these organisms belonged to the genus *Nosema,* and they were named *N. corneum* (*V. corneae*) (Shadduck et al., 1990). *E. hellem* has been grown from corneal tissue and conjunctival scrapings in MDCK cells (Didier et al., 1991a; Hollister et al., 1993). However, species identification of tissue culture-derived ocular microsporidia using SDS-PAGE and Western blot (immunoblot) immunodetection has proved cumbersome and time-consuming (Didier et al., 1991a; Didier et al., 1991c; Visvesvara et al., 1994; Didier et al., 1996a).

Schwartz et al. reported success with immunofluorescent antibody (IFA) techniques against eye-derived microsporidia for species identification with species-specific antisera (Schwartz et al., 1993). However, this approach requires access to tissue culture lines, which produce high numbers of heterologous spores, and laboratory animals to generate the antisera (Fig. 17).

Newer diagnostic techniques have utilized the tools of molecular taxonomy including a PCR methodology that employed specific primers based on rRNA (Vossbrinck et al.,

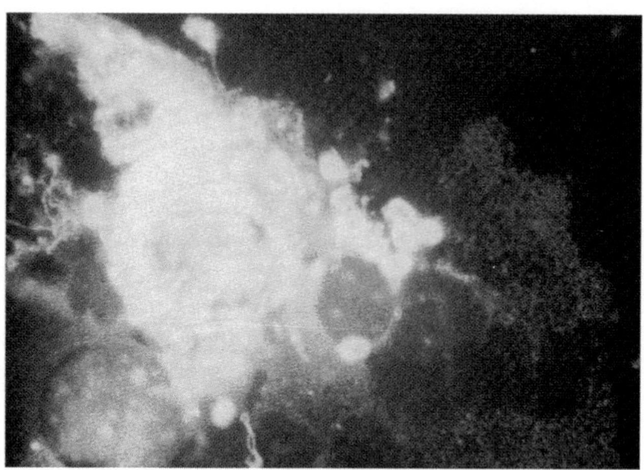

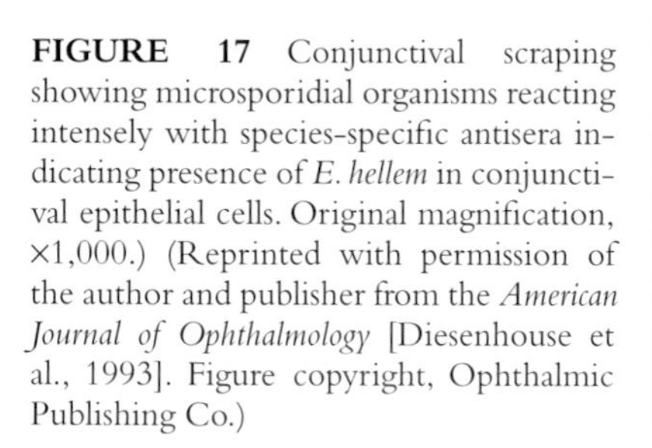

FIGURE 17 Conjunctival scraping showing microsporidial organisms reacting intensely with species-specific antisera indicating presence of *E. hellem* in conjunctival epithelial cells. Original magnification, ×1,000.) (Reprinted with permission of the author and publisher from the *American Journal of Ophthalmology* [Diesenhouse et al., 1993]. Figure copyright, Ophthalmic Publishing Co.)

1993, 1994; Weiss, 1995). Monoclonal antibodies to polar tube proteins (Visvesvara et al., 1994; Keohane et al., 1996) are also being employed for this purpose. Species-specific primers for the ocular microsporidian *E. hellem* (Visvesvara et al., 1994; Didier et al., 1995, 1996a) have been described. Second-step methods following PCR to corroborate or specifically identify the PCR products have been utilized. These procedures include DNA sequencing (Hartskeerl et al., 1993; Vossbrinck et al., 1993; Zhu et al., 1993; Didier et al., 1996a), restriction fragment length polymorphism (Hartskeerl et al., 1993; Vossbrinck et al., 1993; Didier et al., 1995, 1996a), and double-stranded DNA heteroduplex mobility shift analyses (Didier et al., 1996a). In addition, PCR may prove useful in helping to provide an understanding of the phylogenetic relationships between these parasites, as well as in understanding the intraspecies epidemiology and the responses to drug therapy.

SPECIMEN COLLECTION

Ocular specimen collection yielding sufficient sample for diagnosis can be accomplished by a variety of methods including conjunctival scrapings, swab, or biopsy; corneal scraping and biopsy; corneal transplant button; and a whole globe from an enucleation (Ashton and Wirasinha, 1973; Pinnolis et al., 1981; Davis et al., 1990; Friedberg et al., 1990; Lowder et al., 1990; Diesenhouse et al., 1993; Rosberger et al., 1993; Lowder et al., 1996; Shah et al., 1996).

With a slit lamp and either a sterile cotton swab or a platinum spatula, any portion of the ocular surface can be scraped to obtain a superficial conjunctival or corneal sample. Topical anesthesia with proparacaine is usually sufficient. A high yield of organisms is often found both in the conjunctiva and in the involved corneal epithelium. Conjunctival cultures or scrapings are usually obtained from the tarsal surface of the lower lid. The patient is instructed to look up, the lower lid is pulled down, and a moistened applicator (culture) or platinum spatula (scraping) is used to swab back and forth over the tarsal conjunctiva. To obtain corneal epithelial tissue, the spatula is held at 45° to the surface and passed multiple times in the same direction until sufficient material is obtained. The sample can then be smeared on glass slides. If possible, a 10-mm-diameter specimen is obtained. If specific stains and TEM studies will be performed, it is best to collect multiple specimens. The morbidity of a superficial corneal scraping is greater than that of a conjunctival scraping, but full reepithelialization of the cornea usually occurs in a few days.

Corneal biopsy is usually employed in cases of stromal keratitis. It involves removing tissue from the involved corneal stroma. Biopsy is indicated if an unidentified destructive corneal process is progressing despite current therapy, particularly in cases when cultures and scrapings have failed to reveal organisms. Clinicians are reluctant to perform a corneal biopsy because removal of corneal stromal tissue may lead to significant scarring. Biopsy is usually performed in the operating room with a microscope and microsurgical instruments to ensure sampling of the involved tissue while causing minimal destruction of uninvolved tissue. Corneal biopsy is often unnecessary, especially in cases of ocular microsporidia causing keratoconjunctivitis, because the organisms can be readily obtained from more superficial conjunctival or corneal samples.

Ocular tissue samples for light microscopy are generally fixed in absolute methanol. Tissue for immunopathologic examination is generally stored in a saline solution. Although other stains are useful, our laboratory prefers a chromotrope-based modified trichrome stain. Trichrome blue stain has two advantages; it is commercially available in a ready-to-use kit, and it can be applied to cultures recently removed from incubation at 37°C without losing its staining characteristics. In addition, it may be particularly useful in laboratories that lack ultraviolet microscopy equipment (Honore et al., 1996). Generally, microsporidia stain bright pinkish red, are oval in shape, and measure between 0.8 and 1.7 μm with trichrome blue. Bacteria, artifacts, and debris may also stain red, but the size and shape are not typical or the staining is unusual, lacking the internal transparent content of the spores (Shadduck, 1989; Shadduck and Greeley, 1989; Weber et al., 1992; Honore et al., 1996). For TEM examination the tissue is generally fixed in a cacodylate-buffered glutaraldehyde solution, postfixed with 1to 2% osmium tetraoxide, and embedded in Spurr's resin (Didier et al., 1991a, 1991c).

Confocal microscopy is a new technique reported to aid in diagnosis of microsporidial keratitis (Shah et al., 1996). Shah reported success in vivo, with a scanning confocal microscope with 24× contact objective lens and a Nipkow disk in imaging and diagnosing microsporidial keratoconjunctivitis. High contrast intraepithelial opacities within surface corneal epithelial cells were observed (Fig. 18), and diagnosis was confirmed on chromotrope-based Weber stain (Shah et al., 1996). Briefly, the optical design of the confocal microscope is based on the principle of Lukosz (Lukosz, 1966). The modern confocal microscope uses a point light source focused on a small volume within a specimen, and a confocal point detector is used to collect the resulting signal. This technique reduces the

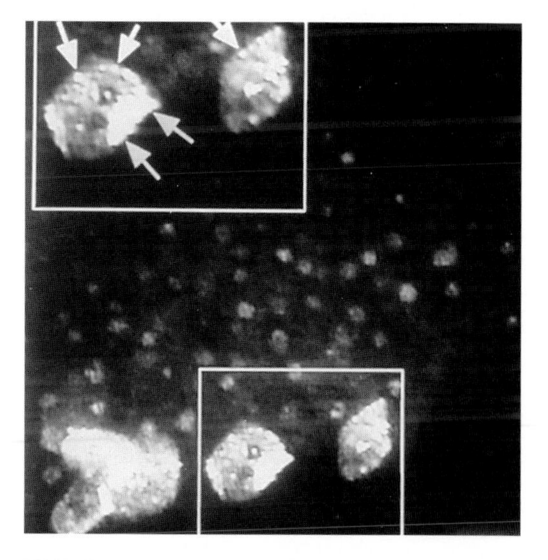

FIGURE 18 In vivo confocal microscopy demonstrates corneal epithelial cells which contain intracellular microsporidial spores. (Inset) Enlargement of the two cells, showing numerous aggregates of small, high contrast intracellular microsporidial spores (arrows). (Micrograph courtesy of Gaurav Shah and Edward Holland.)

amount of out-of-focus signal from above and below the focal plane and contributes to the detected image producing a marked increase in both lateral (*x, y*) and axial (*z*) resolution (Wilson and Shepard, 1984; Wilson, 1986; Petroll et al., 1996). The first scanning confocal microscope, developed by Petran and colleagues, used a modified Nipkow disk containing thousands of optically conjugate (source: detector) pinholes arranged in archimedian spirals (Wilson and Shepard, 1984). Rotation of the disk results in even scanning of the tissue in real time. A more detailed discussion of confocal scanning microscopy is beyond the scope of this chapter and has been provided in great detail by Wilson (Wilson and Shepard, 1984).

TREATMENT

The four reported cases of deep stromal keratitis were treated surgically. One patient underwent enucleation for a blind, painful eye (Pinnolis et al., 1981), and two patients underwent penetrating keratoplasty. In one case vision was restored (Davis et al., 1990), while in the second, vision was initially restored but the graft eventually failed (Ashton and Wirasinha, 1973).

No treatment was available for the early cases of microsporidial keratoconjunctivitis in AIDS patients. Attempts to treat punctate keratitis with intensive lubrication, bandage contact lens, or topical antibiotics proved futile (Friedberg et al., 1990). The medical literature had scattered reports of success with various pharmaceutical agents. In 1985 Ledford and colleagues described a patient with AIDS and myositis who reported symptomatic relief with trimethoprim-sulfisoxazole (Ledford et al., 1985). Trimethoprim had been used with limited success in managing microsporidial diarrhea (Current and Owen, 1989). However, topical trimethoprim-sulfisoxazole has not been beneficial in ocular microsporidial patients as reported by Yee et al. (1991), Diesenhouse et al. (1993) and Rosberger et al. (1993). Several clinical reports have also shown microsporidia to be largely unresponsive to standard topical antibiotic therapy (Davis et al., 1990; Friedberg et al., 1990; Lowder et al., 1990; Diesenhouse et al., 1993; Rosberger et al., 1993).

Successful treatment of keratoconjunctivitis with topical propamidine isethionate (Brolene) has been reported (Metcalfe et al., 1992; McCluskey et al., 1993). Brolene, a diamidine amebicidal drug, was initially tried on an empirical basis because it was shown to be effective against other classes of protozoa, particularly acanthamoeba (Metcalfe et al., 1992). McLuskey and colleagues reported significant subjective improvement with Brolene ointment with no progression of disease in one patient (McCluskey et al., 1993). Metcalfe had similar success, although a recurrence of symptoms occurred after cessation of therapy (Metcalfe et al., 1992). Diesenhouse and coauthors (1993) reported no clinical success with Brolene. Whether homologs of various diamidines, which have shown superior effectiveness in vitro against acanthamoebas, would be successful in eradicating microsporidia is unknown (Perrine et al., 1995).

Five published case reports have described variable therapeutic success in treating microsporidial keratoconjunctivitis with itraconazole, a systemic oral triazole antifungal agent (Yee et al., 1991; Diesenhouse et al., 1993; Wilkins et al., 1994; Didier et al., 1996a; Gritz et al., 1997). In one case, an AIDS patient with cryptococcal meningitis was noted to have coincident resolution of microsporidial keratoconjunctivitis following systemic treatment. However, the response was described as a very gradual improvement over 6 weeks. The authors further suggested that diagnostic and therapeutic scrapings may also have played a role (Yee et al., 1991). Didier et al. (1996a), Gritz et al. (1997), and Wilkins et al. (1994) reported no clinical response to a trial of itraconazole. In two cases reported by Diesenhouse and colleagues (1993), itraconazole was a complete therapeutic failure in one and proved helpful temporarily in the second although the infection recurred on cessation of treatment. In a case of microsporidiosis reported by Rosberger et al. (1993), microsporidial keratoconjunctivitis developed in a patient on maintenance therapy with fluconazole, another closely related oral triazole agent commonly used for cryptococcal disease. Neither itraconazole nor fluconazole has been demonstrated to have specific action against microsporidia in controlled studies.

Thiabendazole, an antihelminthic benzimidazole with larvicidal activity, has been shown to have in vitro activity against other intestinal parasites such as *Giardia* (Edlind et al., 1990). A 0.4% suspension applied topically had no effect in a case of microsporidiosis reported by Diesenhouse et al. (1993).

Albendazole, another oral benzimidazole, is a promising new agent. It blocks cellular division by interfering with microtubule formation through its binding action on β-tubulin, thereby inhibiting microtubule polymerization (Lacey, 1985; Ornstein, 1991). Albendazole is effective against larval and adult stages of cestodes and trematodes (Edwards and Breckenridge, 1988). Therapy with albendazole (400 mg twice daily) has been reported to have limited effectiveness in managing intestinal microsporidiosis caused by *Enterocytozoon bieneusi* (Blanshard et al., 1992; Dieterich et al., 1994; Didier et al., 1996b) but is considered clinically more effective in eliminating *Encephalitzoon intestinalis* infection and symptomatology (Blanshard et al., 1992; Molina et al., 1995; Joste et al., 1996). Albendazole has also been effective in eliminating a disseminated microsporidial infection involving the intestine, urinary tract, nasal mucosa, and conjunctiva in a patient with AIDS (Lecuit et al., 1994). Gritz et al. (1997) reported similar success with albendazole, which caused resolution of symptoms in a patient with ocular and paranasal sinus microsporidial infection and AIDS. Silverstein et al. (1997) described complete resolution of the signs and symptoms of microsporidial conjunctivitis and pneumonitis following treatment with albendazole (400 mg twice daily) in a patient without human immunodeficiency syndrome on chronic steroid therapy.

Albendazole has also been shown to suppress parasite division when tested in vitro against ocular-derived *V. corneae*-infected cell lines (Silveira and Canning, 1995a). *V. cornae* is the only microsporidial isolate identified and isolated as a cause of stromal keratitis (Davis et al., 1990; Silveira and Canning, 1995a, 1995b). The major side effects of albendazole are hepatotoxicity, neutropenia, and alopecia, although they do not occur frequently (Steiger et al., 1990). These effects, however, are reversible on cessation of therapy. Patients undergoing therapy should be monitored by an experienced physician and receive weekly liver function tests and complete blood counts (Gritz et al., 1997).

The most promising treatment for ocular microsporidiosis causing superficial keratoconjunctivitis is topical fumagillin, a crystalline antibiotic isolated from cultures of *Aspergillus*. It was first described in 1951 as having potent in vitro amebicidal activity, particularly against *Entamoeba histolytica* (McCowen et al., 1951). In 1952, fumagillin in a sucrose solution was reported to be effective in controlling microsporidial disease of the adult honeybee caused by the protozoan *Nosema apis* (Katznelson and Jamieson, 1952). That same year, oral fumagillin was tested in 22 male patients hospitalized because of infection with *E. histolytica*. All 22 received various doses of oral fumagillin. Signs of toxicity were few, although two patients who received over 50 mg daily complained of dizziness. Four other patients receiving this dosage complained of anorexia without nausea or vomiting. Eradication of amebae from stool was prompt in the less symptomatic patients, however, one patient with severe amebic dysentery did not respond to therapy (Killough et al., 1952). More recently, Molina and coworkers tested fumagillin (20 mg three times daily) in AIDS patients with *E. bieneusi* infection, but the treatment, which cleared spores from stool and duodenal biopsy specimens, caused severe thrombocytopenia and had to be discontinued (Molina et al., 1997).

The success of Fumidil B, a water-soluble form of fumagillin used commercially to control microsporidial disease in honeybees, led to its compassionate use in several early cases of microsporidial keratoconjunctivitis (Diesenhouse et al., 1993; Rosberger et al., 1993). Diesenhouse and colleagues used a concentration of 3 mg of Fumidil B per ml in six rabbit eyes for 5 days and in the left eye of a coworker for 3 days without signs of redness or irritation. To obtain the 3-mg/ml solution, they added 60 mg of Fumidil B to 20 ml of sterile saline, protected the solution from light, and filtered it through a 0.22-µm cellulose acetate filter. The solution contains 70 µg of fumagillin per ml, has a pH of 6.9, and has an osmolarity of 320 mosM. Initially, the medication was instilled in the first patient's blind left eye hourly while he was awake. After 1 week, symptoms were markedly diminished in the treated eye. Treatment was begun bilaterally, and within 1 week, symptoms abated bilaterally; conjunctival hyperemia and epithelial changes were considerably decreased. Medication was gradually tapered over the next several weeks, but after discontinuation of the fumagillin, the patient had a recurrence of symptoms and clinical findings in one eye. Reinitiation of therapy and tapering to twice daily led to complete absence of clinical disease. Subsequently, several clinical successes with topical fumagillin were reported (Rosberger et al., 1993; Wilkins et al., 1994; Didier et al., 1996a; Lowder et al., 1996; Shah et al., 1996; Gritz et al., 1997). At present, it remains the treatment of choice for isolated ocular microsporidiosis causing keratoconjunctivitis. Optimum dosing schedules and length of treatment are unknown; clinical success has been reported at concentrations of Fumadil B between 3 mg/ml (Diesenhouse et al., 1993) and 10 mg/ml (Rosberger et al., 1993). The discovery of dissemination of microsporidial disease in patients with keratitis suggests that systemic treatment with albendazole should be considered (Schwartz et al., 1992; Didier et al., 1996a).

The mechanism of action of fumagillin has not been clearly defined. The drug probably acts by destroying the parasite and arresting its multiplication within the host cell. There are also data suggesting that it specifically may alter DNA content, inhibit RNA synthesis, or affect the fatty acid metabolism of spore membranes (Hartwig and Przellecka, 1971; Jaronski, 1972; Shadduck, 1980). Some mature spores or sporoblasts may survive after clinically successful treatment, which suggests that they are inhibited rather than being destroyed. Fumagillin is also recognized as a potent angiogenesis inhibitor (Ingber et al., 1990).

A new fumagillin analog, *O*-(chloroacetyl-carbamoyl) (TNP-470), has shown promise for its in vitro antimicrosporidial activity (Didier, 1997). TNP-470 is a potent inhibitor of angiogenesis and inhibits neovascularization in several solid tumor models (Yanase et al., 1993; Figg et al., 1997; Fujimoto et al., 1997). Pharmacokinetic studies have demonstrated TNP-470 concentrations with in vitro activity to be achievable in vivo. The drug can be rapidly cleared from the circulation after a single 1-h infusion, however, there is considerable interpatient variability in drug clearance (Figg et al., 1997). TNP-470 has been well tolerated systemically in AIDS patients with Kaposi's sarcoma. TNP-470 treatments did not accelerate $CD4^+$ T-cell decline and did not cause HIV p24 levels to increase (Pluda et al., 1993). The potential topical or systemic use of this analog of fumagillin against microsporidia infected human tissues is unknown.

SUMMARY

Both isolated, deep stromal keratitis and superficial epithelial keratitis caused by microsporidia are rare. However, superficial keratitis is important, as it may be the initial manifestation of systemic microsporidia infection. Therapy, either topical or systemic, can be visually and pathologically evaluated with minimal morbidity. Why certain species involve ocular tissue has yet to be elucidated.

REFERENCES

Arnesen, K., and K. Nordstoga. 1977. Ocular encephalitozoonosis Nosematosis) in blue foxes. *Acta Ophthalmologica* **55:**641–651.

Ashton, B., and P. Wirasinha. 1973. Encephalitozoonosis (nosematosis) of the cornea. *Br. J. Ophthalmol.* **57:**669–674.

Ashton, N., C. Cook, and F. Clegg. 1976. Encephalitozoonosis (nosematosis) causing bilateral cataract in a rabbit. *Br. J. Ophthalmol.* **60:**618–631.

Blanshard, C., D. Ellis, D. Tovey, S. Dowell, and B. Gazzard. 1992. Treatment of intestinal microsporidiosis with albendazole in patients with AIDS. *AIDS* **6:**311–313.

Bryan, R. 1990. p. Microsporidia, p. 2130–2134. *In* G. Mandell, R. Douglas, and J. Bennett (ed.), *Principles and Practice of Infectious Diseases.* Churchill Livingstone, New York, N.Y.

Bryan, R., A. Cali, R. Owen, and H. Spencer. 1990. Microsporidia: opportunistic pathogens in patients with AIDS. p. 1–26. *In* T. Sun (ed.), *Progress in Clinical Parasitology.* W.W. Norton, New York, N.Y.

Buyukmihci, N., R. Belhorn, J. Hunziker, and J. Clinton. 1977. Encephalitozoon (*Nosema*) infection of the cornea in a cat. *JAVMA* **171:**355–357.

Cali, A., D. Meisler, C. Lowder, R. Lembach, L. Ayers, P. Takvorkian, I. Rutherford, D. Longworth, J. McMaron, and R. Bryan. 1991. Corneal microsporidioses: characterization and identification. *J. Protozool.* **38**(Suppl.):215S–217S.

Cali, A., and R. Owen. 1988. Microsporidiosis, pp. 928–949. *In* A. Balows, W. J. Hausler, and E. Lennette (ed.), *The Laboratory Diagnosis of Infectious Diseases: Principles and Practice.* Springer-Verlag, New York.

Canning, E., and J. Lom. 1986. *The Microsporidia of Vertebrates.* Academic Press, New York, N.Y.

Centers for Disease Control. 1990. Microsporidian keratoconjunctivitis in patients with AIDS. *Morbid. Mortal. Weekly Rep.* **39:**188–189.

Current, W., and R. Owen. 1989. Cryptosporidiosis and microsporidiosis, p. 223–249. *In* M. Farthing and G. Keusch (ed.), *Enteric Infections: Mechanisms, Manifestations and Management.* Chapman and Hall, London.

Davis, R., R. Font, M. Keisler, and J. Shadduck. 1990. Corneal microsporidiosis: a case report including ultrastructural observations. *Ophthalmology* **97:**953–957.

Didier, E. 1997. Effects of albendazole, fumagillin, and TNP-470 on microsporidial replication in vitro. *Antimicrob. Agents Chemother.* **41:**1541–1546.

Didier, E., P. Didier, D. Friedberg, S. Stenson, J. Orenstein, R. Yee, F. Tio, R. Davis, C. Vossbrinck, N. Millichamp, and J. Shadduck. 1991a. Isolation and charcterization of a new human microsporidian, *Encephalitozoon hellum* (n. sp.), from three AIDS patients with keratoconjunctivitis. *J. Infect. Dis.* **163:**617–621.

Didier, E., D. Kotler, D. Dieterich, J. M. Orenstein, A. M. Aldras, R. Davis, D. N. Friedberg, W. K. Gourley, R. Lembach, C. Y. Lowder, D. M. Meisler, I. Rutherford, R. W. Yee, and J. A. Shadduck. 1993. Serological studies in human microsporidiosis. *AIDS* **7:**S8–S11.

Didier, E., L. Rogers, A. Brush, W. S., V. Traina-Dorge, and D. Bertucci. 1996a. Diagnosis of disseminated microsporidian *Encephalitozoon hellem* infection by PCR-Southern analysis and successful treatment with albendazole and fumagillin. *J. Clin. Microbiol.* **34:**947–952.

Didier, E., L. B. Rogers, J. M. Orenstein, M. D. Baker, C. R. Vossbrinck, T. van Gool, R. Hartskeerl, R. Soave, and L. M. Beaudet. 1996b. Characterization of *Encephalitozoon* (*Septata*) *intestinalis* isolates cultured from the nasal mucosa and bronchoalveolar lavage fluids from two AIDS patients. *J. Eukaryot. Microbiol.* **43:**34–43.

Didier, E., J. Shadduck, P. Didier, N. Millichamp, and C. Vossbrinck. 1991b. Studies of ocular microsporidia. *J. Protozool.* **38:**635–638.

Didier, E., M. Vossbrink, L. Baker, L. B. Rogers, D. C. Bertucci, and J. A. Shadduck. 1995. Identification and characterization of three *Encephalitozoon cuniculi* strains. *Parasitology* **111:**411–421.

Didier, P., E. Didier, J. Orenstein, and J. Shadduck. 1991c. Fine structure of a new microsporidian, *Encephalitozoon hellem,* in culture. *J. Protozool.* **38:**502–507.

Diesenhouse, M., L. Wilson, G. Corrent, G. Visvesvara, H. Grossniklaus, and R. Bryan. 1993. Treatment of microsporidial keratoconjunctivitis with topical fumagillin. *Am. J. Ophthalmol.* **115:**293–298.

Dieterich, D., E. Lew, D. Kotler, M. Poles, and J. Orenstein. 1994. Treatment with albendazole for intestinal disease due to *Enterocytozoon bieneusi* in patients with AIDS. *J. Infect. Dis.* **169:**178–183.

Edlind, T., T. Hang, and P. Chakraborty. 1990. Activity of the antihelmintic benzimidazoles against *Giardia lambia* in vitro. *J. Infect. Dis.* **162:**1408.

Edwards, G. and A. Breckenridge. 1988. Clinical pharmacokinetics of antihelminthic drugs. *Clin. Pharmacokinet.* **115:**67–93.

Figg, W., J. Pluda, R. Lush, M. Saville, K. Wyvill, E. Reed, and R. Yarchoan. 1997. The pharmacokinetics of TNP-470, a new angiogenesis inhibitor. *Pharmacotherapy* **17:**91–97.

Friedberg, D., E. Didier, and R. Yee. 1993. Microsporidial keratoconjunctivitis. *Am. J. Ophthalmol.* **116:**380–381.

Friedberg, D., S. Stenson, J. Orenstein, P. Tierno, and N. Charles. 1990. Microsporidial keratoconjunctivitis in acquired immunodeficiency syndrome. *Arch. Ophthalmol.* **108:**504–508.

Fujimoto, J., M. Hori, S. Ichigo, Hirose, H. Sakaguchi, and T. Tamaya. 1997. Plausible novel therapeutic strategy of uterine endometrial cancer with reduction of basic fibroblast growth factor secretion by progestin and O-(chloroacetyl-carbamoyl) fumagillol. *Cancer Lett.* **113:**187–194.

Garvey, M., P. Ambrose, and J. Ulmer. 1995. Topical fumagillin in the treatment of microsporidial keratoconjunctivitis in AIDS. *Ann. Pharmacother.* **29:**872–874.

Gritz, D., D. Holsclaw, R. Neger, J. Whitchr and T. Margolis. 1997. Ocular and sinus microsporidial infection cured with systemic albendazole. *Am. J. Ophthalmol.* **124:**241–243.

Hartskeerl, R., A. Schuitema, T. van Gool, and W. Terpstra. 1993. Genetic evidence for the occurrence of extra-intestinal *Enterocytozoon bieneusi* infections. *Nucleic Acids Res.* **21:**4150.

Hartwig, A., and A. Przellecka. 1971. Nucleic acids in intestine of *Apis mellifica* infected with *Nosema apis* and treated with fumagillin DCH. *J. Invertbr. Pathol.* **18:**331.

Hollister, W., E. Canning, N. Colbourn, A. Curry, and C. Lacey. 1993. Characterization of *Encephalitozoon hellem* (Microspora) isolated from the nasal mucosa of a patient with AIDS. *Parasitology* **107:**351–358.

Honore, P., S. Houze, C. Sarfati, S. Challier, G. Kac, J. Le Bras, and F. Derouin. 1996. Contribution of trichrome blue in the diagnosis of microsporidiosis. *Bull. Soc. Pathol. Exot.* **89:**179–180.

Ingber, D., T. Fujita, S. Kishimoto, K. Sudo, T. Kanamura, H. Brem, and J Folkman (1990). Synthetic analogues of fumagillin that inhibit angiogenesis and suppress tumor growth. *Nature* **348:**555–557.

Jaronski, S. 1972. Cytological evidence for RNA synthesis inhibition by fumagillin. *J. Antibiot.* **25:** 327.

Joste, N., J. Rich, K. Busam, and D. Schwartz. 1996. Autopsy verification of *Encephalitozoon intestinalis* (microsporidiosis) eradication following albendazole therapy. *Arch. Pathol. Lab. Med.* **120:** 199–203.

Katznelson, H., and C. Jamieson. 1952. Control of *Nosema* disease of honeybees with fumagillin. *Science* **115:**70–71.

Keohane, E., P. Takorian, A. Cali, H. Tanowitz, M. Wittner, and L. Weiss. 1996. Identification of a microsporidian polar tube protein reactive monoclonal antibody. *J. Eukaryot. Microbiol.* **43:** 26–31.

Killough, J., G. Magill, and R. Smith. 1952. The treatment of amebiasis with fumagillin. *Science* **115:**71–72.

Lacey, C., A. Clarke, P. Fraser, T. Metcalfe, G. Bonsor, and A. Curry. 1992. Chronic microsporidian infection of the nasal mucosae, sinuses and conjunctivae in HIV disease. *Genitourin. Med.* **68:**179–181.

Lacey, E. 1985. The role of cytoskeletal protein, tubulin, in the mode of action and mechanism of drug resistance to benzimidazoles. *Int. J. Parasitol.* **18:** 855–936.

Lecuit, M., E. Oksenhendler, and C. Sarfati. 1994. Use of albendazole for disseminated microsporidial infection in a patient with AIDS. *Clin. Infect. Dis.* **19:**332–333.

Ledford, D., M. Overman, A. Gonzalo, A. Cali, W. Mester, and R. Lockey. 1985. Microsporidiosis myositis in a patient with the acquired immunodeficiency syndrome. *Ann. Intern. Med.* **102:**628–630.

Levine, N., J. Corliss, F. Cox, G. Deroux, J. Grain, B. Honigberg, G. Leedale, A. Loeblich III, J. Lom, D. Lynn, E. Merinfeld, F. Page, G. Poljansky, V. Sprague, J. Vavra, and F. Wallace. 1980. A newly revised classification of the protozoa. *J. Protozool.* **27:**37–58.

Lowder, C., J. McMahon, D. Meisler, E. Dodds, L. Calabrese, E. Didier, and A. Cali. 1996. Microsporidial keratoconjunctivitis caused by *Septata intestinalis* in a patient with acquired immunodeficiency syndrome. *Am. J. Ophthalmol.* **121:**715–717.

Lowder, C., D. Meisler, J. McMahon, D. Longworth, and I. Rutherford. 1990. Microsporidia infection of the cornea in a man seropositive for human immunodeficiency virus. *Am. J. Ophthalmol.* **109:**242–244.

Lukosz, W. 1966. Optical systems with resolving powers exceeding the classical limit. *J. Opt. Soc. Am.* **57:**1190.

McCluskey, P., P. Goonan, D. Marriott, and A. Field. 1993. Microsporidial keratoconjunctivitis in AIDS. *Eye* **7:**80–83.

McCowen, M., M. Callender, and J. J. Lawles. 1951. Fumagillin (H-3), a new antibiotic with amebicidal properties. *Science* **113:**202–203.

Metcalfe, T., R. Doran, P. Rowlands, A. Curry, and C. Lacey. 1992. Microsporidial keratoconjunctivitis in a patient with AIDS. *Br. J. Ophthalmol.* **76:**177–178.

Molina, J., J. Goguel, C. Sarfati, C. Chastang, I. Desportes-Livage, J.-F. Michiels, C. Maslo, C. Katlama, L. Cotte, C. Leport, F. Raffi, F. Derouin, J. Modai, and the French Microsporidoisis Study Group. 1997. Potential efficacy of fumagillin in intestinal microsporidiosis due to *Enterocytozoon bieneusi* in patients with HIV infection: results of a drug screening study. *AIDS* **11:**1603–1610.

Molina, J., E. Oksenhendler, B. Beauvais, C. Scarfati, A. Jaccard, and F. Dermouin. 1995. Disseminated microsporidiosis due to *Septata in-*

testinalis in patients with AIDS: clinical features and response to albendazole therapy. *J. Infect. Dis.* **171:**245–249.

Orenstein, J. 1991. Microsporidiosis in the acquired immunodeficiency syndrome. *J. Parasitol.* **77:**843–864.

Perrin, T. 1943. Spontaneous and experimental encephalitozoon infection in laboratory animals. *Arch. Pathol.* **36:**559–567.

Perrine, D., J. Chenu, P. Georges, J. Lancelot, C. Saturnino, and M. Robba. 1995. Amoebicidal efficiencies of various diamidines against two strains of *Acanthamoeba polyphagia. Antimicrob. Agents Chemother.* **39:**339–342.

Petroll, W., J. Jester, and H. Cavanaugh. 1996. Quantitative three-dimensional confocal imaging of the cornea in situ and in vivo system design and calibration. *Scanning* **18:**45–49.

Pinnolis, M., P. Egbert, R. Font, and F. Winter. 1981. Nosematosis of the cornea. *Arch. Ophthalmol.* **99:**1044–1047.

Pluda, J., K. Wyvill, Lietzan, D. Figg, Saville, B. Ngyuen, A. Foli, J. Bailey, M. Cooper, S. Broder, and R. Yarchoan. 1993. A phase I trial of TNP-470 (AGM-1470) administered to patients with HIV-associated Kaposi's sarcoma (KS). 1st National Conference on Human Retroviruses and Related Infections, Alexandria, Virginia, Infectious Disease Society of America.

Rijpstra, A., E. Canning, R. Van Ketel, J. Eeftinck Schattenkerk, and J. Laarman. 1988. Use of light microscopy to diagnose small-intestinal microsporidiosis in patients with AIDS. *J. Infect. Dis.* **157:**827–831.

Rosberger, D., Serdarevic, R. Erlandson, R. Bryan, D. Schwartz, G. Visvesvara, and P. Keenan. 1993. Successful treatment of microsporidial keratoconjunctivitis with topical fumagillin in a patient with AIDS. *Cornea* **12:** 261–265.

Schwartz, D., R. Byran, K. Hewan-Lowe, G. Visvesvara, R. Weber, A. Cali, and P. Angritt. 1992. Disseminated microsporidiosis (*Encephalitozoon hellem*) and acquired immunodeficiency syndrome. *Arch. Pathol. Lab. Med.* **116:** 660–668.

Schwartz, D., I. Sobottka, G. Leitch, A. Cali, and G. Visvesvara. 1996. Pathology of microsporidiosis. *Arch. Pathol. Lab. Med.* **120:**173–188.

Schwartz, D., G. Visvesvara, M. Diesenhouse, R. Weber, R. Font, L. Wilson, G. Corrent, O. Serdarevis, D. Rosberger, P. Keenen, H. Grossniklaus, K. Hewan-Lowe, and R. Bryan. 1993. Pathologic features and immunofluorescent antibody demonstration of ocular microsporidiosis (*Encephaliozoon hellem*) in seven patients with acquired immunodeficiency syndrome. *Am. J. Ophthalmol.* **115:**285–292.

Shadduck, J. 1980. Effect of fumagillin on in vitro multiplication of *Encephalitozoon cuniculi. J. Protozool.* **27:**202–208.

Shadduck, J. 1989. Human microsporidiosis and AIDS. *Rev. Infect. Dis.* **11:**203–207.

Shadduck, J., and E. Greeley. 1989. Microsporidia and human infections. *Clin. Microbiol. Rev.* **2:**158–165.

Shadduck, J., R. Meccoli, R. Davis, and R. Font. 1990. Isolation of a microsporidian from a human patient. *J. Infect. Dis.* **162:**773–776.

Shah, G., D. Pfister, L. Probst, P. Ferrieri, and E. Holland. 1996. Diagnosis of microsporidial keratitis by confocal microscopy and chromatrope stain. *Am. J. Ophthalmol.* **121:**89–91.

Silveira, H., and E. Canning. 1995a. In vitro cultivation of the human microsporidium *Vittaforma corneae*: development and effect of albendazole. *Folia Parisitol.* **42:**241–250.

Silveira, H., and E. Canning. 1995b. *Vittaforma corneae* n. comb. for the human microsporidium *Nosema corneum* Shadduck, Meccoli, Davis & Font, 1990, based on its ultrastructure in the liver of experimentally infected athymic mice. *J. Eukaryot. Microbiol.* **42:**158–165.

Silverstein, B., E. J. Cunningham, T. Margolis, V. Cevallos, and I. Wong. 1997. Microsporidial keratoconjunctivitis in a patient without human immunodeficiency virus infection. *Am. J. Ophthalmol.* **124:**395–396.

Sprague, V. 1977. Systemics of microsporidia, p. 1–29. *In* L. Bulla and T. Cheng (ed.), *Comparative Pathobiology.* Plenum Press, New York, N.Y.

Steiger, U., J. Cotting, and J. Reichen. 1990. Albendazole treatment of echinococcus in humans: effects on microsomal metabolism and drug tolerance. *Clin. Pharmacol. Ther.* **47:**347–353.

Visvesvara, G., G. Leitch, A. Da Silva, G. P. Croppo, H. Moura, S. Wallace, S. B. Slemenda, D. A. Schwartz, D. Moss, R. T. Bryan, and N. J. Pieniazek. 1994. Polyclonal and monoclonal antibody and PCR-amplified small-subunit rRNA identification of a microsporidian, *Encephalitozoon hellem,* isolated from an AIDS patient with disseminated infection. *J. Clin. Microbiol.* **32:**2760–2768.

Vossbrinck, C., M. Baker, E. Didier, B. Debrunner-Vossbrinck, and J. Shadduck. 1993. Ribosomal DNA sequences of *Encephalitozoon hellem* and *Encephalitozoon cuniculi*: species identification and phylogenetic construction. *J. Eukaryot. Microbiol.* **40:**354–362.

Weber, R., R. Bryan, R. Owen, C. M. Wilcox, L. Gorelkin, and G. S. Visvesvara. 1992. Improved light-microscopical detection of microsporidia spores in stool and duodenal aspirates. *N. Engl. J. Med.* **326:**161–166.

Weiss, L. 1995. And now microsporidiosis. *Ann. Intern. Med.* **123:**954–956.

Wilkins, J., N. Joshi, T. Margolis, V. Cevallos, and C. Dawson. 1994. Microsporidial keratoconjunctivitis treated successfully with a short course of fumagillin. *Eye* **8:**703–704.

Wilson, T. 1986. Confocal light microscopy. *Ann. N. Y. Acad. Sci.* **483:**416–427.

Wilson, T., and C. Shepard. 1984. *Theory and Practice of Scanning Optical Microscopy.* Academic Press, London, United Kingdom.

Wolf, A., and D. Cowen. 1937. Granulomatous encephalomyelitis due to encephalitozoon (encephalitozoic encephalomyelitis). *Bull. Neurol. Inst. N. Y.* **6:**306–371.

Yanase ,T., M. Tamura, K. Fujita, S. Kodama, and K. Tanaka. 1993. Inhibitory effect of angiogenesis inhibitor TNP-470 on tumor growth and metastasis of human cell lines in vitro and in vivo. *Cancer Res.* **53:**2566–2570.

Yee, R., F. Tio, A. Martinez, K. Held, J. Shadduck, and E. Didier. 1991. Resolution of microsporidial epithelial keratopathy in a patient with AIDS. *Ophthalmology* **98:**196–201.

Zhu, X., H. Wittner, B. Tanowitz, et al. 1993. Small subunit rRNA sequence of *Enterocytozoon bieneusi* and its potential diagnostic role with use of the polymerase chain reaction. *J. Infect. Dis.* **168:**1570–1575.

LABORATORY DIAGNOSIS OF MICROSPORIDIOSIS

Rainer Weber, David A. Schwartz, and Peter Deplazes

10

The most robust and widely practicable technique for the diagnosis of microsporidial infection is light microscopic morphological demonstration of the organisms themselves. The spores, the stages by which microsporidia usually are identified, are small, ranging in size from 1 to 3 μm in most species found in humans, and are not sufficiently stained by the routine staining techniques used in parasitology and pathology laboratories to diagnose other protozoa. Therefore, microscopic visualization of the parasites requires special staining methods and adequate microscopic techniques including sufficient illumination and magnification (Weber et al., 1994a). Demonstration of spores and other developmental stages by fluorescein-labeled monoclonal or polyclonal antibodies has been performed experimentally, but none of these antibody detection procedures has been commercialized.

Rainer Weber, Division of Infectious Diseases and Hospital Epidemiology, Department of Internal Medicine, University Hospital, CH-8091 Zurich, Switzerland. *David A. Schwartz,* Department of Pathology, Grady Memorial Hospital, 80 Butler Street SE, Atlanta, Georgia 30335. *Peter Deplazes,* Institute of Parasitology, University of Zurich, Winterthurerstrasse 266A, CH-8057 Zurich, Switzerland.

Microsporidial ultrastructure is unique and pathognomonic for the phylum, and electron microscopy allows one to distinguish among all microsporidial genera. Nevertheless, morphological features alone do not sufficiently characterize all microsporidial species pathogenic in humans; i.e., characterization of the three *Encephalitozoon* spp., which share most of their morphological features, requires antigenic or molecular analyses which may also reveal subspecific variation.

Nucleic acid-based methods have been developed for diagnostic purposes and species identification and are invaluable for taxonomic classification and phylogenetic analyses (Weiss and Vossbrinck, 1998).

In vitro isolation of microsporidia has no relevance for diagnostic purposes, but it is an important research tool. All three *Encephalitozoon* spp. pathogenic in humans, *Trachipleistophora hominis,* and *Vittaforma corneae* have been isolated with different cell culture systems. Only short-term in vitro propagation has been accomplished with *Enterocytozoon bieneusi* (Visvesvara, 1995b).

Serologic assays have been useful in detecting antibodies to *Encephalitozoon cuniculi* in several species of animals, but validated serologic

The Microsporidia and Microsporidiosis (Murray Wittner, editor; Louis M. Weiss, contributing editor), ©1999 American Society for Microbiology, Washington, D.C.

tests for diagnosis of human microsporidiosis are lacking. Available results suggest that antibody detection may be particularly difficult or unreliable in human immunodeficiency virus (HIV)-infected patients with advanced immunodeficiency.

APPROACH TO DIAGNOSIS IN HUMANS

Current data suggest that microsporidia are opportunistic pathogens capable of causing disease predominantly in HIV-infected persons and are also emerging pathogens causing infection in otherwise immunocompromised hosts including HIV-seronegative recipients of organ transplants (Rabodonirina et al., 1996; Sax et al., 1995) (Table 1). Patients with severe cellular immunodeficiency appear at highest risk for developing microsporidial disease. Therefore, it is prudent to consider microsporidia the etiologic agents when they are detected in clinical specimens of such patients. Preliminary observations indicate that microsporidia may also cause illness in immunocompetent and otherwise healthy persons (Bornay-Llinares et al., 1998; Enriquez et al., 1997a; Raynaud et al., 1998; Sandfort et al., 1994; Sobottka et al., 1995b). Reported human infections are globally dispersed and have been documented in persons from all continents (Weber et al., 1994a).

TABLE 1 Microsporidial species pathogenic in humans and their clinical manifestations

Microsporidial species	Organ manifestations	
	Immunocompromised patients	Immunocompetent persons
Enterocytozoon bieneusi	Chronic diarrhea; wasting syndrome; "AIDS cholangiopathy," cholangitis, acalculous cholecystitis; chronic sinusitis, chronic cough, pneumonitis	Self-limiting diarrhea in adults and children; traveler's diarrhea; asymptomatic carriers
Encephalitozoon hellem	Disseminated infection; keratoconjunctivitis; sinusitis, bronchitis, pneumonia; nephritis, ureteritis, cystitis, prostatitis, urethritis	Not described
Encephalitozoon intestinalis (formerly *Septata intestinalis*)	Chronic diarrhea; cholangiopathy; sinusitis, bronchitis, pneumonitis; nephritis; bone infection	Self-limiting diarrhea; asymptomatic carriers
Encephalitozoon cuniculi	Disseminated infection; keratoconjunctivitis; sinusitis, bronchitis, pneumonia; nephritis; hepatitis; peritonitis; symptomatic and asymptomatic intestinal infection; encephalitis	Not described; two HIV-seronegative children with seizure disorder and presumed *E. cuniculi* infection presumably were immunocompromised
Pleistophora spp.	Myositis	Not described
Trachipleistophora hominis	Myositis; keratoconjunctivitis; sinusitis	Not described
Trachipleistophora anthropophthera	Disseminated infection, encephalitis	Not described
Nosema connori	Disseminated infection	Not described
Nosema ocularum	Not described	Keratitis (corneal stroma infection)
Vittaforma corneae (formerly *Nosema corneum*)	Disseminated infection	Keratitis (corneal stroma infection)
Nosema-like microsporidian[a]	Myositis	Not described
Microsporidium ceylonensis[b]	Not described	Corneal ulcer, keratitis (corneal stroma infection)
Microsporidium africanum[b]	Not described	Corneal ulcer, keratitis (corneal stroma infection)

[a] Species not yet classified.
[b] *Microsporidium* is a collective generic name for microsporidia that cannot be classified because available information is not sufficient.

13 Total

Detection of microsporidial parasites has initially been based on electron microscopic examination of tissue specimens because of the organisms' small size and staining properties. In recent years, initial detection of microsporidia by light microscopic examination has been shown to be sensitive and specific and thus has become routine practice (Kotler et al., 1994; Rijpstra et al., 1988; van Gool et al., 1993; Weber et al., 1992a). Evaluation of patients with suspected microsporidiosis should begin with light microscopic examination of stool specimens and urine or cytological examination of other body fluids. Definitive species identification is made with immunofluorescent staining, electron microscopy, antigenic or biochemical analysis, or molecular analysis. Cytological methods are preferred for monitoring of therapy. When cytological techniques cannot be used or are negative, histologic and electron microscopic tissue examinations are employed (Kotler et al., 1994).

Morphological demonstration of microsporidia in specimens obtained from immunocompromised patients, although sensitive and specific, does not usually allow identification of the organisms to the genus and species level. Epidemiological studies in patients not infected with HIV are lacking, but it is assumed on the basis of case observations that immunocompetent persons may excrete lower numbers of microsporidial spores in feces or urine, and therefore the threshold of current detection procedures may not be sufficiently reliable to detect microsporidia in this group.

Clinical and epidemiological studies using polymerase chain reaction (PCR) methodology to detect and identify microsporidia in clinical samples have demonstrated that this approach is sensitive, specific, and feasible as a standard diagnostic procedure (Weiss and Vossbrinck, 1998). Yet, molecular diagnostic procedures are at present more labor-intensive than microscopic techniques, require special equipment, and are costly (Owen, 1997).

Electron microscopy is at present still the "gold standard" for diagnostic confirmation and species identification although the technique is laborious and relatively insensitive for detection of microsporidia (because only small samples are examined and sampling error may occur) and not all microsporidia pathogenic in humans can be identified to the species level by this technique.

Fresh material (without fixative) may be useful for cell culture if available and for future molecular analysis and/or for type collection.

Intestinal Microsporidiosis

In patients with suspected enteric microsporidiosis, examination of stool specimens is a first step that has a high diagnostic yield in many cases (Didier et al., 1995; van Gool et al., 1990, 1993; Weber et al., 1992a, 1994a). Stool examination is at least as sensitive as examination of biopsy specimens (Verre et al., 1993). It is not known, however, whether excretion of microsporidial spores is intermittent or if the sensitivity of stool examination can be improved by screening more than one specimen. Furthermore, the minimum number of *E. bieneusi* spores that can be detected in stool specimens by routine diagnostic procedures is not defined, and it is unknown if the number of spores shed in active intestinal disease is continually above the threshold of detection. Dilution experiments using stool specimens seeded with *Encephalitozoon intestinalis* spores harvested from tissue culture suggested a lower limit for detecting microsporidia in stool of approximately 5×10^4 spores per ml when light microscopy was used (Didier et al., 1995). Another study showed the threshold of detection of *Encephalitozoon* spores to be between 10^4 and 10^6 spores per ml by microscopic examination, and 10^2 spores per ml with the PCR technique (Rinder et al., 1998, in press).

Intestinal microsporidial spores have also been detected in sediments of duodenal aspirate, bile, or biliary aspirates obtained with endoscopic papillary cannulation, transhepatic catheterization, and during surgical cholecystectomy.

Epidemiological studies have yet to determine the optimal diagnostic approach in HIV-infected patients with chronic diarrhea in whom comprehensive stool examination is negative. Particularly, further studies are needed to assess when and in what order or combination endoscopic procedures should be used to obtain biopsies or intestinal fluid. Although microsporidia tend to be most numerous in the jejunum, examination of duodenal as well as terminal ileal tissue has also resulted in detection of the parasites (Weber et al., 1992c). Microsporidia are rarely found in colonic tissue sections. Colonoscopy may remain an important diagnostic tool because it permits detection of cytomegalovirus colitis, a treatable condition occurring in patients with AIDS that is not diagnosed by stool examination. Extension of this procedure to the distal ileum may provide a reliable means for diagnosing intestinal microsporidiosis, sparing the patient the additional discomfort of upper endoscopy.

Disseminated Microsporidiosis

Microsporidial species that cause systemic infection are best detected in urine sediments, respiratory specimens (including bronchoalveolar lavage fluid, sputum, nasal discharge, nasopharyngeal washings, and sinus aspirates), cerebrospinal fluid, and smears from conjunctival swabs.

Because multiple organ involvement occurs, detection of microsporidia in virtually any tissue or body fluid should prompt a thorough search of other sites and body fluids. Because dual infection due to two different microsporidial species has been observed, characterization of microsporidial isolates obtained from different anatomic sites may be indicated (Deplazes et al., in press).

Keratoconjunctival Microsporidiosis

The diagnosis of ocular infection is important in patients with AIDS because it is often the initial sign of systemic microsporidial infection. The most common clinical finding of ocular microsporidiosis is keratoconjunctivitis. The presence of numerous minute corneal ulcers, termed punctate epithelial keratopathy, is characteristic of this condition and are identified by slit lamp examination. Characteristic microsporidial spores can be demonstrated in the corneal and conjunctival epithelium (Schwartz et al., 1993a; Shah et al., 1996).

Cytological preparations, including scrapings, smears, and nontraumatic swab specimens from the conjunctiva or cornea, are ideal methods for demonstrating microsporidial spores. In patients with suspected ocular microsporidiosis, urine and respiratory secretions also should be examined for microsporidia. It is important to identify an ocular infection to the level of species using antibody-based or molecular methods. *Encephalitozoon* spp. cannot be distinguished by morphological features alone.

Rarely, localized microsporidial infection of the cornea (without systemic dissemination) has also been diagnosed by histologic or electron microscopic examination of corneal biopsies in otherwise healthy immunocompetent persons (Weber et al., 1994a).

COLLECTION, TRANSPORT, AND STORAGE OF SPECIMENS

Microsporidial spores are environmentally highly resistant and, if prevented from drying, can remain stable and infectious for periods of up to several years. They can be stored at -20°C for years.

Spores of enteropathogenic microsporidia can be stained and detected in stool specimens or duodenal aspirates that have been fixed in 10% formalin or in sodium acetate-acetic acid-formalin (SAF), in fresh stool samples, or in biopsy specimens. Spores of microsporidia causing disseminated infection can usually be detected in fresh or fixed urine sediments, in other body fluids including sputum, bronchoalveolar lavage fluid, nasal secretions, cerebrospinal fluid, conjunctival smears, and corneal scrapings, or in tissue. For histologic examination, tissue specimens are fixed in formalin. For electron microscopy, fixation of tissue with glutaraldehyde is preferred.

Fresh material (without fixative), stored at 4°C in physiological saline or culture medium supplemented with antibiotics, may be useful for cell culture if available. Also, fresh material is preferred for molecular analyses. Because mi-

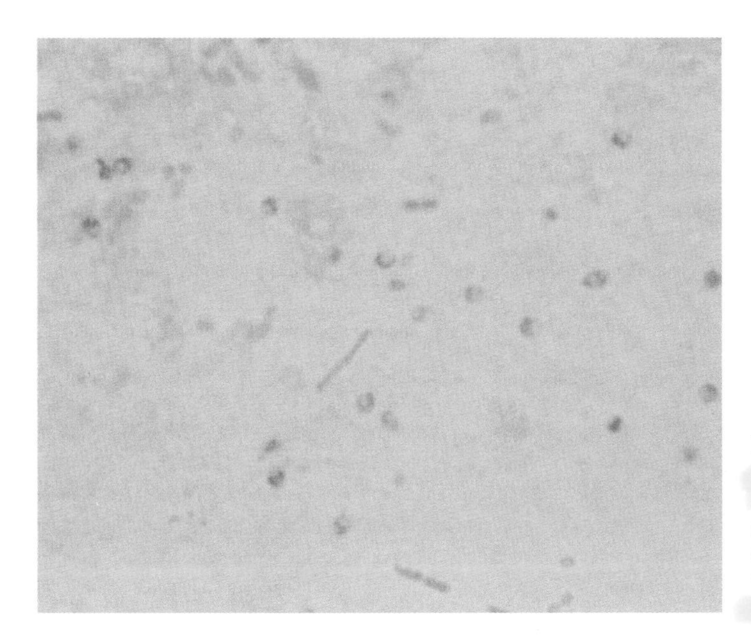

FIGURE 1 Stool specimen from a patient with AIDS and chronic diarrhea, showing pinkish red-stained spores of *Enterocytozoon bieneusi* measuring 0.7 to 1.0 by 1.1 to 1.6 μm. Chromotrope stain was used. Magnification, ×1,000 (oil immersion).

crosporidial spores are highly resistant, fresh (unfixed) clinical specimens can be mailed to laboratories at room temperature.

COPRODIAGNOSIS

Procedures

For examination of stool specimens, the most commonly used stains are chromotrope-based stains (Fig. 1) and chemofluorescent optical brightening agents (Fig. 2). In many laboratories, tests for microsporidia must be specifically requested because the general request for "stool for O & P" (ova and parasites) often does not mean that the specific methods for detecting these organisms are applied. Regardless of which staining technique is employed, the use of positive control material is essential. Detection of microsporidial spores requires adequate illumination and magnification, i.e., ×630 or ×1,000 magnification (oil immersion).

The difference in size between the two major intestinal microsporidia—*Enterocytozoon* and *Encephalitozoon*—should permit the careful observer to make a tentative genus level diagnosis from a stool examination (Weber et al., 1994b). It is critical that intestinal microsporidia be identified to the level of genus because *Encephalitozoon* spp. have a different propensity for dissemination and a different drug sensitivity pattern than *Enterocytozoon* spp. As discussed elsewhere in this book, *E. intestinalis* infection is potentially curable with albendazole. Electron microscopic examination has been successfully used for confirmation of microsporidial spores in feces and to distinguish between *Enterocytozoon* and *Encephalitozoon* spp. (Weber et al., 1994b). However, it is both difficult and time-consuming, and stool electron microscopy is not recommended for routine diagnostic use.

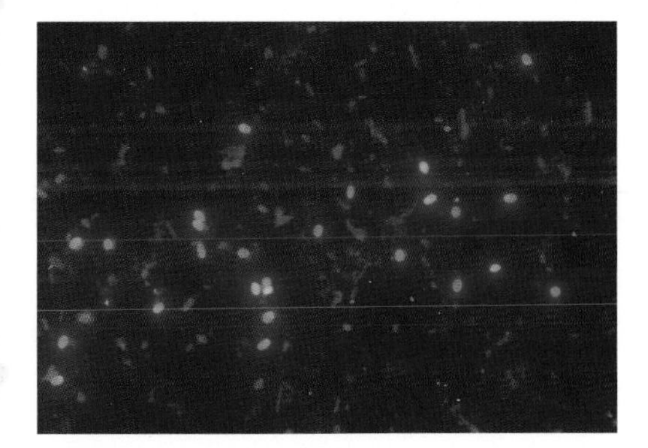

FIGURE 2 Stool specimen from a patient with AIDS and chronic diarrhea, showing microsporidial spores stained with a chemofluorescent optical brightening agent (Fungi-Fluor Kit, Polysciences, Inc., Warrington, Pa.). Magnification, ×1,000 (oil immersion).

Future routine diagnostic techniques are likely to include specific immunofluorescence staining of organisms, antigen tests, and molecular techniques. Immunofluorescent procedures for diagnosis of *Encephalitozoon*-like microsporidial spores are promising. Antisera to *Enterocytozoon* have not yet been developed. Nucleic acid-based methods have been developed for diagnostic purposes and species identification but are still limited to specialized laboratories.

Stool Concentration Techniques

Most of the procedures that have been adapted for concentration of ova and parasites fail to concentrate microsporidial spores in stool specimens, although the formalin-ethyl acetate concentration and different flotation methods remove significant amounts of fecal debris and smears prepared from these concentrates appear easier to read by light microscopic examination when compared with smears from unconcentrated specimens. Such concentration techniques, however, lead to a substantial loss of microsporidial spores and may give false-negative results (Weber et al., 1992a).

New concentration techniques including modifications in centrifugation time and speed have been proposed or are being investigated (van Gool et al., 1994a). No data are available confirming whether such techniques may lead to adequate or improved recovery of spores.

In one study, pretreatment of SAF-fixed stool specimens with 10% KOH (as a mucolytic) prior to a 5-min centrifugation appeared to increase the sensitivity of microscopic detection of microsporidial spores (Carter et al., 1996). This approach appears plausible because when stool samples are examined, microsporidial spores are often found clustered in mucuslike structures. We did not find an advantage to this technique in our routine laboratory when we examined clinical specimens, but a controlled study has yet to be performed.

Performance of Coprodiagnostic Techniques

Epidemiological comparisons of the chromotrope staining technique with methods using chemofluorescent optical brighteners indicated that these tests are robust for routine use. The sensitivity of both methods appeared similarly high when clinical samples of patients with AIDS and diarrhea were examined (DeGirolami et al., 1995; Didier et al., 1995; Ignatius et al., 1997). Some laboratories use both staining techniques because chromotrope stains result in a highly specific visualization of spores, whereas chemofluorescent agents might be more sensitive but may produce false-positive results (Didier et al., 1995).

In experienced parasitological laboratories coprodiagnostic procedures appear to be as reliable as histological examination techniques, although only limited results comparing stool examination with endoscopic evaluation are available (Verre et al., 1993; Weber et al., submitted).

Microscopists may not be accustomed to scanning for structures the size of microsporidia by light microscopy, and the spores are indeed so small that they are easily overlooked if attention is not directed to their detection or when microscopic examination techniques are inappropriate. Nevertheless, when positive control slides are used, after adequate training microscopists are able to reliably use light microscopic stool examination procedures for diagnosis of microsporidiosis. A blinded comparison of microscopic techniques with diagnostic PCR in detecting microsporidia in fecal specimens of HIV-infected patients with diarrhea has shown similar results (Rinder et al., 1998, in press). Nevertheless, other clinical studies using the PCR methodology to detect and identify microsporidia in clinical samples have demonstrated that this approach has the potential for improved sensitivity and specificity (Weiss and Vossbrinck, 1998). It is unknown whether symptomatic patients with or without immunodeficiency, or asymptomatic carriers, always excrete such large numbers of spores that microscopic detection is assured. Therefore, studies using molecular diagnosis are necessary to assess the epidemiological extent of microsporidial infection in humans (Owen, 1997). As shown for other parasites, e.g., cryptosporidia (Weber et al., 1991), the threshold necessary for microscopic detection of enteric pathogens in stool specimens may hamper diagnosis. The limit for detecting spores in stool appears to be approximately 5×10^4 spores per ml (Didier et al., 1995), whereas PCR techniques may detect lower spore concentrations of 10^2 per ml (Rinder et al., 1998, in press).

CYTOLOGY

Microsporidial spores have been detected in urine and other body fluids as summarized in Table 2. Disseminated microsporidial infection often involves multiple organs, and therefore detection of microsporidia in virtually any tissue or body fluid should prompt a thorough search of other sites.

High-speed centrifugation (at least $1,500 \times g$ for 10 min) of body fluids, e.g., urine or duodenal aspirate, may be necessary to concentrate organisms in sediments. For cytological examination of body fluids that do not contain substantial background debris, bacteria, or fungi, staining with Gram stain (microsporidial spores are often gram-variable and stain partially gram-positive or dark reddish), Giemsa stain, or chemofluorescent agents may be useful. Diagnostic confidence may be attained using the chromotrope-based staining technique or fluorescein-tagged monoclonal or polyclonal antibodies, which have been favorably applied in identifying microsporidial spores in duodenal aspirate, urine, bronchoalveolar lavage fluid, sputum, and conjunctival smears.

Urine

High-speed centrifugation of urine at $1,500 \times g$ for 10 min is recommended to concentrate spores in sediments because the number of excreted spores may be low. Urine usually does not contain substantial background debris, bacteria, or fungi, and therefore staining with Gram stain, Giemsa stain, or chemofluorescent agents may be useful. Chromotrope stains (Fig. 3) or immunofluorescence detection procedures have allowed sensitive and specific diagnosis of disseminated microsporidiosis (Visvesvara et al., 1991; Weber et al., 1993a).

Because microsporidial spores may be shed periodically, as is seen in mammalian encephalitozoonosis, repeat examination of single urine specimens or a 24-h urine collection may be necessary for detection of *Encephalitozoon* spores.

Duodenal Aspirate

Microscopic examination of stained smears of centrifuged duodenal aspirate obtained by endoscopy is a highly sensitive technique for diagnosis of intestinal microsporidiosis due to *E. bieneusi* or *E. intestinalis* (Weber et al., 1992a, 1994a). Although there are no epidemiological data evaluating this technique, it is plausible to assume that the diagnostic yield for the examination of duodenal aspirates is probably higher than that of stool examination because of the following reasons: (1) duodenal aspirates do not contain as much background debris as stool, and thus microsporidial spores can be concentrated by high-speed centrifugation, resulting in a higher spore concentration in the sediment, and (2) spores can be better visualized because of the low amount of background debris. The most useful stains for examination of such aspirates include chromotrope stains or chemofluorescent agents.

Respiratory Specimens and Other Body Fluids

Respiratory specimens (including sinonasal washings, sputum, and bronchoalveolar lavage fluid), ascites fluid, cerebrospinal fluid, or other body fluids are prepared using standard cytological techniques and stained with Gram stain (Fig. 4), Giemsa stain (Fig. 5), chromotrope stains (Fig. 6), chemofluorescent agents or immunofluorescence (Fig. 7) techniques, if available (Rijpstra et al., 1988; Sobottka et al., 1995a; Weber et al., 1993a, 1994a, 1997a).

Conjunctival Smears and Corneal Scrapings

The diagnosis of microsporidial conjunctivitis or keratitis can be established by means of a nontraumatic conjunctival swab specimen which is smeared on a slide and stained (Fig. 8) or by examination of conjunctival and/or corneal scrapings or biopsy specimens. Useful staining techniques include chromotrope stains, Gram or Giemsa stain, and fluorescein-tagged

TABLE 2 Sites of detection of microsporidia[a]

Specimen	*Enterocytozoon bieneusi*	*Encephalitozoon intestinalis*	*Encephalitozoon hellem*	*Encephalitozoon cuniculi*	Other microsporidial species
Body fluids or mucosal smears					
Stool	++	++	+	+	–
Duodenal aspirate	++	+	–	–	–
Bile	+	–	–	–	–
Urine	–	+	++	++	*V. corneae*
Sputum, bronchoalveolar lavage fluid	+	+	++	++	–
Nasal secretion (swab), nasal washing	+	+	+	+	*T. hominis*
Conjunctival smear	–	+	+	+	–
Corneal scraping	–	–	+	+	*V. corneae, N. ocularum*
Cerebrospinal fluid	–	–	–	+	–
Tissue					
Duodenum, jejunum, terminal ileum	++	++	–	+	*N. connori*
Colon	+	+	–	–	*N. connori*
Biliary tract, gallbladder	+	+	–	–	–
Liver	+	–	–	–	*T. anthropophthera, N. connori*
Pancreatic duct, pancreas	+	–	–	–	*T. anthropophthera*
Peritoneum	–	–	–	+	–
Bronchial epithelium, sinus, nasal epithelium	+	+	+	+	*N. connori*
Cornea, conjunctiva	–	+	+	+	*V. corneae, N. ocularum, M. ceylonensis, M. africanum, T. hominis*
Kidney	–	+	+	+	*T. anthropophthera, N. connori*
Ureter, urinary bladder, prostate, urethra	–	–	+	+	*N. connori*
Brain	–	+[b]	–	+	*T. anthropophthera*
Muscle	–	–	–	+	*Pleistophora* sp., *T. hominis, Nosema*–like microsporidian
Heart	–	–	–	+	*T. anthropophthera, N. connori*
Lymph nodes	–	–	–	–	*T. anthropophthera*
Spleen	–	–	–	–	*T. anthropophthera*
Bone marrow	–	–	–	–	*T. anthropophthera*
Bone	–	+	–	–	–
Thyroid, parathyroid	–	–	–	–	*T. anthropophthera*
Adrenal glands	–	–	–	+	*N. connori*

[a] –, not reported; +, case report(s) of detection; ++, consistently reliable specimens for detection of microsporidia.
[b] Pituitary gland.

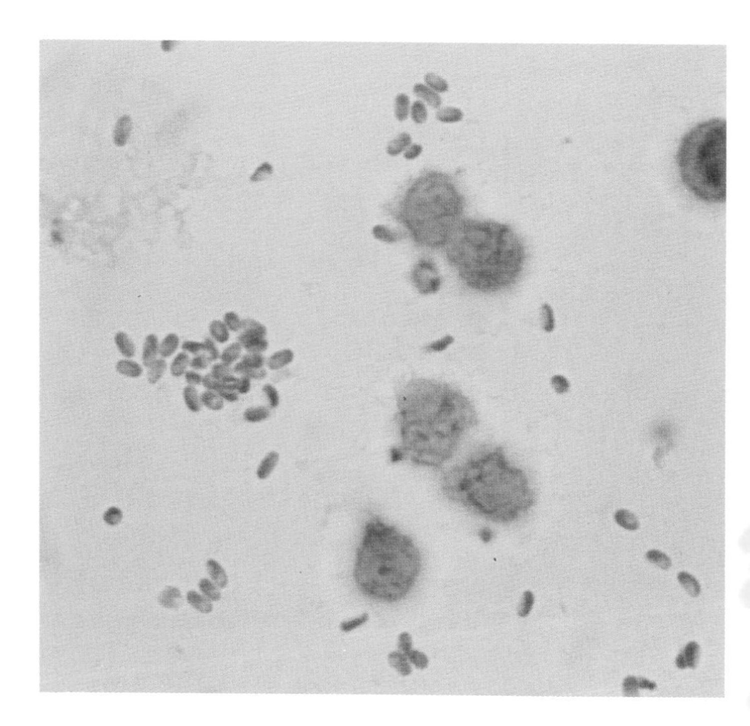

FIGURE 3 Urine sediment from a patient with AIDS and disseminated *Encephalitozoon cuniculi* infection, showing pinkish red-stained microsporidial spores measuring 1.0 to 1.5 by 2.0 to 3.0 μm. Chromotrope stain was used. Magnification, ×1,000 (oil immersion).

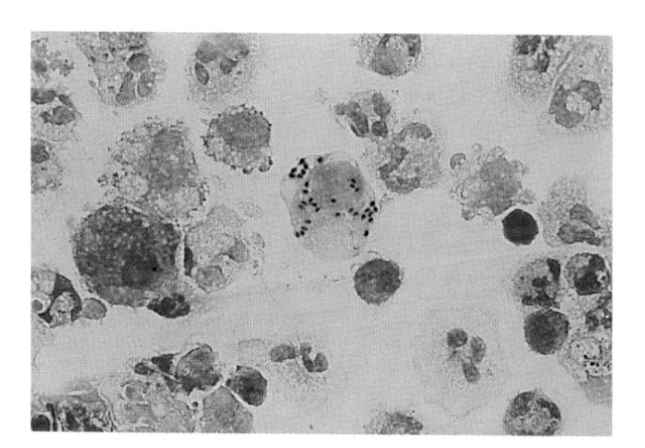

FIGURE 4 Cytospin preparation of bronchoalveolar lavage fluid from a patient with AIDS and intestinal *E. bieneusi* infection, showing intracellular gram-positive microsporidial spores. Gram stain was used. Magnification ×1,000 (oil immersion).

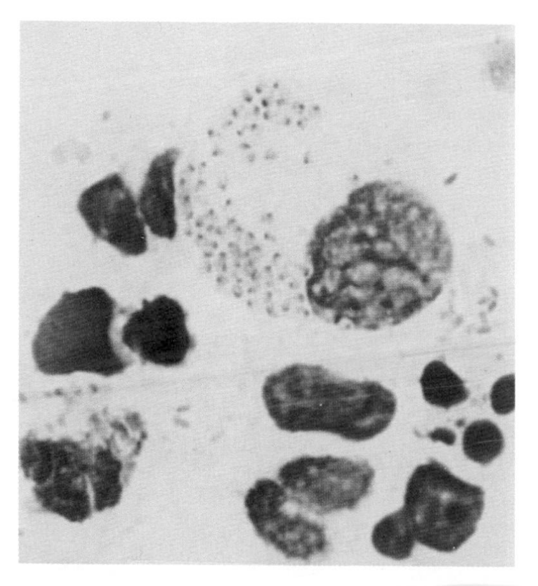

FIGURE 5 Cytospin preparation of bronchoalveolar lavage fluid from a patient with AIDS and intestinal *E. bieneusi* infection, showing intracellular microsporidia. Giemsa stain was used. Magnification, ×1,000 (oil immersion).

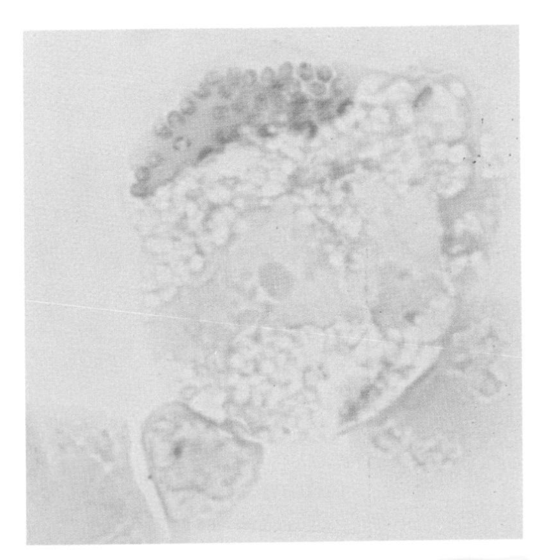

FIGURE 6 Cytospin preparation of cerebrospinal fluid, showing intracellular clusters of pink-stained microsporidial spores measuring 2 to 3 μm. Chromotrope stain was used. Magnification, ×400.

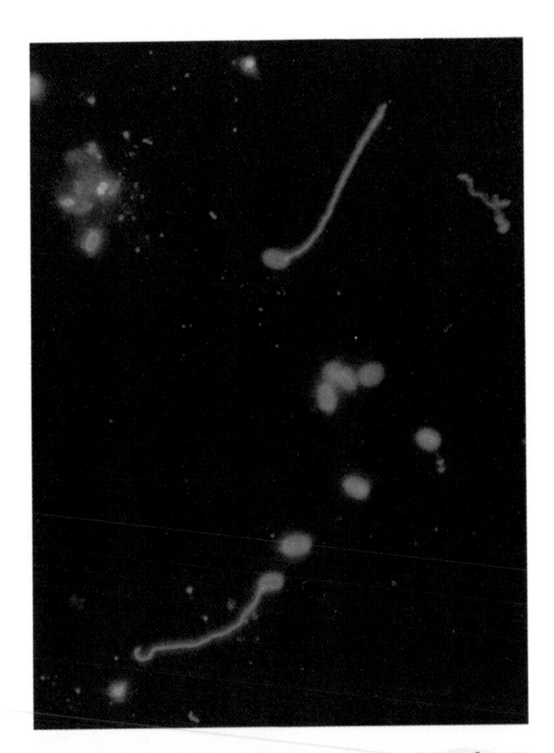

FIGURE 7 Immunofluorescence staining of sputum from a patient with disseminated *E. cuniculi* infection (Weber et al., 1997a). Polyclonal rabbit antibody to *E. cuniculi* was used. Spore walls and extruded polar tubes are seen. Magnification, ×600.

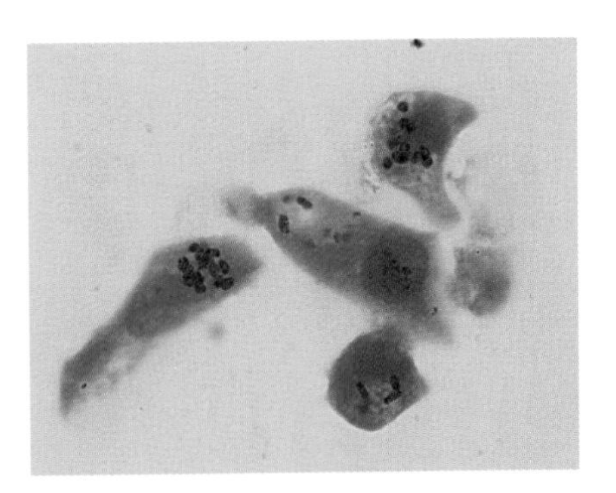

FIGURE 8 Cytological preparation of conjunctival smear with superficial epithelial cells containing numerous gram-positive spores of *Encephalitozoon hellem*. Brown and Hopps stain was used. Magnification, ×1,000.

antisera (Schwartz et al., 1993a; Shah et al., 1996; Weber et al., 1994a). In patients with ocular microsporidiosis, urine and respiratory secretions should also be examined for the presence of microsporidial spores.

MORPHOLOGICAL DETECTION PROCEDURES

Diagnostic techniques are summarized in Table 3, and specific staining procedures are described in the appendix to this chapter.

Chromotrope Staining

The chromotrope-based staining technique for light microscopic examination of stool specimens, duodenal aspirate, and other body fluids, as initially described by Weber and colleagues (1992a, 1993a, 1994a), includes steps similar to those used in the trichrome staining procedure of Wheatley (Wheatley, 1951), which is routinely used for parasitological examination of stool specimens in many laboratories. The chromotrope concentration of the modified staining solution, however, is 10-fold higher and the time the smear is exposed to the staining solution is prolonged.

Smears are prepared with 10 to 20 μl of unconcentrated stool spread very thinly on the slides or with sediments obtained by centrifugation of other body fluids. Microsporidial spores are ovoid and have a specific appearance when stained with chromotrope stains. The

TABLE 3 Diagnostic techniques and identification of microsporidia[a]

Technique	Specimen	Staining technique	Detection procedure[b]	Identification of species[b]	Comments
Light microscopy	Stool	Chromotrope	+	–	Robust, sensitive, specific, readily available
		Chemofluorescence	+	–	Sensitive, other fecal elements (including fungi) are also stained; fluorescent microscope required
		Giemsa	–	–	Difficult to differentiate microsporidial spores from other fecal elements
		Immunofluorescence	(+)	(+)	Commercial availability limits use; anti-*Encephalitozoon* antibodies developed; anti-*Enterocytozoon* antibodies not available
	Urine and other body fluids	Stains as above	+	–	As above
		Gram	+	–	Sensitive; not specific; useful for examination of urine, respiratory, cerebrospinal fluid specimens
	Touch preparation of tissue	Giemsa	+	–	Can visualize intracellular parasites; specific
	Histology	Hematoxylin and eosin	–	–	Sensitivity uncertain; usefulness debated
		Modified Gram (Brown and Brenn, Brown and Hopps)	+	–	Generally recommended
		Warthin-Starry	+	–	Sensitive; promising
		Chromotrope	+	–	Reliable; sensitive
		Chemofluorescence	+	–	Sensitive
		Immunofluorescence	(+)	(+)	Commercial availability limits use

(Continued)

TABLE 3 *(Continued)*

Technique	Specimen	Staining technique	Detection procedure[b]	Identification of species[b]	Comments
	Plastic-embedded sections	Toluidine	+	–	Sensitive
Electron microscopy	Tissue and body fluids		+	+	Gold standard for diagnostic confirmation and species identification; sensitivity may be lower than that of light microscopy
Serology	Serum		–	–	Results of *Encephalitozoon* serology controversial; not useful in immunocompromised patients; *Enterocytozoon* serology not available
Cell culture	Parasites in body fluids and tissue		(+)	–	*Encephalitozoon*, *Nosema*, *Trachipleistophora*, *Vittaforma* can be isolated
Molecular analyses	Body fluids and tissue		(+)	(+)	Results of molecular detection techniques promising; molecular identification of several microsporidial species possible

[a]The table has been modified from Weber et al. (1994a).
[b]+, recommended; (+), useful, but not widely available (research technique or commercially not available); –, not recommended.

spore wall stains bright pinkish-red, some spores appear transparent, and others show a distinct pinkish-red-stained beltlike stripe that girds the spores diagonally or equatorially (Fig. 1). Spores of *E. bieneusi* measure approximately 0.9 by 1.5 µm; spores of *Encephalitozoon* spp. appear consistently and significantly larger, 1.0 to 1.5 by 2.0 to 3.0 µm, than those of *Enterocytozoon* spp. and show similar staining patterns (Weber et al., 1994b). Most background debris counterstains faint green (or blue depending on the staining technique). Other elements, such as yeast and some bacteria, may also stain reddish, but they are distinguished from microsporidial spores by their size, shape, and staining pattern. To control the staining procedures, we recommend the use of fixed control samples with every diagnostic staining procedure.

Several modifications of the original chromotrope staining solution have been proposed, including modifications of the counterstain, described by Ryan and colleagues (1993), and changes in the temperature of the standard chromotrope staining solution and staining time, described by Kokoskin and colleagues (1994) as well as by Didier and colleagues (Didier et al., 1995) (Table 4). The choice of counterstain is a matter of preference and does not influence the contrast with the pink-staining microsporidia. Staining at a temperature of 50°C for 10 min (Kokoskin et al., 1994) or staining at a temperature of 37°C for 30 min (Didier et al., 1995) may improve the detection of microsporidia, as the background may be clearer and spores may stain more intensely.

An acid-fast trichrome stain (Ignatius et al., 1997b) that permits visualization of acid-fast cryptosporidial oocysts as well as microsporidial spores on the same slide, and a quick hot Gram-chromotrope staining technique in which the staining time is reduced to 5 min with the microsporidial spores staining dark violet against a pale-green background (Moura et al., 1997b) (Fig. 9), have also been suggested.

Chemofluorescent Agents

Chemofluorescent optical brightening agents are chitin stains requiring the use of an ultraviolet microscope (van Gool et al., 1993; Vavra and Chalupsky, 1982; Vavra et al., 1993). With the correct wavelength (see the appendix to this chapter) the chitinous wall of the microsporidial spores fluoresces brightly, facilitating the detection of spores (Fig. 2). However, staining is not specific, and small fungal spores that may be present in fecal material along with other fecal elements may also fluoresce. The ability to distinguish the microsporidia requires a certain amount of experience.

Chemofluorescent optical brightening agents include Calcofluor White 2 MR (American Cyanamid Corp., Princeton, N.J.), Fungi-Fluor Kit (Polysciences, Inc., Warrington, Pa.), Fungiqual A (Medical Diagnostics, Kandern, Germany), Cellufluor (Polysciences, Inc., Warrington, Pa.), Uvitex 2B (Ciba Geigy, Basel, Switzerland; not commercially available), and other chemofluorescent stains.

Giemsa Stain

Giemsa staining of stool specimens or body fluids results in a light-blue staining of microsporidia, and sometimes a characteristic darkly stained nucleus can be visualized (van Gool et al., 1990). The very small, blue-stained microsporidia spores, however, are difficult to identify in smears of stool specimens and to differentiate from other fecal elements that also stain blue. Giemsa staining may be useful for light microscopic examination of touch preparations of intestinal biopsies (Rijpstra et al., 1988) or cytological preparations of body fluids (Fig. 5).

Immunodetection

Fluorescein-tagged polyclonal or monoclonal antibodies have been used for detection of microsporidia in clinical samples from humans and animals (Weber et al., 1994a; Weiss and Vossbrinck, 1998). In contrast to histochemical methods, which stain only the wall of sporoblasts or spores, immunofluorescence detection procedures (indirect immunofluorescent antibody test [IFAT]), particularly those that use polyclonal antibodies, visualize spores, intracellular developmental stages, and extruded polar tubes, the latter indicating the viability of the spores and representing a pathognomonic light

TABLE 4 Chromotrope stains and their modifications[a]

Technique	Reference	Modifications[b]	Comments
Chromotrope stain	Weber et al., 1992a	Original stain.	Slide stays in chromotrope staining solution for 90 min.
Trichrome blue stain	Ryan et al., 1993	Modification of the counterstain (aniline blue instead of fast green); different concentration of phosphotungstic acid.	Choice of counterstain is a matter of preference and does not influence the contrast to the pink-staining microsporidia.
Modified chromotrope stain	Kokoskin et al., 1994	Staining temperature increased to 50°C. Staining time reduced to 10 min.	Staining time is reduced. Spores stain more intensely. Equipment to heat staining solution is needed.
Modified chromotrope stain	Didier et al., 1995	Staining temperature increased to 37°C for 30 min. Aniline blue used for counterstain.	As above.
Quick hot Gram-chromotrope stain	Moura et al., 1997	Additional staining step using Gram stain. Chromotrope stain at increased temperature.	Staining time is reduced. Additional staining step. Results are comparable to those for original stain or modifications.
Acid-fast trichrome stain	Ignatius et al., 1997b	Staining steps include acid-fast stain (to detect cryptosporidia) and chromotrope stain. FEA sedimentation technique was used.	Stains two parasites using the same procedure. Requires equipment to measure pH. Recommended sedimentation technique (to concentrate cryptosporidial oocysts) does not concentrate microsporidial spores but may lead to a loss of microsporidial spores (Weber et al., 1992a).
KOH pretreatment and centrifugation	Carter et al., 1996	Pretreatment of SAF-fixed stool specimens with 10% KOH (as a mucolytic) and centrifugation for 5 min.	Not validated in controlled epidemiologic studies. In our experience, there was no improvement in sensitivity.

[a] Laboratory procedures are described in the appendix to this chapter.
[b] FEA, formalin-ethyl acetate; SAF, sodium acetate-acetic acid-formalin.

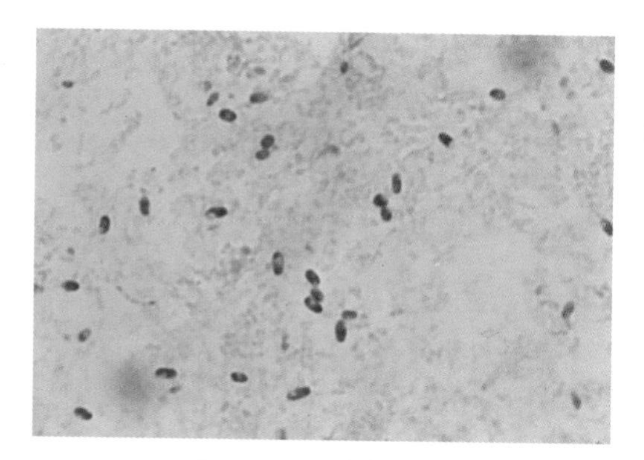

FIGURE 9 Stool specimen from a patient with AIDS and diarrhea, showing dark violet-stained spores of *Encephalitozoon intestinalis* measuring 1.0 to 1.2 by 2.0 to 2.5 µm. Quick hot Gram-chromotrope stain was used. Magnification, ×1,000 (oil immersion). (Reprinted with permission from G. S. Visvesvara, Centers for Disease Control, Atlanta, Ga.)

microscopic feature of all microsporidial species (Fig. 7). Extruded polar tubes have been demonstrated in clinical samples obtained from patients infected with different microsporidia including *E. cuniculi* (in sputum, cerebrospinal fluid, urine, intestinal biopsy specimens [Weber et al., 1997a]), *E. hellem* (in urine, sputum, conjunctival swab specimens [Schwartz et al., 1993a; Weber et al., 1993a]), *E. intestinalis* (in sputum and urine [Beckers et al., 1996]), and *E. bieneusi* (in biopsy material and fluid from the duodenum, but only rarely from spores in fecal samples [Zierdt et al., 1993]).

Polyclonal antibodies with low cross-reactivity between *E. cuniculi* and *E. hellem* have enabled investigators to distinguish these two morphologically identical species (Schwartz et al., 1996; Visvesvara et al., 1991). Cross-reactivity of polyclonal antibodies between *Encephalitozoon* spp. and other microsporidial species has been documented (Aldras et al., 1994; Didier et al., 1995; Niederkorn et al., 1980; Weiss et al., 1992; Zierdt et al., 1993), and has also been used to detect *Enterocytozoon* spores in stool specimens. However, diagnostic application of polyclonal antibodies, especially in fecal samples, was hampered by high levels of background staining and by cross-reactions to yeasts and bacteria (Garcia et al., 1994; Zierdt et al., 1993). In a comparative study with Calcofluor or chromotrope staining, an IFAT that used polyclonal murine antibodies directed against *E. intestinalis* had a lower sensitivity in 55 clinical samples with confirmed *E. intestinalis* or *E. bieneusi* infections and also a lower limit of spore detection (Didier et al., 1995).

Only a few monoclonal antibodies against *Encephalitozoon* spp. have been generated so far (Table 5). An anti-*E. cuniculi* monoclonal antibody directed exclusively against the exospore cross-reacted with spores of *E. hellem, E. intestinalis, Vairimorpha,* and *Nosema* spp. originating from invertebrates but not with spores of *E. bieneusi, V. corneae,* fungi, and bacteria (Enriquez et al., 1997b). This carefully evaluated antibody was shown to be of diagnostic value for the detection of *E. intestinalis* in stool (Fig. 10) and biopsy specimens. However, species-specific diagnosis with specific antibodies would be desirable, especially for the *Encephalitozoon* spp. which are easily detected and identified to genus level with staining methods. Other monoclonal antibodies reacting with *Encephalitozoon* spp. have also been described. So far, only monoclonal antibodies directed against *E. hellem* spores have been proven to be species-specific in the immunofluorescence test (Croppo et al., 1998). The most appropriate application for this test would involve the use of monoclonal antibodies specifically directed against the most prevalent intestinal microsporidian, *E. bieneusi*. Such an assay, however, has not yet been developed.

Histology

Microsporidia spores are easily demonstrated in formalin-fixed and paraffin-embedded tissues with a variety of special staining methods (Kotler et al., 1994; Schwartz et al., 1994a). Tissue Gram staining using either Brown and Hopps or Brown and Brenn modifications (Luna, 1968; Weber et al., 1992a), work well in demonstrating microsporidial spores of all species and is the preferred stain in our laboratory (Weber et al., 1992a). Mature spores generally stain gram-positive (violet/purple), although some variablity in the intensity of

TABLE 5 Murine monoclonal antibodies with diagnostic potential for the detection of microsporidia in humans

Antibody[a]	Specificity as determined by fluorescence antibody test	Comments	Reference(s)
MAbs to *E. hellem:* C12, E9, E11 (IgM directed to spore wall and polar tube)	Cross-reaction with *E. cuniculi, E. intestinalis,* and *E. bieneusi* but not with *V. corneae, Cryptosporidia, Giardia, Trichomonas, Isospora,* yeast, and bacteria	Of diagnostic value for detection of spores in formalin-fixed stool but not evaluated for diagnostic use	Aldras et al., 1994
MAb to *E. cuniculi:* 3B6 (IgG2b directed exclusively to the exospore)	Cross-reaction exclusively with exospore of *E. hellem, E. intestinalis, Vairimorpha,* and *Nosema* spp. from invertebrates; no reactions with *E. bieneusi* and *V. corneae* and with a large number of intestinal bacteria, fungi, and yeasts	High potential for the diagnostic detection of *Encephalitozoon* spp. in fixed material	Enriquez et al., 1997b
MAbs to *E. intestinalis:* Si13 (IgG3 directed against spore wall and early stages of development), Si91 (IgG1 directed exclusively to the extruded polar tube)	No cross-reaction of both antibodies to *E. hellem;* specificity not tested against *E. cuniculi;* some cross-reaction with fecal fungi and bacteria	Diagnostic use not evaluated	Beckers et al., 1996
MAb to *E. hellem:* ED4H10B11/B12 (IgG1 directed against spore wall)	Low cross-reactivity with the polar tube of *E. cuniculi;* no reactivity to *E. intestinalis, E. bieneusi, V. corneae, Cryptosporidia,* or *Giardia*	Can be used in formalin-fixed, paraffin-embedded tissue sections after pretreatment with trypsin or heating in a microwave	Visvesvara et al., 1994; Croppo et al., 1998
MAbs to *E. hellem:* ED4H6B3/A4, ED4H6B1/G4, ED4H10H4/C12, (IgG1 directed against spore wall)	Species-specific for *E. hellem* from cultures or tissue	See above	Croppo et al., 1998

[a] MAbs, monoclonal antibodies.

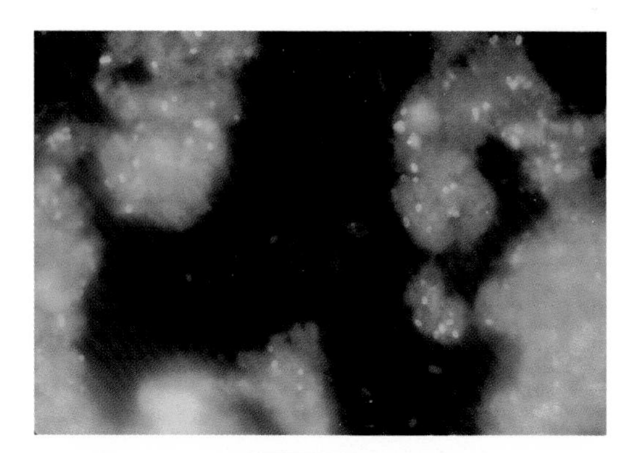

FIGURE 10 Stool specimen from an immunocompetent Mexican patient with diarrhea, showing microsporidial spores stained with a monoclonal antibody to *E. cuniculi* (Mab 3B6) (Enriquez et al., 1997b). The microsporidian was identified as *E. intestinalis,* as described by Bornay-Llinares et al. (1998). Immunofluorescence detection was used. (Reprinted with permission from J. Enriquez, University of Arizona, Tucson, Ariz.)

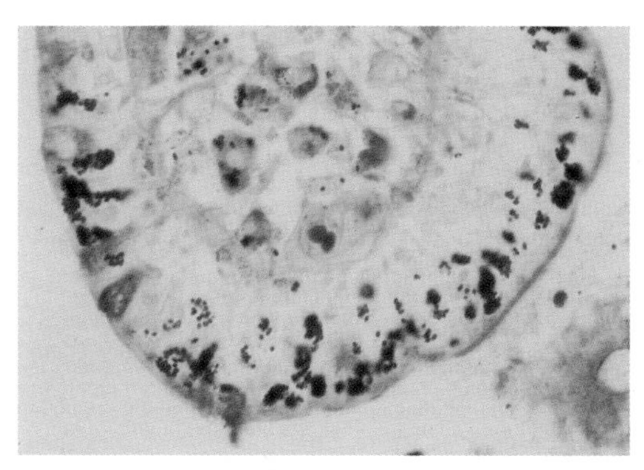

FIGURE 11 Section of duodenal villus tip, showing epithelial cells infected with *E. bieneusi.* Clusters of black-stained spores and plasmodia at various stages of development are shown. Modified Warthin-Starry stain was used. Magnification, ×600. (Reprinted with permission from A. S. Field, St.Vincent's Hospital, Darlinghurst, Australia.)

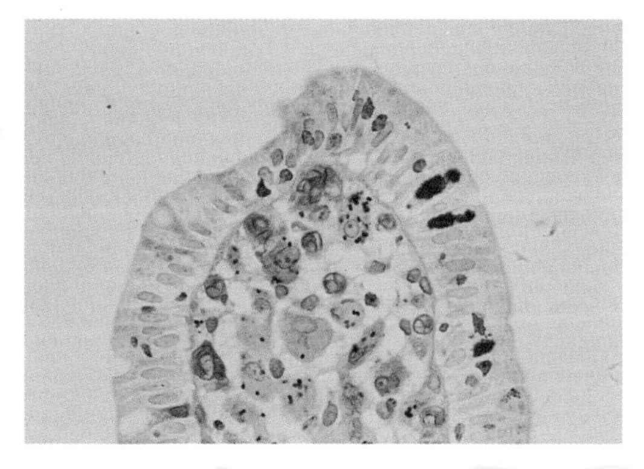

FIGURE 12 Section of duodenal villus tip, showing enterocytes and macrophages infected with black-stained clusters of *E. intestinalis.* Modified Warthin-Starry stain was used. Magnification, ×600. (Reprinted with permission from A. S. Field, St.Vincent's Hospital, Darlinghurst, Australia.)

staining can be noted. Red-staining spores are frequently found and may be immature spores. The spores may appear homogeneously dark-staining in some preparations, but with detailed examination a characteristic spore wall and a circumferential beltlike stripe extending around the equator of the spore can be identified. A modification of tissue Gram staining has been described that incorporates this method with Weber chromotrope staining used in stool examination (Moura et al., 1997). This quick hot Gram-chromotrope method has produced good results in a variety of biopsy and autopsy tissues.

Other useful techniques include silver staining (e.g., modified Warthin-Starry stain) (Field et al., 1993a, 1993b), chromotrope-based staining (Giang et al., 1993), and staining with chemofluorescent agents (Conteas et al., 1996; Vavra et al., 1993). The modified Warthin-Starry stain for microsporidia is promising and allows detection of both mature and developing stages of microsporidia in intestinal tissues (Fig. 11 and 12) (Field et al., 1993a, 1993b). This stain also makes the spores appear larger than their actual size. However, caution must be exercised when using this stain as the specific structural details of microsporidial spores, easily seen with Gram staining, are more difficult to visualize with the silver-based Warthin-Starry stain.

An interesting property of microsporidial spores is their birefringence in paraffin-embedded tissue sections. This characteristic results from the presence of chitin in the endospore layer of microsporidial spores. Birefringence is especially prominent in sections stained with Gram and modified Warthin-Starry stains and can be useful in differentiating microsporidial spores from lysosomes, intracytoplasmic neuroendocrine granules, karyorrhexic debris, and mucin droplets.

In addition to histochemical stains, a wide variety of diagnostic antibodies are available in specialized laboratories for genus- and species-specific identification of microsporidial spores in formalin-fixed, paraffin-embedded tissues (Croppo et al., 1998; Schwartz et al., 1992; Visvesvara et al., 1994).

Formalin fixation with paraffin embedding of biopsy or autopsy tissues is not an optimal method for preservation of specimens for electron microscopy. However, it does not preclude subsequent ultrastructural examination. Microsporidia can be identified to the level of species following removal of tissues from paraffin, followed by postfixation with osmium tetroxide and processing as usual for ultrastructural evaluation (Joste et al., 1996).

Using routine techniques such as hematoxylin and eosin, only highly experienced pathologists have reliably and consistently identified microsporidia in formalin-fixed, paraffin-embedded tissue sections (Orenstein, 1991; Orenstein et al., 1990). Although the periodic acid-Schiff stain is not optimal for initial examination of tissues for microsporidia, spores demonstrate a periodic acid-Schiff-positive polar granule. Giemsa staining is often used for tissue identification of other protozoal infections, but it is not a reliable method for demonstrating microsporidia.

Preparation of ultrathin plastic sections stained with methylene blue-azure II-basic fuchsin (Orenstein et al., 1990) or with toluidine blue may facilitate light microscopic detection of microsporidia.

Electron Microscopy

The identification of microsporidia and their taxonomy have been based primarily on ultrastructural characteristics (Fig. 13). Microsporidial ultrastructure is unique and pathognomonic for the phylum, and ultrastructural features can distinguish among all microsporidial genera (Canning, 1993, 1998). Nevertheless, morphological features alone do not sufficiently characterize all microsporidial species pathogenic in humans. Characterization of the three *Encephalitozoon* spp., which share most of their morphological features, requires antigenic or molecular analyses which may also reveal subspecific variation (Didier et al., 1995).

Although not all microsporidia pathogenic in humans can be identified to the species level by electron microscopy, this technique is still the gold standard for diagnostic confirmation and species identification. Yet, electron microscopy is relatively insensitive for detection of microsporidia because only small samples are examined and sampling error may occur. Major ultrastructural features for species identification are the characteristics of developmental stages of the parasite and the host-parasite interface, which are observed only in tissue specimens, as well as the nuclear configuration and spore morphology.

Electron microscopy has also been successfully used for confirmation or identification of microsporidia in feces and other body fluids, but the technique has a low sensitivity because only a small amount of a sample can be examined (Weber et al., 1994b). Furthermore, electron microscopic examination of stool or urine specimens may reveal only spore ultrastructure. Spore size and ultrastructure, particularly in regard to the configuration of the coiled tubules, differ between *E. bieneusi* (tubules arranged in two rows) (Fig. 14) and *E. intestinalis* (tubules arranged in one row), whereas differentiation of the three *Encephalitozoon* spp. on the basis of spore ultrastructure is not achievable.

NUCLEIC ACID-BASED METHODS

Nucleic acid-based methods have been developed to detect or identify microsporidia. Such methods are invaluable for identification, taxonomic classification, and phylogenetic analyses, but their role in diagnosis has yet to be validated in epidemiological studies. Several PCR assays targeting the RNA genes have been sucessfully used to diagnose and identify microsporidia pathogenic in humans, including *E. bieneusi,* different *Encephalitozoon* spp., *V. corneae,* and *Nosema* spp., as well as microsporidia parasitic in animals

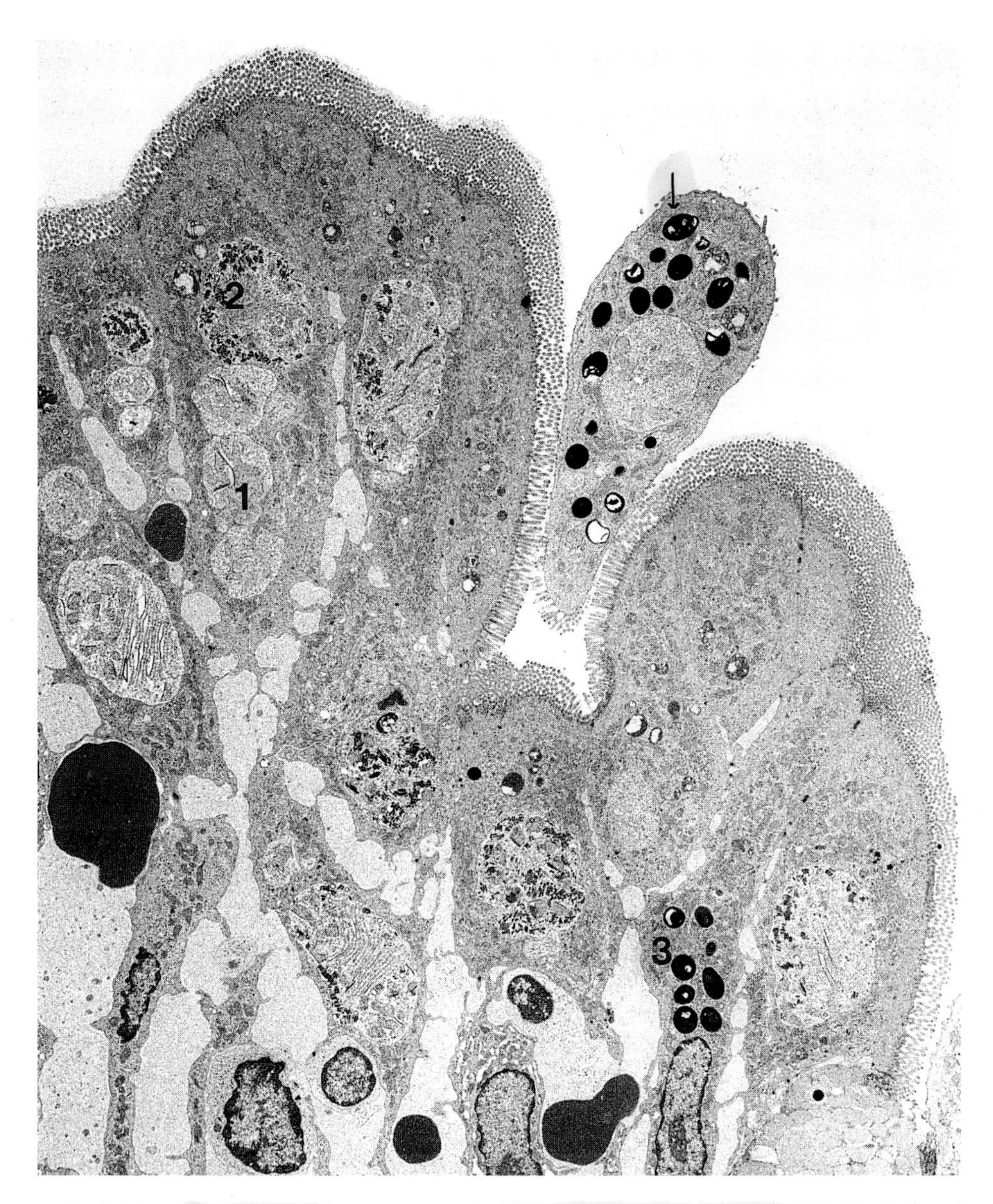

FIGURE 13 Transmission electron micrograph showing duodenal epithelium from an HIV-infected patient infected with *E. bieneusi*. The different developmental stages between the enterocyte nuclei and the microvillus border include a proliferative plasmodium (1), late sporogonial plasmodia (2), and mature spores (3). A sloughing enterocyte containing mature spores is also shown (arrow). Magnification, ×3,700. (Reprinted with permission from M. A. Spycher, University Hospital, Zurich, Switzerland.)

(Vossbrinck et al., 1993; Weiss and Vossbrinck, 1998). Molecular tests have been applied in the examination of microsporidia obtained from cell culture or clinical specimens, including fresh stool, intestinal tissue obtained by endoscopical biopsy, urine, and other body fluids (Coyle et al., 1996; Da Silva et al., 1996, 1997; De Groote et al., 1995; Deplazes et al., 1996a; Talal et al., 1998).

The molecular techniques are described in Chapter 4.

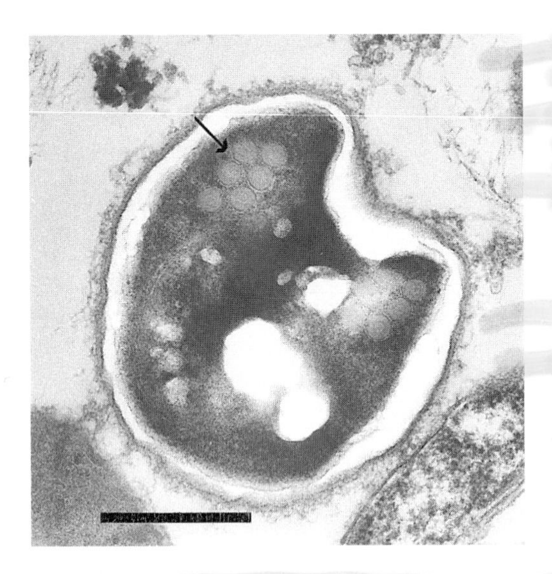

FIGURE 14 Transmission electron micrograph of a microsporidial spore. The polar tubes (arrow) lie in two rows characteristic of *E. bieneusi*. Bar, 0.5 μm. Magnification, ×51,000.

ISOLATION

Isolation of microsporidia has little practical relevance for diagnostic purposes but is of importance for further phenotypic and genetic characterization of isolates obtained from humans and animals as well as for the production of parasite antigens. Especially for microsporidia for which only single isolates have been described so far, e.g., *Nosema*-like microsporidia, isolation and in vitro propagation are of major interest.

Microsporidia cannot be cultivated axenically because of their obligate intracellular development. However, as they can infect a variety of cell types, a wide range of cells lines have been used for successful parasite propagation, including RK-13 (rabbit kidney), MDCK (Madin-Darby canine kidney), and MRC-5 (human embryonic lung fibroblast) cells (Canning et al., 1986; Canning, 1993; Deplazes et al., 1996a; Didier et al., 1991a; Hollister et al., 1996; Shadduck, 1969; Shadduck et al., 1990; Silveira and Canning, 1995; and Visvesvara et al., 1991). A crucial problem of the initial in vitro propagation of new microsporidial isolates is to overcome both the simultaneous contamination of the samples with bacteria, fungi, and viruses as well as the potential cytopathic effect of the samples such as stool, body fluids, or tissue. All specimens should be collected in a sterile culture medium in sterile vessels and transported at 4°C, avoiding secondary microbial contamination which has very often been shown to be resistant to antimicrobial agents in the culture medium. Prior homogenization of the material followed by sieving and centrifugation with Percoll (30 to 50%) allows separation of the spores from debris and fungal spores. Treatment of the purified pellet with 5 mM HCl or 0.02% sodium dodecyl sulfate minimizes bacterial contamination. Growth of bacteria and fungi can be reduced by antibiotics and amphotericin B. Virus infections (e.g., adenovirus, BK virus) can mimic cytopathic effects of microsporidia in cell cultures and impair mass cultivation (Deplazes et al., 1996a; Visvesvara et al., 1996).

Several microsporidial species infecting humans and mammals have been propagated in cell culture, permitting mass production of spores (Didier et al., 1998). *V. corneae* (formerly *Nosema corneum*) was the first microsporidian of human origin isolated from corneal tissue of an immunocompetent patient to be propagated in vitro (Shadduck et al., 1990); a second isolate originated from the urine of a patient with AIDS (Deplazes et al., in press). *E. hellem* was grown from conjunctival and corneal tissue and scrapings (Didier et al., 1991a), as well as from urine, sputum, sinus aspirates, and bronchoalveolar lavage fluid. *E. intestinalis* was isolated from stool specimens, sputum, and urine (Didier et al., 1996; van Gool et al., 1994b). *E. cuniculi* was isolated from urine, respiratory secretions, and cerebrospinal fluid (De Groote et al., 1995; Didier et al., 1996; Weber et al., 1997a), and *T. hominis* from muscle biopsies (Hollister et al., 1996). Various attempts to culture *E. bieneusi* have not achieved long-term cultures, but short-term propagation has been described (Visvesvara et al., 1995b).

Therefore, only limited amounts of *Enterocytozoon* spores purified from stool samples of heavily infected patients have been available for investgative use.

Techniques for culturing and propagating microsporidia are described in Chapter 11.

Several microsporidial species infecting mammals can be maintained in their natural hosts, in immunodeficient mice, or in immunodeficient monkeys, but such approaches are far beyond diagnostic purposes.

SEROLOGY

Serologic assays (including carbon immunoassay [CIA], IFAT, enzyme-linked immunosorbent assay [ELISA], and Western blot [WB]) have been applied in detecting specific IgG antibodies directed to *E. intestinalis* spores in humans and to *E. cuniculi* spores in humans and several animal species. So far, no serologic tests are available for serodiagnosis of the microsporidial species most prevalent in humans, *E. bieneusi,* because this parasite has not been continuously propagated in culture as a prerequisite for antigen production.

Seroepidemiological studies with ELISA and IFAT in human populations have shown high prevalences of spore antigens of *E. cuniculi* in some groups (Bergquist et al., 1984a, 1984b; Hollister and Canning, 1987; Hollister et al., 1991; Singh et al., 1982; WHO Parasitic Diseases Surveillance, 1983). As compared to results for healthy blood donors, seroprevalences of up to 42% have been reported in patients with a history of tropical diseases, from persons living in or who have traveled to tropical countries, and from patients with renal diseases or psychiatric or neurological disorders (Hollister et al., 1991; Singh et al., 1982). WB analysis was used to increase the specificity of the ELISA (Hollister et al., 1991) and demonstrated that sera of patients with a positive ELISA also had a WB banding pattern similar to that of the sera of mice experimentally infected with *E. cuniculi*. This study has shown the diagnostic potential of WB analysis for the identification of microsporidia-specific immune reactions.

A serologic study with ELISA, CIA, and IFAT using spores of *E. intestinalis* has suggested seroprevalences of 8% of 300 Dutch blood donors and 5% of 276 pregnant French women (van Gool et al., 1997). The results of this study are intriguing because, with few exceptions, antibody reactions in the IFAT were found against the extruded polar tube and anchoring disk but not against the spore wall. The authors suggest that their test has a higher specificity than other test systems that detect antibodies directed to the spore wall. In our laboratory, however, an IFAT with spores of *E. cuniculi* always detected antibodies against both spore wall antigens and antigens of the polar tube in humans, farm foxes, monkeys, and rabbits, all with parasitologically proven *E. cuniculi* infections (Deplazes, unpublished data). Cross-reactivity between antigens of the polar tube and of the spore wall of *E. cuniculi, E. hellem, E. intestinalis, E. bieneusi* and microsporidial species of invertebrates have been demonstrated with monoclonal (Aldras et al., 1994; Enriquez et al., 1997a; Croppo et al., 1998) and with polyclonal (Didier et al., 1995; Niederkorn et al., 1980; Weiss et al., 1992; Zierdt et al., 1993) antibodies.

The specificity of all serologic tests used in humans, however, is still unknown, and it is uncertain whether detection of antibodies to *E. cuniculi* or *E. intestinalis* reflects true infections, antigen exposure without establishment of the parasite, cross-reactivity to other microsporidial or nonmicrosporidial microorganisms, or reactions due to polyclonal B cell stimulation, particularly in serum samples of patients with tropical diseases. More studies with sera of immunocompetent subjects with proven microsporidial infections are required to define the diagnostic parameters of these serologic tests. Unfortunately, success with this approach is extremely difficult to achieve because such individuals are difficult to identify because of the assumed intermittent excretion of possibly low numbers of spores. To date, in only one human case, involving a 2-year-old Colombian boy living in Sweden, was the detection of antibodies to *E. cuniculi* in serum accomplished prior to the identification of organisms in urine specimens (Bergquist et al., 1984b).

Furthermore, the feasibility and the interpretation of serologic assays in the detection of antibodies to any parasite infection are notably challenging in the immunocompromised host. Studies have demonstrated a high degree of variability in ELISA and WB for the demonstration of specific antibody responses to microsporidial antigens, suggesting that serologic tests may not become feasible for the diagnosis of microsporidiosis, especially in patients with AIDS (Didier et al., 1991).

MICROSPORIDIA DETECTED IN HUMANS

Enterocytozoon bieneusi

E. bieneusi has been identified in immunodeficient and immunocompetent humans (Desportes et al., 1985; Dobbins and Weinstein, 1985; Modigliani et al., 1985; Sandfort et al., 1994) and in immunocompetent pigs (Deplazes et al., 1996b). Most frequently, the parasite has been found in small intestinal enterocytes of severely immunodeficient HIV-infected patients with chronic diarrhea and weight loss. It has also been infrequently detected in other epithelial cells of patients with AIDS, i.e., in the biliary tree and gallbladder of patients with cholangiopathy, in nonparenchymal liver cells, and in the pancreatic duct, as well as in tracheal, bronchial, and nasal epithelia of patients with lower and upper respiratory tract infections (Weber et al., 1992b; Del Aguila et al., 1997). Systemic *E. bieneusi* has not been documented.

E. bieneusi develops in direct contact with the host cell cytoplasm, unlike members of the genus *Encephalitozoon* which develop within a parasitophorous vacuole (Fig. 13). The proliferative and sporogonial forms are rounded, multinucleate plasmodia measuring up to 6 μm in diameter. Organelles unique to this genus are electron-lucent clefts which are present throughout the life cycle, and electron-dense disks which are formed during sporogony and represent precursors of the polar tube and anchoring disks. Sporoblasts separate from large plasmodial sporonts by invagination of the plasmalemma. The oval spores measure 0.7 to 1.0 by 1.1 to 1.6 μm. The polar tubule has five to seven coils which appear in two rows when seen in transverse section by transmission electron microscopy (Desportes et al., 1985; Cali and Owen, 1990) (Fig. 14).

Ultrastructural features that distinguish *E. bieneusi* from all other microsporidia include nuclear division not immediately followed by cytokinesis, premature development of the tubular apparatus occurring within a multinucleated plasmodium which precedes division into sporoblasts, the occurrence of cleftlike spaces (electron-lucent inclusions) which develop into electron-dense disks that eventually form the polar tubes, spores with five to seven turns of the polar tube arranged in an outer and an inner layer, and elongated nuclei.

Encephalitozoon spp.

Human isolates of three *Encephalitozoon* spp. and *E. cuniculi* of animal origin are morphologically almost identical. In 1991, *E. hellem* was distinguished from *E. cuniculi* on the basis of different protein patterns found by sodium dodecyl sulfate-polyacrylamide gel electrophoresis separation and immunoblots (Didier et al., 1991a). In 1993, *Septata intestinalis* was described and named on the basis of the morphological finding that the intracellular vacuoles containing the parasite showed a unique parasite-secreted fibrillar network surrounding the developing organisms so that the vacuoles appeared septate (Cali et al., 1993; Orenstein et al., 1992a, 1992b). Subsequently, on the basis of phylogenetic analyses, it was proposed that *S. intestinalis* be regarded as a species of the genus *Encephalitozoon*, and it was reclassified as *E. intestinalis* (Hartskeerl et al., 1995). Analyses of the nucleotide sequences of the small-subunit rRNA have confirmed that the three *Encephalitozoon* spp. are indeed distinct organisms (Hartskeerl et al., 1993a; Didier et al., 1995, 1998). *E. cuniculi* has also been found to cause disease in humans (De Groote et al., 1995; Deplazes et al., 1996a; Didier et al., 1996; Mertens et al., 1997; Weber et al., 1997a). Three different strains of *E. cuniculi* (the so-called rabbit, mouse, and canine strains) have been identified phenotypically, by WB analysis of spore antigens, and genetically by random amplification of polymorphic DNA and determination of differences in the rDNA intergenic spacer region (Didier et al., 1995; Mathis et al., 1997).

ENCEPHALITOZOON CUNICULI

E. cuniculi is parasitic in different mammals including rodents, carnivores, primates, and humans (Canning, 1998; Canning et al., 1986). Human infections due to the *E. cuniculi* rabbit strain have been found in European HIV-infected patients (Deplazes et al., 1996a; Hollister et al., 1995; Mathis et al., 1997; Weber et al., 1997a), and infections due to the canine strain in patients originating from the United States and Mexico (Didier et al., 1996; Mathis et al., 1997; Mertens et al., 1997), but no human infections due to the mouse strain have been recognized so far.

E. cuniculi infects macrophages, epithelial cells, vascular endothelial cells, kidney tubule cells, and possibly other cell types and can be found in most tissues, with a predilection, in mammals, for the brain and kidney (Canning et al., 1986). Whether central nervous system infection due to *E. cuniculi* reported in two children not infected with HIV (Bergquist et al., 1984b; Matsubayashi et al., 1959), as well as liver (Terada et al., 1987) and peritoneal infection (Zender et al., 1989) reported in patients with AIDS, were indeed caused by this species, by another *Encephalitozoon*-like species, or possibly by *E. hellem*, is not known because protein and antigenic analyses or molecular studies were not performed in these cases. The spectrum of recognized *E. cuniculi*-associated disease in patients with AIDS includes keratoconjunctivitis, respiratory tract infections, hepatitis, peritonitis, urinary tract infections, and encephalitis (Weber et al., 1997).

E. cuniculi develops intracellularly in a parasitophorous vacuole bounded by a membrane of presumed host cell origin. Nuclei of all stages are unpaired. Meronts divide repeatedly by binary fission, are round to ovoid structures measuring 2 to 6 by 1 to 3 μm, and lie close to the vacuolar membrane. Sporonts appear free in the center of the vacuole and divide into two sporoblasts, which mature into spores. The spores measure about 1.0 to 1.5 by 2.0 to 3.0 μm, and the polar tubule has four to seven coils which appear in a single row. *E. cuniculi* cannot be distinguished from *E. hellem* by ultrastructural methods (Canning, 1998).

ENCEPHALITOZOON HELLEM

E. hellem has been identified in immunodeficient humans and birds to infect macrophages, corneal and conjunctival epithelial cells, nasal epithelium, bronchial epithelium, lining epithelium of the urinary tract, and renal tubular cells (Didier et al., 1991a). *E. hellem* infection in patients with AIDS was observed to be associated with keratoconjunctivitis, sinusitis, bronchiolitis, pneumonitis, nephritis, ureteritis, cystitis, prostatitis, urethritis, hepatitis, peritonitis, and diarrhea (Weber et al., 1994a). Clinical manifestations may vary substantially, ranging from an asymptomatic carrier state to organ failure.

The spores measure about 1.0 to 1.5 by 1.5 to 2.5 μm, and the polar tubule has six to eight coils which appear in a single row (Canning, 1998; Didier et al., 1991a). *E. hellem* is morphologically indistinguishable from *E. cuniculi* by both light and electron microscopy (Fig. 15).

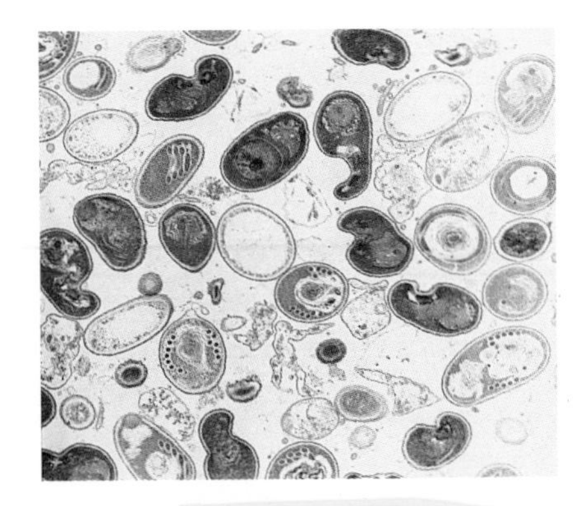

FIGURE 15 Immature and mature spores of *E. hellem* within a parasitophorous vacuole in an infected renal tubular epithelial cell. Note the high concentration of spores within this vacuole. Original magnification, ×19,200.

ENCEPHALITOZOON INTESTINALIS

E. intestinalis has been identified exclusively in humans, both in HIV-infected (Cali et al., 1993; Orenstein et al., 1992a, 1992b), and in immunocompetent persons (Bornay-Llinares et al., 1998; Enriquez et al., 1997a; Flepp et al., 1995; Raynaud et al., 1998). It was first detected in enterocytes and lamina propria macrophages of patients with AIDS and chronic diarrhea. In addition, parasites have been detected in fibroblasts and endothelial cells of the lamina propria, as well as in epithelial cells of the biliary tree, tubular kidney cells, bronchial epithelial cells, and nasal epithelial cells.

E. intestinalis has developmental stages somewhat similar to those of the other two *Encephalitozoon* spp. that infect humans. Meronts proliferate by cellular elongation and development of cytoplasmic invaginations between nuclei. Sporonts also divide by fission. Sporogony occurs in the cytoplasm within a parasitophorous vacuole of host origin (Fig. 16). The proliferative cells of *E. intestinalis* are uni-, bi-, or tetranucleated. Sporogony is tetrasporous. The most obvious diagnostic ultrastructural feature characterizing *E. intestinalis* is the formation of a parasite-secreted fibrillar matrix surrounding the developing organisms. This

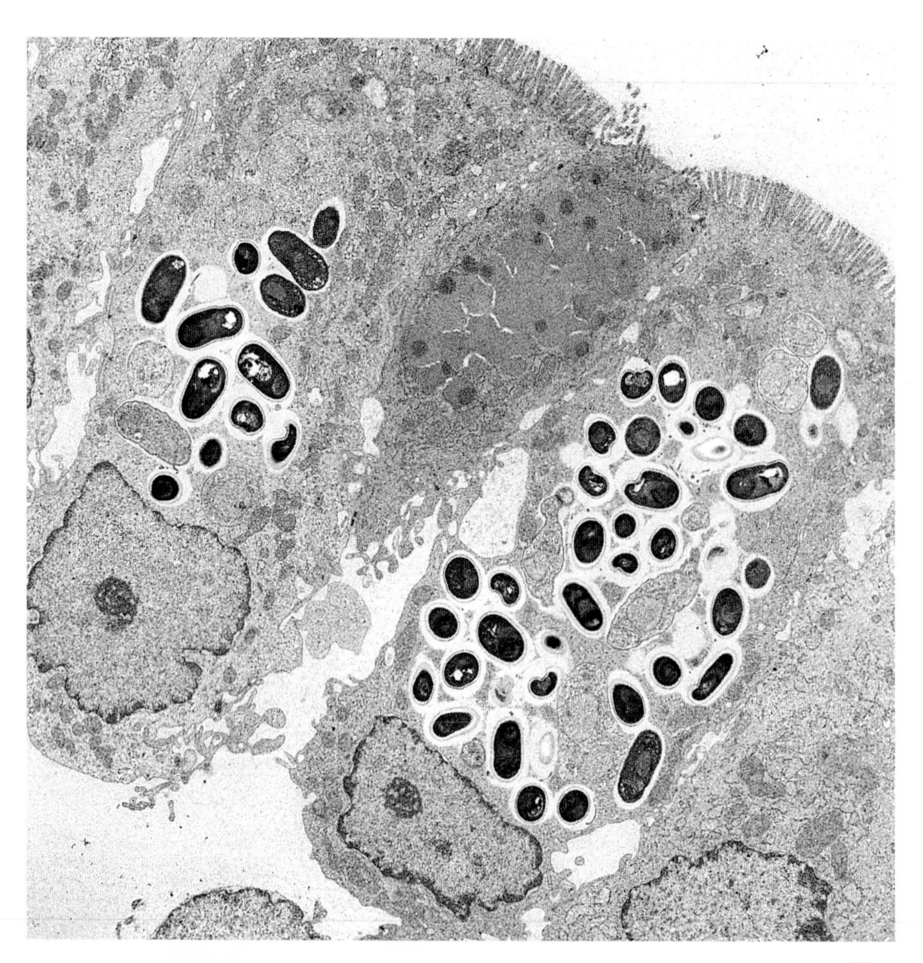

FIGURE 16 Transmission electron micrograph of meronts and developing *E. intestinalis* spores within enterocytes separated by a fibrillar matrix. *E. intestinalis* develops within parasitophorous vacuoles. Magnification, ×5,300. (Reprinted with permission from M. A. Spycher, University Hospital, Zurich, Switzerland.)

matrix gives the parasitophorous vacuole a septated, or honeycombed, appearance. Another unique finding is the presence of tubular appendages, up to 1.2 mm in length and 50 nm in diameter, which originate from the sporont surface and end in a bulbular structure. The mature spores measure 2.0 by 1.2 µm and contain a single row of polar tubules with four to seven coils (Cali et al., 1993).

Pleistophora spp.

Pleistophora spp. are parasitic in insects and in vertebrates, mainly fish (Canning et al., 1986). In humans they have been identified in the muscles of three immunodeficient patients with myositis (Chupp et al., 1993; Grau et al., 1996; Ledford et al., 1985; Macher et al., 1988). *Pleistophora* spp. develop in contact with host cell cytoplasm, forming sporophorous vesicles enclosing spores in variable numbers within a thick, amorphous, parasite-formed coat. Nuclei are unpaired. In merogonic and sporogonic proliferation multinucleate plasmodia are formed. The spores, containing polar tubules with 9 to 12 coils, measure 2.8 by 3.2 to 3.4 µm (Canning, 1998).

Trachipleistophora spp.

T. hominis, identified in muscle biopsies (and additionally in corneal epithelium and nasopharyngeal washings) of patients with AIDS and myositis, forms sporophorous vesicles that do not arise from multinucleate plasmodia. These vesicles, which contain 2 to more than 32 spores, enlarge as the number of spores increases. The nuclei are unpaired in all stages of development. Division of meronts and sporonts is by binary fission. The pyriform spores measure 2.4 by 4.0 µm and have about 11 coils (Field et al., 1996; Hollister et al., 1996).

T. anthropophthera has been identified at autopsy in cerebral, cardiac, renal, pancreatic, thyroid, hepatic, splenic, lymphoid, and bone marrow tissue of patients with AIDS who initially presented with seizures (Yachnis et al., 1996). It is similar to *T. hominis* but appears to be dimorphic, as two different forms of sporophorous vesicles and spores were observed (Vavra et al., 1998).

Nosema spp.

Numerous *Nosema* spp. have been described that are parasitic primarily in invertebrates (Canning, 1993). A case of disseminated *Nosema connori* infection involving almost all tissues examined at autopsy has been reported in an athymic child (Margileth et al., 1973; Sprague, 1974). Keratitis and corneal ulcers due to *N. ocularum* have been observed in otherwise healthy persons (Cali et al., 1991b). A third *Nosema*-like microsporidian, identified in an immunodeficient patient with myositis, has yet to be classified (Cali et al., 1996).

Nosema spp. develop in direct contact with host cell cytoplasm, nuclei are paired (diplokaryotic) (Fig. 17), and sporonts are disporoblastic. The diplokaryotic spores of *N. connori* measure 2.0 to 2.5 by 4.0 to 4.5 µm, and contain polar tubules with approximately 11 coils (Margileth et al., 1973; Sprague, 1974). The spores of *N. ocularum* measure 3 by 5 µm and have polar tubules with 11 to 12 coils (Cali et al., 1991b).

Vittaforma corneae (formerly *N. corneum*)

A new genus, *Vittaforma,* was proposed for *N. corneum* (Shadduck et al., 1990) on the basis of ultrastructural features (Silveira and Canning, 1995). *N. corneum* was previously identified in corneal stroma of a patient with a corneal ulcer (Shadduck et al., 1990). Also, *V. corneae* has been isolated from the urine of an HIV-infected patient with signs and symptoms of urinary tract infection (Deplazes et al., in press).

Development of the parasite takes place in direct contact with the host cell cytoplasm. Sporogony is polysporoblastic, sporonts are ribbon-shaped, and all developmental stages are individually enveloped by a cisterna of host endoplasmic reticulum studded with ribosomes. The spores, containing polar tubules with five to seven coils, are diplokaryotic and measure 1.2 by 3.8 µm.

Uncharacterized Microsporidia

The term *Microsporidium* includes microsporidia that cannot be taxonomically classified because available information is not sufficient (Canning et al., 1986).

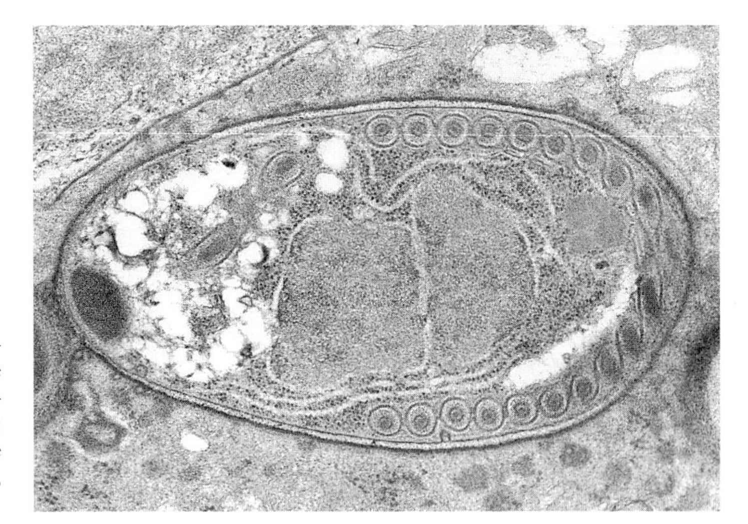

FIGURE 17 Typical ultrastructural appearance of a mature spore of a *Nosema* sp. A diplokaryotic nucleus (with a dividing membrane) and 10 turns of the polar tubule are evident. Original magnification, ×38,000.

Microsporidium ceylonensis spores, measuring 1.5 by 3.5 μm, were detected in the corneal stroma of a patient with a corneal ulcer (Ashton and Wirasinha, 1973). Meronts and sporonts were not seen, and nucleation was not observed. *Microsporidium africanum* spores, measuring 2.5 by 4.5 μm and containing polar tubules with 11 to 13 coils, were detected in the corneal stroma of a patient suffering from a perforated corneal ulcer (Pinnolis et al., 1981). Developmental stages of the parasite were not seen.

PATHOLOGY

Lesions

Aside from studies on ocular microsporidiosis in presumably otherwise healthy persons, detailed histopathologic investigation of human microsporidial infection has been performed only in immunodeficient individuals. In these patients, inflammatory reaction in tissues infected with microsporidia is often minimal or even absent. In contrast, inflammatory reaction in animal encephalitozoonosis is typically an intense diffuse cellular infiltration or granulomatous lesion characterized by infiltration of mononuclear cells including lymphocytes, plasma cells, and macrophages, often around a necrotic center. These lesions may persist after the disappearance of the organism themselves (Canning et al., 1986).

Gastrointestinal Tract

The small intestine is the most common site of human microsporidial infection, and three microsporidial species have been detected at this location: *E. bieneusi* (Desportes et al., 1985; Dobbins and Weinstein, 1985; Modigliani et al., 1985; Orenstein, 1991; Orenstein et al., 1990), *E. intestinalis* (Cali et al., 1993; Orenstein et al., 1992a, 1992b), and rarely *E. cuniculi* (Franzen et al., 1995; Weber et al., 1997a), of which *E. bieneusi* is found in more than 90% of cases. In addition, microscopic examination of feces from an HIV-infected patient with chronic diarrhea revealed *Nosema*-like microsporidial spores. Because these microsporidia were enclosed within striated muscle cells, it was suggested that they had probably been ingested in food and thus represented an incidental finding rather than a true infection (McDougall et al., 1993).

There is pathological evidence that diarrhea due to microsporidiosis in patients with AIDS may be the result of malabsorption (Kotler et al., 1990, 1993; Lambl et al., 1996; Schmidt et al., 1997). In a study on 19 patients with intestinal microsporidiosis, all were found to have reduced absorptive function as demonstrated by D-xylose, vitamin B_{12}, and lactase (Schmidt et al., 1997). Lactase deficiency was found by in situ measurement of levels of duodenal brush border enzymes. Villus height and villus surface areas were also significantly lower. These abnormalities might be a result of the increased apical enterocyte sloughing observed in intestinal microsporidosis biopsy specimens (Fig. 13).

As in the case of other intestinal pathogens, the demonstration of one microsporidial agent in an intestinal biopsy should not exclude a search for other infectious agents. Simultaneous infection of the gastrointestinal tract by *Enterocytozoon* spp. and *E. intestinalis* has been described, although it is a rare occurrence. More commonly seen is a coexistent infection involving microsporidia and *Cryptosporidium, Giardia lamblia,* cytomegalovirus, or other enteric pathogens (Hewan-Lowe et al., 1997; Kotler et al., 1994; Weber et al., 1993b; Sobottka et al., 1998).

INTESTINAL *E. BIENEUSI* INFECTION

In infected individuals, *E. bieneusi* is present throughout the length of the small intestine, in which the proliferative stages and spores are almost always restricted to the enterocytes of the superficial lining (Orenstein et al., 1992c). In the rare cases where spores have been observed in fibrovascular tissues of the lamina propria, no clinical evidence of disseminated disease was noted (Schwartz et al., 1995). Infected tissues usually show no histologic abnormalities specific for *Enterocytozoon,* and active enteritis or ulceration is absent (Schwartz et al., 1992).

Histologic findings may range from virtually normal villus architecture to severe epithelial degeneration (Kotler et al., 1990; Orenstein 1991; Orenstein et al., 1990). In some cases, histopathologic changes have included villus atrophy and fusion, crypt elongation, goblet cell depletion, prominent intraepithelial lymphocytic infiltrates, and enterocyte vesiculation, swelling, or sloughing. In other cases, general preservation of villus architecture in association with minimal (or absent) lamina propria mononuclear cell infiltrates has been noted. One clue that microsporidiosis may be present is based on the observation that infected epithelial cells containing mature spores are often shed intact into the lumen from the underlying basement membrane (Schwartz et al., in press). Recognition of these cells with hematoxylin and eosin staining should prompt the observer to request tissue special staining of replicate sections to look for microsporidial spores.

Because microsporidia are very small, can occur in the absence of noticeable inflammatory tissue reaction, and can be focally distributed, attentive examination and adequate light microscopic magnification, i.e., ×630 or ×1,000 (oil immersion), are required for identification of microsporidia independent of the staining technique used. With Gram staining, preferred by many workers, including ourselves, mature spores of *E. bieneusi* are round to oval, appear smaller than those of *Encephalitozoon* spp., stain gram-positive or gram-variable, and are characteristically located in the cytoplasm of enterocytes between the microvillus border and the nucleus (Fig. 18). Immature stages can stain gram-negative and be more difficult to identify. As in the histologic examination of other microsporidial species, a beltlike stripe can often be seen extending around the equatorial diameter of some *Enterocytozoon* spores. *Enterocytozoon* infection is usually focal, and spores may be very scarce. Several levels should be examined with special stains before a biopsy is confirmed to be negative for microsporidia.

Enterocytozoon spores must be differentiated from enteroendocrine cell granules, which appear similar to spores but are located between

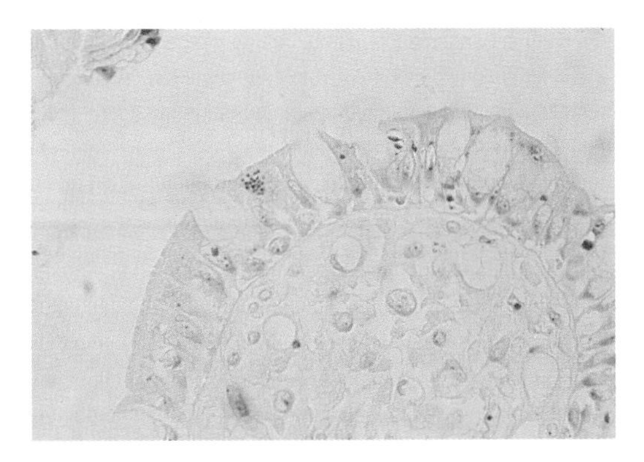

FIGURE 18 Mucosal biopsy specimen, showing duodenal epithelial cells infected with *E. bieneusi.* A cluster of gram-positive spores is located in the supranuclear portion of the cytoplasm. Brown and Brenn stain was used. Magnification, ×630.

the enterocyte nucleus and the basal lamina. Plasmodia often stain lighter than surrounding host cytoplasm and are characterized by a cleft (the electron-lucent inclusion), which is present in all developmental stages of *E. bieneusi*. Non-spore stages can be inferred, even when not visualized, if they indent the superior pole of the enterocyte nucleus toward the microvillus border. Since microsporidia can penetrate and infect host enterocytes only after they emerge from intestinal crypts, the longer enterocytes have been exposed to luminal spores, the more likely they are to be parasitized. Consequently, enterocytes are most heavily parasitized at villus tips, where they are most likely to contain mature spore forms. Within crypts, granules of Paneth cells should not be mistaken for *E. bieneusi* spores, which are not found in crypts, because of the absence of enterocyte penetration by spore polar tubules at this location (Cali and Owen, 1990; Weber et al., 1994a).

Because *E. bieneusi* is most reliably detected in the enterocytes at the tips of villi by both light microscopy and electron microscopy, histologic diagnosis of intestinal microsporidiosis is facilitated by orientating intestinal biopsy specimens prior to fixation. Endoscopic biopsy specimens are folded cut surface to cut surface with the two jaws of a biopsy forceps. The two halves of these specimens can be teased apart with a blunt probe and the cut surface affixed to paper by contact. When the specimen is removed from the fixative, the paper base can be removed or used to orient the specimen in embedding medium so that sections taken perpendicular to the cut edge give crypt-to-tip profiles of villi (Weber et al., 1994a).

INTESTINAL *E. INTESTINALIS* INFECTION

E. intestinalis also infects primarily small intestinal enterocytes (Fig. 16). Like *Enterocytozoon* spp., *E. intestinalis* does not produce any specific endoscopic or tissue abnormality, and inflammatory response as well as intestinal cell injury appears similarly minimal. Nevertheless, small bowel perforation due to *E. intestinalis* infection has also been reported (Soule et al., 1997) . Unlike *E. bieneusi,* however, *E. intestinalis* infects not only enterocytes but also cells in the lamina propria, including endothelial cells, fibroblasts, and macrophages (Cali et al., 1993; Orenstein et al., 1992a, 1992b) (Fig. 19). This observation explains the propensity of this microsporidian to produce disseminated disease. *E. intestinalis* can also infect the large bowel mucosa (Dore et al., 1995). In well-prepared paraffin-embedded sections and in plastic-embedded semithin sections, the parasitophorous vacuole can often be seen surrounding the developing spores and separating them from the host cell cytoplasm (Fig. 19). Thus, the finding of large (up to 2.5 μm) microsporidial spores within a vacuole in intestinal epithelial cells, and also present in lamina propria macrophages and endothelial cells, permits a tentative light microscopic diagnosis of *E. intestinalis* infection.

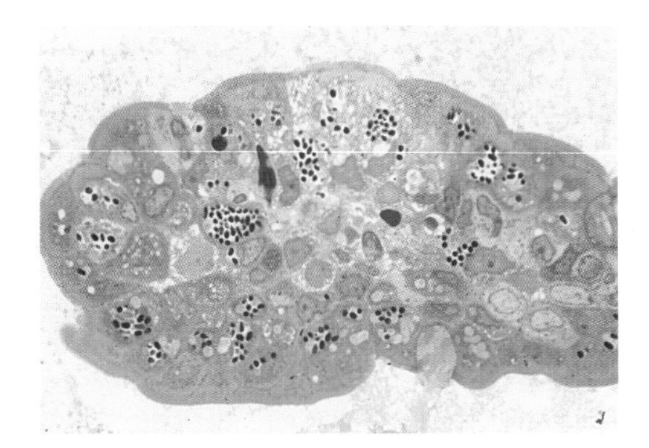

FIGURE 19 *E. intestinalis* infection of the small intestine. The plastic-embedded semithin section shows spores larger than those of *E. bieneusi*. The parasites are not in direct contact with the host cell cytoplasm but instead develop within a parasitophorous vacuole. Toluidine blue stain was used. Magnification, ×400.

Hepatobiliary Tract

BILIARY TRACT

E. bieneusi, and less frequently *E. intestinalis,* can infect the biliary tract (Beaugerie et al., 1992; McWhinney et al., 1991; Pol et al., 1993). Infections of nonparenchymal liver cells, of the epithelium of bile ducts, and, rarely, of the gallbladder have been reported. The majority of these cases involved concurrent infection of the small intestine. It is believed that microsporidia extend into the extrahepatic bile ducts of the liver and pancreas by direct mucosal extension from the duodenum. Infections of the biliary tract may result in sclerosing cholangitis, bile duct dilatation, papillary stenosis, and acalculous cholecystitis (Schwartz et al., 1997).

The discovery of natural *E. bieneusi* infections in SIV-infected captive rhesus monkeys has permitted detailed study of the pathologic features of microsporidial infection of the gallbladder (Mansfield et al., 1997; Schwartz et al., in press). In these animals, the gallbladder is the most commonly infected organ, showing a high parasite burden compared with the pancreas, small and large intestines, and liver. Infected gallbladders are usually abnormal, showing various combinations of epithelial hyperplasia, acute and chronic cholecystitis, and prominent dilatation of mucosal sinusoids. An unusual finding in these gallbladders is the occurrence of prominent shedding of epithelial cells, infected with *E. bieneusi* spores, which are extruded into the lumina of the mucosal sinusoids or gallbladder. This phenomenon of cell shedding is also seen in the intestinal tract of humans infected with *E. bieneusi*.

LIVER

Granulomatous and suppurative hepatitis due to a presumed *E. cuniculi* infection has been described in a patient with AIDS (Terada et al., 1987). Another HIV-infected patient died as a result of liver failure associated with microsporidial infection of the liver, gallbladder, and mediastinal lymph nodes. Autopsy revealed widespread hepatocyte necrosis and extensive infection with an *Encephalitozoon*-like microsporidian (Sheth et al., 1997). In a patient with AIDS and a disseminated *T. anthropophthera* infection, hepatocytes were also infected (Yachnis et al., 1996).

The Eye and Ocular Adnexae

Ocular microsporidial infection has been classified pathologically as either stromal or epithelial, and its pathogenesis varies according to the immune status of the patient (Weber et al., 1994a). In otherwise healthy patients without immunodeficiency, ocular microsporidiosis was associated with keratitis or corneal ulcer, possibly related to prior trauma in some patients. The parasites involved, *V. corneae* (formerly *N. corneum*) (Davis et al., 1990; Shadduck et al., 1990), *N. ocularum* (Cali et al., 1991), *Nosema*-like organisms of the nontaxonomic group *Microsporidium* (Ashton and Wirasinha, 1973; Pinnolis et al., 1981), and unidentified microsporidia (Silverstein et al., 1997), were found deep in the corneal stroma. Their spores were contained within phagocytic cells and lying free between the fibrous layers of corneal lamellae, and a marked inflammatory reaction including mononuclear, neutrophil, and epitheloid infiltration, was present. No systemic microsporidial infection was noted in any of these patients.

In HIV-infected patients, in contrast, microsporidial parasites including *E. cuniculi, E. hellem,* and *E. intestinalis,* were confined to the superficial epithelial cells of the conjunctiva or cornea (Fig. 20 and 21), associated with an inflammatory infiltrate comprised mostly of neutrophils and mononuclear cells (Cali et al., 1991a; Canning et al., 1992; Centers for Disease Control, 1990; Didier et al., 1991a, 1991b; Friedberg et al., 1990; Lowder et al., 1990; Metcalfe et al., 1992; Schwartz et al., 1993a). The inflammatory reaction was generally mild or even absent. These

FIGURE 20 A solitary corneal epithelial cell infected with *E. hellem* from an enucleated globe obtained at autopsy. The spores are gram-positive. Brown and Hopps stain was used. Magnification, ×200.

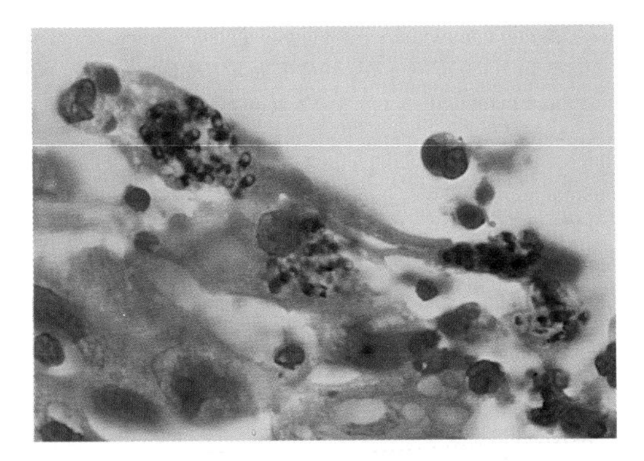

FIGURE 21 Conjunctival biopsy specimen, demonstrating shedding and necrotic conjunctival epithelial cells containing gram-positive microsporidial spores admixed with neutrophils and fibrin from a patient with AIDS and disseminated *E. hellem* infection. Brown and Hopps stain was used. Magnification, ×1,000.

ocular infections occur primarily, if not exclusively, in patients systemically infected with *Encephalitozoon* spp. (Schwartz et al., 1993a).

Respiratory Tract

Upper and lower respiratory tract infections due to microsporidia are associated almost exclusively with disseminated disease produced by all three members of the genus *Encephalitozoon* (Weber et al., 1994a).

PARANASAL SINUS AND NOSE

Upper respiratory microsporidial infections are possibly underdiagnosed in HIV-positive persons. The clinical and pathological features of sinonasal microsporidiosis are nonspecific and include rhinitis, sinusitis, and nasal polyposis (Canning et al., 1992; Hartskeerl et al., 1993a; Josephson et al., 1996; Lacey et al., 1992; Rossi et al., 1996; Weber et al., 1997b). In the majority of patients with sinonasal microsporidosis, infection of other organs is found when appropriate specimens are examined (Fig. 22 and 23).

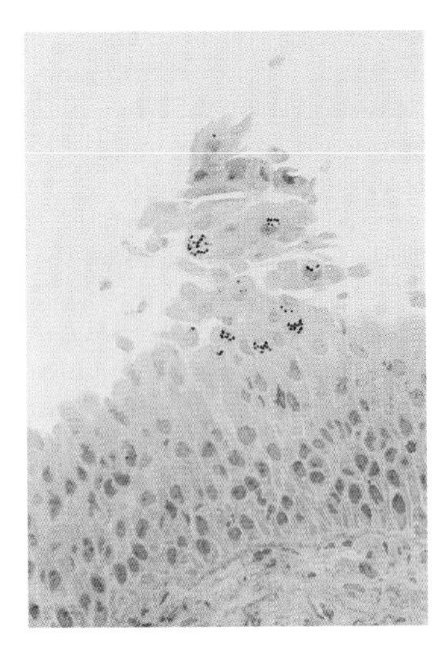

FIGURE 22 Plastic embedded semithin section of a nasal mucosal biopsy specimen, showing epithelial infection with *E. hellem* spores. Toluidine blue stain was used. Magnification, ×400.

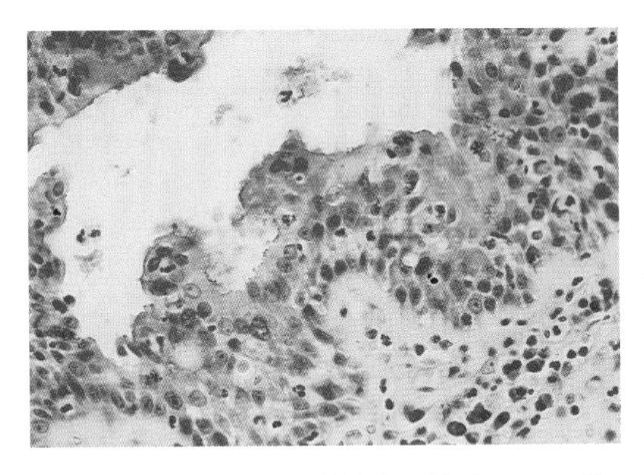

FIGURE 23 Microsporidial sinusitis due to an *Encephalitozoon* sp. Numerous infected epithelial cells can be seen to contain faintly gram-positive clusters of spores. Brown and Hopps stain was used. Magnification, ×400.

TRACHEOBRONCHIAL AND PULMONARY TRACTS

Tracheobronchial and pulmonary microsporidial infections, either with or without symptoms, appear to be a common component of disseminated microsporidiosis (Scaglia et al., 1998; Schwartz et al., 1993; Weber et al., 1993a). A study on AIDS patients with keratoconjunctivitis due to *E. hellem* showed that a significant number had microsporidial spores present in their sputum even though they had no pulmonary symptoms (Schwartz et al., 1993b). It is now recognized that asymptomatic respiratory tract colonization or subclinical pulmonary infection with microsporidiosis can occur, and several patients with AIDS have been described as having *E. hellem* or *E. cuniculi* spores present in sputum or bronchoalveolar lavage fluid and no respiratory symptoms (Scaglia et al., 1998; Weber et al., 1993a). Our understanding of the spectrum of pathologic changes in the lungs of persons infected with microsporidiosis has been limited because of the limited number of patients examined by biopsy or autopsy (Haselton et al., 1996).

Symptomatic pulmonary disease due to *Encephalitozoon* sp. has been described in several patients with AIDS. In the first patient who was autopsied following death due to disseminated *E. hellem* infection, massive numbers of gram-positive microsporidial spores were present diffusely throughout the entire length of the tracheobronchial tree (Fig. 24), extending into terminal bronchioles and associated in some areas with erosive tracheitis, bronchitis, and bronchiolitis (Schwartz et al., 1992). This confluent pattern of microsporidial colonization of the superficial tracheobronchial mucosa, extending into terminal bronchioles, was suggestive of a respiratory mechanism of acquisition. The pulmonary findings from this case were confirmed by subsequent biopsies of patients having pulmonary *E. hellem* infections, which demonstrated bronchiolitis with or without pneumonia (Schwartz et al., 1993b). Microsporidial spores were present in epithelial cells lining bronchi and bronchioles, in neutrophils within the bronchiolar wall, in cells lining the alveoli, in neutrophils, and extracellularly in alveolar spaces. *E. cuniculi* pulmonary infection has been reported in AIDS patients with bronchiolitis and pneumonia (Fig. 25 and 26) (De Groote et al., 1995; Deplazes et al., 1996a).

E. intestinalis can also infect the respiratory tract, in which it is associated with a clinical presentation similar to that of *E. hellem* and *E. cuniculi* (Molina et al., 1995). There are few data available on the pulmonary pathologic findings of *E. intestinalis* infection; however, spores have been described in bronchial epithelial cells (Orenstein et al., 1992a, 1992b).

Rare occurrences of *E. bieneusi* spores in sputum and bronchoalveolar lavage specimens from patients with AIDS and intestinal microsporidiosis have been reported, including one patient with pulmonary symptoms (Del Aguila et al., 1997; Weber et al., 1992b). However, the

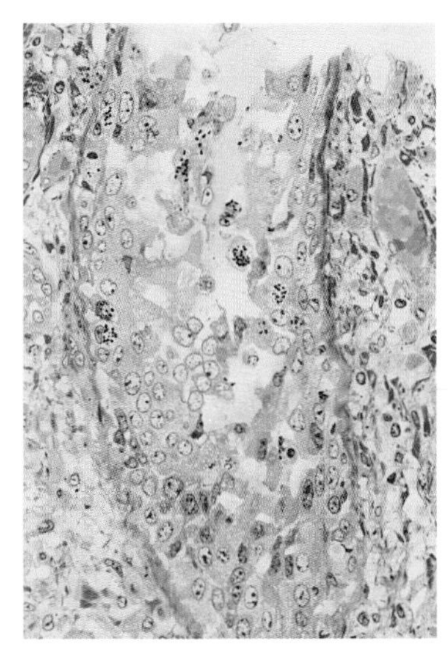

FIGURE 24 Plastic-embedded semithin section of tracheal mucosa with numerous spores of *E. hellem* present in respiratory lining epithelium. Toluidine blue stain was used. Magnification, ×400.

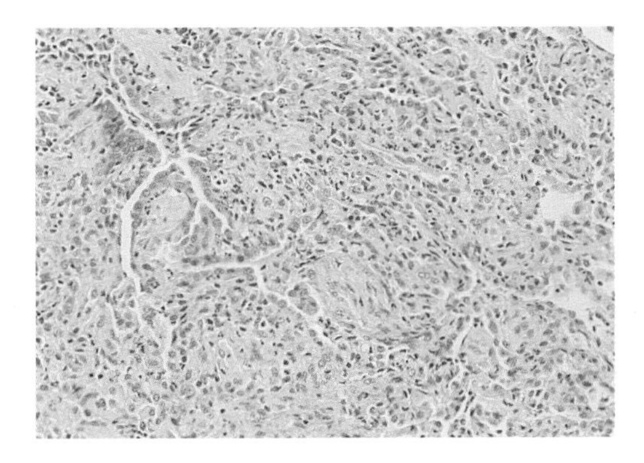

FIGURE 25 Open lung biopsy specimen, showing bronchiolitis and bronchopneumonia due to *E. cuniculi*. Microsporidia cannot be seen in this section, but the identity of this microsporidian was confirmed with both fluorescent antibody and PCR methods. Hematoxylin and eosin stain was used. Magnification, ×200.

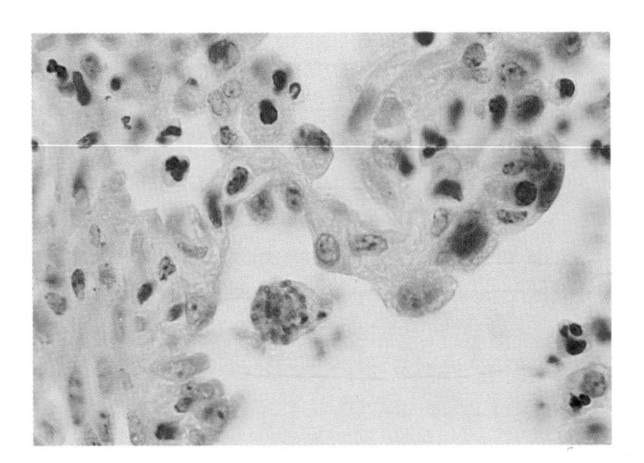

FIGURE 26 Alveolar macrophage containing phagocytosed spores of *E. cuniculi*. Brown and Hopps stain was used. Magnification, ×1,000.

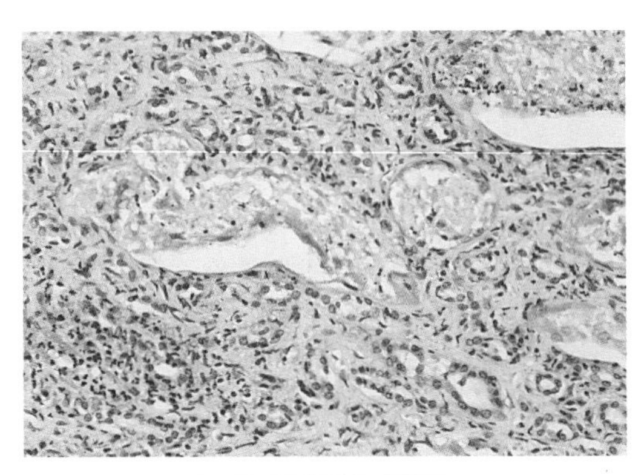

FIGURE 27 Chronic interstitial nephritis in a patient with disseminated *E. hellem* infection. The glomeruli are not involved. Hematoxylin and eosin stain was used. Magnification, ×200.

lung involvement in patients infected with *Enterocytozoon* spp. is probably very unusual. In endoscopically obtained transbronchial biopsy specimens, microsporidial spores are best identified in bronchial or bronchiolar epithelium. In well-oriented sections of tracheobronchial epithelium, spores are concentrated in the supranuclear or subapical region of infected host cells. Microsporidial spores can also be found within the alveolar spaces.

Genitourinary Tract

Urinary tract microsporidiosis appears to be common in patients with *Encephalitozoon* spp. infections (Gunnarsson et al., 1995; Molina et al., 1995). Patients with AIDS and urinary tract infection due to *T. anthropophthera* (Yachnis et al., 1996) and *V. corneae* have also been observed (Deplazes et al., in press).

In patients with AIDS and *Encephalitozoon* infections, simultaneous infection of the eyes, urinary tract, and bronchial tree frequently occurs, although many persons do not have symptoms referable to the kidneys or bladder (Schwartz et al., 1993a; Weber et al., 1994a). An autopsy of a patient with *E. hellem* infection revealed the urinary tract to be the most severely affected system (Schwartz et al., 1992). The kidneys showed a geographic pattern of chronic and granulomatous interstitial nephritis composed of variable numbers of plasma cells, lymphocytes, histiocytes, and neutrophils (Fig. 27). Extensive tubular necrosis was present in advanced cases, and the lumina of many necrotic tubules were filled with amorphous, granular material (Fig. 28). The coalescence of inflammatory cells around necrotic tubules resulted in microabscesses and poorly formed granulomas. With special stains, such as Steiner (Fig. 29) or tissue Gram stain, microsporidial spores were seen to be concentrated in necrotic renal tubules and, to a lesser extent, in the interstitium. The glomeruli were usually spared.

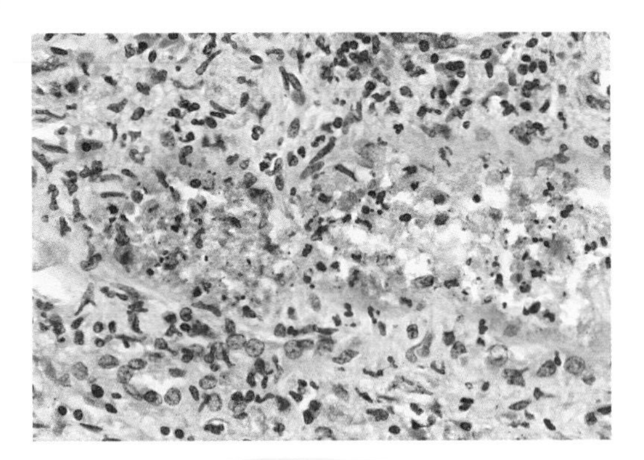

FIGURE 28 Renal tubule with necrosis, infected with *E. hellem*. The granular material filling the lumen represents microsporidial spores and necrotic cellular debris. Hematoxylin and eosin stain was used. Magnification, ×400.

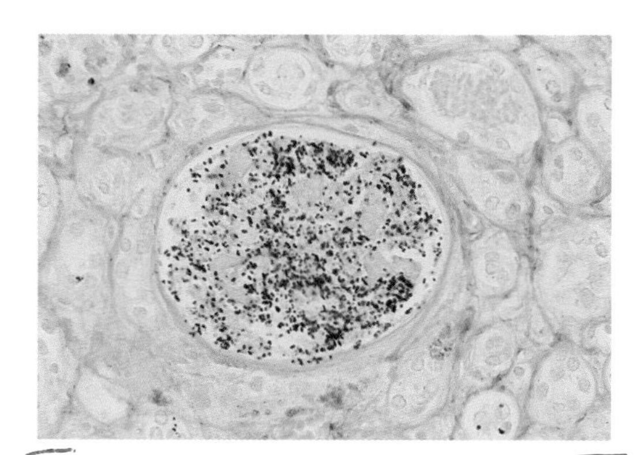

FIGURE 29 The lumen of this renal tubule is filled with *E. hellem* spores. Steiner stain was used. Magnification, ×200.

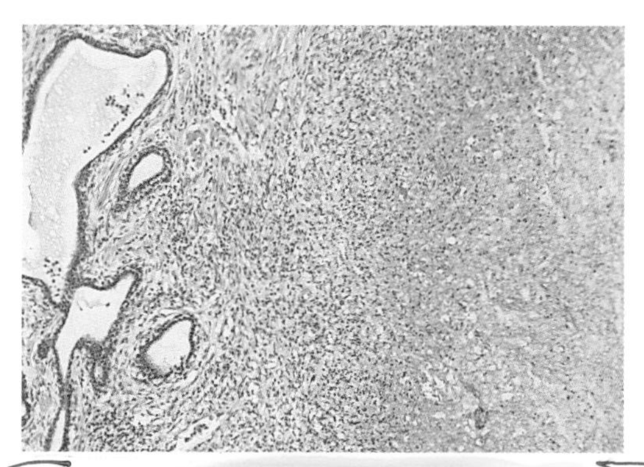

FIGURE 30 Microsporidial prostate abscess. The abscess material contained numerous spores which cannot be seen with routine hematoxylin and eosin staining. Hematoxylin and eosin stain was used. Magnification, ×200.

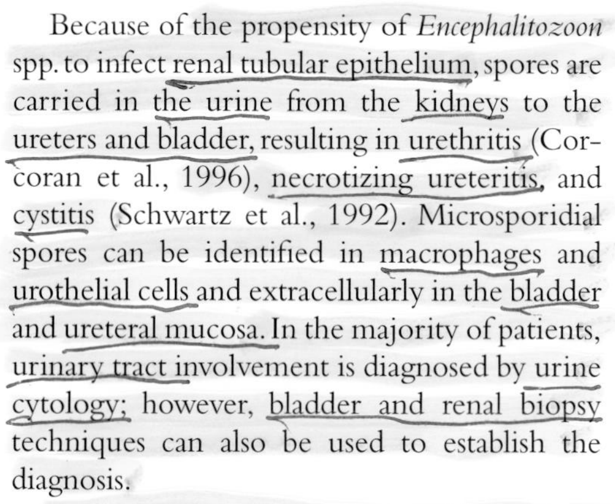

Because of the propensity of *Encephalitozoon* spp. to infect renal tubular epithelium, spores are carried in the urine from the kidneys to the ureters and bladder, resulting in urethritis (Corcoran et al., 1996), necrotizing ureteritis, and cystitis (Schwartz et al., 1992). Microsporidial spores can be identified in macrophages and urothelial cells and extracellularly in the bladder and ureteral mucosa. In the majority of patients, urinary tract involvement is diagnosed by urine cytology; however, bladder and renal biopsy techniques can also be used to establish the diagnosis.

Genital tract infection caused by microsporidia, consisting of a large, central prostatic abscess, has been described in one patient with severe urinary tract involvement due to *E. hellem* (Schwartz et al., 1994a). This abscess probably resulted from extension of an infection of the prostatic urethral mucosa. Spores were present within necrotic abscess material, in adjacent granulation tissue, and in the inflamed prostate gland (Fig. 30). The frequency of prostatic involvement in persons with disseminated microsporidiosis is not known, but involvement of this organ suggests a potential for sexual transmission of the agent.

Lesions of atrophic nephrons, interstitial edema, and fibrosis were found in a patient with AIDS and disseminated *T. anthropophthera* infection. The parasite was detected in tubular and glomerular epithelium, endothelium, and macrophages (Yachnis et al., 1996).

Central Nervous System

Central nervous sytem infection with a microsporidial agent presumed to be *E. cuniculi* was reported in 1959 and 1984 from two children with encephalitis and seizure disorders, but no tissue was available for pathologic examination. Spores consistent with *Encephalitozoon* spp. were found in the cerebrospinal fluid and urine from one child (Matsubayashi et al., 1959) and in urine alone from the other child (Bergquist et al., 1984b). Documented cerebral microsporidiosis has recently been reported in several patients with AIDS.

Disseminated canine strain *E. cuniculi* involving almost all organ systems was identified at the autopsy of one female patient (Mertens et al., 1997). Fluorochrome staining of brain sections revealed free spores in the parenchyma and perivascular spaces and spores within macrophages. There was no evidence of microsporidial infection of astrocytes, oligodendrocytes, neurons, or meningeal cells. A computed tomography scan and magnetic resonance imaging of another patient revealed multiple ringlike contrast-enhanced and micronodular lesions of the hippocampal, mesencephalic, and intracortical regions (Weber et al., 1997). Spores of rabbit strain *E. cuniculi* were identified in the cerebrospinal fluid, as well as in other body fluids, but brain tissue was not available for study.

Two patients with AIDS, a child and an adult, died from disseminated infection due to *T. anthropophthera* (Yachnis et al., 1996). In both patients, central nervous system involvement was present, consisting of multiple ringlike contrast-enhanced gray matter lesions shown by computed tomography scan. Microscopically they contained central areas of necrosis filled with free spores and spore-laden macrophages surrounded by microsporidian-infected astrocytes.

Voluntary Muscles

Myositis due to three different microsporidial agents—a *Pleistophora* sp., *T. hominis,* and a yet unidentified *Nosema*-like microsporidian—has been reported to occur in immunodeficient patients.

Muscle biopsies of three patients with myositis associated with *Pleistophora* infection showed atrophic and degenerating muscle fibers infiltrated by clusters of parasites (Fig. 31). Whereas the inflammatory infiltration was mild in two patients with AIDS (Chupp et al., 1993; Grau et al., 1996), the inflammatory reaction in the HIV-seronegative patient (who had severe cellular immunodeficiency of unknown origin) was intense, involving plasma cells, lymphocytes, and histiocytes (Ledford et al., 1985; Macher et al., 1988).

A biopsy of the deltoid muscle of another patient with AIDS revealed myositis due to a novel microsporidian, *T. hominis* (Field et al., 1996). Paraffin sections showed discrete lesions consisting of a central zone of fibrosis and degenerate skeletal muscle myofibers, surrounded by myofibers of approximately normal size containing a variable number of abutting polygonal sporophorous vesicles. The vesicles had a discernible eosinophilic wall and contained up to 32 spores or spore precursors. Similar parasites were also found in a conjunctival smear and nasopharyngeal washings of this patient, but not in the urine or feces.

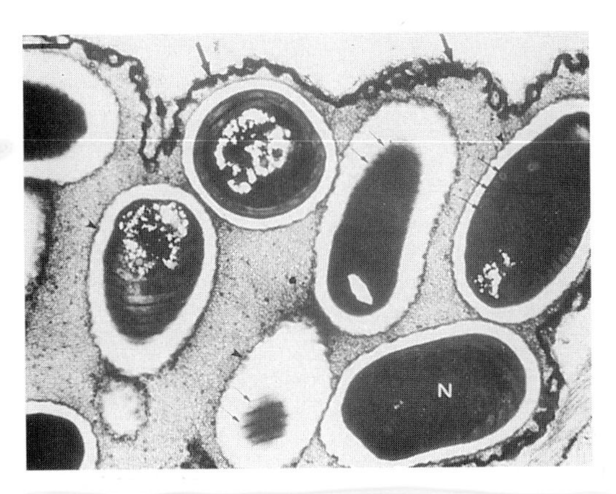

FIGURE 31 Muscle fiber with a sporont of *Pleistophora* containing several sporoblasts. The sporont has a thick, electron dense pansporoblastic membrane (large arrows). Sporoblasts have a nucleus (N), electron-dense exospore (arrowheads), and a coiled polar tubule (small arrows) with 9 to 12 turns. (Reprinted with permission from L. S. Adelman and from J. Alroy, Department of Pathology, Tufts University School of Medicine, Boston, Mass.)

Heart

Autopsy of a patient with AIDS and disseminated *E. cuniculi* infection revealed scattered foci of microsporidian-laden histiocytes and/or myocytes that were sometimes necrotic and associated with inflammation (Mertens et al., 1997). *T. anthropophthera*-infected myocytes surrounding central areas of necrosis were seen in another patient with AIDS (Yachnis et al., 1996). *N. connori* was identified in smooth muscle cells of arteries and myocytes of an athymic child (Margileth et al., 1973).

Spleen, Lymph Node, Bone Marrow, and Other Tissues

Microsporidia have been found in almost every organ system. In addition to the cases described above, *T. anthropophthera* has also been detected in the spleen, lymph nodes, bone marrow, endocrine and exocrine cells of the pancreas, follicular epithelium of the thyroid, and parathyroid epithelium and adipocytes of a patient with AIDS (Yachnis et al., 1996). *N. connori* infection in an athymic child has been reported, involving almost all tissues examined at autopsy (Margileth et al., 1973).

In a patient with cytomegalovirus adrenalitis, *E. cuniculi* spores were also seen in epithelium and endothelial cells of the adrenal glands (Mertens et al., 1997). At the autopsy of a patient with AIDS and a right upper quadrant mass, a 20-cm lobulated inflammatory mass of the omentum was found which contained focal necrosis, nongranulomatous inflammation, and *Encephalitozoon*-like spores (Zender et al., 1989). The species identity of this microsporidian was not confirmed. Rare microsporidial spores were identified in the soft tissues beneath a tongue ulcer in an AIDS patient with disseminated *E. cuniculi* infection (De Groote et al., 1995). *E. intestinalis* infection was found in the mandible of a patient with AIDS (Belcher et al., 1997).

DIAGNOSTIC ASPECTS IN ANIMALS

Microsporidial infections have been described in a large number of mammals, but only *E. cuniculi* infections of rabbits, dogs, and foxes have been of major veterinary medical interest so far. Molecular techniques have been used to compare microsporidial isolates obtained from humans and animals in order to study possible sources of infection and reservoir hosts. Such molecular data have provided strong evidence that the rabbit and the dog strains of *E. cuniculi* are zoonotic parasites (Deplazes et al., 1996a, 1996b; Didier et al., 1996; Hollister et al., 1996; Mathis et al., 1997; Mertens et al., 1997). Furthermore, *E. bieneusi* was detected for the first time in fecal samples from asymptomatic pigs by microscopic and PCR techniques (Deplazes et al., 1996b), and *E. hellem* was found in feces and intestinal contents of different organs of budgerigars (Black et al., 1997) and in a parrot (Suter et al., in press). Although the modes of transmission of *E. bieneusi* and *E. hellem* have yet to be elucidated, these findings have indicated possible zoonotic transmission of these two microsporidial species.

Antemortem diagnosis of *Encephalitozoon* infection in animals can be achieved by detecting specific antibodies by CIA, IFAT, ELISA, and WB (as described in serology), but serology often cannot differentiate between latent infection and microsporidial disease. Microscopic or PCR-based detection of microsporidia in specimens from animals may have reduced diagnostic sensitivity because of the intermittent excretion of spores by the infected animals but is the only reliable means of antemortem diagnosis of *E. bieneusi* in pigs or *E. hellem* in birds.

Postmortem diagnosis in animals has been performed mainly by histologic examination of a variety of tissues by the methods described above. In rabbits a putative diagnosis of encephalitozoonosis was based on granulomatous lesions in the brain, kidney, and other organs (Fig. 32). However, with this procedure

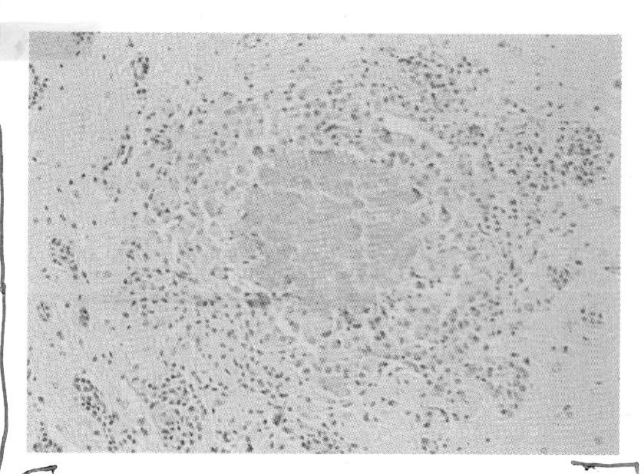

FIGURE 32 Section of brain tissue from a rabbit with torticollis, showing a microgranuloma with a central focus of necrosis but without detectable spores. Hematoxylin and eosin stain was used. Magnification, ×250. (Reprinted with permission from F. Guscetti, University of Zurich, Zurich, Switzerland.)

only heavily infected animals with evident histopathologic findings were considered infected, and the hematoxylin and eosin stain did not allow detection of microsporidial parasites. Immunohistochemical methods (Park et al., 1993; Guscetti et al., personal communication) using polyclonal anti-*Encephalitozoon* antibodies allowed demonstration of the parasite in tissue sections even in locations without severe cellular reactions (Fig. 33).

Rabbits

Serology is a valuable approach in identifying symptomatic as well as asymptomatic *E. cuniculi* infection in rabbits, allowing identification of animal colonies in which sanitation, which interrupts horizontal and vertical transmission, is necessary (Cox and Gallichio, 1978). The initially developed CIA has been replaced by IFAT, ELISA, or WB procedures using spores or soluble spore extracts as antigens (Cox et al., 1979; Müller, 1998; Weiss et al., 1992). After experimental oral infection of rabbits, specific antibodies were demonstrated 3 to 4 weeks before spores were found in the urine (Cox and Gallichio, 1978; Cox et al., 1979). Specific IgG antibodies are detectable by IFAT about 3 to 6 weeks after experimental oral infection, and by ELISA approximately 2 weeks later (Cox et al., 1979; Müller, 1998). Thereafter, specific antibody reactions, often in remarkably high titers, can persist over years in chronically infected asymptomatic animals. The specificity of the serologic tests for *E. cuniculi* infections has been found to be very high in domestic rabbit populations, and no cross-reactions have been found with antibodies due to other infections, e.g., those caused by *Toxoplasma* or *Eimeria.* Positive serology correlated well with the presence of parasites, as isolation from brain tissue was achieved by in vitro cultivation in 19 of 20 seropositive rabbits (Deplazes et al., 1996b).

Because spore excretion is sporadic even in severely symptomatic rabbits with neurological symptoms or nephritis, spore detection is unreliable (Cox and Walden, 1972). Horizontal transmission by ingestion (less likely by inhalation) of spores is thought to be the major mode of infection in rabbit colonies, but intrauterine *E. cuniculi* infections have also been described in rabbits (Cox et al., 1979).

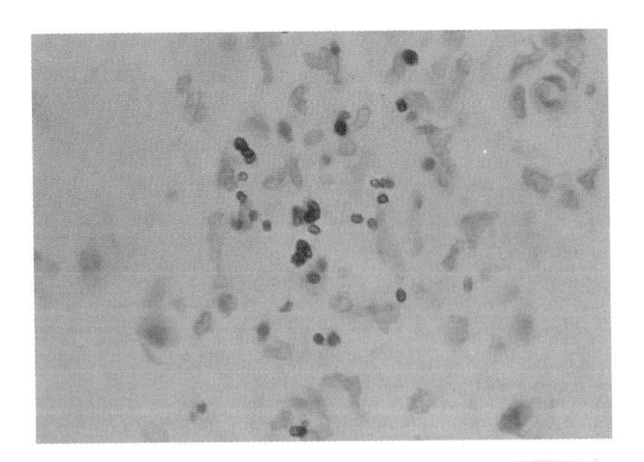

FIGURE 33 Section of brain tissue from an emperor tamarin with lethal neonatal encephalitozoonosis. Immunohistochemical detection of *E. cuniculi* spores in a granuloma was done with a polyclonal rabbit antibody to *E. cuniculi*. Visualization was by the labeled streptavidin-biotin method. Magnification, ×250. (Reprinted with permission from F. Guscetti, University of Zurich, Zurich, Switzerland.)

Monkeys, Foxes, and Dogs

Intrauterine and neonatal infections due to *E. cuniculi* have been reported in farmed blue foxes (*Alopex lagopus*) (Mohn et al., 1974), in domestic dogs in South Africa and the United States (Botha et al., 1979, 1986; Shadduck and Geroulo, 1979), and in nonhuman primates kept in captivity (Canning et al., 1986; Guscetti et al., personal communication; Shadduck and Geroulo, 1979). Severe, disseminated, and lethal *E. cuniculi* infection (Botha et al., 1986) was caused by the dog strain, which has also been found to infect humans (Didier et al., 1995), and by the mouse strain, which infects farmed blue foxes (Mathis et al., 1996). Subclinically infected animals (after intrauterine or peroral infection) can remain infected for years, representing a reservoir of the parasite in this population and possibly also for humans. Therefore, serologic identification and isolation of *Encephalitozoon*-infected animals and their isolation are of practical value in avoiding vertical and horizontal parasite transmission.

Pigs

Enterocytozoon-like spores in fecal samples from three pigs were detected with the chromotrope staining technique, and PCR amplification and sequence analysis proved the presence of *E. bieneusi* in these samples (Deplazes et al., 1996b). In a subsequent study, the PCR technique was used to detect the excretion of *E. bieneusi* spores in feces of pigs from several farms. This procedure was shown to be more sensitive than light microscopy (chromotrope stain), which revealed that only a very few spores were excreted. Spore excretion was not correlated with animal age or clinical symptoms of the pigs. Samples obtained weekly from three naturally infected pigs revealed that the shedding of spores was intermittent or sometimes decreased below the threshold of detection (Breitenmoser et al., in press).

Birds

Microsporidial infections have been diagnosed in psittacine birds in aviaries in the United States and Australia (Canning et al., 1986). Infections due to *Encephalitozoon*-like microsporidia, associated with a high mortality, were detected in affected tissues by electron microscopy. Molecular analyses have shown that *E. hellem* was the etiologic agent of lethal infections in budgerigar chicks in an aviary in the United States (Black et al., 1997). With the same methods, *E. hellem* was also detected in a bird belonging to the family Psittacidae (*Chalopsitta scintillata*) that had been captured in the wild on Aru Island (Indonesia) and died in a quarantine ward in Switzerland (Suter et al., in press). Intestinal infections were detected in all these birds, but no information is available about the diagnostic value of spore detection in feces, and no serologic tests have been developed to identify asymptomatic infections in birds.

APPENDIX: LABORATORY PROCEDURES

Chromotrope Staining (Weber et al., 1992a)

Chromotrope staining solution: Dissolve 6.0 g of chromotrope 2R with 0.15 g of fast green and 0.7 g of phosphotungstic acid in 3 ml of glacial acetic acid (940 ml 96% ethyl alcohol, 62 ml distilled water, 4.5 ml acetic acid). Allow the solution to stand for 30 min, and then mix with 100 ml of distilled water.

Staining procedure: Prepare slides from 10- to 20-μl aliquots of a suspension of unconcentrated liquid stool in 10% formalin (1:3 ratio) very thinly spread over the whole slide. Fix smears in methanol for 5 min, allow to air dry, and stain for 90 min with chromotrope stain at room temperature. Rinse in acid alcohol (4.5 ml of acetic acid, 995.5 ml of 90% ethyl alcohol) for 10 s, and then rinse briefly in 95% alcohol. Dehydrate successively in 95% alcohol for 5 min, in 100% alcohol for 10 min, and in xylene or a xylene substitute for 5 min. View the slide under a light microscopic magnification of ×630 to ×1,000 (oil immersion).

Modified Chromotrope Blue Staining (Ryan et al., 1993)

Use 0.5 g of aniline blue instead of fast green and reduce the amount of phosphotungstic acid to 0.25 g. The concentration of the chromotrope stain and the staining procedure are as described above (Ryan et al., 1993).

Modifications of Staining Temperature of Chromotrope Staining (Kokoskin et al., 1994; Didier et al., 1995)

Stain with chromotrope at 50°C for 10 min (Kokoskin et al., 1994) or at 37°C for 30 min (Didier et al., 1995). Use original or modified chromotrope stains as described above.

Quick Hot Gram-Chromotrope Staining (Moura et al., 1997)

Staining procedure: Prepare a thin smear of the material to be stained and allow it to air dry. Heat-fix the smear three times for 1 s over a low flame or 5 min on a slide warmer at 60°C. Cool to room temperature before staining. Perform Gram staining (without the safranin step). A slight modification of the Gram staining procedure (as described in Gram Stain Kit SG 100, available from Fisher Scientific, or Carr-Scarborough, Decatur, Ga., and for the Boekel Slide Stainer) is used: Dip slides into a staining jar containing gentian violet solution and let stand for 30 s. Rinse off excess stain gently with water. Dip slides into a jar with Gram iodine solution and allow to remain on the slide for 30 s. Remove the iodine solution by gently rinsing with a decolorizer solution. Hold the slide at an angle and add the decolorizer solution dropwise until it flows off the slide colorless. Wash the slide gently with cold water to remove excess decolorizer solution. Perform chromotrope staining. Place the slide in warm (50 to 55°C) chromotrope stain (3.0 g of chromotrope 2R, 0.15 g of fast green, 0.25 g of phosphotungstic acid, 3 ml of acetic acid, 100 ml of distilled water) for at least 1 min. Rinse in 90% acid alcohol (4.5 ml acetic acid, 995.5 ml of 90% ethyl alcohol) for 1 to 3 s. Rinse in 95% ethyl alcohol for 30 s. Rinse twice, 30 s each time, in 100% ethyl alcohol in two different containers and let dry. Mount with Cytoseal 60 (Stephens Scientific, Riverdale, N.J.) and examine.

For formalin-fixed, paraffin-embedded tissue sections, deparaffinize as usual, hydrate in a series of alcohols, and bring the slides to water before performing Gram staining.

Acid-Fast Trichrome Staining (Ignatius et al., 1997b)

Specimen preparation: For concentration of *Cryptosporidium* oocysts, SAF-preserved stool specimens are concentrated by using the formalin-ethyl acetate sedimentation procedure with centrifugation at 650 × *g* for 10 min. A 0.01-ml sediment sample is thinly spread on a glass slide and air-dried. Slides are immediately fixed in methanol for 5 min.

Staining procedure: Fixed slides are covered with a carbol-fuchsin solution (25.0 g of phenol, 500 ml of distilled water, and 25.0 ml of a saturated alcoholic fuchsin solution consisting of 2.0 g of basic fuchsin in 25 ml of 96% ethanol) for 10 min without heating, wash briefly with tap water, decolorize with 0.5% HCI-alcohol, and wash again with tap water. Stain slides immediately for 30 min at 37°C with Didier's trichrome solution. The latter is prepared by dissolving 6.0 g of chromotrope 2R, 0.5 g of aniline blue, and 0.7 bg of phosphotungstic acid in 3 ml of acetic acid at room temperature for 30 min, adding 100 ml of distilled water, and adjusting to pH 2.5 by the addition of 2 N HCI. Then rinse the slides for 10 s with acid alcohol (4.5 ml of acetic acid in 995.5 ml of 90% ethanol), wash for 30 s with 95% ethanol, and examine directly after air drying with oil immersion (magnification, ×1,000).

Calcofluor Staining

Different techniques of chemofluorescence staining, developed by Vávra et al., (1982, 1993) and van Gool et al. (1993) have been described. Because Uvitex 2B as initially used by van Gool et al. is not commercially available, the technique employed by Didier et al., (1995) is described.

Prepare Calcofluor White M2R (Sigma Chemical Co., St. Louis, Mo.) as a 0.5% (wt/vol) solution in Tris-buffered saline (pH 7.2), store at room temperature, and centrifuge at 15,000 × *g* for 2 min to remove precipitates prior to use. Fix smears with methanol and allow to air dry. Stain smears with 1 to 2 drops of Calcofluor solution for 2 to 3 min at room temperature. Rinse with slow-running water and counterstain with 0.1% Evans blue (Sigma catalog no. E-2129) in Tris-buffered saline (pH 7.2) for 1 min at room temperature. Rinse under slow-running water, allow to air dry, seal with Cytoseal 60 (Stephens Scientific), and view under an ultraviolet microscope at a wavelength of 395 to 415 nm (observation light of 455 nm). Organisms appear as bluish-white or turquoise oval halos.

Brown-Brenn Gram Stain for Tissue Staining

Staining procedure: Deparaffinize paraffin-embedded, formalin-fixed sections (4 to 5 μm) and hydrate sections with distilled water. Place slides on a staining rack and cover sections with crystal violet solution for 1 min. Drain off crystal violet solution and rinse with distilled water. Mordant in Gram iodine solution for 5 min. Rinse in distilled water. Blot sections on filter paper. Flood with decolorizer (70% isopropanol [75%] and acetone [25%] are preferred to absolute alcohol or acetone) up to 10 s or until the color stops running. Rinse in distilled water. Stain in basic fuchsin for 1 min (a safranin O solution is thought to produce results that are too red, and thus basic fuchsin is used instead [Luna, 1968]). Dip in acetone to start differentiation. Immediately place slides in 1% picric acid acetone and differentiate until sections are yellowish-pink. Rinse quickly in acetone. Rinse quickly in acetone-xylene. Clear in xylene and mount.

Modified Warthin-Starry Stain for Tissue Staining (Field et al., 1993a, 1993b, 1996)

Staining procedure (from the *Manual of Histologic Staining Methods of the Armed Forces Institute of Pathology* [Luna, 1968] with modifications [Field et al., 1993a, 1993b, 1996]): Cut sections from paraffin blocks into distilled water, including a known positive control for *Helicobacter* or spirochaetes. Rinse slides back and front thoroughly with distilled water. Place sections in freshly prepared 1% silver nitrate (dilute 1:10 from stock 10% silver nitrate with acidified water, pH 4.0) for 30 min at 37°C. Wash in three changes of distilled water. Place sections in the developing solution (15 ml) of water (pH 4.0) and 5% gelatin (mixed and preheated in a 60°C waterbath), 1.5 ml of 10% silver nitrate, and 1.0 ml of 1.5% quinol (added just before developing) in a 60°C waterbath until the sections turn a light golden brown (approximately 2 to 3 min; see below). Check microscopically for the right result. Rinse well in distilled water, then hot running tap water, and then in distilled water. Fix sections in 5% sodium thiosulfate for 3 min. Wash well in water for 2 min. Dehydrate, clear in xylol, and mount in Eukitt immediately to prevent fading.

The major problems that occur with this stain can be minimized (Field et al., 1993a, 1993b, 1996): (1) To prevent a dirty background, (a) do not use an adhesive (e.g., polylysine, gelatin, albumin) since these substances attract the silver stain, leaving a deposit on the slide; (b) ensure that all glassware used, especially glass slides, is clean and free of any contaminants at all stages; (c) use distilled water for making solutions and rinsing glassware; (d) use high-quality reagents; (e) carefully watch slides for the end point of color development since the change in color from yellow to brown background is rapid (between 2 and 3 min) and easily missed. (2) To achieve a bright golden-brown color (a) use acidulated water (pH 4.0) (obtain correct pH with citric acid and do not use an acetic acetate buffer); (b) use fresh working solutions. The quinol (1.5% hydroxyquinone), which should be freshly made or not more than 1 week old, is most critical: development of a yellow tea color in the background indicates oxidation, and the quinol should be discarded. (3) To prevent fading, (a) treat stained sections with 5% sodium thiosulfate, (b) use xylol as a clearing agent, (c) use a reliable mounting medium (Eukitt is recommended), and (d) mount sections immediately after staining.

REFERENCES

Aldras, A. M., J. M. Orenstein, D. P. Kotler, J. A. Shadduck, and E. Didier. 1994. Detection of microsporidia by indirect immunofluorescence antibody test using polyclonal and monoclonal antibodies. *J. Clin. Microbiol.* **32:**608–612.

Ashton, N., and P. A. Wirasinha. 1973. Encephalitozoonosis (nosematosis) of the cornea. *Br. J. Ophthalmol.* **57:**669–674.

Beaugerie, L., M. F. Teilhac, A. M. Deluol, J. Fritsch, P. M. Girard, W. Rozenbaum, Y. LeQuintrec, and F. P. Chatelet. 1992.

Cholangiopathy associated with microsporidia infection of the common bile duct mucosa in a patient with HIV infection. *Ann. Intern. Med.* **117:** 401–402.

Beckers, P. J., G. J. Derks, T. van Gool, F. J. Rietveld, and R. W. Sauerwein. 1996. *Encephalitozoon intestinalis:* specific monoclonal antibodies for laboratory diagnosis of microsporidiosis. *J. Clin. Microbiol.* **34:**282–285.

Belcher, J. W. Jr., S. A. Guttenberg, and B. M. Schmookler. 1997. Microsporidiosis of the mandible in a patient with acquired immunodeficiency syndrome. *J. Oral Maxillofac. Surg.* **55:** 424–426.

Bergquist, R., L. Morfeldt-Mansson, P. O. Pehrson, B. Petrini, and J. Wasserman. 1984a. Antibody against *Encephalitozoon cuniculi* in Swedish homosexual men. *Scand. J. Infect. Dis.* **16:**389–391.

Bergquist, N. R., G. Stintzing, L. Smedman, T. Waller, and T. Andersson. 1984. Diagnosis of encephalitozoonosis in man by serological tests. *Br. Med. J.* **288:**902.

Black, S. S., L. A. Steinohrt, D. C. Bertucci, L. B. Rogers, and E. S. Didier. 1997. *Encephalitozoon hellem* in budgerigars (*Melopsittacus undulatus*). *Vet. Pathol.* **34:**189–198.

Bornay-Llinares, F. J., A. J. da Silva, H. Moura, D. A. Schwartz, G. S. Visvesvara, N. J. Pieniazek, A. Cruz-Lopez, P. Hernandez-Jauregui, and J. Enriquez. 1998. Immunological, microscopic and molecular evidence of *Encephalitozoon (Septata) intestinalis* infection in mammals other than man. *J. Infect. Dis.* **178:**820–826.

Botha, W. S., C. G. Stewart, and A. F. van Dellen. 1986. Observations on the pathology of experimental encephalitozoonosis in dogs. *J. South Afr. Vet. Assoc.* **57:**17–24.

Botha, W. S., A. F. van Dellen, and C. G. Stewart. 1979. Canine encephalitozoonosis in South Africa. *J. S. Afr. Vet. Med. Assoc.* **50:**135–144.

Breitenmoser, A., A. Mathis, E. Bürgi, R. Weber, and P. Deplazes. High prevalence of *Enterocytozoon bieneusi* in swine with 4 genotypes that differ from those identified in humans. *Parasitology,* in press.

Cali, A., D. P. Kotler, and J. M. Orenstein. 1993. *Septata intestinalis* n. g., n. sp., an intestinal microsporidian associated with chronic diarrhea and dissemination in AIDS patients. *J. Protozool.* **40:**101–112.

Cali, A., D. Meisler, C. Y. Lowder, R. Lembach, L. Ayers, P. M. Takvorian, I. Rutherford, D. L. Longworth, J. T. McMahon, and R. T. Bryan. 1991a. Corneal microsporidioses: characterization and identification. *J. Protozool.* **38:** S215–S217.

Cali, A., D. M. Meisler, I. Rutherford, C. Y. Lowder, J. T. McMahon, D. L. Longworth, and R. T. Bryan. 1991b. Corneal microsporidiosis in a patient with AIDS. *Am. J. Trop. Med. Hyg.* **44:**463–468.

Cali, A., and R. L. Owen. 1990. Intracellular development of *Enterocytozoon,* a unique microsporidian found in the intestine of AIDS patients. *J. Protozool.* **37:**145–155.

Cali, A., P. M. Takvorian, S. Lewin, M. Rendel, C. Sian, M. Wittner, and L.M. Weiss. 1996. Identification of a new *Nosema*-like microsporidian associated with myositis in an AIDS patient. *J. Eukaryot. Microbiol.* **43:**S108.

Canning, E. U. 1993. Microsporidia, p. 219–370. *In* J. P. Kreier (ed.) *Parasitic Protozoa,* vol. 6. Academic Press, San Diego, Calif.

Canning, E. U. 1998. Microsporidiosis, p. 609–623. *In* S. R. Palmer, Lord Soulsby, and D. I. H. Simpson (ed.), *Zoonoses: Biology, Clinical Practice, and Public Health Control.* Oxford University Press, Oxford, United Kingdom.

Canning, E. U., A. Curry, C. J. N. Lacey, and D. Fenwick. 1992. Ultrastructure of *Encephalitozoon* sp. infecting the conjunctival, corneal and nasal epithelia of a patient with AIDS. *Eur. J. Protistol.* **28:**226–237.

Canning, E. U., J. Lom, and I. Dykova. 1986. *The Microsporidia of Vertebrates.* Academic Press, New York.

Carter, P. L., D. W. MacPherson, and R. A. McKenzie. 1996. Modified technique to recover microsporidian spores in sodium acetate-acetic acid-formalin-fixed fecal samples by light microscopy and correlation with transmission electron microscopy. *J. Clin. Microbiol.* **34:**2670–2673.

Centers for Disease Control. 1990. Microsporidian keratoconjunctivitis in patients with AIDS. *Morbid. Mortal. Weekly Rep.* **39:**188–189.

Chupp, G. L., J. Alroy, L. S. Adelman, J. C. Breen, and P. R. Skolnik. 1993. Myositis due to *Pleistophora* (microsporidia) in a patient with AIDS. *Clin. Infect. Dis.* **16:**15–21.

Conteas, C. N., T. Sowerby, G. W. Berlin, F. Dahlan, A. Nguyen, R. Porschen, J. Donovan, M. LaRiviere, and J. M. Orenstein. 1996. Fluorescence techniques for diagnosing intestinal microsporidiosis in stool, enteric fluid, and biopsy specimens from acquired immunodeficiency syndrome patients with chronic diarrhea. *Arch. Pathol. Lab. Med.* **120:**847–853.

Corcoran, G. D., J. R. Isaacson, C. Daniels, and P. L. Chiodini. 1996. Urethritis associated with disseminated microsporidiosis: clinical response to albendazole. *Clin. Infect. Dis.* **22:**592–593.

Cox, J. C., and H. A. Gallichio. 1978. Serological and histological studies on adult rabbits with recent, naturally acquired encephalitozoonosis. *Res. Vet. Sci.* **24:**260–261.

Cox, J. C., R. C. Hamilton, and H. D. Attwood. 1979. An investigation of the route and progression of *Encephalitozoon cuniculi* infection in adult rabbits. *J. Protozool.* **26:**260–265.

Cox, J. C., and N. B. Walden. 1972. Presumptive diagnosis of *Nosema cuniculi* in rabbits by immunofluorescence. *Res. Vet. Sci.* **13:**595–597.

Coyle, C. M., M. Wittner, D. P. Kotler, C. Noyer, J. M. Orenstein, H. B. Tanowitz, and L. M. Weiss. 1996. Prevalence of microsporidiosis due to *Enterocytozoon bieneusi* and *Encephalitozoon (Septata) intestinalis* among patients with AIDS-related diarrhea: determination by polymerase chain reaction to the microsporidian small-subunit rRNA gene. *Clin. Infect. Dis.* **23:**1002–1006.

Croppo, J. P., G. S. Visvesvara, G. J. Leitch, S. Wallace, and D. A. Schwartz. 1998. Identification of the microsporidian *Encephalitozoon hellem* using immunoglobulin G monoclonal antibodies. *Arch. Pathol. Lab. Med.* **122:**182–186.

Da Silva, A. J., F. J. Bornay-Llinares, C. A. del Aguila de la Puente, H. Moura, J. M. Peralta, I. Sobottka, D. A. Schwartz, G. S. Visvesvara, S. B. Slemenda, and N. J. Pieniazek. 1997. Diagnosis of *Enterocytozoon bieneusi* (microsporidia) infections by polymerase chain reaction in stool samples using primers based on the region coding for small-subunit ribosomal RNA. *Arch. Pathol. Lab. Med.* **121:**874–879.

Da Silva, A. J., D. A. Schwartz, G. S. Visvesvara, H. Demoura, S. B. Slemenda, and N. J. Pieniazek. 1996. Sensitive PCR diagnosis of infections by *Enterocytozoon bieneusi* (microsporidia) using primers based on the region coding for small subunit rRNA. *J. Clin. Microbiol.* **34:**986–987.

Davis, R. M., R. L. Font, M. S. Keisler, and J. A. Shadduck. 1990. Corneal microsporidiosis. a case report including ultrastructural observations. *Ophthalmology* **97:**953–957.

De Groote, M. A., G. S. Visvesvara, M. L. Wilson, N. J. Pieniazek, S. B. Slemenda, A. J. daSilva, G. J. Leitch, R. T. Bryan, and R. Reves. 1995. Polymerase chain reaction and culture confirmation of disseminated *Encephalitozoon cuniculi* infection in a patient with AIDS: Successful therapy with albendazole. *J. Infect. Dis.* **171:**1375–1378.

DeGirolami, P. C., C. R. Ezratty, G. Desai, A. McCullough, D. Asmuth, C. Wanke, and M. Federman. 1995. Diagnosis of intestinal microsporidiosis by examination of stool and duodenal aspirate with Weber's modified trichrome and Uvitex 2B stains. *J. Clin. Microbiol.* **33:**805–810.

Del Aguila, C., R. Lopez Velez, S. Fenoy, C. Turreientes, J. Cobo, R. Navajas, G. S. Visvesvara, G. P. Croppo, A. J. Da Silva, and N. J. Pieniazek. 1997. Identification of *Enterocytozoon bieneusi* spores in respiratory samples from an AIDS patient with a 2-year history of intestinal microsporidiosis. *J. Clin. Microbiol.* **35:**1862–1866.

Deplazes, P. Unpublished data.

Deplazes, P., A. Mathis, R. Baumgartner, I. Tanner, and R. Weber. 1996a. Immunologic and molecular characteristics of *Encephalitozoon*-like microsporidia isolated from humans and rabbits indicate that *Encephalitozoon cuniculi* is a zoonotic parasite. *Clin. Infect. Dis.* **22:**557–559.

Deplazes, P., A. Mathis, C. Müller, and R. Weber. 1996b. Molecular epidemiology of *Encephalitozoon cuniculi* and first detection of *Enterocytozoon bieneusi* in fecal samples of pigs. *J. Eukaryot. Microbiol.* **43:**93S.

Deplazes, P., A. Mathis, M. van Saanen, A. Iten, R. Keller, I. Tanner, M. Glauser, R. Weber, and E. U. Canning. Dual microsporidial infection due to *Vittaforma corneae* and *Encephalitozoon hellem* in a patient with AIDS. *Clin. Infect. Dis.*, in press.

Desportes, I., Y. Le Charpentier, A. Galian, F. Bernard, B. Cochand-Priollet, A. Lavergne, P. Ravisse, and R. Modigliani. 1985. Occurrence of a new microsporidian: *Enterocytozoon bieneusi* n. g., n. sp., in the enterocytes of a human patient with AIDS. *J. Protozool.* **32:**250–254.

Didier, E. S., P. J. Didier, D. N. Friedberg, S. M. Stenson, J. M. Orenstein, R. W. Yee, F. O. Tio, R. M. Davis, C. Vossbrinck, N. Millichamp, and J. A. Shadduck. 1991a. Isolation and characterization of a new human microsporidian, *Encephalitozoon hellem* (n. sp.), from three AIDS patients with keratoconjunctivitis. *J. Infect. Dis.* **163:**617–621.

Didier, E. S., J. M. Orenstein, A. Aldra, D. Bertucci, L. B. Rogers, and F. A. Janney. 1995. Comparison of three staining methods for detecting microsporidia in fluids. *J. Clin. Microbiol.* **33:**3138–3145.

Didier, E. S., L. B. Rogers, J. M. Orenstein, M. D. Baker, C. R. Vossbrinck, T. van Gool, R. Hartskeerl, R. Soave, and L. M. Beaudet. 1996. Characterization of *Encephalitozoon (Septata) intestinalis* isolates cultured from nasal mucosa and bronchoalveolar lavage fluids of two AIDS patients. *J. Eukaryot. Microbiol.* **43:**34–43.

Didier, E. S., J. A. Shadduck, P. J. Didier, N. Millichamp, and D. R. Vossbrinck. 1991b. Studies on ocular microsporidia. *J. Protozool.* **38:**635–638.

Didier, E. S., K. F. Snowden, and J. A. Shadduck. 1998. Biology of microsporidian species infecting mammals. *Adv. Parasitol.* **40:**284–320.

Didier, E. S., G. S. Visvesvara, M. D. Baker, L. B. Rogers, D. C. Bertucci, M. A. De Groote, and C. R. Vossbrinck. 1996. A microsporidian isolated from an AIDS patient corresponds to *Encephalitozoon cuniculi* III, originally isolated from domestic dogs. *J. Clin. Microbiol.* **34:**2835–2837.

Didier, E. S., C. R. Vossbrinck, M. D. Baker, L. B. Rogers, D. C. Bertucci, and J. A. Shadduck. 1995. Identification and characterization of three *Encephalitozoon cuniculi* strains. *Parasitology* **111:** 411–421.

Dobbins, W. O. III, and W. M. Weinstein. 1985. Electron microscopy of the intestine and rectum in acquired immunodeficiency syndrome. *Gastroenterology* **88:**738–749.

Dore, G. J., D. J. Marriott, M. C. Hing, J. L. Harkness, and A. S. Field. 1995. Disseminated microsporidiosis due to *Septata intestinalis* in nine patients infected with the human immunodeficiency virus: response to therapy with albendazole. *Clin. Infect. Dis.* **21:**70–76.

Enriquez, F. J., A. P. Cruz-Lopez, J. D. Palting, P. Cruz-Lopez, P. Hernandez-Jauregui, C. Tellez, J. Guerrero, and B. Curran. 1997a. Prevalence of microsporidial infections in children and adults with diarrhea, abstr. C-216, p. 158. *In Abstracts of the 97th General Meeting of the American Society for Microbiology.* American Society for Microbiology, Washington, D.C.

Enriquez, F. J., O. Ditrich, J. D. Palting, and K. Smith. 1997b. Simple diagnosis of *Encephalitozoon* sp. infections by using a panspecific antiexospore monoclonal antibody. *J. Clin. Microbiol.* **35:** 724–729.

Field, A., M. Hing, S. Milliken, and D. Marriott. 1993a. Microsporidia in the small intestine of HIV-infected patients: a new diagnostic technique and a new species. *Med. J. Aust.* **158:**390–394.

Field, A. S., D. J. Marriott, and M. C. Hing. 1993b. The Warthin-Starry stain in the diagnosis of small intestinal microsporidiosis in HIV-infected patients. *Folia Parasitol.* (Praha) **40:**261–266.

Field, A. S., D. J. Marriott, S. T. Milliken, B. J. Brew, E. U. Canning, J. G. Kench, P. Darveniza, and J. L. Harkness. 1996. Myositis associated with a newly described microsporidian, *Trachipleistophora hominis,* in a patient with AIDS. *J. Clin. Microbiol.* **34:**2803–2811.

Flepp, M., B. Sauer, R. Lüthy, and R. Weber. 1995. Human microsporidiosis in HIV-seronegative, immunocompetent patients, abstr. LM25, p. 331. *In Abstracts of the 35th Interscience Conference on Antimicrobial Agents and Chemotherapy.* American Society for Microbiology, Washington, D.C.

Franzen, C., D. A. Schwartz, G. S. Visvesvara, A. Müller, A. Schwenk, B. Salzberger, G. Fätkenheuser, G. Mahrle, V. Diehl, and M. Schrappe M. 1995. Disseminated antibody-confirmed *Encephalitozoon cuniculi* with asymptomatic infection of the gastrointestinal tract in a patient with AIDS. *Clin. Infect. Dis.* **21:**1480–1484.

Friedberg, D. N., S. M. Stenson, J. M. Orenstein, P. M. Tierno, and N. C. Charles. 1990. Microsporidial keratoconjunctivitis in acquired immunodeficiency syndrome. *Arch. Ophthalmol.* **108:**504–508.

Garcia, L. S., R. Y. Shimizu, and D. A. Bruckner. 1994. Detection of microsporidial spores in fecal specimens from patients diagnosed with cryptosporidiosis. *J. Clin. Microbiol.* **32:**1739–1741.

Giang, T. T., D. P. Kotler, M. L. Garro, and J. M. Orenstein. 1993. Tissue diagnosis of intestinal microsporidiosis using the chromotrope-2R modified trichrome stain. *Arch. Pathol. Lab. Med.* **117:**1249–1251.

Grau, A., M. E. Valls, J. E. Williams, D. S. Ellis, M. J. Muntane, and C. Nadal. 1996. Miositis por *Pleistophora* en un paciente con SIDA. *Med. Clin.* (Barcelona) **107:**779–781.

Gunnarsson, G., D. Hurlbut, P. C. DeGirolami, M. Federman, and C. Wanke. 1995. Multiorgan microsporidiosis: report of five cases and review. *Clin. Infect. Dis.* **21:**37–44.

Guscetti, F., A. Mathis, J. N. Hatt, and P. Deplazes. Personal communication.

Hartskeerl, R. A., A. R. J. Schuitema, and R. deWachter. 1993b. Secondary structure of the small subunit ribosomal RNA sequence of the microsporidium *Encephalitozoon cuniculi.* Nucl. Acids Res. **21:**1489.

Hartskeerl, R. A., A. R. J. Schuitema, T. van Gool, and J. Terpstra. 1993a. Genetic evidence for the occurrence of extra-intestinal *Enterocytozoon bieneusi* infections. *Nucl. Acids Res.* **21:**4150.

Hartskeerl, R. A., T. van Gool, A. R. Schuitema, E. S. Didier, and W. J. Terpstra. 1995. Genetic and immunological characterization of the microsporidian *Septata intestinalis.* Cali, Kotler and Orenstein, 1993: reclassification to *Encephalitozoon intestinalis. Parasitology* **110:** 277–285.

Hasleton, P. S., D. A. Schwartz, and S. Lucas. 1996. Pulmonary parasitic infections, p. 305–356. *In* P. S. Hasleton (ed.), *Spencer's Pathology of the Lung,* 4th ed. McGraw-Hill, New York.

Hewan-Lowe, K., B. Furlong, M. Sims, and D. A. Schwartz. 1997. Co-infection with *Giardia lam-*

blia and *Enterocytozoon bieneusi* in a patient with AIDS and chronic diarrhea. *Arch. Pathol. Lab. Med.M* **121:**417–422.

Hollister, W. S., and E. U. Canning. 1987. An enzyme-linked immunosorbent assay (ELISA) for detection of antibodies to *E. cuniculi* and its use in determination of infections in man. *Parasitology* **94:**209–219.

Hollister, W. S., E. U. Canning, N. I. Colbourn, and E. J. Aarons. 1995. *Encephalitozoon cuniculi* isolated from the urine of an AIDS patient, which differs from canine and murine isolates. *J. Eukaryot. Microbiol.* **42:**367–372.

Hollister, W. S., E. U. Canning, E. Weidner, A. S. Field, J. Kench, and D. J. Marriot. 1996. Development and ultrastructure of *Trachipleistophora hominis* n. g., n. sp. after in vitro isolation from an AIDS patient and inoculation into athymic mice. *Parasitology* **112:**143–154.

Hollister, W. S., E. U. Canning, and A. Willcox. 1991. Evidence for widespread occurrence of antibodies to *Encephalitozoon cuniculi* (Microspora) in man provided by ELISA and other serological tests. *Parasitology* **102:**33–43.

Ignatius, R., S. Henschel, O. Liesenfeld, U. Mansmann, W. Schmidt, S. Köppe, T. Schneider, W. Heise, U. Futh, E. O. Ricken, H. Hahn, and R. Ullrich. 1997a. Comparative evaluation of modified trichrome and Uvitex 2B stains for detection of low numbers of microsporidial spores in stool specimens. *J. Clin. Microbiol.* **35:**2266–2269.

Ignatius, R., M. Lehmann, K. Miksits, T. Regnath, M. Arvand, E. Engelmann, H. Hahn, and J. Wagner. 1997b. A new acid-fast trichrome stain for simultaneous detection of *Cryptosporidium parvum* and microsporidial species in stool specimens. *J. Clin. Microbiol.* **35:**446–449.

Josephson, G. D., J. Sarlin, J. Reidy, and R. Pincus. 1996. Microsporidial rhinosinusitis: is this the next pathogen to infect the sinuses of the immunocompromised host? *Otolaryngol. Head Neck Surg.* **114:**137–139.

Joste, N., J. D. Rich, K. J. Busam, and D. A. Schwartz. 1996. Autopsy verification of *Encephalitozoon intestinalis* (microsporidiosis) eradication following albendazole therapy. *Arch. Pathol. Lab. Med.* **120:**199–203.

Kokoskin, E., T. W. Gyorkos, A. Camus, L. Cedilotte, T. Purtill, and B. Ward. 1994. Modified technique for efficient detection of microsporidia. *J. Clin. Microbiol.* **32:**1974–1975.

Kotler, D. P., A. Francisco, F. Clayton, J. V. Scholes, and J. M. Orenstein. 1990. Small intestinal injury and parasitic diseases in AIDS. *Ann. Intern. Med.* **113:**444–449.

Kotler, D. P., T. T. Giang, M. L. Garro, and J. M. Orenstein. 1994. Light microscopic diagnosis of microsporidiosis in patients with AIDS. *Am. J. Gastroenterol.* **89:**540–544.

Kotler, D. P., and J. M. Orenstein. 1994. Prevalence of intestinal microsporidiosis in HIV-infected individuals referred for gastroenterological evaluation. *Am. J. Gastroenterol.* **89:**1998–2001.

Kotler, D. P., S. Reka, K. Chow, and J. M. Orenstein. 1993. Effects on enteric parasitoses and HIV infection upon small intestinal structure and function in patients with AIDS. *J. Clin. Gastroenterol.* **16:**10–15.

Lacey, C. J. N., A. Clark, P. Frazer, T. Metcalfe, and A. Curry. 1992. Chronic microsporidian infection in the nasal mucosae, sinuses and conjunctivae in HIV disease. *Genitourin. Med.* **68:** 179–181.

Lambl, B. B., M. Federman, D. Pleskow, and C. A. Wanke. 1996. Malabsorption and wasting in AIDS patients with microsporidia and pathogen-negative diarrhea. *AIDS* **10:**739–744.

Ledford, D. K., M. D. Overman, A. Gonzalo, A. Cali, W. Mester, and R.F. Lockey. 1985. Microsporidiosis myositis in a patient with acquired immunodeficiency syndrome. *Ann. Intern. Med.* **102:**628–630.

Lowder, C.Y., D. M. Meisler, J. T. McMahon, D. L. Longworth, and I. Rutherford. 1990. Microsporidia infection of the cornea in a man seropositive for human immunodeficiency virus. *Am. J. Ophthalmol.* **109:**242–244.

Luna, L. 1968. *Manual of Histologic Staining Methods of the Armed Forces Institute of Pathology,* 3rd ed. McGraw-Hill, New York.

Macher, A. M., R. Neafie, P. Angritt, and S. M. Tuur. 1988. Microsporidial myositis and the acquired immunodeficiency syndrome (AIDS): a four-year follow-up. *Ann. Intern. Med.* **109:**343. (Letter.)

Mansfield, K. G., A. Carville, D. Shvetz, J. MacKey, S. Tzipori, and A. A. Lackner. 1997. Identification of an *Enterocytozoon bieneusi*-like microsporidian parasite in simian-immunodeficiency-virus-inoculated macaques with hepatobiliary disease. *Am. J. Pathol.* **150:**1395–1405.

Margileth, A. M., A. J. Strano, R. Chandra, R. Neafie, M. Blum, and R. M. McCully. 1973. Disseminated nosematosis in an immunologically compromised infant. *Arch. Pathol.* **95:**145–150.

Mathis, A., J. Akerstedt, J. Tharaldsen, O. Odegaard, and P. Deplazes. 1996. Isolates of *Encephalitozoon cuniculi* from farmed blue foxes (*Alopex lagopus*) from Norway differ from isolates from Swiss domestic rabbits (*Oryctolagus cuniculus*). *Parasitol. Res.* **82:**727–730.

Mathis, A., M. Michel, H. Kuster, C. Müller, R. Weber, and P. Deplazes. 1997. Two *Encephalitozoon cuniculi* subtypes of human origine are infectious to rabbits. *Parasitology* **114:**29–35.

Matsubayashi, H., T. Koike, T. Mikata, and S. Hagiwara. 1959. A case of Encephalitozoon-like body infection in man. *Arch. Pathol.* **67:** 181–187.

McDougall, R. J., M. W. Tandy, R. E. Boreham, D. J. Stenzel, and P. J. O'Donoghue. 1993. Incidental finding of a microsporidian parasite from an AIDS patient. *J. Clin. Microbiol.* **31:** 436–439.

McWhinney, P. H. M., D. Nathwani, S. T. Green, J. F. Boyd, and J. A. H. Forrest. 1991. Microsporidiosis detected in association with AIDS-related sclerosing cholangitis. *AIDS* **5:**1394–1395.

Mertens, R. B., E. S. Didier, M. C. Fishbein, D. C. Bertucci, L. B. Rogers, and J. M. Orenstein. 1997. *Encephalitozoon cuniculi* microsporidiosis: infection of the brain, heart, kidneys, trachea, adrenal glands, and urinary bladder in a patient with AIDS. *Mod. Pathol.* **10:**68–77.

Metcalfe, T. W., R. M. L. Foran, P. L. Rowland, A. Curry, and C. J. M. Lacey. 1992. Microsporidial keratoconjunctivitis in a patient with AIDS. *Br. J. Ophthalmol.* **76:**177–178.

Modigliani, R., C. Bories, Y. le Charpentier, M. Salmeron, B. Messing, A. Galian, J. C. Rambaud, A. Lavergene, B. Cochand-Priollet, and I. Desportes. 1985. Diarrhoea and malabsorption in acquired immune deficiency syndrome: a study of four cases with special emphasis on opportunistic protozoan infections. *Gut* **26:** 179–187.

Mohn, S. F., K. Nordstoga, J. Krogsrud, and A. Helgebostad. 1974. Transplacental transmission of *Nosema cuniculi* in the blue fox (*Alopex lagopus*). *Acta Pathol. Microbiol. Scand. Sect. B* **82:**299–300.

Molina, J. M., E. Oksenhendler, B. Beauvais, C. Sarfati, A. Jaccard, F. Derouin, and J. Modai. 1995. Disseminated microsporidiosis due to *Septata intestinalis* in patients with AIDS: clinical features and response to albendazole therapy. *J. Infect. Dis.* **171:** 245–249.

Moura, H., D. A. Schwartz, F. Bornayllinares, F. C. Sodre, S. Wallace, and G. S. Visvesvara. 1997. A new and improved quick-hot chromotrope technique that differentially stains microsporidian spores in clinical samples, including paraffin-embedded tissue sections. *Arch. Pathol. Lab. Med.* **121:**888–893.

Müller, C. 1998. Untersuchungen zur Diagnostik, Biologie und Verbreitung von Microsporidien bei Kaninchen und anderen Tierarten. Veterinary doctoral thesis. University of Zürich, Switzerland.

Niederkorn, J. Y., J. A. Shadduck, and E. Weidner. 1980. Antigenic cross-reactivity among different microsporidian spores as determined by immunofluorescence. *J. Parasitol.* **66:**675–677.

Orenstein, J. M. 1991. Microsporidiosis in the acquired immunodeficiency syndrome. *J. Parasitol.* **77:**843–864.

Orenstein, J. M., J. Chiang, W. Steinberg, P. D. Smith, H. Rotterdam, and D. P. Kotler. 1990. Intestinal microsporidiosis as a cause of diarrhea in human immunodeficiency virus-infected patients: a report of 20 cases. *Hum. Pathol.* **21:**475–481.

Orenstein, J. M., D. T. Dieterich, and D. P. Kotler. 1992a. Systemic dissemination by a newly recognized intestinal microsporidia species in AIDS. *AIDS* **6:**1143–1150.

Orenstein, J. M., M. Tenner, A. Cali, and D. P. Kotler. 1992b. A microsporidian previously undescribed in humans, infecting enterocytes and macrophages, and associated with diarrhea in an acquired immunodeficiency syndrome patient. *Hum. Pathol.* **23:**722–728.

Orenstein, J. M., M. Tenner, and D. P. Kotler. 1992c. Localization of infection by the microsporidian *Enterocytozoon bieneusi* in the gastrointestinal tract of AIDS patients with diarrhea. *AIDS* **6:**195–197.

Owen, R. L. 1997. Polymerase chain reaction of stool: a powerful tool for specific diagnosis and epidemiologic investigation of enteric microsporidia infections. *AIDS* **11:**817–818. (Editorial.)

Park, J. H., K. Ochiai, and C. Itakura. 1993. Direct ABC immunohistochemistry to *Encephalitozoon cuniculi*. *J. Vet. Med. Sci.* **55:**325–328.

Pinnolis, M., P. R. Egbert, R. L. Font, and F. C. Winter. 1981. Nosematosis of the cornea. *Arch. Ophthalmol.* **99:**1044–1047.

Pol, S., C. A. Romania, S. R. Richard, P. Amouyal, I. Desportes-Livage, F. Carnot, J. F. Pays, and P. Berthelot. 1993. Microsporidia infection in patients with the human immunodeficiency virus and unexplained cholangitis. *N. Engl. J. Med.* **328:**95–99.

Rabodonirina, M., M. Bertocchi, I. Desportes-Livage, L. Cotte, H. Levrey, M. A. Piens, G. Monneret, M. Celard, J. F. Mornex, and M. Mojon. 1996. *Enterocytozoon bieneusi* as a cause of chronic diarrhea in a heart-lung transplant recipient who was seronegative for human immunodeficiency virus. *Clin. Infect. Dis.* **23:**114–117.

Raynaud, L., F. Delbac, V. Broussolle, M. Rabodonirina, V. Girault, M. Wallon, G. Cozon,

C. P. Vivares, and F. Peyron. 1998. Identification of *Encephalitozoon intestinalis* in travelers with chronic diarrhea by specific PCR amplification. *J. Clin. Microbiol.* **36:**37–40.

Rijpstra, A. C., E. U. Canning, R. J. van Ketel, J. K. M. Eeftinck Schattenkerk, and J. J. Laarman. 1988. Use of light microscopy to diagnose small-intestinal microsporidiosis in patients with AIDS. *J. Infect. Dis.* **157:**827–831.

Rinder, H., K. Janitschke, H. Aspöck, A. J. Da Silva, P. Deplazes, D. P. Fedorko, C. Franzen, U. Futh, F. Hünger, A. Lehmacher, C. G. Meyer, J.-M. Molina, J. Sandfort, R. Weber, T. Löscher, and the Diagnostic Multicenter Study Group on Microsporidia. 1998. Blinded, externally controlled multicenter evaluation of light microscopy and PCR for detection of microsporidia in stool specimens. *J. Clin. Microbiol.* **36:**1814–1818.

Rossi, R. M., C. Wanke, and M. Federman. 1996. Microsporidian sinusitis in patients with the acquired immunodeficiency syndrome. *Laryngoscope* **106:**966–971.

Ryan, N., G. Sutherland, K. Coughlan, M. Globan, J. Doultree, J. Marshall, R. W. Baird, J. Pedersen, and B. Dwyer. 1993. A new trichrome-blue stain for detection of microsporidial species in urine, stool, and nasopharyngeal specimens. *J. Clin. Microbiol.* **31:** 3264–3269.

Sandfort, J., A. Hannemann, D. Stark, R. L. Owen, and B. Ruf. 1994. *Enterocytozoon bieneusi* infection in an immunocompetent HIV-negative patient with acute diarrhea. *Clin. Infect. Dis.* **19:** 514–516.

Sax, P. E., J. D. Rich, W. S. Pieciak, and Y. M. Trnka. 1995. Intestinal microsporidiosis occurring in a liver transplant recipient. *Transplantation* **60:**617–618.

Scaglia, M., S. Gatti, L. Sacchi, S. Corona, G. Chichino, A. M. Bernuzzi, G. Barbarini, G. P. Croppo, A. J. Da Silva, N. J. Pieniazek, and G. S. Visvesvara. 1998. Asymptomatic respiratory tract microsporidiosis due to *Encephalitozoon hellem* in three patients with AIDS. *Clin. Infect. Dis.* **26:**174–176.

Schmidt, W., T. Schneider, W. Heise, J. D. Schulzke, T. Weinke, R. Ignatius, R. L. Owen, M. Zeitz, E. O. Riecken, and R. Ullrich. 1997. Mucosal abnormalities in microsporidiosis. *AIDS* **11:**1589–1594.

Schwartz, D. A., A. Abou-Elella, C. M. Wilcox, L. Gorelkin, G. S. Visvesvara, S. E. Thompson, S. Hogan, R. Weber, and R. T. Bryan. 1995. The presence of *Enterocytozoon bieneusi* spores in the lamina propria of small bowel biopsies with no evidence of disseminated microsporidiosis. *Arch. Pathol. Lab. Med.* **119:**424–428.

Schwartz, D. A., D. C. Anderson, S. A. Klumpp, and H. M. McClure. 1998. Ultrastructure of atypical (teratoid) developmental stages of *Enterocytozoon bieneusi* in naturally infected rhesus monkeys (*Macaca mulatta*). *Arch. Pathol. Lab. Med.* **122:**423–429.

Schwartz, D. A., and R. T. Bryan. 1997. Microsporidia, p. 61–93. *In* C. R. Horsburgh, Jr., and A. M. Nelson (ed.), *Pathology of Emerging Infections.* ASM Press, Washington, D.C.

Schwartz, D. A., R. T. Bryan, K. O. Hewan-Lowe, G. S. Visvesvara, R. Weber, A. Cali, and P. Angritt. 1992. Disseminated microsporidiosis (*Encephalitozoon hellem*) and acquired immunodeficiency syndrome. *Arch. Pathol. Lab. Med.* **116:** 660–668.

Schwartz, D. A., I. Sobottka, G. J. Leitch, A. Cali, and G. S. Visvesvara. 1996. Pathology of microsporidiosis: emerging parasitic infections in patients with acquired immunodeficiency syndrome. *Arch. Pathol. Lab. Med.* **120:**173–188.

Schwartz, D. A., G. S. Visvesvara, M. C. Diesenhouse, R. Weber, R. L. Font, L. A. Wilson, G. Corrent, D. F. Rosberger, P. J. Keenen, H. Grossniklaus, K. Hewan-Lowe, and R. T. Bryan. 1993a. Ocular pathology of microsporidiosis: role of immunofluorescent antibody for diagnosis of *Encephalitozoon hellem* in biopsies, smears, and intact globes from seven AIDS patients. *Am. J. Opththalmol.* **115:**285–292.

Schwartz, D. A., G. S. Visvesvara, G. J. Leitch, L. Tashjian, M. Pollack, J. Holden, and R. T. Bryan. 1993b. Pathology of symptomatic microsporidial (*Encephalitozoon hellem*) bronchiolitis in AIDS: a new respiratory pathogen diagnosed from lung biopsy, bronchoalveolar lavage, sputum, and tissue culture. *Hum. Pathol.* **24:**937–943.

Schwartz, D. A., G. S. Visvesvara, R. Weber, and R. T. Bryan. 1994a. Male genital tract microsporidiosis and AIDS: prostatic infection with *Encephalitozoon hellem*. *J. Eukaryot. Microbiol.* **41:** 61S.

Schwartz, D. A., G. S. Visvesvara, R. Weber, C. M. Wilcox, and R. T. Bryan. 1994b. Microsporidiosis in HIV positive patients: current methods for diagnosis using biopsy, cytologic, ultrastructural, immunological and tissue culture techniques. *Folia Parasitol.* **41:**91–99.

Shadduck, J. A. 1969. *Nosema cuniculi:* in vitro isolation. *Science* **166:**516–517.

Shadduck, J. A., R. Bendele, and G. T. Robinson. 1978. Isolation of the causative organism of canine encephalitozoonosis. *Vet. Pathol.* **15:**449–460.

Shadduck, J. A., and M. J. Geroulo. 1979. A simple method for the detection of antibodies to *Encephalitozoon cuniculi* in rabbits. *Lab. Anim. Sci.* **29:** 330–334.

Shadduck, J. A., R. A. Meccoli, R. Davis, and R. L. Font. 1990. First isolation of a microsporidian from a human patient. *J. Infect. Dis.* **162:**773–776.

Shah, G. K., D. Pfister, L. E. Probst, P. Ferrieri, and E. Holland. 1996. Diagnosis of microsporidial keratitis by confocal microscopy and the chromatrope stain. *Am. J. Ophthalmol.* **121:** 89–91.

Sheth, S. G., C. Bates, M. Federman, and S. Chopra. 1997. Fulminant hepatic failure caused by microsporidial infection in a patient with AIDS. *AIDS* **11:**553–554. (Letter.)

Silveira, H., and E. U. Canning. 1995. *Vittaforma corneae* n. comb. for the human microsporidium *Nosema corneum* Shadduck, Meccoli, Davis & Font, 1990, based on its ultrastructure in the liver of experimentally infected athymic mice. *J. Eukaryot. Microbiol.* **42:**158–165.

Silverstein, B. E., E. T. Cunningham, Jr., T. P. Margolis, V. Cevallos, and I.G. Wong. 1997. Microsporidial keratoconjunctivitis in a patient without human immunodeficiency virus infection. *Am. J. Ophthalmol.* **124:**395–396.

Singh, M., G. J. Kane, L. Mackinlay, I. Quaki, E. H. Yap, B. C. Ho, L. C. Ho, and L. C. Kim. 1982. Detection of antibodies to *Nosema cuniculi* (Protozoa: Microsporidia) in human and animal sera by the indirect fluorescent antibody technique. *Southeast Asian J. Trop. Med. Public Health* **13:**110–113.

Sobottka, I., H. Albrecht, H. Schäfer, J. Schottelius, G. S. Visvesvara, R. Laufs, and D. A. Schwartz. 1995a. Disseminated *Encephalitozoon (Septata) intestinalis* infection in a patient with AIDS: novel diagnostic approaches and autopsy-confirmed parasitological cure following treatment with albendazole. *J. Clin. Microbiol.* **33:** 2948–2952.

Sobottka, I., H. Albrecht, J. Schottelius, M. Bentfeld, R. Laufs, and D. A. Schwartz. 1995b. Self-limited traveller's diarrhea due to a dual infection with *Enterocytozoon bieneusi* and *Cryptosporidium parvum* in an immunocompetent HIV-negative child. *Eur. J. Clin. Microbiol.* **14:** 919–920.

Sobottka, I., D. A. Schwartz, J. Schottelius, G. S. Visvesvara, N. J. Pieniazek, C. Schmetz, N. P. Kock, R. Laufs, and H. Albrecht. 1998. Prevalence and clinical significance of intestinal microsporidiosis in human immunodeficiency virus-infected patients with and without diarrhea in Germany: a prospective coprodiagnostic study. *Clin. Infect. Dis.* **26:**475–480.

Soule, J. B., A. L. Halverson, R. B. Becker, M. C. Pistole, and J. M. Orenstein. 1997. A patient with acquired immunodeficiency syndrome and untreated *Encephalitozoon (Septata) intestinalis* microsporidiosis leading to small bowel perforation: response to albendazole. *Arch. Pathol. Lab. Med.* **121:**880–887.

Sprague, V. 1974. *Nosema connori* n. sp., microsporidian parasite of man. *Trans. Am. Microsc. Soc.* **93:**400–403.

Suter, C., A. Mathis, R. Hoop, and P. Deplazes. Imported *Encephalitozoon hellem* infection in a wild parrot (*Chalopsitta scintillata*) from Indonesia. *Vet. Rec.*, in press.

Talal, A. H., D. P. Kotler, J. M. Orenstein, and L. M. Weiss. 1998. Detection of *Enterocytozoon bieneusi* in fecal specimens by polymerase chain reaction analysis with primers to the small-subunit rRNA. *Clin. Infect. Dis.* **26:**673–675.

Terada, S., K. R. Reddy, L. J. Jeffers, A. Cali, and E. R. Schiff. 1987. Microsporidian hepatitis in the acquired immunodeficiency syndrome. *Ann. Intern. Med.* **107:**61–62.

van Gool, T., E. U. Canning, and J. Dankert. 1994a. An improved practical and sensitive technique for the detection of microsporidian spores in stool samples. *Trans. R. Soc. Trop. Med. Hyg.* **88:** 189–190.

van Gool, T., E. U. Canning, H. Gilis, M. A. van den Bergh-Weerman, J. K. Eeftinck-Schattenkerk, and J. Dankert. 1994b. *Septata intestinalis* frequently isolated from stool of AIDS patients with a new cultivation method. *Parasitology* **109:**281–289.

van Gool, T., W. S. Hollister, J. Eeftinck Schattenkerk, M. A. Weerman, R. J. van Ketel, P. Reiss, and E. U. Canning. 1990. Diagnosis of *Enterocytozoon bieneusi* microsporidiosis in AIDS patients by recovery of spores from faeces. *Lancet* **336:**697–698. (Letter.)

van Gool, T., F. Snijders, P. Reiss, J. K. M. Eeftinck Schattenkerk, M. A. van den Bergh Weerman, J. F. W. M. Bartelsman, J. J. M. Bruins, E. U. Canning, and J. Dankert. 1993. Diagnosis of intestinal and disseminated microsporidia infections in patients with HIV by a new rapid fluorescence technique. *J. Clin. Pathol.* **46:** 694–699.

van Gool, T., J. C. Vetter, B. Weinmayr, A. Van Dam, F. Derouin, and J. Dankert. 1997. High seroprevalence of *Encephalitozoon* species in immunocompetent subjects. *J. Infect. Dis.* **175:**1020–1024.

Vávra, J., and J. Chalupsky. 1982. Fluorescence staining of microsporidian spores with the brightener Calcofluor White M2R. *J. Protozool.* **29**(Suppl.):503.

Vávra, J., R. Dahbiova, W. S. Hollister, and E. U. Canning. 1993. Staining of microsporidian spores by optical brighteners with remarks on the use of brighteners for the diagnosis of AIDS associated human microsporidioses. *Folia Parasitol.* (Prague) **40:**267–272.

Vávra, J., A. T. Yachnis, J. A. Shadduck, and J. M. Orenstein. 1998. Microsporidia of the genus *Trachipleistophora*—causative agents of human microsporidiosis: description of *Trachipleistophora anthropophthera* n. sp. (Protozoa: Microsporidia). *J. Eukaryot. Microbiol.* **45:**273–283.

Verre, J., D. Marriott, M. C. Hing, A. S. Field, and J. L. Harkness. 1993. Light microscopic detection of microsporidial spores in faeces from HIV-infected patients. AIDS **7**(Suppl. 3): S55–S56.

Visvesvara, G. S., A. J. da Silva, G. P. Croppo, N. J. Pieniazek, G. J. Leitch, D. Ferguson, H. de Moura, S. Wallace, S. B. Slemenda, I. Tyrrell, D. R. Moore, and J. Meador. 1995a. In vitro and serologic and molecular identification of *Septata intestinalis* isolated from urine of a patient with AIDS. *J. Clin. Microbiol.* **33:**930–936.

Visvesvara, G. S., G. J. Leitch, A. J. Da Silva, G. P. Croppo, H. Moura, S. Wallace, S. B. Slemenda, D. A. Schwartz, D. Moss, R. T. Bryan, and N. J. Pieniazek. 1994. Polyclonal and monoclonal antibody and PCR-amplified small-subunit rRNA identification of a microsporidian, *Encephalitozoon hellem,* isolated from an AIDS patient with disseminated infection. *J. Clin. Microbiol.* **32:**2760–2768.

Visvesvara, G. S., G. J. Leitch, H. Moura, S. Wallace, R. Weber, and R. T. Bryan. 1991. Culture, electron microscopy, and immunoblot studies on a microsporidian parasite isolated from the urine of a patient with AIDS. *J. Protozool.* **38:** S105–S111.

Visvesvara, G. S., G. J. Leitch, N. J. Pieniazek, A. J. Da Silva, S. Wallace, S. B. Slemenda, R. Weber, D. A. Schwartz, L. Gorelkin, C. M. Wilcox, and R. T. Bryan. 1995b. Short-term in vitro culture and molecular analysis of *Enterocytozoon bieneusi. J. Eukaryot. Microbiol.* **42:**506–510.

Visvesvara, G. S., G. J. Leitch, S. Wallace, C. Seaba, D. Erdman, and E. P. Ewing, Jr. 1996. Adenovirus masquerading as microsporidia. *J. Parasitol.* **82:**316–319.

Vossbrinck, C. R., M. D. Barker, E. S. Didier, B. A. Debrunner-Vossbrinck, and J. A. Shadduck. 1993. Ribosomal DNA sequences of *Encephalitozoon hellem* and *Encephalitozoon cuniculi:* species identification and phylogenetic construction. *J. Eukaryot. Microbiol.* **40:**354–362.

Weber, R., R. T. Bryan, H. S. Bishop, S. P. Wahlquist, J. J. Sullivan, and D. D. Juranek. 1991. Threshold of detection of *Cryptosporidium* oocysts in human stool specimens: evidence for low sensitivity of current diagnostic methods. *J. Clin. Microbiol.* **29:**1323–1327.

Weber, R., R. T. Bryan, R. L. Owen, C. M. Wilcox, L. Gorelkin, G. S. Visvesvara, and the Enteric Opportunistic Infections Working Group. 1992a. Improved light-microscopical detection of microsporidia spores in stool and duodenal aspirates. *N. Engl. J. Med.* **326:**161–166.

Weber, R., R. T. Bryan, D. A. Schwartz, and R. L. Owen. 1994a. Human microsporidial infections. *Clin. Microbiol. Rev.* **7:**426–461.

Weber, R., P. Deplazes, M. Flepp, A. Mathis, R. Baumann, B. Sauer, H. Kuster, and R. Luthy. 1997a. Cerebral microsporidiosis due to *Encephalitozoon cuniculi* in a patient with human immunodeficiency virus infection. *N. Engl. J. Med.* **336:** 474–478.

Weber, R., M. Flepp, and W. Wichmann. 1997b. Cerebral microsporidiosis due to *Encephalitozoon cuniculi. N. Engl. J. Med.* **337:**640–641. (Letter.)

Weber, R., H. Kuster, R. Keller, T. Bächi, M. A. Spycher, J. Briner, E. Russi, and R. Lüthy. 1992b. Pulmonary and intestinal microsporidiosis in a patient with the acquired immunodeficiency syndrome. *Am. Rev. Respir. Dis.* **146:**1603–1605.

Weber, R., H. Kuster, G. S. Visvesvara, R. T. Bryan, D. A. Schwartz, and R. Lüthy. 1993a. Disseminated microsporidiosis due to *Encephalitozoon hellem:* pulmonary colonization, microhematuria and mild conjunctivitis in a patient with AIDS. *Clin. Infect. Dis.* **17:**415–419.

Weber, R., B. Ledergerber, R. Zbinden, M. Altwegg, G. Pfyffer, M. A. Spycher, J. Briner, M. Opravil, C. Meyenberger, M. Flepp, and the Swiss HIV Cohort Study. Diarrhea and enteric pathogens in HIV infected patients: prospective community based cohort study. Submitted for publication.

Weber, R., A. Müller, M. A. Spycher, M. Opravil, R. Ammann, and J. Briner. 1992c. Intestinal *Enterocytozoon bieneusi* microsporidiosis in an HIV-infected patient: diagnosis by ileocolonoscopic biopsies and long-term follow-up. *Clin. Invest.* **70:**1019–1023.

Weber, R., B. Sauer, R. Lüthy, and D. Nadal. 1993b. Intestinal *Enterocytozoon bieneusi* and *Cryptosporidium* coinfection in an HIV-infected child with chronic diarrhea. *Clin. Infect. Dis.* **17:**480–483.

Weber, R., B. Sauer, M. A. Spycher, P. Deplazes, R. Keller, R. Ammann, J. Briner, and R. Lüthy. 1994b. Detection of *Septata intestinalis* (Microsporidia) in stool specimens, and coprodiagnostic monitoring of successful treatment with albendazole. *Clin. Infect. Dis.* **19:**342-345.

Weiss, L. M., A. Cali, E. Levee, D. Lapplace, H. Tanowitz, D. Simon, and M. Wittner. 1992. Diagnosis of *Encephalitozoon cuniculi* infection by Western blot and the use of cross-reactive antigens for the possible detection of microsporidiosisi in humans. *Am. J. Trop. Med. Hyg.* **47:** 456–462.

Weiss, L. M., and C. R. Vossbrinck. 1998. Microsporidiosis: molecular and diagnostic aspects. *Adv. Parasitol.* **40:**351–395.

Wheatley, W. B. 1951. A rapid staining procedure for intestinal amoebae and flagellates. *Am. J. Clin. Pathol.* **21:**990–991.

WHO Parasitic Diseases Surveillance. 1983. Antibody to *Encephalitozoon cuniculi* in man. *WHO Weekly Epidem. Rec.* **58:**30–32.

Yachnis, A. T., J. Berg, A. Martinez-Salazar, B. S. Bender, L. Diaz, A. M. Rojiani, T. A. Eskin, and J. M. Orenstein. 1996. Disseminated microsporidiosis especially infecting the brain, heart, and kidneys: report of a newly recognized pansporoblastic species in two symptomatic AIDS patients. *Am. J. Clin. Pathol.* **106:**535–543.

Zender, H. O., E. Arrigoni, J. Eckert, and Y. Kapanci. 1989. A case of *Encephalitozoon cuniculi* peritonitis in a patient with AIDS. *Am. J. Clin. Pathol.* **92:**352–356.

Zierdt, C. H., V. J. Gill, and W. S. Zierdt. 1993. Detection of microsporidian spores in clinical samples by indirect fluorescent-antibody assay using whole-cell antisera to *Encephalitozoon cuniculi* and *Encephalitozoon hellem*. *J. Clin. Microbiol.* **31:**3071–3073.

CULTURE AND PROPAGATION OF MICROSPORIDIA

Govinda S. Visvesvara, Hercules Moura, Gordon J. Leitch, and David A. Schwartz

11

Although more than 1,200 species belonging to the approximately 143 genera of the ancient, intracellular, spore-forming, mitochondria-lacking eukaryotic protozoan parasites classified under the phylum Microsporidia (Sprague and Becnel, 1998) are known to infect members of virtually every phylum of the animal kingdom, only a few have been adapted to grow in culture. Before the advent of AIDS, interest in the microsporidia was focused on only a few parasites: (1) *Nosema apis* and *N. bombycis,* which parasitize economically important insects, honeybees and silkworms; (2) *Ameson michaelis,* a parasite of the commercially important blue crab; (3) *Glugea stephani,* a parasite of the economically profitable winter flounder; (4) *Nosema locustae* and *Vairimorpha necatrix,* parasites used in the biological control of agricultural pests; and (5) *Nosema algerae* and *Vavraia culicis,* parasites of medically important insect vectors (Jaronski, 1984). However, interest in propagating certain microsporidia that cause human disease in in vitro culture has intensified recently because several genera (e.g., *Encephalitozoon, Enterocytozoon, Nosema, Pleistophora, Trachipleistophora,* and *Vittaforma*) of microsporidia have been identified during the past decade as opportunistic pathogens of humans, especially patients with AIDS (Schwartz and Bryan, 1997; Weber et al., 1994).

Although *Enterocytozoon bieneusi* is the most frequently identified microsporidian, and is the agent of gastrointestinal (GI) disease leading to diarrhea in patients with AIDS (Schwartz and Bryan, 1997; Weber et al., 1994), all efforts to establish a continuous culture of these parasites have failed so far (Visvesvara et al., 1995b). Since infections due to *Encephalitozoon* spp. are also frequently recognized (Schwartz and Bryan, 1997), several isolates of *E. cuniculi, E. hellem,* and *Encephalitozoon* spp. that cause disseminated microsporidiosis without involving the GI tract have been established in culture (Bocket et al., 1992; Croppo et al., 1997, 1998;

Govinda S. Visvesvara, Division of Parasitic Diseases, M.S.-F/13, National Center for Infectious Diseases, Centers for Disease Control and Prevention, 4770 Buford Highway NE, Atlanta, GA 30341-3724. *Hercules Moura,* Division of Parasitic Diseases, M.S.-F/13, Centers for Disease Control and Prevention, 4770 Buford Highway NE, Atlanta, GA 30341-3724, and Faculdade de Ciências Médicas, Universidade do Estado do Rio de Janeiro and Hospital Evandro Chagas, Instituto Oswaldo Cruz, FIOCRUZ, Rio de Janeiro, Rio de Janeiro, Brazil. *Gordon J. Leitch,* Department of Physiology, Morehouse School of Medicine, Atlanta, Ga. *David A. Schwartz,* Department of Pathology, Emory University and Grady Memorial Hospital, Atlanta, Ga.

The Microsporidia and Microsporidiosis (Murray Wittner, editor; Louis M. Weiss, contributing editor), ©1999 American Society for Microbiology, Washington, D.C.

Deplazes et al., 1996a, 1996b; Desser et al., 1992; Didier et al., 1991, 1995; Furuya et al., 1995; Gatti et al., 1997; Hollister et al., 1993, 1995; Katzwinkel-Wladarsch et al., 1997; Scaglia et al., 1994, 1997, 1998; Visvesvara et al., 1991, 1994; Weber et al., 1997). Similarly, several isolates of *Encephalitozoon intestinalis,* the second most frequently identified microsporidial pathogen causing disseminated microsporidiosis, including GI tract infection, also have been established in culture (del Aguila et al., 1998; Deplazes et al., 1996a; Didier et al., 1995; Doultree et al., 1995; Molina et al., 1995; Visvesvara et al., 1995a). In addition, single isolates of *Nosema corneum* Shadduck et al., 1990 (renamed *Vittaforma corneae* Silveira and Canning, 1995), and *Trachipleistophora hominis* (Hollister et al., 1996) have been maintained in continuous culture. Recently, we have established in our laboratory a continuous culture of an isolate of *Nosema* from a human with an ocular infection (Visvesvara, Moura, et al., unpublished data) and an isolate of *N. algerae* from adult and larval mosquitoes (Moura et al., 1998).

In vitro cultivation of microsporidia, like that of any other parasitic protozoan causing human disease, can be an important adjunct to diagnosis. It is invaluable in studying the biochemistry, physiology, and metabolism of these parasites, as well as in determining their nutritional requirements. Since large numbers of specific stages of these parasites are available, in vitro cultivation can be especially useful for antigen production in the preparation of monoclonal and polyclonal antibodies used in immunologic testing and for vaccine development. Cultures are particularly useful in the in vitro screening of potential pharmaceutical agents, as well as in identifying the susceptible from the resistant isolates. Animal models can be developed to reproduce or simulate the naturally occurring disease so that pathologic processes can be understood. In vitro growth also provides an important means for studying these parasites at the light and ultrastructural levels.

Isolation in culture should always be attempted even when a presumptive diagnosis has been made. This is particularly important for the purpose of establishing a bank of isolates to be used for antigenic, molecular, and biochemical analyses. The isolates will be available to provide material for future studies and will be invaluable in epidemiologic investigations.

IN VITRO CULTURE DURING THE EARLY YEARS (1937 TO 1985)

The first attempt to culture microsporidia was made in 1937 by Trager, who was partially successful in establishing *N. bombycis* infection of a cell culture developed from the ovarian tube lining cells of the silkworm (*Bombyx mori*) (Trager, 1937; Jaronski, 1984). Interest in cultivation of these organisms waned until 1964 when Sen Gupta reported growing *Nosema mesnili,* a parasite of the cabbage worm (*Pieris brassicae*) by explanting gut and fat body tissue of the cabbage worm and allowing infected cells to multiply (Sen Gupta, 1964). Ishihara and Sohi (1966) described the infectivity of *N. bombycis* in cultures of ovarian tissue of the silkworm. In 1968 and 1969 Ishihara reported on the germination and development of *N. bombycis* at 28°C in primary cell cultures of rat, mouse, rabbit, and chicken embryos. Spore formation, however, was found only in rat embryo cells. Over the next few years, several microsporidian species were established in cultures, mostly in insect cell lines, although a few in mammalian cell lines were also established. For example, *N. algerae* was grown in pig kidney cells (Undeen, 1975), and *Nosema disstriae* was grown in hemocytes, imaginal cells, ovarian cell lines, and pupal gonads, as well as silk gland explants of the tent caterpillar (*Malacosoma* sp.) (Sohi and Wilson, 1976; Jaronski, 1984; Wilson and Sohi, 1977). *N. algerae* was also grown in insect cell cultures of *Heliothis zea* (IPLB) incubated at 26°C (Kurtti and Brooks, 1971, 1977; Kurtti et al., 1983), and continuous cultures of *N. bombycis* were established in *Antheraea eucalypti* cell lines (Kawarabata and Ishihara, 1984). Therefore, the period between the 1960s and early 1980s saw a flurry of activity with regard to the in vitro culture of microsporidia (e.g., *N. algerae, N. bombycis, N. apis, N. disstriae, N. eurytremae, N. heliothidis, N. mesnili, A. michaelis,* a *Pleistophora* sp., *V. necatrix,* and *V. culicis*) that parasitize insects, as demonstrated by a large number of reports (Jaronski, 1984). The various microsporidia were grown mostly in explanted insect tissues such as ovarian, midgut, embryo, fat body, gonad, heart, imaginal cell, pupal gonad, and silk gland tissues, as well as in a variety of insect cell culture systems (e.g., TN368 from *Trichoplusia ni*, IZD-Mb-0503 from *Mamestra brassicae*, IPLB-1075 from *H. zea*, ATC15 from *Aedes albopictus*, MOS20 from *A. egypti*, MOS55 from *Anopheles stephensi* and/or *A. gambiae*, ARM from *Armigeres subalbatus*, DMI from *Drosophila melanogaster*, BTC32 from *Triatoma infestans* embryo, and UM-BGE-1,2,4 from *Blattella germanica* embryo), and Xen and XTC6 from *Xenopus*. Furthermore, mammalian cell lines such as canine kidney, canine embryo, rabbit choroid plexus, pig kidney, rat brain, rat embryo, mouse embryo, and chick embryo, Chang liver cells, and even chimpanzee lung-baboon placenta cell cultures were cocultivated with different microsporidia (Jaronski, 1984).

Interest in the culture of microsporidia of mammalian origin was kindled in 1956 when Morris, et al. (Jaronski, 1984) grew *E. cuniculi* of mouse origin in the mouse lymphosarcoma MB III cell line for short periods of time. However, when Shadduck (1969) succeeded in establishing continuous cultivation of *E. cuniculi* of rabbit origin in rabbit kidney (RK) cells, interest in these parasites soon increased (Bismanis, 1970; Jaronski, 1984; Canning and Lom, 1986). Until 1990, *E. cuniculi* was the only microsporidian of mammalian hosts that had been cultivated in vitro, either for short periods or continuously, in a variety of cell lines by a number of researchers. The cell lines used included "finite" lines with a limited life span as well as lines with a continuous life span. The medium used for cultivation consisted of Eagle's minimal essential medium (MEM), Dulbecco's modification of MEM (DMEM), and RPMI 1640 supplemented with either fetal bovine serum (FBS) or fetal calf serum (FCS). Cultures were generally incubated at 37°C in a gas phase with 5% CO_2 or without CO_2 in an ordinary laboratory incubator as static culture systems. In one instance, *E. cuniculi* was grown at 25°C in RCP cells and even at 18°C in fathead minnow (*Timephalis promeles*) fish cells (Bedrnik and Vavra, 1975; Jaronski, 1984). The wide variety of cell lines that have been used to grow *E. cuniculi* have included those of (1) mouse origin, such as mouse fibroblast, mouse lymphosarcoma, embryonic mouse cells, and mouse kidney; (2) hamster origin, such as hamster plasmacytoma, baby hamster kidney, hamster primary glial cells, and hamster kidney; (3) rabbit origin, such as rabbit choroid plexus, weanling albino rabbit, rabbit primary glial cells, rabbit brain, RK, and rabbit fibroblasts; (4) dog origin, such as canine embryo cells, primary canine kidney, canine embryo fibroblasts, primary canine embryo fibroblasts, and canine kidney cells; (5) human origin, such as human uterine cancer cells, Ehrlich cancer cells, human embryonic fibroblasts, human fetal lung cells, and human diploid fetal cells; and (6) miscellaneous origin, such as feline lung, PK cells, ovine choroid plexus cells, bovine kidney cells, primary baboon placental cells, chick embryonic kidney cells, chick embryonic fibroblasts, and fish cells such as fathead minnow cells (Canning and Lom, 1986; Jaronski, 1984). Cultures of *E. cuniculi* were initiated in several different ways: by adding infected tissue explants to cultured cells, by allowing infected cells in explanted cells to grow, by allowing germination of spores in the presence of cells and thereafter infecting cells, and by scraping infected cells from infected cultures and adding them to fresh cell cultures (Canning and Lom, 1986; Jaronski, 1984).

IN VITRO CULTURE FROM 1985 TO THE PRESENT WITH SPECIAL REFERENCE TO HUMAN ISOLATES

Even though the first case of human microsporidiosis was reported in 1959 by Matsubayashi et al. (Schwartz and Bryan, 1997; Weber et al., 1994), nearly 21 years elapsed before successful isolation and in vitro cultivation of a microsporidian of human origin were achieved. Shadduck et al. (1990) reported continuous cultivation of *V. corneae* (*N. corneum*) from a corneal biopsy sample from an immunocompetent person. Since then a number of microsporidia belonging to several genera that cause human infection have been established in culture. Furthermore, the past 8 years have been fruitful for the isolation by microsporidiologists of microsporidial isolates from humans and other animals including the salmonid fish. More than 20 isolates of *E. cuniculi* from humans, rabbits, foxes, mice, and rats (Deplazes et al., 1996a; Thomas et al., 1997); 28 isolates of *E. hellem,* all from humans; 12 isolates of *E. intestinalis,* all from humans; and one isolate each of *V. corneae* (Silveira and Canning, 1995), *T. hominis* (Hollister et al., 1996), and *Nosema* spp. (Moura et al., 1998; Visvesvara et al., unpublished data) have been maintained in continuous culture. However, only short-term culture of *E. bieneusi* has been achieved during this period (Visvesvara et al., 1995b).

INITIATION OF CULTURES FROM PATIENT MATERIALS

The isolates mentioned above have been established in culture from corneal scraping and biopsy samples, urine, sputum, bronchoalveolar lavage (BAL) fluid, feces, duodenal aspirate and biopsy samples, cerebrospinal fluid, and biopsy samples from muscle tissue. Attempts have been made to isolate and culture these organisms, usually by adding the samples to cell culture monolayers. However, in some cases attempts have also been made to grow the microsporidia in explants of tissues obtained from patients. Unfortunately, many published reports state only that the causal agent of microsporidiosis was cultured and do not describe the exact methods adopted for establishing the cultures. We have outlined below the procedures used in our laboratory, and we have also attempted to include the exact methods used by others, if available.

Isolation from Conjunctival Scrapings and Corneal Biopsies

The first microsporidian isolated from a human, *V. corneae* (*N. corneum*) was cultured from a corneal biopsy that was shipped to the laboratory overnight in Hanks' balanced salt solution (HBSS). The corneal pieces were minced, and a portion was treated at 37°C with 0.1% trypsin and 0.25% collagenase for 75 min and then inoculated into three different partially confluent cell lines. These included established cell lines of rabbit corneal epithelium (SIRC) (ATCC60) and MDCK (ATCC34) and a primary cell culture derived from rabbit embryo fibroblasts. Minced pieces of cornea not treated as above were also inoculated into the cell lines separately as noted above. Furthermore, a few bits of cornea were directly explanted onto 24-well plates. All cultures were incubated at 37°C in 5% CO_2 in air. The culture medium used for the inoculated cell lines was Eagle's MEM with 5% FBS and 0.1% gentamicin. The same medium, but with 10 ng of epidermal growth factor per ml, 5 μg of insulin per ml, and 0.1 μg of cholera toxin per ml, was used to grow the corneal explants (Shadduck et al., 1990).

Didier et al. (1991) established continuous cultures of three isolates of *E. hellem* from samples of conjunctival scraping and corneal tissue by adding them to near confluent MDCK (ATCC CCL 34) cell cultures after mincing. The medium consisted of RPMI 1640 supplemented with 5% heat-inactivated FBS and antibiotics. The cultures were incubated at 37°C in 5% CO_2 in air. Didier et al. (1996) also established an isolate of *E. hellem* from conjunctival scrapings from a 37-year-old man with AIDS and disseminated microsporidiosis by first suspending the sample in Tris-buffered saline containing Tween 20 and centrifuging at 400 × *g* for 10 min at room temperature. The pellet was washed once with Tris-buffered saline and suspended in RPMI 1640 medium supplemented with 5% heat-inactivated FBS, 2 mM L-glutamine, penicillin, and streptomycin before being added to a near-confluent RK-13 cell culture. An isolate of a *Nosema* sp. was established in culture by inoculating corneal scrapings and triturated corneal biopsy specimens into human lung fibroblast (HLF) and monkey kidney (E6) cell cultures (Visvesvara, Moura, et al., unpublished data). The medium used was DMEM supplemented with glutamine, 10% heat-inactivated FBS, 50 μg of gentamicin per ml, and 2 μg of amphotericin B per ml, and the cultures were incubated at 37°C without CO_2.

Isolation from Urine

Several isolates of *E. cuniculi, E. hellem,* and *E. intestinalis* have been established in culture from washed and centrifuged sediments of patients' urine. It is preferable to obtain a 24-h sample of urine, as the chances of obtaining a good culture in a relatively short time are enhanced by the availability of large number of spores in the initial inoculum. In our laboratory, the urine sample is usually sedimented at 1,500 × *g* for 20 min and the sediment collected and washed twice in distilled water (to lyse any cells present). The sediment from the final wash is then inoculated in E6 or HLF cell culture along with 50 μg of gentamicin per ml, 1,000 μg of piperacillin per ml, and 5 to 10 μg of amphotericin B per ml. The cell cultures are usually grown in 25-cm^2 Corning flasks containing 10 ml of DMEM and 5 to 10% heat-inactivated FBS, which are incubated at 37°C.

Isolation from Sputum

Several isolates of *E. cuniculi, E. hellem,* and *E. intestinalis* have been established in culture from sputum samples. In our laboratory, sputum samples were mixed with Sputolysin (Calbiochem-Novabiochem, La Jolla, Calif.) and centrifuged as noted. The sediment was washed with 50 ml distilled water, centrifuged, and inoculated into cell cultures as mentioned above.

Isolation from BAL Fluid

BAL samples have also been used to initiate in vitro cultures of *Encephalitozoon*. These samples were usually centrifuged and washed as described above and inoculated into cell cultures.

Isolation from Nasal Sinus and Throat Washing Samples

Encephalitozoon spp. have been isolated from nasal biopsy samples and throat washes. Nasal biopsy specimens were usually minced and inoculated into cell cultures. Throat washes were concentrated by centrifugation, washed by centrifugation, and inoculated into cell cultures as noted above.

Isolation from Duodenal Aspirates and Duodenal Biopsies

An isolate of *E. intestinalis* and several isolates of *E. bieneusi* have also been established in continuous and short-term cultures, respectively. The aspirates were washed and inoculated into cell cultures, whereas the duodenal biopsy material was minced and then inoculated into cell cultures as described above.

Isolation from Stools

Only a few attempts have been made to isolate microsporidia into culture from feces because enteric bacteria and yeast usually overgrow the rich culture medium that is used, which impedes isolation of the fastidious microsporidia, especially *E. bieneusi*. A unique method of isolating these parasites from stool was attempted (van Gool et al., 1994) and resulted in the isolation of only *E. intestinalis*. In this method, stool samples with large numbers of spores obtained from AIDS patients who were biopsy positive for *E. bieneusi* were concentrated by a water-ether sedimentation method. The fecal samples containing spores were then suspended in an antibiotic mixture containing 100 μg each of amoxicillin, vancomycin, and gentamicin per ml and 50 μg of flucytosine per ml placed on a shaker and incubated at 37°C for 18 h. The feces-spore-antibiotic mixture was centrifuged at 1,550 × *g* for 10 min, and the sediment was washed twice with phosphate-buffered saline (PBS), pH 7.2. Monolayers of RK-13 cells were grown on collagen-treated 24.5-mm Transwell membranes (Costar) with a pore size of 0.4 μm placed in six-well cluster dishes containing DMEM (Gibco catalog no. 041-01095) supplemented with 10% heat-inactivated FCS and 2.5 μg of erythromycin per ml. The inoculum consisted of 400 μl of the antibiotic-treated stool mixture which was pipetted into each Transwell membrane, and the dishes were centrifuged at 1,070 × *g* for 30 min in a microtiter plate centrifuge. The pH of the medium changed from 7.0 to 8.0. After centrifugation, the Transwell membranes were gently washed twice with the culture medium, and the culture dishes containing the membranes were incubated for 2 days at 37°C in a CO_2 incubator. The Transwell membranes were again inoculated with the stool-antibiotic mixture as before. The medium in the dishes below the Transwell membranes was replaced every 2 days with fresh medium.

Isolation from Muscle Biopsies

T. hominis was isolated from a muscle biopsy sample and established in continuous culture by Hollister et al. (1996). Spores were liberated from the muscle biopsy specimens first by teasing apart the specimens and then incubating them with 0.25% trypsin (without collagenase) for 15 min at 37°C. The samples were then washed by centrifugation and resuspended in MEM. The medium containing spores and muscle tissue was then layered on 50% Percoll in PBS and centrifuged at 1,000 × *g* for 25 min. The spores obtained were then washed in PBS, the spore number was estimated, and 1×10^5 partially purified spores were added to each well of a 24-well plate containing monolayers of several different types of mammalian cell cultures growing on coverslips. The cell cultures used were the following: MDCK, RK-13, African green monkey kidney (COS-1), rat skeletal myoblasts (L6-C10), and mouse myoblasts (G-7). The medium used for MDCK and RK-13 consisted of MEM with 10% FCS; for COS-1 and L6-C10, it was DMEM with 2 mM L-glutamine and 10% FCS; and for G-7, it was DMEM with 10% FCS and 10% horse serum.

EXAMINATION, MANIPULATION, AND MAINTENANCE OF INOCULATED CELL CULTURES

Inoculated cell cultures should be examined frequently with an inverted microscope preferably equipped with phase-contrast or differential interference-contrast optics. If the host cells are noted to have rounded up and appear to flake off, the culture medium and the flaked-off cells should be poured off into a centrifuge tube and fresh medium containing antibiotics added to the cell culture flasks. The medium in the tubes should then be centrifuged, as described previously, and the sediment washed once with 50 ml of distilled water and reinoculated into the same flask. Thereafter, the culture medium should be replaced once every 24 h for the first week and once every 3 days thereafter. The supernatant medium is always washed, and the sediment reinoculated into the original flasks. In this way the spores released into the culture medium, as well as flaked-off cells containing spores and possibly developing stages of parasites, are concentrated before reinoculation into the original flasks. After 2 to 3 weeks of such manipulation, foci of mammalian cells infected with the parasite will be seen. Once foci of infections appear, it is no longer necessary to reinoculate the sedimented spores. The medium is then poured off into test tubes, and spores are concentrated to establish a spore bank for future use. The flasks are replenished with fresh medium. Two to 3 months after cultivation about 70 to 80% of cells in the monolayers will be infected and appear to be distended with spores. The infected cell culture, especially E6 cell cultures, can be maintained this way for more than 16 months by just removing the old medium and replacing it with fresh medium (Canning and Hollister, 1991; Visvesvara et al., 1991, 1994). If large numbers of spores are needed for experimental work, small areas of the infected cell culture can be scraped and the scraped material inoculated into fresh cell cultures to establish infection and thus expand the number of flasks containing the infected monolayers of the particular microsporidial species that is needed. The infected cell culture can also be expanded by routine subculture after trypsinization. The infected cell culture is split into three flasks after trypsin treatment. For trypsinization, the medium is poured off and the infected culture is rinsed with about 1 ml (for a 25 cm^2 flask) of a trypsin solution containing 0.05% trypsin and 0.53 mM EDTA in Ca^{2+}- and Mg^{2+}-free

HBSS. About 1 ml of fresh trypsin solution is then added to barely cover the cell surface, and the cells are incubated for 1 to 3 min; flasks are gently tapped to detach the cells from the walls of the flask. The trypsin-cell mixture is vigorously pipetted several times, added to 30 ml of fresh medium, and mixed thoroughly; about 10 ml of this mixture is then added to each of three 25-cm^2 flasks. Within 3 to 4 days, monolayers of infected cell cultures will be established in these flasks. The infected host cells, especially those in HLF monolayers, resemble corn on the cob (Fig. 1). In some instances, especially in HLF monolayers, where more than 80 to 90% of cells are infected, the infected cells are usually seen detaching or flaking off after several months of culture. In such old flasks it is not unusual to see patches of broken cells and aggregates of cell debris and spores still attached to the walls of the flasks. In such cases the spores have often extruded their polar tubules and may resemble spermatozoa (Fig. 1).

HARVESTING AND PURIFICATION OF SPORES BY PERCOLL GRADIENT CENTRIFUGATION

For all types of experimental research, especially biochemical work, it is necessary to obtain large numbers of spores free of cell debris and media constituents. To obtain such a clean preparation, spores are released into the culture medium, collected by pouring off the culture medium into 50-ml centrifuge tubes, and sedimented by centrifugation at 1,500 × *g* for 20 min. The supernatant is then removed by suction and the spore sediment washed once with 0.25% sodium dodecyl sulfate in PBS. The sediment is washed once more with PBS and then mixed with an equal volume of Percoll (Sigma or Pharmacia) to obtain 50% Percoll. Percoll, as obtained from the manufacturer, should be stored in the refrigerator. Before use, an appropriate volume is removed and mixed with a 10× salt solution in the ratio of 9 volumes of Percoll to 1 volume of the salt solution to obtain 100% Percoll (Didier, personal communication, 1991). The spore-Percoll mixture is centrifuged at 500 × *g* for 30 min at 4°C. The supernatant is then removed by suction, and the sediment is resuspended in HBSS or PBS and washed by centrifugation as described above. The sediment contains spores free of cell debris.

Percoll density gradient centrifugation has also been used to fractionate different life cycle stages of the cricket microsporidian *Nosema grylli* (Seleznev et al., 1995).

INOCULATION OF TISSUE FRAGMENTS CONTAINING MICROSPORIDIA INTO ESTABLISHED CELL CULTURES VERSUS TISSUE EXPLANTS

Although explants of insect tissue have been employed successfully to obtain cultures of insect microsporidia, the use of human tissue explants has not been fruitful. It appears, therefore, that for the isolation of human microsporidia the tissue should be treated, after teasing or triturating, with trypsin or trypsin and collagenase and then inoculated into cell culture. Listed below are the various microsporidial species that have been established in culture. See also Tables 1 to 4. Methods, as described by the original investigators whenever available, for the in vitro cultivation of the various species of microsporidia of clinical importance, are included. Figures 1 to 6 depict the light and transmission electron microscope (TEM) features of the various microsporidia during their growth in mammalian cell cultures.

Encephalitozoon cuniculi

Currently, 10 isolates of *E. cuniculi* of human origin have been established in culture (Table 1). Of these isolates, 7 originated from Switzerland (Deplazes et al., 1996a, 1996b; Weber et al., 1997) and 1 each originated from the United Kingdom (Hollister et al., 1995), the United States (De Groote et al., 1995), and Spain (del Aguila, personal communication). These isolates of *E. cuniculi* have been cultivated in different types of mammalian cell cultures, including E6 and HLF (De Groote et al., 1995; Croppo et al., 1997); MRC-5 lung

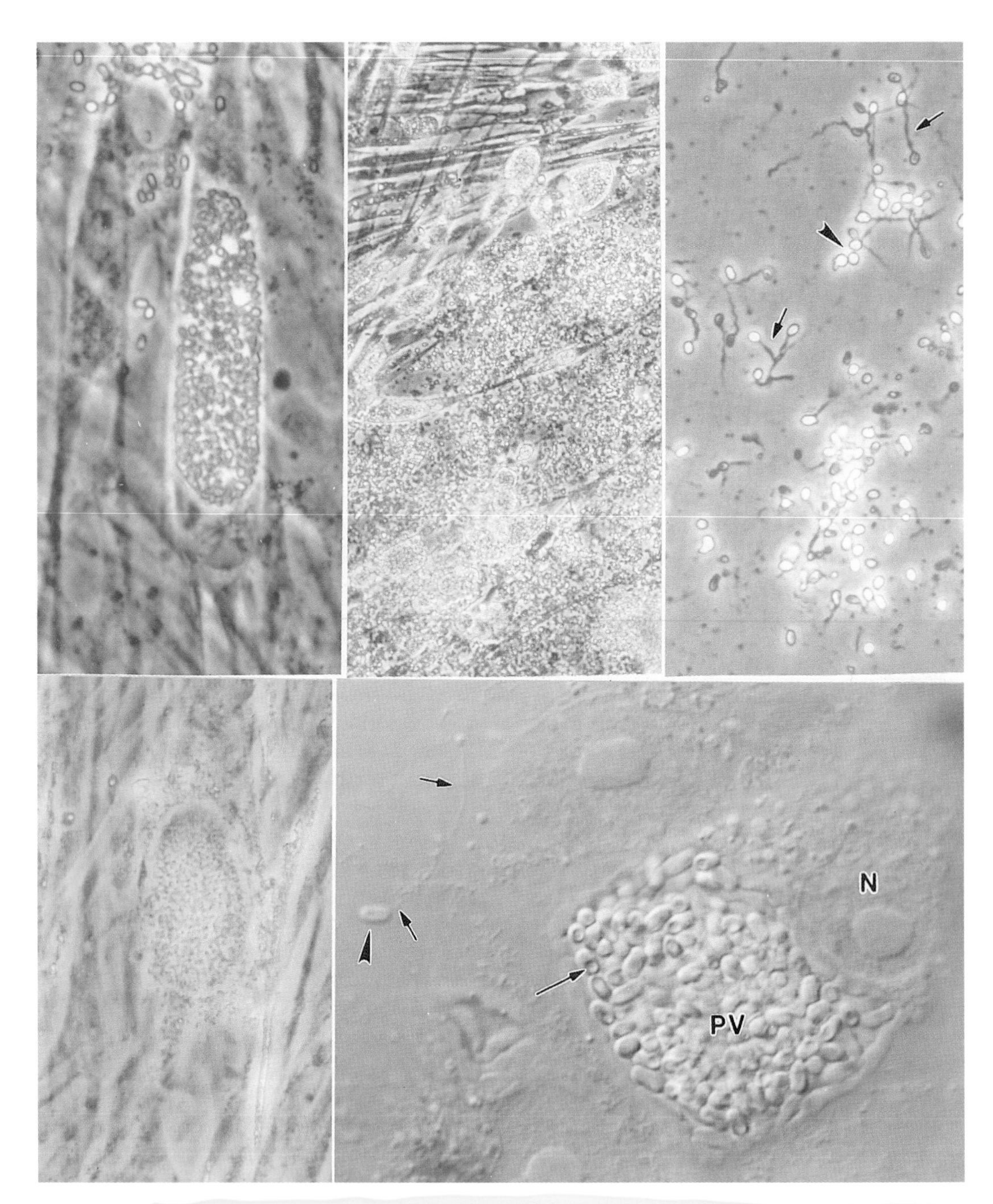

FIGURE 1 Growth of various microsporidia in mammalian cell cultures as seen with a microscope equipped with differential interference contrast (DIC) optics. (Top left) *Encephalitozoon hellem* in human lung fibroblast (HLF) cell culture. A fibroblast cell is distended with developing stages and spores of *E. hellem,* giving the cell a corn-on-the cob appearance. Original magnification, ×600. (Top middle) A low-power (×150) view of the same cell culture showing extensive infection of the monolayer. (Top right) A high-power (×600) view of the same cell culture several months later, exhibiting total destruction of the monolayer. Note the birefringent spores (arrowhead), a number of which have discharged their polar tubules (arrow). (Bottom left) An HLF monolayer inoculated with a *Enterocytozoon bieneusi*-positive duodenal aspirate from an AIDS patient. Note the corn-on-the cob appearance of the cell, but this time it is due to an adenovirus. Original magnification, ×600. (Bottom right) A monkey kidney cell in an E6 monolayer infected with *E. cuniculi.* Note the parasitophorous vacuole (PV) bulging with *E. cuniculi* spores. Some of the spores exhibit a distinct vacuole at one end (long arrow). A spore (arrowhead) has discharged its polar tubule (short arrows). N, host cell nucleus. Original magnification, ×1,250.

TABLE 1 *Encephalitozoon cuniculi:* isolates, origin, and culture conditions

No. of isolates	Country of origin	Sample[a]	Cell line	Medium and serum concentration	Reference(s)
2	United States	Urine, sputum	E6, HLF	MEM, 5 to 10% FBS	De Groote et al., 1995; Croppo et al., 1997
1	United Kingdom	Urine	MDCK	MEM, 10% FBS	Hollister et al., 1995
8	Switzerland	Urine, BAL, CSF	MRC-5	MEM, 5% FBS	Deplazes et al., 1996a, 1996b; Mathis et al., 1996, 1997; Weber et al., 1997
2	Spain	Urine, sputum	E6[b]	MEM, 5 to 10% FBS	del Aguila, personal communication

[a]CSF, cerebrospinal fluid.
[b]No CO_2 (air).

fibroblasts (Deplazes, 1996a; Mathis et al., 1996, 1997); RK-13 cells, (Didier et al., 1996); and the MDCK cell line (Hollister et al., 1995). Careful electron micrographic analysis indicated that the parasites developed within a parasitophorous vacuole (PV) of host origin, which contained all developing stages. Meronts were found attached to the PV, and the cell membranes, on thickening, detached from the PV membrane and developed into sporogonic stages. The sporogonic stages seen were sporonts, sporoblasts, and spores. These stages were usually seen lying free within the PV (Figs. 2 to 5). *E. cuniculi* has been described as disporoblastic, however, data suggest that this parasite also exhibits di-, tri-, tetra-, and even octosporous sporogony (Visvesvara et al., submitted) as depicted in Fig. 3.

The method described above for establishing *E. cuniculi* isolates in culture is generally effective in most cases. However, certain isolates fail to grow in spite of repeated manipulations of the type discussed. For example, we have tried unsuccessfully to establish an isolate from a kidney biopsy of a female patient with nephritis even after 1 year of labor-intensive culture manipulations, although a few spores were periodically seen. Similarly, an isolate that temporarily began to develop in E6 cells after inoculation of a urine sample failed to become established as a continuous culture even though spores were occasionally seen in the supernatant for more than 1 year. In both cases the etiologic agent was *E. cuniculi,* which was determined only after PCR was performed on a few of the spores collected from the supernatant.

Encephalitozoon hellem

Didier et al. (1991) isolated and continuously cultured in vitro three strains of a microsporidian parasite obtained from the corneal tissue of an AIDS patient and from conjunctival scrapings from two other AIDS patients. To establish these isolates in continuous culture, the authors minced the tissue and added it to near confluent monolayers of MDCK cell lines. They used RPMI medium and incubated the cultures at 37°C with 5% CO_2. The electron micrographs obtained for these isolates indicated that morphologically these parasites were similar to *E. cuniculi*. However, biochemical studies, including sodium dodecyl sulfate-polyacrylamide gel electrophoresis analysis, indicated that they differed from *E. cuniculi* and hence they were designated *E. hellem*. Since then as many as 28

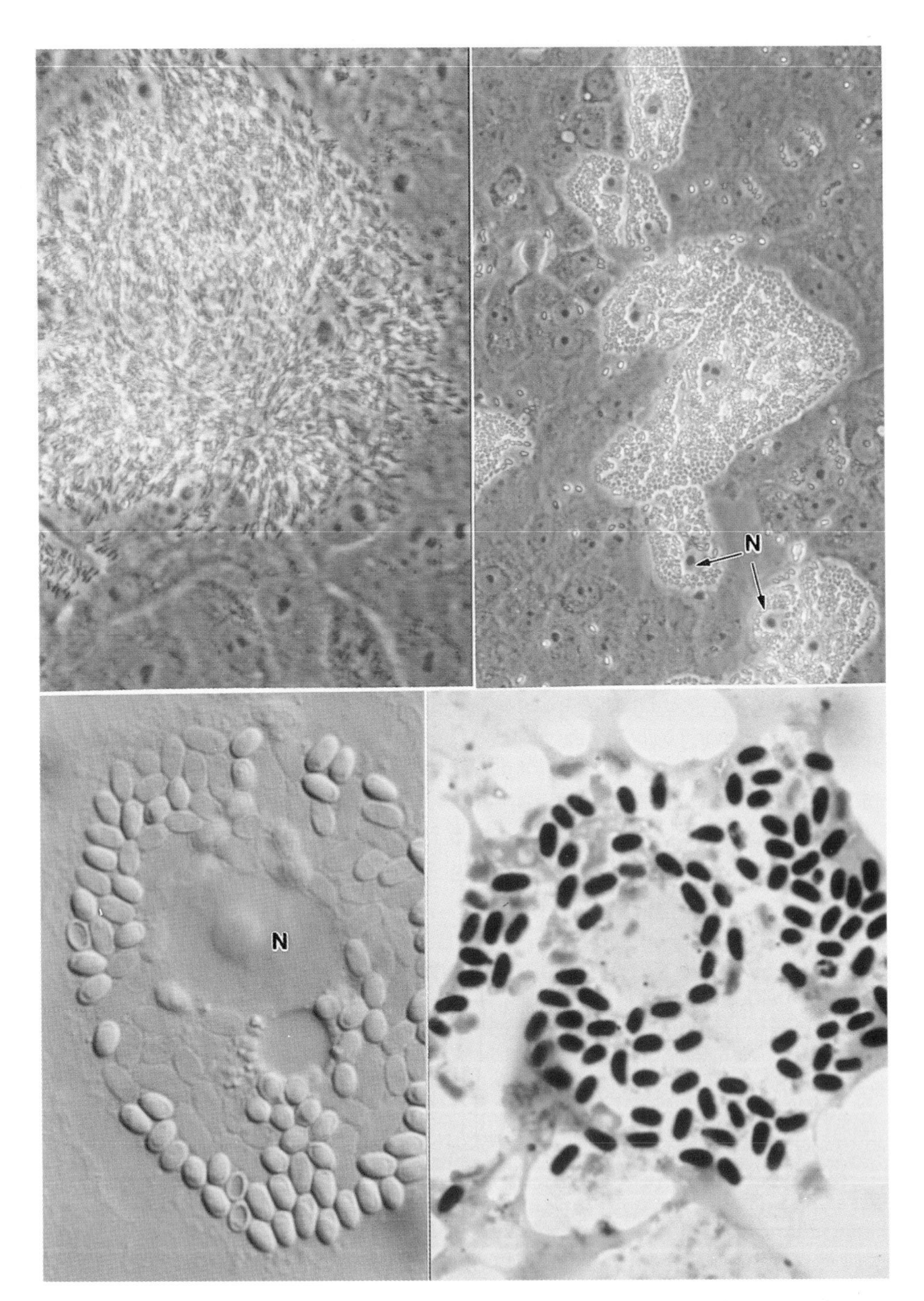

FIGURE 2 (Top left) DIC image of *Vittaforma corneae* growing on an E6 monolayer. Note the centrifugal growth pattern. Original magnification, ×300. (Top right) DIC image of *Nosema algerae,* a mosquito isolate, growing on an E6 monolayer. Note that the cells are distended with spores that are relatively larger than those of *Encephalitozoon* spp. N, host cell nucleus. Original magnification, ×300. (Bottom left) DIC image of *N. algerae,* isolated from the corneal biopsy of an immunocompetent person, growing within an E6 cell. Note spores and developing stages arranged around the host cell nucleus (N). Original magnification, ×1,250. (Bottom right) An E6 monolayer infected with mosquito-derived *N. algerae* stained with the quick hot Gram chromotrope stain. Original magnification, ×1,250.

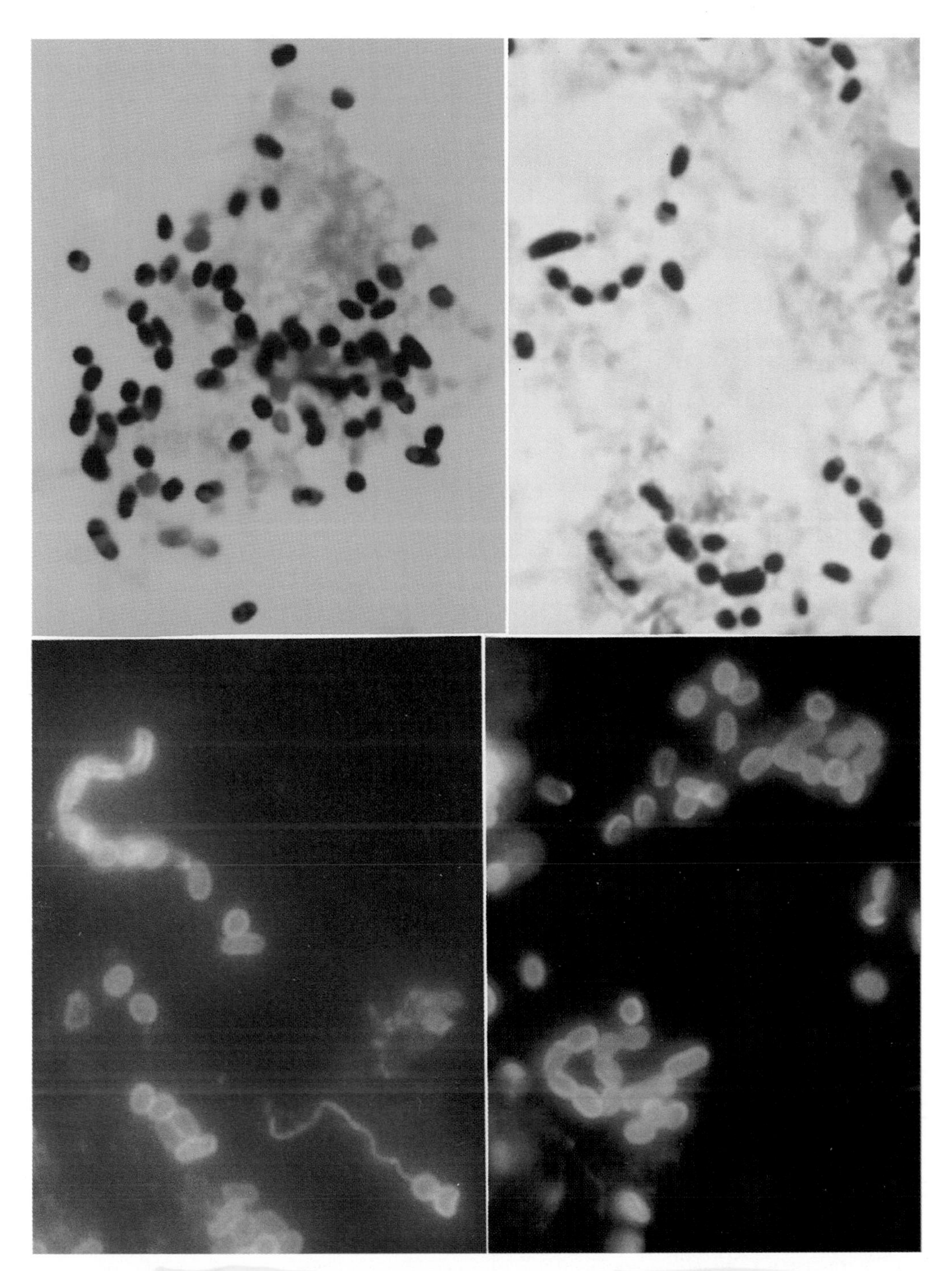

FIGURE 3 Culture smears of *Encephalitozoon hellem* (top left and right) stained with the quick hot Gram chromotrope technique and that of *E. cuniculi* (bottom left and right) after reaction with polyclonal rabbit antiserum in the indirect immunofluorescence test. Note the chains of spores representing polysporoblastic sporogony in both *E. hellem* (top right) and *E. cuniculi* (bottom left and right). Note that a spore still in the chain configuration has already extruded its polar tubule (bottom left). Original magnification, ×1,250.

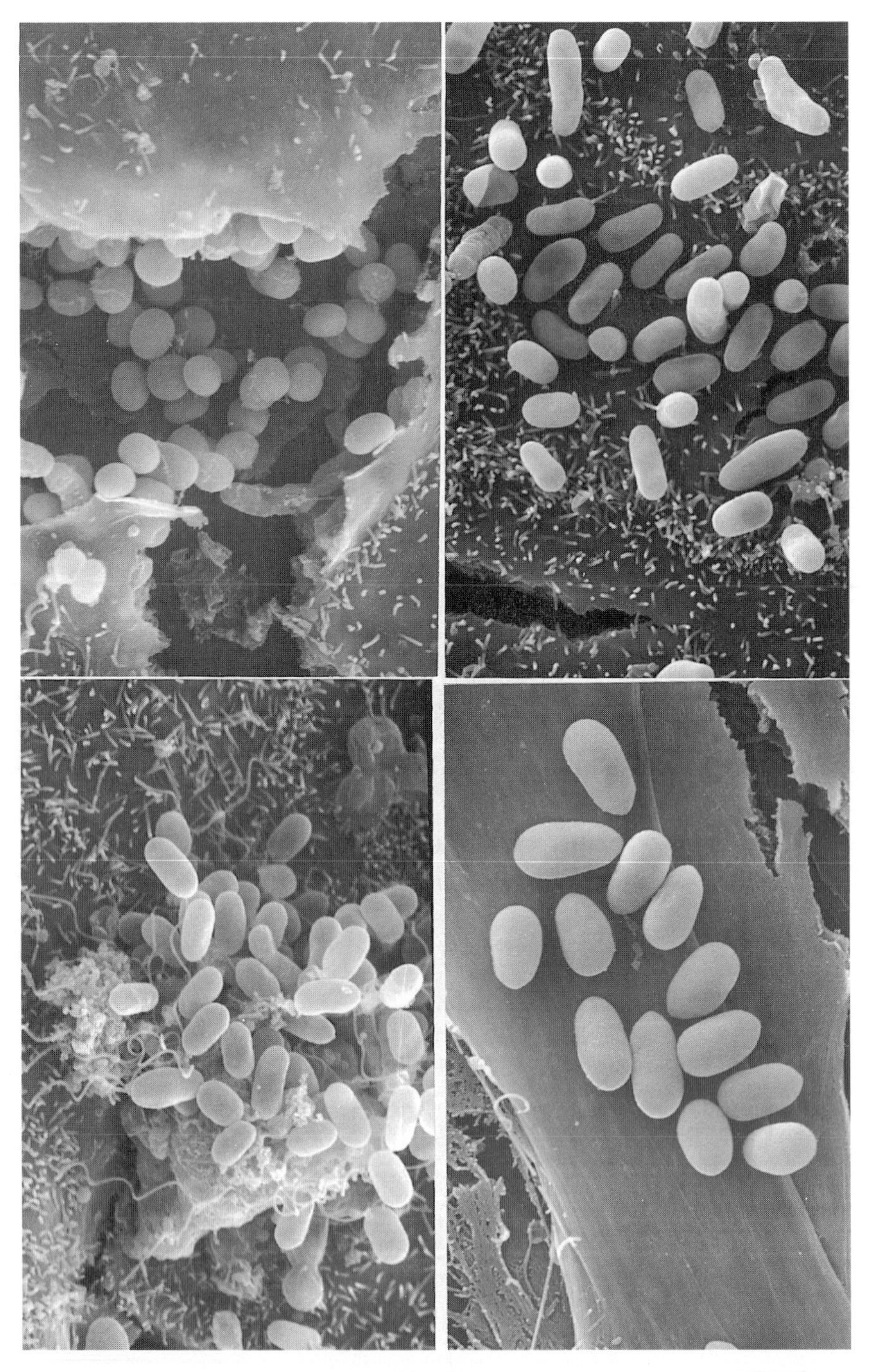

FIGURE 4 Scanning electron micrographs of microsporidia growing in cell cultures. (Top left) *E. cuniculi* in E6 cells. (Top right) *E. hellem* in E6 cells. (Bottom left) *Encephalitozoon intestinalis* in E6 cells. Note that some spores have already discharged their polar tubules. (Bottom right) an isolate of *N. algerae,* from a human cornea, growing in HLF cells. Original magnifications, ×5,000.

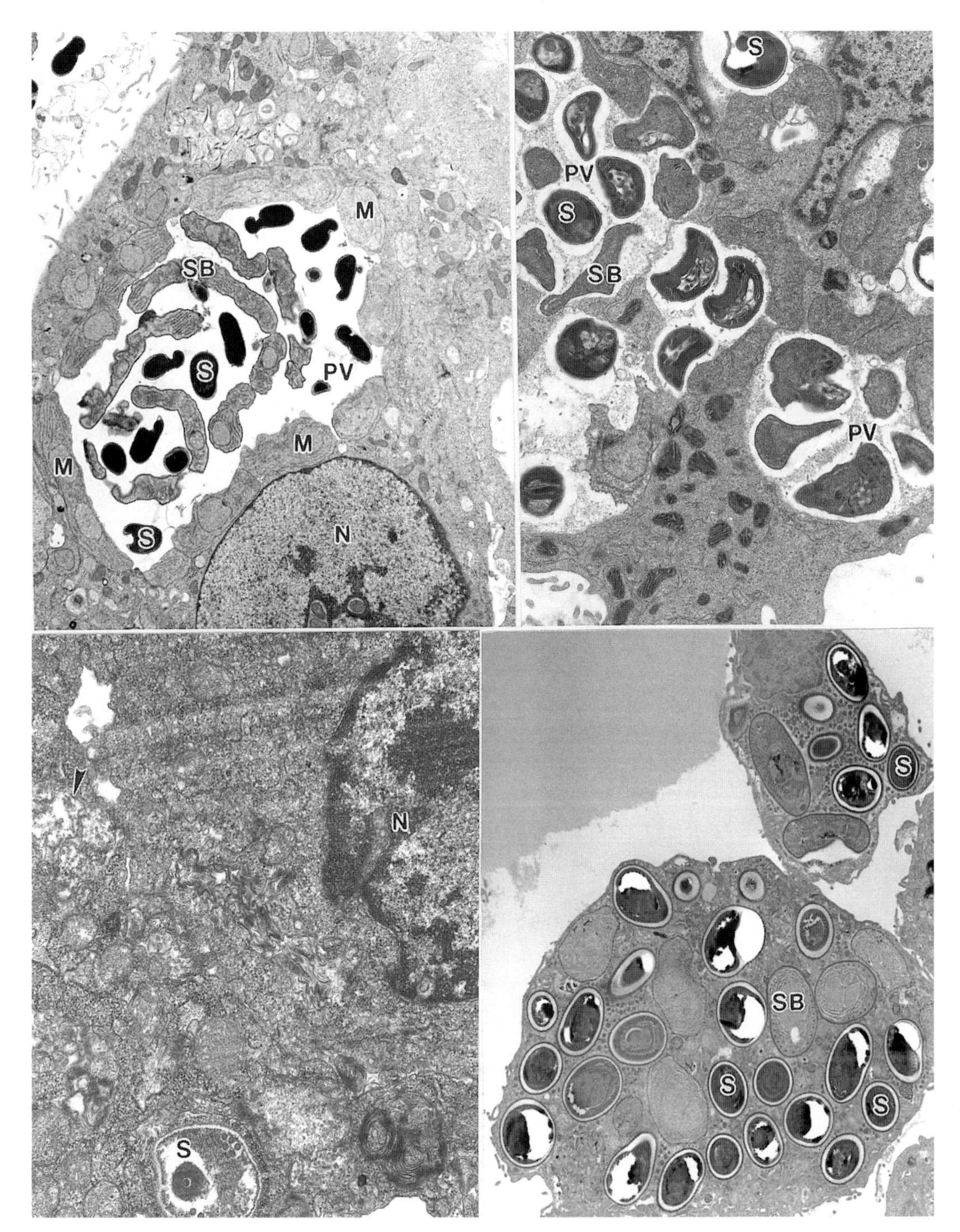

FIGURE 5 Transmission electron micrographs of microsporidia. (Top left) *E. cuniculi.* Original magnification, ×4,000. Note various stages growing inside a parasitophorous vacuole (PV) which is unseptated. (Top right) *E. intestinalis.* Note the various developmental stages within a PV which is septated. Original magnification, ×6,000. (Bottom left) *E. bieneusi.* Note several degenerating developing stages (arrowhead) and a single spore in the cytoplasm. Original magnification, ×10,000. (Bottom right) Mosquito-derived *N. algerae* within an E6 cell. Note several electron-dense spores and a number of developing stages. Original magnification, ×2,500. M, meront; S, spore; SB, sporoblast; N, host cell nucleus.

isolates of *E. hellem* have been established in culture (Table 2): 1 each from the United Kingdom (Hollister et al., 1993b) and Spain (Bornay-Llinares, personal communication); 4 from Switzerland (Deplazes et al., 1996a); 9 from Italy (Gatti et al., 1997; Scaglia et al., 1994, 1997, 1998, unpublished data); and 15 from the United States (Didier et al., 1991, 1996a; Visvesvara et al., 1991, 1994; Visvesvara, Leitch, et al., unpublished data; Croppo et al., 1998).

Various cell lines including E6, HLF, MRC-5, MDCK, RK-13, and fetal bovine lung fibroblasts have been used to culture these parasites. They were grown in tissue culture flasks or on coverslips placed in 24-well Nunc plates. The isolates were established from cornea or conjunctival scrapings (Didier et al., 1991, 1996a); urine (Scaglia et al., 1994; Croppo et al., 1998; Deplazes et al., 1996a; Didier et al., 1996a; Visvesvara et al., 1991); BAL fluid (Croppo et al., 1998; Gatti et al., 1997; Scaglia et al., 1997, 1998; Visvesvara, Leitch, et al., unpublished data); sputum (Visvesvara, Leitch, et al., unpublished data); throat washes (Gatti et al., 1997, Scaglia et al., 1994); and nasal mucosa (Croppo et al., 1998; Hollister et al., 1993; Visvesvara, Leitch, et al., unpublished data).

Specimens are processed as described above and inoculated into E6 or HLF cell cultures. Other methods of treating samples before inoculation into cell cultures include holding of the sample in a solution (VIB) containing glutamine, 0.5% FBS, 1.5% $NaHCO_3$, 500 U of penicillin per ml, 500 μg of streptomycin per ml, 100 μg of gentamicin per ml, 50 μg of neomycin per ml, and 25 μg of amphotericin B per ml for 2 h at 37°C (Scaglia et al., 1994). The samples are then washed and inoculated into cell cultures as noted above. The cell supernatant is removed at least once every 24 h, or earlier if necessary (i.e., if the monolayers appear to flake off). Then the culture flasks are replenished with fresh medium. The supernatant

TABLE 2 *Encephalitozoon hellem:* isolates, origin, and culture conditions

No. of isolates	Country of origin	Sample[a]	Cell line[b]	Medium and serum concentration[c]	Reference
3	United States	Cornea, nasal	MDCK	RPMI 1640, 5% FBS[d]	Didier et al., 1991
1	United States	Urine	E6, HLF	MEM, 5 to 10% FBS[e]	Visvesvara et al., 1991, 1994
1	United Kingdom	Nasal	MDCK	MEM, 10% FCS[f]	Hollister et al., 1993b
1	Italy	Urine, throat wash	Several[g]	EMEM, 10% FBS	Scaglia et al., 1994
4	Switzerland	—	MRC-5	EMEM, 10% FBS	Deplazes et al., 1996
1	United States	Urine, conjunctiva	RK-13	RPMI 1640, 5% FBS	Didier et al., 1996a
1	Italy	BAL	MRC-5, FBF	MEM, 10% FBS	Scaglia et al., 1997
4	Italy	Several[h]	Several[g]	MEM, 10% FBS	Gatti et al., 1997
3	Italy	BAL	MRC-5, FBF	MEM, 10% FBS	Scaglia et al., 1998
5	United States	Several[h]	E6, HLF	MEM, 5% FBS	Croppo et al., 1998
2	United States	Urine, BAL	E6, HLF	MEM, 5% FBS	Visvesvara, Leitch, et al., unpublished data
5	United States	Urine, sputum	E6, HLF	MEM, 5% FBS	Visvesvara, Leitch, et al., unpublished data
1	Spain	BAL	E6, HLF	MEM, 5% FBS	Bornay-Llinares, personal communication

[a]BAL, bronchoalveolar lavage.
[b]HLF, human lung fibroblast; FBF, fetal bovine fibroblast.
[c]FBS, fetal bovine serum; MEM, minimal essential medium; FCS, fetal calf serum; EMEM, Eagle's MEM.
[d]Cultivated in 5% CO_2.
[e]No CO_2 (air).
[f]Fetal calf serum plus 200 IU of penicillin, 200 μg of streptomycin, and 2.5 μg of amphotericin B.
[g]MRC-5, FBF, MDCK, E6.
[h]Urine, sputum, nasal wash, BAL.

is centrifuged, and the sediment reinoculated into the original flasks. In this way we have established 11 isolates in continuous culture.

As stated above, the first indication of a successfully established culture is the appearance of spores in the culture supernatant. Careful microscopic examination under an inverted microscope reveals the presence of foci of infection within the monolayers. As development proceeds, these foci increase in size, making the cells appear distended, resembling corn on the cob (Fig. 1). Electron microscopic analyses reveal the characteristic development of the organisms within a PV, and the PV contains all the developmental stages (Fig. 4 and 5). It is not possible to distinguish *E. hellem* from *E. cuniculi* based on TEM. Additional studies using Western blot or PCR techniques are necessary to identify the species in question. In vitro cultures of these parasites have greatly facilitated study of the taxonomic classification of the microsporidia.

Encephalitozoon intestinalis

E. intestinalis was previously named *Septata intestinalis* but was reclassified as *E. intestinalis* on the basis of its close antigenic and molecular relationships with *Encephalitozoon* spp. (Hartskeerl et al., 1995). *E. intestinalis* has been isolated from urine (del Aguila et al., 1998; Moura et al., unpublished data; Molina et al., 1995; Visvesvara et al., 1995a); stool (van Gool et al., 1994); BAL fluid (Didier et al., 1996b; del Aguila et al., 1998); sputum (del Aguila et al., 1998); and nasal mucosa (Didier et al., 1996b; Doultree et al., 1995; del Aguila et al., 1998). The cell cultures used to propagate these microsporidia include E6 and HLF (del Aguila et al., 1998; Moura et al., unpublished data; Visvesvara et al., 1995a), human embryonic lung (Doultree et al., 1995), monocyte (MDM) (Doultree et al., 1995), MRC-5 (Molina et al., 1995), RK-13, MDCK, I 047, HT-29, and CABO-2 (Didier et al., 1996b) cell lines, and duodenal aspirate and biopsy samples (del Aguila et al., 1998). The medium used consisted of DMEM (del Aguila et al., 1998; Moura et al., unpublished data; Visvesvara et al., 1995a); RPMI 1040 supplemented with heat-inactivated 10% FBS, 2 mM glutamine, penicillin, and streptomycin (Didier et al., 1996b), Eagle's MEM supplemented with 5% FBS, 100 IU of penicillin, and 100 μg of streptomycin per ml (Doultree et al., 1995) (Table 3).

TABLE 3 *Encephalitozoon intestinalis:* isolates, origin and culture conditions

No. of isolates	Country of origin	Sample[a]	Cell line[b]	Media and serum concentration[c]	Reference
7	Netherlands	Stool	RK-13	MEM, 10% FCS[d]	Van Gool et al., 1994
1	United States	Urine	E6, HLF	MEM, 5 to 10% FBS[e]	Visvesvara et al., 1995
3	France	Urine	MRC-5	MEM, 10% FCS	Molina et al., 1995
1	Australia	Nasal	HEL, HMDM	MEM, 5% FCS; Iscove media[f]	Doultree et al., 1995
2	United States	Nasal, BAL	Several[g]	RPMI 1640, 10% FBS	Didier et al., 1996
7	United States	Urine, sputum, intestine	E6, HLF	MEM, 5% FBS	del Aguila et al., 1998
1	Brazil	Urine	E6, HLF	MEM, 5% FBS	Moura et al., unpublished data

[a]Nasal, nasopharyngeal aspirate; BAL, bronchoalveolar lavage; intest, duodenal aspirate/biopsy.
[b]HLF, human lung fibroblast; HEL, human embryonic lung; HMDM, human monocyte-derived macrophage.
[c]MEM, Eagle's minimum essential medium; FBS, fetal bovine serum; FCS, fetal calf serum.
[d]100 μg of erythromycin per ml.
[e]No CO_2 (air).
[f]With 10% human AB-positive serum.
[g]RK13, MDCK, HT-29, Caco-2, Vero, I 047.

It is recommended that patient specimens (urine, BAL fluid, sputum, duodenal aspirate, and biopsy specimens) be treated as described earlier before inoculation into cell cultures. With the methods of inoculation and manipulation previously described, foci of infection began to appear in the inoculated monolayers after about 3 to 4 weeks of incubation. At the same time, clusters of spores were also seen in the culture medium. Smears stained with the quick hot Gram chromotrope also revealed chains of spores, each chain having two, three, four, or even eight spores. Some of the spores had already discharged their polar tubule. TEM revealed the development of parasites within a PV that appeared honeycomb-shaped. On TEM the septation appeared to consist of fine, meshlike, fibrous material (Fig. 4 and 5).

Encephalitozoon sp.

Bocket et al. (1992) reported the isolation and limited multiplication of a microsporidian parasite in an MRC-5 cell line (Table 4). They collected urine from a 23-year-old female drug abuser with a 2-year history of AIDS. The urine was centrifuged at 200 × *g* for 10 min, and 0.2 ml of the sediment was suspended in Eagle's MEM containing 10% FBS, 100 units of penicillin per ml, 100 μg of streptomycin per ml, and 0.25 μg of amphotericin B per ml and stored at 4°C before inoculation into MRC-5, HEp-2, and Vero cell lines. Incubation was at 37°C in a humidified atmosphere in a 5% CO_2 incubator. The culture medium was removed weekly, and the flasks replenished with fresh medium. On light microscopic examination, the authors observed that 20% of the cells had rounded up and that cells were generally enlarged and exhibited cytopathic effects in the intracytoplasmic vesicles. Hemalum eosin and Giemsa staining of the cells revealed the presence of small nucleated basophilic bodies later identified as microsporidia by TEM. Although they did not identify these agents as *Encephalitozoon,* it is apparent from their published electron micrographs that the parasite belonged to the genus *Encephalitozoon* because of the presence of a PV containing developmental stages.

TABLE 4 Isolation and in vitro cultivation of human microsporidia

Organism and no. of isolates	Country of origin	Sample	Cell line	Medium and serum concentration[a]	Reference(s)
Encephalitozoon-like					
1	Canada	Corneal scrapings	MRC-5	MEM, 10% FBS	Desser et al., 1992
1	France	Urine	MRC-5	MEM, 10% FBS	Bocket et al., 1992
1	Japan	Hydatid liver lesion	Liver cells	RPMI 1640, 10% FCS	Furuya, 1995
Enterocytozoon bieneusi					
4	United States	Intestine[b]	E6, HLF	MEM, 5% FBS	Visvesvara et al., 1995b
2	United States	Intestine[b]	E6, HLF	MEM, 10% FBS	Visvesvara et al., 1996
Vittaforma corneae (1)	United States	Corneal biopsy	Several[c]	MEM, 5% FBS	Shadduck et al., 1990; Silveira and Canning, 1995
Trachipleistophora hominis (1)	Australia	Muscle biopsy	Several[d]	DMEM, 10% FCS[e]	Hollister et al., 1996

[a]MEM, minimal essential medium; FBS, fetal bovine serum; FCS, fetal calf serum; DMEM, Dulbecco's modified MEM.
[b]Duodenal aspirate/biopsy.
[c]SIRC, MDCK, MRC-5, XEN, L-929, FHM.
[d]COS-1, RK-13, MDCK, myoblasts G-7 and L6-C10.
[e]With 2 mM L-glutamine.

Since no clear-cut septation of the PV was seen, it was probably either *E. cuniculi* or *E. hellem*.

Desser et al. (1992) also reported the isolation of a microsporidian from corneal scrapings from a 30-year-old man with AIDS and described it as a species of *Encephalitozoon* (Table 4). They inoculated the scrapings into MRC-5 cell lines grown on coverslips in plastic Leighton tubes in Eagle's MEM supplemented with 10% FBS. The development of parasites was observed within a PV as well as inside the cytoplasm. Furthermore, they also noted the presence of chains of sporonts. The authors ascribed the difference in developmental stages from those seen in *E. cuniculi* and *E. hellem* to the way the specimens were processed. In this regard, they processed specimens for TEM on coverslips rather than scraping them from tissue culture flasks. However, sporont chains have been seen in *E. hellem, E. intestinalis,* and *E. cuniculi* (Fig. 3) (Visvesvara et al., in press). Since parasite development occurred within a PV that was not septated, it is quite likely that the parasite described by Desser et al. was either *E. cuniculi* or *E. hellem*.

Furuya et al. (1995) isolated *Encephalitozoon*-like organisms from a human liver lesion caused by larval *Echinococcus multilocularis* (Table 4). They obtained the lesion at surgery by sterile dissection from a patient with alveolar hydatid disease, minced the tissue, and treated it with 0.25% trypsin (Difco) at room temperature for 30 min. Thereafter, the trypsin digests were filtered through a stainless steel mesh (no. 200) to remove cellular debris, and the filtrate washed twice in RPMI 1640 medium by centrifugation at 500 × *g* for 5 min. The sediment was dispersed in RPMI medium containing 500 units of penicillin, 500 μg of streptomycin per ml, 2 mM L-glutamine, 1 mM sodium pyruvate, and 10% FBS, plated out into collagen-coated plastic dishes and incubated at 37°C in a CO_2 incubator. After the tenth passage they noticed a few parasitized cells, which, after repeated subculture, took over the cell culture. Like *E. cuniculi* and *E. hellem,* the parasite grows within an unseptated PV.

Enterocytozoon bieneusi

Although it is the most commonly identified organism in AIDS patients with microsporidiosis, *E. bieneusi* has resisted all efforts of continuous in vitro culture. Van Gool et al. (1994) tried unsuccessfully to culture these organisms from stool samples of patients who were confirmed by biopsy to have *E. bieneusi* microsporidiosis but cultured *E. intestinalis* instead. However, short-term culture of these parasites was achieved (Visvesvara et al., 1995b, 1996) with the same culture system used for the in vitro growth of *Encephalitozoon* spp. (Table 4). Short-term culture, lasting anywhere from 6 weeks to 6 months, was achieved by inoculating duodenal biopsy and aspirates into E6 and HLF monolayers. The medium used was Eagle's MEM supplemented with 5% FBS or with nonessential amino acids, 10 mM *N*-2-hydroxyethylpiperazine-*N*′-2-ethanesulfonic acid (HEPES), 2 mM L-glutamine, 10% FBS, 0.005 μg of epidermal growth factor per ml, 5 μg of transferrin per ml, 5 μg of insulin per ml, 0.005 μg of selenium, 0.0072 μg of hydrocortisone per ml, 50 μg of gentamicin per ml, and 5 μg of amphotericin B per ml. Epidermal growth factor was made up as a 100× stock solution in 0.01% bovine serum albumin, insulin-transferrin-selenium was made up as a 100× stock solution in 0.26% HCl, and hydrocortisone was made up as a 100× stock solution in absolute ethanol. The duodenal aspirates were concentrated by centrifugation and inoculated into cell cultures. The biopsy specimens were macerated and inoculated into the same cultures that received the aspirates. The inoculated cell cultures were incubated in an ordinary laboratory incubator without any special gases, in a CO_2 incubator equipped to obtain 5% CO_2, or in a candle jar. The cultures were treated in exactly the same manner as other established cultures. After several weeks, gram-positive sporelike structures measuring 1 to 1.2 μm long were seen (Visvesvara et al., 1995). TEM revealed various stages of parasites in clusters and in close proximity to one another without intervening spaces (Fig. 6, top). The proliferating stages were characterized by clear spaces interpreted as electron-lucent inclusions, and these

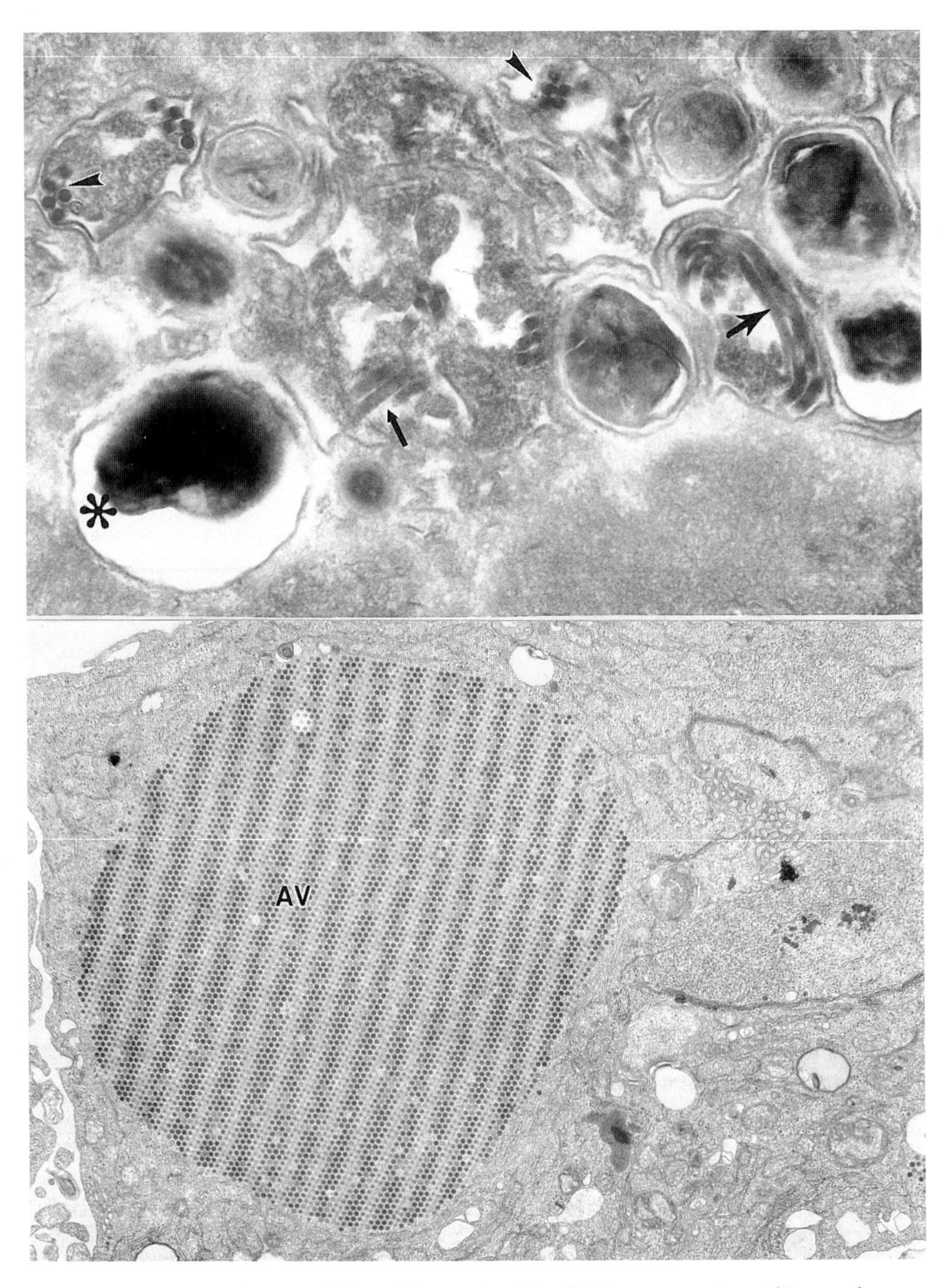

FIGURE 6 (Top) *E. bieneusi* within a disintegrating E6 cell. Note a spore (asterisk), several sporoblasts with polar tubules in cross-sectional (arrowhead), and stacked (small arrow) and coiled (large arrow) profiles. Original magnification, ×12,000. (Reprinted with permission from the Society of Protozoologists.) (Bottom) E6 cell containing a mass of adenoviral particles (AV). Magnification, ×6,000.

stages appeared in close association with the host endoplasmic reticulum and mitochondria. In one instance the host mitochondria appeared to be tightly associated with the membrane of a proliferative stage. Mature spores and sporoblasts with double rows of polar tubule coils were seen (Fig. 5, bottom left, and Fig. 6, top). In several cell cultures adenovirus were also observed (Visvesvara et al., 1996). By light microscopy the cells infected with adenovirions appeared very much like cells distended with spores of microsporidia (Fig. 1, bottom left); however, TEM revealed the presence of adenovirus (Fig. 6, bottom). It is not surprising, therefore, that *E. bieneusi* failed to multiply in these adenovirus contaminated cell cultures.

Nucleospora (Enterocytozoon) salmonis

N. salmonis is a microsporidian parasite primarily of both freshwater and seawater chinook salmon (*Oncorhynchus tshawytscha*). Recently, it has also been found to infect steel head and rainbow trout (*Oncorhynchus mykiss*). In salmonid fish the infection with *N. salmonis* is characterized by intranuclear developmental stages and is associated with the abnormal proliferation of mononuclear leukocytes. This parasite has been cultivated continuously in primary cell cultures of mononuclear leukocytes obtained from uninfected chinook salmon (Table 4). The mononuclear leukocytes are grown in a complex medium consisting of Iscove's modified DMEM supplemented with L-glutamine, sodium pyruvate, MEM nonessential amino acids, concanavalin A, lipopolysaccharide, 2-mercaptoethanol, human recombinant interleukin-2, heat-inactivated FBS, penicillin, and streptomycin. According to these investigators, the optimum pH for growth is 7.0 to 7.2, and the parasites can be maintained in primary culture for up to 60 days with periodic changes of medium. However, continuous culture can be achieved by periodically adding small numbers of infected cells to fresh cultures of mononuclear cells. The parasites can also be cryopreserved and stored as a stabilate (Wongtavatchai et al., 1994).

Vittaforma corneae

Shadduck et al. (1990) isolated a microsporidian from a 45-year-old immunocompetent human with no history of ocular trauma or use of contact lenses who had traveled to the Caribbean and Central America and had lived near a large recreational lake before he developed ocular problems. A corneal biopsy revealed the presence of microsporidial organisms, and later, at the time of corneal transplantation, about one half of the excised cornea was placed in HBSS and shipped overnight to the laboratory. The authors inoculated the corneal samples into three cell lines: rabbit corneal epithelium (SIRC) (ATCC 60); MDCK; and primary rabbit embryo fibroblasts obtained from 14- to 16-day-old rabbit fetuses. They minced the corneal tissue and inoculated a portion into cell cultures. The other portion was treated with 0.1% trypsin and 0.25% collagenase and inoculated into partially confluent cell cultures. A portion was also explanted directly into 24-well plates. The medium used for cell culture was Eagle's MEM supplemented with 5% FBS and 0.1% gentamicin, and the cultures were incubated in a 5% CO_2 atmosphere. After approximately 30 days in culture, foci of infection were seen in SIRC and MDCK cell lines inoculated with trypsin-collagenase-treated corneal tissue. No growth was seen in rabbit embryo cells or corneal tissue directly explanted into culture wells. TEM revealed that the organisms exhibited a diplokaryon resembling those of the genus *Nosema,* and therefore it was classified as a new species, *N. corneum*. On further study, Silveira and Canning (1995a) considered that this parasite was substantially different from *Nosema* and redescribed it as *V. corneae* n. comb. (Silveira and Canning, 1995a). This organism can be grown in E6 and HLF in addition to MDCK and RK-13 cells (Table 4). Its growth characteristics differ substantially from those of the *Encephalitozoon* spp. It grows in a centrifugal formation (Fig. 2, top left), and the spores line up in rows, sometimes completely surrounding the host cell nucleus.

Trachipleistophora hominis

Spores purified from corneal scrapings and muscle biopsies were inoculated into several cell lines (e.g., MDCK, RK-13, COS-1, and L6-C10) and mouse myoblasts established on coverslips in 24-well tissue culture plates containing MEM supplemented with 10% FBS (MDCK, RK-13), or in DMEM with 2 mM glutamine and 10% FBS (COS-1, L6-C10), and DMEM with 10% FBS and 10% horse serum (mouse myoblasts) (Hollister et al., 1996). Muscle biopsy material was teased apart and incubated with 0.25% trypsin for 15 min at 37°C. The suspension was washed twice by centrifugation, and the sediment was layered on 50% Percoll in PBS. After the spores were washed, they were suspended in PBS, their numbers estimated, and 1×10^5 spores were inoculated into several wells of a 24-well culture plate or 1×10^6 spores were inoculated into a petri dish. Although all cell lines supported the growth of the parasites, COS-1 supported growth the best, followed by RK-13 cells (Table 4). All stages of the parasite were obtained, and on the basis of TEM analysis, it was decided that the parasite was different from *Pleistophora* sp. The authors designated it a new genus and species, *T. hominis*.

Nosema spp.

The fact that *Nosema* spp. can also cause infections in humans should not be ignored. A few cases of ocular as well as disseminated *Nosema* sp. infections in immunocompetent persons have been reported (Schwartz and Bryan, 1997; Weber et al., 1994). We are also aware of a recent case of an ocular infection in an immunocompetent patient with keratoconjunctivitis. This organism, isolated from corneal scrapings and a biopsy, was tentatively identified as *N. algerae* and has also been established in in vitro culture (Fig 2, bottom left).

In view of this observation, as well as those on *Nosema* spp., further consideration of *Nosema* spp. (with special reference to *N. algerae*) will be discussed. During the 1990s a number of reports, largely on the cultivation of *Nosema* spp. that infect insects, have been published. For example: *N. locustae* was grown in a fat body cell line derived from *Mythimna convecta* (Khurad et al., 1991); a *Nosema* sp. was cultured in *A. eucalypti* cell cultures that produced two types of spores, one with 10 to 12 coils of the polar tube and the other with three to five coils (Iwano and Ishihara, 1991); *N. bombycis* Y90101 was grown in *A. eucalypti* cell cultures and produced spores that germinated intracellularly and infected other nearby cells (Yasunaga et al., 1994); *N. furnacalis* was grown on *Helicoverpa* cell lines and declined in virulence after 40 transfers (Kurtti et al., 1994); *N. furnacalis* infected and grew well on cell cultures of *H. zea* and produced dimorphic spores (Iwano and Kurtti, 1995). Furthermore, spores of a species of *Nosema* were successfully germinated and allowed to infect a *Spodoptera frugiperda* SF21AEII cell line. It was found that of a number of reagents used to allow extrusion of the polar tubule before infection, Rinaldini's salt solution containing KCl instead of NaCl gave the highest percentage of initial cell infection (Yasunaga et al., 1994). It was also found that growth patterns of *N. pyrausta* and *N. furnacalis* established in *H. zea* cell cultures differed from one another (Sagers et al., 1996).

Nosema algerae

Numerous studies on the physiology and biochemistry of *N. algerae* have been conducted during the past two decades, partly because of the size of its spores (which are about twice as large as those of *Encephalitozoon* spp.) and the ease with which the parasites can be cultivated. Undeen (1975) cultivated *N. algerae* in pig kidney cells at 26 and 35°C. He also inoculated cell cultures maintained at 37 and 38°C, however, no development was seen at 38°C. The organisms invaded the cell line at 37°C but died within 3 days, producing no spores. The parasites, however, were able to invade the cells at 26 and 35°C and produced all developmental stages including mature spores. The spores obtained from in vitro culture were able to infect mosquitoes, indicating their viability and infectivity. In addition, cultures were incubated at 27, 34, and 38°C. No development occurred at 38°C, but the microsporidia grew at the lower temperatures and produced viable spores. Streett et al. (1980) investigated the ability of three different insect cell lines to sustain infection of *N. algerae*. The cell lines used were TN-368, derived from the cabbage looper (*T. ni*), IPLB-1075, derived from the corn earworm (*H. zea*), and IZD-Mb-0503, derived from the cabbage moth (*M. brassicae*). The cell lines were maintained in a complicated growth medium, TNM-FH medium, that had been previously developed for insect cell lines. All three cell lines supported the growth of *N. algerae*, resulting in complete development of the parasite. Growth of the parasites in TNP-368 cells was slow. The production of spores, however, decreased with time and ceased completely by the sixth passage. Spores obtained from all cell cultures, except from TNP-368, were able to infect larvae of *T. ni*. *N. algerae* has also been cultured in embryonic rat brain, *Xenopus* XTC-6 cells, and Chang liver cells (Smith et al., 1982). In our laboratory *N. algerae* was recently established in continuous culture on E6 and HLF cell monolayers at 37°C and 30°C (Moura et al., unpublished data). Adults and larvae of *A. stephensi* obtained from the Walter Reed Army Institute of Research were macerated in a small volume of PBS and washed in PBS. The sediments were inoculated into separate monolayers of E6 and HLF along with antibiotics (gentamicin, imipenem, and amphotericin B) and incubated at 30°C and 37°C. The supernatant medium was removed every day for 3 days and centrifuged, and the sediment was reinoculated into the original flasks. After several days of such culture manipulations, foci of infection were seen in several cells (Fig. 2, top right and bottom right, and Fig. 4, bottom right, and Fig. 5, bottom right. The cells appeared to be distended and were full of developing forms and spores. TEM revealed the presence of diplokaryon in all stages of development, and the spores had 8 to 10 or 11 turns in the polar tube coils. The entire coding region of the parasite was sequenced, and it was found to be identical to that of *N. algerae*.

SPECIMEN STORAGE AND TRANSPORTATION

Since microsporidia culture is a specialized, complicated procedure and one that is not routinely performed in many clinical and hospital laboratories, specimens thought to contain microsporidia have to be sent to laboratories that specialize in research on microsporidia. Therefore, it is necessary to store the samples appropriately before they can be transported. It is fortunate, in a sense that spores, the agents that initiate infection in the appropriate hosts as well as in cell cultures, are generally resistant to many different physical and chemical stimuli in the environment. Furthermore, microsporidian spores, for example, those of *N. algerae,* can be stored at 4°C for prolonged periods of time without any adverse effect on their germination properties (Undeen et al., 1975). It is likely that this feature is characteristic of other microsporidial spores. Spores of culture-grown *Encephalitozoon* spp. have been successfully stored for several weeks at 4°C without a decrease in their infectivity for cultured mammalian cells. We have also received patient specimens from worldwide sources within a few days of their collection. Hollister et al. (1996), of the United Kingdom, also have reported receiving specimens 9 days after they were removed from a patient in Australia and were still successful in infecting cell cultures as well as mice. However, it is suggested that specimens be transported to a specialty laboratory as quickly as possible, preferably within 2 to 3 days. Specimens should also be kept at refrigeration temperatures and shipped with a cold pack so that the chances of contamination and growth of unwanted bacteria and fungi are minimized. It is also recommended that personnel handling patient specimens take appropriate precautions, such as wearing gloves and face masks, while handling these specimens, as they may contain human immunodeficiency virus, hepatitis B, and/or mycobacteria.

CRYOPRESERVATION

Cell cultures infected with microsporidia can be cryopreserved and stored at liquid nitrogen temperatures almost indefinitely. Cryopreservation has many advantages. For example (1) the maintenance of microsporidia in culture is a tedious, labor-intensive procedure, and when an active research program is in abeyance, it is best to store these organisms so that they do not lose their characteristic properties when brought back into culture; (2) cultures may become contaminated and overgrown with unwanted bacteria or fungi and thus may be lost; and (3) continuous cultivation of these parasites for prolonged periods may change their antigenicity, infectivity, and virulence properties. Therefore, it is advisable that, following establishment in culture, and periodically thereafter, the parasites be stored frozen as stabilates. However, cryopreservation also has disadvantages in that the survival levels of many protozoa are very low. It is therefore necessary to identify the appropriate cryoprotectants and methods of cryopreservation and storage that will improve the survival rate. The ideal method is to hold the microsporidia in growth medium containing a cryoprotectant such as dimethyl sulfoxide (DMSO) or glycerol and subject the mixture to cooling at the rate of 1°C per min, preferably in an automatic instrument such as a Linde cell freezer. Such equipment is expensive, however, and may not be available in many laboratories. A relatively simple and inexpensive apparatus, the Cell Freezer (Nalgene Nunc International, Rochester, N.Y.), has been very helpful in freezing many different protozoa, including *Entamoeba histolytica, Giardia lamblia, Trichomonas vaginalis, Acanthamoeba* spp., *Balamuthia mandrillaris, Naegleria* spp., and *Willaertia* spp., as well as microsporidia such as *E. cuniculi, E. hellem, E. intestinalis, V. corneae,* and *Nosema* spp. The Cell Freezer (Nalgene Cryo 1°C Freezing Container, no. 5100-0001) consists of two components, a round plastic shell with a screw-cap closure and a molded plastic insert containing 18 spaces (tube or vial holders) that accept 1-ml freezer vials or tubes. The shell is a receptacle for the cooling agent, isopropanol or ethylene glycol. Approximately 50 ml of the cooling agent is poured into the shell, and the molded plastic insert is put inside the shell. The cell culture in which the microsporidia grow is either scraped or trypsinized and washed once in balanced salt solution. The sediment containing the cell culture and microsporidia is dispersed in DMEM with 10% FBS. It is then mixed with equal volumes of DMEM containing 20% DMSO so that the final concentration of DMSO is 10%. The mixture is then aliquoted in 1-ml volumes to a number of cryopreservation vials, and the vials are placed in the vial holders of the molded plastic. The Cell Freezer is then transferred to a −70°C freezer and held overnight. The Cell Freezer is removed from the −70°C freezer, and the vials are transferred to a liquid nitrogen freezer and stored in the vapor phase. The frozen cultures, now referred to as stabilates, can be stored indefinitely under liquid nitrogen vapor. To establish viable cultures from the frozen stabilates, vials are removed from the freezer and transported to the culture room in a bed of dry ice and immediately placed in a 37°C water bath. The vials are gently swirled in the warm water, and as soon as the culture thaws and becomes liquid, it is transferred to a culture flask with the appropriate cell line and incubated at 37°C. Within a week, foci of infection will be seen in the recipient cell culture. Using this method, we have cryopreserved a number of isolates of *E. cuniculi, E. hellem, E. intestinalis, V. corneae,* and *Nosema* spp. for several years.

APPLICATION AND USES OF CULTURE-GROWN MICROSPORIDIA

One great advantage of obtaining spores from in vitro cultures is the absence of bacterial and fungal contaminants. While these spores are contaminated with cellular components and cell debris, these contaminants can be easily removed by washing and purification by Percoll gradient centrifugation (Didier et al., 1991; Visvesvara et al., 1991). Relatively large numbers of spores can be obtained for many different types of in vitro and in vivo studies including the following: testing the efficacy of pharmaceutical and therapeutic agents; generation of monoclonal and polyclonal antibodies; immunologic studies; animal inoculation studies; and molecular, biochemical, and physiologic studies. For example, that fumagillin strongly inhibits the development of *E. cuniculi* in vitro was determined in cell culture by Shadduck (1980). As a result of this study, patients with ocular microsporidiosis have been treated with this antibiotic with complete resolution of this condition (Diesenhouse et al., 1993; Garvey et al., 1995; Lowder et al., 1996; Rosberger et al., 1993). However, fumagillin is very toxic and cannot be given systemically. Hence research has focused on a relatively less toxic semisynthetic analog of fumagillin, TNP-470, which has been found by Didier (1997) to be very active against several microsporidia in vitro. This finding has been confirmed independently both in vitro and in vivo by Coyle et al. (1998). In vitro testing of the anthelmintic albendazole has also shown that this drug is active against the *Encephalitozoon* spp. (Beauvais et al., 1994; Colbourn et al., 1994; Didier, 1997; Haque et al., 1993). In vitro culture of the microsporidian *E. hellem* also has been helpful in physiologic and biochemical studies dealing with the inhibition of spontaneous and H_2O_2-stimulated polar tubule extrusion by the microfilament disrupter cytochalasin D, the microtubule disrupter demecolcine, the calcium channel blocker nifedipine, and the antifungal agent itraconazole. These studies have suggested that microtubules, microfilaments, and voltage-gated calcium channels play a role in the germination process (Leitch et al., 1993, 1995). Another study directed at determining the calcium and hydrogen ion concentrations in the PV of *E. hellem*-infected cultured cells suggested that the PV membrane is not a significant barrier to ion diffusion between the host cell and parasite (Leitch et al., 1995). In vitro assays with cell cultures infected with *E. cuniculi, E. hellem,* and *E. intestinalis* have also been used to evaluate the efficacy of candidate antiparasitic agents such as nifedipine, metronidazole, and nitric oxide donors, all of which exhibited antimicrosporidial activity (He et al., 1996). Additionally, a method that uses the fluorescent probe calcein and confocal microscopy to study the shape and organization of the PV to determine the viability of various stages of the parasite and to detect drug-induced effects in *Encephalitozoon* spp.-infected E6 cells has also been developed (Leitch et al., 1997).

In vitro culture of microsporidia was instrumental in the development of chemical stains that specifically identified microsporidian spores in complex human specimens such as feces (Moura et al., 1996, 1997; Weber et al., 1992).

In vitro culture of microsporidia is invaluable in the development of monoclonal and polyclonal antibodies for the purpose of identifying the causal agents in patient samples including feces, urine, conjunctival and corneal scrapings, sputum, BAL fluid, nasal wash, and biopsy and autopsy specimens, as well as in identifying species and strains of microsporidia (Aldras et al., 1994; Beckers et al., 1996; Croppo et al., 1997, 1998; Didier et al., 1991, 1996c; Deplazes et al., 1996a; Enriquez et al., 1997; Franzen et al., 1995; Hollister et al., 1993b, 1995; Scaglia et al., 1998; Sobottka et al., 1998; Schwartz et al., 1992, 1993, 1996; Visvesvara et al., 1994, 1995a; Weiss et al., 1992; Zierdt et al., 1993). Culture-grown spores also have been used to infect different species of laboratory animals to produce disease and to study infectivity and replication of isolates as well as pathogenesis and immune mechanisms (Achbarou et al., 1996; Cox et al., 1986; Didier et al., 1994; Didier and Bertucci, 1996; Pang and Shadduck, 1985; Shadduck and Polley, 1978; Silveira et al., 1993; Van Dellen et al., 1989). Additionally, it has been helpful in the delineation of serum antibody responses in patients with microsporidiosis as well as in uninfected controls and in seroepidemiology studies (Bergquist et al., 1984; Didier et al., 1991; Hollister et al., 1991; Visvesvara et al., 1994, 1995a; van Gool et al., 1997).

In vitro propagation of microsporidia and subsequent extraction of DNA have been helpful in designing species-specific PCR primers that can be used to identify microsporidia of a particular species in patient specimens. Furthermore, it has been invaluable in the characterization of strains based on random amplification of polymorphic DNA, as well as by the determination of differences in the rDNA intergenic spacer region, and in the study of molecular phylogeny (Baker et al., 1995; Black et al., 1997; Da Silva et al., 1997; Didier et al., 1995, 1996a, 1996b; Deplazes et al., 1996a, 1996b; De Groote et al., 1995; Edlind et al., 1996; Furuya et al., 1995; Hartskeerl et al., 1995; Hollister et al., 1995; Ombrouck et al., 1996; Scaglia et al., 1994, 1997, 1998; Visvesvara et al., 1994, 1995a, 1995b; Vossbrinck et al., 1993; Zhu et al., 1994).

Infected cell cultures can be processed for electron microscopy to study intracellular development as well as host-parasite interactions at the cellular level (Silveira and Canning, 1995b; Hollister et al., 1996; Visvesvara et al., 1995a, 1996).

CONCLUSIONS

During the past few years, a number of isolates of microsporidia belonging to different genera and species have been isolated from human tissue and body fluids and established in culture. Undoubtedly, they are only a small fraction of the total number of organisms represented in this group. At the present time, we are ignorant of other species of hosts that these human isolates of microsporidia (except for *E. cuniculi*) infect, though we have begun to identify some of them. For example, *E. bieneusi, E. hellem,* and *E. intestinalis* have been identified in domestic animals such as pigs, donkeys, dogs, cows, and goats and in birds, (budgerigars) (Deplazes et al., 1996; Bornay-Llinares et al., 1998; Black et al., 1997). These observations should, therefore, give further impetus to establishing *E. bieneusi* in continuous culture since it is the most prevalent microsporidian in AIDS patients as well as for other animal-derived microsporidial species. Unfortunately, methods used for the cultivation of *Encephalitozoon* spp. have not been successful for *E. bieneusi*. It is likely, therefore, that supplements, including trace elements, growth factors, cytokines, and so on, which are being used to support the growth of mammalian cells in vitro, will be applied to the cultivation of these parasites. Currently, interest in the cultivation of *E. bieneusi* appears to have diminished, probably because of the difficulties in obtaining patient specimens or even the nonavailability of samples containing these parasites. This may reflect the use of multiple agents, including protease inhibitors, for the treatment of AIDS patients.

In vitro cultivation of these parasites over the next decade will certainly continue to be a challenging but fruitful area of study. Since further research will likely include microsporidian surface antigens that may be involved in the invasion of the host cell, identification of specific proteins, and the development of monoclonal antibodies that will neutralize the invading potential of the parasite, in vitro cultivation will remain invaluable. Furthermore, cultured microsporidia are increasingly being used for serodiagnostic studies. In vitro culture is also invaluable in elucidating isolate or strain differences and in molecular epidemiology. Assessment of functional antibodies and cell-mediated protective systems against these microsporidia requires suitable assays that can be determined in a cost-effective manner only by in vitro culture. Eventually, efforts will be made to develop a vaccine, and vaccine efficacy must be determined with intact parasites that can be obtained in large quantities and without the contaminating influences of host components. This objective can be achieved only by in vitro methods. In this regard, long-term continuous culture may cause attenuation of strains that have potential for development of a vaccine. In vitro culture also may provide lymphokine and cytokine assay systems that may prevent host cell invasion by blocking polar tube extrusion.

ACKNOWLEDGMENTS

We are indebted to Sara Wallace for her invaluable help in the maintenance of cultures and cryopreservation of the various protozoan parasites during the past 15 years. We also thank John Goosey and Marcel Belloso for the corneal biopsy specimen.

REFERENCES

Achbarou, A., C. Ombrouck, T. Generagbe, F. Charlotte, L. Renia, I. Desportes-Livage, and D. Mazier. 1996. Experimental model for intestinal human microsporidiosis in interferon gamma receptor knockout mice infected by *Encephalitozoon intestinalis. Parasite Immunol.* **18:**387–392.

Aldras, A. M., J. M. Orenstein, D. P. Cotler, J. A. Shadduck, and E. S. Didier. 1994. Detection of microsporidia by indirect immunofluorescence antibody test using polyclonal and monoclonal antibodies. *J. Clin. Microbiol.* **32:**608–612.

Baker, M. D., C. R. Vossbrinck, E. S. Didier, J. V. Maddox, and J. A. Shadduck. 1995. Small subunit ribosomal DNA phylogeny of various microsporidia with emphasis on AIDS related forms. *J. Eukaryot. Microbiol.* **42:**564–570.

Beauvais, B., C. Sarfati, S. Challier, and F. Derouin. 1994. In vitro model to assess effect of antimicrobial agents on *Encephalitozoon cuniculi. Antimicrob. Agents Chemother.* **38:**2440–2448.

Beckers, P. J. A., G. J. M. M. Derks, T. van Gool, F. J. R. Rietveld, and R. W. Sauerwein. 1996. *Encephalitozoon intestinalis*-specific monoclonal antibodies for laboratory diagnosis of microsporidiosis. *J. Clin. Microbiol.* **34:**282–285.

Bedrník, P., and J. Vávra. 1975. Further observations on the maintenance of *Encephalitozoon cuniculi* in tissue culture. *J. Protozool.* **19** (suppl.): 75.

Bergquist, N. R., G. Stintzing, L. Smedman, T. Waller, and T. Andersson. 1984. Diagnosis of encephalitozoonosis in man by serological tests. *Br. Med. J.* **288:**902.

Bismanis, J. E. 1970. Detection of latent murine nosematosis and growth of *Nosema cuniculi* in cell cultures. *Can. J. Microbiol.* **16:**237–242.

Black, S. S., L. A. Steinohrt, D. C. Bertucci, L. B. Rogers, and E. S. Didier. 1997. *Encephalitozoon hellem* in budgerigars (*Melopsittacus undulatus*). *Vet. Pathol.* **34:**189–198.

Bocket, L., C. H. Marquette, A. Dewilde, D. Hober, and P. Wattre. 1992. Isolation and replication in human fibroblast cell (MRC-5) of a microsporidian from an AIDS patient. *Microb. Pathog.* **12:**187–191.

Bornay-Llinares, F. J. Personal communication.

Bornay-Llinares, F. J., A. J. da Silva, H. Moura, D. A. Schwartz, G. S. Visvesvara, N. J. Pieniazek, A. Cruz-Lopes, P. Herandez-Jauregui, J. Guerrero, and J. Enriquez. 1998. Immunological, microscopic and molecular evidence of natural *Encephalitozoon* (*Septata*) *intestinalis* infection in mammals other than humans. *J. Infect. Dis.* **178:** 820–826.

Canning, E. U., and W. S. Hollister. 1991. In vitro and in vivo investigations of human microsporidia. *J. Protozool.* **38:**631–635.

Canning, E. U., and J. Lom. 1986. *The Microsporidia of Vertebrates.* Academic Press, Inc., New York, N.Y.

Colbourn, N. I., W. S. Hollister, A. Curry, and E. U. Canning. 1994. Activity of albendazole against *Encephalitozoon cuniculi in vitro. Eur. J. Protistol.* **30:**211–220.

Cox, J. V., R. C. Hamilton, D. Pye, and J. W. Edmonds. 1986. The infectivity of *Encephalitozoon cuniculi in vivo* and *in vitro. Z. Parasitenkd.* **72:** 65–72.

Coyle, C., M. Kent, H. B. Tanowitz, M. Wittner, and L. M. Weiss. 1998. TNP-470 is an effective antimicrosporidial agent. *J. Infect. Dis.* **177:**515–518.

Croppo, G. P., G. S. Visvesvara, G. J. Leitch, S. Wallace, and M. A. De Groote. 1997. Western blot and immunofluorescence analysis of a human isolate of *Encephalitozoon cuniculi* established in culture from the urine of a patient with AIDS. *J. Parasitol.* **83:**66–69.

Croppo, G. P., G. S. Visvesvara, G. J. Leitch, S. Wallace, and D. A. Schwartz. 1998. Western blot identification of the microsporidian *Encephalitozoon hellem* using immunoglobulin G monoclonal antibodies. *Arch. Pathol. Lab. Med.* **122:**182–186.

Da Silva, A. J., S. B. Slemenda, G. S. Visvesvara, D. A. Schwartz, C. M. Wilcox, S. Wallace, and N. J. Pieniazek. 1997. Detection of *Septata intestinalis* (microsporidia) Cali et al. 1993 using polymerase chain reaction primers targeting the small subunit ribosomal RNA coding region. *Mol. Diagn.* **2:**47–52.

De Groote, M. A., G. S. Visvesvara, M. L. Wilson, N. J. Pieniazek, S. B. Slemenda, A. J. da Silva, G. J. Leitch, R. T. Bryan, and R. Reves. 1995. Polymerase chain reaction and culture confirmation of disseminated *Encephalitozoon cuniculi* in a patient with AIDS: successful therapy with albendazole. *J. Infect. Dis.* **171:**1375–1378.

del Aguila, C. Personal communication.

del Aguila, C., G. P. Croppo, H. Moura, A. J. Da Silva, G. J. Leitch, D. M. Moss, S. Wallace, S. B. Slemenda, N. J. Pieniazek, and G. S. Visvesvara. 1998. Ultrastructure, immunofluorescence, Western blot, and PCR analysis of eight isolates of *Encephalitozoon* (*Septata*) *intestinalis* established in culture from sputum and urine samples and duodenal aspirates of five patients with AIDS. *J. Clin. Microbiol.* **36:**1201–1208.

Deplazes, P., A. Mathis, C. Müller, and R. Weber. 1996a. Molecular epidemiology of *Encephalitozoon cuniculi* and first detection of *Enterocytozoon bieneusi* in faecal samples of pigs. *J. Eukaryot. Microbiol.* **43:**93S.

Deplazes, P., A. Mathis, R. Baumgartner, I. Tanner, and R. Weber. 1996b. Immunologic and molecular characteristics of *Encephalitozoon*-like microsporidia isolated from humans and rabbits indicate that *Encephalitozoon cuniculi* is a zoonotic parasite. *Clin. Infect. Dis.* **22:**557–559.

Desser, S. S., H. Hong, and Y. J. Yang. 1992. Ultrastructure of the development of a species of *Encephalitozoon* cultured from the eye of an AIDS patient. *Parasitol. Res.* **78:**677–683.

Didier, E. S. 1992. Personal communication.

Didier, E. S. 1997. Effects of albendazole, fumagillin, and TNP-470 and microsporidial replication in vitro. *Antimicrob. Agents Chemother.* **41:**1541–1546.

Didier, E. S., and D. C. Bertucci. 1996. Identification of *Encephalitozoon intestinalis* proteins that induce proliferation of sensitized murine spleen cells. *J. Eukaryot. Microbiol.* **43:**92S.

Didier, E. S., P. J. Didier, D. N. Friedberg, S. M. Stenson, J. M. Orenstein, R. W. Yee, F. O. Tio, R. M. Davis, C. Vossbrinck., N. Millichamp, and J. S. Shadduck. 1991. Isolation and characterization of a new human microsporidian, *Encephalitozoon hellem* (n. sp.), from three AIDS patients with keratoconjunctivitis. *J. Infect. Dis.* **163:**617–621.

Didier, E. S., L. B. Rogers, A. D. Brush, S. Wong, V. Traina-Dorge, and D. Bertucci. 1996a. Diagnosis of disseminated microsporidian *Encephalitozoon hellem* infection by PCR- Southern analysis and successful treatment with albendazole and fumagillin. *J. Clin. Microbiol.* **34:**947–952.

Didier, E. S., L. B. Rogers, J. M. Orenstein, M. D. Baker, C. R. Vossbrinck, T. Van Gool, R. Hartskeerl, R. Soave, and L. M. Beaudet. 1996b. Characterization of *Encephalitozoon (Septata) intestinalis* isolates cultured from nasal mucosa and bronchoalveolar lavage fluids of two AIDS patients. *J. Eukaryot. Microbiol.* **43:**34–43.

Didier, E. S., P. W. Varner, P. J. Didier, A. M. Aldras, N. J. Millichamp, M. Murphey-Corb, R. Bohm, and J. A. Shadduck. 1994. Experimental microsporidiosis in immunocompetent and immunodeficient mice and monkeys. *Folia Parasitol.* **41:**1–11.

Didier, E. S., G. S. Visvesvara, M. D. Baker, L. B. Rogers, D. C. Bertucci, M.A. De Groote, and C. R. Vossbrinck. 1996c. A microsporidian isolated from an AIDS patient corresponds to *Encephalitozoon cuniculi* III, originally isolated from domestic dogs. *J. Clin. Microbiol.* **34:**2835–2837.

Didier, E. S., C. R. Vossbrinck, M. D. Baker, L. B. Rogers, D. C. Bertucci, and J. A. Shadduck. 1995. Identification and characterization of three *Encephalitozoon cuniculi* strains. *Parasitology* **111:** 411–421.

Diesenhouse, M. C., L. A. Wilson, G. F. Corrent, G. S. Visvesvara, H. E. Grossniklaus, and Bryan, R. T. 1993. Treatment of microsporidial keratoconjunctivitis with topical fumagillin. *Am. J. Ophthalmol.* **115:**293–298.

Doultree, J. V., A. L., Maerz, N. J. Ryan, R. W. Baird, E. Wright, S. M. Crowe, and J. A. Marshall. 1995. In vitro growth of the microsporidian *Septata intestinalis* from an AIDS patient with disseminated illness. *J. Clin. Microbiol.* **33:**463–470.

Edlind, T, S. Katiyar, G. S. Visvesvara, and J. Li. 1996. Evolutionary origins of microsporidia and basis for benzimidazole sensitivity: an update. *J. Eukaryot. Microbiol.* **43:**109S.

Enriquez, F. J., O. Ditrich, J. D. Palting, and K. Smith. 1997. Simple diagnosis of *Encephalitozoon* sp. microsporidial infections by using a panspecific antiexospore monoclonal antibody. *J. Clin. Microbiol.* **35:**724–729.

Franzen, C., D. A. Schwartz, G. S. Visvesvara, A. Müller, A. Schwenk, B. Salzberger, G. Fätkenheuer, P. Hartmann, G. Mahrle, V. Diehl, and M. Schrappe. 1995. Immunologically confirmed disseminated, asymptomatic *Encephalitozoon cuniculi* infection of the gastrointestinal tract in a patient with AIDS. *Clin. Infect. Dis.* **21:**1480–1484.

Furuya, K, C. Sato, H. Nagano, N. Sato, and J. Uchino. 1995. *Encephalitozoon*-like organisms in patients with alveolar hydatid disease: cell culture, ultrastructure, histoimmunochemical localization and seroprevalence. *J. Eukaryot. Microbiol.* **42:**518–525.

Gatti, S., L. Sacchi, S. Novati, S. Corona, A. M. Bernuzzi, H. Moura, N. J. Pieniazek, G. S. Visvesvara, and M. Scaglia. 1997. Extraintestinal microsporidiosis in AIDS patients: clinical features and advanced protocols for diagnosis and characterization of the isolates. *J. Eukaryot. Microbiol.* **44:**79S

Garvey, M. J., P. G. Ambrose, and J. L. Ulmer. 1995. Topical fumagillin in the treatment of microsporidial keratoconjunctivitis in AIDS. *Ann. Pharmacother.* **29:**872–874.

Haque, A., W. S. Hollister, A. Wilcox, and E. U. Canning. 1993. The antimicrosporidial activity of albendazole. *J. Invertebr. Pathol.* **62:**171–177.

Hartskeerl, R. A., T. van Gool, A. R. J. Schuitema, E. S. Didier, and W. J. Tepstra. 1995. Genetic and immunological characterization of the microsporidian *Septata intestinalis* Cali, Kotler and Orenstein, 1993: reclassification to *Encephalitozoon intestinalis. Parasitology* **110:**277–285.

He, Q., G. J. Leitch, G. S. Visvesvara, and S. Wallace. 1996. Effects of nifedipine, metronidazole, and nitric oxide donors on spore germination and cell culture infection of the microsporidia *Encephalitozoon hellem* and *Encephalitozoon intestinalis. Antimicrob. Agents Chemother.* **40:**179–185.

Hollister, W. S., E. U. Canning, and N. I. Colbourn. 1993a. A species of *Encephalitozoon* isolated from an AIDS patient: criteria for species differentiation. *Folia Parasitol.* **40:**293–295.

Hollister, W. S., E. U. Canning, N. I. Colbourn, and E. J. Aarons. 1995. *Encephalitozoon cuniculi* isolated from the urine of an AIDS patient, which differs from canine and murine isolates. *J. Eukaryot. Microbiol.* **42:**367–372.

Hollister, W. S., E. U. Canning, N. I. Colbourn, A. Curry, and C. J. N. Lacey. 1993b. Characterization of *Encephalitozoon hellem* (Microspora) isolated from the nasal mucosa of a patient with AIDS. *Parasitology* **107:**351–358.

Hollister, W. S., E. U. Canning, E. Weidner, A. S. Field, J. Kench, and D. J. Marriott. 1996. Development and ultrastructure of *Trachipleistophora hominis* n. g., n. sp. after in vitro isolation from an AIDS patient and inoculation into athymic mice. *Parasitology* **112:**143–154.

Hollister, W. S., E. U. Canning, and A. Wilcox. 1991. Evidence for widespread occurrence of antibodies to *Encephalitozoon cuniculi* (Microspora) in man provided by ELISA and other serological tests. *Parasitology* **102:**33–43.

Ishihara, R. 1968. Growth of *Nosema bombycis* in primary cell cultures of mammalian and chick embryos. *J. Invertebr. Pathol.* **11:**238.

Ishihara, R. 1969. The life cycle of *Nosema bombycis* as revealed in tissue culture cells of *Bombyx mori*. *J. Invertebr. Pathol.* **14:**316–320.

Ishihara, R., and S. S. Sohi. 1966. Infection of ovarian tissue culture of *Bombyx mori* by *Nosema bombycis* spores. *J. Invertebr. Pathol.* **8:**538–540.

Iwano, H., and R. Ishihara. 1991. Dimorphism of spores of *Nosema* spp. in cultured cell. *J. Invertebr. Pathol.* **57:**211–219.

Iwano, H., and T. J. Kurtti. 1995. Identification and isolation of dimorphic spores from *Nosema furnacalis* (Microspora: Nosematidae). *J. Invertebr. Pathol.* **65:**230–236.

Jaronski, S. T. 1984. Microsporidia in cell culture, p. 183–229, *In* K. Maramorosch (ed.). *Advances in Cell Culture,* vol. 18. Academic Press, New York.

Katzwinkel-Wladarsch, S., P. Deplazes, R. Weber, T. Loscher, and H. Rinder. 1997. Comparison of polymerase chain reaction with light microscopy for detection of microsporidia in clinical specimens. *Eur. J. Clin. Microbiol. Infect. Dis.* **16:** 7–10.

Kawarabata, T., and R. Ishihara. 1984. Infection and development of *Nosema bombycis,* Microsporida, Protozoa, in a cell line of *Antheraea eucalypti*. *J. Invertebr. Pathol.* **44:**52–62.

Khurad, A. M., Raina, S. K., and Pandharipande, T. N. 1991. In vitro propagation of *Nosema locustae* using fat body cell line derived from *Mythimna convecta* (Lepidoptera: Noctuidae). *J. Protozool.* **38:** 91S–93S.

Kurtti, T. J., and M. A. Brooks. 1971. Growth of microsporidian parasite in cultured cells of tent caterpillars (*Malacostoma*). *Curr. Topics Microbiol. Immunol.* **55:**204–208.

Kurtti, T. J., and Brooks, M. A. 1977. The rate of development of a microsporidian in moth cell cultures. *J. Invertebr. Pathol.* **29:**126–132.

Kurtti, T. J., S. E. Ross, Y. Liu, and U. G. Munderloh. 1994. In vitro developmental biology and spore production in *Nosema furnacalis* (Microspora: Nosematidae). *J. Invertebr. Pathol.* **63:**188–196.

Kurtti, T. J., R. Tsang, and M. A. Brooks. 1983. The spread of infection by the microsporidian, *Nosema disstriae,* in insect cell lines. *J. Protozool.* **30:**652–657.

Leitch, G. J., Q. He, S. Wallace, and G. S. Visvesvara. 1993. Inhibition of the polar filament extrusion of the microsporidium, *Encephalitozoon hellem,* isolated from an AIDS patient. *J. Eukaryot. Microbiol.* **40:**711–717.

Leitch, G. J., M. Scanlon, A. Shaw, G. S. Visvesvara, and S. Wallace. 1997. Use of a fluorescent probe to assess the activities of candidate agents against intracellular forms of *Encephalitozoon microsporidia*. *Antimicrob. Agents Chemother.* **41:**337–344.

Leitch, G. J., M. Scanlon, G. S. Visvesvara, and S. Wallace. 1995. Calcium and hydrogen ion concentrations in the parasitophorous vacuoles of epithelial cells infected with the microsporidian *Encephalitozoon hellem*. *J. Eukaryot. Microbiol.* **42:**445–451.

Lowder, C. Y., J. T. McMahon, D. M. Meisler, E. M. Dodds, L. H. Calabrese, E. S. Didier, and A. Cali. 1996. Microsporidial keratoconjunctivitis caused by *Septata intestinalis* in a patient with acquired immunodeficiency syndrome. *Am. J. Ophthalmol.* **121:**715–717.

Mathis, A., J. Åkerstedt, J. Tharaldsen, Ø. Ødegaard, and P. Deplazes. 1996. Isolates of *Encephalitozoon cuniculi* from farmed blue foxes (*Alopex lagopus*) from Norway differ from isolates from Swiss domestic rabbits (*Oryctolagus cuniculus*). *Parasitol. Res.* **82:**727–730.

Mathis, A., M. Michel, H. Kuster, C. Muller, R. Weber, and P. Deplazes. 1997. Two *Encephalitozoon cuniculi* subtypes of human origin are infectious to rabbits. *Parasitology* **114:**29–35.

Molina, J. M., E. Oksenhendler, B. Beauvais, C. Sarfati, A. Jaccard, F. Derouin, and J. Modai. 1995. Disseminated microsporidiosis due to *Septata intestinalis* in a patient with AIDS: clinical features and response to albendazole therapy. *J. Infect. Dis.* **171:**245–249.

Moura, H., A. J. Dasilva, I. N. Moura, N. J. Pieniazek, D. A. Schwartz, S. Wallace, R. A.

Wirtz, and G. S. Visvesvara. 1998. Continuous *in vitro* cultivation of *Nosema algerae,* an insect microsporidian, in mammalian cells at 37°C, abstr. X-14, p. 545. *In Abstracts of the 98th General Meeting of the American Society for Microbiology.* American Society for Microbiology Washington, D.C.

Moura, H., J. L. Nunes Da Silva, F. C. Sodré, P. Brasil, K. Wallmo, S. Wahlquist, S. Wallace, G. P. Croppo, and G. S. Visvesvara. 1996. Gram-chromotrope: a new technique that enhances detection of microsporidial spores in clinical samples. *J. Eukaryot. Microbiol.* **43:**94S–95S.

Moura, H., D. A. Schwartz, F. Bornay-Llinares, F. C. Sodré, S. Wallace, and G. S. Visvesvara. 1997. A new and improved "quick-hot Gram-chromotrope" technique that differentially stains microsporidian spores in clinical samples, including paraffin-embedded tissue sections. *Arch. Pathol. Lab. Med.* **121:**888–893.

Ombrouck, C., I. Desportes-Livage, A. Achbarou, and M. Gentilini. 1996. Specific detection of the microsporidian *Encephalitozoon intestinalis* in AIDS patients. *C. R. Acad. Sci.* **319:**39–43.

Pang, V. F., and J. A. Shadduck. 1985. Susceptibility of cats, sheep, and swine to a rabbit isolate of *Encephalitozoon cuniculi. Am. J. Vet. Res.* **46:**1071–1077.

Rosberger, D. F., O. N. Serdarevic, R. A. Erlandson, R. T. Bryan, D. A. Schwartz, G. S. Visvesvara, and P. C. Keenan. 1993. Successful treatment of microsporidial keratoconjunctivitis with topical fumagillin in a patient with AIDS. *Cornea* **12:**261–265.

Sagers, J. B., U. G. Munderloh, and T. J. Kurtti. 1996. Early events in the infection of a *Helicoverpa zea* cell line by *Nosema furnacalis* and *Nosema pyrausta* (Microspora: Nosematidae). *J. Invertebr. Pathol.* **67:**28–34.

Scaglia, M. Personal communication.

Scaglia, M., S. Gatti, L. Sacchi, S. Corona, G. Chichino, A. M. Bernuzzi, G. Barbarini, G. P. Croppo, A. J. Da Silva, N. J. Pieniazek, and G. S. Visvesvara. 1998. Asymptomatic respiratory tract microsporidiosis due to *Encephalitozoon hellem* in three patients with AIDS. *Clin. Infect. Dis.* **26:**174–176.

Scaglia, M., L. Sacchi, G. P. Croppo, A. da Silva, S. Gatti, S. Corona, A. Orani, A. M. Bernuzzi, N. J. Pieniazek, S. B. Slemenda, S. Wallace, and G. S. Visvesvara. 1997. Pulmonary microsporidiosis due to *Encephalitozoon hellem* in a patient with AIDS. *J. Infect.* **34:**119–126.

Scaglia, M., L. Sacchi, S. Gatti, A. M. Bernuzzi, P. D. P. Polver, I. Piacentini, E. Concia, G. P. Croppo, A. J. Da Silva, N. J. Pieniazek, S. B. Slemenda, S. Wallace, G. J. Leitch, and G. S. Visvesvara. 1994. Isolation and identification of *Encephalitozoon hellem* from an Italian AIDS patient with disseminated microsporidiosis. *APMIS* **102:**817–827.

Schwartz, D. A., and R. T. Bryan. 1997. Microsporidia, p. 61–94. *In* C. R. Horsburgh, Jr., and A. M. Nelson (ed.), *Pathology of Emerging Infections.* ASM Press, Washington, D.C.

Schwartz, D. A., R. T. Bryan, K. O. Hewan-Lowe, G. S. Visvesvara, R. Weber, A. Cali, and P. Angritt. 1992. Disseminated microsporidiosis (*Encephalitozoon hellem*) and acquired immunodeficiency syndrome: autopsy evidence for respiratory infection. *Arch. Pathol. Lab. Med.* **116:**660–668.

Schwartz, D. A., I. Sobottka, G. J. Leitch, A. Cali, and G. S. Visvesvara. 1996. Pathology of microsporidiosis: emerging parasitic infections in patients with acquired immunodeficiency syndrome. *Arch. Pathol. Lab. Med.* **120:**173–188.

Schwartz, D. A., G. S. Visvesvara, G. J. Leitch, L. Tashjian, M. Pollack, J. Holden, and R. T. Bryan. 1993. Pathology of symptomatic microsporidial (*Encephalitozoon hellem*) bronchiolitis in the acquired immunodeficiency syndrome: a new respiratory pathogen diagnosed from lung biopsy, bronchoalveolar lavage, sputum, and tissue culture. *Hum. Pathol.* **24:**937–943.

Seleznev, K. V., I. V. Issi, V. V. Dolgikh, G. B. Belostotskaya, O. A. Antonova, and J. J. Sokolova. 1995. Fractionation of different life cycle stages of microsporidia *Nosema grylli* from crickets *Gryllus bimaculatus* by centrifugation in Percoll density gradient for biochemical research. *J. Eukaryot. Microbiol.* **42:**288–292.

Sen Gupta, K. 1964. Cultivation of *Nosema mesnili* Paillot (microsporidia) in vitro. *Curr. Sci.* **33:**407–408.

Shadduck, J. A. 1969. *Nosema cuniculi:* in vitro isolation. *Science* **166:**516–517.

Shadduck, J. A. 1980. Effect of fumagillin on in vitro multiplication of *Encephalitozoon cuniculi. J. Protozool.* **27:**202–208.

Shadduck, J. A., R. A. Meccoli, R. Davis, and R. L. Font. 1990. First isolation of a microsporidian from a human patient. *J. Infect. Dis.* **162:**773–776.

Shadduck, J. A., and M. B. Polley. 1978. Some factors influencing the in vitro infectivity and replication of *Encephalitozoon cuniculi. J. Protozool.* **25:** 491–496.

Shadduck, J. A., W. T. Watson, S. P. Pakes, and A. Cali. 1979. Animal infectivity of *Encephalitozoon cuniculi. J. Parasitol.* **65:**123–129.

Silveira, H., and E. U. Canning. 1995a. *Vittaforma corneae* n. comb. for the human microsporidium

Nosema corneum Shadduck, Meccoli, Davis & Font, 1990, based on its ultrastructure in the liver of experimentally infected athymic mice. *J. Eukaryot. Microbiol.* **42:**158–165.

Silveira, H., and E. U. Canning. 1995b. In vitro cultivation of the human microsporidium *Vittaforma corneae:* development and effect of albendazole. *Folia Parasitol.* **42:**241–250.

Silveira, H., E. U. Canning, and J. A. Shadduck. 1993. Experimental infection of athymic mice with the human microsporidian *Nosema corneum. Parasitology* **107:**489–496.

Smith, J. E., R. J. Barker, and P. F. Lai. 1982. Culture of microsporidia from invertebrates in vertebrate cells. *Parasitology* **85:**427–436.

Sobottka, I., D. A. Schwartz, J. Schottelius, G. S. Visvesvara, N. J. Pieniazek, C. Schmetz, N. P. Kock, R. Laufs, and H. Albrecht. 1998. Prevalence and clinical significance of intestinal microsporidiosis in human immunodeficiency virus-infected patients with and without diarrhea in Germany: a prospective coprodiagnostic study. *Clin. Infect. Dis.* **26:**475–480.

Sodre, F. C., H. Moura, S. Wahlquist, F. J. Bornay-Llinares, and G. S. Visvesvara. 1997. An immunofluorescent (IF) test detects Encephalitozoon intestinalis spores in stool samples. abstr. C-225, p. 159. *In Abstracts of the 97th General Meeting of the American Society for Microbiology.* American Society for Microbiology, Washington, D.C.

Sohi, S. S., and G. G. Wilson. 1976. Persistent infection of *Malacosoma disstria* (Lepidoptera: Lasiocampidae) cell cultures with *Nosema* (*Glugea*) *disstriae* (Microsporida: Nosematidae). *Can. J. Zool.* **54:**336–342.

Sprague, V., and J. J. Becnel. 1998. Note on the name-author-date combination for the taxon Microsporidies Balbiani, 1882, when ranked as a phylum. *J. Invertebr. Pathol.* **71:**91–94

Streett, D. A., D. Ralph, and W. F. Hink. 1980. Replication of *Nosema algerae* in three insect cell lines. *J. Protozool.* **27:**113–117.

Thomas, C., M. Finn, L. Twigg, P. Deplazes, R. C. Thompson. 1997. Microsporidia (*Encephalitozoon cuniculi*) in wild rabbits in Australia. *Aust. Vet. J.* **75:**808–810.

Trager, W. 1937. The hatching of spores of *Nosema bombycis* Nägeli and the partial development of the organism in tissue cultures. *J. Parasitol.* **23:**226–227.

Undeen, A. H. 1975. Growth of *Nosema algerae* in pig kidney cell cultures. *J. Protozool.* **22:**107–110.

Van Dellen, A. F., C. G. Stewart, and W. S. Botha. 1989. Studies of encephalitozoonosis in vervet monkeys (*Cercopithecus pygerythrus*) orally inoculated with spores of *Encephalitozoon cuniculi* isolated from dogs (*Canis familiaris*). *Onderstepoort J. Vet. Res.* **56:**1–22.

Van Gool, T., E. U. Canning, H. Gilis, M. A. van den Bergh Weerman, J. K. M. Eeftinck Schattenkerk, and J. Dankert. 1994. *Septata intestinalis* frequently isolated from stool of AIDS patients with a new cultivation method. *Parasitology* **109:**281–289.

Van Gool, T., J. V. M. Vettre, B. Weinmayer, A. Van Dam, F. Derouin, and J. Dankert. 1997. High seroprevalence of *Encephalitozoon* species in immunocompetent subjects. *J. Infect. Dis.* **175:**1020–1024.

Visvesvara, G. S., A. J. Da Silva, G. P. Croppo, N. J. Pieniazek, G. J. Leitch, D. Ferguson, H. Moura, S. Wallace, S. B. Slemenda, I. Tyrrel, D. F. Moore, and J. Meador. 1995a. In vitro culture and serologic and molecular identification of *Septata intestinalis* isolated from urine of a patient with AIDS. *J. Clin. Microbiol.* **33:**930–936.

Visvesvara, G. S., G. J. Leitch, A. J. Da Silva, G. P. Croppo, H. Moura, S. Wallace, S. B. Slemenda, D. A. Schwartz, D. Moss, R. T. Bryan, and N. J. Pieniazek. 1994. Polyclonal and monoclonal antibody and PCR-amplified small sub-unit rRNA identification of a microsporidian, *Encephalitozoon hellem,* isolated from an AIDS patient with disseminated infection. *J. Clin. Microbiol.* **32:**2760–268.

Visvesvara, G. S., G. J. Leitch, L. Gorelkin, M. C. Wilcox, R. Weber, and R. T. Bryan. 1995b. Short term in vitro culture of *Enterocytozoon bieneusi* from four different patients with AIDS. *J. Eukaryot. Microbiol.* **42:**506–510.

Visvesvara, G. S., G. J. Leitch, H. Moura, S. Wallace, R. Weber, and R. T. Bryan. 1991. Culture, electron microscopy, and immunoblot studies on a microsporidian isolated from the urine of a patient with AIDS. *J. Protozool.* **38:**105S–111S.

Visvesvara, G. S., H. Moura, G. J. Leitch, D. A. Schwartz, A. J. Da Silva, S. Wallace, and N. J. Pieniazek. Unpublished data.

Visvesvara, G. S., G. J. Leitch, and N. J. Pieniazek. *Encephalitozoon cuniculi:* light and electron microscopic evidence for di-, tri-, tetra-, and octasporogonic development with a note on the molecular phylogeny of Encephalitozoonidae. *J. Eukaryot. Microbiol.*, in press.

Visvesvara, G. S., G. J. Leitch, D. A. Schwartz, A. J. Da Silva, S. Wallace, H. Moura, N. J. Pieniazek, and R. T. Bryan. Unpublished data.

Visvesvara, G. S., G. J. Leitch, S. Wallace, C. Seaba, D. Erdman, and E. P. Ewing, Jr. 1996. Adenovirus masquerading as microsporidia. *J. Parasitol.* **82:**316–319.

Vossbrinck, C. R., M. D. Baker, E. S. Didier, B. A. Debrunner-Vossbrinck, and J. A. Shadduck.

1993. Ribosomal DNA sequences of *Encephalitozoon hellem* and *Encephalitozoon cuniculi:* species identification and phylogenetic construction. *J. Eukaryot. Microbiol.* **40:**354–362.

Weber, R., R. T. Bryan, R. L. Owen, C. M. Wilcox, L. Gorelkin, and G. S. Visvesvara. 1992. Improved light-microscopical detection of microsporidia spores in stool and duodenal aspirates. *N. Engl. J. Med.* **326:**161–166.

Weber, R., R. T. Bryan, D. A. Schwartz, and R. Owen. 1994. Human microsporidial infections. *Clin. Microbiol. Rev.* **7:**426–461.

Weber, R., P. Deplazes, M. Flepp, A. Mathis, R. Baumann, B. Sauer, H. Kuster, and R. Luthy. 1997. Cerebral microsporidiosis due to *Encephalitozoon cuniculi* in a patient with human immunodeficiency virus infection. *N. Engl. J. Med.* **336:** 474–478.

Weiss, L. M., A. Cali, E. Levee, D. LaPlace, H. Tanowitz, D. Simon, and M. Wittner. 1992. Diagnosis of *Encephalitozoon cuniculi* by Western blot and the use of cross-reactive antigens for the possible detection of microsporidiosis in humans. *Am. J. Trop. Med. Hyg.* **47:**456–462.

Wilson, G. G., and S. S. Sohi. 1977. Effect of temperature on healthy and microsporidia-infected continuous cultures of *Malacosoma disstria* hemocytes. *Can. J. Zool.* **55:**713–717.

Wongtavatchai, J., P. A. Conrad, and R. P. Hedrick. 1994. In vitro cultivation of the microsporidian *Enterocytozoon salmonis* using a newly developed medium for salmonid lymphocytes. *J. Tissue Cult. Methods* **16:**125–131.

Yasunaga, C., M. Funakoshi, and T. Kawarabata. 1994. Effects of host cell density on cell infection level in *Antheraea eucalypti* (Lepidoptera: Saturniidae) cell cultures persistently infected with *Nosema bombycis* (Microsporida: Nosematidae). *J. Eukaryot. Microbiol.* **41:**133–137.

Zhu, X., M. Wittner, H. B. Tanowitz, A. Cali, and L. M. Weiss. 1994. Ribosomal RNA sequences of *Enterocytozoon bieneusi, Septata intestinalis* and *Ameson michaelis:* phylogenetic construction and structural correspondence. *J. Eukaryot. Microbiol.* **41:**204–209.

Zierdt, C. H., V. J. Gill, and W. S. Zierdt. 1993. Detection of microsporidian spores in clinical samples by indirect fluorescent-antibody assay using whole-cell antisera to *Encephalitozoon cuniculi* and *Encephalitozoon hellem. J. Clin. Microbiol.* **31:**3071–3074.

MICROSPORIDIA IN HIGHER VERTEBRATES

Karen F. Snowden and John A. Shadduck

12

Microsporidial infections have been described in a variety of domestic and wild mammalian hosts and less frequently in avian, amphibian, and reptilian hosts. *Encephalitozoon cuniculi* was first reported in rabbits and laboratory rodents in the early 1920s, and this species remains the most frequently observed organism in nonhuman vertebrates. Several additional microsporidial species have been identified in various hosts, usually as single reports.

MICROSPORIDIOSIS IN RODENTS

Introduction

With the advent of improved care and management techniques for laboratory animals, spontaneous rodent encephalitozoonosis has become a rare disease. Previously, this infection had been considered one of the most common and important sources of interference with experimental results (Innes et al., 1962). Shadduck and Pakes (1971) and Canning and Lom (1986) should be consulted for details. Serological surveys reported prevalence rates as high as 80% in hamsters, 30% in rats, and 85% in guinea pigs (Chalupsky et al., 1979; Gannon, 1980). Colonies of laboratory mice were frequently infected, but without the expression of clinical signs or lesions other than the detection of focal granulomas and parasites in various tissues (Kyo, 1958; Lainson, et al., 1964; Morris et al., 1956; Nelson, 1962). The use of laboratory rodents with clinically silent infections resulted in altered host immune responses (Huldt and Waller, 1974), contamination of transplantable neoplasms with accompanying alterations in the biological behavior of these tumors (Arison et al., 1966; Petri, 1966), and the mistaken attribution of causality in studies on infectious diseases (Shadduck and Pakes, 1971). With modern laboratory animal management techniques, including gnotobiotic derivation, microsporidial infections have largely disappeared, although investigators and laboratory animal veterinarians should remain vigilant especially in situations in which materials are serially transmitted from animal to animal or in colonies derived from wild or poorly maintained founder animals. The availability of reliable serological diagnostic techniques should greatly assist in the detection and elimination of this parasitic disease among laboratory rodents (Gannon, 1980). Table 1

Karen F. Snowden, Department of Veterinary Pathobiology, College of Veterinary Medicine, Texas A&M University, College Station, TX 77843-4467. *John A. Shadduck,* Heska Corporation, 1825 Sharp Point Drive, Fort Collins, CO 80525.

The Microsporidia and Microsporidiosis (Murray Wittner, editor; Louis M. Weiss, contributing editor), ©1999 American Society for Microbiology, Washington, D.C.

TABLE 1 Spontaneous microsporidioses in rodents

Rodent	Parasite[a]	Clinical disease	Lesions	Reference(s)
Mice (immunocompromised)	*Encephalitozoon* sp.	Ascites, death	Peritonitis; nephritis; encephalitis	Bismanis, 1970
Mice	*Encephalitozoon* sp.	None; ascites; retardation of growth of transplantable tumor	Focal granulomas in kidney and brain	Arison et al., 1966; Chalupsky et al, 1979; Gannon, 1980; Innes et al., 1962; Morris et al., 1956; Nelson, 1962
Rats	*E. cuniculi*	None; transplantable tumor contaminated with *E. cuniculi* showed altered growth patterns	Cerebral, renal, or hepatic granulomas	Attwood and Sutton, 1965; Chalupsky et al., 1982; Gannon, 1980; Lainson et al., 1964; Majeed and Zubzidy, 1982; Petri, 1966
Guinea pigs	*E. cuniculi*	None	Cerebral and renal granulomas; vasculitis; nephritis; encephalitis	Chalupsky et al., 1979; Gannon, 1980; Moffatt and Schiefer, 1973; Perrin, 1943
Hamsters	*E. cuniculi*	Contamination of transplantable tumor		Chalupsky et al., 1979; Meiser et al., 1971
Arctic lemmings	*E. cuniculi*	Central nervous system disturbances	Disseminated granulomatous foci	Cutlip and Beal, 1989; Cutlip and Denis, 1993
Muskrats	*E. cuniculi*	None	Cerebral granulomas	Woebeser and Schuh, 1979
Mastomys	*E. cuniculi*	None	Cerebral lesions	Lainson et al., 1964
Voles	*Thelohania apodemi*	None	Organisms in muscle	Doby et al., 1963

[a]Parasite genus and species assigned by morphological criteria only. The possibility that some of these infections were caused by *E. hellem* or *E. intestinalis* cannot be excluded.

summarizes information documenting spontaneous microsporidial infections in various rodent hosts.

Diagnosis

A variety of serological tests have been described for the diagnosis of spontaneous encephalitozoonosis (Cox et al., 1981; Cox and Gallichio, 1977; Cox and Pye, 1975; Gannon, 1980; Kellett and Bywater, 1980). Although most are useful, the most commonly employed test today is either immunofluorescence or enzyme-linked immunosorbent assay. Both techniques are reliable and have shown good correlation with the presence of organisms even in the absence of clinical signs. Experimental data support the conclusion that the presence of antibody constitutes reliable evidence of active, albeit clinically silent, infection. Absence of antibody, however, must be interpreted with caution because, at least in some species, several weeks may be required between initial infection and the appearance of detectable serum antibodies.

Transmission

Both horizontal and vertical transmission of encephalitozoonosis seem to occur. Evidence for vertical transmission includes the presence of organisms in specific-pathogen-free mice delivered by cesarean section and foster-suckled on germ-free rats (Kyo, 1958) and serological positivity in gnotobiotic guinea pigs (Boot et al., 1988).

MICROSPORIDIOSIS IN MICE

So far as is known, spontaneously occurring microsporidiosis in rodents is caused by *E. cuniculi* with the one exception of a single case of *Thelohania apodemi* in wild meadow voles (Doby et al., 1963). However, most species identification has been based on light or electron microscopic morphological features. The possibility that rodents are spontaneously infected with microsporidia other than *E. cuniculi* cannot be excluded. The most frequently reported spontaneous infections in rodents are in laboratory mice, perhaps because of the relatively large number of mice employed in research as opposed to other rodent species. In addition, laboratory mice have been used extensively in the study of experimental microsporidiosis.

Clinical Disease

In mice the disease takes two distinct forms depending on whether the animals are immunologically compromised. In mice that are immunologically intact, clinical signs of spontaneous microsporidiosis are rare. The infection is almost always clinically silent, but in some situations ascites and retardation of growth of contaminated transplantable tumors have been observed (Arison et al., 1966; Morris et al., 1956; Nelson, 1962). Experimentally, mice with fully functional immune systems are resistant to the induction of overt infection with clinical signs, illness, and death (Didier et al., 1994).

The situation is quite different in immunologically compromised mice. Typically, these animals either are treated with immunosuppressive drugs such as corticosteroids (Bismanis, 1970; Ruiz, 1964) or are spontaneously immunologically compromised (Didier et al., 1994) (e.g., *nu/nu* and severe combined immunodeficient mice). In these animals, lethal disease is the usual outcome of either activated spontaneous or experimentally induced infection. Clinically these mice display either massive ascites or a wasting syndrome characterized by inappetance, dehydration, severe weight loss, inactivity, coma, and death. Experimentally, time from infection to death varies according to the organism used and dose delivered, but clinical signs followed by death typically occurred within 2 to 4 weeks after the administration of parasite. Experimental infection can be induced via the peritoneal, oral, or intranasal route of administration. Large numbers of parasites can be recovered from the ascitic fluid. The organisms are typically found in parasitophorous vacuoles in the peritoneal macrophages present in the ascitic fluid.

Ascites = accumulation of serous fluid in the peritoneal cavity.

→ Ascites = common in St Bernards we had

Pathology

With the exception of ascites and occasional hepatosplenomegaly (Morris et al., 1956; Nelson, 1962), there are commonly no obvious gross lesions at necropsy. Microscopically, peritonitis, nephritis, and encephalitis are the commonly occurring lesions in both spontaneously and experimentally infected mice. Additional lesions, however, can be seen in the pancreas, adrenal, lung, liver, and spleen depending on the route of inoculation and the organism employed. Splenomegaly is reported in some cases and has been variously attributed to either lymphoid hyperplasia or diffuse enlargement of the splenic red pulp (Shadduck and Pakes, 1971). Weiser (1965) observed organisms in heart and omentum (Weiser, 1965).

Microscopic lesions consist of focal granulomas typically found in the brain and/or kidney. In the brain lesions are commonly composed of loose aggregations of activated microglia and are accompanied by small numbers of lymphocytes and macrophages (Innes et al., 1962). The classic large epithelioid cells and central caseous necrosis seen in *Encephalitozoon*-induced granulomas of the rabbit brain are rare or nonexistent in mice. There is often slight nonsuppurative diffuse meningitis and occasional perivascular cuffs. Lesions do not have a distinct tissue distribution within the central nervous system. Diffuse lymphocytic interstitial nephritis with occasional poorly formed cortical or medullary granulomas is the typical renal lesion. In addition, enlargement with subsequent rupture and exfoliation of renal tubular epithelial cells is a common finding. The amount and severity of renal epithelial cell loss are directly related to the intensity of the infection. In immunologically intact mice, cerebral and renal lesions generally are mild.

Often it is difficult to detect parasites but, when found, they are present either in the lesions or focally in parasitophorous vacuoles quite distinctly separated from the areas of inflammation. A search for organisms is usually accomplished most successfully by concentrating on tubular epithelial cells in the renal cortex and corticomedullary junction. Spores of *E. cuniculi* are gram-positive, and so the use of properly differentiated tissue Gram stains greatly assists in detecting and identifying parasites (Moller, 1968). This technique is not equally valuable in all experimental infections. *Vittaforma corneae,* for example, stains poorly with the Gram technique, but the spores of all *Encephalitozoon* species studied to date are intensely gram-positive.

In immunologically compromised mice the infections may be massive. Large numbers of organisms accompanied by massive necrosis and moderate to severe lymphocytic inflammatory infiltrates characterize this disease (Fig. 1). In these situations, organisms and lesions may be found in most major organs and tissues. As noted above, parasites and lesions have been detected in the brain, kidney, liver, lung, spleen, adrenal, heart, pancreas, intestine, and both visceral and parietal peritoneum. Classical granulomatous inflammatory foci are rare; however, they may occur if experimentally infected animals are allowed to recover with the aid of antiparasitic drugs.

MICROSPORIDIOSIS IN RATS, GUINEA PIGS, AND HAMSTERS

Spontaneous encephalitozoonosis also has been reported in laboratory rats (Attwood and Sutton, 1965; Majeed and Zubaidy, 1982; Petri, 1966), guinea pigs (Moffatt and Schiefer, 1973), and hamsters (Meiser et al., 1971). Clinical signs and gross lesions are essentially absent in immunologically intact animals, and we are unaware of reports of encephalitozoonosis in these animals resulting from immunosuppression. Histopathologically organisms are identified incidentally, typically in the kidney and brain (Attwood and Sutton, 1965), with hepatic lesions and organisms reported in rats (Canning, 1965; Majeed and Zubaidy, 1982). Lesions consist of focal cerebral, renal, or hepatic granulomas and occasionally diffuse lymphocytic interstitial nephritis (Canning, 1967; Majeed and Zubaidy, 1982). Classic cerebral granulomas are detected more frequently in the rat than in the mouse and consist of a collection of epithelioid cells, central necrosis,

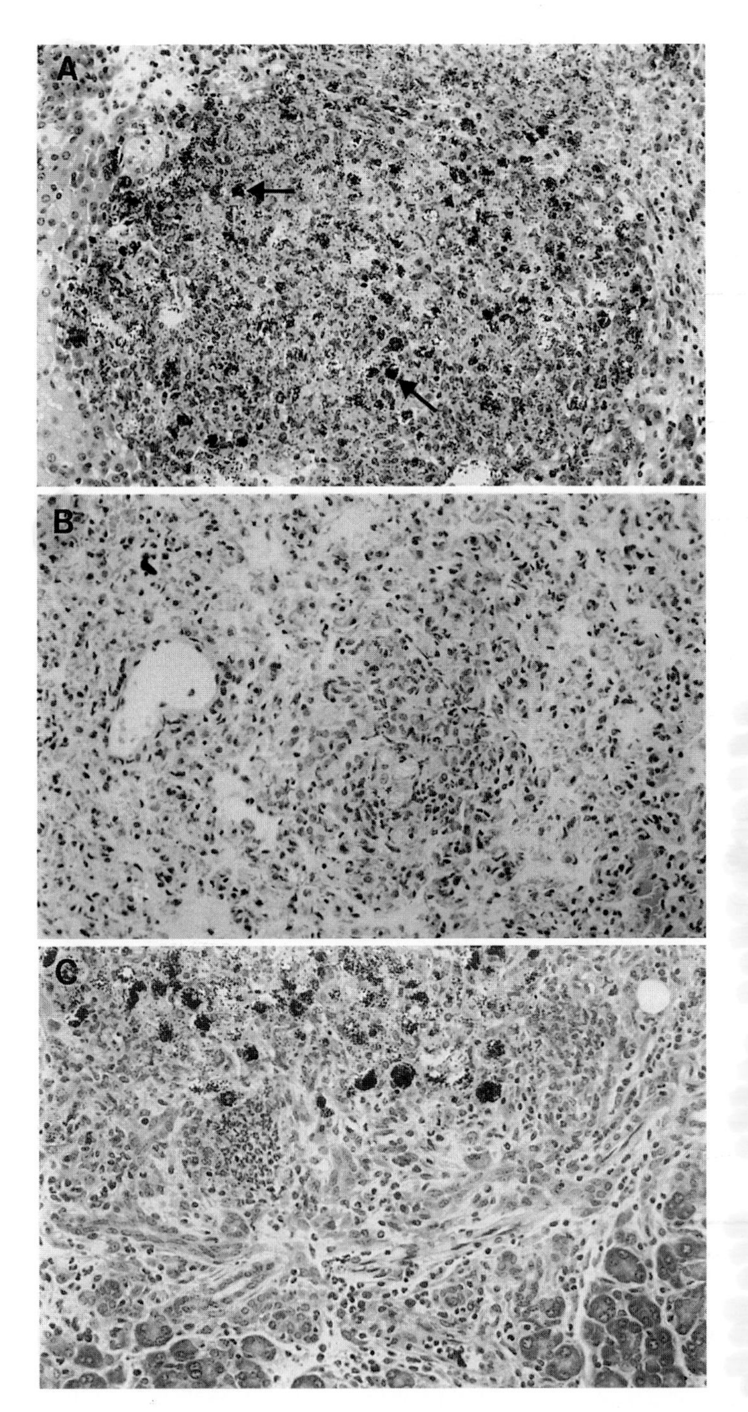

FIGURE 1 Histologic lesions in immunodeficient mouse tissues. (A) Adrenal medulla of an athymic mouse experimentally infected with *Encephalitozoon hellem*. There is massive necrosis of the adrenal medullary cells with collapse of the normal architecture and infiltration of macrophages with a few lymphocytes. Many macrophages and adrenal medullary cells are infected with organisms (arrows). Gram stain was used. (B) Lung of an athymic mouse experimentally infected with *E. hellem*. Moderate interstitial pneumonitis and a granuloma composed of macrophages and lymphocytes characterize this lesion. Hematoxylin and eosin stain was used. (C) Pancreas of an athymic mouse experimentally infected with *E. hellem*. There is massive necrosis and collapse of the exocrine pancreas with masses of neutrophils. Pancreatic exocrine cells and macrophages are filled with parasites. Gram stain was used.

giant cells, and lymphocytic infiltrates at the periphery (Majeed and Zubaidy, 1982). Lesions of the spinal cord have also been described in infected rats. Organisms were reported by Canning in the walls of large blood vessels in a heavily infected rat, and in the same report he described lesions of the lung with parasites in the epithelial cells of the bronchioles and the squamous epithelium of the alveoli (Canning, 1967). Reports on rats describe the contamination of transplantable tumors with alteration in the growth pattern of the tumor but no clinical signs in the infected rats (Chalupsky et al., 1982; Petri, 1966). Other reports mention seropositivity in colonies of laboratory rats without accompanying clinical signs (Gannon, 1980). Similarly, in guinea pigs the infection has been detected serologically and is also seen as an incidental finding during histopathologic examination of these animals (Chalupsky et al., 1979; Gannon, 1980). Microgranulomas with epithelioid cells also are observed in guinea pig brain (Moffatt and Schiefer, 1973). In one animal, vasculitis of the brain was detected, but focal gliosis was the only lesion found in other guinea pigs (Moffatt and Schiefer, 1973).

In hamsters, the existing reports describe contamination of a transplantable tumor line without accompanying clinical signs. Serologically positive hamster colonies have also been reported (Chalupsky et al., 1979).

MICROSPORIDIOSIS IN WILD RODENTS

There are a few reports of microsporidiosis in wild rodents. Arctic lemmings were observed to have clinically detectable disturbances of the central nervous system, including circling and paralysis accompanied by disseminated granulomatous encephalitis with the presence of organisms resembling *Encephalitozoon* (Cutlip and Beall, 1989). Organisms were found in the vascular endothelial cells in many tissues accompanied by minimal or no inflammatory response (Cutlip and Denis, 1993). Lesions also were seen in the spleen, kidney, heart, lung and liver.

In muskrats, there is a single report of cerebral granulomas unaccompanied by detectable clinical signs (Woebeser and Schuh, 1979). There is a single description of wild meadow voles (*Apodemus sylvaticus*) infected with a pansporoblastic microsporidian, *T. apodemi* (Doby et al., 1963; Weiser, 1993). Organisms were detected in smears of the brain as spherical or ovoid colonies with no tissue reaction around them or as free octospores in small granulomata. No clinical signs were observed.

MICROSPORIDIOSIS IN RABBITS

Introduction

Several reports published in the early 1920s described an encephalitis syndrome in the domestic rabbit and recognized microsporidial organisms in histologic lesions (Wright and Craighead, 1922; Goodpasture, 1924; Doerr and Zdansky, 1923). The organism associated with the neurologic disease was named *E. cuniculi* by Levaditi et al. (1923). This scientific name was used for a number of years for the parasites identified in rabbits and other rodent hosts. In 1964 it was suggested that the genus be changed to *Nosema* because of similarities to organisms identified in arthropods, and for a period of time the scientific name *Nosema cuniculi* was widely used (Lainson et al., 1964). In 1971 Shadduck and Pakes (1971) proposed a return to the original nomenclature, and *E. cuniculi* has subsequently been the name used for the rabbit and rodent parasite species.

In addition to encephalitis, chronic renal disease in rabbits due to *E. cuniculi* has also been well characterized (Flatt and Jackson, 1970; Testoni, 1974). Encephalitozoonosis is now widely recognized as a naturally occurring parasite in domestic rabbits. It frequently exists as a chronic, latent infection, and only a percentage of infected animals develop clinical disease.

The prevalence of *Encephalitozoon* infection in rabbits varies widely. On the basis of a histologic examination of tissues, Goodpasture (1924) reported that 10 of 30 rabbits had brain lesions, while Perrin (1943) reported that 0 of 50 rabbits had lesions. In another study, 100 of 2,338 rabbits (4.3%) raised for meat had renal lesions (Flatt and Jackson, 1970). In a survey of 100 New Zealand White rabbits maintained in a research colony, 72% of the animals showed renal lesions, and parasites were identified in 9% (Testoni, 1974). Outbreaks as well as endemic asymptomatic infections can also occur. In one study, 15% of 450 New Zealand White rabbits animals at a large breeding farm developed neurologic disease (Pattison et al., 1971). In contrast to the situation in rodent research colonies where *E. cuniculi* infections have been largely eliminated, it is likely that subclinical infections are still widespread in commercial and pet rabbitries and in research colonies where rabbits are not maintained as specific-pathogen-free animals (Bywater et al., 1980).

Diagnosis

A number of methods have been described for diagnosis of *E. cuniculi* infection, and antibody detection procedures are used most frequently. An India ink immunoreaction assay was developed as a simple method for diagnosis of encephalitozoonosis without the need for special equipment (Waller, 1977). This technique was used widely in the late 1970s as a tool for monitoring both natural infections in rabbitries and experimental infections in research facilities (Bywater and Kellett, 1978a, 1978b, 1979; Kellett and Bywater, 1978; Lyngset, 1980; Waller et al., 1978). The indirect immunofluorescent antibody (IFA) detection test is probably the most commonly employed diagnostic test for rabbits at this time, and several variations of the method have been described (Cox et al., 1972; Cox and Gallichio, 1977; Cox and Pye, 1975; Jackson et al., 1974; Waller et al., 1979). This procedure has been used in serological surveys of laboratory-reared rabbits (Singh et al., 1982) and in experimentally infected animals (Kunstyr et al., 1986; Waller et al., 1978; Wosu et al., 1977). An immunoperoxidase test has also been devised as an alternative to the IFA test when appropriate microscopic facilities are not available (Cox et al., 1981; Gannon, 1978), and a complement fixation method has also been reported (Wosu et al. 1977). Most comparisons among the various tests suggest a similar sensitivity in detecting humoral responses of the rabbit to the parasite (Kellett and Bywater, 1980; Pakes et al., 1984). In rabbits experimentally infected by the intravenous or oral route, significant serological responses were detected 2 to 4 weeks after infection and persisted for at least 3 to 6 months (Cox et al., 1979).

In addition to serological methods, intradermal skin testing for detection of encephalitozoonosis has also been described (Pakes et al., 1972). In one study, intradermal testing showed similar sensitivity to IFA testing in experimentally infected rabbits (Wosu et al., 1977). In another study comparing a variety of serological tests and the intradermal skin test, the latter test correlated best with histologic lesions in experimentally infected rabbits (Pakes et al., 1984).

Active parasite infections can also be diagnosed by detection of the spore form of the parasite in the urine of rabbits. Spores may be directly observed microscopically, particularly after Gram or acid-fast staining of the urine sediment, or indirectly by infecting tissue culture monolayers with urine sediment (Goodman and Garner, 1972; Pye and Cox, 1977). In one study, organisms were excreted in the urine of about 25% of serologically positive animals (Cox and Pye, 1975). In experimentally infected rabbits, spores were shed in high numbers from day 31 to day 63, and shedding continued as long as 98 days after infection (Cox et al., 1979).

Transmission

Experimental infections have been established by administering cultured parasites by the intravenous, oral, intratracheal, intraperitoneal, or intracerebral route (Cox et al., 1979; Pakes et al., 1972; Perrin, 1943; Wosu et al., 1977). It is generally accepted that the most common route of transmission is an oral one since spores are passed in large numbers in the urine of infected rabbits (Cox et al, 1972; Owen and Gannon, 1980). There are conflicting reports concerning the possibility that infections can be transmitted transplacentally (Hunt et al., 1972; Owen and Gannon, 1980; Wilson, 1986). It is probable that the parasite can be transmitted in this manner, but it is unlikely that it is an important means of transmission in natural infections.

The possibility of zoonotic transmission of *E. cuniculi* from rabbits to humans has been discussed in the literature. Bywater (1979) suggested that the organism should not be considered a human pathogen zoonotically transmitted from animal hosts. However, the frequency with which *E. cuniculi* is reported in immunocompromised humans is increasing because of the increase in human immunodeficiency virus infection. With the molecular technique of restriction fragment length polymorphism analysis of the small-subunit rRNA gene (small-subunit rDNA), three of six human isolates and nine rabbit isolates were found to be identical (Deplazes et al., 1996a). The possibility of zoonotic transmission of *E. cuniculi* between rabbits and humans needs further study.

Clinical Disease

Clinical signs of disease are rarely observed, but signs of cerebral dysfunction are occasionally reported in spontaneously infected rabbits (Pattison et al., 1971; Shadduck and Pakes, 1971). Animals have been reported to be dull and inactive and sometimes to have a head tilt. Disease may result in death or continue chronically with a progression of clinical signs including ataxia, paresis, and anterior or posterior paralysis (Goodpasture, 1924; Pattison et al., 1971). Although nephritis has been well documented histologically, clinical signs of renal disease have not generally been reported (Flatt and Jackson, 1970; Testoni, 1974).

Pathology

Few gross lesions have been detected in either asymptomatic or clinically ill rabbits. Occasionally macroscopic changes, including depressed pale areas on the cortical surface of the kidney and adhesion of the renal capsule, have been observed (Flatt and Jackson, 1970; Shadduck and Pakes, 1971; Testoni, 1974).

In spontaneous infections, the most commonly reported histologic lesions are granulomatous and nonsuppurative encephalitis and nephritis (Fig. 2). Characteristic focal lesions in the brain, meninges, and sometimes the spinal cord include areas of central necrosis surrounded by epithelioid cells and a mononuclear infiltrate (Cox and Gallichio, 1978; Goodpasture, 1924; Malherbe and Munday, 1958; Shadduck and Pakes, 1971; Wright and Craighead, 1922). Parasites may or may not be found in the lesions as clusters of intracellular organisms or as small numbers of individual parasites. In severe cases, marked perivascular cuffing with a mononuclear infiltrate has been found in the brain and meninges.

Kidney lesions included mononuclear infiltration and necrosis in both the renal tubules and adjacent interstitial areas (Flatt and Jackson, 1970; Testoni, 1974). In severe cases, the kidney architecture is altered when degenerated renal tubules are replaced by connective tissue in focal areas. Parasites are identified both as groups of intracellular organisms and as individual spores in renal tubular epithelium. Occasionally parasites or granulomatous foci are found in glomeruli or in perivascular areas, and lesions have also been described in the spleen, liver, and myocardium of spontaneously infected animals (Wright and Craighead, 1922). In the acute stage of experimental infections, parasites and nonsuppurative lesions were histologically identified in the lung, kidney, and liver, while lesions were limited to the brain, kidney, and occasionally heart in chronically infected animals (Cox et al., 1979).

MICROSPORIDIOSIS IN WILD RABBITS

Since domestic rabbits are common hosts for *E. cuniculi* infection, several studies have investigated whether microsporidial infections are also found in wild rabbits and hares. A single case of histologically confirmed microsporidial infection in a single wild-caught cottontail rabbit has been reported in the scientific literature (Jungherr, 1955). After serological screening, three wild rabbits (*Oryctolagus cuniculus*) collected in Scotland were reported positive based on the India ink immunoreaction test (Wilson, 1979). Subsequently, sera from 175 wild rabbits were tested by IFA, and no positive animals were identified (Cox and Ross, 1980). In a similar study, no positive animals were found among 823 and 57 rabbits collected in Australia and New Zealand, respectively (Cox et al., 1980).

MICROSPORIDIOSIS IN DOGS

Introduction

Early reports describing neurologic disease in dogs and "parasites" in histologic tissues did not specifically identify microsporidia but mentioned canine distemper virus and/or toxoplasmosis as possible pathogens (Kantorowicz and Lewy, 1923; Perdrau and Pugh, 1930; Peters and Yamagiwa, 1936). The first confirmed reports of microsporidial infections based on parasite morphologic characteristics were in dogs from eastern and southern Africa. In 1952

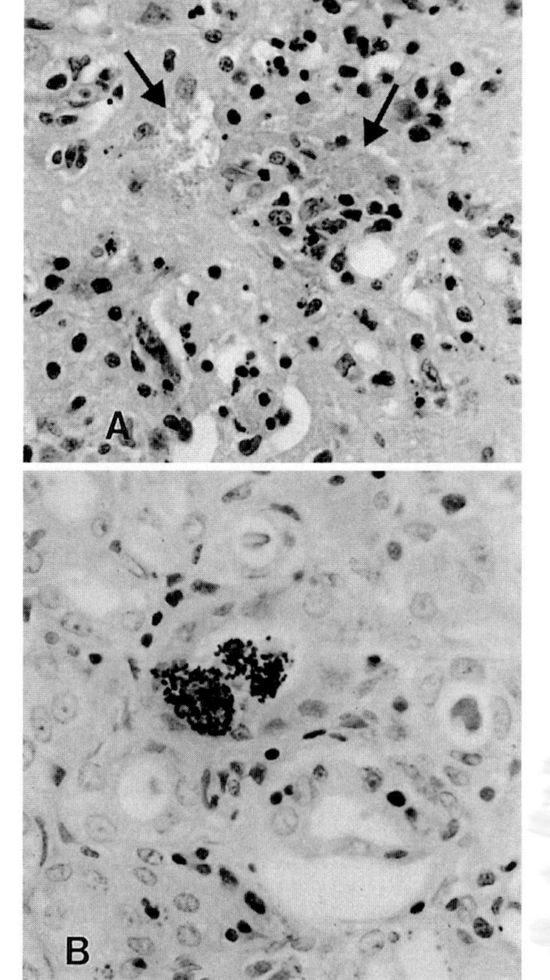

FIGURE 2 Histologic lesions in domestic rabbit tissues. (A) Cerebral cortex from a rabbit spontaneously infected with *Encephalitozoon cuniculi.* Focal necrosis, lymphocytic infiltration, activated microglia, and masses of organisms (arrows) characterize the lesion. Hematoxylin and eosin stain was used. (B) Kidney of a rabbit spontaneously infected with *E. cuniculi.* Masses of parasites are present either free in the lumen or in parasitophorous vacuoles of renal tubular epithelial cells. There is very little interstitial inflammation. Gram stain was used.

Plowright observed an encephalitis-nephritis syndrome in two litters of puppies and identified *E. cuniculi* on the basis of histologic findings (Plowright, 1952; Plowright and Yeoman, 1952). In a 1966 report from the Republic of South Africa, Basson et al. discussed similar clinical and histopathologic features in a litter of puppies but used the nomenclature *Nosema cuniculi* (Basson et al., 1966). Canine encephalitozoonosis has subsequently been found most frequently in South Africa (reviewed by Botha et al., 1979; Stewart and Botha, 1989), but infections have been reported in other geographic locations including England, Germany, the United States, Zimbabwe, and Tanzania (Botha et al., 1979; Van Dellen et al., 1989).

The first reported spontaneous *E. cuniculi* infection of dogs in the United States occurred in a litter of beagle puppies in 1978, and a canine isolate was first propagated in vitro with primary canine kidney cell cultures and canine embryo fibroblast cultures (Shadduck et al., 1978). The following year, *E. cuniculi* was propagated in primary canine kidney cell cultures from puppies from three unrelated litters, and in vitro cultures were successfully established and cryopreserved in a canine kidney (MDCK) cell line (Stewart et al., 1979b).

Historically, the identification of canine isolates of *E. cuniculi* was based on morphologic and ultrastructural characteristics. The species identity and strain of canine microsporidial isolates have been confirmed as *E. cuniculi* with the use of molecular techniques. E. S. Didier et al. (1995, 1996) amplified the internal transcribed spacer (ITS) region of the rRNA gene and used restriction endonuclease digestion and heteroduplex mobility shift analysis or DNA sequencing to compare parasite isolates from rabbit, mouse, human, and dog sources (E. S. Didier et al., 1995, 1996). The human isolate and two dog isolates (type III) were identical and differed from rabbit (type I) and rodent (type II) isolates based on the number of four base pair repeats in the internal transcribed spacer region. Subsequently, a third dog isolate established from one of several infected Maltese puppies was identified as *E. cuniculi* type III (Snowden and Didier, unpublished data). The epidemiologic significance of these observations and the potential for zoonotic transmission of the *E. cuniculi* need to be studied more extensively.

One additional report provides data addressing the issue of zoonotic potential. Vervet monkeys (*Cercopithecus pygerythrus*) were experimentally infected with *E. cuniculi* of dog origin, and subclinical disease was identified histopathologically in monkey tissues, suggesting the possibility of zoonotic transmission *E. cuniculi* (Van Dellen et al., 1989).

Diagnosis

Both serological and parasitologic methods have been used to diagnose microsporidial infections in dogs. An immunofluorescent serodiagnostic test, first described in 1979, has been employed in several serological surveys (Stewart et al., 1979a). In one study, dogs became serologically positive between 32 and 39 days after infection, and 70% of 50 dogs from infected kennels were seropositive (Stewart et al., 1981). In the same report, 18% of 220 random dog sera from South Africa were positive. In another South African study, 12 of 52 dogs with chronic renal disease were positive for antibodies to *E. cuniculi,* while only 2 of 42 control dogs were serologically positive (Stewart et al., 1986b). In contrast, in a more recent survey in Switzerland, 212 healthy dogs were serologically negative, and 102 fecal samples were negative for spores (Deplazes et al., 1996b). In experimental infections, dogs developed antibodies within 7 days of infection and remained seropositive for the duration of the experiment (370 days) (Stewart et al., 1986a).

Detecting parasites in fecal or urine samples is a common diagnostic method in immunocompromised humans, and the technique may also be useful in detecting current infection in dogs. In one study clinically ill puppies shed large numbers of parasite spores in their urine (Stewart et al., 1986b), while in another study no parasites were observed on repeated urinalysis (Botha et al., 1986). Parasites have been histologically identified in the kidney of asymptomatic adult dogs (usually bitch mothers of infected puppies), as well as in clinically ill puppies, and so it is logical that asymptomatic animals may also shed spores (McInnes and Stewart, 1991). In a recent clinical case, two surviving clinically ill puppies and one asymptomatic puppy from a litter of eight Boston terriers shed spores in their urine and feces for an extended period of time (Snowden, unpublished data). The sire and dam of the litter also shed parasites in their urine and feces although they remained subclinically infected.

Transmission

Both oral and transplacental routes of infection have been described. Botha et al. (1979) established infections in four of five dogs after feeding them infected tissues from naturally infected dogs. The experimental dogs developed no clinical disease but demonstrated histopathologic lesions. It is probable that many young puppies develop clinical disease after in utero infection from their asymptomatic dams (McInnes and Stewart, 1991; Plowright, 1952).

Clinical Disease

The most common clinical manifestation of canine microsporidial infection is an encephalitis-nephritis syndrome in puppies from 4 to 10 weeks of age that often results in death (Basson et al., 1966; Botha et al., 1979; Plowright, 1952; Plowright and Yeoman, 1952; Shadduck et al., 1978; Stewart and Botha, 1989; Stewart et al., 1979b). Neurologic signs may include behavior changes, repeated vocalizing, loss of awareness of surroundings, weakness and incoordination progressing to rear limb ataxia, tremors, convulsions, and/or blindness over a period of several days, as well as anorexia and weight loss. Dehydration and mucopurulent ocular discharge and conjunctivitis are sometimes observed (Botha et al., 1979). Clinical disease is most frequently confused with neurologic signs associated with canine distemper virus infections, as well as rabies or systemic toxoplasmosis. Most puppies succumb to the infection over the course of a few days, but some animals slowly recover. Entire litters may become clinically ill, or one or more puppies may display clinical signs of varying severity while their littermates remain asymptomatic. Bitches producing infected litters consistently are asymptomatic (McInnes and Stewart, 1991).

While neurologic disease in puppies is the most frequent clinical scenario, chronic nephritis eventually resulting in renal failure has occasionally been reported in young adult dogs (Botha et al., 1979; Stewart and Botha, 1989). These animals are usually emaciated, and the kidneys may be palpably enlarged or fibrotic and shrunken in size.

Pathology

Limited hematologic or clinical chemistry data have been included in case reports of naturally infected dogs. In one case, no abnormalities were reported on hemograms, urinalysis, or blood urea nitrogen tests performed on naturally infected puppies (Shadduck et al., 1978). In other cases, low hemoglobin and/or hematocrit values were reported, while leukograms were variable (Botha et al., 1979; Szabo and Shadduck, 1987). In one report, serum alkaline phosphatase enzyme levels were consistently elevated (Botha et al., 1979). Impaired renal function, indicated by elevated blood urea nitrogen or serum creatinine levels, has been observed in clinically ill puppies and in asymptomatic bitches (Botha et al., 1979; McInnes and Stewart, 1991). In one study, experimentally infected young adult dogs were monitored for 18 months. Dogs occasionally showed lower packed cell volume (PCV) and modest lymphocytosis during that time period, but the only significant laboratory results were consistently elevated antibody responses to parasite spores (Szabo and Shadduck, 1988).

The carcasses of most puppies are in poor or emaciated condition, often with distended abdomens (Botha et al., 1979; Snowden, unpublished observations). Gross abnormalities are described in the brain and kidney, with lesions in other organs occasionally mentioned. In some puppies with the encephalitis-nephritis syndrome, the brain and meninges showed no macroscopic lesions (Botha et al., 1979;

Plowright, 1952). In other cases, brain congestion or swelling of the gyri was found, as well as edematous distention or hemorrhagic foci on the meninges (Basson et al., 1966; Szabo and Shadduck, 1987). Typically, kidneys are bilaterally enlarged and pale and frequently have radial streaks extending through the renal cortex. Adhesion of the renal capsule, irregular renal surfaces, petechial hemorrhages, and disseminated or focal pinpoint white spots have also been observed (Botha et al., 1979; McInnes and Stewart, 1991; Plowright, 1952; Szabo and Shadduck, 1987). The kidneys of older, chronically infected dogs are shrunken, with irregular cortical surfaces and an adhered capsule, symptoms indicative of chronic renal disease (Botha et al., 1979). Lymphadenopathy, splenomegaly, gastric mucosal petechiation, and disseminated pinpoint white lesions in the hepatic parenchyma have also been found (Basson et al., 1966; Plowright, 1952; Szabo and Shadduck, 1987).

The most consistently described histologic lesions are cerebral vascular and renal tubular abnormalities in puppies (Fig. 3A to C). Lesions are varied and progressive depending on the age of the animal and the duration and severity of clinical disease. In puppies with neurologic clinical signs, vasculitis and perivascular nonsuppurative inflammatory lesions with plasma cells, lymphocytes, and macrophages are consistently demonstrated in various regions of the brain (Botha et al., 1979; McCully et al., 1978; Plowright and Yeoman, 1952; Shadduck et al., 1978; Szabo and Shadduck, 1987; VanDellen et al., 1978). Mild to marked gliosis and multifocal necrosis were also observed in the neuropile. Parasites can be visualized at sites of inflammation and vasculitis, particularly in vascular endothelial cells lining small vessels and sometimes in areas lacking inflammation. Mononuclear cellular infiltrate and thickening of the meninges are also often found in puppies.

Histologic lesions in the kidney are often seen as interstitial lymphoplasmacytic and granulomatous nephritis, often starting at the corticomedullary junction and extending into the cortex (Basson et al., 1966; Botha et al., 1979; McCully et al., 1978; Plowright, 1952; Plowright and Yeoman, 1952; Shadduck et al., 1978; Szabo and Shadduck, 1987). Inflammation and necrosis are often identified in the renal tubules, and parasites are frequently observed in tubular epithelial cells and within tubular lumens and sometimes in glomerular endothelium. In chronic cases in older dogs, renal tubular fibroplasia and dilation or obstruction, sometimes with a plasma cell infiltrate, have been found (Botha et al., 1979). In some cases, Bowman's capsule is thickened and glomeruli are atrophied, illustrating the end-stage renal disease occasionally described in adult dogs.

In some clinically ill dogs, multifocal granulomatous hepatitis has been observed with organisms in hepatocytes and sometimes in vascular endothelium of small to medium-sized vessels (Botha et al., 1979; McCully et al., 1978; Plowright and Yeoman, 1952). Mild multifocal myocarditis was occasionally reported (Botha et al., 1979). Ocular lesions have also been noted, including a lymphoplasmacytic infiltration of the iris (Botha et al., 1979).

MICROSPORIDIOSIS IN BLUE FOXES

Introduction

A series of reports in the 1970s and early 1980s characterized microsporidial infections in farmed blue foxes (*Alopex lagopus*) in the Scandinavian countries (Arnesen and Nordstoga, 1977; Mohn et al., 1974, 1981; Nordstoga and Westbye, 1976). Breeding colonies on fur farms in Nordic countries sometimes suffer heavy animal losses, resulting in significant economic losses to the industry. *E. cuniculi,* also called *N. cuniculi* in some early reports, was the parasite species identified morphologically and ultrastructurally as the cause of two disease syndromes. Young animals develop polyarteritis nodosa, a systemic disease characterized by severe disseminated vasculitis in the small and medium sized arteries of many organs (Nordstoga and Westbye, 1976). This disease may be accompanied by ocular vasculitis and the

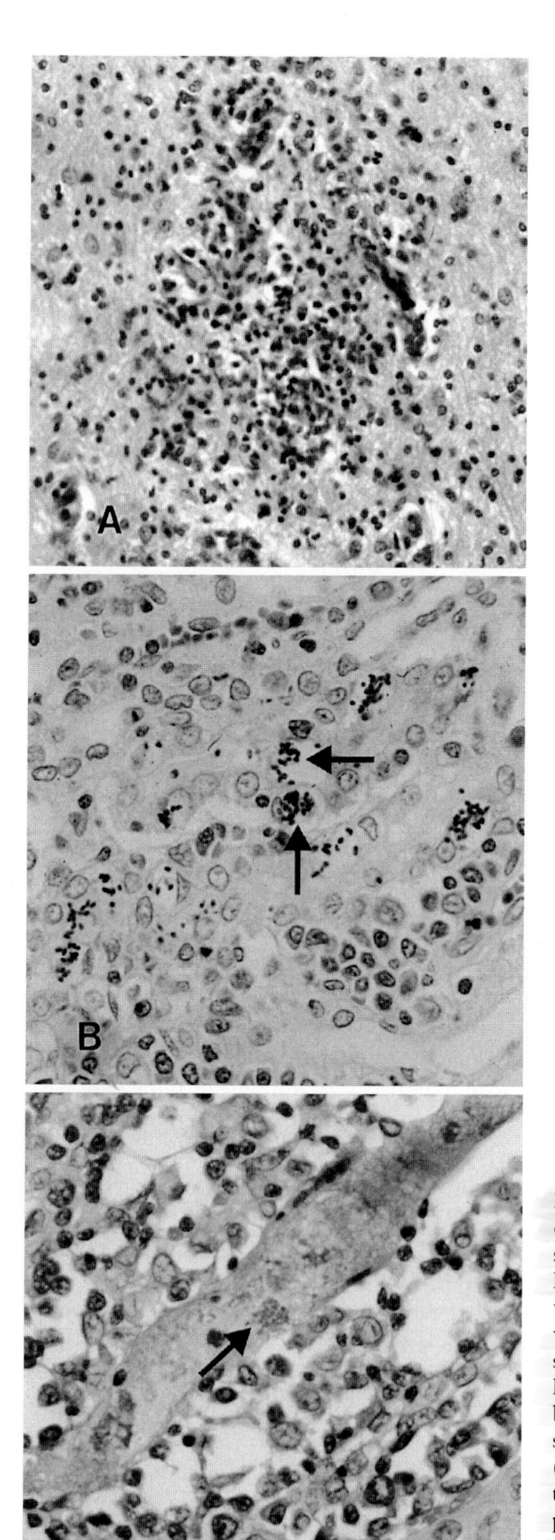

FIGURE 3 Histologic lesions in dog tissues. (A) Brain of a dog spontaneously infected with *E. cuniculi*. There is severe lymphocytic perivascular infiltration with diffuse lymphocytic encephalitis. Hematoxylin and eosin stain was used. (B) Kidney from a dog spontaneously infected with *E. cuniculi*. There is a moderate lymphocytic interstitial infiltrate accompanying severe disruption of tubular architecture. Note the large number of organisms both free in the tubular lumens and contained in parasitophorous vacuoles in the renal tubular epithelial cells (arrows). Gram stain was used. (C) Kidney of a dog spontaneously infected with *E. cuniculi*. There is severe lymphocytic interstitial nephritis with collapse and loss of renal tubules. Tubular epithelial cells are disrupted and heavily infected with organisms (arrow). Hematoxylin and eosin stain was used.

development of lens cataracts, resulting in vision deficits (Arnesen and Nordstoga, 1977).

With the use of molecular methods, analysis of the ITS region of the RNA gene from a number of *E. cuniculi* isolates indicated multiple types or strains of the parasite (Didier et al., 1995). Typically, rabbit, rodent, and human or dog isolates had two, three, or four repeats, respectively, of a 5′-GTTT-3′ sequence, and these patterns were designated types I, II, and III. Three isolates from foxes showed a type II identity, similar to that of most rodent isolates (Mathis et al., 1996). Additional immunoblotting data suggested subtle antigenic and protein differences in the fox isolates.

Diagnosis

Two types of serological testing for specific antibody responses in foxes have been described, including an India ink immunoreaction assay and an indirect IFA test (Mohn, 1982b; Mohn and Odegaard, 1977). Tests were reported to have similar sensitivities, and experimentally infected pups and vixens all became serologically positive. In natural infections, clinically ill pups seroconvert if they live long enough, while asymptomatic puppies may or may not become serologically positive. These data suggest that some pups may not become infected by either the vertical or the lateral mode of transmission even if they associate with infected siblings (Mohn and Nordstoga, 1982).

Transmission

Blue foxes have been experimentally infected with *E. cuniculi* by oral and intrauterine injection routes (Mohn and Nordstoga, 1982; Mohn et al., 1982b; Nordstoga et al., 1978). Vixens rarely develop clinical disease, but they maintain subclinical infections that can be passed transplacentally to offspring as long as a year after initial infection (Mohn et al., 1974; Nordstoga et al., 1978). The major route of infection of young animals that develop clinical disease appears to be transplacental. Typically most or all pups in a litter become ill and die while the vixen remains asymptomatic (Arnesen and Nordstoga, 1977). Lateral transmission among young pups has also been documented in naive pups that seroconverted after being housed with experimentally infected pups who were presumably shedding parasite spores (Mohn and Nordstoga, 1982). Puppies that were orally infected failed to develop clinical disease.

Clinical Disease

Most reports describe systemic disease in fox pups from 4 weeks to 5 months old. In both experimental and natural infections, animals frequently show anorexia and stunted growth, a dull attitude, and sometimes abnormal thirst (Mohn, 1982a; Nordstoga and Westbye, 1976). Various neurologic signs such as ataxia, circling and head tilt, lameness, paresis of the hindlimbs, and convulsions are also observed (Mohn, 1982a; Mohn and Nordstoga, 1982; Nordstoga and Westbye, 1976). Sometimes reduced vision or blindness is also reported.

Pathology

Several hematologic abnormalities are frequently observed in clinically ill and asymptomatic infected pups, including leukocytosis and mild normocytic hypochromic anemia (Mohn, 1982a). Serum electrophoresis shows a slight elevation in total protein values, with a mild decrease in albumin levels but marked hypergammaglobulinemia (Mohn, 1982a; Mohn and Nordstoga, 1975). Clinically ill pups may have elevated urea nitrogen and creatinine values and hypermagnesemia reflecting renal dysfunction (Mohn, 1982a). Experimentally infected vixens showed minimal changes in a variety of hematologic and clinical chemistry tests over a period of months when compared with uninfected females (Mohn et al., 1982a).

Nordstoga and Westbye (1976) reviewed the clinical and pathologic features of a large number of naturally infected pups from multiple fur farms and described the polyarteritis nodosa syndrome. In young animals, gross lesions included prominent distorted coronary arteries with thickened, nodular lesions on the surface of the heart, and sometimes the pericardial sac contained blood or fibrinous exudate. Frequently, the spleen and lymph nodes were enlarged, and the kidneys were pale and enlarged, sometimes with circumscribed pale spots on the external surfaces. Often the meninges were hyperemic or showed multifocal hemorrhages. Similar arterial lesions and renal enlargement were described in some experimentally infected pups, while no gross lesions were identified in others (Mohn and Nordstoga, 1982). Histologic lesions focus on vascular alterations seen most frequently in the brain, cardiac arteries, and eye (Arnesen and Nordstoga, 1977; Mohn and Nordstoga, 1982; Nordstoga, 1972; Nordstoga and Westbye, 1976). In the brain, multifocal gliosis and mononuclear cuffing of blood vessels were the most prominent lesions. Sometimes small vessels also had polymorphonuclear or eosinophilic inflammatory infiltrates. Parasites were frequently identified in vascular endothelium, and the associated meninges were often thickened and infiltrated with mononuclear inflammatory cells.

If animals survived long enough, superficial coronary vessels in the myocardium and arteries of other muscles and organs showed marked fibrocellular thickening characteristic of polyarteritis nodosa disease (Nordstoga, 1972). The fibrotic lesions included mononuclear inflammatory cells and frequently eosinophils forming periarterial granulomas. Mononuclear interstitial nephritis progressing to fibrosis has been observed in the kidney depending on the length of the disease process.

In the eye, arterial lesions affecting the posterior ciliary arteries, the small arteries of the iris, sclera, and retina, have also been described (Arnesen and Nordstoga, 1977). Changes in the lens also caused cataracts, resulting in vision deficits and blindness in chronically infected animals.

In one study, vixens with reproductive problems and subclinical infections had histologic lesions of metritis (Mohn et al., 1982a). The lamina propria of the uterine wall was thickened and infiltrated with mononuclear inflammatory cells.

MICROSPORIDIOSIS IN NONHUMAN PRIMATES

Introduction

Only a few cases of naturally occurring microsporidiosis have been reported in nonhuman primates. Several genera of monkeys have been found to have with microsporidial infections from two different parasite genera. Additionally, experimental infections have been established in three monkey genera for three different parasite species from two genera.

NATURAL INFECTIONS WITH *ENCEPHALITOZOON CUNICULI*

A single case report describes a fatal intestinal infection in an adult female South American titi monkey (*Callicebus moloch cupreus*) (Seibold and Fussell, 1973). Over a 4-week period, the animal became anorexic and lost weight before succumbing to systemic bacterial infection. The most striking histologic feature of the infection was total desquamation of jejunal epithelium which was heavily infected with clusters of microsporidial organisms. The size, morphology, staining, and ultrastructural characteristics were compatible with *E. cuniculi* (then named *N. cuniculi*).

Several reports describe natural *E. cuniculi* infections in squirrel monkeys (*Saimiri sciureus*). In the first case, a 1-month-old orphan male squirrel monkey developed seizures and lost weight over a 1-month period before death (Brown et al., 1973). Histologic findings included meningitis and perivascular cuffing in the brain, as well as multifocal granulomatous lesions in the liver and kidney. Gram-positive organisms were identified in all three tissues. The ultrastructural features of the organisms were compatible with *E. cuniculi*.

In a second case, a stillborn, premature male squirrel monkey was diagnosed with encephalitozoonosis as indicated by histologic lesions (Anver et al., 1972). Granulomatous encephalitis with associated acid-fast organisms was present in all regions of the brain. No lesions were observed in other organs. The ultrastructural features of the organism were compatible with *E. cuniculi*.

In a more recent report, a series of 22 cases of naturally occurring encephalitozoonosis in squirrel monkeys from a breeding colony were described (Zeman and Baskin, 1985). Most of these animals were young, including 9 animals that were less than 4 weeks old. Most animals were asymptomatic, and microsporidial infections were diagnosed histologically. All animals had vasculitis and multifocal granulomas in the brain. Granulomatous lesions with or without parasites were common in the meninges, lung, kidney, and adrenal and were occasionally identified in heart and liver.

A single report describes organisms morphologically identical to *E. cuniculi* that grew in primary cell cultures of baboon (*Papio* sp.) placenta (Shadduck et al., 1979). The donor animal and other members of the primate colony were serologically negative. Parasite-containing materials were preserved before the organisms could be identified, and so no further analysis was possible.

EXPERIMENTAL INFECTIONS WITH *ENCEPHALITOZOON* SPP.

Several attempts have been made to experimentally infect primate hosts. Experimental administration of *E. cuniculi* to vervet monkeys (*Cercopithecus pygerythrus*) resulted in subclinical infections (Van Dellen et al., 1989). Nongravid and late pregnant adult monkeys and infants were infected by the oral or intravenous route with a parasite isolate derived from a natural fatal infection in a dog. Subclinical infection was documented in all the monkeys on the basis of histologic lesions. Granulomatous lesions of varying intensity were identified most consistently in the liver, kidney, and brain and resembled those in dogs. Horizontal and vertical transmission of infection were documented.

Experimental infections with *E. cuniculi* and the closely related *Encephalitozoon hellem* (Didier et al., 1991) have also been established in rhesus macaque monkeys (*Macaca mulatta*) (Didier et al., 1994; P. J. Didier et al., 1996). Both healthy and immunocompetent simian immunodeficiency virus (SIV)-infected monkeys were infected with organisms by the intravenous or oral route. The immunocompetent monkeys developed serological responses to the appropriate parasite species but displayed no clinical signs of disease and rarely shed organisms in their feces or urine. SIV-infected immunocompetent monkeys also seroconverted, but as T-cell levels fell, parasite shedding increased. These monkeys died of wasting disease associated with the viral infection, but microsporidia were not identified histologically in their tissues. A third group of immunocompromised SIV-infected monkeys were infected with *E. hellem* and shed parasites in their urine and feces. These animals succumbed to wasting disease, and animals infected by an intravenous route had disseminated histologic lesions in the liver and kidney.

INFECTIONS WITH *ENTEROCYTOZOON BIENEUSI*

The microsporidial parasite *Enterocytozoon bieneusi,* previously reported only in humans, was recently identified in three species of macaques (*Macaca mulata, M. nemestrina,* and *M. cyclopis*) (Mansfield et al., 1997). The macaques were experimentally infected with the immunosuppressive SIV and had low CD4 T-cell counts. They demonstrated chronic hepatobiliary infection with proliferative cholecystitis. Parasites were identified in the gallbladder and common bile duct and sometimes in the upper small intestine with Gram-stained histologic sections, immunohistochemistry, and in situ hybridization on formalin-fixed, paraffin-embedded tissues. Ultrastructural studies showed organisms with *E. bieneusi* characteristics, a single nucleus and a polar filament with five to seven coils in double rows. Polymerase chain reaction amplification of parasites from frozen tissues and DNA sequencing was used to speciate the parasites. This organism had previously been identified only in humans and is the most common microsporidial infection in AIDS patients.

Therefore, the same research group tried to establish experimental infections in rhesus macaques. SIV-infected monkeys were orally infected with *E. bieneusi* spores from a human patient, and chronic infections were established (Tzipori et al., 1997). Monkeys shed spores for prolonged times, and when the animals were terminated and histologically evaluated, parasites could be detected in the biliary tract and jejunum by in situ hybridization.

MICROSPORIDIOSIS IN OTHER MAMMALIAN HOSTS

Microsporidial infections have been described in several other mammals, usually in single clinical reports.

Natural Infections in Domestic Animal Hosts

A single case report described *E. cuniculi* infection in a near-term draft horse fetus in South Africa (VanRensburg et al., 1991). Significant macro- and microscopic lesions were identified in the kidney. Grossly the kidneys were congested, and the renal cortex had a mottled appearance. Histologic evaluation revealed severe diffuse interstitial lymphoplasmacytic nephritis with numerous microsporidial organisms in glomeruli and tubular epithelial cells. Electron microscopy showed spores with the characteristic *Encephalitozoon* morphology containing a single nucleus and a polar filament with five to seven coils. No lesions or parasites were observed in the placenta. Although the dam had aborted previously and several associated mares had had pregnancy problems that breeding season, microsporidial infections were not detected in other foals or adult horses on biopsy or serologically.

A single case report described microsporidial infection in 1 of 122 kidneys collected from goats routinely slaughtered in India (Khanna and Iyer, 1971). Grossly the kidney had multiple 1- to 2-mm pale spots on the cortical surface and pale radial striations on cut surface, similar to the appearance of gross renal lesions in other hosts (Botha et al., 1979; Plowright and Yeoman, 1952). Focal interstitial nephritis with appropriately sized gram-positive organisms was described histologically, and the parasite was identified as *N. cuniculi* (current nomenclature, *E. cuniculi*).

In a survey in England, serum samples from 40 sheep were tested for antibody responses to *E. cuniculi* (then called *N. cuniculi*) (Singh et al., 1982). Thirty-eight (95%) of the animals were serologically positive with low titers; however, the relevance of these observations has not been further evaluated.

Fecal samples from domestic pigs and a number of other animal hosts were screened for the presence of microsporidian spores (Deplazes et al., 1996b), and spores were detected in samples from four pigs. DNA analysis of the ribosomal intergenic spacer region showed 97% homology with the enteric microsporidian, *E. bieneusi*. This parasite species is the most common microsporidian in immunocompromised humans, and this case was the first report of this parasite in a nonprimate host.

VanRensburg et al. (1971) described neurologic disease in three Siamese cat littermates from South Africa. Kittens demonstrated "spasms and twitching of muscles and depression." Histopathologic examination of one kitten showed disseminated *N. cuniculi* infection (current nomenclature, *E. cuniculi*). The most significant lesions were nonpurulent meningoencephalitis, interstitial nephritis, and systemic vasculitis. Parasites were observed in the parenchyma of the brain, kidney, spleen, and lymph node, and in the tunica media of numerous blood vessels in most organs. This encephalitis-nephritis syndrome is similar to that observed in young dogs (Plowright, 1952).

Another report describes ocular lesions in a cat cornea, with corneal opacities and inflammation of the cornea, conjunctiva, and anterior uvea (Buyukmihci et al., 1977). The organism was identified as *E. cuniculi;* however, the morphologic features (1 by 4 μm in size) and ultrastructural features (15 to 16 coils in the polar filament) reported in the article are not compatible with this parasite species. In a review, Canning and Lom (1986) briefly discussed morphologic and nomenclature discrepancies in the original report and suggested a new name, *Microsporidium buyukmihcii,* based on the morphologic features.

Microsporidial infection does not appear to be common in cats. In a survey in Switzerland, 1 cat of 45 animals with ocular lesions showed a specific antibody reaction (Deplazes et al., 1996b). In the same study, 45 feline fecal samples were found to be negative for parasite spores.

In a recent survey, *Encephalitozoon* spores were identified in fecal samples from 19 of 172 mammalian and 16 of 99 avian hosts by a monoclonal antibody-based immunofluorescent assay (Bornay-Llinares et al., 1998). Ten samples from eight animal hosts were further analyzed by additional staining methods, electron microscopy, and PCR amplification of a portion of the small-subunit rRNA gene. By using these methods, the microsporidia were identified as *E. intestinalis* in a total of six samples from a goat, a cow, a dog, a donkey, and two pigs. This is the first report of this microsporidian species in nonhuman hosts.

Natural Infections in Wild Carnivores

Several reports describe microsporidial infections in a selection of wild carnivores maintained as captive or zoological exhibition animals.

Encephalitozoonosis was diagnosed in two groups of wild dog (*Lycaon pictus*) pups whelped in captivity and maintained at a South African zoological park (VanHeerden et al., 1989). In both groups, 8- or 11-week-old pups developed clinical neurologic disease about 2 weeks after vaccination for canine distemper virus. Puppies demonstrated ataxia and incoordination, paresis, seizures, and excessive vocalization before they succumbed to the infection or were humanely terminated. Abnormal clinical pathologic test results for at least one pup included elevated protein levels in the cerebrospinal fluid, an elevated blood urea nitrogen level, mild hypergammaglobulinemia, and leukocytosis. Gross lesions at the time of necropsy included pustules and papules on inguinal and abdominal skin, cerebral edema, and meningeal congestion, and pups in one group had pulmonary edema, hepatomegaly, enteritis, and gastritis. The most striking histopathologic findings included multifocal mononuclear inflammation associated with *Encephalitozoon* sp.-like organisms in the brainstem, cerebrum, cerebellum, and kidney. The species of microsporidia was not identified, but the morphologic appearance and clinical encephalitis-nephritis presentation strongly resembled that of *E. cuniculi* infection in dogs.

Microsporidial disease caused by organisms morphologically compatible with *E. cuniculi* has been observed in three genera of carnivores at the Prague Zoo (Vávra and Blazek, 1971). Neurologic disease and high mortality were reported in a colony of suricates (meerkats) (*Suricata suricatta,* Carnivora, Viverridae) over a period of years. The major histologic lesions were meningoencephalitis with numerous parasites identified as *N. cuniculi* (current nomenclature, *E. cuniculi*). Encephalitis and nephritis were also reported in two clouded leopards (*Neofelis nebulosa,* Carnivora, Felidae) on the basis of histologic findings.

Microsporidial disease was observed in a litter of Siberian polecats (*Mustela eversmanii satunini,* Carnivora, Mustelidae) maintained at a research laboratory after importation from Siberia (Novilla et al., 1980). Seven of 11 kits died at about 2 weeks of age, several days after the onset of illness, while two littermates died of other causes and two were asymptomatic. Concurrent infections with a *Staphylococcus* bacterial infection and *Hepatozoon* sp. protozoa were identified on autopsy. Microsporidia were histologically identified in the brain, liver, and lung of the kits. The ultrastructure of the organism was compatible with *E. cuniculi.*

In a case report of a single red fox (*Vulpes vulpes*) reared in England, gram-positive intracellular organisms were histologically identified in brain and small intestinal muscle sections (Wilson, 1979). The identity of the parasite was presumed to be *E. cuniculi.* In a serological survey of numerous animal hosts in Switzerland, 86 wild red foxes were serologically negative for antibodies against *E. cuniculi* (Deplazes et al., 1996b).

MICROSPORIDIOSIS IN AVIAN HOSTS

Introduction

Only a few cases of natural microsporidial infections in birds are reported in the scientific literature. These reports describe single animal infections or isolated flock outbreaks, and all cases reported so far are in psittacine birds. Peach-faced lovebirds (*Agapornis roseicollis*) are the most frequently reported hosts (Branstetter and Knipe, 1982; Lowenstine and Petrak, 1980; Norton and Prior, 1994; Novilla and Kwapien, 1978), followed by masked lovebirds (*Agapornis personata*) (Kemp and Kluge, 1975; Powell et al., 1989; Randall et al., 1986). Microsporidiosis has also been observed in a mixed flock of black-masked and Fischer's lovebirds (*Agapornis fischeri*) (Randall et al., 1986), in a flock of budgerigars (*Melopsittacus undulatus*) (Black et al., 1997), in a single double yellow-headed Amazon parrot (*Amazona ochrocephala*) (Poonacha et al., 1985), and in two unrelated eclectus parrots (*Eclectus roratus*) (Pulparampil et al., 1998). In a survey for microsporidia, no spores were detected in fecal samples from 39 birds (Deplazes et al., 1996b), however, the species of the birds were not identified in this report.

Until recently, the species of microsporidia in avian tissues had not been determined, and no parasite species had been categorized as an avian pathogen (Canning and Lom, 1986). In three reports describing infections in lovebirds, the organisms were simply identified as members of the phylum Microspora, and the genus or species of parasite was not defined (Norton and Prior, 1994; Novilla and Kwapien, 1978; Randall et al., 1986). In four reports, the morphologic and ultrastructural description of the parasites suggested that they were members of the genus *Encephalitozoon* (Kemp and Kluge, 1975; Lowenstine and Petrak, 1980; Poonacha et al., 1985; Powell et al., 1989). In another report, Black et al. (1997) identified microsporidia in an outbreak in a commercial pet bird aviary, where death losses occurred in young budgerigar chicks. The molecular Southern blot method was used on DNA extracted from chick tissues, and the organisms were identified as *E. hellem*. By molecular sequencing of a portion of the small-subunit rRNA gene, *E. hellem* has also been identified in archival tissues of two eclectus parrots (Pulparampil et al., 1998). This microsporidian species was first described in 1991 and had previously been identified only in immunocompromised humans (Didier et al., 1991). The transmission, epidemiology, and zoonotic potential of *E. hellem* need to be explored.

Clinical Disease

Limited descriptions of the clinical course of infections in birds are included in the literature. In lovebirds, anorexia and loss of condition typically were observed for several days before birds succumbed to the infection (Norton and Prior, 1994; Novilla and Kwapien, 1978; Randall et al., 1986). In young budgerigar chicks, ill-thrift, dehydration, and diarrhea followed rapidly by death were reported (Black et al., 1997). In the Amazon parrot case, the animal had a 2-week history of decreased appetite and weight loss, diarrhea, and respiratory difficulty and was unresponsive to antibiotic therapy (Poonacha et al., 1985).

Pathology

In a limited number of descriptions of gross lesions in lovebirds, an enlarged liver and/or kidney was reported, sometimes mottled with small, pale foci (Novilla and Kwapien, 1978; Powell et al., 1989; Randall et al., 1986). In the Amazon parrot, an enlarged, congested liver was observed, as well as thickened air sacs containing a yellow-white exudate (Poonacha et al., 1985).

The kidney, liver, and small intestine are the most frequently reported locations of histologic lesions and parasites in several bird species. Less frequently, lesions are also found in spleen, bile duct, and lung. In the kidney, the renal tubules were dilated and often contained parasites; renal tubular epithelium was extensively damaged and frequently contained intracellular parasites (Black et al., 1997; Kemp and Kluge, 1975; Norton and Prior, 1994; Novilla and Kwapien, 1978; Poonacha et al.,

TABLE 2 Spontaneous microsporidioses in amphibians and reptiles

Host	Parasite	Reference(s)	Comment
Bufo bufo (common toad)	*Pleistophora myotrophica*	Canning and Elkan, 1963; Canning et al., 1964; Elkan, 1963	
Bufo vulgaris (B. bufo)	*Pleistophora bufonis*	Guyenot and Ponse, 1926, referenced in Canning et al., 1964	
Bufo marinus (cane toad)	*Alloglugea bufonis*	Paperna and Lainson, 1995	
Rana pipiens (common frog)	Unidentified microsporidian	Schuetz et al., 1978	Parasites affected growth of frog oocytes
Rana temporaria (frog)	*Glugea (Pleistophora?) danilewskyi*	Guyenot and Naville, 1920, referenced in Canning et al., 1964	
Common newt	*Nosema tritoni*	Weiser, 1960, referenced in Canning et al., 1964	
Spenodon punctatus (tuatara [primitive reptile resembling a lizard])	*Pleistophora* sp.	Liu and King, 1971	
Tropidonotus natrix (grass snake)	*Glugea (Pleistophora?) danilewskyi*	Debaisieux, 1919, referenced in Canning et al., 1964	Canning discusses the history of this report

1985; Powell et al., 1989; Randall et al., 1986). Inflammatory cellular infiltrate was not often described in infected kidneys. In the liver, hepatic necrosis was reported with multifocal or periportal damage to hepatocytes with a varying intensity of accompanying inflammation (Novilla and Kwapien, 1978; Powell et al., 1989; Randall et al., 1986). In a smaller number of cases, parasites and lesions were primarily identified in the small intestine. In budgerigar chicks and in a peach-faced lovebird, parasites were localized to the enterocytes of the duodenal mucosa with little associated inflammatory change (Black et. al., 1997; Norton and Prior, 1994). In both cases "megabacteriosis" was reported in the gastrointestinal tract, as well as microsporidial infection. In contrast, in the eclectus parrot cases, parasites were primarily located in the lamina propria of the small intestine rather than in the more superficial epithelium (Puparampil et al., 1998, submitted).

The ultrastructural characteristics of microsporidian parasites identified in all reported avian cases were similar (Black et al., 1997; Branstetter and Knipe, 1982; Kemp and Kluge, 1975; Norton and Prior, 1994; Powell et al., 1989). Each approximately 1- to 3-μm oval spore contained a single nucleus surrounded by a typical electron-dense outer wall and an inner electron-lucent wall. The lamellar polarplast and coiled polar filament with five to seven turns were characteristic of the genus *Encephalitozoon.*

MICROSPORIDIOSIS IN AMPHIBIANS AND REPTILES

Only a few cases of naturally occurring microsporidiosis, including several different parasite genera, are reported in amphibian or reptile hosts. A brief summary of references appears in Table 2.

REFERENCES

Anver, M. R., N. W. King, and R. D. Hunt. 1972. Congenital encephalitozoonosis in a squirrel monkey (*Saimiri sciureus*). *Vet. Pathol.* **9:**475–480.

Arison, R. N., J. A. Cassaro, and M. P. Pruss. 1966. Studies on a murine ascites-producing agent and its effect on tumor development. *Cancer Res.* **26:** 1915–1929.

Arnesen, K., and K. Nordstoga. 1977. Ocular encephalitozoonosis (nosematosis) in blue foxes. *Acta Ophthalmol.* **55:**641–651.

Attwood, H. D., and R. D. Sutton. 1965. *Encephalitozoon granulomata* in rats. *J. Pathol. Bacteriol.* **89:** 735–738.

Basson, P. A., R. M. McCully, and W. E. J. Warnes. 1966. Nosematosis: report of a canine case in the Republic of South Africa. *J. S. Afr. Vet. Med. Assoc.* **37:**3–9.

Bismanis, J. E. 1970. Detection of latent murine nosematosis and growth of *Nosema cuniculi* in cell cultures. *Can. J. Microbiol.* **16:**237–242.

Black, S. S., L. A. Steinohrt, D. C. Bertucci, L. B. Rogers, and E. S. Didier. 1997. *Encephalitozoon hellem* in budgerigars (*Melopsittacus undulatus*). *Vet. Pathol.* **34:**189–198.

Boot, R., F. VanKnapen, B. C. Kruijt, and H. C. Walvoort. 1988. Serological evidence for *Encephalitozoon cuniculi* infection (nosemiasis) in gnotobiotic guinea pigs. *Lab. Anim.* **22:**337–342.

Bornay-Llinares, F. J., A. J. da Silva, H. Moura, D. A. Schwartz, G. S. Visvesvara, N. J. Pieniazek, A. Cruz-Lopez, P. Hernandez-Jauregui, J. Guerrero, and F. J. Enriquez. 1998. Immunologic, microscopic and molecular evidence of *Encephalitozoon intestinalis (Septata intestinalis)* infection in mammals other than man. *J. Infect. Dis.* **178:**820–826.

Botha, W. S., I. C. Dormehl, and D. J. Goosen. 1986. Evaluation of kidney function in dogs suffering from canine encephalitozoonosis by standard clinical pathological and radiopharmaceutical techniques. *J. S. Afr. Vet. Med. Assoc.* **57:**79–86.

Botha, W. S., A. F. Van Dellen, and C. G. Stewart. 1979. Canine encephalitozoonosis in South Africa. *J. S. Afr. Vet. Med. Assoc.* **50:**135–144.

Branstetter, D. G. and S. M. Knipe. 1982. Microsporan infection in the lovebird *Agapornis roseicollis*. *Micron* **13:**61–62.

Brown, R. J., D. K. Hinkle, W. P. Trevethan, J. L. Kupper, and A. E. McKee. 1973. Nosematosis in a squirrel monkey (*Saimiri sciureus*). *J. Med. Primatol.* **2:**114–123.

Buyukmihci, N., R. W. Bellhorn, J. Hunziker, and J. Clinton. 1977. *Encephalitozoon* (*Nosema*) infection of the cornea in a cat. *J. Am. Vet. Med. Assoc.* **171:**355–357.

Bywater, J. E. C. 1979. Is encephalitozoonosis a zoonosis? *Lab. Anim.* **13:**149–151.

Bywater, J. E. C., and B. S. Kellett. 1978a. *Encephalitozoon cuniculi* antibodies in a specific- pathogen-free rabbit unit. *Infect. Immun.* **21:**360–364.

Bywater, J. E. C., and B. S. Kellett. 1978b. The eradication of *Encephalitozoon cuniculi* from a specific pathogen-free rabbit colony. *Lab. Anim. Sci.* **28:**402–404.

Bywater, J. E. C., and B. S. Kellett. 1979. Humoral immune response to natural infection with *Encephalitozoon cuniculi* in rabbits. *Lab. Anim.* **13:** 293–297.

Bywater, J. E. C., B. S. Kellett, and T. Waller. 1980. *Encephalitozoon cuniculi* antibodies in commercially available rabbit antisera and serum reagents. *Lab. Anim.* **14:**87–89.

Canning, E. U. 1965. An unusually heavy natural infection of *Nosema cuniculi* (Levaditi et al.) in a laboratory rat. *Trans. R. Soc. Trop. Med. Hyg.* **59:**371.

Canning, E. U. 1967. Vertebrates as hosts to microsporidia, with special reference to rats infected with *Nosema cuniculi*. *J. Protozool.* **2:**197–205.

Canning, E. U., and E. Elkan. 1963. Microsporidiosis in *Bufo bufo*. *Parasitology* **53:**11p. (Abstract.)

Canning, E. U., E. Elkan, and P. I. Trigg. 1964. *Pleistophora myotrophica* spec. nov., causing high mortality in the common toad *Bufo bufo* L., with notes on the maintenance of *Bufo* and *Xenopus* in the laboratory. *J. Protozool.* **11:**157–166.

Canning, E. U., and J. Lom. 1986. *The Microsporidia of Vertebrates.* Academic Press, New York.

Chalupsky, J., J. Vávra, and P. Bedrnik. 1979. Encephalitozoonosis in laboratory animals: a serological survey. *Folia Parasitol.* **26:**1–8.

Chalupsky, J., J. Vávra, Z. Zuskova, M. Melka, and P. Bedrnik. 1982. The occurrence of *Encephalitozoon cuniculi* (EC) in Yoshida rat ascites sarcoma (YAS) and the attempt to eliminate EC from this tumor. *J. Protozool.* **29:**504.

Cox, J. C., and H. A. Gallichio. 1977. An evaluation of indirect immunofluorescence in the serological diagnosis of *Nosema cuniculi* infection. *Res. Vet. Sci.* **22:**50–52.

Cox, J. C., and H. A. Gallichio. 1978. Serological and histological studies on adult rabbits with recent, nautrally acquired encephalitozoonosis. *Res. Vet. Sci.* **24:**260–261.

Cox, J. C., R. C. Hamilton, and H. D. Attwood. 1979. An investigation of the route and progression of *Encephalitozoon cuniculi* infection in adult rabbits. *J. Protozool.* **26:**260–265.

Cox, J. C., R. Horsburgh, and D. Pye. 1981. Simple diagnostic test for antibodies to *Encephalitozoon cuniculi* based on enzyme immunoassay. *Lab. Anim.* **15:**41–43.

Cox, J. C., and D. Pye. 1975. Serodiagnosis of nosematosis by immunofluorescence using cell-culture-grown organisms. *Lab. Anim.* **9:**297–304.

Cox, J. C., D. Pye, J. W. Edmonds, and R. Shepherd. 1980. An investigation of *Encephalitozoon cuniculi* in the wild rabbit. *J. Hyg.* **84:**295–300.

Cox, J. C., and J. Ross. 1980. A serological survey of *Encephalitozoon cuniculi* infection in the wild rabbit in England and Scotland. *Res. Vet. Sci.* **28:** 396.

Cox, J. C., N. B. Walden, and R. C. Nairn. 1972. Presumptive diagnosis of *Nosema cuniculi* in rabbits by immunofluorescence. *Res. Vet. Sci.* **13:**595–597.

Cutlip, R. C., and C. W. Beall. 1989. Encephalitozoonosis in arctic lemmings. *Lab. Anim. Sci.* **39:** 331–333.

Cutlip, R. C., and E. D. Denis. 1993. Retrospective study of diseases in a captive lemming colony. *J. Wildl. Dis.* **29:**620–622.

Deplazes, P., A. Mathis, R. Baumgartner, I. Tanner, and R. Weber. 1996a. Immunologic and molecular characteristics of *Encephalitozoon*-like microsporidia isolated from humans and rabbits indicate that *Encephalitozoon cuniculi* is a zoonotic parasite. *Clin. Infect. Dis.* **22:**557–559.

Deplazes, P., A. Mathis, C. Muller, and R. Weber. 1996b. Molecular epidemiology of *Encephalitozoon cuniculi* and first detection of *Enterocytozoon bieneusi* in faecal samples of pigs. *J. Eukaryot. Microbiol.* **43:**93S.

Didier, E. S., P. J. Didier, D. N. Friedberg, S. M. Stenson, J. M. Orenstein, R. W. Yee, F. O. Tio, R. M. Davis, C. Vossbrinck, N. Millichamp, and J. A. Shadduck. 1991. Isolation and characterization of a new human microsporidian, *Encephalitozoon hellem* (n. sp.), from three AIDS patients with keratoconjunctivitis. *J. Infect. Dis.* **163:**617–621.

Didier, E. S., P. W. Varner, P. J. Didier, A. M. Aldras, N. J. Millichamp, M. Murphey-Corb, R. Bohm, and J. A. Shadduck. 1994. Experimental microsporidiosis in immunocompetent and immunodeficient mice and monkeys. *Folia Parasitol.* **41:**1–11.

Didier, E. S., G. S. Visvesvara, M. D. Baker, L. B. Rogers, D. D. Bertucci, M. A. DeGroote, and C. R. Vossbrinck. 1996. A microsporidian isolated from an AIDS patient corresponds to *Encephalitozoon cuniculi* III, originally isolated from domestic dogs. *J. Clin. Microbiol.* **34:**2835–2837.

Didier, E. S., C. R. Vossbrinck, M. D. Baker, L. B. Rogers, D. C. Bertucci, and J. A. Shadduck. 1995. Identification and characterization of three *Encephalitozoon cuniculi* strains. *Parasitology* **111:** 411–421.

Didier, P. J., E. S. Didier, and M. Murphey-Corb. 1996. Experimental microsporidiosis in immunodeficient monkeys. *FASEB J.* **9:**A5614. (Abstract.)

Doby, J. M., A. Jeannes, and B. Rault. 1963. *Thelohania apodemi* n. sp., première microsporidie du genre *Thelohania* observée chez un mammifère. *C. R. Acad. Sci.* **257:**248–251.

Doerr, R., and E. Zdansky. 1923. Zur aetiologie der encephalitis epidemica. *Schweiz. Med. Wochenschr.* **53:**349–351.

Elkan, E. 1963. A microsporidium affecting the common toad (*Bufo bufo* L.). *Br. J. Herpetol.* **3:**89.

Flatt, R. E., and S. J. Jackson. 1970. Renal nosematosis in young rabbits. *Pathol. Vet.* **7:**492–497.

Gannon, J. 1978. The immunoperoxidase test diagnosis of *Encephalitozoon cuniculi* in rabbits. *Lab. Anim.* **12:**125–127.

Gannon, J. 1980. A survey of *Encephalitozoon cuniculi* in laboratory animal colonies in the United Kingdom. *Lab. Anim.* **14:**91–94.

Goodman, D. G., and F. M. Garner. 1972. A comparison of methods for detecting *Nosema cuniculi* in rabbit urine. *Lab. Anim. Sci.* **22:**568–572.

Goodpasture, E. W. 1924. Spontaneous encephalitis in rabbits. *J. Infect. Dis.* **34:**428–432.

Huldt, G., and T. Waller. 1974. Accidental nosematosis in mice with impaired immunological competence. *Acta Pathol. Microbiol. Scand. Sect. B* **82:**451–452.

Hunt, R. D., N. W. King, and H. L. Foster. 1972. Encephalitozoonosis: evidence for vertical transmission. *J. Infect. Dis.* **126:**212–214.

Innes, J. R. M., W. Zeman, J. K. Frenkel, and G. Borner. 1962. Occult endemic encephalitozoonosis of the central nervous system of mice. *J. Neuropathol. Exp. Neurol.* **21:**519–533.

Jackson, S., R. F. Solorzano, and D. Middleton. 1974. Indirect fluorescent antibody test for *Nosema cuniculi* (*Encephalitozoon*) in rabbits. *J. Am. Vet. Med. Assoc.* **164:**176.

Jungherr, E. 1955. *Encephalitozoon* encephalomyelitis in a rabbit. *J. Am. Vet. Med. Assoc.* **127:**518. (Abstract.)

Kantorowicz, R., and F. M. Lewy. 1923. Neueparasitologische und patholgisch-anatomische Befunde der nervosen Staupe der Hunde. *Arch. Wiss. Prakt. Tierheilkd.* **49:**137–157.

Kellett, B. S., and J. E. C. Bywater. 1978. A modified india-ink immunoreaction for the detection of encephalitozoonosis. *Lab. Anim.* **12:**59–60.

Kellett, B. S., and J. E. C. Bywater. 1980. The indirect india-ink immunoreaction for detection of antibodies to *Encephalitozoon cuniculi* in rat and mouse serum. *Lab. Anim.* **14:**83–86.

Kemp, R. L., and J. P. Kluge. 1975. *Encephalitozoon* sp. in the blue-masked lovebird, *Agapornis personata* (Reichenow): first confirmed report of microsporidian infection in birds. *J. Protozool.* **22:**489–491.

Khanna, R. S., and P. K. R. Iyer. 1971. A case of *Nosema cuniculi* infection in a goat. *Indian J. Med.* **59:**993–995.

Kunstyr, I., L. Lev, and S. Naumann. 1986. Humoral antibody response of rabbits to experimental infection with *Encephalitozoon cuniculi*. *Vet. Parasitol.* **21:**223–232.

Kyo, Y. 1958. *Encephalitozoon* parasitemia: observations on the infestation of *Encephalitozoon* in mice, with special note on its parasitemia. *Nisshin Igaku* **45:** 500–504.

Lainson, R., P. C. C. Garnham, R. Killick-Kendrick, and R. G. Bird. 1964. Nosematosis, a microsporidial infection of rodents and other animals, including man. *Br. Med. J.* **2:**470–472.

Levaditi, C., S. Nicolau, and R. Schoen. 1923. L'étiologie de l'encephalite. *C. R. Acad. Sci.* **177:** 985–988.

Liu, S.-K., and F. W. King. 1971. Microsporidiosis in the tuatara. *J. Am. Vet. Med. Assoc.* **159:**1578–1582.

Lowenstine, L. J., and M. L. Petrak. 1980. Microsporidiosis in two peach-faced lovebirds, p. 365–368. *In* R. J. Montali and G. Migaki (ed.), *Comparative Pathology of Zoo Animals: Proceedings of Symposium of National Zoological Park, Smithsonian Institution.* Smithsonian Institution Press, Washington, D.C.

Lyngset, A. 1980. A survey of serum antibodies to *Encephalitozoon cuniculi* in breeding rabbits and their young. *Lab. Anim. Sci.* **30:**558–561.

Majeed, S. K., and A. J. Zubaidy. 1982. Histopathological lesions associated with *Encephalitozoon cuniculi* (nosematosis) infection in a colony of Wistar rats. *Lab. Anim.* **16:**244–247.

Malherbe, H., and V. Munday. 1958. *Encephalitozoon cuniculi* infection of laboratory rabbits and mice in South Africa. *J. S. Afr. Vet. Med. Assoc.* **29:**241–246.

Mansfield, K. G., A. Carville, D. Shvetz, J. MacKey, S. Tzipori, and A. A. Lackner. 1997. Identification of an *Enterocytozoon bieneusi*-like microsporidian parasite in simian immunodeficiency virus-inoculated macaques with hepatobiliary disease. *Am. J. Pathol.* **150:**1395–1405.

Mathis, A., J. Akerstedt, J. Tharaldsen, O. Odegaard, and P. Deplazes. 1996. Isolates of *Encephalitozoon cuniculi* from farmed blue foxes (*Alopex lagopus*) from Norway differ from isolates from Swiss domestic rabbits (*Oryctolagus cuniculus*). *Parasitol. Res.* **82:**727–730.

McCully, R. M., A. F. Van Dellen, P. A. Basson, and J. Lawrence. 1978. Observations on the pathology of canine microsporidiosis. *Onderstepoort J. Vet. Res.* **45:**75–92.

McInnes, E. F., and C. G. Stewart. 1991. The pathology of subclinical infection of *Encephalitozoon cuniculi* in canine dams producing pups with overt encephalitozoonosis. *J. S. Afr. Vet. Med. Assoc.* **62:**51–54.

Meiser, J., V. Kinzel, and P. Jirovec. 1971. Nosematosis as an accompanying infection of plasmacytoma ascites in Syrian golden hamsters. *Pathol. Microbiol.* **37:**249–269.

Moffatt, R. E., and B. Schiefer. 1973. Microsporidiosis (encephalitozoonosis) in the guinea pig. *Lab. Anim. Sci.* **23:**282–284.

Mohn, S. F. 1982a. Experimental encephalitozoonosis in the blue fox. *Acta Vet. Scand.* **23:**503–514.

Mohn, S. F. 1982b. Encephalitozoonosis in the blue fox: comparison between the India-ink immunoreaction and the indirect fluorescent antibody test in detecting *Encephalitozoon cuniculi* antibodies. *Acta Vet. Scand.* **23:**99–126.

Mohn, S. F., T. Landsverk, and K. Nordstoga. 1981. Encephalitozoonosis in the blue fox: morphological identification of the parasite. *Acta Pathol. Microbiol. Scand.* **89:**117–122.

Mohn, S. F., and K. Nordstoga. 1975. Electrophoretic patterns of serum proteins in blue foxes with special reference to changes associated with nosematosis. *Acta Vet. Scand.* **16:**297–306.

Mohn, S. F., and K. Nordstoga. 1982. Experimental encephalitozoonosis in the blue fox: neonatal exposure to the parasite. *Acta Vet. Scand.* **23:**344–360.

Mohn, S. F., K. Nordstoga, and I. W. Dishington. 1982a. Experimental encephalitozoonosis in the blue fox: clinical, serological, and pathological examinations of vixens after oral and intrauterine inoculation. *Acta Vet. Scand.* **23:**490–502.

Mohn, S. F., K. Nordstoga, J. Krogsrud, and A. Helgebostad. 1974. Transplacental transmission of *Nosema cuniculi* in the blue fox (*Alopex lagopus*). *Acta Pathol. Microbiol. Scand.* **82:**299–300.

Mohn, S. F., K. Nordstoga, and O. M. Moller. 1982b. Experimental encephalitozoonosis in the blue fox: transplacental transmission of the parasite. *Acta Vet. Scand.* **23:**211–220.

Mohn, S. F., and O. A. Odegaàrd. 1977. The indirect fluorescent antibody test (IFAT) for the detection of *Nosema cuniculi* antibodies in the blue fox (*Alopex lagopus*). *Acta Vet. Scand.* **18:**290–292.

Moller, T. 1968. A survey on toxoplasmosis and encephalitozoonosis in laboratory animals. *Z. Versuchstierk. Bd.* **10:**27–38.

Morris, J. A., J. M. C. McCowen, and R. E. Blount. 1956. Ascites and hepatosplenomegaly in mice associated with protozoan-like cytoplasmic structures. *J. Infect. Dis.* **98:**306–311.

Nelson, J. B. 1962. An intracellular parasite resembling a microsporidian associated with ascites in swiss mice. *Proc. Soc. Exp. Biol. Med.* **109:**714–717.

Nordstoga, K. 1972. Nosematosis in blue foxes. *Nord. Vet. Med.* **24:**21–24.

Nordstoga, K., S. F. Mohn, J. Aamdal, and A. Helgebostad. 1978. Nosematosis (encephalitozoonosis) in a litter of blue foxes after intrauterine injection of nosema spores. *Acta Vet. Scand.* **19:** 150–152.

Nordstoga, K., and K. Westbye. 1976. Polyarteritis nodosa associated with nosematosis in blue foxes. *Acta Pathol. Microbiol. Scand. Sect. A* **84:**291–296.

Norton, J. H., and H. C. Prior. 1994. Microsporidiosis in a peach-faced lovebird (*Agapornis roseicollis*). *Aust. Vet. J.* **71:**23–24.

Novilla, M. N., J. W. Carpenter, and R. P. Kwapien. 1980. Dual infection of Siberian polecats with

Encephalitozoon cuniculi and *Hepatozoon mustelis* n. sp., p. 353–363. *In* R. J. Montali and G. Migaki (ed.), *Comparative Pathology of Zoo Animals: Proceedings of a Symposium of National Zoological Park, Smithsonian Institution.* Smithsonian Institution, Washington, D.C.

Novilla, M. N., and R. P. Kwapien. 1978. Microsporidian infection in the pied peach-faced lovebird (*Agapornis roseicollis*). *Avian Dis.* **22:**198–204.

Owen, D. G., and J. Gannon. 1980. Investigation into the transplacental transmission of *Encephalitozoon cuniculi* in rabbits. *Lab. Anim.* **14:**35–38.

Pakes, S. P., J. A. Shadduck, D. B. Feldman, and J. A. Moore. 1984. Comparison of tests for the diagnosis of spontaneous encephalitozoonosis in rabbits. *Lab. Anim. Sci.* **34:**356–359.

Pakes, S. P., J. A. Shadduck, and R. G. Olsen. 1972. A diagnostic skin test for encephalitozoonosis (nosematosis) in rabbits. *Lab. Anim. Sci.* **22:**870–877.

Paperna, I., and R. Lainson. 1995. *Alloglugea bufonis* nov. gen., nov. sp. (Microsporea: Glugeidae), a microsporidian of *Bufo marinus* tadpoles and metamorphosing toads (Amphibia: Anura) from Amazonian Brazil. *Dis. Aquat. Org.* **23:**7–16.

Pattison, M., F. G. Clegg, and A. L. Duncan. 1971. An outbreak of encephalomyelitis in broiler rabbits caused by *Nosema cuniculi. Vet. Rec.* **88:**404–405.

Perdrau, J. R., and L. P. Pugh. 1930. The pathology of disseminated encephalomyelitis of the dog (the "nervous form of canine distemper"). *J. Pathol. Bacteriol.* **33:**79–91.

Perrin, T. L. 1943. Spontaneous and experimental *Encephalitozoon* infection in laboratory animals. *Arch. Pathol.* **36:**559–567.

Peters, G., and S. Yamagiwa. 1936. *Arch. Wiss. Prakt. Tierheilkd.* **70:**138.

Petri, M. 1966. The occurrence of *Nosema cuniculi* (*Encephalitozoon cuniculi*) in the cells of transplantable, malignant ascites tumours and its effect upon tumour and host. *Acta Pathol. Microbiol. Scand.* **66:**13–30.

Plowright, W. 1952. An encephalitis-nephritis syndrome in the dog probably due to congenital *Encephalitozoon infection. J. Comp. Pathol.* **62:**83–93.

Plowright, W., and G. Yeoman. 1952. Probable *Encephalitozoon* infection of the dog. *Vet. Rec.* **64:** 381–383.

Poonacha, K. B., P. D. William, and R. D. Stamper. 1985. Encephalitozoonosis in a parrot. *J. Am. Vet. Med. Assoc.* **186:**700–702.

Powell, S., K. Tang, F. Chandler, D. Parks, and C. Hood. 1989. Microsporidiosis in a lovebird. *J. Vet. Diagn. Invest.* **1:**69–71.

Pulparampil, N., D. Graham, D. Phalen, and K. Snowden. 1998. *Encephalitozoon hellem* infection in eclectus parrots (*Eclectus roratus*): identification from archival tissues. *J. Eukaryot. Microbiol.* **45:**651–655.

Pye, D., and J. C. Cox. 1977. Isolation of *Encephalitozoon cuniculi* from urine samples. *Lab. Anim.* **11:**233–234.

Randall, C. J., S. Lees, R. J. Higgins, and N. H. Harcourt-Brown. 1986. Microsporidian infection in lovebirds (*Agapornis* spp.). *Avian Pathol.* **15:** 223–231.

Ruiz, A. 1964. Ueber die spontaninfektion mit *Encephalitozoon cuniculi* bei weissen maeusen. *Rev. Biol. Trop.* **12:**225–227.

Schuetz, A. W., K. Selman, and D. Samson. 1978. Alterations in growth, function and composition of *Rana pipiens* oocytes and follicles associated with microsporidian parasites. *J. Exp. Zool.* **204:**81–94.

Seibold, H. R., and E. N. Fussell. 1973. Intestinal microsporidiosis in *Callicebus moloch. Lab. Anim. Sci.* **23:**115–118.

Shadduck, J. A., R. Bendele, and G. T. Robinson. 1978. Isolation of the causative organism of canine encephalitozoonosis. *Vet. Pathol.* **15:**449–460.

Shadduck, J. A., G. Kelsoe, and R. J. Helmke. 1979. A microsporidian contaminant of a nonhuman primate cell culture: ultrastructural comparison with *Nosema connori. J. Parasitol.* **65:**185–188.

Shadduck, J. A., and S. P. Pakes. 1971. Spontaneous diseases of laboratory animals which interfere with biomedical research: encephalitozoonosis and toxoplasmosis. *Am. J. Pathol.* **64:**657–674.

Singh, M., G. J. Kane, L. Mackinlay, I. Quaki, Y. E. Hian, H. B. Chuan, H. L. Chuen, and L. K. Chew. 1982. Detection of antibodies to *Nosema cuniculi* (Protozoa: Microsporidia) in human and animal sera by the indirect fluorescent antibody technique. *Southeast Asian J. Trop. Med. Public Health* **13:**110–113.

Snowden, K. Unpublished data.

Snowden, K., and E. S. Didier. Unpublished data.

Stewart, C. G., and W. S. Botha. 1989. Canine encephalitozoonosis. *Zimbabwe Vet. J.* **20:**89–93.

Stewart, C. G., W. S. Botha, and A. F. Van Dellen. 1979a. The prevalence of *Encephalitozoon* antibodies in dogs and an evaluation of the indirect fluorescent antibody test. *J. S. Afr. Vet. Med. Assoc.* **50:** 169–172.

Stewart, C. G., M. G. Collett, and H. Snyman. 1986a. The immune response in a dog to *Encephalitozoon cuniculi* infection. *Onderstepoort J. Vet. Res.* **53:**35–37.

Stewart, C. G., F. Reyers, and H. Snyman. 1986b. The relationship in dogs between primary renal disease and antibodies to *Encephalitozoon cuniculi. J. S. Afr. Vet. Med. Assoc.* **59:**19–21.

Stewart, C. G., A. F. Van Dellen, and W. S. Botha. 1979b. Canine encephalitozoonosis in kennels and the isolation of *Encephalitozoon* in tissue culture. *J. S. Afr. Vet. Med. Assoc.* **50:**165–168.

Stewart, C. G., A. F. Van Dellen, and W. S. Botha. 1981. Antibodies to a canine isolate of *Encephalitozoon* in various species. *S. Afr. J. Sci.* **77:**572.

Szabo, J. R. and J. A. Shadduck. 1987. Experimental encephalitozoonosis in neonatal dogs. *Vet. Pathol.* **24:**99–108.

Szabo, J. R., and J. A. Shadduck. 1988. Immunologic and clinicopathologic evaluation of adult dogs inoculated with *Encephalitozoon cuniculi.* J. Clin. Microbiol. **26:**557–562.

Testoni, F. J. 1974. Enzootic renal nosematosis in laboratory rabbits. *Aust. Vet. J.* **50:**159–163.

Tzipori, S., A. Carville, G. Widmer, D. Kotler, K. Mansfield, and A. Lackner. 1997. Transmission and establishment of a persistent infection of *Enterocytozoon bieneusi,* derived from a human with AIDS, in simian immunodeficiency virus-infected rhesus monkeys. *J. Infect. Dis.* **175:**1016–1020.

Van Dellen, A. F., W. S. Botha, J. Boomker, and W. E. J. Warnes. 1978. Light and electron microscopical studies on canine encephalitozoonosis: cerebral vasculitis. *Onderstepoort J. Vet. Res.* **45:**165–186.

Van Dellen, A. F., C. G. Stewart, and W. S. Botha. 1989. Studies of encephalitozoonosis in vervet monkeys (*Cercopithecus pygerythrus*) orally inoculated with spores of *Encephalitozoon cuniculi* isolated from dogs (*Canis familiaris*). *Onderstepoort J. Vet. Res.* **56:**1–22.

VanHeerden, J., N. Bainbridge, R. E. J. Burroughs, and N. P. J. Kriek. 1989. Distemper-like disease and encephalitozoonosis in wild dogs (*Lycaon pictus*). *J. Wildl. Dis.* **25:**70–75.

VanRensburg, I. B. J., and J. L. duPlessis. 1971. Nosematosis in a cat: a case report. *J. S. Afr. Vet. Med. Assoc.* **42:**327–331.

VanRensburg, I. B. J., D. H. Volkmann, J. T. Soley, and C. G. Stewart. 1991. *Encephalitozoon* infection in a still-born foal. *J. S. Afr. Vet. Med. Assoc.* **62:**130–132.

Vávra, J., and K. Blazek. 1971. Nosematosis in carnivores. *J. Parasitol.* **57:**923–924.

Waller, T. 1977. The india-ink immunoreaction: a method for the rapid diagnosis of encephalitozoonosis. *Lab. Anim.* **11:**93–97.

Waller, T., A. Lyngset, and B. Morein. 1979. Diagnosis of encephalitozoonosis: a simple method for collection of rabbit blood. *Lab. Anim.* **13:**207–208.

Waller, T., B. Morein, and E. Fabiansson. 1978. Humoral immune response to infection with *Encephalitozoon cuniculi* in rabbits. *Lab. Anim.* **12:** 145–148.

Weiser, J. 1965. *Nosema muris* n. sp., a new microsporidian parasite of the white mouse (*Mus musculus* L.). *J. Protozool.* **12:**78–83.

Weiser, J. 1993. Early experiences with microsporidia of man and mammals. *Folia Parasitol.* **40:**257–260.

Wilson, J. M. 1979. *Encephalitozoon cuniculi* in wild european rabbits and a fox. *Res. Vet. Sci.* **26:**114.

Wilson, J. M. 1986. Can *Encephalitozoon cuniculi* cross the placenta? *Res. Vet. Sci.* **40:**138.

Wobeser, G., and J. C. L. Schuh. 1979. Microsporidal encephalitis in muskrats. *J. Wildl. Dis.* **15:** 413–417.

Wosu, N. J., R. G. Olsen, J. A. Shadduck, A. Koestner, and S. P. Pakes. 1977. Diagnosis of experimental encephalitozoonosis of rabbits by complement fixation. *J. Infect. Dis.* **135:**944–948.

Wosu, N. J., J. A. Shadduck, S. P. Pakes, J. K. Frenkel, K. S. J. Todd, and J. D. Conroy. 1977. Diagnosis of encephalitozoonosis in experimentally infected rabbits by intradermal and immunofluorescence tests. *Lab. Anim. Sci.* **27:**210–216.

Wright, J. H., and E. M. Craighead. 1922. Infectious motor paralysis in young rabbits. *J. Exp. Med.* **36:**135–141.

Zeman, D. H., and G. B. Baskin. 1985. Encephalitozoonosis in squirrel monkeys (*Saimiri sciureus*). *Vet. Pathol.* **22:**24–31.

FISH MICROSPORIDIA

Ross W. Shaw and Michael L. Kent

13

About 100 species and 14 genera of the phylum Microsporidia Balbiani, 1882, have been described from fishes, and several cause severe disease. Microsporidia of fishes are widely distributed by both host species and geographic location. Whereas most fish microsporidia are host specific, at least at the genus level, a few (e.g., *Glugea stephani* Hagenmuller 1899) show broad host specificity (Canning and Lom, 1986). Comprehensive lists of fish hosts and microsporidian species appear in Canning and Lom (1986), Lom and Dyková (1992), and Dyková (1995). A general overview of economic importance, immunology, and treatment is followed by a review of the most important genera. Features of pathology and transmission relevant to individual species are included under the genera section. For additional review of transmission please refer to Chapter 3.

The ability of some microsporidia to impact negatively on fish health is well known. Increased mortality in wild fishes such as rainbow smelt (*Osmerus mordax*) (Haley, 1952; Nepszy et al., 1978), young gizzard shad (*Dorsoma cepedianum*) (Putz, 1969; Price, 1982), and freshwater salmonids (Putz et al., 1965; Urawa, 1989) have been attributed to microsporidial infections. In wild fish, declines of entire commercial fisheries have been attributed to microsporidiosis (Mann, 1954; Sindermann, 1966; Ralphs and Matthews, 1986). For example, *Pleistophora macrozoarcides* Nigrelli 1946 has been implicated in the collapse of the North American ocean pout (*Macrozoarces americanus*) fishery (Fischthal, 1944; Sandholzer et al., 1945; Sheehy et al., 1974). Declines in the rainbow smelt fishery of New Hampshire were due in part to infection by *Glugea hertwigi* Weissenberg, 1911 (Haley, 1954). Disfiguring of fish by large cysts or liquefaction of muscle can decrease catch value (Nigrelli, 1946; Grabda, 1978; Egidius and Soleim, 1986; Pulsford and Matthews, 1991). Microsporidiosis can also cause indirect mortality through starvation and reduced growth rates (Matthews and Matthews, 1980; Figueras et al., 1992). Sprengel and Lüchtenberg (1991) found that impairment of swimming ability in European smelt (*Osmerus eperlanus*) infected by *Pleistophora ladogensis* Voronin, 1978, may contribute to higher mortality due to predation. Wilkund et al. (1996)

Ross W. Shaw, Department of Zoology, University of British Columbia, Vancouver, British Columbia V6T 1Z4, Canada. *Michael L. Kent,* Department of Fisheries and Oceans, Pacific Biological Station, Nanaimo, British Columbia V9R 5K6, Canada.

The Microsporidia and Microsporidiosis (Murray Wittner, editor; Louis M. Weiss, contributing editor), ©1999 American Society for Microbiology, Washington, D.C.

reported that *Pleistophora mirandellae* Vaney and Conte, 1901, which infects gonads, decreased the reproductive capacity in roach (*Rutilus rutilus*). Selective mortality in juvenile stocks can further reduce reproductive capacity. For example, *Microgemma hepaticus* Ralphs and Matthews, 1986, infects primarily the liver of juvenile grey mullet (*Chelon labrosus*) and may contribute to mortality (Ralphs and Matthews, 1986).

An increase in aquaculture on a worldwide basis has led to the description of several new microsporidia and outbreaks of microsporidiosis in recent years. Compared to their wild counterparts, cultured fish are particularly susceptible to microsporidial infections because of high stocking densities. Extremely high mortality has occurred in ornamental fishes as a result of infection by *Glugea* spp. (Lom et al., 1995). *Pleistophora hyphessobryconis* Shäperclaus, 1941 (Fig. 1), well known to fish hobbyists as the source of neon tetra (*Paracheirodon inessi*) disease, is one of the most common parasites of aquarium fishes (Canning and Lom, 1986; Lom et al., 1989). Massive mortality of neon tetras held in aquariums is still not uncommon. Morphological abnormalities and mortality due to microsporidiosis in ornamental zebra danio (*Brachydanio rerio*) have also been reported (Kinkelin, 1980). The zebra fish is now a very important laboratory animal for genetic research, and thus large colonies of this fish are maintained at several universities in Europe and North America. We have also observed this microsporidial infection in zebra fish kept in these colonies. The primary site of infection is the central nervous system, where it is often associated with chronic inflammatory changes (Fig. 2). In addition, we have seen infections and associated severe inflammation in the skeletal muscle surrounding vertebrae and in the kidney. *Heterosporis* spp. can infect cichlids (*Pseudocrenilabrus multicolor*) (Lom et al., 1989), ornamental Siamese fighting fish (*Betta splendens*) (Lom et al., 1993), and angelfish (*Pterophyllum scalare*) (Michel et al., 1989), causing emaciation and distress or severe pathological changes in muscle tissue.

Microsporidia are also important in cultured food fishes. Hatcheries raising salmonids for enhancement of wild stocks or for profit have experienced epizootics of microsporidia (Putz, 1969; Hauck, 1984; Urawa and Awakura, 1994). Mortality and disease have occurred at both freshwater (Markey et al., 1994) and marine (Kent et al., 1989) sites where fish are grown to market size. Chronic low mortality in sea bream (*Sparus arurata*), for example, has been caused by a *Pleistophora* sp. that infects the muscle (Abela et al., 1996). Culture of Japanese eels (*Anguilla japonica*) and ayu (*Plecoglossus altivelis*) is often plagued by infections with *Heterosporis anguillarum* (Hoshina, 1951) (syn. *Pleistophora*

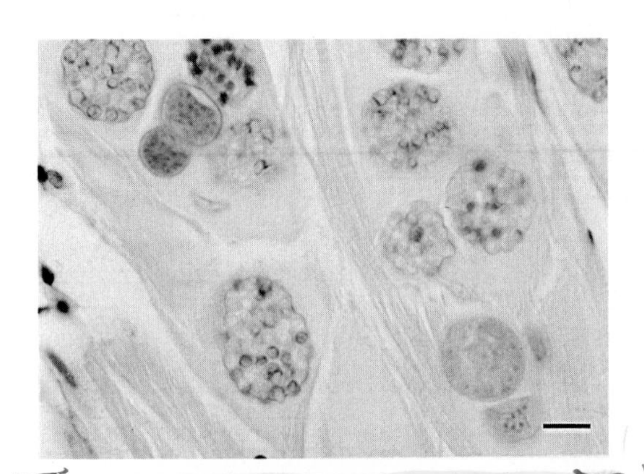

FIGURE 1 *Pleistophora hyphessobryconis* infection in skeletal muscle of a neon tetra (*Paracheirodon inessi*). Scale bar, 14 μm.

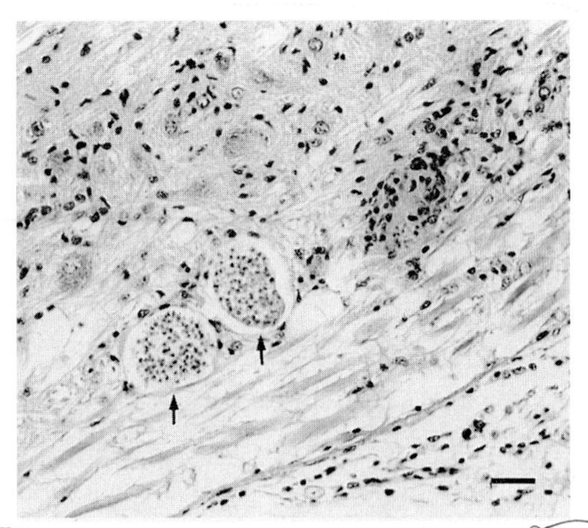

FIGURE 2 A *Microsporidium* sp. infecting the central nervous system of a zebra fish (*Brachydanio rerio*). Arrows denote xenomas. Scale bar, 35 μm.

anguillarum) and *Glugea plecoglossi* Takahashi and Egusa, 1977 (Awakura, 1974; Kano and Fukui, 1982; Kim et al., 1996), respectively. Production of minnows for bait is often impacted by infections with *P. ovariae* Summerfelt, 1964, which damage the ovaries of fish and thereby reduce fecundity (Nagel and Hoffman, 1977).

Early investigations of fish microsporidia included some observations of host response (Drew, 1910; Debaisieux, 1920; Nigrelli, 1946). Researchers during this time period occasionally confused host reparative events (i.e., phagocytosis) with early parasite development stages (Canning and Lom, 1986). Presently, host-parasite relationships of fish microsporidia can be broadly classified into two groups: those associated with xenoma formation (e.g., *Glugea, Ichthyosporidium, Jirovecia, Loma, Microfilum, Microgemma, Nosemoides, Spraguea,* and *Tetramicra*) (Fig. 3), and those associated with non-xenoma-forming microsporidia (e.g., *Nucleospora, Heterosporis, Pleistophora,* and *Thelohania*) (Fig. 4). The term "xenoma" was developed from the early works of Chatton (1920) and Weissenberg (1922). Cells infected by certain species of microsporidia become transformed and hypertrophied, resulting in a unique host cell-parasite complex. The host cell alters its structure and size, becoming physiologically integrated with the parasite. Cytoplasmic contents are replaced by the parasite, and the surface of the cell, or xenoma, becomes modified for increased absorption. Microvilluslike structures form in cells infected by *Microsporidium cotti* Chatton and Courier, 1923, *Ichthyosporidium,* and *Tetramicra,* while *Glugea* and *Loma* spp. utilize numerous pinocytotic vesicles (Canning and Lom, 1986). Hypertrophied cells can reach a size of 400 to 500 μm or larger and are visible macroscopically as white cysts (Matthews and Matthews, 1980; Ralphs and Matthews, 1986) (Fig. 5). Eventually the cell is totally destroyed by the parasite. Spores released when a xenoma ruptures are easily identified by a large posterior vacuole in wet mounts and histological sections (Fig. 6).

FIGURE 4 *Pleistophora hippoglossoideos* in the musculature of a plaice (*Hippoglossoides platessoides*). (Reprinted with permission of the author and publisher from Möller and Möller, 1986.)

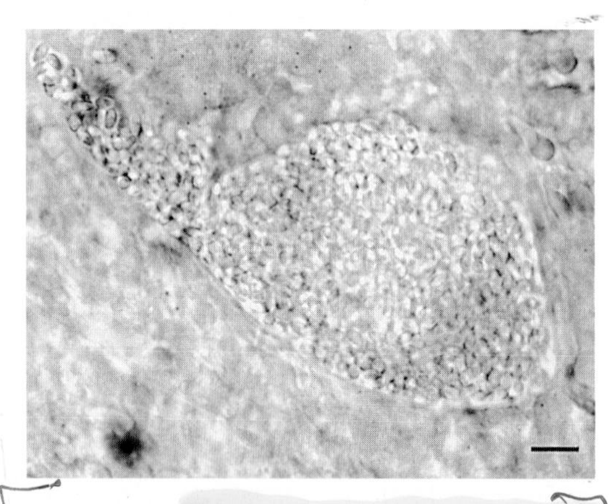

FIGURE 3 Wet mount of a ruptured xenoma filled with spores of an undescribed *Loma* sp. from the gill of a ling cod (*Ophiodon elongatus*). Scale bar, 10 μm.

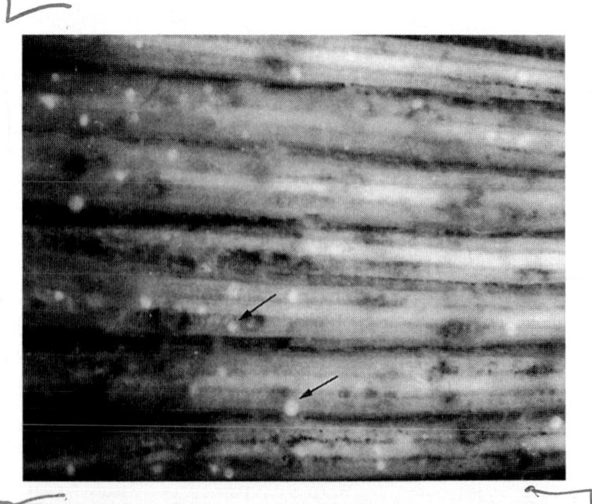

FIGURE 5 Heavy infection of *Loma salmonae* (white xenomas) in fresh gill of a wild spawning sockeye salmon (*Oncorhynchus nerka*) from Babine Lake, British Columbia, Canada. (Reprinted with permission from Mark Higgins.)

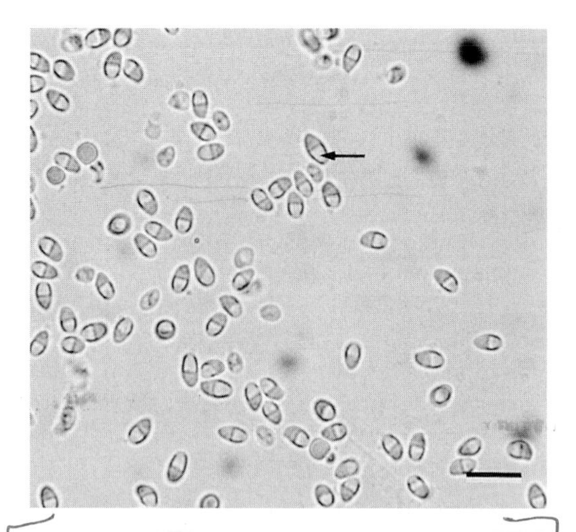

FIGURE 6 Wet mount of spores from an undescribed microsporidian from mountain whitefish (*Prosopium williamsoni*). Arrow indicates large posterior vacuole. Scale bar, 10 μm.

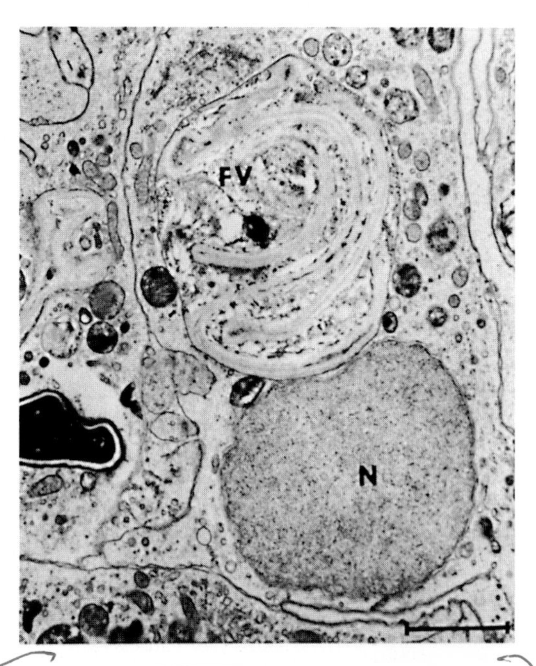

FIGURE 7 Electron micrograph of phagocytic cell with prominent nucleus (N) and remnants of *Pleistophora ovariae* spore walls within a food vacuole (FV). Scale bar, 2 μm. (Reprinted with permission of the author and publisher from Canning and Lom, 1986.)

IMMUNOLOGY

The specific and nonspecific defense systems of fish are well reviewed by Kaattari and Piganelli (1996), Secombes (1996), and Dalmo et al. (1997). In comparison to the situation in viral and bacterial pathogens, the response of fish to microsporidial infections is relatively unexplored.

Phagocytosis of spores by macrophages has been described for several microsporidian species (Dyková and Lom, 1980; Canning and Lom, 1986; Pulsford and Matthews, 1991; Kim et al., 1996). Macrophages actively ingest and break down spores (Fig. 7), playing a crucial role in the host defense mechanism. The spore coat contains an outer protein layer and an inner chitin layer. Spores appear to be digested from the inside out, indicating that fish macrophages possess a wide repository of chitinases (Dyková and Lom, 1980; Canning and Lom, 1986). Microsporidia are not without their own defenses, as spores may prevent phagosome-lysosome fusion. Weidner and Sibley (1985) found that anionic components on the spores of *G. hertwigi* can form ionic bonds with the phagosome membrane, increasing rigidity and preventing fusion with a lysosome. It is not surprising that some microsporidia have evolved this protective mechanism, as has occurred with other intracellular parasites such as *Chlamydia psittaci* (Wyrick and Brownridge, 1978), and *Toxoplasma gondii* Nicolle and Manceaux, 1908 (Jones and Hirsch, 1972).

Suppression of the host inflammatory response by microsporidia has also been documented (Dyková and Lom, 1978, 1980). Laudan et al. (1986a, 1986b, 1987, 1989) found that spores of *G. stephani,* when phagocytized, suppressed immunoglobulin levels of winter flounder (*Pleuronectes americanus*) by stimulating macrophages to release prostaglandin and/or leukotrienes. This finding was substantiated by injecting serum from infected fish into healthy fish, which resulted in immunosuppression of healthy fish. Administering indomethacin after spore exposure also failed to produce immunosuppression in flounder. *G. stephani* impairs the host's humoral response to other infectious agents and interferes with initiation, as well as with other levels, of the response. After *G. stephani* becomes established, the host's immune system can recover.

It has been suggested that *Tetramicra brevifilum* Matthews and Matthews, 1980, can immunosuppress turbot (*Scophthalmus maximus*) by using a mechanism similar to that found in *G. stephani* (Leiro et al., 1994). Figueras et al. (1992) found that the serum agglutinin titers of naturally infected fish were low, although this did not increase their susceptibility to vibriosis (*Vibrio anguillarum*). These authors also isolated a serum factor from infected fish, concluding that further study was necessary to determine if the factor was an immunomodulator or only a by-product of the parasite. Leiro et al. (1993) injected fish intraperitoneally with whole spores or a crude extract of *T. brevifilum*. At 30 days after injection they gave the fish a second immunization of whole spores. They found that whole spores lead to higher antibody production initially, but that after 30 days fish responded more to crude extract antigens. Low seropositive values were obtained from enzyme-linked immunosorbent assay studies on infected fish, leading them to suggest that this was indirect evidence of immunosuppression.

Humoral and/or cellular response of a fish may also be impaired by infection with *Nucleospora salmonis* Hedrick, Groff and Baxa 1991 (Wongtavatchai et al., 1995b) or *Heterosporis anguillarum* (Hung et al., 1997). Serum factors affecting T and B cells directly may be secreted by these parasites, or parasites may stimulate release of mediators from macrophages. Leiro et al. (1996a) found that *Glugea caulleryi* Van den Berghe, 1940, showed considerable homology of spore surface antigens to those of *T. brevifilum*. However, *G. caulleryi* did not suppress the humoral immune response of turbot but rather stimulated it during first and subsequent exposures. In general, caution should be exercised in assuming that lymphopenia is caused directly by a concurrent microsporidium infection because this condition may be caused by a wide variety of stressors (Barton and Iwama, 1991; Figueras et al., 1992).

Immune responses by fish are often weak or nonexistent when microsporidia evade detection (e.g., by forming xenomas or simply by being intracellular). However, during initial infection, and especially during rupture of xenomas (see section on *Glugea* pathology) or pseudocysts (i.e., *Pleistophora*-type infections), microsporidia are vulnerable to attack by the immune system of fish. Although antibody production in response to infection does occur (Buchmann et al., 1992; Leiro et al., 1993; Hung et al., 1996, 1997), it may not play a protective role (Kim et al., 1996). Resistance to reinfection in recovered fish occurs in *Loma salmonae* Morrison and Sprague, 1981 (syn. *Pleistophora salmonae* Putz, Hoffman, and Dunbar, 1965) (Speare et al., 1998b).

Different strains of the same fish species can exhibit varying degrees of susceptibility. For example, we have observed marked differences in susceptibility to *L. salmonae* between different strains of chinook salmon (*Oncorhynchus tshawytscha*) (Shaw et al., 1996). Phagocytes may play a primary role in preventing initial and subsequent infection by microsporidia. General phagocytic activity, measured by percent phagocytosis (PP) (the percentage of cells showing phagocytic activity) and the phagocytic index (PI) (the mean number of objects ingested per phagocyte), usually increases in the presence of serum. It is well known that opsonization by antibody and/or complement of nonself organisms enhances phagocytosis in fish (Griffin, 1983; Sakai, 1984; Scott et al., 1985; Olivier et al., 1986; Waterstrat et al., 1991; Matsuyama et al., 1992; Rose and Levine, 1992; Pedrera et al., 1993; Leiro et al., 1996b). However, Leiro et al. (1996b) found that serum did not increase phagocytosis of *G. caulleryi* or *T. brevifilum* by turbot macrophages. Neither did macrophages from immunized turbot ingest significantly more spores than those from nonimmunized fish. They investigated this further by treating spores with sodium *m*-periodate to block the binding of macrophages to surface-borne sugars on spores. These spores were less effectively ingested, suggesting that macrophages recognize sugars on the surface of the microsporidia. Lectinophagocytosis has also been shown in tilapia (*Oreochromis spilurus*) by

Saggers and Gould (1989). Leiro et al. (1996b) concluded that microsporidia may modify the phagocytic response in fish. We have found that the PP and PI of *L. salmonae* spores by macrophages differ significantly between Atlantic salmon (*Salmo salar*) which is completely resistant to *L. salmonae,* and the very susceptible chinook salmon (Shaw, unpublished data).

Temperature can also affect the ability of a fish host to mount an immune response, and phagocytic activity of macrophages is dependent on temperature (Finn and Nielson, 1971; Leiro et al., 1995). Development of fish-infecting microsporidia can also depend on ambient temperature (Awakura, 1974; Olson, 1981; Speare et al., 1998a). For example, Speare et al. (1998a) were unable to detect *L. salmonae* infections in rainbow trout (*Oncorhynchus mykiss*) held at 10°C, although fish held at 14.5°C developed infections.

The health of fish has often been used as a biological indicator of environmental conditions (Kent and Fournie, 1993). Parasitic burdens can have a significant impact on mortality in wild fishes and in some cases may be related to environmental conditions (Overstreet, 1993). There have been very few studies dealing with the effects of anthropogenic contamination on microsporidial infection in fish. For example, Barker et al. (1994) found that the abundance of *G. stephani* in winter flounder was significantly higher at a pulp mill effluent site where pollutants accumulated in the sediment. In fish from polluted waters, compared to those from unpolluted waters, Barker et al. (1994) reported larger and more varied cysts forming in visceral organs, in addition to the intestine, which is considered the normal site of infection (Takvorian and Cali, 1981). Chronic exposure to pulp mill effluent may have further immunosuppressed the flounder, allowing *G. stephani* to proliferate.

TREATMENT

Several drugs have been used to treat microsporidian infections in fish, mostly on an experimental basis. Fumagillin, an antimicrobial agent developed for treating *Nosema apis* Zander, 1902, infections in honeybees, is the drug most widely used to treat microsporidiosis in fishes. The drug apparently acts by inhibiting RNA synthesis (Jaronski, 1972). Kano et al. (1982) reported that fumagillin was effective against the microsporidian *H. anguillarum* in eels (*Anguilla japonica*). Since this first report on treating microsporidiosis in fish with fumagillin, the drug has been used against *N. salmonis* infections in chinook salmon (Hedrick et al., 1991b) and *L. salmonae* infections in chinook salmon (Kent and Dawe, 1994). Fumagillin has also been used successfully to control several myxosporean diseases in fish (e.g., whirling disease, proliferative kidney disease, and spherosporosis) (Molnár et al., 1987; Hedrick et al., 1988; Székely et al., 1988; Laurén et al., 1989; Wishkovsky et al., 1990; Yokoyama et al., 1990; El-Matbouli and Hoffmann, 1991; Sitjà-Bobadilla and Alvarez-Pellitero, 1992; Higgins and Kent, 1996).

Various concentrations of the drug were employed in these studies, and on the basis of these reports, 3 to 10 mg of fumagillin per kg of fish per day is the recommended dose for salmonids. Higher concentrations or prolonged treatment (e.g., 30 to 60 days) may cause anorexia, poor growth, anemia, renal tubule degeneration, and atrophy of hematopoietic tissues in salmonids (Laurén et al., 1989; Wishkovsky et al., 1990). Lower doses may be efficacious for some microsporidial infections in fish (e.g., Hedrick et al. [1991b] controlled *N. salmonis* infections with oral fumagillin treatment at 1 mg of drug per kg of fish per day for 4 weeks). We reported positive results for controlling *L. salmonae* infections at 10 mg of drug per kg of fish per day with a 4 week treatment (Kent and Dawe, 1994). Our recent experiments with the drug demonstrated that infections can be controlled with lower doses (2 or 4 mg of drug per kg of fish per day).

Fumagillin is not heat-stable. Therefore, it is recommended that the feed be coated with the drug instead of the drug being incorporated into the feed during milling. The drug is not very soluble in water but is very soluble in alcohol. In most studies, fumagillin was mixed with alcohol and sprayed on the feed; then the feed was coated with oil.

An analog of fumagillin, TNP-470 (Takeda Chemical Industries, Ltd.), is an potent antiangiogenesis agent (Kusaka et al., 1994). Furthermore, in laboratory studies TNP-470 has been shown to be effective against the mammalian microsporidia pathogens *Encephalitozoon intestinalis, Vittaforma corneae, E. cuniculi,* and *E. hellem*. Therefore, we tested the efficacy of this drug in controlling *L. salmonae* and *N. salmonis* in experimentally infected salmon. Fish treated orally at 1.0 mg or 0.1 mg of drug per kg per fish per day for 6 weeks showed markedly reduced infections by both microsporidia even at the low dose. No significant toxic side effects were associated with the treatment. We also found that treating fish at this low dose prevented infections in fish naturally exposed to the PKX myxosporean.

The effects of a systemic triazinone, Toltrazuril (Bayer AG), on *Glugea anomala* Moniez 1887 infections in sticklebacks (*Gasterosteus aculeatus*) have been investigated at the ultrastructure level (Schmahl and Mehlhorn, 1989; Schmahl et al., 1990). Bath exposure of the drug was found to cause destruction of all life stages of the parasite, but the overall effects of this drug in reducing the prevalence or intensity of microsporidian infection in fish have not been reported.

Awakura and Kurahashi (1967) reported that amprolium inhibited merogony in *Microsporidium takedai* administered at 0.06% of body weight per day for up to 48 days, but the treatment was associated with toxic side effects. There are a few other reports of treatment of fish microsporidia. For example, Andodi and Frank (1969) (cited in Canning and Lom, 1986) claimed that they cured *P. hyphessobryconis* infections in aquarium fish by adjusting the pH to 7.5 to 8.0 and introducing ozone at a rate of 1 mg/h/100 L. Nagel and Summerfelt (1977) reported that nitrofurazone reduced infections of *Pleistophora ovariae* in golden shiners (*Notemigonus chrysoleucas*) when used in oral treatments at 2.2 to 3.3 g of drug per kg feed.

SIGNIFICANT MICROSPORIDIAN DISEASES

Xenoma-Forming Genera

GLUGEA

Species of *Glugea* are among the most intensively studied fish microsporidia. They infect the submucosal intestinal cells of a variety of wild and cultured fish. Disease in the food fish sand smelt (*Atherina boyeri*), pike perch (*Stizostedion lucioperca*), and a bait fish (*D. cepedianum*), can be caused by *Glugea atherinae* Berrebi, 1978, *G. luciopercae* Dogiel and Bykhowsky, 1939, and *G. cepedianae* Putz, Hoffman, and Dunbar, 1965, respectively (Price 1982; Canning and Lom, 1986). However, most research has focused on *G. anomala* in sticklebacks (*Gasterosteus* and *Pungitius* spp.), *G. hertwigi* in smelts, *G. plecoglossi* in ayu, and *G. stephani* in flatfishes (*Pleuronectes* spp. and *Platessa* spp.). *G. anomala* represents classical developmental and pathological features (Weissenberg, 1967, 1968; Dyková and Lom, 1980), while mortality of fish has been documented in the other species (Olson, 1976; Nepszy et al., 1978; Cali et al., 1986; Kim et al., 1996).

G. anomala and *G. plecoglossi* develop in almost all body organs of the host, forming large (2 to 3 mm), white xenomas which can cause serious injury to organs (Fig. 8). *G. hertwigi* (Fig. 9) and *G. stephani* (Fig. 10) primarily infect the subepithelial connective tissue of the intestine, although in heavy infections *G. hertwigi* can

FIGURE 8 Large cysts of *Glugea anomala* in a stickleback (*Gasterosteus aculeatus*). (Reprinted with permission of the author and publisher from Möller and Möller, 1986.)

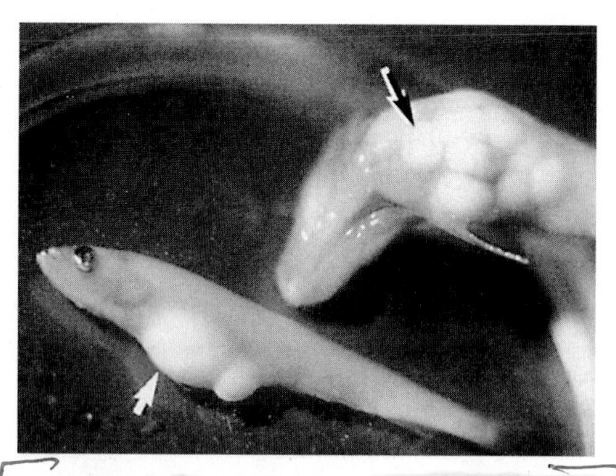

FIGURE 9 Abdominal cysts of *Glugea hertwigi* (arrows) in rainbow smelts (*Osmerus mordax*). (Reprinted with permission of the author and publisher from Noga, 1992.)

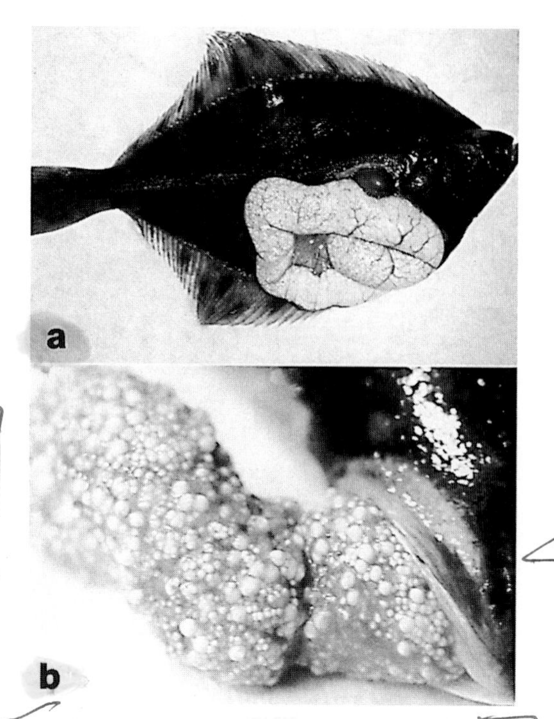

FIGURE 10 *Glugea stephani* in a winter flounder (*Pleuronectes americanus*). (a) Heavy infection of cysts in the intestine; (b) close-up showing pebbled, chalky white appearance of intestine in the same individual. (Reprinted with the permission of the author and publisher from Möller and Möller, 1986.)

spread to all other organs, even the skeletal muscle and gills (Canning and Lom, 1986).

Pathology. *Glugea* spp., like *Loma* and *Tetramicra,* form xenomas that can be classified by stage of development (Dyková and Lom, 1980). Early or young xenomas have a cytoplasm filled uniformly with developmental stages of the parasite (Fig. 11a). As they grow, the developmental stages occupy the center of the xenoma surrounded by peripheral host cytoplasm. For *Glugea,* maturing xenomas exhibit a light, refractile wall and peripheral host cell cytoplasm with spores in the center. A fully developed xenoma is filled with spores and a few host cell components (Fig. 11b). The host's reaction to xenoma formation varies depending on the stage of xenoma development, the tissue infected, and the age and species of the host. However, some generalizations can be made. The stages of host reaction described below were originally developed by Dyková and Lom (1978) for *G. anomala, G. aculeatus, G. hertgiwi, G. plecoglossi,* and *G. stephani.* Further study and clarification was made by Dyková and Lom (1980) and Canning and Lom (1986).

Weakly Reactive Stage. Early or very young xenomas begin to exert pressure atrophy on the surrounding tissue as they grow. The host responds by local proliferation of connective tissue and collagen fibers, forming a concentric layer of connective tissue around the xenoma. The weakly reactive stage is also seen with fully developed xenomas.

Productive Stage. Mature xenomas elicit prominent inflammation (Fig. 11c). Inflammation includes influx of fibroblasts and macrophages. Proliferative inflammation accompanies changes in the xenoma wall, although it is not clear if this is a causal relationship (Dyková et al., 1980; Canning and Lom, 1986). The wall of the xenoma swells, fibroblasts appear within it, and the wall ruptures and begins to disappear. In addition to fibroblasts, the formation of a capsule composed of eosinophilic granular cells around the xenoma was reported by Reimschuessel et al. (1987). *Ichthyosporidium* spp. elicit a slightly different

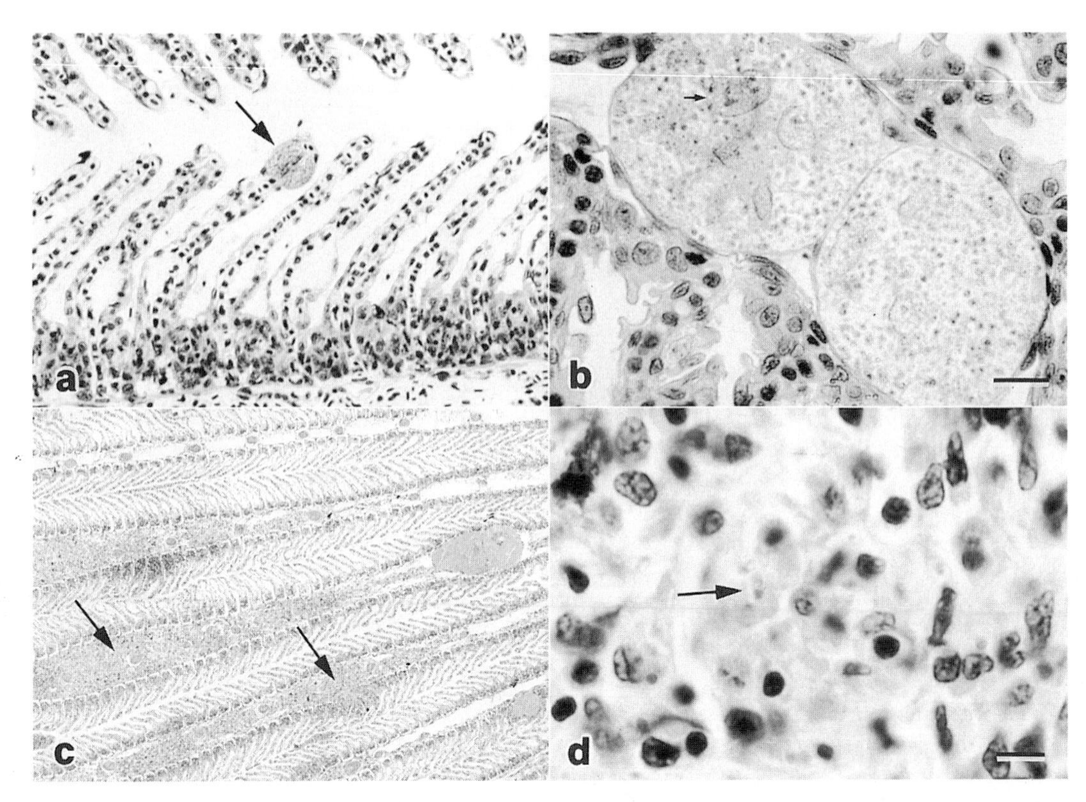

FIGURE 11 Various stages of microsporidian development represented by *Loma salmonae* in coho salmon (*Oncorhynchus kisutch*) gills. (a) Young xenoma in secondary lamellae (arrow). (b) Mature xenoma with few remnants of host cell nuclei (arrow). Scale bar, 45 μm. (c) Proliferative inflammatory response (arrows) associated with infection in the primary lamellae. (d) Granuloma resulting from a ruptured xenoma with a few remaining spores (arrow) in macrophages. Scale bar, 10 μm.

inflammatory response characterized by epithelioid cells and histiocytes. Epithelioid cells encasing the xenoma are arranged with their long axis perpendicular to the xenoma wall. Destruction or rupture of the xenoma wall is accompanied by dystrophic changes that occur with the appearance of periodic acid-Schiff-positive substances and Ca^{2+} within the mass of spores. The xenoma is then replaced by granulomatous tissue (Fig. 11d).

Granuloma Involution Stage. After the xenoma ruptures, spores are released and ingested by macrophages. The granulation tissue matures, and the granuloma gradually recedes. Fibrous connective tissue that has developed during the productive stage undergoes hyalinization, and the lesion is slowly resolved. Although the xenoma can be completely eliminated, functional restitution of heavily infected organs is not possible (Dyková et al., 1980).

Glugea spp. can cripple, disfigure, debilitate, and kill fishes (Sindermann, 1990; Dyková, 1995). These microsporidia can also contribute to decreased fecundity and retarded growth (Chen and Power, 1972; McVicar, 1975). The large cysts of *G. anomala* create serious tissue injury due to pressure atrophy, although sticklebacks appear to survive even heavy infections (Canning and Lom, 1986). Host response follows the stages described in this section under "Pathology." *G. plecoglossi* has low organ specificity, although development and host tissue reactions are similar to those of *G. anomala*

(Dyková et al., 1980). The prevalence of *G. hertwigi* in smelt can reach 100% in some lakes, and epizootics with intensities of 250 xenomas per fish have occurred (Anenkova-Khlopina, 1920; Haley, 1952, 1953, 1954; Bogdanova, 1957; Petrushevski and Shulman, 1958; Delisle, 1965; Nepszy and Dechtiar, 1972; Canning and Lom, 1986). Delisle (1972) estimated that in Lake Erie alone, 10 million fish per year were lost as a result of *G. hertwigi* infections. The intestinal epithelium can disintegrate, resulting in general septicemia, intoxication, and death (Canning and Lom, 1986). The fecundity of smelt can also be severely affected by the parasite (Sindermann, 1963). Infections by *G. stephani* occlude the intestinal lumen, disrupt its integrity, and contribute to emaciation or even death of the host (Cali et al., 1986). Xenomas replace the intestinal wall, forming a rigid layer up to 4 mm thick which has a chalk-white, pebbled appearance (Fig. 10b). Fish can die, even from low level infections, if xenomas form in the mucosa and rupture into the epithelial lining. In contrast, Cali et al. (1986) showed that heavily infected fish may survive if xenomas form on the serosal side of the intestinal tract. Olson (1976) reported collecting emaciated starry flounder (*Platichthys stellatus*) from Yaquina Bay, Oreg., infected with *G. stephani*. Cali et al. (1986) also observed *G. stephani* in moribund winter flounder of Sandy Hook Bay, N.J. Heavy infections may be fatal in plaice (*Pleuronectes platessa*) held in aquariums or at fish farms (Bückmann, 1952; McVicar, 1975). McVicar (1975) suggested that *G. stephani* may be highly contagious at a farm site.

Transmission. *Glugea* spp., like most fish microsporidia, are transmitted directly via ingestion. These species have been transmitted by intraperitoneal injection or via a crustacean paratenic host (e.g., *Daphnia* spp., brine shrimp, or amphipods) (Weissenberg, 1921, 1968; McVicar, 1975; Olson, 1976; Kim et al., 1996). Predation may disperse spores of *G. hertwigi*, or cannibalism may lead to infection in other smelt (Haley, 1954; Delisle, 1972). Olson (1976, 1981) found that spores of *G. stephani* passed directly through the crustacean digestive tract and that infections heavier than intraperitoneal injection resulted when a crustacean paratenic host was used. He suggested that amphipods may represent a natural route of transmission for *G. stephani*.

The target host cell of *Glugea* spp. can vary. *G. anomala* may target a migratory mesenchyme cell, such as a macrophage or histocyte (Weissenberg, 1968), whereas *G. stephani* may infect a neutrophil (Canning and Lom, 1986). The formation of xenomas free within the body cavity by intraperitoneal injection suggested to McVicar (1975) that *G. stephani* infects macrophages.

Seasonal and geographic variations in the prevalence of *G. stephani* have been related to water temperature (Olson, 1976, 1981; Takvorian and Cali, 1981, 1984). Olson (1981) found that development of the parasite was arrested at 10°C and resumed when water temperatures were raised to 15°C or higher.

LOMA

Loma is another important xenoma-forming genus of microsporidia. Xenomas of *Loma* spp., unlike those of *Glugea,* develop asynchronously with various developmental stages located throughout the xenoma. There are nine described species of *Loma,* and this genus includes some species noted for infecting fishes of commercial importance (i.e., cod and salmon). *Loma branchialis* Nemeczek, 1911 (syn. *Nosema branchialis* Nemeczek, 1911, *Glugea branchialis* Nemeczek, 1911, and *Loma morhua* Morrison and Sprague, 1981) has been described from several members of the family Gadidae: *Gadus aeglefinus, G. callarias, G. morhua marisalbi, G. morhua kildinensis, G. morhua,* haddock (*Melanogrammus aeglefinus*), and rockling (*Enchelyopus cimbrius*) (Bazikalova, 1932; Dogiel, 1936; Fantham et al., 1941; Shulman and Shulman-Albova, 1953; Morrison and Sprague, 1981a, 1981b; Morrison and Marryatt, 1986). *Loma camerounensis* Fomena, Coste, and Bouix, 1992, infects tilapia (*Oreochromis niloticus*), which are widely cultured in West Africa (Fomena et al., 1992). Morrison and Sprague (1981c, 1983) described *Loma fontinalis* Morrison and Sprague, 1983, from brook trout

(*Salvelinus fontinalus*) in Canada. *Loma salmonae* is well known as a serious pathogen in farmed salmonids (*Oncorhynchus* spp.) in the Pacific Northwest (Kent et al., 1989; Kent, 1992; Shaw et al., 1997). This parasite may have been introduced into Europe with rainbow trout and coho (*Oncorhynchus kisutch*) transported from Californian stocks (Poynton, 1986). Descriptions of *Loma* spp. from salmonids such as chinook, coho, and rainbow trout (Awakura et al., 1982; Hauck, 1984; Mora, 1988; Speare et al., 1989; Magor, 1987; Gandhi et al., 1995) are likely *L. salmonae* on the basis of host, parasite morphology, and site of infection.

All these species except *L. camerounensis* infect endothelial cells, causing formation of xenomas throughout vascularized organs (e.g., heart, kidney, spleen, liver) but primarily in the gills (Fig. 5). *L. camerounensis* forms xenomas in connective tissue of the gut submucosa.

Species descriptions for *Loma* have been problematic. Most earlier researchers designated new parasite species based on a new host, geographic location, and slight differences in spore morphology. Morrison and Sprague (1981a, 1981b, 1981c; 1983) described *Loma morhua* and *L. fontinalis,* recognizing that these species may be conspecific to *L. branchialis* and *L. salmonae,* respectively (Canning and Lom, 1986). Analysis of the rDNA of *Loma* spp. and transmission studies are beginning to clarify the taxonomy of this genus (Docker et al., 1997a; Shaw et al., 1997). Shaw et al. (1997) described *L. embiotocia* Shaw, Kent, Docker, Brown, Devlin, and Adamson, 1997, in shiner perch (*Cymatogaster aggregata*) with all these techniques. They found that the internal transcribed spacer region of *Loma embiotocia* differed significantly from the internal inscribed spacer region of *L. salmonae* but not within isolates of each species. Furthermore, they were unable to transmit *L. salmonae* to shiner perch and thus assigned the shiner perch parasite as a new species, *L. embiotocia.* In our laboratory we have been unable to transmit *L. embiotocia* to either chinook or *Loma* spp. isolated from Pacific cod (*Gadus macrocephalus*), or ling cod (*Ophiodon elongatus*) to chinook. This finding suggests that *Loma* spp. found in marine hosts (Kent et al., 1998) may be distinct species or that all are at least distinct from *L. salmonae.*

Wales and Wolf (1955) found 75% of wild yearling rainbow trout were infected with *L. salmonae* in California. We have observed *L. salmonae* to be widespread in wild Pacific salmon populations in both fresh and marine habitats in British Columbia, Canada (Shaw, unpublished data). Epizootics of *L. salmonae* have been recorded in Japan, the Pacific Northwest, the eastern United States, and Europe (Wales and Wolf, 1955; Hauck, 1984; Bekhti and Bouix, 1985; Canning and Lom, 1986; Kent et al., 1989; Bruno et al., 1995). In 1955 Wales and Wolf first described transport and grow-out of farmed fish as being more difficult because of *L. salmonae.* Since that time the parasite has continued to cause large economic losses to chinook farms in British Columbia. Recently we have also noted high mortality in wild spawning sockeye of the Babine Lake system, British Columbia, due to heavy *L. salmonae* infections.

In our laboratory we also found that *L. salmonae* infected all seven species of *Oncorhynchus* and brook trout. Speare et al. (1998a) were unable to infect brook trout, although their fish may have developed resistance from a previous exposure to *L. salmonae* or *L. fontinalis* (Speare, personal communication). Atlantic salmon, Arctic char (*Salvelinus alpinus*), herring (*Clupea pallasi*), prickly sculpins (*Cottus asper*), sticklebacks, and guppies (*Poecilia* sp.) are resistant to experimental infection by *L. salmonae* (Kent et al., 1995; Shaw, unpublished data). *L. salmonae,* therefore, displays host specificity, with the main hosts being species of the genus *Oncorhynchus.*

Pathology. Cod infected with *L. branchialis* exhibit no obvious clinical signs (Morrison and Sprague, 1981a). However, xenomas in the gills reached 1.2 mm in size, resulting in distortion and displacement of gill tissue and blood vessels (Kabata, 1959; Morrison and

Sprague, 1981a). Pathological changes caused by these infections are similar to those described for *G. anomala,* except that Morrison (1983) found that spores remained in the center of granulomas with both phagocytes and fibroblasts undergoing coagulative necrosis. The impact on host mortality is unknown, although infections may affect metabolism negatively through decreased respiratory efficiency (Morrison and Sprague, 1981a).

Gross clinical signs of *L. salmonae* infection may include darkening of the tail or body; lethargy; gill pallor; petechial hemorrhaging in the gills, skin, and fins; ascites; hemorrhagic pyloric ceca; and white cysts (xenomas) on the gills (Hauck, 1984; Kent, 1992; Markey et al., 1994; Bruno et al., 1995). Reduced growth, impaired swimming efficiency, and increased mortality of young chinook have also been noted (Hauck, 1984). The growth rate appears to be reduced during xenoma formation in rainbow trout (Speare, personal communication). Hauck (1984) provides a comprehensive pathological description of a *L. salmonae* systemic infection which includes necrosis of cartilage and musculature, occlusion of arteries, pericarditis of the bulbus arteriosus, and hyperplasia of gill and heart tissues. The effects of occlusion are seen when xenomas extend through the tunicae media and adventitia of arteries. Xenomas and/or free spores in the heart can result in mural emboli, hyperplasia of ventricular and atrial tissue, and an oligocythemic condition. Gill xenomas can cause subacute to chronic vasculitis, perivasculitis, vascular thrombosis, and hyperplasia of gill tissue (Kent et al., 1989; Speare et al., 1989). As with other xenoma-forming species, the inflammatory response and associated tissue damage are more severe after xenomas have ruptured (Kent et al., 1989).

Fomena et al. (1992) did not describe the pathology of *L. camerounensis* except to note a high prevalence (94%) of the parasite and that large xenomas could protrude into the intestinal cavity. Negative impacts on fish health were also not described by Morrison and Sprague (1981c, 1983) for *L. fontinalis.* It is likely that the pathology is similar to that of *L. salmonae.*

Transmission. Spores of *Loma* spp. released from xenomas can be infectious when ingested. Spores may be liberated directly from the gills and urine into the external environment (Hauck, 1984). Autoinfection may occur via spores released from ruptured xenomas being transported to other tissues. Morrison (1983) proposed this phenomenon for *L. branchialis,* and Shaw et al. (1998) substantiated it for *L. salmonae* by experimentally inducing infection by injecting purified spores intravascularly. Hauck (1984) suggested that infection may occur by direct phagocytic uptake of spores by the gills. However, placement of infectious spores directly on gills of chinook did not result in infection (Shaw et al., 1998). Salmon may also be infected experimentally by anal gavage, per os, or by intravascular, intramuscular, or intraperitoneal injection. Poynton (1986) and Kent et al. (1995) suggested that *L. salmonae* is transmitted from fish to fish within a freshwater or marine net pen. Fish have been infected by cohabitation with carrier fish in a flow-through system, suggesting that direct fish-to-fish transmission occurs within a net pen (Shaw et al., 1998). The infection is prevalent in wild ocean-caught salmonids (Kent et al., 1998), which may be a source of infection for marine fish farms. However, infections are probably maintained and perpetuated at these farms by stocking of parasite-free smolts at the same sites with older, infected fish. Chinook can remain infected for extended periods (i.e., >80 days at 10°C), and purified spores of *L. salmonae* are viable between 52 and 100 days in both fresh- and saltwater (Shaw, unpublished data). Epizootics of *L. salmonae* have occurred at freshwater hatcheries (Hauck, 1984; Magor, 1987), indicating that smolts may also be infected before they are transferred to seawater farms.

Development of immunity to *L. salmonae* occurs in rainbow trout. Speare et al. (1998b) found that fish resolved experimental infections by 10 weeks after initial exposure and were resistant to experimental reinfection. Our preliminary observations suggest that the same phenomenon can occur in chinook salmon.

The structure and development of xenomas formed by *Loma* spp. have been studied extensively. However, very little is known about the route the parasite takes after entering the gut of the host. It is likely that sporoplasms from extruded spores are injected into epithelial cells of the alimentary canal. Shaw et al. (1998) observed intracellular structures, possible presporogonic stages, progressing though intestinal epithelial cells and into the lamina propria. Markey et al. (1994) described unidentified intracellular structures that preceded xenoma formation in all infected tissues. These structures may have represented earlier stages of the parasite. Work is currently under way using in situ hybridization with the rDNA probe (Docker et al., 1997a) to clarify the route of early *L. salmonae* infection (Speare, personal communication).

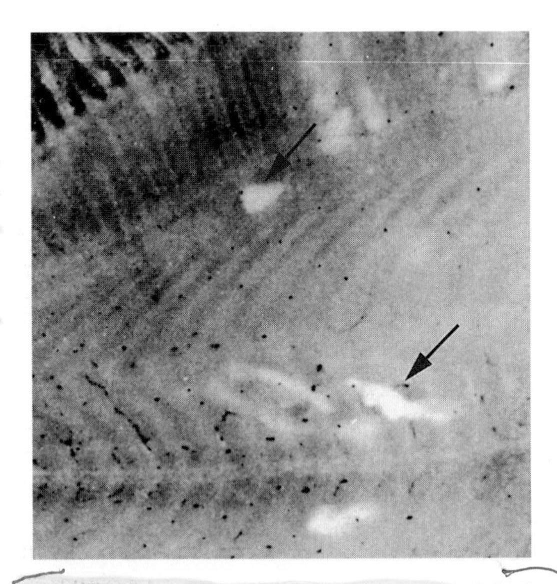

FIGURE 12 *Tetramicra brevifilum* (arrows) in skeletal musculature of a turbot (*Scophthalmus maximus*). (Reprinted with permission of the publisher from Matthews and Matthews, 1980.)

TETRAMICRA

T. brevifilum, another xenoma-forming microsporidian, has been noted to cause significant disease in cultured and wild fish. The host fish, turbot, is becoming more widely cultured in countries such as Spain, and the parasite represents a significant threat to turbot culture (Figueras et al., 1992). *T. brevifilum* infects connective tissue of the skeletal musculature (Fig. 12), and in heavy infections xenomas can be found in the intestine, kidney, liver, and spleen (Estévez et al., 1992; Figueras et al., 1992). Matthews and Matthews (1980) first described this parasite in wild turbot collected off the north coast of Cornwall, United Kingdom. The parasite was then isolated in turbot culture in Galicia, Spain, during 1990 by Figueras et al. (1992). Xenomas lack the multilaminate layer characterizing those of *Glugea* spp. However, xenomas of *T. brevifilum* may reach 1.5 mm in diameter with additional adhesions between xenomas creating composite cysts (Matthews and Matthews, 1980; Dyková and Figueras, 1994). Spores of this species contain a conspicuous inclusion in the sporoplasm and posterior vacuole, which is unique among fish-infecting microsporidia (Lom and Dyková, 1992).

Pathology. Turbot with heavy infections of *T. brevifilum* exhibit erratic swimming behavior, general tissue swelling, dorsal darkening, overproduction of mucus, muscle liquefaction, visible cysts, and chronic low mortality. A 50% reduction in growth rate and a jellylike consistency of muscle in infected stocks contribute to decreased marketability (Figueras et al., 1992). The impairment of swimming ability could contribute to decreased feeding rates in cultured stocks and to increased predation and starvation in wild stocks (Matthews and Matthews, 1980).

The presence of such large xenomas creates pressure atrophy consistent with trauma in the perimysium, leading to the displacement of muscle fibers and loss of fiber attachment to myocommata (Matthews and Matthews, 1980). Canalization in connective tissue of the spinal column and localized hemorrhaging, fibrosis, and necrosis of infected tissues occur with this infection (Matthews and Matthews, 1980; Estévez et al., 1992). Rupture of xenomas results in general cellular infiltration and collagen deposition, and spores are phagocytized by macrophages and subsequently destroyed. Myodegeneration by vacuolization of the sarcoplasm and separation of myofibrils can create liquefaction of musculature. As in *Pleistophora*-type infections, this damage is thought to be caused by substances secreted by the parasite (Dyková and Lom, 1980). Estévez et al. (1992) noted that degeneration of the muscularis mucosa occurred but to a lesser extent than in skeletal muscles. Fish that survive infection may recover fully (Estévez et al., 1992). Figueras et al. (1992) and Leiro et al. (1993) investigated the effects of *T. brevifilum* on the immune status of fish (see "Immunology" above).

Transmission. Matthews and Matthews (1980) were able to transmit *T. brevifilum* by intramuscular injection but not per os. No cross transmission to controls was observed by these researchers during the 7 weeks fish were held at 15°C. They suggested that phagocytes may be involved in the life cycle of *T. brevifilum* by acting as a transport mechanism within the host, which could transport the parasite through the endothelium, or acting as sites for development themselves. Phagocytes might become directly infected by a sporoplasm or during phagocytosis of meronts in the lamina propria of the intestine. Estévez et al. (1992) observed free spores within the lamina propria of the intestine.

Figueras et al. (1992) were unable to infect fish intraperitoneally or by waterborne exposure. No microsporidia were found in the food fed to cultured turbot, leading them to conclude that the fish were infected by aquatic crustaceans such as copepods, decapod larvae, or mysids. In their study infection corresponded with a drop in temperature, possibly reflecting stress induced in the fish. The death of the host and the release of spores directly into seawater and/or cannibalism of moribund fish may be important in transmission of *T. brevifilum*. Frequent mortality in the juvenile population could ensure transmission of the parasite to other fish (Matthews and Matthews, 1980).

Non-Xenoma-Forming Genera

PLEISTOPHORA

Pleistophora, like *Glugea,* causes great damage in a wide variety of fish species. These parasites usually invade skeletal muscles, replacing the sarcoplasm and destroying the cells. In general, they are associated with musculature destruction or liquefaction, deformity, and production of tumorlike masses in fish (Pulsford and Matthews, 1991). Some species, such as *Pleistophora mirandellae* and *P. ovariae,* invade and destroy oocytes. *Pleistophora* spp. are characterized by diffuse infections, with only a few species such as *P. senegalensis* Faye, Toguebaye, and Bouix, 1990, and *P. hippoglossoideos* Bosanquet, 1910, forming xenomas (Morrison et al., 1984; Faye et al., 1990). More than 30 species of *Pleistophora* have been described. Of these, 8 infect fishes of economic importance: *P. ehrenbaumi* Reichenow, 1929, in wolffish (*Anarhichas* spp.), *P. finisterrensis* Leiro, Ortega, Iglesias, Estévez, and Sanmartin, 1996, in blue whiting (*Micromesistius poutassou*), *P. hippoglossoideos* in plaice (*Hippoglossoides platessoides*); *P. hyphessobryconis* in ornamental fishes, *P. macrozoarcides* in ocean pout, *P. mirandellae* in European cyprinids, *P. ovariae* in bait fishes (family Cyprinidae), and *P. senegalensis* in sea bream. *H. anguillarum* (formerly *P. anguillarum*) is discussed with other species of *Heterosporis.*

Muscle fibers of wolffish (*Anarhichas lupus* and *A. minor*) are infected by *P. ehrenbaumi,* and up to 10% of wolffish in the waters of Iceland may be infected (Meyer, 1952). The parasite creates tumorlike swellings up to 8 by 15 by 4 cm, which can make the commercial catches unfit for consumption (Egidius and Soleim,

1986). Wolffishes are promising species for aquaculture (Wiseman and Brown, 1996), and thus this parasite may ultimately become a problem in the rearing of these fishes.

Blue whiting collected off northwest Spain have been infected by *P. finisterrensis.* Leiro et al. (1996c) found that 5% of fish harbored infections in the hypoaxial musculature, in which infective 3- to 6-mm foci were located. They did not associate any significant morbidity with the infection.

The type host for *P. hippoglossoideos* is *H. platessoides* (syn. *H. limandoides, Drepanopsetta hippoglossoides*) (Bosanquet, 1910; Kabata, 1959; Canning and Lom, 1986). Canning and Lom (1986) and Lom and Dyková (1992) also included sole (*Solea solea*) as a host. Muscles of the fins, walls of the visceral cavity, and somatic musculature of fish may be infected (Fig. 4). Cystlike structures are visible externally and can be up to 2.5 by 10 mm in some cases, making the fish unfit for eating (Canning and Lom, 1986; Dyková, 1995). Morrison et al. (1984) provide a redescription of this species.

Originally imported from wild fishes of the upper Amazon River, *P. hyphessobryconis* is now distributed worldwide in many families of freshwater tropical fishes, infecting over 16 species (Lom and Dyková, 1992; Dyková, 1995). It invades skeletal muscle (Fig. 1), and heavy infections can occur in connective tissue of the ovaries, intestinal epithelium, skin, and renal tubules. Large cysts (2 mm) can form, and spores can concentrate in the subcutaneous tissue and skin (Canning and Lom, 1986). Heavy losses occur in culture ponds of ornamental species infected by *P. hyphessobryconis,* with few successful treatments available (see "Treatment").

Skeletal muscles of ocean pout infected with *P. macrozoarcides* can contain tumorlike masses up to 8 cm in size which exude a puslike fluid when cut (Canning and Lom, 1986). These pseudocysts increase with the age and size of fish and have been blamed for the 1940 collapse of the ocean pout fishery (Fischthal, 1944; Sandholzer et al., 1945) as a result of difficulties in marketing fish infected with *P. macrozoarcides* to consumers (Sheehy et al., 1974).

P. mirandellae (syn. *P. longifilis* Schuberg, 1910, *P. oolytica* Weiser, 1949) infects oocytes of common European cyprinids such as bleak (*Alburnus alburnus*), barbel (*Barbus barbus*), and roach (*Rutilus rutilus*) (Dyková, 1995). It has also been found in common pike (*Esox lucius*) (Maurand et al., 1988). Macroscopic white lesions appear in the ovary, and an infection of 10 to 20% of follicles can significantly decrease the fecundity of fish (Lom and Dyková, 1992). Wiklund et al. (1996) reported that roach in the Archipelago Sea (Finland) infected with *P. mirandellae* likely experienced decreased fecundity as a result of ovary degeneration and destruction.

Golden shiners and fathead minnows (*Pimephales promelas*) are the main hosts of *P. ovariae* (Fig. 13). Summerfelt and Warner (1970) found an overall prevalence of 48% in shiners from U.S. bait-minnow hatcheries they surveyed. The lowest prevalence was found in young fish, with the infection rate increasing up to 79% for older fish. The prevalence then dropped off, possibly as a result of selective mortality. Although this parasite was widespread in golden shiners, hatcheries were able to obtain eggs from young fish not yet damaged by *P. ovariae* (Nagel and Hoffman, 1977). The organism may cause a 40% decrease in fecundity (Summerfelt 1964).

Sea bream is an important species in the markets of Senegal. Faye et al. (1990) described *P. senegalensis* from the muscularis of this host's

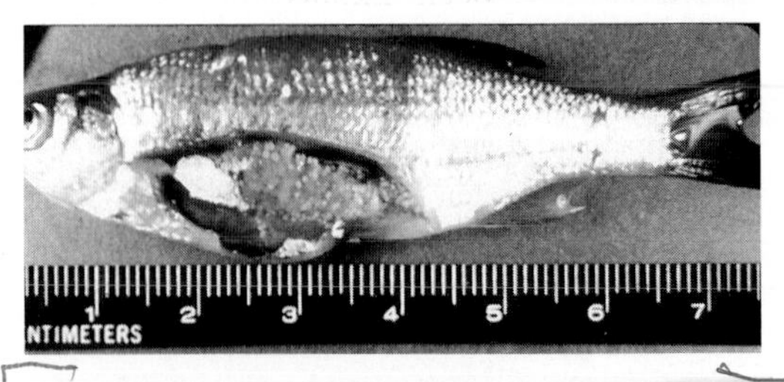

FIGURE 13 *Pleistophora ovariae* in a golden shiner (*Notemigonus chrysoleucas*). (Reprinted with permission from G. Hoffman.)

intestinal wall, but no significant pathological changes were associated with infection.

Pathology. In non-xenoma-forming genera such as *Pleistophora,* infected cell contents are replaced without inducing prominent cell hypertrophy. These microsporidia infect muscle cells or oocytes, and the host reaction is often minimal. Minor inflammatory cell infiltration into the myosepta can take place as the parasite develops. Fusion of infected muscle fibers can also occur with some parasite species (e.g., *Pleistophora ehrenbaumi* and *P. macrozoarcidis*). Eventually all the muscle cell contents are replaced and the cell is destroyed, liberating mature spores. The host then responds with an influx of macrophages that phagocytize and digest the spores, but tissue regeneration is limited (Dyková and Lom, 1980; Pulsford and Matthews, 1991). Invaded muscle fibers are often not isolated by granuloma formation, as is the case in xenoma-forming species. However, granulomas can develop as a result of heavy infection by certain species (e.g., *P. macrozoarcidis* and *P. hyphessobryconis*). *P. hyphessobryconis* is unique in that it forms discrete islets of degenerated sarcoplasm in muscle that directly abuts intact myofibrils. These islets result from substances secreted by the parasite (Dyková and Lom, 1980).

Pathological changes have not been described for *P. ehrenbaumi* infections. However, fish can become extremely emaciated, suggesting that such infections can be fatal (Egidius and Soleim, 1986; Lom and Dyková, 1992). Mortality due to *P. hippoglossoideos* in smaller individuals has been implied but not confirmed (Morrison et al., 1984). Large cysts of *P. hippoglossoideos* can cause compression and distortion of tissues, but no great damage results (Kabata 1959; Canning and Lom 1986). Morrison et al. (1984) observed influx of phagocytes and engulfment of spores in infected muscle, and some nodules were also encapsulated with fibrous tissue.

Fish infected with *P. hyphessobryconis* show anomalous behavior and movement, fading of color, and the appearance of grayish-white patches in the muscle. Emaciation, scoliosis, kyphosis, and bristled scales are also characteristic of the infection (Canning and Lom, 1986). Granulomas can form in the mesenteries or viscera with an envelope of connective tissue cells. Canning and Lom (1986) found connective tissue encapsulation was better at limiting spread of the parasite than phagocytosis by macrophages. Heavy infections can cause high mortality within 14 days (Thieme, 1954, cited in Canning and Lom, 1986). Destruction of the host muscle begins with a unique halo formation around the parasite, which consists of disrupted smooth endoplasmic reticulum, free ribosomes, and myofibrils (Canning and Lom, 1986; Dyková, 1995). Infections can lead to atrophy of the testes (parasitic castration) and liver, although these organs do not have to be heavily infected (Lom and Dyková, 1992).

Canning and Lom (1986) describe the pathology of *P. macrozoarcides* infections. The parasite can form small, whitish cylinders in muscle fibers during early stages of infection, but large tumor-like masses result when the host encapsulates several infected fibers with concentrically arranged connective tissue. The center of these "pseudocysts" contains free spores. The muscle turns brownish, and muscle fibers are hyalinized and destroyed. The host response follows that described earlier for other non-xenoma-forming species.

Both *P. mirandellae* and *P. ovariae* infect oocytes of fish, although the pathogenesis of these infections differs. *P. mirandellae* infects before the zona radiata forms, developing in the yolk and replacing it with spores. A significant proliferative granulomatous inflammatory response can occur, with destruction of infected oocytes and spores (Canning and Lom, 1986; Dyková, 1995). Infected connective tissue of the seminiferous tubules can also create hypertrophy of local epithelial cells, leading to decreased fecundity (Schuberg, 1910). In contrast, inflammation and encapsulation of infected oocytes is not associated with *P. ovariae* infection. The inflammatory reaction has not been well described for this species, although Summerfelt and Warner (1970) provide a brief description of its pathogenesis. Infected ova become mottled with white spots and streaks, each representing a mass of ovarian stroma and spores. Atresia of heavily infected ova occurs, followed by the coalescing of spores in a stroma of zona radiata and yolk. Hyperplasia of the follicular epithelium occurs and then a collapse of the zona radiata, and an influx of phagocytes that destroy the oocyte. Atretic follicles of the ovary are invaded by fibroblasts, and some fibrosis can be seen. Infected ova develop a stroma of connective tissue and increase in size, as reflected by postspawning fish having heavier ovaries than healthy fish. Parasitic castration can be pronounced, reducing fecundity by 37% or more (Canning and Lom, 1986). Curiously, infected fish are often larger than healthy spawning fish, because of reduced commitment of nutrients to egg production (Summerfelt and Warner, 1970).

Transmission. As with many microsporidia, death of the host may liberate free spores of *Pleistophora* spp. to infect the next host. Spores may also be released with urine, or directly from the skin, of fish infected with *P. hyphessobryconis* (Canning and Lom, 1986). Tetras and goldfish have been infected per os, and intramuscular injections are helpful in maintaining *P. hyphessobryconis* in the laboratory (Lom, 1969; Canning and Lom, 1986). Autoinfection, in which spores produced within the host hatch and form secondary infections, has been proposed for *P. hyphessobryconis*. Nigrelli (1946) proposed that infection of large areas of the host by *P. macrozoarcides* could be explained by autoinfection. It is likely that autoinfection occurs in some *Pleistophora* spp., but this finding has yet to be substantiated. Summerfelt (1972) infected golden shiners with *P. ovariae* per os and suggested transovarial (vertical) transmission after finding the microsporidian in 5% of 38 blastulas he examined. Canning and Lom (1986) did not consider this observation definitive evidence for transovarial transmission.

Interestingly, Leiro et al. (1994) infected turbot per os with *P. finisterrensis* from blue whiting. Outbreaks of *P. finisterrensis* may occur in turbot culture in the future, as raw blue whiting are often used as feed.

HETEROSPORIS

Three species of *Heterosporis* infect fish: *Heterosporis finki* Schubert, 1969, *H. schuberti* Lom, Dyková, Körting, and Flinger, 1989, and *H. anguillarum*. The first two infect ornamental fishes, whereas the latter is an important pathogen in eel culture.

H. finki infects the musculature of angelfish (*Pterophyllum scalare*) (Schubert, 1969; Michel et al., 1989), a popular aquarium fish. *H. schuberti* has been described from the musculature of *Pseudocrenilabrus multicolor* (family Cichlidae) and *Ancistrus cirrhosus* (family Loricariidae) (Lom et al., 1989). Both species are characterized by formation of a sporophorocyst, which is a thick, dense envelope of parasitic origin inside the sarcoplasm (Fig. 14) (Lom et al., 1993). The sporophorocyst contains only parasite stages, unlike a xenoma which contains host cell cytoplasm and nuclei. *H. anguillarum* development resembles that of *Pleistophora* spp. (see species under this genus).

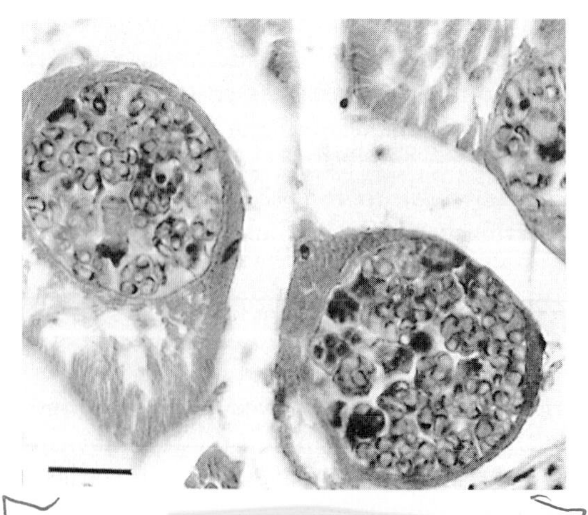

FIGURE 14 Sporophorocysts of *Heterosporis finiki* in the musculature of an angelfish (*Pterophyllum scalare*). Scale bar, 20 μm. (Reprinted with permission of the publisher from Michel et al., 1989.)

Pathology. Michel et al. (1989) noted that few fish exhibited clinical signs during an outbreak of *H. finki* at an angelfish farm in France. However, some fish were emaciated and exhibited lesions up to 5 mm in diameter. Liquefaction of infected striated muscle was also seen, although internal organs appeared normal. Lom et al. (1989) described signs of distress, emaciation, and up to 95% mortality among fish infected with *H. schuberti*. Sporophorocysts provoked only moderate cellular infiltration. However, rupture of these sporophorocysts resulted in the formation of aggregates of macrophages containing phagocytized spores in the mesenteries, intestine, and kidney. A more severe reaction in *B. splendens* infected by a *Heterosporis* sp. was described by Lom et al. (1993). In these fish granulomatous myositis was characterized by a central area of lesions containing spores, and recovery of muscle function was deemed impossible.

The growth of Japanese eels can be retarded and their market value can be decreased by *H. anguillarum* infection. Eels mount an extensive inflammatory response when cysts rupture, followed by a fibroblastic response. Deformities and muscle liquefaction can result from such infections (T'sui and Wang, 1988).

Transmission. Little information is available on the transmission of *Heterosporis* spp. Michel et al. (1989) hypothesized that clinically normal adult angelfish transferred *H. finki* to juveniles or that the infection was introduced with a food source. Infection in eels is possible by oral administration of *H. anguillarum* spores (Kano and Fukui, 1982) or by waterborne exposure (T'sui et al., 1988).

MICROSPORIDIUM

The collective group *Microsporidium* contains many species that infect fish, some of which may not even belong to the phylum Microsporidia. For example, *Microsporidium pseudotumefaciens* Pflugfelder, 1952, infects common freshwater aquarium fishes in the genera *Brachydanio, Colisa, Lebistes, Gambusia, Platypoecilus, Molliensia,* and *Xiphophorus.* It was thought to cause considerable mortality but has been reported only once (Pflugfelder, 1952, cited in Canning and Lom, 1986). Canning and Lom (1986) transferred this parasite to the group *Microsporidium* but pointed out that its life cycle did not fit that of a microsporidian.

Both Becko disease in yellowtail (*Seriola quinqueradiata*) caused by *Microsporidium seriolae* Egusa 1982 (syn. *Pleistophora* sp. Ghittino 1974), and *M. takedai* Awakura 1974 infection in freshwater salmonids affect muscle and can be pathogenic in commercially important fish in Japan (Awakura, 1974; Egusa, 1982). *M. takedai* has been recorded from various salmonid hosts and geographic locations in Japan (Takeda, 1933; Awakura et al., 1966; Kubota, 1967; Funahashi et al., 1973; Awakura, 1974, 1978; Vyalova, 1984; Vyalova and Voronin, 1987). Hosts for *M. takedai* include chinook, chum (*Oncorhynchus keta*), pink (*O. gorbuscha*), kokanee (*O. nerka*), and masu (*O. masu*) salmon, white-spotted char (*Salvelinus leuconaenis*), brown trout (*Salmo trutta*), Dolly Varden (*S. malma*), and rainbow trout. Kokanee salmon and rainbow trout are particularly susceptible to infection (Urawa and Awakura, 1994). Urawa (1989) found that hatchery-reared masu salmon smolts were very susceptible to *M. takedai* infection when released into rivers. Susceptibility is age dependent, with young fish being highly susceptible (Awakura, 1974). The prevalence of *M. takedai* is also seasonal, peaking during the months of highest water temperature. Urawa (1989) found that prevalence ranged from 80 to 100% in wild and hatchery-reared salmon from 1982 to 1984.

Pathology. Beko disease is characterized by cyst formation in the muscle. Small and large depressions on the lateral surface of yellowtail can indicate the presence of *M. seriolae* cysts (Canning and Lom, 1986). Spores can be found within cysts bounded by a fibrous host membrane (Egusa, 1982). The parasite causes muscle liquefaction, which undoubtedly contributes to lower marketability of infected yellowtail.

M. takedai also forms cysts, which are located in the heart during chronic infections, and in the trunk, fin, jaw, eye, throat, and gullet muscles during acute infections (Fig. 15) (Awakura, 1974). Cysts of *M. takedai* are spindle shaped (2 to 3 mm wide and 3 to 6 mm long) and lack a xenoma wall (Awakura, 1974; Urawa and Awakura, 1994). Both acute and chronic infections, with associated mortality, have been recorded from wild and hatchery-reared fishes. Chronic infections are characterized by cyst formation causing extreme hypertophy, multiplication of connective tissue, vacuolization, and deformation of the heart (Fig. 15a) (Urawa, 1989; Urawa and Awakura, 1994). Acute infections include cyst formation in general skeletal musculature that can result in considerable necrosis. Acute infections have also been associated with extremely high mortality in hatchery fishes (Awakura, 1965; Urawa and Awakura, 1994). Urawa (1989) observed a strong correlation between the condition of fish and the intensity of *M. takedai* infection. Hosts mount a typical inflammatory response to the parasite, and formations of granulomas are observed (Awakura, 1974).

Transmission. Little information is available on the transmission of *M. seriolae.* Awakura (1974) transmitted *M. takedai* per os, and by waterborne infection and suggested that paratenic hosts such as rotifers (*Euchlanis* spp.), and glochidia larvae of mussels (*Margaritifera* spp.) may play a role in transmission. However, Canning and Lom (1986) noted that the organisms he described in potential paratenic hosts were not microsporidia. Temperature plays a significant role in the transmission and development of *M. takedai*. Development may be arrested at 8°C and retarded between 11 and 15°C (Awakura, 1974). A successful control strategy has been to rear smolts below 15°C (Urawa and Awakura, 1994), and fish that have recovered from infection can develop immunity for up to 1 year (Awakura and Kurahashi, 1967).

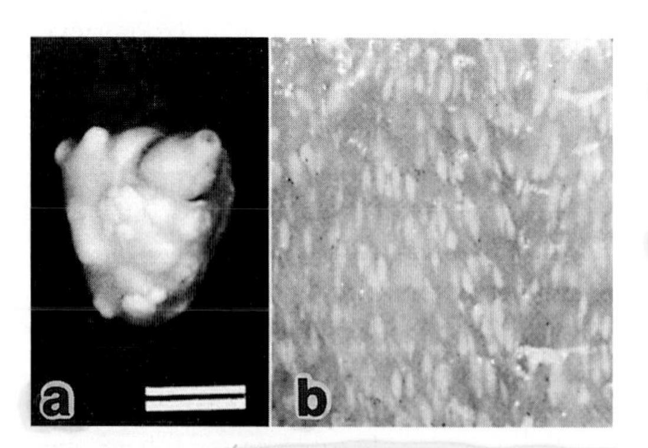

FIGURE 15 (a) *Microsporidium takedai* in the heart of a sockeye salmon (*Oncorhynchus nerka*). Scale bar, 4 mm. (Reprinted with permission of the author and publisher from Canning and Lom, 1986.) (b) *M. takedai* in musculature of a rainbow trout (*Oncorhynchus mykiss*). (Reproduced with permission of the authors and publisher from Lom and Dyková, 1992.)

NUCLEOSPORA

N. salmonis is a remarkable microsporidium that causes disease in salmonid fishes. It infects the nuclei of hemoblasts, particularly lymphoblasts or plasmablasts. Infected cells exhibit massive proliferation and are immature, suggestive of neoplasia. This microsporidium was first observed in pen-reared chinook in Washington State, where it was associated with anemia (Elston et al., 1987). The parasite has also been reported in freshwater-reared chinook, kokanee, and steelhead trout (Morrison et al., 1990; Hedrick et al., 1990, 1991b). The infection is common in caged-reared chinook salmon in British Columbia (Kent, 1998) and in Atlantic salmon in Chile (Bravo, 1996). Intranuclear microsporidia very similar to *N. salmonis* have been observed in Atlantic lumpfish (*Cyclopterus lumpus*) (Mullins et al., 1994) and Atlantic halibut *Hippoglossus hippoglossus* (Nilsen et al., 1995), and these organisms were tentatively assigned to the genus *Enterocytozoon*.

The intranuclear microsporidian of salmonids was originally described as *N. salmonis* by Hedrick et al. (1991a) but was described shortly thereafter as *Enterocytozoon salmonis* by Chilmonczyk et al. (1991). rDNA sequence comparison of *N. salmonis* and *Enterocytozoon bieneusi* does not provide compelling evidence to suppress the nominal genus *Nucleospora*. Although intrageneric sequence comparisons among microsporidia are limited, *N. salmonis* and *E. bieneusi* (at 20.1% genetic divergence in the 16S and 28S genes) show greater differences than congeneric species examined to date (Docker et al., 1997b).

Differences in host and site of infection also support the separation of *Nucleospora* from *Enterocytozoon*. *Nucleospora* develops within the nuclei of fishes, whereas *Enterocytozoon* infects the cytoplasm of enterocytes in humans. *Nucleospora* spp. from fishes differ from *E. bieneusi* in that they possess spores with 8 to 12 turns in the polar filament, whereas those of *E. bieneusi* have only 5 or 6. On the basis of these distinctions and the morphological characteristics available to date, the intranuclear microsporidian found in both Atlantic lumpfish and Atlantic halibut should also be transferred to the genus *Nucleospora*.

Despite the reestablishment of *Nucleospora* and *Enterocytozoon* as separate genera, they are more closely related to each other than they are to other microsporidian genera. Docker et al. (1997b) therefore proposed that *Nucleospora* be retained in the family Enterocytozoonidae. *N. salmonis* exhibits many of the distinctive characteristics of the family Enterocytozoonidae (Cali and Owen, 1990; Desportes-Livage et al., 1996). In other microsporidia, polar tubes do not form until sporonts divide and the plasmalemma has thickened, whereas *Nucleospora* and *Enterocytozoon* form polar tube precursors (i.e., electron-dense disks) prior to plasmodial division and thickening of the sporoginal plasmalemma (Chapter 3). In addition, both *Nucleospora* and *Enterocytozoon* are polysporous but do not form sporophorous vesicles or pansporoblastic membranes and do not possess diplokarya in any stage of development. Nevertheless, after more detailed comparisons, Desportes-Livage et al. (1996) concluded that the morphology and development of these microsporidia are less closely related than originally supposed.

Nucleospora infections are usually associated with a concurrent neoplastic condition involving massive lymphoproliferation, known as plasmacytoid leukemia (PL), in chinook salmon in British Columbia (Kent et al., 1990). The actual cause of PL is controversial. Laboratory transmission studies indicated that *N. salmonis* may not be the primary cause of all cases of PL; i.e., PL was transmitted with tissue homogenates or filtrates in the absence of the microsporidian (Kent and Dawe, 1990; Newbound and Kent, 1991; Kent and Dawe, 1993). Furthermore, Eaton and Kent (1992) described a retrovirus from fish with PL.

Our most convincing studies suggesting a viral etiology for PL were conducted with tissues from fish with no sign of *N. salmonis* infection, which was often the case in pen-reared salmon that we examined in the late 1980s and early 1990s. However, in essentially all cases that we have investigated in recent years, and in studies from other countries, *N. salmonis* is consistently observed in the proliferating plasmablasts. Therefore, it is possible that these morphologically similar lymphoproliferative disorders are actually different diseases caused by two different agents. Studies with fumagillin (Hedrick et al., 1991b) and its analog TNP-470 (Higgins et al., 1998) support the microsporidian hypothesis. In these studies, both the parasite and the lesions were prevented by treatment with these antimicrosporidian compounds, which is in contrast to our earlier experiments with fumagillin in the control of PL (Kent and Dawe, 1993). The microsporidian has been maintained in lymphocyte cultures, and soluble fractions of these cultures stimulate uninfected cells to proliferate (Wongtavatchai et al., 1995a).

Sensitive and specific polymerase chain reaction tests have been developed for the detection of *N. salmonis* based on rDNA sequence from the small-subunit region (Barlough et al., 1995) or ITS region (Docker et al., 1997b).

Pathology. Heavily infected fish are anemic, with a packed blood cell volume as low as 5%. Typical of severe anemia in fish, affected salmon exhibit prominent pallor of the gills.

Histological examination of affected fish reveals a proliferation of plasmablasts in essentially every organ, including the kidney, spleen, liver, intestine, pancreas and associated mesenteric fat, meninges, heart, skeletal muscle, skin, and eye (Hedrick et al., 1990; Morrison et al., 1990). In histological sections, the plasmablasts contain large, often deeply clefted or lobated nuclei and prominent nucleoli. They have a moderate amount of finely granular, eosinophilic, or amphophilic cytoplasm, and many of these cells are mitotically active. The eye, spleen, and kidney are the primary organs affected (Fig. 16).

In the eye, there is massive infiltration of plasmablasts into the periorbital connective tissue and ocular muscles. Kidneys exhibit prominent hyperplasia of the interstitium as a result of proliferation of the plasmablasts. Thickened basement membranes in the capillaries of the glomerulus are often observed. In severely affected fish, there is a perivascular infiltration of the plasmablasts within the liver, and the cells often proliferate within the sinusoids.

The pericardium of the ventricle, atrium, and bulbus arteriosus of the heart may be infiltrated by plasmablasts, which form a thick cellular capsule that surrounds the heart. The cells may also infiltrate the endocardium, particularly in the bulbus arteriosus, and plasmablasts occur throughout the vascular sinuses of the heart. When the lower intestine is affected, there is massive proliferation of the plasmablasts in the lamina propria and in the submucosa, resulting in expansion of the intestinal villi.

The microsporidian is very small and is identified by careful examination of nuclei of hemoblasts in histological sections (Fig. 17a), in Gram-stained imprints (Fig. 17b), or by electron microscopy (Fig. 18). In tissue sections stained with hematoxylin and eosin, the parasites appear as eosinophilic spherical bodies (2 to 4 μm) in host cell nuclei, surrounded by a rim of basophilic host cell chromatin. Warthin-Starry stain combined with hematoxylin and eosin enhances detection of the parasite in tis-

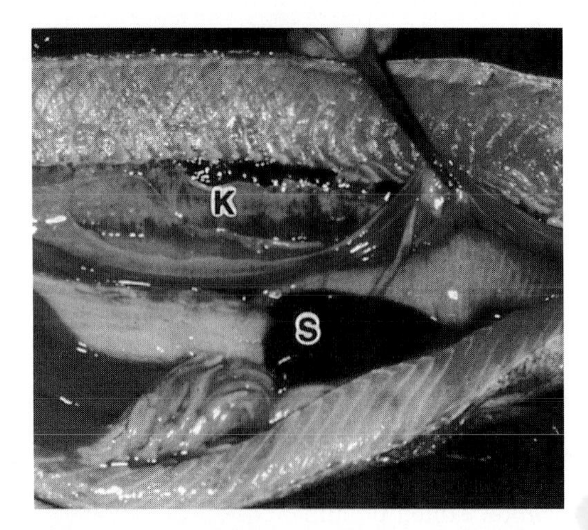

FIGURE 16 Renosplenomegaly in a chinook salmon (*Oncorhynchus tshawytscha*) infected with *Nucleospora salmonis.* K, kidney; S, spleen.

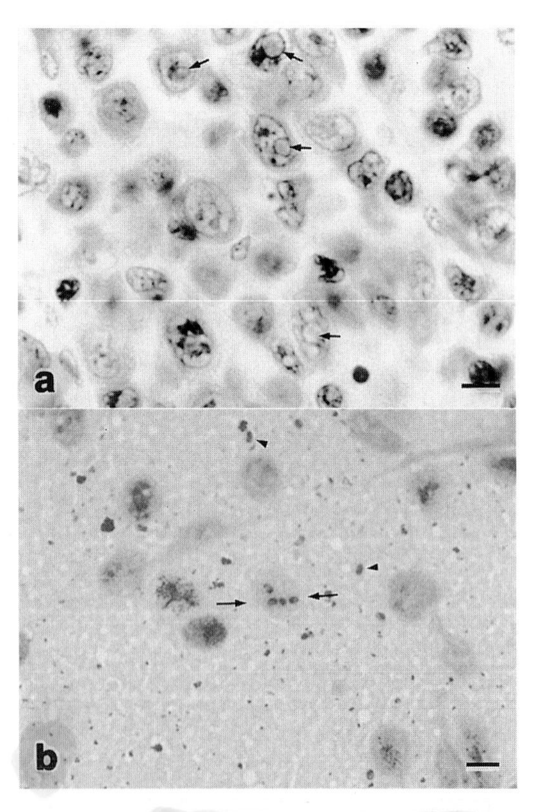

FIGURE 17 (a) *Nucleospora salmonis* (arrows) in the nuclei of hemoblasts from the retrobulbar tissue of the eye of a chinook salmon (*O. tshawytscha*) in histological section. Scale bar, 5 μm. (b) Gram stain of the same tissue, showing four spores in nucleus remnant (arrows) and free spores (arrowheads). Scale bar, 5 μm.

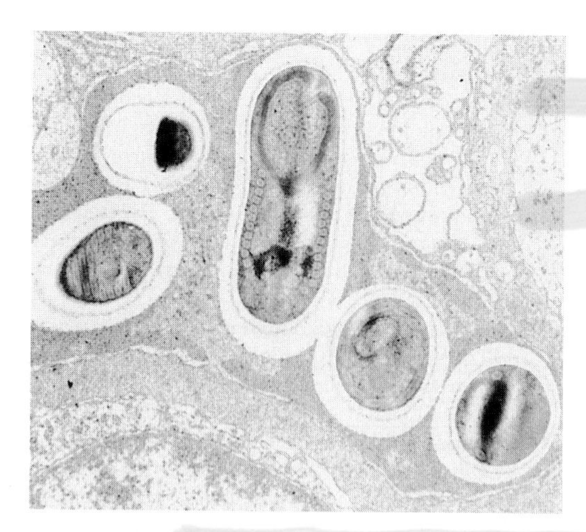

FIGURE 18 Electron micrograph of *N. salmonis* in the nucleus of a hemoblast.

sue sections (Kent et al., 1995). In these preparations, prespore stages stain brown or black, and spores stain dark black. In one outbreak, the infection in pen-reared Atlantic salmon was associated with multiple focal lesions appearing mainly on the head. Histologically the lesions were characterized by massive fibroplasia in which the nuclei of proliferating fibrocytes were infected by the parasite.

Transmission. *N. salmonis* is transmitted by cohabitation or by feeding infected tissues to fish in freshwater (Baxa-Antonio et al., 1992). We repeated these findings in our laboratory but were unable to transmit the infection by cohabitation in seawater.

CONCLUSIONS

In recent years there has been increased research interest in fish microsporidia along with an increase in the economic importance of microsporidia in fish culture. This has lead to a better understanding of the pathogenesis and immune response of the host to the parasite. Furthermore, advances are being made in the development of effective chemotherapeutants for these microsporidia. In the past, fish microsporidia were assigned to specific taxa based mostly on morphological characteristics. Unfortunately, few useful morphological characteristics are present in fish-infecting microsporidia. The use of molecular systematics (Chapter 4), along with interhost transmission experiments, has provided a better understanding of the taxonomic relationships of fish microsporidia, and we suggest that these tools be employed (when possible) as new taxa are proposed.

REFERENCES

Abela, M., J. Brinch-Iversen, J. Tanti, and A. L. Breton. 1996. Occurrence of a new histozoic microsporidian (Protozoa, Microspora) in cultured gilt head sea bream *Sparus aurata* L. *Bull. Eur. Assoc. Fish Pathol.* **16:**196–199.

Anenkova-Khlopina, N. P. 1920. Contribution to the study of parasitic diseases of *Osmerus eperlanus. Izv. Otd. Rybovad. Nauch-prom. Issled.* 1 and 2.

Awakura, T. 1965. Studies on the *Pleistophora* disease of salmonid fishes. I. Observation on the state of the occurrence and the cause in Chitose River in Hokkaido. *Sci. Rep. Hokkaido Fish Hatchery* **20:**1–27.

Awakura, T. 1974. Studies on the microsporidian infection in salmonid fishes. *Sci. Rep. Hokkaido Fish Hatchery* **29:**1–96.

Awakura, T. 1978. A new epizootic of a microsporidiosis of salmonids in Hokkaido, Japan. *Fish Pathol.* **13:**17–18.

Awakura, T., and S. Kurashashi. 1967. Studies on the *Pleistophora* disease of salmonid fish. III. On prevention and control of the disease. *Sci. Rep. Hokkaido Fish Hatchery* **22:**51–68.

Awakura, T., S. Kurashashi, and H. Matsumoto. 1966. Studies on the *Pleistophora* disease of salmonid fishes. II. Occurrence of the microsporidian disease in a new district. *Sci. Rep. Hokkaido Fish Hatchery* **21:**1–12.

Awakura, T., M. Tanaka, and M. Yoshimizu. 1982. Studies of masu salmon, *Oncorhynchus masou.* IV. *Loma* sp. (Protozoa: Microsporea) found in the gills. *Sci. Rep. Hokkaido Fish Hatchery* **37:**49–54.

Barker, D. E., R. A. Khan, and R. Hooper. 1994. Bioindicators of stress in winter flounder, *Pleuronectes americanus,* captured adjacent to a pulp and paper mill in St. George's Bay, Newfoundland. *Can. J. Fish. Aquat. Sci.* **51:**2203–2209.

Barlough, J. E., T. S. McDowell, A. Milani, L. Bigornia, S. B. Slemenda, N. J. Pieniazek, and R. P. Hedrick. 1995. Nested polymerase chain reaction for detection of *Enterocytozoon salmonis* genomic DNA in chinook salmon *Oncorhynchus tshawytscha. Dis. Aquat. Org.* **23:**17–23.

Barton, B. A., and G. K. Iwama. 1991. Physiological changes in fish from stress in aquaculture with emphasis on the response and effects of corticosteroids, p. 3–26. *In* M. Faisal and F. H. Hetrick (ed.), *Annual Review of Fish Diseases.* Pergamon Press, New York, N.Y.

Baxa-Antonio, D., J. M. Groff, and R. P. Hedrick. 1992. Experimental horizontol transmission of *Enterocytozoon salmonis* to chinook salmon, *Oncorhynchus tshawytscha. J. Protozool.* **39:**699–702.

Bazikalova, A. 1932. Additions to parasitology of Murmansk fishes, p.136–153. *In* S. I. Mittel'mana (ed.), *Sbor nauchno-prom. Rab. Murman.* CNIRC, Moscow, Russia.

Bekhti, M., and G. Bouix. 1985. *Loma salmonae* (Putz, Hoffman et Dunbar, 1965) et *Loma diplodae* n. sp., microsporidies parasites de branchies de poissons téléostéens: implantation et données ultrastructurales. *Protistologica* **21:**47–59.

Bogdanova, E. A. 1957. The microsporidian *Glugea hertwigi* Weissenberg in the stint (*Osmerus eperlanus m. spirinchus*) from Lake Ylyna-yarvi, p. 1–328. *In* G. K. Petrushevskii (ed.), *Parasites and Diseases of fish.* Izvestiya Vsesoyuznoga Nauchno-Issledovatelskogo Instituta Ozernogo I Rechnogo Rybnogo Khozyaistva, Leningrad, Russia.

Bosanquet, W. C. 1910. Brief notes on two Myxosporidian organisms (*Pleistophora hippoglossoideos,* n. sp. and *Myxidium mackiei* n. sp.). *Zool. Anz.* **35:** 434–439.

Bravo, S. 1996. *Enterocytozoon salmonis* in Chile. *Fish Health Sect. Am. Fish. Soc. Newsl.* **24**(1):12–13.

Bruno, D. W., R. O. Collins, and C. M. Morrison. 1995. The occurrence of *Loma salmonae* (Protozoa: Microspora) in farmed rainbow trout, *Oncorhynchus mykiss* Walbaum, in Scotland. *Aquaculture* **133:**341–344.

Buchmann, K., K. Ogawa, and C. F. Lo. 1992. Immune reponse of the Japanese eel (*Anguilla japonica*) against major antigens from the microsporean *Pleistophora anguillarum* Hoshina, 1951. *Gyobyo Kenkyu* **27:**157–161.

Bückmann, A. 1952. Infektionen mit *Glugea stephani* und mit *Vibrio anguillarum* bei Schollen (*Pleuronectes platessa* L.). *Kurze Mitt. Fishbiol. Abt. Max Planck Inst. Meeresbiol.* **1:**1–7.

Cali, A., and R. L. Owen. 1990. Intracellular development of *Enterocytozoon,* a unique microsporidian found in the intestine of AIDS patients. *J. Protozool.* **37:**145–155.

Cali, A., P. M. Takvorian, J. J. Ziskowski, and T. K. Sawyer. 1986. Experimental infection of American winter flounder (*Pseudopleuronectes americanus*) with *Glugea stephani* (Microsporida). *J. Fish Biol.* **28:**199–206.

Canning, E. U., and J. Lom. 1986. *The Microsporidia of Vertebrates.* Academic Press, New York, N.Y.

Chatton, E. 1920. Un complexe xéno-parasitaire morphologique et physiologique *Neresheimeria paradoxa* chez *Fritillaria pellucida. C. R. Acad. Sci.* **171:**55–57.

Chen, M., and G. Power. 1972. Infection of the American smelt in Lake Ontario and Lake Erie with the microsporidian parasite *Glugea hertwigi* (Weissenberg). *Can. J. Zool.* **50:**1183–1188.

Chilmonczyk, S., W. T. Cox, and R. P. Hedrick. 1991. *Enterocytozoon salmonis* n. sp.: an intranuclear microsporidium from salmonid fish. *J. Protozool.* **38:**264–269.

Dalmo, R. A., K. Ingebrigtsen, and J. Bøgwald. 1997. Non-specific defence mechanisms in fish, with particular reference to the reticuloendothelial system (RES). *J. Fish Dis.* **20:**241–273.

Debaisieux, P. 1920. Etudes sure les microsporidies. IV. *Glugea anomala* Monz. *Cellule* **30:**215–245.

Delisle, C. E. 1965. A study on the mass mortality of the stunted smelt population, *Osmerus eperlanus mordax* at Heney Lake, Gatineau Co., Quebec, p. 25–27. In *Proceedings of the 10th Technical Session of the Ontario Research Foundation.*

Delisle, C. E. 1972. Variations mensuelles de *Glugea hertwigi* (Sporozoa: Microsporidia) chez différents tissues et organes de l'éperlan adulte dulcicole et consequences de cette infection sur une mortalité massive annuelle de ce poisson. *Can. J. Zool.* **50:** 1589–1600.

Desportes-Livage, I., S. Chilmonczyk, R. Hedrick, C. Ombrouck, D. Monge, I. Maiga, and M. Gentilini. 1996. Comparative development of two microsporidian species: *Enterocytozoon bieneusi* and *Enterocytozoon salmonis,* reported from AIDS patients and salmonid fish, respectively. *J. Eukaryot. Microbiol.* **43:**49–60

Docker, M. F., R. H. Devlin, J. Richard, and M. L. Kent. 1997a. Sensitive and specific polymerase chain reaction assay for detection of *Loma salmonae* (Microsporea). *Dis. Aquat. Org.* **29:**13–20.

Docker, M. F., M. L. Kent, D. M. L. Hervio, L. Weiss, A. Cali, and R. H. Devlin. 1997b. Ribosomal DNA sequence of *Nucleospora salmonis* Hedrick, Groff, and Baxa, 1991 (Microsporea: Enterocytozoonidae): implications for phylogeny and nomenclature. *J. Eukaryot. Microbiol.* **44:**55–60.

Dogiel, V. A. 1936. Parasites of cod from the relic lake Mogilny. *Trans. Leningrad Univ. Biol.* **7:**123–133.

Drew, G. H. 1910. Some notes on parasitic and other diseases of fish, 2nd series. *Parasitol.* **3:**54–62.

Dyková, I. 1995. Phylum Microspora, p. 149–179. *In* P. T. K. Woo (ed.), *Fish Diseases and Disorders,* vol. 1. CAB International, Wallingford, United Kingdom.

Dyková, I., and A. Figueras. 1994. Histopathological changes in turbot *Scophthalmus maximus* due to a histophagous ciliate. *Dis. Aquat. Org.* **18:**5–9.

Dyková, I., and J. Lom. 1978. Tissue reaction of the three-spined stickleback *Gasterosteus aculeatus* L. to infection with *Glugea anomala* (Moniez, 1887). *J. Fish Dis.* **1:**83–90.

Dyková, I., and J. Lom. 1980. Tissue reactions to microsporidian infections in fish. *J. Fish Dis.* **3:** 265–283.

Dyková, I., J. Lom, and S. Egusa. 1980. Tissue reaction to *Glugea plecoglossi* infection in its natural host, *Plecoglossus altivelis. Folia Parasitol.* **27:**213–216.

Eaton, W. D., and M. L. Kent. 1992. A retrovirus in chinook salmon (*Oncorhynchus tshawytscha*) with plasmacytoid leukemia and evidence for the etiology of the disease. *Cancer Res.* **52:**6496–6500.

El-Matbouli, M., and R. W. Hoffmann. 1991. Prevention of experimentally induced whirling disease in rainbow trout *Oncorhynchus mykiss* by fumagillin. *Dis. Aquat. Org.* **10:**109–113.

Egidius, E., and Ø. Soleim. 1986. *Pleistophora ehrenbaumi,* a microsporidian parasite in wolffish: *Anarhichas lupus. Bull. Eur. Assoc. Fish Pathol.* **6:** 13–15.

Egusa, S. 1982. A microsporidian species from yellowtail juveniles, *Seriola quinqueradiata,* with "Beko" disease. *Fish Pathol.* **16:**187–192.

Elston, R. A., M. L. Kent, and L. H. Harrell. 1987. An intranuclear microsporidium associated with acute anemia in chinook salmon, *Oncorhynchus tshawytscha. J. Protozool.* **34:**274–277.

Estévez, J., R. Iglesias, J. Leiro, F. M. Ubeira, and M. L. Sanmartin. 1992. An unusual site of infection by a microsporean in the turbot *Scophthalmus maximus. Dis. Aquat. Org.* **13:**139–142.

Fantham, H. B., A. Porter, and L. R. Richardson. 1941. Some microsporidia found in certain fishes and insects in eastern Canada. *Parasitology* **33:**186–208.

Faye, N., B. S. Toguebaye, and G. Boix. 1990. Ultrastructure and development of *Pleistophora senegalensis* sp. nov. (Protozoa, Microspora) from the gilt-head sea bream, *Sparus aurata* L. (Teleost, Sparidae) from the coast of Senegal. *J. Fish Dis.* **13:**179–192.

Figueras, A., B. Novoa, M. Santarém, E. Martínez, J. M. Alvarez, A.E. Toranzo, and I. Dyková. 1992. *Tetramicra brevifilum,* a potential threat to farmed turbot *Scophthalmus maximus. Dis. Aquat. Org.* **14:**127–135.

Finn, J. P., and N. O. Nielson. 1971. The effect of temperature variation on the inflammatory response of rainbow trout. *J. Pathol.* **105:**257–268.

Fischthal, J. H. 1944. Observations on a sporozoan parasite of the eelpout, *Zoarces anguillaris,* with an evaluation of candling methods for its detection. *J. Parasitol.* **30:**35–36.

Fomena, A., F. Coste, and G. Bouix. 1992. *Loma camerounensis* sp. nov. (Protozoa: Microsporida) a parasite of *Oreochromis niloticus* Linnaeus, 1757 (Teleost: Cichlidae) in fish-rearing ponds in Melen, Yaoundé, Cameroon. *Parasitol. Res.* **78:**201–208.

Funahashi, N., T. Miyazaki, and S. S. Kubota. 1973. Histopathological studies on microsporidian infection in young rainbow trout. *Fish Pathol.* **8:**64–67.

Gandhi, S., L. Locatelli, and S. W. Feist. 1995. Occurrence of *Loma* sp. (Microsporidia) in farmed rainbow trout (*Oncorhynchus mykiss*) at a site in South West England. *Bull. Eur. Assoc. Fish Pathol.* **15:**58–60.

Grabda, J. 1978. Studies on parasitic infestation of blue whiting (*Micromesistius* sp.) with respect to fish utilization for consumption. *Acta Ichthol. Piscatoria* **8:**29–38.

Griffin, B. R. 1983. Opsonic effect of rainbow trout (*Salmo gairdneri*) antibody on phagocytosis of *Yersinia ruckeri* by trout leukocytes. *Dev. Comp. Immunol.* **7:**253–255.

Haley, A. J. 1952. Preliminary observations on a severe epidemic of microsporidiosis in the smelt, *Osmerus mordax* (Mitchell). *J. Parasitol.* **38:**183.

Haley, A. J. 1953. Observations on a protozoan infection in the freshwater smelt, p. 7. *In Proceedings of the 32nd Annual Session of the New Hampshire Academy of Science.*

Haley, A. J. 1954. Microsporidian parasite, *Glugea hertwigi,* in American smelt from the Great Bay region, New Hampshire. *Trans. Am. Microsc. Soc.* **83:** 84–90.

Hauck, A. K. 1984. A mortality and associated tissue reactions of chinook salmon, *Oncorhynchus tshawytscha* (Walbaum), caused by the microsporidan *Loma* sp. *J. Fish Dis.* **7:**217–229.

Hedrick, R. P., J. M. Groff, and D. V. Baxa. 1991a. Experimental infections with *Nucleospora salmonis* n. g. n. sp.: an intranuclear microsporidium from chinook salmon (*Oncorhynchus tshawytscha*). *Fish Health Sect. Am. Fish. Soc. Newsl.* **19:**5.

Hedrick, R. P., J. M. Groff, and D. V. Baxa. 1991b. Experimental infections with *Enterocytozoon salmonis* Chilmonczyk, Cox, Hedrick (Microsporea): an intranuclear microsporidium from chinook salmon *Oncorhynchus tshawytscha. Dis. Aquat. Org.* **10:**103–108.

Hedrick, R. P., J. M. Groff, P. Foley, and T. McDowell. 1988. Oral adminsitration of fumagillin DCH protects chinook salmon *Oncorhynchus tshawytscha* from experimentally induced proliferative kidney disease. *Dis. Aquat. Org.* **4:**165–168.

Hedrick, R. P., J. M. Groff, T. S. McDowell, M. Willis, and W. T. Cox. 1990. Hematopoietic intranuclear microsporidian infections with features of leukemia in chinook salmon *Oncorhynchus tshawytscha. Dis. Aquat. Org.* **8:**189–197.

Higgins, M. J., M. L. Kent, J. D. W. Moran, L. M. Weiss, and S. C. Dawe. 1998. Efficacy of the fumagillin analog TNP-470 for *Nucleospora salmonis* and *Loma salmonae* infections in chinook salmon *Oncorhynchus tshawytscha*. *Dis. Aquat. Org.* **34:**45–49.

Higgins, M. J., and M. L. Kent. 1996. Field trials with fumagillin for the control of proliferative kidney disease in coho salmon. *Prog. Fish Cult.* **58:**268–272.

Hung, H. W., C. F. Lo, C. C. Tseng, and G. H. Kou. 1996. Humoral immune response of Japanese eel *Anguilla japonica* Temminck & Schlegel, to *Pleistophora anguillarum* Hoshina, 1951 (Microspora). *J. Fish Dis.* **19:**243–250.

Hung, H. W., C. F. Lo, C. C. Tseng, and G. H. Kou. 1997. Antibody response of glass eels, *Anguilla japonica* Temminck and Schlegel, to *Pleistophora anguillarum* Hosina (Microspora) infection. *J. Fish Dis.* **20:**237–239.

Jaronski, S. T. 1972. Cytochemical evidence for RNA synthesis inhibition by fumagillin. *J. Antibiot.* **6:**327–331.

Jones, T. O., and H. G. Hirsch. 1972. The interaction between *Toxoplasma gondii* and mammalian cells. II. The absence of lysosomal fusion with phagocytic vacuoles containing living parasites. *J. Exp. Med.* **136:**1173–1194.

Kaattari, S. L., and J. D. Piganelli. 1996. The specific immune system:humoral defense, p. 207–254. *In* G. Iwama and T. Nakanishi (ed.), *The Fish Immune System: Organism, Pathogen, and Environment.* Academic Press, San Diego, Calif.

Kabata, Z. 1959. On two little-known microsporidia of marine fishes. *Parasitology* **49:**309–315.

Kano, T., and H. Fukui. 1982. Studies on *Pleistophora* infection in eel, *Anguilla japonica.* I. Experimental induction of microsporidiosis and fumigillin efficacy. *Fish Pathol.* **16:**193–200.

Kano, T., T. Okauchi, and H. Fukui. 1982. Studies on *Pleistophora* infection in eel, *Anguilla japonica.* II. Preliminary tests for application of fumagillin. *Fish Pathol.* **17:** 107–114.

Kent, M. L. 1992. Diseases of seawater netpen-reared salmonid fishes in the Pacific Northwest. *Can. Spec. Publ. Fish. Aquat. Sci.* 116.

Kent, M. L., and S. C. Dawe. 1990. Transmission of a plasmacytoid leukemia of chinook salmon *Oncorhynchus tshawytscha.* *Cancer Res.* **50**(Suppl.): 5679S–5681S.

Kent, M. L., and S. C. Dawe. 1993. Further evidence for a viral etiology in plasmacytoid leukemia of chinook salmon. *Dis. Aquat. Org.* **15:**115–121

Kent, M. L., and S. C. Dawe. 1994. Efficacy of fumagillin DCH against experimentally induced *Loma salmonae* (Microsporea) infections in chinook salmon *Oncorhynchus tshawytscha. Dis. Aquat. Org.* **20:**231–233.

Kent, M. L., S. C. Dawe, and D. J. Speare. 1995. Transmission of *Loma salmonae* (Microsporea) to chinook salmon in sea water. *Can. Vet. J.* **36:** 98–101.

Kent, M. L., D. G. Elliott, J. M. Groff, and R. P. Hedrick. 1989. *Loma salmonae* (Protozoa: Microspora) infections in seawater reared coho salmon *Oncorhynchus kisutch. Aquaculture* **80:**211–222.

Kent, M. L., and J. W. Fournie. 1993. Importance of marine fish diseases: an overview, p.1–24. *In* J. A. Couch and J. W. Fournie (ed.), *Advances in Fisheries Science Pathobiology of Marine and Estuarine Organisms.* CRC Press, Boca Raton, Fla.

Kent, M. L., J. M. Groff, G. S. Traxler, J. G. Zinkl, and J. W. Bagshaw. 1990. Plasmacytoid leukemia in seawater reared chinook salmon *Oncorhynchus tshawytscha. Dis. Aquat. Org.* **8:**199–209.

Kent, M. L. 1998. Protozoa and myxozoa, p. 49–67. *In* M. L. Kent and T. T. Poppe (ed.), *Diseases of Seawater Netpen-Reared Salmonid Fishes.* Pacific Biological Station, Nanaimo, British Columbia, Canada.

Kent, M. L., V. Rantis, J. W. Bagshaw, and S. C. Dawe. 1995. Enhanced detection of *Enterocytozoon salmonis* (Microspora), an intranuclear microsporean of salmonid fishes, with the Warthin-Starry stain combined with hematoxylin and eosin. *Dis. Aquat. Org.* **23:**235–237.

Kent, M. L., G. S. Traxler, D. Kieser, J. Richard, S. C. Dawe, R. W. Shaw, G. Prosperi-Porta, J. Ketchenson, and T. P. T. Evelyn. 1998. Survey of salmonid pathogens in ocean-caught fishes in British Columbia, Canada. *J. Aquat. Anim. Health* **10:**211–219.

Kim, J. H., H. Yokoyama, K. Ogawa, S. Takahashi, and H. Wakabayashi. 1996. Humoral immune response of ayu, *Plecoglossus altivelis,* to *Glugea plecoglossi* (Protozoa: Microspora). *Fish Pathol.* **31:** 215–220.

Kinkelin, P. D. 1980. Occurrence of a microsporidian infection in zebra danio *Brachydanio rerio* (Hamilton-Buchanan). *J. Fish Dis.* **3:**71–73.

Kubota, S. S. 1967. Pathological studies on the regressive processes of cardiac muscle found on the X-disease, and the additional discussion on the etiology of the disease. *J. Fac. Fish. Prefect. Univ. Mie* **7:**199–209.

Kusaka, M., K. Sudo, E. Matsutani, Y. Kozai, S. Marui, T. Fujita, D. Ingber, and J. Folkman. 1994. Cytostatic inhibition of endothelial cell growth by the angiogenesis inhibitor TNP-470 (AGM-1470). *Br. J. Cancer* **69:**212–216.

Laudan, R., J. S. Stolen, and A. Cali. 1986a. Immunoglobulin levels of the winter flounder *Pseudopleuronectes americanus* and summer flounder *Paralichthys dentarus* injected with the microsporidian parasite *Glugea stephani. Dev. Comp. Immunol.* **10:**331–340.

Laudan, R., J. S. Stolen, and A. Cali. 1986b. The immune response of a marine teleost, *Pseudopleuronectes americanus* (winter flounder) to the protozoan parasite *Glugea stephani*. *Vet. Immunol. Immunopathol.* **12:**403–412.

Laudan, R., J. S. Stolen, and A. Cali. 1987. The immunomodulating effect of the microsporidan *Glugea stephani* on the humoral response and immunoglobulin levels in winter flounder, *Pseudopleuronectes americanus. J. Fish Biol.* **31:**155–160.

Laudan, R., J. S. Stolen, and A. Cali. 1989. The effect of the microsporidia *Glugea stephani* on the immunoglobulin levels of juvenile and adult winter flounder (*Pseudopleuronectes americanus*). *Dev. Comp. Immunol.* **13:**35–41.

Laurén, D. J., A. Wishkovsky, J. M. Groff, R. P. Hedrick, and D. E. Hinton. 1989. Toxicity and pharmocokinetics of the antibiotic fumagillin in yearling rainbow trout (*Salmo gairdneri*). *Toxicol. Appl. Pharmacol.* **98:**444–453.

Leiro, J. J. Estevez, M. T. Santamarina, M. L. Sanmartin, and F. M. Ubeira. 1993. Humoral immune response of turbot, *Scophthalmus maximus* (L.), to antigens from *Tetramicra brevifilum* Matthews & Matthews, 1980 (Microspora). *J. Fish Dis.* **16:**577–584.

Leiro, J., J. Estévez, F. M. Ubeira, M. T. Santamarina, and M. L. Sanmartin. 1994. Serological relationships between two microsporidian parasites of fish. *Aquaculture* **125:**1–9.

Leiro, J., M. Ortega, J. Estévez, M. T. Santamarina, M. L. Sanmartin, and F. M. Ubeira. 1996a. The humoral immune response of turbot, *Scophthalmus maximus* L., to spore-surface antigens of microsporidian parasites. *Vet. Immunol. Immunopathol.* **55:**235–242.

Leiro, J., M. Ortega, J. Estévez, F. M. Ubeira, and M. L Sanmartin. 1996b. The role of opzonization by antibody and complement in in vitro phagocytosis of microsporidian parasites by turbot spleen cells. *Vet. Immunol. Immunopathol.* **51:**201–210.

Leiro, J., M. Ortega, R. Iglesias, J. Estévez, and M. L. Sanmartin. 1996c. *Pleistophora finisterrensis* n. sp., a microsporidian parasite of blue whiting *Micromesistius poutassou. Syst. Parasitol.* **34:**163–170.

Leiro, J., M. I. G. Siso, M. Ortega, M. T. Santamarina, and M. L. Sanmartin. 1995. A factorial experimental design for investigation of the effects of temperature, incubation time, and pathogen-to-phagocyte ratio on in vitro phagocytosis by turbot adherent cells. *Comp. Biochem. Physiol.* **112C:**215–220.

Lom, J. 1969. Experimental transmission of a microsporidian, *Pleistophora hyphessobryconis,* by intramuscular transplantation. *J. Protozool.* **16**(Suppl.)**:**17.

Lom, J. and I. Dyková. 1992. *Developments in Aquaculture and Fisheries Science,* vol. 26. *Protozoan Parasites of Fishes.* Elsevier, Amsterdam, The Netherlands.

Lom, J., I. Dyková, W. Körting, and H. Klinger. 1989. *Heterosporis schuberti* n. sp., a new microsporidian parasite of aquarium fish. *Eur. J. Protistol.* **25:**129–135.

Lom, J., I. Dyková, K. Tonguthai, and S. Chinabut. 1993. Muscle infection due to *Heterosporis* sp. in the Siamese fighting fish, *Betta splendens* Regan. *J. Fish Dis.* **16:**513–516.

Lom, J., E. J. Noga, and I. Dyková. 1995. Occurrence of a microsporean with characteristics of *Glugea anomala* in ornamental fish of the family Cyprinodontidae. *Dis. Aquat. Org.* **21:**239–242.

Magor, B. G. 1987. First report of *Loma* sp. (Microsporida) in juvenile coho salmon (*Oncorhynchus kisutch*) from Vancouver Island, British Columbia. *Can. J. Zool.* **65:**751–752.

Mann, H. 1954. Die wirtschaftliche Bedeutung von Krankheiten bei Seefischen. *Fischwirt. Bremerhaven* **6:**38–39.

Markey, P. T., V. S. Blazer, M. S. Ewing, and K. M. Kocan. 1994. *Loma* sp. in salmonids from the Eastern United States: associated lesions in rainbow trout. *J. Aquat. Anim. Health* **6:**318–328.

Matsuyama, H., T. Yano, T. Yamakawa, and M. Nakao. 1992. Opsonic effect of the third complement component (C3) of carp (*Cyprinus carpio*) on phagocytosis by neutrophils. *Fish Shellfish Immunol.* **2:**68–78.

Matthews, R. A. and B. F. Matthews. 1980. Cell and tissue reactions of turbot *Scophthalmus maximus* (L.) to *Tetramicra brevifilum* gen. n., sp. n. (Microspora). *J. Fish Dis.* **3:**495–515.

Maurand, J., C. Loubés, C. Gasc, J. Pelletier, and J. Barral. 1988. *Pleistophora mirandellae* Vaney and Conte, 1901, a microsporidian parasite in cyprinid fish of rivers in Hérault: taxonomy and histopathology. *J. Fish Dis.* **11:**251–259.

McVicar, A. H. 1975. Infection of plaice *Pleuronectes platessa* L. with *Glugea* (*Nosema*) *stephani* (Hagenmüller 1899) (Protozoa: Microsporidia) in a fish farm and under experimental conditions. *J. Fish Biol.* **7:**611–619.

Meyer, A. 1952. Veränderung des Fleisches beim Katfish. *Fischereiwelt,* 57–58.

Michel, C., J. Maurand, C. Loubés, S. Chilmonczyk, and P. Kinkelin. 1989. *Heterosporis finki,* a microsporidian parasite of the angel fish *Pterophyllum scalare:* pathology and ultrastructure. *Dis. Aquat. Org.* **7:**103–109.

Möller, H. and K. Möller. 1986. *Diseases and Parasites of Marine Fishes.* Verlag Möller, Kiel, Germany.

Molnár, K., F. Baska, and C. Székely. 1987. Fumagillin, an efficacious drug against renal sphaerosporosis of the common carp *Cyprinus carpio. Dis. Aquat. Org.* **2:**187–190.

Mora, J. A. L. 1988. Prospeccion de los principales generos parasitarios que se encuentran en salmonidos provenientes de tres pisciculturas de agua dulce de la provincia de Llanquihue (X region) Chile. Veterinary thesis. Universidad Austral de Chile, Valdivia, Chile.

Morrison, C. M. 1983. The distribution of the microsporidian *Loma morhua* in the tissues of the cod *Gadus morhua* L. *Can. J. Zool.* **61:**2155–2161.

Morrison, C. M., and V. Marryatt. 1986. Further observations on *Loma morhua* Morrison & Sprague, 1981. *J. Fish Dis.* **9:**63–67.

Morrison, C. M., V. Marryatt, and B. Gray. 1984. Structure of *Pleistophora hippoglossoideos* Bosanquet in the American plaice *Hippoglossoides platessoides* (Fabricius). *J. Parasitol.* **70:**412–421.

Morrison, C. M., and V. Sprague. 1981a. Electron microscopical study of a new genus and new species of microsporida in the gills of Atlantic cod *Gadus morhua* L. *J. Fish Dis.* **4:**15–32.

Morrison, C. M., and V. Sprague. 1981b. Light and electron microscope study of microsporida in the gill of haddock, *Melanogrammus aeglefinus* (L.). *J. Fish Dis.* **4:**179–184.

Morrison, C. M., and V. Sprague. 1981c. Microsporidian parasites in the gills of salmonid fishes. *J. Fish Dis.* **4:**371–386.

Morrison, C. M., and V. Sprague. 1983. *Loma salmonae* (Putz, Hoffman and Dunbar, 1965) in the rainbow trout, *Salmo gairdneri* Richarson, and *L. fontinalis* sp. nov. (Microsporida) in the brook trout, *Salvelinus fontinalis* (Mitchill). *J. Fish Dis.* **6:** 345–353.

Morrison, J. K., E. MacConnell, P. F. Chapman, and R. L. Westgard. 1990. A microsporidium-induced lymphoblastosis in chinook salmon *Oncorhynchus tshawytscha*. *Dis. Aquat. Org.* **8:**99–104.

Mullins, J. E., M. Powell, D. J. Speare, and R. Cawthorn. 1994. An intranuclear microsporidian in lumpfish *Cyclopterus lumpus*. *Dis. Aquat. Org.* **20:**7–13.

Nagel, M. L., and G. L. Hoffman. 1977. A new host for *Pleistophora ovariae* (Microsporida). *J. Parasitol.* **63:**160–162.

Nagel, M. L., and R. C. Summerfelt. 1977. Nitrofurazone for control of the microsporidan parasite *Pleistophora ovariae* in golden shiners. *Prog. Fish Cult.* **39:**18–23.

Nepszy, S. J., J. Budd, and A. O. Dechtiar. 1978. Mortality of young-of-the-year rainbow smelt (*Osmerus mordax*) in Lake Erie associated with the occurrence of *Glugea hertwigi*. *J. Wildl. Dis.* **14:** 233–239.

Nepszy, S. J., and A. O. Dechtiar. 1972. Occurrence of *Glugea hertwigi* in Lake Erie rainbow smelt (*Psmerus mordax*) and associated mortality of adult smelt. *J. Fish. Res. Board Can.* **29:**1639–1641.

Newbound, G. C., and M. L. Kent. 1991. Experimental interspecies transmission of plasmacytoid leukemia in salmonid fishes. *Dis. Aquat. Org.* **10:** 159–166.

Nilsen, F., A. Ness, and A. Nylund. 1995. Observations on an intranuclear microsporidian in lymphoblasts from farmed Atlantic halibut larvae (*Hippoglossus hippoglossus* L.). *J. Eukaryot. Microbiol.* **42:**131–135.

Nigrelli, R. F. 1946. Studies on the marine resources of southern New England. V. Parasites and diseases of the ocean pout, *Macrozoarces americanus*. *Bull. Bingham Oceanogr. Coll.* **9:**187–221.

Noga, E. J. 1996. *Fish Disease Diagnosis and Treatment*. Mosby-Year Book, St. Louis, Mo.

Olivier, G., C. A. Eaton, and N. Campbell. 1986. Interaction between *Aeromonas salmonicida* and peritoneal macrophages of brook trout (*Salvelinus fontinalis*). *Vet. Immunol. Immunopathol.* **12:**223–234.

Olson, R. E. 1976. Laboratory and field studies on *Glugea stephani* (Hagenmuller), a microsporidan parasite of pleuronectid flatfishes. *J. Protozool.* **23:** 158–164.

Olson, R. E. 1981. The effect of low temperature on the development of microsporidan *Glugea stephani* in English sole (*Parophrys vetulus*). *J. Wildl. Dis.* **17:** 559–562.

Overstreet, R. M. 1993. Parasitic diseases of fishes and their relationship with toxicants and other environmental factors, p. 111–156. *In* J. A. Couch and J. W. Fournie (ed.), *Advances in Fisheries Science Pathobiology of Marine and Estuarine Organisms*. CRC Press, Boca Raton, Fla.

Pedrera, I. M., A. B. Rodríguez, G. M. Salido, and C. Barriga. 1993. Phagocytic process of head kidney granulocytes of tench (*Tinca tinca*, L.). *Fish Shellfish Immunol.* **3:**411–421.

Petrushevski, G. K., and S. S. Shulman. 1958. Parasitic diseases in fish in water reservoirs of the USSR, p. 301–320. *In* G. K. Petrushevski and Y. I. Polyanski (ed.), *Parasitology of Fishes*. Leningrad University Press, Leningrad, Russia.

Poynton, S. L. 1986. Distribution of the flagellate *Hexamita salmonis*, Moore, 1922 and the microsporidian *Loma salmonae* Putz, Hoffman and Dunbar, 1965 in brown trout, *Salmo trutta* L., and the rainbow trout, *Salmo gairdner* Richardson, in the River Itchen (U.K.) and three of its fish farms. *J. Fish Biol.* **29:**417–429.

Price, R. L. 1982. Incidence of *Pleistophora cepedianae* (Microsporida) in gizzard shad (*Dorosoma cepedianum*) of Carlyle Lake, Illinois. *J. Parasitol.* **68:** 1167–1168.

Pulsford, A., and R. A. Matthews. 1991. Macrophages and giant cells associated with a microsporidian parasite causing liquefaction of the skeletal muscle of the Norway pout, *Trisopterus esmarkii* (Nilsson). *J. Fish Dis.* **14:**67–78.

Putz, R. E. 1969. Parasites of freshwater fishes. II. Protozoa 1. Microsporidia of fishes. *Fish Dis. Leaflet* **20.**

Putz, R. E., G. L. Hoffman, and C. E. Dunbar. 1965. Two new species of *Pleistophora* (Microsporidea) from North American fish with a synopsis of microsporidea of freshwater and euryhaline fishes. *J. Protozool.* **12:**228–236.

Ralphs, J. R., and R. A. Matthews. 1986. Hepatic microsporidiosis of juvenile grey mullet, *Chelon labrosus* (Risso), due to *Microgemma hepaticus* gen. nov. sp. nov. *J. Fish Dis.* **9:**225–242.

Reimschuessel, R., R. O. Bennett, E. B. May, and M. M. Lipsky. 1987. Eosinophilic granular cell response to a microsporidian infection in a sergeant major fish, *Abudefduf saxatilis* (L.). *J. Fish Dis.* **10:**319–322.

Rose, A. S., and R. P. Levine. 1992. Complement-mediated opsonization and phagocytosis of *Renibacterium salmoninarum. Fish Shellfish Immunol.* **2:** 223–240.

Saggers, B. A., and M. L. Gould. 1989. The attachment of microorganisms to macrophages isolated from tilapia *Oreochromis spilurus* Gunther. *J. Fish Biol.* **35:**287–294.

Sakai, D. K. 1984. Opzonization by fish antibody and complement in the immune phagocytosis by peritoneal exudate cells isolated from salmonid fishes. *J. Fish Dis.* **7:**29–38.

Sandholzer, L. A., T. Nostrand, and L. Young. 1945. Studies on an ichthyosporidian-like parasite of ocean pout (*Zoarces anguillaris*). *U.S. Fish Wild. Serv. Spec. Sci. Rep.* **31:**1–12.

Schmahl, G., and H. Mehlhorn. 1989. Treatment of fish parasites. 6. Effects of sym-triazinone (Toltrazuril) on developmental stages of *Glugea anomala* Moniez, 1887 (Microsporidia): a light and electron microscopic study. *Eur. J. Protistol.* **24:** 252–259.

Schmahl, G., A. E. Toukhy, and F. A. Ghaffer. 1990. Transmission electron microscopic studies on the effects of toltrazuril on *Glugea anomala,* Moniez, 1887 (Microsporidia) infecting the three-spinded stickleback *Gasterosteus aculeatus. Parasitol. Res.* **76:**700–706.

Schuberg, A. 1910. Über Mikrosporidien aus dem Hoden der Barbe und durch sie verursachte Hypertrophie der Kerne. *Arb. Kaiserl. Gesundh.* (Berlin) **33:**401–434.

Schubert, G. 1969. Ultracytologische Untersuchungen an der Spore der Mikrosporidienart, *Heterpsoris finki* gen. n., sp.n. *Z. Parasitenkd.* **32:**59–79.

Scott, A. L., W. A. Rogers, and P. H. Klesius. 1985. Chemiluminescence by peripheral blood phagocytes from channel catfish: function of opsonin and temperature. *Dev. Comp. Immunol.* **9:**241–250.

Secombes, C. J. 1996. The nonspecific immune system: cellular defenses, p. 63–103. *In* G. Iwama and T. Nakanishi (ed.), *The Fish Immune System: Organism, Pathogen, and Environment,* Academic Press, San Diego, Calif.

Shaw, R. W. Unpublished data.

Shaw, R. W., M. L. Kent, and M. L. Adamson. 1996. Variation in susceptibility of three chinook salmon (*Oncorhynchus tshawystcha*) strains to *Loma salmonae* (Protozoa: Microspora), p. 109. Abstr. Joint Meet. Am. Soc. Parasitol. Soc. Protozool. 1996, Tucson, Ariz.

Shaw, R. W., M. L. Kent, and M. L. Adamson. 1998. Modes of transmission of *Loma salmonae* (Microsporidia). *Dis. Aquat. Org.* **33:**151–156.

Shaw, R. W., M. L. Kent, M. F. Docker, A. M. V. Brown, R. H. Devlin, and M. L. Adamson. 1997. A new species of *Loma* (Microsporea) in shiner perch (*Cymatogaster aggregata*). *J. Parasitol.* **83:**296–301.

Sheehy, D. J., M. P. Sissenwine, and S. B. Saila. 1974. Ocean pout parasites. *Mar. Fish. Rev.* **36:**29–33.

Shulman, S. S., and R. E. Shulman-Albova. 1953. *Parasites of White Sea Fishes.* Academy of Sciences Publishing House, USSR.

Sindermann, C. J. 1963. Disease in marine populations. *Trans. N. Am. Wildl. Nat. Res. Conf.* **28:**336–356.

Sindermann, C. J. 1966. Diseases of marine fishes. *Adv. Mar. Biol.* **4:**1–89.

Sindermann, C. J. 1990. *Principal Diseases of Marine Fish and Shellfish,* 2nd ed., vol. 1. Academic Press, San Diego, Calif.

Sitjà-Bobadilla, A., and P. Alvarez-Pellitero. 1992. Effect of fumagillin treatment on sea bass *Dicentrarchus labrax* parasitized by *Sphaerospora testicularis* (Myxosporea: Bivalvulida). *Dis. Aquat. Org.* **14:**171–178.

Speare, D. J. Personal communication.

Speare, D. J., G. J. Arsenault, and M. A. Boute. 1998a. Evaluation of rainbow trout as a model species for studying the pathogenesis of the branchial microsporidian *Loma salmonae. Contemp. Top. Lab. Anim. Sci.* **37:**55–58.

Speare, D. J., H. J. Beaman, S. R. M. Jones, R. J. F. Markham, and G. J. Arsenault. 1998b. Induced resistance of rainbow trout to gill disease associated with the microsporidian gill parasite *Loma salmonae. J. Fish Dis.* **21:**93–100.

Speare, D. J., J. Brackett, and H. W. Ferguson. 1989. Sequential pathology of the gills of coho salmon with a combined diatom and microsporidian gill infection. *Can. Vet. J.* **30:**571–575.

Sprengel, G., and H. Lüchtenberg. 1991. Infection by endoparasites reduces maximum swimming speed of European smelt *Osmerus eperlanus* and European eel *Anguilla anguilla*. *Dis. Aquat. Org.* **11:** 31–35.

Summerfelt, R. C. 1964. A new microsporidian parasite from the golden shiner, *Notemigonus crysoleucas. Trans. Am. Fish. Soc.* **93:**6–10.

Summerfelt, R. C. 1972. Studies on the transmission of *Pleistophora ovariae,* the ovary parsite of the golden shiner (*Notemigonus crysoleucas*), p. 1–19. *In* Final Report, Project 4-66-R. National Marine Fisheries Service.

Summerfelt, R. C., and M. C. Warner. 1970. Geographical distribution and host parasite relationships of *Pleistophora ovariae* (Microsporida, Nosematidae) in *Notemigonus crysoleucas. J. Wildl. Dis.* **6:**457–465.

Székely, C., K. Molnár, and F. Baska. 1988. Efficacy of fumagillin against *Myxidium giardi* Cépéde, 1906 infection of the European eel (*Anguilla anguilla*): new observations of imported class eels. *Acta Vet. Hung.* **36:**239–246.

Takeda, S. 1933. A new disease of rainbow trout. *Keison Iho* **5:**1–9.

Takvorian, P. M., and A. Cali. 1981. The occurrence of *Glugea stephani* (Hagenmuller, 1899) in American winter flounder, *Pseudopleuronectes americanus* (Walbaum) from the New York-New Jersey lower bay complex. *J. Fish Biol.* **18:**491–501.

Takvorian, P. M., and A. Cali. 1984. Seasonal prevalence of the microsporidan, *Glugea stephani* (Hagenmuller), in winter flounder, *Pseudopleuronectes americanus* (Walbaum), from the New York-New Jersey lower bay complex. *J. Fish Biol.* **24:**655–663.

T'sui, W. H., and C. H. Wang. 1988. On the *Pleistophora* infection in eel. I. histopathology, ultrastructure and development of *Pleistophora anguillarum* in eel, *Anguilla japonica. Bull. Inst. Zool. Acad. Sin.* **27:**159–166.

T'sui, W. H., C. H. Wang, and C. F. Lo. 1988. On the *Pleistophora* infection in eel. II. the development of *Pleistophora anguillarum* in experimentally infected elvers, *Anguilla japonica. Bul. Inst. Zool. Acad. Sin.* **27:**249–258.

Urawa, S. 1989. Seasonal occurrence of *Microsporidium takedai* (Microsporida) infection in masu salmon, *Oncorhynchus masou,* from the Chitose River. *Physiol. Ecol. Jpn Spec.* **1:**587–598.

Urawa, S., and T. Awakura. 1994. Protozoan diseases of freshwater fishes in Hokkaido. *Sci. Rep. Hokkaido Fish Hatchery* **48:**47–58.

Vyalova, G. P. 1984. Microsporidian infection in pink salmon. *Fish Res. Inland Waters* **10:**9–11.

Vyalova, G. P., and V. N. Voronin. 1987. Microsporidiosis of salmonids from Sakhalin: distribution and dynamics of the infection. *Parazitologia* **21:**553–558.

Wales, J. H, and H. Wolf. 1955. Three protozoan diseases of trout in California. *Calif. Fish Game* **41:** 183–187.

Waterstrat, P. R., A. J. Ainsworth, and G. Capley. 1991. In vitro responses of channel catfish, *Ictalurus punctatus,* neutrophils to *Edwardsiella ictaluri. Dev. Comp. Immunol.* **15:**53–63.

Weidner, E., and L. D. Sibley. 1985. Phagocytized intracellular microsporidian blocks phagosome acidification and phagosome-lysosome fusion. *J. Protozool.* **32:**311–317.

Weissenberg, R. 1921. Zür Wirtsgewebsableitung des Plasmakorpers der *Glugea anomala* Cysten. *Arch. Protistenk.* **42:**400–421.

Weissenberg, R. 1922. Mikrosporidien und Chlamydozoen als Zellparasiten von Fischen. *Verh. Deutsch. Zool. Ges.* **27:**41–43.

Weissenberg, R. 1967. Contribution to the study of the intracellular development of the microsporidium *Glugea anomala* in the fish *Gasterosteus aculeatus. J. Protozool.* **14:**28–29.

Weissenberg, R. 1968. Intracellular development of the microsporidian *Glugea anomala* Moniez in hypertrophying migratory cells of the fish *Gasterosteus aculeatus* L., an example of the formation of "xenoma tumors." *J. Protozool.* **15:**44–57.

Wiklund, T., L. Lounasheimo, J. Lom, and G. Bylund. 1996. Gonadal impairment in roach *Rutilus rutilus* from Finnish coastal areas of the northern Baltic sea. *Dis. Aquat. Org.* **26:**163–171.

Wiseman, D. L., and J. A. Brown. 1996. Early growth and survival of larval striped wolffish (*Anarhichas lupus*): a behavioural approach. *Bull. Aquac. Assoc. Can.* **1:**12–13.

Wishkovsky, A., J. M. Groff, D. J. Lauren, R. J. Toth, and R. P. Hedrick. 1990. Efficacy of fumagillin against proliferative kidney disease and its side effects in rainbow trout (*Oncorhynchus mykiss*) fingerlings. *Fish Pathol.* **25:**141–146.

Wongtavatchai, J, P. A. Conrad, and R. P. Hedrick. 1995a. In vitro characteristics of the microsporidian: *Enterocytozoon salmonis. J. Eukaryot. Microbiol.* **42:**401–405.

Wongtavatchai, J, P. A. Conrad, and R. P. Hedrick. 1995b. Effect of the microsporidian *Enterocytozoon salmonis* on the immune response of chinook salmon. *Vet. Immunol. Immunopathol.* **48:** 367–374.

Wyrick, R. B., and E. A. Brownridge. 1978. Growth of *Chlamydia psittaci* in macrophages. *Infect. Immun.* **19:**1054–1060.

Yokoyama, H., K. Ogawa, and H. Wakabayashi. 1990. Chemotherapy with fumagillin and toltrazuril against kidney enlargement disease of goldfish caused by the myxosporean *Hoferellus carassii. Fish Pathol.* **25:**157–163.

MICROSPORIDIA IN INSECTS

James J. Becnel and Theodore G. Andreadis

14

The earliest record of a microsporidian parasite can probably be attributed to Gluge (1838), who reported *Glugea anomala* (Moniez, 1887) from a fish host. However, the first named species of microsporidia was *Nosema bombycis* Naegeli, 1857, from the silkworm (*Bombyx mori*). Pasteur, in his landmark studies, established that a protozoan parasite (later determined to be *N. bombycis* by Balbiani [1882] and Stempell [1909]) was the etiological agent of pebrine or silkworm disease. His careful investigations proved that this parasite was transmitted from adult to progeny via the egg (transovarial transmission) and by the ingestion of spores. On the basis of this information, he developed preventative methods that essentially saved the silkworm industry worldwide. Pasteur's investigations provided the foundation for all subsequent studies on entomogenous microsporidia. Many of these investigations have been directed at exploiting microsporidia as manipulated biological control agents or to cure beneficial insects of disease. Both areas have received considerable attention and have required an in-depth understanding of the biology of the microsporidium as well as its relationship to the host. Perhaps no other group has been more carefully studied than the entomogenous species, and this work has provided much of our basic knowledge on the microsporidia.

Currently, almost half of the 143 described genera of microsporidia have an insect as the type host. Most of these genera are distinguished on the basis of descriptions of one sporulation sequence and one spore type with no information on transmission. The dogma "one spore, one species" was first questioned by Hazard and Weiser (1968). They conducted transmission studies with *Parathelohania anophelis* in mosquitoes and discovered that two functionally and morphologically distinctive spore types were part of the same life cycle. This was the first study in which one microsporidian species produced two spore types, and it had important taxonomic implications (Sprague, 1976). Two more recent discoveries with taxonomic implications have also resulted from careful transmission studies. One of these was the revelation that some species of *Amblyospora* required a copepod intermediate host in order to complete the life cycle (Sweeney et al., 1985; Andreadis, 1985a). The other was the discovery that some species of the genera *Nosema, Vairimorpha,* and *Edhazardia* produced a previously unknown binucleate spore early in the infection process that functioned to spread the parasite to other host tissues (Iwano and Ishihara,

James J. Becnel, USDA/ARS, Center for Medical, Agricultural, and Veterinary Entomology, Gainesville, Florida 32604. *Theodore G. Andreadis,* Connecticut Agricultural Experiment Station, New Haven, Connecticut 06504.

The Microsporidia and Microsporidiosis (Murray Wittner, editor; Louis M. Weiss, contributing editor), ©1999 American Society for Microbiology, Washington, D.C.

1989; de Graaf et al., 1994a, 1994b; Johnson et al., 1997; Solter et al., 1997; Solter and Maddox, 1998). The discovery of currently unrecognized sporulation sequences for other genera of microsporidia from insects will unquestionably have profound taxonomic and biological implications for all groups of microsporidia.

It is the intent of this chapter to highlight the biological and life cycle features of entomogenous microsporidia and provide some basic information on their taxonomic distribution. Because of the large volume of information on entomogenous microsporidia, we have selected representative species and studies for the discussions in this chapter. For additional information, the following references are suggested: Kudo (1924), Weiser (1961, 1977), Bulla and Cheng (1976, 1977), Larsson (1986a, 1988), Beyer and Issi (1986), Canning and Lom (1986), Brooks (1988), Sprague et al. (1992), and Undeen and Vávra (1997).

PATHOLOGY

There is wide variation in the types of pathologies that can occur in insects as a result of infection with microsporidia. Generally, microsporidian infections (microsporidioses) are classified as chronic (slow-acting, progressing seriousness) and rarely as acute (fast acting, short duration). The various signs and symptoms associated with microsporidiosis in insects range from obvious (patent) tissue manifestations to abnormal developmental and behavioral changes.

Organism Level

Microsporidia have exploited practically all tissues of all stages of the insect host. Infected insects often exhibit external as well as internal changes as a result of development of the microsporidium. The external effects can be expressed as changes in color, size, form, or behavior as compared to healthy individuals.

The two most commonly infected insect tissues from which microsporidia have been reported are the fat body and midgut epithelium. In many soft-bodied aquatic insects with transparent cuticles, infections are often indicated by the porcelain-white appearance of the midgut cells or the fat body in the thoracic and abdominal segments. These segments may be distended and deformed as a result of the accumulation of large masses of spores (Fig. 1 and 2). Infections in the midgut are normally restricted to the gastric ceca (Fig. 3) and/or Malphigian tubules. In insects with more heavily scleratized cuticles, signs of fat body infections are less obvious, and hosts must usually be dissected to detect the parasite. In these cases, fat body tissue is often lobate and porcelain-white in appearance as compared to healthy tissue (Fig. 4 to 7).

Overt signs of infection in larvae may include the formation of dark, melanized spots or areas on the cuticle that often appear puffy in comparison to healthy individuals (Fig. 8) and are probably due to host defense reactions. Infected larvae and pupae are typically stunted and can exhibit dramatic differences in size compared to healthy individuals of the same age (Fig. 9 and 10). Infected adults are often smaller and may have deformed wings or exhibit lethargic activity. Symptoms of infection in adults, however, are often more subtle and are expressed as general effects on fitness, including a shortened life span and/or a reduction in fertility (fecundity plus natality).

Some microsporidia may also cause systemic infections, particularly in the later stages of infection. In such cases virtually all tissues are eventually invaded, and death of the host follows. With certain polymorphic microsporidia another condition can occur in which sequential successions of different tissues are infected during the course of parasite development. For example, in *Amblyospora* spp. infections spread from the gastric ceca (gut) to the oenocytes as the host matures from one larval molt to the next or from larva to pupa to adult (see "Life Cycles").

Cellular Level

The development of all insect-parasitic microsporidia is restricted to the cytoplasm of the host cell. Invasion of the nucleus is rare and is thought to be inadvertent. Because the parasite

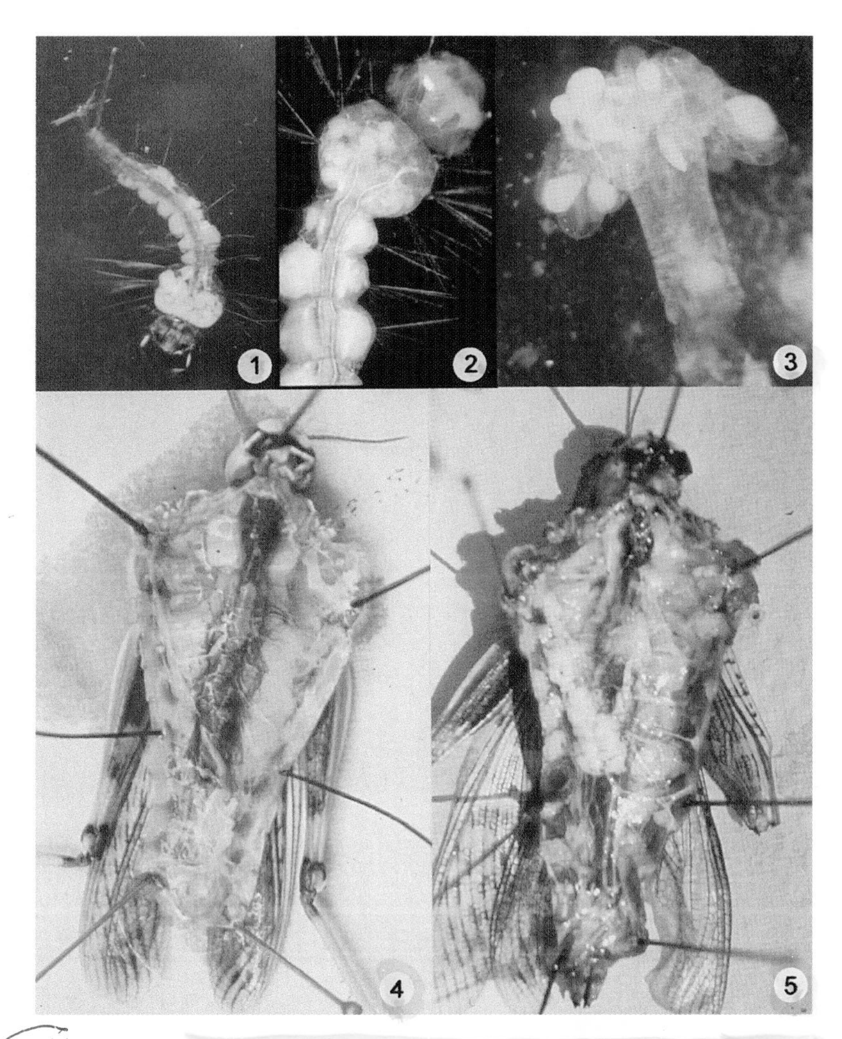

FIGURES 1–5 Gross pathology of insects with microsporidian infections. **FIGURE 1** *Culex salinarius* larva exhibiting fat body infection with *Amblyospora salinaria*. Magnification, ×4.6. **FIGURE 2** *Aedes aegypti* larva exhibiting fat body infection with *Edhazardia aedis*. Magnification, ×6.6. **FIGURE 3** Gastric ceca of a *Culiseta inornata* larva infected with *Polydispyrenia caecorum*. White cysts are cells filled with spores. Magnification, ×13.2. **FIGURE 4** Dissected *Locusta migratoria capito* exhibiting healthy, yellowish-appearing fat body. Magnification, ×1.3. **FIGURE 5** Dissected *L. migratoria capito* infected with *Johenrea locustae*, demonstrating infected fat body tissue appearing as white cysts. Magnification, ×1.2. Reprinted from Lange et al. (1996) with permission from the authors and publisher (Fig. 4 and 5).

develops intracellularly, host defense reactions are usually limited to infections when cell integrity has been compromised. The most typical defense reaction is melanization of spores that have been released into the hemolymph (Fig. 11 and 12). Cell and nuclear hypertrophy are common features of infection. The former presumably results from rapid multiplication of the parasite to completely fill the cytoplasm with stages of the microsporidium, as demonstrated in the midgut epithelium (Fig. 13 to 16) and in oenocytes (Fig. 17).

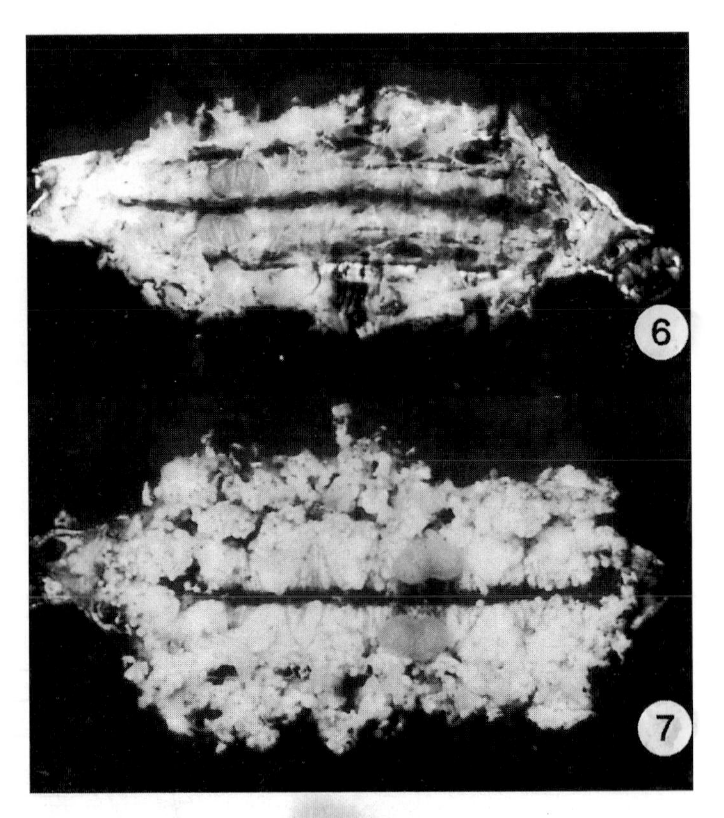

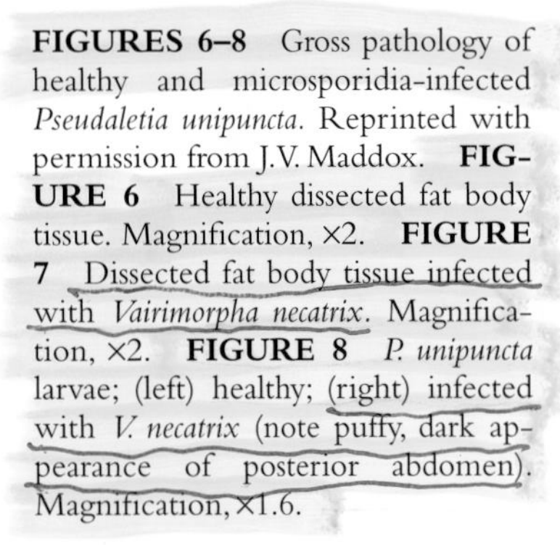

FIGURES 6–8 Gross pathology of healthy and microsporidia-infected *Pseudaletia unipuncta.* Reprinted with permission from J.V. Maddox. **FIGURE 6** Healthy dissected fat body tissue. Magnification, ×2. **FIGURE 7** Dissected fat body tissue infected with *Vairimorpha necatrix*. Magnification, ×2. **FIGURE 8** *P. unipuncta* larvae; (left) healthy; (right) infected with *V. necatrix* (note puffy, dark appearance of posterior abdomen). Magnification, ×1.6.

Specialized relationships between microsporidia and the host at the cellular level have been termed xenomas (Canning and Lom, 1986). Weiser (1976) distinguished two main types of xenomas in insects, syncytial and neoplastic. A syncytial xenoma results from the fusion of infected host cells to form large, multinucleated plasmodia. The nuclei of these plasmodia do not increase in number but are generally hypertrophied. An example of this type of xenoma is found in the fat body of the blackfly (*Odagmia ornata*) infected with *Amblyospora bracteata*. Neoplastic xenomas are characterized by two types of parasite-induced

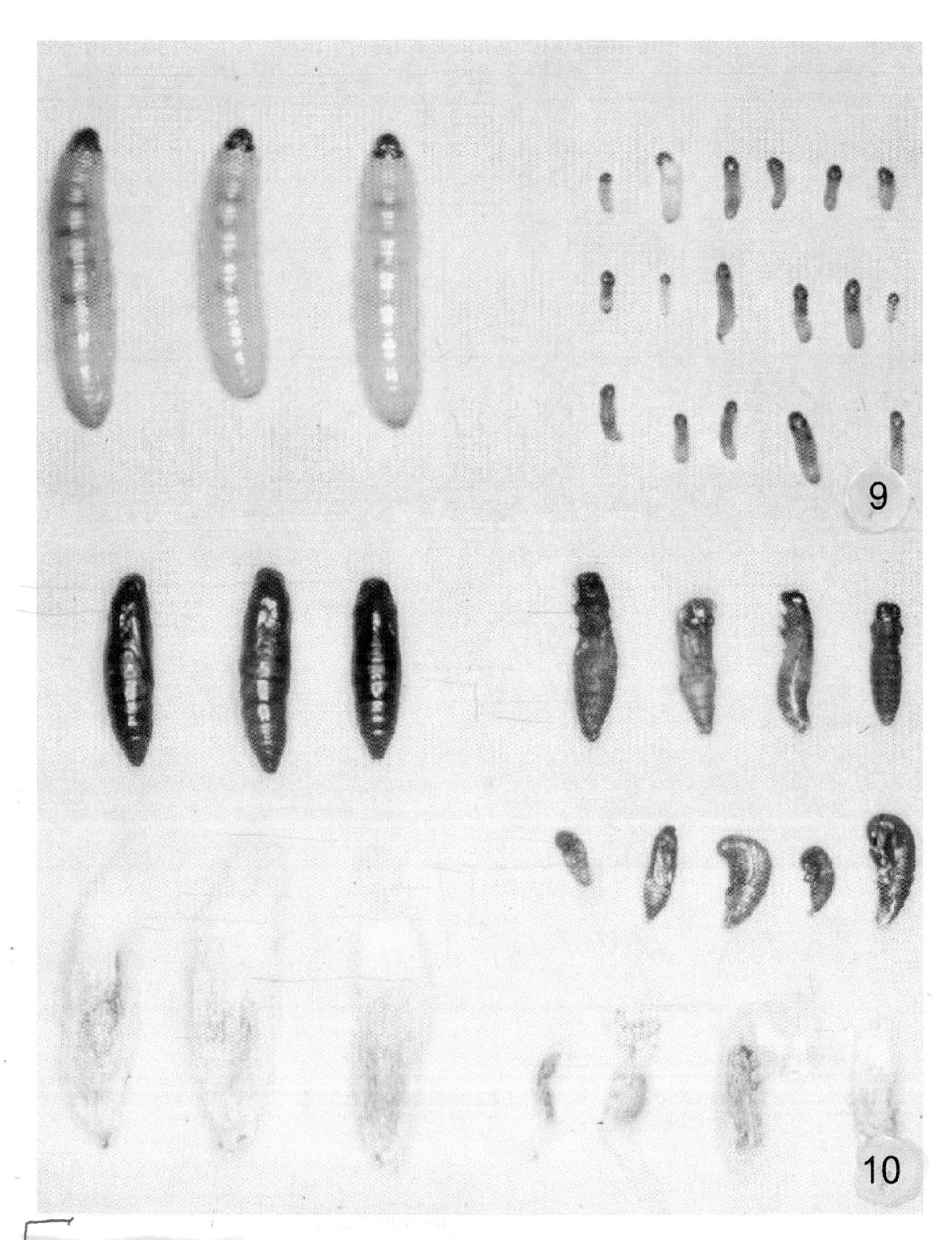

FIGURES 9 AND 10 Navel orangeworm (*Amyelois transitella*) larvae infected with *Vairimorpha heterosporium*. Reprinted with permission from W. Kellen, Horticultural Crops Research Laboratory, USDA, ARS, Fresno, Calif. **FIGURE 9** Large, healthy control larvae on left, and stunted, infected larvae on right. Four weeks after infection. **FIGURE 10** Pupae infected with *V. heterosporium*. Large, healthy control pupae at left, and stunted, deformed, infected pupae on right.

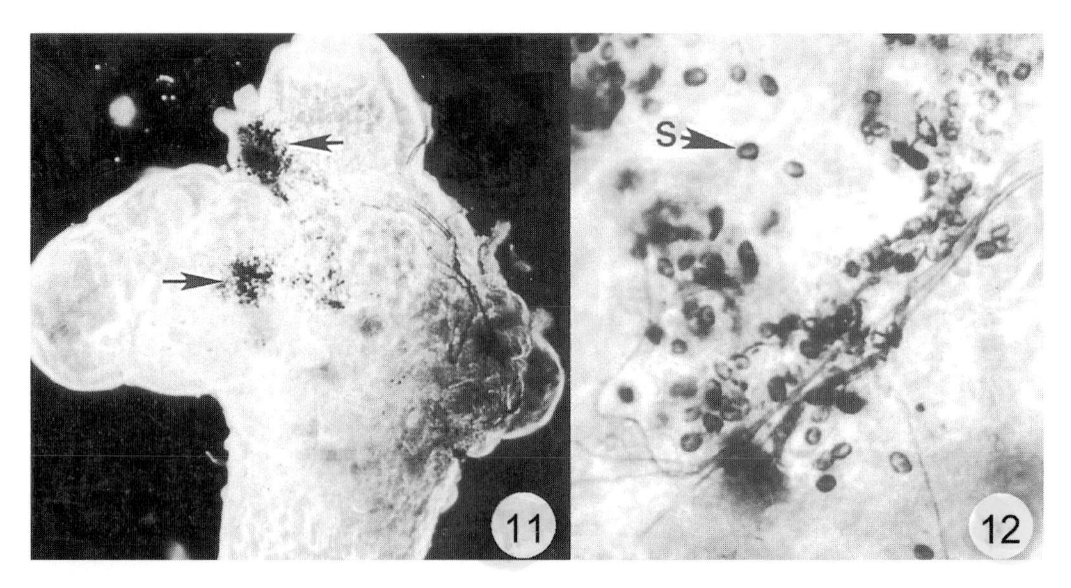

FIGURES 11 AND 12 Gastric ceca of *Aedes albopictus,* demonstrating immune reaction to *E. aedis* infection. Reprinted with permission from M. A. Johnson. **FIGURE 11** Whole gastric cecum, exhibiting melanization of *E. aedis* primary binucleate spores (arrows). Magnification, ×40. **FIGURE 12** High magnification of infected area showing melanized spores. Magnification, ×250. S, spores.

increases (hyperplasia) in the number of infected host cells. In one type, the infected cells divide such that each parasite is contained in a cell called a "xenocyte." The number of cells in the infected tissue can increase to 10 to 30 times that of the original number found in the fat body of blackflies infected with *Janacekia debaisieuxi* (Weiser, 1976). The other type of neoplastic xenoma is the *Glugea*-type cyst commonly reported from fish (Canning and Lom, 1986). In this case, the host cell nucleus is hypertrophic and often becomes branched or fragmented into numerous separate nuclei. The cell surface of the infected cell can become greatly modified, and the xenoma can become very large, up to 13 mm in size. This type of xenoma is relatively rare in insect microsporidia, but a new species from the migratory locust (*Locusta migratoria capito*) has been described with a *Glugea*-type xenoma (Lange et al., 1996). The xenoma of *Johenrea locustae* (Fig. 18 and 19) is composed primarily of fat body cells with the wall limited by a basement membrane beneath which there is a region of collagen fibers (Fig. 20). This and other types of xenomas apparently provide benefits to both the parasite and the host. The parasite has a suitable environment in which to develop protected from host defense reactions. The host restricts the parasite to certain areas, preventing spread to other more essential tissues. This is, however, a very poorly studied area in which there is little information on the physiological and immunological aspects of xenoma formation and function.

Impact on Hosts

Most insect microsporidia induce sublethal effects on hosts, resulting in reduced fertility, shortened longevity, and a loss of vigor (Brooks, 1988). Because of the chronic nature of infections caused by many entomogenous microsporidia, effects on the host are commonly evaluated by measuring the impact on fitness. The individual or cumulative effects of lower survival, reduced longevity, and fertility can be used to measure fitness. Reductions in pupal weight, fecundity, and adult longevity of host

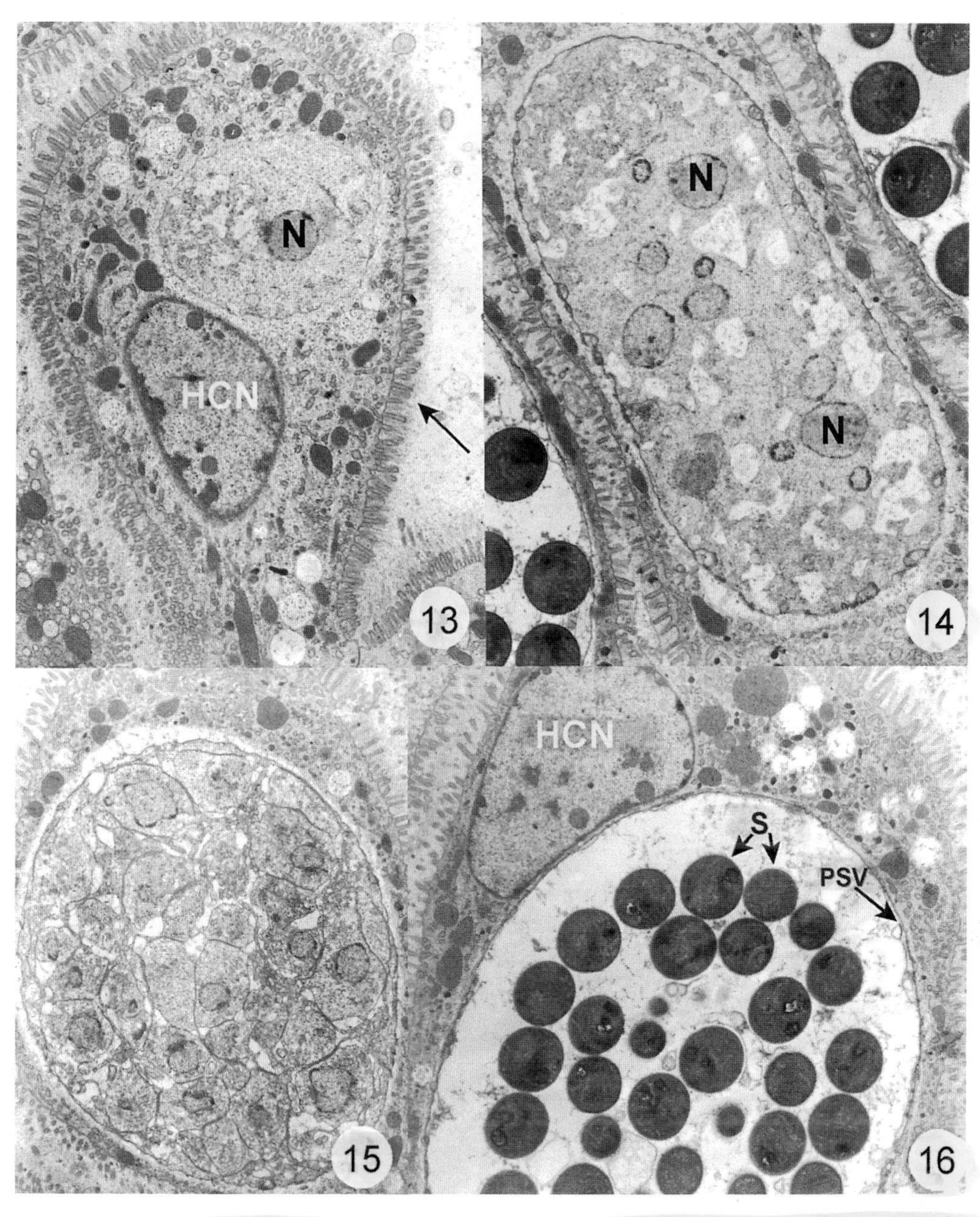

FIGURES 13–16 Electron micrographs of *Nolleria pulicis* demonstrating developmental cycle within midgut epithelium cells of *Ctenocephalides felis.* **FIGURE 13** Early plasmodium at onset of nuclear division and microvilli (arrow). Magnification, ×5,600. **FIGURE 14** Multinucleate plasmodium. Magnification, ×6,400. **FIGURE 15** Multinucleate sporogonial plasmodium in early stages of multiple division by vacuolation. Magnification, ×4,250. **FIGURE 16** Mature spores within polysporophorous vesicle. Magnification, ×5,900. Reprinted from Beard et al. (1990) with permission from the authors and publisher. HCN, host cell nucleus; N, nucleus; PSV, polysporophorous vesicle; S, spores.

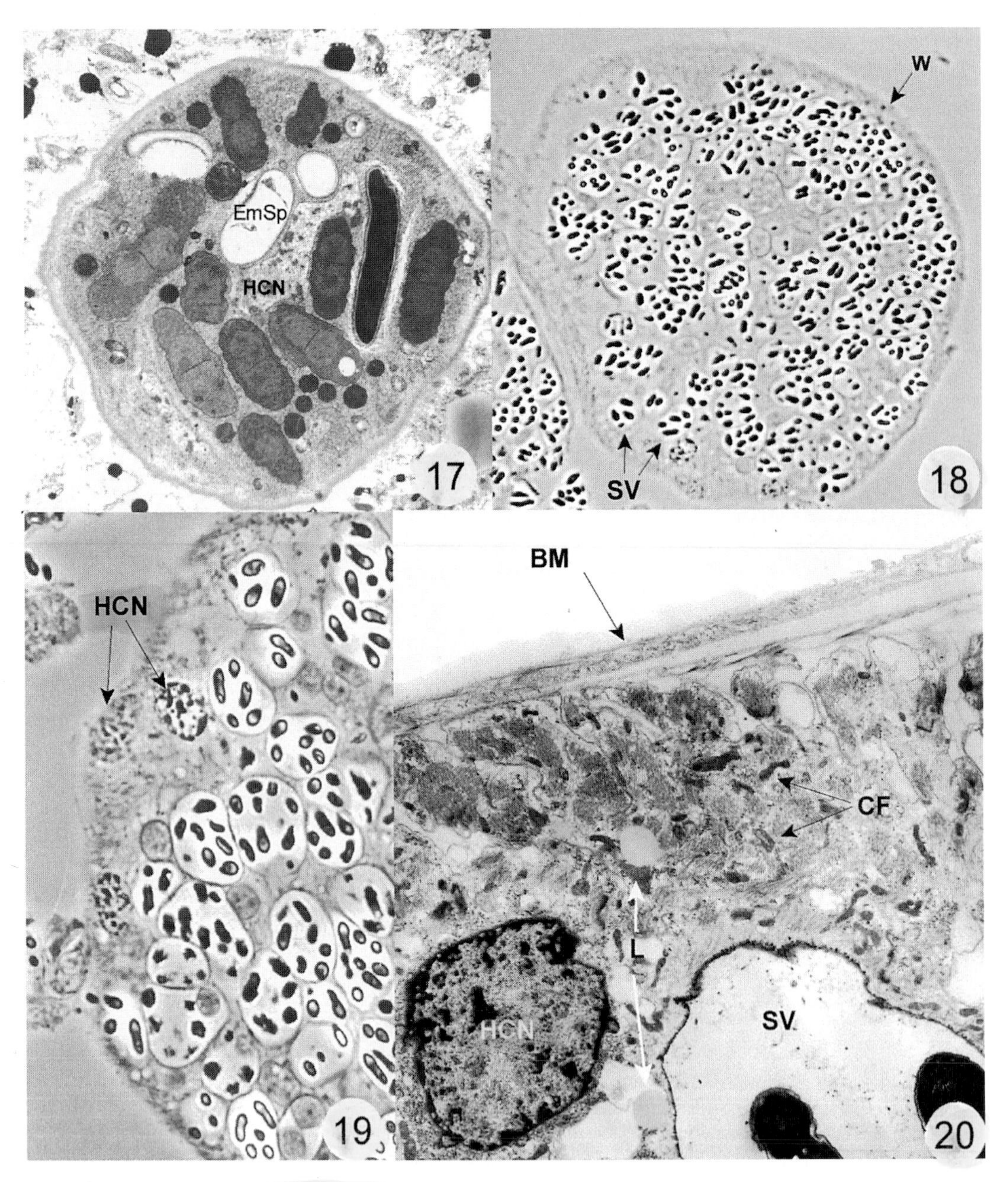

FIGURES 17–20 Microsporidia-infected host cells. **FIGURE 17** Oenocyte in adult *Aedes triseriatus* infected with *Pilosporella chapmani*. Magnification, ×4,400. **FIGURE 18** Xenoma of *J. locustae*. Magnification, ×250. **FIGURE 19** Higher magnification of *J. locustae* xenoma demonstrating multiple host cell nuclei. Magnification, ×350. **FIGURE 20** Higher magnification of xenoma wall of *J. locustae*. Magnification, ×6,500. Reprinted from Becnel et al. (1986) (Fig. 17) and Lange et al. (1996) (Fig. 18 to 20) with permission of the authors and publishers. BM, basement membrane; CF, collagen fibrils; EmSP, empty spore; HCN, host cell nucleus; L, lipid; SV, sporophorous vesicle; W, wall.

insects have been documented for many microsporidia, including *Nosema pyrausta* (Windels et al., 1976), *N. heliothidis* (Gaugler and Brooks, 1975), *N. fumiferanae* (Wilson, 1977), and *Endoreticulatus schubergi* (Wilson, 1984). *Nosema muscidifuracis,* a parasite of the parasitoid *Muscidifurax raptor,* seriously reduced the effectiveness of this commercially produced biological control agent of muscoid flies (Becnel and Geden, 1994). Adults of *M. raptor* infected with *N. muscidifuracis* live half as long and produce about 10% as many progeny as uninfected parasitoids (Geden et al., 1995). In addition, 100% of the progeny from infected females are infected. In a study on *Edhazardia aedis,* a pathogen of *Aedes aegypti,* it was estimated that the reproductive capacity of the female mosquito was reduced 98.2% as compared to that of uninfected individuals (Becnel et al., 1995). The reasons for reduced fecundity and egg hatch in infected insects are not known but have been related to destruction of the ovaries or depletion of nutritional reserves by the parasite (Gaugler and Brooks, 1975; Brooks, 1988).

Disease Management

The chronic infections caused by microsporidia are a significant problem for all types of beneficial insects from honeybees to biological control agents such as parasitoids. Colonies must be constantly monitored and screened for the presence of disease, and when disease is detected, remedial actions must be implemented immediately. This procedure can often be as simple as selecting uninfected progeny (the Pasteur method) followed by good sanitation procedures (Undeen and Vávra, 1997). The most successful drug therapy has been the use of fumagillin (Fumidil B) to control *Nosema apis* in colonies of honeybees (Goetze and Zeutzschel, 1959; Gochnauer et al., 1975; Furgala and Mussen, 1978; Brooks, 1988). While Fumadil B has been effective against *N. apis* and some other species of microsporidia, it may be ineffective or provide only temporary suppression of other entomogenous microsporidia (Brooks, 1988). Newer drugs, such as albendozole (Haque et al., 1993), have not been evaluated for their effectiveness against entomogenous microsporidia. Heat therapy has also been used successfully in at least two cases to reduce the prevalence of microsporidiosis (Raun, 1961; Geden et al., 1995). This method exploits the higher heat tolerance of the host compared to that of the parasite. While this can be an effective method, host survival at the temperatures required to eliminate the parasite (47°C for 30 to 45 min) is poor (Geden et al., 1995).

TRANSMISSION

Horizontal Transmission

The major and minor pathways of transmission for insect microsporidia are presented in Fig. 21. The most common method of transmission is through direct oral ingestion of infectious spores found in food or liquid within the insect's immediate environment (soil, water, plants) (Canning, 1970; Andreadis, 1987). Transmission through oral ingestion is dependent on the acquisition of a sufficient quantity of spores, as well as appropriate conditions (pH, enzymes) within the host's alimentary tract to facilitate spore germination and successful entry into susceptible host tissues (Tanada, 1976). A detailed discussion of the infective process is presented in Chapter 7.

Contamination of the environment with spores is typically achieved when an infected host dies or when spores are released in fecal excrement. In general, spores of enteric microsporidia (i.e., those that infect the alimentary tract) are almost always disseminated in feces (Weiser, 1961), and they may be passed out throughout the lifetime of an infected host. In microsporidia that infect fat body tissue or produce systemic infections, however, spore release into the environment can take place only with the disintegration of infected tissues following the death of the host. This is commonly observed in most aquatic dipteran hosts, especially mosquitoes (Becnel, 1994; Lucarotti and Andreadis, 1995).

Pathogen dissemination and subsequent oral transmission may also be effected through regurgitation of spores and through secretion of

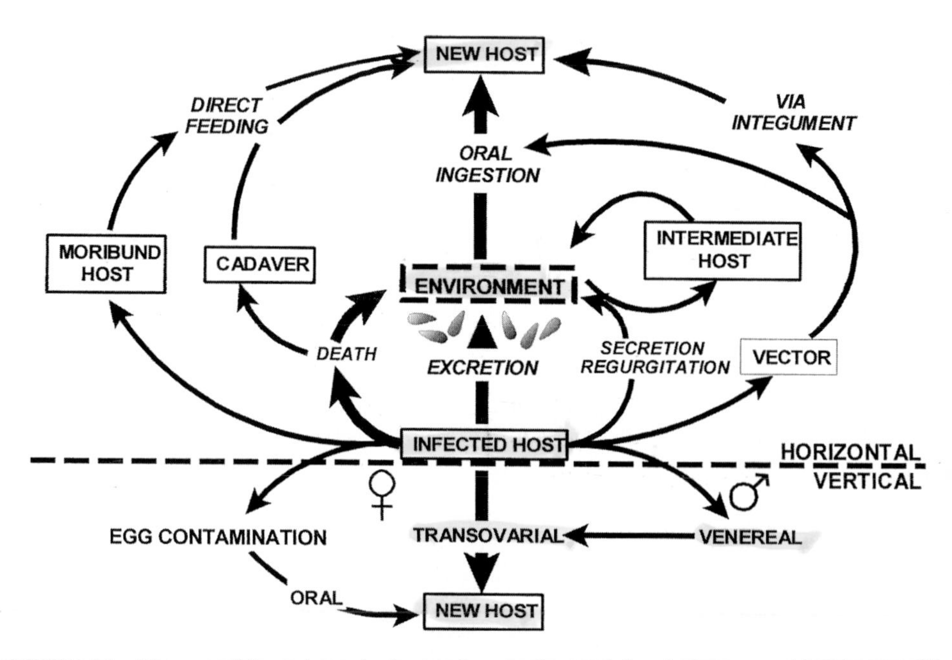

FIGURE 21 Diagram of the major and minor pathways of transmission for insect parasitic microsporidia.

spores in larval silk when these glands are heavily infected. The former has been reported with *N. fumiferanae* in the spruce budworm (*Choristoneura fumiferana*) (Thomson, 1958; Wilson, 1982), and the latter with an undescribed *Nosema* sp. in the gypsy moth (*Lymantria dispar*) (Jeffords et al., 1987), both lepidopteran hosts.

Direct insect-to-insect transmission via oral ingestion of microsporidian spores occurs through cannibalistic feeding on weak or moribund infected individuals or by feeding on infected cadavers (Brooks, 1988). According to Kramer (1976), cannibalism probably occurs to some extent in all gregarious insect hosts for microsporidia. It is reported to be the primary method of horizontal transmission of *Nosema locustae* in rangeland grasshopper populations (Henry, 1972) and has been observed in Indian meal moth larvae (*Plodia interpunctella*) (Lepidoptera) infected with *Vairimorpha plodiae* (Kellen and Lindegren, 1971) and in flour beetles (*Tribolium confusum* and *T. castaneum*) (Coleoptera) infected with *Nosema whitei* (Watson, 1979).

Microsporidia can also be transmitted by the ovipositional activities of hymenopterous parasitoids. Brooks (1973, 1993) has extensively reviewed this subject. In most instances, entry into the insect host is through the integument via intrahemocelic inoculation of spores on the parasitoid's contaminated ovipositor. This vector-mediated mode of transmission is generally thought to be mechanical (i.e., the microsporidium neither develops nor multiplies within the parasitoid), and contamination of the adult parasitoid vector typically occurs when uninfected females initially oviposit in infected hosts. According to Brooks (1973, 1993), most evidence for mechanical transmission of microsporidia by parasitoids in nature is presumptive and is based primarily on a close association between the prevalence of microsporidian infection and the prevalence of parasitism in the insect host. However, this mode of transmission has been documented in laboratory studies involving at least three different pathogen-parasitoid-host complexes. The parasitoid is first allowed to oviposit in an infected host prior to exposure to an uninfected host: (1) a *Nosema* sp. transmitted by *Cotesia marginiventris* to the lawn armyworm (*Spodoptera mauritia acronytoides*) (Lepidoptera) (Laigo and Tamashiro, 1967); (2) a

Vairimorpha sp. transmitted by *Microplitis croceipes* to the corn earworm (*Helicoverpa zea*) (Lepidoptera) (Hamm et al., 1983); and (3) *Nosema epilachnae* transmitted by *Pediobius foveolatus* to the Mexican bean beetle (*Epilachna varivestis*) (Coleoptera) (Own and Brooks, 1986).

In some instances, female parasitoids that develop in infected hosts may also become infected with the host microsporidium and serve as vectors when they subsequently oviposit in an uninfected host. This phenomenon has been thoroughly documented in the transmission of *N. pyrausta* to larvae of the European corn borer (*Ostrinia nubilalis*) (Lepidoptera) by the braconid parasitoid *Macrocentrus grandii* (Andreadis, 1980; Siegel et al., 1986b). However, it is unclear whether actual transmission during oviposition is biological (via direct inoculation of spores) or mechanical (via a contaminated ovipositor).

Certain parasitoids that are directly susceptible to their host's microsporidia may further transmit the infection to their own progeny transovarially, provided they survive to the adult stage (Brooks, 1993). The extent to which this mode of transmission occurs in nature is unknown, but under laboratory conditions, transovarial transmission to parasitoid progeny through several generations has been reported with at least five species: *Nosema campoletidis* (*Campoletis aeneoviridis*) (Brooks, 1973; McNeil and Brooks, 1974), *N. heliothides* (*Campoletis sonorensis*) (Brooks and Cranford, 1972; Brooks, 1973), *Nosema mesnili* (*Cotesia glomerata*) (Blunck, 1954; Issi and Maslennikova, 1966), *N. pyrausta* (*M. grandii*) (Siegel et al., 1986b), and *Nosema varivestis* (*P. foveolatus*) (Own and Brooks, 1986).

Social hymenopterous insects may also serve as vectors of microsporidia by feeding infectious spores to larval brood. This transmission mode occurs in *Burenella dimorpha,* a parasite of the tropical fire ant (*Solenopsis geminata*) (Jouvenaz et al., 1981). Adult ants cannibalize ruptered infected pupae but do not ingest spores. Instead, the spores and particulate food are diverted to the infrabuccal cavity where they are formed into a pellet. This pellet is then fed to fourth instar larvae, which is the only host stage that is vulnerable to infection since younger larvae and adults are fed only liquid food.

In some microsporidian parasites of mosquitoes, namely, *Amblyospora* and *Parathelohania* spp., an indirect method of horizontal transmission involving obligatory development in an intermediate copepod (Crustacea: Copepoda) host has been described (Andreadis, 1985a; Sweeney et al., 1985; Avery and Undeen, 1990; Sweeney et al., 1990; Becnel, 1992; White et al., 1994; Becnel and Andreadis, 1998). In this pathway, copepods acquire infection through oral ingestion of spores discharged into the aquatic environment following the death of diseased mosquito larvae. A second morphologically distinct spore is formed in the copepod. This spore is similarly released into the water with the death of the copepod host and serves to orally reinfect mosquito larvae. The details of this unique developmental cycle are described in "Life Cycles."

Vertical Transmission

Vertical transmission, defined here as the direct transfer of infection from parent to progeny (Fine, 1975), is a major pathway of transmission for many microsporidian parasites of insects. For *Amblyospora* spp. that infect mosquitoes, it is the principal means by which the microsporidium is maintained from one host generation to the next (Kellen et al., 1965; Chapman et al., 1966; Andreadis and Hall 1979a; Becnel, 1994), and it represents the single most important adaptation for survival that has evolved within this group (Lucarotti and Andreadis, 1995). In other host insects, vertical transmission augments horizontal routes of infection and facilitates the persistence of the microsporidium when host densities are low or when no susceptible stages of the insect host are present (Anderson and May, 1981).

In most host insects, vertical transmission is maternally mediated; i.e., it occurs entirely through the female line (Fine, 1975). This form of transmission in which the microsporidium is passed from one generation to the next by way of the egg is commonly referred to as transovum transmission. It can occur in two distinct ways depending on whether passage of the microsporidium occurs within the ovary (transovarially) or on the surface of the egg.

When transmission is transovarial, the microsporidium gains entry to the egg within the female host via infection of the ovaries and associated reproductive structures. This route appears to be the most common mode of vertical transmission of microsporidia in insects, but the precise mechanism by which the microsporidium gains entry into the egg is not entirely understood. Among polymorphic microsporidia that parasitize dipteran hosts, such as *Amblyospora, Culicospora,* and *Edhazardia,* infection is generally thought to occur by direct inoculation of individual oocytes with sporoplasms from specially produced spores (Andreadis and Hall, 1979a; Andreadis, 1983; Becnel et al., 1987, 1989; Becnel, 1994). These spores are intimately associated with host oocytes, and their resulting sporoplasms have been observed within nurse cells of oocytes prior to oviposition (Andreadis and Hall, 1979a; Andreadis, 1983). With monomorphic microsporidia such as *Nosema,* which parasitize coleopteran, lepidopteran, and hymenopteran hosts, however, ovarian infection appears to be achieved by the incorporation of vegetative or spore stages into the developing eggs during oogenesis as demonstrated by the presence of these stages within late-stage embryos and eggs (Kramer, 1959a; Brooks, 1968; Kellen and Lindegren, 1971, 1973b; Bauer and Pankratz, 1993; Becnel and Geden, 1994). In this case, no special spores are produced; the same spores that function in oral transmission are involved in transovarial transmission. In a study on *Nosema otiorrhynchi* transmission in *Otiorrhynchus ligustici,* Weiser (1958) has reported that schizonts penetrate the interior of the oocyte where they develop into secondary centers of infection. Kellen and Lindegren (1973b) indicated that *N. plodiae* schizonts invade and multiply in nurse cells of *P. interpunctella* before passing into associated oocytes.

In another distinct form of transovarial transmission, larval progeny of an infected female can become infected through oral consumption of spores found within the yolk at, or shortly after, eclosion from the egg. This has been observed in two lepidopteran hosts, the winter moth (*Operophthera brumata*) infected with *Orthosomella operophterae* (Canning, 1982), and *C. fumiferana* infected with *N. fumiferanae* (Thomson, 1958). According to Canning (1982), this mechanism represents an important adaptation by the microsporidium that ensures that hosts do not succumb to infection while still within the egg and thus defeats the purpose for which transovarial transmission has evolved.

Transovum transmission can also occur when spores from feces, anal hairs, or ovarian connective tissue contaminate the external surface of the egg shell and are consumed by host larvae at eclosion. Infection can usually be eliminated in the progeny through surface sterilization of eggs (Undeen and Vávra, 1997). Transmission via surface-contaminated eggs occurs far less frequently than transovarial transmission but has been documented with *Nosema algerae* in *Anopheles stephensi* (Alger and Undeen, 1970; Canning and Hulls, 1970) and with *N. pyrausta* in *O. nubilalis* (Kramer, 1959b).

Paternally mediated vertical transmission has been observed with microsporidia in insects but is apparently rare, having been reported from only two species, *N. fumiferanae* (Thomson, 1958) and *N. plodiae* (Kellen and Lindegren, 1971). In both instances, transmission was thought to be effected by venereal transfer of the microsporidium from the male to the female host during mating, with subsequent transfer to the egg.

HOST SPECIFICITY

In Vitro Studies

Insect microsporidia have been studied in tissue culture systems derived from either established cell lines or tissue explants. Many of these studies have been conducted in tissue cultures of the natural host to investigate basic developmental cycles of the pathogen difficult to study in the whole organism. Other studies have examined the ability to establish and grow entomogenous microsporidia in cells from hosts of different origins to resolve possible safety concerns. In the 1980s Jaronski (1984) and Brooks (1988) provided excellent reviews of all the microsporidia from insects that had been studied in in vitro systems. At that time, eight species of microsporidia from insects had been successfully grown in cell culture, including *N. algerae, N. apis, N. bombycis, N. disstriae, N. heliothidis, N. mesnili, Vairimorpha necatrix,* and *Vavraia culicis.* Since then, a number of additional species have been studied. *Cystosporogenes operophterae,* a parasite of *O. brumata,* was cultured in a *Spodoptera frugiperda* cell line (Xie, 1988). A *Nosema* sp. from the lawn grass cutworm (*Spodoptera depravata*) was cultured in an *Antheraea eucalypti* cell line (Iwano and Ishihara, 1989). *Nosema furnacalis,* a parasite of *Ostrinia furnacalis,* was placed in continuous culture in a *H. zea* cell line (Kurtti et al., 1994). And a *Vairimorpha* sp. (identified as NIS M12) was successfully grown in four different insect cell lines (Inoue et al., 1995).

A number of microsporidia have been grown in tissue cultures derived from insects other than the natural host. The ability of *N. disstriae* to grow in tissue cultures of three different insects was reviewed by Brooks (1988). In another study, Inoue et al. (1995) found that a *Vairimorpha* sp. grew and developed in four different lepidopteran cell lines; however, growth rates of the parasite differed in the various lines, and a persistent infection was maintained only in a *S. frugiperda* cell line.

Without question, *N. algerae* has demonstrated the greatest ability to grow in a wide variety of cells from both invertebrate and vertebrate hosts. This species (isolated from a mosquito host) has been grown in a number of different dipteran and lepidopteran cell lines (for an extensive review, see Brooks, 1988). *N. algerae* has also been grown in pig kidney cell cultures but only at temperatures below 37°C (Undeen, 1975). It has yet to be demonstrated that any entomogenous microsporidium can grow at or above 37°C. This appears to be a major physiological difference between microsporidia that parasitize insects and those that parasitize homeothermic vertebrates.

Growth of microsporidia in cells other than the natural host has been interpreted to indicate a low degree of in vitro specificity (Brooks, 1988). It appears, however, that infectivity or susceptibility of a cell line to a microsporidium may not necessarily reflect specificity. This seems particularly true when one considers normal growth and persistence as opposed to atypical development in an alien tissue culture system (Kurtti et al., 1994; Inoue et al., 1995).

Tissue culture studies are responsible for an important discovery regarding the developmental cycles of microsporidia. In 1989, Iwano and Ishihara studied the early development of a *Nosema* sp. in cultured cells. They demonstrated that 35 h after inoculation binucleated spores were formed and germinated spontaneously, a form of autoinfection. A study of ultrastructural features verified that two spore morphs were present, one at 36 h and another at 72 h. The early spore differed from the later one in the morphology of the spore wall and polar filament. This type of spore dimorphism was later demonstrated for *N. bombycis* in cell culture (Iwano and Ishihara, 1991a) and in the midgut epithelium of the natural host, *B. mori* (Iwano and Ishihara, 1991b). Since then, an early (or primary) spore that germinates spontaneously in tissues of the host has been reported for *N. apis* (de Graaf et al., 1994a, 1994b), *E. aedis* (Johnson et al., 1997), several microsporidia of the gypsy moth (Solter et al., 1997), and *V. necatrix* (Solter and Maddox, 1998). This discovery has been a fundamental one for the microsporidia and provides a possible explanation for the spread of microsporidia in tissue culture systems and in the insect host.

In Vivo Studies

The specificity of many entomogenous microsporidia has been investigated principally to determine the host range of potential target pests and safety for nontarget organisms. Brooks (1988) has reviewed this subject and discussed a number of entomogenous microsporidia considered potential microbial control agents. Two of these, *N. algerae* and *N. locustae,* are notable because of the extensive specificity studies conducted against a phylogenetically diverse group of hosts.

N. locustae is the only microsporidian parasite registered by the Environmental Protection Agency as a microbial insecticide, and it has been used in the control of grasshoppers. It infects about 90 different species of grasshoppers but is not infectious for the honeybee (*Apis mellifera*) or two lepidopterans, *H. zea* and *Agrotis ipsilon* (see Table 36 in Brooks, 1988). Extensive testing was conducted against mammals, birds, and fish without any significant effects (see Table 37 in Brooks, 1988). The specificity of this pathogen, therefore, seems to be restricted to the Orthoptera. Conversely, *N. algerae* has been transmitted orally to species from four different families of insects and two species of trematodes (see Table 31 in Brooks, 1988). It has also been transmitted by injection to species from six orders of insects, a crustacean (Undeen and Maddox, 1973), and a mouse (Undeen and Alger, 1976). Additional studies conducted with athymic mice injected with *N. algerae* have verified development for up to 96 days in the tail (Alger et al., 1980) and up to 49 days in the tail and foot (Trammer et al., 1997). Alger et al. (1980) concluded that development of *N. algerae* was limited first by temperature and second by the immune system. Additionally, mice fed on by adult mosquitoes heavily infected with *N. algerae* were uninfected (Alger et al., 1980). A human accidentally injected with spores of *N. algerae,* tested positive for antibodies to *N. algerae,* but a volunteer who allowed infected mosquitoes to blood-feed on her arm did not (Alger et al., 1980). It was concluded that *N. algerae* posed no risk to warm-blooded animals.

Several studies have compared the physiological host specificity (laboratory) with the ecological host specificity (field) for several microsporidia from terrestrial and aquatic systems. One study was conducted to determine the potential risks posed by exposing native North American Lepidoptera to exotic microsporidia. Forty-nine North American species of Lepidoptera were fed five species of microsporidia isolated from European populations of the gypsy moth (*Lymantria dispar*) (Solter et al., 1997), and three categories of responses were noted: (1) refractory, (2) atypical infection, and (3) heavy infection. One pathogen, an *Endoreticulatus* sp., produced extensive infection in two-thirds of the hosts tested and was deemed a high risk for introduction. The other four microsporidia produced heavy infections in 2 to 19% of nontarget hosts. The remaining nontargets developed atypical infections. This information was used to rank the different species of microsporidia for their safety to nontargets when released as biocontrol agents against the gypsy moth in North America.

Amblyospora connecticus is a pathogen of the mosquito *Aedes cantator* and the intermediate host *Acanthocyclops vernalis* (Andreadis, 1988b) (see "Life Cycles"). Twenty mosquito species from five genera were exposed to *A. connecticus* to determine susceptibility and the ability to complete the life cycle of the microsporidium via transovarial transmission (Andreadis, 1989). Four species of *Aedes* were susceptible to *A. connecticus* and underwent normal vegetative development but produced binucleate spores only in *Aedes epactius.* However, *A. connectius* was not transmitted transovarially in any of the alternate hosts, demonstrating a high degree of specificity for *A. cantator.*

Studies on the susceptibility and specificity of mosquitoes to *E. aedis,* a pathogen of *A. aegypti* (Becnel et al., 1989), were conducted by Becnel and Johnson (1993) and Andreadis (1994a). Twenty species of mosquitoes representing eight genera were exposed to *E. aedis.* Eight species developed infections and produced binucleate spores in the adult mosquito. Transovarial transmission, however, was not successful in any of the alternate mosquito hosts. These studies have verified that while *E. aedis* can infect a variety of mosquitoes, it is specific for *A. aegypti,* the only host in which it is transmitted transovarially.

ENVIRONMENTAL PERSISTENCE

Very little is known about the natural persistence of insect-parasitic microsporidia in the external environment. Most studies have focused on the impact of temperature, moisture, and solar radiation on spore survival, and with a few exceptions, this research has been limited to laboratory investigations of species that infect economically important terrestrial insects. Thorough reviews of this subject can be found in Kramer (1970, 1976), Maddox (1973, 1977), and Brooks (1980, 1988). On the basis of available data found therein, the following conclusions have been made: (1) spore survival varies from species to species; (2) spores of most species of microsporidia do not persist in the general environment for more than 1 year; (3) spore survival is greater when spores are bound in feces or dried cadavers or maintained in aqueous media at temperatures slightly above freezing (0 to 6°C); (4) drying is harmful, and most microsporidian spores do not survive more than 2 to 3 months at room temperature; (5) drying is immediately lethal to species from aquatic hosts; (6) temperatures above 35°C greatly reduce spore viability; (7) unprotected spores cannot withstand exposure to direct sunlight for more than several days, and in bright sunlight complete inactivation may occur within minutes or hours; (8) any substrate that offers some protection from direct sunlight and provides a constant source of moisture extends spore longevity; and (9) the presence of other microorganisms is usually detrimental. According to Kramer (1976), the main threat to the survival of microsporidian spores is slow dehydration, which can be brought about by low humidity at moderate to high temperatures or by freezing, as water is lost from the spore through the thinner portions of the exospore at the apical pole.

LIFE CYCLES

The life cycles of entomogenous microsporidia range from relatively simple to extremely complex. Microsporidia described as having a simple life cycle are usually characterized by a single sporulation sequence involving only one host individual. The spores formed in the originally infected host are directly infectious orally to other individuals of the same host species to repeat the cycle. This concept has required modification with the discovery that some species (previously believed to have only one sporulation sequence) have now been determined to have a sporulation sequence early in development. This early or primary spore apparently functions to spread the parasite to other host cells (autoinfection). Three species, *Endoreticulatus fidelis, N. apis,* and *V. necatrix,* have been selected as examples of entomogenous microsporidia that complete their life cycle in one host individual. Microsporidia with complex life cycles usually have multiple sporulation sequences involving more than one generation of the host and sometimes an intermediate host. These are represented by the polymorphic microsporidia of aquatic Diptera and will be described for two important species, *A. connecticus* and *E. aedis.*

Endoreticulatus fidelis

As an example of a microsporidian species that is uninucleate throughout its development, *E. fidelis* has only one sporulation sequence and requires only one host individual to complete the life cycle (Hostounsky and Weiser, 1975). The host for this pathogen is the potato beetle (either *Leptinotarsa undecimlineata,* as reported by Hostounsky and Weiser [1975] or *L. decemlineata* and *L. juncta,* as reported by Brooks et al. [1988]). The following information on the life cycle of *E. fidelis* is based on the report by Brooks et al. (1988), and details are illustrated in Fig. 22 to 36. Infections in the larval stages of the beetle are initiated when uninucleate spores are ingested by larvae feeding on contaminated potato foliage. The uninucleate sporoplasms are inoculated directly into the midgut epithelium where the entire developmental sequence occurs. All stages of schizogony and sporulation (sporogony plus sporogenesis) develop within a parasitophorous vacuole derived from the cisterna of the host endoplasmic reticulum (Fig. 33 and 34). Schizogony (presporogonic reproduction of uninucleate stages) occurs by repeated binary division of uninucleate schizonts or multiple fission of spherical or moniliform plasmodia (Fig. 23 to 29) and is the

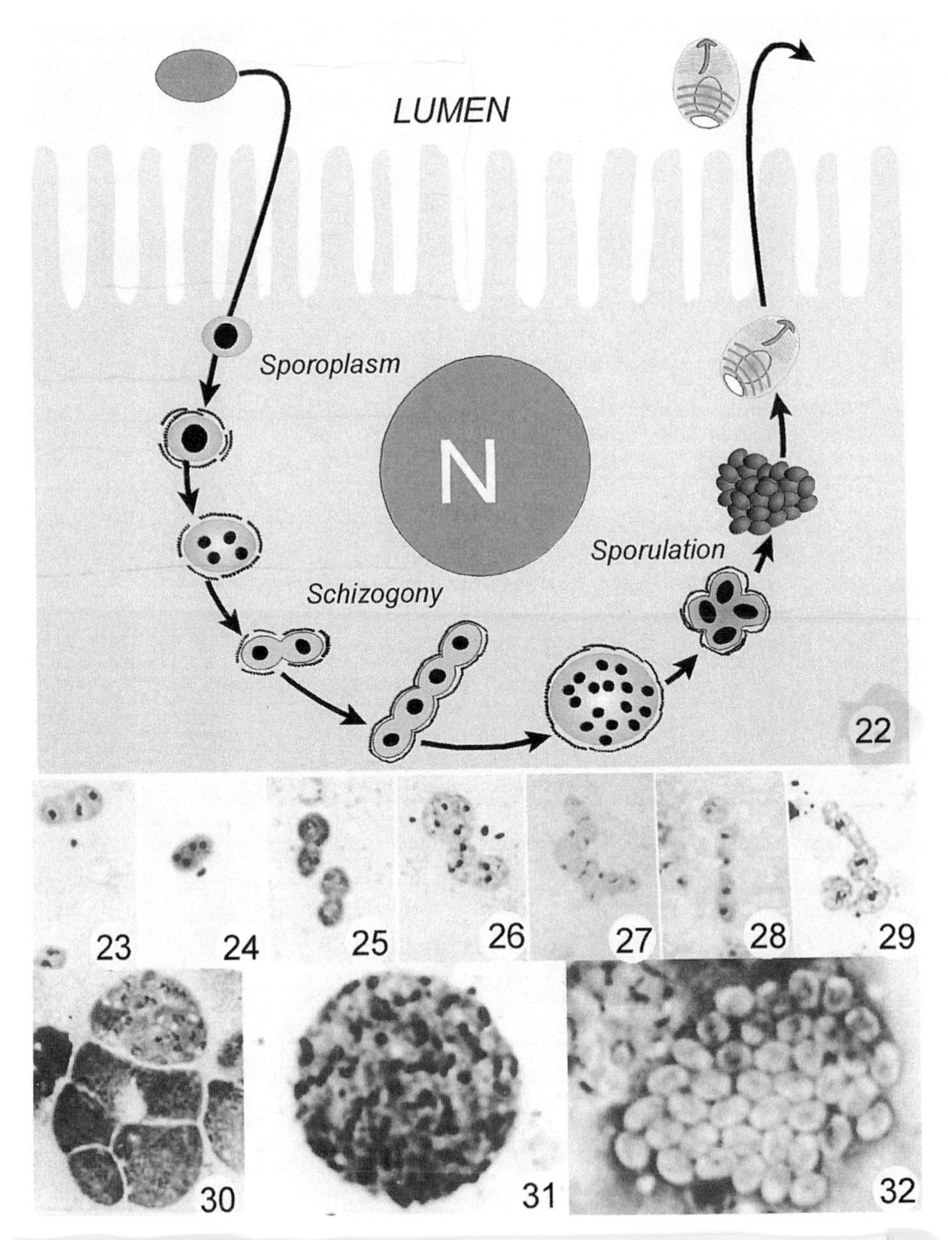

FIGURE 22 Diagram of the life cycle of *Endoreticulatus fidelis* in a larval midgut epithelial cell of *Leptinotarsa decemlineata*. **FIGURES 23–32** Developmental stages of *E. fidelis* as seen in Giemsa-stained smears. Reprinted from Brooks et al. (1988) with permission from the authors and publisher. **FIGURES 23 AND 24** Haplokaryotic schizonts. **FIGURE 25** Schizonts undergoing binary fission. **FIGURES 26–29** Ribbonlike schizonts, some dividing by budding or multiple fission. **FIGURE 30** Sporogonial plasmodia exhibiting evidence of division by plasmotomy. **FIGURE 31** Late-stage sporogonial plasmodium undergoing multiple fission to form sporoblasts. **FIGURE 32** Group of spores. All magnifications, ×1,400.

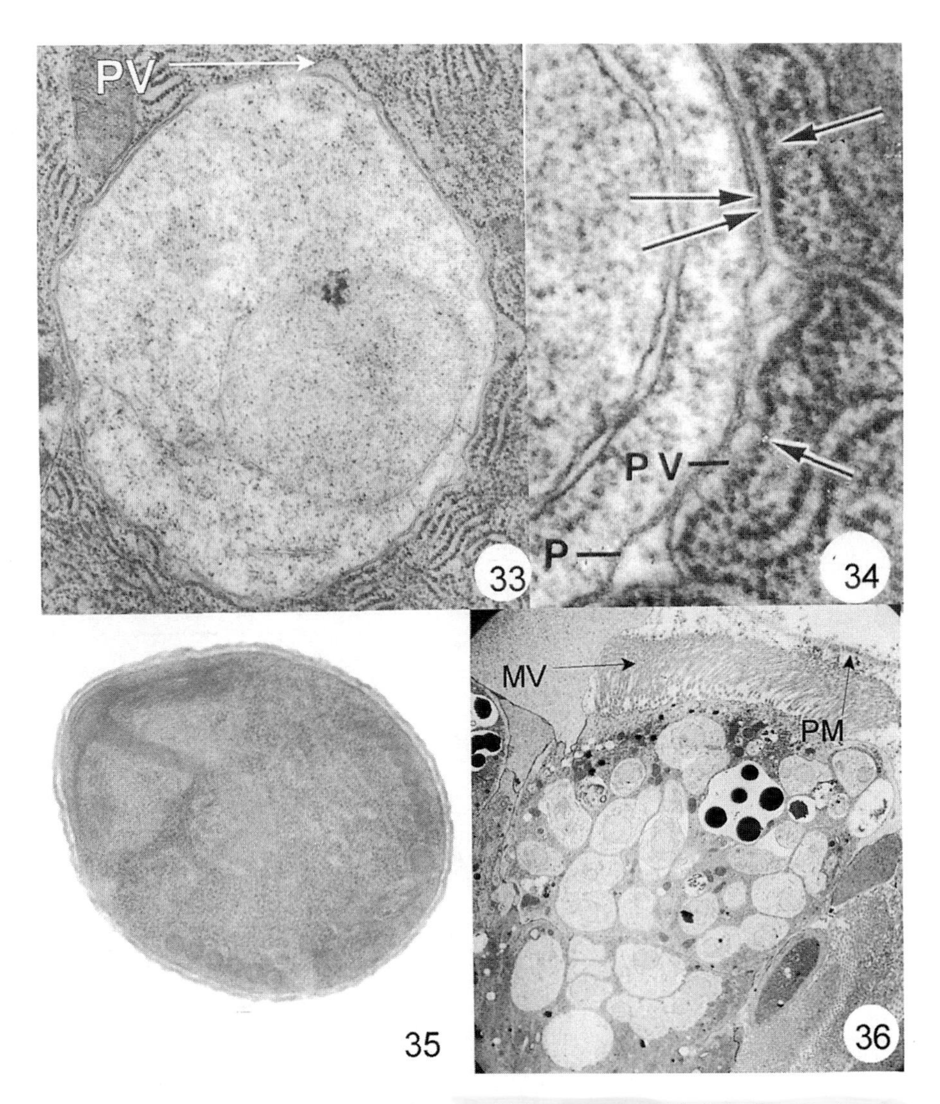

FIGURES 33–36 Electron micrographs of *E. fidelis* in midgut epithelial cells of *L. decemlineata*. **FIGURE 33** Schizont in parasitophorous vacuole (PV) derived from cisternae of host cell rough endoplasmic reticulum. Magnification, ×34,600. **FIGURE 34** Higher magnification of a schizont plasmalemma-parasitophorous vacuole interface. Double arrows indicate inner membrane without ribosomes closely associated with the plasmalemma, and single arrows indicate parasitophorous vacuole studded with ribosomes. Magnification, ×89,000. **FIGURE 35** Uninucleate spore. Magnification, ×42,300. **FIGURE 36** Infected midgut epithelial cell of *E. fidelis.* Magnification, ×2,500. Reprinted from Brooks et al. (1988) with permission from the authors and publisher (Fig. 33–35). MV, microvilli; PV, parasitophorous vacuole; PM, peritrophic membrane; P, plasmalemma.

main multiplicative phase of the parasite. At some point, schizonts transform into cells that give rise to sporogonial plasmodia (Fig. 30 and 31). These plasmodia are distinguished by withdrawal of the sporont plasmalemma from the parasitophorous vacuole. Sporogony is by multiple fission of sporogonial plasmodia to produce sporoblasts. Variable numbers of uninucleate spores (up to about 40) are formed within the parasitophorous vacuole (Fig. 32

and 35). Infected midgut epithelial cells (Fig. 36) rupture, releasing spores that are passed out of the host, resulting in contamination of the environment by the frass. *E. fidelis* is found in both larvae and adults of the potato beetle. Detrimental effects of the parasite are manifested by the premature mortality of larvae and adults and a pronounced effect on the reproductive capability of adults.

Nosema apis

An economically important parasite of the honeybee is *N. apis.* It is representative of microsporidia that possess diplokarya throughout their development and complete their life cycle in one host individual (Fig. 37 to 39). The following information is primarily based on observations by Fries (1993) and de Graaf et al. (1994a, 1994b). Adult honeybees ingest binucleate spores probably (in part) as a result of comb cleaning. Binucleate spores are stimulated to germinate in the gut lumen and deposit the sporoplasm directly into midgut epithelial cells. The binucleate sporoplasm grows in size and matures into the first meront. Approximately 24 h after infection the first nuclear division occurs, followed by binary fission to initiate the merogonial cycle. Multiplication is typically by binary fission of diplokaryotic cells but can also involve multiple fission of paucinucleate plasmodia. Approximately 48 h after infection, some stages transform into sporonts which divide once by binary fission to form two sporoblasts (de Graaf et al., 1994a). This event results in a primary binucleate spore characterized by a thin spore wall and a short polar filament which germinates spontaneously within the cytoplasm of the epithelial cells (Fig. 38). It is believed that this mechanism (autoinfection) serves to spread the parasite to adjacent epithelial cells. Other meronts within the cell continue to multiply and after a num-

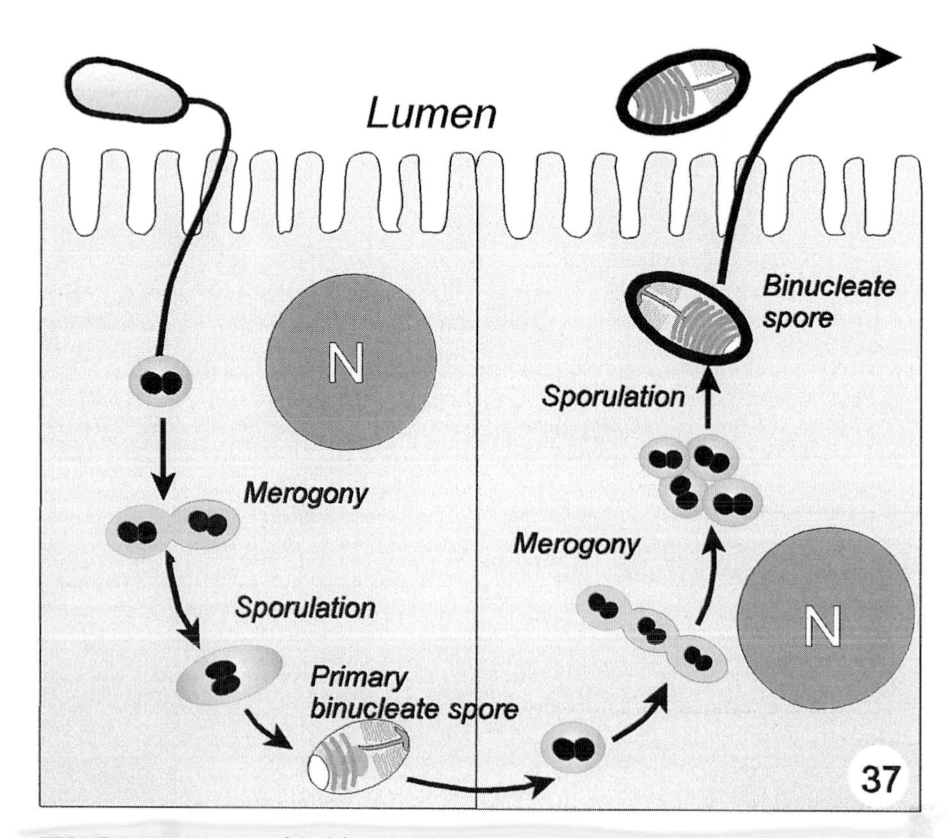

FIGURE 37 Diagram of the life cycle of *Nosema apis* in *Apis mellifera* midgut epithelial cells.

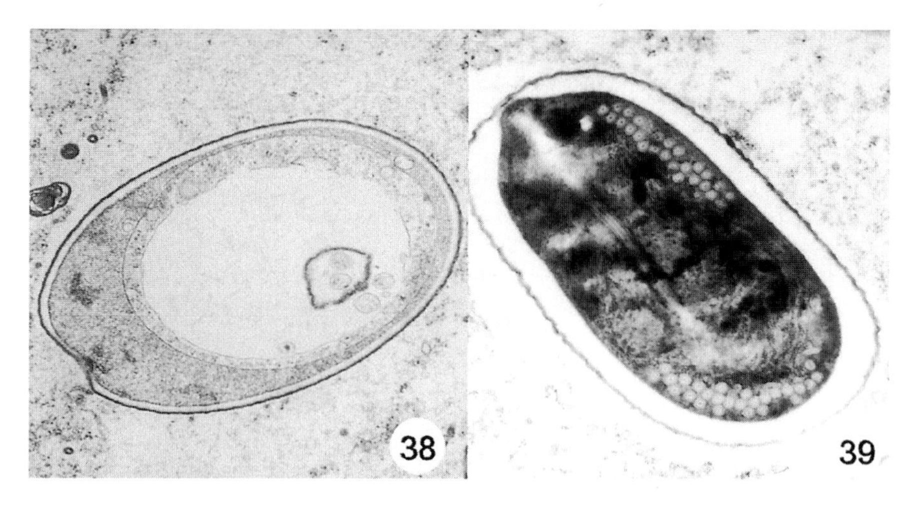

FIGURES 38–39 Electron micrographs of *N. apis* spores formed in *Apis mellifera*. Reprinted with permission from I. Fries. **FIGURE 38** Germinated primary binucleate spore of *N. apis* characterized by a thin spore wall and a large vacuole. Magnification, ×17,700. **FIGURE 39** Binucleate spore (environmental spore) of *N. apis* characterized by a thick spore wall and a long polar filament. Magnification, ×17,000.

ber of divisions enter into a second sporulation sequence. Diplokaryotic sporonts divide once to produce two sporoblasts which mature into spores. The binucleate spores are thick-walled, with a long polar filament (Fig. 39), and are produced in large numbers. The infected epithelial cells are filled with spores and eventually rupture, releasing the spores into the gut lumen. The spores are voided with the frass and contaminate the bee's environment until they are ingested by a new host individual.

Vairimorpha necatrix

Another life cycle variation involves microsporidian species with more than one sporulation sequence in the same host individual. This life cycle involves sporonts that can enter one of two different sporulation sequences. *V. necatrix* is an example of this type of life cycle and requires only one host individual to complete development (Fig. 40). A third sporulation sequence early in the infection process has also been described (Solter and Maddox, 1998). The following account is a summary of information from Maddox et al. (1981) and Solter and Maddox (1998, in press).

A susceptible host for *V. necatrix* is the fall armyworm (*S. frugiperda*), which becomes infected by ingesting spores while feeding on foliage. Germination of spores results in the deposition of sporoplasms into midgut epithelial cells. The sporoplasm matures and undergoes at least one replication by binary fission. At between 30 and 72 h, these cells transform into diplokaryotic sporonts which divide once to produce two sporoblasts. The resulting primary binucleate spores are thin-walled and have a large posterior vacuole (Fig. 40). These spores apparently germinate spontaneously at maturation, releasing sporoplasms that become established in the larval fat body. A second merogonial sequence produces large numbers of diplokaryotic stages, primarily by binary fission. Diplokaryotic sporonts then enter one of two sporulation sequences. One sequence is disporous, produces binucleate spores of the *Nosema* type, and always occurs first (Fig. 41). The other sequence occurs later and involves meiosis to produce uninucleate spores (meiospores) (Fig. 42) in groups of eight contained within a sporophorous vesicle (Fig. 40). The fat body literally becomes a sac filled with spores, with most infected individuals dying as larvae. The environment becomes contaminated with spores, providing a source of inoculum for new host individuals.

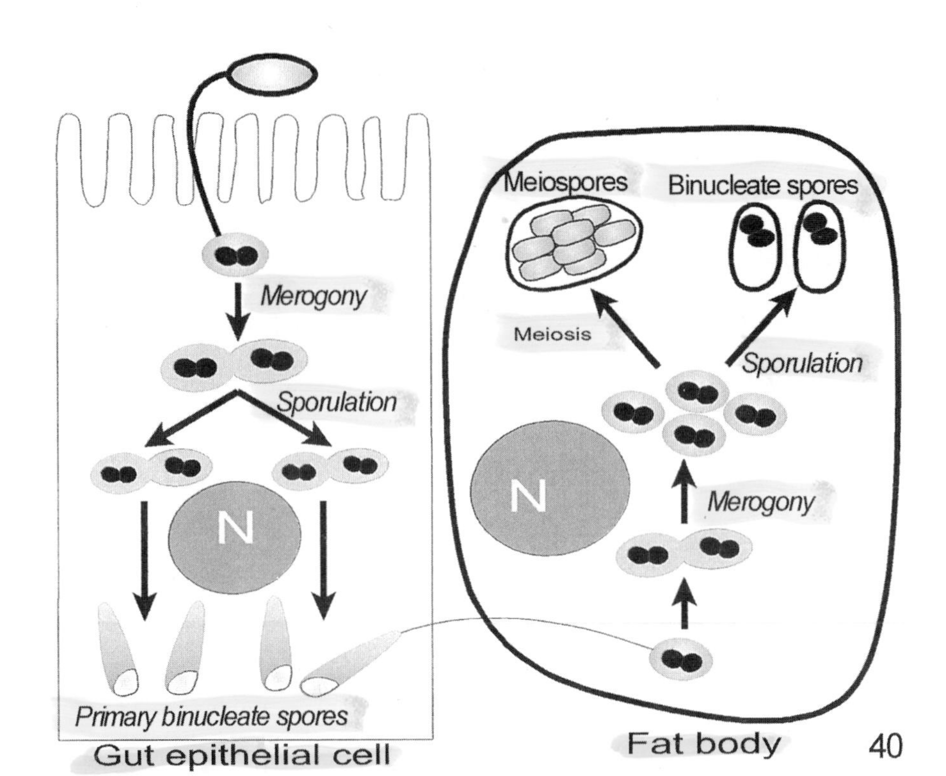

FIGURE 40 Diagram of the life cycle of *V. necatrix* in a *Spodoptera frugiperda* midgut epithelial cell and fat body.

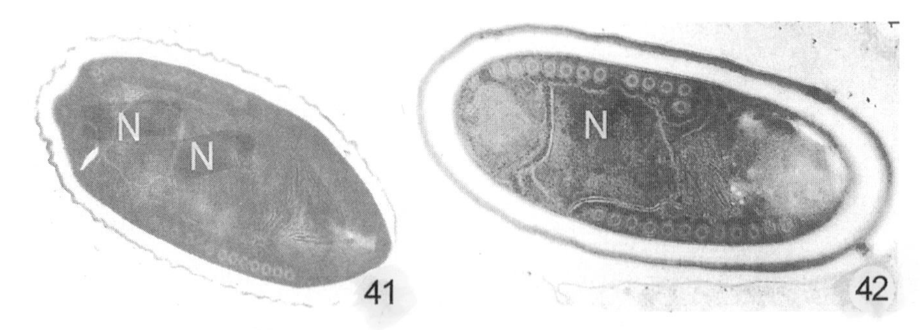

FIGURES 41 AND 42 Electron micrographs of *V. necatrix* spores. Reprinted from Moore et al. (1992) with permission from the authors and publisher. **FIGURE 41** Binucleate spore (environmental spore) of *V. necatrix* characterized by a thick spore wall and a long polar filament. Magnification, ×8,700. **FIGURE 42** Meiospore of *V. necatrix*. Magnification, ×19,600. N, nucleus.

Polymorphic Microsporidia of Aquatic Diptera

The polymorphic or heterosporous microsporidia that infect aquatic Diptera, especially mosquitoes, possess the most complex life cycles yet described for any microsporidia. These life cycles include asexual (schizogony, merogony, and sporogony) and sexual (karyogamy, gametogenesis, and plasmogamy) reproduction, the formation of multiple spore types (each with a specific function), and both vertical (transovarial) and horizontal transmission. Most species require two successive host generations to complete their life cycle, and at least three genera, *Amblyospora, Hyalinocysta,* and *Parathelohania,* require obligatory development in an intermediate copepod host. Detailed studies on these microsporidia have revealed numerous variations on the life cycle particular for each species and host (Hazard and Oldacre, 1975; Andreadis, 1990a, 1990b; Sweeney and Becnel, 1991; Becnel, 1994). However, it is beyond the scope of this chapter to describe all these cycles in detail. Therefore, representative life histories of the two most extensively studied genera, *Amblyospora* and *Edhazardia,* will be presented here. Details on the life cycle and ultrastructure of other notable polymorphic genera can be found in the following references: *Culicospora* (Hazard et al., 1985; Becnel et al., 1987; Becnel, 1994); *Culicosporella* (Hazard and Savage, 1970; Hazard et al., 1984; Becnel and Fukuda, 1991); *Hazardia* (Hazard and Fukuda, 1974; Hazard et al., 1985); *Parathelohania* (Hazard and Weiser, 1968; Avery and Undeen, 1990).

Amblyospora connecticus

The genus *Amblyospora* represents the largest group of microsporidia that infect natural populations of mosquitoes. To date, more than 90 species or isolates have been described worldwide from 79 different species of mosquitoes in 8 genera (for a partial host/species list, see Andreadis, 1994b). All species of *Amblyospora,* as far as we know, are transovarially transmitted by adult female mosquitoes and undergo obligatory development in an intermediate copepod host as a prerequisite to horizontal transmission (Andreadis, 1985a; Sweeney et al., 1985, 1990; Becnel, 1992; White et al., 1994; Becnel and Andreadis, 1998). A representative life cycle of one species, *A. connecticus,* as it occurs in the brown saltmarsh mosquito (*Aedes cantator*) and the copepod *Acanthocyclops vernalis* (Andreadis 1983, 1985a, 1988a, 1988b, 1990b), is described below and in Fig. 43 to 74.

The microsporidium is horizontally transmitted to larval mosquitoes following oral ingestion of extracellular haploid spores (Fig. 46 and 64) that are released into the water with the death of infected copepods. Spores germinate

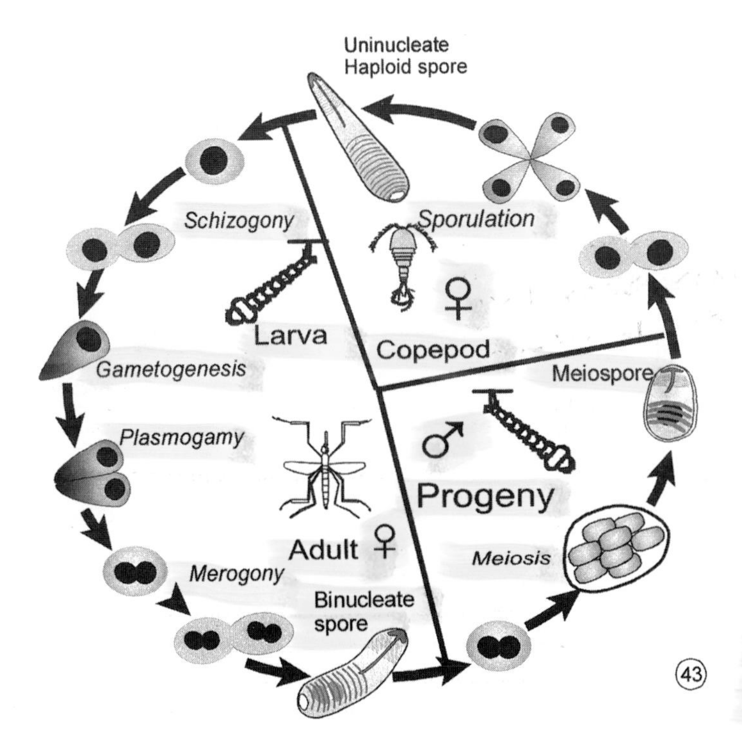

FIGURE 43 Diagram of the life cycle of *Amblyospora connecticus* in *Acanthocyclops vernalis* and *Aedes cantator.*

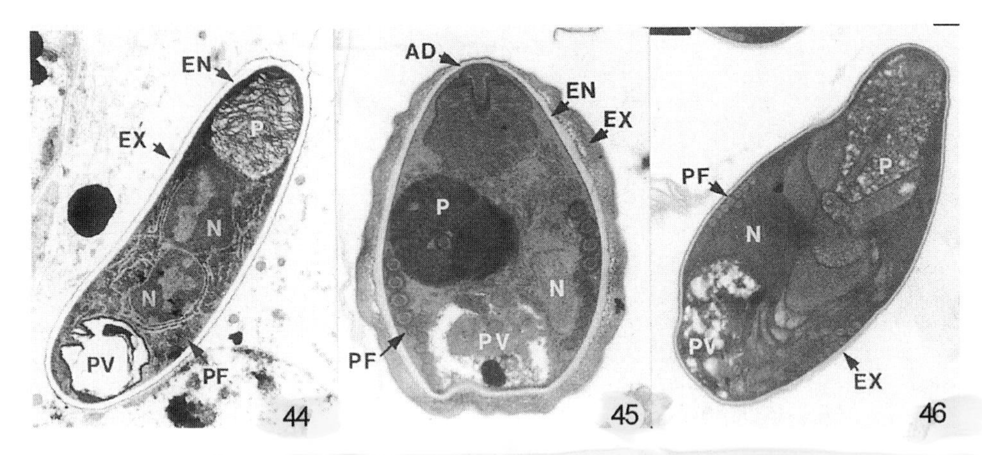

FIGURES 44–46 Longitudinal section of spores of *A. connecticus.* **FIGURE 44** Binucleate spore from a female mosquito, *A. cantator.* Magnification, ×7,500. **FIGURE 45** Meiospore from a fourth instar larva of *A. cantator.* Magnification, ×11,200. **FIGURE 46** Haploid spore from a copepod, *Acanthocyclops vernalis.* Magnification, ×10,600. AD, anchoring disk; EN, endospore; EX, exospore; P, polaroplast; PF, polar filament; PV, posterior vacuole; N, nucleus.

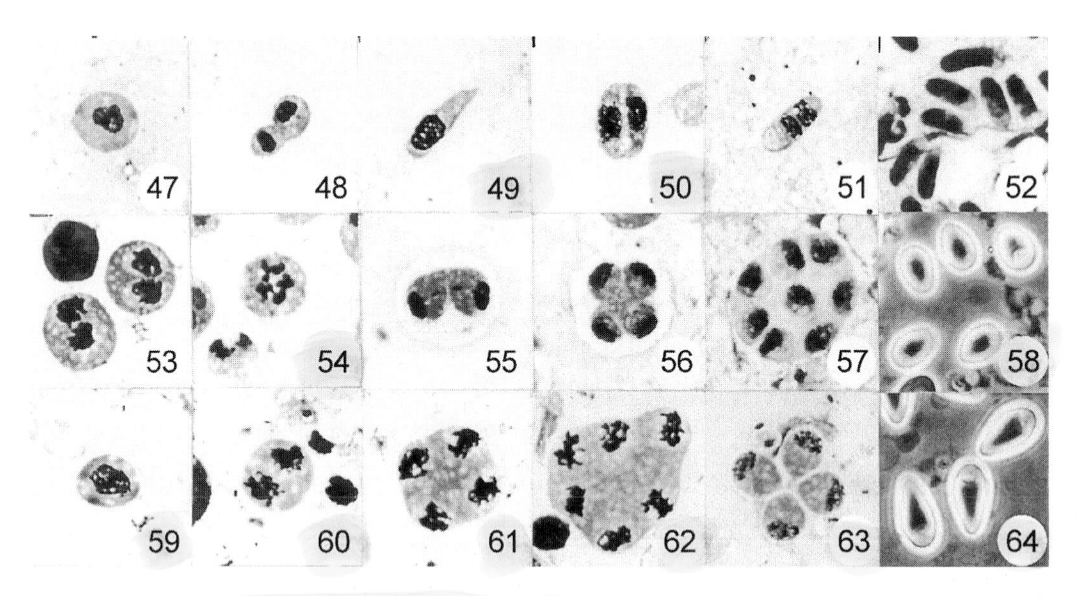

FIGURES 47–64 Life cycle stages of *A. connecticus* as seen in Giemsa-stained smears. **FIGURES 47–52** Stages in larval (Fig. 47–50) and adult (Fig. 51 and 52) *A. cantator* following oral ingestion of copepod spore. **FIGURE 47** Uninucleate schizont. **FIGURE 48** Dividing schizont. **FIGURE 49** Gamete. **FIGURE 50** Gametes undergoing plasmogamy. **FIGURE 51** Sporoblast. **FIGURE 52** Binucleate spores. **FIGURES 53–58** Stages in larval *A. cantator* following transovarial transmission. **FIGURE 53** Diplokaryotic sporont. **FIGURE 54** Sporont undergoing meiosis. **FIGURE 55** Binucleate sporont. **FIGURE 56** Quadrinucleate sporont. **FIGURE 57** Eight sporoblasts within sporophorous vesicle. **FIGURE 58** Live meiospores (phase-contrast). **FIGURES 59–64** Stages in *A. vernalis* following ingestion of meiospores. **FIGURE 59** Early sporont. **FIGURE 60** Binucleate sporont. **FIGURES 61 and 62** Sporogonial plasmodia. **FIGURE 63** Sporoblasts. **FIGURE 64** Live spores (phase-contrast). All magnifications, ×950.

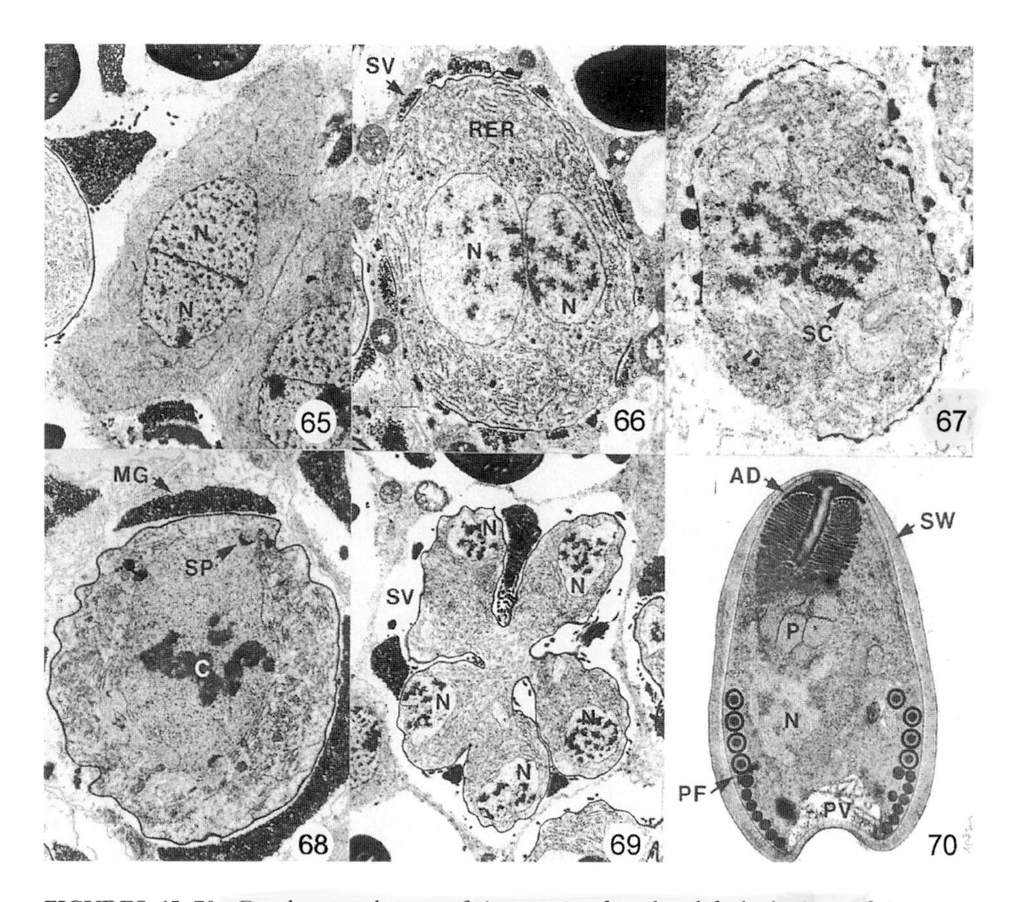

FIGURES 65–70 Developmental stages of *A. connecticus* from larval fat body tissue of *A. cantator.* **FIGURE 65** Diplokaryotic meront. Magnification, ×5,000. **FIGURE 66** Binucleate sporont. Magnification, ×5,200. **FIGURE 67** Binucleate sporont during prophase of meiosis. Magnification, ×6,000. **FIGURE 68** Sporont at metaphase I. Magnification, ×5,400. **FIGURE 69** Multinucleated sporont within sporophorous vesicle. Magnification, ×3,700. **FIGURE 70** Sporoblast. Magnification, ×9,600. AD, anchoring disk; C, chromosomes; MG, metabolic granules; N, nucleus; P, polaroplast; PF, polar filament; PV, posterior vacuole; SC, synaptonemal complex; SP, spindle plaque; SV, sporophorous vesicle; RER, rough endoplasmic reticulum; SW, spore wall.

within the lumen of the larval gut and invade epithelial cells of the midgut and gastric cecum (Fig. 47 and 72) via injection of the sporoplasm through the evaginating polar tube. After a brief period of multiplication by binary fission (Fig. 48), the microsporidium spreads to muscle tissue and oenocytes (Fig. 73) where it undergoes a sexual phase of development involving gametogenesis (Fig. 49) and plasmogamy (cytoplasmic fusion) (Fig. 50) of uninucleate gametes, thereby restoring itself to the diplokaryotic condition (Fig. 51). Infected oenocytes become hypertrophied (Fig. 73), but there is no overt pathology associated with infection. Larval hosts develop normally and emerge as apparently healthy adults. The microsporidium undergoes limited multiplication in the adult female and sporulates when the female mosquito acquires a blood meal. Sporulation coincides with maturation of the ovaries and is stimulated by the secretion of host reproductive hormones, specifically 20-hydroxyecdysone

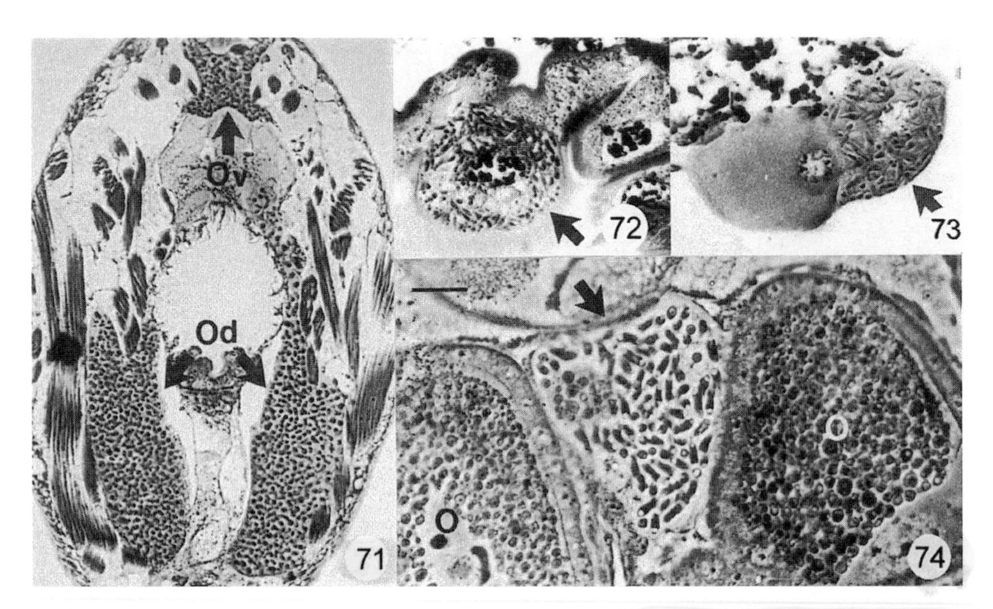

FIGURES 71–74 Histological sections of *A. connecticus* infections in copepod and mosquito hosts. **FIGURE 71** Sagittal section of a female *A. vernalis* copepod, showing infection within the median ovary (Ov) and paired lateral oviducts (Od). Magnification, ×150. **FIGURE 72** Infected epithelial cell (arrow) of the gastric cecum from a fourth instar *A. cantator* larva. Magnification, ×250. **FIGURE 73** Infected oenocyte cell (arrow) from a fourth instar *A. cantator* larva. Magnification, ×340. **FIGURE 74** Infected oenocyte containing binucleate spores (arrow) from the ovaries of an adult female *A. cantator.* O, host oocyte. Magnification, ×920.

(Lord and Hall, 1983). This mechanism results in the formation of binucleate spores (Fig. 44 and 52) which are typically found lying between the ovarioles and oocytes within the ovariale sheath (Fig. 74). Spores germinate within 1 to 2 days, infect the ovaries via the nurse cells, and are thereby responsible for transovarial transmission of the microsporidium to the F_1 generation progeny.

Parasite development in larval progeny is dimorphic and dependent to some degree on the host sex. In certain females, the microsporidium infects the oenocytes and undergoes a simple developmental sequence during which it undergoes limited multiplication by binary and multiple fission (merogony). These larvae show no adverse effects and develop normally to adulthood. When these females are mated with healthy males, the pathogen is transmitted transovarially following a blood meal, as in the previous generation. This pathway of continuous transovarial transmission through female progeny can sustain *A. connecticus* for several generations. However, some degree of horizontal transmission must occur, as these microsporidia are not transmitted with 100% efficiency and there is no assistance from males (i.e., paternally-mediated transmission). This generality is true for all species of *Amblyospora* (Andreadis and Hall, 1979b; Andreadis, 1985b; Sweeney et al., 1988, 1989).

In other female and usually in all male progeny that hatch from infected eggs, the microsporidium invades fat body tissue and exhibits an entirely different developmental sequence. It initially undergoes a proliferative merogony (Fig. 53 and 65) during which infection spreads throughout the fat body of the larval host. Since no spores are formed at this time, it is presumed that diplokaryotic meronts are responsible for cell-to-cell transmission, perhaps as a result of cell fusion. Diplokaryotic sporonts (Fig. 54 and 66) are subsequently formed and undergo meiosis and a prolonged sporulation sequence (Fig. 55 to 57 and 67 to 69) resulting in the production of tens of thousands of haploid spores in groups of eight which are commonly referred to as meiospores (Fig. 45, 58 and 70). Infections typically kill the larval host during the fourth stadium by destroying normal fat body function and depleting larvae of essential reserves. Meiospores produced in these mosquito larvae are orally infectious to female stages of *A. vernalis.* Following ingestion and subsequent spore germination, the microsporidium infects copepod ovarian tissue (Fig. 71) and undergoes repeated schizogony followed by polysporoblastic sporogony (Fig. 59 to 63) and the formation of thousands of haploid spores (Fig. 46 and 64). This process ultimately kills the copepod, permitting the release of spores into the water where they can be ingested by mosquito larvae to complete the cycle. *A. connecticus* thereby persists by surviving in one of two living hosts throughout most of its life cycle.

Notable variations on this life cycle reported for other species of *Amblyospora* include (1) binucleate spore formation in adult female mosquitoes without a blood meal (Sweeney et al., 1989), (2) meiospore development in all F_1 progeny following transovarial transmission (Lord et al., 1981; Sweeney et al., 1990), (3) benign oenocytic infections in all F_1 females following transovarial transmission (Kellen and Wills, 1962; Andreadis and Hall, 1979b), and (4) the susceptibility of both male and female copepods (Sweeney et al., 1990). Further specific host-parasite relationships related to tissue specificity and the expression of infection in each sex can be found in Kellen et al. (1965, 1966), Chapman et al. (1966), Anderson (1968), and Hazard and Oldacre (1975).

Edhazardia aedis

Edhazardia is a monotypic genus represented by *E. aedis,* a highly infectious, virulent parasite of the yellow fever mosquito (*A. aegypti*) (Hembree, 1982; Hembree and Ryan, 1982). It is a polymorphic (heterosporous) species that has four different sporulation sequences and, like *Amblyospora,* is transmitted both vertically and horizontally. However, unlike that of *Amblyospora,* its life cycle does not involve an intermediate host. Details of its life cycle as characterized by Becnel et al. (1989) and Johnson et al. (1997) are illustrated in Fig. 75 to 107 and described below.

E. aedis is horizontally transmitted to larval mosquitoes via oral ingestion of uninucleate lanceolate spores (Fig. 88 and 106) released into the aquatic environment with the death of transovarially infected larvae. Spores readily germinate within the lumen of the midgut and initially infect epithelial cells of the gastric ceca. Here the microsporidium undergoes a limited asexual multiplicative phase (schizogony) (Fig. 76 and 77) followed by gametogenesis (Fig. 78). This process results in the formation of uninucleate, pyriform gametes possessing a distinctive double-membraned papilla or nipplelike structure on the plasmalemma (Fig. 94). Gametes subsequently undergo plasmogamy (Fig. 79 and 95) to form diplokaryotic stages (Fig. 96) which then develop into small, binucleate spores (Fig. 80 and 104). These primary spores, which have only recently been discovered, germinate quickly and are responsible for the dissemination of *E. aedis* to other host tissues, most significantly oenocytes. This portion of the life cycle is completed typically within 120 h of ingestion of the lanceolate spore. Although some variation may occur, most lightly to moderately infected larvae develop to adulthood in which *E. aedis* exhibits a second asexual multiplicative phase (merogony). This development takes place within host oenocytes that circulate within the hemocoel and move to areas surrounding the ovaries in female hosts. Sporulation (Fig. 81, 82, and 97) ensues after the female takes a blood meal, and this event results in the production of a binucleate spore (Fig. 83 and 105). This spore, often called the transovarial spore, is responsible for infection of the ovaries and subsequent transmission to the filial generation. In addition to being functionally distinct, the transovarial spore is larger and more oblong and possesses a longer polar filament and a smaller posterior vacuole than the early spore

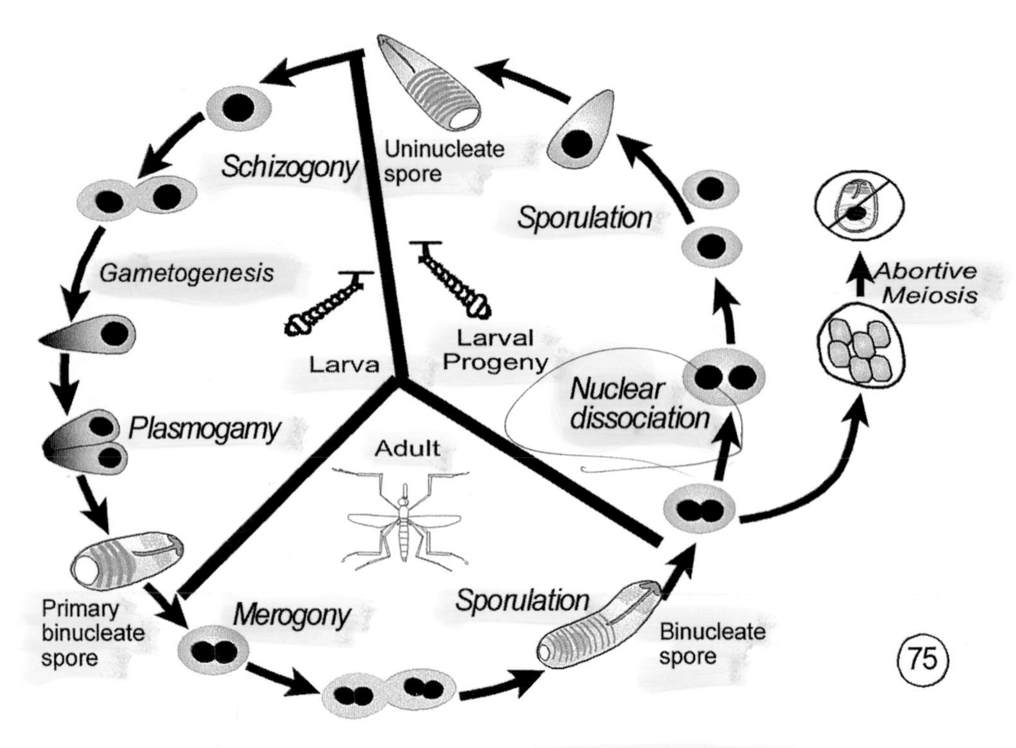

FIGURE 75 Diagram of the life cycle of *E. aedis* in *A. aegypti.*

(Fig. 104 and 105). As in *Amblyospora,* this portion of the life cycle causes no acute pathology. However, infected females exhibit reduced fecundity, longevity (Becnel et al. 1995), and blood-feeding success (Koella and Agnew, 1997).

In larval progeny of the filial generation, *E. aedis* invades fat body tissue and undergoes a third merogony (Fig. 84 and 98), following which the diplokaryotic phase (dihaplophase) of the life cycle ends by one of two different processes, meiosis or nuclear dissociation. The meiotic sequence (Fig. 89 to 92) is similar to that occurring in *Amblyospora,* but it usually aborts and rarely forms meiospores (Fig. 93 and 107). In the predominate nuclear dissociation sequence, the two members of the diplokaryon separate (Fig. 85 and 99) and undergo cytokinesis to form two independent haploid cells. These cells then undergo a sporogonial sequence (Fig. 86, 87, 100 to 103) to form large numbers of uninucleate spores (Fig. 88 and 106). This process results in death of the larval host and the ensuing liberation of infectious uninucleate spores into the aquatic environment where they may be ingested by other susceptible mosquito larvae to complete the cycle. These spores are thus analogous to those produced in the intermediate copepod host with species of *Amblyospora.* This shift in the mechanism of haplosis by nuclear dissociation rather than by meiosis to form orally infectious spores in *E. aedis* and other closely related species such as *Culicospora magna* may represent an example of regressive evolution in which these microsporidia are evolving away from a two-host system, as seen in *Amblyospora* to a simpler, more efficient one-host system (Becnel, 1994). This hypothesis is supported by molecular phylogenetic analyses of the small-subunit rDNA sequences of these microsporidia (Baker et al., 1998; Vossbrinck et al., 1998), which suggest that mosquitoes and their microsporidial parasites have coevolved.

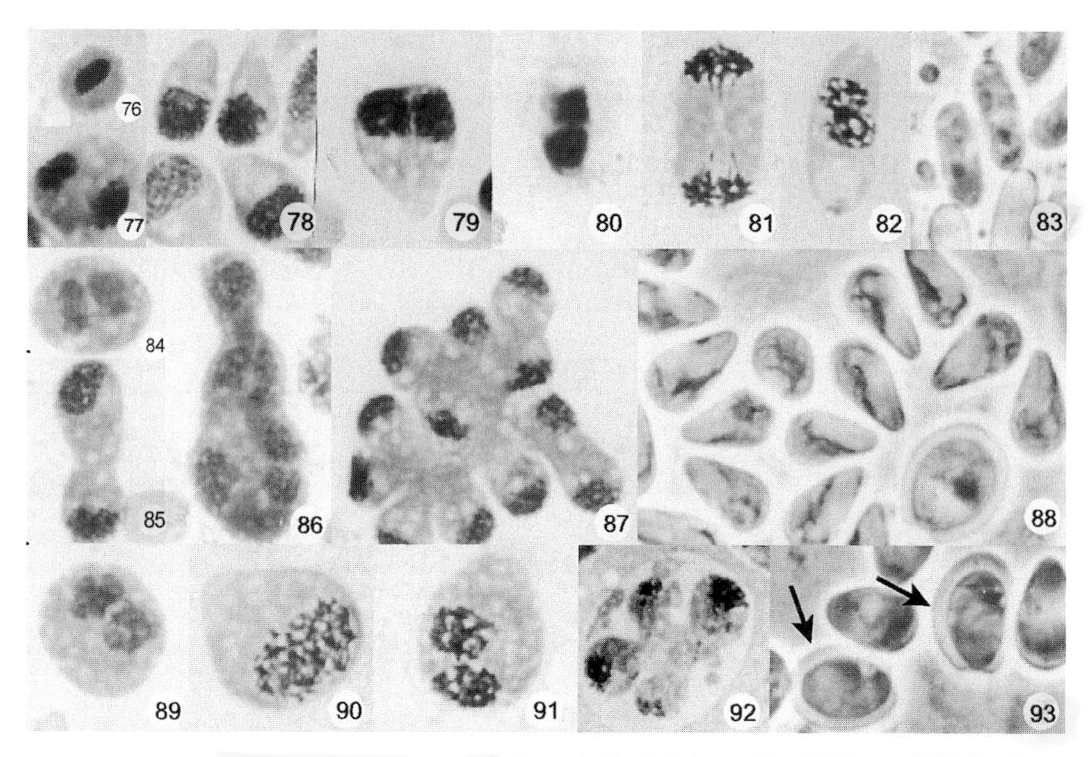

FIGURES 76–93 Life cycle stages of *E. aedis* as seen in Giemsa-stained smears. Reproduced from Becnel et al. (1989) with permission from the publisher. **FIGURES 76–80** Sporulation sequence of *E. aedis* in *A. aegypti* after horizontal transmission producing the primary binucleate spore. **FIGURE 76** Uninucleate sporoplasm in gastric cecum. **FIGURE 77** Schizont dividing. **FIGURE 78** Gametes. **FIGURE 79** Paired gametes undergoing plasmogamy. **FIGURE 80** Primary binucleate spore. **FIGURES 81–83** Sporulation sequence of *E. aedis* in *A. aegypti* after horizontal transmission producing binucleate (transovarial) spores. **FIGURE 81** Diplokaryotic sporont undergoing binary fission. **FIGURE 82** Binucleate sporoblast. **FIGURE 83** Binucleate (transovarial) spore (phase-contrast). **FIGURES 84–88** Sporulation sequence of *E. aedis* producing the uninucleate pyriform spore from nuclear dissociation in the filial generation. **FIGURE 84** Diplokaryotic meront. **FIGURE 85** Sporont undergoing cytokinesis after nuclear dissociation. **FIGURE 86** Sporogonial plasmodium. **FIGURE 87** Sporogonial plasmodium dividing into uninucleate sporoblasts. **FIGURE 88** Uninucleate thin-walled pyriform spores from filial host larva (phase-contrast). **FIGURES 89–93** Sporulation sequence of *E. aedis* that involves meiosis to produce the meiospores in the filial generation. **FIGURE 89** Diplokaryotic meront. **FIGURE 90** Zygote or early sporont derived from a diplokaryotic meront. **FIGURE 91** Sporont. **FIGURE 92** Tetranucleate sporogonial plasmodium. **FIGURE 93** Meiospores (arrows, phase-contrast). All magnifications, ×1,700.

GENERA OF INSECT-PARASITIC MICROSPORIDIA

In 1924 Kudo reported 14 genera and a total of 178 species of microsporidia including some unnamed and doubtful ones. At that time, there were approximately 75 named species from insects. Weiser (1963a) noted that there were more than 200 species of microsporidia with insects as hosts. Sprague (1977) recognized 44 genera and a total of approximately 750 species of microsporidia with about 150 unnamed. These included approximately 380 species from insects, about 90 of which were unnamed. Currently, there are 143 genera of microsporidia, 69 of which have insects as type hosts. Approximately 600 of the 1,000 named species of microsporidia have been reported from insects (the exact number of described species of microsporidia is difficult to ascertain, estimates ranging from approximately 900 to 1300).

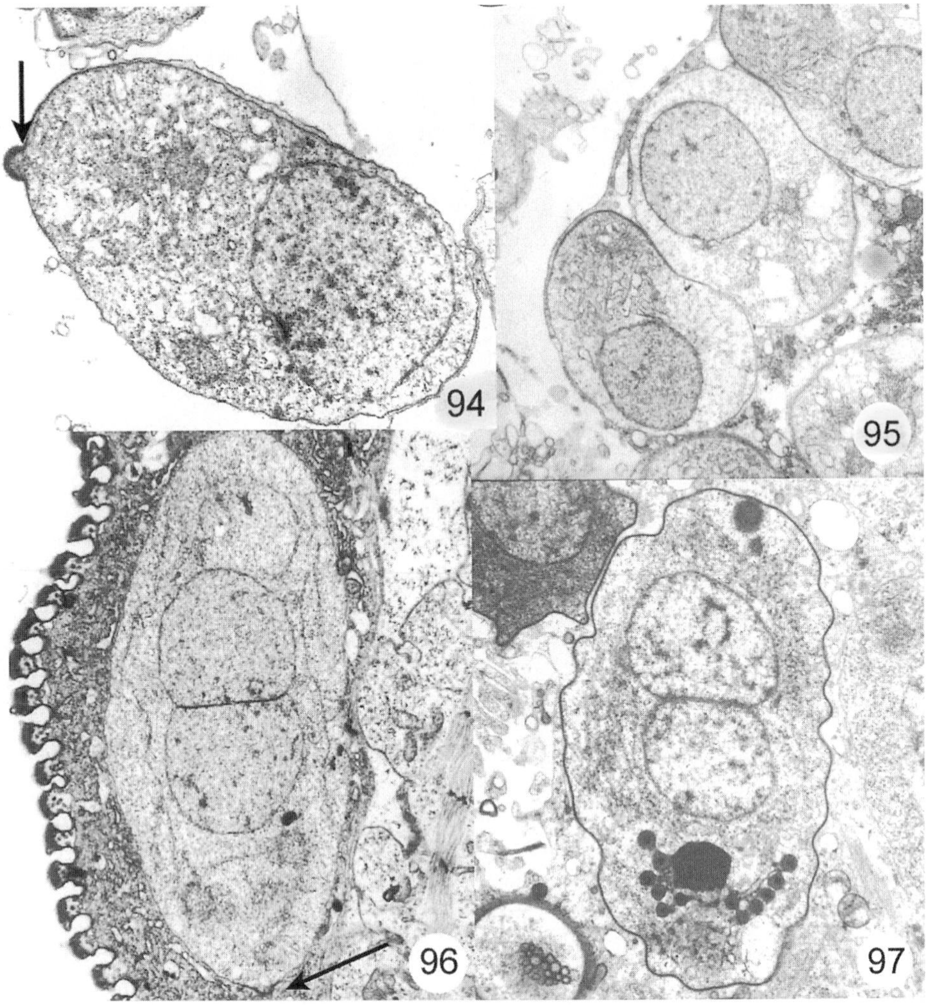

94 95 96 97

FIGURES 94–97 Electron micrographs of *E. aedis* in *A. aegypti* after horizontal transmission. Reprinted from Becnel et al. (1989) with permission from the publisher. **FIGURE 94** Gamete with papilla (arrow). Magnification, ×11,200. **FIGURE 95** A pair of gametes at the beginning of plasmogamy. Magnification, ×6,400. **FIGURE 96** Diplokaryotic meront. The arrow indicates a papilla at the apex of the cell. Magnification, ×7,600. **FIGURE 97** Diplokaryotic sporoblast. Magnification, ×9,900.

Microsporidia have been described from all insect orders. The genera of microsporidia with an insect as the type host are presented by insect order in Table 1. It is remarkable that 42 of the 69 entomogenous genera have been described from dipteran hosts; if aquatic insects are combined (Diptera, Ephemeroptera, Trichoptera, and Odonata), this number becomes 52 of the 69 entomogenous genera. There are several possible explanations for this apparent generic diversity in aquatic insects. Perhaps it is simply due to the relative ease of detecting microsporidian infections through the clear cuticles of aquatic insects. Another likely reason is that the many described species of *Nosema* and *Vairimorpha* (mostly from terrestrial insects)

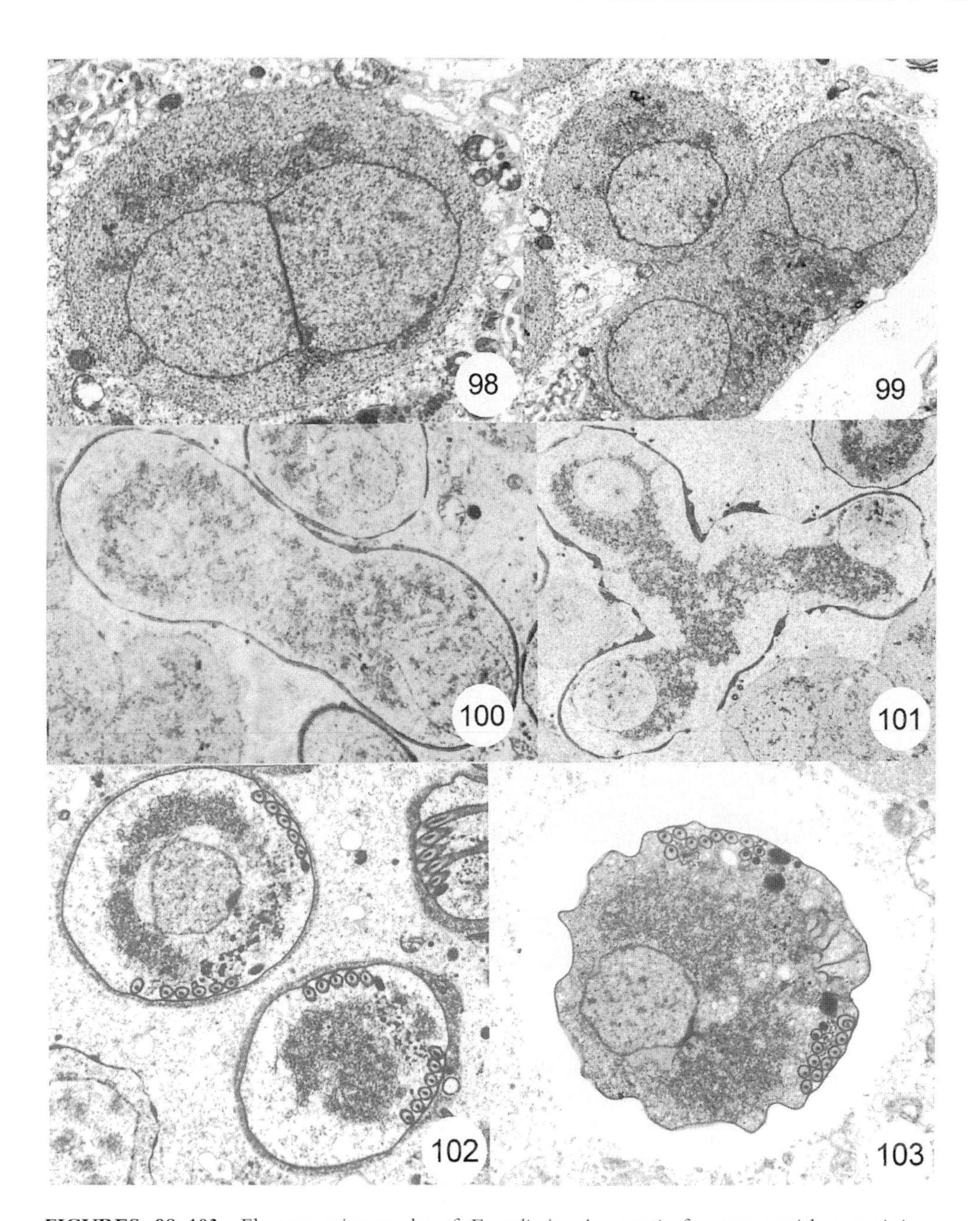

FIGURES 98–103 Electron micrographs of *E. aedis* in *A. aegypti* after transovarial transmission. Reprinted from Becnel et al. (1989) with permission from the publisher. **FIGURE 98** Diplokaryotic meront. Magnification, ×8,300. **FIGURE 99** Dissociation of diplokaryon. Magnification, ×7,650. **FIGURE 100** Dividing schizont. Magnification, ×5,300. **FIGURE 101** Sporogonial plasmodium in process of sporogony. Magnification, ×3,700. **FIGURE 102** Early sporoblasts. Magnification, ×5,800. **FIGURE 103** Sporoblast isolated in a sporophorous vesicle. Magnification, ×7,650.

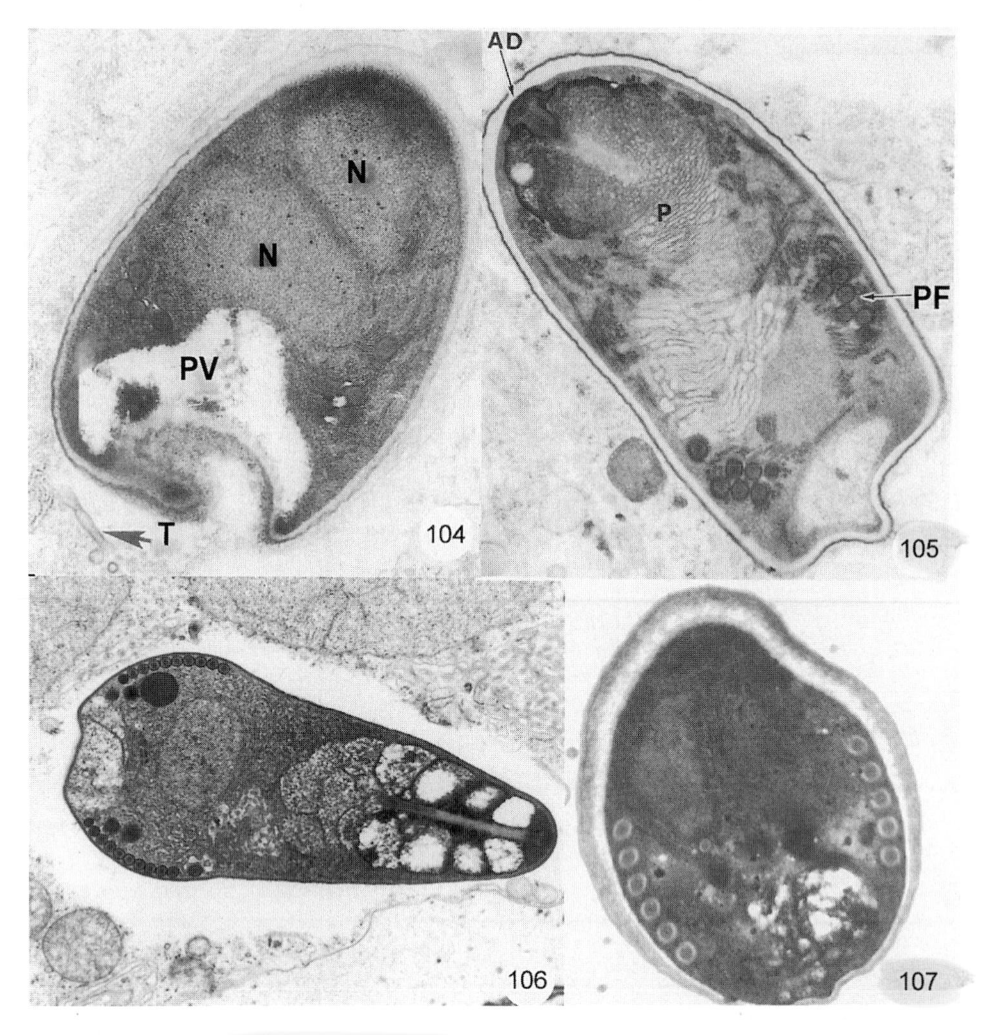

FIGURES 104–107 Electron micrographs of the four different spore types of *E. aedis* in *A. aegypti*. **FIGURE 104** Primary binucleate spore. N, nucleus; PV, posterior vacuole; T, tubules. Magnification, ×16,000. **FIGURE 105** Binucleate (transovarial) spore. AD, anchoring disk; PF, polar filament; P, polaroplast. Magnification, ×15,400. **FIGURE 106** Haploid, uninucleate (environmental) spore. Magnification, ×7,700. **FIGURE 107** Meiospore. Magnification, ×12,700. Reprinted from Johnson et al. (1997) with permission from the authors and publisher (Fig. 104) and from Becnel et al. (1989) with permission from the publisher (Fig. 105–107).

represent an assemblage of many genera, as indicated by molecular studies (Baker et al., 1994). Alternatively, the criteria used to establish the genera of microsporidia from aquatic insects may not truly reflect phylogenetic diversity but adaptations to specific habitats and host systems. Hopefully, additional molecular studies will help resolve some of these fundamental taxonomic questions.

The genera having insects as type hosts are listed below. Diagnostic information is primarily restricted to the features of sporulation and the spore, with the addition of distinguishing life cycle characteristics when available. The type host and species are given followed by comments on distribution and other matters deemed of importance. Space is not available to provide comprehensive information on these

TABLE 1 Distribution by insect order of microsporidia with insects as type hosts

Insect order	No. of microsporidian genera	Microsporidian genus or genera
Diptera	42	*Amblyospora, Bohuslavia, Campanulospora, Caudospora, Chapmanium, Coccospora, Cristulospora, Culicospora, Culicosporella, Cylindrospora, Edhazardia, Evlachovaia, Flabelliforma, Golbergia, Hazardia, Helmichia, Hessea, Hirsutusporos, Hyalinocysta, Janacekia, Merocinta, Napamichum, Neoperezia, Octosporea, Parathelohania, Pegmatheca, Pernicivesicula, Pilosporella, Polydispyrenia, Ringueletium, Scipionospora, Semenovaia, Spherospora, Spiroglugea, Striatospora, Systenostrema, Toxoglugea, Toxospora, Trichoctosporea, Tricornia, Vavraia, Weiseria*
Coleoptera	5	*Anncaliia, Canningia, Chytridiopsis, Endoreticulatus, Ovavesicula*
Ephemeroptera	5	*Geusia, Mitoplistophora, Stempellia, Telomyxa, Trichoduboscqia*
Lepidoptera	4	*Cystosporogenes, Nosema, Orthosomella, Vairimorpha*
Trichoptera	3	*Episeptum, Issia, Tardivesicula*
Orthoptera	2	*Heterovesicula, Johenrea*
Odonata	2	*Nudispora, Resiomeria*
Siphonaptera	2	*Nolleria, Pulcispora*
Collembola	1	*Auraspora*
Thysanura	1	*Buxtehudea*
Hymenoptera	1	*Burenella*
Isoptera	1	*Duboscqia*
Total	69	

genera, but such information can be found elsewhere (Canning and Lom, 1986; Larsson, 1986a, 1988; Weiser, 1977; Sprague, 1977; Sprague et al., 1992). Also listed in this section are a few genera for which insects are not the type hosts. This is because many species have been reported to occur in insects (such as *Thelohania, Pleistophora*) or because there are possible links to insect microsporidia (such as *Trichotuzetia*).

***Amblyospora* Hazard & Oldacre, 1975**

Diagnosis: Three sporulation sequences resulting in the production of three morphologically and functionally distinctive spore types characterize this genus. A binucleate spore is formed in oenocytes of the adult female and functions to infect the developing oocytes. In progeny, meiospores are formed in groups of eight in the fat body. These spores are infectious to an obligate intermediate host which in all cases thus far has been a copepod. A second type of uninucleate spore is formed in the ovaries of the copepod host. The large, lanceolate spore is responsible for infecting a new generation of mosquitoes, leading to formation of the binucleate spore in the adult mosquito to complete the life cycle.

Type species: *Amblyospora californica* (Kellen & Lipa, 1960) Hazard & Oldacre, 1975

Type host: *Culex tarsalis* (Diptera: Culicidae)

Comments: This is a very large genus, with most species isolated from Diptera (about 90, mostly mosquitoes) and about 5 species from Trichoptera reported. An intermediate host has been verified for approximately 10 species. Most species are described from the larval stages of the host on the basis of the one sporulation sequence that involves meiosis and produces spores in groups of eight. This is the first microsporidian genus for which an intermediate host has been documented.

***Ameson* Sprague, 1977**

Diagnosis: Presporulation stages are diplokaryotic. Only one sporulation sequence is known that involves polysporoblastic sporogony. Sporulation ends with the production of binucleate spores which are small and oval. Muscle is the main site of infection. Crabs can be infected by exposure to infected tissue, which indicates that transmission is either by mouth or via the gills.

Type species: *Ameson michaelis* (Sprague, 1970)

Type host: *Callinectes sapidus* (Decapoda: Portunidae)

Comments: Only one species from an insect (Diptera: Tabanidea) has been described. The type species for this genus is a marine decapod, and placement of an insect microsporidium in this genus is questionable.

Anncaliia Issi, Krylova & Nicolaeva, 1993

Diagnosis: Presporulation stages are diplokaryotic. There is only one sporulation sequence of the *Nosema* type. Sporogony is disporoblastic, with the production of oval, binucleate spores. This genus is distinguished by the production of tubular secretions during development. Transmission is presumably by mouth, with infections occurring primarily in the fat body.

Type species: *Anncaliia meligethi* Issi, Krylova & Nicolaeva, 1993

Type host: *Meligethes aeneus* (Coleoptera: Nitidulidae)

Comments: The type host is a beetle. The authors transferred *N. varivestis* (Brooks et al., 1985), a parasite of the Mexican bean beetle, into this genus on the basis of the presence of tubular secretions.

Auraspora Weiser & Purrini, 1980

Diagnosis: Presporulation stages are diplokaryotic. The group is characterized by two sporulation sequences: One is a *Nosema*-like sequence producing pyriform spores that are thin-walled and diplokaryotic. The other, described as anomalous, produces thick-walled, uninucleate spores within a sporophorous vesicle. Infections occur in the male gonad.

Type species: *Auraspora canningae* Weiser & Purrini, 1980

Type host: *Lepidocyrtus lignorum* (Collembola: Entomobryidae)

Comments: Monotypic. The type host is a springtail.

Bacillidium Janda, 1928

Diagnosis: Presporulation stages are diplokaryotic. Only one sporulation sequence is known that begins with diplokaryotic sporonts. Sporogony is disporoblastic, producing large, rod-shaped, binucleate spores. Transmission is by mouth, and the main sites of infection are lymphocytes and nephridia.

Type species: *Bacillidium criodrili* Janda, 1928

Type host: *Criodrilus lacuum* (Annelida: Oligochaeta)

Comments: Two species from insects (one from a chironomid and one from the Thysanura [silverfish] have been reported, but the type host is not an insect.

Bohuslavia Larsson, 1985

Diagnosis: Presporulation stages are diplokaryotic. Only one sporulation sequence is known. Diplokaryotic sporonts apparently undergo meiosis to produce uninucleate sporoblasts in groups of 8 (typical) or 16 and eventually 8 or 16 spores in a persistent sporophorous vesicle. Infections are found in the fat body of larvae.

Type species: *Bohuslavia asterias* (Weiser, 1963b) Larsson, 1985

Type host: *Endochironomus* sp. (Diptera: Chironomidae)

Comments: Monotypic. This is one of many genera known with only a *Thelohania*-like sporulation sequence.

Burenella Jouvenaz & Hazard, 1978

Diagnosis: Diplokaryotic stages lead to two distinctive sporulation sequences in the host. One sequence is *Nosema*-like and produces spores in the hypodermis. The other occurs in the fat body and is similar to sporulation in the genus *Thelohania,* producing uninucleate spores in groups of eight and presumably involving meiosis. Transmission is by mouth via the binucleate spores. The role of the uninucleate spores unknown.

Type species: *Burenella dimorpha* Jouvenaz & Hazard, 1978

Type host: *Solenopsis geminata* (Hymenoptera: Formicidae)

Comments: Monotypic. Isolated from a fire ant.

Buxtehudea Larsson, 1980

Diagnosis: Presporulation stages are absent. Only one sporulation sequence is known. Development is uninucleate with polysporoblastic sporogony, each sporont producing 50 to 100 sporoblasts. Uninucleate spores are spherical and are contained within a parasitophorous vesicle. Infection is restricted to the midgut epithelium with transmission apparently by mouth.

Type species: *Buxtehudea scaniae* Larsson, 1980

Type host: *Petrobius brevistylis* (Thysanura: Machilidae)

Comments: Monotypic. Possibly two spore types are present, one with a short polar filament and little or no endospore, and the other with a longer polar filament and a thick endospore (Sprague et al., 1992).

Campanulospora Issi, Radischcheva & Dolzhenko, 1983

Diagnosis: Only one sporulation sequence is known. Sporonts are diplokaryotic, disporoblastic sporogony producing binucleate spores that are elongate-ovoid. Tissue infected is the midgut epithelium (?).

Type species: *Campanulospora denticulata* Issi, Radischcheva & Dolzhenko, 1983

Type host: *Delia floralis* (Diptera: Muscidae)

Comments: Monotypic. Isolated from a muscoid fly. The description is in Russian, and additional details are unavailable.

Canningia Weiser, Wegensteiner & Zizka, 1995

Diagnosis: Only one sporulation sequence is known. All stages are uninucleate and occur in the hyaloplasm of the host cell. Sporulation produces small, uninucleate spores which are elongate-ovoid to tubular. The parasite is transmitted both orally and transovarially. It infects primarily the fat body and Malpighian tubules, but the infection may become systemic.

Type species: *Canningia spinidentis* Weiser, Wegensteiner & Zizka, 1995

Type host: *Pityokteines spinidens* (Coleoptera: Scolytidae)

Comments: Monotypic. Isolated from a bark beetle.

Caudospora Weiser, 1946

Diagnosis: Only one sporulation sequence is known. All stages are diplokaryotic. Octosporoblastic sporogony produces binucleate spores without the presence of a sporophorous vesicle. A taillike posterior appendage that is an extension of the exospore characterizes the spore. The mechanism of transmission is unknown. Infections are found in the fat body of larvae.

Type species: *Caudospora simulii* Weiser, 1946

Type host: *Simulium hirtipes* (Diptera: Simuliidae)

Comment: Approximately five other species have been reported, all from blackflies.

Chapmanium Hazard & Oldacre, 1975

Diagnosis: Presporulation stages are diplokaryotic. Only one sporulation sequence is known. Diplokaryotic sporonts undergo octosporoblastic sporogony (perhaps meiosis is involved) to produce eight spores within a persistent, fusiform sporophorous vesicle. Infections occur in the larval fat body.

Type species: *Chapmanium cirritus* Hazard & Oldacre, 1975

Type host: *Corethrella brakeleyi* (Diptera: Chaoboridae)

Comments: Three species from insects have been described: two from Diptera and one from Hemiptera.

Chytridiopsis Schneider, 1884

Diagnosis: Presporulation stages are presumably absent. Only one sporulation sequence is known and occurs in either a thick-walled cyst or in a membrane-bound vacuole. Uninucleate spores are small and spherical, with reduced organelles and little or no endospore. The typical polaroplast is lacking but has been replaced by a structure with a honeycomb appearance. Infections are found in the intestinal epithelium of the adult.

Type species: *Chytridiopsis socius* Schneider, 1884

Type host: *Blaps mortisaga* (Coleoptera: Tenebrionidae)

Comments: Five species from beetles and one from a caddis fly. They are examples of what have been called primitive microsporidia because of the reduced development of spore organelles and apparent lack of presporulation development.

Coccospora Kudo, 1925

Diagnosis: Presporulation stages are diplokaryotic. Only one sporulation sequence is known. Sporonts are diplokaryotic, and sporogony is octosporoblastic and probably involves meiosis to produce eight spores in a subpersistent sporophorous vesicle. Spores are spherical and rather small. Infections are found in adipose tissue of the larvae, but there is no information on transmission.

Type species: *Coccospora micrococcus* (Léger & Hesse, 1921) Kudo, 1925

Type host: *Tanypus setiger* (Diptera: Chironomidae)

Comment: There are three species, all from chironomids.

Cougourdella Hesse, 1935

Diagnosis: Only one sporulation sequence is known. Spores are gourd-shaped, large, and uninucleate. Infections have been reported in the adipose tissue, muscle, and gonad.

Type species: *Cougourdella magna* Hesse, 1935

Type host: *Megacyclops viridis* (Copepoda: Cyclopidae)

Comments: The type host is not an insect. The original description was inadequate in several respects. Three species from Trichoptera have been described on the basis of the presence of gourd-shaped spores.

Cristulospora Khodzhaeva & Issi, 1989

Diagnosis: Presporulation stages are diplokaryotic. Two sporulation sequences are known. One is in the adult female mosquito and produces binucleate spores. The other is octosporoblastic (presumably involving meiosis) and produces eight spores in a sporophorous vesicle. Infections occur in the larval fat body and presumably in the oenocytes of adults. This genus is distinguished by "appendages in the shape of magnificent plumes" at the poles of each meiospore.

Type species: *Cristulospora sherbani* Khodzhaeva & Issi, 1989

Type host: *Culex modestus* (Diptera: Culicidae)

Comments: Monotypic. Authors considered this new genus similar to *Amblyospora*.

Culicospora Weiser, 1977

Diagnosis: Presporulation stages are uninucleate and diplokaryotic. Two sporulation sequences are involved in the life cycle. One produces an oval-cylindrical binucleate spore in oenocytes of the adult female mosquito which infects the developing oocyte. The second sequence in the fat body of progeny involves diplokaryotic sporonts which undergo nuclear dissociation to form uninucleate sporoblasts. Spores are large and lanceolate, and each is contained in a delicate sporophorous vesicle.

Type species: *Culicospora magna* (Kudo, 1920) Weiser, 1977

Type host: *Culex pipiens* (Diptera: Culicidae)

Comments: The type species is from a mosquito, but there is evidence that the type host may have been misidentified and is actually *Culex restuans* (Sprague et al., 1992). One additional species, from a blackfly, has been described. The uninucleate spore in larvae is remarkably similar to the spore formed in the copepod intermediate host for *A. californica*.

Culicosporella Weiser, 1977

Diagnosis: Three sporulation sequences are involved in the life cycle, each initiated by a diplokaryotic sporont. A small, oblong-ovoid binucleate spore is formed in adult female mosquitoes and is presumably responsible for transovarial transmission. In the fat body of progeny, two distinctive sporulation sequences occur. Multiple fission of large diplokaryotic sporogonial plasmodia results in the formation of large, lanceolate binucleate spores, each within a delicate sporophorous vesicle. These spores are orally infectious to mosquito larvae. The other sequence involves meiosis which aborts during the process and rarely results in meiospores.

Type species: *Culicosporella lunata* (Hazard & Savage, 1970) Weiser, 1977

Type host: *Culex pilosus* (Diptera: Culicidae)

Comments: Monotypic. The lanceolate binucleate spore is similar morphologically to the uninucleate spores formed in the copepod intermediate host of *A. californica* except for nucleation.

Cylindrospora Issi & Voronin, 1986

Diagnosis: Presporulation stages are diplokaryotic. Only one sporulation sequence is known. The sporont is a diplokaryotic cell with sporogony octosporoblastic and perhaps involves meiosis. Uninucleate spores are bacilliform with a short, funnel-shaped polar filament and a reduced endospore. Infections are found in the larval fat body.

Type species: *Cylindrospora chironomi* Issi & Voronin, 1986

Type host: *Chironomus plumosus* (Diptera: Chironomidae)

Comment: Only two known species are known, both from chironomids.

Cystosporogenes Canning, Barker, Nicholas & Page, 1985

Diagnosis: Only one sporulation sequence is known. All stages are uninucleate throughout development. Multinucleate sporogoninal plasmodia divide by budding to produce 8, 12, or 16 spores in a vesicle of uncertain origin. Spores are small with a thin, rugose exospore and a thin endospore. Infections occur primarily in the silk glands but can spread to many other tissues.

Type species: *Cystosporogenes operophterae* (Canning, 1960) Canning Barker, Nicholas & Page, 1985

Type host: *Orthosoma brumata* (Lepidoptera: Geometridae)

Comment: Additional species from a dipteran and a hymenopteran have been described.

Duboscqia Perez, 1908

Diagnosis: Presporulation stages are uninucleate and diplokaryotic. One sporulation sequence results in 16 spores within a sporophorous vesicle. Development probably involves diplokaryotic cells that undergo meiosis, but interpretations differ. Infections occur in the fat body. The parasite and host form a conspicuous xenoma.

Type species: *Duboscqia legeri* Perez, 1908

Type host: *Reticulitermes lucifugus* (Isoptera: Rhinotermitidae)

Comment: One additional species has been reported from a mosquito on the basis of formation of 16 spores in a sporophorous vesicle (Sweeney et al., 1993). This placement should be reevaluated when the type is examined by modern methods.

Edhazardia Becnel, Sprague & Fukuda, 1989

Diagnosis: This is the only microsporidium for which four sporulation sequences have been documented. One produces a small, thin-walled binucleate spore in the gastric ceca of larval mosquitoes and is responsible for autoinfection. A second, larger binucleate spore develops in the oenocytes of the female adult and functions to infect the developing oocytes. In the fat body of infected progeny, two sporulation sequences occur. One (the primary sequence) begins with nuclear dissociation of a diplokaryotic cell to produce uninucleate sporonts which undergo multiple division to produce uninucleate pyriform spores, each within a delicate sporophorous vesicle. The other sequence involves meiosis of a diplokaryotic sporont which usually aborts prior to the formation of meiospores.

Type species: *Edhazardia aedis* (Kudo 1930) Becnel, Sprague & Fukuda, 1989

Type host: *Aedes aegypti* (Diptera: Culicidae)

Comments: Monotypic. This genus has the characteristic *Amblyospora*-like life cycle features of transovarial and horizontal transmission but without the involvement of an intermediate host.

Endoreticulatus Brooks, Becnel & Kennedy, 1988

Diagnosis: Only one sporulation sequence is known. All stages are uninucleate throughout development. Moniliform or irregular sporogonial plasmodia divide by multiple fission to produce up to 50 sporoblasts. Spores are small and ovocylindrical and are contained within a parasitophorous vesicle formed by the host endoplasmic reticulum. Infections are restricted to the midgut epithelium, and transmission occurs by mouth.

Type species: *Endoreticulatus fidelis* (Hostounsky & Weiser, 1975) Brooks, Becnel & Kennedy, 1988

Type host: *Leptinotarsa undecemlineata* (Coleoptera: Chrysomelidae)

Comment: Type host is the potato beetle, with one additional species (from a lepidopteran) described.

Episeptum Larsson, 1986

Diagnosis: Only one sporulation sequence is known. All stages are uninucleate throughout development. Tetrasporoblastic sporogony produces small, ovoid, uninucleate spores. Infections are restricted to the larval fat body.

Type species: *Episeptum inversum* Larsson, 1986b

Type host: *Holocentropus picicornis* (Trichoptera: Polycentropodidae)

Comment: There are three species in this genus, all from caddis flies.

Evlachovaia Issi, 1986

Diagnosis: Two sporulation sequences have been reported, one that produces oval binucleate spores and one that produces short, oval, uninucleate spores within a sporophorous vesicle.

Type species: *Evlachovaia chironomi* Voronin & Issi, 1986

Type host: *Chironomus plumosus* (Diptera: Chironomidae)

Comment: The original description is brief and perhaps does not satisfy the criteria of availability.

Flabelliforma Canning, Killick-Kendrick & Killick-Kendrick, 1991

Diagnosis: Only one sporulation sequence is known. All stages are uninucleate throughout development. Multiple division of sporogonial plasmodia produces as many a 32 ovoid, uninucleate spores contained within a delicate polysporophorous vesicle. Infections are restricted to the midgut and transmission occurs by mouth.

Type species: *Flabelliforma montana* Canning, Killick-Kendrick & Killick-Kendrick, 1991

Type host: *Phlebotomus ariasi* (Diptera: Psychodidae)

Comment: Two species have been described, both from moth flies.

Geusia Rühl & Korn, 1979

Diagnosis: Information is meager. One sporulation sequence produces oval, uninucleate (?) spores.

Type species: *Geusia gamocysti* Rühl & Korn, 1979

Type host: *Gamocystis ephemerae* (Gregarinida: Gregarinidae) in *Ephemera danica* (Ephemeroptera: Ephemeridae)

Comments: Monotypic. This genus is mentioned here only because it has been reported from a gregarine parasite of a mayfly.

Golbergia Weiser, 1977

Diagnosis: Presporulation stages are diplokaryotic. Two sporulation sequences occur in the larval fat body. One sequence begins with diplokaryotic sporogonial plasmodia which divide by multiple fission to produce thick-walled, binucleate spores. The binucleate spore is flattened at the narrow end, and the exospore is ornamented at the broad end with ridges and naillike protrusions. The other sequence involves uninucleate sporonts which divide by multiple fission to produce uninucleate spores in sporophorous vesicles with 4, 8, 12, or 16 spores. Spores are pyriform and thin-walled. Transmission occurs by mouth.

Type species: *Golbergia spinosa* (Golberg, 1971) Weiser, 1977

Type host: *Culex pipiens* (Diptera: Culicidae)

Comments: Monotypic. This genus is very similar to *Hazardia* morphologically and difficult to distinguish from it.

Gurleya Doflein, 1898

Diagnosis: Only one sporulation sequence is known. Sporogony is tetrasporoblastic, resulting in four uninucleate, pyriform spores contained within a sporophorous vesicle. Infections are found in hypodermal tissue.

Type species: *Gurleya tetraspora* Doflein, 1898

Type host: *Daphnia maxima* (Cladocera: Daphniidae)

Comments: The type host is a microcrustacean, but about 11 species in insects from Diptera, Ephemeroptera, Isoptera, Lepidoptera, Odonata, and Trichoptera have been described. Placement of these species in this genus has been based primarily on the presence of four spores within a sporophorous vesicle.

Hazardia Weiser, 1977

Diagnosis: This genus is characterized by three sporulation sequences occurring primarily in the fat body of larval mosquitoes. The first sequence results in the production of small, oval binucleate spores. The second sequence involves diplokaryotic sporonts which divide by binary fission to produce lanceolate, thick-walled, binucleate spores with a rugose exospore. The third sequence (the most common) involves uninucleate sporonts which form sporogonial plasmodia that divide by multiple fission, producing 2 to 16 spores (usually 8). These uninucleate spores are pyriform and thin-walled. Transmission is by mouth.

Type species: *Hazardia milleri* (Hazard & Fukuda, 1974) Weiser, 1977

Type host: *Culex pipiens quinquefasciatus* (Diptera: Culicidae)

Comments: Monotypic. The distinctions between this genus and *Golbergia* are in need of resolution.

Helmichia Larsson, 1982

Diagnosis: Presporulation stages are diplokaryotic. One sporulation sequence is known. Diplokaryotic sporonts undergo meiosis followed by octosporoblastic sporogony which results in the formation of eight spores within a sporophorous vesicle. Uninucleate spores are rod-shaped and have a short polar filament without coils. Infections are restricted to the fat body of larvae.

Type species: *Helmichia aggregata* Larsson, 1982

Type host: *Endochironomus* sp. (Diptera: Chironomidae)

Comment: Two species from chironomids and one from a blackfly have been described.

Hessea Ormières & Sprague, 1973

Diagnosis: Presporulation stages are diplokaryotic. Two sporulation sequences are known. The parasite is contained within a thick-walled vesicle of host and parasite origin. Diplokaryotic sporonts divide by multiple fission to form either binucleate or uninucleate spores. Spores are generally teratological, with most organelles reduced. Infections are restricted to the gut epithelium of larvae.

Type species: *Hessea squamosa* Ormières & Sprague, 1973

Type host: *Sciara* sp. (Diptera: Lycoriidae)

Comments: Monotypic. The type host is a fungus gnat.

Heterovesicula Lange, Macvean, Henry & Streett, 1995

Diagnosis: Presporulation stages are diplokaryotic. This genus is characterized by two concurrent sporulation sequences occurring in the fat body and Malpighian tubules of the adult. One involves large, diplokaryotic plasmodia which transform into chains of sporonts. Sporogony is disporous and results in variable numbers of binucleate spores within a sporophorous vesicle. Binucleate spores are ovocylindrical with a rugose, layered exospore. The other sequence involves uninucleate sporonts produced as a result of nuclear dissociation of a diplokaryon. Sporogony is octosporoblastic, producing eight slightly pyriform uninucleate spores within a sporophorous vesicle. Transmission occurs by mouth.

Type species: *Heterovesicula cowani* Lange, Macvean, Henry & Streett, 1995

Type host: *Anabrus simplex* (Orthoptera: Tettigoniidae)

Comments: Monotypic. This genus is similar to several microsporidia (such as *Vairimorpha* and *Burenella*) in that there are two concurrent sporulation sequences producing binucleate and uninucleate spores. It is distinctive because each spore type is formed within a sporophorous vesicle.

Hirsutusporos Batson, 1983

Diagnosis: Presporulation stages are diplokaryotic. Only one sporulation sequence is known. All stages occur in the hyaloplasm of the host cell. A diplokaryotic sporont divides by binary fission to produce two sporoblasts. Binucleate spores are broadly oval and are characterized by exospore ornamentation in the form of a posterior tuft of large and small filamentous appendages extending toward the anterior and becoming closely packed tubercles. Infections are restricted to the fat body of larvae.

Type species: *Hirsutusporos austrosimulii* Batson, 1983

Type host: *Austrosimulium* sp. (Diptera: Simuliidae)

Comment: Monotypic.

Hyalinocysta Hazard & Oldacre, 1975

Diagnosis: Presporulation stages are diplokaryotic. Only one sporulation sequence is known. Meiosis of diplokaryotic sporonts is followed by octosporoblastic sporogony. Uninucleate spores are ovoid and are contained within a sporophorous vesicle. Infections are restricted to the fat body of larvae.

Type species: *Hyalinocysta chapmani* Hazard & Oldacre, 1975

Type host: *Culiseta melanura* (Diptera: Culicidae)

Comments: One additional species from a blackfly has been described. Recently completed studies (Andreadis, unpublished data) have demonstrated the involvement of an intermediate copepod host in the life cycle of *H. chapmani*.

Issia Weiser, 1977

Diagnosis: Presporulation stages are diplokaryotic. Only one sporulation sequence is known. The sporont is diplokaryotic, and sporogony is disporoblastic. A binucleate ovoid spore and two spores adhere to one another within a sporophorous vesicle. Infections are found in the fat body of larvae.

Type species: *Issia trichopterae* (Weiser, 1946) Weiser, 1977

Type host: *Plectrocnemia geniculata* (Trichoptera: Polycentropidae)

Comments: Data are fragmentary. In addition to the type species, one species from a mosquito has been described.

Janacekia Larsson, 1983

Diagnosis: Presporulation stages are diplokaryotic. Only one sporulation sequence is known. The sporont is a diplokaryotic cell which undergoes meiosis and octosporoblastic sporogony. Eight spores are contained within a sporophorous vesicle. Spores are uninucleate and oval, and the exospore is covered with short, thick tubules. Infections are restricted to the larval fat body.

Type species: *Janacekia debaisieuxi* (Jírovec, 1943) Larsson, 1983

Type host: *Simulium maculatum* (Diptera: Simuliidae)

Comment: One species from a crane fly (Ptychopteridae) and one from a beetle have been described.

Johenrea Lange, Becnel, Razafindratiana, Przybyszewski & Razafindrafara, 1996

Diagnosis: Only one sporulation sequence is known. All stages are uninucleate throughout development, and a conspicuous neoplastic xenoma is formed. Sporogonial plasmodia with 16 nuclei divide by plasmotomy to form tetranucleate plasmodia. Two divisions follow to produce 16 spores within a sporophorous vesicle. Sometimes 8 or 32 spores are observed within a sporophorous vesicle. The uninucleate spores are elongate-ovoid and orally infectious. Infections are mainly restricted to the fat body.

Type species: *Johenrea locustae* Lange, Becnel, Razafindratiana, Przybyszewski & Razafindrafara, 1996

Type host: *Locusta migratoria capito* (Orthoptera: Acrididae)

Comments: Monotypic. This microsporidium is only one of a few from arthropods to have a neoplastic xenoma of the *Glugea* type commonly found in microsporidia from fish.

Larssoniella Weiser & David, 1997

Diagnosis: Only one sporulation sequence is known. All stages are apparently uninucleate and occur in the hyaloplasm of the host cell. Sporonts are identified by the presence of tufts of tubules arising from the plasmalemma. Tufts remain attached to the exospore of young spores at the posterior end. Mature uninucleate spores are elongate-ovoid without the tufts of tubules. Infections are found in the silk gland, Malpighian tubules, and gonads.

Type species: *Larssoniella resinellae* Weiser & David, 1997

Type host: *Petrova resinella* (Lepidoptera: Tortricidae)

Comment: Monotypic.

Merocinta Pell & Canning, 1993

Diagnosis: Presporulation stages are diplokaryotic. Two sporulation sequences have been reported. One sequence forms binucleate spores in adults which are responsible for transovarial transmission. The other occurs in the gut tissue of larvae. Diplokaryotic cells undergo nuclear dissociation to form uninucleate cells, and these cells divide repeatedly to form 40 to 60 uninucleate spores within a parasitophorous vesicle. Uninucleate spores are oval and small and have a short polar filament.

Type species: *Merocinta davidii* Pell & Canning, 1993

Type host: *Mansonia africana* (Diptera: Culicidae)

Comment: Monotypic.

Microsporidium Balbiani, 1884

Comments: This is a collective group intended by Sprague (1977) "as a temporary expedient when species (both previously named ones and new ones) can not readily be assigned to genera." There are approximately 45 species of microsporidia from 7 orders of insects assigned to this genus.

Mitoplistophora Codreanu, 1966

Diagnosis: Only one sporulation sequence is known. Sporulation ends with 2, 8, 16, 32, 48, or 64 spores in a triangular sporophorous vesicle. The spore is pyriform, but other details are unknown. Infections are found in the fat body of nymphs.

Type species: *Mitoplistophora angularis* Codreanu, 1966

Type host: *Ephemera danica* (Ephemeroptera: Ephemeridae)

Comment: Monotypic.

Napamichum Larsson, 1990

Diagnosis: Presporulation stages are diplokaryotic. One sporulation sequence is known. The sporont is a diplokaryotic cell which undergoes meiosis and octosproblastic sporogony to produce eight spores within a sporophorous vesicle. Spores are uninucleate and pyriform and are formed in the fat body of larvae.

Type species: *Napamichum dispersus* (Larsson, 1984) Larsson, 1990a

Type host: *Endochironomus* sp. (Diptera: Chironomidae)

Comments: Basically another example of a *Thelohania*-like sporulation sequence. Two additional species have been described, one from a chironomid and another from an aquatic mite.

Neoperezia Issi & Voronin, 1979

Diagnosis: Presporulation stages are diplokaryotic. Only one sporulation sequence is known. The sporont is a diplokaryotic cell with octosporoblastic sporogony (typical) resulting in eight spores, likely in a sporophorous vesicle. Spores are uninucleate and elongate-ovoid and are joined together in pairs. Infections are found in the larval fat body.

Type species: *Neoperezia chironomi* Issi & Voronin, 1979

Type host: *Chironomus plumosus* (Diptera: Chironomidae)

Comment: Monotypic.

Nolleria Beard, Butler & Becnel, 1990

Diagnosis: Presporulation stages are presumably absent. One sporulation sequence is known. The sporont is a uninucleate cell. Sporogony involves multiple division by vacuolation of a large plasmodium, producing 150 to 200 spores within an envelope composed of both host and parasite components. The spore is uninucleate, spherical, and rather small with many organelles reduced. Infections are found in the midgut epithelium of the adult flea.

Type species: *Nolleria pulicis* Beard, Butler & Becnel, 1990

Type host: *Ctenocephalides felis* (Siphonaptera: Pulicidae)

Comments: Monotypic. This genus is another example of what have been called primitive microsporidia because of the reduced development of spore organelles and apparent lack of presporulation development.

Nosema Naegeli, 1857

Diagnosis: Presporulation stages are diplokaryotic (some reports indicate that uninucleate stages occur [Sprague et al., 1992]). Two sporulation sequences are known, both with diplokaryotic sporonts and disporoblastic sporogony. The primary (early) sporulation sequence produces a thin-walled, binucleate spore with a short polar filament and a large posterior vacuole. This spore apparently germinates spontaneously to spread the parasite within the host (autoinfection) (Iwano and Ishihara, 1991a, 1991b). The second sporulation sequence produces a binucleate spore with a thicker spore wall and a longer polar filament and apparently functions to infect a new host. Infections are systemic, and transmission can occur either transovarially or by mouth.

Type species: *Nosema bombycis* Naegeli 1857

Type host: *Bombyx mori* (Lepidoptera: Bombycidae)

Comments: This genus has more than 150 described species representing at least 12 orders of insects. The type species is responsible for pébrine disease in silkworms and was the subject of work by Pasteur and his discovery that this disease is caused by a microbial pathogen. The discovery of an early sporulation sequence that produces spores to spread the infection in the host has also been found to occur in N. apis (de Graaf et al., 1994a, 1994b), *V. necatrix* (Solter and Maddox, 1998), and *E. aedis* (Johnson et al., 1997).

Nudispora Larsson, 1990

Diagnosis: Presporulation stages are diplokaryotic. Only one sporulation sequence is known. The sporont is a diplokaryotic cell which undergoes meiosis and octosporoblastic sporogony. Eight uninucleate, ovoid spores are produced in the larval fat body without a sporophorous vesicle.

Type species: *Nudispora biformis* Larsson, 1990b

Type host: *Coenagrion hastulatum* (Odonata: Coenagrionidae)

Comments: Monotypic. Sporulation sequence of the *Thelohania* type but without the formation of a sporophorous vesicle.

Octosporea Flu, 1911

Diagnosis: Presporulation stages are diplokaryotic. Only one sporulation sequence is known. The sporont is a diplokaryotic cell, and sporogony is usually octosporoblastic. Diplokaryotic sporogonial plasmodia divide by multiple fission to produce eight binucleate spores within a sporophorous vesicle. Binucleate spores are ovocylindrical and orally infectious. Infections are found in the midgut epithelium of larvae and adults.

Type species: *Octosporea muscaedomesticae* Flu, 1911

Type host: *Musca domestica* (Diptera: Muscidae)

Comments: The type host is the common housefly, but about 13 species from Diptera, Ephemeroptera, Hemiptera, Lepidoptera, and Collembola have been reported.

Orthosomella Canning, Wigley & Barker, 1991

Diagnosis: Only one sporulation sequence is known. All developmental stages are uninucleate and are in direct contact with the host cell hyaloplasm. Sporogony involves moniliform sporogonial plasmodia with 2, 4, 8, or 12 nuclei which divide to produce uninucleate spores. Uninucleate spores are elongate-ellipsoid and are formed in the silk gland, gut, and other tissues of the larva and adult. Transmission is thought to be transovarial.

Type species: *Orthosomella operophterae* (Canning, 1960) Canning, Wigley & Barker, 1991

Type host: *Operophtera brumata* (Lepidoptera: Geometridae)

Comment: Two species have been described, both from the Lepidoptera.

Ovavesicula Andreadis & Hanula, 1987

Diagnosis: Presporulation stages are diplokaryotic. Only one sporulation sequence is known. A diplokaryotic cell undergoes nuclear dissociation followed by four nuclear divisions without cytokinesis to form a plasmodium with 32 nuclei. Plasmodia break up, eventually producing 32 uninucleate spores within a thick-walled polysporophorous vesicle. Uninucleate spores are spheroid to ovoid and small and are formed in the Malpighian tubules of larvae.

Type species: *Ovavesicula popilliae* Andreadis & Hanula, 1987

Type host: *Popillia japonica* (Coleoptera: Scarabaeidae)

Comment: Monotypic.

Parathelohania Codreanu, 1966

Diagnosis: Presporulation stages are diplokaryotic. Two sporulation sequences are involved in the life cycle. One produces a binucleate spore in the oenocytes of the adult female mosquito which functions to infect the developing oocyte. The second sequence occurs in the fat body of progeny and involves diplokaryotic sporonts which undergo meiosis and octosporoblastic sporogony to produce eight uninucleate spores within a sporophorous vesicle. Meiospores are ovoid and are characterized by a posterior extension of the spore wall referred to as a "bottleneck."

Type species: *Parathelohania legeri* (Hesse, 1904) Codreanu, 1966

Type host: *Anopheles maculipennis* (Diptera: Culicidae)

Comments: Members of this genus are almost exclusively parasites of anopheline mosquitoes, and there are about 14 described species. A third sporulation sequence has been demonstrated for *P. anophelis,* which produces uninucleate spores in a copepod intermediate host (Avery and Undeen, 1990). This observation suggests that the life cycles of species in this genus are similar to that of *Amblyospora.*

Pegmatheca Hazard & Oldacre, 1975

Diagnosis: Presporulation stages are diplokaryotic. Only one sporulation sequence is known. The sporont is a diplokaryotic cell which undergoes meiosis. Sporogony is octosporoblastic, resulting in eight uninucleate spores within a sporophorous vesicle. These vesicles can remain tied together after sporogenesis is complete. Uninucleate spores are oval and are formed in the fat body of larvae.

Type species: *Pegmatheca simulii* Hazard & Oldacre, 1975

Type host: *Simulium tuberosum* (Diptera: Simuliidae)

Comments: One additional species, from a trichopteran, has been described. This genus provides a further example of a *Thelohania*-like sporulation sequence.

Perezia Léger & Duboscq, 1909

Diagnosis: One sporulation sequence is known. Diplokaryotic cells in early development become uninucleate in an unknown manner.

Moniliform sporogonial plasmodia divide to eventually produce ovoid, uninucleate spores.

Type species: *Perezia lankesteriae* Léger & Duboscq, 1909

Type host: *Lankesteria ascidiae* (Gregarinida)

Comments: The type host is a gregarine parasite of a marine tunicate. About eight species from insects have been described: five from Lepidoptera and one each from Coleoptera, Hymenoptera, and Orthoptera. It is questionable whether the species described from insects can be accommodated in this genus.

***Pernicivesicula* Bylén & Larsson, 1994**

Diagnosis: Presporulation stages are diplokaryotic. Only one sporulation sequence is known. The sporont is a diplokaryotic cell which undergoes meiosis. Sporogony is octosporoblastic, producing eight uninucleate sporoblasts. Multiples of eight spores are formed in a polysporophorous vesicle composed of host- and parasite-derived membranes. Uninucleate spores are slender and rod-shaped with an uncoiled polar filament. Infections are found in the fat body of larvae.

Type species: *Pernicivesicula gracilis* Bylén & Larsson, 1994

Type host: *Pentaneurella* sp. (Diptera: Chironomidae)

Comment: Monotypic.

***Pilosporella* Hazard & Oldacre, 1975**

Diagnosis: Presporulation stages are diplokaryotic. There are two sporulation sequences, one in larvae and one in adults. In larvae, the sporont is a diplokaryotic cell which undergoes meiosis. Sporogonial plasmodia are moniliform, and sporogony is octosporoblastic. Eight subspherical uninucleate spores are contained within a sporophorous vesicle. Infections are found in the fat body. In adults, binucleate spores are formed in oenocytes and are responsible for transovarial transmission.

Type species: *Pilosporella fishi* Hazard & Oldacre, 1975

Type host: *Wyeomyia vanduzeei* (Diptera: Culicidae)

Comment: Much of the information about this genus is based on *P. chapmani* (the only other species in this genus) as reported by Becnel et al. (1986).

***Pleistophora* Gurley, 1893**

Diagnosis: Only one sporulation sequence is known. All stages are uninucleate throughout development. Sporogonial plasmodia with up to 200 nuclei divide by repeated segmentation. Polysporophorous vesicle with the envelope composed of both host and parasite components contains uninucleate, ovoid spores. Infections are typically found in striated muscle. Some species have been demonstrated to be infectious by mouth.

Type species: *Pleistophora typicalis* Gurley, 1893

Type host: *Cottus scorpius* (*Myoxocephalus scorpius*) (Perciformes: Cottidae)

Comments: The type host is a fish. This genus has been a sort of catchall for species that are uninucleate throughout and produce large numbers of spores in a polysporophorous vesicle. Numerous species (more than 50) have been described and are primarily from the insect groups Blattaria, Coleoptera, Diptera, Lepidoptera, and Orthoptera. New genera such as *Vavraia* and *Polydispyrenia* (Canning and Hazard 1982), *Cystosporogenes* (Canning et al., 1985), and *Endoreticulatus* (Brooks et al., 1988) have been created for some of the species in insects. Most, however, are yet to be reassigned. Investigators will encounter many species assigned to the genus "*Plistophora.*" Labbe (1899) amended *Pleistophora* to *Plistophora* in violation of the code. Sprague (1977) transferred many of the *Plistophora* spp. to *Pleistophora* and has provided additional details on this subject.

***Polydispyrenia* Canning & Hazard, 1982**

Diagnosis: Presporulation stages are diplokaryotic. Only one sporulation sequence is known. Sporogonial plasmodia divide into diplokaryotic sporonts within a vesicle composed of both host and parasite components. The sporont undergoes meiosis and octospoblastic sporogony, producing multiples of eight uninucleate spores in a polysporophorous vesicle. Infections are found in the larval fat body.

Type species: *Polydispyrenia simulii* (Lutz & Splendore, 1908) Canning & Hazard, 1982

Type host: *Simulium venustum* (Simulium pertinax) (Diptera: Simuliidae)

Comment: Species from the Culicidae are also known.

***Pulcispora* Vedmed, Krylova & Issi, 1991**

Diagnosis: Presporulation stages are diplokaryotic. Only one sporulation sequence is known. The sporont is a diplokaryotic cell which (presumably) undergoes meiosis and octosporoblastic sporogony. Uninucleate, oval-cylindrical spores (8 to 32) are produced within a polysporophorous vesicle that was formed prior to sporogony.

Type species: *Pulcispora xenopsyllae* Vedmed, Krylova & Issi, 1991

Type host: *Xenopsylla hirtipes* (Siphonaptera: Pulicidae)

Comments: Monotypic. The description was in Russian, and details are from a brief English summary.

***Pyrotheca* Hesse, 1935**

Diagnosis: Only one sporulation sequence is known. Sporogony is tetrasporoblastic, producing uninucleate spores in groups of four (sporophorous vesicle assumed). Spores are large and shaped like a powder horn. Infections are found in adipose tissue.

Type species: *Pyrotheca cyclopis* (Leblanc, 1930) Poisson, 1953

Type host: *Cyclops albidus* (Copepoda: Cyclopidae)

Comment: One species from a trichopteran (caddis fly) has been described.

***Resiomeria* Larsson, 1986**

Diagnosis: Presporulation stages are diplokaryotic. Only one sporulation sequence is known. The sporont is a diplokaryotic cell which undergoes meiosis and octosporoblastic sporogony. Eight rod-shaped, uninucleate spores are contained within a sporophorous vesicle. Infections are found in the fat body of larvae.

Type species: *Resiomeria odonatae* Larsson, 1986c

Type host: *Aeshna grandis* (Odonata: Aeshnidae)

Comments: Monotypic. This genus provides another example of a *Thelohania*-like sporulation sequence.

***Ringueletium* Garcia, 1990**

Diagnosis: Presporulation stages are diplokaryotic. Only one sporulation sequence is known with all stages in the hyaloplasm of the host cell. Diplokaryotic sporogonial plasmodia divide by multiple division to produce eight binucleate spores. Spores are oval and characterized by uniformly distributed filamentous appendages attached to the exospore. Infections are found in the fat body of larvae.

Type species: *Ringueletium pillosa* Garcia, 1990

Type host: *Gigantodax rufidulum* (Diptera: Simuliidae)

Comment: Monotypic.

***Scipionospora* Bylén & Larsson, 1996**

Diagnosis: Presporulation stages are diplokaryotic. Only one sporulation sequence is known. The sporont is a diplokaryotic cell which undergoes two successive divisions without cytokinesis to form a tetranucleate sporogonial plasmodium. The plasmodium divides by budding into four binucleate sporoblasts which form four spores within a sporophorous vesicle. Binucleate spores are rod-shaped with an uncoiled polar filament. Infections are found in the fat body of larvae.

Type species: *Scipionospora tetraspora* (Léger & Hesse, 1922) Bylén & Larsson, 1996

Type host: *Tanytarsus* sp. (Diptera: Chironomidae)

Comment: Monotypic.

***Semenovaia* Voronin & Issi, 1986**

Diagnosis: Two sporulation sequences have been reported. One produces 16 uninucleate, ovoid spores. The other produces 2 ovoid, binucleate spores.

Type species: *Semenovaia chironomi* Voronin & Issi, 1986, in Issi, 1986

Type host: *Chironomus plumosus* (Diptera: Chironomidae)

Comments: Monotypic. The description was brief and had few details.

Spherospora Garcia, 1991

Diagnosis: Presporulation stages are diplokaryotic. Only one sporulation sequence is known. The sporont is a diplokaryotic cell which undergoes meiosis and octosporoblastic sporogony. Eight uninucleate, spherical spores are contained within a delicate sporophorous vesicle.

Type species: *Spherospora andinae* Garcia, 1991

Type host: *Gigantodox chilense* (Diptera: Simuliidae)

Comments: Monotypic. This genus provides yet another example of a *Thelohania*-like sporulation sequence.

Spiroglugea Léger & Hesse, 1924

Diagnosis: Only one sporulation sequence is known. Sporogony is octosporoblastic, producing eight spirilform spores in groups (a sporophorous vesicle is presumed). Infections are found in the fat body of larvae.

Type species: *Spiroglugea octospora* (Léger & Hesse, 1922) Léger & Hesse, 1924

Type host: *Ceratopogon* sp. (Diptera: Ceratopogonidae)

Comment: Monotypic.

Stempellia Léger & Hesse, 1910

Diagnosis: Presporulation stages are diplokaryotic. Two sporulation sequences have been reported. The sporont is a diplokaryotic cell. One sequence involves tetranucleate sporonts which produce four pyriform, uninucleate spores within a sporophorous vesicle. The other sequence involves octonucleate sporonts producing eight uninucleate, ovoid spores within a sporophorous vesicle. Infections are found in the fat body of larvae.

Type species: *Stempellia mutabilis* Léger & Hesse, 1910

Type host: *Ephemera vulgata* (Ephemeroptera: Ephemeridae)

Comment: Additional species from Diptera, Coleoptera, and Isoptera have been described.

Striatospora Issi & Voronin, 1986, in Issi, 1986

Diagnosis: Presporulation stages are diplokaryotic. Only one sporulation sequence is known. The sporont is a diplokaryotic cell. Sporogony is octosporoblastic and presumably involves meiosis. Uninucleate spores are short and cylindrical, with the exospore ornamented with longitudinal rows of electron-dense material.

Type species: *Striatospora chironomi* Issi & Voronin, 1986, in Issi, 1986

Type host: *Chironomus plumosus* (Diptera: Chironomidae)

Comments: Monotypic. The description of the type is brief, but another example of a *Thelohania*-like sporulation sequence appears to be involved.

Systenostrema Hazard & Oldacre, 1975

Diagnosis: Only one sporulation sequence is known. The sporont is a diplokaryotic cell. Sporogony is octosporoblastic and presumably involves meiosis. Eight uninucleate, pyriform spores are contained within a sporophorous vesicle. Infections are found in the fat body of larvae.

Type species: *Systenostrema tabani* Hazard & Oldacre, 1975

Type host: *Tabanus lineola* (Diptera: Tabanidae)

Comments: The type host is a deerfly, and one species from a midge larva has been described. Two other species from Odonata have been described.

Tardivesicula Larsson & Bylén, 1992

Diagnosis: Only one sporulation sequence is known. All stages are uninucleate throughout development. Sporogonial plasmodia undergo multiple division to produce 16 to 32 sporoblasts. Uninucleate spores are shaped like stout rods contained within a polysporophorous vesicle. Infections are found in the fat body of larvae.

Type species: *Tardivesicula duplicata* Larsson & Bylén, 1992

Type host: *Limnephilus centralis* (Trichoptera: Limnephilidae)

Comment: Monotypic.

***Telomyxa* Léger & Hesse, 1910**

Diagnosis: Only one sporulation sequence is known. All stages are uninucleate throughout development. Sporogony of the uninucleate sporont is disporoblastic. Uninucleate spores occur in pairs with a common exospore and are contained within a sporophorous vesicle. Infected tissue is the fat body of larvae.

Type species: *Telomyxa glugeiformis* Léger & Hesse, 1910

Type host: *Ephemera vulgata* (Ephemeroptera: Ephemeridae)

Comments: There are two other described species, one from a dipteran and one from a caddis fly. One additional species from a semiaquatic beetle has been described, although its inclusion in this genus is questioned (Larsson, 1988).

***Thelohania* Henneguy, 1892**

Diagnosis: Presporulation stages are diplokaryotic. Only one sporulation sequence is known. The sporont is a diplokaryotic cell which undergoes meiosis and octosporoblastic sporogony. Uninucleate spores are pyriform and are contained within a sporophorous vesicle to form packets of eight spores. Infected tissue is primarily muscle.

Type species: *Thelohania giardi* Henneguy, 1892, in Henneguy and Thélohan, 1892

Type host: *Crangon vulgaris* (Decapoda: Crangonidae)

Comments: The type species is a marine shrimp, but many species from insects have been described because of the sporulation sequence that produces packets of eight spores. Over 60 species from insects in the orders Diptera, Collembola, Ephemeroptera, Hemiptera, Lepidoptera, Hymenoptera, Odonata, and Trichoptera have been described.

***Toxoglugea* Léger & Hesse, 1924**

Diagnosis: Only one sporulation sequence is known. Presporulation stages are probably diplokaryotic. Sporogony is octosporoblastic and probably involves meiosis. Uninucleate spores are small and in the form of U-shaped rods. Eight spores are contained within a sporophorous vesicle. Infected tissue is the fat body of larvae.

Type species: *Toxoglugea vibrio* (Léger & Hesse, 1922) Léger & Hesse, 1924

Type host: *Ceratopogon* sp. (Diptera: Ceratopogonidae)

Comment: Species described have been mostly from Diptera, but some from Plecoptera, Odonata, Hemiptera, and Homoptera have been described.

***Toxospora* Voronin, 1993**

Diagnosis: Only one sporulation sequence is known. Presporulation presumably involves diplokaryotic stages. The sporont is a diplokaryotic cell. Sporogony is octosporoblastic and results in eight uninucleate sporoblasts contained in a sporophorous vesicle. Uninucleate spores are in the form of U-shaped rods. Infections are found in the fat body of larvae.

Type species: *Toxospora volgae* Voronin, 1993

Type host: *Corynoneura* sp. (Diptera: Chironomidae)

Comments: Monotypic. The distinction between this genus and *Toxoglugea* is unclear.

***Trichoctosporea* Larsson, 1994**

Diagnosis: Presporulation stages are diplokaryotic. Only one sporulation sequence is known. The sporont is a diplokaryotic cell which undergoes meiosis and octosporoblastic sporogony to produce eight sporoblasts within a sporophorous vesicle. Uninucleate spores are ovoid and ornamented, with as many as five fibrous projections which are extensions of the exospore. Infected tissue is the fat body of larvae.

Type species: *Trichoctosporea pygopellita* Larsson, 1994

Type host: *Aedes vexans* (Diptera: Culicidae)

Comments: Monotypic. This genus has the general features of the *Amblyospora*-like sporulation sequence that produces meiospores in mosquito larvae. It is distinguished from these forms by the ornamentation of the spore.

Trichoduboscqia Léger, 1926

Diagnosis: Presporulation stages are probably diplokaryotic. Only one sporulation sequence is known. Sporogony involves multiple division of sporogonial plasmodia typically with 16 nuclei but often 32 and rarely 8, 12, 20, or 24. Uninucleate spores are elongate-pyriform and contained within a polysporophorous vesicle. The vesicle is characterized by two to six appendages with a core of exospore material. Infected tissue is the larval fat body.

Type species: *Trichoduboscqia epeori* Léger, 1926

Type host: *Epeorus torrentium* (Ephemeroptera: Heptageniidae)

Comment: Monotypic.

Trichotuzetia Vávra, Larsson & Baker, 1997

Diagnosis: All developmental stages are uninucleate. Only one sporulation sequence is known. Sporogonial plasmodia divide by binary division to produce uninucleate sporoblasts. Uninucleate spores are pyriform and are individually contained within a sporophorous vesicle. The gonads of both sexes are initially infected but, with the exception of the muscle and gut, infection eventually becomes systemic.

Type species: *Trichotuzetia guttata* Vávra, Larsson & Baker, 1997

Type host: *Cyclops vicinus* (Copepoda: Cyclopidae)

Comments: Monotypic. This genus is mentioned here because molecular analysis has indicated that *T. guttata* is most closely related to a group of microsporidia that includes *Amblyospora californica*.

Tricornia Pell & Canning, 1992

Diagnosis: Presporulation stages are diplokaryotic. Only one sporulation sequence is known. The sporont is a diplokaryotic cell which undergoes meiosis and octosporoblastic sporogony, producing eight sporoblasts. Spores are uninucleate and oval and are contained in a sporophorous vesicle. The spore wall is characterized by two hornlike projections on the anterior end and one on the posterior end. Infected tissue is the larval fat body.

Type species: *Tricornia muhezae* Pell & Canning, 1992

Type host: *Mansonia africana* (Diptera: Culicidae)

Comments: Monotypic. Sporulation sequences resemble those of *Amblyospora* and *Parathelohania* spp. found in the larval mosquito host.

Tuzetia Maurand, Fize, Fenwick & Michel, 1971

Diagnosis: All developmental stages are uninucleate. Only one sporulation sequence is known. Sporogonial plasmodia divide by multiple division to produce sporoblasts. Uninucleate spores are pyriform, and each is contained in a sporophorous vesicle. Infections are found in the fat body, reproductive organs, and muscle.

Type species: *Tuzetia infirma* (Kudo, 1921) Maurand, Fize, Fenwick & Michel, 1971

Type host: *Cyclops albidus* (Copepoda: Cyclopidae)

Comments: The type host is a microcrustacean. There are about three species in insects of the order Ephemeroptera.

Unikaryon Canning, Lai & Lie, 1974

Diagnosis: Only one sporulation sequence is known. All stages are uninucleate and occur in the hyaloplasm of the host cell. Sporogony occurs by binary division. Uninucleate spores are small and pyriform.

Type species: *Unikaryon piriformis* Canning, Lai & Lie, 1974

Type host: *Echinostoma audyi* (Digenea: Echinostomatidae)

Comment: Species in insects have been primarily described as being from flower and bark beetles.

Vairimorpha Pilley, 1976

Diagnosis: Three sporulation sequences are known. The first occurs in midgut cells soon after infection and produces binucleate spores. The spores germinate spontaneously and deposit a binucleate sporoplasm into fat body cells where two different sporulation sequences take place. In one sequence, diplokaryotic sporonts divide by binary fission (disporous) to produce two diplokaryotic sporoblasts. Binucleate spores are oblong and are in direct contact with the cytoplasm. The other sequence begins with diplokaryotic sporonts and involves meiosis and octosporoblastic sporogony to produce eight sporoblasts within a sporophorous vesicle. Uninucleate spores are ovoid. The binucleate spores are orally infectious to new host individuals. The role of the meiospores is unknown.

Type species: *Vairimorpha necatrix* (Kramer, 1965) Pilley, 1976

Type host: *Pseudaletia unipunctata* (Lepidoptera: Noctuidae)

Comments: Mainly a parasite of Lepidoptera, but species from Hymenoptera and Diptera have also been described.

Vavraia Weiser, 1977

Diagnosis: All stages are uninucleate. Only one sporulation sequence is known. Sporogonial plasmodia divide by multiple division, producing polysporophorous vesicles with sporoblasts of 8, 16, and 32 commonly and 64 rarely. Uninucleate spores are ovoid, with the anchoring disk positioned off-center. Sporophorous vesicles are thick, with contributions from both the host and parasite. Infected tissue is Malpighian tubules and various other organs depending on the host.

Type species: *Vavraia culicis* (Weiser, 1947) Weiser, 1977

Type host: *Culex pipiens* (Diptera: Culicidae)

Comments: The type species has been reported from many different mosquito hosts. One species from a trichopteran has been described.

Weiseria Doby & Saguez, 1964

Diagnosis: Presporulation stages are probably diplokaryotic. Only one sporulation sequence is known. Sporogonial plasmodia have 16 to 22 diplokarya which divide by multiple division, usually to produce 16 sporoblasts. Binucleate spores are pyriform, and the posterior regions of the exospore have ridges. Infected tissue is the fat body of larvae.

Type species: *Weiseria laurenti* Doby & Saguez, 1964

Type host: *Prosimulium inflatum* (Diptera: Simuliidae)

Comments: Monotypic. Similar to *Caudospora* and mainly differs in the ornamentation of the exospore.

ACKNOWLEDGMENT

We are indebted to Margaret (Peg) Johnson for her assistance throughout the preparation of the manuscript.

REFERENCES

Alger, N. E., and A. H. Undeen. 1970. The control of a microsporidian, *Nosema* sp., in an anopheline colony by an egg rinsing technique. *J. Invertebr. Pathol.* **15:**321–327.

Alger, N. E., J. V. Maddox, and J. A. Shadduck. 1980. *Nosema algerae:* Infectivity and immune response in normal and nude mice. Mimeograph. VBC/80.778. World Health Organization, Rome.

Anderson, J. F. 1968. Microsporidia parasitizing mosquitoes collected in Connecticut. *J. Invertebr. Pathol.* **11:**440–445.

Anderson, R. M., and R. M. May. 1981. The population dynamics of microparasites and their invertebrate hosts. *Philos. Trans. R. Soc. London B* **291:**451–524.

Andreadis, T. G. Unpublished data.

Andreadis, T. G. 1980. *Nosema pyrausta* infection in *Macrocentrus grandii,* a braconid parasite of the European corn borer, *Ostrinia nubilalis. J. Invertebr. Pathol.* **35:**229–233.

Andreadis, T. G. 1983. Life cycle and epizootiology of *Amblyospora* sp. (Microspora: Amblyosporidae) in the mosquito, *Aedes cantator. J. Protozoology.* **30:**509–518.

Andreadis, T. G. 1985a. Experimental transmission of a microsporidian pathogen from mosquitoes to an alternate copepod host. *Proc. Natl. Acad. Sci. USA* **82:**5574–5577.

Andreadis, T. G. 1985b. Life cycle, epizootiology, and horizontal transmission of *Amblyospora* sp. (Microspora: Amblysoporidae) in a univoltine mosquito, *Aedes stimulans. J. Invertebr. Pathol.* **46:**31–46.

Andreadis, T. G. 1987. Horizontal transmission of *Nosema pyrausta* (Microsporida: Nosematidae) in the European corn borer, *Ostrinia nubilalis* (Lepidoptera: Pyralidae). *Environ. Entomol.* **16:**1124–1129.

Andreadis, T. G. 1988a. Comparative susceptibility of the copepod *Acanthocyclops vernalis* to a microsporidian parasite, *Amblyospora connecticus,* from the mosquito *Aedes cantator. J. Invertebr. Pathol.* **52:**73–77.

Andreadis, T. G. 1988b. *Amblyospora connecticus* sp. nov. (Microsporida: Amblyosporidae): horizontal transmission studies in the mosquito *Aedes cantator* and formal description. *J. Invertebr. Pathol.* **52:** 90–101.

Andreadis, T. G. 1989. Host specificity of *Amblyospora connecticus,* a polymorphic microsporidian parasite of *Aedes cantator. J. Med. Entomol.* **26:**140–145.

Andreadis, T. G. 1990a. Polymorphic microsporidia of mosquitoes: potential for biological control. *New Direct. Biol. Control* **112:**175–188.

Andreadis, T. G. 1990b. Epizootiology of *Amblyospora connecticus* (Microsporida) in field populations of the saltmarsh mosquito, *Aedes canator,* and the cyclopoid copepod, *Acanthocyclops vernalis. J. Protozool.* **37:**174–182.

Andreadis, T. G. 1994a. Host range tests with *Edhazardia aedis* (Microsporida: Culicosporidae) against northern Nearctic mosquitoes. *J. Invertebr. Pathol.* **64:**46–51.

Andreadis, T. G. 1994b. Ultrastructural characterization of meiospores of six new species of *Amblyospora* (Microsporida: Amblyosporidae) from northern *Aedes* (Diptera: Culicidae) mosquitoes. *J. Eukaryot. Microbiol.* **41:**147–154.

Andreadis, T. G., and D. W. Hall. 1979a. Development, ultrastructure, and mode of transmission of *Amblyospora* sp. (Microspora) in the mosquito. *J. Protozool.* **26:**444–452.

Andreadis, T. G., and D. W. Hall. 1979b. Significance of transovarial infections of *Amblyospora* sp. (Microspora: Thelohaniidae) in relation to parasite maintenance in the mosquito *Culex salinarius. J. Invertebr. Pathol.* **34:**152–157.

Andreadis, T. G., and J. L. Hanula. 1987. Ultrastructural study and description of *Ovavesicula popilliae* n. g., n. sp. (Microsporida: Pleistophoridae) from the Japanese beetle, *Popilla japonica* (Coleoptera: Scarabaeidae). *J. Protozool.* **34:**15–21.

Avery, S. W., and A. H. Undeen. 1990. Horizontal transmission of *Parathelohania anophelis* to the copepod *Microcyclops varicans,* and the mosquito, *Anopheles quadrimaculatus. J. Invertebr. Pathol.* **56:**98–105.

Baker, M. D., C. R. Vossbrinck, J. J. Becnel, and T. G. Andreadis. 1998. Phylogeny of *Amblyospora* (Microsporida: Amblyosporidae) and related genera based on small subunit ribosomal DNA data: a possible example of host parasite cospeciation. *J. Invertebr. Pathol.* **71:**199–206.

Baker, M. D., C. R. Vossbrinck, J. V. Maddox, and A. H. Undeen. 1994. Phylogenetic relationships among *Vairimorpha* and *Nosema* species (Microspora) based on ribosomal RNA sequence data. *J. Invertebr. Pathol.* **64:**100–106.

Balbiani, G. 1882. Sur les microsporidies ou psorospermies des articules. *C. R. Acad. Sci.* **95:**1168–1171.

Balbiani, G. 1884. *Leçons sur les Sporozoaires.* Octave Doin, Paris.

Batson, B. S. 1983. A light and electron microscopic study of *Hirusutusporos austrosimulii* gen. n., sp. n. (Microspora: Nosematidae), a parasite of *Austrosimulium* sp. (Diptera: Simuliidae) in New Zealand. *Protistologica* **19:**263–280.

Bauer, L. S., and H. S. Pankratz. 1993. *Nosema scripta* n. sp. (Microsporida: Nosematidae), a microsporidian parasite of the cottonwood leaf beetle, *Chrysomela scripta* (Coleoptera: Chrysomelidae). *J. Eukaryot. Microbiol.* **40:**135–141.

Beard, B., J. F. Butler, and J. J. Becnel. 1990. *Nolleria pulicis* n. gen., n. sp. (Microsporida: Chytridiopsidae), a microsporidian parasite of the cat flea *Ctenocephalides felis* (Siphonaptera: Pulicidae). *J. Protozool.* **37:**90–99.

Becnel, J. J. 1992. Horizontal transmission and subsequent development of *Amblyospora californica* (Microsporida: Amblyosporidae) in the intermediate and definitive hosts. *Dis. Aquat. Org.* **13:**17–28.

Becnel, J. J. 1994. Life cycles and host-parasite relationships of microsporidia in culicine mosquitoes. *Folia Parasitol.* **41:**91–96.

Becnel, J. J., and T. G. Andreadis. 1998. *Amblyospora salinaria* n. sp. (Microsporidia: Amblyosporidae), parasite of *Culex salinarius* (Diptera: Culicidae): its life cycle stages in an intermediate host. *J. Invertebr. Pathol.* **71:**258–262.

Becnel, J. J., and T. Fukuda. 1991. Ultrastructure of *Culicosporella lunata* (Microsporidia: Culicosporellidae fam. n.) in the mosquito *Culex pilosus* (Diptera: Culicidae) with new information on the development cycle. *Eur. J. Protistol.* **26:**319–329.

Becnel, J. J., J. J. Garcia, and M. A. Johnson. 1995. *Edhazardia aedis* (Microspora: Culicosporidae) effects on the reproductive capacity of *Aedes aegypti* (Diptera: Culicidae). *J. Med. Entomol.* **32:**549–553.

Becnel, J. J., and C. J. Geden. 1994. Description of a new species of microsporidia from *Muscidifurax raptor* (Hymenoptera: Pteromalidae), a pupal parasitoid of muscoid flies. *J. Eukaryot. Microbiol.* **41:**236–243.

Becnel, J. J., E. I. Hazard, and T. Fukuda. 1986. Fine structure and development of *Pilosporella chapmani* (Microspora: Thelohaniidae) in the mosquito, *Aedes triseriatus* (Say). *J. Protozool.* **33:**60–66.

Becnel, J. J., E. I. Hazard, T. Fukuda, and V. Sprague. 1987. Life cycle of *Culicospora magna* (Kudo, 1920) (Microsporida, Culicosporidae) in *Culex restuans* Theobold with special reference to sexuality. *J. Protozool.* **34:**313–322.

Becnel, J. J., and M. A. Johnson. 1993. Mosquito host range and specificity of *Edhazardia aedis* (Microspora: Culicosporidae). *J. Am. Mosq. Control Assoc.* **9:**269–274.

Becnel, J. J., V. Sprague, T. Fukuda, and E. I. Hazard. 1989. Development of *Edhazardia aedis* (Kudo, 1930) n. g., n. comb. (Microsporida: Amblyosporidae) in the mosquito *Aedes aegypti* (L.) (Diptera: Culicidae). *J. Protozool.* **36:**119–130.

Beyer, T. V., and I. V. Issi. 1986. *Microsporidia*. Protozoology no. 10. Academy of Sciences of the USSR, Leningrad.

Blunck, H. 1954. Mikrosporidien bei *Pieris brassicae* L.: ihren Parasiten und Hyperparasiten. *Z. Angew. Entomol.* **36:**316–333.

Brooks, W. M. 1968. Transovarian transmission of *Nosema heliothidis* in the corn earworm, *Heliothis zea*. *J. Invertebr. Pathol.* **11:**510–512.

Brooks, W. M. 1973. Protozoa: Host-parasite-pathogen interrelationships. *Misc. Publ. Entomol. Soc. Am.* **9:**105–111.

Brooks, W. M. 1980. Production and efficacy of protozoa. *Biotech. Bioeng.* **22:**1415–1440.

Brooks, W. M. 1988. Entomogenous protozoa. p.1–149. *In* C. M. Ignoffo (ed.), *Handbook of Natural Pesticides,* Vol. 5, CRC Press, Boca Raton, Fla.

Brooks, W. M. 1993. Host-parasitoid-pathogen interactions. p. 231–272. *In* N. Beckage, S. Thompson, and B. Federici (ed.), *Parasites and Pathogens of Insects 2*. Academic Press, New York, N.Y.

Brooks, W. M., J. J. Becnel, and G. G. Kennedy. 1988. Establishment of *Endoreticulatus* n. g. for *Pleistophora fidelis* (Hostounsky & Weiser, 1975) (Microsporida: Pleistophoridae) based on the ultrastructure of a microsporidium in the Colorado potato beetle, *Leptinotarsa decemlineata* (Say) (Coleoptera: Chrysomelidae). *J. Protozool.* **35:** 481–488.

Brooks, W. M., and J. D. Cranford. 1972. Microsporidoses of the hymenopterous parasites, *Campoletis sonorensis* and *Cardiochiles nigriceps,* larval parasites of *Heliothis* species. *J. Invertebr. Pathol.* **20:**77–94.

Brooks, W. M., E. I. Hazard, and J. J. Becnel. 1985. Two new species of *Nosema* (Microsporida, Nosematidae) from the Mexican bean beetle *Epilachna varivestis* (Coleoptera, Coccinellidae). *J. Protozool.* **32:**525–534.

Bulla, L. A., and T. C. Cheng (ed.). 1976. *Comparative Pathobiology,* vol. 1. Plenum Press, New York, N.Y.

Bulla, L. A., and T. C. Cheng (ed.). 1977. *Comparative Pathobiology,* vol. 2. Plenum Press, New York, N.Y.

Bylén, E. K. C., and J. I. R. Larsson. 1994. Ultrastructural study and description of *Pernicivesicula gracilis* gen. et sp. nov. (Microspora, Pereziidae), a rod-shaped microsporidium of midge larvae, *Pentaneurella* sp. (Diptera, Chironomidae), in Sweden. *Eur. J. Protistol.* **30:**139–150.

Bylén, E. K. C., and J. I. R. Larsson. 1996. Ultrastructural study and description of *Mrazekia tetraspora* Leger & Hesse, 1922 and transfer to a new genus *Scipionospora* n.g. (Microspora, Caudosporidae). *Eur. J. Protistol.* **32:**104–115.

Canning, E. U. 1960. Two new microsporidian parasites of the winter moth, *Operophtera brumata* (L.). *J. Parasitol.* **46:**755–763.

Canning, E. U. 1971. Transmission of Microsporida, p. 415–424. Proceedings of the 4th International Colloquium on Insect Pathology.

Canning, E. U. 1982. An evaluation of protozoal characteristics in relation to biological control of pests. *Parasitology* **84:**119–149.

Canning, E. U., R. J. Barker, J. P. Nicholas, and A. M. Page. 1985. The ultrastructure of three microsporidia from winter moth, *Operophtera brumata* (L.), and the establishment of a new genus *Cystosporogenes* n. g. for *Pleistophora operophterae* (Canning, 1960). *Syst. Parasitol.* **7:**213–225.

Canning, E. U., and E. I. Hazard. 1982. Genus *Pleistophora Gurley,* 1893: an assemblage of at least three genera. *J. Protozool.* **29:**39–49.

Canning, E. U., and R. H. Hulls. 1970. A microsporidan infection of *Anopheles gambiae* Giles, from Tanzania: interpretation of its mode of transmission and notes on *Nosema* infection in mosquitoes. *J. Protozool.* **17:**532–539.

Canning, E. U., R. Killick-Kendrick, and M. Killick-Kendrick. 1991a. A new microsporidian parasite, *Flabelliforma montana* n. g., n. sp., infecting *Phlebotomus ariasi* (Diptera: Psychodidae) in France. *J. Invertebr. Pathol.* **57:**71–81.

Canning, E. U., P. F. Lai, and K. J. Lie. 1974. Microsporidian parasites of trematode larvae from aquatic snails in West Malaysia. *J. Protozool.* **21:**19–25.

Canning, E. U., and J. Lom. 1986. *The Microsporidia of Vertebrates.* Academic Press, London.

Canning, E. U., P. J. Wigley, and R. J. Barker. 1991b. *Orthosomella* nomen novum for the junior homonym *Orthosoma* Canning, Wigley and Barker, 1983. *J. Invertebr. Pathol.* **58:**464.

Chapman, H. C., D. B. Woodard, W. R. Kellen, and T. B. Clark. 1966. Host-parasite relationships of *Thelohania* associated with mosquitoes in Louisiana (Microsporidia: Nosematidae). *J. Invertebr. Pathol.* **8:**452–456.

Codreanu, R. 1966. On the occurrence of spore or sporont appendages in the microsporidia and their taxonomic significance, p. 602–603. *In Proceedings of the 1st International Congress on Parasitology.* Pergamon Press, New York, N.Y.

de Graaf, D. C., H. Raes, G. Sabbe, P. H. de Rycke, and F. J. Jacobs. 1994a. Early development of *Nosema apis* (Microspora: Nosematidae) in the midgut epithelium of the honeybee (*Apis mellifera*). *J. Invertebr. Pathol.* **63:**74–81.

de Graaf, D. C., H. Raes, and F. J. Jacobs. 1994b. Spore dimorphism in *Nosema apis* (Microsporida: Nosematidae) developmental cycle. *J. Invertebr. Pathol.* **63:**92–94.

Doby, J. M., and F. Saguez. 1964. *Weiseria,* genre nouveau de microsporidies et *Weiseria laurenti* n. sp., parasite de larves de *Prosimulium inflatum* Davies, 1957 (Dipteres: Paranematoceres). *C. R. Acad. Sci.* **259:**3614–3617.

Doflein, F. 1898. Studien zur Naturgeschichte der Protozoen. III. Ueber Myxosporidien. *Zool. Jahrb. Abt. Anat.* **11:**281–350.

Fine, P. E. M. 1975. Vectors and vertical transmission: an epidemiologic perspective. *Ann. N.Y. Acad. Sci.* **266:**173–194.

Flu, P. 1911. Studien über die im Darm der Stubenfliege, *Musca domestica,* vorkommenden protozöaren Gebilde. *Centralbl. Bakteriol. Infect. Hyg. Abt. I. Orig.* **57:**522–535.

Fries, I. 1993. *Nosema apis:* a parasite in the honey bee colony. *Bee World* **74:**5–19.

Furgala, B., and Mussen, E. C. 1978. Protozoa. p. 1–62. *In* R. A. Morse, (ed.), *Honey Bee Pests, Predators, and Diseases.* Comstock, Ithaca, N.Y.

Garcia, J. J. 1990. Un nuevo microsporidio patogeno de larvas de simulidos (Diptera: Simulidae) *Ringueletium pillosa* gen. et sp. nov. (Microspora: Caudosporidiae). *Neotropica* **36:**111–122.

Garcia, J. J. 1991. Estudios sobre el cicly de vida y ultrastructura de *Spherospora andinae* gen. et sp. nov. (Microspora: Thelohaniidae), un nuevo microsporidio de simulidos neotropicales. *Neotropica* **37:**15–23.

Gaugler, R. R., and W. M. Brooks. 1975. Sublethal effects of infection by *Nosema heliothidis* in the corn earworm, *Heliothis zea. J. Invertebr. Pathol.* **26:**57–63.

Geden, C. J., S. J. Long, D. A. Rutz, and J. J. Becnel. 1995. *Nosema* disease of the parasitoid *Muscidifurax raptor* (Hymenoptera: Pteromalidae): prevalence, patterns of transmission, management, and impact. *Biol. Control* **5:**607–614.

Gluge, G. 1838. Notice sur quelques points d'anatomie pathologique comparée, suivie de quelques observations sur la structure des branchies dans épinoches. *Bull. Acad. R. Belg.* **5:**771–772.

Gochnauer, T. A., B. Furgala, and H. Shimanuki. 1975. Diseases and enemies of the honey bee, p. 1–615. *In The Hive and the Honey Bee.* C. Dadant and Sons, Hamilton, Ill.

Goetze, G. and B. Zeutzschel. 1959. *Nosema* disease of honeybees, and its control with drugs: review of research work since 1954. *Bee World* **40:**217.

Golberg, A. M. 1971. Mikrosporidiozy komarov *Culex pipiens* L. *Med. Parasitol. Parazitol. Bolezni* **2:** 204–207.

Gurley, R. 1893. On the classification of the Myxosporidia, a group of protozoan parasites infecting fishes. *Article 10 Bull. U.S. Fish Comm. for 1891* **11:**407–420.

Haque, M. A., W. S. Hollister, A. Willcox, and E. U. Canning. 1993. The antimicrosporidial activity of albendazole. *J. Invertebr. Pathol.* **62:**171–177.

Hamm, J. J., D. A. Nordlund, and B. G. Mullinix, Jr. 1983. Interaction of the microsporidium *Vairimorpha* sp. with *Microplitis croceipes* (Cresson) and *Cotesia marginiventris* (Cresson) (Hymenoptera: Braconidae), two parasitoids of *Heliothis zea* (Boddie) (Lepidoptera: Noctuidae). *Environ. Entomol.* **12:**1547–1550.

Hazard, E. I., and T. Fukuda. 1974. *Stempellia milleri* sp. n. (Microsporida: Nosematidae) in the mosquito *Culex pipiens quinquifasciatus* Say. *J. Protozool.* **21:**497–504.

Hazard, E. I., T. Fukuda, and J. J. Becnel. 1984. Life cycle of *Culicosporella lunata* (Hazard & Savage, 1970) Weiser, 1977 (Microspora) as revealed in the light microscope with a redescription of the genus and species. *J. Protozool.* **31:**385–391.

Hazard, E. I., T. Fukuda, and J. J. Becnel. 1985. Gametogenesis and plasmogamy in certain species of Microspora. *J. Invertebr. Pathol.* **46:**63–69.

Hazard, E. I., and S. W. Oldacre. 1975. Revision of Microsporidia (Protozoa) close to *Thelohania,* with descriptions of one new family, eight new genera and thirteen new species. *U.S. Dept. Agric. Tech. Bull.* **1530:**1–104.

Hazard, E. I., and K. E. Savage. 1970. *Stempellia lunata* sp. n. (Microsporida: Nosematidae) in larvae of the mosquito *Culex pilosus* collected in Florida. *J. Invertebr. Pathol.* **15:**49–54.

Hazard, E. I., and J. Weiser. 1968. Spores of *Thelohania* in adult female *Anopheles:* development and transovarial transmission, and redescriptions of *T. legeri* Hesse and *T. obesa* Kudo. *J. Protozool.* **15:** 817–823.

Hembree, S. 1982. Dose-response studies of a new species of per os and vertically transmittable microsporidian pathogens of *Aedes aegypti* from Thailand. *Mosquito News* **42:**55–61.

Hembree, S. C., and J. R. Ryan. 1982. Observations on the vertical transmission of a new microsporidian pathogen of *Aedes aegypti* from Thailand. *Mosquito News* **42:**49–54.

Henneguy, F., and P. Thélohan. 1892. Myxosporidies parasites des muscles chez quelques crustacés décapodes. *Ann. Microgr.* **4:**617–641.

Henry, J. 1972. Epizootiology of infections by *Nosema locustae* Canning (Microsporida: Nosematidae) in grasshoppers. *Acrida* **1:**111–120.

Hesse, E. 1904. *Thelohania legeri* n. sp., microsporidie nouvelle, parasite des larves d'*Anopheles maculipennis* Meig. *C. R. Soc. Biol.* **57:**570–571.

Hesse, E. 1935. Sur quelques microsporidies parasites de *Megacyclops viridis* Jurine. *Arch. Zool. Exp. Gen.* **75:**651–661.

Hostounsky, Z., and J. Weiser. 1975. *Nosema polygrammae* sp. n. and *Plistophora fidelis* sp. n. (Microsporidia, Nosematidae) infecting *Polygramma undecimlineata* (Coleoptera: Chrysomelidae) in Cuba. *Vest. Cesk. Spol. Zool.* **39:**104–110.

Inoue, S., S. Yokota, C. Yasunaga, M. Funakoshi, T. Kawarabata, and S. Hayasaka. 1995. Continuous culture of *Vairimorpha* sp. Nis m12 (Microsporida: Protozoa) in insect cell lines. *J. Seric. Sci. Jpn.* **64:**515–522.

Issi, I. V. 1986. Microsporidia as a phylum of parasitic protozoa. *Protozoology* **10:**6–136.

Issi, I. V., S. V. Krylova, and V. M. Nicolaeva. 1993. The ultrastructure of the microsporidium *Nosema meligethi* and establishment of the new genus *Ancaliia*. *Parazitologiya* **27:**127–133.

Issi, I. V., and V. A. Maslennikova. 1966. Role of *Apanteles glomeratus* (Hymenoptera, Braconidae) in the transmission of *Nosema polyvora* (Protozoa, Microsporidia). *Entomol. Obozr.* **45:**494–499.

Issi, I. V., D. F. Radischcheva, and V. T. Dolzhenko. 1983. Microsporidia of flies of genus *Delia* (Diptera, Muscidae), harmful to farm crops. *Trans. Vses. Nauchno-Issled. Inst. Zashch. Rast.* **55:**3–9.

Issi, I. V., and V. N. Voronin. 1979. The contemporary state of the problem on bispore genera of microsporidians. *Parazitologiya* **13:**150–158.

Iwano, H., and R. Ishihara. 1989. Intracellular germination of spores of a *Nosema* sp. immediately after their formation in cultured cell. *J. Invertebr. Pathol.* **54:**125–127.

Iwano, H., and R. Ishihara. 1991a. Dimorphism of spores of *Nosema* spp. in cultured cell. *J. Invertebr. Pathol.* **57:**211–219.

Iwano, H., and R. Ishihara. 1991b. Dimorphic development of *Nosema bombycis* spores in gut epithelium of larvae of the silkworm, *Bombyx mori*. *J. Seric. Sci. Jpn.* **60:**249–256.

Janda, V. 1928. Über Microorganismen aus der Leibeshöhle von *Criodrilus lacuum* Hoffm. und eigenartige Neubildungen in der Körperwand dieses Tieres. *Arch. Protistenkd.* **63:**84–93.

Jaronski, S. T. 1984. Microsporida in cell culture. *Adv. Cell Cult.* **3:**183–229.

Jeffords, M. R., J. V. Maddox, and K. W. O'Hayer. 1987. Microsporidian spores in gypsy moth larval silk: a possible route of horizontal transmission. *J. Invertebr. Pathol.* **49:**332–333.

Jírovec, O. 1943. Revision der in *Simulium*-larven parasitierenden Mikrosporidien. *Zool. Anz.* **142:** 173–179.

Johnson, M. A., J. J. Becnel, and A. H. Undeen. 1997. A new sporulation sequence in *Edhazardia aedis* (Microsporidia: Culicosporidae), a parasite of the mosquito *Aedes aegypti* (Diptera: Culicidae). *J. Invertebr. Pathol.* **70:**69–75.

Jouvenaz, D. P., and E. I. Hazard. 1978. New family, genus, and species of microsporida (Protozoa: Microsporida) from the tropical fire ant, *Solenopsis geminata* (Fabricius) (Insecta: Formicidae). *J. Protozool.* **25:**24–29.

Jouvenaz, D. P., C. S. Lofgren, and G. E. Allen. 1981. Transmission and infectivity of spores of *Burenella dimorpha* (Microsporida: Burenellidae). *J. Invertebr. Pathol.* **37:**265–268.

Kellen, W. R., H. C. Chapman, T. B. Clark, and J. E. Lindegren. 1965. Host-parasite relationships of some *Thelohania* from mosquitoes (Nosematidae: Microsporidia). *J. Invertebr. Pathol.* **7:**161–166.

Kellen, W. R., H. C. Chapman, T. B. Clark, and J. E. Lindegren. 1966. Transovarian transmission of some *Thelohania* (Nosematidae: Microsporidia) in mosquitoes of California and Louisiana. *J. Invertebr. Pathol.* **8:**355–359.

Kellen, W. R., and J. E. Lindegren. 1971. Modes of transmission of *Nosema plodia* Kellen and Lindegren, a pathogen of *Plodia interpunctella* (Hübner). *J. Stored Prod. Res.* **7:**31–34.

Kellen, W. R., and J. E. Lindegren. 1973a. *Nosema invadens* sp. n. (Microsporida: Nosematidae), a pathogen causing inflammatory response in Lepidoptera. *J. Invertebr. Pathol.* **21:**293–300.

Kellen, W. R., and J. E. Lindegren. 1973b. Transovarian transmission of *Nosema plodiae* in the Indian meal moth *Plodia interpunctella*. *J. Invertebr. Pathol.* **21:**248–254.

Kellen, W. R., and J. J. Lipa. 1960. *Thelohania californica* n. sp., a microsporidian parasite of *Culex tarsalis* Coquillett. *J. Insect Pathol.* **2:**1–12.

Kellen, W. R., and W. Wills. 1962. The transovarian transmission of *Thelohania californica* Kellen and Lipa in *Culex tarsalis* Coquillett. *J. Insect Pathol.* **4:**321–326.

Khodzhaeva, L. F., and I. V. Issi. 1989. New genus of microsporidiae, *Cristulospora* gen. n. (Amblyosporidae), with three new species from blood-sucking mosquitoes from the Uzbekistan. *Parazitologiya* **23:**140–145.

Koella, J. C., and P. Agnew. 1997. Blood-feeding success of the mosquito *Aedes aegypti* depends on the transmission route of its parasite *Edhazardia aedis*. *Oikos* **78:**311–316.

Kramer, J. P. 1959a. On *Nosema heliothidis* Lutz and Splendor, a microsporidian parasite of *Heliothis zea* (Boddie) and *Heliothis virescens* (Fabricius) (Lepidoptera, Phalaenidae). *J. Invertebr. Pathol.* **1:**297–303.

Kramer, J. P. 1959b. Some relationships between *Perezia pyraustae* Paillot (Sporozoa, Nosematidae) and *Pyrausta nubilalis* (Hubner) (Lepidoptera, Pyralidae). *J. Insect Pathol.* **1:**25–33.

Kramer, J. P. 1965. *Nosema necatrix* sp. n. and *Thelohania diazoma* sp. n., microsporidians from the armyworm *Pseudaletia unipuncta* (Haworth). *J. Invertebr. Pathol.* **7:**117–121.

Kramer, J. P. 1970. Longevity of microsporidian spores with special reference to *Octosporea muscaedomesticae* Flu. *Acta Protozool.* **8:**127–135.

Kramer, J. P. 1976. The extra-corporeal ecology of microsporidia, p.127–136. *In* L. A. Bulla and T. C. Cheng (ed.), *Comparative Pathobiology,* vol. 1. Plenum Press, New York, N.Y.

Kudo, R. 1920. On the structure of some microsporidian spores. *J. Parasitol.* **6:**178–182.

Kudo, R. 1921. Microsporidia parasitic in copepods. *J. Parasitol.* **7:**137–143.

Kudo, R. 1924. A biologic and taxonomic study of the Microsporidia. *Ill. Biol. Monogr.* **9:**268.

Kudo, R. 1925. Microsporidia. *Science* **61:**366.

Kudo, R. 1930. Studies on microsporidia parasitic in mosquitoes. VIII. On a microsporidian, *Nosema aedis* nov. spec., parasitic in a larva of *Aedes aegypti* of Puerto Rico. *Arch. Protistenkd.* **69:**23–28.

Kurtti, T. J., S. E. Ross, Y. Liu, and U. G. Munderloh. 1994. In vitro developmental biology and spore production in *Nosema furnacalis* (Microspora: Nosematidae). *J. Invertebr. Pathol.* **63:**188–196.

Labbe, A. 1899. Sporozoa. p. v–xiii, 1–180. *In* Butschi, O. (ed.), *Das Tierrich.* Friedländer und Sohn, Berlin, Germany.

Laigo, F. M., and M. Tamashiro. 1967. Interactions between a microsporidian pathogen of the lawn armyworm and the hymenopterous parasite *Apanteles marginiventris. J. Invertebr. Pathol.* **9:**546–554.

Lange, C. E., J. J. Becnel, E. Razafindratiana, J. Przybyszewski, and H. Razafindrafara. 1996. *Johenrea locustae* n. g., n. sp. (Microspora: Glugeidae): a pathogen of migratory locusts (Orthoptera: Acrididae: Oedipodinae) from Madagascar. *J. Invertebr. Pathol.* **68:**28–40.

Lange, C. E., C. M. Macvean, J. E. Henry, and D. A. Streett. 1995. *Heterovesicula cowani* n. g., n. sp. (Heterovesiculidae n. fam.), a microsporidian parasite of Mormon crickets, *Anabrus simplex* Haldeman, 1852 (Orthoptera: Tettigoniidae). *J. Eukaryot. Microbiol.* **42:**552–558.

Larsson, J. I. R. 1980. Insect pathological investigations on Swedish Thysanura. II. A new microsporidian parasite of *Petrobius brevistylis* (Microcoryphia, Machilidae): description of the species and creation of two new genera and a new family. *Protistologica* **16:**85–101.

Larsson, J. I. R. 1982. Cytology and taxonomy of *Helmichia aggregata* gen. et sp. nov. (Microspora, Thelohaniidae), a parasite of *Endochironomus* larvae (Diptera, Chironomidae). *Protistologica* **18:**355–370.

Larsson, J. I. R. 1983. A revisionary study of the taxon *Tuzetia* Maurand, Fize, Fenwick and Michel, 1971, and related forms (Microspora, Tuzetiidae). *Protistologica* **19:**323–355.

Larsson, J. I. R. 1984. Ultrastructural study and description of *Chapmanium dispersus* n. sp. (Microspora, Thelohaniidae) a microsporidian parasite of *Endochironomus* larvae (Diptera, Chironomidae). *Protistologica* **20:**547–563.

Larsson, J. I. R. 1985. On the cytology, development and systematic position of *Thelohania asterias* Weiser, 1963, with creation of the new genus *Bohuslavia* (Microspora, Thelohaniidae). *Protistologica* **21:**235–248.

Larsson, J. I. R. 1986a. Ultrastructure, function, and classification of microsporidia. *Prog. Protistol.* **1:** 325–390.

Larsson, J. I. R. 1986b. Ultracytology of a tetrasporoblastic microsporidium of the caddis fly *Holocentropus picicornis* (Trichoptera, Polycentropodidae), with a description of *Episeptum inversum* gen. et sp. nov. (Microspora, Gurleyidae). *Arch. Protistenkd.* **131:**257–280.

Larsson, J. I. R. 1986c. Ultrastructural investigation of two microsporidia with rod-shaped spores, with descriptions of *Cylindrospora fasciculata* sp. nov. and *Resiomeria odonatae* gen. et sp. nov. (Microspora: Thelohaniidae). *Protistologica* **22:**379–398.

Larsson, J. I. R. 1988. Identification of microsporidian genera: a guide with comments on the taxonomy. *Arch. Protistenkd.* **136:**1–37.

Larsson, J. I. R. 1990a. Description of a new microsporidium of the water mite *Limnochares aquatica* and establishment of the new genus *Napamichum* (Microspora, Thelohaniidae). *J. Invertebr. Pathol.* **55:**152–161.

Larsson, J. I. R. 1990b. On the cytology and taxonomic position of *Nudispora biformis* n. g., n. sp. (Microspora, Thelohaniidae), a microsporidian parasite of the dragon fly *Coenagrion hastulatum* in Sweden. *J. Protozool.* **37:**310–318.

Larsson, J. I. R. 1994. *Trichoctosporea pygopellita* gen. et sp. nov. (Microspora, Thelohaniidae), a microsporidian parasite of the mosquito *Aedes vexans* (Diptera, Culicidae). *Arch. Protistenkd.* **144:**147–161.

Larsson, J. I. R., and E. K. C. Bylén. 1992. *Tardivesicula duplicata* gen. et sp. nov. (Microspora, Duboscqiidae), a microsporidian parasite of the caddis fly *Limnephilus centralis* (Trichoptera, Limnephilidae) in Sweden. *Eur. J. Protistol.* **28:**25–36.

Leblanc, L. 1930. Deux microsporidies nouvelles des copepodes: *Gurleya cyclopis* et *Plistophora cyclopis. Ann. Soc. Sci. Brux. Ser. B* **59:**272–275.

Léger, L. 1926. Une microsporidie nouvelle à sporontes èpineux. *C. R. Acad. Sci.* **182:**727–729.

Léger, L., and O. Duboscq. 1909. *Perezia lankesterle,* n. g., n. sp., microsporidie parasite de *Lankesteria ascidae* (Ray-Lank). *Arch. Zool. Exp. Gen. Ser.* **5,1, N. et R.:**89–93.

Léger, L., and E. Hesse. 1910. Cnidosporidies des larves d' éphémères. *C. R. Acad. Sci.* **150:**411–414.

Léger, L., and E. Hesse. 1921. Microsporidies a spores spheriques. *C. R. Acad. Sci.* **173:**1419–1421.

Léger, L., and E. Hesse. 1922. Microsporidies bactériformes et essai de sytematique du groupe. *C. R. Acad. Sci.* **174:**327–330.

Léger, L., and E. E. Hesse. 1924. Microsporidies nouvelles parasites des animaux d'eau dounce. *Trav. Lab. Hydrobiol. Pisc. Univ. Grenoble* **14:**49–56.

Lord, J. C., and D. W. Hall. 1983. Sporulation of *Amblyospora* (Microspora) in female *Culex salinarius:* induction by 20-hydroxyecdysone. *Parasitol.* **87:** 377–383.

Lord, J. C., D. W. Hall, and E. A. Ellis. 1981. Life cycle of a new species of *Amblyospora* (Microspora: Amblyosporidae) in the mosquito *Aedes taeniorhynchus. J. Invertebr. Pathol.* **37:**66–72.

Lucarotti, C. J., and T. G. Andreadis. 1995. Reproductive strategies and adaptations for survival among obligatory microsporidian and fungal parasites of mosquitoes: a comparative analysis of *Amblyospora* and *Coelomomyces. J. Am. Mosquito Control Assoc.* **11:**111–121.

Lutz, A., and A. Splendore. 1908. Ueber pebrine und verwandte Mikrosporidien. Zweite Mitteilung. *Zentbl. Bakteriol. Parasitenkd. Infektionskr. Hyg. Abt. 1 Orig.* **46:**311–315.

Maddox, J. V. 1973. The persistence of the Microsporida in the environment. *Misc. Publ. Entomol. Soc. Am.* **9:**99–104.

Maddox, J. V. 1977. Stability of entomopathogenic Protozoa: environmental stability of microbial insecticides. *Misc. Publ. Entomol. Soc. Am.* **10:**3–18.

Maddox, J. V., W. M. Brooks, and J. R. Fuxa. 1981. *Vairimorpha necatrix* a pathogen of agricultural pests: potential for pest control. p. 587–594. *In* H. D. Burges (ed.), *Microbial Control of Pests and Plant Diseases 1970–1980.* Academic Press, New York, N.Y.

Maurand, J., A. Fize, B. Fenwick, and R. Michel. 1971. Etude au microscope électronique de *Nosema infirmum* Kudo, 1921, microsporidie parasite d'un copépode cyclopoïde; création du genre nouveau *Tuzetia* a propos de cette espece. *Protistologica* **7:**221–225.

McNeil, J. N., and W. M. Brooks. 1974. Interactions of the hyperparasitoids *Catolaccus aeneoviridis* [Hym.: Pteromalidae] and *Spilochalcis side* [Hym.: Chalcididae] with the microsporidians *Nosema heliothidis* and *N. campoletidis. Entomophaga* **19:**195–204.

Moniez, R. 1887. Observations pour la revision des Microsporidies. *C. R. Acad. Sci.* **104:**1312–1314.

Moore, C. B., and W. M. Brooks. 1992. An ultrastructural study of *Vairimorpha necatrix* (Microspora, Microsporida) with particular reference to episporontal inclusions during octosporogony. *J. Protozool.* **39:** 392–398.

Naegeli, C. 1857. Über die neue Krankheit der Seidenraupe und verwandte Organismen. *Botan. Zeit.* **15:**760–761.

Ormières, R., and V. Sprague. 1973. A new family, new genus, and new species allied to the Microsporida. *J. Invertebr. Pathol.* **21:**224–240.

Own, O. S., and W. M. Brooks. 1986. Interactions of the parasite *Pediobius foveolatus* (Hymenoptera: Eulophidae) with two *Nosema* spp. (Microsporida: Nosematidae) of the Mexican bean beetle (Coleoptera: Coccinellidae). *Environ. Entomol.* **15:** 32–39.

Pell, J. K., and E. U. Canning. 1992. Ultrastructure of *Tricornia muhezae* n. g., n. sp. (Microspora, Thelohaniidae), a parasite of *Mansonia africana* (Diptera, Culicidae) from Tanzania. *J. Protozool.* **39:**242–247.

Pell, J. K., and E. U. Canning. 1993. Ultrastructure and life cycle of *Merocinta davidii* gen. et sp. nov., a dimorphic microsporidian parasite of *Mansonia africana* (Diptera, Culicidae) from Tanzania. *J. Invertebr. Pathol.* **61:**267–274.

Pérez, C. 1908. Sur *Duboscqia legeri,* microsporidie nouvelle parasite de *Termes lucifugus* et sur la classification des microsporidies. *C. R. Soc. Biol.* **65:** 631–633.

Pilley, B. M. 1976. A new genus, *Vairimorpha* (Protozoa: Microsporida), for *Nosema necatrix* Kramer 1965: pathogenicity and life cycle in *Spodoptera exempta* (Lepidoptera: Noctuidae). *J. Invertebr. Pathol.* **28:**177–183.

Poisson, R. 1953. Ordre des Microsporidies. p. 1042–1070. *In* P. P. Grasse (ed.), *Traite de Zoologie,* vol. 1. Masson et Cie, Paris, France.

Raun, E. S. 1961. Elimination of microsporidiosis in laboratory-reared European corn borers by the use of heat. *J. Insect Pathol.* **3:**446–448.

Rühl, H., and H. Korn. 1979. Ein Mikrosporidier, *Geusia gamocysti* n. gen., n. sp. als Hyperparsit bei *Gamocystis ephemerae. Arch. Protistenkd.* **121:**349–355.

Schneider, A. 1884. Sur le développement du *Stylorhynchus longicollis. Arch. Zool. Exp. Gen.* **2:**1–36.

Siegel, J. P., J. V. Maddox, and W. G. Ruesink. 1986a. Lethal and sublethal effects of *Nosema pyrausta* on the European corn borer (*Ostrinia nubilalis*) in central Illinois. *J. Invertebr. Pathol.* **48:** 167–173.

Siegel, J. P., J. V. Maddox, and W. G. Ruesnik. 1986b. Impact of *Nosema pyrausta* on a braconid,

Macrocentrus grandii, in central Illinois. *J. Invertebr. Pathol.* **47:**271–276.

Solter, L. F., and J. V. Maddox. 1998. Timing of an early sporulation sequence of microsporidia in the genus *Vairimorpha* (Microsporidia: Burenellidae). *J. Invertebr. Pathol.* **72:**323–329.

Solter, L. F., J. V. Maddox, and M. L. McManus. 1997. Host specificity of Microsporidia (Protista: Microspora) from European populations of *Lymantria dispar* (Lepidoptera: Lymantriidae) to indigenous North American Lepidoptera. *J. Invertebr. Pathol.* **69:**135–150.

Sprague, V. 1970. Some protozoan parasites and hyperparasites in marine decapod crustacea, p. 416–430. *In* S. F. Snieszko (ed.), *American Fisheries Society Special Publication 5. A Symposium on Diseases of Fishes and Shellfishes.* American Fisheries Society, Washington, D.C.

Sprague, V. 1976. Implications of dimorphism in taxonomy of the Microsporidia, p. 454–457. *In Proceedings of the 1st Colloquium on Invertebrate Pathology, Kingston, Ontario, Canada.*

Sprague, V. 1977. Systematics of the Microsporidia, p. 1–510. *In* L. A. Bulla and T. C. Cheng (ed.), *Comparative Pathobiology 2.* Plenum Press, New York, N.Y.

Sprague, V., J. J. Becnel, and E. I. Hazard. 1992. Taxonomy of phylum Microspora. *Crit. Rev. Microbiol.* **18:**285–395.

Stempell, W. 1909. Über *Nosema bombycis* Nageli. *Arch. Protistenkd.* **16:**281–358.

Sweeney, A. W., and J. J. Becnel. 1991. Potential of microsporidia for the biological control of mosquitoes. *Parasitol. Today* **7:**217–220.

Sweeney, A. W., S. L. Doggett, and G. Gullick. 1989. Laboratory experiments on infection rates of *Amblyospora dyxenoides* in the mosquito *Culex annulirostris. J. Invertebr. Pathol.* **53:**83–92.

Sweeney, A. W., S. L. Doggett, and R. G. Piper. 1990. Host specificity studies of *Amblyospora indicola* and *Amblyospora dyxenoides* (Microspora: Amblyosporidae) in mosquitoes and copepods. *J. Invertebr. Pathol.* **56:**415–418.

Sweeney, A. W., S. L. Doggett, and R. G. Piper. 1993. Life cycle of a new species of *Duboscquia* (Microsporida: Thelohaniidae) infecting the mosquito *Anopheles hilli* and an intermediate copepod host, *Apocyclops dengizicus. J. Invertebr. Pathol.* **62:** 137–146.

Sweeney, A. W., E. I. Hazard, and M. F. Graham. 1985. Intermediate host for an *Amblyospora* sp. (Microspora) infecting the mosquito *Culex annulirostris. J. Invertebr. Pathol.* **46:**98–102.

Sweeney, A. W., E. I. Hazard, and M. F. Graham. 1988. Life cycle of *Amblyospora dyxenoides* sp. nov. in the mosquito *Culex annulirostris* and the copepod *Mesocyclops albicans. J. Invertebr. Pathol.* **51:** 46–57.

Tanada, Y. 1976. Epizootiology and microbial control, p. 247–280. *In* L. A. Bulla and T. C. Cheng (ed.), *Comparative Pathobiology,* vol. 2. Plenum Press, New York.

Thomson, H. M. 1958. Some aspects of the epidemiology of a microsporidian parasite of the spruce budworm, *Choristoneura fumiferana* (Clem.). *Can. J. Zool.* **36:**309–316.

Trammer, T., F. Dombrowski, M. Doehring, W. A. Maier, and H. M. Seitz. 1997. Opportunistic properties of *Nosema algerae* (Microspora), a mosquito parasite, in immunocompromised mice. *J. Eukaryot. Microbiol.* **44:**258–262.

Undeen, A. H., 1975. Growth of *Nosema algerae* in pig kidney cell cultures. *J. Protozool.* **22:**107–110.

Undeen, A. H., and N. E. Alger. 1976. *Nosema algerae:* Infection of the white mouse by a mosquito parasite. *Exp. Parasitol.* **40:**86–88.

Undeen, A. H., and J. V. Maddox. 1973. The infection of nonmosquito hosts by injection with spores of the microsporidian *Nosema algerae. J. Invertebr. Pathol.* **22:**258–265.

Undeen, A. H., and J. Vávra. 1997. Research methods for entomopathogenic Protozoa. p. 117–151. *In* L. L. Lacey (ed.), *Manual of Techniques on Insect Pathology.* Academic Press, San Diego, Calif.

Vávra, J., J. I. R. Larsson, and M. D. Baker. 1997. Light and electron microscopic cytology of *Trichotuzetia guttata* gen. et sp. n. (Microspora, Tuzetiidae), a microsporidian parasite of *Cyclops vicinus* Uljanin, 1875 (Crustacea, Copepoda). *Arch. Protistenkd.* **147:**293–306.

Vávra, J., and A. H. Undeen. 1970. *Nosema algerae* n. sp. (Cnidospora, Microsporida) a pathogen in a laboratory colony of *Anopheles stephensi* Liston (Diptera: Culicidae). *J. Protozool.* **17:**240–249.

Vedmed, A. I., S. V. Krylova, and I. V. Issi. 1991. The *Pulicispora xenopsyllae* new genus new species microsporidium from fleas of the genus *Xenopsylla. Parazitologiya* **25:**13–19.

Voronin, V. N. 1993. The microsporidium *Toxospora volgae* gen. n., sp. n. from chironomidae larvae of the genus *Corynoneura. Parazitologiya* **27:**148–154.

Vossbrinck, C. R., T. G. Andreadis, and B. A. Debrunner-Vossbrinck. 1998. Verification of intermediate hosts in the life cycles of microsporidia by small subunit rDNA sequencing. *J. Eukaryot. Microbiol.* **45:**290–292.

Watson, P. L. 1979. The biology and computer simulation of the population dynamics of *Tribolium confusum* (order Coleoptera, family Tenebrionidae) with an introduced pathogen, *Nosema whitei* (order Microsporidia, family Nosematidae). Ph.D. dissertation. University of Illinois, Urbana-Champaign.

Weiser, J. 1946. The microsporidia of insect larvae. *Vest. Cesk. Spol. Zool.* **10:**245–272.

Weiser, J. 1947. Klíč k určování Mikrosporidií. *Acta Soc. Sci. Nat. Moravicae* **18:**1–64.

Weiser, J. 1958. Transovariale Ubertragung der *Nosema otiorrhynchi* W. *Vest. Cesk. Spol. Zool.* **22:**10–12.

Weiser, J. 1961. Die Mikrosporidien als Parasiten der Insekten. *Monogr. Angew. Entomol.* **17:**1–149.

Weiser, J. 1963a. Sporozoan infections. p. 291–334. *In* E. A. Steinhaus (ed.), *Insect Pathology: An Advanced Treatise,* vol. 2. Academic Press, New York, N.Y.

Weiser, J. 1963b. Zur Kenntnis der Mikrosporidien aus *Chironomiden Larven.* III. *Zool. Anz.* **170:**226–230.

Weiser, J. 1976. The *Pleistophora debaisieuxi* xenoma. *Z. Parasitenkd.* **48:**263–270.

Weiser, J. 1977. Contribution to the classification of microsporidia. *Vest. Cesk. Spol. Zool.* **41:**308–321.

Weiser, J., and L. David. 1997. A light and electron microscopic study of *Larssoniella resinellaen.* n. gen., n. sp. (Microspora, Unikaryonidae), a parasite of *Petrova resinella* (Lepidoptera, Tortricidae) in Central Europe. *Arch. Prostistenkd.* **147:**405–410.

Weiser, J., and K. Purrini. 1980. Seven new microsporidian parasites of springtails (Collembola) in the Federal Republic of Germany. *Z. Parasitenkd.* **62:**75–84.

Weiser, J., R. Wegensteiner, and Z. Zizka. 1995. *Canningia spinidentis* gen. et sp. n. (Protista: Microspora), a new pathogen of the fir bark beetle *Pityokteines spinidens. Folia Parasitol.* **42:**1–10.

White, S. E., T. Fukuda, and A. H. Undeen. 1994. Horizontal transmission of *Amblyospora opacita* (Microspora: Amblyosporidae) between the mosquito, *Culex territans,* and the copepod, *Paracyclops fimbriatus* Chiltoni. *J. Invertebr. Pathol.* **63:**19–25.

Wilson, G. G. 1977. The effects of feeding microsporidian (*Nosema fumiferanae*) spores to naturally infected spruce budworm (*Choristoneura fumiferana*). *Can. J. Zool.* **55:**249–250.

Wilson, G. G. 1982. Preliminary observations of life stages of a microsporidian parasite in the white pine weevil *Pissodes strobi. J. Protozool.* **29:**484.

Wilson, G. G. 1984. The transmission and effects of *Nosema fumiferanae* and *Pleistophora schubergi* (Microsporida) on *Choristoneura fumiferana* (Lepidoptera: Tortricidae). *Proc. Entomol. Soc. Ont.* **115:** 71–75.

Windels, M. B., H. C. Chiang, and B. Furgala. 1976. Effects of *Nosema pyrausta* on pupa and adult stages of the European corn borer *Ostrinia nubilalis. J. Invertebr. Pathol.* **27:**239.

Xie, W. D. 1988. Propagation of *Cystosporogenes operophterae,* a microsporidian parasite of the winter moth, *Operophtera brumata,* in a *Spodoptera frugiperda* cell line. *Parasitology* **97:**229–239.

EPIDEMIOLOGY OF MICROSPORIDIOSIS

Ralph T. Bryan and David A. Schwartz

15

Over the past decade or so, the microsporidia have risen from relatively obscure organisms to well-recognized human pathogens. This decade was also witness to the discovery of several genera and species of microsporidia that are "new to science." The preceding chapters contain numerous examples of our rapidly expanding understanding of the many factors involved in the interplay between these organisms and human disease—including advances in clinical management, improved diagnostic methods, meticulous histopathologic descriptions, and a rapidly changing molecular taxonomy. Despite this progress, the epidemiology of human microsporidiosis is still poorly understood. The scientific advances that have occurred, however, have provided some tantalizing clues for epidemiologists interested in studying this condition.

This chapter will focus on what is known regarding the epidemiology of human microsporidiosis and what we may be poised to better understand in the very near future. Topics to be covered include prevalence and geographic distribution, case demographics and populations at risk, and potential modes of transmission.

Ralph T. Bryan, National Center for Infectious Diseases, Centers for Disease Control and Prevention, Albuquerque, NM 87110. *David A. Schwartz,* Department of Pathology, Emory University School of Medicine, Grady Memorial Hospital, Atlanta, GA 30335.

PREVALENCE AND GEOGRAPHIC DISTRIBUTION

Reliable estimates of prevalence, in the strict sense of the term, are not available, partly because studies based on truly random samples have rarely, if ever, been conducted for microsporidiosis and partly because systematic, routine state or federally sponsored surveillance for this condition is not in place anywhere in the world. In the literature reports of prevalence generally refer to highly selected population groups such as AIDS patients with chronic diarrhea and have relied on noncontrolled, nonrandom study designs, including prospective cohort studies, retrospective specimen and chart reviews, and limited surveys using primarily convenience samples. Also problematic has been the lack of consistency in terms of diagnostic approaches and case selection criteria. These factors, in conjunction with the multiple species and diverse clinical manifestations of human microsporidial infections,

The Microsporidia and Microsporidiosis (Murray Wittner, editor; Louis M. Weiss, contributing editor), ©1999 American Society for Microbiology, Washington, D.C.

make it very difficult to understand the true burden of disease attributable to microsporidiosis.

Despite these drawbacks, investigations conducted worldwide reveal reasonably consistent results for some patient populations. Most of what is now known about human microsporidiosis can be attributed to experience with patients infected with human immunodeficiency virus (HIV). Since 1985 and the first recognition of *Enterocytozoon bieneusi* as an AIDS-associated opportunistic pathogen (Desportes et al., 1985), several hundred patients with chronic diarrhea attributed to this organism have been reported. Worldwide, between 1989 and 1998, at least 25 published studies on *E. bieneusi* in AIDS patients with chronic diarrhea have reported prevalences ranging from 4 to 50%. Combined, these studies have evaluated over 2,400 patients and confirmed about 375 cases of *E. bieneusi* infections—for an overall prevalence of approximately 15%. These observations suggest that we can reasonably expect that *E. bieneusi* accounts for a significant proportion of the chronic diarrheal disease observed in patients with AIDS. Although closer scrutiny is needed, there appear to be no consistent trends, or variations, in prevalence based on country of origin or other demographic characteristics (Bryan et al., 1991; Bryan 1995a, 1995b; Weber et al., 1994; Coyle et al., 1996; Van Gool et al., 1995; Brasil et al., 1996; Weitz et al., 1995; Voglino et al., 1996; Drobniewski et al., 1995; Kyaw et al., 1997).

Although human disease associated with microsporidia has been reported predominantly from developed nations in North America, western Europe, and Australia, the recognition of human microsporidiosis is on the rise worldwide (Fig. 1). Persons suffering from a variety of microsporidial disease manifestations have been identified from all continents except Antarctica. With new interest in these organisms and improved diagnostic approaches, case recognition in the Third World is becoming increasingly common. Human infections are now well-documented in several African nations (Aoun et al., 1997; Bretagne et al., 1993; Cegielski et al., unpublished data; Drobniewski et al., 1995; Hautvast et al., 1997; Lucas et al., 1989; Maiga et al., 1997; Van Gool et al., 1995), Southeast Asia (Morakote et al., 1995), and South America (Brasil et al., 1996; Wuhib et al., 1994; Weitz et al., 1995).

SEROLOGIC STUDIES

Surveys for antibodies to microsporidia in human sera have focused exclusively on human exposure to *Encephalitozoon* species. Studies published prior to 1991 have been summarized by Bryan and colleagues (Bryan et al., 1991; Singh et al., 1982; Bergquist et al., 1984; Hollister and Canning, 1987). The frequency of positive specimens ranged from 0 to 38% in the various population groups evaluated. These studies, however, lacked clinical or pathologic correlations and failed to produce any definitive epidemiologic conclusions. Since that time, only a few serologic investigations have been performed. Whereas earlier studies targeted (presumably) *Encephalitozoon cuniculi*, recent publications have reported results based on human immunoreactivity to *E. intestinalis* and *E. hellem* as well. In three studies published in 1997, human sera from France, the Netherlands, Slovakia, and the Czech Republic were variably screened for antibodies to *Encephalitozoon* spp. with techniques such as enzyme-linked immunosorbent assay, indirect immunofluorescent antibody testing, counterimmunoelectrophoresis, and complement fixation. Van Gool and colleagues found antibodies to *E. intestinalis* among pregnant French women (5%) and Dutch blood donors (8%) (van Gool et al., 1997). HIV-infected Czech patients were screened by Pospisilova and colleagues, who found that 5.3% were seropositive to *E. cuniculi* while 1.3% were seropositive to *E. hellem;* all (presumably HIV-negative) blood donors that were screened were seronegative for both species (Pospisilova et al., 1997). In Slovakia, Cislakova and colleagues found that 5 of 1,998 (5.1%) slaughterhouse workers were seropositive for *Encephalitozoon* spp. (not specified) but that 92 forestry workers, 22 dog breeders, and 150 blood donors were all

FIGURE 1 Geographic distribution of human microsporidiosis. Shaded regions indicate countries with confirmed human cases. The figure is based on a literature review as of May 1998.

seronegative (Cislakova et al, 1997). These more recent studies suffer from many of the same limitations as those of earlier years, including the use of convenience rather than random sampling techniques, but do suggest that human infections with *Encephalitozoon* spp. may be more common than previously recognized.

CASE DEMOGRAPHICS AND POPULATIONS AT RISK

Serious illness due to infections with the various human-infecting microsporidia appears to occur predominantly in adults suffering from immunosuppression, especially that associated with HIV/AIDS (Bryan, 1995a, 1995b; Weber et al., 1994). Human microsporidiosis, however, is being increasingly recognized in persons with other forms of immunosuppression. Infections with *E. bieneusi*, for example, have been confirmed in one HIV-negative liver transplant recipient and one heart-lung transplant recipient (Rabodonirina et al., 1996; Sax et al., 1995). Cotte and colleagues also noted a number of *E. bieneusi* infections in bone marrow, heart-lung, liver, and kidney transplant recipients during a possible outbreak of intestinal microsporidiosis in France in 1995 (Cotte et al., 1998). *E. cuniculi* was recently identified in a kidney-pancreas transplant recipient (Goodman et al., unpublished data) and a case of pulmonary microsporidial infection (presumably an *Encephalitozoon* sp.) was confirmed at autopsy in a patient with chronic myeloid leukemia who had undergone bone marrow transplantation (Kelkar et al., 1997). Transplant recipients, however, do not appear to be the only HIV-negative, immunosuppressed populations at risk for microsporidiosis. Silverstein and colleagues described a patient who developed chronic bilateral microsporidial keratoconjunctivitis while taking systemic prednisone (20 mg per day) for severe asthma (Silverstein et al., 1997). Although the infecting species was not reported, organisms pictured were consistent with *Encephalitozoon* and the patient responded successfully to therapy with albendazole. In yet another case of non-HIV-associated immunosuppression, Wanke and colleagues observed a patient with *E. bieneusi*-associated diarrhea who had an unexplained decreased CD4 cell count and a history of cryptococcal meningitis (Wanke et al., 1996). Interestingly, this patient also responded successfully to therapy with albendazole. With these various reports in mind, it is reasonable to anticipate that the recognition of microsporidiosis in association with other, non-HIV-associated, forms of immunosuppression will continue to increase.

Although most recognized cases of human microsporidiosis are associated with some form of immunosuppression, reports describing microsporidial infections in HIV-negative immunocompetent patients are also increasing (Table 1). These reports have included instances of *E. bieneusi* infections in travelers as well as in adult and child residents of various tropical countries (Albrecht et al., 1997; Bryan et al., 1997; Cegielski et al., unpublished data; Cotte et al., 1998; Hautvaust et al., 1997; Maiga et al., 1997; Sandfort et al., 1994; Sobottka et al., 1995; Wanke et al., 1996). Also, Raynaud and colleagues have reported the first association of

TABLE 1 Non-HIV-associated microsporidiosis: *E. bieneusi* self-limited diarrhea in immunocompetent hosts

Patient[a]	Location	Epidemiologic history	Duration of diarrhea	Yr reported
26 yo, M	Germany	Travel to Middle East	2 wk	1994
3 yo, F	Germany	Travel to Turkey	6 wk	1995
26 yo, F	Switzerland	Nurse	2 wk	1995
42 yo, M	United States	HIV-negative; CD4 cell count, 500	3 mo	1996

[a]yo, years old; M, male; F, female.

E. intestinalis with chronic diarrhea in immunocompetent travelers (Raynaud et al., 1998). It is interesting to speculate, therefore, that intestinal microsporidiosis may be an underappreciated cause of travelers' diarrhea that warrants further investigation. Heightened awareness of this disease by clinicians caring for travelers returning from tropical destinations should help to increase recognition and, ultimately, our understanding of the epidemiology of this disease.

POTENTIAL MODES OF TRANSMISSION

Zoonotic Infections

ENCEPHALITOZOON CUNICULI

Microsporidial infections in a wide variety of invertebrates and vertebrates were well recognized prior to the initial report of human infection in 1959. *E. cuniculi* was first described in a rabbit with motor paralysis by Wright and Craighead in 1922 (Wright et al., 1922). Because it was the first mammalian microsporidian to be grown in culture, it is the most extensively studied microsporidian infecting mammals. It has a broad host range, and a partial listing of potential hosts includes the rabbit, mouse, rat, muskrat, guinea pig, hamster, ground shrew, fox, cat, leopard, puma, squirrel monkey, rhesus monkey, baboon, goat, sheep dog, pig, and horse (Canning and Hollister, 1987). *E. cuniculi* was for many years the sole member of its genus. Many of the initial descriptions of microsporidial infection of humans, both prior to and during the AIDS pandemic, have been due to *Encephalitozoon*-like organisms. These isolates microscopically resembled *E. cuniculi,* and thus they were believed to be due to this microsporidian. However, the isolation from three AIDS patients in 1991 of a morphologically identical, but antigenically and biochemically distinct, species of *Encephalitozoon, E. hellem,* raised the possibility that some previously reported human infections had resulted from this newly discovered agent and not from *E. cuniculi* (Didier et al., 1991). Additional confirmation of the importance of *E. hellem* as a human pathogen was obtained in 1993, when Schwartz and colleagues identified *E. hellem,* but not *E. cuniculi,* using antibody-based methods from the eyes of seven patients with AIDS and keratoconjunctivitis (Schwartz et al., 1997). The existence of *E. cuniculi* infections in humans was in doubt until 1995 when disseminated *E. cuniculi* was confirmed in an AIDS patient with both nucleic acid-based and tissue culture methods (De Groote et al., 1995). Since then, additional cases of *E. cuniculi* in AIDS patients have been confirmed by antibody-based and molecular methods (Didier et al., 1996; Deplazes et al., 1996a; Mathis et al., 1997).

Because of differences in the parasite distribution and lesions seen in naturally identified *E. cuniculi* infections of animals, Weiser believed that strain differences might exist (Weiser, 1964, 1965). In 1995 Didier and colleagues found that *E. cuniculi* isolates obtained from rabbits, dogs, and mice could be differentiated on the basis of subtle differences in protein profiles in the range of 54 to 59 kDa with sodium dodecyl sulfate-polyacrylamide gel electrophoresis and antigenic analysis, and differences in the 5′-GTTT-3′ repeats present in the internal transcribed spacer region of the rDNA identified by polymerase chain reaction (PCR) methods (Didier et al., 1995a). They designated these isolates strain I (rabbits and one mouse), strain II (mice only), and strain III (domestic dogs only). Using similar methods, Didier and colleagues examined an isolate of *E. cuniculi* obtained from a patient with AIDS and disseminated microsporidiosis. They found that the rDNA internal transcribed spacer region of this isolate was identical to that of *E. cuniculi* strain III which had been isolated only from domestic dogs. Although the patient owned a dog, no microsporidia were detected in the pet's urine (Didier et al., 1996).

Additional evidence is accumulating that human infections with *E. cuniculi* may result from animal sources. Deplazes and colleagues (1996a) reported that *E. cuniculi* strain I, originally described from rabbits, has been identified as infecting humans in Switzerland. Four

E. cuniculi isolates from HIV-infected patients living in Switzerland were indistinguishable by immunological and molecular methods from isolates obtained from Swiss rabbits (Mathis et al., 1997). None of these patients owned a rabbit or remembered having been exposed to a rabbit as an adult. However, one patient grew up on a farm where rabbits and other animals were raised, and another two patients kept a variety of pets.

Natural infections of nonhuman primates with an *Encephalitozoon* sp. have been described. In a primate colony housing 250 squirrel monkeys (*Saimiri sciureus*), immunofluorescence and dot enzyme-linked immunosorbent assay techniques revealed serum antibodies to *E. cuniculi* at least once in 179 monkeys, and three or more times in 56 monkeys (Shadduck and Baskin, 1989). The animals had organisms and granulomatous infection in a variety of organs including the brain, kidney, lung, adrenal gland, and liver (Zeman and Baskin, 1985). Nonhuman primates are also susceptible to experimental infection with several microsporidian agents that also infect humans, including *E. cuniculi* (Didier et al., 1995b), *E. hellem* (Didier et al., 1995b), and *E. bieneusi* (see below).

The source of human microsporidial infections and the mechanism(s) of transmission are still not clear. Although too few microsporidial infections of humans have been analyzed to provide firm evidence of zoonotic transmission, it appears likely that at least some cases of human infection are acquired from animals (Table 2).

TABLE 2 Potential zoonotic reservoirs for human infection

Animal host	Microsporidian
Rabbit	*Encephalitozoon cuniculi*
Dog	*Encephalitozoon cuniculi*
Parakeet[a]	*Encephalitozoon hellem*
Parrot[b]	*Encephalitozoon hellem*
Pig	*Enterocytozoon bieneusi*
Monkey[c]	*Enterocytozoon bieneusi*
Dog, pig, goat, cattle	*Encephalitozoon intestinalis*

[a]Budgerigar (*Melopsittacus undulatus*).
[b]*Eclectus roratus*.
[c]Rhesus macaque (*Macaca mulatta*).

ENCEPHALITOZOON HELLEM

Several reports based on light and electron microscopy have described *Encephalitozoon* spp. infection of psittacine birds, and considerable interest has been raised regarding the possible role of these animals as potential reservoirs of human infection. Following the discovery of *E. hellem* in 1991 by Didier and colleagues (Didier et al., 1991), efforts to isolate this agent from nonhuman sources proved unsuccessful, and for several years it was believed that the organism was confined to human hosts. However, in 1997 an *Encephalitozoon* sp. was reported to cause fatal microsporidial disease in young birds at a budgerigar, or parakeet (*Melopsittacus undulatus*), aviary in Mississippi. The birds, chicks from 1 to 2 weeks of age, had heavy microsporidial infection of the intestines and multifocal microsporidial hepatic necrosis. Southern blotting and PCR methods confirmed the etiologic agent as *E. hellem* (Black et al., 1997). This occurrence was followed by a report in 1998 of *E. hellem* infection in a different avian host, an eclectus parrot (*Eclectus roratus*). Microsporidia were identified in this bird from liver and kidney sections submitted following autopsy and confirmed as *E. hellem* by PCR techniques (Pulparampil et al., 1998). *E. hellem* has not been identified thus far from nonhuman mammals, but it is possible that mammalian isolates previously believed to be *E. cuniculi* by morphological methods are, in fact, *E. hellem*. Although some patients with AIDS who have developed *E. hellem* infection keep companion animals and birds (Schwartz et al., 1992), there has been no documentation of concurrent infection of both humans and their companion animals with this microsporidian.

ENCEPHALITOZOON (FORMERLY *SEPTATA*) *INTESTINALIS*

E. intestinalis is the most recent addition to the genus *Encephalitozoon*. It was first identified in

1993 from patients with AIDS who developed a diarrhea/wasting syndrome and was initially named *Septata intestinalis* because of its characteristic ultrastructural appearance featuring septations between spores and developing stages (Cali et al., 1993). Unlike the other major intestinal microsporidian infecting humans, *E. bieneusi*, *E. intestinalis* has the capability to disseminate beyond the intestinal tract and cause systemic and potentially fatal disease. Humans remained the only recognized host of this agent until 1998, when Bornay-Llinares and colleagues described *E. intestinalis* in stool from a variety of mammals in Mexico. Spores of *E. intestinalis* were identified in stool specimens from five different animals—donkey, pig, dog, cow and goat—collected during an epidemiological survey in two rural villages in Pueblo State. Light microscopic examination of stool showed clusters of microsporidian spores, morphologically consistent with an *Encephalitozoon* spp., within the cytoplasm of shed epithelial cells. In stool samples with a high parasite burden, there was ultrastructural evidence of an *Encephalitozoon*-type infection. The identity of the microsporidia as *E. intestinalis* was confirmed with an anti-*E. intestinalis* antiserum and PCR (Bornay-Llinares et al., 1998).

Prior to the AIDS pandemic, an intestinal microsporidian was identified in 1973 at necropsy from a female *Callicebus moloch* monkey that had been housed in a primate colony (Seibold and Fussell, 1973). Although the morphological features of this agent were somewhat unclear, it was believed to resemble an *Encephalitozoon* sp. on the basis of light and electron microscopy. In retrospect, it is possible that this agent might have been the then-undiscovered microsporidian *E. intestinalis.*

ENTEROCYTOZOON BIENEUSI

Since its initial description in 1985 as a cause of diarrhea and wasting syndrome in patients with AIDS, *E. bieneusi* has been found to be the most prevalent microsporidial infection of humans. However, its potential environmental source(s) continue to elude the scientific community. *E. bieneusi* was first identified in a nonhuman host in 1996 when Deplazes and colleagues described spores from immature and adult pigs in Switzerland (Deplazes et al., 1996b). Their identity was confirmed as *E. bieneusi* by DNA analysis. Following this report, there were independent reports from two geographically disparate primate colonies in 1997 describing naturally acquired *E. bieneusi* infections occurring in captive macaques (Anderson et al., 1997; Mansfield et al., 1997). All the animals had been experimentally infected with simian immunodeficiency virus (SIV) prior to their diagnosis as having microsporidiosis at the time of necropsy. Although most infected macaques were *Macaca mulatta,* a few *M. cyclopis* and *M. nemestrina* also had *E. bieneusi* infection. An interesting feature of the pathology of these primates was that the majority of the parasite burden resided in the gallbladder and not, as has been described from human infections, in the small intestine (Mansfield et al., 1997; Schwartz et al., 1998a). More recently, Mansfield and colleagues have found *E. bieneusi* infections in immunocompetent, non-SIV infected macaques from the same primate colony (Mansfield, personal communication; Carville et al., 1997). The environmental source for both human and nonhuman primate infections remains unknown.

In addition to the occurrence of natural infections of macaques with *Enterocytozoon,* this agent has been experimentally transmitted from a human with AIDS and intestinal microsporidiosis to SIV-infected macaques via oral inoculation with concentrated spores (Tzipori et al., 1997). Infected animals began shedding spores within 1 week after inoculation and continued until euthanized 7 to 8 months later. Microscopic examination of the intestine and hepatobiliary tract revealed that the parasite burden was so sparse that *E. bieneusi* could be detected only with the use of in situ hybridization techniques.

Waterborne Transmission

Despite the presence of more than 1,000 species of microsporidia throughout the biosphere, potential environmental sources of human-infecting microsporidia have been poorly researched. Other intestinal protozoa, such as *Giardia, Cyclospora,* and *Cryptosporidium,* are commonly acquired by ingestion of contaminated water (or food), suggesting that the same may be true for *E. bieneusi* and *E. intestinalis.* Although data supporting this contention remain limited, recent observations are beginning to provide a stronger scientific foundation for the theory of waterborne transmission of some forms of human microsporidiosis.

Microsporidia have been isolated from ditch water in a mosquito larval habitat in Florida (Avery and Undeen, 1987). Several genera were identified, including two that were potentially infective to humans, *Pleistophora* sp. and *Nosema* sp. The first identification of a pathogenic human microsporidian from a natural water source was reported in 1997 (Sparfel et al., 1997). Spores of *E. bieneusi,* the most prevalent microsporidian infecting humans, were identified with a novel filtration/PCR method in samples of environmental surface water from the Seine River in France. Other microsporidian species potentially infective to humans, including *Pleistophora* spp. and *Vittaforma corneae,* were identified in water from the Loire and Seine Rivers by this technique.

Several groups of investigators, working separately in different parts of the world, have found that water contact may be an independent risk factor for intestinal microsporidiosis. A case-control study was conducted in clinics serving HIV-infected persons in Massachusetts and Texas to identify environmental risk factors for acquiring intestinal infection (Watson et al., 1996). Twelve patients with intestinal microsporidiosis and 54 uninfected controls were enrolled. Risk factors for acquiring microsporidiosis in HIV-infected persons included swimming in rivers, ponds, and lakes and drinking unfiltered tap water. In 1997, another prospective unmatched case-control study of risk factors for acquiring intestinal microsporidiosis was conducted among HIV-infected patients (<200 CD4 cells per mm^3) in France (Hutin et al., 1997). The only two factors associated with microsporidial infection were swimming in a pool and male homosexuality. These investigators suggested that the mode of transmission of intestinal microsporidiosis is fecal-oral, including waterborne and person-to-person transmission.

In 1998 Cotte and colleagues reported an outbreak of intestinal microsporidiosis in HIV-infected persons which occurred in mid-1995 in France (Cotte et al., 1998). In a retrospective study of 1,453 patients (978 HIV-infected, 395 non-HIV-infected) who underwent stool examinations for microsporidia in Lyon, France, between April 1993 and December 1996, a cluster of microsporidiosis cases was noted to occur from May to November 1995. There were 200 infected patients identified during this outbreak versus 138 "endemic" cases found outside the outbreak period. Both HIV-infected and non-HIV-infected persons were found to have intestinal microsporidiosis in each period. Variations in the geographic distribution of the affected patients during the outbreak were suggestive of an association between the municipal water distribution system and occurrence of the outbreak. However, none of the typical markers for fecal contamination of water were identified during and outside the outbreak period.

In another study to address the potential for waterborne transmission, Enriquez and colleagues assessed the prevalence of *E. intestinalis* in two rural agricultural communities in central Mexico (Enriquez et al., 1998). This cross-sectional survey revealed that 15 (21.4%) of 70 households had at least one member who was infected with this organism. They also found that the use of untreated water from an indoor faucet and the use of community wells as a primary water source were significantly more common in households with an *E. intestinalis*-infected resident. Further, the use of a private well as the primary household water source appeared to be protective.

Two studies have failed to show an association of water with intestinal microsporidiosis. In Brazil, Wuhib and colleagues examined stool specimens from 166 HIV-infected persons to determine the prevalence of intestinal parasites (Wuhib et al., 1994). In addition to their finding that cryptosporidiosis and microsporidiosis were the most common intestinal parasites, cryptosporidial, but not microsporidial, infection was found to be associated with the rainy season. This study, however, did not examine the specific forms of water contact (swimming, drinking, humidifiers) that were observed as risk factors in the previously discussed investigations. Conteas and colleagues examined 8,439 stool specimens from HIV-infected persons in southern California from 1993 to 1996 and found no seasonal association with either recreational water use or seasonal contamination of the water supply (Conteas et al., 1998a).

In summary, although data supporting waterborne transmission are accumulating, some uncertainty remains. It is likely, however, that waterborne transmission of human microsporidiosis does indeed occur. Further studies are needed to confirm a definitive link between human infections and contaminated water sources before appropriate public health initiatives are considered.

Airborne Transmission and Respiratory Tract Acquisition

The occurrence of upper and lower respiratory tract infections suggests that microsporidiosis can be acquired by inhalation or transmitted via aerosolized infected materials. In the mouse model of microsporidiosis, oral, intranasal, and intratracheal application of *E. cuniculi* leads to disseminated infections (Cox et al., 1979; Wicher et al., 1991). Transmission via the aerosol route has been especially implicated as a likely route of infection in some of the reported cases of *E. hellem* infection. The initial evidence for respiratory acquisition was based on the unexpected autopsy findings of prominent respiratory epithelial infection with *E. hellem*, extending confluently from the proximal trachea down to the terminal bronchioles, in the first described patient with both AIDS and disseminated microsporidiosis (Schwartz et al., 1992). This superficial pattern of respiratory epithelial infection was similar to that seen in other infections that have a respiratory route of acquisition. Since then many additional cases of microsporidial respiratory infection have been reported (Schwartz et al., 1993b; Lucas et al., 1996). Several cases of *E. hellem* infection limited to the paranasal sinuses have been described in which intestinal infection was notably absent (Albrecht and Schwartz, unpublished data; Dunand et al., 1997; Scaglia et al., 1998), providing indirect evidence for primary respiratory tract infection. Many patients with *Encephalitozoon* spp. infection have initial symptoms of sinonasal disease, suggesting an upper respiratory portal of entry in these cases.

The presence of *Encephalitozoon* spp. spores in respiratory secretions as well as in urine has led to the suggestion that ocular infections with these agents may be acquired by external autoinoculation, perhaps by contaminated fingers (Schwartz et al., 1993a). Ingestion or inhalation of spore-laden urine contaminating animal cages is an established means of *E. cuniculi* transmission in rabbits and other laboratory animals, suggesting that comparable forms of environmental exposure could lead to human infection. As *E. hellem* has recently been isolated from bird feces, it is conceivable that this agent is acquired through inhalation of aerosolized fecal material containing infective spores, a route of transmission that is well documented for *Cryptococcus neoformans*.

Direct Person-to-Person Transmission

Extensive genitourinary tract involvement is typical of disseminated infections due to all *Encephalitozoon* spp. The identification of prostatic abscess (Schwartz et al., 1994) and urethral infection (Birthistle et al., 1996; Corcoran et al., 1996; Schwartz et al., 1992) in some male patients raises the possibility that person-to-person transmission by sexual means may at least be anatomically feasible.

The identification of concurrent microsporidial infection in homosexual male sexual partners has drawn attention to the potential of person-to-person transmission. In one partner pair infection involving cohabitating men who were both infected with microsporidia, the sexual partner of a patient with intestinal microsporidiosis was diagnosed with microsporidial urethritis that did not resolve until he abstained from unprotected sexual intercourse and treatment with albendazole was initiated (Birthistle et al., 1996). The patient's good response to albendazole therapy and the organ systems affected together seem to indicate infection with an *Encephalitozoon* sp., but the etiologic agent was not further characterized by the authors. They concluded that sexual transmission of microsporidia was the most probable explanation for the patient's urethritis.

Four additional cohabitating homosexual male partner pairs have been found to have concurrent microsporidiosis (Schwartz et al., 1998b). Three of these four sets of men were concurrently infected with HIV. In the fourth partner pair, one partner was HIV-infected and the other was not. In all four pairs the partners were alive and infected with microsporidiosis concurrently with one another, were infected with the same microsporidian species, and were shedding spores in either feces or body fluids. Two partner pairs were infected with *E. hellem,* one pair with *E. bieneusi,* and the HIV-discordant pair was infected with *E. intestinalis.*

Other authors have reported that microsporidiosis is more prevalent among their homosexual patients compared to other risk groups (Voglino et al., 1996). Two prospective case-control studies have shown that male homosexuality (Hutin et al., 1998) and having an HIV-infected cohabitant (Watson et al., 1996) are risk factors for acquiring intestinal microsporidiosis. Interestingly, in the mouse, rectal installation of *E. cuniculi* has been shown to cause disseminated infection (Fuentealba et al., 1992).

Although the circumstantial evidence seems strong, the documentation of partner infections does not prove the sexual transmissibility of microsporidia. It is conceivable that partner infections are the result of fecal-oral, urinary-oral, or common-source transmission. The well-documented presence of viable, infective spores in multiple body fluids and excreta is consistent with all these potential modes of transmission.

PREVENTION

Microsporidian spores can survive, and remain infective, in the environment for extended periods of time. In vitro data evaluating the potential efficacy of preventive measures have been published only for *E. cuniculi* (Waller, 1980). These experiments indicate that spores can survive in the environment for months to years depending on humidity and temperature. Even in a typical dry hospital environment (22°C) spores can survive for at least a month. Exposure to recommended working concentrations of most disinfectants for 30 min, boiling for 5 min, and autoclaving at 120°C was reported to kill the spores. Freezing may not be an effective means of disinfection, as it has been possible to grow *E. hellem* successfully after storing it at –70°C for months. Because *E. intestinalis* has only recently been sustained in culture, and *E. bieneusi* has been grown successfully only in short-term cultures, the potential infectivity of both parasites as well as the efficacy of preventive measures have not been evaluated.

For these reasons, data supporting effective preventive strategies are limited. However, the presence of infective spores in body fluids and feces suggests that body substance precautions in health care and other institutional settings, general attention to hand washing, and other personal hygiene measures, may be helpful in preventing primary infections. Meticulous hand washing may be of particular importance in preventing ocular infection, which may occur as a result of inoculation of conjunctival surfaces by fingers contaminated with respiratory fluids or urine. Whether respiratory precautions might be efficacious for persons documented to have spores in sputum or other respiratory secretions is unknown. Similarly, precautions pertinent to environmental or zoonotic exposure to microsporidia are as yet undefined (Bryan, 1995a).

Whether primary prophylaxis with antimicrobials is possible remains to be determined. Cotrimoxazole and atovaquone (for *Pneumocystis carinii,* and *Toxoplasma gondii*) and azithromycin (for mycobacteria) are effective or are currently under investigation for prophylactic treatment of other opportunistic infections in patients with AIDS. It is possible that the use of these agents could potentially also reduce the incidence of microsporidiosis. *E. bieneusi* infection, however, has been detected in patients receiving clindamycin, TMP-SMX, dapsone, pyrimethamine, and itraconazole. In a prospective evaluation of a series of 76 patients in Hamburg, 3 of 22 patients (14%) treated with TMP-SMX for *P. carinii* prophylaxis developed *E. bieneusi* infection as compared to 11 of 54 patients (20%) who were treated with inhalational pentamidine or did not receive any prophylaxis ($p > 0.2$) (Albrecht et al., 1995). Even though these numbers are small, it seems that the currently used prophylactic dose of 960 mg/day will offer no or, at best, minimal protection from *E. bieneusi* infection. Four out of five patients with disseminated *E. intestinalis* infection reported by Molina et al. (1995) also received TMP-SMX prior to diagnosis, indicating that it is also ineffective in preventing *E. intestinalis* infection. Future trials evaluating prophylactic strategies involving drugs with potential antimicrosporidial activity should attempt to address this question. Finally, although its efficacy as a therapeutic agent for *Encephalitozoon* infections is well-confirmed (Molina et al., 1995; Dore et al., 1995; De Groote et al., 1995; Corcoran et al., 1996), rigorous clinical trials using albendazole for primary or secondary prophylaxis have yet to be performed.

One of the most exciting observations in preventing microsporidiosis in persons with AIDS has been the role of highly active antiretroviral therapy in raising CD4 lymphocyte counts and heightening immunity. Several small studies have now shown that administration of a protease inhibitor together with one or two nucleoside analogs has resulted in remission of diarrheal symptoms of most patients with microsporidia-associated diarrhea (Goguel et al., 1997; Foudraine et al., 1998; Conteas et al., 1998b). In some patients, the stool was persistently negative for microsporidia spores for up to 10 months after treatment. Although highly active antiretroviral therapy is a promising treatment regimen for prevention of microsporidiosis in patients with AIDS, long-term follow-up is clearly needed to assess whether these patients are actually free of the intracellular stages of microsporidia or are temporarily in remission with an inactive infection.

A high standard of personal hygiene seems mandatory for patients infected with microsporidia. HIV-positive persons are advised to avoid raw eggs and undercooked meat in order to avoid exposure to *Salmonella* spp. and *T. gondii*. Because *E. cuniculi* can infect eggs and consumable animals such as rabbits, this general recommendation should be reinforced (Reetz et al., 1994). In addition, infected patients should be warned that sexual transmission of microsporidiosis cannot be excluded. Cohabitating sexual partners of infected patients should be offered screening for microsporidiosis regardless of their HIV status.

REFERENCES

Albrecht, H., and D. A. Schwartz. 1998. Unpublished data.

Albrecht, H., and I. Sobottka. 1997. *Enterocytozoon bieneusi* infection in patients who are not infected with human immunodeficiency virus. *Clin. Infect. Dis.* **25:**344.

Albrecht, H., I. Sobottka, H. J. Stellbrink, and H. Greten. 1995. Does the choice of *Pneumocystis carinii* prophylaxis influence the prevalence of *Enterocytozoon bieneusi* microsporidiosis in AIDS-patients? *AIDS* **9:**302–303.

Anderson, D. C., S. A. Klumpp, A. J. Da Silva, N. J. Pieniazek, H. M. McClure, and D. A. Schwartz. 1997. Naturally-acquired *Enterocytozoon bieneusi* (Microsporida) hepatobiliary infection in rhesus monkeys with simian immunodeficiency virus (SIV): a possible animal model of disease, abstr. K-120b, p. 349. *In Abstracts of the 37th Interscience Conference on Antimicrobial Agents and Chemotherapy*. American Society for Microbiology, Washington, D.C.

Aoun, K., A. Bouratbine, A. Datry, S. Biligui, and R. Ben Ismail. 1997. Presence of intestinal microsporidia in Tunisia: apropos of 1 case. *Bull. Soc. Pathol. Exot.* **90:**176.

Avery, S. W., and A. H. Undeen. 1987. The isolation of microsporidia and other pathogens from concentrated ditch water. *J. Am. Mosquito Control Assoc.* **3:**54–58.

Bergquist, R., L. Morfeldt-Mansson, P. O. Pherson, B. Petrini, and J. Wasserman. 1984. Antibody against *Encephalitozoon cuniculi* in Swedish homosexual men. *Scand. J. Infect. Dis.* **16:**389–391.

Birthistle, K., P. Moore, and P. Hay. 1996. Microsporidia: a new sexually transmissible cause of urethritis. *Genitourin. Med.* **72:**445.

Black, S. S., L. A. Steinohrt, D. C. Bertucci, L. B. Rogers, and E. S. Didier. 1997. *Encephalitozoon hellem* in budgerigars (*Melopsittacus undulatus*). *Vet. Pathol.* **34:**189–198.

Bornay-Llinares, F. J., A. J. Da Silva, H. Moura, D. A. Schwartz, G. S. Visvesvara, N. J. Pieniazek, A. Cruz-Lopez, P. Hernandez-Jauregui, and J. Enriquez. 1998. Immunological, microscopic and molecular evidence of *Encephalitozoon* (*Septata*) *intestinalis* infection in mammals other than man. *J. Infect. Dis.* **178:**820–826.

Brasil, P., F. C. Sodre, T. Cuzzi-Maya, M. C. Gutierrez, H. Mattos, and H. Moura. 1996. Intestinal microsporidiosis in HIV-positive patients with chronic unexplained diarrhea in Rio de Janeiro, Brazil: diagnosis, clinical presentation, and follow-up. *Rev. Inst. Med. Trop. Sao Paulo.* **38:**97–102.

Bretagne, S., F. Foulet, W. Alkassoum, J. Fleury-Feith, and M. Develoux. 1993. Prévalence des spores d'*Enterocytozoon bieneusi* dans les selles de patients sidéens et d'enfants africains non infectés par le VIH. *Bull. Soc. Path. Exot.* **86:**351–357.

Bryan, R. T. 1995a. Microsporidiosis as an AIDS-related opportunistic infection. *Clin. Infect. Dis.* **21**(Suppl.):S62–S65.

Bryan, R. T. 1995b. Microsporidia, p. 2513–2524. *In* G. L. Mandell, J. E. Bennett, and R. Dolin (ed.), *Principles and Practice of Infectious Diseases,* 4th ed. Churchill Livingstone, New York.

Bryan, R. T., A. Cali, R. L. Owen, and H. C. Spencer. 1991. Microsporidia: opportunistic pathogens in patients with AIDS, p. 1–26. *In* T. Sun (ed.), *Progress in Clinical Parasitology,* vol. 2. Field and Wood, Philadelphia.

Bryan, R. T., R. Weber, and D. A. Schwartz. 1997. Microsporidiosis in persons without HIV. *Clin. Infect. Dis.* **24:**534–535. (Letter.)

Cali, A., D. P. Kotler, and J. M. Orenstein. 1993. *Septata intestinalis* n. g., n. sp., an intestinal microsporidian associated with chronic diarrhea and dissemination in AIDS patients. *J. Eukaryot. Microbiol.* **40:**101–112.

Canning, E. U., and W. S. Hollister. 1987. Microsporidia of mammals: widespread pathogens or opportunistic curiosities? *Parasitol. Today* **3:**267–273.

Carville, A., K. Mansfield, K. C. Lin, J. McKay, L. Chalifoux, and A. Lackner. 1997. Genetic and ultrastructural characterization of *Enterocytozoon bieneusi* in simian immunodeficiency virus infected and immunocompetent rhesus macaques. *Vet. Pathol.* **34:**515.

Cegielski, J. P., Y. R. Ortega, S. McKee, J. F. Madden, L. Gaido, D. A. Schwartz, K. Manji, A. F. Jorgensen, S. E. Miller, U. P. Pulipaka, A. E. Msengi, D. H. Mwakyusa, C. R. Sterling, and L. B. Reller. 1998. Unpublished data.

Cislakova, L., H. Prokopcakova, M. Stefkovic, and M. Halanova. 1997. *Encephalitozoon cuniculi,* clinical and epidemiologic significance: results of a preliminary serologic study in humans. *Epidemiol. Mikrobiol. Imunol.* **46:**30–33.

Conteas, C. N., O. G. W. Berlin, M. J. Lariviere, S. S. Pandhumas, C. E. Speck, R. Porschen, and T. Nakaya. 1998a. Examination of the prevalence and seasonal variation of intestinal microsporidiosis in the stools of persons with chronic diarrhea and human immunodeficiency virus infection. *Am. J. Trop. Med. Hyg.* **58:**559–561.

Conteas, C. N., O. G. W. Berlin, C. E. Speck, S. S. Pandhumas, M. J. Lariviere, and C. Fu. 1998b. Modification of the clinical course of intestinal microsporidiosis in acquired immunodeficiency syndrome patients by immune status and anti-human immunodeficiency virus therapy. *Am. J. Trop. Med. Hyg.* **58:**555–558.

Corcoran, G. D., J. R. Isaacson, C. Daniels, and P. L. Chiodini. 1996. Urethritis associated with disseminated microsporidiosis: clinical response to albendazole. *Clin. Infect. Dis.* **22:**592–593.

Cotte, L., M. Rabodonirina, C. Raynal, F. Chapuis, M. A. Piens, and C. Trepo. 1998. Outbreak of intestinal microsporidiosis in HIV-infected and non-infected patients, abstr. 483. *In Abstracts of the 5th Conference on Retroviruses and Opportunistic Infections.*

Cox, J. C., R. C. Hamilton, and H. D. Attwood. 1979. An investigation of the route and progression of *Encephalitozoon cuniculi* in adult rabbits. *J. Protozool.* **26:**260–265.

Coyle, C. M., M. Wittner, D. P. Kotler, C. Noyer, J. M. Orenstein, H. B. Tanowitz, and L. M. Weiss. 1996. Prevalence of microsporidiosis due to *Enterocytozoon bieneusi* and *Encephalitozoon intestinalis* among patients with AIDS-related diarrhea: determination by polymerase chain reaction to the microsporidian small-subunit rRNA gene. *Clin. Infect. Dis.* **23:**1002–1006.

De Groote, M. A., G. S. Visvesvara, M. L. Wilson, N. J. Pieniazek, S. B. Slemenda, A. J. Da Silva, G. J. Leitch, R. T. Bryan, and R. Reves. 1995. Polymerase chain reaction and culture confirmation of disseminated *Encephalitozoon cuniculi* in a patient with AIDS: successful therapy with albendazole. *J. Infect. Dis.* **171:**1375–1378.

Deplazes, P., A. Mathis, R. Baumgartner, I. Tanner, and R. Weber. 1996a. Immunologic and molecular characteristics of *Encephalitozoon*-like microsporidia isolated from humans and rabbits indicate that *Encephalitozoon cuniculi* is a zoonotic parasite. *Clin. Infect. Dis.* **22:**557–559.

Deplazes, P., A. Mathis, C. Muller, and R. Weber. 1996b. Molecular epidemiology of *Encephalitozoon cuniculi* and first detection of *Enterocytozoon bieneusi* in faecal samples of pigs. *J. Eukaryot. Microbiol.* **43:**93S.

Desportes, I., Y. Le Charpentier, A. Galian, F. Bernard, B. Cochand-Priollet, A. Lavergne, P. Ravisse, and R. Modigliani. 1985. Occurrence of a new microsporidian: *Enterocytozoon bieneusi* n. g., n. sp., in the enterocytes of a human patient with AIDS. *J. Protozool.* **32:**250–254.

Didier, E. S., P. J. Didier, D. N. Friedberg, S. M. Stenson, J. M. Orenstein, R. W. Yee, F. W. Tio, R. M. Davis, C. Vossbrinck, N. Millichamp, and J. A. Shadduck. 1991. Isolation and characterization of a new human microsporidian, *Encephalitozoon hellem* (n. sp.), from three AIDS patients with keratoconjunctivitis. *J. Infect. Dis.* **163:**617–621.

Didier, E. S., G. S. Visvesvara, M. D. Baker, L. B. Rogers, D. C. Bertucci, M. A. DeGroote, and C. R. Vossbrinck. 1996. A microsporidian isolated from an AIDS patient corresponds to *Encephalitozoon cuniculi* III, originally isolated from domestic dogs. *J. Clin. Microbiol.* **34:**2835–2837.

Didier, E. S., C. R. Vossbrinck, M. D. Baker, L. B. Rogers, D. C. Bertucci, and J. A. Shadduck. 1995a. Identification and characterization of three *Encephalitozoon cuniculi* strains. *Parasitology* **111:** 411–421.

Didier, P. J., E. S. Didier, and M. Murphey-Corb. 1995b. Experimental microsporidiosis in immunodeficient monkeys, *FASEB J.* **9:**A967, abstr. 5614.

Dore, G. J., D. J. Marriott, M. C. Hing, J. L. Harkness, and A. S. Field. 1995. Disseminated microsporidiosis due to *Septata intestinalis* in nine patients infected with the human immunodeficiency virus: response to therapy with albendazole. *Clin. Infect. Dis.* **21:**70–76.

Drobniewski, F., P. Kelly, A. Carew, B. Ngwenya, N. Luo, C. Pankhurst, and M. Farthing. 1995. Human microsporidiosis in African AIDS patients with chronic diarrhea. *J. Infect. Dis.* **171:**515–516.

Dunand, V. A., S. M. Hammer, R. Rossi, M. Poulin, M. A. Albrecht, J. P. Doweiko, P. C. DeGirolami, E. Coakley, E. Piessens, and C. A. Wanke. 1997. Parasitic sinusitis and otitis in patients infected with human immunodeficiency virus: report of five cases and review. *Clin. Infect. Dis.* **25:**267–272

Enriquez, F. J., D. Taren, A. Cruz-López, M. Muramoto, J. D. Palting, and P. Cruz. 1998. Prevalence of intestinal encephalitozoonosis in Mexico. *Clin. Infect. Dis.* **26:**1227–1229.

Foudraine, N. A., G. J. Weverling, T. van Gool, M. T. L. Roos, F. de Wolf, P. P. Koopmans, P. J. van den Broek, P. L. Meenhorst, R. van Leeuwen, J. M. A. Lange, and P. Reiss. 1998. Improvement of chronic diarrhoea in patients with advanced HIV-1 infection during potent antiretroviral therapy. *AIDS* **12:**35–41.

Fuentealba, I. C., N. T. Mahoney, J. A. Shadduck, J. Harvill, V. Wicher, and K. Wicher. 1992. Hepatic lesions in rabbits infected with *Encephalitozoon cuniculi* administered per rectum. *Vet. Pathol.* **29:**536–540.

Goguel, J., C. Katlama, C. Sarfati, C. Maslo, C. Leport, and J. M. Molina. 1997. Remission of AIDS-associated intestinal microsporidiosis with highly active antiretroviral therapy. *AIDS* **11:** 1658–1659.

Goodman, H., R. T. Bryan, and D. A. Schwartz. 1998. Unpublished data.

Hautvast, J. L. A., J. J. M. Tolboom, T. J. M. M. Derks, P. Beckers, and R. W. Sauerwein. 1997. Asymptomatic intestinal microsporidiosis in a human immunodeficiency virus-seronegative, immunocompetent Zambian child. *Pediatr. Infect. Dis. J.* **16:**415–416.

Hollister, W. S., and E. U. Canning. 1987. An enzyme-linked immunosorbent assay (ELISA) for detection of antibodies to *E. cuniculi* and its use in determination of infections in man. *Parasitology* **94:**209–219.

Hutin, Y. J. F., M.-N. Sombardier, O. Liguory, C. Sarfati, F. Derouin, J. Modaï, and J.-M. Molina. 1998. Risk factors for intestinal microsporidiosis in patients with human immunodeficiency virus infection: a case-control study. *J. Infect. Dis.* **178:**904–907.

Kelkar, R., P. S. Sastry, S. S. Kulkarni, T. K. Saikia, P. M. Parikh, and S. H. Advani. 1997. Pulmonary microsporidial infection in a patient with CML undergoing allogenic marrow transplant. *Bone Marrow Transplant.* **19:**179–182.

Kyaw, T., A. Curry, V. Edwards-Jones, J. Craske, and B. K. Mandal. 1997. The prevalence of *Enterocytozoon bieneusi* in acquired immunodeficiency syndrome (AIDS) patients from the northwest of England: 1992–1995. *Br. J. Biomed. Sci.* **54:**186–191.

Lucas, S. B., L. Papadaki, C. Conlon, N. Sewankambo, R. Goodgame, and D. Serwadda. 1989. Diagnosis of intestinal microsporidiosis in patients with AIDS. *J. Clin. Pathol.* **42:**885–890.

Lucas, S. B., D. A. Schwartz, and P. A. Hasleton. 1996. Pulmonary parasitic diseases, p. 316–319. *In* P.S. Hasleton (ed.), *Spencer's Pathology of the Lung*, 5th ed. McGraw-Hill, New York.

Maiga, I., O. Doumba, M. Dembele, H. Traore, I. Desportes-Livage, I. Hilmarsdottir, E. Giboyau, L. Maiga, L. Kassambara, Y. el Fakhry, A. Datry, M. Gentilini, and E. Pichard. 1997. Human intestinal microsporidiosis in Bamako (Mali): the presence of *Enterocytozoon bieneusi* in HIV seropositive patients. *Sante* **7:**257–262.

Mansfield, K. G. 1998. Personal communication.

Mansfield, K. G., A. Carville, D. Shvetz, J. MacKey, S. Tzipori, and A. A. Lackner. 1997. Identification of an *Enterocytozoon bieneusi*-like microsporidian parasite in simian-immunodeficiency-virus-inoculated macaques with hepatobiliary disease. *Am. J. Pathol.* **150:**1395–1405.

Mathis, A., M. Michel, H. Kuster, C. Muller, R. Weber, and P. Deplazes. 1997. Two *Encephalitozoon cuniculi* strains of human origin are infectious to rabbits. *Parasitology* **1146:**29–35.

Molina, J. M., E. Oksenhendler, B. Beauvais, C. Sarfati, A. Jaccard, F. Derouin, and J. Modaï. 1995. Disseminated microsporidiosis due to *Septata intestinalis* in patients with AIDS: clinical features and response to albendazole therapy. *J. Infect. Dis.* **171:**245–249.

Morakote, N., P. Siriprasert, S. Piangjai, P. Vitayasai, B. Tookyan, and P. Uparanukraw. 1995. *Microsporidium* and *Cyclospora* in human stools in Chiang Mai, Thailand. *Southeast Asian J. Trop. Med. Public Health* **26:**799–800.

Pospisilova, Z., O. Ditrich, M. Stankova, and P. Kodym. 1997. Parasitic opportunistic infections in Czech HIV-infected patients: a prospective study. *Cent. Eur. J. Public Health* **5:**208–213.

Pulparampil, N., D. Graham, D. Phalen, and K. Snowden. 1998. *Encephalitozoon hellem* infection in an eclectus parrot: zoonotic potential? p. 64. *In* Proceedings of the International Conference on Emerging Infectious Diseases, Atlanta, Ga.

Rabodonirina, M., M. Bertocchi, I. Desportes-Livage, L. Cotte, H. Levrey, M. A. Piens, G. Monneret, M. Celard, J. F. Mornex, and M. Mojon. 1996. *Enterocytozoon bieneusi* as a cause of chronic diarrhea in a heart-lung transplant recipient who was seronegative for human immunodeficiency virus. *Clin. Infect. Dis.* **23:**114–117.

Raynaud, L., F. Delbac, V. Broussolle, M. Rabodonirina, V. Girault, M. Wallon, G. Cozon, C. P. Vivares, and F. Peyron. 1998. Identification of *Encephalitozoon intestinalis* in travelers with chronic diarrhea by specific PCR amplification. *J. Clin. Microbiol.* **36:**37–40.

Reetz, J. 1994. Natürliche Übertragung von Mikrosporidien (*Encephalitozoon cuniculi*) über das Hühnerei. *Tierärztl. Prax.* **22:**147–150.

Sandfort, J., A. Hannerman, H. Gelderblom, K. Stark, R. L. Owen, and B. Ruf. 1994. *Enterocytozoon bieneusi* infection in an immunocompetent patient who had acute diarrhea and who was not infected with the human immunodeficiency virus. *Clin. Infect. Dis.* **19:**514–516.

Sax, P. E., J. D. Rich, W. S. Pieciak, and Y. M. Trnka. 1995. Intestinal microsporidiosis occurring in a liver transplant recipient. *Transplantation.* **60:**617–618.

Scaglia, M., S. Gatti, L. Sacchi, S. Corona, G. Chichino, A. M. Bernuzzi, G. Barbarini, G. P. Croppo, A. J. Da Silva, N. J. Pieniazek, and G. S. Visvesvara. 1998. Asymptomatic respiratory tract microsporidiosis due to *Encephalitozoon hellem* in three patients with AIDS. *Clin. Infect. Dis.* **26:**174–176.

Schmidt, W., T. Schneider, W. Heise, J. D. Schulzke, T. Weinke, R. Ignatius, R. L. Owen, M. Zeitz, E. O. Riecken, and R. Ullrich. 1997. Mucosal abnormalities in microsporidiosis. *AIDS* **11:**1589–1594.

Schwartz, D. A., D. C. Anderson, S. A. Klumpp, and H. M. McClure. 1998a. Ultrastructure of atypical (teratoid) sporogonial stages of *Enterocytozoon bieneusi* (Microsporidia) in naturally infected rhesus monkeys (*Macacca mulatta*). *Arch. Pathol. Lab. Med.* **122:**423–429.

Schwartz, D. A, R. T. Bryan, K. O. Hewan-Lowe, G. S. Visvesvara, R. Weber, A. Cali, and P. Angritt. 1992. Disseminated microsporidiosis (*Encephalitozoon hellem*) and acquired immunodeficiency syndrome: autopsy evidence for respiratory acquisition. *Arch. Pathol. Lab. Med.* **116:**660–668.

Schwartz, D. A., M. Flepp, G. S. Visvesvara, R. T. Bryan, and R. Weber. 1998b. Unpublished data.

Schwartz, D. A., G. S. Visvesvara, M. C. Diesenhouse, R. Weber, R. L. Font, L. A. Wilson, G. Corrent, D. F. Rosberger, P. C. Keenen, H. E. Grossniklaus, K. O. Hewan-Lowe, and R. T. Bryan. 1993a. Ocular pathology of microsporidiosis: role of immunofluorescent antibody for diagnosis of *Encephalitozoon hellem* in biopsies, smears, and intact globes from seven AIDS patients. *Am. J. Ophthalmol.* **115:**285–292.

Schwartz, D. A., G. S. Visvesvara, G. J. Leitch, L. S. Tashjian, M. Pollack, and R. T. Bryan. 1993b. Pathology of symptomatic microsporidial (*Encephalitozoon hellem*) bronchiolitis in AIDS: a new respiratory pathogen diagnosed by biopsy, bronchoalveolar lavage, sputum cytology, and tissue culture. *Hum. Pathol.* **24:**937–943.

Schwartz, D. A., G. S. Visvesvara, R. Weber, and R. T. Bryan. 1994. Male genital tract microsporidiosis and AIDS: Prostatic abscess due to *Encephalitozoon hellem. J. Eukaryot. Microbiol.* **41:** 61S.

Seibold, H. R., and E. N. Fussell. 1973. Intestinal microsporidiosis in *Callicebus moloch. Lab. Anim. Sci.* **23:**115–118.

Shadduck, J. A., and G. Baskin. 1989. Serologic evidence of *Encephalitozoon cuniculi* infection in a colony of squirrel monkeys (*Saimiri sciureus*). *Lab. Anim. Sci.* **39:**328–330.

Silverstein, B. E., E. T. Cunningham, T. P. Margolis, V. Cevallos, and I. G. Wong. 1997. Microsporidial keratoconjunctivitis in a patient without human immunodeficiency virus infection. *Am. J. Ophthalmol.* **124:**395–396.

Singh, M., G. J. Kane, L. MacKinlay, I. Quaki, E. H. Yap, B. C. Ho, L. C. Ho, and L. C. Kim. 1982. Detection of antibodies to *Nosema cuniculi* (protozoa: microsporidia) in human and animal sera by the indirect fluorescent antibody technique. *Southeast Asian J. Trop. Med. Public Health* **13:** 110–113.

Sobottka, I., H. Albrecht, J. Schottelius, C. Schmetz, M. Bentfield, R. Laufs, and D. A. Schwartz. 1995. Self-limited traveller's diarrhea due to a dual infection with *Enterocytozoon bieneusi* and *Cryptosporidium parvum* in an immunocompetent HIV-negative child. *Eur. J. Clin. Microbiol. Infect. Dis.* **14:**919–920.

Sparfel, J. M., C. Sarfati, O. Ligoury, B. Caroff, N. Dumoutier, B. Gueglio, E. Billaud, F. Raffi, J. M. Molina, M. Miegeville, and F. Derouin. 1997. Detection of microsporidia and identification of *Enterocytozoon bieneusi* in surface water by filtration followed by specific PCR. *J. Eukaryot. Microbiol.* **44:**78S.

Tzipori, S., A. Carville, G. Widmer, D. Kotler, K. Mansfield, and A. Lackner. 1997. Transmission and establishment of a persistent infection of *Enterocytozoon bieneusi,* derived from a human with AIDS, in simian immunodeficiency virus-infected rhesus monkeys. *J. Infect. Dis.* **175:**1016–1020.

Van Gool, T., E. Luderhoff, K. J. Nathoo, C. F. Kiire, J. Dankert, and P. R. Mason. 1995. High prevalence of *Enterocytozoon bieneusi* infections among HIV-positive individuals with persistent diarrhea in Harare, Zimbabwe. *Trans. R. Soc. Trop. Med. Hyg.* **89:**478–480.

Van Gool, T., J. C. M. Vetter, B. Weinmayr, A. Van Dam, F. Derouin, and J. Dankert. 1997. High seroprevalence of *Encephalitozoon species* in immunocompetent subjects. *J. Infect. Dis.* **175:**1020–1024.

Voglino, M. C., G. Donelli, P. Rossi, A. Ludovisi, V. Rinaldi, F. Goffredo, R. Paloscia, and E. Pozio. 1996. Intestinal microsporidiosis in Italian individuals with AIDS. *Ital. J. Gastroenterol.* **28:** 381–386.

Waller, T. 1980. Sensitivity of *Encephalitozoon cuniculi* to various temperatures, disinfectants and drugs. *Lab. Anim. Sci.* **13:**277–280.

Wanke, C. A., P. DeGirolami, and M. Federman. 1996. *Enterocytozoon bieneusi* infection and diarrheal disease in patients who were not infected with human immunodeficiency virus: case report and review. *Clin. Infect. Dis.* **23:**816–818.

Watson, D. A. R., D. Asmuth, and C. A. Wanke. 1996. Environmental risk factors for acquisition of microsporidia in HIV-infected persons. *Clin. Infect. Dis.* **23:**903.

Weber, R., R. T. Bryan, D. A. Schwartz, and R. L. Owen. 1994. Human microsporidial infections. *Clin. Microbiol. Rev.* **7:**426–461.

Weiser, J. 1964. On the taxonomic position of the genus *Encephalitozoon* Levaditi, Nicolau & Schoen, 1923 (Protozoa: Microsporidia). *Parasitology* **54:** 749–751.

Weiser, J. 1965. Microsporidian infections of mammals and the genus *Encephalitozoon. Int. Congr. Parasitol.* **1:**445–446.

Weitz, J. C., R. Botehlo, and R. T. Bryan. 1995. Microsporidiosis in patients with chronic diarrhea and AIDS, in HIV asymptomatic patients and in patients with acute diarrhea. *Rev. Med. Chile* **123:** 849–856. (In Spanish.)

Wicher, V., R. E. Baughn, C. Fuentealba, J. A. Shadduck, F. Abbruscato, and K. Wicher. 1991. Enteric infection with an obligate intracellular parasite, *Encephalitozoon cuniculi,* in an experimental model. *Infect. Immun.* **59:**2225–2231.

Wright, J. H. and E. M. Craighead. 1922. Infectious motor paralysis in young rabbits. *J. Exp. Med.* **36:**135–140.

Wuhib, T., T. M. J. Silva, R. D. Newman, L. S. Garcia, M. L. D. Pereira, C. S. Chaves, S. P. Wahlquist, R. T. Bryan, R. L. Guerrant, A. Q. de Sousa, T. R. B. S. de Queiroz, and C. L. Sears. 1994. Cryptosporidial and microsporidial infections in human immunodeficiency virus-infected patients in northeastern Brazil. *J. Infect. Dis.* **170:**494–497.

Zemen, D. H. and G. B. Baskin. 1985. Encephalitozoonosis in squirrel monkeys (*Saimiri sciureus*). *Vet. Pathol.* **22:**24–31.

APPENDIX: CHECKLIST OF AVAILABLE GENERIC NAMES FOR MICROSPORIDIA WITH TYPE SPECIES AND TYPE HOSTS

Victor Sprague and James J. Becnel

This list includes all the generic names that have come to our attention and that are deemed to have met the criteria of availability as defined by the Code of Zoological Nomenclature.

1. *Abelspora* Azevedo, 1987. Type species *Abelspora portucalensis* Azevedo, 1987. Type host *Carcinus maenas* (Linnaeus) Leach, 1814 (Decapoda: Portunidae).

2. *Agglomerata* Larsson & Yan, 1988. Type species *Agglomerata sidae* (Jírovec, 1942) Larsson & Yan, 1988. Type host *Sida crystallina* (O. F. Mueller, 1785) (Cladocera: Sididae).

3. *Agmasoma* Hazard & Oldacre, 1975. Type species *Agmasoma penaei* (Sprague, 1950) Hazard & Oldacre, 1975. Type host *Penaeus setiferus* (Linnaeus) (Decapoda: Penaeidae).

4. *Alfvenia* Larsson, 1983. Type species *Alfvenia nuda* Larsson, 1983. Type host *Acanthocyclops vernalis* Fisher (Copepoda: Cyclopidae).

5. *Alloglugea* Paperna & Lainson, 1995. Type species *Alloglugea bufonis* Paperna & Lainson, 1995. Type host *Bufo marinus* Linnaeus (Anura: Bufonidae).

6. *Amblyospora* Hazard & Oldacre, 1975. Type species *Amblyospora californica* (Kellen & Lipa, 1960) Hazard & Oldacre, 1975. Type definitive host *Culex tarsalis* Coquillett (Diptera: Culicidae). Type intermediate host *Mesocyclops leukarti* (Claus, 1875) (Copepoda: Cyclopidae).

7. *Ameson* Sprague, 1977. Type species *Ameson michaelis* (Sprague, 1970) Sprague, 1977. Type host *Callinetes sapidus* (Rathbun, 1896) (Decapoda: Portunidae).

8. *Amphiacantha* Caullery & Mesnil, 1914. Type species *Amphiacantha longa* Caullery & Mesnil, 1914. Type host *Ophioidina elongata* Ming. "or related species" (Gregarinida) parasite of *Lumbriconereis tingens* (Polychaeta: Eunicidae).

9. *Amphiamblys* Caullery & Mesnil, 1914. Type species *Amphiamblys capitellidis* (Caullery & Mesnil, 1897) Caullery & Mesnil, 1914. Type host *Ancora* sp. (Gregarinida) parasite of *Capitellides giardi* (Polychaeta: Eunicidae).

10. *Anncaliia* Issi, Krylova & Nicolaeva, 1993. Type species *Anncaliia melagethi* (Issi & Radishcheva, 1979) Issi, Krylova & Nicolaeva, 1993. Type host *Meligethes aenus* (Coleoptera: Nitidulidae).

Victor Sprague, Chesapeake Biological Laboratory, Solomons, MD 20688. *James J. Becnel,* USDA/ARS Center for Medical, Agricultural, and Veterinary Entomology, Gainesville, FL 32604.

11. *Auraspora* Weiser & Purrini, 1980. Type species *Auraspora canningae* Weiser & Purrini, 1980. Type host *Lepidocyrtus lignorum* Fabricius, 1781 (Collembola: Entomobryidae).

12. *Bacillidium* Janda, 1928. Type species *Bacillidium criodrilli* Janda, 1928. Type host *Criodrilus lacuum* Hoffmeister (Oligochaeta).

13. *Baculea* Loubes & Akbarieh, 1978. Type species *Baculea daphniae* Loubes & Akbarieh, 1978. Type host *Daphnia pulex* (de Geer, 1778) (Cladocera: Daphniidae).

14. *Berwaldia* Larsson, 1981. Type species *Berwaldia singularis* Larsson, 1981. Type host *Daphnia pulex* (de Geer, 1778) (Cladocera: Daphniidae).

15. *Binucleospora* Bronnvall & Larsson, 1995. Type species *Binucleospora elongata* Bronnvall & Larsson, 1985. Type host *Candona* sp. (Ostracoda: Cyprididae).

16. *Bohuslavia* Larsson, 1985. Type species *Bohuslavia asterias* (Weiser, 1963) Larsson, 1985. Type host *Endochironomus* sp. (Diptera: Chironomidae).

17. *Brachiola* Cali, Takvorian & Weiss, 1998. Type species *Brachiola vesicularum* Cali, Takvorian & Weiss, 1998, in Cali et al., 1998. Type host *Homo sapiens* Linnaeus (Primates: Hominidae).

18. *Burenella* Jouvenaz & Hazard, 1978. Type species *Burenella dimorpha* Jouvenaz & Hazard, 1978. Type host *Solenopsis geminata* (Fabricius) (Isoptera: Formicidae).

19. *Burkea* Sprague, 1977. Type species *Burkea gatesi* (Putyorac & Tourret, 1963) Sprague, 1977. Type host *Pheretima hawayana* (Oligochaeta: Megascolecidae) selected here from two hosts mentioned.

20. *Buxtehudea* Larsson, 1980. Type species *Buxtehudea scaniae* Larsson, 1980. Type host *Petrobius brevistylis* Carpenter, 1913 (Thysanura: Machilidae).

21. *Campanulospora* Issi, Radischcheva & Dolzhenko, 1983. Type species *Campanulospora denticulata* Issi, Radischcheva & Dolzhenko, 1983. Type host *Delia floralis* Fall. (Diptera: Muscidae).

22. *Canningia* Weiser, Wegensteiner & Zizka, 1995. Type species *Canningia spinidentis* Weiser, Wegensteiner & Zizka, 1995. Type host *Pityokteines spinidens* Rtt. (Coleoptera: Scolytidae).

23. *Caudospora* Weiser, 1946. Type species *Caudospora simulii* Weiser, 1946. Type host *Simulium hirtipes* (Fries, 1824) (Diptera: Simuliidae).

24. *Caulleryetta* Dogiel, 1922. Type species *Caulleryetta mesnili* Dogiel, 1922. Type host *Selenidium* sp. (Gregarinida: Schizocystidae) parasite of *Travisia forbesi* (Polychaeta).

25. *Chapmanium* Hazard & Oldacre, 1975. Type species *Chapmanium cirritus* Hazard & Oldacre, 1975. Type host *Corethrella brakeleyi* (Coquillett) (Diptera: Chaoboridae).

26. *Chytridioides* Trégouboff, 1913. Type species *Chytridioides schizophylli* Trégouboff, 1913. Type host *Schizophyllum mediterraneus* Latzel (Diplopoda).

27. *Chytridiopsis* Schneider, 1884. Type species *Chytridiopsis socius* Schneider, 1884. Type host *Blaps mortisaga* Linnaeus (Coleoptera: Tenebrionidae).

28. *Ciliatosporidium* Foissner & Foissner, 1995. Type species *Ciliatosporidium platyophryae* Foissner & Foissner, 1995. Type host *Platyophrya terricola* (Foissner, 1987) Foissner & Foissner, 1995 (Ciliophora: Colpodea).

29. *Coccospora* Kudo, 1925. Replacement name for *Cocconema* Léger & Hesse, 1921, preoccupied. Type species *Coccospora micrococcus* (Léger & Hesse, 1921) Kudo, 1925. Type host *Tanypus setiger* Kieffer (Diptera: Chironomidae).

30. *Cougourdella* Hesse, 1935. Type species *Cougourdella magna* Hesse, 1935. Type host *Megacyclops viridis* Jurine (Copepoda: Cyclopidae).

31. *Cristulospora* Khodzhaeva & Issi, 1989. Type species *Cristulospora sherbani* Khodzhaeva & Issi, 1989. Type host *Culex modestus* Ficalbi (Diptera: Culicidae).

32. *Cryptosporina* Hazard & Oldacre, 1975. Type species *Cryptosporina brachyfila* Hazard & Oldacre, 1975. Type host *Piona* sp. (Arachnida: Hygrobatinae).

33. *Culicospora* Weiser, 1977. Type species *Culicospora magna* (Kudo, 1920) Weiser, 1977.

Type host (?) *Culex pipiens* Linnaeus (Diptera: Culicidae).

34. *Culicosporella* Weiser, 1977. Type species *Culicosporella lunata* (Hazard & Savage, 1970) Weiser, 1977. Type host *Culex pilosus* (Dyar & Knab, 1906) (Diptera: Culicidae).

35. *Cylindrospora* Issi & Voronin, 1986. Type species *Cylindrospora chironomi* Issi & Voronin, 1986, in Issi, 1986. Type host *Chironomus plumous* Linnaeus (Diptera: Chironomidae).

36. *Cystosporogenes* Canning, Barker, Nicholas & Page, 1985. Type species *Cystosporogenes operophterae* (Canning, 1960) Canning, Barker, Nicholas & Page, 1985. Type host *Operophtera brumata* Linnaeus (Lepidoptera: Geometridae).

37. *Desportesia* Issi & Voronin, 1986. Type species *Desportesia lauberi* (Desportes & Theodorides, 1979) Issi & Voronin, 1986, in Issi, 1986. Type host *Lecudina* sp. (Gregarinida: Lecudinidae) parasite of unidentified marine annelid (Echiurida).

38. *Duboscqia* Perez, 1908. Type species *Duboscqia legeri* Perez, 1908. Type host *Termes lucifugus* = *Reticulitermes lucifugus* Rossi (Isoptera: Rhinotermitidae).

39. *Edhazardia* Becnel, Sprague & Fukuda, 1989. Type species *Edhazardia aedis* (Kudo, 1930) Becnel, Sprague & Fukuda, 1989. Type host *Aedes aegypti* Linnaeus (Diptera: Culicidae).

40. *Encephalitozoon* Levaditi, Nicolau & Schoen, 1923. Type species *Encephalitozoon cuniculi* Levaditi, Nicolau & Schoen, 1923. Type host is a rabbit (Lagomorpha: Leporidae).

41. *Endoreticulatus* Brooks, Becnel & Kennedy, 1988. Type species *Endoreticulatus fidelis* (Hostounsky & Weiser, 1975) Brooks, Becnel & Kennedy, 1988. Type host *Leptinotarsa undecimlineata* Stal. (Coleoptera: Chrysomelidae).

42. *Enterocytozoon* Desportes, Le Charpentier, Galian, Bernard, Cochand-Priollet, Lavergne, Ravisse & Modigliani, 1985. Type species *Enterocytozoon bieneusi* Desportes, Le Charpentier, Galian, Bernard, Cochand-Priollet, Lavergne, Ravisse & Modigliani, 1985. Type host *Homo sapiens* Linnaeus (Primates: Hominidae).

43. *Episeptum* Larsson, 1986. Type species *Episeptum inversum* Larsson, 1986. Type host *Holocentropus picicornis* (Stevens, 1836) (Trichoptera: Polycentropidae).

44. *Evlachovaia* Voronin & Issi, 1986. Type species *Evlachovaia chironomi* Voronin & Issi, 1986, in Issi, 1986. Type host *Chironomus plumosus* (Linnaeus) (Diptera: Chironomidae).

45. *Flabelliforma* Canning, Killick-Kendrick & Killick-Kendrick, 1991. Type species *Flabelliforma montana* Canning, Killick-Kendrick & Killick-Kendrick, 1991. Type host *Phlebotomus ariasi* Tonnoir, 1921 (Diptera: Psychodidae).

46. *Geusia* Rühl & Korn, 1979. Type species *Geusia gamocysti* Rühl & Korn, 1979. Type host *Gamocystis ephemerae* Frantzius, 1848 (Gregarinida: Gregarinidae), parasite of *Ephemera danica* (Ephemeroptera: Ephemeridae).

47. *Glugea* Thélohan, 1891. Type species *Glugea anomala* (Moniez, 1887) Gurley, 1893. Type host *Gasterosteus aculeatus* Linnaeus (Gasterosteiformes: Gasterosteidae).

48. *Glugoides* Larsson, Ebert, Vávra & Voronin, 1996. Type species *Glugoides intestinalis* (Chatton, 1907) Larsson, Ebert, Vávra & Voronin, 1996. Type host *Daphnia magna* Straus, 1820 (Cladocera: Daphniidae), selected here from two hosts mentioned.

49. *Golbergia* Weiser, 1977. Junior objective synonym *Diffingeria* Issi, 1979. Type species *Golbergia spinosa* (Golberg, 1971) Weiser, 1977. Type host *Culex pipiens* Linnaeus (Diptera: Culicidae).

50. *Gurleya* Doflein, 1898. Type species *Gurleya tetraspora* Doflein, 1898. Type host *Daphnia maxima* (Cladocera: Daphniidae).

51. *Gurleyides* Voronin, 1986. Type species *Gurleyides biformis* Voronin, 1986. Type host *Ceriodaphnia reticulata* Jurine (Cladocera: Daphniidae).

52. *Hazardia* Weiser, 1977. Type species *Hazardia milleri* (Hazard & Fukuda, 1974) Weiser, 1977. Type host *Culex pipiens quinquefasciatus* Say, 1823 (Diptera: Culicidae).

53. *Helmichia* Larsson, 1982. Type species *Helmichia aggregata* Larsson, 1982. Type host *Endochironomus* sp. (Diptera: Chironomidae).

54. *Hessea* Ormières & Sprague, 1973. Type species *Hessea squamosa* Ormières & Sprague, 1973. Type host *Sciara* sp. (Diptera: Lycoriidae).

55. *Heterosporis* Schubert, 1969. Type species *Heterosporis finki* Schubert, 1969. Type host *Pterophyllum scalare* (Curs & Valens, 1831) (Perciformes: Cichlidae).

56. *Heterovesicula* Lange, Macvean, Henry & Streett, 1995. Type species *Heterovesicula cowani* Lange, Macvean, Henry & Streett, 1995. Type host *Anabrus simplex* Haldeman, 1852 (Orthoptera: Tettigoniidae).

57. *Hirsutusporos* Batson, 1983. Type species *Hirsutusporos austrosimulii* Batson, 1983. Type host *Austrosimulium* sp. (Diptera: Simuliidae).

58. *Holobispora* Voronin, 1986. Type species *Holobispora thermocyclopis* Voronin, 1986. Type host *Thermocyclops orthonoides* (Sars) (Copepoda: Cyclopidae).

59. *Hrabyeia* Lom & Dykova, 1990. Type species *Hrabyeia xerkophora* Lom & Dykova, 1990. Type host *Nais christinae* Kasparzak, 1973 (Oligochaeta: Naididae).

60. *Hyalinocysta* Hazard & Oldacre, 1975. Type species *Hyalinocysta chapmani* Hazard & Oldacre, 1975. Type host *Culiseta melanura* Coquillett, 1902 (Diptera: Culicidae).

61. *Ichthyosporidium* Caullery & Mesnil, 1905. Type species *Ichthyosporidium giganteum* (Thélohan, 1895) Swarczewsky, 1914. Type host *Crenilabrus melops* Linnaeus (Pisces).

62. *Inodosporus* Overstreet & Weidner, 1974. Type species *Inodosporous spraguei* Overstreet & Weidner, 1974. Type host *Palaemonetes pugio* Holthius, 1949 (Decapoda: Palaemonidae).

63. *Intexta* Larsson, Steiner & Bjørnson, 1997. Type species *Intexta acarivora* Larsson, Steiner & Bjørnson, 1997. Type host *Tyrophagus putrescentiae* (Schrank, 1781) (Acarina: Acaridae)

64. *Issia* Weiser, 1977. Type species *Issia trichopterae* (Weiser, 1946) Weiser, 1977. Type host *Plectrocnemia geniculata* (Tricoptera: Polycentropidae).

65. *Janacekia* Larsson, 1983. Type species *Janacekia debaisieuxi* (Jírovec, 1943) Larsson, 1983. Type host *Simulium maculatum* Meig. (Diptera: Simuliidae).

66. *Jirovecia* Weiser, 1977. Type species *Jirovecia caudata* (Léger & Hesse, 1916) Weiser, 1977. Type host *Tubifex tubifex* Mueller (Oligochaeta: Tubificidae).

67. *Jiroveciana* Larsson, 1980. Type species *Jiroveciana limnodrilli* (Jírovec, 1940) Larsson, 1980. Type host *Limnodrilus missionicus* (Oligochaeta: Tubificidae).

68. *Johenrea* Lange, Becnel, Razafindratiana, Przybyszewski & Razafindrafara, 1996. Type species *Johenrea locustae* Lange, Becnel, Razafindratiana, Przybyszewski & Razafindrafara, 1996. Type host *Locusta migratoria capito* (Saussure, 1884) (Orthoptera: Acridae).

69. *Kinorhynchospora* Adrianov & Rybakov, 1991. Type species *Kinorhynchospora japonica* Adrianov & Rybakov, 1991. Type host *Kinorhynchus yushini* Adrianov, 1989 (Homalorhagida: Pycnophyidae).

70. *Lanatospora* Voronin, 1986. Type species *Lanatospora macrocyclopis* (Voronin, 1977) Voronin, 1986. Type host *Macrocyclops albidus* Jurine (Copepoda: Cyclopidae).

71. *Larssonia* Vidtman & Sokolova, 1994. Type species *Larssonia obtusa* (Moniez, 1887) Vidtman & Sokolova, 1994. Type host *Daphnia pulex* de Geer (Cladocera: Daphniidae).

72. *Larssoniella* Weiser & David, 1997. Type species *Larssoniella resinellae* Weiser & David, 1997. Type host *Petrova resinella* (Linnaeus) (Lepidoptera: Tortricidae).

73. *Loma* Morrison & Sprague, 1981. Type species *Loma morhua* Morrison & Sprague, 1981. Type host *Gadus morhua* Linnaeus (Gadiformes: Gadidae).

74. *Mariona* Stempell, 1909. Type species *Mariona marionis* (Thélohan, 1895) Stempell, 1909. Type host *Ceratomyxa coris* Georgevitch, 1916 (Bivalvulida: Ceratomyxidae), parasite of *Coris julis* Linnaeus (Pisces).

75. *Marssoniella* Lemmermann, 1900. Type species *Marssoniella elegans* Lemmermann, 1900. Type host *Cyclops strenuus* Fischer, 1851 (Copepoda: Cyclopidae).

76. *Merocinta* Pell & Canning, 1993. Type species *Merocinta davidii* Pell & Canning, 1993. Type host *Mansonia africana* (Theobald) (Diptera: Culicidae).

77. *Metchnikovella* Caullery & Mesnil, 1897. Type species *Metchnikovella spionis* Caullery & Mesnil, 1897. Type host *Polyrhabdina brasili* Caullery & Mesnil (Gregarinida: Lecudinidae) parasite of *Spio martinensis* Mesnil (Polychaeta: Spionidae).

78. *Microfilum* Faye, Toguebaye & Bouix, 1991. Type species *Microfilum lutjani* Faye, Toguebaye & Bouix, 1991. Type host *Lutjanus fulgens* (Valenciennes, 1830) (Perciformes: Lutjanidae).

79. *Microgemma* Ralphs & Matthews, 1986. Type species *Microgemma hepaticus* Ralphs & Matthews, 1986. Type host *Chelon labrosus* (Risso) (Pisces).

80. *Microsporidium* Balbiani, 1884. Not an available name sensu stricto but used under the provisions of the Code (see Glossary) as the legitimate name of a collective group. Useful as a provisional generic name if an author desires to record an unidentified species or to form a binomen and establish a new species while there is indecision about the genus.

81. *Microsporidyopsis* Schereschewsky, 1925. Type species *Microsporidyopsis nereidis* Schereschewsky, 1925. Type host *Dolyocystis* sp. (Gregarinida) parasite of *Nereis parallelogramma* Claparede (Polychaeta: Nereidae).

82. *Mitoplistophora* Codreanu, 1966. Type species *Mitoplistophora angularis* Codreanu, 1966. Type host *Ephemera danica* (Ephemeroptera: Ephemeridae).

83. *Mrazekia* Léger & Hesse, 1916. Type species *Mrazekia argoisi* Léger & Hesse, 1916. Type host *Asellus aquaticus* Linnaeus (Isopoda: Asellidae).

84. *Myxocystis* Mrazek, 1897. Type species *Myxocystis ciliata* Mrazek, 1897. Type host *Limnodrilus claparedianus* Ratzel (Oligochaeta: Tubificidae).

85. *Nadelspora* Olson, Tiekotter & Reno, 1994. Type species *Nadelspora canceri* Olson, Tiekotter & Reno, 1994. Type host *Cancer magister* Dana, 1852 (Decapoda: Cancridae).

86. *Napamichum* Larsson, 1990. Type species *Napamichum dispersus* (Larsson, 1984) Larsson, 1990a. Type host *Endochironomus* sp. (Diptera: Chironomidae).

87. *Nelliemelba* Larsson, 1983. Type species *Nelliemelba boeckella* (Milner & Mayer, 1982) Larsson, 1983. Type host *Boeckella triarticulata* (Thomson) (Copepoda: Calinoidea).

88. *Neoperezia* Issi & Voronin, 1979. Type species *Neoperezia chironomi* Issi & Voronin, 1979. Type host *Chironomus plumosus* Linnaeus (Diptera: Chironomidae).

89. *Neonosemoides* Faye, Toguebaye & Bouix, 1996. Type species *Neonosemoides tilapiae* (Sakiti & Bouix, 1987) Faye, Toguebaye & Bouix, 1996. Type host *Tilapia guineensis* (Perciformes: Cichlidae).

90. *Nolleria* Beard, Butler & Becnel, 1990. Type species *Nolleria pulcis* Beard, Butler & Becnel, 1990. Type host *Cteocephalides felis* (Boche, 1833) (Siphonaptera: Pulicidae).

91. *Norlevinea* Vávra, 1984. Type species *Norlevinea daphniae* Vávra, 1984. Type host *Daphnia longispina* O. F. Mueller (Cladocera: Daphniidae).

92. *Nosema* Naegeli, 1857. Type species *Nosema bombycis* Naegeli, 1857. Type host *Bombyx mori* Linnaeus (Lepidoptera: Bombycidae).

93. *Nosemoides* Vinckier, 1975. Type species *Nosemoides vivieri* (Vinckier, Devauchelle & Prensier, 1970) Vinckier, 1975. Type host *Lecudina linei* Vinckier, 1975 (Gregarinida: Monocystidae), parasite of *Lineus viridis* (Fabricius) (Heteronemertea: Lineidae).

94. *Nucleospora* Docker, Kent & Devlin, 1996. Type species *Nucleospora salmonis* (Chilmonczyk, Cox & Hedrick, 1991) Docker, Kent & Devlin, 1996. Type host *Orcorhynchus tshawytscha* (Walbalm) (Salmoniformes: Salmonidae).

95. *Nudispora* Larsson, 1990. Type species *Nudispora biformis* Larsson, 1990b. Type host *Coenagrion hastulatum* Charpentier, 1925 (Odonata: Coenagrionidae).

96. *Octosporea* Flu, 1911. Type species *Octosporea muscaedomesticae* Flu, 1911. Type host *Musca domestica* Linnaeus (Diptera: Muscidae).

97. *Oligosporidium* Codreanu-Bălcescu, Codreanu & Traciuc, 1981. Type species *Oligosporidium arachnicolum* (Codreanu-Bălcescu, Codreanu & Traciuc, 1978) Codreanu-Bălcescu, Codreanu & Traciuc, 1981. Type host *Xysticus cambridgei* (Aranea: Thomisidae).

98. *Ordospora* Larsson, 1997. Type species *Ordospora colligata* Larsson, 1997. Type host *Daphnia magna* Straus, 1820 (Cladocera: Daphniidae).

99. *Ormieresia* Vivares, Bouix & Manier, 1977. Type species *Ormieresia carcini* Vivares, Bouix & Manier, 1977. Type host *Carcinus mediterraneus* Czerniavsky, 1884 (Decapoda: Portunidae).

100. *Orthosomella* Canning, Wigley & Barker, 1991. Replacement name for *Orthosoma* Canning, Wigley & Barker, 1983, preoccupied. Type species *Orthosomella operophterae* (Canning, 1960) Canning, Wigley & Barker, 1991. Type host *Operophtera brumata* (Linnaeus) (Lepidoptera: Geometridae).

101. *Orthothelohania* Codreanu & Bălcescu-Codreanu, 1974. Type species *Orthothelohania octospora* (Henneguy, 1892, in Henneguy & Thélohan, 1892, sensu Pixel-Goodrich, 1920) Codreanu & Bălcescu-Codreanu, 1974. Type host *Palaemon serratus* (Pennant, 1777) (Decapoda: Palaemonidae).

102. *Ovavesicula* Andreadis & Hanula, 1987. Type species *Ovavesicula popilliae* Andreadis & Hanula, 1987. Type host *Popilla japonica* Newman (Coleoptera: Scarabaeidae).

103. *Parathelohania* Codreanu, 1966. Type species *Parathelohania legeri* (Hesse, 1904a, 1904b) Codreanu, 1966. Type host *Anopheles maculipennis* Meigen, 1818 (Diptera: Culicidae).

104. *Pegmatheca* Hazard & Oldacre, 1975. Type species *Pegmatheca simulii* Hazard & Oldacre, 1975. Type host *Simulium tuberosum* Lindstrom, 1911 (Diptera: Simuliidae).

105. *Perezia* Léger & Duboscq, 1909. Type species *Perezia lankesteriae* Léger & Duboscq, 1909. Type host *Lankesteria ascidiae* (Lankester, 1872) (Gregarinida: Diplocystidae), parasite of *Ciona intestinalis* (Linnaeus) (Dictyobranchia: Ascidiidae).

106. *Pernicivesicula* Bylén & Larsson, 1994. Type species *Pernicivesicula gracilis* Bylén & Larsson, 1994. Type host *Pentaneurella* sp. Fittkau & Murray, 1983 (Diptera: Chironomidae).

107. *Pilosporella* Hazard & Oldacre, 1975. Type species *Pilosporella fishi* Hazard & Oldacre, 1975. Type host *Wyeomyia vanduzeei* Dyar & Knab, 1906 (Diptera: Culicidae).

108. *Pleistophora* Gurley, 1893. Type species *Pleistophora typicalis* Gurley, 1893. Type host *Cottus scorpius* = *Myoxocephalus scorpius* (Linnaeus) (Perciformes: Cottidae).

109. *Pleistophoridium* Codreanu-Bălcescu & Codreanu, 1982. Type species *Pleistophoridium hyperparasiticum* (Codreanu-Bălcescu & Codreanu, 1976) Codreanu-Bălcescu & Codreanu, 1982. Type host *Enterocystis rhithrogenae* M. Codreanu, 1940 (Gregarinida: Monocystidae), parasite of *Rhithrogena semicolorata* (Curt, 1834) (Ephemeroptera).

110. *Polydispyrenia* Canning & Hazard, 1982. Type species *Polydispyrenia simulii* (Lutz & Splendore, 1908) Canning & Hazard, 1982. Type host *Simulium venustum* Say = *Simulium pertinax* Kollar (Diptera: Simuliidae).

111. *Pseudopleistophora* Sprague, 1977. Type species *Pseudopleistophora szollosii* Sprague, 1977. Type host *Armandia brevis* (Polychaeta).

112. *Pulcispora* Vedmed, Krylova & Issi, 1991. Type species *Pulcispora xenopsyllae* Vedmed, Krylova & Issi, 1991. Type host *Xenopsylla hirtipes* (Siphonaptera: Pulicidae).

113. *Pyrotheca* Hesse, 1935. Type species *Pyrotheca cyclopis* (Leblanc, 1930) Poisson, 1953. Type host *Cyclops albidus* Jurine, 1820 (Copepoda: Cyclopidae).

114. *Rectispora* Larsson, 1990. Type species *Rectispora reticulata* Larsson, 1990c. Type host *Pomatothrix hammonienis* (Michaelson, 1901) (Oligochaeta: Tubificidae).

115. *Resiomeria* Larsson, 1986. Type species *Resiomeria odonatae* Larsson, 1986b. Type host *Aeshna grandis* (Odonata: Aeshnidae).

116. *Ringueletium* Garcia, 1990. Type species *Ringueletium pilosa* Garcia, 1990. Type host *Gigantodox rufidulum* Wigodzinsky & Coscaron (Diptera: Simuliidae).

117. ***Scipionospora*** **Bylén & Larsson, 1996.** Type species *Scipionospora tetraspora* (Léger & Hesse, 1922) Bylén & Larsson, 1996. Type host *Tanytarsus* sp. Léger & Hesse, 1922 (Diptera: Chironomidae).

118. ***Semenovaia*** **Voronin & Issi, 1986.** Type species *Semenovaia chironomi* Voronin & Issi, 1986, in Issi, 1986. Type host *Chironomus plumosus* (Linnaeus) (Diptera: Chironomidae).

119. ***Septata*** **Cali, Kotler & Orenstein, 1993.** Type species *Septata intestinalis* Cali, Kotler & Orenstein 1993. Type host *Homo sapien* Linnaeus (Primates: Hominidae).

120. ***Spherospora*** **Garcia, 1991.** Type species *Spherospora andinae* Garcia, 1991. Type host *Gigantodox chilense* (Philippi) (Diptera: Simuliidae).

121. ***Spiroglugea*** **Léger & Hesse, 1924.** Replacement name for *Spironema* Léger & Hesse, 1922, preoccupied. Junior objective synonyms *Spirospora* Kudo, 1925, and *Spirillonema* Wenyon, 1926. Type species *Spiroglugea octospora* (Léger & Hesse, 1922) Léger & Hesse, 1924. Type host *Ceratopogon* sp. (Diptera: Ceratopogonidae).

122. ***Spraguea*** **Weissenberg, 1976.** Type species *Spraguea lophii* (Doflein, 1898) Weissenberg, 1976. Type host *Lophius piscatorius* (Lophiiformes: Lophiidae).

123. ***Steinhausia*** **Sprague, Ormières & Manier, 1972.** Type species *Steinhausia mytilovum* (Field, 1924) Sprague, Ormières & Manier, 1972. Type host *Mytilus edulis* Linnaeus (Pelecypoda: Mitilidae).

124. ***Stempellia*** **Léger & Hesse, 1910.** Type species *Stempellia mutabilis* Léger & Hesse, 1910. Type host *Ephemera vulgata* Linnaeus (Ephemeroptera: Ephemeridae).

125. ***Striatospora*** **Issi & Voronin, 1986.** Type species *Striatospora chironomi* Issi & Voronin, 1986, in Issi, 1986. Type host *Chironomus plumosus* (Linnaeus) (Diptera: Chironomidae).

126. ***Systenostrema*** **Hazard & Oldacre, 1975.** Type species *Systenostrema tabani* Hazard & Oldacre, 1975. Type host *Tabanus lineola* Fabricius (Diptera: Tabanidae).

127. ***Tardivesicula*** **Larsson & Bylén, 1992.** Type species *Tardivesicula duplicata* Larsson & Bylén, 1992. Type host *Limnephilus centralis* (Curtis, 1884) (Trichoptera: Limnephilidae).

128. ***Telomyxa*** **Léger & Hesse, 1910.** Type species *Telomyxa glugeiformis* Léger & Hesse, 1910. Type host *Ephemera vulgata* Linnaeus (Ephemeroptera: Ephemeridae).

129. ***Tetramicra*** **Matthews & Matthews, 1980.** Type species *Tetramicra brevifilum* Matthews & Matthews, 1980. Type host *Scophthalmus maximus* (Linnaeus) (Pleuronectiformes: Bothridae).

130. ***Thelohania*** **Henneguy, 1892.** Type species *Thelohania giardi* Henneguy, 1892, in Henneguy & Thélohan, 1892. Type host *Crangon vulgaris* (Decapoda: Crangonidae).

131. ***Toxoglugea*** **Léger & Hesse, 1924.** Replacement name for *Toxonema* Léger & Hesse, 1922, preoccupied. Junior objective synonym *Toxospora* Kudo, 1925. Type species *Toxoglugea vibrio* (Léger & Hesse, 1922) Léger & Hesse, 1924. Type host *Ceratopogon* sp. (Diptera: Ceratopogonidae).

132. ***Toxospora*** **Voronin, 1993.** Type species *Toxospora volgae* Voronin, 1993. Type host *Corynoneura* sp. (Diptera: Chironomidae).

133. ***Trachipleistophora*** **Hollister, Canning, Weidner, Field, Kench & Marriott, 1996.** Type species *Trachipleistophora hominis* Hollister, Canning, Weidner, Field, Kench & Marriott, 1996. Type host *Homo sapiens* Linnaeus (Primates: Hominidae).

134. ***Trichoctosporea*** **Larsson, 1994.** Type species *Trichoctosporea pygopellita* Larsson, 1994. Type host *Aedes vexans* (Meig.) (Diptera: Culicidae).

135. ***Trichoduboscqia*** **Léger, 1926.** Type species *Trichoduboscqia epeori* Léger, 1926. Type host *Epeorus torrentium* Eat. (Ephemeroptera: Heptageniidae).

136. ***Trichotuzetia*** **Vávra, Larsson & Baker, 1997.** Type species *Trichotuzetia guttata* Vávra, Larsson & Baker, 1997. Type host *Cyclops vicinus* Uljanin, 1875 (Copepoda: Cyclopidae).

137. ***Tricornia*** **Pell & Canning, 1992.** Type species *Tricornia muhezae* Pell & Canning, 1992. Type host *Mansonia africana* (Theobald) (Diptera: Culicidae).

138. *Tuzetia* Maurand, Fize, Fenwik & Michel, 1971. Type species *Tuzetia infirma* (Kudo, 1921) Maurand, Fize, Fenwick & Michel, 1971. Type host *Cyclops albidus* (Jurine, 1820) (Copepoda: Cyclopidae).

139. *Unikaryon* Canning, Lai & Lie, 1974. Type species *Unikaryon piriformis* Canning, Lai & Lie, 1974. Type host *Echinostoma audyi* Umathevy (Digenea: Echinostomidae), parasite of *Lymnaea rubiginosa* (Pulmonata: Lymnaeidae).

140. *Vairimorpha* Pilley, 1976. Type species *Vairimorpha necatrix* (Kramer, 1965) Pilley, 1976. Type host *Pseudaletia unipunctata* (Hayworth) (Lepidoptera: Noctuidae).

141. *Vavraia* Weiser, 1977. Type species *Vavraia culicis* (Weiser, 1947) Weiser, 1977. Type host *Culex pipiens* Linnaeus (Diptera: Culicidae).

142. *Vittaforma* Silveira & Canning, 1995. Type species *Vittaforma corneae* (Shadduck, Meccoli, Davis & Font, 1990) Silveira & Canning, 1995. Type host *Homo sapiens* Linnaeus (Primates: Hominidae).

143. *Weiseria* Doby & Saguez, 1964. Type species *Weiseria laurenti* Doby & Saguez, 1964. Type host *Prosimulium inflatum* (Davies) (Diptera: Simuliidae).

144. *Wittmannia* Czaker, 1997. Type species *Wittmannia antarctica* Czaker, 1997. Type host *Kantharella antiarctica* Czaker, 1997 (Mesozoa: Dicyemida: Kantharellidae), parasite of *Pareledone turqueti* Joubin, 1905 (Cephalopoda).

ACKNOWLEDGMENTS

We are indebted to Margaret (Peg) Johnson for assistance in preparing the manuscript.

This is contribution 3122 from the University of Maryland Center for Environmental Science.

REFERENCES

Adrianov, A. V., and A. V. Rybakov. 1991. *Kinorhynchospora japonica* gen. n., sp. n. (Microsporidia) from the intestinal epithelium of *Kinorhynchus yushini* (Homalorhagida, Pycnophyidae). *Zool. Zh.* **70:**5–11.

Andreadis, T. G., and J. L. Hanula. 1987. Ultrastructural study and description of *Ovavesicula popilliae* n. g., n. sp. (Microsporida: Pleistophoridae) from the Japenese beetle, *Popilla japonica* (Coleoptera: Scarabaeidae). *J. Protozool.* **34:**15–21.

Azevedo, C. 1987. Fine structure of the microsporidan *Abelspora portucalensis* gen. n., sp. n. (Microsporida) parasite of the hepatopancreas of *Carcinus maenas* (Crustacea, Decapoda). *J. Invertebr. Pathol.* **49:**83–92.

Balbiani, G. 1884. *Leçons sur les Sporozoaires.* Octave Doin, Paris.

Batson, B. S. 1983. A light and electron microscopic study of *Hirsutusporos austrosimulii* gen. n., sp. n., (Microspora: Nosematidae), a parasite of *Austrosimulium* sp. (Diptera: Simuliidae) in New Zealand. *Protistologica* **19:**263–280.

Beard, B., J. F. Butler, and J. J. Becnel. 1990. *Nolleria pulicis* n. gen., n. sp. (Microsporida: Chytridiopsidae), a microsporidian parasite of the cat flea *Ctenocephalides felis* (Siphonaptera: Pulicidae). *J. Protozool.* **37:**90–99.

Becnel, J. J., V. Sprague, T. Fukuda, and E. I. Hazard. 1989. Development of *Edhazardia aedis* (Kudo, 1930) n. g., n. comb. (Microsporida: Amblyosporidae) in the mosquito *Aedes aegypti* (L.) (Diptera: Culicidae). *J. Protozool.* **36:**119–130.

Bronnvall, A. M., and J. I. R. Larsson. 1995. Description of *Binucleospora elongata* gen. et sp. nov. (Microspora, Caudosporidae), a microsporidian parasite of ostracods of the genus *Candona* (Crustacea, Cyprididae) in Sweden. *Eur. J. Protistol.* **31:**63–72.

Brooks, W. M., J. J. Becnel, and G. G. Kennedy. 1988. Establishment of *Endoreticulatus* n. g. for *Pleistophora fidelis* (Hostounsky & Weiser, 1975) (Microsporida: Pleistophoridae) based on the ultrastructure of a microsporidium in the Colorado potato beetle, *Leptinotarsa decemlineata* (Say) (Coleoptera: Chrysomelidae). *J. Protozool.* **35:** 481–488.

Bylén, E. K. C., and J. I. R. Larsson. 1994. Ultrastructural study and description of *Pernicivesicula gracilis* gen. et sp. nov. (Microspora, Pereziidae), a rod-shaped microsporidium of midge larvae, *Pentaneurella* sp. (Diptera, Chironomidae), in Sweden. *Eur. J. Protistol.* **30:**139–150.

Bylén, E. K. C., and J. I. R. Larsson. 1996. Ultrastructural study and description of *Mrazekia tetraspora* Léger & Hesse, 1922 and transfer to a new genus *Scipionospora* n.g. (Microspora, Caudosporidae). *Eur. J. Protistol.* **32:**104–115.

Cali, A., D. P. Kotler, and J. M. Orenstein. 1993. *Septata intestinalis* n. g., n. sp., an intestinal microsporidian associated with chronic diarrhea and dissemination in AIDS patients. *J. Eukaryot. Microbiol.* **40:**101–112.

Cali, A., P. M. Takvorian, S. Lewin, M. Rendel, C. Sian, M. Wittner, H. B. Tanowitz, E. Keo-

hane, and L. M. Weiss. 1998. *Brachiola vesicularum,* a new AIDS microsporidium. *J. Eukaryot. Microbiol.* **45:**240–251.

Canning, E. U. 1960. Two new microsporidian parasites of the winter moth, *Operophtera brumata* (L.). *J. Parasitol.* **46:**755–763.

Canning, E. U., R. J. Barker, J. P. Nicholas, and A. M. Page. 1985. The ultrastructure of three microsporidia from the winter moth, *Operophtera brumata* (L.), and the establishment of a new genus *Cystosporogenes* n. g. for *Pleistophora operophterae* (Canning, 1960). *Syst. Parasitol.* **7:**213–225.

Canning, E. U., and E. I. Hazard. 1982. Genus *Pleistophora* Gurley, 1893: an assemblage of at least three genera. *J. Protozool.* **29:**39–49.

Canning, E. U., R. Killick-Kendrick, and M. Killick-Kendrick. 1991. A new microsporidian parasite, *Flabelliforma montana* n. g., n. sp., infecting *Phlebotomus ariasi* (Diptera, Psychodidae) in France. *J. Invertebr. Pathol.* **57:**71–81.

Canning, E. U., P. F. Lai, and K. J. Lie. 1974. Microsporidian parasites of trematode larvae from aquatic snails in West Malaysia. *J. Protozool.* **21:**19–25.

Canning, E. U., P. J. Wigley, and R. J. Barker. 1983. The taxonomy of three species of microsporidia (Protozoa: Microspora) from an oakwood population of winter moths *Operophtera brumata* (L.) (Lepidoptera: Geometridae). *Syst. Parasitol.* **5:**147–159.

Canning, E. U., P. J. Wigley, and R. J. Barker. 1991. *Orthosomella* nomen novum for the junior homonym *Orthosoma* Canning, Wigley and Barker, 1983. *J. Invertebr. Pathol.* **58:**464.

Caullery, M., and F. Mesnil. 1897. Sur un type nouveau (*Metchnikovella* n. g.) d'organism parasites des gregarines. *C. R. Acad. Sci.* **125:**787–790.

Caullery, M., and F. Mesnil. 1905. Recherches sur les Haplosporidies. *Arch. Zool. Exp. Gen.* **4:**101–181.

Caullery, M., and F. Mesnil. 1914. Sur les Metchnikovellidae et autres protistes parásites des grégarines d'annélides. *C. R. Soc. Biol.* **77:**527–532.

Chatton, E. 1907. Revue des parasites et des commensaux des Cladoceres: observations sur des formes nouvelles ou peu connues. *C. R. Assoc. Fr. Avan. Sci.* 797-811.

Chilmonczyk, S., W. T. Cox, and R. P. Hedrick. 1991. *Enterocytozoon salmonis* n. sp.: an intranuclear microsporidium from salmonid fish. *J. Protozool.* **38:**264–269.

Codreanu, R. 1966. On the occurrence of spore or sporont appendages in the microsporidia and their taxonomic significance, p. 602–603. *In Proceedings of the 1st International Congress on Parasitology.* Pergamon Press, New York, N.Y.

Codreanu, R., and D. Bălcescu-Codreanu. 1974. On the morphology and ultrastructure of the microsporidian *Thelohania octospora* Henneguy, 1892, parasitic in the prawn *Palaemon serratus* (Pennant) 1777 from the Atlantic French coast: need for a revision of its taxonomic status, p. 15–16. *In Proceedings of the 3rd International Congress on Parasitology.* Egermann Druckereigesellschaften, Vienna, Austria.

Codreanu-Bălcescu, D., and R. Codreanu. 1976. L'ultrastructure d'une Microsporidie hyperparasite d'une grégarine d'Ephémère. *J. Protozool.* **23:**8A.

Codreanu-Bălcescu, D., and R. Codreanu. 1982. Sur la position taxonomique d'apres les caracteres ultrastructuraux d'une Microsporidie hyperparasite d'une Grégarine. *J. Protozool.* **29:**515.

Codreanu-Bălcescu, D., R. Codreanu, and E. Traciuc. 1978. Ultrastructural aspects of a microsporidian parasitizing the ovaries of Araneid. Abstr. VIth Int. Colloq. Invertbr. Pathol., Prague, abstr. 21.

Codreanu-Bălcescu, D., R. Codreanu, and E. Traciuc. 1981. Ultrastructural data on a microsporidian infesting the ovaries of an araneid. *J. Invertebr. Pathol.* **37:**28–33.

Czaker, R. 1997. *Wittmannia antarctica* n. g., n. sp. (Nosematidae), a new hyperparasite in the Antarctic dicyemid mesozoan *Kantharella antarctica*. *J. Eukaryot. Microbiol.* **44:**438–446.

Desportes, I., Y. Le Charpentier, A. Galian, F. B. Bernard, Cochand-Priollet, A. Lavergne, F. Ravisse, and R. Modigliani. 1985. Occurrence of a new microsporidian: *Enterocytozoon bieneusi* n. g., n. sp., in the enterocytes of a human patient with AIDS. *J. Protozool.* **32:**250–254.

Desportes, I., and J. Theodorides. 1979. Étude ultrastructurale d'*Amphiamblys laubieri* n. sp. (Microsporidie, Metchnikovellidae) parasite d'une gregarine (*Lecudina* sp.) d'un Echiurien abyssal. *Protistologica* **15:**435–457.

Doby, J. M., and F. Saguez. 1964. *Weiseria,* genre nouveau de microsporidies et *Weiseria laurenti* n. sp., parasite de larves de *Prosimulium inflatum* Davies, 1957 (Dipteres: Paranematoceres). *C. R. Acad. Sci.* **259:**3614–3617.

Docker, M. F., M. L. Kent, and R. H. Devlin. 1996. Ribosomal DNA sequence of *Nucleospora salmonis* (Microsporea): implications for phylogeny and nomenclature, abstr. 92, p. 102. *In Program and Abstracts of the Meeting of the American Society of Parasitologists and Society of Protozoologists.*

Doflein, F. 1898. Studien zur Naturgeschichte der Protozoen. III. Ueber Myxosporidien. *Zool. Jahrb. Abt. Anat.* **11:**281–350.

Dogiel, V. 1922. Sur un noveau genre de *Metchnikovellidae. Ann. Inst. Pasteur* **36:**574–577.

Faye, N., B. S. Toguebaye, and G. Bouix. 1991. *Microfilum lutjani* n. g. n. sp. (Protozoa: Microsporida), a gill parasite of the golden African snapper *Lutjanus fulgens* (Valenciennes, 1830) (Teleost Lutjanidae): developmental cycle and ultrastructure. *J. Protozool.* **38:**30–40.

Faye, N., B. S. Toguebaye, and G. Bouix. 1996. Ultrastructure and development of *Neonosemoides tilapiae* (Sakiti and Bouix, 1987) n. g., n. comb. (Protozoa, Microspora) from African cichlid fish. *Eur. J. Protistol.* **32:**320–326.

Field, I. A. 1924. Biology and economic value of the sea mussel *Mytilus edulis. Bull. U.S. Bur. Fish.* 1921–22 **38:**127–259.

Flu, P. 1911. Studien über die im Darm der Stubenfliege, *Musca domestica,* vorkommenden protozoaren Gebilde. *Zentbl. Bakteriol. Infektionskr. Hyg. Abt. I Orig.* **57:**522–535.

Foissner, I., and W. Foissner. 1995. *Ciliatosporidium platyophryae* nov. gen., nov. spec. (Microspora incerta sedis), a parasite of *Platyophrya terricola* (Ciliophora, Colpodea). *Eur. J. Protistol.* **31:**248–259.

Garcia, J. J. 1990. Un nuevo microsporidio patogeno de larvas de simulidos (Diptera: Simuliidae) *Ringueletium pillosa* gen. et sp. nov. (Microspora: Caudosporidiae). *Neotropica* **36:** 111–122.

Garcia, J. J. 1991. Estudios sobre el cicly de vida y ultrastructura de *Spherospora andinae* gen. et sp. nov. (Microspora. Thelohaniidae), un nuevo microsporidio de simulidos neotropicales. *Neotropica* **37:**15-23.

Golberg, A. M. 1971. Mikrosporidiozy komarov *Culex pipiens* L. *Med. Parasitol. Parazit. Bolezni* **2:** 204–207.

Gurley, R. 1893. On the classification of the Myxosporidia, a group of protozoan parasites infecting fishes. Art. 10. *Bull. U.S. Fish Comm.* 1891 **11:**407–420.

Hazard, E. I., and T. Fukuda. 1974. *Stempellia milleri* sp. n. (Microsporida: Nosematidae) in the mosquito *Culex pipiens quinquefasciatus* Say. *J. Protozool.* **21:**497–504.

Hazard, E. I., and S. W. Oldacre. 1975. Revision of Microsporidia (Protozoa) close to *Thelohania,* with descriptions of one new family, eight new genera and thirteen new species. *U.S. Dept. Agric. Tech. Bull.* **1530:**1–104 .

Hazard, E. I., and K. E. Savage. 1970. *Stempellia lunata* sp. n. (Microsporida: Nosematidae) in larvae of the mosquito *Culex pilosus* collected in Florida. *J. Invertebr. Pathol.* **15:**49–54.

Henneguy, F., and P. Thélohan. 1892. Myxosporidies parasites des muscles chez quelques crustacés décapodes. *Ann. Microgr.* **4:**617–641.

Hesse, E. 1904a. *Thelohania legeri* n. sp., microsporidie nouvelle, parasite des larves d'*Anopheles maculipennis* Meig. *C. R. Soc. Biol.* **57:**570–571.

Hesse, E. 1904b. Sur le développement de *Thelohania legeri* Hesse. *C. R. Soc. Biol.* **57:**571–572.

Hesse, E. 1935. Sur quelques microsporidies parasites de *Megacyclops viridis* Jurine. *Arch. Zool. Exp. Gen.* **75:**651–661.

Hollister, W. S., E. U. Canning, E. Weidner, A. S. Field, J. Kench, and D. J. Marriott. 1996. Development and ultrastructure of *Trachipleistophora hominis* n. g., n. sp. after in vitro isolation from an AIDS patient and inoculation into athymic mice. *Parasitology* **112:**143–154.

Hostounsky, Z., and J. Weiser. 1975. *Nosema polygrammae* sp. n. and *Plistophora fidelis* sp. n. (Microsporidia, Nosematidae) infecting *Polygramma undecimlineata* (Coleoptera: Chrysomelidae) in Cuba. *Vest. Cesk. Spol. Zool.* **39:**104–110.

Issi, I. V. 1979. Novyi rod mikrosporidii *Diffengeria (=Weiseria) spinosa* (Goldberg, 1971) iz komarov *Culex pipiens* (Diptera, Culicidae), p. 91–97. In *Sistematika i ekologia sporovinkov i knidosporidii,* vol. 78. Leningrad, Russia.

Issi, I. V. 1986. Microsporidia as a phylum of parasitic protozoa. *Protozoology* **10:**6–136.

Issi, I. V., S. V. Krylova, and V. M. Nicolaeva. 1993. The ultrastructure of the microsporidium *Nosema meligethi* and establishment of the new genus *Ancaliia. Parazitologiya* **27:**127. (In Russian with English summary.)

Issi, I. V., and D. F. Radishcheva. 1979. Microsporidiosis diseases of beetles, pests of Cruciferae crops in the Leningrad Region Natural control . *Tr. Vses. Nauchno–Issled. Inst. Zashch. Rast.* **46:**19–23. (In Russian with English summary.)

Issi, I. V., D. F. Radishcheva, and V. T. Dolzhenko. 1983. Microsporidia of flies of genus *Delia* (Diptera, Muscidae), harmful to farm crops. *Tr. Vses. Nauchno-Issled. Inst. Zashch. Rast.* **55:**3–9. (In Russian with English summary.)

Issi, I. V., and V. N. Voronin. 1979. The contemporary state of the problem on bispore genera of microsporidians. *Parazitologiya* **13:**150–158. (In Russian with English summary.)

Janda, V. 1928. Über Microorganismen aus der Leibeshöhle von *Criodrilus lacuum* Hoffm. und eigenartige Neubildungen in der Körperwand dieses Tieres. *Arch. Protistenkd.* **63:**84–93.

Jírovec, O. 1940. Zur Kenntnis einiger in Oligochäten parasitierenden Protisten. I. *Arch. Protistenkd.* **94:** 80–92.

Jírovec, O. 1942. Zur Kenntnis einiger Cladoceren-Parasiten. II. *Zool. Anz.* **140:**129–133.

Jírovec, O. 1943. Revision der in *Simulium* Larven parasitierenden Mikrosporidien. *Zool. Anz.* **142:** 173–179.

Jouvenaz, D. P., and E. I. Hazard. 1978. New family, genus, and species of microsporida (Protozoa: Microsporida) from the tropical fire ant, *Solenopsis* *geminata* (Fabricius) (Insecta: Formicidae). *J. Protozool.* **25:**24–29.

Kellen, W. R., and J. J. Lipa. 1960. *Thelohania californica* n. sp., a microsporidian parasite of *Culex tarsalis* Coquillet. *J. Invertebr. Pathol.* **2:**1–12.

Khodzhaeva, L. F., and I. V. Issi. 1989. New genus of microsporidiae, *Cristulospora* gen. n. (Amblyosporidae), with three new species from bloodsucking mosquitoes from the Uzbekistan. *Parazitologiya* **23:**140–145. (In Russian with English summary.)

Kramer, J. P. 1965. *Nosema necatrix* sp. n. and *Thelohania diazoma* sp. n., microsporidians from the armyworm *Pseudaletia unipuncta* (Haworth). *J. Invertebr. Pathol.* **7:**117–121.

Kudo, R. 1920. On the structure of some microsporidian spores. *J. Parasitol.* **6:**178–182.

Kudo, R. 1921. Microsporidia parasitic in copepods. *J. Parasitol.* **7:**137–143.

Kudo, R. 1925. Microsporidia. *Science* **61:**366.

Kudo, R. 1930. Studies on microsporidia parasitic in mosquitoes. VIII. On a microsporidian, *Nosema aedis* nov. spec., parasitic in a larva of *Aedes aegypti* of Puerto Rico. *Arch. Protistenkd.* **69:**23–28.

Lange, C. E., J. J. Becnel, E. Razafindratiana, J. Przybyszewski, and H. Razafindrafara. 1996. *Johenrea locustae* n. g., n. sp. (Microspora: Glugeidae): a pathogen of migratory locusts (Orthoptera: Acrididae: Oedipodinae) from Madagascar. *J. Invertebr. Pathol.* **68:**28–40.

Lange, C. E., C. M. Macvean, J. E. Henry, and D. A. Streett. 1995. *Heterovesicula cowani* n. g., n. sp. (Heterovesiculidae n. fam.), a microsporidian parasite of Mormon crickets, *Anabrus simplex* Haldeman, 1852 (Orthoptera: Tettigoniidae). *J. Eukaryot. Microbiol.* **42:**552–558.

Larsson, J. I. R. 1980. Insect pathological investigations on Swedish Thysanura. II. A new microsporidian parasite of *Petrobius brevistylis* (Microcoryphia, Machilidae): description of the species and creation of two new genera and a new family. *Protistologica* **16:**85–101.

Larsson, J. I. R. 1981. A new microsporidium *Berwaldia singularis* gen. et sp. nov. from *Daphnia pulex* and a survey of microsporidia described from Cladocera. *Parasitology* **83:**325–342.

Larsson, J. I. R. 1982. Cytology and taxonomy of *Helmichia aggregata* gen. et sp. nov. (Microspora, Thelohaniidae), a parasite of *Endochironomus* larvae (Diptera, Chironomidae). *Protistologica* **18:**355–370.

Larsson, J. I. R. 1983. A revisionary study of the taxon *Tuzetia* Maurand, Fize, Fenwick and Michel, 1971, and related forms (Microspora, Tuzetiidae). *Protistologica* **19:**323–355.

Larsson, J. I. R. 1984. Ultrastructural study and description of *Chapmanium dispersus* n. sp. (Microspora, Thelohaniidae) a microsporidian parasite of *Endochironomus* larvae (Diptera, Chironomidae). *Protistologica* **20:**547–563.

Larsson, J. I. R. 1985. On the cytology, development and systematic position of *Thelohania asterias* Weiser, 1963, with creation of the new genus *Bohuslavia* (Microspora, Thelohaniidae). *Protistologica* **21:**235–248.

Larsson, J. I. R. 1986a. Ultracytology of a tetrasporoblastic microsporidium of the caddisfly *Holocentropus picicornis* (Trichoptera, Polycentropodidae), with description of *Episeptum inversum* gen. et sp. nov. (Microspora, Gurleyidae). *Arch. Protistenkd.* **131:**257–280.

Larsson, J. I. R. 1986b. Ultrastructural investigation of two microsporidia with rod-shaped spores, with descriptions of *Cylindrospora fasciculata* sp. nov. and *Resiomeria odonatae* gen. et sp. nov. (Microspora, Thelohaniidae). *Protistologica* **22:**379–398.

Larsson, J. I. R. 1990a. Description of a new microsporidium of the water mite *Limnochares aquatica* and establishment of the new genus *Napamichum* (Microspora, Thelohaniidae). *J. Invertebr. Pathol.* **55:**152–161.

Larsson, J. I. R. 1990b. On the cytology and taxonomic position of *Nudispora biformis* n. g., n. sp. (Microspora, Thelohaniidae), a microsporidian parasite of the dragon fly *Coenagrion hastulatum* in Sweden. *J. Protozool.* **37:**310–318.

Larsson, J. I. R. 1990c. *Rectispora reticulata* gen. et sp. nov. (Microspora, Bacillidiidae), a new microsporidian parasite of *Pomatothrix hammoniensis* (Michaelsen, 1901) (Oligochaeta, Tubificidae). *Eur. J. Protistol.* **26:**55–64.

Larsson, J. I. R. 1994. *Trichoctosporea pygopellita* gen. et sp. nov. (Microspora, Thelohaniidae), a microsporidian parasite of the mosquito *Aedes vexans* (Diptera, Culicidae). *Arch. Protistenkd.* **144:**147–161.

Larsson, J. I. R., and E. K. C. Bylén. 1992. *Tardivesicula duplicata* gen. et sp. nov. (Microspora, Duboscqiidae), a microsporidian parasite of the caddis fly *Limnephilus centralis* (Trichoptera, Limnephilidae) in Sweden. *Eur. J. Protistol.* **28:**25–36.

Larsson, J. I. R., D. Ebert, J. Vávra, and V. N. Voronin. 1996. Redescription of *Pleistophora intestinalis* Chatton, 1907, a microsporidian parasite of *Daphnia magna* and *Daphnia pulex,* with establishment of the new genus *Glugoides* (Microspora, Glugeidae). *Eur. J. Protistol.* **32:**251–261.

Larsson, J. I. R., M. Y. Steiner, and S. Bjørnson. 1997. *Intexta acarivora* gen. et sp. n. (Microspora: Chytridiopsidae): ultrastructural study and description of a new microsporidian parasite of the forage mite *Tyrophagus putrescentiae* (Acari: Acaridae). *Acta Protozool.* **36:**295–304.

Larsson, J. I. R., and N. D. Yan. 1988. The ultrastructural cytology and taxonomy of *Duboscqui sidae,* with establishment of the new genus *Agglomerata* gen. nov. *Arch. Protistenkd.* **135:**271–288.

Leblanc, L. 1930. Deux microsporidies nouvelles des copepodes: *Gurleya cyclopis* et *Plistophora cyclopis. Ann. Soc. Sci. Brux. Ser. B* **59:**272–275.

Léger, L. 1926. Une microsporidie nouvelle à sporontes èpineux. *C. R. Acad. Sci.* **182:**727–729.

Léger, L., and O. Duboscq. 1909. *Perezia lankesteriae,* n. g., n. sp., microsporidie parasite de *Lankesteria ascidae* (Ray-Lank). *Arch. Zool. Exp. Gen. Ser.* **5,1, N. et R.:**89–93.

Léger, L., and E. Hesse. 1910. Cnidosporidies des larves d' éphémères. *C. R. Acad. Sci.* **150:**411–414.

Léger, L., and E. Hesse. 1916. *Mrazekia,* genre nouveau de microsporidies a spore tubuleuses. *C. R. Soc. Biol.* **79:**345–348.

Léger, L., and E. Hesse. 1921. Microsporidies a spores spheriques. *C. R. Acad. Sci.* **173:**1419–1421.

Léger, L., and E. Hesse. 1922. Microsporidies bactériformes et essai de sytematique du groupe. *C. R. Acad. Sci.* **174:**327–330.

Léger, L., and E. Hesse. 1924. Microsporidies nouvelles parasites des animaux d'eau douce. *Trav. Lab. Hydrobiol. Pisc. Univ. Grenoble* **14:**49–56.

Lemmermann, E. 1900. Beiträge zur Kenntnis der Planktonalgen. VII. *Ber. Dtsch. Bot. Ges.* **18:**135–143.

Levaditi, C., S. Nicolau, and R. Schoen. 1923. Nouvelles donnees sur l'*Encephalitozoon cuniculi. C. R. Soc. Biol.* **89:**1157–1162.

Lom, J., and I. Dykova. 1990. *Hrabyeia xerkophora* n. gen. n. sp., a new microsporidian with tailed spores from the oligochaete *Nais christinae* Kasparzak, 1973. *Eur. J. Protistol.* **25:**243–248.

Loubes, C., and M. Akbarieh. 1978. Étude ultrastructurale de la microsporidie *Baculaea daphniae* n. g., n. sp., parasite de l'épithélium intestinal de *Daphnia pulex* Leydig, 1860 (Crustace, Cladocere). *Protistologica* **14:**23–38.

Lutz, A., and A. Splendore. 1908. Ueber Pebrine und verwandte Mikrosporidien. Zweite Mitteilung. *Centralbl. Bakteriol. Parasitenkd. Infekt. Hyg. Abt. 1. Orig.* **46:**311–315.

Matthews, R. A., and B. F. Matthews. 1980. Cell and tissue reactions of turbot *Scophthalmus maximus* (L.) to *Tetramicra brevifilum* gen. n., sp. n. (Microspora). *J. Fish Dis.* **3:**495–515.

Maurand, J., A. Fize, B. Fenwick, and R. Michel. 1971. Étude au microscope électronique de *Nosema infirmum* Kudo, 1921, microsporidie parasite d'un copépode cyclopoïde: création du genre nouveau *Tuzetia* a propos de cette espece. *Protistologica* **7:**221–225.

Milner, R. J., and J. A. Mayer. 1982. *Tuzetia boeckella* sp. nov. (Protozoa: Microsporida), a parasite of *Boeckella triarticulata* (Copepoda: Calanoidea) in Australia. *J. Invertebr. Pathol.* **39:**174–184.

Moniez, R. 1887. Observations pour la revision des Microsporidies. *C. R. Acad. Sci.* **104:**1312–1314.

Morrison, C. M., and V. Sprague. 1981. Electron microscopical study of a new genus and new species of microsporida in the gills of Atlantic cod *Gadus morhua* L. *J. Fish Dis.* **4:**15–32.

Mrazek, A. 1897. Über eine neue Sporozoenform aus *Limnodrilus. Sitzungsber. Bohm. Ges. Wiss. Math.-Naturwiss. Cl.* **8:**1-5.

Naegeli, C. 1857. Über die neue Krankheit der Seidenraupe und verwandte Organismen. *Bot. Zeit.* **15:**760–761.

Olson, R. E., K. L. Tiekotter, and P. W. Reno. 1994. *Nadelspora canceri* n. g., n. sp., an unusual microsporidian parasite of the Dungeness crab, *Cancer magister. J. Eukaryot. Microbiol.* **41:**349–359.

Ormières, R., and V. Sprague. 1973. A new family, new genus, and new species allied to the Microsporida. *J. Invertebr. Pathol.* **21:**224–240.

Overstreet, R. M., and E. Weidner. 1974. Differentiation of microsporidian spore tails in *Inodosporus spraguei* gen. et sp. n. *Z. Parasitenkd.* **44:**169–186.

Paperna, I., and R. Lainson. 1995. *Alloglugea bufonis* nov. gen., nov. sp. (Microsporea: Glugeidae), a microsporidian of *Bufo marinus* tadpoles and metamorphosing toads (Amphibia: Anura) from Amazonian Brazil. *Dis. Aquat. Org.* **23:**7–16.

Pell, J. K., and E. U. Canning. 1992. Ultrastructure of *Tricornia muhezae* n. g., n. sp. (Microspora, Thelohaniidae), a parasite of *Mansonia africana* (Diptera, Culicidae) from Tanzania. *J. Protozool.* **39:**242–247.

Pell, J. K., and E. U. Canning. 1993. Ultrastructure and life cycle of *Merocinta davidii* gen. et sp. nov., a dimorphic microsporidian parasite of *Mansonia africana* (Diptera, Culicidae) from Tanzania. *J. Invertebr. Pathol.* **61:**267–274.

Pérez, C. 1908. Sur *Duboscqia legeri,* microsporidie nouvelle parasite de *Termes lucifugus* et sur la classification des microsporidies. *C. R. Soc. Biol.* **65:** 631–633.

Pilley, B. M. 1976. A new genus, *Vairimorpha* (Protozoa: Microsporida), for *Nosema necatrix* Kramer 1965: Pathogenicity and life cycle in *Spodoptera exempta* (Lepidoptera: Noctuidae). *J. Invertebr. Pathol.* **28:**177–183.

Pixell-Goodrich, H. L. M. 1920. The spore of *Thelohania. Arch. Zool. Exp. Gen. N. & R.* **59:**17–19.

Poisson, R. 1953. Ordre des Microsporidies, p. 1042-1070. *In* P. P. Grasse (ed.), *Traité de Zoologie,* vol. 1. Masson et Cie, Paris, France.

Puytorac, P., and M. Tourret. 1963. Étude de kystes d'origine parasitaire (Microsporidies ou Grégarines) sur la paroi interne du corps des Vers

Megasclocidae. *Ann. Parasitol. Hum. Comp.* **38:** 861–874.

Ralphs, J. R., and R. A. Matthews. 1986. Hepatic microsporidiosis of juvenile grey mullet, *Chelon labrosus* (Risso), due to *Microgemma hepaticus* gen. nov. sp. nov. *J. Fish Dis.* **9:**225–242.

Rühl, H. and H. Korn. 1979. Ein Mikrosporidier, *Geusia gamocysti* n. gen., n. sp. als Hyperparasit bei *Gamocystis ephemerae. Arch. Protistenkd.* **121:**349–355.

Schereschewsky, H. 1925. La famille Mechnikovellidae (C. & M.) et la plaie quelle occupe dans le systéme des protistes. *Arch. Russ. Protistenkd. Moscow* **3:**137–145.

Schneider, A. 1884. Sur le développement du *Stylorhynchus longicollis. Arch. Zool. Exp. Gen.* **2:**1–36.

Schubert, G. 1969. Ultracytologische Untersuchungen an der Spore der Mikrosporidienart, *Heterosporis finki,* gen. n., sp. n. *Z. Parasitenkd.* **32:** 59–79.

Shadduck, J. A., R. A. Meccoli, R. Davis, and R. L. Font. 1990. Isolation of a microsporidian from a human patient. *J. Infect. Dis.* **162:**773–776.

Silveira, H., and E. U. Canning. 1995. In vitro cultivation of the human microsporidium *Vittaforma corneae:* development and effect of albendazole. *Folia Parasitol.* **42:**241–250.

Sprague, V. 1950. Notes on three microsporidian parasites of decapod Crustacea of Louisiana coastal waters. *Occ. Pap. Mar. Lab. La. State Univ.* **5:**1–8.

Sprague, V. 1970. Some protozoan parasites and hyperparasites in marine decapod Crustacea. *Am. Fish. Soc. Spec. Publ.* **5:**416–430.

Sprague, V. 1977. Systematics of the Microsporidia. *In* L. A. Bulla and T. C. Cheng (ed.), *Comparative Pathobiology,* vol. 2. Plenum Press, New York, N.Y.

Sprague, V., R. R. Ormières, and J. F. Manier. 1972. Creation of a new genus and a new family in the Microsporida. *J. Invertebr. Pathol.* **20:**228–231.

Stempell, W. 1909. Über *Nosema bombycis* Naegeli. *Arch. Protistenkd.* **16:**281–358.

Swarczewsky, B. 1914. Über den Lebenscyclus einiger Haplosporidien. *Arch. Protistenkd.* **33:**49–108.

Thélohan, P. 1891. Sur deux sporozoaires nouveaux, parasites des poissons. *C. R. Acad. Sci.* **112:**168–171.

Thélohan, P. 1895. Recherches sur les Myxosporidies. *Bull. Sci. Fr. Belg.* **26:**100–394.

Trégouboff, G. 1913. Sur un Chytridiopside nouveau, *Chytridioides schizophylli* n. g., n. sp., parasite de l'intestin de *Schizophyllum mediterraneum Latzel. Arch. Zool. Exp. Gen. N. et R.* **52:**25–31.

Vávra, J. 1984. *Norlevinea* n. g., a new genus for *Glugea daphniae* (Protozoa: Microspora), a parasite of *Daphina longispina* (Crustacea: Phyllopoda). *J. Protozool.* **31:**508–513.

Vávra, J., J. I. R. Larsson, and M. D. Baker. 1997. Light and electron microscopic cytology of *Trichotuzetia guttata* gen. et sp. n. (Microspora, Tuzetiidae), a microsporidian parasite of *Cyclops vicinus* Uljanin, 1875 (Crustacea, Copepoda). *Arch. Protistenkd.* **147:**293–306.

Vedmed, A. I., S.V. Krylova, and I.V. Issi. 1989. The microsporidium *Pulicipora xenopsyllae* gen. n., sp. n. from fleas of the genus *Xenopsylla. Parazitologiya* **25:**13–19. (In Russian with English summary.)

Vidtmann, S. S., and Y.V. Sokolova. 1994. The description of the new genus *Larssonia* gen. n. based on the ultrastructural analysis of microsporidium (*Pleistophora*) *obtrusa* from *Daphnia pulex* (Cladocera). *Parazitologiya* **28:**202–213. (In Russian with English summary.)

Vinckier, D. 1975. *Nosemoides* gen. n. *N. vivier* (Vinckier, Devauchelle and Prensier, 1970) comb. nov. (Microsporidie): etude de la différenciation sporoblastique et genèse des différentes structures de la spore. *J. Protozool.* **22:**170–184.

Vinckier, D., G. Devauchelle, and G. Prensier. 1970. *Nosema vivieri* n. sp. (Microsporidae, Nosematidae) hyperparasite d'une grégarine vivant dans le caelome d'une némerte. *C. R. Acad. Sci.* **270:**821–823.

Vivares, C. P., G. Bouix, and J. F. Manier. 1977. *Ormieresia carcini* gen. n., sp. n., microsporidie de crabe Mediterraneen, *Carcinus mediterraneus* Czerniavsky, 1884: cycle evolutif et étude ultrastructurale. *J. Protozool.* **24:**83–94.

Voronin, V. N. 1977. Microsporidians (Protozoa, Microsporidia) of Entomostraca from waters of the Leningrad region. *Parazitologiya* **11:**505–512. (In Russian.)

Voronin, V. N. 1986. The microsporidia of crustaceans. *Protozoology (Leningrad)* **10:**137–166. (In Russian.)

Voronin, V. N. 1993. The microsporidium *Toxospora volgae* gen. n., sp. n. from chironomidae larvae of the genus *Corynoneura. Parazitologiya* **27:**148–154. (In Russian with English summary.)

Weiser, J. 1946. The microsporidia of insect larvae. *Vest. Cesk. Zool. Spol.* **10:**245–272.

Weiser, J. 1947. Klíč k určování Mikrosporidií. *Acta Soc. Sci. Nat. Morav.* **18:**1–64.

Weiser, J. 1963. Zur Kenntnis der Mikrosporidien aus *Chironomiden* Larven. III. *Zool. Anz.* **170:**226–230.

Weiser, J. 1977. Contribution to the classification of microsporidia. *Vest. Cesk. Spol. Zool.* **41:**308–321.

Weiser, J., and L. David. 1997. A light and electron microscopic study of *Larssoniella resinellae* n. gen., n. sp. (Microspora, Unikaryonidae), a parasite of *Petrova resinella* (Lepidoptera, Tortricidae) in Central Europe. *Arch. Protistenkd.* **147:**405–410.

Weiser, J., and K. Purrini. 1980. Seven new microsporidian parasites of springtails (Collembola) in the Federal Republic of Germany. *Z. Parasitenkd.* **62:**75–84.

Weiser, J., R. Wegensteiner, and Z. Zizka. 1995. *Canningia spinidentis* gen. et sp. n. (Protista: Microspora), a new pathogen of the fir bark beetle *Pityokteines spinidens. Folia Parasitol.* **42:**1–10.

Weissenberg, R. 1976. Microsporidian interactions with host cells. p. 203-237. *In* L. A. Bulla and T. C. Cheng (ed.), *Comparative Pathobiology,* vol. 2. Plenum Press, New York, N.Y.

Wenyon, C. M. 1926. *Protozoology,* vol. 1, p. 734–755. William Wood and Co., New York, N.Y.

GLOSSARY

Victor Sprague and James J. Becnel

Kudo (1924) presented the first glossary for the microsporidia. It contained only 30 entries but served well for nearly the next half-century, a period during which technological limitations permitted very little progress in the acquisition of detailed knowledge of these minute organisms. Eventually, the application of new technologies, especially electron microscopy, ushered in a period of revolutionary advancement in the knowledge of the morphology and life cycle of the microsporidia. Two of the early publications on electron microscope observations are especially noteworthy for having contributed very significantly to knowledge of the cytology of microsporidia and for introducing much of the basic terminology. Huger (1960) studied *Nosema locustae* Canning, 1953, revealing for the first time, in much detail but somewhat indistinctly, the accurate internal structure of a microsporidian spore. Vávra (1965) studied four species and made a contribution of monumental importance, presenting illustrations unexcelled in quality and revealing for the first time many cytological details of various developmental stages. Soon there was an expanding flow of studies involving the use of electron microscopy and other new technologies. For a period of time, authors necessarily introduced many new terms when publishing the new information, and readers at that time were generally unsure about how to interpret many of the terms and to select the most suitable of the many synonyms. Vávra and Sprague (1976) published the next glossary, one containing 135 entries and many synonyms. This work was essentially a comprehensive list of terms that had been used with the microsporidia, not a list of selected terms well defined and purported to be acceptable for general usage.

Sprague et al. (1992), in a comprehensive taxonomic study, published the next and last glossary, one containing only 47 entries and prepared mainly to clarify the terms used elsewhere in the same publication. Many of the terms listed in previously published glossaries, especially that by Vávra and Sprague, are now obsolete, and some are inaccurately defined.

We have attempted to prepare a glossary that includes most of the basic terms that are

Victor Sprague, Chesapeake Biological Laboratory, Solomons, MD 20688. *James J. Becnel,* USDA/ARS, Center for Medical, Agricultural and Veterinary Entomology, Gainesville, FL 32604.

The Microsporidia and Microsporidiosis (Murray Wittner, editor; Louis M. Weiss, contributing editor), ©1999 American Society for Microbiology, Washington, D.C.

currently in general use and have meanings that are more or less agreed upon. These terms relate primarily to morphology, life cycle, and parasite-host cell relations. Terms that relate primarily to such specialized fields as biochemistry, molecular biology, and disease are not included. Most of the basic terms can be further clarified by consulting two publications of Vávra (1976a, 1976b).

Acapsulate. Without a capsule. Refers particularly to a xenoma that lacks a capsule. Term newly adopted by the authors.

Alternation of generations. "Biol. the alternation in the life cycle of an organism of forms produced in a different manner, esp. the alternation of sexual with asexual generations." Dictionary definition. More specifically in the microsporidia, the alternation of a series of dihaplophasic (i.e., sexual) generations in the life cycle with a series of haplophasic (i.e., asexual) generations. Alternation of generations is a familiar concept for many plant and animal species. First proposed for a microsporidian species by Becnel et al. (1987) in *Culicospora magna* (Kudo, 1920).

Anchoring disk. A complex membrane, demonstrated by electron microscopy and appearing as a vertical section of an umbrella, continuous with the outer covering of the polar filament and presumed to anchor the everted filament to the spore during discharge of the sporoplasm. Coined by Vávra (1971).

Anterior end. Refers particularly to the end of the spore from which the polar filament becomes everted. Paraphrase of Kudo (1924).

Autoinfection. Transmission of an infection from one cell to another in the same host individual, in the same or different tissue, by inoculation with a sporoplasm ejected from an autoinfective spore. Term of undetermined origin.

Autoinfective spore. One of a class of spores designated by function. A spore whose specific function is to inject its sporoplasm into another host cell of the same or a different tissue. Probably quite common in the microsporidia. Term of undetermined origin.

Binary division. Dividing of a cell into two nearly equal daughter individuals. Common term of undetermined origin.

Binucleate. Having two haploid nuclei. Useful in characterizing a spore without reference to whether it has a diplokaryon or two nuclei from a dissociated diplokaryon or two nuclei that were never associated as a diplokaryon.

Budding. Dividing of an individual, uninucleate or multinucleate, into two individuals of distinctly unequal size. Common term of undetermined origin.

Capsulate. Having a capsule. Used here only with reference to a xenoma covered with a capsule, the classic example being the *Glugea* xenoma. Newly applied to the microsporidia.

Capsule. Literally, diminutive of *capsa,* Latin for box. A common and very general term that has been applied to a great variety of unrelated covering devices, including several in the microsporidia. Weissenberg (1968) applied the term to a wall that surrounds the *Glugea* xenoma, and some other authors have used it in the same sense. It can be reserved for this structure on a xenoma because covering devices that appear elsewhere now have special names.

Chromosome cycle. "Haploidy and diploidy in the successive phases of the life cycle and the processes of change, if any, from one number of chromosome sets to another. As used here, it has no reference to specific numbers of chromosomes" (Sprague et al., 1992).

Cistellae. Flattened vesicles of uniform size and clustered around the polar filament in the genus *Burkea.* Thought to be homologous with the polaroplast of most genera. Coined by Sprague et al. (1992).

Dihaplophase. The part of a microsporidian life cycle in which each individual is diplokaryotic. Coined by Naville (1931).

Dihaplophasic. Diplokaryotic.

Dimorphosporous. Having two distinctly different morphological types of spores during the life cycle. Used by Sprague et al. (1992) as a replacement for the ambiguous term "dimorphic."

Diplokaryon. A pair of haploid nuclei more or less intimately associated and functioning together as a diploid nucleus. Probably first applied to microsporidia by Debaissieux and Gastaldi (1919). "Dikaryon" of some authors, "... the genetic and physiological equivalent of the diploid and differs from a diplophase primarily in the competence of the haploid nuclei to act independently. Nuclear fusion is ultimately required for meiosis to occur. . . ." (Raper and Flexer, 1970).

Diplokaryotic. Having one or more diplokarya or pertaining to the association of two nuclei in a diplokaryon. Dihaplophasic.

Diplophase. The succession of stages in a life cycle that have a diploid nucleus. Limited to the zygote in microsporidia.

Diplosis. "Doubling of the chromosome number." (Sharp, 1934).

Disporoblastic. Pertaining to a sporont or to sporogony that produces two sporoblasts. Modification by Kudo (1924).

Distal. "Situated away from the point of attachment of origin." Dictionary definition. Here recommended for use, rather than "posterior," in reference to the polar filament and parts of the polaroplast.

Endospore. The chitinous inner spore coat. Term used in the past and with a different meaning for Sporozoa. Adopted by Vávra (1968) for microsporidia.

Exospore. The proteinaceous outer spore coat. Proposed by Vávra (1968).

Extracorporeal spore. A functional designation for a spore that lives for some time outside the body of its host before transmitting its sporoplasm to another host individual. New term, inspired by Kramer's (1976) treatise on "The Extra-Corporeal Ecology of Microsporidia."

Extrusion apparatus. A complex of spore organelles that injects the sporoplasm into a host cell. "It is composed of polar sac, polar aperture, polaroplast, polar tube and posterior vacuole." Term coined by Weidner (1972) who gave a detailed account of microsporidian invasion into host cells.

Gametogony. "Formation of gametes, often by schizogony" (Levine, 1971). First demonstrated for the microsporidia by Hazard et al. (1985) near the end of haplophase and designated "gametogenesis." Use of the latter term with microsporidia is inappropriate because it usually implies gamete production accompanied by meiosis.

Golgi apparatus. Vávra first demonstrated the Golgi apparatus in microsporidia (Vávra, 1965) and discussed the structure and function in detail (Vávra, 1976a). "The . . . structure has little resemblance to stacks of flattened saccules of typical Golgi." It often appears in cytoplasm of microsporidia as one or more groups of small, opaque vesicles limited by a single membrane and situated in a homogeneous matrix more dense than the surrounding cytoplasm. During sporogenesis, it becomes conspicuous and plays a primary role in elaborating the extrusion apparatus.

Haplophase. The part of the life cycle in which all the nuclei are haploid. Haploid phase. First applied to the microsporidia by Naville (1931). "A haploid phase is considered to occur whenever the immediate products of meiosis [haplosis, either by meiosis or nuclear dissociation in microsporidia] are competent to multiply vegetatively" (Raper and Flexer, 1970).

Haplophasic. "Having only unpaired (haploid) nuclei" (Sprague et al., 1992).

Haplosis. Reduction of the chromosome number from diploid to haploid. Paraphrase of Sharp (1934).

Heterosporous. "Having more than one kind of spore." Dictionary definition. Term used by Becnel et al. (1989) to replace the ambiguous terms "dimorphic," "trimorphic," "polymorphic," etc.

Interface. "A surface regarded as the common boundary of two bodies or spaces." Dictionary definition. "In microsporida the boundary is in some cases the outer surface of the parasite plasmalemma and in other cases an envelope external to that plasmalemma" (Sprague et al., 1992).

Interfacial envelope. An envelope of any composition or origin that is situated between the plasmalemma of the parasite and the cytoplasm of the host cell. Paraphrase of Sprague et al. (1992).

Karyogamy. Fusion of two haploid nuclei to form a synkaryon.

Life cycle. The complete sequence (or series of sequences) of morphological patterns within the cyclic development of an organism. Although in higher animals it starts with the zygote and ends with the production of a new generation of gametes, in microsporidia the situation is more complicated. In species with diplokarya the life cycle seems logically to start with formation of the first diplokaryotic cell (meront) by plasmogamy and nuclear association. In species without diplokarya the logical starting point is less evident. In microsporidia generally, it is convenient, if not otherwise logical, to treat the sporoplasm that infects a new host individual as the starting point of the life cycle.

Meiospore. A spore produced in a sporulation sequence that was accompanied by meiosis.

Merogonial plasmodium. A presporogonic individual with more than two diplokarya, rarely more than eight, as the result of delayed cytokinesis during merogony.

Merogony. (Gk. *meros,* part + Gk. *gonei,* production). An indeterminate series of binary divisions of diplokaryotic cells (meronts), sometimes with delayed cytokinesis and production of transitory paucinucleate plasmodia. First applied to a microsporidian species, *Thelohania chaetogastris* Schroeder, 1909, by the author of this binomen and incorrectly attributed to Stempell (1902) who adopted "meront" (1901, 1902, 1909) but not "merogony." Term treated by Schroeder as a synonym of "schizogony" and further ignored, a practice followed by Kudo (1924) and most other authors for many years thereafter. Sharply distinguished by the present authors.

Meront. (Gk. *meros,* part + Gk. *ont,* being). Introduced by Stempell (1901) for some presporogonic stages, variable in size, that were observed in *Thelohania muelleri* (Pfeiffer, 1894). Etymologically, the term suggests unequal division or budding, this being the reason that Stempell gave for adopting it. One can now infer that the smaller cells were uninucleate individuals in a schizogonic and/or a gametogonic sequence, leaving only the larger, diplokaryotic, cells to bear the original name "meronts." It is now logical and expedient to restrict this term to a diplokaryotic cell. The etymological incongruity, probably having rarely been recognized, presents no problem; it can easily give way to practical considerations and be ignored.

Merozoite. In the Apicomplexa, a certain vermiform and motile individual with an apical complex, so named in distinction from a morphologically similar "sporozoite." A misnomer for any individual in the microsporidia and listed here because during the past 40 years several authors have borrowed it and misapplied it to daughter meronts. It has a verbal resemblance to some individuals in the microsporidia but no counterpart. Application of the term "merozoite" to a diplokarytic cell in the microsporidia is not only inappropriate but is also misleading because it reflects a misconception of merogony in these organisms.

Mictosporoblastic. (Gk. *miktos,* mixed, + sporoblastic). Pertaining to a sporont or to sporogony that produces a variable number of sporoblasts. Modification of Kudo (1924).

Morphological type of spore. Any of a number of kinds of normal spores that exhibit conspicuously different morphological characters, the most important being the number of nuclei. Some other conspicuous characters are shape of the spore and structure of the extrusion apparatus. The different types are yet to be clearly distinguished, and descriptive terms for them remain to be standardized. Used by Sprague et al. (1992).

Mucocalyx. (*muco* + Gk. *kalyx,* husk or covering). A thick layer of mucous material covering the spores of some species and believed to be a flotation device. Coined by Vávra in Vávra and Sprague (1976). Described in detail by Vávra and Barker (1980).

Multiple division. Simultaneous division of a plasmodium into as many cells as there were nuclei in the parent body. Common and rarely defined term of undetermined origin.

Nuclear association. Pairing of two haploid nuclei following plasmogamy, at the end of the haplophase, to form a diplokaryon. A form of diplosis in the microsporidia. (Becnel et al., 1987).

Nuclear dissociation. Separation of the two members of a diplokaryon to form two independent haploid nuclei. One of two methods of haplosis in microsporidia, the other being meiosis (Becnel et al., 1987).

Octosporoblastic. Pertaining to a sporont or to sporogony that produces eight sporoblasts. Modification of Kudo (1924).

Pansporoblast. Coined by Gurley (1893) for use with both myxosporidia and microsporidia (treated as two distinct orders under a subclass). The term was defined in words that do not apply well to the latter order. The meaning of "pansporoblast" for the microsporidia becomes clear when we note that it was proposed "in distinction from the sporoblasts which result from the segmentation of the pansporoblast." Clearly, for the microsporidia the term "pansporoblast" may accurately be defined as an individual that divides into sporoblasts. "Pansporoblast" is an exact synonym of "sporont" as these were originally used with the microsporidia. Both terms have been frequently misused, particularly by their application to any or every stage in a sporulation sequence.

Pansporoblast(ic) membrane. Coined by Gurley (1893) for an interfacial envelope produced by the pansporoblast.

Paramural body. A saclike invagination of the plasmalemma of a dividing individual, containing a whorl of tubular structures, adhering to the cleavage furrow, and participating in the cleavage process. Adopted by Vávra (1976c) to replace the junior synonym "scindosome" of Vávra (1975). Previously reported only in plant cells and attributed by Vávra (1976c) to Marchant and Robards (1968).

Parasite-host cell relations. All interrelated morphological characters and all physiological interactions between a parasite and its host cell.

Parasitophorous vacuole. As defined by a dictionary, a vacuole is "a cavity within a cell, often containing a watery fluid or secretion." Self-evidently, a parasitophorous vacuole contains parasites. The vacuole is enveloped by a membrane which is of host cell origin.

Plasmodium. "A syncytium, especially in the case of Protista" (Wilson, 1925). A plasmodium with relatively few nuclei, as a merogonial plasmodium with about eight diplokarya, is often characterized as "paucinucleate." A plasmodium with a relatively large number of nuclei has been characterized as "plurinucleate."

Plasmogamy. Cytoplasmic fusion of gametes as opposed to nuclear fusion. First applied to microsporidia by Hazard et al. (1985) for cytoplasmic fusion of gametes without karyogamy. "Plastogamy" of Mercier (1909).

Plasmotomy. Dividing of a plasmodium into smaller plasmodia. Coined by Doflein (1898).

Polar cap. A chromophilic body at the anterior end of the spore contained within the polar sac. Term proposed independently by Hiller (1959) and Vávra (1959). Found independently by Huger (1960) and Vávra (1959) to be periodic acid-Schiff positive. Has many synonyms.

Polar filament. A filiform organelle inside the spore, attached anteriorly, extending backward and typically coiling several times just inside the spore wall. Quite variable, in many ways, according to species. Part of the extrusion apparatus that everts and inoculates the sporoplasm into a host cell.

Polaroplast. A complex of smooth membranes surrounding the base of the polar filament and arranged in layers or vesicles at right angles to it. Quite variable according to species. The term was coined by Huger (1960) who first demonstrated this organelle.

Polar sac. The inflated central part of the anchoring disk containing the polar cap. Proposed by Petri and Schiodt (1966).

Polar tube. Used as early as 1972 by Weidner instead of the usual term "polar filament" because this organelle was found to be tubular when everted. Later (1976) Weidner found evidence that it "is assembled as a tube within the spore." Some authors, perhaps feeling that a structure may be both filamentous and tubular, use the terms interchangeably. Petri (1969) referred to a "filament (filamentous tube)."

Polar vesicles. Introduced by Youssef and Hammond (1971) for a group of vesicles near the spindle plaque in *Nosema apis* Zander. The vesicles were later observed in many species, but their significance is unknown.

Polysporoblastic. Pertaining to a sporont or to sporogony that produces many sporoblasts. Modification of Kudo (1924).

Posterior body. Proposed by Weiser and Zizka (1974, 1975) for a "spongioid structure" in the posterior vacuole of immature spores which authors generally interpret as the Golgi apparatus in the final stages of elaborating the extrusion apparatus. Vávra (1976a, Fig. 29 and text) apparently saw only the abstract (Weiser and Zizka, 1974), misunderstood the application of this term, and adopted it for a Golgi secretion body to replace its many other names ("electron-dense body," "amorphous body," "metachromatic granule," "refringent granule," etc.). Since the terms "posterior body" and "posterosome," an alternative offered by Weiser and Zizka (1975), have become ambiguous, a means of avoiding confusion in their application is needed. One way is to reject them in favor of more specific terms (such as "Golgi apparatus" and "Golgi secretion").

Posterior end. Refers to the end of a spore opposite that from which the polar filament becomes everted. Paraphrase of Kudo (1924).

Posterior vacuole. A large, clear area in the posterior part of most spores, usually conspicuous in living spores, stained spores, and spores viewed with the electron microscope. Believed by some authors to be bounded by a membrane. Formed during sporogenesis, probably with participation of the Golgi apparatus.

Posterosome. Introduced by Weiser and Zizka (1975) as an alternative to the term "posterior body."

Schizogony. Coined by Schaudinn, fide Minchin (1922), "to denote the non-sexual cycle" in one of the Sarcodina which has alternation of generations. Has frequently been used indiscriminately for any presporogonic division in the microsporidia. Since alternation of generations is now known in microsporidia, it is expedient to follow the original use of this term by restricting its application to division by haplophasic individuals.

Schizont. Coined by Schaudinn, fide Minchin (1922), to denote an individual that undergoes schizogony.

Scindosome. Coined by Vávra (1975) and rejected by Vávra (1976c) as a junior synonym for "paramural body."

Secondary infective form. Term used by Ishihara (1969) for extracellular forms of *N. bombycis* Naegeli, 1857, found in tissue culture and resembling sporoplasms. Found by Iwano and Ishihara (1991) to be sporoplasms discharged by recently discovered and morphologically identifiable spores that develop early in the life cycle and function as autoinfective spores.

Spindle plaque. An electron-dense, laminated body lying on the nuclear membrane at the site where the intranuclear microtubules are attached. Coined by Moens and Rapport (1971) for a similar body in the yeast *Saccharomyces cerevisiae* (Hansen). Adopted by Vávra (1977), who considered it the most appropriate of several terms that have been applied to the spindle terminus in microsporidia.

Spore. A stage in development that has a structure especially adapted to its function of transmitting the infection to another cell in the same or another host individual. It has three basic components, the sporoplasm, the extrusion apparatus, and the spore envelope. All specialized organelles originate from the sporoblast by cytoplasmic differentiation. Paraphrase of Vávra (1976a).

Spore morphogenesis. Sporogenesis. Transformation of a sporoblast into a spore.

Sporoblast. The product of sporogony. "A cell which develops directly into a spore" (Kudo, 1924).

Sporogenesis. Spore morphogenesis. Transformation of a sporoblast into a spore. "During spore morphogenesis the Golgi assumes a central role in the elaboration of various spore organelles" (Vávra, 1976a).

Sporogonial plasmodium. Multinucleate body that divides into sporoblasts. Introduced by Tuzet et al. (1971) as "plasmode sporogonial." Not to be confused with the term "sporont."

Sporogony. Dividing of a sporont or a sporogonial plasmodium into sporoblasts.

Sporont. A cell, uninucleate or diplokaryotic, that divides directly or with the intervention of a plasmodial stage into sporoblasts. Not to be confused with "sporogonial plasmodium." An old term, adopted for the microsporidia by Stempell (1901) and recognized by the same author (1902) as a synonym of "pansporoblast."

Sporophorous vesicle. A sac that contains spores. Used by Gurley (1893) with particular reference to a pansporoblast (sporont) membrane which, after becoming separated from the host cell with its contents, remains intact for an indefinite time as a sac that contains spores. It is now known that a sporophorous vesicle is not necessarily produced by the sporont.

Sporoplasm. Defined by Naville (1931) as the generative portion of the spore. Designated by Debaisieux (1928) as the "germe" of the spore. Some sporoplasms have a single haploid nucleus, some have two separate haploid nuclei, and some have a diplokaryon. See Vávra (1976a) for a general consideration.

Sporulation. Production of spores. Usually sporogony followed by spore morphogenesis.

Sporulation plasmodium. A plasmodium that divides into young spores rather than sporoblasts, internal spore organelles having already been formed before the onset of division. Example: *Enterocytozoon*. New term.

Subpersistent. "More or less persistent." Dictionary definition. Used by Gurley (1893) to indicate that sporophorous vesicles differ in the time period during which they remain intact.

Synaptonemal complex. A particular configuration of chromosomes in the prophase of meiosis that is demonstrable with electron microscopy as a group of three parallel elements, a very thin central element, and two broader lateral elements. Its demonstration is generally accepted as proof of sexuality in the organism. Discovered for microsporidia independently by Vávra (1976a) in *Janacekia debaisieuxi* (Jirovec, 1943) and Loubes et al. (1976) in *Gurleya chironomi* Loubes & Maurand, 1975.

Tetrasporoblastic. Pertaining to a sporont or to sporogony that produces four sporoblasts. Modification of Kudo (1924).

Transovarial spore. A functional designation for a particular spore, so called "because it is responsible for the infection of the filial generation within the ovaries" (Johnson et al., 1997).

Vacuolar membrane. An envelope that encloses the contents of a vacuole, meaning here a parasitophorous vacuole. The membrane is of host cell origin but may be formed in different ways. The following are examples: The vacuolar membrane in a cell infected with *Abelspora portucalensis* is formed by small vesicles in the host cell cytoplasm that make contact with the sporont and coalesce (Azevedo, 1987). The vacuolar membrane in a cell infected with *Encephalitozoon intestinalis* is a phagocytic membrane formed by internalization of a portion of host cell plasmalemma in a phagocytic response to the stimulus of a spore starting to evert its polar filament (Magaud et al., 1997). Term of uncertain origin. Newly applied to the microsporidia.

Vacuole. See Parasitophorous vacuole.

Vesicle. Defined by a dictionary as "a little sac or cyst."

Xenoma. "Xenon," "xenone," "xenom," and "xenoma" are consecutive spellings used by Weissenberg (1922a, 1922b, 1949, 1968), who coined the term. Synonym of "xenoparasitic complex." Chatton (1920) characterized the complex as an "intimate" association. "Integrated" may be more descriptive for the situation in microsporidia because a complete morphological and physiological integration of parasite and host cell seems to be generally regarded as an essential character of this group.

Authors tend to apply the term "xenoma" only to a complex that exhibits conspicuous or spectacular features. Perhaps the highly integrated condition is in itself the essence of a xenoma, in which case this term may logically apply to every parasite-host cell combination in the microsporidia. Whatever the theoretical situation may be, it seems premature to define the extent of the applicability of the term "xenoma" at this time.

Xenoparasitic complex. "Complexe xéno-parasitaire" of Chatton (1920). An association of parasite and host cell so intimate that it is morphologically and physiologically an indivisible unit, essentially an autonomous organism. First proposed for the intimate association between a parasitic dinoflagellate and a tunicate host cell, but Chatton also called attention to similar associations between some microsporidia and the infected host cells. Synonym of "xenoma."

ACKNOWLEDGMENTS

We are indebted to Margaret (Peg) Johnson for assistance in preparing the manuscript.

This is contribution 3123 from the University of Maryland Center for Environmental Science.

REFERENCES

Azevedo, C. 1987. Fine structure of the microsporidian *Abelspora portucalensis* gen. n., sp. n. (Microsporidia) parasite of the hepatopancreas of *Carcinus maenas* (Crustacea, Decapoda). *J. Invertebr. Pathol.* **49:**83–92.

Becnel, J. J., E. I. Hazard, T. Fukuda, and V. Sprague. 1987. Life cycle of *Culicospora magna* (Kudo, 1920) (Microsporida, Culicosporidae) in *Culex restuans* Theobold with special reference to sexuality. *J. Protozool.* **34:**313–322.

Becnel, J. J., V. Sprague, T. Fukuda, and E. I. Hazard. 1989. Development of *Edhazardia aedis* (Kudo, 1930) n. g., n. comb. (Microsporida: Amblyosporidae) in the mosquito *Aedes aegypti* (L.) (Diptera: Culicidae). *J. Protozool.* **36:**119–130.

Chatton, E. 1920. Sur un complexe xeno-parasitaire morphologique et physiologique, *Neresheimeria catenara* chez *Fritillaria pellucida*. *C. R. Acad. Sci.* **171:**55–57.

Debaisieux, P. 1928. Etudes cytologiques sur quelques microsporidies. *Cellule* **38:**389–450.

Debaisieux, P., and L. Gastaldi. 1919. Les microsporidies parasites des larves de *Simulium*. II. *Cellule* **30:**185–213.

Doflein, F. 1898. Studien zur Naturgeschichte der Protozoen. III. Ueber Myxosporidien. *Zool. Jahrb. Abt. Anat.* **11:**281–350.

Gurley, R. 1893. On the classification of the myxosporidia, a group of protozoan parasites infecting fishes. *Bull. U.S. Fish Commission* **11:**407–420.

Hazard, E. I., T. Fukuda, and J. J. Becnel. 1985. Gametogenesis and plasmogamy in certain species of Microspora. *J. Invertebr. Pathol.* **46:**63–69.

Hiller, S. R. 1959. The morphology and life cycle of *Plistophora scatopsi* sp. n., a microsporidian parasitic in the mid-gut of *Scatopse notata* Mg. (Diptera: Nematocera). *Parasitology* **49:**464–472.

Huger, A. 1960. Electron microscope study on the cytology of a microsporidian spore by means of ultrathin sections. *J. Insect Pathol.* **2:**84–105.

Ishihara, R. 1969. The life cycle of *Nosema bombycis* as revealed in tissue culture cells of *Bombyx mori*. *J. Invertebr. Pathol.* **14:**316–320.

Iwano, H., and R. Ishihara. 1991. Dimorphic development of *Nosema bombycis* spores in gut epithelium of larvae of the silkworm, *Bombyx mori*. *J. Seric. Sci. Jpn.* **60:**249–256.

Johnson, M. A., J. J. Becnel, and A. H. Undeen. 1997. A new sporulation sequence in *Edhazardia aedis* (Microsporidia: Culicosporidae), a parasite of the mosquito *Aedes aegypti* (Diptera: Culicidae). *J. Invertebr. Pathol.* **70:**69–75.

Kramer, J. P. 1976. The extra-corporeal ecology of microsporidia, p. 127–135. *In* L. A. Bulla, Jr., and T. C. Cheng (ed.), *Comparative Pathobiology*, vol. 1. Plenum Press, New York, N.Y.

Kudo, R. 1924. A biologic and taxonomic study of the Microsporidia. *Ill. Biol. Monogr.* **9:**1–269.

Levine, N. D. 1971. Uniform terminology for the protozoan subphylum Apicomplexa. *J. Protozool.* **18:**352–355.

Loubes, C., J. Maurand, and V. Rousset-Galangau. 1976. Présence de complexes synaptonématiques dans le cycle biologique de *Gurleya chironomi* Loubes et Maurand, 1975: un argument en faveur d'une sexualité chez les Microsporidies? *C. R. Acad. Sci.* **282:**1025–1027.

Magaud, A., A. Achbarou, and I. Desportes-Livage. 1997. Cell invasion by the microsporidium *Encephalitozoon intestinalis*. *J. Eukaryot. Microbiol.* **44:**81S.

Mercier, L. 1909. Contribution a l'etude de la sexualite chez les Myxosporidies et chez les Microsporidies. *Mem. Acad. R. Belg. Cl. Sci.* **2:**1–52.

Minchin, E. A. 1922. *An Introduction to the Study of the Protozoa*. Edward Arnold, London.

Moens, P. B., and E. Rapport. 1971. Spindles, spindle plaques, and meiosis in the yeast *Saccharomyces cerevisiae* (Hansen). *J. Cell Biol.* **50:** 344–361.

Naville, A. 1931. Les sporozoaires (cycles chromosomiques sexualite). *Mem. Soc. Phys. Hist. Nat.* **41:**1–223.

Petri, M. 1969. Studies on *Nosema cuniculi* found in transplantable ascites tumours with a survey of microsporidiosis in mammals. *Acta Pathol. Microbiol. Scand.* **204:**1–91.

Petri, M., and T. Schiödt. 1966. On the ultrastructure of *Nosema cuniculi* in the cells of the Yoshida rat ascites sarcoma. *Acta Pathol. Microbiol. Scand.* **66:**437–446.

Raper, J. R., and Flexer, A. S. 1970. The road to diploidy with emphasis on a detour. *Symp. Soc. Gen. Microbiol.* **20:**401–432.

Sharp, L. W. 1934. *Introduction to Cytology.* McGraw-Hill Book Co. New York, N.Y.

Sprague, V., J. J. Becnel, and E. I. Hazard. 1992. Taxonomy of phylum Microspora. *Crit. Rev. Microbiol.* **18:**285–395.

Stempell, W. 1901. Zur Entwicklung von *Plistophora mulleri* (L. Pfr.). *Zool. Anz.* **24:**157–158.

Stempell, W. 1902. Ueber *Thelohania mulleri* (L. Pfr.). *Zool. Jahrb. Abt. Morphol.* **16:**235–272.

Stempell, W. 1909. Ueber *Nosema bombycis* Nageli. *Arch. Protistenkd.* **16:**281–358.

Tuzet, O., J. Maurand, J. A. Fize, R. Michel, and B. Fenwick. 1971. Proposition d'un nouveau cadre systematique pour les genres de Microsporidies. *C. R. Acad. Sci. Paris* **272:**1268–1271.

Vávra, J. 1959. Beitrag zur Cytologie einiger Mikrosporidien. *Vest. Cesk. Spol. Zool.* **23:**347–350.

Vávra, J. 1965. Étude au microscope electronique de la morphologie du developpement de quelques Microsporidies. *C. R. Acad. Sci. Paris* **261:**3467–3470.

Vávra, J. 1968. Ultrastructural features of *Caudospora simulii* Weiser (Protozoa, Microsporidia). *Folia Parasitol.* (Prague) **15:**1–9.

Vávra, J. 1971. Ultrahistochemical detection of carbohydrates in microsporidian spores. *J. Protozool.* **18**(Suppl.):47. (Abstract 179.)

Vávra, J. 1975. Scindosome: a new microsporidian organelle. *J. Protozool.* **22:**69A. (Abstract 208.)

Vávra, J. 1976a. Structure of the Microsporidia, p. 1–85. *In* L. A. Bulla, Jr., and T. C. Cheng (ed.), *Comparative Pathobiology,* vol. 1. Plenum Press, New York, N.Y.

Vávra, J. 1976b. Development of the Microsporidia. p. 87–109. *In* L. A. Bulla, Jr., and T. C. Cheng, (ed.), *Comparative Pathobiology,* vol. 1. Plenum Press, New York, N.Y.

Vávra, J. 1976c. The occurrence of paramural bodies in microsporidia. *J. Protozool.* **23:**21A. (Abstract 63.)

Vávra, J. 1977. The microsporidian mitotic apparatus. *J. Protozool.* **24:**17A. (Abstract 115.)

Vávra, J., and R. J. Barker. 1980. The microsporidian mucocalyx as seen in the scanning electron microscope. *Folia Parasitol.* **27:**19–21.

Vávra, J., and Sprague, V. 1976. Glossary for the Microsporidia. p. 341–363. *In* L. A. Bulla, Jr., and T. C. Cheng, (ed.), *Comparative Pathobiology,* vol. 1. Plenum Press, New York, N.Y.

Weidner, E. 1972. Ultrastructural study of microsporidian invasion into cells. *Z. Parasitenkd.* **40:** 227–242.

Weidner, E. 1976. The microsporidian spore invasion tube: the ultrastructure, isolation, and characterization of the protein comprising the tube. *J. Cell Biol.* **71:**23–34.

Weiser, J., and Z. Zizka. 1974. Stages in sporogony of *Pleistophora debaisieuxi* (Microsporidia). *J. Protozool.* **21:**477. (Abstract 229.)

Weiser, J., and Z. Zizka. 1975. Stages in sporogony of *Plistophora debaisieuxi* Jìrovec (Microsporidia). *Acta Protozool.* **14:**185–194.

Weissenberg, R. 1922a. Mikrosporidien, Myxosporidien und Chlamydozoen als Zellparasiten von Fischen. *Verh. Dtsch. Zool. Ges.* **27:**41–43.

Weissenberg, R. 1922b. Fremddienliche Reaktionen beim intrazellulalren Parasitismus, ein Beitrag zur Kenntnis gallenahnlicher Bildungen in Teirkorper. *Verh. Dtsch. Zool. Ges.* **27:**96–98.

Weissenberg, R. 1949. Cell growth and cell transformation induced by cellular parasites. *Anat. Rec.* **103:**517–518.

Weissenberg, R. 1968. Intracellular development of the microsporidian *Glugea anomala* Moniez in hypertrophying migratory cells of the fish *Gasterosteus aculeatus* L., an example of the formation of "xenoma" tumors. *J. Protozool.* **15:**44–57.

Wilson, E. B. 1925. *The Cell in Development and Heredity.* The Macmillan Co., New York, N.Y.

Youssef, N., and D. M. Hammond. 1971. The fine structure of the developmental stages of the microsporidian *Nosema apis* Zander. *Tissue Cell* **3:** 283–294.

INDEX